MONOGRAPHS AND RESEARCH NOTES IN MATHEMATICS

Elements of Quasigroup Theory and Applications

Victor Shcherbacov

Principal Researcher
Institute of Mathematics and Computer Science
Academy of Sciences
Moldova

CRC Press is an imprint of the
Taylor & Francis Group, an **informa** business
A CHAPMAN & HALL BOOK

MONOGRAPHS AND RESEARCH NOTES IN MATHEMATICS

Published Titles

Actions and Invariants of Algebraic Groups, Second Edition, Walter Ferrer Santos and Alvaro Rittatore

Analytical Methods for Kolmogorov Equations, Second Edition, Luca Lorenzi

Application of Fuzzy Logic to Social Choice Theory, John N. Mordeson, Davender S. Malik and Terry D. Clark

Blow-up Patterns for Higher-Order: Nonlinear Parabolic, Hyperbolic Dispersion and Schrödinger Equations, Victor A. Galaktionov, Enzo L. Mitidieri, and Stanislav Pohozaev

Bounds for Determinants of Linear Operators and Their Applications, Michael Gil′

Complex Analysis: Conformal Inequalities and the Bieberbach Conjecture, Prem K. Kythe

Computational Aspects of Polynomial Identities: Volume l, Kemer's Theorems, 2nd Edition Alexei Kanel-Belov, Yakov Karasik, and Louis Halle Rowen

A Concise Introduction to Geometric Numerical Integration, Fernando Casas and Sergio Blanes

Cremona Groups and Icosahedron, Ivan Cheltsov and Constantin Shramov

Delay Differential Evolutions Subjected to Nonlocal Initial Conditions Monica-Dana Burlică, Mihai Necula, Daniela Roşu, and Ioan I. Vrabie

Diagram Genus, Generators, and Applications, Alexander Stoimenow

Difference Equations: Theory, Applications and Advanced Topics, Third Edition Ronald E. Mickens

Dictionary of Inequalities, Second Edition, Peter Bullen

Elements of Quasigroup Theory and Applications, Victor Shcherbacov

Finite Element Methods for Eigenvalue Problems, Jiguang Sun and Aihui Zhou

Introduction to Abelian Model Structures and Gorenstein Homological Dimensions Marco A. Pérez

Iterative Methods without Inversion, Anatoly Galperin

Iterative Optimization in Inverse Problems, Charles L. Byrne

Line Integral Methods for Conservative Problems, Luigi Brugnano and Felice Iavernaro

Lineability: The Search for Linearity in Mathematics, Richard M. Aron, Luis Bernal González, Daniel M. Pellegrino, and Juan B. Seoane Sepúlveda

Modeling and Inverse Problems in the Presence of Uncertainty, H. T. Banks, Shuhua Hu, and W. Clayton Thompson

Monomial Algebras, Second Edition, Rafael H. Villarreal

Published Titles Continued

Nonlinear Functional Analysis in Banach Spaces and Banach Algebras: Fixed Point Theory Under Weak Topology for Nonlinear Operators and Block Operator Matrices with Applications, Aref Jeribi and Bilel Krichen

Partial Differential Equations with Variable Exponents: Variational Methods and Qualitative Analysis, Vicenţiu D. Rădulescu and Dušan D. Repovš

A Practical Guide to Geometric Regulation for Distributed Parameter Systems Eugenio Aulisa and David Gilliam

Reconstruction from Integral Data, Victor Palamodov

Signal Processing: A Mathematical Approach, Second Edition, Charles L. Byrne

Sinusoids: Theory and Technological Applications, Prem K. Kythe

Special Integrals of Gradshteyn and Ryzhik: the Proofs – Volume I, Victor H. Moll

Special Integrals of Gradshteyn and Ryzhik: the Proofs – Volume II, Victor H. Moll

Stochastic Cauchy Problems in Infinite Dimensions: Generalized and Regularized Solutions, Irina V. Melnikova

Submanifolds and Holonomy, Second Edition, Jürgen Berndt, Sergio Console, and Carlos Enrique Olmos

Symmetry and Quantum Mechanics, Scott Corry

The Truth Value Algebra of Type-2 Fuzzy Sets: Order Convolutions of Functions on the Unit Interval, John Harding, Carol Walker, and Elbert Walker

Forthcoming Titles

Groups, Designs, and Linear Algebra, Donald L. Kreher

Handbook of the Tutte Polynomial, Joanna Anthony Ellis-Monaghan and Iain Moffat

Microlocal Analysis on R^n and on NonCompact Manifolds, Sandro Coriasco

Practical Guide to Geometric Regulation for Distributed Parameter Systems, Eugenio Aulisa and David S. Gilliam

CRC Press
Taylor & Francis Group
6000 Broken Sound Parkway NW, Suite 300
Boca Raton, FL 33487-2742

Printed on acid-free paper
Version Date: 20170215

International Standard Book Number-13: 978-1-4987-2155-4 (Hardback)

Visit the Taylor & Francis Web site at
http://www.taylorandfrancis.com

and the CRC Press Web site at
http://www.crcpress.com

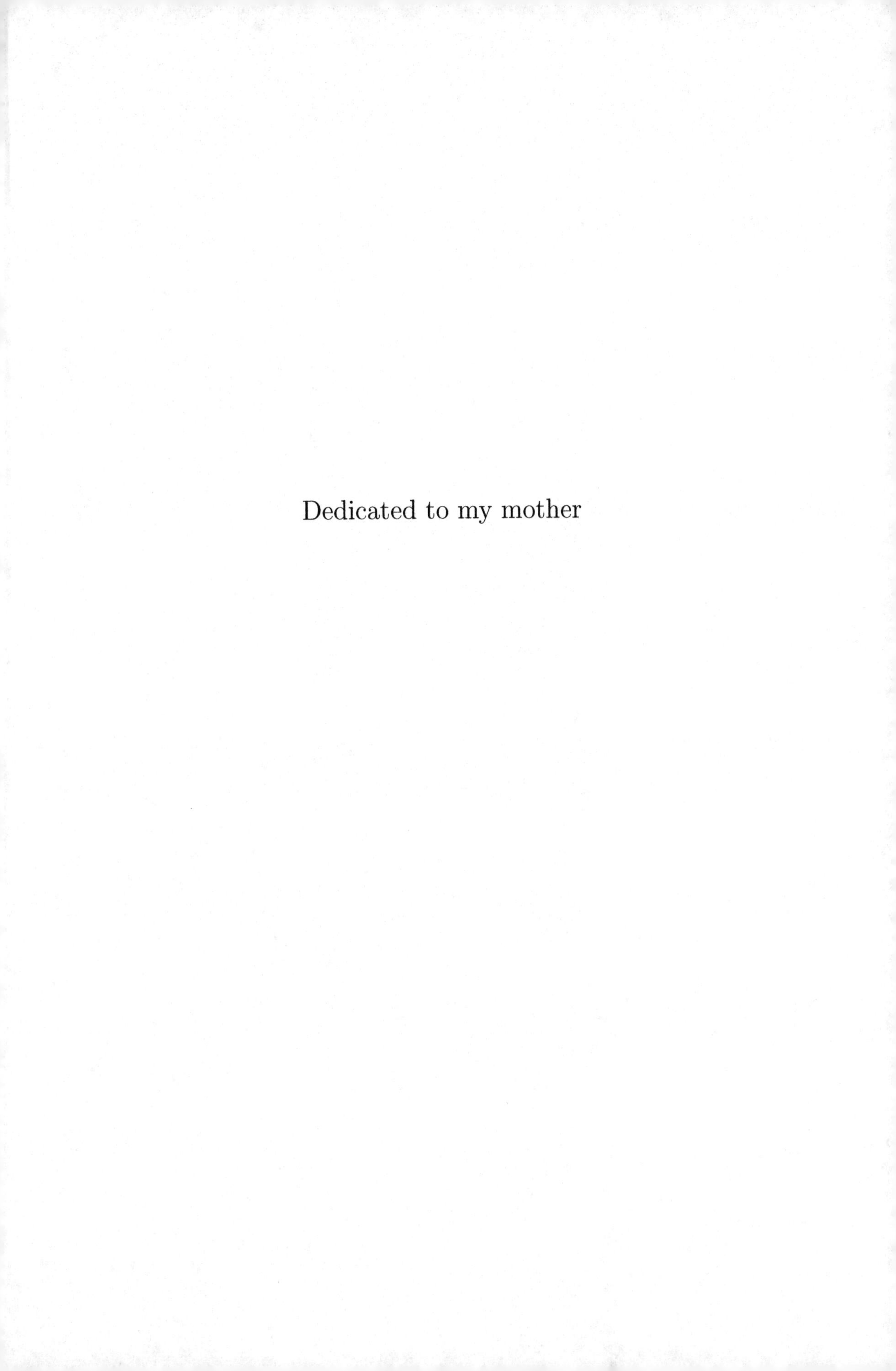

Dedicated to my mother

Contents

Foreword xv

List of Figures xix

List of Tables xxi

I Foundations 1

1 Elements of quasigroup theory 3

1.1 Introduction . . . 5
1.1.1 The role of definitions . . . 5
1.1.2 Sets . . . 5
1.1.3 Products and partitions . . . 5
1.1.4 Maps . . . 6
1.2 Objects . . . 8
1.2.1 Groupoids and quasigroups . . . 8
1.2.2 Parastrophy: Quasigroup as an algebra . . . 11
1.2.2.1 Parastrophy . . . 11
1.2.2.2 Middle translations . . . 12
1.2.2.3 Some groupoids . . . 13
1.2.2.4 Substitutions in groupoid identities . . . 15
1.2.2.5 Equational definitions . . . 15
1.2.3 Some other definitions of e-quasigroups . . . 18
1.2.4 Quasigroup-based cryptosystem . . . 20
1.2.5 Identity elements . . . 21
1.2.5.1 Local identity elements . . . 21
1.2.5.2 Left and right identity elements . . . 21
1.2.5.3 Loops . . . 21
1.2.5.4 Identity elements of quasigroup parastrophes . . . 22
1.2.5.5 The equivalence of loop definitions . . . 22
1.2.5.6 Identity elements in some quasigroups . . . 23
1.2.5.7 Inverse elements in loops . . . 24
1.2.6 Multiplication groups of quasigroups . . . 24
1.2.7 Transversals: "Come back way" . . . 26
1.2.8 Generators of inner multiplication groups . . . 27
1.3 Morphisms . . . 29
1.3.1 Isotopism . . . 29
1.3.2 Group action . . . 32
1.3.3 Isotopism: Another point of view . . . 33
1.3.4 Autotopisms of binary quasigroups . . . 34
1.3.5 Automorphisms of quasigroups . . . 38

1.3.6 Pseudo-automorphisms and G-loops 38
1.3.7 Parastrophisms as operators 42
1.3.8 Isostrophism 43
1.3.9 Autostrophisms 45
1.3.9.1 Coincidence of quasigroup parastrophes 45
1.3.10 Inverse loops to a fixed loop 46
1.3.11 Anti-autotopy 48
1.3.12 Translations of isotopic quasigroups 48
1.4 Sub-objects 51
1.4.1 Subquasigroups: Nuclei and center 51
1.4.1.1 Sub-objects 51
1.4.1.2 Nuclei 52
1.4.1.3 Center 52
1.4.2 Bol and Moufang nuclei 52
1.4.3 The coincidence of loop nuclei 54
1.4.3.1 Nuclei coincidence and identities 56
1.4.4 Quasigroup nuclei and center 57
1.4.4.1 Historical notes 57
1.4.4.2 Quasigroup nuclei 57
1.4.4.3 Quasigroup center 58
1.4.5 Regular permutations 58
1.4.6 A-nuclei of quasigroups 59
1.4.7 A-pseudo-automorphisms by isostrophy 60
1.4.8 Commutators and associators 62
1.5 Congruences 63
1.5.1 Congruences of quasigroups 63
1.5.1.1 Congruences in universal algebra 63
1.5.1.2 Normal congruences 64
1.5.2 Quasigroup homomorphisms 66
1.5.3 Normal subquasigroups 67
1.5.4 Normal subloops 68
1.5.5 Antihomomorphisms and endomorphisms 68
1.5.6 Homotopism 70
1.5.7 Congruences and isotopism 71
1.5.8 Congruence permutability 72
1.6 Constructions 72
1.6.1 Direct product 73
1.6.2 Semidirect product 74
1.6.3 Crossed (quasi-direct) product 75
1.6.4 n-Ary crossed product 76
1.6.5 Generalized crossed product 76
1.6.6 Generalized singular direct product 76
1.6.7 Sabinin's product 78
1.7 Quasigroups and combinatorics 79
1.7.1 Orthogonality 79
1.7.1.1 Orthogonality of binary operations 79
1.7.1.2 Orthogonality of n-ary operations 81
1.7.1.3 Easy way to construct n-ary orthogonal operations 82
1.7.2 Partial Latin squares: Latin trades 83
1.7.3 Critical sets of Latin squares, Sudoku 85
1.7.4 Transversals in Latin squares 86

1.7.5 Quasigroup prolongations: Combinatorial aspect 87
1.7.5.1 Bruck-Belousov prolongation 87
1.7.5.2 Belyavskaya prolongation 90
1.7.5.3 Algebraic approach 91
1.7.5.4 Prolongation using quasicomplete mappings 93
1.7.5.5 Two-step mixed procedure 95
1.7.5.6 Brualdi problem 96
1.7.5.7 Contractions of quasigroups 97
1.7.6 Orthomorphisms 97
1.7.7 Neo-fields and left neo-fields 98
1.7.8 Sign of translations 99
1.7.9 The number of quasigroups 100
1.7.10 Latin squares and graphs 101
1.7.11 Orthogonal arrays 105

2 Some quasigroup classes **107**

2.1 Definitions of loop and quasigroup classes 108
2.1.1 Moufang loops, Bol loops, and generalizations 110
2.1.2 Some linear quasigroups 113
2.2 Classical inverse quasigroups 115
2.2.1 Definitions and properties 115
2.2.2 Autotopies of LIP- and IP-loops 119
2.2.3 Moufang and Bol elements in LIP-loops 123
2.2.4 Loops with the property $I_l = I_r$ 125
2.3 Medial quasigroups 126
2.3.1 Linear forms: Toyoda theorem 126
2.3.2 Direct decompositions: Murdoch theorem 130
2.3.3 Simple quasigroups 133
2.3.4 Examples 133
2.4 Paramedial quasigroups 135
2.4.1 Kepka-Nemec theorem 135
2.4.2 Antiendomorphisms 136
2.4.3 Direct decomposition 138
2.4.4 Simple paramedial quasigroups 140
2.4.5 Quasigroups of order 4 141
2.5 CMLs and their isotopes 142
2.5.1 CMLs 142
2.5.2 Distributive quasigroups 144
2.6 Left distributive quasigroups 146
2.6.1 Examples, constructions, orders 146
2.6.2 Properties, simple quasigroups, loop isotopes 148
2.7 TS-quasigroups 149
2.7.1 Constructions, loop isotopes 149
2.7.2 2-nilpotent TS-loops 151
2.7.3 Some properties of TS-quasigroups 152
2.8 Schröder quasigroups 152
2.9 Incidence systems and block designs 154
2.9.1 Introduction 154
2.9.2 3-nets and binary quasigroups 156
2.9.3 On orders of finite projective planes 156

2.9.4 Steiner systems 157
2.9.5 Mendelsohn design 161
2.9.6 Spectra of quasigroups with 2-variable identities 161
2.10 Linear quasigroups 164
2.10.1 Introduction 164
2.10.2 Definitions 165
2.10.3 Group isotopes and identities 166
2.10.4 Nuclei, identities 169
2.10.5 Parastrophes of linear quasigroups 170
2.10.6 On the forms of n-T-quasigroups 172
2.10.7 (m, n)-Linear quasigroups 173
2.11 Miscellaneous 176
2.11.1 Groups with triality 177
2.11.2 Universal properties of quasigroups 179
2.11.3 Alternative and various conjugate closed quasigroups and loops . . . 180

3 Binary inverse quasigroups **183**

3.1 Definitions 183
3.1.1 Definitions of "general" inverse quasigroups 183
3.2 (r, s, t)-Inverse quasigroups 188
3.2.1 Elementary properties and examples 188
3.2.2 Left-linear quasigroups which are (r, s, t)-inverse 190
3.2.3 Main theorems 194
3.2.4 Direct product of (r, s, t)-quasigroups 197
3.2.5 The existence of (r, s, t)-inverse quasigroups 201
3.2.6 WIP-quasigroups 202
3.2.7 Examples of WIP-quasigroups 205
3.2.8 Generalized balanced parastrophic identities 206
3.2.9 Historical notes 207

4 A-nuclei of quasigroups **209**

4.1 Preliminaries 210
4.1.1 Isotopism 210
4.1.2 Quasigroup derivatives 212
4.1.2.1 G-quasigroups 213
4.1.2.2 Garrison's nuclei in quasigroups 214
4.1.2.3 Mixed derivatives 214
4.1.3 Set of maps 215
4.2 Garrison's nuclei and A-nuclei 216
4.2.1 Definitions of nuclei and A-nuclei 216
4.2.2 Components of A-nuclei and identity elements 219
4.2.3 A-nuclei of loops by isostrophy 222
4.2.4 Isomorphisms of A-nuclei 222
4.2.5 A-nuclei by some isotopisms 225
4.2.6 Quasigroup bundle and nuclei 226
4.2.7 A-nuclei actions 228
4.2.8 A-nuclear quasigroups 230
4.2.9 Identities with permutation and group isotopes 232
4.3 A-centers of a quasigroup 235

4.3.1 Normality of A-nuclei and autotopy group 235
4.3.2 A-centers of a loop 239
4.3.3 A-centers of a quasigroup 241
4.4 A-nuclei and quasigroup congruences 243
4.4.1 Normality of equivalences in quasigroups 243
4.4.2 Additional conditions of normality of equivalences 244
4.4.3 A-nuclei and quasigroup congruences 248
4.4.4 A-nuclei and loop congruences 251
4.4.5 On loops with nucleus of index two 253
4.5 Coincidence of A-nuclei in inverse quasigroups 254
4.5.1 (α, β, γ)-inverse quasigroups 254
4.5.2 λ-, ρ-, and μ-inverse quasigroups 256
4.6 Relations between a loop and its inverses 257
4.6.1 Nuclei of inverse loops in Belousov sense 257
4.6.2 LIP- and AAIP-loops 258
4.6.3 Invariants of reciprocally inverse loops 259
4.6.3.1 Middle Bol loops 259
4.6.3.2 Some invariants 259
4.6.3.3 Term-equivalent loops 260
4.6.4 Nuclei of loops that are inverse to a fixed loop 261

II Theory 263

5 On two Belousov problems 265

5.1 The existence of identity elements in quasigroups 265
5.1.1 On quasigroups with Moufang identities 265
5.1.2 Identities that define a CML 272
5.2 Bruck-Belousov problem 274
5.2.1 Introduction 274
5.2.2 Congruences of quasigroups 276
5.2.3 Congruences of inverse quasigroups 282
5.2.4 Behavior of congruences by an isotopy 283
5.2.5 Regularity of quasigroup congruences 284

6 Quasigroups which have an endomorphism 287

6.1 Introduction 287
6.1.1 Parastrophe invariants and isostrophisms 291
6.2 Left and right F-, E-, SM-quasigroups 293
6.2.1 Direct decompositions 297
6.2.2 F-quasigroups 301
6.2.3 E-quasigroups 307
6.2.4 SM-quasigroups 310
6.2.5 Finite simple quasigroups 311
6.2.6 Left FESM-quasigroups 312
6.2.7 CML as an SM-quasigroup 314
6.3 Loop isotopes 315
6.3.1 Left F-quasigroups 315
6.3.2 F-quasigroups 321
6.3.3 Left SM-quasigroups 321

6.3.4 Left E-quasigroups . . . 322

7 Structure of n-ary medial quasigroups 327

7.1 On n-ary medial quasigroups . . . 327
7.1.1 n-ary quasigroups: Isotopy and translations . . . 327
7.1.2 Linear n-ary quasigroups . . . 329
7.1.3 n-Ary medial quasigroups . . . 330
7.1.4 Homomorphisms of n-ary quasigroups . . . 331
7.1.5 Direct product of n-ary quasigroups . . . 334
7.1.6 Multiplication group of n-ary T-quasigroup . . . 335
7.1.7 Homomorphisms of n-ary linear quasigroups . . . 336
7.1.8 n-Ary analog of Murdoch theorem . . . 339
7.2 Properties of n-ary simple T-quasigroups . . . 344
7.2.1 Simple n-ary quasigroups . . . 344
7.2.2 Congruences of linear n-ary quasigroups . . . 345
7.2.3 Simple n-T-quasigroups . . . 348
7.2.4 Simple n-ary medial quasigroups . . . 349
7.3 Solvability of finite n-ary medial quasigroups . . . 354

8 Automorphisms of some quasigroups 357

8.1 On autotopies of n-ary linear quasigroups . . . 357
8.1.1 Autotopies of derivative groups . . . 358
8.1.2 Automorphisms of n-T-quasigroups . . . 363
8.1.3 Automorphisms of some quasigroup isotopes . . . 368
8.1.4 Automorphisms of medial n-quasigroups . . . 370
8.1.5 Examples . . . 372
8.2 Automorphism groups of some binary quasigroups . . . 374
8.2.1 Isomorphisms of IP-loop isotopes . . . 374
8.2.2 Automorphisms of loop isotopes . . . 377
8.2.3 Automorphisms of LD-quasigroups . . . 379
8.2.4 Automorphisms of isotopes of LD-quasigroups . . . 381
8.2.5 Quasigroups with transitive automorphism group . . . 383
8.3 Non-isomorphic isotopic quasigroups . . . 383

9 Orthogonality of quasigroups 387

9.1 Orthogonality: Introduction . . . 388
9.1.1 Squares and Latin squares . . . 388
9.1.2 m-Tuples of maps and its product . . . 389
9.1.3 m-Tuples of maps and groupoids . . . 390
9.1.4 τ-Property . . . 392
9.1.5 Definitions of orthogonality . . . 395
9.1.6 Orthogonality in works of V.D. Belousov . . . 397
9.1.7 Product of squares . . . 398
9.2 Orthogonality and parastroph orthogonality . . . 398
9.2.1 Orthogonality of left quasigroups . . . 399
9.2.2 Orthogonality of quasigroup parastrophes . . . 400
9.2.3 Orthogonality in the language of quasi-identities . . . 402
9.2.4 Orthogonality of parastrophes in the language of identities . . . 403

9.2.5 Spectra of some parastroph orthogonal quasigroups 405
9.3 Orthogonality of linear and alinear quasigroups 407
9.3.1 Orthogonality of one-sided linear quasigroups 408
9.3.2 Orthogonality of linear and alinear quasigroups 412
9.3.3 Orthogonality of parastrophes 416
9.3.4 Parastrophe orthogonality of T-quasigroups 420
9.3.5 (12)-parastrophe orthogonality 422
9.3.6 totCO-quasigroups 425
9.4 Nets and orthogonality of the systems of quasigroups 426
9.4.1 k-nets and systems of orthogonal binary quasigroups 426
9.4.2 Algebraic (k, n)-nets and systems of orthogonal n-ary quasigroups . 427
9.4.3 Orthogonality of n-ary quasigroups and identities 428
9.5 Transformations which preserve orthogonality 429
9.5.1 Isotopy and (12)-isostrophy 430
9.5.2 Generalized isotopy 431
9.5.3 Gisotopy and orthogonality 433
9.5.4 Mann's operations 434

III Applications 437

10 Quasigroups and codes 439

10.1 One check symbol codes and quasigroups 439
10.1.1 Introduction 439
10.1.2 On possibilities of quasigroup codes 443
10.1.3 TAC-quasigroups and n-quasigroup codes 445
10.1.4 5-n-quasigroup codes 450
10.1.5 Phonetic errors 451
10.1.6 Examples of codes 452
10.2 Recursive MDS-codes 458
10.2.1 Some definitions 459
10.2.2 Singleton bound 459
10.2.3 MDS-codes 460
10.2.4 Recursive codes 460
10.2.5 Gonsales-Couselo-Markov-Nechaev construction 462
10.2.6 Orthogonal quasigroups of order ten 464
10.2.7 Additional information 465
10.3 On signs of Bol loop translations 466

11 Quasigroups in cryptology 471

11.1 Introduction 472
11.1.1 Quasigroups in "classical" cryptology 473
11.2 Quasigroup-based stream ciphers 474
11.2.1 Introduction 474
11.2.2 Modifications and generalizations 475
11.2.3 Further development 476
11.2.4 Some applications 478
11.2.5 Additional modifications of Algorithm 1.69 478
11.2.6 n-Ary analogs of binary algorithms 479
11.2.7 Further development of Algorithm 11.10 481

11.3 Cryptanalysis of some stream ciphers 482
11.3.1 Chosen ciphertext attack 482
11.3.2 Chosen plaintext attack 482
11.4 Combined algorithms 483
11.4.1 Ciphers based on the systems of orthogonal n-ary operation 483
11.4.2 Modifications of Algorithm 11.14 483
11.4.3 Stream cipher based on orthogonal system of quasigroups 485
11.4.4 T-quasigroup-based stream cipher 485
11.4.5 Generalization of functions of Algorithm 11.16 487
11.4.6 On quasigroup-based cryptcode 488
11.4.6.1 Code part 488
11.4.6.2 Cryptographical part 489
11.4.6.3 Decoding 490
11.4.6.4 Resistance 491
11.4.6.5 A code-crypt algorithm 491
11.4.7 Comparison of the power of the proposed algorithms 491
11.5 One-way and hash functions 492
11.5.1 One-way function 492
11.5.2 Hash function 493
11.6 Secret-sharing schemes 494
11.6.1 Critical sets 494
11.6.2 Youden squares 495
11.6.3 Reed-Solomon codes 496
11.6.4 Orthogonality and secret-sharing schemes 496
11.7 Some algebraic systems in cryptology 497
11.7.1 Inverse quasigroups in cryptology 498
11.7.2 Some groups in cryptology 498
11.7.2.1 El Gamal cryptosystem 499
11.7.2.2 De-symmetrization of Algorithm 1.69 499
11.7.2.3 RSA and GM cryptosystems 500
11.7.2.4 Homomorphic encryption 501
11.7.2.5 MOR cryptosystem 501
11.7.3 El Gamal signature scheme 502
11.7.4 Polynomially complete quasigroups in cryptology 503
11.7.5 Cryptosystems which are based on row-Latin squares 504
11.7.6 Non-binary pseudo-random sequences over Galois fields 505
11.7.7 Authentication of a message 505
11.7.8 Zero-knowledge protocol 506
11.7.9 Hamming distance between quasigroups 507
11.7.10 Generation of quasigroups for cryptographical needs 507

A Appendix **509**

A.1 The system of German banknotes 509
A.2 Outline of the history of quasigroup theory 510
A.3 On 20 Belousov problems 512

References **517**

Index **567**

Foreword

This book is based on lectures which the author gave for graduate students of Charles University (Prague, Czech Republic) in autumn 2003 [779].

We hope this book will assist young (and not so young) mathematicians who would like to start their own research on the topic of quasigroup theory and especially on its applications.

In Chapters 1–4 a sufficiently elementary introduction to the theory of quasigroups and main classes of quasigroups is given. In Chapters 5–9 some results obtained mainly in the last twenty years in the branch of "pure" quasigroup theory are presented. In Chapters 10 and 11 information on applications of quasigroups in code theory and cryptology is collected.

Therefore it is possible to divide this book into three parts: Foundations, Theory, and Applications.

Chapter 1 "Elements of Quasigroup Theory" gives a short, somewhat "elementary" from the point of view of an "advanced" algebraist, and in some places only the outlined, introduction to quasigroup theory. In this chapter we try to demonstrate to the reader that there is no essentially big difference between binary and n-ary objects and that the "work" with an n-ary object is quite similar to the work with a binary object.

As it seems to the author, this introductory chapter is written in the spirit of Belousov quasigroup school: that is, attention is given especially to algebraic questions concerning quasigroup theory.

The author hopes that many questions which mainly concern loop theory will be discussed in more detail in other books.

We hope that this chapter will be comprehensible to undergraduate students (both for independent study and also for study with a teacher).

In writing this chapter, the author has tried also to take into consideration possible needs and interests of engineers, mathematicians who are non-algebraists and physicists who would like to use quasigroups in coding theory, cryptology, physics or in some other "suitable places" for application of quasigroups.

Notice, the theory of quasigroups is in the process of rapid growth and even the foundations of quasigroup theory are being changed quite quickly.

Most applications of quasigroups are connected with the fact that quasigroups often have some kind of inverse property and/or that they, like semigroups, are a generalization of the concept of a group.

Readers can find more detailed introductions to quasigroup theory in the books of V.D. Belousov [72, 80], H.O. Pflugfelder [685], and J.D.H. Smith [819]. See, also, [173, 201, 865]. Smooth quasigroups and loops are studied in [654, 727], and topological algebraic systems, including topological quasigroups, in [205].

Good introductions to the theory of Latin squares and applications can be found in [240, 490, 559, 86].

In Chapter 2 "Some Quasigroup Classes" information about the most researched quasigroup classes is included. A big part of this information is well known but there are some new results, too.

Chapter 3 is based on information from joint articles of A.D. Keedwell and

V.A. Shcherbacov concerning properties of some new classes of inverse quasigroups [486, 487, 488, 489]. There exists a sufficient basis to expect that quasigroups with these inverse properties will be applicable in cryptology.

In Chapter 4 relatively new "autotopical" approach to the nuclei and center of a quasigroup is presented.

Chapter 5 contains solutions of two problems from V.D. Belousov's book [72].

In Chapter 6 information on properties (direct decompositions, simple objects, loop isotopes) of finite left and right F-, E-, and SM-quasigroups is presented. The solutions of "1a" Belousov problem [72] and two problems of Kinyon and Phillips [520] are obtained as corollaries.

Chapter 7 contains a theorem that is a generalization of the classical Murdoch Theorem [646] on the structure of binary finite medial quasigroups. Simple n-ary medial quasigroups are also described.

In Chapter 8 information on automorphisms of binary and n-ary (mainly linear) quasigroups is given.

Chapter 9 contains a quite new approach and results concerning orthogonality of binary quasigroups. See, also [643, 213]. Various applications of the property of orthogonality of quasigroups and groupoids are described in books by J. Denes and A.D. Keedwell [240, 241, 243].

In Chapter 10 an n-ary quasigroup approach to codes with one check symbol is developed. This part is based on joint articles of G.L. Mullen and V.A. Shcherbacov [641, 642]. Information on quasigroup-based recursive MDS codes is mainly taken from [371].

Results on applications of quasigroups in cryptology are presented in Chapter 11.

The Appendix contains short historical data on quasigroup theory and information about the 20 Belousov problems [72] (formulations, information about solutions and names of solvers).

All theorems, corollaries, lemmas, remarks and formulas are identified by two numbers, the first of which specifies the chapter number. In compiling of the list of references, we tried to use mainly readily available sources. To make reading of this book more comfortable, we sometimes repeat the definitions and basic facts. We have put a few open problems in both explicit and implicit forms.

Acknowledgments. First of all, the author expresses profound gratitude to Professor A.D. Keedwell, whom he considers one of his scientific teachers. He thanks Professors Alberto Marini, Aleš Drápal and J.D. Phillips for their help. The author's work was partially supported by the grants FRVS 2733/2003, MSM 113200007, CRDF-MRDA grant and the grant 08.820.08.08 RF.

The work on this book was continued during his visit at Central European University (September, 2010–February, 2011). The author thanks the head of Mathematical Department of CEU, Prof. Gheorghe Morosanu, for his hospitality and help.

He also thanks his colleagues from the Algebra and Topology Department of the Institute of Mathematics and Computer Science of the Academy of Sciences of Republic Moldova, especially Prof. Galina Belyavskaya and Tatiana Verlan, for their help and fruitful discussions. He also thanks the staff of Shevchenko Transnistria State University.

He also thanks his family and relatives.

Victor A. Shcherbacov
Chişinău

This book was approved for publication by the Scientific Council of the Institute of Mathematics and Computer Science of the Academy of Sciences of Moldova.

2010 Mathematics Subject Classification.
Primary: 20N05 (Loops, quasigroups);
20N15 (n-ary systems).
Secondary:
20N02 (Sets with a single binary operation (groupoids));
05B15 (Orthogonal arrays, Latin squares, Room squares)
94B05 (Linear codes, general);
94A60 (Cryptography);
94A62 (Authentication and secret sharing).

Reviewer: Vladimir Arnautov, Doctor Habilitatus in Physics and Mathematics, Academician of the Academy of Sciences of Republic Moldova.

List of Figures

1.1 Graph of quasigroup $(Q, \circ)$ and the respective Latin square. 102
1.2 Graph of quasigroup $(Q, \cdot)$ and of the respective Latin square. 103
1.3 Graph of middle translations of the group Z_4. 104
1.4 Graph of middle translations of Klein group K_4. 105

2.1 The Fano plane. 160

9.1 4-net of order four. 426

List of Tables

1.1 Translations of quasigroup parastrophes. 13
1.2 Local identity elements of parastrophic quasigroups. 22
1.3 Translations of quasigroup isostrophes. 49
1.4 Connections between components of A-nuclei by isostrophy. 61
1.5 Connections between components of A-pseudo-automorphisms by isostrophy. 62

4.1 Components of loop nuclei at isostrophy. 222

10.1 Error types and their frequencies [752]. 440

Part I

Foundations

Chapter 1

Elements of quasigroup theory

1.1	Introduction	5
1.1.1	The role of definitions	5
1.1.2	Sets	5
1.1.3	Products and partitions	5
1.1.4	Maps	6
1.2	Objects	8
1.2.1	Groupoids and quasigroups	8
1.2.2	Parastrophy: Quasigroup as an algebra	11
1.2.2.1	Parastrophy	11
1.2.2.2	Middle translations	12
1.2.2.3	Some groupoids	13
1.2.2.4	Substitutions in groupoid identities	15
1.2.2.5	Equational definitions	15
1.2.3	Some other definitions of e-quasigroups	18
1.2.4	Quasigroup-based cryptosystem	20
1.2.5	Identity elements	21
1.2.5.1	Local identity elements	21
1.2.5.2	Left and right identity elements	21
1.2.5.3	Loops	21
1.2.5.4	Identity elements of quasigroup parastrophes	22
1.2.5.5	The equivalence of loop definitions	22
1.2.5.6	Identity elements in some quasigroups	23
1.2.5.7	Inverse elements in loops	24
1.2.6	Multiplication groups of quasigroups	24
1.2.7	Transversals: "Come back way"	26
1.2.8	Generators of inner multiplication groups	27
1.3	Morphisms	29
1.3.1	Isotopism	29
1.3.2	Group action	32
1.3.3	Isotopism: Another point of view	33
1.3.4	Autotopisms of binary quasigroups	34
1.3.5	Automorphisms of quasigroups	38
1.3.6	Pseudo-automorphisms and G-loops	38
1.3.7	Parastrophisms as operators	42
1.3.8	Isostrophism	43
1.3.9	Autostrophisms	45
1.3.9.1	Coincidence of quasigroup parastrophes	45
1.3.10	Inverse loops to a fixed loop	46
1.3.11	Anti-autotopy	48
1.3.12	Translations of isotopic quasigroups	48
1.4	Sub-objects	51

1.4.1 Subquasigroups: Nuclei and center 51
1.4.1.1 Sub-objects 51
1.4.1.2 Nuclei 52
1.4.1.3 Center 52
1.4.2 Bol and Moufang nuclei 52
1.4.3 The coincidence of loop nuclei 54
1.4.3.1 Nuclei coincidence and identities 56
1.4.4 Quasigroup nuclei and center 57
1.4.4.1 Historical notes 57
1.4.4.2 Quasigroup nuclei 57
1.4.4.3 Quasigroup center 58
1.4.5 Regular permutations 58
1.4.6 A-nuclei of quasigroups 59
1.4.7 A-pseudo-automorphisms by isostrophy 60
1.4.8 Commutators and associators 62
1.5 Congruences 63
1.5.1 Congruences of quasigroups 63
1.5.1.1 Congruences in universal algebra 63
1.5.1.2 Normal congruences 64
1.5.2 Quasigroup homomorphisms 66
1.5.3 Normal subquasigroups 67
1.5.4 Normal subloops 68
1.5.5 Antihomomorphisms and endomorphisms 68
1.5.6 Homotopism 70
1.5.7 Congruences and isotopism 71
1.5.8 Congruence permutability 72
1.6 Constructions 72
1.6.1 Direct product 73
1.6.2 Semidirect product 74
1.6.3 Crossed (quasi-direct) product 75
1.6.4 n-Ary crossed product 76
1.6.5 Generalized crossed product 76
1.6.6 Generalized singular direct product 76
1.6.7 Sabinin's product 78
1.7 Quasigroups and combinatorics 79
1.7.1 Orthogonality 79
1.7.1.1 Orthogonality of binary operations 79
1.7.1.2 Orthogonality of n-ary operations 81
1.7.1.3 Easy way to construct n-ary orthogonal operations 82
1.7.2 Partial Latin squares: Latin trades 83
1.7.3 Critical sets of Latin squares, Sudoku 85
1.7.4 Transversals in Latin squares 86
1.7.5 Quasigroup prolongations: Combinatorial aspect 87
1.7.5.1 Bruck-Belousov prolongation 87
1.7.5.2 Belyavskaya prolongation 90
1.7.5.3 Algebraic approach 91
1.7.5.4 Prolongation using quasicomplete mappings 93
1.7.5.5 Two-step mixed procedure 95
1.7.5.6 Brualdi problem 96

1.7.5.7	Contractions of quasigroups	97
1.7.6	Orthomorphisms	97
1.7.7	Neo-fields and left neo-fields	98
1.7.8	Sign of translations	99
1.7.9	The number of quasigroups	100
1.7.10	Latin squares and graphs	101
1.7.11	Orthogonal arrays	104

1.1 Introduction

1.1.1 The role of definitions

In this section we adopt some remarks from the books [334, 422, 296].

In mathematics one should strive to avoid ambiguity. A very important ingredient of mathematical creativity is the ability to formulate useful definitions, ones that will lead to interesting results.

Every definition is understood to be an "if and only if"type of statement, even though it is customary to suppress the only if. Thus one may define: "A triangle is isosceles if it has two sides of equal length", really meaning that a triangle is isosceles if and only if it has two sides of equal length.

The basic importance of definitions to mathematics is also a structural weakness because not every concept used can be defined.

1.1.2 Sets

A set is well-defined collection of objects. We summarize briefly some of the things about sets we shall take for granted.

1. A set S is made up of elements, and if a is one of these elements, we shall denote this fact by $a \in S$. The order of elements in the set S is not taken into consideration, i.e., $S = \{a, b\} = \{b, a\}$.

2. There is exactly one set with no elements. It is the empty set and is denoted by $\varnothing$.

3. We may describe a set either by giving a characterizing property of the elements, such as "the set of all members of the United State Senate", or by listing the elements, for example, $\{1, 3, 4\}$ is a set.

4. A set is well defined, meaning that if S is a set and a is some object, then either a is definitely in S, denoted by $a \in S$, or a is definitely not in S, denoted by $a \notin S$. Thus one should never say "Consider the set S of some positive numbers", for it is not definite whether $2 \in S$ or $2 \notin S$.

1.1.3 Products and partitions

Definition 1.1. The Cartesian product of sets $S_1, S_2, \ldots, S_n$ is the set of all ordered n-tuples $(a_1, a_2, \ldots, a_n)$, where $a_i \in S_i$.

The Cartesian product is denoted by either $S_1 \times S_2 \times \cdots \times S_n$ or by

$$\prod_{i=1}^{n} S_i.$$

If $S_1 = S_2 = \cdots = S_n = S$, then we have $S \times S \times \cdots \times S = S^n = \{(a_1, a_2, \ldots, a_n) \mid a_i \in S\}$ (the n-th power of the set S).

A binary relation on a set Q is any subset of the set $Q \times Q$, a ternary relation is any subset of the set $Q \times Q \times Q$, and an n-ary relation is any subset of the set Q^n.

If φ and ψ are binary relations on Q, then their product is defined in the following way: $(a, b) \in \varphi \circ \psi$ if there is an element $c \in Q$ such that $(a, c) \in \varphi$ and $(c, b) \in \psi$. If φ is a binary relation on Q, then $\varphi^{-1} = \{(y, x) \mid (x, y) \in \varphi\}$. The operation of the product of binary relations is associative [211, 678, 813, 711].

Definition 1.2. A partition of a set is a decomposition of the set into cells such that every element of the set is in exactly one of the cells. Two cells (or sets) having no elements in common are disjoint.

Let $a \sim b$ denote that a is in the same cell as b for a given partition of a set containing both a and b. Clearly the following properties are always satisfied: $a \sim a$; if $a \sim b$, then $b \sim a$; if $a \sim b$ and $b \sim c$, then $a \sim c$, i.e., if a is in the same cell as b and b is in the same cell as c, then a is in the same cell as c.

Theorem 1.3. *Let S be a nonempty set and let $\sim$ be a relation between elements of S that satisfies the following properties:*

1. *(Reflexive) $a \sim a$ for all $a \in S$.*
2. *(Symmetric) If $a \sim b$, then $b \sim a$.*
3. *(Transitive) If $a \sim b$ and $b \sim c$, then $a \sim c$.*

Then $\sim$ yields a natural partition of S, where $\bar{a} = \{x \in S \mid x \sim a\}$ is the cell containing a for all $a \in S$. Conversely, each partition of S gives rise to a natural relation $\sim$ satisfying the reflexive, symmetric, and transitive properties if $a \sim b$ is defined to mean that $a \in \bar{b}$.

Definition 1.4. A relation $\sim$ on a set S satisfying the reflexive, symmetric, and transitive properties is an equivalence relation on S. Each cell $\bar{a}$ in the natural partition given by an equivalence relation is an equivalence class.

1.1.4 Maps

One of the truly universal concepts that arises in almost every part of mathematics is that of a function or a mapping from one set to another. One can safely say that there is no part of mathematics where the notion arises or plays a central role.

The definition of a function from one set to another can be given in a formal way in terms of a subset of the Cartesian product of these sets (i.e., in terms of binary relations). Instead, here, we shall give an informal and admittedly nonrigorous definition of a mapping (function) from one set to another.

Definition 1.5. Let S, T be sets; a function or mapping f from S to T is a rule that assigns a unique element $t \in T$ to each element $s \in S$.

Definition 1.6. The mapping $f : S \to T$ is onto or surjective if every $t \in T$ is the image of some $s \in S$ under f; that is, if and only if, given $t \in T$, there exists at least one $s \in S$ such that $t = f(s)$.

Definition 1.7. A mapping $f : S \to T$ is said to be one-to-one (written 1-1) or injective if for $s_1 \neq s_2$ in S, $f(s_1) \neq f(s_2)$ in T. Equivalently, f is 1-1 if $f(s_1) = f(s_2)$ implies $s_1 = s_2$.

Definition 1.8. A mapping $f : S \to T$ is said to be a 1-1 correspondence or bijection if f is both 1-1 and onto (i.e., f is both injective and surjective).

Remark 1.9. Mainly we shall use the following order of multiplication (of composition) of maps: $(\alpha\beta)(x) = \alpha(\beta(x))$, where α, β are maps. Cases when other orders of multiplication of maps are used, are also specified.

Lemma 1.10. *Let $f : A \to B$ be a map of non-empty sets. Then*

(i) the map f is injective if and only if there exists a map $g : B \to A$ such that $gf = 1_A$;
(ii) the map f is surjective if and only if there exists a map $g : B \to A$ such that $fg = 1_B$;
(iii) the map f is bijective if and only if there exists a map $g : B \to A$ such that $gf = 1_A$ and $fg = 1_B$ [312, 1. Proposition].

Here 1_A denotes the identity map of the set A. It is clear that any identity map is bijective.

Definition 1.11. Let $f : A \to B$ be a map. A left inverse map to the map f is a map $g : B \to A$ such that $gf = 1_A$. A right inverse map to the map f is a map $h : B \to A$ such that $fh = 1_B$. A left and right inverse map is called an inverse map.

It is known that if a map has an inverse, it is unique. In general this is not true for left or right inverse maps. See Example 1.13.

Lemma 1.12. *1. Let $\varphi : A \to B$ be a map of non-empty sets. Then φ is injective if and only if φ has a left inverse map ${}^{-1}\varphi$.*

2. Let $\psi : A \to B$ be a map of non-empty sets. Then ψ is surjective if and only if ψ has a right inverse map ψ^{-1};

3. Let $f : A \to B$ be a map of non-empty sets. Then f is bijective if and only if f has an inverse map f^{-1} [312, 2. Proposition].

Example 1.13. Let $A = \{1, 2\}$, $B = \{a, b, c\}$, $\varphi = \{(1, a), (2, b)\}$. It is clear that the mapping φ is a 1-1 mapping of A into B. The left inverse mappings to the injection φ are, for example, the following surjections ($B \to A$): ${}^{-1}\varphi_1 = \{(a, 1), (b, 2), (c, 1)\}$; ${}^{-1}\varphi_2 = \{(a, 1), (b, 2), (c, 2)\}$.

Indeed, ${}^{-1}\varphi_1\varphi(1) = {}^{-1}\varphi_1(\varphi(1)) = {}^{-1}\varphi_1(a) = 1$, ${}^{-1}\varphi_1\varphi(2) = {}^{-1}\varphi_1(b) = 2$; ${}^{-1}\varphi_2\varphi(1) = {}^{-1}\varphi_1(a) = 1$, ${}^{-1}\varphi_2\varphi(2) = {}^{-1}\varphi_2(b) = 2$. Therefore ${}^{-1}\varphi_1\varphi = {}^{-1}\varphi_2\varphi = 1_A$.

Let $A = \{5, 7, 9\}$, $B = \{t, q\}$, $\psi = \{(5, t), (7, q), (9, t)\}$. The mapping ψ is a surjective mapping of A onto B. It is easy to check that the mapping $\psi^{-1} = \{(t, 5), (q, 7)\}$ is an injective mapping. Indeed, $\psi(\psi^{-1}(t)) = \psi(5) = t$, $\psi(\psi^{-1}(q)) = \psi(7) = q$. Therefore $\psi\psi^{-1} = 1_B$.

Lemma 1.14. *Let μ and ν be some maps of a non-empty set Q. If the product $\mu\nu$ of the maps is a bijective map of the set Q, then the map ν is injective and the map μ is surjective.*

Proof. If $\mu\nu = \alpha$, where α is a bijection of the set Q, then $(\alpha^{-1}\mu)\nu = 1_Q$. Then by Lemma 1.10, the map $\alpha^{-1}\mu$ is surjective and the map ν is injective. Therefore, the map μ is surjective, since the map α^{-1} is a bijective map. □

Recall, any mapping of a non-empty set Q into itself can be considered as a binary relation defined on the set Q. In this case, multiplication of mappings is a special case of the operation of multiplication of binary relations.

Lemma 1.15. *If Q is a finite set, then:*

1. any injective ($x \neq y \Rightarrow \varphi x \neq \varphi y$) map φ on this set ($\varphi(Q) \subseteq Q$) is a bijective map;

2. any surjective map ψ of this set into itself ($\psi(Q) = Q$) is a bijective map [538, Theorem 3, p. 45].

Proof. Case 1. First of all we re-write implication ($x \neq y \Rightarrow \varphi x \neq \varphi y$) in the following equivalent form:

$$\varphi x = \varphi y \Rightarrow x = y. \tag{1.1}$$

We should prove that the map φ is surjective. In other words we should prove that for any $x \in Q$ there exists an element $x' \in Q$ such that $\varphi(x') = x$.

Let $\varphi^k(x) = \varphi(\varphi^{k-1})(x)$, $k = 0, 1, \dots,$ be a sequence of elements of the set Q. Since the set Q is finite, then there exist integers m, n, $n < m$, such that $\varphi^m(x) = \varphi^n(x)$. Applying n times the relation (1.1) to the last equality, we obtain the following equality $\varphi^{m-n}(x) = x$, $\varphi(\varphi^{m-n-1}(x)) = x$. Denote the element $\varphi^{m-n-1}(x)$ by the symbol x' and finally obtain the following equality $\varphi(x') = x$.

Case 2. If the map ψ is surjective, then by Lemma 1.12 there exists a right inverse map ψ^{-1} such that $\psi\psi^{-1} = 1_Q$, i.e., $\psi\psi^{-1} = \varepsilon$, where ε denotes identity permutation of the set Q. The fact that the map ψ^{-1} is injective follows from Lemma 1.14. By Case 1 of this Lemma the mapping ψ^{-1} is bijective. Therefore the mapping ψ is bijective, too. □

We shall consider the set $A(Q)$ of all bijections of the set Q onto itself. The set $A(Q)$ forms a group relative to "usual" multiplication of these mappings. This group is called the symmetric group of degree n and it will be denoted by S_n, if the set Q has finite number n of elements. Elements of the group S_n are called permutations of the set Q. In the investigation of finite groups and quasigroups, the group S_n and its subgroups play a central role.

An n-ary operation defined on a non-empty set Q is a map $A : Q^n \longrightarrow Q$ such that $D(A) = Q^n$, i.e., this map is defined for any n-tuple. The number n is called the "arity" of operation A. If the element c corresponds to the n-tuple $(b_1, b_2, \dots, b_n)$, then we shall write this fact in the following form $A(b_1, b_2, \dots, b_n) = c$, or in the form $A : (b_1, b_2, \dots, b_n) \longmapsto c$.

If $n = 2$, then the operation A is called a binary operation, if $n = 1$, then the operation A is called a unary operation. If $n = 0$, then the operation A is called a *nullary operation*. A nullary operation is a fixation of an element from the basic set Q [184].

Exercise 1.16. On a finite set of order n there exist $n^{(n^k)}$ k-ary operations. For example, on the set of order ten, there exist 10^{1000} ternary operations.

1.2 Objects

1.2.1 Groupoids and quasigroups

A binary groupoid (G, A) is a non-empty set G together with a binary operation A.

Many different symbols can be used to denote binary operations, for example, $\circ, \star, \cdot$. Thus, we can write $x \circ y$ instead of $A(x, y)$, and $x \star y$ instead of $B(x, y)$.

An n-ary groupoid (G, A) is a non-empty set G together with an n-ary operation A.

There exists a bijection (1-1 correspondence) between the set of all binary (n-ary, for fixed n) operations defined on a set Q and the set of all groupoids, defined on the set Q. Indeed, $A \longleftrightarrow (Q, A)$.

A sequence $x_m, x_{m+1}, \dots, x_n$, where m, n are natural numbers and $m \leqslant n$, will be denoted by x_m^n, and a sequence $x, \dots, x$ (k times) will be denoted by $\overline{x}^k$. The expression $\overline{1,n}$ designates the set $\{1, 2, \dots, n\}$ of natural numbers [77].

We shall say that the operations A and B, which are defined on a set Q, coincide, if $A(a_1^n) = B(a_1^n)$ for all $a_i \in Q$, $i \in \overline{1,n}$.

The order of an n-ary groupoid (Q, A) is cardinality $|Q|$ (sometimes denotation $\bar{\bar{Q}}$ is used) of the carrier set Q. An n-ary groupoid $(Q, \cdot)$ is said to be finite if its order is finite.

It is possible to represent a finite n-ary groupoid (Q, A) (theoretically of any, but practically, only not a very big size) as a set of $(n+1)$-tuples $(a_1, a_2, \dots, a_n, A(a_1^n))$.

In the binary case it is possible to define a groupoid as a set of triplets. For example, the set

$$(Q, \cdot) = \{(a, a, a), (a, b, a), (a, c, b), (b, a, b), (b, b, c), (b, c, a), (c, a, c), (c, b, a), (c, c, b)\}$$

defines a groupoid with the following multiplication table:

$\cdot$	a	b	c
a	a	a	b
b	b	c	a
c	c	a	b

where, for example, the triple (a, c, b) defines the element $b = a \cdot c$, which appears at the intersection of the row headed by a and the column headed by c. This table is called the *Cayley table* of the groupoid $(Q, \cdot)$, where $Q = \{a, b, c\}$.

Note 1.17. Usually it is supposed that elements of carried set Q are arranged. So the groupoid $(Q, \circ)$, which is defined with the help of the following Cayley table,

$\circ$	b	c	a
b	c	a	b
c	a	b	c
a	a	b	a

is equal (as a set of triplets) to the groupoid $(Q, \cdot)$, but $(Q, \cdot) \neq (Q, *)$, where the groupoid $(Q, *)$ has the following Cayley table:

$*$	b	c	a
b	a	a	b
c	b	c	a
a	c	a	b

A groupoid $(Q, \circ)$ is called a *right quasigroup* if, for all $a, b \in Q$, there exists a unique solution $x \in Q$ to the equation $x \circ a = b$.

Example 1.18. Examples of right quasigroups. All columns are permutations of the set {0, 1, 2}.

$\circ$	0	1	2
0	0	0	2
1	1	1	1
2	2	2	0

$*$	0	1	2
0	0	1	2
1	1	0	1
2	2	2	0

A groupoid $(Q, \circ)$ is called a *left quasigroup* if, for all $a, b \in Q$, there exists a unique solution $y \in Q$ to the equation $a \circ y = b$.

Example 1.19. Examples of left quasigroups. All rows are permutations of the set $\{0, 1, 2\}$.

$\circ$	0	1	2
0	1	0	2
1	0	1	2
2	2	1	0

$*$	0	1	2
0	0	1	2
1	2	0	1
2	0	1	2

A left and right quasigroup is a *quasigroup.*

Example 1.20. Example of quasigroup.

$*$	0	1	2
0	1	0	2
1	0	2	1
2	2	1	0

We give a main definition of a quasigroup:

Definition 1.21. A binary groupoid $(Q, \circ)$ is called a quasigroup if for any ordered pair $(a, b) \in Q^2$ there exist unique solutions $x, y \in Q$ to the equations $x \circ a = b$ and $a \circ y = b$ [72].

Often this definition is called *existential.* Some equational (using identities) quasigroup definitions are given below (Definition 1.54).

From Definition 1.21 it follows, that any two elements from the triple $(a, b, a \circ b)$ specify the third element in a unique way. Indeed, for any elements a, b there exists a unique element $a \circ b$. This follows from the definition of operation $\circ$.

Elements $a, a \circ b$ determine the third element in a unique way since there exists a unique solution to the equation $a \circ y = b$.

Elements $b, a \circ b$ determine the third element in unique way since there exists a unique solution to the equation $x \circ a = b$.

Therefore, it is convenient to define an n-ary quasigroup in the following manner.

Definition 1.22. An n-ary groupoid (Q, A) with n-ary operation A such that in the equality $A(x_1, x_2, \dots, x_n) = x_{n+1}$ the fact of knowing any n elements of the set $\{x_1, x_2, \dots, x_n, x_{n+1}\}$ uniquely specifies the remaining one element is called an n-ary quasigroup [77].

If we put $n = 2$, then we obtain one more definition of a binary quasigroup.

Example 1.23. Let $Q = \{0, 1, 2, 3, 4\}$. Using the operations of addition $(+)$ and multiplication $(\cdot)$ modulo 5, we define a 4-ary quasigroup (Q, f) in the following way: $f(x_1^4) = 2 \cdot x_1 + x_2 + 4 \cdot x_3 + 3 \cdot x_4$.

Exercise 1.24. Construct a Cayley table of the quasigroup defined in Example 1.23 or list all 5-tuples that define this 4-ary quasigroup.

Definition 1.25. Let $(G, \cdot)$ be a groupoid and let a be a fixed element in G. Translation maps L_a (left) and R_a (right) are defined by $L_a x = a \cdot x$, $R_a x = x \cdot a$ for all $x \in G$.

Translations play a prominent role in quasigroup theory.

Example 1.26. Example of a quasigroup and its left and right translations.

$\circ$	a	b	c
a	a	c	b
b	b	a	c
c	c	b	a

For this quasigroup we have the following left and right translations:

$$L_a^\circ = \begin{pmatrix} a & b & c \\ a & c & b \end{pmatrix}; \quad L_b^\circ = \begin{pmatrix} a & b & c \\ b & a & c \end{pmatrix}; \quad L_c^\circ = \begin{pmatrix} a & b & c \\ c & b & a \end{pmatrix};$$
$$R_a^\circ = \begin{pmatrix} a & b & c \\ a & b & c \end{pmatrix}; \quad R_b^\circ = \begin{pmatrix} a & b & c \\ c & a & b \end{pmatrix}; \quad R_c^\circ = \begin{pmatrix} a & b & c \\ b & c & a \end{pmatrix}.$$

We can express these permutations in the cycle form: $L_a^\circ = (b\,c)$; $L_b^\circ = (a\,b)$; $L_c^\circ = (a\,c)$; $R_a^\circ = \varepsilon$; $R_b^\circ = (a\,c\,b)$; $R_c^\circ = (a\,b\,c)$.

Remark 1.27. Usually the symbol of binary operation in the record of translation is omitted, i.e., it is written L_a instead of L_a°.

Notice if, for example,

$$L_a = \begin{pmatrix} a & b & c \\ c & a & b \end{pmatrix}, \text{then } L_a^{-1} = \begin{pmatrix} c & a & b \\ a & b & c \end{pmatrix} = \begin{pmatrix} a & b & c \\ b & c & a \end{pmatrix},$$

$L_a L_a^{-1} = L_a^{-1} L_a = \varepsilon$. We shall denote identity permutation by ε.

It is easy to see that in the Cayley table of a quasigroup $(Q, \cdot)$ each row and each column is a permutation of the set Q. So we may give the following definition of a quasigroup.

Definition 1.28. A groupoid $(Q, \cdot)$ is called a quasigroup if the maps $L_a : G \longrightarrow Q$, $R_a : Q \longrightarrow Q$ are bijections for all $a \in Q$ [685].

Lemma 1.29. *1. The statements "the equation $x \circ a = b$ has a unique solution for all $a, b \in Q$" and "R_a is bijection of the set Q for any $a \in Q$" are equivalent.*

2. Similarly, the conditions "the equation $a \circ y = b$ has a unique solution for all $a, b \in Q$" and "L_a is bijection of the set Q for any $a \in Q$" are equivalent.

Proof. Case 1."$\Longrightarrow$." Fix the element a and write equality $x \circ a = b$ in the following form: $R_a^\circ x = b$. Variable x takes all values in the set Q. Therefore, we have a mapping of the set Q into itself.

From the fact that equation $x \circ a = b$ has a unique solution, it follows that the mapping R_a° is injective ($(x \neq y) \rightarrow (R_a^\circ x \neq R_a^\circ y)$).

Since for any element $b \in Q$, there exists an element $x \in Q$ such that $R_a^\circ x = b$, we have that this mapping is surjective, and therefore, is a bijection of the set Q.

"$\Longleftarrow$." If translation R_a° is a bijection, then $x = (R_a^\circ)^{-1} b$.

Case 2 is proved similarly. □

Definition 1.30. An unbordered Cayley table of a finite quasigroup is called a Latin square.

Quasigroup triples have interesting combinatorial properties [571, 721].

1.2.2 Parastrophy: Quasigroup as an algebra

1.2.2.1 Parastrophy

Definition 1.31. From Definition 1.22 it follows that with a given binary quasigroup (Q, A) it is possible to associate $(3! - 1)$ others, so-called parastrophes of quasigroup (Q, A):

1. $A(x_1, x_2) = x_3 \Longleftrightarrow$
2. $A^{(12)}(x_2, x_1) = x_3 \Longleftrightarrow$
3. $A^{(13)}(x_3, x_2) = x_1 \Longleftrightarrow$
4. $A^{(23)}(x_1, x_3) = x_2 \Longleftrightarrow$
5. $A^{(123)}(x_2, x_3) = x_1 \Longleftrightarrow$
6. $A^{(132)}(x_3, x_1) = x_2$

[845, p. 230], [72, p. 18].

Notice, Cases 5 and 6 are "(12)-parastrophes" of Cases 3 and 4, respectively. Therefore

$$A^{\sigma}(x_i, x_j) = x_k \Leftrightarrow A(x_{\sigma^{-1}i}, x_{\sigma^{-1}j}) = x_{\sigma^{-1}k}, \tag{1.2}$$

where $\sigma \in S_3$, $i, j, k \in \{1, 2, 3\}$, the numbers i, j, k are different in pairs, and $x_i, x_j, x_k \in Q$.

In the language of triplets, we have $A^{\sigma}(x_1, x_2, x_3) = A(x_{\sigma^{-1}1}, x_{\sigma^{-1}2}, x_{\sigma^{-1}3})$, i.e., in order to have the set of triplets of quasigroup A^{σ} we permute elements of any triplet by the "σ^{-1}-rule" in the set of triplets of operation A.

Then $A^{\sigma}(x_{\sigma 1}, x_{\sigma 2}) = x_{\sigma 3} \Leftrightarrow A(x_{\sigma^{-1}\sigma 1}, x_{\sigma^{-1}\sigma 2}) = x_{\sigma^{-1}\sigma 3} \Leftrightarrow A(x_1, x_2) = x_3$. For example, $A^{(132)}(x_1, x_2) = x_3 \Leftrightarrow A(x_{(123)1}, x_{(123)2}) = x_{(123)3}$. That is, $A^{(132)}(x_1, x_2) = x_3 \Leftrightarrow A(x_2, x_3) = x_1$.

Usually the operation $A^{(12)}$ is denoted as "$*$", the operation $A^{(13)}$ is denoted as "/", and the operation $A^{(23)}$ is denoted as "\". Thus, if $x \cdot y = z$, then $y * x = z$, $x \backslash z = y$ and $z / y = x$. Sometimes [72, 80] it denoted ${}^{(13)}A$ instead of $A^{(13)}$, and so on. In [786] operation $A^{(123)}$ is denoted by //, operation $A^{(133)}$ is denoted by \\. Other denotations of these operations are used in [310].

The following convenient designation of quasigroup parastrophes (Belousov's designation [700]) is also used [80, 85]:

$$\begin{aligned}
A^{(12)} &= A^{*} = {}^{s}A, \\
A^{(13)} &= {}^{-1}A = {}^{l}A, \\
A^{(23)} &= A^{-1} = {}^{r}A, \\
A^{(123)} &= {}^{-1}(A^{-1}) = {}^{l}({}^{r}A) = {}^{\alpha}A = {}^{s}({}^{l}A), \\
A^{(132)} &= ({}^{-1}A)^{-1} = {}^{r}({}^{l}A) = {}^{\beta}A = {}^{s}({}^{r}A).
\end{aligned}$$

Note 1.32. The concept of parastrophy, especially (12)-parastrophy, can be extended to groupoids.

1.2.2.2 Middle translations

We defined left and right translations of a groupoid and, therefore, of a quasigroup. But for quasigroups it is possible to define a third kind of translation, namely, middle translations. If P_a is a middle translation of a quasigroup $(Q, \cdot)$, then $x \cdot P_a x = a$ for all $x \in Q$ [75].

Suppose that quasigroups (Q, A^{σ}) and (Q, A^{δ}) are parastrophes of a quasigroup (Q, A). Any translation of a quasigroup (Q, A^{σ}) can be expressed as a translation of a quasigroup (Q, A^{δ}) or as its inverse translation. For example, $R^{*}_a x = x * a = a \cdot x = L_a x$. Then $R^{*}_a = L_a$, $R^{*} = L$, $R^{(12)} = L$. Or

$$P^{\backslash}_a x = y \Leftrightarrow x \backslash y = a \Leftrightarrow x \cdot a = y \Leftrightarrow R_a x = y.$$

Then $P^{\backslash}_a = R_a$, $P^{\backslash} = R$, $P^{(23)} = R$.

The following table shows, for each kind of translation, the equivalent one in each of the (six) parastrophes of a quasigroup $(Q, \cdot)$. In fact, Table 1.1 is a rewritten form of results on three kinds of translations from [75]. See also [298, 786].

From Table 1.1 it follows, for example, that $R^{(132)} = L^{-1} = L^{(23)} = P^{(13)} = (R^{-1})^{(12)} = (P^{-1})^{(123)}$.

Using Table 1.1 it is possible to construct parastrophes of a quasigroup $(Q, \cdot)$. In order to construct (12)-parastrophe of a quasigroup $(Q, \cdot)$ we can write any row of the Cayley table of quasigroup $(Q, \cdot)$ as a corresponding column.

Table 1.1: Translations of quasigroup parastrophes.

	ε	(12)	(13)	(23)	(123)	(132)
R	R	L	R^{-1}	P	P^{-1}	L^{-1}
L	L	R	P^{-1}	L^{-1}	R^{-1}	P
P	P	P^{-1}	L^{-1}	R	L	R^{-1}
R^{-1}	R^{-1}	L^{-1}	R	P^{-1}	P	L
L^{-1}	L^{-1}	R^{-1}	P	L	R	P^{-1}
P^{-1}	P^{-1}	P	L	R^{-1}	L^{-1}	R

Example 1.33.

$$\begin{array}{c|ccc} \circ & a & b & c \\ \hline a & a & c & b \\ b & b & a & c \\ c & c & b & a \end{array} \quad \Longrightarrow \quad \begin{array}{c|ccc} \overset{(12)}{\circ} & a & b & c \\ \hline a & a & b & c \\ b & c & a & b \\ c & b & c & a \end{array}$$

In order to construct the (23)-parastrophe of a quasigroup $(Q, \cdot)$ we can replace any row of the Cayley table of quasigroup $(Q, \cdot)$ with the corresponding inverse row.

Example 1.34.

$$\begin{array}{c|ccc} \circ & a & b & c \\ \hline a & a & b & c \\ b & b & c & a \\ c & c & a & b \end{array} \quad \Longrightarrow \quad \begin{array}{c|ccc} \overset{(23)}{\circ} & a & b & c \\ \hline a & a & b & c \\ b & c & a & b \\ c & b & c & a \end{array}$$

In order to construct the (13)-parastrophe of a quasigroup $(Q, \cdot)$ we can replace any column of the Cayley table of quasigroup $(Q, \cdot)$ with the corresponding inverse column.

In order to construct the (123)-parastrophe of a quasigroup $(Q, \cdot)$ we can replace any row of the Cayley table of quasigroup $(Q, \cdot)$ with the corresponding inverse column.

In order to construct the (132)-parastrophe of a quasigroup $(Q, \cdot)$ we can replace any column of the Cayley table of quasigroup $(Q, \cdot)$ with the corresponding inverse row.

Exercise 1.35. Find all parastrophes of the quasigroups given in Examples 1.33 and 1.34.

1.2.2.3 Some groupoids

The class of binary quasigroups is close to the classes of binary left (right) cancellation (division) groupoids. See, for example, Lemma 1.47 and Theorem 1.58. It is known that a homomorphic image of an existential quasigroup may be only division groupoid and not a quasigroup [43].

Definition 1.36. A groupoid $(G, \cdot)$ is called a left cancellation groupoid, if the following implication is fulfilled: $a \cdot x = a \cdot y \Rightarrow x = y$ for all $a, x, y \in G$, i.e., the translation L_a is an injective map for any $a \in G$.

Definition 1.37. A groupoid $(G, \cdot)$ is called a right cancellation groupoid, if the following implication is fulfilled: $x \cdot a = y \cdot a \Rightarrow x = y$ for all $a, x, y \in G$, i.e., the translation R_a is an injective map for any $a \in G$ [462].

Definition 1.38. A groupoid $(G, \cdot)$ is called a cancellation groupoid if it is both a left and right cancellation groupoid.

Remark 1.39. In this text the terms *"cancellation groupoid"* and *"cancellative groupoid"* are synonymous.

Example 1.40. Let $x \circ y = 2x + 4y$ for all $x, y \in \mathbb{Z}$, where $(\mathbb{Z}, +, \cdot)$ is the ring of integers. The reader can check that $(\mathbb{Z}, \circ)$ is a cancellation groupoid.

Example 1.41. Let $x \circ y = 2x + 3y$ for all $x, y \in \mathbb{Z}$, where $(\mathbb{Z}, +, \cdot)$ is the ring of integers. The reader can check that $(\mathbb{Z}, \circ)$ is a cancellation groupoid.

Definition 1.42. A groupoid $(G, \cdot)$ is said to be a left (right, resp.) division groupoid if L_a (R_a, resp.) is surjective for every $a \in G$; it is said to be a division groupoid if it is both a left and right division groupoid [462].

We can give an equivalent definition of a division groupoid.

Definition 1.43. A groupoid $(G, \cdot)$ is called a division groupoid, if the equations $a \cdot x = b$ and $y \cdot a = b$ have solutions (not necessarily unique solutions) for any ordered pair of elements $a, b \in Q$.

Example 1.44. Let $x \circ y = 2 \cdot x + [y/3]$ for all $x, y \in \mathbb{Z}$, where $(\mathbb{Z}, +, \cdot)$ is the ring of integers, $[y/3] = n$, if $y = 3n$, or $y = 3n + 1$, else $y = 3n + 2$, where $n \in \mathbb{Z}$. It is possible to check that $(\mathbb{Z}, \circ)$ is a right cancellation and left division groupoid.

Example 1.45. Let $x \circ y = 1 \cdot x + [y/3]$ for all $x, y \in \mathbb{Z}$, where $(\mathbb{Z}, +, \cdot)$ is the ring of integers. It is possible to check that $(\mathbb{Z}, \circ)$ is a right quasigroup and a left division groupoid.

Example 1.46. Let $x \circ y = x^2 + y^3$ for all $x, y \in \mathbb{C}$, where $(\mathbb{C}, +, \cdot)$ is the field of complex numbers. The reader can check that $(\mathbb{C}, \circ)$ is a division groupoid.

Lemma 1.47. *A division cancellation groupoid $(Q, \cdot)$ is a quasigroup.*

Proof. The proof follows from definitions and Lemma 1.29. □

Theorem 1.48. *A finite cancellation groupoid $(G, \cdot)$ is a quasigroup; a finite division groupoid $(G, \cdot)$ is a quasigroup.*

Proof. By Lemma 1.15, if G is a finite set, then any injective ($x \neq y \Rightarrow \varphi x \neq \varphi y$) map φ on this set ($\varphi(G) \subseteq G$) is a bijective map and any surjective map ψ of this set into itself ($\psi(G) = G$) is a bijective map, too.

Since any left and right translation of a cancellation groupoid $(G, \cdot)$ is an injective map, then in the case when the set G is finite, we have that $(G, \cdot)$ is a quasigroup.

Similarly, since any left and right translation of a division groupoid $(G, \cdot)$ is a surjective map, then in the case when the set G is finite, we have that any division groupoid is a quasigroup. □

Theorem 1.49. *A finite left (right) cancellation groupoid is a left (right) quasigroup and a finite left (right) division groupoid is a left (right) quasigroup.*

Proof. The proof is similar to the proof of Theorem 1.48 and we omit it. □

1.2.2.4 Substitutions in groupoid identities

Here we give non-formal, but, we hope, relatively clear information from universal algebra about substitutions in identities. See [184] for more details. Recall that any algebraic operation is a function defined on a non-empty set Q, therefore any algebraic operation is a mapping.

Suppose that $(Q, \circ)$ is a binary groupoid. Define the set $Q \circ Q$ in the following way: $Q \circ Q = \{x \circ y \mid \text{for all}\, x, y \in Q\}$.

Definition 1.50. Let $(Q, \circ)$ be a binary groupoid. The following property

$$Q \circ Q = Q \tag{1.3}$$

of the operation $\circ$ is called *surjective.*

From Definition 1.42 it follows that the operation of left (right) division groupoid $(Q, \cdot)$ satisfies the property (1.3).

Example 1.51. The left cancellation groupoid from Example 1.40 satisfies the property $\mathbb{Z} \circ \mathbb{Z} \subsetneq \mathbb{Z}$ and it is not surjective. The cancellation groupoid from Example 1.41 satisfies the surjective property (1.3) since $g.c.d.(2; 3) = 1$ and from standard information on linear Diophantine equations [919], it follows that the equation $2x + 3y = t$ has solution for any $t \in \mathbb{Z}$.

An algebra (Q, F) will be called surjective, if any n-ary ($n \geqslant 1$) operation $f \in F$ has the surjective property, i.e., $f(Q, Q, \dots, Q) = Q$.

In surjective binary groupoid $(Q, \cdot)$ any element $c \in Q$ can be represented as a product of two elements, say, a and b, and any variable x can be presented as a product of variables, say, y and z. Surjective medial groupoids are defined and studied in [462].

Notice, "inverse procedure" is admissible in any identity of any algebra. We have in mind that the replacement (the substitution) of, say, variable x, by suitable term T [917] in an identity of an algebra in all occurrences of an individual variable, is possible in the majority of "usual" cases.

It is clear that operation $\circ$ of quasigroup $(Q, \circ)$ has a surjective property. Similarly, any operation of quasigroup $(Q, \cdot, /, \backslash)$ (see below) has a surjective property.

1.2.2.5 Equational definitions

We recall, an algebra (or algebraic structure) is a set A together with a collection of operations defined on A [184, 212].

Bates and Kiokemeister [43] discovered that a class of quasigroups which is defined using equations (Definition 1.21) is not closed relative to homomorphic images. In the general case the quasigroup homomorphic image is only a division groupoid (see Lemma 1.280).

The problem of finding definitions of a quasigroup class such that this class is closed relative to homomorphic images was solved by Garret Birkhoff. He has defined a quasigroup using three binary operations and six identities [149, 151]. Quasigroups defined in this way often are called *equational quasigroups.* From standard universal algebraic facts (many of which also were discovered by Garret Birkhoff), it follows that class of equational quasigroups is a variety and it is closed relative to homomorphic images [151, 184].

Garrett Birkhoff [149, 151] has defined an equational quasigroup as an algebra with three binary operations $(Q, \cdot, /, \backslash)$ that satisfies the following six identities:

$$x \cdot (x \backslash y) = y, \tag{1.4}$$

$$(y/x) \cdot x = y, \tag{1.5}$$

$$x\backslash(x \cdot y) = y, \tag{1.6}$$

$$(y \cdot x)/x = y, \tag{1.7}$$

$$x/(y\backslash x) = y, \tag{1.8}$$

$$(x/y)\backslash x = y. \tag{1.9}$$

Remark 1.52. In [819] the identities (1.4)–(1.7) are called respectively (SL), (SR), (IL), (IR), since these identities guarantee that the left (L) and right (R) translations of an algebra $(Q, \cdot, /, \backslash)$ relative to the operation "$\cdot$" are surjective (S) or injective (I) mappings of the set Q.

Following this logic we can denote identity (1.8) by (SP) and identity (1.9) by (IP) since these identities guarantee that middle translations (P) are respectively surjective and injective mappings relative to the operation "$\cdot$" [786]. Indeed using Table 1.1 we see that $L^{/}_x = P^{-1}_x$.

The following lemma is well known:

Lemma 1.53. *In algebra $(Q, \cdot, \backslash, /)$ with the identities (1.4)–(1.7), the identities (1.8) and (1.9) are true [819, 779, 786].*

Proof. From the identity (1.7) $(y \to x, x \to x\backslash y)$ we have

$$(x \cdot (x\backslash y))/(x\backslash y) = x. \tag{1.10}$$

But by the identity (1.4), $x \cdot (x\backslash y) = y$. Thus from the identity (1.10) we obtain $y/(x\backslash y) = x$, i.e., we obtain (up to renaming of variables) the identity (1.8).

We can re-write identity (1.6) in the following form:

$$(x/y)\backslash((x/y) \cdot y) = y. \tag{1.11}$$

By the identity (1.5) $(x/y) \cdot y = x$. Thus from equality (1.11) we have $(x/y)\backslash x = y$, i.e., we obtain the identity (1.9). □

Therefore the following Evans equational definition of a quasigroup is usually used [307].

Definition 1.54. [150, 151, 307]. A groupoid $(Q, \cdot)$ is called a quasigroup if, on the set Q, there exist operations "$\backslash$" and "$/$" such that in the algebra $(Q, \cdot, \backslash, /)$ identities (1.4)–(1.7) are fulfilled.

Algebra $(Q, \cdot, \backslash, /)$ with identities (1.4)–(1.7) is called *e-quasigroup* (probably from the words equational quasigroup) [151, 80]. Below we shall also name an algebra $(Q, \cdot, \backslash, /)$ with identities (1.4)–(1.7) as a quasigroup $(Q, \cdot, \backslash, /)$.

A.I. Maltsev [590, 591] has called this algebra a *primitive quasigroup*. Definitions of quasigroup with the help of quasi-identities can be found in [590, 786].

Note 1.55. The identities (1.4) and (1.5) guarantee the existence of solutions of equations $a \cdot y = b$ and $x \cdot a = b$, while the identities (1.6) and (1.7) guarantee the uniqueness of solutions in these equations. See below.

Theorem 1.56. *Definitions 1.21 and 1.54 are equivalent.*

Proof. (Definition 1.21 ⇒ Definition 1.54). Let $(Q, \cdot)$ be a quasigroup. Since for every pair of elements $a, b \in Q$ there exists a unique element x such that $a \cdot x = b$, we can associate with this equation an operation on the set Q, namely $a \cdot x = b \leftrightarrow a\backslash b = x$. If we substitute the last expression in the equality $a \cdot x = b$, then we obtain $a \cdot (a\backslash b) = b$ for all $a, b \in Q$. We obtained identity (1.4) from Definition 1.54.

Similarly, $y \cdot a = b \leftrightarrow b/a = y$, $(b/a) \cdot a = b$ for all $a, b \in Q$ and we obtain identity (1.5).

Identities (1.6) and (1.7) follow from the definitions of the operations $\backslash$ and $/$. Indeed, $x\backslash(x \cdot y) = y \leftrightarrow x \cdot y = x \cdot y$, $(y \cdot x)/x = y \leftrightarrow y \cdot x = y \cdot x$.

(Definition 1.54 $\Rightarrow$ Definition 1.21). Let $(Q, \cdot, /, \backslash)$ be an algebra with three binary operations such that in the algebra identities (1.4), (1.5), (1.6) and (1.7) hold.

We need to prove the existence and the uniqueness of solutions to the equations $a \cdot x = b$ and $y \cdot a = b$.

(Existence). Let $x = a\backslash b$. Then $a \cdot x = a(a\backslash b) \overset{(1.4)}{=} b$. Similarly, if $y = b/a$, then $y \cdot a = (b/a) \cdot a \overset{(1.5)}{=} b$.

(Uniqueness). Suppose that there exist two solutions x_1 and x_2 of equation $a \cdot x = b$, i.e., $a \cdot x_1 = b$ and $a \cdot x_2 = b$. Then $x_1 = a\backslash b$ and further we have

$$x_1 = a\backslash b = a\backslash(ax_2) \overset{(1.6)}{=} x_2.$$

Similarly, if $y_1 \cdot a = b$ and $y_2 \cdot a = b$, then $y_1 = b/a$,

$$y_1 = b/a = (y_2 \cdot a)/a \overset{(1.7)}{=} y_2.$$

□

It is possible to rewrite identities (1.4)–(1.7) in the language of translations in the following form:

$$L_x^{(\cdot)} L_x^{(\backslash)} y = y, \tag{1.12}$$

$$R_x^{(\cdot)} R_x^{(/)} y = y, \tag{1.13}$$

$$L_x^{(\backslash)} L_x^{(\cdot)} y = y. \tag{1.14}$$

$$R_x^{(/)} R_x^{(\cdot)} y = y. \tag{1.15}$$

Remark 1.57. Using Table 1.1 we can re-write identities (1.12)–(1.15) as follows:

$$L_x L_x^{-1} y = y, \tag{1.16}$$

$$R_x R_x^{-1} y = y, \tag{1.17}$$

$$L_x^{-1} L_x y = y, \tag{1.18}$$

$$R_x^{-1} R_x y = y. \tag{1.19}$$

Taking into consideration Lemma 1.14 and equalities (1.12)–(1.15) we can say that identities (1.4)–(1.7) guarantee that in a quasigroup $(Q, \cdot)$ any left and right translation is a bijection of the set Q.

In the language of translations, identities (1.9) and (1.8) take the following forms, respectively, $R_x^{(\backslash)} L_x^{(/)} y = y$, $L_x^{(/)} R_x^{(\backslash)} y = y$. From Table 1.1 it follows that $L_x^{(/)} = P_x^{-1}$, $R_x^{(\backslash)} = P_x$.

Algebras with identities (1.4)–(1.9) are discussed in [786, 798]. The following "equivalence" theorem is proved.

Theorem 1.58. *1. A groupoid $(Q, \cdot)$ is a left division groupoid if and only if in algebra $(Q, \cdot, \backslash)$ the identity (1.4) holds true.*

2. A groupoid $(Q, \cdot)$ is a right division groupoid if and only if in algebra $(Q, \cdot, /)$ the identity (1.5) holds true.

3. *A groupoid $(Q, \cdot)$ is a left cancellation groupoid if and only if in algebra $(Q, \cdot, \backslash)$ the identity (1.6) holds true.*

4. *A groupoid $(Q, \cdot)$ is a right cancellation groupoid if and only if in algebra $(Q, \cdot, /)$ the identity (1.7) holds true [786].*

The results about groupoids which satisfy some modifications of associative law are presented in [706, 707].

1.2.3 Some other definitions of e-quasigroups

The identities (1.8) and (1.9) play an important role in the following definitions of e-quasigroups.

Lemma 1.59. *In algebra $(Q, \cdot, \backslash, /)$ the identity (1.4) follows from the identities (1.5) and (1.8).*

Proof. If in identity (1.5) we change the variable y by the variable x, and variable x by the symbol (term) $y\backslash x$, then

$$(x/(y\backslash x)) \cdot (y\backslash x) = x. \tag{1.20}$$

By the identity (1.8) $x/(y\backslash x) = y$. Therefore we can rewrite the identity (1.20) as follows:

$$y \cdot (y\backslash x) = x. \tag{1.21}$$

Then we obtain the identity (1.4). □

Lemma 1.60. *In algebra $(Q, \cdot, \backslash, /)$ the identity (1.7) follows from the identities (1.6) and (1.8).*

Proof. We can re-write the identity (1.8) in the following form:

$$(x \cdot y)/(x\backslash(x \cdot y)) = x. \tag{1.22}$$

By the identity (1.6), $x\backslash(x \cdot y) = y$. Therefore the identity (1.22) takes the form $(x \cdot y)/y = x$ and it coincides with the identity (1.7). □

Lemma 1.61. *In algebra $(Q, \cdot, \backslash, /)$ the identity (1.6) follows from the identities (1.7) and (1.9).*

Proof. We can re-write the identity (1.9), $(x \to (x \cdot y), y \to x)$, in the following form:

$$((x \cdot y)/y)\backslash(x \cdot y) = y. \tag{1.23}$$

By the identity (1.7), $(x \cdot y)/y = x$. Therefore the identity (1.23) takes the form: $x\backslash(x \cdot y) = y$ and it coincides with the identity (1.6). □

Lemma 1.62. *In algebra $(Q, \cdot, \backslash, /)$ the identity (1.5) follows from the identities (1.4) and (1.9).*

Proof. We can re-write the identity (1.4) in the following form:

$$(x/y) \cdot ((x/y)\backslash x) = x. \tag{1.24}$$

By the identity (1.9), $(x/y)\backslash x = y$. Therefore the identity (1.24) takes the form $(x/y) \cdot y = x$ and it coincides with the identity (1.5). □

Theorem 1.63. *An algebra* $(Q, \cdot, \backslash, /)$ *with the identities (1.5), (1.6) and (1.8) is a quasigroup [797, 692].*

Proof. The proof follows from Lemmas 1.59 and 1.60. □

Theorem 1.64. *An algebra* $(Q, \cdot, \backslash, /)$ *with the identities (1.4), (1.7) and (1.9) is a quasigroup [797, 692].*

Proof. The proof follows from Lemmas 1.61 and 1.62. □

Theorem 1.65. *(1) If the* $(1\,3)$*-parastrophe of a groupoid* (Q, A) *is a groupoid, then* (Q, A) *is a right quasigroup.*

(2) If the $(2\,3)$*-parastrophe of a groupoid* (Q, A) *is a groupoid, then* (Q, A) *is a left quasigroup.*

(3) If the $(1\,2\,3)$*-parastrophe of a groupoid* (Q, A) *is a groupoid, then* (Q, A) *is a quasigroup.*

(4) If the $(1\,3\,2)$*-parastrophe of a groupoid* (Q, A) *is a groupoid, then* (Q, A) *is a quasigroup [489].*

(5) If the $(1\,2)$*-parastrophe of a left quasigroup* (Q, A) *is a groupoid, then* (Q, A) *is a quasigroup.*

(6) If the $(1\,2)$*-parastrophe of a right quasigroup* (Q, A) *is a groupoid, then* (Q, A) *is a quasigroup.*

(7) If the $(2\,3)$*-parastrophe of a right quasigroup* (Q, A) *is a groupoid, then* (Q, A) *is a quasigroup.*

(8) If the $(2\,3)$*-parastrophe of a left quasigroup* (Q, A) *is a groupoid, then* (Q, A) *is a quasigroup [786].*

Proof. Case 1. The main idea is the following: in the $(1\,3)$-parastrophe of a groupoid (Q, A) and in groupoid (Q, A), right translations are inverse in pairs. See Table 1.1. □

The definition of middle ternary relation for groupoids (an analogue of middle quasigroup translation) is given in [786].

Theorem 1.66. *A groupoid* $(Q, \cdot)$ *is a quasigroup if and only if all middle translations of* $(Q, \cdot)$ *are bijective maps of the set* Q.

Proof. Let a, b be a pair of fixed elements of the set Q. Since translation P_b of groupoid $(Q, \cdot)$ is a bijective map, then we have that for the element a of set Q there exists a unique element $x \in Q$ such that $P_b\, a = x$, i.e., $a \cdot x = b$. Since all middle translations of $(Q, \cdot)$ are bijective maps, then for any fixed elements $a, b \in Q$ there exists a unique element x such that equation $a \cdot x = b$ has a unique solution.

If a translation P_b is a bijective map, then translation P_b^{-1} also is a bijective map. Further we have that for any fixed element a of the set Q there exists a unique element $y \in Q$ such that $P_b^{-1} a = y$, i.e., $y \cdot a = b$. Therefore, equation $y \cdot a = b$ has a unique solution in $(Q, \cdot)$ for any fixed elements $a, b \in Q$.

Converse. From Table 1.1 it follows that middle translations of a quasigroup are bijective mappings. See, also, [75]. □

Some other definitions of a quasigroup are given in [590, 819, 786].

Example 1.67. Define binary groupoid $(\mathbb{Z}, *)$ $x * y = x + 2y$ for all $x, y \in \mathbb{Z}$, where $(\mathbb{Z}, +, \cdot)$ is the ring of integers.

It is easy to check that groupoid $(\mathbb{Z}, *)$ is a right quasigroup with the left cancellative property.

From Theorem 1.58 it follows that groupoid $(\mathbb{Z}, *)$ satisfies the identities (1.5), (1.6), and (1.7).

We prove that groupoid $(\mathbb{Z}, *)$ satisfies the identity (1.9). If $x * y = x + 2y$, then the set of solutions of the equation $a * y = b$ is described in the following way: $y = a\backslash b = \frac{b-a}{2}$. It is clear that a solution of this equation exists only for any pair of elements of equal parity from the set of integers.

Solution of equation $x * a = b$ has the following form $x = b/a = b - 2a$. Then

$$(a/b)\backslash a = (a - 2b)\backslash a = \frac{a - a + 2b}{2} = b.$$

Therefore $(a/b)\backslash a = b$ for all $a, b \in \mathbb{Z}$, groupoid $(\mathbb{Z}, *)$ satisfies identities (1.5), (1.6), (1.7), and (1.9) and it is not a quasigroup.

Remark 1.68. In Example 1.67 both the sub-term a/b and the term $(a/b)\backslash a = b$ are defined for all $a, b \in Q$.

1.2.4 Quasigroup-based cryptosystem

References [601, 602] proposed using quasigroups for secure encryption.

A quasigroup $(Q, \cdot)$ and its (23)-parastrophe $(Q, \backslash)$ satisfy the following identities $x\backslash(x \cdot y) = y$, $x \cdot (x\backslash y) = y$ (identities (1.4) and (1.6)). It is proposed to use this property of the quasigroups to construct a stream cipher.

Algorithm 1.69. Let A be a non-empty alphabet, k be a natural number, $u_i, v_i \in A$, $i \in \{1, ..., k\}$. Define a quasigroup $(A, \cdot)$. It is clear that the quasigroup $(A, \backslash)$ is defined in a unique way. We take a fixed element l ($l \in A$), which is called a leader.

Let $u_1 u_2 ... u_k$ be a k-tuple of letters from the alphabet A. The authors propose the following ciphering procedure $v_1 = l \cdot u_1, v_i = v_{i-1} \cdot u_i$, $i = 2, ..., k$. Therefore we obtain the following ciphertext $v_1 v_2 \dots v_k$.

The enciphering algorithm is constructed in the following way: $u_1 = l\backslash v_1, u_i = v_{i-1}\backslash v_i, i = 2, ..., k$.

We shall name this algorithm the Markovski algorithm.

In this encryption procedure, a private key for both sender and receiver is quasigroup $(A, \cdot)$ and leader element l. Therefore this is a cryptosystem with symmetric keys. The authors claim that this cipher is resistant to the brute force attack (exhaustive search) and to the statistical attack (in many languages some letters appear together more frequently, than other letters).

Example 1.70. Let alphabet A consist of the letters a, b, c. We construct the following quasigroup $(A, \cdot)$:

$\cdot$	a	b	c
a	b	c	a
b	c	a	b
c	a	b	c

Then quasigroup $(A, \backslash)$ has the following Cayley table:

$\backslash$	a	b	c
a	c	a	b
b	b	c	a
c	a	b	c

Let $l = a$ and the open text is $u = b\,b\,c\,a\,a\,c\,b\,a$. Then the ciphertext is $v = c\,b\,b\,c\,a\,a\,c\,a$. Applying the decoding function to v we get $b\,b\,c\,a\,a\,c\,b\,a = u$.

Remark 1.71. In this construction n-ary quasigroups and their parastrophes can also be used. See Algorithm 11.10 or [684]. Notice, row-Latin squares can be used by construction of a cryptosystem with non-symmetric keys. See Section 11.7.5.

1.2.5 Identity elements

1.2.5.1 Local identity elements

Definition 1.72. An element $f(b)$ of a quasigroup $(Q, \cdot)$ is called a left local identity element of an element $b \in Q$, if $f(b) \cdot b = b$. Therefore $f(b) = b/b$, or $f(b) = R_b^{-1}b$.

An element $e(b)$ of a quasigroup $(Q, \cdot)$ is called a right local identity element of an element $b \in Q$, if $b \cdot e(b) = b$. Therefore $e(b) = b\backslash b$, or $e(b) = L_b^{-1}b$.

An element $s(b)$ of a quasigroup $(Q, \cdot)$ is called a middle local identity element of an element $b \in Q$, if $b \cdot b = s(b)$. Therefore $s(b) = L_b b = R_b b$ [764, 767].

In a quasigroup there exists a unique left (right, middle) local identity element of any fixed element a. Indeed, if $f(b) \cdot a = f(a) \cdot a$, then $f(b) = f(a)$. And so on.

Exercise 1.73. Construct a quasigroup with at least two left (right, middle) local identity elements.

1.2.5.2 Left and right identity elements

Definition 1.74. An element $f \in Q$ $(e \in Q)$ is a left (right) identity element for quasigroup $(Q, \cdot)$ means that $f = f(x)$ for all $x \in Q$ (respectively, $e = e(x)$ for all $x \in Q$). An element $s \in Q$ is a middle identity element for quasigroup $(Q, \cdot)$ means that $s = s(x)$ for all $x \in Q$.

Definition 1.75. An element f is a left identity element of a quasigroup $(Q, \cdot)$ means that $f(x) = f$ for all $x \in Q$, i.e., all left local elements of quasigroup $(Q, \cdot)$ coincide.

An element e is a right identity element of a quasigroup $(Q, \cdot)$ means that $e(x) = e$ for all $x \in Q$, i.e., all right local elements of quasigroup $(Q, \cdot)$ coincide.

An element e is an identity element of a quasigroup $(Q, \cdot)$ means that $e(x) = f(x) = e$ for all $x \in Q$, i.e., all left and right local elements of quasigroup $(Q, \cdot)$ coincide.

Lemma 1.76. *1. In a quasigroup $(Q, \cdot)$ there exists a unique left identity element.*

2. In a quasigroup $(Q, \cdot)$ there exists a unique right identity element.

3. In a quasigroup $(Q, \cdot)$ there exists a unique identity element.

Proof. Case 1. If $f_1 x = f_2 x = x$ for all $x \in Q$, then $f_1 = f_2$. Case 2. If $xe_1 = xe_2 = x$ for all $x \in Q$, then $e_1 = e_2$. Case 3. The proof follows from Cases 1 and 2. □

1.2.5.3 Loops

Definition 1.77. A quasigroup $(Q, \cdot)$ with a left identity element $f \in Q$ is called a *left loop*.

A quasigroup $(Q, \cdot)$ with a right identity element $e \in Q$ is called a *right loop*.

A quasigroup $(Q, \cdot)$ with a middle identity element $s \in Q$ is called a *unipotent quasigroup*.

A quasigroup $(Q, \cdot)$ with an identity element $e \in Q$ is called a *loop*.

Example 1.78. We give an example of a non-associative unipotent loop.

$\circ$	0	1	2	3	4
0	0	1	2	3	4
1	1	0	3	4	2
2	2	4	0	1	3
3	3	2	4	0	1
4	4	3	1	2	0

Unipotent loops are studied in [504].

Define in a quasigroup $(Q, \cdot)$ the following mappings: $f : x \mapsto f(x)$, where $f(x) \cdot x = x$; $e : x \mapsto e(x)$, where $x \cdot e(x) = x$; $s : x \mapsto s(x)$, where $s(x) = x \cdot x$.

Remark 1.79. In a left loop $|f(Q)| = 1$, in a right loop $|e(Q)| = 1$, in a unipotent quasigroup $|s(Q)| = 1$.

1.2.5.4 Identity elements of quasigroup parastrophes

Connections between different kinds of local identity elements in various parastrophes of a quasigroup $(Q, \cdot)$ are given in the following table [504, 764, 767].

Table 1.2: Local identity elements of parastrophic quasigroups.

	ε	(12)	(13)	(23)	(123)	(132)
f	f	e	s	f	e	s
e	e	f	e	s	s	f
s	s	s	f	e	f	e

In Table 1.2, for example, $s^{(123)} = f^{(\cdot)}$.

Definition 1.80. A quasigroup $(Q, \cdot)$ with identity $x \cdot x = x$ is called an *idempotent quasigroup*. An element x with this property is called *idempotent element*.

Remark 1.81. It is easy to see that any parastrophe of an idempotent quasigroup is an idempotent quasigroup. In an idempotent quasigroup, the mappings e, f, s are identity permutations of the set Q. Moreover, if one of these three mappings is an identity permutation, then all other ones are identity permutations, too.

1.2.5.5 The equivalence of loop definitions

Definition 1.82. A quasigroup $(Q, \cdot, /, \backslash)$ with identity

$$x/x = y/y \tag{1.25}$$

is called a left loop.

Lemma 1.83. *A quasigroup $(Q, \cdot)$ is a left loop if and only if quasigroup $(Q, \cdot, /, \backslash)$ is a left loop.*

Proof. Suppose that quasigroup $(Q, \cdot)$ has the left identity element, i.e., there exists an element f such that the equality $f \cdot z = z$ is true for any $z \in Q$. We can re-write equality

$f \cdot z = z$ in the following form: $z/z = f$ for any $z \in Q$. Therefore $x/x = y/y = f$ and the following identity is true $x/x = y/y$.

Suppose that quasigroup $(Q, \cdot, /, \backslash)$ satisfies the identity (1.25). Therefore we have $(x/x) \cdot z \overset{(1.25)}{=} (z/z) \cdot z \overset{(1.5)}{=} z$. Thus quasigroup $(Q, \cdot)$ has the left identity element $f = x/x$. The uniqueness follows from Lemma 1.76. □

Recall identity (1.5) is true in any right division groupoid (Theorem 1.58). Therefore in any right division groupoid with the identity (1.25) there exists at least one left identity element.

Definition 1.84. A quasigroup $(Q, \cdot, /, \backslash)$ with identity

$$x\backslash x = y\backslash y \tag{1.26}$$

is called a right loop.

Exercise 1.85. Prove the analogue of Lemma 1.83 for right loops.

Definition 1.86. A quasigroup $(Q, \cdot, /, \backslash)$ with identity

$$x\backslash x = y/y \tag{1.27}$$

is called a loop.

Theorem 1.87. *Quasigroup $(Q, \cdot)$ is a loop if and only if quasigroup $(Q, \cdot, /, \backslash)$ is a loop [591, p. 97].*

Proof. A quasigroup $(Q, \cdot)$ is a loop means that there exists an element, say 1, such that the following equalities $1 \cdot x = x$ and $y \cdot 1 = y$ are true for any $x, y \in Q$. Therefore $x/x = 1$, $y\backslash y = 1$, and $x/x = y\backslash y$ for all $x, y \in Q$.

Let $x\backslash x = y/y$ in quasigroup $(Q, \cdot, /, \backslash)$. Prove the existence of the left and right identity element in quasigroup $(Q, \cdot)$. We have: $(x\backslash x) \cdot z \overset{(1.27)}{=} (z/z) \cdot z \overset{(1.5)}{=} z$; $z \cdot (x/x) \overset{(1.27)}{=} z \cdot (z\backslash z) \overset{(1.4)}{=} z$.

The uniqueness of the identity element follows from Lemma 1.76. □

1.2.5.6 Identity elements in some quasigroups

Lemma 1.88. *[819, Proposition 1.3]. A nonempty quasigroup $(Q, \cdot, /, \backslash)$ is a loop if and only if it satisfies the "slightly associative identity"*

$$x(y/y) \cdot z = x \cdot (y/y)z. \tag{1.28}$$

Proof. If $(Q, \cdot, /, \backslash)$ is a loop with identity element e relative to the operation " $\cdot$ ", then $y/y = e$ and identity (1.28) follows.

Conversely, if identity (1.28) is true, then setting $z = y$ we obtain $x(y/y) \cdot y = x \cdot (y/y)y = xy$ (since by identity (1.5) $(y/y)y = y$). Further we have $x(y/y) \cdot y = xy$. After cancellation from the right we have $x(y/y) = x$, $x\backslash x = y/y$. □

Corollary 1.89. *A quasigroup $(Q, \cdot)$ with identity*

$$(x \cdot y) \cdot z = x \cdot (y \cdot z) \tag{1.29}$$

(identity of associativity) has the identity element, i.e., this quasigroup is a group [549].

Proof. In identity associativity we can put $y := y/y$ and after this apply Lemma 1.88. □

Notice that in [72] V.D. Belousov posed the following problem (Problem 18): "From what identities, that are true in a quasigroup $Q(\cdot)$, does it follow that the quasigroup $Q(\cdot)$ is a loop?" More detailed information on Belousov problems is given in Chapter 5.

The following identities are called Moufang identities: $x(y \cdot xz) = (xy \cdot x)z$, $(zx \cdot y)x = z(x \cdot yx)$, $yx \cdot zy = y(xz \cdot y)$, $yx \cdot zy = (y \cdot xz)y$.

Theorem 1.90. *A quasigroup $(Q, \cdot)$ with any of Moufang identities is a loop [545, 794].*

From Theorem 1.90 and the well-known result that in a loop all Moufang identities are equivalent [173, 72], it follows that in a quasigroup all Moufang identities are equivalent.

Definition 1.91. A quasigroup with identity $xy \cdot uv = xu \cdot yv$ is called *medial*, and a quasigroup with equality $yz \cdot x = yf(x) \cdot zx$ is called *right F-quasigroup* [80].

Problem 1.1. It is easy to see that in loops $1 \cdot ab = 1a \cdot 1b$. Describe quasigroups with the property $f(ab) = f(a)f(b)$ for all $a, b \in Q$, where $f(a)$ is a left local element of element a.

Medial and right F-quasigroups have this property [80].

The identity $x(y \cdot xz) = (x \cdot yx)z$ is called the left Bol identity. A loop with the left Bol identity is called a left Bol loop.

Lemma 1.92. *A quasigroup $(Q, \cdot)$ with the left Bol identity has a right identity element.*

Proof. Indeed, if we put $z := e_x$ in identity $x(y \cdot xz) = (x \cdot yx)z$, then we obtain $x \cdot yx = (x \cdot yx)e_x$ for all $x, y \in Q$. □

Notice, there exist right loops with the left Bol identity:

*	0	1	2
0	0	2	1
1	1	0	2
2	2	1	0

1.2.5.7 Inverse elements in loops

Let $(Q, \circ)$ be a loop. Let I_r, I_l be the following maps: $I_r(x) = x^{-1}$, where $x \circ x^{-1} = 1$; $I_l(x) = {}^{-1}x$, where ${}^{-1}x \circ x = 1$, for all $x \in Q$.

Lemma 1.93. *In any loop $(Q, \circ)$, $I_l = I_r^{-1}$.*

Proof. In a loop $(Q, \circ)$ we have $x \circ I_r x = 1$, i.e., $P_1 x = I_r x$, $I_r = P_1$. Similarly we have $I_l x \circ x = 1$, i.e., $P_1^{-1} x = I_l x$, $I_l = P_1^{-1}$. Therefore $I_l = I_r^{-1}$ in any loop. □

1.2.6 Multiplication groups of quasigroups

Let $(Q, \cdot)$ be a quasigroup. With every element $a \in Q$ it is possible to associate left (L_a), right (R_a) and middle (P_a) translations. These translations are some permutations of the set Q. They can be considered as elements of the symmetric group S_Q.

With any quasigroup $(Q, \cdot)$ we can associate the sets of all left translations $(\mathbb{L})$, right translations $(\mathbb{R})$, and middle translations $(\mathbb{P})$. Denote the groups generated by all left, right, and middle translations of a quasigroup $(Q, \cdot)$ as $LM(Q, \cdot)$, $RM(Q, \cdot)$, and $PM(Q, \cdot)$, respectively.

The group generated by all left and right translations of a quasigroup $(Q, \cdot)$ is called (following A.A. Albert [10, 11]), a multiplication group of quasigroup. Usually this group is denoted by $M(Q, \cdot)$. A multiplication group of a quasigroup plays an important role by the study of quasigroups, especially in the study of loops.

By $FM(Q, \cdot)$ we shall denote a group generated by sets $\mathbb{L}, \mathbb{R}, \mathbb{P}$ of a quasigroup $(Q, \cdot)$. Information about the group $FM(Q, \cdot)$ in the case when $(Q, \cdot)$ is an IP-loop (a Moufang loop, a group) is presented in [758].

We can see on a quasigroup $(Q, \cdot)$ of a finite order n, as on a set $\mathbb{T}$ of permutations of the group S_n with the following property: if $\alpha, \beta \in \mathbb{T}$ and there exists an element $x \in Q$ such that $\alpha^{-1}\beta x = x$, then $\alpha = \beta$.

Definition 1.94. A set $\mathbb{T}$ of permutations on a finite set Q is called sharply transitive, if for any pair of elements $a, b \in Q$ there exists exactly one permutation $\alpha \in \mathbb{T}$ such that $\alpha a = b$.

Sets of all left, right and middle translations of a quasigroup $(Q, \cdot)$ give us examples of sharply transitive sets of permutations on the set Q.

We shall use the following fact about the properties of groups that act transitively on a set Q: if a group G acts transitively on a set Q, then stabilizers of any two elements $a, b \in Q$ are isomorphic subgroups and $|G| = |St(a)| \cdot |Q|$ [471].

Multiplication groups play an important role in quasigroup theory for the following reason: groups are algebraic objects which are much better known than quasigroups. Often it is possible to express some quasigroup properties in the language of corresponding groups.

For example, the theory of normality of quasigroups, and especially of loops, can be expressed in terms of corresponding subgroups of its multiplication group [10, 11].

Properties of multiplication groups of quasigroups and loops are described in many articles, see, for example, [280, 268, 269]. In [268] the following result is proved. Let W be a free loop with a basis $X \neq \varnothing$. Then the left multiplication group $LM(W)$ is a free group of infinite rank and a Frobenius permutation group [435]. Notice, for the group $M(W)$ this fact is not true [268, 280]. Multiplication groups of free and free commutative quasigroups are also described in [360, 361].

Let $(Q, \cdot)$ be a quasigroup. The group $I_h = \{\alpha \in M(Q, \cdot) | \alpha h = h\}$ is called an inner mapping group of a quasigroup $(Q, \cdot)$ relative to element $h \in Q$. The group I_h is a stabilizer of element h by action of the group $M(Q, \cdot)$ on the set Q.

In the loop case the group $I_1(Q, \cdot) = I(Q, \cdot)$ is usually studied, where 1 is the identity element of a loop $(Q, \cdot)$.

It is possible to define the "inner mapping groups" for the groups $LM(Q, \cdot)$, $RM(Q, \cdot)$, $PM(Q, \cdot)$, $FM(Q, \cdot)$ of a quasigroup $(Q, \cdot)$, namely, it is possible to define the groups $LI_h(Q, \cdot)$, $RI_h(Q, \cdot)$, $PI_h(Q, \cdot)$, $FI_h(Q, \cdot)$ for any element $h \in Q$.

Example 1.95. We use quasigroup from Example 1.20. It is easy to check that $L_0 = (01)$, $L_1 = (12)$, $L_2 = (02)$, $LM(Q, \cdot) = \{\varepsilon, (0\,1), (0\,2), (1\,2), (0\,1\,2), (0\,2\,1)\}$, $LI_0(Q, \cdot) = \{\varepsilon, (1\,2)\}$, $LI_1(Q, \cdot) = \{\varepsilon, (0\,2)\}$, $LI_2(Q, \cdot) = \{\varepsilon, (0\,1)\}$, $LM(Q, \cdot) \cong S_3$, $LI_0(Q, \cdot) \cong Z_2$.

Of course, it is possible to define "inner mapping groups" for other "multiplication groups" of a quasigroup $(Q, \cdot)$.

Since all the above listed multiplication groups of a quasigroup $(Q, \cdot)$ act transitively on the set Q, inner mapping groups relative to different elements of the set Q are isomorphic.

For example, $PI_h(Q, \cdot) \cong PI_g(Q, \cdot)$, $FI_h(Q, \cdot) \cong FI_g(Q, \cdot)$ and so on. If $|Q|$ is finite, then $|M(Q, \cdot) = |Q||I_h(\cdot)|$ and so on.

Exercise 1.96. Construct various multiplication groups and inner multiplication groups for a quasigroup of order 4 or (and) 5.

1.2.7 Transversals: "Come back way"

Definition 1.97. Suppose we have a partition of a set Q into cosets A_i. The set of representatives $\{a_i \mid a_i \in A_i\}$ (a unique element from any coset) is called a transversal of the set Q relative to the partition A_i.

The term "transversal" is used in combinatorics. See, for example, [409, Chapter V].

Let $(Q, \cdot)$ be a quasigroup. It is clear that $LI_h(Q, \cdot) \subseteq LM(Q, \cdot)$, and that a unique left translation lies in any left coset class of the group $LM(Q, \cdot)$ by the group $LI_h(Q, \cdot)$. Therefore we have the following partition of the group $LM(Q, \cdot)$ through the group $LI_h(Q, \cdot)$ (for brevity, we omit the expression $(Q, \cdot)$)

$$\begin{aligned} LM = a_1 \cdot LI_h \sqcup a_2 \cdot LI_h \sqcup a_3 \cdot LI_h \sqcup \dots a_n \cdot LI_h \sqcup \dots = \\ L_{a_1} \cdot LI_h \sqcup L_{a_2} \cdot LI_h \sqcup L_{a_3} \cdot LI_h \sqcup \dots L_{a_n} \cdot LI_h \sqcup \dots \end{aligned} \tag{1.30}$$

It is clear that the situation with other multiplication groups and corresponding inner multiplication groups is similar.

Notice, the ordered set of translations $\{L_{a_1}, \dots, L_{a_n}, \dots\}$ defines the Cayley table of quasigroup $(Q, \cdot)$ in a unique way.

Example 1.98. We continue Example 1.95. We have $LM = L_0 \cdot LI_0 \sqcup L_1 \cdot LI_0 \sqcup L_2 \cdot LI_0$, where $LI_0 = \{\varepsilon, (12)\}$, $L_0 \cdot LI_0 = \{(01), (012)\}$, $L_1 \cdot LI_0 = \{\varepsilon, (12)\}$, $L_2 \cdot LI_0 = \{(02), (021)\}$.

This situation can be generalized in the following way. Let $(G, \cdot)$ be a group, and $(H, \cdot)$ be its subgroup. A complete system T of representatives of the left cosets aH, $a \in G$ is called a left transversal in group $(G, \cdot)$ to subgroup $(H, \cdot)$.

In other words, from any coset $a_i \cdot H$ we take only one element, for example, element a_i. Thus $G = a_1 \cdot H \sqcup a_2 \cdot H \sqcup a_3 \cdot H \sqcup \dots a_n \cdot H \sqcup \dots = \sqcup a_i \cdot H$ [471], and $T = \{a_1, a_2, \dots, a_n, \dots\}$ is a left transversal.

Define operation $*$ on the set T in the following way:

$$a * b = a \cdot b \pmod{H}. \tag{1.31}$$

Lemma 1.99. *Groupoid $(T, *)$ is a left quasigroup, i.e., equation $a * x = b$ has a unique solution for any $a, b \in T$.*

Proof. Indeed, left translations L^*_d, $d \in T$, defined by the equality (1.31), correspond to permutations (this fact is easy to check or see [407, Theorem 5.3.1. (a)], [471, 12.2.1 Theorem]) by the following action of the element d on the set of cosets $d(a_j \cdot H) = d \cdot a_j \cdot H$, where $a_j \cdot H \in \sqcup a_i \cdot H$. □

R. Baer, by definition of transversal, supposed that $1 \in T$, where 1 is the identity element of the group G [24, 25]. In this case the left quasigroup has the identity element 1, i.e., $1 * x = x * 1 = x$ for all $x \in T$.

Notice, the set of left translations of the groupoid $(T, *)$ is a subset of the set of permutations S of the group G by its subgroup H (permutation presentation of the group G by its subgroup H) [407, Theorem 5.3.1. (a)], [471, 12.2.1 Theorem]).

It is known [407, 471] that the set S forms a group relative to standard multiplication of permutations and this group is isomorphic to the group $G/(ker\, H)$, where $kerH$ is the largest normal subgroup of the group G, which is contained in H.

Lemma 1.100. *If $L^*_a = L^*_b$ in the left quasigroup $(T, *)$, then $a = b$.*

Proof. If $L^*_a x = L^*_b x$ for all $x \in T$, then, using equality (1.31), we have $a \cdot x = b \cdot x \pmod{H}$, $x = a^{-1}bx \pmod{H}$, $a^{-1}b \in H$, $a = b \pmod{H}$, i.e., $a = b$ in left quasigroup $(T, *)$. □

Corollary 1.101. *In the Cayley table of the left quasigroup $(T, *)$, all rows are pairwise different.*

Proof. The proof follows from Lemma 1.100. □

The conditions when a transversal T is a loop transversal, i.e., $(T, *)$ is a loop, are given in [24, 25, 658].

Example 1.102. Let $G = S_3 = \langle a, b \,|\, a^3 = b^2 = (ab)^2 = 1 \rangle$, $S_3 = \{1, a, a^2, b, ab, a^2b\}$, $H = \langle b \rangle$. We have the following set of left cosets: $H = \{1, b\}$, $aH = \{a, ab\}$, $a^2H = \{a^2, a^2b\}$. Elements $\{b, ab, a^2\}$ form the left transversal T. We can construct a Cayley table of the left quasigroup $(T, *)$.

$*$	b	ab	a^2
b	b	a^2	ab
ab	ab	b	a^2
a^2	a^2	b	ab

If we denote b as 1, ab as 2, a^2 as 3, then we obtain the following Cayley table:

$*$	1	2	3
1	1	3	2
2	2	1	3
3	3	1	2

In [556, Theorem 3.4] it is proved that every right quasigroup with identity element can be embedded as a right transversal in a group which is universal in some sense.

Transversals are studied in [449, 450, 447, 448]. These objects are used in construction of codes [429, 433]. Loop transversal is based on a loop and some its subloop. Such transversals are introduced and studied in [553, 554].

1.2.8 Generators of inner multiplication groups

We start this subsection from the following Belousov theorem, which is a generalization of the corresponding Bruck theorem, which was proved for loop case [167]. Taking into consideration the importance of these theorems for the development of quasigroups and especially loop theory, we decided to name these theorems for their discoverers.

Theorem 1.103. *Belousov theorem. In a quasigroup $(Q, \cdot)$*

$$I_h(Q, \cdot) = \langle R_{a,b}, L_{a,b}, T_a \,|\, a, b \in Q \rangle,$$

where $R_{a,b} = R^{-1}_{a \bullet b} R_b R_a$, $h(a \bullet b) = (ha)b$, $L_{a,b} = L^{-1}_{a \circ b} L_a L_b$, $(a \circ b)h = a(bh)$, $T_a = L^{-1}_{\sigma a} R_a$, $\sigma = R_h^{-1} L_h$ [72].

Theorem 1.104. *Bruck theorem. In a loop $(Q, \cdot)$ with identity element* 1

$$I_1(Q, \cdot) = \langle R_{a,b}, L_{a,b}, T_a \,|\, a, b \in Q \rangle,$$

where $R_{a,b} = R^{-1}_{ab} R_b R_a$, $L_{a,b} = L^{-1}_{ab} L_a L_b$, $T_a = L_a^{-1} R_a$ [167].

Theorem 1.103 and Theorem 1.104 are proved in [72, 167] using induction. Theorem 1.103 and Theorem 1.104 play important roles in definition of normal subloops and subquasigroups.

It is possible to prove analogs of these theorems using Theorem 1.106. This plan, for example, was realized in [772]. Namely, in Theorem 1.106 induction is hidden. Properties of the group $FM(Q, \cdot)$ and generators of the group $FI_h(Q, \cdot)$ are given in [75].

Theorem 1.105. *In a quasigroup* $(Q, \cdot)$

$$FI_h(Q, \cdot) = \langle L_{a,b}, T_{a,b}, P_{a,b} \,|\, a, b \in Q \rangle,$$

where $L_{a,b} = L^{-1}_{a \circ b} L_a L_b$*,* $(a \circ b)h = a(bh)$*,* $T_{a,b} = L^{-1}_{a \star b} R_a L_b$*,* $(a \star b)h = bh \cdot a$*,* $P_{a,b} = L^{-1}_{a \blacklozenge b} P_a L_b$*,* $(a \blacklozenge b)h = a/(bh)$*.*

Proof. Firstly we note that group $FM(Q, \cdot)$ of a quasigroup $(Q, \cdot)$ acts transitively on the set Q. Really, for any fixed elements $a, b \in Q$ there exists an element c, such that $L_c a = b$ since the equation $x \cdot a = b$ has a unique solution in a quasigroup $(Q, \cdot)$ for any fixed elements $a, b \in Q$.

A stabilizer of element h in group FM of a quasigroup $(Q, \cdot)$ is group FI_h. Moreover, we can write

$$FM(Q, \cdot) = L_{f(h)}(FI_h) \sqcup L_{a_1}(FI_h) \sqcup L_{a_2}(FI_h) \sqcup \cdots, \tag{1.32}$$

where $f(h) \cdot h = h$.

In [471, 14.3.1. Theorem] the following Otto Schreier Theorem is proved:

Theorem 1.106. *Let* T *be a set of generators of a group* G*,* H *be a subgroup of the group* G*,* $f : u \to \overline{u}$ *be a function such that any element* $u \in G$ *corresponds to a fixed element* $\bar{u}$ *from the set* uH*, and* S *be a set of selected representatives (one element from any coset). Then* $H = \langle \overline{ps}^{-1} ps \mid s \in S, p \in T \rangle$*.*

We apply this theorem for obtaining a generator set of the group FI_h of a quasigroup $(Q, \cdot)$. The set $T = \mathbb{L} \cup \mathbb{R} \cup \mathbb{P}$ is a generator set of the group $FM(Q, \cdot)$ of a quasigroup $(Q, \cdot)$.

Let a set of selected representatives from every left coset of subgroup FI_h of a group $FM(Q, \cdot)$ be the following set: $S = \{L_a \mid a \in Q\}$. Therefore we defined the "selecting" function f.

Finally, we specify elements of the form $\overline{ps}^{-1}$. They must have the form L_z^{-1}. We can find the element z knowing elements p and s.

(i) Let $p = L_a$, $s = L_b$. We need to specify element z such that $L_a L_b h = L_z h$, $a(bh) = zh$. We have $z = R_h^{-1}(a \cdot bh)$. We can re-write the last relation as a binary operation with variables $a, b \in Q$ in the following form $(a \circ b)h = (a \cdot bh)$.

(ii) Let $p = R_a$, $s = L_b$. We need to specify element z such that $R_a L_b h = L_z h$, $(bh)a = zh$. Further we have $z = R_h^{-1}((bh)a)$. We can re-write the last relation as a binary operation with variables $a, b \in Q$ in the following form $(a \star b)h = (bh \cdot a)$.

(iii) Let $p = P_a$, $s = L_b$. We need to specify element z such that $P_a L_b h = L_z h$, $P_a(bh) = zh$. Further we have $zh \cdot bh = a$, $zh = a/bh$. We can re-write the last relation as a binary operation with variables $a, b \in Q$ in the following form $(a \blacklozenge b)h = a/(bh)$. □

Corollary 1.107. *In a quasigroup* $(Q, \cdot)$*,*

$$I_h(Q, \cdot) = \langle L_{a,b}, T_{a,b} \,|\, a, b \in Q \rangle,$$

where $L_{a,b} = L^{-1}_{a \circ b} L_a L_b$*,* $(a \circ b)h = a(bh)$*,* $T_{a,b} = L^{-1}_{a \star b} R_a L_b$*,* $(a \star b)h = bh \cdot a$*.*

Proof. The proof follows from Theorem 1.105. □

Corollary 1.108. *In a loop* $(Q, \cdot)$*,*

$$\begin{aligned} FI_1(Q, \cdot) &= \langle L_{a,b}, T_{a,b}, P_{a,b} \,|\, a, b \in Q \rangle, \\ I_1(Q, \cdot) &= \langle L_{a,b}, T_{a,b} \,|\, a, b \in Q \rangle, \end{aligned} \tag{1.33}$$

where $L_{a,b} = L^{-1}_{a \cdot b} L_a L_b$*,* $T_{a,b} = L^{-1}_{b \cdot a} R_a L_b$*,* $P_{a,b} = L^{-1}_{a/b} P_a L_b$*.*

Proof. It is sufficient to put $h = 1$ in conditions of Theorem 1.105, where the symbol 1 denotes identity element of loop $(Q, \cdot)$. □

Notice, $\langle L_{a,b}, T_{a,b} \mid a, b \in Q\rangle \subseteq \langle R_{a,b}, L_{a,b}, T_a \mid a, b \in Q\rangle$. Indeed,

$$(L_{ba}^{-1} R_{ba})(R_{ba}^{-1} R_a R_b)(R_b^{-1} L_b) = L_{ba}^{-1} R_a L_b.$$

The proof of inverse inclusion is more complicated.

Lemma 1.109. *In a quasigroup* $(Q, \cdot)$

$$LI_h(Q, \cdot) = \langle L_{a,b} \mid a, b \in Q\rangle\,,$$

where $L_{a,b} = L_{a\circ b}^{-1} L_a L_b$, $(a \circ b)h = a(bh)$ *[816, 772].*

Proof. The proof follows from Theorem 1.105. □

Exercise 1.110. Construct two more triples of generators of the group FI_h of a quasigroup $(Q, \cdot)$, $h \in Q$.

Lemma 1.111. *In a loop* $(Q, \cdot)$,

$$PI_h(Q, \cdot) = \langle P_{a,b} \mid a, b \in Q\rangle\,,$$

where $P_{a,b} = P_{a\backslash b}^{-1} P_a P_b$ *[767].*

Proof. The proof is similar to the proof of Theorem 1.105. □

The following results are known: a permutation group G of a set Q is the multiplication group of some quasigroup if and only if there is a loop $(Q, +)$ and permutations f and g of Q such that $\langle Mlt(Q, +), f, g\rangle = G$ [438]; any Hamiltonian group (finite non-abelian groups with only normal subgroups [407, p. 213]) cannot be a multiplication group of a loop [438].

Problem 1.2. Describe groups that can be or cannot be a multiplication group of a loop (of a quasigroup). We hope our readers will be able to generalize this problem on other "multiplication groups" of quasigroups and loops (or left quasigroups).

We notice, there are many articles in which properties of quasigroups (or loops) are studied with various conditions on their various inner multiplication groups. Mainly, these are the articles of T. Kepka, A. Drapal, M. Niemenmaa and their pupils and followers.

1.3 Morphisms

1.3.1 Isotopism

We start from a traditional definition of isotopism (of isotopy).

Definition 1.112. n-Ary groupoid (G, f) is an isotope of an n-ary groupoid (G, g) (*in other words* (G, f) *is an isotopic image of* (G, g)), if there exist permutations $\mu_1, \mu_2, \dots, \mu_n$, μ of the set G such that

$$f(x_1, x_2, \dots, x_n) = \mu^{-1} g(\mu_1 x_1, \dots, \mu_n x_n) \tag{1.34}$$

for all $x_1, \dots, x_n \in G$. We can also write this fact in the form $(G, f) = (G, g)T$ where $T = (\mu_1, \mu_2, \dots, \mu_n, \mu)$. The ordered $(n + 1)$-tuple T is called *isotopy* of n-ary groupoids.

If in equality (1.34) $f = g$, then $(n+1)$-tuple $(\mu_1, \mu_2, \dots, \mu_n, \mu)$ of permutations of the set G is called an *autotopy of n-groupoid* (Q, f). The last component of an autotopy of an n-groupoid is called a *quasiautomorphism* (by analogy with binary case).

A set of all autotopies of a groupoid (Q, f) forms the group of autotopies relative to the usually defined operation on this set: if $T_1 = (\mu_1, \mu_2, \dots, \mu_n, \mu)$ and $T_2 = (\nu_1, \nu_2, \dots, \nu_n, \nu)$ are autotopies of groupoid (Q, f), then $T_1T_2 = (\mu_1\nu_1, \mu_2\nu_2, \dots, \mu_n\nu_n, \mu\nu)$ is an autotopy of groupoid (Q, f). Autotopy group of a groupoid (Q, f) will be denoted as $\mathfrak{T}(Q, f)$.

If in (1.34) $\mu_1 = \mu_2 = \dots = \mu_n = \mu$, then groupoids (Q, f) and (Q, g) are isomorphic.

At last, if in (1.34) the n-ary operations f and g are equal and $\mu_1 = \mu_2 = \dots = \mu_n = \mu$, then we obtain an *automorphism of groupoid* (Q, f), i.e., a permutation μ of the set Q is called an automorphism of an n-groupoid (Q, f) if for all $x_1, \dots, x_n \in Q$ the following relation is fulfilled: $\mu f(x_1, \dots, x_n) = f(\mu_1 x_1, \dots, \mu_n x_n)$. We denote by $Aut(Q, f)$ the automorphism group of an n-ary groupoid (Q, f). If $n = 2$, we obtain the following definition of isotopism.

Definition 1.113. Binary groupoid $(G, \circ)$ is an isotopic image of a binary groupoid $(G, \cdot)$, if there exist permutations α, β, γ of the set G such that $x \circ y = \gamma^{-1}(\alpha x \cdot \beta y)$.

We list some properties of isotopisms. As usual, ε denotes the identity permutation.

Lemma 1.114. *If (α, β, γ) is an isotopism, then*

$$(\alpha, \beta, \gamma) = (\alpha, \beta, \varepsilon) * (\varepsilon, \varepsilon, \gamma) = (\alpha, \varepsilon, \varepsilon) * (\varepsilon, \beta, \varepsilon) * (\varepsilon, \varepsilon, \gamma) = \\ (\varepsilon, \beta, \varepsilon) * (\varepsilon, \varepsilon, \gamma) * (\alpha, \varepsilon, \varepsilon)$$

and so on.

Lemma 1.114 helps to construct a Cayley table of isotopic images of a finite quasigroup (groupoid).

Lemma 1.115. *If $(Q, \circ) = (Q, \cdot)(\alpha, \varepsilon, \varepsilon)$, i.e., $x \circ y = \alpha x \cdot y$ for all $x, y \in Q$, then $L^{\circ}_x = L^{\cdot}_{\alpha x}$ for all $x \in Q$.*

If $(Q, \circ) = (Q, \cdot)(\varepsilon, \beta, \varepsilon)$, i.e., $x \circ y = x \cdot \beta y$ for all $x, y \in Q$, then $R^{\circ}_y = R^{\cdot}_{\beta y}$ for all $y \in Q$.

If $(Q, \circ) = (Q, \cdot)(\varepsilon, \varepsilon, \gamma)$, i.e., $x \circ y = \gamma^{-1}(x \cdot y)$ for all $x, y \in Q$, then $P^{\circ}_z = P^{\cdot}_{\gamma z}$ for all $z \in Q$.

Proof. Case 3. We can re-write equality $(Q, \circ) = (Q, \cdot)(\varepsilon, \varepsilon, \gamma)$ in the form $x \circ y = \gamma^{-1}(x \cdot y)$ for all $x, y \in Q$. Therefore, if $x \circ y = \gamma^{-1}(x \cdot y) = z$, then $P^{\cdot}_z x = y$, $P^{\circ}_{\gamma^{-1}z} x = y$, $P^{\cdot}_z = P^{\circ}_{\gamma^{-1}z}$. □

Lemma 1.115 helps to find the Cayley table of isotopic images of a groupoid in the following way: if we have isotopy (α, β, γ), then we permute rows by the rule $L^{\circ}_x = L^{\cdot}_{\alpha x}$, after this we permute columns by the rule $R^{\circ}_y = R^{\cdot}_{\beta y}$, and finally we rename elements in the Cayley table by the following rule: if $x \cdot y = a$, then $x \circ y = \gamma^{-1} a$. As it follows from Lemma 1.114, we can change the order of execution of steps 1, 2, and 3.

Example 1.116. Let $T = ((1234), (12)(34), (123))$. Let a quasigroup $(Q, \cdot)$ have the following Cayley table:

·	1	2	3	4
1	2	1	3	4
2	3	2	4	1
3	4	3	1	2
4	1	4	2	3

If we apply isotopy $((1234), \varepsilon, \varepsilon)$ to this quasigroup (it changes the rows), then we obtain the following Cayley table

$*$	1	2	3	4
1	1	4	2	3
2	2	1	3	4
3	3	2	4	1
4	4	3	1	2

Further, if we apply the isotopy $(\varepsilon, (12)(34), \varepsilon)$ to the obtained quasigroup (we change the order of columns in the previous Cayley table), then we have quasigroup:

$\circ$	1	2	3	4
1	4	1	3	2
2	1	2	4	3
3	2	3	1	4
4	3	4	2	1

Finally, with the help of isotopy $(\varepsilon, \varepsilon, (123))$ $(\gamma^{-1} = (132))$, we rename elements inside the last Cayley table:

$\circ$	1	2	3	4
1	4	3	2	1
2	3	1	4	2
3	1	2	3	4
4	2	4	1	3

Definition 1.117. An isotopism of the form $(\alpha, \beta, \varepsilon)$ is called a principal isotopism.

Usually we shall write the fact that groupoids (Q, A) and (Q, B) are isotopic in this form: $(Q, A) \sim (Q, B)$

Lemma 1.118. *Any isotopism up to isomorphism is a principal isotopism.*

Proof. Suppose that (Q, A) and (Q, B) are isotopic groupoids. If $(Q, B) = (Q, A)\ (\alpha, \beta, \gamma)$, then $(Q, B)(\gamma^{-1}, \gamma^{-1}, \gamma^{-1}) = (Q, A)(\alpha\gamma^{-1}, \beta\gamma^{-1}, \varepsilon)$. Thus $(Q, C) = (Q, A)\ (\alpha\gamma^{-1}, \beta\gamma^{-1}, \varepsilon)$, where $(Q, C) = (Q, B)\ (\gamma^{-1}, \gamma^{-1}, \gamma^{-1})$. □

Definition 1.119. Isotopy of the form $(R_a^{-1}, L_b^{-1}, \varepsilon)$, where L_b, R_a are left and right translations of a quasigroup $(Q, \cdot)$, is called LP-isotopy (loop isotopy) [72, 80].

Theorem 1.120. *Any LP-isotope of a quasigroup $(Q, \cdot)$ is a loop.*

Proof. Prove that quasigroup $(Q, \circ)$, where $x \circ y = R_a^{-1}x \cdot L_b^{-1}y$, is a loop. Let $1 = b \cdot a$. If we take $x = 1$, then $1 \circ y = R_a^{-1}ba \cdot L_b^{-1}y = R_a^{-1}R_a b \cdot L_b^{-1}y = b \cdot L_b^{-1}y = L_b L_b^{-1}y = y$.

If we take $y = 1$, then we have $x \circ 1 = R_a^{-1}x \cdot L_b^{-1}ba = R_a^{-1}x \cdot a = R_a R_a^{-1}x = x$. Element 1 is the identity element of the quasigroup $(Q, \circ)$. □

Lemma 1.121.

1. *If $(Q, \circ) = (Q, \cdot)(\alpha, \beta, \varepsilon)$ and $(Q, \circ)$ is a loop, then there exist elements $a, b \in Q$ such that $\alpha = R_a^{-1}$, $\beta = L_b^{-1}$, where $R_a x = x \cdot a$, $L_b x = b \cdot x$ ([80], Lemma 1.1).*
2. *If $(Q, \circ) = (Q, \cdot)(\alpha, \varepsilon, \gamma)$ and $(Q, \circ)$ is a unipotent left loop, then there exist elements $a, b \in Q$ such that $\alpha = P_a^{-1}$, and $\gamma = L_b$.*
3. *If $(Q, \circ) = (Q, \cdot)(\varepsilon, \beta, \gamma)$ and $(Q, \circ)$ is a unipotent right loop, then there exist elements $a, b \in Q$ such that $\beta = P_a$, and $\gamma = R_b$.*

Proof. 1. Let $x \circ y = \alpha x \cdot \beta y$. If $x = 1$, then we have $1 \circ y = y = \alpha 1 \cdot \beta y$. Therefore $L_{\alpha 1}\beta = \varepsilon$, $\beta = L_{\alpha 1}^{-1}$. If we take $y = 1$, then we have $x \circ 1 = x = \alpha x \cdot \beta 1$, $R_{\beta 1}\alpha = \varepsilon$, $\alpha = R_{\beta 1}^{-1}$.

2. Let $x \circ y = \gamma^{-1}(\alpha x \cdot y)$. If $x = 1$, then we have $y = \gamma^{-1}(\alpha 1 \cdot y)$. Therefore $\gamma y = L_{\alpha 1} y$. If $x = y$, then $1 = \gamma^{-1}(\alpha \cdot x)$, $\gamma 1 = \alpha x \cdot x$, $P_{\gamma 1}^{-1} x = \alpha x$. In order to obtain the claimed in this case, we denote $\alpha 1$ by b, and $\gamma 1$ by a.

3. Let $x \circ y = \gamma^{-1}(x \cdot \beta y)$. If $y = 1$, then we have $x = \gamma^{-1}(x \cdot \beta 1)$. Therefore $\gamma x = R_{\beta 1} x$. If $x = y$, then $1 = \gamma^{-1}(x \cdot \beta x)$, $\gamma 1 = x \cdot \beta x$, $P_{\gamma 1} x = \beta x$. In order to obtain the claimed in conditions of the lemma, we denote $\beta 1$ by b, and $\gamma 1$ by a.

□

Lemma 1.122. *(i) If $(Q, \cdot)$ is a binary quasigroup, L_a, R_b are some of its left and right translations, $\varphi \in Aut(Q, \cdot)$, then $\varphi L_a = L_{\varphi a}\varphi$, $\varphi R_b = R_{\varphi b}\varphi$.*

(ii) If $(Q, +)$ is a group, then $L_a R_b = R_b L_a$, $L_a^{-1} = L_{-a}$, $R_a^{-1} = R_{-a}$.

(iii) If $(Q, +)$ is a group, then $R_d = L_d I_d$, where I_d is the inner automorphism of the group $(Q, +)$, i.e., $I_d x = -d + x + d$ for all $x \in Q$.

Proof. (i) We have $\varphi L_a x = \varphi(a \cdot x) = \varphi a \cdot \varphi x = L_{\varphi a}\varphi x$, $\varphi R_b x = \varphi(x \cdot b) = \varphi x \cdot \varphi b = R_{\varphi b}\varphi x$.

(ii) $L_a R_b x = a + (x + b) = (a + x) + b = R_b L_a x$. $L_a^{-1} = L_{-a}$ since $L_a^{-1} L_a x = x = -a + a + x = L_{-a} L_a x$.

(iii) $R_d x = x + d = d - d + x + d = L_d I_d x$. □

Theorem 1.123. *Generalized Albert theorem. If $(Q, \circ) = (Q, \cdot)(\alpha, \beta, \varepsilon)$, $(Q, \cdot)$ is a group, and $(Q, \circ)$ is a loop, then $(Q, \circ)$ is a group isomorphic to group $(Q, \cdot)$ [80, 10, 11, 72, 548, 685, 779].*

Proof. By Lemma 1.121 $\alpha = R_a^{-1}$, $\beta = L_b^{-1}$. However in a group $R_a^{-1} = R_{a^{-1}}$, $L_b^{-1} = L_{b^{-1}}$ (Lemma 1.122).

Therefore $x \circ y = R_a^{-1} x \cdot L_b^{-1} y = x a^{-1} \cdot b^{-1} y = x(a^{-1} b^{-1}) y$. Denote the element $a^{-1} b^{-1}$ as c. Then $(x \circ y) \cdot c = (x \cdot c) \cdot (y \cdot c)$, $R_c(x \circ y) = R_c x \cdot R_c y$.

Hence $(Q, \circ) \cong (Q, \cdot)$. If $a \cdot b = 1$, then $(Q, \circ) = (Q, \cdot)$. □

Exercise 1.124. Find the form of isotopy between quasigroups from Example 1.34.

1.3.2 Group action

We recall some definitions from [334, 471, 911].

Definition 1.125. A group G acts on a set M if for any pair of elements (g, m), $g \in G, m \in M$, an element $(gm) \in M$ is defined. Moreover, $g_1(g_2(m)) = (g_1 g_2)m$ and $em = m$ for all $m \in M$, $g_1, g_2 \in G$. Here e is the identity element of the group G.

The set $Gm = \{gm \mid g \in G\}$ is called an orbit of element m. For every m in M, we define the stabilizer subgroup of m as the set of all elements in G that fix m: $G_m = \{g \mid gm = m\}$.

Theorem 1.126. *Let x and y be two elements in M, and let g be a group element such that $y = g(x)$. Then the two stabilizer groups G_x and G_y are related by $G_y = gG_x g^{-1}$ [911].*

Proof. By definition, $h \in G_y$ if and only if $h(g(x)) = g(x)$. Applying g^{-1} to both sides of this equality yields $(g^{-1}hg)(x) = (g^{-1}g)(x) = x$; that is, $g^{-1}hg \in G_x$. □

The aforesaid gives us that the stabilizers of elements in the same orbit are conjugate to each other. Thus, one can associate a conjugacy class of a subgroup of G (i.e., the set of all conjugates of the subgroup) to each orbit.

The orbits of any two elements of the set M coincide or are not intersected. Then the set M is divided into a set of non-intersected orbits. In other words, if we define a binary relation $\sim$ on the set M as:

$$m_1 \sim m_2 \text{ if and only if there exists } g \in G \text{ such that } m_2 = gm_1,$$

then $\sim$ is an equivalence relation on the set M.

Every orbit is an invariant subset of M on which G acts transitively. The action of G on M is transitive if and only if all elements are equivalent, meaning that there is only one orbit.

A partition θ of the set M on disjoint subsets $\theta(x)$, $x \in M$ is called a partition on blocks relative to the group G, if for any $\theta(a)$ and any $g \in G$ there exists a subset $\theta(b)$ such that $g\theta(a) = \theta(b)$. It is obvious that there exist trivial partitions of the set M, namely, partitions into one-element blocks and partitions into unique blocks.

If there does not exist a partition of the set M into non-trivial blocks, then the group G is called primitive.

Definition 1.127. The action of G on M is referred to as follows:

1. Faithful (or effective), if for any two distinct $g, h \in G$ there exists an $x \in M$ such that $g(x) \neq h(x)$; or equivalently, if for any $g \neq e \in G$ there exists an $x \in M$ such that $g(x) \neq x$. Intuitively, different elements of G induce different permutations of M;

2. Free (or semiregular), if for any two distinct $g, h \in G$ and all $x \in M$ we have $g(x) \neq h(x)$; or equivalently, if $g(x) = x$ for some x, then $g = e$;

3. Regular (or simply transitive), if it is both transitive and free; this is equivalent to saying that for any two x, y in M there exists precisely one g in G such that $g(x) = y$. In this case, M is known as a principal homogeneous space for G or as a G-torsor [911].

1.3.3 Isotopism: Another point of view

Here we present another point of view on the concept of isotopism using the concept of action of a group.

Definition 1.128. An ordered $(n+1)$-tuple of permutations (bijections) of a set G is called an *isotopism (an isotopy)*.

Lemma 1.129. *Set $\mathcal{T}$ of all isotopisms of a set Q forms the group $S_Q^{n+1} = S_Q \times S_Q \times \dots \times S_Q$, which is the direct sum of $(n+1)$ copies of the group S_Q relative to the following operation (componentwise multiplication of $(n+1)$-tuples): $(\mu_1, \mu_2, \dots, \mu_{n+1}) * (\nu_1, \nu_2, \dots, \nu_{n+1}) = (\mu_1\nu_1, \mu_2\nu_2, \dots, \mu_{n+1}\nu_{n+1})$.*

Let $\mathcal{G}$ be a class of all n-ary groupoids (arity is fixed) defined on a set Q. By $\mathcal{Q}$ we denote a quasigroup class defined on the set Q. Define the action of elements of the group $\mathcal{T}$ on classes $\mathcal{G}, \mathcal{Q}$ in the following way: if (Q, f) is n-ary groupoid, $T = (\nu_1, \nu_2, \dots, nu_n, n_{n+1}) \in \mathcal{T}$, then $(Q, f)T = \nu_{n+1}^{-1} f(\nu_1 x_1, \nu_2 x_2, \dots, \nu_n x_n)$ for all $x_1, x_2, \dots, x_n$.

Theorem 1.130. *(i) $\mathcal{G}\mathcal{T} = \mathcal{G}$, (ii) $\mathcal{Q}\mathcal{T} = \mathcal{Q}$.*

Proof. (i). If (Q, f) is an n-ary groupoid, $T = (\nu_1, \nu_2, \dots, nu_n, n_{n+1}) \in \mathcal{T}$, then $(Q, f)T$ defines some other n-ary groupoid (Q, g) since the operation $g(x_1^n) = \nu_{n+1}^{-1} f(\nu_1 x_1, \nu_2 x_2, \dots, \nu_n x_n)$ is defined for all $x_1, \dots, x_n \in Q$.

(ii). Prove this theorem for the binary case. For n-ary ($n > 2$) the proof is similar. Let $(Q, \cdot)$ be a quasigroup and $T = (\alpha, \beta, \gamma)$ – an isotopy. Prove that operation $x \circ y = \gamma^{-1}(\alpha x \cdot \beta y)$ is a quasigroup operation. From (i) it follows that $(Q, \circ)$ is a binary groupoid.

For any fixed element x, the map L_x° is a permutation of the set Q, since $L_x^\circ y = \gamma^{-1} L_{\alpha x}^{\cdot} \beta y$ and the product of permutations is a permutation. Similarly, the map $R_y^\circ x = \gamma^{-1} R_{\beta y}^{\cdot} \alpha x$ is a permutation. Taking into consideration Definition 1.28, we conclude that groupoid $(Q, \circ)$ is a quasigroup. □

Corollary 1.131. *Any isotope of a left (right) quasigroup $(Q, \circ)$ is a left (right) quasigroup.*

Remark 1.132. Researches of quasigroup classes closed relative to all isotopisms or isotopisms of a fixed kind (for example, LP-isotopisms) form a direction in quasigroup theory [72, 401, 400, 869, 870, 866].

Quasigroup isotopism has a relatively clear geometrical [76] and automata theory [400] interpretation.

1.3.4 Autotopisms of binary quasigroups

Theorem 1.133. *If n-ary quasigroups (Q, f) and (Q, g) are isotopic with isotopy T, i.e., $(Q, f) = (Q, g)T$, then $Avt\,(Q, f) = T^{-1} Avt\,(Q, g) T$ [77].*

Proof. Quasigroups (Q, f) and (Q, g) are in one orbit (they are isotopic) by action of the group $\mathcal{T}$ on the set $\mathcal{Q}$ of all quasigroups of a fixed arity n. Autotopy groups of these quasigroups are stabilizers of elements (Q, f) and (Q, g) by this action. It is known that stabilizers of elements of a set S from one orbit by action of a group G on the set S are isomorphic [471, 334], moreover, they are conjugate subgroups of the group S_Q^3. □

Automorphisms and automorphism groups of some binary and n-ary quasigroups are studied in many articles, see, for example, [646, 461, 462, 736, 760, 769, 290, 772, 832, 444, 524, 525, 600].

It is clear that any automorphism is an autotopy with equal components. So, if we know the structure of autotopies of a "good" n-ary quasigroup (Q, f) and the form of isotopy T, then we have a possibility to obtain information on autotopies and automorphisms of n-ary quasigroup $(Q, g) = (Q, f)T$.

This observation was used by the study of automorphism groups of quasigroups isotopic to groups in [762].

In the binary case, Theorem 1.133 allows us (up to isomorphism) to reduce the study of autotopy group of a quasigroup to the study of autotopy group of an LP-isotope of this quasigroup, i.e., to the study of autotopy group of a loop.

Definition 1.134. An autotopism (sometimes we shall refer to autotopism as autotopy) is an isotopism of a quasigroup $(Q, \cdot)$ into itself, i.e., a triple (α, β, γ) of permutations of the set Q is an autotopy if the equality $x \cdot y = \gamma^{-1}(\alpha x \cdot \beta y)$ is fulfilled for all $x, y \in Q$.

We denote the set of all autotopies of a quasigroup $(Q, \cdot)$ as $Avt\,(Q, \cdot)$. It is clear that the defined on this set (on the set $Avt\,(Q, \cdot)$) operation $\star$ of autotopies multiplication

$$(\alpha_1, \beta_1, \gamma_1) \star (\alpha_2, \beta_2, \gamma_2) = (\alpha_1\alpha_2, \beta_1\beta_2, \gamma_1\gamma_2)$$

is a group operation. Then we have a possibility to speak on the group $(Avt(Q, \cdot), \star)$. This group has more than one denotation. The following denotation $Top(Q, \cdot)$ of the group of autotopism of a quasigroup $(Q, \cdot)$ will be also used.

Lemma 1.135. *The set of all the first (second, third) components of autotopies of a quasigroup $(Q, \cdot)$ forms a group.*

We shall denote the group of all the first components of autotopisms of a quasigroup $(Q, \cdot)$ as $AC_1(Q, \cdot)$ (the letters "A"and "C"are initial letters of the words autotopy component), of all the second components as $AC_2(Q, \cdot)$, and the set of all the third ones as $AC_3(Q, \cdot)$. The group $AC_3(Q, \cdot)$ is called the group of quasiautomorphisms of a quasigroup $(Q, \cdot)$.

Definition 1.136. The third component of any autotopism is called a *quasiautomorphism.*

Lemma 1.137. *If T is a quasigroup autotopy, then its two components uniquely determine the third one.*

Proof. If $(\alpha_1, \beta, \gamma)$ and $(\alpha_2, \beta, \gamma)$ are autotopies, then $(\alpha_2^{-1}, \beta^{-1}, \gamma^{-1})$ is an autotopy and $(\alpha_1\alpha_2^{-1}, \beta\beta^{-1}, \gamma\gamma^{-1}) = (\alpha_1\alpha_2^{-1}, \varepsilon, \varepsilon)$ is an autotopy too. We can re-write the last form of autotopy in this form: $\alpha_1\alpha_2^{-1}x \cdot y = x \cdot y$, then $\alpha_1 = \alpha_2$.

If $(\varepsilon, \varepsilon, \gamma_1\gamma_2)$ is an autotopy, then we have $x \cdot y = \gamma_1\gamma_2^{-1}(x \cdot y)$. If we put in the last equality $y = e(x)$, then we obtain $x = \gamma_1\gamma_2^{-1}x$ for all $x \in Q$, i.e., $\gamma_1 = \gamma_2$. □

Corollary 1.138. *If two components of a quasigroup autotopy are identity mappings, then the third component is also an identity mapping.*

Proof. The proof follows from Lemma 1.137 because in any quasigroup there exists identity autotopy $(\varepsilon, \varepsilon, \varepsilon)$. □

A stronger result than Lemma 1.137 is proved by I.V. Leakh [580].

Theorem 1.139. *Leakh theorem. Any autotopy $T = (\alpha_1, \alpha_2, \alpha_3)$ of a quasigroup $(Q, \circ)$ is uniquely defined by its autotopy component α_i, $i \in \{1, 2, 3\}$, and by element $b = \alpha_j a$, where a is any fixed element of set Q, $i \neq j$ [580].*

Proof. Case 1. $i = 1, j = 2$. If we have autotopies $(\alpha, \beta_1, \gamma_1)$ and $(\alpha, \beta_2, \gamma_2)$ such that $\beta_1 a = \beta_2 a = b$, then we have $\alpha x \circ \beta_1 a = \gamma_1(x \circ a)$ and $\alpha x \circ \beta_2 a = \gamma_2(x \circ a)$. Since the left sides of the last equalities are equal, then we have $\gamma_1(x \circ a) = \gamma_2(x \circ a)$, $\gamma_1 R_a x = \gamma_2 R_a x$, $\gamma_1 = \gamma_2$ and by Lemma 1.137, $\beta_1 = \beta_2$.

Case 2. $i = 1, j = 3$. Suppose there exist autotopies $(\alpha, \beta_1, \gamma_1)$ and $(\alpha, \beta_2, \gamma_2)$ such that $\gamma_1 a = \gamma_2 a = b$ for some fixed element $a \in Q$. Since $(Q, \circ)$ is a quasigroup, then for any element $x \in Q$ there exists a unique element $x' \in Q$ such that $x \circ x' = a$. Using the concept of middle quasigroup translation we can re-write the last equality in the form $P_a x = x'$ and say that P_a is a permutation of the set Q.

For all pairs x, x' we have $\alpha x \circ \beta_1 x' = \gamma_1(x \circ x') = b$ and $\alpha x \circ \beta_2 x' = \gamma_2(x \circ x') = b$. Since the right sides of the last equalities are equal, we have $\alpha x \circ \beta_1 x' = \alpha x \circ \beta_2 x'$, $\beta_1 x' = \beta_2 x'$ for all $x' \in Q$. The variable x' takes all values from the set Q since P_a is a permutation of the set Q. Therefore $\beta_1 = \beta_2$ and by Lemma 1.137, $\gamma_1 = \gamma_2$.

All other cases are proved in a similar way as Cases 1 and 2. □

Lemma 1.140. *If $(Q, \cdot)$ is a loop, then its autotopy has the form*

$$(\alpha, \beta, \gamma) = (R_b^{-1}, L_a^{-1}, \varepsilon)(\gamma, \gamma, \gamma), \tag{1.35}$$

where $\gamma 1 = a \cdot b$, γ is some bijection of the set Q.

Proof. Let $T = (\alpha, \beta, \gamma)$ be an autotopy of a loop $(Q, \cdot)$, i.e., $\alpha x \cdot \beta y = \gamma(x \cdot y)$. If we put $x = 1$, then we obtain $\alpha 1 \cdot \beta y = \gamma y$, $\gamma = L_{\alpha 1}\beta$, $\beta = L_{\alpha 1}^{-1}\gamma$. If we put $y = 1$, then, by analogy, we obtain, $\alpha = R_{\beta 1}^{-1}\gamma$. Then $T = (R_{\beta 1}^{-1}\gamma, L_{\alpha 1}^{-1}\gamma, \gamma) = (R_b^{-1}, L_a^{-1}, \varepsilon)(\gamma, \gamma, \gamma)$, where $\beta 1 = b$, $\alpha 1 = a$. If we put $x = y = 1$, then $\alpha 1 \cdot \beta 1 = \gamma 1$. □

Corollary 1.141. *If $(Q, \cdot)$ is a loop, then any its autotopy has the form*

$$(\alpha, \beta, \gamma) = (\gamma, \gamma, \gamma)(R_b, L_a, \varepsilon), \tag{1.36}$$

where γ is some bijection of the set Q.

Proof. We use Lemma 1.140. We have

$$\begin{aligned}(\alpha_1, \beta_1, \gamma_1) = T^{-1} = (\alpha^{-1}, \beta^{-1}, \gamma^{-1}) \overset{(1.35)}{=} (\gamma^{-1}, \gamma^{-1}, \gamma^{-1})(R_b, L_a, \varepsilon) = \\ (\gamma^{-1}, \gamma^{-1}, \gamma^{-1})(R_b, L_a, \varepsilon).\end{aligned} \tag{1.37}$$

□

Remark 1.142. In some articles, loop autotopy presented in the form (1.36) is called crypto-automorphism. See for example [6].

Theorem 1.143. *The order of autotopy group of a finite quasigroup Q of order n is a divisor of the number $n! \cdot n$.*

Proof. The proof follows from Theorem 1.133 (we can prove the loop case), Lemma 1.137 and Lemma 1.140 (we can take the second and the third components of loop autotopy). □

Example 1.144. The order of autotopy group of the group $Z_2 \times Z_2$ is equal to $4 \cdot 4 \cdot 6 = 4! \cdot 4$, i.e., in this case the order of autotopy group is equal to the upper bound.

There exist quasigroups (loops) with identity autotopy group [245], i.e., $|Avt(Q, \cdot)| = 1$.

Example 1.145. We give examples of such loops of order seven and nine. In this case the autotopy group is of minimal order. The proof is based on the analyses of cycle structure of quasigroup (loop) translations.

·	1	2	3	4	5	6	7
1	1	2	3	4	5	6	7
2	2	1	7	6	4	5	3
3	3	6	1	2	7	4	5
4	4	5	2	1	3	7	6
5	5	7	4	3	6	2	1
6	6	3	5	7	2	1	4
7	7	4	6	5	1	3	2

·	1	2	3	4	5	6	7	8	9
1	1	2	3	4	5	6	7	8	9
2	2	3	1	8	6	7	5	9	4
3	3	1	2	9	7	5	6	4	8
4	4	5	6	7	9	8	1	3	2
5	5	6	4	2	1	9	8	7	3
6	6	4	5	3	8	1	9	2	7
7	7	8	9	5	3	2	4	6	1
8	8	9	7	1	4	3	2	5	6
9	9	7	8	6	2	4	3	1	5

Lemma 1.146. *For any loop autotopy (α, β, γ) the following equality is true*

$$(\alpha, \beta, \gamma) = (R_b^{-1}, R_b^{-1}, R_b^{-1})(\varepsilon, R_b, R_b)(L_a, \varepsilon, L_a)(L_a^{-1}, L_a^{-1}, L_a^{-1})(\gamma, \gamma, \gamma), \tag{1.38}$$

where $a = \alpha 1$, $b = \beta 1$.

Proof. In order to check equality (1.38) it is sufficient to multiply factors in the right side of this equality and to compare the obtained result with the right side of equality (1.35). □

Remark 1.147. Notice in the conditions of Lemma 1.146 that $\gamma 1 = \alpha 1 \cdot \beta 1$. Comparison of Lemma 1.146 and Theorem 1.148 demonstrates that in the decomposition of autotopies of a loop and a group there are triples corresponding to the elements of the left and right nucleus.

We can obtain more detailed information on autotopies of a group, and since autotopy groups of isotopic quasigroups are isomorphic, on autotopies of quasigroups that are group isotopes.

Theorem 1.148. *Any autotopy of a group $(Q, +)$ can be decomposed in the following product of autotopies:*

$$(L_a\delta, R_b\delta, L_aR_b\delta) = (L_a, \varepsilon, L_a)(\varepsilon, R_b, R_b)(\delta, \delta, \delta), \tag{1.39}$$

where L_a is a left translation of the group $(Q, +)$, R_b is a right translation of this group, and δ is an automorphism of the group $(Q, +)$ [80].

Proof. Let $T = (\alpha, \beta, \gamma)$ be an autotopy of a group $(Q, +)$, i.e., for all $x, y \in Q$ the following equality

$$\alpha x + \beta y = \gamma(x + y) \tag{1.40}$$

is true.

If in equality 1.40 we put $x = y = 0$, then we obtain $\alpha 0 + \beta 0 = \gamma 0$.

If in equality 1.40 we put only $x = 0$, then $\alpha 0 + \beta y$, $L_{\alpha 0}\beta = \gamma$, $\beta = L_{-\alpha 0}\gamma$.

If in equality 1.40 we put only $y = 0$, then $\alpha x + \beta 0 = \gamma x$, $R_{\beta 0}\alpha = \gamma$, $\alpha = R_{-\beta 0}\gamma$.

Now we can re-write equality 1.40 in this form: $R_{-\beta 0}\gamma x + L_{-\alpha 0}\gamma y = \gamma(x + y)$, i.e., $\gamma x - \beta 0 - \alpha 0 + \gamma y = \gamma(x + y)$. Denote $-\beta 0 - \alpha 0$ as c, and it is easy to conclude that $-c = \alpha 0 + \beta 0$. From the last equality we have $\gamma x + c + \gamma y + c = \gamma(x + y) + c$, i.e., $R_c\gamma$ is an automorphism of the group $(Q, +)$.

Let $\theta = R_c\gamma$. Then $\gamma = R_{-c}\theta$, $\alpha x = R_{-\beta 0}\gamma x = R_{-\beta 0}R_{-c}\theta x = \theta x + \alpha 0 + \beta 0 - \beta 0 = \theta x + \alpha 0 = \alpha 0 - \alpha 0 + \theta x + \alpha 0 = L_{\alpha 0}I_{\alpha 0}\theta x$, where $I_{\alpha 0}x = -\alpha 0 + x + \alpha 0$ is an inner automorphism of the group $(Q, +)$.

Similarly,

$$\beta x = L_{-\alpha 0}\gamma x = L_{-\alpha 0}R_{-c}\theta x = -\alpha 0 + \theta x + \alpha 0 + \beta 0 = R_{\beta 0}I_{\alpha 0}\theta x.$$

We can also write the permutation γ in the following form: $\gamma x = \theta x + \alpha 0 + \beta 0 = \alpha 0 - \alpha 0 + \theta x + \alpha 0 + \beta 0 = L_{\alpha 0}R_{\beta 0}I_{\alpha 0}\theta x$.

If we rename $\alpha 0$ as a, $\beta 0$ as b, and $I_{\alpha 0}\theta$ as δ, then we obtain the following form of any autotopy of a group $(Q, +)$:

$$(L_a\delta, R_b\delta, L_aR_b\delta).$$

□

Corollary 1.149. *1. If $L_a\delta = L_aR_b\delta$, then $R_b = \varepsilon$. 2. If $R_b\delta = L_aR_b\delta$, then $L_a = \varepsilon$. 3. If $L_a\delta = R_b\delta$, then $a \in C(Q, +)$.*

Proof. 3. We have $a + \delta x + a + \delta y = a + a + \delta x + \delta y$, $\delta x + a = a + \delta x$ for all $x \in Q$. □

Corollary 1.150. *Any group quasiautomorphism has the form $L_d\,\varphi$, where $\varphi \in Aut(Q, +)$ [769].*

Proof. We have $L_aR_b\,\delta x = a + \delta x + b = a + b - b + \delta x + b = L_{a+b}I_b\,\delta x = L_d\,\varphi$, where $d = a + b$, $\varphi = I_b\delta$, $I_b\,x = -b + x + b$. □

Theorem 1.151. *[580]. If $(Q,+)$ is a group, then*

$$Avt(Q,+) \cong ((Q,+) \times (Q,+)) \leftthreetimes Aut(Q,+).$$

Remark 1.152. For the loop case, similar results are given in Theorem 4.107: in any loop $(Q,\cdot)$ we have $(N_l \times N_r) \leftthreetimes Aut(Q,\cdot) \subseteq Avt(Q,\cdot)$; $(N_l \times N_m) \leftthreetimes Aut(Q,\cdot) \subseteq Avt(Q,\cdot)$; $(N_r \times N_m) \leftthreetimes Aut(Q,\cdot) \subseteq Avt(Q,\cdot)$. It is clear that Theorem 1.151 follows from Theorem 4.107.

Cycle structure of autotopisms of quasigroups and Latin squares is studied in [313, 314, 315, 857]. Principal loop-isotopes of a finite quasigroup are studied in [176].

The form of any autotopy of a quasigroup (of a loop) often plays an important role in the study of properties of these quasigroups. The form of autotopy of any IP-loop is described in [173, 72].

1.3.5 Automorphisms of quasigroups

Theorem 1.153. *If α is an automorphism of a quasigroup (Q,A), then α is an automorphism of any parastrophe of this quasigroup.*

Proof. From equality $A(\alpha,\alpha,\alpha) = A$ we have $(A(\alpha,\alpha,\alpha))^{(13)} = A^{(13)}$. Using Lemma 1.175 further we have $A^{(13)}(\alpha,\alpha,\alpha) = A^{(13)}$.

Similarly, applying Lemma 1.175 to equality $(A(\alpha,\alpha,\alpha))^{(23)} = A^{(23)}$ we obtain $A^{(23)}(\alpha,\alpha,\alpha) = A^{(23)}$. And so on. □

Lemma 1.154. *1. If $x \cdot y = \alpha x * y$, where $(Q,*)$ is an idempotent quasigroup, α is a permutation of the set Q, then $Aut(Q,\cdot) = C_{Aut(Q,*)}(\alpha) = \{\tau \in Aut(Q,*) \mid \tau\alpha = \alpha\tau\}$, in particular, $Aut(Q,\cdot) \subseteq Aut(Q,*)$.*

2. If $x \cdot y = x * \beta y$, where $(Q,*)$ is an idempotent quasigroup, β is a permutation of the set Q, then $Aut(Q,\cdot) = C_{Aut(Q,*)}(\beta) = \{\tau \in Aut(Q,*) \mid \tau\beta = \beta\tau\}$, in particular, $Aut(Q,\cdot) \subseteq Aut(Q,*)$ ([600], Corollary 12).

Proof. 1. We give a sketch of the proof. If $\varphi \in Aut(Q,\cdot)$, then $\varphi(x \cdot y) = \varphi(\alpha x * y) = \varphi x \cdot \varphi y = \alpha\varphi x * \varphi y$. If $y = \alpha x$, then $\varphi\alpha x = \alpha\varphi x * \varphi\alpha x$, $\varphi\alpha = \alpha\varphi$, $\varphi(\alpha x * y) = \varphi\alpha x * \varphi y$.

2. The proof of Case 2 is similar to the proof of Case 1. □

Quasigroups that have an identity automorphism group and a non-identity autotopy group are constructed in Corollary 8.30. Information on automorphism groups of some loops is given in [526]. In Chapter 8 is given information on automorphisms of some quasigroups.

1.3.6 Pseudo-automorphisms and G-loops

In this subsection we follow [80, 685]. The following definition belongs to R.H. Bruck (loop case) [169] and Belousov (quasigroup case) [72]. In [72, p. 45] Belousov names right pseudo-automorphism in "our" sense as a left pseudo-automorphism.

Definition 1.155. 1. A bijection θ of a set Q is called a right pseudo-automorphism of a quasigroup $(Q,\cdot)$ if there exists at least one element $c \in Q$ such that $\theta x \cdot (\theta y \cdot c) = \theta(x \cdot y) \cdot c$ for all $x, y \in Q$, i.e.,

$$(\theta, R_c\theta, R_c\theta) \tag{1.41}$$

is an autotopy of a quasigroup $(Q,\cdot)$. The element c is called a companion of θ.

2. A bijection θ of a set Q is called a left pseudo-automorphism of a quasigroup $(Q, \cdot)$ if there exists at least one element $c \in Q$ such that $(c \cdot \theta x) \cdot \theta y = c \cdot \theta(x \cdot y)$ for all $x, y \in Q$, i.e.,
$$(L_c\theta, \theta, L_c\theta) \tag{1.42}$$
is an autotopy of a quasigroup $(Q, \cdot)$. The element c is called a companion of θ.

We can give the following:

Definition 1.156. A bijection θ of a set Q such that $\theta 1 = 1$ is called a middle pseudo-automorphism of a loop $(Q, \cdot)$, if there exists at least one element $c \in Q$ such that $(\theta x \cdot c)\theta y = \theta(x \cdot (c \cdot y))$ for all $x, y \in Q$, i.e.,
$$(R_c\theta, \theta L_c^{-1}, \theta) \tag{1.43}$$
is an autotopy of a quasigroup $(Q, \cdot)$. The element c is called a companion of θ.

Another definition of middle pseudo-automorphism in loop case is given in [386]. See also [580, 287, 546].

Definition 1.157. Let $(Q, \cdot)$ be a loop. If $\gamma \in S_Q$ and $c \in Q$ satisfy
$$\gamma(xy) = (\gamma x/(c\backslash 1))(c\backslash \gamma y)$$
for all $x, y \in Q$, then γ is called a middle pseudo-automorphism with companion c.

Example 1.158. In a group, any element c is a companion of any right pseudo-automorphism (in fact, of any automorphism) θ. This follows from Theorem 1.148, because the triplet (ε, R_c, R_c) is an autotopy of this group.

In any loop, the set of companions of right pseudo-automorphism ε (of identity pseudo-automorphism) coincides with its right nucleus (see Definition 1.211).

We can deduce Bruck's concept of right (left) loop pseudo-automorphism from a formally more general concept of right (left) A-pseudo-automorphisms. It is clear that the following definition is not a unique possible generalization of the concept of pseudo-automorphism on the quasigroup case.

Definition 1.159. 1. A bijection α of a set Q is called a right A-pseudo-automorphism of a quasigroup $(Q, \cdot)$, if there exists a bijection β of the set Q such that the triple (α, β, β) is an autotopy of quasigroup $(Q, \cdot)$.

2. A bijection β of a set Q is called a left A-pseudo-automorphism of a quasigroup $(Q, \cdot)$, if there exists a bijection α of the set Q such that the triple (α, β, α) is an autotopy of quasigroup $(Q, \cdot)$.

3. A bijection γ of a set Q is called a middle A-pseudo-automorphism of a quasigroup $(Q, \cdot)$, if there exists a bijection α of the set Q such that the triple (α, α, γ) is an autotopy of quasigroup $(Q, \cdot)$.

We shall also name as a right (left, middle) A-pseudo-automorphism the autotopical triplet that corresponds to the permutation α (β, γ).

Remark 1.160. In [386] autotopies are used by definition of pseudo-automorphisms. Autotopies with two equal components are used by definitions of loop nuclei. See below. The idea to give Definition 1.159 was born after reading the work [834]. Geometrical approach to pseudo-automorphisms is realized in [186].

Example 1.161. If $(Q,+)$ is a group, then autotopies, which correspond to left, right, and middle A-pseudo-automorphisms have the following forms, respectively: $(L_a\theta, \theta, L_a\theta)$, $(\theta, R_b\theta, R_b\theta)$, and $(L_c\theta, L_c\theta, L_{c+c}\theta)$, where $c+x=x+c$ for all $x \in Q$. The last triplet we can re-write in the following form: $(L_{-c}\delta, L_{-c}\delta, \delta)$, where $\delta = L_{c+c}\theta$.

Theorem 1.162. *Sets of all the first, second, and third components of right (left, middle) A-pseudo-automorphisms of a quasigroup $(Q,\cdot)$, and sets of right (left, middle) A-pseudo-automorphisms of a quasigroup $(Q,\cdot)$ form groups relative to operation of multiplication of these A-pseudo-automorphisms as autotopisms of the quasigroup $(Q,\cdot)$.*

Proof. If the triplets $(\alpha_1, \beta_1, \beta_1)$ and $(\alpha_2, \beta_2, \beta_2)$ are autotopisms of quasigroup $(Q,\cdot)$, then the product of these triplets is an autotopism of quasigroup $(Q,\cdot)$ and is also a right A-pseudo-automorphism of $(Q,\cdot)$.

If the triplet (α, β, β) is an autotopism of quasigroup $(Q,\cdot)$, then the triplet $(\alpha^{-1}, \beta^{-1}, \beta^{-1})$ is also an autotopism and therefore is a right A-pseudo-automorphism of $(Q,\cdot)$.

The proofs for the sets of left and middle A-pseudo-automorphisms, for the sets of the first, second, and third components of right, left, and middle A-pseudo-automorphisms are similar and we omit them. □

Following tradition [72] (see also Note 1.248) we shall denote the above listed groups using the letter Π with various indexes as follows: ${}_1\Pi_l^A$, ${}_2\Pi_l^A$, ${}_3\Pi_l^A$, ${}_1\Pi_m^A$, ${}_2\Pi_m^A$, ${}_3\Pi_m^A$, ${}_1\Pi_r^A$, ${}_2\Pi_r^A$, and ${}_3\Pi_r^A$. The letter A in the right upper corner means that this is an autotopical pseudo-automorphism. For example, ${}_2\Pi_r^A$ denotes the group of second components of right A-pseudo-automorphisms of a quasigroup $(Q,\cdot)$.

Notice, from Definition 1.159 we have that ${}_2\Pi_r^A = {}_3\Pi_r^A$; ${}_1\Pi_l^A = {}_3\Pi_l^A$; ${}_1\Pi_m^A = {}_2\Pi_m^A$.

The following theorem demonstrates that existence of a right (left) pseudo-automorphism is a sufficiently strong condition.

Theorem 1.163. *1. If a quasigroup $(Q,\cdot)$ has a right pseudo-automorphism θ with companion c, then $(Q,\cdot)$ has a right identity element $e = \theta^{-1}f(c)$, where $f(c)$ is the left local identity element of the element c.*

2. If a quasigroup $(Q,\cdot)$ has a left pseudo-automorphism θ with companion c, then $(Q,\cdot)$ has a left identity element $f = \theta^{-1}e(c)$, where $e(c)$ is the right local identity element of the element c.

Proof. Case 1. From $\theta x\cdot(\theta y\cdot c) = \theta(x\cdot y)\cdot c$ with $y = \theta^{-1}f(c)$ we have $\theta x\cdot c = \theta(x\cdot\theta^{-1}f(c))\cdot c$, $x = x\cdot\theta^{-1}f(c)$, i.e., $\theta^{-1}f(c)$ is the right identity element of the quasigroup $(Q,\cdot)$, i.e., any quasigroup with at least one right non-trivial pseudo-automorphism is a right loop.

Case 2. From $(c\cdot\theta x)\cdot\theta y = c\cdot\theta(x\cdot y)$ with $x = \theta^{-1}e(c)$ we have $c\cdot\theta y = c\cdot\theta(\theta^{-1}e(c)\cdot y)$, $y = \theta^{-1}e(c)\cdot y$, i.e., the element $\theta^{-1}e(c)$ is the left identity element of the quasigroup $(Q,\cdot)$, i.e., any quasigroup with at least one left non-trivial pseudo-automorphism is a left loop. □

Corollary 1.164. *If $(Q,\cdot)$ is a loop and θ is its right (left) pseudo-automorphism, then $\theta 1 = 1$.*

Proof. If θ is a right pseudo-automorphism, then in a loop with identity element 1 we have: $1 = \theta^{-1}f(1) = \theta^{-1}1$, $\theta 1 = 1$.

If θ is a left pseudo-automorphism, then in a loop with identity element 1 we have: $1 = \theta^{-1}e(1) = \theta^{-1}1$, $\theta 1 = 1$. □

Lemma 1.165. *1. In a right loop $(Q,\cdot)$ with the right identity element e, any right A-pseudo-automorphism is a right pseudo-automorphism.*

2. *In a left loop $(Q, \cdot)$ with the left identity element f, any left A-pseudo-automorphism is a left pseudo-automorphism.*

Proof. Case 1. From equality $\alpha x \cdot \beta y = \beta(x \cdot y)$ by $y = e$ we have $R_{\beta e}\alpha x = \beta x$, and we obtain the following autotopy triplet: $(\alpha, R_{\beta e}\alpha, R_{\beta e}\alpha)$. By Definition 1.155 component α of the autotopy $(\alpha, R_{\beta e}\alpha, R_{\beta e}\alpha)$ is a right pseudo-automorphism with companion βe.

Case 2. From equality $\alpha x \cdot \beta y = \alpha(x \cdot y)$ by $x = f$ we have $L_{\alpha f}\beta y = \alpha y$, and we obtain the following autotopy triplet: $(L_{\alpha f}\beta, \beta, L_{\alpha f}\beta)$. By Definition 1.155 component β of the autotopy $(L_{\alpha f}\beta, \beta, L_{\alpha f}\beta)$ is a right pseudo-automorphism with companion αf. □

Remark 1.166. From Lemma 1.165 it follows that the set of left (right) pseudo-automorphisms of a loop $(Q, \cdot)$ forms a group relative to component-wise multiplication. If $(L_a\theta, \theta, L_a\theta)$ and $(L_b\delta, \delta, L_b\delta)$ are left pseudo-automorphisms, then $(L_a\theta, \theta, L_a\theta)(L_b\delta, \delta, L_b\delta) = (L_c\theta\delta, \theta\delta, L_c\theta\delta)$, where $c = a \cdot \theta b$. Indeed $L_a\theta L_b\delta 1 = a \cdot \theta b$.

Taking into consideration the Leakh Theorem (Theorem 1.139) we can identify the group of left pseudo-automorphisms of a loop $(Q, \cdot)$ with the set of pairs (a, θ), where $(L_a\theta, \theta, L_a\theta)$ is a left pseudoautomorphism of the loop $(Q, \cdot)$, and with the following multiplication of these pairs,

$$(a, \theta)(b, \delta) = (a \cdot \theta b, \theta\delta).$$

Definition 1.167. G-loop is a loop which is isomorphic to all its loop isotopes (LP-isotopes) [173, 685].

This definition has a geometric origin. R.H. Bruck [173, p. 60] cited as an unsolved problem the problem of description of G-loops. Geometrically, the Bruck problem was solved by Barlotti and Strambach [34].

Belousov solved this problem (see [72, Theorem 3.8], [685, p. 82] for details) and proved the following:

Theorem 1.168. *A loop $(L, \cdot)$ is a G-loop if and only if every element $x \in L$ is a companion of some right and some left pseudo-automorphism of $(L, \cdot)$ [72].*

By solving this Bruck problem, Belousov used the concept of left (right) derivative operation of a quasigroup (in fact, derivative operation is defined using a special kind of quasigroup isotopy). Moreover, in some sense Belousov constructed G-loops theory, which is equivalent but alternative to the Bruck theory of G-loops, because Belousov has swapped definition, and characterizing theorem in the theory of G-loops (Definition 1.167 and Theorem 1.168). This change lets him to define left and right G-quasigroups (see Chapter 4 Section 4.1 for details).

Examples, constructions, and the theory of G-loops are given in [72, Chapter X], [65]. In recent years, the theory of G-loops has developed very actively. New classes of G-loops are discovered. For example, any CC-loop (conjugacy closed loops are loops that satisfy the following identities: $xy \cdot z = xz \cdot (z\backslash(yz))$ and $z \cdot yx = ((zy)/z) \cdot zx$), Buchsteiner loop (a loop with identity $x\backslash(xy \cdot z) = (y \cdot zx)/x$ [223]), and Cheban loop ($x(xy \cdot z) = (y \cdot zx)x$) [695] is a G-loop. See also [38, 39, 374, 546].

Example 1.169. We give an example of a Buchsteiner loop that is constructed using Mace

4 [614]. Notice, any quasigroup with Buchsteiner identity is a loop [482].

$*$	0	1	2	3	4	5
0	0	1	2	3	4	5
1	1	2	3	5	0	4
2	2	4	5	1	3	0
3	3	0	4	2	5	1
4	4	5	1	0	2	3
5	5	3	0	4	1	2

Using concepts of left, right, middle A-pseudo-automorphisms it is possible to give the following:

Definition 1.170. 1. A quasigroup $(Q, \cdot)$ is called a right GA-quasigroup, if the group ${}_2\Pi_r^A$ (or the group ${}_3\Pi_r^A$) is transitive on the set Q.

2. A quasigroup $(Q, \cdot)$ is called a left GA-quasigroup, if the group ${}_1\Pi_l^A$ (or the group ${}_3\Pi_l^A$) is transitive on the set Q.

3. A quasigroup $(Q, \cdot)$ is called a middle GA-quasigroup, if the group ${}_1\Pi_m^A$ (or the group ${}_2\Pi_m^A$) is transitive on the set Q.

4. A right and left GA-quasigroup is called GA-quasigroup.

From Lemma 1.165 it follows that a GA-loop is a "standard" G-loop. From Example 1.161 it follows that any group is a left and right GA-quasigroup, and any commutative group is a middle GA-quasigroup.

Problem 1.3. Research left, right, and middle GA-quasigroups.

1.3.7 Parastrophisms as operators

Similar to isotopisms, we can consider quasigroup parastrophisms as operators that act on the class of all quasigroups. Recall, we have 3! operators in the binary case and $(n+1)!$ operators in the n-ary case.

It is clear that all parastrophisms of a quasigroup $(Q, \cdot)$ relative to the operation of its composition form the group of parastrophism PRS. It is easy to see that $PRS \cong S_3 \cong D_3 \cong \langle l, r, s \mid l^2 = r^2 = s^2 = lrls = rlrs = \varepsilon \rangle$, where S_3 means the symmetric group defined on 3 elements and D_3 means the dihedral group of order six [471, 407, 924]. Notice $(lr)^3 = (lr)lrlr = (sl)lrlr = s(rl)r = ssrr = \varepsilon$. For the convenience of readers, we give the Cayley table of the group S_3 in terms of elements l, r, s.

$*$	ε	l	r	s	lr	rl
ε	ε	l	r	s	lr	rl
l	l	ε	lr	rl	r	s
r	r	rl	ε	lr	s	l
s	s	lr	rl	ε	l	r
lr	lr	s	l	r	rl	ε
rl	rl	r	s	l	ε	lr

It is clear that there exist $15 = \frac{6 \cdot 5}{2}$ unordered pairs of parastrophes of a quasigroup (Q, A) such that in any pair, the parastrophes are different.

Lemma 1.171. *Action of the group PRS on the set of all unordered pairs of parastrophes of a quasigroup (Q, A) divide this set in the following four orbits [700]:*

I. $(A, {}^rA), ({}^lA, {}^{lr}A), ({}^{rl}A, {}^sA)$;
II. $(A, {}^lA), ({}^rA, {}^{rl}A), ({}^sA, {}^{lr}A)$;
III. $(A, {}^sA), ({}^rA, {}^{lr}A), ({}^lA, {}^{rl}A)$;
IV. $({}^lA, {}^rA), (A, {}^{lr}A), ({}^rA, {}^sA), ({}^{lr}A, {}^{rl}A), (A, {}^{rl}A), ({}^lA, {}^sA)$.

Proof. It is possible to use direct calculations using the Cayley table of the group S_3 in terms of elements l, r, s. For example, $s(A, {}^rA) = ({}^sA, {}^{sr}A) = ({}^sA, {}^{rl}A) = ({}^{rl}A, {}^sA)$. □

Lemma 1.172. *The set of all unordered pairs of parastrophes of a quasigroup (Q, A) relative to the action of group $\langle \varepsilon, s\rangle \cong Z_2$ is shared in the following orbits:*

I. $(A, {}^rA), ({}^sA, {}^{rl}A)$; *II.* $(A, {}^lA), ({}^sA, {}^{lr}A)$; *III.* $(A, {}^sA)$;
IV. $(A, {}^{lr}A), ({}^sA, {}^lA)$; *V.* $(A, {}^{rl}A), ({}^sA, {}^rA)$;
VI. $({}^lA, {}^{lr}A)$; *VII.* $({}^rA, {}^{rl}A)$;
VIII. $({}^lA, {}^rA), ({}^{lr}A, {}^{rl}A)$; *IX.* $({}^rA, {}^{lr}A), ({}^lA, {}^{rl}A)$.

Proof. The proof is similar to the proof of Lemma 1.171. □

1.3.8 Isostrophism

Isostrophy of a quasigroup is a transformation that is a combination of parastrophy and isotopy. Therefore, the isostrophic image of a quasigroup (Q, A) is an isotopic image of its parastrophe (Definition 1.174).

It is possible to define an isostrophic image of a quasigroup (Q, A) as a parastrophic image of its isotopic image. Therefore, there exists a possibility to define isostrophy in at least two ways. Here we use the first possibility.

Sade [733] names this transformation *paratopy.* Isostrophism (synonym of isostrophy) has clear geometrical (net) motivation [72, 685, 34]. See also [580]. In fact isostrophism is net isomorphism [72].

We define the action of parastrophy on an isotopic triplet in the following way: if $T = (\alpha_1, \alpha_2, \alpha_3)$ is an isotopy, σ is a parastrophy of a quasigroup (Q, A), then we shall denote by T^σ the triple $(\alpha_{\sigma^{-1}1}, \alpha_{\sigma^{-1}2}, \alpha_{\sigma^{-1}3})$.

Remark 1.173. It is also possible to define the action of a parastrophy σ on the triple $T = (\alpha_1, \alpha_2, \alpha_3)$ in the following way: $T^\sigma = (\alpha_{\sigma 1}, \alpha_{\sigma 2}, \alpha_{\sigma 3})$.

Definition 1.174. A quasigroup (Q, B) is an isostrophic image of a quasigroup (Q, A) if there exists a collection of permutations $(\sigma, (\alpha_1, \alpha_2, \alpha_3)) = (\sigma, T)$, where $\sigma \in S_3$, $T = (\alpha_1, \alpha_2, \alpha_3)$ and $\alpha_1, \alpha_2, \alpha_3$ are permutations of the set Q such that

$$B(x_1, x_2) = A(x_1, x_2)(\sigma, T) = \alpha_3^{-1} A(\alpha_1 x_{\sigma^{-1}1}, \alpha_2 x_{\sigma^{-1}2}) \tag{1.44}$$

for all $x_1, x_2 \in Q$ [788]. Here $x_3 = B(x_1, x_2)$.

A collection of permutations $(\sigma, (\alpha_1, \alpha_2, \alpha_3)) = (\sigma, T)$ will be called an *isostrophy* of a quasigroup (Q, A).

Often an isostrophy (σ, T) is called a σ-isostrophy T or an isostrophy of type σ. We can re-write the equality from Definition 1.174 in the form $A^\sigma T = B$, where $T = (\alpha_1, \alpha_2, \alpha_3)$.

Probably R. Artzy was the first to give an algebraic definition of isostrophy [20]. For n-ary quasigroups the concept of isostrophy is studied in [77]. Some generalizations of concepts of isotopy and parastrophy for n-ary relations and some algebraic systems are given in [954].

Relations between parastrophy and isotopy are reflected in the following classical lemma:

Lemma 1.175. *In a quasigroup (Q, A): (i) $(AT)^\sigma = A^\sigma T^\sigma$; (ii) $(T_1T_2)^\sigma = T_1^\sigma T_2^\sigma$, where T is an isotopy and σ is a parastrophy of quasigroup (Q, A) [72].*

Proof. (i) We have $AT = (\alpha_1 x_1, \alpha_2 x_2, \alpha_3 x_3)$, where $x_3 = A(x_1, x_2)$. By the definition of parastrophy (equality 1.2) we have

$$(A(x_1, x_2, x_3))^\sigma = A^\sigma(x_1, x_2, x_3) = A(x_{\sigma^{-1}1}, x_{\sigma^{-1}2}, x_{\sigma^{-1}3}).$$

Then we have

$$(AT)^\sigma = (A(\alpha_1 x_1, \alpha_2 x_2, \alpha_3 x_3))^\sigma = A^\sigma(\alpha_1 x_1, \alpha_2 x_2, \alpha_3 x_3) = \\ A(\alpha_{\sigma^{-1}1} x_{\sigma^{-1}1}, \alpha_{\sigma^{-1}2} x_{\sigma^{-1}2}, \alpha_{\sigma^{-1}3} x_{\sigma^{-1}3}) = A^\sigma T^\sigma.$$

(ii) Let $T_1 = (\beta_1, \beta_2, \beta_3)$, $T_2 = (\gamma_1, \gamma_2, \gamma_3)$. Then we have

$$(T_1T_2)^\sigma = (\beta_1\gamma_1, \beta_2\gamma_2, \beta_3\gamma_3)^\sigma = \\ (\beta_{\sigma^{-1}1}\gamma_{\sigma^{-1}1}, \beta_{\sigma^{-1}2}\gamma_{\sigma^{-1}2}, \beta_{\sigma^{-1}3}\gamma_{\sigma^{-1}3}) = T_1^\sigma T_2^\sigma.$$

□

Definition 1.176. Let (σ, T) and (τ, S) be some isostrophisms, $T = (\alpha_1, \alpha_2, \alpha_3)$, $S = (\beta_1, \beta_2, \beta_3)$. Define the following operation of multiplication (successive application of maps) of a pair of isostrophisms:

$$(\sigma, T)(\tau, S) = (\sigma\tau, T^\tau S). \tag{1.45}$$

We can write the equality $B = A(\sigma\tau, T^\tau S)$ in more detail: (the right record of maps)

$$B(x_1, x_2, x_3) = A(x_{1(\tau^{-1}\sigma^{-1})}\alpha_{1\tau^{-1}}\beta_1, x_{2(\tau^{-1}\sigma^{-1})}\alpha_{2\tau^{-1}}\beta_2, x_{3(\tau^{-1}\sigma^{-1})}\alpha_{3\tau^{-1}}\beta_3). \tag{1.46}$$

If we use the left record of maps, then

$$B(x_1, x_2, x_3) = A(\beta_1\alpha_{\tau^{-1}1}x_{\sigma^{-1}(\tau^{-1}1)}, \beta_2\alpha_{\tau^{-1}2}x_{\sigma^{-1}(\tau^{-1}2)}, \beta_3\alpha_{\tau^{-1}3}x_{\sigma^{-1}(\tau^{-1}3)}). \tag{1.47}$$

Lemma 1.177. *If $(\sigma, T) = (\sigma, (\alpha_1, \alpha_2, \alpha_3))$ is an isostrophy, then*

$$(\sigma, T)^{-1} = (\sigma^{-1}, (T^{-1})^{\sigma^{-1}}) = (\sigma^{-1}, (\alpha_{\sigma 1}^{-1}, \alpha_{\sigma 2}^{-1}, \alpha_{\sigma 3}^{-1})). \tag{1.48}$$

Proof. Indeed, $(\sigma, T)^{-1} = (\sigma^{-1}, (T^{-1})^{\sigma^{-1}})$, $(T^{-1})^{\sigma^{-1}} = (\alpha_1^{-1}, \alpha_2^{-1}, \alpha_3^{-1})^{\sigma^{-1}} = (\alpha_{\sigma 1}^{-1}, \alpha_{\sigma 2}^{-1}, \alpha_{\sigma 3}^{-1})$. Then

$$(\sigma, T)(\sigma^{-1}, (T^{-1})^{\sigma^{-1}}) = (\sigma\sigma^{-1}, T^{\sigma^{-1}}(T^{-1})^{\sigma^{-1}}) = (\varepsilon, \varepsilon), \\ (\sigma^{-1}, (T^{-1})^{\sigma^{-1}})(\sigma, T) = (\sigma^{-1}\sigma, T^{-1}T) = (\varepsilon, \varepsilon).$$

□

Lemma 1.178. *Let (σ, ε) and (ε, S) be some isostrophisms. Then*

$$(\sigma, \varepsilon)(\varepsilon, S) = (\sigma, S). \tag{1.49}$$

Proof. The proof directly follows from Definition 1.176. □

Therefore any isostrophism can be considered as a consecutive application of parastrophism and isotopism (here the order is important).

Example 1.179. We construct an isostrophic image of quasigroup $(Q, \circ)$ by isostrophy $((\sigma),(T)) = ((23),((12),(23),(123)))$ of quasigroup of order 3 from Example 1.34 in two steps. In the first step we pass from initial quasigroup $(Q, \circ)$ to its (23)-parastrophe and after this we pass from quasigroup $(Q, \overset{(23)}{\circ})$ to its isotope of the form $T = ((12),(23),(123))$ (see Example 1.116 for more details).

$\circ$	a	b	c
a	a	b	c
b	b	c	a
c	c	a	b

$\Longrightarrow$

$\overset{(23)}{\circ}$	a	b	c
a	a	b	c
b	c	a	b
c	b	c	a

$\Longrightarrow$

$(\overset{(23)}{\circ}, T)$	a	b	c
a	b	a	c
b	c	b	a
c	a	c	b

1.3.9 Autostrophisms

Definition 1.180. If in Definition 1.174 $(Q, B) = (Q, A)$, then we obtain the definition of quasigroup autostrophy (autostrophism) [72].

Autostrophisms and the autostrophy group of a quasigroup have been studied in the following articles and books: [77, 76, 580, 685, 619, 203]. In [203] the following result is proved: if $n \to \infty$, then almost all quasigroups of order n have an identity autostrophy group.

Exercise 1.181. Prove that the set of all autostrophies of a quasigroup $(Q, \cdot)$ forms a group relative to the operation defined by formula (1.45).

Lemma 1.182. *If quasigroup $(Q, \circ)$ is an isostrophic image of quasigroup $(Q, \cdot)$ with an isostrophy T, i.e., $(Q, \circ) = (Q, \cdot)T$, then $Aus(Q, \circ) = T^{-1}Aus(Q, \cdot)T$.*

Proof. The proof, in fact, repeats the proof of Lemma 1.4 from [80]. Let $S \in Aus(Q, \circ)$. Then $(Q, \cdot)T = (Q, \cdot)TS$, $(Q, \cdot) = (Q, \cdot)TST^{-1}$,

$$TAus(Q, \circ)T^{-1} \subseteq Aus(Q, \cdot). \tag{1.50}$$

If $(Q, \circ) = (Q, \cdot)T$, then $(Q, \circ)T^{-1} = (Q, \cdot)$, expression (1.50) takes the form $T^{-1}Aus(Q, \cdot)T \subseteq Aus(Q, \circ)$,

$$TAus(Q, \circ)T^{-1} \supseteq Aus(Q, \cdot). \tag{1.51}$$

Comparing (1.50) and (1.51) we obtain $Aus(Q, \circ) = T^{-1}Aus(Q, \cdot)T$. □

Notice, if we apply "group action ideology" to the definition of isostrophism (as it was done in Subsection 1.3.3), then Lemma 1.182 is a variant of Theorem 1.126.

Remark 1.183. Isotopism is used in cryptology [395]. It is clear that there is a sense to try to use isostrophism instead of isotopism.

1.3.9.1 Coincidence of quasigroup parastrophes

The following lemma is well known. See, for example, [845, 18].

Lemma 1.184. *Quasigroup $(Q, \cdot)$ coincides with its:*

1. *$(12) = s$ parastrophe if and only if the identity $xy = yx$ is true (identity of commutativity);*

2. *(13) = l parastrophe if and only if the identity $xy \cdot y = x$ is true (right semi-symmetric identity);*

3. *(23) = r parastrophe if and only if the identity $x \cdot xy = y$ is true (left semi-symmetric identity);*

4. *(123) = sl parastrophe (left record of action of operators, i.e., ${}^{sl}A = {}^{s}({}^{l}A)$) if and only if the identity $y \cdot xy = x$ is true (semi-symmetric identity);*

5. *(132) = sr parastrophe (left record of action of operators) if and only if the identity $xy \cdot x = y$ is true (semi-symmetric identity).*

Proof. Case 1. In this case quasigroup $(Q, \cdot)$ has autostrophy $T = ((12), (\varepsilon, \varepsilon, \varepsilon))$. If we apply autostrophy T to a triple $(x, y, x \cdot y)$, then we obtain the triple $(y, x, x \cdot y)$. Therefore we have identity $yx = xy$. It is clear that the inverse is also true.

Cases 2, 3 are proved in a similar way.

Case 4. In this case quasigroup $(Q, \cdot)$ has autostrophy $T = ((123), (\varepsilon, \varepsilon, \varepsilon))$. If we apply autostrophy T to a triple $(x, y, x \cdot y)$, then we obtain the triple $(y, x \cdot y, x)$. Therefore we have identity $y \cdot xy = x$. It is clear that the inverse is also true.

Case 5 is proved in a similar way. □

Corollary 1.185. *In a quasigroup $(Q, \cdot)$ identities $y \cdot xy = x$ and $xy \cdot x = y$ are equivalent.*

Proof. If quasigroup $(Q, \cdot)$ has autostrophy $T = ((123)(\varepsilon, \varepsilon, \varepsilon))$, then it has autostrophy $T^2 = ((132)(\varepsilon, \varepsilon, \varepsilon))$ and vice versa, since $(T^2)^2 = T$. □

Lemma 1.186. *In a quasigroup $(Q, \cdot)$ all identities of the set $A = \{x \cdot xy = y, xy = yx, xy \cdot y = x, x \cdot yx = y\}$ result from any two identities of the set A.*

Proof. The proof is based on Lemma 1.184 and the following property of the group S_3: any two elements of the set $\{l, r, s, sl\}$ generate the group S_3. □

1.3.10 Inverse loops to a fixed loop

In [65, 72] V.D. Belousov defined inverse loops to a fixed loop (Q, A) with the same identity element in the following way: $A^{\lambda}(x, y) = x/Iy$; $A^{\rho}(x, y) = I^{-1}x\backslash y$, where $x \cdot Ix = 1$ for all $x \in Q$. The main purpose of this definition is to provide that any "almost pure" parastrophe of a loop is also a loop. In [685, Chapter II] the similar loop transformations are named as isostrophies. See also [21].

In [72, p. 19] it is noticed that $A^{\lambda^2} = A$. Taking this condition into consideration we can prove the following:

Lemma 1.187. *Any inverse loop to a fixed loop $(Q, \cdot)$ has the following form, where $I^2 = \varepsilon$.*

1. $(Q, \lambda) = (Q, \cdot)((1\,3), (\varepsilon, I, \varepsilon))$, $x \lambda y = x/Iy$;

2. $(Q, \rho) = (Q, \cdot)((2\,3), (I, \varepsilon, \varepsilon))$, $x \rho y = Ix\backslash y$;

3. $(Q, *) = (Q, \cdot)((1\,2), (I, I, I))$, $x * y = I(Iy \cdot Ix)$;

4. $(Q, \rho^*) = (Q, \cdot)((1\,2\,3), (I, \varepsilon, I))$, $x \rho^* y = I(Iy\backslash x)$;

5. $(Q, \lambda^*) = (Q, \cdot)((1\,3\,2), (\varepsilon, I, I))$, $x \lambda^* y = I(y/Ix)$;

Proof. In the language of isostrophic images we can rewrite Belousov's inverse loop operations in the following way $(Q, \lambda) = (Q, \cdot)((1\,3), (\varepsilon, I, \varepsilon))$, $(Q, \rho) = (Q, \cdot)((2\,3), (I^{-1}, \varepsilon, \varepsilon))$, respectively.

The property $A^{\lambda^2} = A$ we can rewrite in the following way $((1\,3), (\varepsilon, I, \varepsilon))^2 = (\varepsilon, (\varepsilon, I^2, \varepsilon))$, i.e., the triple $(\varepsilon, I^2, \varepsilon)$ is an autotopy of loop (Q, A). Then by Corollary 1.138 $I^2 = \varepsilon$, $I = I^{-1}$.

After this, Cases 1–5 are easy. □

Corollary 1.188. *Any loop $(Q, \cdot)$ with the property $I^2 = \varepsilon$ has no more than 5 loops that are inverse to it.*

Proof. It is easy to calculate that $\langle((1\,3), (\varepsilon, I, \varepsilon)), ((2\,3), (I, \varepsilon, \varepsilon))\rangle \cong S_3$. □

Information on loops with the property $I^2 = \varepsilon$ is in Section 2.2. See also [72, 777].

Moreover, these six loops form the system of reciprocally inverse loops. Notice, if $I^2 \neq \varepsilon$, then any loop can contain even an infinite number of loops that are inverse to it.

Say that a loop $(Q, \cdot)$ satisfies the *antiautomorphic inverse property* (AAIP) if $I^2 = \varepsilon$ and $I(xy) = IyIx$ for all $x, y \in Q$.

Corollary 1.189. *Let $(Q, \cdot)$ be a loop with $I^2 = \varepsilon$.*

1. *$(Q, \cdot) = (Q, \cdot)((1\,3), (\varepsilon, I, \varepsilon))$, $x \cdot y = x/Iy$ if and only if $(Q, \cdot)$ is an RIP-loop;*
2. *$(Q, \cdot) = (Q, \cdot)((2\,3), (I, \varepsilon, \varepsilon))$, $x \cdot y = Ix\backslash y$ if and only if $(Q, \cdot)$ is an LIP-loop;*
3. *$(Q, \cdot) = (Q, \cdot)((1\,2), (I, I, I))$, $x \cdot y = I(Iy \cdot Ix)$ if and only if $(Q, \cdot)$ is an antiautomorphic inverse property loop (AAIP-loop);*
4. *$(Q, \cdot) = (Q, \cdot)((1\,2\,3), (I, \varepsilon, I))$, $x \cdot y = I(Iy\backslash x)$ if and only if $I = \varepsilon$, $y \cdot xy = x$;*
5. *$(Q, \cdot) = (Q, \cdot)((1\,3\,2), (\varepsilon, I, I))$, $x \cdot y = I(y/Ix)$ if and only if $I = \varepsilon$, $y \cdot xy = x$.*

Proof. Cases 1–3 are easy.

Case 4. If $x = 1$, then $y = I(Iy\backslash 1)$. From equality $x \cdot Ix = 1$ we have $x\backslash 1 = Ix$. Therefore $y = I(Iy\backslash 1) = I^3 y$. But $I^2 = \varepsilon$. Then $I = \varepsilon$ and we obtain $x \cdot y = y\backslash x$, $y \cdot xy = x$. It is easy to see that from the last equality by $x = 1$ we have $y^2 = 1$.

Converse. It is easy to see.

Case 5 is proved in a way similar to Case 4. Prove only that in quasigroups the identities $y \cdot xy = x$ and $yx \cdot y = x$ are equivalent. From identity $y \cdot xy = x$ we have $L_y R_y = \varepsilon$, further $R_y L_y = \varepsilon$, $yx \cdot y = x$. □

It is possible to define the following inverse loops to a fixed loop $(Q, \cdot)$ with the same identity element.

Definition 1.190. 1. $(Q, \lambda) = (Q, \cdot)((1\,3), (\varepsilon, I, \varepsilon))$, $x\lambda y = x/Iy$;

2. $(Q, \rho) = (Q, \cdot)((2\,3), (I^{-1}, \varepsilon, \varepsilon))$, $x\rho y = I^{-1}x\backslash y$;

3. $(Q, *) = (Q, \cdot)((1\,2), (\varepsilon, \varepsilon, \varepsilon))$, $x * y = y \cdot x$;

4. $(Q, \rho^*) = (Q, \cdot)((1\,2\,3), (\varepsilon, I^{-1}, \varepsilon))$, $x\rho^* y = I^{-1}y\backslash x$;

5. $(Q, \lambda^*) = (Q, \cdot)((1\,3\,2), (I, \varepsilon, \varepsilon))$, $x\lambda^* y = y/Ix$.

Notice, the fulfilment of condition $I^2 = \varepsilon$ in Definition 1.190 is not necessary.

Exercise 1.191. Check that all isostrophic images of a loop $(Q, \cdot)$ from Lemma 1.187 and Definition 1.190 are loops.

1.3.11 Anti-autotopy

Here we give definitions of some weakly examined objects.

Definition 1.192. An autostrophy of the form $((12), (\alpha, \beta, \gamma))$ is called an anti-autotopy [83].

Definition 1.193. An algebraic structure consisting of a non-empty set $(H, [\,])$ with a ternary operation denoted by $[\,]$ which satisfies the para-associative law

$$[[a, b, c], d, e] = [a, [b, c, d], e] = [a, b, [c, d, e]] \text{ for all } a, b, c, d, e \in H$$

and the identity law

$$[a, a, x] = [x, a, a] = x \text{ for all } a, x \in H$$

is called a heap [902, 906].

A group can be regarded as a heap under the operation $[x, y, z] = xy^{-1}z$.

Theorem 1.194. *The set of quasigroup anti-autotopies forms a heap relative to the following operation* $[T_1, T_2, T_3] = T_1T_2^{-1}T_3$*, where* T_1, T_2, T_3 *are some antiautotopies [83].*

Theorem 1.195. *The set of quasigroup autostrophies of the form* $((13), (\alpha, \beta, \gamma))$ *forms a heap relative to composition* $[T_1, T_2, T_3] = T_1T_2^{-1}T_3$*, where* T_1, T_2, T_3 *are some* (13)*-autostrophies.*

The set of quasigroup autostrophies of the form $((23), (\alpha, \beta, \gamma))$ *forms a heap relative to standard* $[T_1, T_2, T_3] = T_1T_2^{-1}T_3$*, where* T_1, T_2, T_3 *are some* (23)*-autostrophies.*

Proof. The proof is standard and we omit it. □

Theorem 1.196. *Any antiautotopy of a group* $(Q, +)$ *has the following form:*

$$T = (L_a, R_b, L_aR_b)\tilde{\theta}, \tag{1.52}$$

where L_a *and* R_b *are left and right translations of group* $(Q, +)$*, and* $\tilde{\theta}$ *is an antiautomorphism of the group* $(Q, +)$ *[879].*

Definition 1.197. An autostrophy of a quasigroup $(Q, \cdot)$ of the form $((\sigma), (\alpha, \beta, \gamma))$, where $\sigma \in S_3$, $\alpha, \beta, \gamma \in S_Q$ will be called: left σ-A-nucleus, if $\beta = \varepsilon$; right σ-A-nucleus, if $\alpha = \varepsilon$; middle σ-A-nucleus, if $\gamma = \varepsilon$; left σ-A-pseudo-automorphism, if $\alpha = \gamma$; right σ-A-pseudo-automorphism, if $\beta = \gamma$; and middle σ-A-pseudo-automorphism, if $\alpha = \beta$.

1.3.12 Translations of isotopic quasigroups

Lemma 1.198. *If quasigroups* $(Q, \cdot)$ *and* $(Q, \circ)$ *are isotopic, i.e.,* $x \circ y = \gamma^{-1}(\alpha x \cdot \beta y)$ *for all* $x, y \in Q$ *and some fixed permutations* α, β, γ *of the set set* Q*, then:*

$$\begin{array}{lll} L^{\circ}_x = \gamma^{-1}L^{\cdot}_{\alpha x}\beta, & R^{\circ}_y = \gamma^{-1}R^{\cdot}_{\beta y}\alpha, & P^{\circ}_z = \beta^{-1}P^{\cdot}_{\gamma z}\alpha, \\ (L^{\circ}_x)^{-1} = \beta^{-1}(L^{\cdot}_{\alpha x})^{-1}\gamma, & (R^{\circ}_y)^{-1} = \alpha^{-1}(R^{\cdot}_{\beta y})^{-1}\gamma, & (P^{\circ}_z)^{-1} = \alpha^{-1}(P^{\cdot}_{\gamma z})^{-1}\beta. \end{array}$$

Proof. If $x \circ y = \gamma^{-1}(\alpha x \cdot \beta y)$, then $L^{\circ}_x y = \gamma^{-1}L^{\cdot}_{\alpha x}\beta y$, i.e., $L^{\circ}_x = \gamma^{-1}L^{\cdot}_{\alpha x}\beta$. Similarly $R^{\circ}_y = \gamma^{-1}R^{\cdot}_{\beta y}\alpha$.

Let $x \circ y = \gamma^{-1}(\alpha x \cdot \beta y) = z$. Then $P^{\circ}_z x = y$, $P^{\cdot}_{\gamma z}\alpha x = \beta y$, $P^{\circ}_z = \beta^{-1}P^{\cdot}_{\gamma z}\alpha$.

Therefore $(L^{\circ}_x)^{-1} = \beta^{-1}(L^{\cdot}_{\alpha x})^{-1}\gamma$, $(R^{\circ}_y)^{-1} = \alpha^{-1}(R^{\cdot}_{\beta y})^{-1}\gamma$, $(P^{\circ}_z)^{-1} = \alpha^{-1}(P^{\cdot}_{\gamma z})^{-1}\beta$. □

Table 1.3: Translations of quasigroup isostrophes.

	(ε, T)	$((12), T)$	$((13), T)$
R°	$\gamma^{-1}R^{\cdot}_{\beta y}\alpha$	$\gamma^{-1}L^{\cdot}_{\alpha y}\beta$	$\alpha^{-1}(R^{\cdot}_{\beta y})^{-1}\gamma$
L°	$\gamma^{-1}L^{\cdot}_{\alpha x}\beta$	$\gamma^{-1}R^{\cdot}_{\beta x}\alpha$	$\alpha^{-1}(P^{\cdot}_{\gamma x})^{-1}\beta$
P°	$\beta^{-1}P^{\cdot}_{\gamma z}\alpha$	$\alpha^{-1}(P^{\cdot}_{\gamma z})^{-1}\beta$	$\beta^{-1}(L^{\cdot}_{\alpha z})^{-1}\gamma$
$(R^{\circ})^{-1}$	$\alpha^{-1}(R^{\cdot}_{\beta y})^{-1}\gamma$	$\beta^{-1}(L^{\cdot}_{\alpha y})^{-1}\gamma$	$\gamma^{-1}R^{\cdot}_{\beta y}\alpha$
$(L^{\circ})^{-1}$	$\beta^{-1}(L^{\cdot}_{\alpha x})^{-1}\gamma$	$\alpha^{-1}(R^{\cdot}_{\beta x})^{-1}\gamma$	$\beta^{-1}P^{\cdot}_{\gamma x}\alpha$
$(P^{\circ})^{-1}$	$\alpha^{-1}(P^{\cdot}_{\gamma z})^{-1}\beta$	$\beta^{-1}P^{\cdot}_{\gamma z}\alpha$	$\gamma^{-1}L^{\cdot}_{\alpha z}\beta$
	$((23), T)$	$((123), T)$	$((132), T)$
R°	$\beta^{-1}P^{\cdot}_{\gamma y}\alpha$	$\alpha^{-1}(P^{\cdot}_{\gamma y})^{-1}\beta$	$\beta^{-1}(L^{\cdot}_{\alpha y})^{-1}\gamma$
L°	$\beta^{-1}(L^{\cdot}_{\alpha x})^{-1}\gamma$	$\alpha^{-1}(R^{\cdot}_{\beta x})^{-1}\gamma$	$\beta^{-1}P^{\cdot}_{\gamma x}\alpha$
P°	$\gamma^{-1}R^{\cdot}_{\beta z}\alpha$	$\gamma^{-1}L^{\cdot}_{\alpha z}\beta$	$\alpha^{-1}(R^{\cdot}_{\beta z})^{-1}\gamma$
$(R^{\circ})^{-1}$	$\alpha^{-1}(P^{\cdot}_{\gamma y})^{-1}\beta$	$\beta^{-1}P^{\cdot}_{\gamma y}\alpha$	$\gamma^{-1}L^{\cdot}_{\alpha y}\beta$
$(L^{\circ})^{-1}$	$\gamma^{-1}L^{\cdot}_{\alpha x}\beta$	$\gamma^{-1}R^{\cdot}_{\beta x}\alpha$	$\alpha^{-1}(P^{\cdot}_{\gamma x})^{-1}\beta$
$(P^{\circ})^{-1}$	$\alpha^{-1}(R^{\cdot}_{\beta z})^{-1}\gamma$	$\beta^{-1}(L^{\cdot}_{\alpha z})^{-1}\gamma$	$\gamma^{-1}R^{\cdot}_{\beta z}\alpha$

Taking into consideration Table 1.1 and Lemmas 1.178 and 1.198 we can establish transformations of quasigroup translations at any isostrophy. In Table 1.3 $T = (\alpha, \beta, \gamma)$, $(Q, \circ) = (Q, \cdot)(\sigma, (\alpha, \beta, \gamma))$. If, for example, $\sigma = (1\,3)$, then $L^{\circ}_x = \alpha^{-1}(P^{\cdot}_{\gamma x})^{-1}\beta$.

There exist interesting connections between the multiplication groups of a quasigroup and its LP-isotopes. Some of these relations are very well known; some are given in [799, 438].

Lemma 1.199. *Suppose that $(Q, \circ)$ and $(Q, \cdot)$ are isotopic quasigroups such that $x \circ y = \alpha x \cdot \beta y$ for all $x, y \in Q$. Then*

$$L^{\circ}_x y = L^{\cdot}_{\alpha x}\beta y, R^{\circ}_y x = R^{\cdot}_{\beta y}\alpha x, \tag{1.53}$$

$LM(Q, \circ) = \langle L_x\beta \mid x \in Q\rangle$, $RM(Q, \circ) = \langle R_x\alpha \mid x \in Q\rangle$.

Proof. The proof follows from the definitions of translations. □

Corollary 1.200. *If $(Q, \circ)$ and $(Q, \cdot)$ are loops such that $x \circ y = \alpha x \cdot \beta y$ for all $x, y \in Q$, then $LM(Q, \circ) = LM(Q, \cdot)$ [173, p. 57].*

Proof. By Lemma 1.121 there exist elements $a, b \in Q$ such that $\alpha = R_a^{-1}$, $\beta = L_b^{-1}$. Thus by Lemma 1.199 $L^{\circ}_x y = L^{\cdot}_{\alpha x}(L_b^{-1})^{\cdot}$, $LM(Q, \circ) \subseteq LM(Q, \cdot)$. Similarly $LM(Q, \cdot) \subseteq LM(Q, \circ)$, therefore $LM(Q, \cdot) = LM(Q, \circ)$. □

Theorem 1.201. *If $(Q, \circ)$ is a loop, $(Q, \cdot)$ is a quasigroup such that $x \circ y = \alpha x \cdot \beta y$ for all $x, y \in Q$, then the following statements are true:*

1. *$LM(Q, \circ) = \langle L_x L_y^{-1} \mid x, y \in Q\rangle$.*
2. *$LM(Q, \cdot) = \langle LM(Q, \circ), \beta\rangle$.*
3. *If $\beta \in Aut(Q, \circ)$, then $LM(Q, \cdot) = LM(Q, \circ)\,\langle\beta\rangle$.*
4. *If $\beta \in Aut(Q, \circ)$, then $LM(Q, \circ) \trianglelefteq LM(Q, \cdot)$.*
5. *If $\beta \in Aut(Q, \cdot)$, then $LM(Q, \circ) \trianglelefteq LM(Q, \cdot)$.*

6. *If* $\beta \in Aut(Q, \circ)$, $LM(Q, \circ) \cap \langle \beta \rangle = \varepsilon$, *then* $LM(Q, \cdot) \cong LM(Q, \circ) \rtimes \langle \beta \rangle$.

7. *If* $\beta \in Aut(Q, \circ)$, $(Q, \circ)$ *is an abelian group, then* $LM(Q, \cdot) \cong (Q, \circ) \rtimes \langle \beta \rangle$.

Proof. Case 1. From Lemmas 1.121 and 1.199 it follows that $LM(Q, \circ) = \langle L_x L_b^{-1} \mid x \in Q \rangle$ for some fixed element $b \in Q$. Suppose that $(Q, *)$ is one more isotope of quasigroup $(Q, \cdot)$. Then $LM(Q, *) = \langle L_x L_c^{-1} \mid x \in Q \rangle$ for some fixed element $c \in Q$.

It is clear that the loops $(Q, \circ)$ and $(Q, *)$ are principally isotopic and we can apply Corollary 1.200. Therefore properties of the group $LM(Q, \circ)$ do not depend on the choice of the element b. Then $LM(Q, \circ) = \langle L_x L_y^{-1} \mid x, y \in Q \rangle$.

Case 2. From Lemma 1.199 and the fact that $\beta = L_b^{-1}$ it follows $L_x^{\circ} \in LM(Q, \cdot)$. Therefore $\langle LM(Q, \circ), \beta \rangle \subseteq LM(Q, \cdot)$.

From equality (1.53) we have $L^{\cdot}_{\alpha x} = L_x^{\circ} \beta^{-1}$. Then $LM(Q, \cdot) \subseteq \langle LM(Q, \circ), \beta \rangle$, $LM(Q, \cdot) = \langle LM(Q, \circ), \beta \rangle$.

Case 3. From Case 2 it follows that $LM(Q, \circ) \langle \beta \rangle \subseteq LM(Q, \cdot)$. By Lemma 1.122

$$\beta L_x^{\circ} = L_{\beta x}^{\circ} \beta; \; \beta (L_x^{\circ})^{-1} = (L_{\beta x}^{\circ})^{-1} \beta \tag{1.54}$$

for any left translation of loop $(Q, \circ)$. Therefore any element of the group $\langle LM(Q, \circ), \beta \rangle$ has the form $\overline{L} \beta^k$, where $\overline{L} \in LM(Q, \circ)$, k is an integer.

From equalities $L^{\cdot}_{\alpha x} = L_x^{\circ} \beta^{-1}$ and (1.54) we have $LM(Q, \cdot) \subseteq LM(Q, \circ) \langle \beta \rangle$.

Case 4. Suppose that $\overline{L}_1 \in LM(Q, \circ)$. Then we have

$$(\overline{L} \beta^k)^{-1} \overline{L}_1 (\overline{L} \beta^k) = \beta^{-k} \overline{L}^{\,-1} \overline{L}_1 \overline{L} \beta^k \overset{1.54}{=} \overline{L}_2 \in LM(Q, \circ). \tag{1.55}$$

Thus $LM(Q, \circ) \trianglelefteq LM(Q, \cdot)$.

Case 5. From equality $\beta L^{\cdot}_x = L^{\cdot}_{\beta x} \beta$ we have $\beta L_x^{\circ} \overset{1.53}{=} \beta L^{\cdot}_{\alpha x} \beta = L^{\cdot}_{\beta \alpha x} \beta^2 = L^{\circ}_{\alpha^{-1} \beta \alpha x} \beta$. Further proof is similar to the proof of Case 4.

Case 6. The proof follows from Cases 3, 4 and the definition of semidirect product.

Case 7. In this case $LM(Q, \circ) = (Q, \circ)$, $(Q, \circ) \cap \langle \beta \rangle = \varepsilon$. Indeed, if $L_a x = \beta x$, then by $x = 0$ we have $L_a 0 = \beta 0 = 0$, $a = 0$, and $\beta = \varepsilon$. Further we can apply the results of Case 6. □

Similar results are true for right multiplication groups and multiplication groups of a quasigroup and a loop which is isotopic to this quasigroup [438, 799].

We formulate the following

Theorem 1.202. *If* $(Q, \circ)$ *is an isotope of the form* $(R_a^{-1}, L_b^{-1}, \varepsilon)$ *of a quasigroup* $(Q, \cdot)$, *i.e.,* $x \circ y = R_a^{-1} x \cdot L_b^{-1} y$ *for all* $x, y \in Q$, *then*

$$\begin{array}{l} LM(Q, \cdot) \cong \langle LM(Q, \circ), L_b^{-1} \rangle, \\ RM(Q, \cdot) \cong \langle RM(Q, \circ), R_a^{-1} \rangle, \\ M(Q, \cdot) \cong \langle M(Q, \circ), R_a^{-1}, L_b^{-1} \rangle. \end{array}$$

If permutation L_b^{-1} *is an automorphism of the loop* $(Q, \circ)$, *then*

$$LM(Q, \circ) \trianglelefteq LM(Q, \cdot), LM(Q, \cdot) = LM(Q, \circ) \langle L_b^{-1} \rangle.$$

If permutation R_a^{-1} *is an automorphism of the loop* $(Q, \circ)$, *then*

$$RM(Q, \circ) \trianglelefteq RM(Q, \cdot), RM(Q, \cdot) = RM(Q, \circ) \langle R_a^{-1} \rangle.$$

If permutations L_b^{-1} *and* R_a^{-1} *are automorphisms of the loop* $(Q, \circ)$, *then*

$$M(Q, \cdot) = M(Q, \circ) \langle R_a^{-1}, L_b^{-1} \rangle.$$

If $(Q, \circ)$ is an abelian group, L_b^{-1} and R_a^{-1} are its automorphisms, then

$$\begin{array}{l} LM(Q,\cdot) = (Q,\circ)\leftthreetimes\langle L_b^{-1}\rangle, \\ RM(Q,\cdot) = (Q,\circ)\leftthreetimes\langle R_a^{-1}\rangle, \\ M(Q,\cdot) = (Q,\circ)\leftthreetimes\langle R_a^{-1}, L_b^{-1}\rangle \end{array}$$

[438, 799].

1.4 Sub-objects

1.4.1 Subquasigroups: Nuclei and center

1.4.1.1 Sub-objects

Definition 1.203. Let $(Q, \cdot)$ be a quasigroup. A subset H of the set Q we shall name a *sub-object* of $(Q, \cdot)$, if $(H, \cdot)$ is closed relative to the operation $\cdot$, i.e., if $a, b \in H$, then $a \cdot b \in H$.

In fact, Definition 1.203 is a definition of a subgroupoid of a quasigroup $(Q, \cdot)$.

Lemma 1.204. *1. Any sub-object of a quasigroup $(Q, \cdot)$ is a cancellation groupoid.*

2. Any sub-object of a quasigroup $(Q, \cdot, \backslash, /)$ is a subquasigroup.

3. Any subquasigroup of a quasigroup $(Q, \cdot)$ is a subquasigroup in $(Q, \cdot, \backslash, /)$ and vice versa, any subquasigroup of a quasigroup $(Q, \cdot, \backslash, /)$ is a a subquasigroup in $(Q, \cdot)$.

Proof. 1. If $a, b, c \in H$, then from $a \cdot b = a \cdot c$ follows $b = c$, since $(H, \cdot) \subseteq (Q, \cdot)$. Similarly from $b \cdot a = c \cdot a$ follows $b = c$.

2. Exercise.

3. Exercise. See, for example, [72, 685, 591, 184]. □

Example 1.205. In the additive group of integers $(Z, +)$ the set of natural numbers N forms a cancellative groupoid (cancellative semigroup [211, 583, 864]) $(N, +)$ relative to the operation "+".

Corollary 1.206. *Any sub-object of a finite quasigroup $(Q, \cdot)$ is a subquasigroup.*

Proof. Any finite cancellation groupoid is a quasigroup. □

Lemma 1.207. *If $(Q, \cdot)$ is a quasigroup, $(H, \cdot)$ is its subquasigroup, $a, b \in H$, then $(H, \cdot)T$ is a subloop of the loop $(Q, \cdot)T$, where T is an isotopy of the form $(R_a^{-1}, L_b^{-1}, \varepsilon)$.*

Proof. We have that $R_a|_H, L_b|_H$ are translations of $(H, \cdot)$, since $a, b \in H$. □

Theorem 1.208. *A countable quasigroup with at least three members is necessarily isotopic to a quasigroup which has no proper subquasigroups [495].*

"This theorem serves to answer affirmatively a question raised by J. Denes and A.D. Keedwell [240]. D.A. Robinson, MR0509347 (80b:20093)". Additional information about this topic is given in [443].

Notice, it is known that any countable n-ary groupoid (it is infinite) can be embedded into a simple n-ary groupoid generated by one element and having no non-identity automorphism [561].

1.4.1.2 Nuclei

Definition 1.209. Let $(Q,\cdot)$ be a groupoid and let $a \in Q$. The element a is a left (middle, right) nuclear element in $(Q,\cdot)$ means that $L_{ax} = L_a L_x \Leftrightarrow ax \cdot y = a \cdot xy$ ($L_{xa} = L_x L_a \Leftrightarrow xa \cdot y = x \cdot ay$, $R_{xa} = R_a R_x \Leftrightarrow y \cdot xa = yx \cdot a$) for all $x, y \in Q$ [685, 346].

Definition 1.210. Let $(Q,\cdot)$ be a groupoid and let $a \in Q$. The element a is a left (middle, right) nuclear element in $(Q,\cdot)$ means that $L_a R_y x = R_y L_a x \Leftrightarrow a \cdot xy = ax \cdot y$ ($L_x R_y a = R_y L_x a \Leftrightarrow x \cdot ay = xa \cdot y$, $R_a L_y x = L_y R_a x \Leftrightarrow yx \cdot a = y \cdot xa$) for all $x, y \in Q$ [685, 346].

An element a is nuclear in groupoid $(Q,\cdot)$ means that a is a left, right, and middle nuclear element in $(Q,\cdot)$.

Definition 1.211. Let $(Q,\cdot)$ be a groupoid. The left nucleus N_l (middle nucleus N_m, right nucleus N_r) of $(Q,\cdot)$ is the set of all left (middle, right) nuclear elements in $(Q,\cdot)$ and the nucleus is given by $N = N_l \cap N_r \cap N_m$ [685, 346, 10, 173].

In other words

$$N_l = \{a \in Q \mid a \cdot xy = ax \cdot y,\ x, y \in Q\},$$
$$N_m = \{a \in Q \mid xa \cdot y = x \cdot ay,\ x, y \in Q\},$$
$$N_r = \{a \in Q \mid xy \cdot a = x \cdot ya,\ x, y \in Q\}.$$

Theorem 1.212. *Let $(Q,\cdot)$ be a groupoid. If N_l (N_m, N_r) is nonempty, then N_l (N_m, N_r) is a subgroupoid of $(Q,\cdot)$.*

Proof. Let $L_{ax} = L_a L_x$, $L_{bx} = L_b L_x$ for all $x \in Q$. Prove that $L_{ab\cdot x} = L_{ab} L_x$. We have $L_{ab\cdot x} = L_{L_{ab}x} = L_{L_a L_b x} = L_{a\cdot bx} = L_a L_{bx} = L_a L_b L_x = L_{ab} L_x$. □

1.4.1.3 Center

The concept of the center is well known in universal algebra [184].

Definition 1.213. The center of a groupoid $(G,\cdot)$ is the following set $Z = N \cap C$, where $C = \{a \in G \mid a \cdot x = x \cdot a\ \forall x \in G\}$ and N is the nucleus of $(G,\cdot)$.

The center of a groupoid $(G,\cdot)$ can also be defined in the following way: $Z = \{a \in N \mid L_a = R_a\}$ [685, p. 19].
$(Z,\cdot)$ is an abelian subgroup of the loop $(Q,\cdot)$ [10, 173].

Theorem 1.214. *A.A. Albert. The center of a loop $(Q,\cdot)$ is isomorphic to the center of the group $M(Q,\cdot)$ [10, 11], [685, p. 25].*

Exercise 1.215. Prove that for a loop (even for a groupoid) $Z = N_l \cap N_r \cap C = N_l \cap N_m \cap C = N_m \cap N_r \cap C$. Prove that for a commutative loop (even for a commutative groupoid) $N_l = N_r$.

1.4.2 Bol and Moufang nuclei

In this section we give information mainly from the articles and Ph.D. dissertation of I.A. Florja [328, 327]. Various kinds of Moufang elements are defined and researched in [691]. Loop identities that can be obtained by a nuclear identification are studied in [275, 276]. Bol and Moufang elements for the quasigroup case are studied in [328]. Here we give information only for the loop case.

Using the language of autotopies it is possible to define various other "elements" almost for all Bol-Moufang type identities (see Definition 2.1).

Definition 1.216. An element a of a loop $(Q, \cdot)$ is called a middle Moufang element, if the following equality is true

$$ax \cdot ya = (a \cdot xy)a \quad (1.56)$$

for all $x, y \in Q$.

Lemma 1.217. *An element a of a loop $(Q, \cdot)$ is a middle Moufang element if and only if the triple $(L_a, R_a, R_a L_a)$ is an autotopy of the loop $(Q, \cdot)$.*

Lemma 1.218. *If equality (1.56) is true in a loop $(Q, \cdot)$, then the following equality is also true $ax \cdot ya = a(xy \cdot a)$.*

Proof. Indeed, from equality (1.56) by $x = 1$ we have that $a \cdot ya = ay \cdot a$ for all $x, y \in Q$, i.e.,

$$L_a R_a = R_a L_a. \quad (1.57)$$

Therefore $(a \cdot xy)a = a(xy \cdot a)$. □

Corollary 1.219. *An element a of a loop $(Q, \cdot)$ is a middle Moufang element if and only if the triple $(L_a, R_a, L_a R_a)$ is an autotopy of the loop $(Q, \cdot)$.*

Definition 1.220.

1. An element a of a loop $(Q, \cdot)$ is called a left Moufang element, if the following equality is true

$$(ax \cdot a)y = a(x \cdot ay) \quad (1.58)$$

for all $x, y \in Q$.

2. An element a of a loop $(Q, \cdot)$ is called a right Moufang element, if the following equality is true

$$x(a \cdot ya) = (xa \cdot y)a \quad (1.59)$$

for all $x, y \in Q$.

We can reformulate Definition 1.220 in the following form:

Lemma 1.221.

1. *An element a of a loop $(Q, \cdot)$ is a left Moufang element if and only if the triple $(R_a L_a, L_a^{-1}, L_a)$ is an autotopy of the loop $(Q, \cdot)$.*

2. *An element a of a loop $(Q, \cdot)$ is a right Moufang element if and only if the triple $(R_a^{-1}, L_a R_a, R_a)$ is an autotopy of the loop $(Q, \cdot)$.*

Definition 1.222. The set of all middle Moufang elements of a loop $(Q, \cdot)$ is called a middle Moufang nucleus M_m. The set of all left Moufang elements of a loop $(Q, \cdot)$ is called a left Moufang nucleus M_l. The set of all right Moufang elements of a loop $(Q, \cdot)$ is called a right Moufang nucleus M_r. Moufang nucleus M is defined as follows: $M = M_l \cap M_r \cap M_m$ [691].

Definition 1.223.

1. An element a of a loop $(Q, \cdot)$ is called a left Bol element, if the following equality is true

$$a(x \cdot ay) = (a \cdot xa)y \quad (1.60)$$

for all $x, y \in Q$.

2. An element a of a loop $(Q, \cdot, /, \backslash)$ is called a middle Bol element, if the following equality is true

$$a((xy)\backslash a) = (a/y)(x\backslash a) \quad (1.61)$$

for all $x, y \in Q$.

3. An element a of a loop $(Q, \cdot)$ is called a right Bol element, if the following equality is true
$$(ya \cdot x)a = y(ax \cdot a) \tag{1.62}$$
for all $x, y \in Q$.

Lemma 1.224. *1. An element a of a loop $(Q, \cdot)$ is a left Bol element if and only if the triple $(L_a R_a, L_a^{-1}, L_a)$ is an autotopy of the loop $(Q, \cdot)$.*

2. An element a of a loop $(Q, \cdot, /, \backslash)$ is a middle Bol element if and only if the triple $(P_a^{-1}, P_a, L_a P_a)$ is an anti-autotopy of the loop $(Q, \cdot, /, \backslash)$.

3. An element a of a loop $(Q, \cdot)$ is a right Bol element if and only if the triple $(R_a^{-1}, R_a L_a, R_a)$ is an autotopy of the loop $(Q, \cdot)$.

Proof. It is sufficient to write equalities 1.60, 1.61, and 1.62 in autotopic form. □

Definition 1.225. The set of all middle Bol elements of a loop $(Q, \cdot)$ is called a middle Bol nucleus B_m. The set of all left Bol elements of a loop $(Q, \cdot)$ is called a left Bol nucleus B_l. The set of all right Bol elements of a loop $(Q, \cdot)$ is called a right Bol nucleus B_m. Bol nucleus B is defined as follows: $B = B_l \cap B_r \cap B_m$.

1.4.3 The coincidence of loop nuclei

The information presented below is mainly taken from [327].

Theorem 1.226. *Left nucleus N_l and right nucleus N_r of a loop $(Q, \cdot)$ coincide if and only if any element of these nuclei is a middle Moufang element.*

Proof. Suppose $N_l = N_r$. If $a \in N_l$, then the triplet (L_a, ε, L_a) is an autotopy of the loop $(Q, \cdot)$. If $a \in N_r$, then the triplet (ε, R_a, R_a) is an autotopy of the loop $(Q, \cdot)$.

Thus the triplet $(L_a, \varepsilon, L_a)(\varepsilon, R_a, R_a) = (L_a, R_a, R_a L_a)$ is also an autotopism of the loop $(Q, \cdot)$. By Lemma 1.217 the element a is a middle Moufang element.

Converse. Let $a \in N_l$, i.e., (L_a, ε, L_a) is an autotopy. Since the element a is a middle Moufang element, i.e., triplet $(L_a, R_a, R_a L_a)$ is an autotopy of the loop $(Q, \cdot)$, then $(L_a, R_a, R_a L_a)(L_a^{-1}, \varepsilon, L_a^{-1}) = (\varepsilon, R_a, R_a)$. Therefore,
$$N_l \subseteq N_r. \tag{1.63}$$

Let $a \in N_r$, i.e., (ε, R_a, R_a) is an autotopy. Since the element a is a Moufang element, i.e., triplet $(L_a, R_a, R_a L_a)$ is an autotopy of the loop $(Q, \cdot)$, then $(\varepsilon, R_a^{-1}, R_a^{-1})(L_a, R_a, R_a L_a) = (L_a, \varepsilon, L_a)$. Therefore,
$$N_r \subseteq N_l. \tag{1.64}$$
Thus $N_l = N_r$. □

Lemma 1.227. *Left nucleus N_l and middle nucleus N_m of a loop $(Q, \cdot)$ coincide if and only if any element of these nuclei is a left Bol element.*

Proof. Suppose $N_l = N_m$. If $a \in N_l$, then the triplet (L_a, ε, L_a) is an autotopy of the loop $(Q, \cdot)$. If $a \in N_m$, then the triplet $(R_a, L_a^{-1}, \varepsilon)$ is an autotopy of the loop $(Q, \cdot)$.

The triplet $(R_a, L_a^{-1}, \varepsilon)(L_a, \varepsilon, L_a) = (R_a L_a, L_a^{-1}, L_a)$ is an autotopism of the loop $(Q, \cdot)$. By Lemma 1.224 the element a is a Bol element.

Converse. Let $a \in N_l$, i.e., (L_a, ε, L_a) is an autotopy. Since the element a is a Bol element, i.e., triplet (R_aL_a, L_a^{-1}, L_a) is an autotopy of the loop $(Q, \cdot)$, then $(R_aL_a, L_a^{-1}, L_a)(L_a^{-1}, \varepsilon, L_a^{-1}) = (R_a, L_a^{-1}, \varepsilon)$. Therefore,

$$N_l \subseteq N_m. \tag{1.65}$$

Let $a \in N_m$, i.e., $(R_a, L_a^{-1}, \varepsilon)$ is an autotopy. Since the element a is a Bol element, then $(R_a^{-1}, L_a, \varepsilon)(R_aL_a, L_a^{-1}, L_a) = (L_a, \varepsilon, L_a)$. Therefore,

$$N_m \subseteq N_l, \tag{1.66}$$

and, finally, $N_l = N_m$. □

Lemma 1.228. *Right nucleus N_r and middle nucleus N_m of a loop $(Q, \cdot)$ coincide if and only if any element of these nuclei is a right Bol element.*

Proof. We can use the following equality that follows from Lemma 1.224 $(R_a^{-1}, R_aL_a, R_a) = (\varepsilon, R_a, R_a)(R_a^{-1}, L_a, \varepsilon)$. Further proof is similar to the proof of Lemma 1.227 and we omit it. □

Notice, it is possible to formulate other conditions of coincidence of loop nuclei.

Lemma 1.229.

1. *Right nucleus N_r and middle N_m of a loop $(Q, \cdot)$ coincide if and only if any element of these nuclei is a right Moufang element.*
2. *Left nucleus N_l and middle N_m of a loop $(Q, \cdot)$ coincide if and only if any element of these nuclei is a left Moufang element.*

Proof. Case 1. Suppose $N_r = N_m$. If $a \in N_r$, then the triplet (ε, R_a, R_a) is an autotopy of the loop $(Q, \cdot)$. If $a \in N_m$, then the triplet $(R_a^{-1}, L_a, \varepsilon)$ is an autotopy of the loop $(Q, \cdot)$.

Therefore the triplet $(R_a^{-1}, L_a, \varepsilon)(\varepsilon, R_a, R_a) = (R_a^{-1}, L_aR_a, R_a)$ is an autotopism of the loop $(Q, \cdot)$. By Lemma 1.221 the element a is a middle-right element.

Converse. Let $a \in N_r$, i.e., (ε, R_a, R_a) is an autotopy. Since the element a is a middle-right element, i.e., triplet (R_a^{-1}, L_aR_a, R_a) is an autotopy of the loop $(Q, \cdot)$, then $(R_a^{-1}, L_aR_a, R_a)(\varepsilon, R_a^{-1}, R_a^{-1}) = (R_a^{-1}, L_a, \varepsilon)$. Therefore,

$$N_r \subseteq N_m. \tag{1.67}$$

Let $a \in N_m$, i.e., $(R_a, L_a^{-1}, \varepsilon)$ is an autotopy. Since the element a is a middle-right element, then $(R_a, L_a^{-1}, \varepsilon)(R_a^{-1}, L_aR_a, R_a) = (\varepsilon, R_a, R_a)$. Therefore,

$$N_m \subseteq N_r, \tag{1.68}$$

and, finally, $N_r = N_m$.

Case 2 is proved similarly to Case 1. □

Theorem 1.230. *Nuclei N_l, N_m, and N_r of a loop $(Q, \cdot)$ coincide if and only if any element of the nucleus N is simultaneously a middle Moufang and a left Bol element [327].*

Proof. The proof follows from Lemmas 1.226 and 1.227. □

Using Theorem 1.226 and Lemmas 1.227, 1.228, and 1.229 the reader can formulate other versions of Theorem 1.230.

1.4.3.1 Nuclei coincidence and identities

Equalities (1.60), 1.62), (1.56), and (1.59) help to obtain identities which guarantee the coincidence of some nuclei in loops. Notice that the equalities mentioned above are non-unique, which guarantees the coincidence of some loop nuclei.

Lemma 1.231. *If the triple* $(L_a^{-1}, R_a, R_a L_a^{-1})$ *is an autotopy of the loop* $(Q, \cdot, /, \backslash)$*, or, equivalently, if equality*

$$x \cdot ya = (a\backslash(ax \cdot y))a \tag{1.69}$$

is true for all $x, y \in Q$*, and all* $a \in N_l$*, then* $N_l = N_r$*, and vice versa, if* $N_l = N_r$ *in some loop* $(Q, \cdot, /, \backslash)$*, then equality (1.69) is true in this loop for all* $x, y \in Q$ *and all* $a \in N_l$*.*

Proof. The proof is similar to the proof of Lemma 1.229 and we omit it. □

Remark 1.232. Following [89] (see also [511]) equality (1.69) can be called a partial identity.

Suppose that we have an identity. If we fix some variables (i.e., every occurrence of the same variable we replace by one fixed element) in this identity, then we obtain an object which we shall name the *retract* of this identity. It is clear, if in algebra an identity is true, then in this algebra any retract of this identity is true, too [184].

Example 1.233. Suppose in a loop $(Q, \cdot, /, \backslash, 1)$ the following identity is true for all $x, y, t, z \in Q$

$$x \cdot yt = (t\backslash((t \cdot x) \cdot (z \cdot (z \cdot y))))t. \tag{1.70}$$

If we put in identity (1.70) $z = 1$, then we obtain the following identity $x \cdot yt = (t\backslash(tx \cdot y))t$ which is a retract of identity (1.70).

It is clear that in this case equality $x \cdot ya = (a\backslash(ax \cdot y))a$ is true for all $x, y \in Q$ and some fixed elements $a \in Q$. Fulfillment of the last equality "guarantee", that in the loop $(Q, \cdot, /, \backslash, 1)$, $N_l = N_r$.

Example 1.234. We give an example of a non-associative loop of order six that satisfies identity (1.70).

*	0	1	2	3	4	5
0	0	1	2	3	4	5
1	1	0	3	2	5	4
2	2	4	0	5	1	3
3	3	5	4	0	2	1
4	4	3	5	1	0	2
5	5	2	1	4	3	0

This example was constructed using Mace 4 [614].

In the next lemma, identity (1.60) (respectively, (1.62), (1.56), (1.58), (1.59), (1.69)) means that corresponding equality (1.60) (respectively, (1.62), (1.56), (1.58), (1.59), (1.69)) is true for all x, y, a.

Lemma 1.235. *If identity (1.56), or identity (1.69) is a retract of an identity that is true in a loop* $(Q, \cdot, /, \backslash)$*, then* $N_l = N_r$*;*

if identity (1.60) or identity (1.58) is a retract of an identity that is true in a loop $(Q, \cdot, /, \backslash)$*, then* $N_l = N_m$*;*

if identity (1.62) or identity (1.59) is a retract of an identity that is true in a loop $(Q, \cdot, /, \backslash)$*, then* $N_r = N_m$*.*

Proof. It is easy to see. □

1.4.4 Quasigroup nuclei and center

1.4.4.1 Historical notes

G.N. Garrison [346] was the first to define the concept of a quasigroup nucleus. In implicit form the definition of loop nucleus is given in [10]. A nucleus (left, right, or middle) "measures" how far a quasigroup is from a group. Unfortunately, if a quasigroup $(Q, \cdot)$ has a non-trivial left (right, middle) Garrison's nucleus, then this quasigroup is a left loop (a right loop, a loop, respectively) [685, p. 17]. Therefore many authors tried to diffuse (to generalize) the concept of nucleus on the "proper" quasigroup case.

Various definitions of groupoid, quasigroup and loop nuclei were given and researched by A. Sade, R.H. Bruck, V.D. Belousov, P.I. Gramma, A.A. Gvaramiya, M.D. Kitoroagá, G.B. Belyavskaya, H.O. Pflugfelder and many other mathematicians [167, 731, 511, 173, 380, 72, 399, 75, 128, 529, 89, 578, 110, 104, 685, 779, 489].

1.4.4.2 Quasigroup nuclei

M.D. Kitoroagé [529] gives the following definition of nuclei of a quasigroup $(Q, \cdot)$:

$$N_l(h) = \{a \in Q \mid ax \cdot y = a \cdot L_h^{-1}(hx \cdot y)\ \forall x, y \in Q\},$$

$$N_m(h) = \{a \in Q \mid R_h^{-1}(xa) \cdot y = x \cdot L_h^{-1}(ay),\ \forall x, y \in Q\},$$

$$N_r(h) = \{a \in Q \mid y \cdot xa = R_h^{-1}(y \cdot xh) \cdot a,\ \forall x, y \in Q\}.$$

Let $(Q, \cdot)$ be a quasigroup. We recall that the inner mapping group $\mathbb{I}_h(Q, \cdot)$ of a quasigroup $(Q, \cdot)$ is generated by the following permutations: $R_{x,y}, L_{x,y}, T_x$, where $R_{x,y} = R^{-1}_{x \bullet y} R_y R_x$, $h(x \bullet y) = (hx)y$, $L_{x,y} = L^{-1}_{x \circ y} L_x L_y$, $(x \circ y)h = x(yh)$, $T_x = L^{-1}_{\sigma x} R_x$, $\sigma = R_h^{-1} L_h$.

The permutation $R_{x,y}$ will be denoted as $R^h_{x,y}$, the permutation $L_{x,y}$ will be denoted as $L^h_{x,y}$, and the permutation T_x will be denoted as T^h_x.

G.B. Belyavskaya gives the following definitions.

Definition 1.236. A left h-nucleus N_l^h of a quasigroup $(Q, \cdot)$ is called the maximal subset H of the set Q such that

(1) $R^h_{x,y} a = a \cdot e(h)$ for all $x, y \in Q$, $a \in H$;
(2) $H \cdot e(h) = H$ [110, 104].

Definition 1.237. A right h-nucleus N_r^h of a quasigroup $(Q, \cdot)$ is called the maximal subset H of the set Q such that

(1) $L^h_{x,y} a = f(a) \cdot a$ for all $x, y \in Q$, $a \in H$;
(2) $f(h) \cdot H = H$ [110, 104].

Definition 1.238. A middle h-nucleus N_m^h of a quasigroup $(Q, \cdot)$ is called the maximal subset H of the set Q such that

(1) $L_h^{-1}(\sigma a \cdot x) \cdot y = x \cdot L_h^{-1}(a \cdot y)$ for all $x, y \in Q$, $a \in H$;
(2) $\sigma \cdot H = H$, where $\sigma \cdot H = \{\sigma a \mid a \in H\}$ [105, 133].

Later G.B. Belyavskaya gave the other definition of middle nucleus of a quasigroup $(Q, \cdot)$.

Definition 1.239. A middle h-nucleus N_m^h of a quasigroup $(Q, \cdot)$ is called the maximal subset H of the set Q such that

(1) $C_{x,y} a = \sigma a$ for all $x, y \in Q$, $a \in H$, where $C_{x,y} = R_y^{-1} L_h L_x^{-1} R_y R_h^{-1} L_x$;
(2) $\sigma \cdot H = H$, where $\sigma \cdot H = \{\sigma a \mid a \in H\}$ [110].

We notice, that in the last definition we can re-write equality (1) in the following form $x \cdot (L_h^{-1}(\sigma a \cdot y)) = (R_h^{-1}(x \cdot a)) \cdot y$.

Problem 1.4. Are the last two definitions of middle nucleus equivalent?

1.4.4.3 Quasigroup center

P.I. Gramma [380] defined the center of a quasigroup $(Q, \cdot)$ as a set C of all elements $a \in Q$ such that $R^h_{x,y}a = a$, $L^h_{x,y}a = a$, $T^h_x a = a$ for all $x, y \in Q$ and a fixed element $h \in Q$.

Later G.B. Belyavskaya [104, 106, 108] and J.D.H. Smith [201, 813, 819] gave a more general definition of center of a quasigroup.

J.D.H. Smith gave this definition in the language of universal algebra (Congruence Theory). J.D.H. Smith defined central congruence in a quasigroup. The center of a quasigroup is a coset class of central congruence. Notice that any congruence of quasigroup Q defines a subquasigroup of Q^2 [813].

For her definition of a quasigroup center, G.B. Belyavskaya used mainly the concept of the set invariant relative to some subgroups of the inner multiplication group of a quasigroup [104]. The G.B. Belyavskaya and J.D.H. Smith definitions are close.

G.B. Belyavskaya gives the following definition of center of a quasigroup.

Definition 1.240. The maximal subset H of the set Q that consists from elements $a \in Q$ such that

(1) $R^h_{x,y}a = a \cdot e(a)$, $L^h_{x,y}a = f(h) \cdot a$, $T^h_x a = \sigma_h^{-1} a$ for all $x, y \in Q$, $a \in H$;

(2) $H \cdot e(h) = f(h) \cdot H = H$, $h \in R_h^{-1}a \cdot H$, for all $a \in H$,

where $R_h^{-1}a \cdot H = \{R_h^{-1}a \cdot b \,|\, b \in H\}$ is called a center Z_h of a quasigroup $(Q, \cdot)$ [104, 105, 110].

Definition 1.241. Quasigroup $(Q, \cdot)$ is a T-quasigroup if and only if there exists an abelian group $(Q, +)$, its automorphisms φ and ψ and a fixed element $a \in Q$ such that $x \cdot y = \varphi x + \psi y + a$ for all $x, y \in Q$ [667, 506]. A T-quasigroup with the additional condition $\varphi\psi = \psi\varphi$ is called medial.

A quasigroup is central, if it coincides with its center.

Theorem 1.242. *Belyavskaya theorem. A quasigroup is central (in the Belyavskaya and Smith sense) if and only if it is a T-quasigroup [108].*

Later G.B. Belyavskaya gave her "nuclear and central" theory in the language of quasigroup commutators and associators [109, 110, 111].

1.4.5 Regular permutations

The concept of regular permutations is very close to the concepts of autotopy and nucleus.

Definition 1.243. A permutation λ (ρ) of a set Q is called a left (right) regular permutation for a quasigroup $(Q, \cdot)$, if $\lambda x \cdot y = \lambda^\star(xy)$ $(x \cdot \rho y = \rho^\star(xy))$ for all $x, y \in Q$. A permutation φ is middle regular if $\varphi x \cdot y = x \cdot \varphi^\star y$ [72, 667].

A quasigroup $(Q, \cdot)$ is called $\mathcal{L}$-transitive (respectively $\mathfrak{R}$-transitive, Φ-transitive), if group $\mathcal{L}$ ($\mathfrak{R}$, Φ) of all left (right, middle) regular permutations is transitive on the set Q.

Theorem 1.244. *$\mathcal{L}$-transitivity and $\mathcal{L}^\star$-transitivity (respectively $\mathcal{R}$ and $\mathcal{R}^\star$-transitivity, Φ and $\Phi^\star$-transitivity) are equivalent properties for a quasigroup [72, 493].*

V.D. Belousov discovered connections of quasigroup nuclei with the groups of regular mappings of quasigroups [59, 72, 80, 491, 488, 489, 500]. In [791] some facts on quasigroup nuclei using this connection are given.

1.4.6 A-nuclei of quasigroups

More detailed information on A-nuclei is given in Chapter 4.

Definition 1.245. The set of all autotopisms of the form $(\alpha, \varepsilon, \gamma)$ of a quasigroup $(Q, \circ)$, where ε is the identity mapping, is called the *left autotopy nucleus* (left A-nucleus) of the quasigroup $(Q, \circ)$.

Similarly, the sets of autotopisms of the forms $(\alpha, \beta, \varepsilon)$ and $(\varepsilon, \beta, \gamma)$ form the *middle and right A-nuclei* of $(Q, \circ)$.

We shall denote these three sets of mappings by N_l^A, N_m^A and N_r^A respectively.

Definition 1.245 and the results stated and proved in Lemma 1.247 and Theorem 1.249 are not entirely new. Various versions of these facts and definitions can be found in the quasigroup literature [59, 491, 493, 529, 110, 512].

We recall, in [59, 491], that A-nuclei are called as groups of left, right and middle regular permutations respectively.

Notice, autotopies of the form $(L_a L_h^{-1}, \varepsilon, L_{a\cdot e(h)} L_h^{-1})$, where $a \in Q$ (in fact $a \in N_l$ in Belyavskaya sense) correspond to the elements of left Belyavskaya nucleus [104]. Recall that isotopic quasigroups have isomorphic groups of autotopisms (Theorem 1.133) or [72].

Lemma 1.246. *Isotopic quasigroups have isomorphic autotopy nuclei.*

Proof. The proof follows from Theorem 1.133. □

Lemma 1.247. *The first components of the autotopisms of any subgroup $K(N_l^A)$ of N_l^A themselves form a group $K_1(N_l^A)$.*

Proof. Let $(\alpha_1, \varepsilon, \gamma_1), (\alpha_2, \varepsilon, \gamma_2) \in K(N_l^A)$. Then $\alpha_1 x \circ y = \gamma_1(x \circ y)$ and $\alpha_2 x \circ y = \gamma_2(x \circ y)$ for all $x, y \in Q$. From the first equation, $\alpha_1(\alpha_2 x) \circ y = \gamma_1(\alpha_2 x \circ y) = \gamma_1\gamma_2(x \circ y)$ by virtue of the second equation. Thus, $(\alpha_1\alpha_2, \varepsilon, \gamma_1\gamma_2) \in K(N_l^A)$.

Let $x = \alpha_1^{-1}u$. Then, $\alpha_1 x \circ y = \gamma_1(x \circ y) \Rightarrow u \circ y = \gamma_1(\alpha_1^{-1}u \circ y) \Rightarrow \alpha_1^{-1}u \circ y = \gamma_1^{-1}(u \circ y)$, so $(\alpha_1^{-1}, \varepsilon, \gamma_1^{-1}) \in K(N_l^A)$.

Clearly, $(\varepsilon, \varepsilon, \varepsilon) \in K(N_l^A)$. Hence, ε is the first component and, if α_1 and α_2 are the first components, so are $\alpha_1\alpha_2$ and α_1^{-1}. The result follows. □

In a similar way, we may prove that the third components of the autotopisms of $K(N_l^A)$, the first and second components of the autotopisms of a subgroup $K(N_m^A)$ of N_m^A, and the second and third components of the autotopisms of a subgroup $K(N_r^A)$ of N_r^A form groups $K_3(N_l^A)$, $K_1(N_m^A)$, $K_2(N_m^A)$, $K_2(N_r^A)$, $K_3(N_r^A)$.

Note 1.248. From Lemma 1.247 it follows that the first, second and third components of N_l^A, N_m^A and N_r^A each form groups. For brevity, we shall denote these nine groups by ${}_1N_l^A$, ${}_2N_l^A$, ${}_3N_l^A$, ${}_1N_m^A$, ${}_2N_m^A$, ${}_3N_m^A$, ${}_1N_r^A$, ${}_2N_r^A$ and ${}_3N_r^A$.

From Lemma 1.247 it follows that N_l^A, N_m^A and N_r^A are also subgroups of the autotopy group of a quasigroup $(Q, \circ)$ relative to usual multiplication of triplets.

Theorem 1.249. *In loop $(Q, \circ)$ with identity element 1,*
$(\alpha, \varepsilon, \gamma) \in Avt(Q, \circ) \Longleftrightarrow \alpha = \gamma = L_{\alpha 1}$ *and* $\alpha 1 \in N_l$,
$(\varepsilon, \beta, \gamma) \in Avt(Q, \circ) \Longleftrightarrow \beta = \gamma = R_{\beta 1}$ *and* $\beta 1 \in N_r$,
$(\alpha, \beta, \varepsilon) \in Avt(Q, \circ) \Longleftrightarrow \alpha = R_{\beta 1}$, $\beta = L_{\beta 1}^{-1}$ *and* $\beta 1 \in N_m$.

Proof. Case 1. From equality

$$\alpha x \circ y = \gamma(x \circ y) \tag{1.71}$$

by $y = 1$ it follows that $\alpha = \gamma$, by $x = 1$ we obtain $\gamma = L_{\alpha 1} = L_a$. Therefore from equality (1.71) we have

$$(a \circ x) \circ y = a \circ (x \circ y), \tag{1.72}$$

i.e., $a \in N_l$. The converse is obvious.

The proof of Case 2 is similar to the proof of Case 1 and we omit it.

The proof of Case 3 is given in Theorem 4.45. □

Then in the loop case, Belousov's concept of regular permutations [72, 80, 685] coincides with the concept of the A-nucleus.

Corollary 1.250. *In any loop, the left (right, middle) A-nucleus is isomorphic to the left (right, middle) nucleus.*

Proof. From Theorem 1.249 it follows that any element of the left A-nucleus of a loop has the form $(L_{\alpha 1}, \varepsilon, L_{\alpha 1})$, where $\alpha 1 \in N_l$. The following map $\xi : L_{\alpha 1} \mapsto \alpha 1$ is the required isomorphism.

In a similar way, the following isomorphisms are constructed: $N_r^A \cong N_r$, $N_m^A \cong N_m$. □

Components of loop A-nuclei are subgroups of some multiplication groups.

Corollary 1.251. *In any loop the following inclusions are true:* ${}_1N_l^A = {}_3N_l^A \subseteq LM \subseteq M$, ${}_2N_r^A = {}_3N_r^A \subseteq RM \subseteq M$, ${}_1N_m^A \subseteq RM \subseteq M$, ${}_2N_m^A \subseteq LM \subseteq M$.

Proof. The proof follows from Theorem 1.249. □

A similar result is true for the quasigroup case. See Corollary 4.48.

There exist some relations between components of the A-nuclei and local identity elements in any quasigroup. See Chapter 4 or [791].

We recall, a group G acts on the set Q transitively if for any elements $a, b \in Q$ there exists an element $\gamma \in G$ such that $\gamma a = b$ [471].

Definition 1.252. If at least one component of the A-nuclei of a quasigroup $(Q, \cdot)$ acts on the set Q transitively, then $(Q, \cdot)$ will be called an *A-nuclear quasigroup.*

Definition 1.252 is a generalization of corresponding definition from [72, p. 32].

Theorem 1.253. *A quasigroup is A-nuclear if and only if it is a group isotope [791, 792].*

1.4.7 A-pseudo-automorphisms by isostrophy

In this section we find connections between components of the A-nuclei of a quasigroup $(Q, \cdot)$ and its isostrophic images of the form $(Q, \circ) = (Q, \cdot)((\sigma)(\alpha, \beta, \gamma))$, where $\sigma \in S_3$, $\alpha, \beta, \gamma \in S_Q$.

We omit the symbol of autotopy nuclei (the symbol A) and put the symbols of binary operations "$\circ$" and "$\cdot$" in its place respectively. Denote isostrophy $(\sigma, (\alpha, \beta, \gamma))$ by (σ, T) for all cases.

Lemma 1.254. *If quasigroup* $(Q, \circ)$ *is an isostrophic image of quasigroup* $(Q, \cdot)$ *with an isostrophy* $S = ((12), T)$*, i.e.,* $(Q, \circ) = (Q, \cdot)S$*, then* $N_l(Q, \circ) = S^{-1}N_r(Q, \cdot)S$*,* ${}_1N_l(Q, \circ) = \alpha^{-1}\,{}_2N_r(Q, \cdot)\alpha$*,* ${}_3N_l(Q, \circ) = \gamma^{-1}\,{}_3N_r(Q, \cdot)\gamma$.

Proof. The proof repeats the proof of Lemma 1.182. Let $K \in N_l(Q, \circ)$. Then $(Q, \cdot)S = (Q, \cdot)SK$, $(Q, \cdot) = (Q, \cdot)SKS^{-1}$,

$$SN_l(Q, \circ)S^{-1} \subseteq N_r(Q, \cdot), \tag{1.73}$$

since

$$((12), (\alpha, \beta, \gamma))\,(\varepsilon, ({}_1N_l^\circ, \varepsilon, {}_3N_l^\circ))\,((12), (\beta^{-1}, \alpha^{-1}, \gamma^{-1})) =$$
$$(\varepsilon, (\varepsilon, \alpha\, {}_1N_l^\circ \alpha^{-1}, \gamma\, {}_3N_l^\circ \gamma^{-1})).$$

If $(Q, \circ) = (Q, \cdot)S$, then $(Q, \circ)S^{-1} = (Q, \cdot)$, expression (1.73) takes the form

$$S^{-1}N_r(Q, \cdot)S \subseteq N_l(Q, \circ). \tag{1.74}$$

Indeed,

$$((12), (\beta^{-1}, \alpha^{-1}, \gamma^{-1}))\,(\varepsilon, (\varepsilon, {}_2N_r^\cdot, {}_3N_r^\cdot))\,((12), (\alpha, \beta, \gamma)) =$$
$$(\varepsilon, (\alpha^{-1}\,{}_2N_r^\cdot\alpha,\ \varepsilon,\ \gamma^{-1}\,{}_3N_r^\cdot\gamma)).$$

We can rewrite expression (1.74) in this form

$$N_r(Q, \cdot) \subseteq SN_l(Q, \circ)S^{-1}. \tag{1.75}$$

Comparing (1.73) and (1.75) we obtain $N_r(Q, \cdot) = SN_l(Q, \circ)S^{-1}$. Therefore $\alpha\,{}_1N_l^\circ\alpha^{-1} = {}_2N_r^\cdot$, ${}_1N_l^\circ = \alpha^{-1}\,{}_2N_r^\cdot\alpha$, ${}_3N_l^\circ = \gamma^{-1}\,{}_3N_r^\cdot\gamma$. □

All other analogs of Lemma 1.254 are proved in a similar way.

In Table 1.4, connections between components of the A-nuclei of a quasigroup $(Q, \cdot)$ and its isostrophic images of the form $(Q, \circ) = (Q, \cdot)((\sigma)(\alpha, \beta, \gamma))$, where $\sigma \in S_3$, $\alpha, \beta, \gamma \in S_Q$ are collected.

Table 1.4: Connections between components of A-nuclei by isostrophy.

	(ε, T)	$((12), T)$	$((13), T)$	$((23), T)$	$((132), T)$	$((123), T)$
${}_1N_l^\circ$	$\alpha^{-1}\,{}_1N_l^\cdot\alpha$	$\alpha^{-1}\,{}_2N_r^\cdot\alpha$	$\alpha^{-1}\,{}_3N_l^\cdot\alpha$	$\alpha^{-1}\,{}_1N_m^\cdot\alpha$	$\alpha^{-1}\,{}_2N_m^\cdot\alpha$	$\alpha^{-1}\,{}_3N_r^\cdot\alpha$
${}_3N_l^\circ$	$\gamma^{-1}\,{}_3N_l^\cdot\gamma$	$\gamma^{-1}\,{}_3N_r^\cdot\gamma$	$\gamma^{-1}\,{}_1N_l^\cdot\gamma$	$\gamma^{-1}\,{}_2N_m^\cdot\gamma$	$\gamma^{-1}\,{}_1N_m^\cdot\gamma$	$\gamma^{-1}\,{}_2N_r^\cdot\gamma$
${}_2N_r^\circ$	$\beta^{-1}\,{}_2N_r^\cdot\beta$	$\beta^{-1}\,{}_1N_l^\cdot\beta$	$\beta^{-1}\,{}_2N_m^\cdot\beta$	$\beta^{-1}\,{}_3N_r^\cdot\beta$	$\beta^{-1}\,{}_3N_l^\cdot\beta$	$\beta^{-1}\,{}_1N_m^\cdot\beta$
${}_3N_r^\circ$	$\gamma^{-1}\,{}_3N_r^\cdot\gamma$	$\gamma^{-1}\,{}_3N_l^\cdot\gamma$	$\gamma^{-1}\,{}_1N_m^\cdot\gamma$	$\gamma^{-1}\,{}_2N_r^\cdot\gamma$	$\gamma^{-1}\,{}_1N_l^\cdot\gamma$	$\gamma^{-1}\,{}_2N_m^\cdot\gamma$
${}_1N_m^\circ$	$\alpha^{-1}\,{}_1N_m^\cdot\alpha$	$\alpha^{-1}\,{}_2N_m^\cdot\alpha$	$\alpha^{-1}\,{}_3N_r^\cdot\alpha$	$\alpha^{-1}\,{}_1N_l^\cdot\alpha$	$\alpha^{-1}\,{}_2N_r^\cdot\alpha$	$\alpha^{-1}\,{}_3N_l^\cdot\alpha$
${}_2N_m^\circ$	$\beta^{-1}\,{}_2N_m^\cdot\beta$	$\beta^{-1}\,{}_1N_m^\cdot\beta$	$\beta^{-1}\,{}_2N_r^\cdot\beta$	$\beta^{-1}\,{}_3N_l^\cdot\beta$	$\beta^{-1}\,{}_3N_r^\cdot\beta$	$\beta^{-1}\,{}_1N_l^\cdot\beta$

If $(Q, \circ) = (Q, \cdot)((123)(\alpha, \beta, \gamma))$, then from Table 1.4 we have ${}_2N_r^A(Q, \circ) = \beta^{-1}\,{}_1N_m^A\beta(Q, \cdot)$. Therefore, if $\rho \in {}_2N_r^A(Q, \circ)$, then there exists an element $\mu \in {}_1N_m^A(Q, \cdot)$ such that $\rho = \beta^{-1}\mu\beta$ and vice versa, if $\mu \in {}_1N_m^A(Q, \cdot)$, then there exists an element $\rho \in {}_2N_r^A(Q, \circ)$ such that $\rho = \beta^{-1}\mu\beta$.

Connections between components of A-pseudo-automorphisms of a quasigroup $(Q, \cdot)$ and its isostrophic images of the form $(Q, \circ) = (Q, \cdot)((\sigma)(\alpha, \beta, \gamma))$, where $\sigma \in S_3$, $\alpha, \beta, \gamma \in S_Q$ are similar to connections between components of A-nuclei (Table 1.4).

Table 1.5: Connections between components of A-pseudo-automorphisms by isostrophy.

	(ε, T)	$((12), T)$	$((13), T)$	$((23), T)$	$((132), T)$	$((123), T)$
${}_1\Pi_l^{\circ}$	$\alpha^{-1}\,{}_1\Pi_l^{\cdot}\alpha$	$\alpha^{-1}\,{}_2\Pi_r^{\cdot}\alpha$	$\alpha^{-1}\,{}_3\Pi_l^{\cdot}\alpha$	$\alpha^{-1}\,{}_1\Pi_m^{\cdot}\alpha$	$\alpha^{-1}\,{}_2\Pi_m^{\cdot}\alpha$	$\alpha^{-1}\,{}_3\Pi_r^{\cdot}\alpha$
${}_2\Pi_l^{\circ}$	$\beta^{-1}\,{}_2\Pi_l^{\cdot}\beta$	$\beta^{-1}\,{}_1\Pi_r^{\cdot}\beta$	$\beta^{-1}\,{}_2\Pi_l^{\cdot}\beta$	$\beta^{-1}\,{}_3\Pi_m^{\cdot}\beta$	$\beta^{-1}\,{}_3\Pi_m^{\cdot}\beta$	$\beta^{-1}\,{}_1\Pi_r^{\cdot}\beta$
${}_3\Pi_l^{\circ}$	$\gamma^{-1}\,{}_3\Pi_l^{\cdot}\gamma$	$\gamma^{-1}\,{}_3\Pi_r^{\cdot}\gamma$	$\gamma^{-1}\,{}_1\Pi_l^{\cdot}\gamma$	$\gamma^{-1}\,{}_2\Pi_m^{\cdot}\gamma$	$\gamma^{-1}\,{}_1\Pi_m^{\cdot}\gamma$	$\gamma^{-1}\,{}_2\Pi_r^{\cdot}\gamma$
${}_1\Pi_r^{\circ}$	$\alpha^{-1}\,{}_1\Pi_r^{\cdot}\alpha$	$\alpha^{-1}\,{}_2\Pi_l^{\cdot}\alpha$	$\alpha^{-1}\,{}_3\Pi_m^{\cdot}\alpha$	$\alpha^{-1}\,{}_1\Pi_r^{\cdot}\alpha$	$\alpha^{-1}\,{}_2\Pi_l^{\cdot}\alpha$	$\alpha^{-1}\,{}_3\Pi_m^{\cdot}\alpha$
${}_2\Pi_r^{\circ}$	$\beta^{-1}\,{}_2\Pi_r^{\cdot}\beta$	$\beta^{-1}\,{}_1\Pi_l^{\cdot}\beta$	$\beta^{-1}\,{}_2\Pi_m^{\cdot}\beta$	$\beta^{-1}\,{}_3\Pi_r^{\cdot}\beta$	$\beta^{-1}\,{}_3\Pi_l^{\cdot}\beta$	$\beta^{-1}\,{}_1\Pi_m^{\cdot}\beta$
${}_3\Pi_r^{\circ}$	$\gamma^{-1}\,{}_3\Pi_r^{\cdot}\gamma$	$\gamma^{-1}\,{}_3\Pi_l^{\cdot}\gamma$	$\gamma^{-1}\,{}_1\Pi_m^{\cdot}\gamma$	$\gamma^{-1}\,{}_2\Pi_r^{\cdot}\gamma$	$\gamma^{-1}\,{}_1\Pi_l^{\cdot}\gamma$	$\gamma^{-1}\,{}_2\Pi_m^{\cdot}\gamma$
${}_1\Pi_m^{\circ}$	$\alpha^{-1}\,{}_1\Pi_m^{\cdot}\alpha$	$\alpha^{-1}\,{}_2\Pi_m^{\cdot}\alpha$	$\alpha^{-1}\,{}_3\Pi_r^{\cdot}\alpha$	$\alpha^{-1}\,{}_1\Pi_l^{\cdot}\alpha$	$\alpha^{-1}\,{}_2\Pi_r^{\cdot}\alpha$	$\alpha^{-1}\,{}_3\Pi_l^{\cdot}\alpha$
${}_2\Pi_m^{\circ}$	$\beta^{-1}\,{}_2\Pi_m^{\cdot}\beta$	$\beta^{-1}\,{}_1\Pi_m^{\cdot}\beta$	$\beta^{-1}\,{}_2\Pi_r^{\cdot}\beta$	$\beta^{-1}\,{}_3\Pi_l^{\cdot}\beta$	$\beta^{-1}\,{}_3\Pi_r^{\cdot}\beta$	$\beta^{-1}\,{}_1\Pi_l^{\cdot}\beta$
${}_3\Pi_m^{\circ}$	$\gamma^{-1}\,{}_3\Pi_m^{\cdot}\gamma$	$\gamma^{-1}\,{}_3\Pi_m^{\cdot}\gamma$	$\gamma^{-1}\,{}_1\Pi_r^{\cdot}\gamma$	$\gamma^{-1}\,{}_2\Pi_l^{\cdot}\gamma$	$\gamma^{-1}\,{}_1\Pi_r^{\cdot}\gamma$	$\gamma^{-1}\,{}_2\Pi_l^{\cdot}\gamma$

Theorem 1.255. *Isostrophic quasigroups have isomorphic components of A-nuclei and isomorphic components of A-pseudo-automorphisms.*

Proof. For A-nuclei the proof follows from Table 1.4. The proof for components of A-pseudo-automorphisms is similar and we omit it. □

Lemma 1.256. *In a commutative loop $(Q, \cdot)$ any left pseudo-automorphism θ with a companion a is a right pseudo-automorphism with the same companion and vice versa.*

Proof. In the language of autostrophisms, commutativity means that the following set of permutations $((1\,2)(\varepsilon, \varepsilon, \varepsilon))$ is an autostrophy of loop $(Q, \cdot)$. Suppose that $((\varepsilon)(L_a\theta, \theta, L_a\theta))$ is a left pseudo-automorphism of loop $(Q, \cdot)$. Further we have $((1\,2)(\varepsilon, \varepsilon, \varepsilon))((\varepsilon)(L_a\theta, \theta, L_a\theta))((1\,2)(\varepsilon, \varepsilon, \varepsilon)) = ((\varepsilon)(\theta, L_a\theta, L_a\theta)) = ((\varepsilon)(\theta, R_a\theta, R_a\theta))$ because in a commutative loop, $L_a = R_a$. □

Problem 1.5. Research quasigroups with various σ-A-nuclei and σ-A-pseudo-automorphisms.

1.4.8 Commutators and associators

Commutators $[x, y]$ in a quasigroup $(Q, \cdot)$ can be defined using the following equality $xy = yx \cdot [x, y]$, and associators $[x, y, z]$ using the equality $xy \cdot z = (x \cdot yz)\,[x, y, z]$ (Definition 2.14), where $x, y, z \in Q$. It is easy to see that there exist other definitions of associators. See, for example, Definition 2.162.

Commutators and associators play an important role in the quasigroup, and especially in loop theory. Concepts of the quasigroup commutator and associator in some sense are dual, like in group theory, to the concepts of the center and the nuclei in quasigroup. We refer the reader to the following articles and books [173, 60, 109, 220, 842].

Notice, objects which are similar to nuclei and associators can be defined for many identities in quasigroup $(Q, \cdot)$.

For example, left nucleus D_l for a left distributive identity in a quasigroup $(Q, \cdot)$ can be defined as follows: $D_l(Q, \cdot) = \{a \in Q \mid a \cdot xy = ax \cdot ay \; \forall x, y \in Q\}$.

The left distributant $[x, y, z]$ can be defined as follows: $x \cdot yz = (xy \cdot xz)[x, y, z]$; the right "alternant" $\lfloor x, y \rfloor$ can be defined using right alternative identity in the following way: $xy \cdot y = (x \cdot yy)\lfloor x, y \rfloor$, and so on. It is clear that it is possible to define objects which are dual to Moufang and Bol nuclei.

For the structure theory of quasigroups and loops it is useful, if the above mentioned nuclei, centers, and their dual objects are normal sub-objects in a quasigroup or loop.

1.5 Congruences

1.5.1 Congruences of quasigroups

1.5.1.1 Congruences in universal algebra

Definition 1.257. Let A be an algebra of type $\mathbb{F}$ and let θ an equivalence. Then θ is a congruence on A if θ satisfies the following compatibility property: For each n-ary function symbol $f \in \mathbb{F}$ and elements $a_i, b_i \in A$, if $a_i \theta b_i$ holds for $1 \leqslant i \leqslant n$, then $f^A(a_1, \dots a_n)\theta f^A(b_1, \dots b_n)$ holds [184].

"The compatibility property is an obvious condition for introducing an algebraic structure on the set of equivalence classes A/θ, an algebraic structure which is inherited from the algebra A" [184].

Here we suppose that a quasigroup $(Q, \cdot)$ in the sense of Definition 1.22 ("existential quasigroup") also is an algebra. For binary groupoids and binary quasigroups of the form $(Q, \cdot)$, i.e., with one binary operation in the signature, we can write Definition 1.257 in the following form.

Definition 1.258. An equivalence θ is a congruence of a groupoid $(Q, \cdot)$, if the following implication is true for all $x, y, w, z \in Q$: $x\theta y \wedge w\theta z \Longrightarrow (x \cdot w)\,\theta\,(y \cdot z)$ [212].

It is well known that Definition 1.258 is equivalent to the following:

Definition 1.259. An equivalence θ is a congruence of a binary groupoid $(Q, \cdot)$, if the following implications are true for all $x, y, z \in Q$: $x\theta y \Longrightarrow (z \cdot x)\theta(z \cdot y), x\theta y \Longrightarrow (x \cdot z)\theta(y \cdot z)$.

In the language of translations, implications from Definition 1.259 can be written in the following form: $x\theta y \Longrightarrow L_z x\theta L_z y, x\theta y \Longrightarrow R_z x\theta R_z y$.

Exercise 1.260. Prove the equivalence of Definitions 1.258 and 1.259 for binary groupoids.

Example 1.261. We give an example of a quasigroup in which the equivalence θ ($\theta(0) = \{0, 2\}$, $\theta(1) = \{1, 3\}$) satisfies the implication $x\theta y \rightarrow (ux)v\theta(uy)v$ and it does not satisfy the implication $x\theta y \rightarrow ux\theta uy$ for all $x, y, u, v \in Q$, because from relation $0\theta 2$ it follows that $(0 \cdot 0)\theta(0 \cdot 2)$, $0\,\theta\,1$. It is clear that $(0, 1) \notin \theta$.

$\cdot$	0	1	2	3
0	0	2	1	3
1	2	0	3	1
2	1	3	0	2
3	3	1	2	0

Any algebra has trivial congruences, namely, the diagonal $\hat{Q} = \{(q, q) \,|q \in Q\}$ and universal $Q \times Q$ congruence.

Notice, there exists a quasigroup $(Q, \cdot)$ in the sense of Definition 1.22 ("existential quasigroup") and its congruence θ in the sense of Definition 1.257 such that Q/θ ("an algebraic structure which is inherited from the algebra A") is a division groupoid which, in general, is not a quasigroup [43, 782].

Therefore for quasigroups, congruence and homomorphism theory has a little non-standard character from the point of view of the universal-algebraic approach [184]. In order to avoid some inconveniences for quasigroups, the concept of normal congruence is introduced.

1.5.1.2 Normal congruences

Definition 1.262. A congruence θ of a quasigroup $(Q,\cdot)$ is *normal*, if the following implications are true for all $x, y, z \in Q$: $(z\cdot x)\theta(z\cdot y) \Longrightarrow x\theta y, (x\cdot z)\theta(y\cdot z) \Longrightarrow x\theta y$ [72, 80].

Using the language of quasigroup translations (Table 1.1), we can rewrite implications from Definition 1.262 in the following form: $x\theta y \Longrightarrow L_z^{-1}x\theta L_z^{-1}y, x\theta y \Longrightarrow R_z^{-1}x\theta R_z^{-1}y$, or, using Table 1.1, in the following form: $x\theta y \Longrightarrow (z\backslash x)\theta(z\backslash y), x\theta y \Longrightarrow (x/z)\,\theta\,(y/z)$.

Definition 1.263. An equivalence θ is a congruence of a quasigroup $(Q,\cdot,/,\backslash)$, if the following implications are true for all $x, y, z \in Q$:

$$\begin{aligned} &x\theta y \Longrightarrow (z\cdot x)\theta(z\cdot y), x\theta y \Longrightarrow (x\cdot z)\theta(y\cdot z),\\ &x\theta y \Longrightarrow (z/x)\theta(z/y), x\theta y \Longrightarrow (x/z)\theta(y/z),\\ &x\theta y \Longrightarrow (z\backslash x)\theta(z\backslash y), x\theta y \Longrightarrow (x\backslash z)\theta(y\backslash z), \end{aligned} \tag{1.76}$$

or, using the language of quasigroup translations (Table 1.1), we can rewrite implications (1.76) in the following form

$$\begin{aligned} &x\theta y \Longrightarrow L_z x\theta L_z y, x\theta y \Longrightarrow R_z x\theta R_z y,\\ &x\theta y \Longrightarrow P_z^{-1}x\theta P_z^{-1}y, x\theta y \Longrightarrow R_z^{-1}x\theta R_z^{-1}y,\\ &x\theta y \Longrightarrow L_z^{-1}x\theta L_z^{-1}y, x\theta y \Longrightarrow P_z x\theta P_z y. \end{aligned} \tag{1.77}$$

Definition 1.264. If θ is a binary relation on a set Q, α is a permutation of the set Q, and from $x\theta y$ it follows that $\alpha x\theta\alpha y$ for all $(x, y) \in \theta$, then we shall say that the permutation α is a *semi-admissible or stable* permutation relative to the binary relation θ and binary relation θ is semi-admissible (or stable) relative to the permutation α.

Definition 1.265. If θ is a binary relation on a set Q, α is a permutation of the set Q and from $x\theta y$ it follows that $\alpha x\theta\alpha y$ and $\alpha^{-1}x\theta\alpha^{-1}y$ for all $(x, y) \in \theta$, then we shall say that the permutation α is an *admissible* permutation relative to the binary relation θ and binary relation θ is admissible relative to the permutation α [72].

Thus any congruence of quasigroup $(Q,\cdot)$ is semi-admissible relative to any left and right quasigroup translation, and any normal congruence of quasigroup $(Q,\cdot)$ is admissible relative to any left and right quasigroup translation.

Lemma 1.266. *Let θ be a normal congruence of a quasigroup $(Q,\cdot)$. The following implications hold true:*

1. *If $a\theta b$, $c\theta d$, then $L_a^{-1}c\,\theta\,L_b^{-1}d$;*
2. *If $a\theta b$, $c\theta d$, then $R_a^{-1}c\,\theta\,R_b^{-1}d$.*

Proof. 1. If $ac\,\theta\,bd$ and $a\theta b$, then $c\theta d$. Indeed, if $a\theta b$, then $ac\theta bc$. If $ac\,\theta\,bc$ and $ac\theta bd$, then $bc\theta bd$, and, finally, $c\theta d$. In other words, if $L_a c\,\theta\,L_b d$ and $a\theta b$, then $c\theta d$.

Since $a\theta b$ and θ is a normal quasigroup congruence, we have $c\,\theta\,d \Longleftrightarrow L_a L_a^{-1}c\,\theta\,L_b L_b^{-1}d \Longrightarrow L_a^{-1}c\,\theta\,L_b^{-1}d$.

2. If $ac\,\theta\,bd$ and $c\theta d$, then $a\theta b$. Indeed, if $c\theta d$, then $ac\theta ad$. If $ac\,\theta\,bd$ and $ac\theta ad$, then $bd\theta ad$, and, finally, $a\theta b$. In other words, if $R_c a\,\theta\,R_d b$ and $c\theta d$, then $a\theta b$.

Since $a\theta b$ and θ is a normal quasigroup congruence, we have $c\,\theta\,d \Longleftrightarrow R_a R_a^{-1}c\,\theta\,R_b R_b^{-1}d \Longrightarrow R_a^{-1}c\,\theta\,R_b^{-1}d$. ☐

Lemma 1.267. *Any normal congruence of a quasigroup $(Q, \cdot)$ is admissible relative to any left, right, and middle translation [75].*

Proof. The fact that any normal quasigroup congruence is admissible relative to any left and right quasigroup translation follows from Definitions 1.259 and 1.262.

Let θ be a normal congruence of a quasigroup $(Q, \cdot)$. Prove the following implication

$$a\theta b \Longrightarrow P_c a \,\theta\, P_c b. \tag{1.78}$$

Implication (1.78) is equivalent to the last implication from the list (1.76):

$$a\theta b \Longrightarrow (a\backslash c)\,\theta\,(b\backslash c). \tag{1.79}$$

If $a\backslash c = k$, then $a \cdot k = c$. Similarly if $b\backslash c = m$, then $b \cdot m = c$. We can rewrite implication (1.79) in the following form

$$\text{if } a\theta b \text{ and } (a \cdot k)\,\theta\,(b \cdot m), \text{ then } k\,\theta\, m. \tag{1.80}$$

Using Lemma 1.266 Case 1 we can say that implication (1.80) is fulfilled.

Prove the following implication

$$a\theta b \Longrightarrow P_c^{-1} a \,\theta\, P_c^{-1} b. \tag{1.81}$$

Implication (1.81) is equivalent to the following implication from the list (1.76):

$$a\theta b \Longrightarrow (c/a)\,\theta\,(c/b). \tag{1.82}$$

If $c/a = k$, then $k \cdot a = c$. Similarly, if $c/b = m$, then $m \cdot b = c$. We can rewrite implication (1.82) in the following form:

$$\text{if } a\theta b \text{ and } (k \cdot a)\,\theta\,(m \cdot b), \text{ then } k\,\theta\, m. \tag{1.83}$$

Using Lemma 1.266 Case 2 we can say that implication (1.83) holds true. □

From Lemma 1.267 it follows that any element of the group $FM(Q, \cdot)$ of a quasigroup $(Q, \cdot)$ is admissible relative to any normal congruence of the quasigroup $(Q, \cdot)$.

Corollary 1.268. *If θ is a normal quasigroup congruence of a quasigroup $(Q, \cdot)$, then θ is a normal congruence of any parastrophe of $(Q, \cdot)$ [75].*

Proof. The proof follows from Lemma 1.267 and Table 1.1. □

Corollary 1.269. *If θ is a normal congruence of a loop Q, then it is a normal congruence of any inverse loop to loop Q.*

Proof. This follows from the fact that any normal congruence of a loop Q is invariant relative to any left, right, middle translation of Q (Lemma 1.267) and the fact that $I = P_1$. □

Remark 1.270. From Corollary 1.269 it follows that lattices of normal congruences of a loop Q and any inverse loop to loop Q are equal.

One of the most important properties of e-quasigroup $(Q, \cdot, \backslash, /)$ is the following one:

Lemma 1.271. *1. Any congruence of a quasigroup $(Q, \cdot, \backslash, /)$ is a normal congruence of quasigroup $(Q, \cdot)$;*

2. any normal congruence of a quasigroup $(Q, \cdot)$ is a congruence of quasigroup $(Q, \cdot, \backslash, /)$ [151, 591, 72, 80].

Proof. 1. Follows from Definitions 1.263 and 1.265. 2. It is possible to use definitions and Lemma 1.267. □

Lemma 1.272. *In finite quasigroup $(Q, \cdot)$ any congruence is normal.*

Proof. Recall, any left and right translation of a quasigroup $(Q, \cdot)$ is a bijection of the set Q. Since Q has the finite order, then for any left translation L_a there exists a natural number m such that $L_a^m = \varepsilon$. Then $L_a^{-1} = L_a^{m-1}$. For right translation the proof is similar. □

Definition 1.273. A quasigroup $(Q, \cdot)$ with identities $xy = yx$, $x \cdot xy = y$, $xy \cdot y = x$ is called a *TS-quasigroup.*

See Lemma 1.186 for more details.

Exercise 1.274. In any TS-quasigroup, any congruence is normal. In a group $(G, \cdot, ^{-1}, 1)$ any congruence is normal.

Important objects in quasigroup theory (similar to other algebraic theories) are simple quasigroups. In some sense, simple quasigroups play a role in quasigroup theory that is similar to the role of primes in number theory.

Definition 1.275. A quasigroup $(Q, \cdot)$ is *simple* if its only normal congruences are the diagonal congruence $\hat{Q} = \{(q, q) \,|\, q \in Q\}$ and universal congruence $Q \times Q$.

Example 1.276. Any quasigroup of prime order is simple.

Definition 1.277. A quasigroup $(Q, \cdot)$ is *strongly simple* if its only congruences are the diagonal $\hat{Q} = \{(q, q) \,|\, q \in Q\}$ and universal $Q \times Q$.

Let θ be a normal congruence of a quasigroup $(Q, \cdot)$. Let $\theta(a) = \{x \,|\, x\theta a\}$ be a coset class of an element a. We list some properties of $\theta(a)$:

- if $b \in \theta(a)$, then $\theta(b) = \theta(a)$; $\theta(a) = \theta(b)$ or $\theta(a) \cap \theta(b) = \varnothing$;
- $a \cdot \theta(b) = \theta(a \cdot b) = (\theta a) \cdot b$; $\theta(a) \cdot \theta(b) = \theta(a \cdot b)$;
- in the finite case $|\theta(a)| = |\theta(b)|$ for all a, b; and
- in the finite case the order of any coset class $\theta(x)$ divides the order of the set Q.

1.5.2 Quasigroup homomorphisms

Definition 1.278. If $(Q, \cdot)$ and $(H, \circ)$ are groupoids, h is a single-valued mapping of Q into H such that $h(x_1 \cdot x_2) = hx_1 \circ hx_2$, then h is called a *homomorphism (a multiplicative homomorphism)* of $(Q, \cdot)$ into $(H, \circ)$ and the set $\{hx \,|\, x \in Q\}$ is called a *homomorphic image* of $(Q, \cdot)$ under h [685].

In case $(Q, \cdot) = (H, \circ)$ a homomorphism is also called an *endomorphism* and an isomorphism is referred to as an automorphism. For the n-ary case see Definition 7.16.

There exists the well-known connection between quasigroup homomorphisms and congruences. Let h be a homomorphism of a quasigroup $(Q, \cdot)$ onto a groupoid $(H, \circ)$. Then h induces a congruence $Ker\, h = \theta$ (the kernel of h) in the following way: $x\,\theta\,y$ if and only if $h(x) = h(y)$ [80, 685].

If θ is a normal congruence of a quasigroup $(Q, \cdot)$, then θ determines natural homomorphism h $(h(a) = \theta(a))$ of $(Q, \cdot)$ onto some quasigroup $(Q', \circ)$ by the rule: $\theta(x) \circ \theta(y) = \theta(x \cdot y)$, where $\theta(x), \theta(y), \theta(x \cdot y) \in Q/\theta$ [80, 685].

Define an operation $\circ$ on the set $\bar{Q}$ of all cosets of normal congruence θ ($\bar{Q} = \{\theta(a) \,|\, a \in Q\}$) in the following way: $\theta(a) \circ \theta(b) = \theta(a \cdot b)$. It is possible to check that $(\bar{Q}, \circ)$ is a quasigroup. A map $\varphi : a \longrightarrow \bar{a} = \theta(a)$ is a homomorphism of a quasigroup $(Q, \cdot)$ onto quasigroup $(\bar{Q}, \circ)$.

The quasigroup $(\bar{Q}, \circ)$ is called a factor-quasigroup of a quasigroup $(Q, \cdot)$ by the normal congruence θ. The $(\bar{Q}, \circ)$ is denoted as Q/θ.

Theorem 1.279. *If h is a homomorphism of a quasigroup $(Q, \cdot)$ onto a quasigroup $(H, \circ)$, then h determines a normal congruence θ on $(Q, \cdot)$ such that $Q/\theta \cong (H, \circ)$, and vice versa, a normal congruence θ induces a homomorphism from $(Q, \cdot)$ onto $(H, \circ) \cong Q/\theta$. ([80], [685], I.7.2 Theorem).*

Lemma 1.280. *1. Any homomorphic image of a quasigroup $(Q, \cdot)$ is a division groupoid [43, 173].*

2. Any homomorphic image of a quasigroup $(Q, \cdot, \backslash, /)$ is a quasigroup [184, 591].

Proof. 1. Let $h(a), h(b) \in h(Q)$. We demonstrate that the solution of the equation $h(a) \circ x = h(b)$ lies in $h(Q)$. Consider the equation $a \cdot y = b$. Denote the solution of this equation by c. Then $h(c)$ is solution of the equation $h(a) \circ x = h(b)$. Indeed, $h(a) \circ h(c) = h(a \cdot c) = h(b)$.

For equation $x \cdot h(a) = h(b)$ the proof is similar.

2. Exercise. See [72, 80, 685, 184, 591]. □

1.5.3 Normal subquasigroups

Definition 1.281. A subquasigroup $(H, \cdot)$ of a quasigroup $(Q, \cdot)$ is called normal (or normal divisor) if H is a coset of a normal congruence θ and we shall denote this fact as follows: $H \trianglelefteq Q$.

Lemma 1.282. *A coset $H = \theta(h)$ of a normal congruence θ of a quasigroup $(Q, \cdot)$ is a subquasigroup if and only if $h \;\theta\; h^2$ [72, 80].*

The proof is similar to the proof of Proposition 7.96.

Corollary 1.283. *If an equivalence class $\theta(h)$ of a normal congruence θ of a quasigroup $(Q, \cdot)$ contains an idempotent element, then $\theta(h)$ is a normal subquasigroup of $(Q, \cdot)$.*

Theorem 1.284. *[460]. A subset H of a quasigroup Q is a coset class of normal congruence if and only if*

1. *$\mathbb{I}_k H \subseteq H$ for any fixed element $k \in H$;*
2. *if $(a/h)b = c$ and any two of the elements a, b, c belong to H, then the third one belongs to H, too.*

Theorem 1.285. *A subquasigroup H of a quasigroup Q is normal if and only if $\mathbb{I}_k H \subseteq H$ for any fixed element $k \in H$ [72].*

Theorem 1.286. *Any subquasigroup of a T-quasigroup is normal [506, Theorem 43].*

As a corollary we obtain the following:

Lemma 1.287. *Any subquasigroup $(H, \cdot)$ of a medial quasigroup $(Q, \cdot)$ is normal [506].*

Definition 1.288. A congruence θ of a quasigroup $(Q, \cdot)$ is called regular if it is uniquely defined by its coset $\theta(a)$, and the coset $\theta(a)$ of a congruence θ is called regular if it is a coset of only one congruence.

Remark 1.289. In [589] A.I. Maltsev has given necessary and sufficient conditions for the fact that a normal complex K of an algebraic system A is a coset of only one congruence, i.e., K is a coset of only one congruence of a system A. From the results of A.I. Maltsev [589], see, also, [782], it follows that normal subquasigroup A is a coset class of only one normal congruence of the quasigroup Q. Therefore in the finite quasigroup $(Q, \cdot)$ its congruence is regular. See Section 5.2 for more details.

1.5.4 Normal subloops

Theorem 1.290. *Let $(Q, \cdot)$ be a loop with the identity element 1 and N be a normal subloop, i.e., $N = \theta(1)$ for a normal congruence θ. Then*

(1) $x \cdot N = N \cdot x$ for all $x \in Q$;
(2) $x \cdot (y \cdot N) = (x \cdot y) \cdot N$, $(N \cdot x) \cdot y = N \cdot (x \cdot y)$ for all $x, y \in Q$;
(3) $(x \cdot N) \cdot (y \cdot N) = (x \cdot y) \cdot N$ for all $x, y \in Q$ [167, 72].

Theorem 1.291. *Let $(Q, \cdot)$ be a loop with the identity element 1 and N be a subloop. Then N is a normal subloop of the loop $(Q, \cdot)$, if*

(1) $x \cdot N = N \cdot x$ for all $x \in Q$;
(2) $x \cdot (y \cdot N) = (x \cdot y) \cdot N$, $(N \cdot x) \cdot y = N \cdot (x \cdot y)$ for all $x, y \in Q$ [167, 72].

A.A. Albert studied normal subloops of a loop $(L, \cdot)$ using normal subgroups of the group $M(L, \cdot)$ [10, 11].

A loop in which every subloop is normal is called Hamiltonian. Properties of Hamiltonian loops are studied in [662].

Exercise 1.292. Let $(Q, \cdot)$ be a loop with the identity element 1 and N be a subloop. Then N is a normal subloop of the loop $(Q, \cdot)$, if

(1) $x \cdot (y \cdot N) = (x \cdot y) \cdot N$,
(2) $(x \cdot y) \cdot N = (x \cdot N) \cdot y$ for all $x, y \in Q$. See Corollary 1.108.

Remark 1.293. There exists a correspondence between generators of an inner multiplication group of a loop (Theorem 1.104), normal subloops of a loop (Theorem 1.291), nuclei and A-nuclei of a loop (Definition 1.243) and some kinds of loop autotopy (Definition 1.134).

1.5.5 Antihomomorphisms and endomorphisms

Definition 1.294. If $(Q, \cdot)$ and $(H, \circ)$ are binary quasigroups, h is a single-valued mapping of Q into H such that $h(x_1 \cdot x_2) = hx_2 \circ hx_1$, then h is called an *antihomomorphism (a multiplicative antihomomorphism)* of $(Q, \cdot)$ into $(H, \circ)$ and the set $\{hx \mid x \in Q\}$ is called an *antihomomorphic image* of $(Q, \cdot)$ under h.

Remark 1.295. It is easy to see that with any antihomomorphism h we can associate a homomorphism $\mathbf{h}$ in the following way: $h(x_1 \cdot x_2) = hx_2 \circ hx_1$ if and only if $\mathbf{h}(x_1 \cdot x_2) = \mathbf{h}x_1 * \mathbf{h}x_2$, where $(H, *)$ is (12)-parastrophe of the quasigroup $(H, \circ)$.

We have used the same letter h in various type faces (fonts) because a single valued mapping h of the set Q into the set H in both cases is the same.

In case $(Q, \cdot) = (H, \circ)$ an antihomomorphism is also called an *antiendomorphism*, and an antiisomorphism is referred to as an anti-automorphism.

Exercise 1.296. Denote by the letter I the following anti-automorphism of a group $(Q,+)$: $I(x) = -x$ for any $x \in Q$. It is well known that $I^2 = \varepsilon$ [471]. Any anti-automorphism $\overline{\psi}$ of the group $(Q,+)$ can be represented in the form $\overline{\psi} = I\psi$, where $\psi \in Aut(Q,+)$.

Lemma 1.297. *Let h be an antihomomorphism of a quasigroup $(Q,\cdot)$ onto a groupoid $(H,\circ)$. Then h induces a congruence $Ker\, h = \theta$ (the kernel of h) in the following way: $x\,\theta\, y$ if and only if $h(x) = h(y)$.*

Proof. It is easy to see that θ is an equivalence. Prove that the equivalence θ is a congruence. Rewrite implication $(a)\,\theta\,(b) \longrightarrow (c{\cdot}a)\,\theta\,(c{\cdot}b)$ in the following form: $h(a) = h(b) \longrightarrow h(c{\cdot}a) = h(c\cdot b)$. The last implication is equivalent to the following: $h(a) = h(b) \longrightarrow h(a)\circ h(c) = h(b)\circ h(c)$. It is clear that the last implication is true. In a similar way the implication $(a)\,\theta\,(b) \longrightarrow (a\cdot c)\,\theta\,(b\cdot c)$ is proved. □

For antiendomorphisms the situation is more interesting. Compare Corollary 1.298 with Lemma 1.280.

Corollary 1.298. *1. If h is an endomorphism of a quasigroup $(Q,\cdot)$, then $(hQ,\cdot)$ is a subquasigroup of $(Q,\cdot)$ [789, 788].*

2. If h is an antiendomorphism of a quasigroup $(Q,\cdot)$, then $(hQ,\cdot)$ is a subquasigroup of $(Q,\cdot)$.

Proof. 1. We re-write the proof from ([80], p. 33) for a more general case. Prove that $(hQ,\cdot)$ is a subquasigroup of quasigroup $(Q,\cdot)$. Let $h(a)$, $h(b) \in h(Q)$. We demonstrate that the solution of equation $h(a)\cdot x = h(b)$ lies in $h(Q)$. Consider the equation $a\cdot y = b$. Denote the solution of this equation by c, i.e., $y = c$. Then $h(c)$ is a solution of equation $h(a)\cdot x = h(b)$. Indeed, $h(a)\cdot h(c) = h(a\cdot c) = h(b)$.

It is easy to see, that this is a unique solution. Indeed, if $h(a){\cdot}c_1 = h(b)$, then $h(a){\cdot}h(c) = h(a)\cdot c_1$. Since $h(a), h(c), c_1$ are elements of quasigroup $(Q,\cdot)$, then $h(c) = c_1$. For equation $x\cdot h(a) = h(b)$ the proof is similar.

2. Prove that $(hQ,\cdot)$ is a subquasigroup of quasigroup $(Q,\cdot)$. Let $h(a)$, $h(b) \in h(Q)$. We demonstrate that the solution of equation $h(a)\cdot x = h(b)$ lies in $h(Q)$. Consider the equation $y\cdot a = b$. Denote the solution of this equation by c, i.e., $y = c$. Then we have $h(c\cdot a) = h(a)\cdot h(c) = h(b)$. Then the element $h(c)$ is a solution of equation $h(a)\cdot x = h(b)$.

Prove that this is a unique solution. Indeed, if $h(a){\cdot}c_1 = h(b)$, then $h(a){\cdot}h(c) = h(a){\cdot}c_1$. Since $h(a), h(c), c_1$ are elements of quasigroup $(Q,\cdot)$, then $h(c) = c_1$.

For equation $x\cdot h(a) = h(b)$ the proof is similar. □

Notice that Case 1 of Corollary 1.298 follows from Theorem 1.279.

Remark 1.299. It is possible to give the following proof of Corollary 1.298. The $(hQ,\cdot)$ is a cancellation groupoid, since it is a sub-object of the quasigroup $(Q,\cdot)$ (Lemma 1.204). From the other side $(hQ,\cdot)$ is a division groupoid, since it is a homomorphic image of $(Q,\cdot)$ (Lemma 1.280). Therefore, by Lemma 1.47, $(hQ,\cdot)$ is a subquasigroup of the quasigroup $(Q,\cdot)$.

Corollary 1.300. *If h is an endomorphism of a quasigroup $(Q,\cdot)$, then h is an endomorphism of the quasigroups $(Q,*)$, $(Q,/)$, $(Q,\backslash)$, $(Q,//)$, $(Q,\backslash\backslash)$, i.e., from $h(x{\cdot}y) = h(x){\cdot}h(y)$ we obtain the following:*

1. $h(x*y) = h(x)*h(y)$. 2. $h(x/y) = h(x)/h(y)$.
3. $h(x\backslash y) = h(x)\backslash h(y)$. 4. $h(x//y) = h(x)//h(y)$.
5. $h(x\backslash\backslash y) = h(x)\backslash\backslash h(y)$.

Proof. From Lemma 1.298 we have that $(hQ, \cdot)$ is a subquasigroup of $(Q, \cdot)$.

Case 1. If we pass from the quasigroup $(Q, \cdot)$ to quasigroup $(Q, *)$, then subquasigroup $(hQ, *)$ of the quasigroup $(Q, *)$ will correspond to the subquasigroup $(hQ, \cdot)$. Indeed, any subquasigroup of the quasigroup $(Q, \cdot)$ is closed relative to parastrophe operations $*, /, \backslash, //, \backslash\backslash$ of the quasigroup $(Q, \cdot)$. Further we have $h(x * y) = h(y \cdot x) = h(y) \cdot h(x) = h(x) * h(y)$.

Case 2. If we pass from the quasigroup $(Q, \cdot)$ to quasigroup $(Q, /)$, then subquasigroup $(hQ, /)$ of the quasigroup $(Q, /)$ will correspond to the subquasigroup $(hQ, \cdot)$.

Let $z = x/y$, where $x, y \in Q$. Then from definition of the operation $/$ it follows that $x = zy$. Then $h(x) = h(z)h(y)$, $h(x/y) = h(z) = h(x)/h(y)$ ([591], p. 96, Theorem 1).

The remaining cases are proved in similar way. □

Lemma 1.301. *Let $(Q, \cdot)$ be a quasigroup. If f is an endomorphism of $(Q, \cdot)$, then $f(e(x)) = e(f(x))$, $f(s(x)) = s(f(x))$ for all $x \in Q$;*

if e is an endomorphism of $(Q, \cdot)$, then $e(f(x)) = f(e(x))$, $e(s(x)) = s(e(x))$ for all $x \in Q$;

if s is an endomorphism of $(Q, \cdot)$, then $s(f(x)) = f(s(x))$, $s(e(x)) = e(s(x))$ for all $x \in Q$ (Lemma 2.4. from [520]).

Proof. We shall use Corollary 1.300. If f is an endomorphism, then $f(e(x)) = f(x\backslash x) = f(x)\backslash f(x) = e(f(x))$, $f(s(x)) = f(x) \cdot f(x) = s(f(x))$.

If e is an endomorphism, then $e(f(x)) = e(x)/e(x) = f(e(x))$, $e(s(x)) = e(x) \cdot e(x) = s(e(x))$.

If s is an endomorphism, then $s(f(x)) = s(x)/s(x) = f(s(x))$, $s(e(x)) = s(x)\backslash s(x) = s(e(x))$. □

Denote by $Aaut(Q, +)$ the set of all anti-automorphisms of a quasigroup $(Q, +)$.

Lemma 1.302. *1. The product of two anti-automorphisms of a quasigroup $(Q, +)$, say $\overline{\varphi}$ and $\overline{\psi}$, is an automorphism of $(Q, +)$.*

2. *If φ is an automorphism of a quasigroup $(Q, +)$, and $\overline{\psi}$ is its anti-automorphism, then $\varphi\overline{\psi}$, $\overline{\psi}\varphi$ are some anti-automorphisms of the quasigroup $(Q, +)$.*
3. *$I\varphi = \varphi I$.*
4. *$I\overline{\varphi} = \overline{\varphi}I$.*
5. *$(\overline{\varphi})^{-1} = \overline{\varphi^{-1}} = I\varphi^{-1}$.*
6. *Denote by J_a the inner automorphism of a group $(Q, +)$, i.e., $J_a x = a + x - a$ for all $x \in Q$. If $\varphi \in Aut(Q, +)$, then $\varphi J_a = J_{\varphi a}\varphi$, $\varphi J_a^{-1} = J_{\varphi a}^{-1}\varphi$.*
7. *If $J_a \in Inn(Q, +)$ and $I\varphi \in Aaut(Q, +)$, then $I\varphi J_a = J_{\varphi a} I\varphi$, i.e., $\overline{\varphi}J_a = J_{\varphi a}\overline{\varphi}$.*

Proof. It is easy to check. See [793]. □

1.5.6 Homotopism

Definition 1.303. A quasigroup $(Q, \cdot)$ is a homotopic image of a quasigroup $(P, \circ)$ if there exist surjective maps $\alpha, \beta, \gamma \in Q^P$ ($\alpha \in Q^P \Leftrightarrow \alpha : P \to Q$) such that $\gamma(x \circ y) = \alpha x \cdot \beta y$ for all $x, y \in Q$ [10].

If $\alpha = \beta = \gamma$, then homotopy (homotopism) is a homomorphism.

Theorem 1.304. *A quasigroup $(P, \circ)$ is homotopic to a quasigroup $(Q, \cdot)$ if and only if a LP-isotope of the quasigroup $(P, \circ)$ is homotopically mapped on an LP-isotope of the quasigroup $(Q, \cdot)$.*

Proof. Let $\varphi : (Q, \cdot) \longrightarrow (Q', \circ)$ be a homomorphism of a quasigroup $(Q, \cdot)$ on a quasigroup $(Q', \circ)$. Define in $(Q, \cdot)$ the following binary relation: $a \sim b \leftrightarrow \varphi a = \varphi b$. The binary relation $\sim$ is an equivalence relation. Moreover, $\varphi(a \cdot c) = \varphi a \circ \varphi c$, $\varphi(b \cdot c) = \varphi b \circ \varphi c$. So, if $a \sim b \leftrightarrow \varphi a = \varphi b$, then

$$\varphi(ac) = \varphi a \circ \varphi c = \varphi b \circ \varphi c = \varphi(ab),$$

i.e., $ac \sim bc$. Similarly, $a \sim b \rightarrow ca \sim cb$. Then, $\sim$ is a congruence on $(Q, \cdot)$. The following implications are fulfilled too:

(1) $ac \sim bc \longrightarrow a \sim b$,

(2) $ca \sim cb \longrightarrow a \sim b$.

We prove (1).

$$ac \sim bc \rightarrow \varphi(ac) = \varphi(bc) \Longleftrightarrow \varphi a \circ \varphi c = \varphi b \circ \varphi c.$$

Since $(Q', \circ)$ is a quasigroup, then $\varphi a = \varphi b \leftrightarrow a \sim b$. □

1.5.7 Congruences and isotopism

Definition 1.305. An isotopy (α, β, γ) is admissible relative to a binary relation θ, if θ is admissible relative to permutations α, β, γ.

In [72, p. 59] the following key lemma is proved:

Lemma 1.306. *Let θ be a normal congruence of a quasigroup $(Q, \cdot)$. If a quasigroup $(Q, \circ)$ is isotopic to $(Q, \cdot)$ and the isotopy (α, β, γ) is admissible relative to θ, then θ is also a normal congruence in $(Q, \circ)$.*

Lemma 7.63 is an n-ary analog of Lemma 1.306. We denote by $nCon(Q, \cdot)$ the set of all normal congruences of a quasigroup $(Q, \cdot)$. We give the following corollary of Lemma 1.306.

Corollary 1.307. *If $(Q, \cdot)$ is a quasigroup, $(Q, +)$ is a loop of the form $x + y = R_a^{-1}x \cdot L_b^{-1}y$ for all $x, y \in Q$, then $nCon(Q, \cdot) \subseteq nCon(Q, +)$.*

Proof. If θ is a normal congruence of a quasigroup $(Q, \cdot)$, then, since θ is admissible relative to the isotopy $T = (R_a^{-1}, L_b^{-1}, \varepsilon)$, θ is also a normal congruence of a loop $(Q, +)$. □

Remark 1.308. It is easy to see, if $x + y = R_a^{-1}x \cdot L_b^{-1}y$, then $x \cdot y = R_a x + L_b y$. If in conditions of Corollary 1.307, in addition, we suppose, that the isotopy $T^{-1} = (R_a, L_b, \varepsilon)$ is admissible relative to any normal congruence of the loop $(Q, +)$, then we obtain the following equality $nCon(Q, \cdot) = nCon(Q, +)$.

Corollary 1.309. *Let $(Q, \cdot) = (Q, +)(\alpha, \beta, \varepsilon)$, where $(Q, +)$ is a loop, $\alpha, \beta \in S_Q$. If $(Q, +)$ does not contain normal subloops admissible relative to permutations α, β, then quasigroup $(Q, \cdot)$ is simple.*

Proof. The proof follows from Lemmas 1.306 and 1.121, Corollary 1.307. □

By the study of linear quasigroups the following theorem is helpful.

Theorem 1.310. *Let $(Q, +)$ be an IP-loop, $x \cdot y = (\varphi x + \psi y) + c$, where $\varphi, \psi \in Aut(Q, +)$, $a \in C(Q, +)$, θ be a normal congruence of $(Q, +)$. Then θ is a normal congruence of $(Q, \cdot)$ if and only if $\varphi\,|_{Ker\,\theta}$, $\psi\,|_{Ker\,\theta}$ are automorphisms of $Ker\,\theta$ [667, 506, 767, 769].*

Simple quasigroups without subquasigroups that have identity automorphism groups are constructed in [443] on the basis of groups using prolongation construction (see Section 1.7).

Lemma 1.311. *There exists a simple quasigroup without subquasigroups that have an identity autotopy group.*

Proof. This follows from Example 1.145 (the loop from this example has order 7, therefore it is simple as its isotope), Kepka's result (Theorem 1.208), and behavior of autotopy groups by isotopy (Theorem 1.133). □

1.5.8 Congruence permutability

In [589] A.I. Maltsev proved that in an e-quasigroup $(Q, \cdot, /, \backslash)$ all congruences are permutable.

Any congruence of e-quasigroup $(Q, \cdot, /, \backslash)$ is a normal congruence of quasigroup $(Q, \cdot)$ and vice versa, any normal congruence of quasigroup $(Q, \cdot)$ is a congruence of e-quasigroup $(Q, \cdot, /, \backslash)$ [591, 786]. Moreover, it is proved that normal congruence of a quasigroup $(Q, \cdot)$ is permutable with any quasigroup congruence [151].

In [886] an example of non-permutable congruences of a quasigroup $(Q, \cdot)$ is constructed. See [589, 336] for additional information on permutability of quasigroup congruences.

We give a sketched proof of the following well-known fact [589, 813, 184]. We follow [813].

Lemma 1.312. *Normal quasigroup congruences commute in pairs.*

Proof. We shall use Lemma 1.266. Let θ_1 and θ_2 be normal congruences of a quasigroup $(Q, \cdot)$. Then $a(\theta_1 \circ \theta_2)b$ means that there exists an element $c \in Q$ such that $a\theta_1 c$ and $c\theta_2 b$.

Further we have

$$\begin{array}{ll} a\theta_2 a, & a\theta_2 a \\ c\theta_2 b, & L_c^{-1}c\theta_2 L_c^{-1}b \\ b\theta_2 b, & L_c^{-1}b\theta_2 L_c^{-1}b. \end{array}$$

Then

$$R^{-1}_{L_c^{-1}c}a \cdot L_c^{-1}b\,\theta_2\,R^{-1}_{L_c^{-1}b}a \cdot L_c^{-1}b = a.$$

From relations

$$\begin{array}{ll} a\theta_1 a, & a\theta_1 a \\ a\theta_1 c, & L_c^{-1}a\theta_1 L_c^{-1}c \\ b\theta_1 b, & L_c^{-1}b\theta_1 L_c^{-1}b \end{array}$$

we obtain

$$b = R^{-1}_{L_c^{-1}a}a \cdot L_c^{-1}b\,\theta_1\,R^{-1}_{L_c^{-1}c}a \cdot L_c^{-1}b.$$

Therefore, $a(\theta_2 \circ \theta_1)b$. □

1.6 Constructions

For quasigroup theory the problem of construction of a quasigroup is important. A good overview of existing methods of construction of quasigroups and loops is in [201].

Non-strictly speaking we can divide algebraic quasigroup constructions into two subclasses.

(i) We take more "simple" objects (groups, rings residues modulo k, algebras, fields) and some quit "natural" operations (addition, multiplication, subtraction, taking of a linear form etc.) or morphisms (isotopy, parastrophy, isostrophy, crossed isotopy or generalized isotopy (Section 9.5.2)) and construct quasigroups and loops in "general" or with some necessary properties.

For example, using some ring construction H. Zassenhaus constructed a Bol loop of order 8 [685, p. 113], V.I. Onoi constructed a left distributive quasigroup that is not isotopic to a left Bol loop [670], A.S. Basarab constructed generalized Moufang loops [36], and A.M. Cheban constructed some loops that are named for him [195, 693].

(ii) Direct constructions, when as "bricks" (or as "adobes") for this construction quasigroups are taken, i.e., in these constructions quasigroups (or groupoids) are used for construction of quasigroups.

In the next subsections we give some constructions of the second kind. A good overview of various quasigroup constructions is presented in [201].

1.6.1 Direct product

A direct product of quasigroups is a standard algebraic construction.

Definition 1.313. If $(A, \cdot)$, $(B, \circ)$ are binary quasigroups, then their *(external) direct product* $(Q, *) = (A, \cdot) \times (B, \circ)$ is the set of all ordered pairs (a, b), where $a \in A$, $b \in B$, and where the operation in $(Q, *)$ is defined component-wise, that is, $(a * b) = (a_1 \cdot a_2, b_1 \circ b_2)$, where $a_1, a_2 \in A$, $b_1, b_2 \in B$.

Direct products of quasigroups are studied in many articles and books, see, for example, [212, 813, 667, 460, 103, 107]. The concept of a direct product of quasigroups was used already in [646]. In the group case it is possible to find these definitions, for example, in [471, 422].

In [151, 184, 813, 819] there is a definition of the (internal) direct product of Ω-algebras. We recall that any quasigroup is an Ω-algebra.

Let U and W be equivalence relations on a set A, let $U \vee W = \{(x, y) \in A^2 \mid \exists\ n \in N,\ \exists\ t_0, t_1, \dots, t_{2n} \in A, x = t_0 U t_1\ W\ t_2\ U \dots U t_{2n-1} W t_{2n} = y\}$. $U \vee W$ is an equivalence relation on A called the join of U and W.

If U and W are equivalence relations on A for which $U \circ W = W \circ U$, then $U \circ W = U \vee W$, U and W are said to commute [813].

If A is an Ω-algebra and U, W are congruences on A, then $U \vee W$, and $U \cap W$ are also congruences on A.

Definition 1.314. If U and W are congruences on the algebra A which commute and for which $U \cap W = \hat{A} = \{(a, a) \mid \forall\, a \in A\}$, then the join $U \circ W = U \vee W$ of U and W is called a *direct product* $U \sqcap W$ *of* U *and* W [151, 813, 819].

The following theorem establishes the connection between concepts of the internal and external direct product of Ω-algebras.

Theorem 1.315. *An Ω-algebra A is isomorphic to a direct product of Ω-algebras B and C with isomorphism φ, i.e., $\varphi : A \to B \times C$, if and only if there exist such congruences U and W of A that $A^2 = U \sqcap W$ ([813, p. 16], [819]).*

In the next lemma, $C(Q)$ means the center of a loop Q.

Lemma 1.316. *If a loop Q is isomorphic to the direct product of the loops A and B, then $C(Q) \cong C(A) \times C(B)$.*

Proof. It is easy to check. □

Lemma 1.317. *If a quasigroup Q is isomorphic to the direct product of quasigroups A and B and quasigroup Q fulfills an identity, then quasigroups A and B also fulfill this identity.*

Proof. From the definition of the direct product it follows that there exists a homomorphism of quasigroup Q on quasigroup A. Similarly, there exists a homomorphism of quasigroup Q on quasigroup B. It is well known [151, 591, 184] that if a quasigroup satisfies an identity, then its homomorphic image also satisfies this identity. □

1.6.2 Semidirect product

There exist various approaches to the concept of a semidirect product of quasigroups [728, 727, 180, 908]. We give an "external" definition of the semidirect product of quasigroups by an analogy with group case [471]. The main principle is that any semidirect product is a direct (Cartesian) product as a set [908].

In a non-formal sense, Schreier extension of quasigroups is an inverse "operation" to the "operation" of the semidirect product of quasigroups. Here we omit description of this kind of quasigroup extension. See [318, 654] for details.

Definition 1.318. Suppose that Q is a quasigroup, A is a normal subquasigroup of Q (i.e., $A \trianglelefteq Q$), and B is a subquasigroup of Q. A quasigroup Q is the semidirect product of quasigroups A and B, if there exists a homomorphism $h : Q \longrightarrow B$ which is the identity on B and whose kernel is A, i.e., A is a coset class of the normal congruence $Ker\, h$. We shall denote this fact as follows $Q \cong A \rtimes B$ [909].

In the group case, information about this construction is in [471, 909].

We give a construction of a semidirect product from [718]. Let $(G, +)$ and $(H, \cdot)$ be groups, whose identity elements are denoted by 0 and e respectively; let $AutH$ be the automorphism group of $(H, \cdot)$ with the identity automorphism denoted by ε; let $\theta : G \rightarrow AutH$. For $g \in G$ and $h \in H$, let $\theta(g)h$ be that element in H which is obtained when $\theta(g)$ acts on h. Now let B be the Cartesian product $B = G \times H$ and define

$$(x, a) \circ (y, b) = (x + y, \theta(y)a \cdot b) \tag{1.84}$$

for all $(x, a), (y, b) \in B$. It is easy to see that $(B, \circ)$ is a quasigroup, quasigroup $(B, \circ)$ is a loop if and only if $\theta(0) = \varepsilon$, and loop $(B, \circ)$ is a Bol loop if and only if

$$\theta(y + z + y) = \theta(y)\theta(z)\theta(y) \tag{1.85}$$

for all $y, z \in G$.

Lemma 1.319. *If a quasigroup Q is the semidirect product of quasigroups A and B, $A \trianglelefteq Q$, then there exists an isotopy T of Q such that QT is a loop and $QT \cong AT \rtimes BT$.*

Proof. If we take isotopy of the form $(R_a^{-1}, L_a^{-1}, \varepsilon)$, where $a \in A$, then we have that QT is a loop and AT is its normal subloop (Lemma 1.306, Remark 1.267). Further we have that BT is a loop since $BT \cong QT/AT$. Therefore BT is a subloop of the loop QT, since the set B is a subset of the set Q. □

Corollary 1.320. *If a quasigroup Q is the direct product of quasigroups A and B, then there exists an isotopy $T = (T_1, T_2)$ of Q such that $QT \cong AT_1 \times BT_2$ is a loop.*

Proof. The proof follows from Lemma 1.319. □

Lemma 1.321. *1. If a left loop $(Q,\cdot)$ with the form $x \cdot y = x + \psi y$, where $(Q,+)$ is a group, $\psi \in Aut(Q,+)$, is the semidirect product of a normal subgroup $(H,\cdot) \trianglelefteq (Q,\cdot)$ and a subgroup $(K,\cdot) \subseteq (Q,\cdot)$, $H \cap K = \{0\}$, then $(Q,\cdot) = (Q,+)$.*

2. If a right loop $(Q,\cdot)$ with the form $x \cdot y = \varphi x + y$, where $(Q,+)$ is a group, $\varphi \in Aut(Q,+)$, is the semidirect product of a normal subgroup $(H,\cdot) \trianglelefteq (Q,\cdot)$ and a subgroup $(K,\cdot) \subseteq (Q,\cdot)$, $H \cap K = \{0\}$, then $(Q,\cdot) = (Q,+)$.

Proof. 1. Since $(Q,\cdot)$ is the semidirect product of a normal subgroup $(H,\cdot)$ and a subgroup $(K,\cdot)$, then we can write any element a of the loop $(Q,\cdot)$ in a unique way as a pair $a = k \cdot h$, where $k \in (K,\cdot)$, $h \in (H,\cdot)$.

We notice, $\psi k = k$, $\psi h = h$, since $(K,\cdot)$, $(H,\cdot)$ are subgroups of the left loop $(Q,\cdot)$. Indeed, from associativity of K, $(k_1 \cdot k_2) \cdot k_3 = k_1 \cdot (k_2 \cdot k_3)$ for all $k_1, k_2, k_3 \in K$ we have $k_1 + \psi k_2 + \psi k_3 = k_1 + \psi k_2 + \psi^2 k_3$, $k_3 = \psi k_3$ for all $k_3 \in K$. By analogy $h = \psi h$ for all $h \in H$.

Further we have $\psi a = \psi(k \cdot h) = \psi(k + \psi h) = \psi k + \psi^2 h = k + \psi h = k \cdot h = a$, $\psi = \varepsilon$, $(Q,\cdot) = (Q,+)$.

2. This case is proved similarly to Case 1. □

Example 1.322. A quasigroup $(Z_9, \circ)$, $x \circ y = x + 4 \cdot y$, where $(Z_9, +, \cdot)$ is the ring of residues modulo 9, $Z_9 = \{0, 1, 2, 3, 4, 5, 6, 7, 8\}$, demonstrates that some restrictions in Lemma 1.321 are essential.

Here we give a relatively general method of construction of a quasigroup (of a loop) using two groups [67]. Let $(M,\cdot)$, $M = \{a, b, c, \dots\}$, be a group and $(N,+)$, $N = \{x, y, z, \dots\}$ be a commutative group. On the set $Q = M \times N$ define multiplication as follows:

$$(a, x) \circ (b, y) = (ab,\ t_{a,b} + \varphi_b^{-1} x + y), \tag{1.86}$$

where $t_{a,b}$ is an element of the group $(N,+)$ which depends on the elements $a, b \in M$, φ_b^{-1} is an automorphism of the group $(N,+)$ which depends on the element $b \in M$. It is easy to check that $(Q,\circ)$ is a quasigroup. If, additionally,

$$\varphi_1 = 1, t_{a,1} = t_{1,a} = 0, \tag{1.87}$$

where 1 is identity element of the group $(M,\cdot)$ and 0 is identity element of the group $(N,+)$, then $(Q,\circ)$ is a loop.

1.6.3 Crossed (quasi-direct) product

We give a construction of a crossed (quasi-direct in terminology of [943]) product of quasigroups [166, 165, 72].

Let $(P,\cdot)$ be a quasigroup which is defined on a set $P = \{u, v, w, \dots\}$. Let Σ be a system of quasigroups defined on a set Q. For any ordered pair of elements $u, v \in P$ we put in correspondence an operation $A \in \Sigma$, i.e., we define a map $\delta : P^2 \longrightarrow \Sigma$.

On the set $M = P \times Q$ we define the following operation $\circ$:

$$(u, a) \circ (v, b) = (v \cdot v, A_{u,v}(a, b)), \tag{1.88}$$

where $u, v \in P, a, b \in Q$. Therefore $(M, \circ)$ is a groupoid.

Theorem 1.323. *If $(P,\cdot)$ is a quasigroup, Σ is a system of quasigroups defined on a set Q, then $(M,\circ)$ is a quasigroup. $(M,\circ)$ is a loop if and only if $(P,\cdot)$ is a loop and there exists an element $c \in Q$ such that*

$$A_{u,1}(a,e) = A_{1,v}(e,a) = a$$

for all $a \in Q$, $u,v \in P$. In this case the identity element of the loop $(M\circ)$ is the pair $(1,e)$.

Remark 1.324. If $\Sigma = \{(Q,A)\}$, then $(M,\circ) \cong (P,\cdot) \times (Q,A)$.

Applications of the binary crossed product of quasigroups are in [60, 943, 228].

1.6.4 n-Ary crossed product

In [157] the crossed product of n-ary quasigroups is used to construct n-ary irreducible quasigroups of finite composite orders.

Let (P,f) be an n-ary quasigroup which is defined on a set P. Let Σ be a system of n-ary quasigroups defined on a set Q.

For any n-tuple of elements $(u_1,u_2,\dots,u_n) \in P^n$ we put in correspondence an operation $g \in \Sigma$, i.e., we define a map $\delta : P^n \longrightarrow \Sigma$.

On the set $M = P \times Q$ we define the following operation h:

$$h((u_1,a_1),(u_2,a_2),\dots,(u_n,a_n)) = (f(u_1,u_2,\dots,u_n), g(a_1,a_2,\dots,a_n)), \tag{1.89}$$

where $u_i \in P, a_j \in Q$, n-ary operation $g \in \Sigma$ corresponds to the n-tuple $(u_1,u_2,\dots,u_n)$, i.e., $g = \delta(u_1,u_2,\dots,u_n) \in \Sigma$.

In [157] it is proved that n-ary groupoid (M,h) is an n-ary quasigroup. In this article the crossed product of n-ary quasigroups is called a twist of an n-ary quasigroup f.

Notice that in [157] by the proof of irreducibility of constructed n-quasigroups the Belousov-Sandik D_{ij} criteria [93] of reducibility of n-ary quasigroups is used.

1.6.5 Generalized crossed product

There is the definition of generalized crossed product of two systems of quasigroups [124]. In this article there are some applications of this construction and some combinatorial comments. Denote by $P = \{u,v,w,\dots\}$ and $Q = \{a,b,c,\dots\}$ some nonempty sets. By Σ_1 and Σ_2 we denote the systems of binary quasigroups defined on the sets P and Q, respectively. To any element a from Q, we put in correspondence a quasigroup $A_a \in \Sigma_1$, and to any pair of elements $u,v \in P$ we put in correspondence a quasigroup $B_{u,v} \in \Sigma_2$. On the set $M = P \times Q$ we define the following operation

$$C((u,a),(v,b)) = (A_{B_{u,v}(a,b)}(u,v), B_{u,v}(a,b)).$$

In [124] it is proved that (M,C) is a quasigroup.

1.6.6 Generalized singular direct product

The singular direct product is defined by Sade [729, 732]. The text below is taken from Stein's review (MR0138575 (25 #2019)) on the article [732].

Let $(Q,\circ)$ be a groupoid with a subgroupoid $(P,\circ)$. Let P' be the complement of P in Q. Assume that $P \circ P' \cup P' \circ P \subset P'$ (note that if $(Q,\circ)$ is a quasigroup, then this condition is automatically satisfied).

On P' we define the groupoid $(P',\otimes)$, not necessarily related to $\circ$. Let $(V,\odot)$ be a groupoid.

Then the singular direct product QV is the groupoid $(P \cup (P' \times V), \oplus)$ defined on the set $P \cup (P' \times V)$ by the conditions

$$p_1 \oplus p_2 = p_1 \circ p_2, \quad \text{if } p_1, p_2 \in P; \tag{1.90}$$

$$(p', v) \oplus p = (p' \circ p, v), \quad \text{if } p \in P, p' \in P', v \in V; \tag{1.91}$$

$$p \oplus (p', v) = (p \circ p', v), \quad \text{if } p \in P, p' \in P', v \in V; \tag{1.92}$$

$$(p_1', v) \oplus (p_2', v) = \begin{cases} p' \circ p_2', & \text{if } p_1' \circ p_2' \in P, \\ (p_1' \circ p_2', v), & \text{if } p_1' \circ p_2' \in P'; \end{cases} \tag{1.93}$$

$$(p_1', v_1) \oplus (p_2', v_2) = (p_1' \otimes p_2', v_1 \odot v_2), \quad \text{if } v_1 \neq v_2. \tag{1.94}$$

If P is void and $\otimes$ is $\circ$, then QV reduces to the ordinary direct product.

The singular direct product is significant in the theory of quasigroups for two reasons: (I) If Q, P, P' are quasigroups and V is an idempotent quasigroup, then QV is a quasigroup; (II) If $Q^*, P^*, P^{*\prime}, V^*$ satisfy (I) and are orthogonal to the respective quasigroups without asterisks, then Q^*V^* is orthogonal to QV.

In particular, if there are idempotent orthogonal quasigroups of orders $q + 1$ and v, and orthogonal quasigroups of order q, then there are orthogonal quasigroups of order $1 + qv$. For $q = 3$ and $v = 7$ this method produces orthogonal quasigroups of order 22. It can also yield results similar to Theorem 8(i) of Bose, Shrikhande and Parker *[Canad. J. Math. 12 (1960), 189–203]*. From Stein's review (MR0138575 (25 #2019)).

In [568] quasigroups satisfying the identity $x(xy) = yx$ are constructed using a singular direct product. Identities of the form $w(x, y) = t(x, y)$ which are preserved by the singular direct product are described in [570].

A generalized singular direct product is a combination of singular direct product and a portion of a crossed (quasi-direct) product [569]. We follow [226].

If in equality (1.94) we change a fixed operation $\otimes$ by the set of operations $\otimes_{v_1,v_2}$, i.e., if equality (1.94) takes the form

$$(p_1', v_1) \oplus (p_2', v_2) = (p_1' \otimes_{v_1,v_2} p_2', v_1 \odot v_2) \quad \text{if } v_1 \neq v_2, \tag{1.95}$$

then we obtain the definition of a generalized singular direct product.

Example 1.325. [226]. We take the following quasigroup $(Q, \circ)$, $Q = \{0, 1, 2, 3\}$, with one-element subquasigroup $(P, \circ)$, $P = \{0\}$, $P' = \{1, 2, 3\}$, $(V, \odot)$ is an idempotent quasigroup, and $p_1' \otimes_{v_1,v_2} p_2' = (-p_1' - p_2' + 2v_1 + v_2 - 1)_3 + 1$. The expression $(-p_1' - p_2' + 2v_1 + v_2 - 1)_3$ means $(-p_1' - p_2' + 2v_1 + v_2 - 1) \pmod 3$. A pair (p', v) will be denoted by $p' + 3v$.

$\circ$	0	1	2	3
0	0	1	2	3
1	2	3	0	1
2	3	2	1	0
3	1	0	3	2

$\odot$	0	1	2
0	0	2	1
1	2	1	0
2	1	0	2

$\oplus$	0	1	2	3	4	5	6	7	8	9
0	0	1	2	3	4	5	6	7	8	9
1	2	3	0	1	8	7	9	6	5	4
2	3	2	1	0	7	9	8	5	4	6
3	1	0	3	2	9	8	7	4	6	5
4	5	9	8	7	6	0	4	2	1	3
5	6	8	7	9	5	4	0	1	3	2
6	4	7	9	8	0	6	5	3	2	1
7	8	5	4	6	3	2	1	9	0	7
8	9	4	6	5	2	1	3	8	7	0
9	7	6	5	4	1	3	2	0	9	8

If $p_1{}' = 2$, $v_1 = 0$, $p_2{}' = 3$, $v_2 = 1$, then $(2,0) \oplus (3,1) = ((-2-3+2\cdot 0+1-1)_3 + 1, 0 \odot 1) = ((-5)_3+1, 2) = (1+1, 2) = (2,2) = 2+3\cdot 2 = 8$. Notice in this case $(p_1{}', v_1) = (2,0) = 2+3\cdot 0 = 2$, $(p_2{}', v_2) = 3+3\cdot 1 = 6$. Finally we have $2 \oplus 6 = 8$.

The singular direct product and its generalization are used intensively in combinatorics [644], by construction of Schröder quasigroups [572].

1.6.7 Sabinin's product

The detailed description of L.V. Sabinin's product of a left loop and a group is in [723, 727]. Here left loop means a groupoid $(Q,\cdot)$ in which equation $a\cdot x = b$ has a unique solution for any fixed elements $a, b \in Q$ and in this groupoid there exists an element 1 such that $1\cdot x = x$ for all $x \in Q$. It is easy to see that in any left loop (even in any left quasigroup) any left translation is a bijective mapping of the set Q.

Below we mainly follow H. Pflugfelder's report from (MR0340461 (49 #5216)). The group denoted by $as_l(Q)$ and called the associant of Q is the inner mapping group of the group $LM(Q,\cdot) \subseteq S_Q$.

Transassociant $(H,\circ)$ of $(Q,\cdot)$ is a subgroup of S_Q such that $as_l(Q) \subseteq H \subseteq S_Q$, and $m_q h = (L_{h(q)}^{-1}) \circ h \circ L_q \circ h^{-1} \in H$ for all $q \in Q$ and $h \in H$.

Here $h(q)$ is the image of q under h, and L_a means left translation of left quasigroup $(Q,\cdot)$.

A semiproduct of $(Q,\cdot)$, and its transassociant $(H,\circ)$, $Q \circledast H$, is a set $Q \times H$ with the following composition:

$$(q_1,h_1)(q_2,h_2) = (q_1\cdot h_1(q_2), l(q_1,h_1(q_2)) \circ m_{q_2}(h_1) \circ h_1 \circ h_2),$$

where $l(a,b) = L_{ab}^{-1} \circ L_a \circ L_b$, $a, b \in Q$.

$Q \circledast H$ is a group that is transitive on Q and determines on Q a structure of uniform space [723, 727].

On the other hand, for every uniform space there can be defined a loop structure in such a way that the semiproduct of this loop with its transassociant is isomorphic to the fundamental group of the uniform space [727].

A generalization of Sabinin's construction is given in [180, 724]. Many geometrical and combinatorial ways of constructing quasigroups are described in [228].

1.7 Quasigroups and combinatorics

We know that any finite quasigroup is a Latin square. Sometimes general theorems about block designs are helpful in the theory of Latin squares and, automatically, in the theory of finite quasigroups [722, 213, 850, 918, 152, 572]. Very close connections exist between Steiner triple systems (combinatorial object) and TS-quasigroups [409, 721]. See Section 2.6 for more details.

1.7.1 Orthogonality

In this section we give an outline of the concept of orthogonality. For more details see [240, 241, 559], and Chapter 9 of this book.

Definition 1.326. A Latin square of order (of size) m is an arrangement of m symbols $x_1, \dots, x_m$ into m rows and m columns such that no row and no column contains any of the symbols $x_1, \dots, x_m$ twice [240, 241].

Infinite Latin squares are also defined and studied. See for example [423].

1.7.1.1 Orthogonality of binary operations

Definition 1.327. Two $n \times n$ Latin squares are called orthogonal if when one is superimposed upon the other, every ordered pair of symbols $x_1, x_2, \dots, x_m$ occurs once in the resulting square [240].

At the present time orthogonality is one of the most "applicable" quasigroup properties.

Leonard (Leonhard in Latin) Euler conjectured that there do not exist orthogonal Latin squares for orders $n = 6, 10, 14, 18, \dots$. Euler's conjecture is true for $n = 6$ [882, 883, 240] but it is not true for all natural numbers $n = 4k + 2$, $k \geqslant 2$ [677, 160, 161, 240].

A short disproof of Euler's conjecture concerning orthogonal Latin squares is in [953, 228]. In this disproof, concepts of quasigroup transversal, quasigroup prolongation and singular direct product are used.

Example 1.328. We give an example of a pair of orthogonal quasigroups (Latin squares) of order 10 [559]. In any cell the first number is from the first quasigroup (Latin square), and the second number is from the second quasigroup (Latin square).

$\cdot, \star$	0	1	2	3	4	5	6	7	8	9
0	12	23	31	46	59	64	78	87	95	00
1	74	42	27	09	61	58	85	90	33	16
2	51	14	45	67	08	80	93	22	76	39
3	07	71	10	38	83	92	44	56	12	65
4	35	57	73	82	94	11	06	49	60	28
5	20	05	52	91	77	36	19	63	48	84
6	43	30	04	55	26	79	62	18	81	97
7	89	98	66	24	32	03	50	75	17	41
8	68	86	99	70	15	47	21	34	02	53
9	96	69	88	13	40	25	37	01	54	72

The elegant construction of a pair of orthogonal Latin squares of order $(12k + 10)$ (including order 10) is in [722, Chapter 7]. See also Example 10.74.

We recall that an unbordered (i.e., without the first row and the first column) Cayley table of any finite quasigroup is a Latin square (Definition 1.30) and vice versa, any Latin square defines a quasigroup.

Definition 1.329. Binary groupoids (Q, A) and (Q, B) are called orthogonal if the system of equations

$$\begin{cases} A(x, y) = a \\ B(x, y) = b \end{cases}$$

has a unique solution (x_0, y_0) for any fixed pair of elements $a, b \in Q$. We shall denote this fact as follows: $(Q, A) \perp (Q, B)$.

Example 1.330. Example of a pair of orthogonal groupoids of order 6.

A	1	2	3	4	5	6
1	1	1	1	1	1	1
2	2	2	2	2	2	2
3	3	3	3	3	3	3
4	4	4	4	4	4	4
5	5	5	5	5	5	5
6	6	6	6	6	6	6

B	1	2	3	4	5	6
1	1	2	3	4	5	6
2	1	2	3	4	5	6
3	1	2	3	4	5	6
4	1	2	3	4	5	6
5	1	2	3	4	5	6
6	1	2	3	4	5	6

Example 1.331. Quasigroups $(Z_{11}, \star)$ and $(Z_{11}, \circ)$, where $x \star y = x + 2 \cdot y$, $x \circ y = 3 \cdot x + y$ for all $x, y \in Z_{11}$, $(Z_{11}, +, \cdot)$ is the ring of residues modulo 11, are orthogonal. See Section 9.3 for details.

Definition 1.332. A Latin square for which an orthogonal Latin square exists is called a basis square [596].

We suppose that Latin squares of order (of size) n from the following two theorems are defined on the set $\overline{1, n}$.

Theorem 1.333. *If in the Latin square L of order $(4n + 2)$ the square formed by the first $(2n + 1)$ rows and the first $(2n + 1)$ columns contains fewer than $(n + 1)$ elements which are different from elements of the set $\{1, 2, \ldots, 2n + 1\}$, then L is not a basis square [596].*

Theorem 1.334. *If in the $(4n + 1)$-sided Latin square L the square formed by the first $2n$ rows and the first $2n$ columns contains fewer than $n/2$ elements different from the elements of the set $\{1, 2, \ldots, 2n\}$, then L is not a basis square [596].*

Denote by $N(q)$ the number of mutually (in pairs) orthogonal Latin squares of order q.

Theorem 1.335. $N(q) \leqslant (q - 1)$;

if q is prime, then $N(q) = (q - 1)$;

$N(q_1 q_2) \geqslant \min\{N(q_1), N(q_2)\}$, *in particular, if $q = q_1 \ldots q_t$ is the canonical decomposition of q, then $N(q) \geqslant \min\{q_1 - 1, \ldots, q_t - 1\}$;*

$N(q) \geqslant q^{10/143} - 2$;

$N(q) \geqslant 3$, *if $q \notin \{2, 3, 6, 10\}$;*

$N(q) \geqslant 6$ *whenever $n > 90$;*

$N(q) \geqslant n^{10/148}$ *for sufficiently large q. [240, 241, 595, 944, 559, 420, 371, 147].*

Problem 1.6. Find the triple of mutually orthogonal Latin squares (of MOLS) of order 10 or prove that such a triple does not exist.

Notice, there exist three mutually orthogonal Latin squares of order 14 [889] and three orthogonal Latin cubes (three orthogonal 3-ary quasigroups) of order 10 [13, 14, 311].

SOMAs (this is the acronym for simple orthogonal multi-array) are studied in [696, 822]. However, SOMAs had been studied earlier by Bailey [28] as a special class of semi-Latin squares used in the design of experiments. The SOMAs of type $(1,\dots,1)$ (that is, SOMAs coming from the superposition of MOLS) are called Trojan squares in [28]. See Example 1.328.

Latin squares C and D of order n defined on the same set of symbols are called pseudo-orthogonal if any two rows $c_i \in C$ and $d_j \in D$ have exactly one common symbol.

Example 1.336. There exist non-orthogonal and pseudo-orthogonal (first pair), orthogonal but non-pseudo-orthogonal (second pair), and orthogonal and pseudo-orthogonal (third pair) Latin squares [156].

$$\begin{matrix} 1 & 2 & 3 \\ 3 & 1 & 2 \\ 2 & 3 & 1 \end{matrix} \qquad \begin{matrix} 1 & 3 & 2 \\ 2 & 1 & 3 \\ 3 & 2 & 1 \end{matrix} \qquad \begin{matrix} 1 & 2 & 3 \\ 2 & 3 & 1 \\ 3 & 1 & 2 \end{matrix} \qquad \begin{matrix} 1 & 2 & 3 \\ 3 & 1 & 2 \\ 2 & 3 & 1 \end{matrix} \qquad \begin{matrix} 1 & 2 & 3 \\ 2 & 3 & 1 \\ 3 & 1 & 2 \end{matrix} \qquad \begin{matrix} 1 & 3 & 2 \\ 2 & 1 & 3 \\ 3 & 2 & 1 \end{matrix}$$

In article [156] pseudo-orthogonal Latin squares are applied for computation of an intersection number of complete l-partite graphs.

1.7.1.2 Orthogonality of n-ary operations

We give a classical definition of orthogonality of n-ary operations [57, 96].

Definition 1.337. n-ary groupoids (Q, f_1), (Q, f_2), $\dots$, (Q, f_n) are called orthogonal, if for any fixed n-tuple $a_1, a_2, \dots, a_n$ the following system of equations

$$\begin{cases} f_1(x_1, x_2, \dots, x_n) = a_1 \\ f_2(x_1, x_2, \dots, x_n) = a_2 \\ \dots \\ f_n(x_1, x_2, \dots, x_n) = a_n \end{cases} \tag{1.96}$$

has a unique solution.

If the set Q is finite, then any system of n orthogonal n-ary groupoids (Q, f_i) $i \in \overline{1,n}$ defines a permutation of the set Q^n and vice versa [74, 96, 57]. Therefore if $|Q| = q$, then there exist $(q^n)!$ systems of n-ary orthogonal groupoids defined on the set Q.

Definition 1.337 is possible to use in the case when the set Q is infinite.

Example 1.338. Operations $A_1(x_1, x_2, x_3) = 1 \cdot x_1 + 0 \cdot x_2 + 0 \cdot x_3$, $A_2(x_1, x_2, x_3) = 0 \cdot x_1 + 1 \cdot x_2 + 0 \cdot x_3$, $A_3(x_1, x_2, x_3) = 0 \cdot x_1 + 0 \cdot x_2 + 1 \cdot x_3$ defined over the field R of real numbers (or over a finite field) are orthogonal, since the system

$$\begin{cases} 1 \cdot x_1 + 0 \cdot x_2 + 0 \cdot x_3 = a_1 \\ 0 \cdot x_1 + 1 \cdot x_2 + 0 \cdot x_3 = a_2 \\ 0 \cdot x_1 + 0 \cdot x_2 + 1 \cdot x_3 = a_3 \end{cases}$$

has a unique solution for any fixed 3-tuple $(a_1, a_2, a_3) \in R^3$.

Definition 1.339. n-ary groupoids (Q, f_1), (Q, f_2), $\dots$, (Q, f_k) $(2 \leqslant k \leqslant n)$ given on a set Q of order m are called orthogonal if the system of equations (1.96) has exactly m^{n-k} solutions for any k-tuple $a_1, a_2, \dots, a_k$, where $a_1, a_2, \dots, a_k \in Q$ [114].

If $k = n$, then from Definition 1.339 we obtain standard Definition 1.337. There exist various generalizations of the definition of orthogonality of n-ary operations. Generalizations of the concept of orthogonality are given in [835, 836]. Orthogonality of partial algebraic systems and d-cubes is studied in [855, 851].

The definition of orthogonality of binary systems has a rich and long history [240]. About the n-ary case see, for example, [309].

1.7.1.3 Easy way to construct n-ary orthogonal operations

In the following example, a sufficiently convenient and general way for construction of systems of orthogonal n-ary groupoids is given.

Example 1.340. Define operations $A_1(x_1, x_2, x_3)$, $A_2(x_1, x_2, x_3)$, $A_3(x_1, x_2, x_3)$ over the set $M = \{0, 1, 2\}$ in the following way. Take all 27 triplets $K = \{(R_i, S_i, T_i) \mid R_i, S_i, T_i \in M, i \in \overline{1, 27}\}$ in any fixed order and put

$$A_1(0,0,0) = R_1, A_1(0,0,1) = R_2, A_1(0,0,2) = R_3, \ldots, A_1(2,2,2) = R_{27},$$
$$A_2(0,0,0) = S_1, A_2(0,0,1) = S_2, A_2(0,0,2) = S_3, \ldots, A_2(2,2,2) = S_{27},$$
$$A_3(0,0,0) = T_1, A_3(0,0,1) = T_2, A_3(0,0,2) = T_3, \ldots, A_3(2,2,2) = T_{27}.$$

The operations A_1, A_2 and A_3 form a system of orthogonal operations. If we take these 27 triplets in different order, then we obtain another system of orthogonal 3-ary groupoids.

This way gives a possibility to easily construct an inverse system B of orthogonal n-ary operations to a fixed system A of orthogonal n-ary operations. Recall that an inverse system means that $B(A(x_1^n)) = x_1^n$, $x_i \in Q$.

Example 1.341. We give an example of a ternary quasigroup (Q, A) of order 4 [77, p. 115]. In some sense this quasigroup is non-trivial since it is not an isotope of 3-ary group (Q, f) with the form $f(x_1^3) = x_1 + x_2 + x_3$, where $(Q, +)$ is a binary group of order 4. Recall, there exist two groups of order 4, namely cyclic group Z_4 and Klein group $Z_2 \times Z_2$. Any binary quasigroup of order 4 is a group isotope [10, 11].

A_0	0	1	2	3
0	0	1	2	3
1	1	2	3	0
2	2	3	0	1
3	3	0	1	2

A_1	0	1	2	3
0	1	0	3	2
1	0	1	2	3
2	3	2	1	0
3	2	3	0	1

A_2	0	1	2	3
0	2	3	0	1
1	3	0	1	2
2	0	1	2	3
3	1	2	3	0

A_3	0	1	2	3
0	3	2	1	0
1	2	3	0	1
2	1	0	3	2
3	0	1	2	3

Notice that $A(0,1,2) = A_0(1,2) = 3$, $A(2,3,2) = A_2(3,2) = 3$. Moreover, $A(0,1,x) = A(2,3,x)$ for any $x \in Q$. Then the actions of translations $T(0,1,-)$ and $T(2,3,-)$ on the set $\{0, 1, 2, 3\}$ are equal.

Example 1.342. We give an example of three orthogonal ternary groupoids that are defined on the four-element set $\{0, 1, 2, 3\}$. A multiplication table of the first groupoid (in fact, of a quasigroup) is given in Example 1.341. Below we give multiplication tables of two other 3-ary groupoids.

B_0	0	1	2	3
0	3	0	1	3
1	0	2	3	0
2	1	2	1	3
3	1	1	2	2

B_1	0	1	2	3
0	2	1	1	0
1	2	3	3	0
2	0	2	1	3
3	0	0	3	1

B_2	0	1	2	3
0	1	2	0	0
1	2	0	3	1
2	0	2	3	2
3	3	2	1	1

B_3	0	1	2	3
0	3	3	2	2
1	0	1	2	1
2	0	2	0	3
3	3	1	0	3

C_0	0	1	2	3
0	3	1	2	0
1	2	1	1	2
2	0	1	0	1
3	3	1	2	3

C_1	0	1	2	3
0	1	2	1	3
1	1	2	3	1
2	0	2	2	0
3	1	3	1	1

C_2	0	1	2	3
0	3	3	0	0
1	2	1	0	1
2	3	3	2	0
3	3	0	2	3

C_3	0	1	2	3
0	2	1	0	0
1	2	0	2	3
2	3	3	2	0
3	2	0	0	3

From the formula $(q^n)!$ it follows that there exist $(4^3)! = 64!$ orthogonal systems of 3-ary groupoids over a set of order 4.

There exist generalizations of the concept of orthogonality both in binary and n-ary ($n > 2$) cases [125, 836, 647]. In [100, 101, 102, 241] the concept of partial orthogonality of binary quasigroups (r-orthogonality) is studied. On application of this concept in code theory see [240].

Remark 1.343. r-Orthogonality was not studied much in the n-ary case.

From a cryptographical point of view, orthogonality of n-ary operations is one of the most important concepts, since n n-ary operations defined on a set Q give us a permutation (a bijective map) on the set Q^n and this system of operations gives us "a come back way."

1.7.2 Partial Latin squares: Latin trades

See [213][Chapter II], [258] for more details. We start from the definition of a partial Latin square.

Definition 1.344. An $n \times n$ array of cells that are either empty or filled with symbols that occur at most once in each column is called a partial column Latin square.

An $n \times n$ array of cells that are either empty or filled with symbols that occur at most once in each row is called a partial row Latin square.

An $n \times n$ array of cells that are either empty or filled with symbols that occur at most once in each row and column is called a partial Latin square.

Notice that each triple $(r, c, s) \in P$ corresponds to the entry of symbol s in row r and column c of the array. It is possible to re-write Definition 1.344 in the following way.

Definition 1.345. A partial Latin square P is a set of triples from $R \times C \times S$, where R, C, and S are the sets of rows, columns, and elements, respectively, satisfying

$$(r, c, s) \in L \Rightarrow \{(r', c, s), (r, c', s), (r, c, s')\} \cap L = \varnothing \text{ for } \quad \text{all } r' \neq r, c' \neq c, s' \neq s. \quad (1.97)$$

In other words, any two elements of any triple of a partial Latin square P in a unique way define the last element.

Example 1.346. Example of a partial Latin square.

*	0	1	2
0	0	1	2
1	1	0	
2	2		

We give an example of a partial quasigroup (a partial Latin square) which cannot be completed to a quasigroup (to a Latin square).

Exercise 1.347. Construct a partial Latin square which is completable to at most one Latin square.

Theorem 1.348. *If P is a partial Latin square of size $n \times n$ with at most $n-1$ non-empty cells, then P can be completed to a Latin square of order n [811, 213].*

The concept of partial Latin square permits the following generalization.

Definition 1.349. Let r be a positive integer and S have cardinality nr. We say that L is an r-semi partial Latin square of order n, or $L \in PLS_r(n)$, if $|L(i,j)| \leqslant r$ for each $i, j \in \overline{1,n}$ and L is both column- and row-Latin [544].

From Definition 1.349 it follows that "usual" partial Latin squares are 1-semi partial Latin squares.

The next step in the study of partial Latin squares (partial quasigroups and groupoids) was made by T. Kepka and A. Drapal [277, 278] by introducing of the concept of Latin trade. The theory of Latin trades developed sufficiently fast and now there are many articles devoted to Latin trades. See, for example, [188, 190, 191, 192, 396]. In [191] it is noticed that Latin trades are connected with critical sets in a Latin square [475]. Some information on critical sets in Latin squares is given in Subsection 1.7.3.

Latin trade can be defined in the following way.

Definition 1.350. Let T be a partial Latin square and L be a Latin square with $T \subseteq L$. We say that T is a Latin trade if there exists a partial Latin square T' with $T' \cap T = \emptyset$ such that $(L \backslash T) \cup T'$ is a Latin square.

The more descriptive definition of Latin trade is given in [188, 192].

Definition 1.351. A Latin bitrade (W, B) is a pair of non-empty partial Latin squares for which:

$$\begin{aligned} &(r,c,s) \in W \Rightarrow \exists r' \neq r, c' \neq c; s' \neq s : \{(r',c,s),(r,c',s),(r,c,s')\} \in B, \text{ and} \\ &(r,c,s) \in B \Rightarrow \exists r' \neq r, c' \neq c; s' \neq s : \{(r',c,s),(r,c',s),(r,c,s')\} \in W. \end{aligned} \tag{1.98}$$

W and B are called Latin trades [396].

From Definition 1.351 it follows that two trades from one bitrade have coinciding sets of non-empty cells, in any fixed row they have coinciding sets of elements in filled cells, and in any fixed column they also have coinciding sets of elements in filled cells.

Example 1.352. We give examples of Latin bitrades (W, B) [479, 192].

0_2		2_3	3_0
1_0	0_3	3_1	
2_1	3_2		1_3
	2_0	1_2	0_1

1_2		2_1				
	1_2		2_1			
2_1		3_2	1_3			
	2_1	1_3	3_2			

A spherical Latin trade is a partial Latin square which is associated with a face 2-colorable triangulation of the sphere. See, for example, [396] for details.

Example 1.353. [396]. The n-cycle (W_n, B_n) is a spherical Latin bitrade of size $2n$ that has 2 rows, n columns and n symbols. It is best represented as a pair of partial Latin squares:

0	1	2	...	$(n-2)$	$(n-1)$
1	2	3	...	$(n-1)$	0

1	2	3	...	$(n-1)$	0
0	1	2	...	$(n-2)$	$(n-1)$

The 2-cycle is the smallest Latin bitrade (size 4), and is also known as an intercalate.

1.7.3 Critical sets of Latin squares, Sudoku

Definition 1.354. A critical set C in a Latin square L of order n is a set $C = \{(i; j; k) \mid i, j, k \in \{1, 2, \dots, n\}\}$ with the following two properties:

(1) L is the only Latin square of order n which has symbols k in cell (i, j) for each $(i; j; k) \in C$;

(2) no proper subset of C has the property (1) [559].

In other words, a critical set of a fixed Latin square L is a partial Latin square which is uniquely completable to L. There exist Latin squares which have more than one critical set.

Example 1.355. The critical sets A and B given below can both be completed uniquely to the same Latin square L [479].

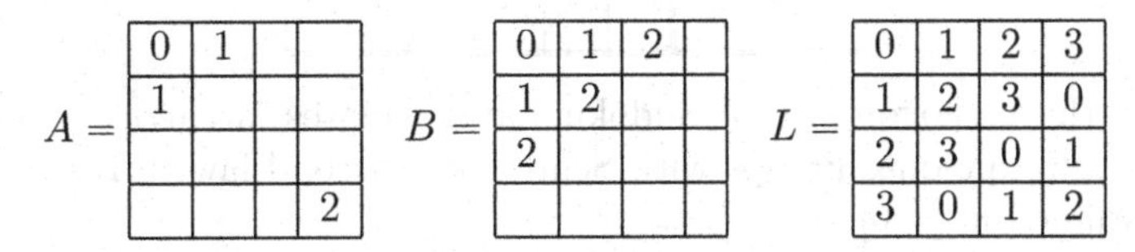

$$A = \begin{array}{|c|c|c|c|}\hline 0 & 1 & & \\ \hline 1 & & & \\ \hline & & & \\ \hline & & & 2 \\ \hline\end{array} \quad B = \begin{array}{|c|c|c|c|}\hline 0 & 1 & 2 & \\ \hline 1 & 2 & & \\ \hline 2 & & & \\ \hline & & & \\ \hline\end{array} \quad L = \begin{array}{|c|c|c|c|}\hline 0 & 1 & 2 & 3 \\ \hline 1 & 2 & 3 & 0 \\ \hline 2 & 3 & 0 & 1 \\ \hline 3 & 0 & 1 & 2 \\ \hline\end{array}$$

A critical set of a fixed Latin square L is called minimal if it is of the smallest possible order. In Example 1.355 the set A is a minimal critical set.

Example 1.356. Critical set C of Latin square K is of minimal order [479].

$$C = \begin{array}{|c|c|c|c|}\hline & 1 & & \\ \hline 1 & & & 2 \\ \hline & & 0 & \\ \hline 3 & & & \\ \hline\end{array} \quad K = \begin{array}{|c|c|c|c|}\hline 0 & 1 & 2 & 3 \\ \hline 1 & 0 & 3 & 2 \\ \hline 2 & 3 & 0 & 1 \\ \hline 3 & 2 & 1 & 0 \\ \hline\end{array}$$

For some natural numbers n there exist Latin squares of order n which have minimal critical sets of different order [479, 617]. For $n = 4$ this fact follows from Examples 1.355 and 1.356.

Definition 1.357. A Latin square L of order n^2 defined on the set of natural numbers $S = \overline{1, n^2}$ is called a Sudoku Latin square, if it consists of n disjoint sub-squares, each of which consists of all elements of the set S [484]. In Sudoku literature [617], a Sudoku Latin square is called a grid.

Example 1.358. We give an example of a Sudoku Latin square of order four. The author thanks Dr. Oleg Chalbash for this example.

$$\begin{array}{|c|c||c|c|}\hline 1 & 2 & 3 & 4 \\ \hline 3 & 4 & 1 & 2 \\ \hline\hline 4 & 3 & 2 & 1 \\ \hline 2 & 1 & 4 & 3 \\ \hline\end{array}$$

Definition 1.359. A critical set of Sudoku Latin square is called Sudoku (Sudoku square, Sudoku puzzle).

Usually Sudoku is represented as a square table with some empty cells. Non-empty cells are called clues [617].

Example 1.360. We give an example of a Sudoku square of order nine (standard size). It is proved that the number of clues (17) for the corresponding grid is minimal [617].

			8		1			
							4	3
5								
				7		8		
							1	
	2			3				
6							7	5
		3	4					
			2			6		

Research on the orthogonality of Sudoku Latin squares is carried out in [484, 485]. Sudoku and palindromic quasigroups with Schroeder's second law (9.14) and Stein's third law (9.12) are studied in [820].

Using heavy computer calculations it is proved that there is no 16-clue Sudoku square of order nine [617]. An overview of some results about critical sets in Latin squares is given in Section 11.6.

1.7.4 Transversals in Latin squares

We give a traditional definition of transversal in a Latin square.

Definition 1.361. A transversal of a Latin square of order n is a set of n cells, one in each row, one in each column, and such that no two of the cells contain the same symbol [240, p. 28].

In [240] it is written: The notion of transversal was first introduced by Euler under the title *formule directrix* [306].

Example 1.362. Transversal elements in a Latin square and the respective quasigroup (the group Z_3) are marked in bold.

$$\begin{array}{ccc} 1 & 2 & \mathbf{3} \\ \mathbf{2} & 3 & 1 \\ 3 & \mathbf{1} & 2 \end{array} \qquad \begin{array}{c|ccc} * & 1 & 2 & 3 \\ \hline 1 & 1 & 2 & \mathbf{3} \\ 2 & \mathbf{2} & 3 & 1 \\ 3 & 3 & \mathbf{1} & 2 \end{array}$$

Example 1.363. We give an example of three disjoint transversals (n-transversals) in the Latin square from Example 1.362 (italic, bold and underlined) and the orthogonal mate to this Latin square.

$$\begin{array}{ccc} \mathit{1} & \underline{2} & \mathbf{3} \\ \mathbf{2} & \mathit{3} & \underline{1} \\ \underline{3} & \mathbf{1} & \mathit{2} \end{array} \qquad \begin{array}{ccc} \mathit{1} & \underline{2} & \mathbf{3} \\ \mathbf{3} & \mathit{1} & \underline{2} \\ \underline{2} & \mathbf{3} & \mathit{1} \end{array}$$

It is quite easy to understand that the following theorem is true [487, 903].

Theorem 1.364. *A Latin square of order n has an orthogonal mate if and only if it has a decomposition into n disjoint transversals.*

Remark 1.365. For all $n > 3$ there exists a quasigroup $(Q, \cdot)$ of order n such that $(Q, \cdot)$ does not have more than $(n-3)$ disjoint transversals [904].

We give a more general definition of transversal.

Definition 1.366. A k-transversal of a Latin square of order n is a set of n cells, one in each row, one in each column, and such that maximum k cells contain different symbols [903].

Therefore an n-transversal of a Latin square is "usual" transversal in sense of Definition 1.361. In [240] a k-transversal is called a partial transversal. In [129, 100, 101] a k-transversal is called a chain.

A generalization of Theorem 1.364 on a disjoint set of k-transversals is in [101].

The problem of the number of k-transversals in a finite quasigroup is researched in many articles [903, 128, 129].

It is clear that any finite quasigroup (any Latin square) has exactly n disjoint 1-transversals. If a square of order n has exactly n disjoint 1-transversals, then this square is a Latin square. The same is true for finite quasigroups.

A middle translation in the respective quasigroup corresponds to any 1-transversal in a Latin square. In Example 1.367, "light-gray" 1-transversal corresponds to a middle translation P_2.

Example 1.367. The following Latin square has 1-, 2-, 3-, 4- and 5-transversals.

1	2	3	4	5
2	1	4	5	3
3	5	1	2	4
4	3	5	1	2
5	4	2	3	1

1	2	3	4	5
2	1	4	5	3
3	5	1	2	4
4	3	5	1	2
5	4	2	3	1

1	2	3	4	5
2	1	4	5	3
3	5	1	2	4
4	3	5	1	2
5	4	2	3	1

1	2	3	4	5
2	1	4	5	3
3	5	1	2	4
4	3	5	1	2
5	4	2	3	1

1	2	3	4	5
2	1	4	5	3
3	5	1	2	4
4	3	5	1	2
5	4	2	3	1

Theorem 1.368. *A cyclic group of composite odd order n has k-transversal for any k with the exception of $k = n-1$.*

A cyclic group of even order n has k-transversal for any k with the exception of $k = n$.

A cyclic group of prime order p has k-transversal for any k with the exception of $k = 2$ and $k = n-1$ [129, Theorem 3, p. 56]

Exercise 1.369. Construct 1-, 3- and 5-transversals of the cyclic group of order five Z_5.

1.7.5 Quasigroup prolongations: Combinatorial aspect

1.7.5.1 Bruck-Belousov prolongation

Quasigroup prolongation is quite a natural way of constructing a finite quasigroup of order $n + k$ $(k \leqslant n)$ from a quasigroup of order n. We start from some definitions. Mainly we follow [240, 72, 71, 247, 60].

R.H. Bruck proposed to use transversals (n-transversals) for prolongation of a quasigroup [166]. We give the Belousov construction of quasigroup prolongation. If transversal elements are situated on the main diagonal, we obtain the Bruck construction.

Algorithm 1.370. We prolong the Latin square and quasigroup $(Q, *)$ of order 3 to the Latin square and the respective quasigroup $(Q', \star)$ of order 4 in the following way.

We add an additional column from the right and an additional row below, transpose all marked (transversal) elements in their fixed order in these new cells and fill all remaining empty cells by the symbol "4."

$$\begin{array}{ccc} 1 & 2 & 3 \\ 2 & 3 & 1 \\ 3 & 1 & 2 \end{array} \rightarrow \begin{array}{cccc} 1 & 2 & \square & 3 \\ \square & 3 & 1 & 2 \\ 3 & \square & 2 & 1 \\ 2 & 1 & 3 & \square \end{array} \rightarrow \begin{array}{cccc} 1 & 2 & 4 & 3 \\ 4 & 3 & 1 & 2 \\ 3 & 4 & 2 & 1 \\ 2 & 1 & 3 & 4 \end{array}$$

Example 1.371. It is easy to see that the initial Latin square from Algorithm 1.370 has more than one transversal. We "isolate" three disjoint transversals in the initial Latin square using italic and bold texts. See Example 1.363.

$$\begin{array}{ccc} \mathit{1} & \mathbf{2} & 3 \\ 2 & \mathit{3} & \mathbf{1} \\ \mathbf{3} & 1 & \mathit{2} \end{array}$$

There exists a possibility to generalize Algorithm 1.370 and to make a quasigroup prolongation using $2, 3, \ldots, n$ disjoint transversals [228, 947]. We demonstrate generalization of Algorithm 1.370 in the following examples.

Example 1.372. We extend the Latin square of order 3 from Example 1.371 to a Latin square of order 5 in the following way. See also Example 10.71.

Step 1. We add two additional columns from the right and two additional rows below, and transpose all marked (transversal) elements in their fixed order in these new cells.

$$\begin{array}{ccccc} \square & \square & 3 & \mathit{1} & \mathbf{2} \\ 2 & \square & \square & \mathit{3} & \mathbf{1} \\ \square & 1 & \square & \mathit{2} & \mathbf{3} \\ \mathit{1} & \mathit{3} & \mathit{2} & \square & \square \\ \mathbf{3} & \mathbf{2} & \mathbf{1} & \square & \square \end{array}$$

Step 2. Fill all transversal cells empty after transposition with the symbols "4, 5."In transversal cells of a fixed transversal we put the same element.

$$\begin{array}{ccccc} \mathit{4} & \mathbf{5} & 3 & \mathit{1} & \mathbf{2} \\ 2 & \mathit{4} & \mathbf{5} & \mathit{3} & \mathbf{1} \\ \mathbf{5} & 1 & \mathit{4} & \mathit{2} & \mathbf{3} \\ \mathit{1} & \mathit{3} & \mathit{2} & \square & \square \\ \mathbf{3} & \mathbf{2} & \mathbf{1} & \square & \square \end{array}$$

Step 3. In the remaining right bottom empty square we put a quasigroup of order 2 defined on the set $\{4, 5\}$.

$$\begin{array}{ccccc} \mathit{4} & \mathbf{5} & 3 & \mathit{1} & \mathbf{2} \\ 2 & \mathit{4} & \mathbf{5} & \mathit{3} & \mathbf{1} \\ \mathbf{5} & 1 & \mathit{4} & \mathit{2} & \mathbf{3} \\ \mathit{1} & \mathit{3} & \mathit{2} & 5 & 4 \\ \mathbf{3} & \mathbf{2} & \mathbf{1} & 4 & 5 \end{array}$$

Modification of Step 1. We can change the order of colored rows and/or columns, for example, in the following way:

4	**5**	3	*1*	**2**
2	*4*	**5**	*3*	**1**
5	1	*4*	*2*	**3**
3	**2**	**1**	5	4
1	*3*	*2*	4	5

Remark 1.373. It is clear that by prolongation of Latin squares and quasigroups we can situate additional columns and rows not only from the right and in the bottom of the initial Latin square, but in any other suitable place.

Example 1.374. We prolong the Latin square of order 3 to a Latin square of order 6 in the following way.

Step 1. We add three additional columns from the right and three additional rows below, and transpose in these new cells all marked (transversal) elements in their fixed order.

□	□	□	*1*	**2**	3
□	□	□	*3*	**1**	2
□	□	□	*2*	**3**	1
1	*3*	*2*	□	□	□
3	**2**	**1**	□	□	□
2	1	3	□	□	□

Step 2. Fill all transversal cells remaining empty after transposition with the symbols "4, 5, 6 "in their "old transversal order," i.e., we put the symbol 5 in all empty cells of transversal with "bold" elements.

6	**5**	4	*1*	**2**	3
4	*6*	**5**	*3*	**1**	2
5	4	*6*	*2*	**3**	1
1	*3*	*2*	□	□	□
3	**2**	**1**	□	□	□
2	1	3	□	□	□

Step 3. In the remaining right bottom empty square we put any quasigroup of order 3 defined on the set $\{4, 5, 6\}$.

6	5	4	1	2	3
4	6	5	3	1	2
5	4	6	2	3	1
1	3	2	4	5	6
3	2	1	5	6	4
2	1	3	6	4	5

In a more formalized manner, the algorithm which is described in Examples 1.372 and 1.374 is given in [371]. In this article many MDS codes are constructed using this algorithm. Also a pair of orthogonal quasigroups of order ten is constructed [371, 953]. See also Section 10.2.

1.7.5.2 Belyavskaya prolongation

G.B. Belyavskaya proposed modification of Belousov's algorithm [99, 97, 98].

Algorithm 1.375. We add an additional column from the right and an additional row below, transpose in these new cells all marked (transversal) elements except one (in our example element **2**) in their fixed order and fill in all remaining empty cells except one (with coordinates $(n+1, n+1)$) the symbol "4". The cell with coordinates $(n+1, n+1)$ is filled by the not transposed transversal element.

$$\begin{array}{ccc} 1 & 2 & 3 \\ \mathbf{2} & 3 & 1 \\ 3 & 1 & 2 \end{array} \quad \rightarrow \quad \begin{array}{cccc} 1 & 2 & \square & 3 \\ \mathbf{2} & 3 & 1 & \square \\ 3 & \square & 2 & 1 \\ \square & 1 & 3 & \square \end{array} \quad \rightarrow \quad \begin{array}{cccc} 1 & 2 & 4 & 3 \\ \mathbf{2} & 3 & 1 & 4 \\ 3 & 4 & 2 & 1 \\ 4 & 1 & 3 & \mathbf{2} \end{array}$$

Algorithm 1.376. Generalized Belyavskaya algorithm. It is possible to generalize the Belyavskaya algorithm using more than one disjoint transversal.

Example 1.377. We prolong the Latin square of order 3 to the Latin square of order 5 using the Belyavskaya prolongation algorithm (Algorithm 1.375) that simultaneously is applied to two transversals.

We add two additional columns from the right and two additional rows below, and transpose in these new cells all marked (transversal) elements in their fixed order with the exception of one element in any transversal. We take element 1 in "italic" transversal and take element 1 in "bold" transversal. It is not obligatory that in "italic" and "bold" transversals we take equal "exceptional" elements.

Step 1.

$$\begin{array}{ccccc} \mathit{1} & \square & 3 & \square & \mathbf{2} \\ 2 & \square & \mathbf{1} & \mathit{3} & \square \\ \square & 1 & \square & \mathit{2} & \mathbf{3} \\ \square & \mathit{3} & \mathit{2} & \square & \square \\ \mathbf{3} & \mathbf{2} & \square & \square & \square \end{array}$$

Step 2. Fill all transversal cells remaining empty after transposition with the symbols "4, 5." In transversal cells of a fixed transversal we put the same element.

In the bottom of main diagonal write elements "4, 5." Unfortunately, direct generalization of the Belyavskaya algorithm is not possible.

$$\begin{array}{ccccc} \mathit{1} & 4 & 3 & \mathit{5} & \mathbf{2} \\ 2 & \mathit{5} & \mathbf{1} & \mathit{3} & 4 \\ 4 & 1 & \mathit{5} & \mathit{2} & \mathbf{3} \\ \mathit{5} & \mathit{3} & \mathit{2} & 4 & 1 \\ \mathbf{3} & \mathbf{2} & 4 & 1 & \mathit{5} \end{array}$$

Modification of Step 1. We can change the order of the last two (marked) rows and/or columns. For example, we changed the 4th and 5th rows. In this case we obtain "a more direct" generalization of Belyavskaya prolongation.

$$\begin{array}{ccccc} \mathit{1} & 4 & 3 & \mathit{5} & \mathbf{2} \\ 2 & \mathit{5} & \mathbf{1} & \mathit{3} & 4 \\ 4 & 1 & \mathit{5} & \mathit{2} & \mathbf{3} \\ \mathbf{3} & \mathbf{2} & 4 & 1 & \mathit{5} \\ \mathit{5} & \mathit{3} & \mathit{2} & 4 & 1 \end{array}$$

Example 1.378. Using the generalized Belyavskaya algorithm we extend the Latin square of order 3 to Latin square of order 6 in the following way.

Step 1. We add three additional columns from the right and three additional rows below, transpose in these new cells all marked (transversal) elements in their fixed order with the exception of the element 3 from "italic" transversal, element 3 from "bold" transversal and element 1 from the third transversal.

□	□	□	*1*	**2**	3
□	*3*	□	□	**1**	2
3	1	□	*2*	□	□
1	□	*2*	□	□	□
□	**2**	**1**	□	□	□
2	□	3	□	□	□

Step 2. Fill all transversal cells remaining empty after transposition with the symbols "4, 5, 6" in their "old transversal order," i.e., we put the symbol 4 in all empty cells of "italic" transversal, the symbol 5 in all empty cells of "bold" transversal and so on.

4	**5**	6	*1*	**2**	3
6	*3*	**5**	*4*	**1**	2
3	1	*4*	*2*	**5**	6
1	*4*	*2*	□	□	□
5	**2**	**1**	□	□	□
2	6	3	□	□	□

Step 3. The remaining right bottom empty square we should complete in order to obtain a quasigroup. The bottom part of the main diagonal we fill with the elements $3, 3, 1$, because namely these elements remain in transversals.

In this case the cell $(5, 4)$ can be filled only by the element 6. The remaining is clear for any fan of Sudoku.

4	5	6	1	2	3
6	3	5	4	1	2
3	1	4	2	5	6
1	4	2	3	6	5
5	2	1	6	3	4
2	6	3	5	4	1

Remark 1.379. As it seems to us, the Belyavskaya algorithm permutes elements of a Latin square better than the Bruck-Belousov algorithm. This statement requires further investigation.

1.7.5.3 Algebraic approach

We give an algebraic variant of Algorithm 1.375 and some other prolongation algorithms [72, 247]. In [443], quasigroup prolongation is used for construction of quasigroups with an identity automorphism group, without non-trivial normal congruences and subquasigroups. See also [561].

Definition 1.380. Let $(Q, \cdot)$ be a finite quasigroup and σ be a mapping of the set Q. We can construct the mapping $\overline{\sigma}$ in the following way:

$$\overline{\sigma}x = x \cdot \sigma x \quad \text{for all} \quad x \in Q. \tag{1.99}$$

The mapping $\overline{\sigma}$ is called a *conjugated mapping* to the mapping σ.

A mapping σ is *quasicomplete*, if σ is a permutation of a set Q and $\overline{\sigma}(Q)$ contains all elements of Q except one. In this case there exists an element $a \in Q$, called special, such that $a = \overline{\sigma}x_1 = \overline{\sigma}x_2$ for some $x_1, x_2 \in Q$, $x_1 \neq x_2$.

If the mappings σ and $\overline{\sigma}$ are permutations of a set Q, then we say that σ is *complete*. A quasigroup having at least one complete mapping is called *admissible*.

For instance, if σ is the identity mapping, then $\overline{\sigma}x = x^2$; if $\sigma x = a$ for all $x \in Q$, then $\overline{\sigma}x = R_a x$. Any idempotent quasigroup is admissible. In this case $\sigma = \overline{\sigma} = \varepsilon$.

The elements x, σx and $\overline{\sigma}x$ define triplets $(x, \sigma x, \overline{\sigma}x)$ in a Latin square and in the corresponding quasigroup $(Q, \cdot)$ in the following way: elements x and σx define coordinates of the cell in which the element $\overline{\sigma}x = x \cdot \sigma x$ is situated.

It is easy to see that any transversal of a Latin square defines complete permutation in a corresponding quasigroup and vice versa, any complete permutation defines a transversal. Similarly, quasicomplete mapping defines an $(n-1)$-transversal (Definition 1.366).

Example 1.381. We find the mappings $\overline{\sigma}$ and σ for the quasigroup $(Q, *)$ from Example 1.362.

If $x = 1$, then we have $\overline{\sigma}1 = \overline{3} = 1 \cdot \sigma 1$. Therefore in this case $\overline{\sigma}1 = \overline{3}$ and $\sigma 1 = 3$, since the element $\overline{3}$ lies in the third column. Indeed, $1 \cdot 3 = \overline{3}$.

If $x = 2$, then we have $\overline{\sigma}2 = \overline{2} = 2 \cdot \sigma 2$. Therefore in this case $\overline{\sigma}2 = \overline{2}$ and $\sigma 2 = 1$, since the element $\overline{2}$ lies in the first column. Indeed, $2 \cdot 1 = \overline{2}$.

If $x = 3$, then we have $\overline{\sigma}3 = \overline{1} = 3 \cdot \sigma 3$. Therefore in this case $\overline{\sigma}3 = \overline{1}$ and $\sigma 3 = 2$, since the element $\overline{1}$ lies in the second column. Indeed, $3 \cdot 2 = \overline{1}$.

Therefore

$$\sigma = \begin{pmatrix} 1 & 2 & 3 \\ 3 & 1 & 2 \end{pmatrix}; \quad \overline{\sigma} = \begin{pmatrix} 1 & 2 & 3 \\ 3 & 2 & 1 \end{pmatrix}.$$

Lemma 1.382. *If a quasigroup $(Q, \cdot)$ satisfies condition $\overline{\sigma}x = x \cdot \sigma x$ for all $x \in Q$, then in its isotope $(Q, \circ) = (Q, \cdot)(\alpha, \beta, \gamma)$ the following equality is true*

$$x \circ \beta^{-1}\sigma\alpha x = \gamma^{-1}\overline{\sigma}\alpha x. \tag{1.100}$$

Proof. If $(Q, \circ) = (Q, \cdot)(\alpha, \beta, \gamma)$, i.e., if $x \circ y = \gamma^{-1}(\alpha x \cdot \beta y)$, then $x \cdot y = \gamma(\alpha^{-1}x \circ \beta^{-1}y)$.

If we put $y = \sigma x$, then $x \cdot \sigma x = \gamma(\alpha^{-1}x \circ \beta^{-1}\sigma x) = \overline{\sigma}x$, $\alpha^{-1}x \circ \beta^{-1}\sigma x = \gamma^{-1}\overline{\sigma}x$, $x \circ \beta^{-1}\sigma\alpha x = \gamma^{-1}\overline{\sigma}\alpha x$. □

Lemma 1.383. *If a quasigroup $(Q, \cdot)$ is admissible, then its isostrophic image is also admissible.*

Proof. Taking into consideration that any isostrophy of a quasigroup $(Q, \cdot)$ is a composition of isotopy and parastrophy, we divide our proof into two parts.

We prove that admissibility is invariant relative to any isotopy.

It is clear that the mappings $\beta^{-1}\sigma\alpha$ and $\gamma^{-1}\overline{\sigma}\alpha$ from equation (1.100) are permutations of the set Q.

We prove that admissibility is invariant relative to any parastrophy.

Operation "/". From equality $\overline{\sigma}x = x \cdot \sigma x$ passing to the operation "/" we have $x = \overline{\sigma}x / \sigma x$, $\overline{\sigma}^{\,-1}x = x / \sigma\overline{\sigma}^{\,-1}x$. It is clear that the mappings $\sigma\overline{\sigma}^{\,-1}$ and $\overline{\sigma}^{\,-1}$ are permutations of the set Q.

Operation "\". From equality $\overline{\sigma}x = x \cdot \sigma x$ passing to the operation "\" we have $\sigma x = x \backslash \overline{\sigma}x$.

For other parastrophes of quasigroup $(Q, \cdot)$ the proof is similar. □

Corollary 1.384. *If a quasigroup $(Q, \cdot)$ is admissible, then $(Q, \cdot)$ is isotopic to an idempotent quasigroup $(Q, \circ)$.*

Proof. If we put, for example, in equality (1.100) $\beta = \sigma$, $\gamma = \overline{\sigma}$, $\alpha = \varepsilon$, then we obtain that quasigroup $(Q, \circ)$ is idempotent. □

R. H. Bruck has pointed out that every finite group of odd order is isotopic to an idempotent quasigroup [166]. Therefore it has a transversal.

Definition 1.385. [226]. A quasigroup $(Q, \cdot)$ is called TA-quasigroup if for all $c, x, y \in Q$:

$$(c \cdot x) \cdot y = (c \cdot y) \cdot x \Rightarrow x = y, \tag{1.101}$$

$$x \cdot y = y \cdot x \Rightarrow x = y. \tag{1.102}$$

If the quasigroup only fulfills the first implication, it is called a weak totally anti-symmetric (WTA-quasigroup).

There are TA-quasigroups of order n for all $n \neq 2; 6$ [226].

TA-quasigroups are used for construction of codes with one check symbol [227]. See Chapter 10 for details.

Lemma 1.386. *A finite WTA-quasigroup $(Q, \cdot)$ is admissible, i.e., it has a transversal.*

Proof. We follow [226]. Suppose that in equality (1.99) $\overline{\sigma} = \varepsilon$, i.e., $\sigma x = x \backslash x$. The existence of transversal will be proved, if we demonstrate that the map σ is a permutation of the set Q. Since Q is finite, it is sufficient to demonstrate that the map σ is injective (see Lemma 1.15).

Suppose that $\sigma(i) = \sigma(j)$. Put in quasi-identity (1.101) $c := i$, $x := \sigma(i) = i \backslash i = \sigma(j) = j \backslash j$, and $y := i \backslash j$ (i.e., $i \cdot x = i$, $j \cdot x = j$, and $i \cdot y = j$). Then we have

$$(i \cdot x) \cdot y = i \cdot y = j = j \cdot x = (i \cdot y) \cdot x \tag{1.103}$$

and hence, $x = y$ resp. $\sigma(i) = i \backslash i = i \backslash j$, which implies $i = j$. Thus, we can see that σ is injective and also a permutation. □

1.7.5.4 Prolongation using quasicomplete mappings

I.I. Derienko and W.A. Dudek propose prolongation construction of a quasigroup using quasicomplete mappings [247]. This construction is a generalization of the Belyavskaya algorithm on quasicomplete mappings.

Algorithm 1.387. We start from a quasigroup and a quasicomplete mapping, act as in the Belyavskaya algorithm but in the cell with coordinates $(n + 1, n + 1)$ we write the element $Q \backslash \overline{\sigma} Q$.

Example 1.388. We take the following quasigroup (Example 1.116),

$\cdot$	1	2	3	4
1	**2**	1	3	4
2	3	2	**4**	1
3	4	**3**	1	2
4	1	4	2	**3**

and the following mapping

$$\sigma = \begin{pmatrix} 1 & 2 & 3 & 4 \\ 1 & 3 & 2 & 4 \end{pmatrix},$$

then

$$\overline{\sigma} = \begin{pmatrix} 1 & 2 & 3 & 4 \\ 2 & 4 & 3 & 3 \end{pmatrix},$$

σ is a quasicomplete mapping and $Q \backslash \overline{\sigma} Q = \{1\}$. Using the Derienko-Dudek algorithm we obtain:

*	1	2	3	4	5
1	□	1	3	4	**2**
2	3	2	□	1	**4**
3	4	□	1	2	**3**
4	1	4	2	**3**	□
5	**2**	**3**	**4**	□	□

→

*	1	2	3	4	5
1	5	1	3	4	**2**
2	3	2	5	1	**4**
3	4	5	1	2	**3**
4	1	4	2	**3**	5
5	**2**	**3**	**4**	5	*1*

There exists a possibility, similar to the Bruck-Belousov algorithm, to generalize the Derienko-Dudek algorithm in the spirit of Yamamoto.

Example 1.389. We further use Example 1.116. It is clear that the following Latin square has four quasi-complete mappings. We shall use "italic" and "bold" quasi-complete mappings for the prolongation of this Latin square.

2	**1**	3	4
3	2	*4*	1
4	*3*	1	**2**
1	4	**2**	*3*

Step 1. We add two columns and two rows and transpose elements of the "italic" and "bold" quasi-complete maps.

□	□	3	4	*2*	**1**
□	2	□	1	*4*	**3**
4	□	1	□	*3*	**2**
1	4	**2**	*3*	□	□
2	*3*	*4*	□	□	□
3	**1**	□	**2**	□	□

Step 2. We fill empty "italic" cells in the initial Latin square with the number 5 and "bold" cells with the number 6. We fill bottom part of main diagonal with the elements 1 and 4 respectively.

5	**6**	3	4	*2*	**1**
6	2	*5*	1	*4*	**3**
4	*5*	1	**6**	*3*	**2**
1	4	**2**	*3*	□	□
2	*3*	*4*	□	1	□
3	**1**	□	**2**	□	4

Step 3. Finally we complement the obtained partial Latin square to complete the Latin square. It is easy to see that it is possible to do this in a unique way.

5	6	3	4	2	1
6	2	5	1	4	3
4	5	1	6	3	2
1	4	2	3	6	5
2	3	4	5	1	6
3	1	6	2	5	4

1.7.5.5 Two-step mixed procedure

Suppose that a quasigroup of order n has two disjoint transversals. At the first step we can expand this quasigroup to a quasigroup of order $n+1$ using the Bruck-Belousov or Belyavskaya algorithm. After this procedure our second transversal passes in complete or quasicomplete mapping (i.e., in n- or $(n-1)$-transversal) and we can prolong the obtained quasigroup using either the Bruck-Belousov or Belyavskaya algorithm, or the Derienko-Dudek algorithm (if we have obtained quasicomplete mapping).

Example 1.390. We start from the well-known Latin square of order 3.

1	2	**3**
2	*3*	1
3	**1**	*2*

At the first step we use "bold" transversal and the Belyavskaya algorithm (see Algorithm 1.375). We leave in its place element 2 and obtain the following Latin square.

1	2	4	3
2	*3*	1	4
3	4	*2*	1
4	1	3	**2**

In this case "italic" transversal passes in $(n-1)$-transversal

$$\sigma = \begin{pmatrix} 1 & 2 & 3 & 4 \\ 1 & 2 & 3 & 4 \end{pmatrix},$$

$$\overline{\sigma} = \begin{pmatrix} 1 & 2 & 3 & 4 \\ 1 & 3 & 2 & 2 \end{pmatrix}.$$

The map σ is a quasicomplete mapping and $Q\backslash\overline{\sigma}Q = \{4\}$. Using the Derienko-Dudek algorithm we obtain

5	2	4	3	1
2	5	1	4	3
3	4	5	1	2
4	1	3	2	5
1	3	2	5	4

1.7.5.6 Brualdi problem

Definition 1.391. A quasigroup that has a quasi-complete mapping is called a *quasi-admissible quasigroup.* A quasigroup with all idempotent elements except one element is called an $(n-1)$*-idempotent quasigroup.*

Notice that any $(n-1)$-idempotent quasigroup is quasi-admissible ($\sigma = \varepsilon$).

Lemma 1.392. *If a quasigroup* $(Q, \cdot)$ *is quasi-admissible, then* $(Q, \cdot)$ *is isotopic to an* $(n-1)$*-idempotent quasigroup* $(Q, \circ)$.

Proof. If we put $\beta = \sigma$, $\alpha = \varepsilon$ in equality (1.100), then we obtain the following equality

$$x \circ x = \gamma^{-1}\overline{\sigma}x. \tag{1.104}$$

Define quasi-inverse permutation δ^{-1} to the mapping $\overline{\sigma}$ in the following way. If $\overline{\sigma}x = z$, then $\delta^{-1}z = x$. Therefore $\delta^{-1}\overline{\sigma}x = \delta^{-1}z = x$ in this case.

If $\overline{\sigma}x = z$ and $\overline{\sigma}y = z$, then $\delta^{-1}z = x$ and $\delta^{-1}w = y$, where $\{w\} = Q\setminus\overline{\sigma}Q$. In this case $\delta^{-1}\overline{\sigma}x = \delta^{-1}z = x$, and $\delta^{-1}y = w$.

If we put $\gamma^{-1} = \delta^{-1}$ in equality (1.104), then we obtain that quasigroup $(Q, \circ)$ is $(n-1)$-idempotent. I.e., $x \circ x = x$ for all $x \in Q$ except the element y, for that $y \circ y = w$. □

Example 1.393. We take the following quasigroup (Example 1.388).

$\cdot$	1	2	3	4
1	**2**	1	3	4
2	3	2	**4**	1
3	4	**3**	1	2
4	1	4	2	**3**

and the following isotopy $(\varepsilon, \beta, \gamma)$, where

$$\beta = \begin{pmatrix} 1 & 2 & 3 & 4 \\ 1 & 3 & 2 & 4 \end{pmatrix}, \quad \gamma = \begin{pmatrix} 1 & 2 & 3 & 4 \\ 2 & 4 & 3 & 1 \end{pmatrix}.$$

It is clear that $\gamma^{-1} = (1\,4\,2)$. We obtain the following $(n-1)$-idempotent quasigroup:

$\circ$	1	2	3	4
1	1	3	4	2
2	3	2	1	4
3	2	4	3	1
4	4	1	2	3

Problem 1.7. R.A. Brualdi conjectured that any $n \times n$ Latin square possesses a sequence of $k \geqslant n-1$ distinct elements selected from different rows and different columns [240, p. 103].

In other words, each finite quasigroup has at least one complete or quasi-complete mapping.

Taking into consideration Corollary 1.384 and Lemma 1.392 we can reformulate a Brualdi problem in the following way: any finite quasigroup is isotopic to idempotent or $(n-1)$-idempotent quasigroup.

Remark 1.394. From Theorem 1.368 (Belyavskaya-Russu Theorem) it follows that for finite cyclic groups, the Brualdi conjecture is true.

1.7.5.7 Contractions of quasigroups

The procedure which is inverse to prolongation of quasigroups is called contraction of quasigroups. Procedures of contraction of quasigroups make, from a quasigroup of size n, a quasigroup of size $(n-1)$ or $(n-2)$. It is clear that any procedure of prolongation has its proper "inverse" procedure of contraction. See [98, 97, 246] for details.

1.7.6 Orthomorphisms

Definition 1.395. A bijective mapping $\varphi : g \rightarrowtail \varphi(g)$ of a finite group $(G, \cdot)$ onto itself is called an orthomorphism if the mapping $\theta : g \rightarrowtail \theta(g)$, where $\theta(g) = g^{-1}\varphi(g)$ is again a bijective mapping of G onto itself. The orthomorphism is said to be in canonical form if $\varphi(1) = 1$, where 1 is the identity element of $(G, \cdot)$.

Example 1.396. Any group $(G, \cdot)$ of odd order has an orthomorphism. Indeed, the mapping $\varphi : x \mapsto x{\cdot}x$ is a permutation of the group $(G, \cdot)$ and the mapping $\theta = \varepsilon$ is also a permutation of the set Q.

Remark 1.397. The concept of an orthomorphism was first introduced explicitly in [466] but the equivalent concept of complete mapping dates back to H.B. Mann [594]. If $(G, \cdot)$ is a group (IP-loop), then we can re-write equality $\theta(g) = g^{-1}\varphi(g)$ in the following form; $g \cdot \theta(g) = \varphi(g)$. In the last equality the map θ is a complete mapping and φ is conjugate mapping to the mapping θ (Definition 1.380). Therefore any orthomorphism φ is a conjugate mapping to a complete mapping.

One more way. We can re-write equality $\theta(g) = Ig \cdot \varphi(g)$ in the following form: $\theta I(g) = g \cdot \varphi I(g)$, where $Ig = g^{-1}$ for all $g \in G$. It is clear that in any group $I^2 = \varepsilon$. Indeed, suppose that $Ig \cdot g = 1$. Then $I^2 g(Ig \cdot g) = I^2 g$, $(I^2 g \cdot Ig)g = g$, $I^2 = \varepsilon$.

Lemma 1.398. *If a group $(G, \cdot)$ of a finite order n has at least one complete mapping (orthomorphism, transversal, n-transversal), then it has n disjoint complete mappings (orthomorphisms, transversals, n-transversals).*

Proof. Suppose that the mapping φ is complete, i.e., it satisfies equality $x \cdot \varphi x = \psi x$ for all $x \in G$ and the mapping ψ is a permutation of the set G. Then we have $xa \cdot \varphi(xa) = \psi(xa)$ for all $a \in G$. By associativity further we have $x \cdot a\varphi(xa) = \psi(xa)$. Rewrite the last equality using translations as follows: $x \cdot L_a\varphi R_a x = \psi R_a x$. We proved that the mapping $L_a\varphi R_a$ is also complete for any $a \in G$, since the mapping ψR_a is a permutation of the set G as the product of two permutations.

Prove that the complete mappings obtained in such way are disjoint in pairs. Suppose the contrary, that there exists an element $c \in G$ such that $\psi R_a c = \psi R_b c$, for some $a, b \in G$, $a \neq b$. Then we have $\psi R_a c = \psi R_b c$, $R_a c = R_b c$, $c \cdot a = c \cdot b$, $a = b$. Our supposition is false, and the mappings $L_a\varphi R_a$ and $L_b\varphi R_b$ are disjoint.

The proofs for the orthomorphisms and transversals (n-transversals) are similar and we omit them. □

Lemma 1.399. *If a finite commutative group $(G, \cdot)$ has an orthomorphism φ, then the product of all the elements of $(G, \cdot)$ in any order is equal to the identity element.*

Proof. We can write any element a of the group $(G, \cdot)$ as the product $b^{-1} \cdot \varphi b$. Since the map φ is a permutation of the set G, then there exists a unique element c such that $\varphi c = b$, there exists a unique element d such that $d^{-1} = (\varphi b)^{-1}$. Therefore $b^{-1} \cdot \varphi c \cdot \varphi b \cdot d^{-1} = 1$. And so on. □

Lemma 1.400. *A cyclic group of even order Z_{2k} does not have any orthomorphism [240].*

Proof. In the group $Z_{2k} = \langle a \mid a^{2k} = 1 \rangle$ the element a^k has the order two. Therefore any product (the group Z_{2k} is commutative) of all the elements of Z_{2k} is equal to $1 \cdot (a \cdot a^{-1}) \cdot (a^2 \cdot a^{-2}) \ldots (a^{k-1} \cdot a^{-(k-1)}) \cdot a^k = a^k$. Thus by Lemma 1.399 the group Z_{2k} does not have any orthomorphism. □

The historical significance of this concept as applied to a finite group G is that it guarantees the existence of, and can be used to construct, an orthogonal mate for the Latin square formed by the multiplication table (Cayley table) of G. Notice, both groups of order ten (the cyclic group Z_{10} and dihedral group D_{10}, [407, 471]) do not have any orthomorphism.

Example 1.401. We continue Example 1.363. Using complete mapping $\theta = (1\,3\,2)$ we construct "italic" transversal of the group Z_3 in the following way: in the cell with coordinates $(1, \theta(1))$ we write the element $1 * \theta(1)$, in the cell with coordinates $(2, \theta(2))$ we write the element $2 * \theta(2)$, and so on. The conjugate mapping to the mapping θ is the following mapping: $\varphi = (1\,3)$. Via equality $x \cdot L_a \theta R_a x = \varphi R_a x$ we obtain two more transversals (light-gray and blue). Construction of the orthogonal mate is given in Example 1.363.

Using orthomorphism $\varphi = (1\,3)$ we construct the respective (red or gray) transversal as follows: in the cell $(1^{-1}, \varphi 1) = (1, 3)$ we write the element $1 * 3 = 3 = \theta 1$, in the cell $(2^{-1}, \varphi 2) = (3, 2)$ we write the element $2^{-1} * \varphi 2 = 3 * 2 = 1 = \theta 2$, and so on. Notice that conjugate mapping to the orthomorphism $\varphi = (1\,3)$ is the mapping $\theta = (1\,3\,2)$. It is clear that "italic" (middle-gray) and underlined transversals coincide (see Remark 1.397).

*	1	2	3
1	1	2	*3*
2	*2*	3	1
3	3	*1*	2

*	1	2	3
1	**1**	2	3
2	2	**3**	1
3	3	1	**2**

*	1	2	3
1	1	2	<u>3</u>
2	<u>2</u>	3	1
3	3	<u>1</u>	2

Concepts of near complete mapping and near orthomorphism of a group are also given and researched [478, 50]. The notion of near-complete mapping is close to the notion of quasicomplete mapping (Definition 1.380).

1.7.7 Neo-fields and left neo-fields

"The concept of a neofield was first introduced and developed by L. J. Paige [675] in 1949. Paige hoped to use this structure as the co-ordinate system for a finite projective plane and so to construct new planes, possibly of non-prime power order" [478].

A left neo-field $(N, +, \cdot)$ of order n consists of a set N of n symbols on which two binary operations "+" and "·" are defined such that $(N, +)$ is a loop, with identity element, say 0. $(N\backslash\{0\}, \cdot)$ is a group and the operation "·" distributes from the left over "+". (That is, $x \cdot (y + z) = x \cdot y + x \cdot z$ for all $x, y, z \in N$.) If the right distributive law also holds, the structure is called a neo-field.

A left neofield (or neofield) whose multiplication group is a group $(G, \cdot)$ is said to be based on that group. Clearly, every left neofield based on an abelian group is a neofield. Also, a neofield, the operation of addition of which satisfies the associative law, is a field.

"Most useful in practice are cyclic neofields: that is, neofields whose multiplication group is cyclic. Such neofields have been investigated in considerable details by D.F. Hsu in his book [428]" [478].

We give the following construction of left neofields [478, Theorem N]. "Let $(G, \cdot)$ be a finite group with identity element 1 which possesses an orthomorphism ψ (in canonical form). Let 0 be a symbol not in the set G. Define $N = G \cup \{0\}$. Then $(N, \oplus, \cdot)$ is a left neofield, where we define $\psi(w) = 1 \oplus w$ for all $w \neq 0, 1$ and $\psi(0) = 1 \oplus 0 = 1$, $\psi(1) = 1 \oplus 1 = 0$. Also, $x \oplus y = x(1 \oplus x^{-1}y)$ for $x \neq 0$, $0 \oplus y = y$ and $0 \cdot x = 0 = x \cdot 0$ for all $x \in N$ [478]."

Notice, $x \oplus x = x(1 \oplus x^{-1}x) = x \cdot 0 = 0$. Therefore any loop $(Q, \oplus)$ satisfies the identity $x \oplus x = 0$. Taking into consideration Example 1.396 and the above given construction, we can say that there exist loops with the identity $x \cdot x = 1$ of any even order. See also [442].

Example 1.402. It is easy to see that there does not exist any canonical orthomorphism over the group Z_3 and any orthomorphism over the group Z_4. It is possible to use the group $Z_2 \times Z_2$, i.e., the Klein group, but here, in order to construct a neo-field of order six we shall use the group Z_5 which is defined on the set {1, 2, 3, 4, 5} with the identity element 1. We take the following orthomorphism $\psi = (2\,3\,5\,4)$ and construct the following neo-field $(N, \oplus, \cdot)$ of order six.

$\cdot$	0	1	2	3	4	5
0	0	0	0	0	0	0
1	0	1	2	3	4	5
2	0	2	3	4	5	1
3	0	3	4	5	1	2
4	0	4	5	1	2	3
5	0	5	1	2	3	4

$\oplus$	0	1	2	3	4	5
0	0	1	2	3	4	5
1	1	0	3	5	2	4
2	2	5	0	4	1	3
3	3	4	1	0	5	2
4	4	3	5	2	0	1
5	5	2	4	1	3	0

We give the sketched proof of left distributivity of neo-field $(N, \oplus, \cdot)$. Non-formally speaking we can write the operation $\oplus$ in the following form: $x \oplus y = x \cdot \psi(x^{-1}y)$ (see Remark 2.130). Therefore $x \cdot (y \oplus z) = x(y \cdot \psi(y^{-1}z)) = xy \cdot \psi(y^{-1}z)$, $(x \cdot y) \oplus (x \cdot z) = xy \cdot \psi(y^{-1}x^{-1} \cdot xz) = xy \cdot \psi(y^{-1}z)$.

Notice, $Aut(N, \oplus) \cong \langle a, b \mid a^5 = b^4 = 1, b^{-1}ab = a^2 \rangle$, $Aut(N, \oplus) \cong Z_5 \rtimes Z_4$, $|Aut(N, \oplus)| = 20$.

In [243, 242] some cryptological applications of neo-fields and left neo-fields are described.

1.7.8 Sign of translations

Let $\mathbf{L} = \{L_a \mid a \in Q\}$, $\mathbf{R} = \{R_a \mid a \in Q\}$, $\mathbf{I} = \{I_a \mid a \in Q\}$ be the sets of all left, right and middle translations of a quasigroup $(Q, \cdot)$, where $L_a x = ax$, $R_a x = xa$, $x \cdot I_a x = a$, respectively.

Under the sign function we mean homomorphism of the symmetric group S_n onto the group Z_2 of order 2, $Z_2 = \{1, -1\}$. If $\alpha \in S_n$ is a product of the even number of cycles of length 2 (2-cycles), then $sgn\,\alpha = 1$. If α is a product of the odd number of 2-cycles, then $sgn\,\alpha = -1$.

We shall mention some properties of the sign function. Let $\alpha, \beta, \gamma \in S_n$. Then $sgn(\alpha\beta) = sgn\,\alpha \cdot sgn\,\beta = sgn\,\beta \cdot sgn\,\alpha = sgn(\beta\alpha)$, $sgn(\alpha(\beta\gamma)) = sgn((\alpha\beta)\gamma)$, because the associative and commutative identities hold in the group Z_2.

Let $(Q, \cdot)$ be a finite quasigroup of order n. We use the known notions $sgn\,\mathbf{L} = \prod_{i=1}^{n} sgn(L_{a_i})$, $sgn\,\mathbf{R} = \prod_{i=1}^{n} sgn(R_{a_i})$, $sgn\,\mathbf{I} = \prod_{i=1}^{n} sgn(I_{a_i})$, where $a_i \in Q$; moreover let's define $tsgn\,Q = \langle sgn\,\mathbf{L}, sgn\,\mathbf{R}, sgn\,\mathbf{I} \rangle$.

A loop $(Q, \cdot)$ with identity $x(y \cdot xz) = (xy \cdot x)z$ is called a Moufang loop; a loop with identity $x(y \cdot xz) = (x \cdot yx)z$ is called a left Bol loop. We shall consider only left Bol loops and shall call them Bol loops omitting the word "left" for short.

Theorem 1.403. *Let Q be a finite Bol loop.*

- *If $|\,Q\,| = 4k$, then $tsgn\,Q = \langle 1, 1, 1 \rangle$;*

- *if* $|Q| = 4k+1$, *then* $tsgn\,Q = \langle 1,1,1\rangle$;
- *if* $|Q| = 4k+2$, *then* $tsgn\,Q = \langle -1,-1,-1\rangle$

[599].

Corollary 1.404. *Let Q be a finite Moufang loop.*

- *If* $|Q| = 4k$, *then* $tsgn\,Q = \langle 1,1,1\rangle$;
- *if* $|Q| = 4k+1$, *then* $tsgn\,Q = \langle 1,1,1\rangle$;
- *if* $|Q| = 4k+2$, *then* $tsgn\,Q = \langle -1,-1,-1\rangle$;
- *if* $|Q| = 4k+3$, *then* $tsgn\,Q = \langle 1,1,-1\rangle$

[599].

Let Q be a quasigroup. Denote by $sgn\,Q$ the following product $sgn\,Q = \mathrm{sgn}\mathbf{L}\cdot sgn\,\mathbf{R}\cdot sgn\,\mathbf{I}$.

Theorem 1.405. *Let Q be a finite quasigroup.*

- *If* $|Q| = 4k$, *then* $sgn\,Q = 1$;
- *if* $|Q| = 4k+1$, *then* $sgn\,Q = 1$;
- *if* $|Q| = 4k+2$, *then* $sgn\,Q = -1$;
- *if* $|Q| = 4k+3$, *then* $sgn\,Q = -1$.

[454, 950].

Theorem 1.405 has deep combinatorial sense. See also [431, 288].

1.7.9 The number of quasigroups

Number N of all binary quasigroups of order n is more than $n!(n-1)!(n-2)!\dots 2!1!$ (i.e., this is the lower bound of number N) [240]. See, also, [656, 349, 618, 619, 203]. An explicit formula for the number of all binary quasigroups of order n is in [755].

Definition 1.406. An $n \times n$ permutation matrix is a $(0,1)$-matrix of order n, each row and column of which contains exactly one nonzero element.

Definition 1.407. Let $m \leqslant n$ be integers, and $S_n(m)$ be the set of all m-permutations of the elements in the set $I_n = 1,2,...,n$.

For any $m \times n$ real matrix $A = (a_{ij})$ we define the permanent $PerA$ of the matrix A in the following way:

$$PerA = \sum_{(i_1,\dots,i_m)\in S_n(m)} (a_{1,i_1}a_{2,i_2}\dots a_{m,i_m}).$$

For the special case $m = n$ we have

$$perA = \sum_{(i_1,\dots,i_n)\in S_n(n)} (a_{1,i_1}a_{2,i_2}\dots a_{n,i_n}).$$

Then $PerA$ is a sum of products of the elements of the matrix A.

Example 1.408. If A is a 2×2 matrix consisting of natural (or integer, rational, real, complex) numbers, or of the elements of a ring, then

$$per \begin{pmatrix} a_{11} & a_{12} \\ a_{21} & a_{22} \end{pmatrix} = (a_{11} \cdot a_{22}) + (a_{12} \cdot a_{21}).$$

See [626] for details.

Using the classical inclusion-exclusion principle, the authors establish an explicit formula for the number of Latin squares of order n, and therefore of quasigroups of order n as well:

$$L_n = n! \sum_{A \in B_n} (-1)^{\sigma_0(A)} \binom{perA}{n}$$

or

$$L_n = \sum_{A \in B_n} (-1)^{\sigma_0(A)} (perA)^n,$$

where B_n is the set of $n \times n$ $(0,1)$ matrices, $\sigma_0(A)$ is the number of zero elements in the matrix A and $perA$ is the permanent of the matrix A.

The authors write: "We point out that although our result gives a simple and explicit formula for the number of Latin squares this formula does not provide an efficient algorithm for computing the value of L_n, since the number of terms in the formula is exponential on n. Thus a different type of formula which is efficient in practical computation is still desirable."

Approximation of the number of finite n-ary quasigroups is given in [543, 702]. The upper bound for this number is given in [573]. Many formulae for the number of Latin rectangles are given in [856].

The number of Latin squares (of order $\leqslant 7$) related to autotopisms is computed in [315] using the Gröbner basis and the program Singular.

1.7.10 Latin squares and graphs

Taking into account that any Latin square is the inner part of Cayley table of a quasigroup, we see that graphs of Latin squares are directly connected with graphs of corresponding quasigroups. There exist various ways to define graphs (including colored graphs) on a Latin square [559, 240, 419, 531].

Definition 1.409. A graph Γ is a pair (V, E), where V is a non-empty set and $E \subseteq V \times V$, i.e., E is a binary relation defined on the set V.

Usually elements of the set V are called vertices, and elements of the set E are called edges of graph Γ. If the graph Γ is a non-oriented graph, then the binary relation E is symmetric.

Recall, we can consider any element of a groupoid $(G, \circ)$ as a triplet. It is clear that the same is true for the elements of a Latin square and respective quasigroup.

Definition 1.410. Define the graph of a groupoid $(G, \cdot)$ in the following way. Any triplet is a graph vertex. If a pair of triplets has at least one equal coordinate, then the corresponding pair is connected with a common edge.

The definition of a graph of respective square (the inner part of Cayley table of groupoid $(G, \cdot)$) is similar.

Example 1.411. In quasigroup $(Q, \circ)$

$\circ$	1	2	3
1	1	2	3
2	2	3	1
3	3	1	2

we have $(Q, \circ) = \{(1, 1, 1), (2, 3, 1), (3, 2, 1), (1, 2, 2), (2, 1, 2), (3, 3, 2), (1, 3, 3), (2, 2, 3), (3, 1, 3)\}$.

We denote triplet (1, 1, 1) by the cipher 1, triplet (2, 3, 1) by the cipher 2, triplet (3, 2, 1) by the cipher 3, and so on. Then we obtain the following graph Γ of quasigroup $(Q, \circ)$ and of the respective Latin square.

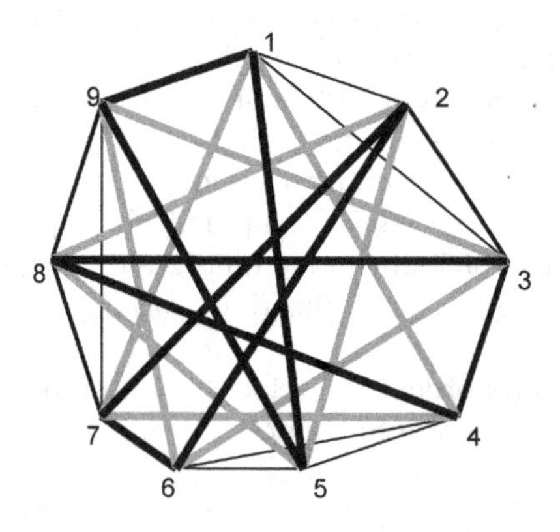

Figure 1.1: Graph of quasigroup $(Q, \circ)$ and the respective Latin square.

Example 1.412. In quasigroup $(Q, \cdot)$

$\cdot$	1	2	3
1	1	2	3
2	3	1	2
3	2	3	1

we have $(Q, \cdot) = \{(1, 1, 1), (1, 2, 2), (1, 3, 3), (2, 1, 3), (2, 2, 1), (2, 3, 2), (3, 1, 2), (3, 2, 3), (3, 3, 1)\}$. We denote triplet (1, 1, 1) by the cipher 1, triplet (1, 2, 2) by the cipher 2, triplet (1, 3, 3) by the cipher 3, and so on. We construct graph Γ_1 of quasigroup $(Q, \cdot)$. See Figure 1.2.

It is easy to see that graphs Γ and Γ_1 coincide, if we do not take into consideration the colors of edges. It is clear that graph Γ has a high degree of symmetry. Properties of this graph can be described using the following definitions.

The degree (or valency) of a vertex of a graph is the number of edges incident to the vertex, with loops counted twice [892].

Definition 1.413. A regular graph is a graph where each vertex has the same number of neighbors; i.e., every vertex has the same degree or valency [892].

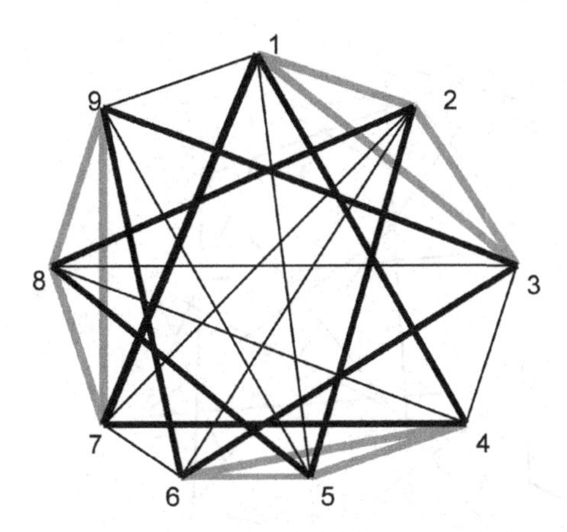

Figure 1.2: Graph of quasigroup $(Q, \cdot)$ and of the respective Latin square.

Graphs from Figures 1.1 and 1.2 are regular and have degree $k = 3(n-1) = 6$.

An adjacent vertex of a vertex v in a graph is a vertex that is connected to v by an edge. In graph Γ (Figure 1.2) vertices $2, 3, 4, 5, 7, 9$ are adjacent to vertex 6.

The neighborhood of a vertex v in a graph G is the induced subgraph of G consisting of all vertices adjacent to v and all edges connecting two such vertices [931].

Definition 1.414. A graph G is said to be strongly regular if there are also integers λ and μ such that: every two adjacent vertices have λ common neighbors; every two non-adjacent vertices have μ common neighbors. A graph of this kind is sometimes said to be an $SRG(v, k, \lambda, \mu)$ [938].

Graph 1.1 has the following parameters: $\lambda = 3$, $\mu = 6$.

Definition 1.415. A Latin square graph is a strongly regular graph $\Gamma = SRG(L)$ with parameters $v = n^2$ (number of vertex), $k = 3(n-1)$, $\lambda = n$, $\mu = 6$ [419].

Theorem 1.416. *A strongly regular graph Γ with the parameters $v = n^2$, $k = 3(n-1)$, $\lambda = n$, $\mu = 6$ is a Latin square graph, provided that $n > 23$.*

See [172, 168, 159, 240] for details.

Define the following translation graphs of a groupoid and the respective square.

Definition 1.417. The left (right, middle) translation graph of a groupoid $(G, \cdot)$ we define in the following way: any triplet is a graph vertex; if a pair of triplets has the first (second, third, respectively) equal coordinate, then the respective pair of vertices is connected with a common edge.

Definitions of translation graphs of respective squares (inner part of Cayley table of groupoid $(G, \cdot)$) are similar. Definitions that are similar to Definition 1.417 are given in [240, 419].

In Figures 1.1 and 1.2 edges of left translation graphs are light-gray, of right translation graphs are black and wide, and edges of middle translation graphs are black and narrow.

We define the group Z_4 as the set of triplets $\{(0,0,0), (0,1,1), (0,2,2), (0,3,3), (1,0,1), (1,1,2), (1,2,3), (1,3,0), (2,0,2), (2,1,3), (2,2,0), (2,3,1), (3,0,3), (3,1,0), (3,2,1), (3,3,2)\}$ and denote these triplets by the numbers $1, \ldots, 16$ in their natural order. In Figure 1.3 the middle translation graph of the group Z_4 is presented.

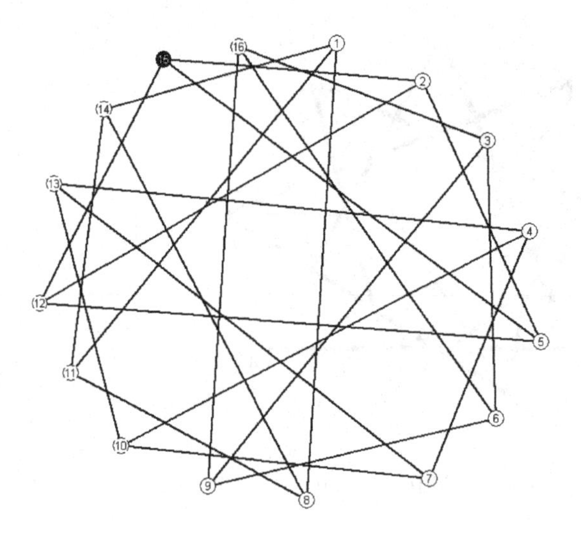

Figure 1.3: Graph of middle translations of the group Z_4.

Graph 1.3 consists of four disjoint subgraphs with the following vertices: {1, 8, 11, 14}, {2, 5, 12, 15}, {3, 6, 9, 16}, {4, 7, 10, 13}.

We define Klein group K_4 as the set of triplets $\{(0,0,0), (0,1,1), (0,2,2), (0,3,3), (1,0,1), (1,1,0), (1,2,3), (1,3,2), (2,0,2), (2,1,3), (2,2,0), (2,3,1), (3,0,3), (3,1,2), (3,2,1), (3,3,0)\}$ and denote these triplets by the numbers $1, \ldots, 16$ in their natural order. The middle translation graph of the group K_4 is presented in Figure 1.3.

This middle translation graph of the group K_4 consists of four disjoint subgraphs with the following vertices: {1, 6, 11, 16}, {2, 5, 12, 15},{3, 8, 9, 14}, {4, 7, 10, 13}.

Definition of oriented linear graph Γ with the coloring of edges is given in [19]. "Let G be a groupoid. From G one can derive an oriented linear graph Γ with the coloring of edges as follows: the vertices of Γ are the elements of G; if $g_1 g_2 = g_3$, then there is a directed edge from g_1 to g_3, colored by g_2. Graph Γ is called the "Cayley diagram of G" (MR0132798 (24 #A2634)). It is easy to see that the Cayley diagram of a quasigroup reflects the structure of right translations of this quasigroup.

"Using Cayley diagrams, the author provides proofs of the known results: (1) if all principal isotopes of a loop G are commutative, then G is a group (see Theorem 4.100); (2) a loop, all principal isotopes of which have the right inverse property, is a Moufang one" (MR0132798 (24 #A2634)).

Automorphism groups of Latin square graphs are studied in [419, 531].

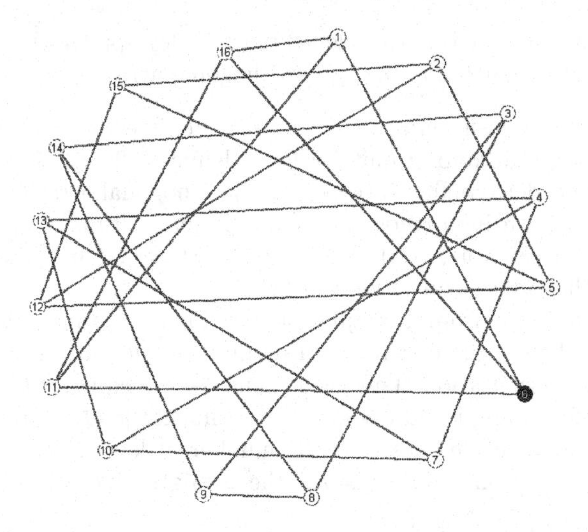

Figure 1.4: Graph of middle translations of Klein group K_4.

1.7.11 Orthogonal arrays

Orthogonal arrays generalize the idea of mutually orthogonal Latin squares in a tabular form. Much more information about arrays is given in [27, 850].

Definition 1.418. A t-(v, k, λ) orthogonal array ($t \leqslant k$) is a $v^t \times k$ array (this array has k columns) the entries of which are chosen from a set X with v points such that in every subset of t columns of the array, every t-tuple of points of X appears in exactly λ rows [932].

Example 1.419. It is easy to see that the quasigroup from Example 1.411 and the quasigroup from Example 1.412 are orthogonal. In the following orthogonal array, the columns I, II, and III set quasigroup $(Q, \circ)$, and the columns I, II, and IV set quasigroup $(Q, \cdot)$.

I	*II*	*III*	*IV*
1	1	1	1
1	2	2	2
1	3	3	3
2	1	2	3
2	2	3	1
2	3	1	2
3	1	3	2
3	2	1	3
3	3	2	1

In this array, the first coordinates of quasigroup triplets form column I, second coordinates form column II, third coordinates of triplets of quasigroup $(Q, \circ)$ form column III, and third coordinates of triplets of quasigroup $(Q, \cdot)$ form column IV.

In this example we have $X = \{1, 2, 3\}$, $v = 3$. Further we have $t = 2$, $k = 4$, $\lambda = 1$, since in any two columns any ordered pair of elements (2-tuple) appears exactly one time ($\lambda = 1$). Therefore we have constructed a 2-$(3, 4, 1)$ orthogonal array.

It is clear that every one of quasigroups $(Q, \circ)$ and $(Q, \cdot)$ defines a 2-$(3, 3, 1)$ orthogonal array. Moreover any binary quasigroup of finite order v defines a 2-$(v, 3, 1)$ array and any n-ary quasigroup of a finite order v defines n-$(v, n+1, 1)$ orthogonal array.

Presentation of quasigroups using orthogonal arrays appears in many articles devoted to quasigroup theory. See, for example, [571]. A 2-$(v, k, 1)$ orthogonal array is equivalent to a set of $k-2$ mutually orthogonal Latin squares of order v.

Note, we can consider the array from Example 1.419 as a permutation of the set $X \times X$, for example columns I and II form the first row of a permutation θ and columns III and IV form the second row of this permutation. This is the cause of the equality $k = 2t$.

In many applications (for example, in design of experiments [860]) parameters of the orthogonal array have the following names: v is the number of levels, k is the number of factors, λv^t is the number of experimental runs, t is the strength, and λ is the index.

Definition 1.420. An orthogonal array is simple if it does not contain any repeated rows. An orthogonal array is linear if X is a finite field of order q, q is a prime power, and the rows of the array form a subspace of the vector space.

Every linear orthogonal array is simple [932, 850]. Orthogonal systems of t t-ary quasigroups of order v (we suppose that $v > t$) form t-$(v, 2t, 1)$ orthogonal arrays.

Exercise 1.421. Permutation of columns in any quasigroup array 2-$(v, 3, 1)$ (three columns) gives $6 = 3!$ quasigroups that are parastrophes of the initial quasigroup. How many quasigroups can be obtained by permutation of columns in: the array of Example 1.419 $(4! \cdot 2 = 48)$ and the array of n orthogonal n-ary quasigroups $((2n!) \cdot n)$?

Note, columns of the array from Example 1.419 can be considered as code words of a code $\mathfrak{C}_1$, similarly rows of this array form a code $\mathfrak{C}_2$. See Chapter X for additional information on codes.

Exercise 1.422. Find the Hamming distance (Definition 10.54) between code words of code $\mathfrak{C}_1$ and of code $\mathfrak{C}_2$.

Chapter 2

Some quasigroup classes

2.1	Definitions of loop and quasigroup classes	108
	2.1.1 Moufang loops, Bol loops, and generalizations	110
	2.1.2 Some linear quasigroups	113
2.2	Classical inverse quasigroups	115
	2.2.1 Definitions and properties	115
	2.2.2 Autotopies of LIP- and IP-loops	119
	2.2.3 Moufang and Bol elements in LIP-loops	123
	2.2.4 Loops with the property $I_l = I_r$	125
2.3	Medial quasigroups	126
	2.3.1 Linear forms: Toyoda theorem	126
	2.3.2 Direct decompositions: Murdoch theorem	130
	2.3.3 Simple quasigroups	133
	2.3.4 Examples	133
2.4	Paramedial quasigroups	135
	2.4.1 Kepka-Nemec theorem	135
	2.4.2 Antiendomorphisms	136
	2.4.3 Direct decomposition	138
	2.4.4 Simple paramedial quasigroups	140
	2.4.5 Quasigroups of order 4	141
2.5	CMLs and their isotopes	142
	2.5.1 CMLs	142
	2.5.2 Distributive quasigroups	144
2.6	Left distributive quasigroups	146
	2.6.1 Examples, constructions, orders	146
	2.6.2 Properties, simple quasigroups, loop isotopes	148
2.7	TS-quasigroups	149
	2.7.1 Constructions, loop isotopes	149
	2.7.2 2-nilpotent TS-loops	151
	2.7.3 Some properties of TS-quasigroups	152
2.8	Schröder quasigroups	152
2.9	Incidence systems and block designs	154
	2.9.1 Introduction	154
	2.9.2 3-nets and binary quasigroups	156
	2.9.3 On orders of finite projective planes	156
	2.9.4 Steiner systems	157
	2.9.5 Mendelsohn design	161
	2.9.6 Spectra of quasigroups with 2-variable identities	161
2.10	Linear quasigroups	164
	2.10.1 Introduction	164
	2.10.2 Definitions	165
	2.10.3 Group isotopes and identities	166
	2.10.4 Nuclei, identities	169

2.10.5 Parastrophes of linear quasigroups 170
2.10.6 On the forms of n-T-quasigroups 172
2.10.7 (m, n)-Linear quasigroups 173
2.11 Miscellaneous 176
2.11.1 Groups with triality 177
2.11.2 Universal properties of quasigroups 179
2.11.3 Alternative and various conjugate closed quasigroups and loops 180

In this chapter we give information on basic properties of the most known quasigroup classes. The languages of identities and linear forms are used in order to define quasigroup classes.

2.1 Definitions of loop and quasigroup classes

We give or repeat definitions of some "popular" classes of quasigroups and loop.

Definition 2.1. Identities that involve three variables, two of which appear once on both sides of the equation and one of which appears twice on both sides are called Bol-Moufang type identities.

Various properties of Bol-Moufang type identities in quasigroups and loops are studied in [319, 694, 219, 7].

It is easy to see that many of the quasigroups and loops defined below satisfy Bol-Moufang type identities with permutations.

Definition 2.2. A quasigroup $(Q, \cdot)$ is

1. *medial*, if $xy \cdot uv = xu \cdot yv$ for all $x, y, u, v \in Q$;
2. *paramedial*, if $xy \cdot uv = vy \cdot ux$ for all $x, y, u, v \in Q$;
3. *left distributive*, if $x \cdot yz = xy \cdot xz$ for all $x, y, z \in Q$;
4. *left alternative*, if $x \cdot xy = xx \cdot y$ for all $x, y \in Q$;
5. *right distributive*, if $xy \cdot z = xy \cdot xz$ for all $x, y, z \in Q$;
6. *right alternative*, if $x \cdot yy = xy \cdot y$ for all $x, y \in Q$;
7. *flexible* (with elasticity [869]) if $x \cdot yx = xy \cdot x$ for all $x, y \in Q$ [869, 870];
8. *distributive*, if it is left and right distributive;
9. *alternative*, if it is left and right distributive;
10. *A-quasigroup* relative to an element h, if the group of all inner permutations relative to h is a subgroup of the group of automorphisms of $(Q, \cdot)$ [528];
11. *idempotent*, if $x \cdot x = x$ for all $x \in Q$;
12. *semi-symmetric*, if $x \cdot yx = y$ for all $x \in Q$ [734, 735, 709];
13. *Schröder*, if $(x \cdot y) \cdot (y \cdot x) = x$ for all $x \in Q$ [552];

14. *unipotent*, if there exists an element $a \in Q$ such that $x \cdot x = a$ for all $x \in Q$;

15. *left semi-symmetric*, if $x \cdot xy = y$ for all $x, y \in Q$;

16. *Stein quasigroup*, if $x \cdot xy = yx$ for all $x, y \in Q$ [845, 85];

17. *C_3-quasigroup*, if $x \cdot (x \cdot xy) = y$ for all $x, y \in Q$ [141, 213, 866];

18. *Moufang quasigroup*, if $(xy \cdot z)y = x(y(e_y z \cdot y))$ for all $x, y, z \in Q$ [328, 88];

19. *Moufang quasigroup*, if $y(x \cdot yz) = ((y \cdot x f_y)y)z$ for all $x, y, z \in Q$ [328, 88];

20. *left Bol quasigroup*, if $x(y \cdot xz) = R^{-1}_{e_x}(x \cdot yx) \cdot z$ for all $x, y, z \in Q$ [328, 326];

21. *right Bol quasigroup*, if $(yx \cdot z)x = yL^{-1}_{f_x}(xz \cdot x)$ for all $x, y, z \in Q$ [328, 326];

22. *left CI-quasigroup*, if $x(xy \cdot z) = R^{-1}_{e_x}(x \cdot xz) \cdot y$ for all $x, y, z \in Q$ [331];

23. *right CI-quasigroup*, if $(x \cdot yz)z = y \cdot L^{-1}_{f_x}(xz \cdot z)$ for all $x, y, z \in Q$ [331];

24. *TS-quasigroup*, if $x \cdot xy = y, xy = yx$ for all $x, y \in Q$;

25. *left F-quasigroup*, if $x \cdot yz = xy \cdot e(x)z$ for all $x, y, z \in Q$, where $x \cdot e(x) = x$ for all x;

26. *right F-quasigroup*, if $xy \cdot z = xf(z) \cdot yz$ for all $x, y, z \in Q$, where $f(x) \cdot x = x$ for all x;

27. *F-quasigroup*, if it is left and right F-quasigroup;

28. *left SM-quasigroup*, if $s(x) \cdot yz = xx \cdot yz = xy \cdot xz$ for all $x, y, z \in Q$, where $s(x) = x \cdot x$ for all x;

29. *right SM-quasigroup*, if $zy \cdot s(x) = zx \cdot yx$ for all $x, y, z \in Q$;

30. *SM-quasigroup*, if it is left and right SM-quasigroup;

31. *left E-quasigroup*, if $x \cdot yz = f(x)y \cdot xz$ for all $x, y, z \in Q$, where $f(x) \cdot x = x$ for all x;

32. *right E-quasigroup*, if $zy \cdot x = zx \cdot ye(x)$ for all $x, y, z \in Q$, where $x \cdot e(x) = x$ for all x;

33. *E-quasigroup*, if it is left and right E-quasigroup [520].

Definition 2.3. A loop $(Q, \cdot)$ is

1. *automorphic (or A-loop)* if the group of all inner permutations is a subgroup of the group of automorphisms of $(Q, \cdot)$ [174, 455];

2. a RIF-loop if it is an IP-loop with the property that $J\theta = \theta J$ for all $\theta \in I_1(Q, \cdot)$ [517];

3. an ARIF-loop if it is a flexible loop ($x \cdot yx = xy \cdot x$) satisfying the following equations: $R_x R_{(yxy)} = R_{(xyx)} R_y$, $L_x L_{(yxy)} = L_{(xyx)} L_y$ [517];

4. *automorphic inverse loop(or AI-loop)* if $I(x \cdot y) = Ix \cdot Iy$ for all $x, y \in Q$, where $x \cdot Ix = 1$;

5. *left Bol loop*, if the identity $x(y \cdot xz) = (x \cdot yx)z$ is true;

6. *right Bol loop*, if the identity $(zx \cdot y)x = z(xy \cdot x)$ is true;

7. *Bruck loop*, if it is left (right) Bol loop with the automorphic inverse property: $(x \cdot y)^{-1} = x^{-1} \cdot y^{-1}$ for all $x, y \in Q$;

8. *middle Bol loop*, if in the respective primitive loop $(Q, \cdot, /, \backslash)$ the identity $x(yz\backslash x) = (x/z)(y\backslash x)$ is true;

9. *CC-loop* (conjugacy closed loop) if the identities $xy \cdot z = xz \cdot (z\backslash(yz))$ and $z \cdot yx = ((zy)/z) \cdot zx$ are true or, equivalently, if for any $x, y \in Q$ the product $L_x^{-1}L_yL_x$ is a left translation, the product $R_x^{-1}R_yR_x$ is a right translation;

10. *Buchsteiner loop*, if the identity $x\backslash(xy \cdot z) = (y \cdot zx)/x$ is true [179];

11. *Osborn loop*, if the identity $xy \cdot (\theta_x z \cdot x) = (x \cdot yz) \cdot x$ is true, where θ_x is a permutation of the set Q which depends on x; if in the corresponding primitive loop $(Q, \cdot, /, \backslash)$ the identity $((1/x)\backslash y) \cdot zx = x(yz \cdot x)$ is true;

12. *left Cheban loop*, if the identity $x(xy \cdot z) = yx \cdot xz$ is true [195];

13. *Cheban loop*, if the identity $x(xy \cdot z) = (y \cdot zx)x$ is true [195];

14. *generalized Moufang loop*, if $x(yz \cdot x) = I(I^{-1}y \cdot I^{-1}x) \cdot zx$ for all $x, y, z \in Q$, where $x \cdot Ix = 1$ for all $x \in Q$, or, equivalently, if $(x \cdot yz) \cdot x = xy \cdot I^{-1}(Ix \cdot Iz)$ [36];

15. *Moufang loop*, if the identity $x(y \cdot xz) = (xy \cdot x)z$ is true (left Moufang identity);

16. *Moufang loop*, if the identity $(zx \cdot y)x = z(x \cdot yx)$ is true (right Moufang identity);

17. *Moufang loop*, if the identity $x(yz \cdot x) = xy \cdot zx$ is true (middle Moufang identity);

18. *commutative Moufang loop (CML)*, if the identity $xx \cdot yz = xy \cdot xz$ is true;

19. *left M-loop*, if the equality $x \circ (y \circ z) = (x \circ (y \circ I\varphi x)) \circ (\varphi x \circ z)$ is true for all $x, y, z \in Q$, where $Ix = x^{-1}$, $x \circ x^{-1} = 1$, φ is a map of the set Q [80, p. 109].

20. *right M-loop*, if the equality $(y \circ z) \circ x = (y \circ \psi x) \circ ((I^{-1}\psi x \circ z) \circ x)$ is true for all $x, y, z \in Q$, where $Ix = x^{-1}$, $x \circ x^{-1} = 1$, ψ is a map of the set Q [80, p. 109].

21. *M-loop*, if it is left and right *M*-loop [80, p. 109];

22. Ring alternative loop *(RA loop)* is a loop whose loop ring over some commutative, associative ring with a unity and of characteristic different from 2, is alternative but not associative [373].

2.1.1 Moufang loops, Bol loops, and generalizations

In [72, p. 77] Moufang quasigroups are defined using two identities: $y(x \cdot yz) = ((y \cdot xf_y)y)z$ and $(xy \cdot z)y = x(y(e_y z \cdot y))$ for all $x, y, z \in Q$. In [328, Lemma 6.4, p. 57], [88] it is proved that for quasigroups these two identities are equivalent. Prover [615] can also be used for the proof.

We give examples of Moufang quasigroups.

$*$	0	1	2
0	1	0	2
1	0	2	1
2	2	1	0

$\circ$	0	1	2	3
0	1	0	2	3
1	0	1	3	2
2	2	3	0	1
3	3	2	1	0

$\star$	0	1	2	3	4
0	1	0	3	2	4
1	0	2	1	4	3
2	3	1	4	0	2
3	2	4	0	3	1
4	4	3	2	1	0

Lemma 2.4. *A left and right Bol quasigroup is a Moufang quasigroup.*

Proof. The proof follows from results presented in [328]. It is possible to use Prover [615]. □

Example 2.5. The following quasigroup is a left Bol quasigroup which is not a right Bol quasigroup.

*	0	1	2	3
0	1	0	3	2
1	2	1	0	3
2	0	3	2	1
3	3	2	1	0

Lemma 2.6. *If any loop isotope of an LIP-quasigroup $(Q, \cdot)$ is an LIP-loop, then quasigroup $(Q, \cdot)$ is a left Bol quasigroup.*

If any loop isotope of an RIP-quasigroup $(Q, \cdot)$ is an RIP-loop, then quasigroup $(Q, \cdot)$ is a right Bol quasigroup.

If any loop isotope of an IP-quasigroup $(Q, \cdot)$ is an IP-loop, then quasigroup $(Q, \cdot)$ is a Moufang quasigroup.

Proof. Case 1. See [328, p. 11]. Case 2 is symmetric to Case 1. Case 3, see [77, p. 55, Theorem 2.2]. □

In [891] the following identities with parameter θ (with additional unary operation) are studied:

$$((\theta x)y \cdot z)x = \theta x \cdot (y \cdot zx), \tag{2.1}$$

$$(xy \cdot z)\theta x = x \cdot (y \cdot z\theta x). \tag{2.2}$$

It is proved that any loop $(Q, \cdot)$ with identity (2.1) or (2.2) is a Moufang loop [891].

RA-loops are studied in many articles. See for example [198, 199, 457, 218]. Many Bol-Moufang identities are listed in [319]. See also [7, 219, 679, 694]. The equivalence of left, right, and middle Moufang identities from Definition 2.3 is a well-known fact [173, 72, 685]. The quasigroup case is studied in [545, 794]. See also Chapter 5.

Remark 2.7. From Moufang identity $x(y \cdot xz) = (xy \cdot x)z$ by $y = 1$ we have $x \cdot xz = xx \cdot z$. This identity is called the *left alternative identity.*

From Moufang identity $(zx \cdot y)x = z(x \cdot yx)$ by $y = 1$ we have $zx \cdot x = z \cdot xx$. This identity is called the *right alternative identity.*

From Moufang identity $x(yz \cdot x) = xy \cdot zx$ by $z = 1$ we have $x \cdot yx = xy \cdot x$. This identity is called the identity of *flexibility.* Therefore we have the following Moufang identity $(x \cdot yz)x = xy \cdot zx$.

We give the example of the Moufang loop [526].

Example 2.8. Define Cayley-Dickson loops $(C_n, \circ)$ inductively as follows: $C_0 = \{\pm(1)\}$, $C_1 = \{\pm(1,0), \pm(1,1)\}$, $C_n = \{\pm(x,0), \pm(x,1) \mid x \in C_{n-1}\}$, $n \in N$.

Conjugates of the elements of C_n are $x^* = -x$ for $x \in C_n \backslash \{1, -1\}$, $1^* = 1, (-1)^* = -1$. In C_{n+1} we have $(x,0)^* = (x^*, 0), (x,1)^* = (-x, 1)$.

Operation $\circ$ is defined in the following way

$$\begin{aligned} (x,0) \circ (y,0) &= (x \cdot y, 0), \\ (x,0) \circ (y,1) &= (y \cdot x, 1), \\ (x,1) \circ (y,0) &= (x \cdot y^*, 1) \\ (x,1) \circ (y,1) &= (-y^* \cdot x, 0). \end{aligned} \tag{2.3}$$

Then $(C_0, \cdot) \cong Z_2$, $(C_1, \circ) \cong Z_4$, $(C_2, \circ) \cong H_8$ (the quaternion group), $(C_3, \circ) \cong O_{16}$ (the octonion loop (Moufang)).

A great theorem that in fact founded quasigroup (loop) theory is the following Moufang theorem.

Theorem 2.9. *If in a Moufang loop* $(Q, \cdot)$ $ab \cdot c = a \cdot bc$ *for some* $a, b, c \in Q$*, then the elements* a, b, c *generate a subgroup of loop* $(Q, \cdot)$.

See the proof in [173, 169, 72, 685, 283]. A loop is *diassociative* if every pair of its elements generates a subgroup.

From Theorem 2.9 follows a very important corollary:

Corollary 2.10. *Any Moufang loop* $(Q, \cdot)$ *is diassociative.*

In any Moufang $(Q, \cdot)$ loop $N_l = N_r = N_m = N \trianglelefteq Q$ [72, Lemma 6.2, p. 94].

Similar to finite groups [317], any finite Moufang loop of odd order is solvable [350, 351].

Every Moufang loop of order less than 12 is necessarily a group and, up to isomorphism, there exists exactly one Moufang loop of order 12 that is not associative [202].

Example 2.11. We give a Cayley table of a non associative Moufang loop of order 12 [202, 685].

*	0	1	2	3	4	5	6	7	8	9	10	11
0	0	1	2	3	4	5	6	7	8	9	10	11
1	1	0	3	2	5	4	9	8	7	6	11	10
2	2	3	0	1	6	9	4	10	11	5	7	8
3	3	2	1	0	7	8	10	4	5	11	6	9
4	4	5	6	9	0	1	2	11	10	3	8	7
5	5	4	7	8	1	0	11	2	3	10	9	6
6	6	7	4	10	2	11	0	1	9	8	3	5
7	7	6	5	11	3	10	8	9	1	0	2	4
8	8	9	11	5	10	3	7	6	0	1	4	2
9	9	8	10	4	11	2	1	0	6	7	5	3
10	10	11	9	6	8	7	3	5	4	2	0	1
11	11	10	8	7	9	6	5	3	2	4	1	0

Exercise 2.12. Find the nucleus of the Moufang loop of order 12 given in Example 2.11.

Finite simple Moufang loops are constructed by Paige [676]. A remarkable connection between the theory of Moufang loops and the theory of groups was established in the articles [265, 389, 390, 181]. The proof that Paige has constructed all finite simple Moufang loops is given in [566]. Automorphism groups of simple Moufang loops are researched in [652].

Theorem 2.13. *(Grishkov-Zavarnitsine Theorem). The order of any subloop* H *of a finite Moufang loop* M *is a divisor of the order of the loop (Lagrange property) [393].*

For finite Moufang loops, an analog of the first Sylow theorem giving a criterion for the existence of a p-Sylow subloop is proved in [394]. Also the maximal order of p-subloops in the Moufang loops that do not possess p-Sylow subloops is founded. For further progress see in [337, 338].

Here we give some results from the Ph.D. dissertation of Alexandr Savel'evich Basarab [36]. Any generalized Moufang loop is a WIP-loop [36]. We denote a generalized Moufang loop by $(Q, \cdot)$, its nucleus we denote by N. Let G' be the group generated by the set $\langle L_x^{-1} R_x \mid x \in Q \rangle$. Let $C \subseteq N$ be a set of elements c, any of which satisfies the following equality

$$a \cdot c = ab \cdot \varphi I \varphi^{-1} b, \tag{2.4}$$

where $a, b \in Q$, and pseudoautomorphism $\varphi \in G'$. Recall that in a generalized Moufang loop any left pseudoautomorphism is also its right pseudoautomorphisms [36].

Subnucleus N' is defined in the following way: $N' = \langle C \rangle$. In a generalized Moufang loop $(Q, \cdot)$ we have $N' \trianglelefteq Q$. In a Moufang loop, $N' = \{\varepsilon\}$ [36].

Definition 2.14. Let $(Q, \cdot)$ be a quasigroup. The solution of equation $ab \cdot c = (a \cdot bc) \cdot x$ is called an associator of elements a, b, c [173]. We shall denote it as $[a, b, c]$.

Theorem 2.15. *Basarab theorem. If in a generalized Moufang loop the associator of elements a, b, c is in N', i.e., $[a, b, c] \in N'$, then these elements generate a subloop such that the associator of any three elements from this subloop is in N' [36].*

Bases of Bol loops and Bol quasigroups theory were given by D.A. Robinson [715, 716]. R. P. Burn proved that the number 8 is the smallest possible order of a Bol loop which is not a group, and that there exist exactly six non-isomorphic Bol loops of order 8 [182, 183].

Proper (non-Moufang) infinite simple Bol loops (two series) are constructed in [513]. Proper finite simple Bol loops are constructed by Gabor Nagy [649, 651]. Minimal order of constructed loops is equal to 24.

Properties of some left Bol and Bruck loops are researched in [46, 45]. Useful information about Bol loops is presented in [333]. Embeddings of Bol loops in groups are studied in [717].

Example 2.16. Hans Zassenhaus construction. We denote the ring of integers modulo 2 by $(R, +, \cdot)$. The $(B, \circ)$ such that $B = R \times R \times R$ and the operation $(\circ)$ defined by $(a, b, c) \circ (m, n, p) = (a + m, b + n, c + p + b \cdot m \cdot n)$ is a Bol loop of order 8 [685].

Example 2.17. Let H be a subgroup of a non-abelian group G and let $A = H \times G$. For $(h_1, g_1), (h_2, g_2) \in A$ define

$$(h_1, g_1) \circ (h_2, g_2) = (h_1 h_2, h_2 g_1 h_2^{-1} g_2). \tag{2.5}$$

Then $(A, \circ)$ is a Bol loop [838, 451].

Notice, that the construction defined by equality (2.5) is similar to the construction of semidirect product in groups [471]. Generalization of this construction is given in [718] (see Section 1.6.2) in order to prove that there exist Bol loops whose nuclei are not normal. See also [248]. Constructions of Bol loops are also given in [375, 200].

In [46], the authors prove that a finite Bruck loop is the direct product of a Bruck loop of odd order with either a solvable Bruck loop of 2-power order or a product of loops related to the groups $PSL_2(q)$, $q = 9$ or $q \geqslant 5$ is a Fermat prime. As corollaries, they obtain versions of Sylow's, Lagrange's and Hall's theorems for these loops.

In some articles [744] the Bruck loop is defined as a left Bol loop with the following Bruck identity: $x(yy \cdot x) = (xy)^2$.

2.1.2 Some linear quasigroups

An n-ary quasigroup of the form

$$\gamma g(x_1, x_2, \dots, x_n) = \gamma_1 x_1 + \gamma_2 x_2 + \dots + \gamma_n x_n,$$

where $(Q, +)$ is a group and $\gamma, \gamma_1, \dots, \gamma_n$ are some permutations of the set Q, is called *n-ary group isotope* (Q, g). Of course this equality (often as the analogous equalities that will appear later in these lectures) is true for all $x_1, x_2, \dots, x_n \in Q$.

An n-ary quasigroup of the form

$$g(x_1, x_2, \dots, x_n) = \alpha_1 x_1 + \alpha_2 x_2 + \dots + \alpha_n x_n + a = \sum_{i=1}^{n} \alpha_i x_i + a, \tag{2.6}$$

where $(Q,+)$ is a group, $\alpha_1,\dots,\alpha_n$ are some automorphisms of the group $(Q,+)$, and the element a is some fixed element of the set Q, is called *linear n-ary quasigroup* (Q,g) (over group $(Q,+)$).

A linear quasigroup over an abelian group is called a *n-T-quasigroup* [871]. The theory of binary T-quasigroups was developed by T. Kepka and P. Nemec [667, 506].

Definition 2.18. A quasigroup $(Q,\cdot)$ with identities $x\cdot xy = y$, $xy = yx$ such that any of its three elements generate a medial subquasigroup is called a CH-quasigroup (Manin quasigroup).

Remark 2.19. From Definition 2.2 (or 1.273) it follows that any CH-quasigroup can be defined as a trimedial TS-quasigroup.

It is possible to obtain any CH-quasigroup $(Q,\circ)$ with the help of the following construction: $x\circ y = (-x-y)+d$, where the element d is an element from the center of CML $(Q,+)$ and $(-x)+x = 0$ [593].

If $(Q,+)$ is a commutative Moufang loop, then quasigroup $(Q,\cdot)$ of the form $x\cdot y = -x-y$, where $x-x = 0$ for all $x\in Q$, is a TS-quasigroup [685]. Indeed, $x\cdot y = -x-y = -y-x = y\cdot x$, $x\cdot xy = -x-(-x-y) = -x+y+x = -x+x+y = y$.

Lemma 2.20. *1. A left F-loop is a group. 2. A right F-loop is a group. 3. A loop $(Q,\cdot)$ is left semimedial if and only if it is a commutative Moufang loop. 4. A loop $(Q,\cdot)$ is right semimedial if and only if it is a commutative Moufang loop. 5. A left E-loop $(Q,\cdot)$ is a commutative group. 6. A right E-loop $(Q,\cdot)$ is a commutative group.*

Proof. Case 1. From $x\cdot yz = xy\cdot e(x)z$ we have $x\cdot yz = xy\cdot z$. Case 2. From $xy\cdot z = xf(z)\cdot yz$ we have $xy\cdot z = x\cdot yz$.

Case 3. We use the proof from [72, p. 99]. Let $(Q,\cdot)$ be a left semimedial loop. If $y=1$, then we have $x^2\cdot z = x\cdot xz$. If $z=1$, then $x^2y = xy\cdot x$. Then $x\cdot xy = xy\cdot x$. If we denote xy by y, then we obtain that $xy = yx$, i.e., the loop $(Q,\cdot)$ is commutative.

Converse. Using commutative identity we can write the Moufang identity (Case 4 of Definition 2.3) as follows: $x(x\cdot yz) = xy\cdot xz$. Applying a left alternative identity to the last identity we obtain $xx\cdot yz = xy\cdot xz$.

Case 4. For the proof of Case 3 it is possible to use "mirror" principles. We give the direct proof. Let $(Q,\cdot)$ be a right semimedial loop, i.e., $zy\cdot x^2 = zx\cdot yx$ for all $x,y,z\in Q$. If $y=1$, then we have $z\cdot x^2 = zx\cdot x$.

If $z=1$, then $yx^2 = x\cdot yx$. Then $zx\cdot x = x\cdot zx$. If we denote zx by z, then we obtain that $zx = xz$, i.e., the loop $(Q,\cdot)$ is commutative. Moreover, we have $x^2\cdot yz = zy\cdot x^2$, $xy\cdot zx = xz\cdot yx$.

It is clear that a commutative Moufang loop is right semimedial. See Case 3.

Case 5. From $x\cdot yz = f(x)y\cdot xz$ we have $x\cdot yz = y\cdot xz$. From the last identity by $z=1$ we obtain $x\cdot y = y\cdot x$. Therefore we can re-write identity $x\cdot yz = y\cdot xz$ in the following form: $yz\cdot x = y\cdot zx$.

Case 6 is proved in a similar way to Case 5. □

Medial quasigroups, distributive quasigroups, T-quasigroups, CH-quasigroups, F-quasigroups are linear quasigroups in the sense of the following definition:

Definition 2.21. A quasigroup $(Q,\cdot)$ is called linear over a loop $(Q,+)$ if there exist automorphisms $\varphi,\psi\in Aut(Q,+)$ and an element c (usually c is an element from a nucleus of the loop) such that $x\cdot y = (\varphi x+\psi y)+c$ for all $x,y\in Q$.

See more details in the next sections.

2.2 Classical inverse quasigroups

2.2.1 Definitions and properties

Almost all the well-known kinds of quasigroups and loops such as LIP-, IP-, WIP- and CI-loops and quasigroups are included among the classes of quasigroup which have some kind of inverse property. Most recently, (r, s, t)-inverse quasigroups were defined as a generalization of various kinds of "crossed-inverse" property quasigroups and loops: in particular, they generalize CI-, WIP- and m-inverse loops [472].

Definition 2.22. 1) A quasigroup $(Q, \circ)$ has the *left inverse-property ($(Q, \circ)$ is an LIP-quasigroup)* if there exists a permutation $x \to \lambda x$ of the set Q such that

$$\lambda x \circ (x \circ y) = y \tag{2.7}$$

for all $x, y \in Q$ [72];

2) a quasigroup $(Q, \circ)$ has the *right inverse-property ($(Q, \circ)$ is an RIP-quasigroup)* if there exists a permutation $x \to \rho x$ of the set Q such that

$$(x \circ y) \circ \rho y = x \tag{2.8}$$

for all $x, y \in Q$ [72];

3) a quasigroup $(Q, \circ)$ has the *inverse property* (is an *IP-quasigroup* [72, 685]) if both (2.7) and (2.8) hold. If $(Q, \circ)$ is a loop, then $\lambda = \rho$ and we write $x\lambda = x\rho = x^{-1}$. However, in the case when $(Q, \circ)$ is a quasigroup, $x\lambda \neq x\rho$ is possible (see Remark 2.32).

4) A quasigroup $(Q, \circ)$ has the *weak-inverse-property ($(Q, \circ)$ is a WIP-quasigroup)* if there exists a permutation $x \to Jx$ of the set Q such that

$$x \circ J(y \circ x) = Jy \tag{2.9}$$

for all $x, y \in Q$ [25, 487, 847];

5) a quasigroup $(Q, \circ)$ has the *crossed-inverse-property ($(Q, \circ)$ is a CI-quasigroup)* if there exists a permutation $x \to Jx$ of the set Q such that

$$(x \circ y) \circ Jx = y \tag{2.10}$$

for all $x, y \in Q$ [17, 487];

6) a quasigroup $(Q, \circ)$ has the *m-inverse-property* if there exists a permutation $x \to Jx$ of the set Q such that

$$J^m(x \circ y) \circ J^{m+1}x = J^m y \tag{2.11}$$

for all $x, y \in Q$ [472, 486];

7) a quasigroup $(Q, \circ)$ has the *(r, s, t)-inverse-property (i.e., $(Q, \circ)$ is an (r, s, t)-quasigroup)* if there exists a permutation $x \to Jx$ of the set Q such that

$$J^r(x \circ y) \circ J^s x = J^t y \tag{2.12}$$

for all $x, y \in Q$ [487, 488].

There exist potential applications of m-inverse quasigroups with long inverse cycles in cryptography [242].

Example 2.23. Loop $(Q, \circ)$ with the following Cayley table

$\circ$	1	2	3	4	5
1	1	2	3	4	5
2	2	1	4	5	3
3	3	4	5	1	2
4	4	5	2	3	1
5	5	3	1	2	4

is a $(0, 5, 1)$-inverse, $(2, 1, 3)$-inverse, and $(4, 3, 5)$-inverse loop with respect to permutation $J = (1\,2)(3\,4\,5)$ and this loop is a $(0, 1, 5)$-inverse, $(2, 3, 1)$-inverse and $(4, 5, 3)$-inverse loop with respect to permutation $J = (1\,2)(3\,5\,4)$.

Example 2.24. Loop $(Q, \cdot)$ with the following Cayley table

$\cdot$	1	2	3	4	5	6
1	1	2	3	4	5	6
2	2	1	4	3	6	5
3	3	4	5	6	1	2
4	4	3	6	5	2	1
5	5	6	2	1	4	3
6	6	5	1	2	3	4

is (1, 2, 5)-inverse loop with respect to permutation $J = (1\,2)(3\,6\,4\,5)$, and is (3, 6, 3)-inverse loop with respect to permutation $J = (3\,6\,4\,5)$.

Examples 2.23 and 2.24 were found using computer calculations [254].

Lemma 2.25. *1. In any LIP-quasigroup the permutation λ is unique.*

2. In any RIP-quasigroup the permutation ρ is unique.

3. In any CI-quasigroup the permutation J is unique.

Proof. Case 1. If in quasigroup $(Q, \cdot)$ the equalities $\lambda_1 x(xy) = y$ and $\lambda_2 x(xy) = y$ are true, then we have $\lambda_1 x(xy) = \lambda_2 x(xy)$, $\lambda_1 x = \lambda_2 x$.

Cases 2 and 3 are proved in a similar way. □

The situation is different for WIP-, m-inverse quasigroups and loops and its generalizations. See Examples 2.27 and 2.28. These examples were found using Mace 4 [614].

Example 2.26. Example of a loop $(Q, *)$ with the weak inverse property relative to permutation $J = (0\,4)$ which is not a WIP-loop relative to permutation I, where $x * I(x) = 0$ for all $x \in Q$ (see Definition 2.29).

$*$	0	1	2	3	4
0	0	1	2	3	4
1	1	4	0	2	3
2	2	3	4	0	1
3	3	0	1	4	2
4	4	2	3	1	0

Example 2.27. Example of a loop with a weak inverse property in the sense of Definition

2.22 Case 4, in which $J_1 = (1\,2)(3\,4) = I$, $J_2 = (0\,1)(2\,3)$. This loop is also a 3-inverse loop in the sense of Definition 2.22 Case 6 with $J_1 = (1\,2)(3\,4)$ and $J_2 = (0\,1)(2\,3)$.

*	0	1	2	3	4
0	0	1	2	3	4
1	1	3	0	4	2
2	2	0	4	1	3
3	3	4	1	2	0
4	4	2	3	0	1

Example 2.28. Example of a 4-inverse loop, in which $J_1 = (2\,4)(3\,5) = I$, $J_2 = (0\,1)(2\,5)(3\,4)$.

*	0	1	2	3	4	5
0	0	1	2	3	4	5
1	1	0	3	2	5	4
2	2	3	4	5	0	1
3	3	2	5	4	1	0
4	4	5	0	1	2	3
5	5	4	1	0	3	2

Therefore we need to give the following:

Definition 2.29. A loop $(Q,+)$ has the *weak-inverse-property* (*$(Q,+)$ is a WIP-loop*) if for all $x, y \in Q$

$$x + I(y + x) = Iy, \tag{2.13}$$

where $x + Ix = 0$.

A loop $(Q,+)$ has the *m-inverse-property* if for all $x, y \in Q$

$$I^m(x + y) + I^{m+1}x = I^m y, \tag{2.14}$$

where $x + Ix = 0$ [472].

In a loop $(Q,\cdot)$ we define element x^{-1} as an element with the property that $x \cdot x^{-1} = 1$, and element ${}^{-1}x$ as ${}^{-1}x \cdot x = 1$ for all $x \in Q$. See Lemma 1.93.

Lemma 2.30. *In any LIP-quasigroup, $\lambda^2 = \varepsilon$. In any RIP-quasigroup, $\rho^2 = \varepsilon$.*

Proof. Case 1. We have $\lambda^2 x(\lambda x(x \cdot y)) = xy$, $\lambda^2 x(\lambda x(x \cdot y)) = \lambda^2 x \cdot y$. Therefore, $\lambda^2 = \varepsilon$.

Case 2 is proved in a similar way. □

Lemma 2.31. *1. In any LIP-loop $(Q,\cdot)$, ${}^{-1}x = x^{-1}$. 2. In any RIP-loop $(Q,\cdot)$, ${}^{-1}x = x^{-1}$.*

Proof. Case 1. By definition of ${}^{-1}x \cdot (x \cdot x^{-1}) = {}^{-1}x \cdot 1 = {}^{-1}x$. Applying LIP-property we have ${}^{-1}x \cdot (x \cdot x^{-1}) = x^{-1}$. Therefore ${}^{-1}x = x^{-1}$.

Case 2 is proved in a similar way to Case 1. □

Remark 2.32. V.D. Belousov [80] gave an example of an IP-quasigroup which satisfies both 2.7 and 2.8 but with $\lambda \neq \rho$.

Let$(G,+)$ be a finite abelian group. Define a quasigroup $(Q,\cdot)$ on the ordered pairs of G by $(a_i, b_i) \cdot (a_j, b_j) = (a_i + a_j, b_j - b_i)$. Then $(Q,\cdot)$ is both an LIP-quasigroup with $\lambda(a,b) = (-a,-b)$ and a RIP-quasigroup with $\rho(a,b) = (-a,b)$.

We have $\lambda(a_i,b_i)\cdot[(a_i,b_i)\cdot(a_j,b_j)] = (-a_i,-b_i)\cdot(a_i+a_j, b_j-b_i) = [-a_i+(a_i+a_j), (b_j - b_i) - (-b_i)] = (a_j, b_j)$ and

$[(a_j,b_j)\cdot(a_i,b_i)]\cdot\rho(a_i,b_i) = (a_j + a_i, b_i - b_j)\cdot(-a_i,b_i)$
$= [(a_j + a_i) - a_i, b_i - (b_i - b_j)] = (a_j, b_j)$.

Lemma 2.33. *Any Moufang loop is an IP-loop.*

Proof. The proof is taken from [685]. From Moufang identity $x(yz \cdot x) = xy \cdot zx$ by $x = z^{-1}$ we have $z^{-1}(yz \cdot z^{-1}) = z^{-1}y \cdot zz^{-1}$, $z^{-1}(yz \cdot z^{-1}) = z^{-1}y$, $yz \cdot z^{-1} = y$. Therefore, any Moufang loop has the right inverse property.

From Moufang identity $(x \cdot yz)x = xy \cdot zx$ (Remark 2.7) by $x = {}^{-1}y$ we have $({}^{-1}y \cdot yz) \cdot {}^{-1}y = {}^{-1}yy \cdot (z \cdot {}^{-1}y)$, ${}^{-1}y \cdot yz = z$. Therefore, any Moufang loop has the left inverse property. □

We give some properties of an IP-quasigroup $(Q, \cdot)$ [72].

Theorem 2.34. *1.* $\rho^2 = \lambda^2 = \varepsilon$.

2. $(y \cdot \rho x) \cdot x = y$, $x \cdot (\lambda x \cdot y) = y$.

3. $a \cdot x = b \Rightarrow x = \lambda a \cdot b$, $y \cdot a = b \Rightarrow y = b \cdot \rho a$.

4. $\rho(x \cdot y) = \lambda y \cdot \lambda x$.

5. $L_{\lambda a} = L_a^{-1}$, $R_{\rho a} = R_a^{-1}$.

6. $\rho R_a \lambda = L_a^{-1}$, $\lambda L_a \rho = R_a^{-1}$. $\rho L_a \lambda = R_{\lambda a}$, $\lambda R_a \rho = L_{\rho a}$.

Proof. Case 1 is proved in Lemma 2.30. Cases 2 and 3 follow from Case 1.

4. If $x \cdot y = z$, then $y = \lambda x \cdot z$, $\lambda x = y \cdot \rho z$ and, finally, $\rho z = \lambda y \cdot \lambda x$. But $z = x \cdot y$. Therefore, $\rho(x \cdot y) = \lambda y \cdot \lambda x$.

5. From equality $\lambda x(xy) = y$ we have $L_{\lambda x} L_x y = y$. Then $L_{\lambda x} = L_x^{-1}$.

6. We have $\rho R_a \lambda x = \rho(\lambda x \cdot a) \overset{4}{=} \lambda a \cdot \lambda^2 x = \lambda a \cdot x = L_{\lambda a} x = L_a^{-1} x$. □

Corollary 2.35. *In any IP-loop: 1.* $IR_aI = L_a^{-1}$. *2.* $IL_aI = R_a^{-1}$. *3.* $R_a = P_aI$. *4.* $L_a = IP_{a^{-1}}$.

Proof. It is easy to see that in any IP-loop $P_1 = I$. Cases 1 and 2 are easy.

Case 3. Let $P_a x = y$. Then $x \cdot y = a$, $y = x^{-1} \cdot a$, $P_a x = R_a I x$.

Case 4. From equality $P_a x = x^{-1} \cdot a$ using Case 4 of Theorem 2.34 we have $P_a x = (a^{-1} \cdot x)^{-1} = IL_{a^{-1}}$. □

Lemma 2.36. *The* $(2\,3)$*-parastrophe of LIP-quasigroup* $(Q, \cdot)$ *is isotopic to* $(Q, \cdot)$*; the* $(1\,3)$*-parastrophe of RIP-quasigroup* $(Q, \cdot)$ *is isotopic to* $(Q, \cdot)$.

Proof. If $(x \cdot y) \cdot \rho y = x$, then $x/y = x \cdot \rho y$. If $\lambda x \cdot (x \cdot y) = y$, then $x\backslash y = \lambda x \cdot y$. □

Theorem 2.37. *Any parastrophe of IP-quasigroup* (Q, A) *is isotopic to* (Q, A).

Proof. From Lemma 2.36 we have

$$A^{(13)} = A(\varepsilon, \rho, \varepsilon), \tag{2.15}$$

$$A^{(23)} = A(\lambda, \varepsilon, \varepsilon). \tag{2.16}$$

Using Lemmas 1.175 and 2.36 further we have
$(A^{(13)})^{(23)} = (A(\varepsilon, \rho, \varepsilon))^{(23)}$, $A^{(123)} = A^{(23)}(\varepsilon, \varepsilon, \rho) \overset{2.16}{=} A(\lambda, \varepsilon, \varepsilon)(\varepsilon, \varepsilon, \rho) = A(\lambda, \varepsilon, \rho)$,

$$A^{(123)} = A(\lambda, \varepsilon, \rho). \tag{2.17}$$

$(A^{(23)})^{(13)} = (A(\lambda, \varepsilon, \varepsilon))^{(13)}$, $A^{(132)} = A^{(13)}(\varepsilon, \varepsilon, \lambda) \overset{2.15}{=} A(\varepsilon, \rho, \varepsilon)(\varepsilon, \varepsilon, \lambda) = A(\varepsilon, \rho, \lambda)$,

$$A^{(132)} = A(\varepsilon, \rho, \lambda). \tag{2.18}$$

$(A^{(132)})^{(23)} = (A(\varepsilon, \rho, \lambda))^{(23)}$, $A^{(12)} \overset{2.16}{=} A(\lambda, \varepsilon, \varepsilon)(\varepsilon, \lambda, \rho) = A(\lambda, \lambda, \rho)$,

$$A^{(12)} = A(\lambda, \lambda, \rho). \tag{2.19}$$

□

Remark 2.38. For groups (even for IP-loops) the equality $A^{(12)} = A(\lambda, \lambda, \rho)$ is transformed in the following equality $y \cdot x = (x^{-1} \cdot y^{-1})^{-1}$.

2.2.2 Autotopies of LIP- and IP-loops

Corollary 2.39. *1. If (α, β, γ) is an autotopy of LIP-quasigroup (Q, A), then the triplet $(\lambda\alpha\lambda, \gamma, \beta)$ is an autotopy of quasigroup (Q, A).*

2. If (α, β, γ) is an autotopy of RIP-quasigroup (Q, A), then the triplet $(\gamma, \rho\beta\rho, \alpha)$ is an autotopy of quasigroup (Q, A).

3. If (α, β, γ) is an autotopy of IP-quasigroup (Q, A), then the triplets $(\gamma, \rho\beta\rho, \alpha)$, $(\lambda\alpha\lambda, \gamma, \beta)$, $(\lambda\gamma\lambda, \alpha, \rho\beta\rho)$, $(\beta, \rho\gamma\rho, \lambda\alpha\lambda)$ and $(\lambda\beta\lambda, \lambda\alpha\lambda, \rho\gamma\rho)$ are autotopies of quasigroup (Q, A).

Proof. We shall use equalities (2.15)–(2.19).

Case 1. If $A = A(\alpha, \beta, \gamma)$, then $A^{(23)} = (A(\alpha, \beta, \gamma))^{(23)}$, $A(\lambda, \varepsilon, \varepsilon) = A(\lambda, \varepsilon, \varepsilon)(\alpha, \gamma, \beta)$, $A = A(\lambda\alpha\lambda, \gamma, \beta)$ since $\lambda = \lambda^{-1}$.

Case 2. If $A = A(\alpha, \beta, \gamma)$, then $A^{(13)} = (A(\alpha, \beta, \gamma))^{(13)}$, $A(\varepsilon, \rho, \varepsilon) = A(\varepsilon, \rho, \varepsilon)(\gamma, \beta, \alpha)$, $A = A(\varepsilon, \rho, \varepsilon)(\gamma, \beta, \alpha)(\varepsilon, \rho^{-1}, \varepsilon)$, $A = A(\gamma, \rho\beta\rho^{-1}, \alpha) = A(\gamma, \rho\beta\rho, \alpha)$, since $\rho = \rho^{-1}$.

Case 3. Any IP-quasigroup is an LIP-quasigroup and an RIP-quasigroup. Therefore further we have: If $A = A(\alpha, \beta, \gamma)$, then $A^{(123)} = (A(\alpha, \beta, \gamma))^{(123)}$, $A(\lambda, \varepsilon, \rho) = A(\lambda, \varepsilon, \rho)(\gamma, \alpha, \beta)$, $A = A(\lambda\gamma\lambda, \alpha, \rho\beta\rho)$.

If $A = A(\alpha, \beta, \gamma)$, then $A^{(132)} = (A(\alpha, \beta, \gamma))^{(132)}$, $A(\varepsilon, \rho, \lambda) = A(\varepsilon, \rho, \lambda)(\beta, \gamma, \alpha)$, $A = A(\beta, \rho\gamma\rho, \lambda\alpha\lambda)$.

If $A = A(\alpha, \beta, \gamma)$, then $A^{(12)} = (A(\alpha, \beta, \gamma))^{(12)}$, $A(\lambda, \lambda, \rho) = A(\lambda, \lambda, \rho)(\beta, \alpha, \gamma)$, $A = A(\lambda\beta\lambda, \lambda\alpha\lambda, \rho\gamma\rho)$. □

Lemma 2.40. *1. In any LIP-quasigroup the following equality is true $AC_2 = AC_3$, i.e., the groups of second and third components of the group of all autotopisms of an LIP-quasigroup coincide.*

2. In any RIP-quasigroup equality $AC_1 = AC_3$ is true.

3. In any IP-quasigroup equality $AC_1 = AC_2 = AC_3$ is true.

Proof. The proof follows from Corollary 2.39. □

Remark 2.41. R.H. Bruck [173] has defined LIP-quasigroup $(Q, \cdot)$ (RIP-quasigroup) giving the condition that λ (ρ) is a map of the set Q because it is not very complicated to demonstrate that in the quasigroup case the map λ (ρ) is a bijection of the set Q.

Indeed, if $\lambda x_1 = \lambda x_2$, then $\lambda x_1(x_1 \cdot y) = \lambda x_1(x_2 \cdot y)$, $x_1 = x_2$. Therefore, λ is an injective map. By Theorem 2.34, Case 1, $\lambda\lambda = \varepsilon$. Then by Lemma 1.14 the map λ is surjective. Thus the map λ is a bijective map, i.e., it is a permutation of the set Q.

Many properties of LIP-quasigroups and LIP-loops are described in the articles and Ph.D. dissertation of I. A. Florja (Ivan Arkhipovich Florja) [328, 327]. In an LIP-loop any Moufang element is a Bol element. Information on properties of Moufang elements is given in [72, 691].

Theorem 2.42. *If (α, β, γ) is an autotopy of an LIP-loop $(Q, \cdot)$, then the element $\alpha 1 = a$ is a left Bol element [328, Lemma 4.1], [327].*

Proof. From equality

$$\gamma(xy) = \alpha x \cdot \beta y \tag{2.20}$$

by $x = 1$ we have $\gamma = L_{\alpha 1}\beta$, $\beta = L_{\alpha 1}^{-1}\gamma = L_a^{-1}\gamma$,

$$(\alpha, \beta, \gamma) = (\alpha, L_a^{-1}\gamma, \gamma). \tag{2.21}$$

From Corollary 2.39, Case 1, it follows that the triple $(I\alpha I, \gamma, L_a^{-1}\gamma)$ is an autotopy of loop $(Q, \cdot)$, where $x \cdot Ix = 1$ for all $x \in Q$.

Further we have that the following triple

$$\begin{aligned}&(I\alpha I, \gamma, L_a^{-1}\gamma)(\alpha, L_a^{-1}\gamma, \gamma)^{-1} = \\&(I\alpha I, \gamma, L_a^{-1}\gamma)(\alpha^{-1}, \gamma^{-1}L_a, \gamma^{-1}) = \\&(I\alpha I\alpha^{-1}, L_a, L_a^{-1})\end{aligned}$$

is an autotopy of loop $(Q, \cdot)$, i.e., the following equality

$$I\alpha I\alpha^{-1}x \cdot L_a y = L_a^{-1}(xy) \tag{2.22}$$

is true for all $x, y \in Q$. From equality (2.22) by $y = 1$ we obtain that $I\alpha I\alpha^{-1}x \cdot a = L_a^{-1}(x)$, $I\alpha I\alpha^{-1} = R_a^{-1}L_a^{-1}$. Therefore the triple $(R_a^{-1}L_a^{-1}, L_a, L_a^{-1})^{-1} = (L_a R_a, L_a^{-1}, L_a)$ is an autotopy of loop $(Q, \cdot)$. The latter means that the element $a = \alpha 1$ is a left Bol element (Lemma 1.224). □

Example 2.43. We give an example of a LIP-loop with a unique left Bol element.

*	0	1	2	3	4	5
0	2	3	5	0	1	4
1	3	4	0	1	5	2
2	4	5	3	2	0	1
3	0	1	2	3	4	5
4	5	0	1	4	2	3
5	1	2	4	5	3	0

Here the identity element is equal to 3. This element is a unique left Bol element, because in this loop $(0 \cdot 40)1 \neq 0(40 \cdot 1); (1 \cdot 21)4 \neq 1(21 \cdot 4); (2 \cdot 42)5 \neq 2(42 \cdot 5); (4 \cdot 04)5 \neq 4(04 \cdot 5); (5 \cdot 15)2 \neq 5(15 \cdot 2)$. The autotopy group of this loop consists of the following elements:

$\{(\varepsilon, \varepsilon, \varepsilon), ((0\ 1), (0\ 4)(1\ 3)\ (2\ 5), (0\ 4)(1\ 3)(2\ 5)),$

$((0\ 5)(1\ 4), (0\ 4)(2\ 3), (0\ 4)(2\ 3)), ((4\ 5), (0\ 4)(1\ 2)\ (3\ 5), (0\ 4)(1\ 2)\ (3\ 5)),$

$((0\ 4\ 1\ 5), (1\ 2\ 5\ 3), (1\ 2\ 5\ 3)), ((0\ 1)\ (4\ 5), (1\ 5)\ (2\ 3), (1\ 5)\ (2\ 3)),$

$((0\ 5\ 1\ 4), (1\ 3\ 5\ 2), (1\ 3\ 5\ 2)), ((0\ 4)\ (1\ 5), (0\ 4)\ (1\ 5), (0\ 4)\ (1\ 5))\}.$

It is clear that all autotopisms of the loop $(Q, *)$ are right pseudo-automorphisms. Theorem 2.42 is true since $\alpha 3 = 3$ for any first component α of any autotopy of loop $(Q, *)$. Let

$$a = ((0\,4\,1\,5), (1\,2\,5\,3), (1\,2\,5\,3)),$$
$$b = ((0\,4)(1\,5), (0\,4)(1\,5), (0\,4)(1\,5)).$$

It is easy to check that $bab = a^{-1}$, $Avt(Q, *) \cong D_4$, where D_4 is the dihedral group of order 8. The automorphism group of this loop has the order two and consists of the following elements: $\{(\varepsilon, \varepsilon, \varepsilon), ((0\,4)(1\,5), (0\,4)(1\,5), (0\,4)(1\,5))\}$.

Lemma 2.44. *If (β, α, γ) is an autotopy of an RIP-loop $(Q, \circ)$, then the element $\alpha 1 = a$ is a right Bol element.*

Proof. Any *RIP*-loop $(Q, \circ)$ is a $(1\,2)$-parastrophe of an *LIP*-loop $(Q, \cdot)$, i.e., $(Q, \circ) = (Q, \cdot)((1\,2), (\varepsilon, \varepsilon, \varepsilon))$. The connections between autotopies of isostrophic (in partial case, of parastrophic) quasigroups are given in Lemma 1.182.

Using this lemma we have that autotopy (β, α, γ) of loop $(Q, \circ)$ passes in the autotopy (α, β, γ) of loop $(Q, \cdot)$.

From Theorem 2.42 it follows that if (α, β, γ) is an autotopy of loop $(Q, \cdot)$, then the triple $(L_a R_a, L_a^{-1}, L_a)$, where $\alpha 1 = a$, is also an autotopy of loop $(Q, \cdot)$. Coming back from the loop $(Q, \cdot)$ to the loop $(Q, \circ)$ we find the form of the autotopy $(L_a R_a, L_a^{-1}, L_a)$ in the loop $(Q, \circ)$:

$$((1\,2), (\varepsilon, \varepsilon, \varepsilon))(\varepsilon, (L_a R_a, L_a^{-1}, L_a))((1\,2), (\varepsilon, \varepsilon, \varepsilon)) = (\varepsilon, (L_a^{-1}, L_a R_a, L_a)).$$

Using Table 1.1 we pass from translations of the loop $(Q, \cdot)$ to translations of loop $(Q, \circ)$ and obtain that triple $(R_a^{-1}, R_a L_a, R_a)$ is an autotopy of loop $(Q, \circ)$. Therefore the element a is a right Bol element of this loop. □

Theorem 2.45. *Let (α, β, γ) be an autotopy of an LIP-loop $(Q, \cdot)$ and let $\beta 1 = b$ be a left Bol element. Then there exists left pseudoautomorphism τ such that the following equality is true:*

$$(\alpha, \beta, \gamma) = (L_k \tau, \tau, L_k \tau)(L_b^{-1} R_b^{-1}, L_b, L_b^{-1}), \tag{2.23}$$

where $k = b \cdot ab$, $a = \alpha 1$ [328, Theorem 4.4], [327].

Proof. If element b is a left Bol element, then $(L_b R_b, L_b^{-1}, L_b)$ is an autotopy of loop $(Q, \cdot)$ (Lemma 1.224). Then $(L_b R_b, L_b^{-1}, L_b)(\alpha, \beta, \gamma) = (\alpha_1, \tau, \gamma_1)$ is also an autotopy, where $\tau = L_b^{-1}\beta$. Then $\tau 1 = L_b^{-1}\beta 1 = L_b^{-1} b = 1$.

If we put $y = 1$ in equality $\gamma_1(xy) = \alpha_1 x \cdot \tau y$, then $\gamma_1 = \alpha_1 = L_b R_b \alpha$, $L_b R_b \alpha(xy) = L_b R_b \alpha x \cdot \tau y$. If we put $x = 1$ in the last equality, then we obtain $L_b R_b \alpha y = (b \cdot ab) \cdot \tau y = L_k \tau y$, where $k = b \cdot ab$. Therefore we have the following equality $L_k \tau(xy) = L_k \tau x \cdot \tau y$. Then $(L_k \tau, \tau, L_k) \in Avt(Q, \cdot)$ and we obtain the equality (2.23). □

Corollary 2.46. *In a left Bol loop any autotopy has the form (2.23) [715].*

Proof. In any left Bol loop any element is a left Bol element. □

Lemma 2.47. *1. In IP-loop $(Q, \cdot)$ any left Bol element is a middle Moufang element and vice versa, any middle Moufang element is a left Bol element.*

2. In IP-loop $(Q, \cdot)$ any right Bol element is a middle Moufang element and vice versa, any middle Moufang element is a right Bol element.

3. In IP-loop $(Q, \cdot)$ any left Moufang element is a middle Moufang element and vice versa, any middle Moufang element is a left Moufang element.

4. In IP-loop $(Q, \cdot)$ any right Moufang element is a middle Moufang element and vice versa, any middle Moufang element is a right Moufang element.

Proof. Case 1. An element a is a left Bol element if and only if the triplet (L_aR_a, L_a^{-1}, L_a) is an autotopy of loop $(Q, \cdot)$ (Lemma 1.224).

Applying Corollary 2.39, Case 2, to this triple further we have that the triple $(L_a, IL_a^{-1}I, L_aR_a) = (L_a, R_a, L_aR_a)$ is also an autotopy of $(Q, \cdot)$. We recall that equality $IL_a^{-1}I = R_a$ is true by Corollary 2.35, Case 2. By Corollary 1.219 the element a is a middle Moufang element.

Converse. Applying Corollary 2.39, Case 2 to triplet (L_a, R_a, L_aR_a) we have that triple $(L_aR_a, IR_aI, L_a) = (L_aR_a, L_a^{-1}, L_a)$ is also an autotopy of $(Q, \cdot)$.

Case 2. An element a is a right Bol element if and only if the triplet (R_a^{-1}, R_aL_a, R_a) is an autotopy of loop $(Q, \cdot)$ (Lemma 1.224).

Applying Corollary 2.39, Case 1 to this triplet further we have that the triple $(IR_a^{-1}I, R_a, R_aL_a) = (L_a, R_a, R_aL_a)$ is also an autotopy of $(Q, \cdot)$. We recall that equality $IR_a^{-1}I = L_a$ is true by Corollary 2.35, Case 2. By Lemma 1.217 the element a is a middle Moufang element. The converse is also true.

Case 3. An element a is a left Moufang element if and only if the triple (R_aL_a, L_a^{-1}, L_a) is an autotopy of loop $(Q, \cdot)$ (Lemma 1.221).

Applying Corollary 2.39, Case 2 to this triplet further we have that the triple $(L_a, IL_a^{-1}I, R_aL_a) = (L_a, R_a, R_aL_a)$ is also an autotopy of $(Q, \cdot)$. We recall that equality $IL_a^{-1}I = R_a$ is true by Corollary 2.35, Case 2. By Lemma 1.217 the element a is a middle Moufang element.

Converse. Applying Corollary 2.39, Case 2 to triplet (L_a, R_a, R_aL_a) we have that triple $(R_aL_a, IR_aI, L_a) = (R_aL_a, L_a^{-1}, L_a)$ is also an autotopy of $(Q, \cdot)$.

The proof of Case 4 is similar to the proof of Case 2 and we omit it. □

Corollary 2.48. *For IP-loop $(Q, \cdot)$ the following statements are equivalent:*

1. *An element a is a left Bol element, the triple (L_aR_a, L_a^{-1}, L_a) is an autotopy of $(Q, \cdot)$.*
2. *An element a is a right Bol element, the triple (R_a^{-1}, R_aL_a, R_a) is an autotopy of $(Q, \cdot)$.*
3. *An element a is a middle Moufang element, the triples (L_a, R_a, R_aL_a) and (L_a, R_a, L_aR_a) are autotopies of $(Q, \cdot)$.*
4. *An element a is a left Moufang element, the triple (R_aL_a, L_a^{-1}, L_a) is an autotopy of $(Q, \cdot)$.*
5. *An element a is a right Moufang element, the triple (R_a^{-1}, L_aR_a, R_a) is an autotopy of $(Q, \cdot)$.*

Proof. The equivalence follows from Lemma 2.47. □

Theorem 2.49. *Any autotopy (α, β, γ) of IP-loop $(Q, \cdot)$ is decomposable in the product of two autotopies:*

$$(\alpha, \beta, \gamma) = (L_k\tau, \tau, L_k\tau)(L_b^{-1}R_b^{-1}, L_b, L_b^{-1}), \tag{2.24}$$

where $a = \alpha 1$, $b = \beta 1$, $k = b \cdot ab$, τ is a permutation of the set Q such that $\tau 1 = 1$ [72, 80].

Proof. Let (α, β, γ) be an autotopy of IP-loop $(Q, \cdot)$. From Lemma 2.44 and Corollary 2.48 it follows that the element $\beta 1 = b$ is a left Bol element. Therefore we can apply Theorem 2.45 which gives us the claimed result. From Corollary 1.164 it follows that $\tau 1 = 1$. □

Theorem 2.49 can also be formulated in the following form.

Theorem 2.50. *Any autotopy (α, β, γ) of an IP-loop is decomposable in the product of two autotopies:*

$$(\alpha, \beta, \gamma) = (L_b, R_b, L_bR_b)(\theta, R_c\theta, R_c\theta), \tag{2.25}$$

where element b is a middle Moufang element and θ is a right pseudo-automorphism with companion c [173], [80, Theorem 2.5].

Proof. The proof is similar to the proof of Theorem 2.49. □

We give a partial case of Theorem 8.5 from [328].

Theorem 2.51. *Any autotopy of IP-quasigroup $(Q, \cdot)$ which has at least one idempotent element d can be presented in the following form:*

$$(\alpha, \beta, \gamma) = (\tau, L_{\tau d}^{-1} R_k \tau R_d L_d, R_k \tau R_d)(L_a, R_a \lambda \rho, R_{e_a} L_a R_a), \tag{2.26}$$

where $\alpha d = a$, $\beta d = b$, $k = d(b\rho(a) \cdot d)$.

In [328] autotopies of IP-quasigroup with a non-empty distributant are studied.

2.2.3 Moufang and Bol elements in LIP-loops

Theorem 2.52. *If any loop, that is isotopic to an IP-quasigroup $(Q, \cdot)$, is an IP-loop, then $(Q, \cdot)$ satisfies the following identity $(xy \cdot z)y = x(y(e_y z \cdot y))$ [80, Theorem 2.2].*

Theorem 2.53. *If any loop, that is isotopic to a loop $(Q, \cdot)$, is an IP-loop, then $(Q, \cdot)$ is a Moufang loop [72].*

Theorem 2.54. *If any loop, that is isotopic to a loop $(Q, \cdot)$, is an LIP-loop, then $(Q, \cdot)$ is a left Bol loop [72].*

Notice, up to isomorphism it is sufficient to research only LP-isotopes of loop $(Q, \cdot)$.

Theorem 2.55. *In IP-loop $(Q, \cdot)$ the set M of all middle Moufang elements forms a Moufang subloop $(M, \cdot)$ of the loop $(Q, \cdot)$ [173].*

Proof. We follow [328]. If the element b is a middle Moufang element, then the triple $(L_b, R_b, L_b R_b)$ is an autotopy of loop $(Q, \cdot)$. Then the triple $(L_b, R_b, L_b R_b)^{-1}$ is also an autotopy. Using equality (1.57) and Theorem 2.34, Case 5 we have $(L_b, R_b, L_b R_b)^{-1} = (L_b^{-1}, R_b^{-1}, R_b^{-1} L_b^{-1}) = (L_b^{-1}, R_b^{-1}, L_b^{-1} R_b^{-1}) = (L_{b^{-1}}, R_{b^{-1}}, L_{b^{-1}} R_{b^{-1}})$, $b^{-1} \in M$.

Let $a, b \in M$. Taking into consideration Corollary 2.48 we have that the product

$$\begin{gathered}(R_a L_a, L_a^{-1}, L_a)(L_b, R_b, L_b R_b)(L_a^{-1} R_a^{-1}, L_a, L_a^{-1})(L_a, R_a, L_a R_a) = \\ (R_a L_a L_b L_a^{-1} R_a^{-1} L_a, L_a^{-1} R_b L_a R_a, L_a L_b R_b L_a^{-1} L_a R_a) = \\ (R_a L_a L_b R_a^{-1}, L_a^{-1} R_b L_a R_a, L_a L_b R_b R_a) = (\alpha, \beta, \gamma)\end{gathered} \tag{2.27}$$

is an autotopy of loop $(Q, \cdot)$. In (2.27) we have used the fact that $L_a R_a = R_a L_a$ for any middle Moufang element a.

Further we have:

$\alpha x = R_a L_a L_b R_a^{-1} x = R_a L_a L_b R_{a^{-1}} x = (a(b \cdot x a^{-1}))a \overset{(1.56)}{=} ab \cdot (x a^{-1})a \overset{RIP}{=} ab \cdot x = L_{ab} x$;

$\beta x = L_a^{-1} R_b L_a R_a x = a^{-1}((a \cdot xa)b) \overset{(1.60)}{=} a^{-1}(a(x \cdot ab)) \overset{LIP}{=} x \cdot ab = R_{ab} x$.

We find the third component of autotopy using the fact that any two components of autotopy define the third one in a unique way (Lemma 1.137). Therefore we have: $L_{ab} x \cdot R_{ab} y = \gamma(xy)$. From the last equality by $x = 1$ we have $L_{ab} R_{ab} y = \gamma y$. Then $(\alpha, \beta, \gamma) = (L_{ab}, R_{ab}, L_{ab} R_{ab})$, $a \cdot b \in M$.

In an IP-loop, equation $x \cdot a = b$ has a solution in the form $x = ba^{-1}$. From the proved above it follows: if $a, b \in M$, then $ba^{-1} \in M$. Similarly, equation $a \cdot y = b$ has the following solution $y = a^{-1} b$. If $a, b \in M$, then $a^{-1} b \in M$. In subloop $(M, \cdot)$ Moufang identities are true. □

Using an argument similar to the arguments used in proving Theorem 2.55 it is possible to prove the following:

Theorem 2.56. *If in a loop $(Q, \cdot)$ any middle Moufang element is a left Bol element, then the set of all middle Moufang elements forms subloop $(M, \cdot)$ of the loop $(Q, \cdot)$ [328, Theorem 4.6].*

Corollary 2.57. *If in a loop $(Q, \cdot)$ any middle Moufang element is a left Moufang element, then the set of all middle Moufang elements forms subloop $(M, \cdot)$ of the loop $(Q, \cdot)$.*

Proof. The proof follows from Corollary 2.48 and Theorem 2.56. □

Corollary 2.58. *In any LIP-loop $(Q, \cdot)$ middle Moufang nucleus M_m is a subloop of the loop $(Q, \cdot)$ [328, Corollary 4.3].*

If in a loop $(Q, \cdot)$ any left Moufang element is also a right Moufang element, then in this loop the set M_l (and M_r) forms subloop $(M_l, \cdot)$ of the loop $(Q, \cdot)$ [328, Corollary 4.4].

In any loop $(Q, \cdot)$ Moufang nucleus M is a subloop of the loop $(Q, \cdot)$ [691].

Proof. The proof follows from Corollary 2.57. □

Theorem 2.59. *Florja Theorem. In any LIP-loop $(Q, \cdot)$ nucleus N is a normal subloop of the middle Moufang nucleus M_m which is a subloop of the loop $(Q, \cdot)$ [328].*

LIP-loop $(Q, *)$ in which left Bol elements do not form a subloop [328, p. 31]:

*	1	2	3	4	5	6	7	8
1	1	2	3	4	5	6	7	8
2	2	1	6	7	8	3	4	5
3	3	5	1	6	2	4	8	7
4	4	7	8	1	6	5	2	3
5	5	8	4	3	1	7	6	2
6	6	3	2	8	7	1	5	4
7	7	4	5	2	3	8	1	6
8	8	7	6	5	4	2	3	1

This loop is not an *RIP*-loop because $(8 * 5) * 5 = 6$. In this loop the elements 2 and 5 are left Bol elements, but its product, the element $2 * 5 = 8$, is not a left Bol element.

AAIP-loop $(Q, \circ)$ in which middle Moufang elements do not form a subloop [691]:

∘	0	1	2	3	4	5	6	7	8	9	10	11
0	0	1	2	3	4	5	6	7	8	9	10	11
1	1	0	3	2	5	4	7	6	10	11	8	9
2	2	4	0	5	1	3	8	9	6	7	11	10
3	3	5	1	4	0	2	9	8	11	10	6	7
4	4	2	5	0	3	1	10	11	7	6	9	8
5	5	3	4	1	2	0	11	10	9	8	7	6
6	6	7	8	10	9	11	0	1	2	3	4	5
7	7	6	9	11	8	10	1	0	3	2	5	4
8	8	10	6	7	11	9	2	4	0	5	1	3
9	9	11	7	6	10	8	4	2	5	0	3	1
10	10	8	11	9	6	7	3	5	1	4	0	2
11	11	9	10	8	7	6	5	3	4	1	2	0

Elements 1 and 2 are middle Moufang elements, but their product $1 \circ 2 = 3$ is not a middle Moufang element.

2.2.4 Loops with the property $I_l = I_r$

We give information on loops with the property $I_l = I_r$. Recall, in any loop $I_l = I_r^{-1}$ (Lemma 1.93). It is not difficult to understand that commutative loops, LIP-loops, Bol loops, and Bruck loops lie in this loop class. The results about loops with this property are in [3], [72, Chapter XI]; results about groupoids with this property are given in [512, p. 22].

Lemma 2.60. *A loop $(Q, \cdot)$ has the property $I_l = I_r$ for all $x \in Q$ if and only if the following implication is true: if $x \cdot y = 1$, then $y \cdot x = 1$ for all $x, y \in Q$.*

Proof. If $x \cdot y = 1$, then $y = I_r x$, but $y = I_r x = I_l x$ and we obtain that $y \cdot x = 1$.

Converse. From relation $x \cdot y = 1$ it follows that $y = I_r x$ and from relation $y \cdot x = 1$ it follows that $y = I_l x$. □

Lemma 2.61. *The implication "if $(x \cdot y) \cdot z = 1$, then $x \cdot (y \cdot z) = 1$ for all $x, y, z \in Q$" defines WIP-loops in the class of all loops [72, p. 87].*

Proof. In a loop $(Q, \cdot)$ equality $(x{\cdot}y){\cdot}z = 1$ is equivalent to the following equality: $z = I(x{\cdot}y)$; equality $x{\cdot}(y{\cdot}z) = 1$ is equivalent to the following ones: $y{\cdot}z = Ix$. Therefore $y{\cdot}I(x{\cdot}y) = Ix$.

Converse. From equality $(x \cdot y) \cdot z = 1$ we have that $z = I(x \cdot y)$. Using the weak inverse property further we have that $y \cdot z = Ix$, where $z = I(x \cdot y)$, and finally we have got that $x \cdot (y \cdot z) = 1$. □

Therefore it is possible to call a WIP-loop an associative loop relative to the identity element.

From this point of view it is possible to call a loop with the property $I_l = I_r$ a commutative loop relative to the identity element of this loop.

We remember a definition of middle translation $P_a : P_a x = y \Leftrightarrow x \cdot y = a \Leftrightarrow x \cdot P_a x = a$. Then $P_a^{-1} x = y \Leftrightarrow y \cdot x = a \Leftrightarrow P_a^{-1} x \cdot x = a$.

Lemma 2.62. *A loop $(Q, \cdot)$ is a loop with the property $I_l x = I_r x$ if and only if $(P_1)^2 = \varepsilon$.*

Proof. We can write $I_l x \cdot x = 1$ as $P_1^{-1} x \cdot x = 1$, $P_1^{-1} x = I_l 1$, and relation $x \cdot I_r x = 1$ as $P_1 x = I_r x$. Condition $I_l x = I_r x$ we can re-write in the form $P_1 = P_1^{-1}$, i.e., $(P_1)^2 = \varepsilon$.

Converse. If $(P_1)^2 = \varepsilon$, then $P_1 = P_1^{-1}$, $I_l x = I_r x$. □

Corollary 2.63. *In the loop $(Q, \cdot)$ the following conditions are equivalent: (i) $I_l = I_r$; (ii) $I_l^2 = \varepsilon$; (iii) $I_r^2 = \varepsilon$.*

Proof. We can use Lemma 1.93. □

We summarize the results of this subsection in the following:

Theorem 2.64. *In a loop $(Q, \cdot)$ with the identity element 1 the following conditions are equivalent:*

(a) $(x^{-1})^{-1} = x$ *for all* $x \in Q$; (b) ${}^{-1}x = x^{-1}$ *for all* $x \in Q$;
(c) ${}^{-1}({}^{-1}x) = x$ *for all* $x \in Q$; (d) $(I_1)^2 = \varepsilon$;
(e) *in the* (13)*-parastrophe* $(Q, \circ)$ *of the loop* $(Q, \cdot)$ *for all* $x \in Q$ *the following conditions are fulfilled: (i)* $x \circ 1 = x$; *(ii)* $x \circ x = 1$; *(iii)* $1 \circ (1 \circ x) = x$;
(f) *the following conditions are fulfilled in the* (13)*-parastrophe* $(Q, \circ)$ *for all* $x \in Q$: *(i)* $R_1^{\circ} = \varepsilon$; *(ii)* $R_x^{\circ} x = 1$; *(iii)* $(L_1^{\circ})^2 = \varepsilon$;
(g) *such implication is true: if* $x \cdot y = 1$, *then* $y \cdot x = 1$ *for all* $x, y \in Q$.

2.3 Medial quasigroups

Taking into consideration the quite general nature of quasigroups it is clear that many quasigroup classes are defined and will be defined. But probably the most important of all quasigroup classes is the class of medial quasigroups.

2.3.1 Linear forms: Toyoda theorem

The Toyoda theorem determined ways of development of quasigroup theory and related systems for many years. This theorem was generalized by many authors in various directions. In this section some generalizations of the Toyoda theorem are also given.

The crucial Toyoda theorem [72, 80, 646, 890, 166] says that

Theorem 2.65. *Every medial quasigroup* $(Q, \cdot)$ *can be presented in the form:*

$$x \cdot y = \varphi x + \psi y + a, \tag{2.28}$$

where $(Q, +)$ *is an abelian group,* φ, ψ *are automorphisms of* $(Q, +)$ *such that* $\varphi\psi = \psi\varphi$, $x, y \in Q$, *and* a *is some fixed element from the set* Q.

Expression $\varphi x + \psi y + a$ we shall name the *linear form* of quasigroup $(Q, \cdot)$.

In view of the Toyoda theorem the theory of medial quasigroups is very close to the theory of abelian groups but it is not exactly the theory of abelian groups. For example, the easily proved fact for abelian groups, that every simple abelian group is finite, was proved for medial quasigroups only in 1977 [589].

Medial quasigroups as well as other classes of quasigroups isotopic to groups give us a possibility to construct quasigroups with preassigned properties. Often these properties can be expressed in the language of properties of groups and components of isotopy.

We recall, any quasiautomorphism γ of a group $(Q, +)$ has the form $L_a^+\beta$, where $a \in Q$, $\beta \in Aut(Q, +)$ (Corollary 1.150). See also [72], [80, p. 24]. Obviously $\beta 0 = 0$, where, as usual, 0 denotes the identity element of $(Q, +)$.

There exist many various proofs of the Toyoda theorem. We give not the shortest, but one of the most known and elementary proofs of this theorem. Notice, in [830] conditions when a quasigroup is isotopic to a group, to an abelian group, conditions of left, right linearity of group isotopes are given. On the basis of this information, the Toyoda theorem is proved in a very short and elegant manner.

Medial quasigroups (as well as any other quasigroup class) can be divided into 2 classes: 1) quasigroups that have one or more idempotent elements; 2) quasigroups that have no idempotent elements.

Theorem 2.66. *Conditions (i) and (ii) are equivalent:*

(i) $(Q, \cdot)$ *is a medial quasigroup that has idempotent element* 0;

(ii) there exists an abelian group $(Q, +)$ *with the identity element* 0 *and two commuting automorphisms* α, β *such that* $x \cdot y = \alpha x + \beta y + a$ *for all* $x, y \in Q$, *where* $-a \in (\alpha + \beta - \varepsilon)Q$.

Proof. (i) $\Longrightarrow$ (ii). LP-isotope $(R_0^{-1}, L_0^{-1}, \varepsilon)$ of quasigroup $(Q, \cdot)$ is a loop $(Q, +)$ with the identity element $0 \cdot 0 = 0$, i.e., $x + y = R_0^{-1}x \cdot L_0^{-1}y$ [72]. Then $x \cdot y = R_0 x + L_0 y, R_0 0 = 0, L_0 0 = 0$. Let $R_0 = \alpha, L_0 = \beta$. Therefore $x \cdot y = \alpha x + \beta y$. So we can rewrite the medial identity in terms of the operation $+$ in the following way:

$$\alpha(\alpha x + \beta y) + \beta(\alpha u + \beta v) = \alpha(\alpha x + \beta u) + \beta(\alpha y + \beta v). \tag{2.29}$$

If we take $x = u = v = 0$ in (2.29), then we obtain $\alpha\beta y = \beta\alpha y$, i.e.,

$$\alpha\beta = \beta\alpha. \tag{2.30}$$

By $u = v = 0$ in (2.29) we have $\alpha(\alpha x + \beta y) = \alpha^2 x + \beta\alpha y =^{(2.30)} \alpha^2 x + \alpha\beta y$. Therefore $\alpha \in Aut(Q, +)$.

If we substitute $x = y = 0$ in (2.29), then $\beta(\alpha u + \beta v) = \alpha\beta u + \beta^2 v =^{(2.30)} \beta\alpha u + \beta^2 v$, $\beta \in Aut(Q, +)$.

By $x = v = 0$ the equality (2.29) takes the form $\alpha\beta y + \beta\alpha u = \alpha\beta u + \beta\alpha y$. Since $\alpha\beta = \beta\alpha$, we have $\alpha\beta y + \alpha\beta u = \alpha\beta u + \alpha\beta y$. Therefore $(Q, +)$ is a commutative loop.

Let $v = 0$ in the relation (2.29). Since $\alpha, \beta \in Aut(Q, +), \alpha\beta = \beta\alpha$, further we obtain $(\alpha^2 x + \alpha\beta y) + \alpha\beta u = (\alpha^2 x + \alpha\beta u) + \alpha\beta y$.

Then we have $(\alpha\beta y + \alpha^2 x) + \alpha\beta u = \alpha\beta y + (\alpha^2 x + \alpha\beta u)$, since $(Q, +)$ is a commutative loop. From the last equality we have that $(Q, +)$ is associative. Therefore $(Q, +)$ is an abelian group. It is easy to see that $0 \in (\alpha + \beta - \varepsilon)Q$.

(ii) $\Longrightarrow$ (i).

If conditions (ii) are fulfilled, then it is easy to check that medial identity holds. Indeed, $\alpha(\alpha x + \beta y + a) + \beta(\alpha u + \beta v + a) + a = \alpha(\alpha x + \beta u + a) + \beta(\alpha y + \beta v + a) + a$, $\alpha^2 x + \alpha\beta y + \alpha a + \beta\alpha u + \beta^2 v + \beta a + a = \alpha^2 x + \alpha\beta u + \alpha a + \beta\alpha y + \beta^2 v + \beta a + a$, $\alpha\beta y + \alpha\beta u = \alpha\beta u + \alpha\beta y$, $0 = 0$.

A quasigroup of this kind has at least one idempotent element. Indeed, let $-a = \alpha d + \beta d - d$, i.e., $\alpha d + \beta d = d - a$. Then $d \cdot d = \alpha d + \beta d + a = d - a + a = d$. □

From the proof of Theorem 2.66 we have the following:

Corollary 2.67. *Any medial quasigroup $(Q, \cdot)$ with an idempotent element 0 can be presented in the form: $x \cdot y = \alpha x + \beta y$, where $(Q, +)$ is an abelian group with the identity element 0 and α, β are commuting automorphisms of the group $(Q, +)$.*

Remark 2.68. Equivalence of conditions (i) and (ii) of Theorem 2.66 is possible to deduce from the results of book [462] (3.1.4. Proposition).

Lemma 2.69. *A medial quasigroup $(Q, \circ)$ of the form $x \circ y = \alpha x + \beta y$ over an abelian group $(Q, +)$ is:*

1. *idempotent, distributive, if $\alpha + \beta = \varepsilon$;*

2. *unipotent, if $\alpha + \beta = 0$;*

3. *isotope of medial idempotent quasigroup $(Q, \cdot)$ of the form $x \cdot y = (\alpha + \beta)^{-1}(\alpha x + \beta y)$, if $\alpha + \beta \neq \varepsilon$, $\alpha + \beta \neq 0$.*

Proof. 1. It is easy to check.

2. If we suppose, that $(\alpha + \beta)x = 0$ for all $x \in Q$, where 0 denotes the zero element of the group $(Q, +)$, then $x \circ x = \alpha x + \beta x = (\alpha + \beta)x = 0$ for all $x \in Q$.

3. If $\alpha + \beta \neq 0$, $\alpha + \beta \neq \varepsilon$, then there exists an element μ of the group $Aut(Q, +)$ such that $\mu(\alpha + \beta) = \varepsilon$, i.e., $\mu = (\alpha + \beta)^{-1}$. Therefore, $x \cdot x = (\alpha + \beta)^{-1}(\alpha x + \beta x) = (\alpha + \beta)^{-1}(\alpha + \beta)x = x$ for all $x \in Q$.

   Quasigroup $(Q, \cdot)$ is medial ([506], Theorem 25). We repeat the proof of Theorem 25: since $\mu\alpha + \mu\beta = \varepsilon$, we have $\mu\alpha\mu\beta = \mu\alpha(\varepsilon - \mu\alpha) = \mu\alpha - (\mu\alpha)^2 = (\varepsilon - \mu\alpha)\mu\alpha = \mu\beta\mu\alpha$. □

Theorem 2.70. *Conditions (i) and (ii) are equivalent:*

(i) $(Q, \cdot)$ *is a medial quasigroup without idempotent elements;*

(ii) there exists an abelian group $(Q, +)$*, its automorphisms* α*,* φ*,* $\alpha\varphi = \varphi\alpha$*, an element* $a \in Q$*,* $-a \notin (\alpha + \varphi - \varepsilon)Q$ *such that* $x \cdot y = \alpha x + \varphi y + a$ *for all* $x, y \in Q$*.*

Proof. (i) $\Longrightarrow$ (ii). By proving this implication we follow in general the book [80]. Let us consider an LP-isotope $(Q, +)$ of a medial quasigroup $(Q, \cdot)$ of the form: $x+y = R^{-1}_{r(0)}x \cdot L^{-1}_0 y$, where $0 \cdot r(0) = 0$, i.e., $r(0)$ is a right local identity element of the element 0. This LP-isotope $(Q, +)$ is a loop with the identity element $0 \cdot r(0) = 0$. Denote $R_{r(0)}$ by α and L_0 by β. We remark that $R_{r(0)}0 = 0$, then $\alpha 0 = 0$.

Using our notations we can write the medial identity in the following form:

$$\alpha(\alpha x + \beta y) + \beta(\alpha u + \beta v) = \alpha(\alpha x + \beta u) + \beta(\alpha y + \beta v). \tag{2.31}$$

By $x = 0, y = \beta^{-1}0$ from (2.31) we have

$$\beta(\alpha u + \beta v) = \alpha\beta u + \beta(\alpha\beta^{-1}0 + \beta v). \tag{2.32}$$

Therefore the permutation β is a quasiautomorphism of the loop $(Q, +)$.

By $u = 0, v = \beta^{-2}0$ in (2.31) we have

$$\alpha(\alpha x + \beta y) = \alpha(\alpha x + \beta 0) + \beta(\alpha y + \beta^{-1}0) \tag{2.33}$$

and we obtain that the permutation α is a quasiautomorphism of the loop $(Q, +)$.

If we use equalities (2.32) and (2.33) in (2.31), then we have

$$\begin{aligned}(\alpha R_{\beta 0}\alpha x + \beta R_{\beta^{-1}0}\alpha y) + (\alpha\beta u + \beta L_{\alpha\beta^{-1}0}\beta v) = \\ (\alpha R_{\beta 0}\alpha x + \beta R_{\beta^{-1}0}\alpha u) + (\alpha\beta y + \beta L_{\alpha\beta^{-1}0}\beta v).\end{aligned} \tag{2.34}$$

If we change in equality (2.34) the element x to the element $\alpha^{-1}R^{-1}_{\beta 0}\alpha^{-1}x$, the element y to $\alpha^{-1}R^{-1}_{\beta^{-1}0}\beta^{-1}y$, the element u to the element $\beta^{-1}\alpha^{-1}u$, and the element v to the element $\beta^{-1}L^{-1}_{\alpha\beta^{-1}0}\beta^{-1}v$, then we have

$$(x + y) + (u + v) = (x + \beta R_{\beta^{-1}0}\alpha\beta^{-1}\alpha^{-1}u) + (\alpha\beta\alpha^{-1}R^{-1}_{\beta^{-1}0}\beta^{-1}y + v).$$

If we take $u = 0$ in the last equality, then we have

$$(x + y) + v = (x + \beta R_{\beta^{-1}0}\alpha\beta^{-1}0) + (\alpha\beta\alpha^{-1}R^{-1}_{\beta^{-1}0}\beta^{-1}y + v). \tag{2.35}$$

If we take $v = 0$ in (2.35), then we obtain

$$x + y = (x + r) + \alpha\beta\alpha^{-1}R^{-1}_{\beta^{-1}0}\beta^{-1}y,$$

where $r = \beta R_{\beta^{-1}0}\alpha\beta^{-1}0$ is a fixed element of the set Q.

If we change $x + y$ in equality (2.35) by the right side of the last equality, then we have

$$((x + r) + \alpha\beta\alpha^{-1}R^{-1}_{\beta^{-1}0}\beta^{-1}y) + v = (x + r) + (\alpha\beta\alpha^{-1}R^{-1}_{\beta^{-1}0}\beta^{-1}y + v).$$

From the last equality it follows that the loop $(Q, +)$ is associative, i.e., is a group.

Since α is a quasiautomorphism of the group and $\alpha 0 = 0$, we have that the permutation α is an automorphism of the group $(Q, +)$. The permutation β has the form $\beta = R_a\varphi$, where $\varphi \in Aut(Q, +)$.

Then we can rewrite the medial identity in the form $\alpha^2x+\alpha\varphi y+\alpha a+\varphi\alpha u+\varphi^2v+\varphi a+a = \alpha^2x+\alpha\varphi u+\alpha a+\varphi\alpha y+\varphi^2v+\varphi a+a$ and, after cancellation in the last equality, we obtain

$$\alpha\varphi y + \alpha a + \varphi\alpha u = \alpha\varphi u + \alpha a + \varphi\alpha y. \tag{2.36}$$

From the last equality by $u = 0$ we have $\alpha\varphi y + \alpha a = \alpha a + \varphi\alpha y$ and by $y = 0$ we have $\alpha a + \varphi\alpha u = \alpha\varphi u + \alpha a$. Using these last equalities we can rewrite equality (2.36) in the form $\alpha a + \varphi\alpha y + \varphi\alpha u = \alpha a + \varphi\alpha u + \varphi\alpha y$. Hence $\varphi\alpha y + \varphi\alpha u = \varphi\alpha u + \varphi\alpha y$, $(Q, +)$ is an abelian group.

Then from equality $\alpha\varphi y + \alpha a = \alpha a + \varphi\alpha y$ it follows that $\alpha\varphi y = \varphi\alpha y$. Therefore $x \cdot y = \alpha x + \varphi y + a$, where $(Q, +)$ is an abelian group and α, φ are automorphisms of $(Q, +)$ such that $\alpha\varphi = \varphi\alpha$.

Now we must only demonstrate that the element $-a \notin (\alpha + \varphi - \varepsilon)Q$. Let us suppose the inverse. Let medial quasigroup $(Q, \cdot)$ have an idempotent element, for example, let $u \cdot u = u$. Then $\alpha u + \varphi u + a = u$, therefore $-a = \alpha u + \varphi u - u = (\alpha + \varphi - \varepsilon)u$, hence $-a \in (\alpha + \varphi - \varepsilon)Q$. We received a contradiction. Our assumption is not true. Hence, if medial quasigroup $(Q, \cdot)$ has no idempotent element, then $-a \notin (\alpha + \varphi - \varepsilon)Q$.

(ii) $\Longrightarrow$ (i). If conditions (ii) are true, then it is easy to check that medial identity holds. Indeed, $\alpha(\alpha x + \varphi y + a) + \varphi(\alpha u + \varphi v + a) + a = \alpha(\alpha x + \varphi u + a) + \varphi(\alpha y + \varphi v + a) + a$, $\alpha^2x+\alpha\varphi y+\alpha a+\varphi\alpha u+\varphi^2v+\varphi a+a = \alpha^2x+\alpha\varphi u+\alpha a+\varphi\alpha y+\varphi^2v+\varphi a+a$, $\alpha\varphi y+\alpha\varphi u = \alpha\varphi u + \alpha\varphi y$, $0 = 0$.

If $-a \notin (\alpha + \varphi - \varepsilon)Q$, then the quasigroup $(Q, \cdot)$ has no idempotent element. Indeed, if the medial quasigroup has an idempotent element u, then $\alpha u + \varphi u + a = u$, therefore $-a = \alpha u + \varphi u - u = (\alpha + \varphi - \varepsilon)u$, hence $-a \in (\alpha + \varphi - \varepsilon)Q$. We received a contradiction.

Hence, if $-a \notin (\alpha + \varphi - \varepsilon)Q$, then the quasigroup $(Q, \cdot)$ has no idempotent element. □

Remark 2.71. From the Toyoda theorem it follows that *any* medial quasigroup is LP-isotopic to an abelian group. From the Albert theorem it follows that for any medial quasigroup the corresponding abelian group is unique up to an isomorphism. A similar conclusion follows from Theorems 2.66 and 2.70.

Remark 2.72. Medial identity is the partial case of the following functional equation of bisymmetry

$$(x\varphi_1 y)\varphi_2(z\varphi_3 t) = (x\varphi_4 z)\varphi_5(y\varphi_6 t), \tag{2.37}$$

where $\varphi_1, \dots, \varphi_6$ are the symbols of binary operations. This equation is solved in the case when $\varphi_1, \dots, \varphi_6$ are quasigroup operations. See [4, 68, 640] for details. This equation is key to many "medial" problems.

The functional equation of the form

$$(x\varphi_1 y)\varphi_2(z\varphi_1 t) = (x\varphi_2 z)\varphi_1(y\varphi_2 t) \tag{2.38}$$

is researched in [66]. Equation (2.38) is true, if, for example, φ_1 is an operation of a medial quasigroup (Q, φ_1) and φ_2 is an operation of any parastrophe of (Q, φ_1) [66, Theorem 3].

Results about medial groupoids are presented in many articles and books. We suggest the Jezek and Kepka book [462]. Sholander showed that every medial cancellation groupoid can be embedded into a medial quasigroup [803]. The dual assertion that every medial division groupoid is a homomorphic image of a medial quasigroup is proved in [459].

2.3.2 Direct decompositions: Murdoch theorem

The following theorem on the structure of finite medial quasigroups was proved by D.C. Murdoch. We give the Murdoch theorem in a slightly modernized form [784]. This theorem is a starting point for a few similar theorems.

For a quasigroup $(Q, \cdot)$ we define the map s: $s(x) = x \cdot x$ for all $x \in Q$. As usual, $s^2(x) = s(s(x))$ and so on. For any medial quasigroup $(Q, \cdot)$ the map s is an endomorphism of this quasigroup, indeed, $s(xy) = xy \cdot xy = xx \cdot yy = s(x) \cdot s(y)$.

Define in a finite medial quasigroup $(Q, \cdot)$ the following chain

$$Q \supset s^1(Q) \supset s^2(Q) \supset \dots \supset s^m(Q) \supset \dots \tag{2.39}$$

When chain (2.39) becomes stable, that there exists a finite number m such that $s^m(Q) = s^{m+1}(Q) = s^{m+2}(Q) \dots$. In this case we shall say that endomorphism s has the order m.

A map h of a non-empty set Q into itself is called a *zero map*, if $|h(Q)| = 1$.

Lemma 2.73. *The endomorphism s of a medial quasigroup $(Q, \cdot)$ is a zero endomorphism if and only if $(Q, \cdot)$ is a unipotent quasigroup, $x \cdot y = \varphi x - \varphi y + g$, $(Q, +)$ is an abelian group, and $\varphi \in Aut(Q, +)$.*

Proof. $(\Rightarrow)$ By the Toyoda theorem (Theorem 2.65) there exists an abelian group $(Q, +)$ and two of its automorphisms φ and ψ such that $x \cdot y = \varphi x + \psi y + g$. From conditions of this lemma, $s(0) = g$. Since the map s is a zero endomorphism of $(Q, \cdot)$, we have $s(x) = g$ for all $x \in Q$. Thus $s(x) = \varphi x + \psi x + g = g$, $\varphi x + \psi x = 0$ for all $x \in Q$, $\psi = -\varphi$.

$(\Leftarrow)$ In the quasigroup $(Q, \cdot)$ with the form $x \cdot y = \varphi x - \varphi y + g$ we have $s^{\cdot} x = x \cdot x = \varphi x - \varphi x + g = g$ for all $x \in Q$. Then the map $s^{\cdot}$ is a zero antiendomorphism of the quasigroup $(Q, \cdot)$. □

Corollary 2.74. *If a medial quasigroup $(Q, \cdot)$ with the form $x \cdot y = \varphi x - \varphi y + g$ has zero endomorphism s, then quasigroup $(Q, \cdot)$ is isomorphic to quasigroup $(Q, \circ)$ with the form $x \circ y = \varphi x - \varphi y$.*

Proof. Let $x \circ y = L_{-g}(L_g x \cdot L_g y)$, $x, y \in Q$. It is clear that $(Q, \circ) \cong (Q, \cdot)$. Quasigroup $(Q, \circ)$ has the following form

$$x \circ y = -g + \varphi(g + x) - \varphi(g + y) + g = \varphi g + \varphi x - \varphi g - \varphi y = \varphi x - \varphi y.$$

□

Theorem 2.75. *If the endomorphism s of a medial quasigroup $(Q, \cdot)$ with the form $x \cdot y = \varphi x + \psi y + g$ is a permutation of the set Q, then*

(i) the map $\varphi + \psi$ is an automorphism of $(Q, +)$;

(ii) quasigroup $(Q, \circ)$ of the form $x \circ y = s^{-1}(x \cdot y) = (\varphi + \psi)^{-1}\varphi x + (\varphi + \psi)^{-1}\psi y$ is a distributive medial quasigroup;

(iii) the map $s^{\cdot}$ is an automorphism of quasigroup $(Q, \circ)$.

Proof. (i) By the Toyoda theorem (Theorem 2.65), $x \cdot y = \varphi x + \psi y + g$. Then $s(x) = \varphi x + \psi x + g$, $s = L_g(\varphi + \psi)$. Since the map s is a permutation of the set Q, L_g is a permutation of the set Q, then the map $\varphi + \psi$ is a permutation and an endomorphism simultaneously, i.e., it is an automorphism of the group $(Q, +)$, $s^{-1} = (\varphi + \psi)^{-1} L_{-g}$.

(ii) Thus

$$x \circ y = s^{-1}(\varphi x + \psi y + g) = (\varphi + \psi)^{-1}(\varphi x + \psi y + g - g) =$$
$$(\varphi + \psi)^{-1}\varphi x + (\varphi + \psi)^{-1}\psi y.$$

Notice, $x \circ x = (\varphi+\psi)^{-1}\varphi x + (\varphi+\psi)^{-1}\psi x = (\varphi+\psi)^{-1}(\varphi+\psi)x = \varepsilon x = x$. Then $(Q, \circ)$ is an idempotent T-quasigroup.

It is well known that any idempotent T-quasigroup is a medial distributive quasigroup [667, 506]. Indeed, it is clear that $(\varphi+\psi)^{-1}\varphi+(\varphi+\psi)^{-1}\psi = \varepsilon$. Denote expression $(\varphi+\psi)^{-1}\varphi$ by α and expression $(\varphi+\psi)^{-1}\psi$ by β. If $\alpha = \varepsilon - \beta$, then $\alpha\beta = (\varepsilon - \beta)\beta = \beta - \beta^2 = \beta(\varepsilon - \beta) = \beta\alpha$. Therefore by Theorem 2.65, $(Q, \circ)$ is a medial quasigroup. Further we have $xx \cdot yz = x \cdot yz = xy \cdot xz$, $xy \cdot zz = xy \cdot z = xz \cdot yz$.

(iii) Prove that the map s is an automorphism of $(Q, \circ)$, i.e., $s(x \circ y) = s(x) \circ s(y)$. We have

$$s(x \circ y) = s(s^{-1}(x \cdot y)) = \varphi x + \psi y + g \tag{2.40}$$

$$\begin{aligned} & s(x) \circ s(y) = (L_g(\varphi + \psi)x) \circ (L_g(\varphi + \psi)y) = \\ & (\varphi + \psi)^{-1}\varphi(L_g(\varphi + \psi)x) + (\varphi + \psi)^{-1}\psi(L_g(\varphi + \psi)y) = \\ & (\varphi + \psi)^{-1}\varphi(g + \varphi x + \psi x) + (\varphi + \psi)^{-1}\psi(g + \varphi y + \psi y) = \\ & (\varphi + \psi)^{-1}(\varphi g + \varphi^2 x + \varphi\psi x) + (\varphi + \psi)^{-1}(\psi g + \psi\varphi y + \psi^2 y) = \\ & (\varphi + \psi)^{-1}(\varphi g + \varphi^2 x + \varphi\psi x + \psi g + \psi\varphi y + \psi^2 y) \overset{(\varphi\psi = \varphi\psi)}{=} \\ & (\varphi + \psi)^{-1}((\varphi + \psi)g + (\varphi + \psi)\varphi x + (\varphi + \psi)\psi y) = \\ & (\varphi + \psi)^{-1}(\varphi + \psi)(\varphi x + \psi y + g) = \varphi x + \psi y + g. \end{aligned} \tag{2.41}$$

Since the right sides of equalities (2.44) and (2.45) are equal we obtain that $s^{\cdot}(x \circ y) = s^{\cdot}(x) \circ s^{\cdot}(y)$. □

Theorem 2.76. *1. Any finite medial quasigroup $(Q, \cdot)$ has the following structure*

$$(Q, \cdot) \cong (A, \circ) \times (B, \cdot),$$

where $(A, \circ)$ is a quasigroup with a unique idempotent element; $(B, \cdot)$ is an isotope of a distributive medial quasigroup $(B, \star)$, $x \cdot y = s(x \star y)$, $x \star y = \varphi x + \psi y$, $s = L_g^+\alpha$, $g \in B$, $\alpha \in Aut(B, +)$, $s = L_g^+\alpha \in$, $\varphi\alpha\varphi = \psi\alpha\psi$.

Proof. The proof of this theorem mainly repeats the proofs of the similar theorems [784, Theorem 6], [789, 796].

If the map s is a permutation of the set Q, then by Theorem 2.75 quasigroup $(Q, \cdot)$ is an isotope of a distributive quasigroup.

If $s(Q) = k$, where k is a fixed element of the set Q, then quasigroup $(Q, \cdot)$ is a unipotent quasigroup, $(Q, \cdot) \cong (Q, \circ)$, where $x \circ y = \varphi x - \varphi y$, $(Q, +)$ is an abelian group, $\varphi \in Aut(Q, +)$ (Lemma 2.73).

We suppose that $s^m = s^{m+1}$, where $m \geqslant 1$. From Corollary 1.298 it follows that $s^m(Q, \cdot) = (B, \cdot)$ is a subquasigroup of quasigroup $(Q, \cdot)$. It is clear that $(B, \cdot)$ is a medial quasigroup in which the map $\overline{s} = s|_{s^m(Q)}$ is a permutation of the set $B \subset Q$. In other words $s(B) = B$.

Define binary relation δ on quasigroup $(Q, \cdot)$ by the following rule: $x\delta y$ if and only if $s^m(x) = s^m(y)$. By Theorem 1.279 the relation δ is a congruence of quasigroup $(Q, \cdot)$. From Lemma 1.272 it follows that in a finite quasigroup any congruence is normal. Therefore, δ is a normal congruence.

It is known that any subquasigroup of a T-quasigroup is normal ([506, Theorem 43]). Then the subquasigroup $(B, \cdot)$ of quasigroup $(Q, \cdot)$ is normal. By Remark 1.289 this subquasigroup defines exactly one normal congruence. Denote this congruence by the letter ρ.

It is known ([72, pages 56–57], [80, 685]) that any coset $\rho(a)$ of a normal congruence ρ of a quasigroup $(Q,\cdot)$ can be presented in the form $a \cdot B$, where B is a normal subquasigroup of quasigroup $(Q,\cdot)$ and $B = \rho(b)$ for some $b \in Q$. Indeed, we can take into consideration the following equalities $a \cdot \rho(b) = \rho(a \cdot b) = \rho(a) \cdot b$ that are true for any normal congruence ρ of a quasigroup $(Q,\cdot)$.

Taking into consideration Remark 1.289 we can say that any normal subquasigroup B of a quasigroup Q defines in a unique way normal congruence ρ by the rule: $x\rho y$ if and only if $B \cdot x = B \cdot y$, i.e., for any $b_1 \in B$ there exists exactly one element $b_2 \in B$ such that $b_1 \cdot x = b_2 \cdot y$ and vice versa, for any $b_2 \in B$ there exists exactly one element $b_1 \in B$ such that $b_1 \cdot x = b_2 \cdot y$.

Prove that $\delta \cap \rho = \hat{Q} = \{(x,x) | \, \forall\, x \in Q\}$. From reflexivity of relations δ, ρ it follows that $\delta \cap \rho \supseteq \hat{Q}$.

Let $(x,y) \in \delta \cap \rho$, i.e., let $x\ \delta\ y$ and $x\ \rho\ y$, where $x, y \in Q$. Using the definitions of relations δ, ρ we have $s^m(x) = s^m(y)$ and $(B,\cdot) \cdot x = (B,\cdot) \cdot y$. Then there exist $a, b \in B$ such that $a \cdot x = b \cdot y$. Applying the map s^m to both sides of the last equality we obtain $s^m(a) \cdot s^m(x) = s^m(b) \cdot s^m(y)$, $s^m(a) = s^m(b)$, $a = b$, since the map $s^m|_B$ is a permutation of the set B. If $a = b$, then from equality $a \cdot x = b \cdot y$ we obtain $x = y$.

We recall that in finite case all congruences are normal and permutable (Lemmas 1.272 and 1.312).

From the Definition 1.314 it follows that in order to prove the existence of the direct decomposition of quasigroup $(Q,\cdot)$ we should prove that $\delta \circ \rho = Q \times Q$.

Let a, c be any fixed elements of the set Q. We will prove the equality if it is shown that there exists an element $y \in Q$ such that $a\delta y$ and $y\rho c$.

From the definition of congruence δ we have that condition $a\delta y$ is equivalent to equality $s^m(a) = s^m(y)$. From the definition of congruence ρ it follows that condition $y\rho c$ is equivalent to the following condition: $y \in \rho(c) = B \cdot c$.

We prove equality $\delta \circ \rho = Q \times Q$, if we demonstrate that there exists an element $y \in B \cdot c$ such that $s^m(a) = s^m(y)$. Such element y exists since $s^m(B \cdot c) = s^m(B) \cdot s^m(c) = B = s^m(Q)$, if the map s^m is an endomorphism of quasigroup $(Q,\cdot)$.

Therefore $\rho \circ \delta = Q \times Q = \delta \circ \rho$, $\delta \cap \rho = \hat{Q}$ and we can use Theorem 7.36. Now we can say that quasigroup $(Q,\cdot)$ is isomorphic to the direct product of a quasigroup $(Q,\cdot)/\delta \cong (B,\cdot)$ (Theorem 1.279) and a quasigroup $(Q,\cdot)/\rho \cong (A,\circ)$.

The medial identity is true in quasigroup $(B,\cdot)$ since $(B,\cdot) \subseteq (Q,\cdot)$. If the quasigroups $(Q,\cdot)$ and $(B,\cdot)$ are medial quasigroups, $(Q,\cdot) \cong (A,\circ) \times (B,\cdot)$, then $(A,\circ)$ also is a medial quasigroup (Lemma 1.317).

Prove that the quasigroup $(A,\circ) \cong (Q,\cdot)/(B,\cdot)$, where $s^m(Q,\cdot) = (B,\cdot)$, has a unique idempotent element.

We can identify elements of quasigroup $(Q,\cdot)/(B,\cdot)$ with cosets of the form $B \cdot c$, where $c \in Q$.

From properties of quasigroup $(A,\circ)$ we have that $s^m(A) = a$, where the element a is a fixed element of the set A that corresponds to the coset class B.

Further, taking into consideration the properties of endomorphism s of the quasigroup $(A,\circ)$, we obtain $s^{m+1}A = s(s^m A) = s(a) = a$. Therefore $s(a) = a$, i.e., the element a is an idempotent element of quasigroup $(A,\circ)$.

Prove that there exists exactly one idempotent element in quasigroup $(A,\circ)$. Suppose that there exists an element c of the set A such that $c \circ c = c$, i.e., such that $s(c) = c$. Then we have $s^m(c) = c = a$, since $s^m(A) = a$.

The fact that $(B,\cdot)$ is an isotope of a distributive quasigroup $(B,\star)$ follows from Theorem 2.75. □

It is clear that Theorem 2.76 reduces study of the structure of finite medial quasigroups to the study of the structure of finite medial unipotent and idempotent quasigroups. Notice, the quasigroup $(A, \circ)$ is unipotently solvable.

We notice that for any unipotent quasigroup $(Q, \cdot)$ with idempotent element e, we have $s(Q) = e$, and for any idempotent quasigroup $(Q, \cdot)$ we have $s = \varepsilon$. Therefore, in these cases we cannot say anything about the structure of medial unipotent and medial idempotent quasigroup using the endomorphism s.

Remark 2.77. In Theorem 2.76 the fact that the map s is an endomorphism of a medial quasigroup is used. It is possible to prove analogs of Theorem 2.76 using the fact that in any medial quasigroup $(Q, \cdot)$ the maps e and f $(x \cdot e(x) = x, f(x) \cdot x = x$ for all $x \in Q)$ are endomorphisms of quasigroup $(Q, \cdot)$ [646]. See Theorem 6.27 for details. We recall that any medial quasigroup is an F-quasigroup.

2.3.3 Simple quasigroups

As has been mentioned above, simple medial quasigroups were described by J. Ježek and T. Kepka in [589]. Ježek and Kepka proved the very important fact that any simple medial quasigroup is finite. We give the Ježek-Kepka theorem in the following form [783].

Theorem 2.78. *If a medial quasigroup $(Q, \cdot)$ of the form $x \cdot y = \alpha x + \beta y + a$ over an abelian group $(Q, +)$ is simple, then*

1. *the group $(Q, +)$ is the additive group of a finite Galois field $GF(p^k)$;*

2. *the group $\langle \alpha, \beta \rangle$ is the multiplicative group of the field $GF(p^k)$ in the case $k > 1$ and the group $\langle \alpha, \beta \rangle$ is any subgroup of the group $Aut(Z_p, +)$ in the case $k = 1$;*

3. *the quasigroup $(Q, \cdot)$ in the case $|Q| > 1$ can be a quasigroup from one of the following disjoint quasigroup classes:*

   (a) *$\alpha + \beta = \varepsilon, a = 0$; in this case the quasigroup $(Q, \cdot)$ is an idempotent quasigroup;*

   (b) *$\alpha + \beta = \varepsilon$ and $a \neq 0$; in this case the quasigroup $(Q, \cdot)$ does not have any idempotent element, and the quasigroup $(Q, \cdot)$ is isomorphic to the quasigroup $(Q, *)$ with the form $x * y = \alpha x + \beta y + 1$ over the same abelian group $(Q, +)$;*

   (c) *$\alpha + \beta \neq \varepsilon$; in this case the quasigroup $(Q, \cdot)$ has exactly one idempotent element, and the quasigroup $(Q, \cdot)$ is isomorphic to the quasigroup $(Q, \circ)$ of the form $x \circ y = \alpha x + \beta y$ over the group $(Q, +)$.*

Exercise 2.79. List all simple medial quasigroups over Galois field $GF(2^2)$.

2.3.4 Examples

It is well known that the direct product of medial idempotent quasigroups is an idempotent quasigroup and a similar situation takes place for unipotent quasigroups.

Lemma 2.80. *If $(Q, \cdot)$ is a medial quasigroup such that $(Q, \cdot) = (Q_1, \cdot_1) \times (Q_2, \cdot_2)$ and the forms of quasigroups $(Q, \cdot)$, $(Q_1, \cdot_1)$ and $(Q_2, \cdot_2)$ are defined over the groups $(Q, +)$, $(Q_1, +_1)$ and $(Q_2, +_2)$, respectively, then*

$$(Q, +) \cong (Q_1, +_1) \times (Q_2, +_2).$$

Proof. It is not very difficult to prove this lemma independently. See also [784]. □

Example 2.81. There exist directly irreducible finite idempotent medial quasigroups and finite unipotent medial quasigroups.

Proof. We denote by $(Z_9, +)$ the additive group of residues modulo 9. The quasigroup $(Z_9, \circ)$ of the form $x \circ y = 2 \cdot x + 8 \cdot y$ is a medial idempotent quasigroup, and quasigroup $(Z_9, *)$ of the form $x * y = 1 \cdot x + 8 \cdot y$ is a medial unipotent quasigroup.

These quasigroups are not simple. Indeed, if $Q = \{0, 3, 6\}$, then $(Q, \circ) \lhd (Z_9, \circ)$ and $(Q, *) \lhd (Z_9, *)$.

These quasigroups are directly irreducible. If we suppose that these quasigroups are directly reducible, then by Lemma 2.69 the group $(Z_9, +)$ is reducible into the direct product of subgroups of order 3. As is well known [471], it is not true. □

Taking into consideration Lemma 2.80 we can say that in a simple medial quasigroup $(Q, \cdot)$ its only subquasigroups are one-element subquasigroups and the quasigroup $(Q, \cdot)$.

Remark 2.82. We notice that in general there exist non-simple medial quasigroups with only trivial subquasigroups. For example, the quasigroup $(Z_9, \diamond)$ with the form $x \diamond y = 2 \cdot x + 8 \cdot y + 1$, where $(Z_9, +)$ is the additive group of residues modulo 9, is a non-simple quasigroup without proper subquasigroups.

But the situation is better for medial idempotent and medial unipotent quasigroups, since these quasigroups contain idempotent elements.

It is known ([72, p. 57], [80, p. 41]) that if θ is a normal congruence of a quasigroup $(Q, \cdot)$ and there exists an idempotent element e of the quasigroup $(Q, \cdot)$, then the equivalence class $\theta(e)$ forms a normal subquasigroup $(\theta(e), \cdot)$ of the quasigroup $(Q, \cdot)$.

We can summarize our remarks in the following:

Lemma 2.83. *In an idempotent medial quasigroup or in a unipotent medial quasigroup $(Q, \cdot)$, any normal congruence θ contains at least one equivalence class $\theta(e)$ such that $(\theta(e), \cdot)$ is a normal subquasigroup of the quasigroup $(Q, \cdot)$.*

Proof. Idempotent quasigroups and unipotent quasigroups contain idempotent elements. □

Definition 2.84. We shall say that a quasigroup $(Q, \cdot)$ is *solvable* if there exists the following finite chain of quasigroups

$$Q/Q_1, Q_1/Q_2, \dots, Q_m/Q_{m+1},$$

where the quasigroup Q_{i+1} is a maximal normal subquasigroup of the quasigroup Q_i and m is the minimal number such that $|Q_m/Q_{m+1}| = 1$.

Remark 2.85. Definition 2.84 differs from the definition of solvability of groups [471].

Lemma 2.86. *Any finite medial idempotent quasigroup $(Q, \cdot)$ is solvable and any quasigroup Q_i/Q_{i+1} is a finite simple medial idempotent quasigroup.*

Proof. The proof follows from Lemma 2.83 and the fact that the quasigroup $(Q, \cdot)$ is finite. □

Lemma 2.87. *Any finite medial unipotent quasigroup $(Q, \cdot)$ is solvable and any quasigroup Q_i/Q_{i+1} is a finite simple medial unipotent quasigroup.*

Proof. The proof is similar to the proof of Lemma 2.86. □

Theorem 2.88. *Any finite medial quasigroup $(Q, \cdot)$ is isomorphic to the direct product of a medial unipotently solvable quasigroup $(Q_1, \circ)$ and a principal isotope of a medial idempotent quasigroup $(Q_2, *)$, where the quasigroups $(Q_i, \circ)/(Q_{i+1}, \circ)$ and $(Q_2, *)$ are solvable for all admissible values of index i, $\gamma \in Aut(Q_2, *)$.*

Lemma 2.89. *A quasigroup $(Q, \cdot)$ of the form $x \cdot y = \alpha x + \beta y$ is isomorphic to a quasigroup $(Q, *)$ of the form $x * y = \gamma x + \delta y$, where $\alpha, \beta, \gamma, \delta$ are automorphisms of an abelian group $(Q, +)$, if and only if there exists an automorphism ψ of the group $(Q, +)$ such that $\psi\alpha = \gamma\psi$, $\psi\beta = \delta\psi$.*

Proof. The proof of this lemma, in fact, repeats the proof of the similar theorem from [760] and we omit it. □

It is easy to see that Lemma 2.89 is true for medial idempotent quasigroups and for medial unipotent quasigroups.

Example 2.90. We denote by $(Z_{16}, +)$ the additive group of residues modulo 16. The quasigroup $(Z_{16}, \circ)$ of the form $x \circ y = 3 \cdot x + 15 \cdot y + 1$ is isomorphic to the quasigroup $(Q, *)$ of the form $x * y = 3 \cdot x + 15 \cdot y$.

This follows from Theorem 2.66. Furthermore, the quasigroup $(Q, *)$ is a quasigroup with the unique idempotent element 0, the quasigroup $(Q, *)$ is a unipotently solvable quasigroup of degree 4, since $s^4(Q) = s^5(Q)$.

Example 2.91. Let $(Z_{16}, +)$ be the additive group of residues modulo 16. The quasigroup $(Z_{16}, *)$ of the form $x * y = 1 \cdot x + 15 \cdot y$ is a solvable unipotent quasigroup of degree 3.

A number of non-isomorphic medial quasigroups, mainly of prime power order, is studied in [832, 523, 841].

2.4 Paramedial quasigroups

2.4.1 Kepka-Nemec theorem

The following identity

$$xy \cdot uv = vy \cdot ux \tag{2.42}$$

is called the *paramedial* identity. Properties of quasigroups with paramedial identity (paramedial quasigroups) are similar to the properties of medial quasigroups.

Theorem 2.92. *(Kepka-Nemec theorem [667]). Every paramedial quasigroup $(Q, \cdot)$ can be presented in the form:*

$$x \cdot y = \varphi x + \psi y + g, \tag{2.43}$$

where $(Q, +)$ is an abelian group, $x, y \in Q$, φ, ψ are automorphisms of $(Q, +)$ such that $\varphi\varphi = \psi\psi$, and g is some fixed element of the set Q.

Remark 2.93. In [667] Kepka and Nemec proved much more general theorem than Theorem 2.92.

Example 2.94. We can define a paramedial quasigroup using a finite ring $(R, +, \cdot)$ of residues modulo n. In this case the automorphisms φ and ψ correspond to pairs of numbers k, l of the ring $(R, +, \cdot)$ such that $gcd(k, n) = gcd(l, n) = 1$ and $k^2 \equiv l^2 \pmod{n}$.

For a quasigroup $(Q, \cdot)$ we define the map s: $s(x) = x \cdot x$ for all $x \in Q$. As usual, $s^2(x) = s(s(x))$ and so on.

2.4.2 Antiendomorphisms

Notice that the fact that a quasigroup has an endomorphism often plays a determining role by the study of the structure of this quasigroup [646, 784, 789, 788]. In this section we apply an endomorphic (or, more exactly, antiendomorphic) approach to the study of finite paramedial quasigroups.

Lemma 2.95. *For any paramedial quasigroup $(Q, \cdot)$ the map s is an antiendomorphism of this quasigroup [667, 506].*

Proof. Indeed, $s(xy) = xy \cdot xy = yy \cdot xx = s(y) \cdot s(x)$. □

Corollary 2.96. *For any paramedial quasigroup $(Q, \cdot)$ the map s^2 is an endomorphism of this quasigroup.*

Proof. Indeed, $s^2(xy) = s(s(xy)) = s(xy \cdot xy) = s(yy \cdot xx) = s(s(y) \cdot s(x)) = s^2(x) \cdot s^2(y)$. □

Corollary 2.97. *For any paramedial quasigroup $(Q, \cdot)$ the map s^{2n+1}, $n \in \mathbb{N}$, is an antiendomorphism of this quasigroup, and for any paramedial quasigroup $(Q, \cdot)$ the map s^{2n}, $n \in \mathbb{N}$, is an endomorphism of this quasigroup.*

Proof. This follows from Lemma 2.95 and Corollary 2.96. □

Theorem 2.98. *The antiendomorphism s of a paramedial quasigroup $(Q, \cdot)$ is a zero antiendomorphism if and only if $(Q, \cdot)$ is a unipotent quasigroup, $x \cdot y = \varphi x - \varphi y + g$, $(Q, +)$ is an abelian group, $\varphi \in Aut(Q, +)$.*

Proof. ($\Rightarrow$) By the Kepka-Nemec theorem (Theorem 2.92) there exists an abelian group $(Q, +)$ and two of its automorphisms φ and ψ such that $x \cdot y = \varphi x + \psi y + g$. From conditions of the theorem, $s(0) = g$. Since the map s is a zero antiendomorphism of $(Q, \cdot)$ we have $s(x) = g$ for all $x \in Q$. Thus $s(x) = \varphi x + \psi x + g = g$, $\varphi x + \psi x = 0$ for all $x \in Q$, $\psi = -\varphi$.

($\Leftarrow$) In the quasigroup $(Q, \cdot)$ with the form $x \cdot y = \varphi x - \varphi y + g$ we have $s^{\cdot} x = x \cdot x = \varphi x - \varphi x + g = g$ for all $x \in Q$. Then the map $s^{\cdot}$ is a zero antiendomorphism of the quasigroup $(Q, \cdot)$. □

Corollary 2.99. *If a paramedial quasigroup $(Q, \cdot)$ with the form $x \cdot y = \varphi x - \varphi y + g$ has zero antiendomorphism s, then quasigroup $(Q, \cdot)$ is isomorphic to quasigroup $(Q, \circ)$ with the form $x \circ y = \varphi x - \varphi y$.*

Proof. Let $x \circ y = L_{-g}(L_g x \cdot L_g y)$, $x, y \in Q$. It is clear that $(Q, \circ) \cong (Q, \cdot)$. Quasigroup $(Q, \circ)$ has the following form

$$x \circ y = -g + \varphi(g + x) - \varphi(g + y) + g = \varphi g + \varphi x - \varphi g - \varphi y = \varphi x - \varphi y.$$

□

Theorem 2.100. *The antiendomorphism s of a paramedial quasigroup $(Q, \cdot)$ is the identity permutation of the set Q if and only if $(Q, \cdot)$ is a medial distributive commutative quasigroup with the form $x \cdot y = \varphi x + \varphi y$, where $(Q, +)$ is an abelian group, $\varphi \in Aut(Q, +)$.*

Proof. ($\Rightarrow$) By the Kepka-Nemec theorem (Theorem 2.92) there exists an abelian group $(Q, +)$ and two of its automorphisms φ and ψ such that $x \cdot y = \varphi x + \psi y + g$. Since $s = \varepsilon$, we have $s(0) = \varphi 0 + \psi 0 + g = 0$, $g = 0$.

Therefore $s^{\cdot}(x) = \varphi x + \psi x = x$ for all $x \in Q$, $\psi = \varepsilon - \varphi$. Then $\varphi^2 = (\varepsilon - \varphi)^2$, $\varphi^2 = \varepsilon - \varphi - \varphi + \varphi^2$, $\varphi + \varphi = \varepsilon$. But $\varphi + \psi = \varepsilon$. Therefore $\varphi = \psi$, $x \cdot y = \varphi x + \varphi y$, and quasigroup $(Q, \cdot)$ is a medial distributive commutative quasigroup.

($\Leftarrow$) In a medial distributive commutative quasigroup $(Q, \cdot)$ the map s is the identity antiendomorphism. □

Lemma 2.101. *Let $(Q, \circ)$ be a distributive medial quasigroup with the form $x \circ y = \varphi x + \psi y$, $\alpha \in Aut(Q, +)$. A quasigroup $(Q, \cdot)$ of the form $x \cdot y = L_a^+ \alpha(x \circ y) = \alpha\varphi x + \alpha\psi y + a$ is a paramedial quasigroup if and only if $\psi = \alpha^{-1}\varphi\alpha$.*

Proof. Since quasigroup $(Q, \cdot)$ is a paramedial quasigroup, we have $\alpha\varphi\alpha\varphi = \alpha\psi\alpha\psi$, $\varphi\alpha\varphi = \psi\alpha\psi$. But $\psi = \varepsilon - \varphi$, since $(Q, \circ)$ is a medial distributive quasigroup. Then $\varphi\alpha\varphi = (\varepsilon - \varphi)\alpha(\varepsilon - \varphi)$, $\varphi\alpha\varphi = \alpha - \alpha\varphi - \varphi\alpha + \varphi\alpha\varphi$, $\alpha = \alpha\varphi + \varphi\alpha$, $\varepsilon = \varphi + \alpha^{-1}\varphi\alpha$. But $\varepsilon = \varphi + \psi$. Therefore $\psi = \alpha^{-1}\varphi\alpha$, $\alpha\psi = \varphi\alpha$.

It is easy to check that the converse also is true. □

Theorem 2.102. *If the antiendomorphism s of a paramedial quasigroup $(Q, \cdot)$ with the form $x \cdot y = \varphi x + \psi y + g$ is a permutation of the set Q, then*

(i) the map $\varphi + \psi$ is an automorphism of $(Q, +)$;

(ii) quasigroup $(Q, \circ)$ of the form $x \circ y = s^{-1}(x \cdot y) = (\varphi + \psi)^{-1}\varphi x + (\varphi + \psi)^{-1}\psi y$ is a distributive medial quasigroup;

(iii) the map s is an antiautomorphism of quasigroup $(Q, \circ)$.

Proof. (i) By the Kepka-Nemec theorem $x \cdot y = \varphi x + \psi y + g$. Then $s(x) = \varphi x + \psi x + g$, $s = L_g(\varphi + \psi)$. Since the map s is a permutation of the set Q, L_g is a permutation of the set Q, then the map $\varphi + \psi$ is an automorphism of the group $(Q, +)$, $s^{-1} = (\varphi + \psi)^{-1}L_{-g}$.

(ii) Thus

$$x \circ y = s^{-1}(\varphi x + \psi y + g) = (\varphi + \psi)^{-1}(\varphi x + \psi y + g - g) = (\varphi + \psi)^{-1}\varphi x + (\varphi + \psi)^{-1}\psi y.$$

Notice, $x \circ x = (\varphi + \psi)^{-1}\varphi x + (\varphi + \psi)^{-1}\psi x = (\varphi + \psi)^{-1}(\varphi + \psi)x = \varepsilon x = x$. Then $(Q, \circ)$ is an idempotent T-quasigroup. It is well known that any idempotent T-quasigroup is a medial distributive quasigroup [667, 506].

(iii) Prove that the map s is an antiautomorphism of $(Q, \circ)$, i.e., $s(y \circ x) = s(x) \circ s(y)$. We have

$$s(y \circ x) = s(s^{-1}(y \cdot x)) = \varphi y + \psi x + g. \tag{2.44}$$

$$\begin{aligned} s(x) \circ s(y) &= (L_g(\varphi + \psi)x) \circ (L_g(\varphi + \psi)y) = \\ &(\varphi + \psi)^{-1}\varphi(L_g(\varphi + \psi)x) + (\varphi + \psi)^{-1}\psi(L_g(\varphi + \psi)y) = \\ &(\varphi + \psi)^{-1}\varphi(g + \varphi x + \psi x) + (\varphi + \psi)^{-1}\psi(g + \varphi y + \psi y) = \\ &(\varphi + \psi)^{-1}(\varphi g + \varphi^2 x + \varphi\psi x) + (\varphi + \psi)^{-1}(\psi g + \psi\varphi y + \psi^2 y) = \\ &(\varphi + \psi)^{-1}(\varphi g + \varphi^2 x + \varphi\psi x + \psi g + \psi\varphi y + \psi^2 y) \overset{(\varphi^2 = \psi^2)}{=} \\ &(\varphi + \psi)^{-1}(\varphi g + \psi^2 x + \varphi\psi x + \psi g + \psi\varphi y + \varphi^2 y) = \\ &(\varphi + \psi)^{-1}((\varphi + \psi)g + (\varphi + \psi)\psi x + (\varphi + \psi)\varphi y) = \\ &(\varphi + \psi)^{-1}(\varphi + \psi)(\psi x + \varphi y + g) = \varphi y + \psi x + g. \end{aligned} \tag{2.45}$$

Since the right sides of equalities (2.44) and (2.45) are equal we obtain that $s(y \circ x) = s(x) \circ s(y)$. □

Remark 2.103. Detailed information on the structure of medial paramedial quasigroups in which the map s is a permutation, is in [506] (Lemma 22, Theorem 23).

Corollary 2.104. *Quasigroup $(Q, \circ)$ from Theorem 2.102 is paramedial if and only if $\varphi(\varphi + \psi)^{-1}\varphi = \psi(\varphi + \psi)^{-1}\psi$.*

Proof. The quasigroup $(Q, \circ)$ is paramedial if and only if $(\varphi + \psi)^{-1}\varphi(\varphi + \psi)^{-1}\varphi = (\varphi + \psi)^{-1}\psi(\varphi+\psi)^{-1}\psi$. The last equality is equivalent with the following: $(\varphi+\psi)(\varphi+\psi)^{-1}\varphi(\varphi+\psi)^{-1}\varphi = (\varphi + \psi)(\varphi + \psi)^{-1}\psi(\varphi + \psi)^{-1}\psi$, $\varphi(\varphi + \psi)^{-1}\varphi = \psi(\varphi + \psi)^{-1}\psi$. □

2.4.3 Direct decomposition

In this section we prove an analog of the Murdoch theorem [646] on the structure of finite medial quasigroups. See also [784, 789]. Direct decompositions of some paramedial quasigroups and abelian groups, which are connected with these quasigroups, are studied in [506].

Define in a finite paramedial quasigroup $(Q, \cdot)$ the following chain

$$Q \supset s^1(Q) \supset s^2(Q) \supset \dots \supset s^m(Q) \supset \dots \qquad (2.46)$$

When chain (6.48) becomes stable, there exists a finite number m such that $s^m(Q) = s^{m+1}(Q) = s^{m+2}(Q) \dots$. In this case we say that antiendomorphism s has the order m.

Taking into consideration Corollary 1.298 we can say that such a chain exists in any paramedial quasigroup. It is clear that chain (6.48) becomes stable in any finite paramedial quasigroup.

Theorem 2.105. *1. Any finite paramedial quasigroup $(Q, \cdot)$ has the following structure*

$$(Q, \cdot) \cong (A, \circ) \times (B, \cdot),$$

where $(A, \circ)$ is a quasigroup with a unique idempotent element; $(B, \cdot)$ is an isotope of a distributive medial quasigroup $(B, \star)$ $x \cdot y = s(x \star y)$, $x \star y = \varphi x + \psi y$, $s = L_a^+\alpha$, $a \in B$, $\alpha \in Aut(B, +)$, $\varphi\alpha\varphi = \psi\alpha\psi$.

Proof. The proof of this theorem in fact repeats the proof of Theorem 2.76.

If the map s is a permutation of the set Q, then by Theorem 2.102 quasigroup $(Q, \cdot)$ is an isotope of a distributive quasigroup.

If $s(Q) = k$, where k is a fixed element of the set Q, then quasigroup $(Q, \cdot)$ is a unipotent quasigroup, $(Q, \cdot) \cong (Q, \circ)$, where $x \circ y = \varphi x - \varphi y$, $(Q, +)$ is an abelian group, $\varphi \in Aut(Q, +)$ (Theorem 2.98).

We suppose that $s^m = s^{m+1}$, where $m \geqslant 1$. From Corollary 1.298 it follows that $s^m(Q, \cdot) = (B, \cdot)$ is a subquasigroup of quasigroup $(Q, \cdot)$. It is clear that $(B, \cdot)$ is a paramedial quasigroup in which the map $\overline{s} = s|_{s^m(Q)}$ is a permutation of the set $B \subset Q$. In other words $s(B) = B$.

Define the binary relation δ on quasigroup $(Q, \cdot)$ by the following rule: $x\delta y$ if and only if $s^m(x) = s^m(y)$.

By Lemma 1.297 (if s^m is an antiendomorphism) or by Theorem 1.279 (if s^m is an endomorphism) the relation δ is a congruence of quasigroup $(Q, \cdot)$.

From Lemma 1.272 it follows that in a finite quasigroup any congruence is normal. Therefore, δ is a normal congruence.

It is known that any subquasigroup of a T-quasigroup is normal ([506], Theorem 43). Then the subquasigroup $(B, \cdot)$ of quasigroup $(Q, \cdot)$ is normal. By Remark 1.289 this subquasigroup defines exactly one normal congruence. Denote this congruence by the letter ρ.

It is known ([72, pages 56–57], [80, 685]) that any coset $\rho(a)$ of a normal congruence ρ of a quasigroup $(Q, \cdot)$ can be presented in the form $a \cdot B$, where B is a normal subquasigroup of quasigroup $(Q, \cdot)$ and $B = \rho(b)$ for some $b \in Q$. Indeed, we can take into consideration the following equalities $a \cdot \rho(b) = \rho(a \cdot b) = \rho(a) \cdot b$, which are true for any normal congruence ρ of a quasigroup $(Q, \cdot)$.

Taking into consideration Remark 1.289 we can say that any normal subquasigroup B of a quasigroup Q defines in a unique way normal congruence ρ by the rule: $x\rho y$ if and only if $B \cdot x = B \cdot y$, i.e., for any $b_1 \in B$ there exists exactly one element $b_2 \in B$ such that $b_1 \cdot x = b_2 \cdot y$ and vice versa, for any $b_2 \in B$ there exists exactly one element $b_1 \in B$ such that $b_1 \cdot x = b_2 \cdot y$.

Prove that $\delta \cap \rho = \hat{Q} = \{(x, x) | \forall x \in Q\}$. From reflexivity of relations δ, ρ it follows that $\delta \cap \rho \supseteq \hat{Q}$.

Let $(x, y) \in \delta \cap \rho$, i.e., let $x\ \delta\ y$ and $x\ \rho\ y$, where $x, y \in Q$. Using the definitions of relations δ, ρ we have $s^m(x) = s^m(y)$ and $(B, \cdot) \cdot x = (B, \cdot) \cdot y$. Then there exist $a, b \in B$ such that $a \cdot x = b \cdot y$. Applying the map s^m to both sides of the last equality we obtain $s^m(a) \cdot s^m(x) = s^m(b) \cdot s^m(y)$, $s^m(a) = s^m(b)$, $a = b$, since the map $s^m|_B$ is a permutation of the set B. If $a = b$, then from equality $a \cdot x = b \cdot y$ we obtain $x = y$.

We notice that in the finite case all congruences are normal and permutable (Lemmas 1.272 and 1.312).

From Definition 1.314 it follows that in order to prove the existence of the direct decomposition of quasigroup $(Q, \cdot)$ we should prove that $\delta \circ \rho = Q \times Q$.

Let a, c be any fixed elements of the set Q. We will prove the equality if it is shown that there exists an element $y \in Q$ such that $a\delta y$ and $y\rho c$.

From the definition of congruence δ we have that condition $a\delta y$ is equivalent to equality $s^m(a) = s^m(y)$. From the definition of congruence ρ it follows that condition $y\rho c$ is equivalent to the following condition: $y \in \rho(c) = B \cdot c$.

We prove equality $\delta \circ \rho = Q \times Q$, if we demonstrate that there exists an element $y \in B \cdot c$ such that $s^m(a) = s^m(y)$. Such element y exists since $s^m(B \cdot c) = s^m(B) \cdot s^m(c) = B = s^m(Q)$, if the map s^m is an endomorphism of quasigroup $(Q, \cdot)$.

If the map s^m is an antiendomorphism of quasigroup $(Q, \cdot)$, then we have $s^m(B \cdot c) = s^m(c) \cdot s^m(B) = B = s^m(Q)$. Thus the element y exists and in this case, too.

Therefore $\rho \circ \delta = Q \times Q = \delta \circ \rho$, $\delta \cap \rho = \hat{Q}$ and we can use Theorem 7.36. Now we can say that quasigroup $(Q, \cdot)$ is isomorphic to the direct product of a quasigroup $(Q, \cdot)/\delta \cong (B, \cdot)$ (Theorem 1.279) and a quasigroup $(Q, \cdot)/\rho \cong (A, \circ)$.

Paramedial identity is true in quasigroup $(B, \cdot)$ since $(B, \cdot) \subseteq (Q, \cdot)$. If the quasigroups $(Q, \cdot)$ and $(B, \cdot)$ are paramedial quasigroups, $(Q, \cdot) \cong (A, \circ) \times (B, \cdot)$, then $(A, \circ)$ is also a paramedial quasigroup (Lemma 1.317).

Prove that the quasigroup $(A, \circ) \cong (Q, \cdot)/(B, \cdot)$, where $s^m(Q, \cdot) = (B, \cdot)$, has a unique idempotent element.

We can identify elements of quasigroup $(Q, \cdot)/(B, \cdot)$ with cosets of the form $B \cdot c$, where $c \in Q$.

From properties of quasigroup $(A, \circ)$ we have that $s^m(A) = a$, where the element a is a fixed element of the set A that corresponds to the coset class B.

Further, taking into consideration the properties of endomorphism s of the quasigroup $(A, \circ)$, we obtain $s^{m+1}A = s(s^m A) = s(a) = a$. Therefore $s(a) = a$, i.e., the element a is an idempotent element of quasigroup $(A, \circ)$.

Prove that there exists exactly one idempotent element in quasigroup $(A, \circ)$. Suppose that there exists an element c of the set A such that $c \circ c = c$, i.e., such that $s(c) = c$. Then we have $s^m(c) = c = a$, since $s^m(A) = a$.

The fact that $(B, \cdot)$ is an isotope of a distributive quasigroup $(B, \star)$, follows from Theorem 2.102. □

Taking into consideration Theorem 2.105 we can formulate an analog of Theorem 2.88 for finite paramedial quasigroups.

Theorem 2.106. *Any finite paramedial quasigroup* $(Q,\cdot)$ *is isomorphic to the direct product of a paramedial unipotently solvable quasigroup* $(Q_1,\circ)$ *with a unique idempotent element and a quasigroup* $(Q_2,*)$*, where* $(Q_2,*)$ *is an isotope of the form* $(\varepsilon,\varepsilon,\gamma)$ *of a medial distributive quasigroup* $(Q_2,\star)$*,* $x*y=\gamma(x\star y)$*,* $x\star y=\varphi x+\psi y$*,* $(Q_2,+)$ *is abelian group corresponding to quasigroup* $(Q_2,\star)$*,* $\gamma=L_a^+\alpha$*,* $a\in Q_2$*,* $\alpha\in Aut(Q_2,+)$*,* $\varphi\alpha\varphi=\psi\alpha\psi$*.*

2.4.4 Simple paramedial quasigroups

Here we give a description of finite simple paramedial quasigroups.

We recall that any antihomomorphism of a quasigroup defines a congruence (Lemma 1.297) which is normal in the finite case (Lemma 1.272).

Then a necessary condition of the simplicity of a finite paramedial quasigroup $(Q,\cdot)$ is the condition that the antiendomorphism s is a permutation of the set Q (i.e., antiendomorphism with zero kernel) or it is a zero antiendomorphism.

This condition is not sufficient. For example, in any distributive quasigroup the map s is a permutation [789, 788], but it is easy to construct a non-simple distributive quasigroup. Finite simple T-quasigroups are researched in [767, 769].

We shall use the following:

Theorem 2.107. *Let* $(Q,\cdot)$ *be a finite T-quasigroup with the form* $x\cdot y=\varphi x+\psi y+a$*, where* $(Q,+)$ *is an abelian group. Then the quasigroup* $(Q,\cdot)$ *is simple if and only if:* $(Q,+)\cong\oplus_{i=1}^{n}(Z_p)_i$ *for some prime p; the group* $\langle\varphi,\psi\rangle$ *is an irreducible two-generated subgroup of the group* $GL(n,p)$ *in the case* $n\geqslant 2$*.*

Theorem 2.107 is true when $(Q,+)$ is a finite-generated commutative Moufang loop ([769], [767, Theorem 2]). Theorem 7.81 is an n-ary analog of Theorem 2.107.

Theorem 2.108. *Finite paramedial quasigroup* $(Q,\cdot)$ *with the form* $x\cdot y=\varphi x+\psi y+c$ *over an abelian group* $(Q,+)$ *is simple if and only if*

1. $(Q,+)\cong\oplus_{i=1}^{n}(Z_p)_i$*;*

2. *the group* $\langle\varphi,\psi\rangle$ *is an irreducible subgroup of the group* $GL(n,p)$ *if* $n\geqslant 2$*, and the group* $\langle\varphi,\psi\rangle$ *is any subgroup of the group* $Aut(Z_p,+)$ *if* $k=1$*;*

3. *the quasigroup* $(Q,\cdot)$ *if* $|\,Q\,|>1$ *can be a quasigroup from one of the following disjoint quasigroup classes:*

   (a) *the map s is a zero antiendomorphism; in this case* $\psi=-\varphi$*, quasigroup* $(Q,\cdot)$ *is a medial unipotent quasigroup, and quasigroup* $(Q,\cdot)$ *is isomorphic to quasigroup* $(Q,\circ)$ *with the form* $x\circ y=\varphi x-\varphi y$ *over the group* $(Q,+)$*;*

   (b) *the map s is an identity permutation; in this case* $c=0$*,* $\varphi=\psi,\varphi+\varphi=\varepsilon$*,* $(Q,\cdot)$ *is a paramedial medial commutative distributive quasigroup;*

   (c) *the map s is a non-identity permutation; quasigroup* $(Q,\circ)$ *of the form* $x\circ y=(\varphi+\psi)^{-1}\varphi x+(\varphi+\psi)^{-1}\psi y$ *is a* $(\varphi+\psi)$*-simple medial distributive quasigroup,* $(\varphi+\psi)\in Aut(Q,+)$*.*

Proof. ($\Rightarrow$) Cases 1 and 2 follow from Theorem 2.107.

From Lemma 1.297 it follows that if a paramedial quasigroup $(Q,\cdot)$ is simple, then in this case the map s is a permutation of the set Q, or it is a zero antiendomorphism.

Case 3, (a) follows from Theorem 2.98.

Case 3, (b) follows from Theorem 2.100.

Case 3, (c) follows from Theorem 2.102.

($\Leftarrow$) Any quasigroup mentioned in Case 3 with the properties mentioned in Cases 1 and 2 is simple and paramedial. $\square$

The problem "Classify the finite simple paramedial quasigroups" is proposed by Jaroslav Jezek and Tomas Kepka at Loops '03, Prague 2003 [907].

2.4.5 Quasigroups of order 4

We shall use the following:

Remark 2.109. If a map f of a quasigroup $(Q,\cdot)$ has the form $L_a\xi$, where ξ is an antiendomorphism of quasigroup $(Q,\cdot)$, then $\xi L_a x = \xi(a\cdot x) = \xi x\cdot \xi a = R_{\xi a}\xi x$. Thus $f^2 = L_a R_{\xi a}\xi^2$ and so on. Therefore, the map f^k is a zero map if and only if the map ξ^k is a zero map.

There are two abelian groups of order 4: additive group Z_4 of residues modulo 4 and elementary abelian 2-group $Z_2\oplus Z_2$.

Let $Z_4 = \{0,1,2,3\}$. Then $Aut\, Z_4 = \{\varepsilon, I\}$, where $I = (13)$. Notice, often the automorphism I is denoted by the sign "–".

The following triplets define paramedial quasigroups of order 4 over the group Z_4:

1) Case $\varphi = \psi$. We have such triplets $(\varepsilon,\varepsilon,\overline{0,3}), (I,I,\overline{0,3})$. Here expression $\overline{0,3}$ denotes the set of integers $\{0,1,2,3\}$.

It is clear that any triplet from the first series defines the group Z_4.

Any quasigroup from the second series is a medial paramedial quasigroup in which the antiendomorphism s has the form $L_i^+\xi$, where $i \in \{0,1,2,3\}$ the map ξ is a multiplication of any quasigroup element x by the number $-2 \equiv 2 \pmod 4$.

Using Remark 2.109 we obtain that any quasigroup from the second series is an unipotently solvable quasigroup of degree 2.

2) Case $\varphi^2 = \psi^2$ and $\varphi \neq \psi$. We have the following cases: $(\varepsilon, I,\overline{0,3}), (I,\varepsilon,\overline{0,3})$. Any quasigroup from the last triplets has zero antiendomorphism s since $I \equiv -$. Therefore in this case we can use Corollary 2.99.

We denote elements of the group $Z_2\oplus Z_2$ as follows: $\{(0;0),(1;0),(0;1),(1;1)\}$. Then the group $Aut(Z_2\oplus Z_2)$ consists of the following automorphisms

$$\begin{pmatrix}1&0\\0&1\end{pmatrix},\begin{pmatrix}1&0\\1&1\end{pmatrix},\begin{pmatrix}1&1\\0&1\end{pmatrix},\begin{pmatrix}0&1\\1&0\end{pmatrix},\begin{pmatrix}1&1\\1&0\end{pmatrix},\begin{pmatrix}0&1\\1&1\end{pmatrix}.$$

Denote these automorphisms in the following way: ε, φ_2, φ_3, φ_4, φ_5, φ_6, respectively.

It is known that $Aut(Z_2\oplus Z_2) \cong S_3$ [407, 471].

1) Case $\varphi = \psi$ is obvious and we omit it.

2) Case $\varphi^2 = \psi^2$ and $\varphi \neq \psi$.

We notice $\varphi_2^2 = \varphi_3^2 = \varphi_4^2 = \varepsilon, \varphi_5^2 = \varphi_6, \varphi_6^2 = \varphi_5$. Then the conditions $\varphi^2 = \psi^2$ and $\varphi \neq \psi$ are fulfilled for the following automorphisms: $\varepsilon, \varphi_2, \varphi_3$ and φ_4.

The following triplets define paramedial quasigroups of order 4 with the property $\varphi \neq \psi$ over the group $Z_2\oplus Z_2$:

$(\varepsilon,\varphi_2,\overline{0,3}), (\varphi_2,\varepsilon,\overline{0,3}), (\varepsilon,\varphi_3,\overline{0,3}), (\varphi_3,\varepsilon,\overline{0,3}), (\varepsilon,\varphi_4,\overline{0,3}), (\varphi_4,\varepsilon,\overline{0,3})$,
$(\varphi_2,\varphi_3,\overline{0,3}), (\varphi_2,\varphi_4,\overline{0,3}), (\varphi_4,\varphi_2,\overline{0,3}), (\varphi_3,\varphi_2,\overline{0,3}), (\varphi_3,\varphi_4,\overline{0,3}), (\varphi_4,\varphi_3,\overline{0,3})$.

Any quasigroup from the first row is a medial paramedial quasigroup. Any of these quasigroups is a unipotently solvable quasigroup of degree 2.

Quasigroups from the second row are simple paramedial quasigroups since any pair of elements of the set $\{\varphi_2,\varphi_3,\varphi_4\}$ generates the group $Aut(Z_2\oplus Z_2) \cong S_3$ [407, 471].

Moreover, none these quasigroups are medial, since automorphisms $\varphi_2, \varphi_3, \varphi_4$ are not permutable in pairs relative to the operation of multiplication. Therefore all these quasigroups are "from" Case 3, (c) of Theorem 2.108.

2.5 CMLs and their isotopes

2.5.1 CMLs

Structure theory of commutative Moufang loops was mainly developed in the works of R.H. Bruck [173]. A good overview of the theory of commutative Moufang loops is in [815]. Here we give only an outline of the theory of CML and distributive quasigroups. This topic merits a separate book.

We recall, a loop $(Q, +)$ with the left SM-identity

$$(x + x) + (y + z) = (x + y) + (x + z) \tag{2.47}$$

is called commutative Moufang loop (CML for short). In Section 5.1.2 other identities that define the CML are given. In some of Kepka's articles the identity (2.47) is called LWA-identity [492].

A commutative Moufang loop in which any element has the order 3 is called a 3-CML. Taking into consideration the importance of the following lemma in CML structure theory, we give it with the proof.

Lemma 2.110. *In a commutative Moufang loop $(Q, +)$ the map $\delta : x \mapsto 3x$ is the central endomorphism [173, 72].*

Proof. In a CML $(Q, +)$ we have $n(x + y) = nx + ny$ for any natural number n since by the Moufang theorem [638, 173, 72] a CML is diassociative (any two elements generate an associative subgroup). Therefore the map δ is an endomorphism. See [521] for many details on commutative diassociative loops.

The proof of centrality of the endomorphism δ is standard [685, 72, 173, 501].

We can re-write the left SM-identity (2.47) as the existence of an autotopy of the form $(L_x, L_x, L_{x+x}) = (L_x, L_x, L_xL_x)$ for all $x \in Q$ in the corresponding quasigroup (in the corresponding loop).

Since any CML is an IP-loop, then $I^2 = \varepsilon$, $I(x \cdot y) = Iy \cdot Ix$ (Theorem 2.34), where $I(x) = -x$. In a CML the map I is an automorphism, since any CML is commutative. Thus $IL_a = L_{-a}I$ for any left translation L_a. From Corollary 2.39 (Case (132)) it follows that (L_x, IL_xL_xI, IL_xI) is also an autotopy of the loop $(Q, +)$. Then $(L_x, IL_xL_xI, IL_xI) = (L_x, L_{-x}L_{-x}, L_{-x})$.

If the triple (L_x, L_x, L_xL_x) is an autotopism of a CML, then the triple $(L_{-x}, L_{-x}, L_{-x}L_{-x})$ is also an autotopism of CML. Then

$$(L_x, L_{-x}L_{-x}, L_{-x})(L_{-x}, L_{-x}, L_{-x}L_{-x}) = (\varepsilon, L_{-x}L_{-x}L_{-x}, L_{-x}L_{-x}L_{-x})$$

is an autotopism. Taking into consideration that any CML has the property of diassociativity, we have $L_{-x}L_{-x}L_{-x} = L_{-3x}$.

Then $(\varepsilon, L_{-3x}, L_{-3x})$ is an autotopism, i.e., $y + (-3x + z) = -3x + (y + z)$. Since a CML $(Q, +)$ is commutative, then $(-3x + z) + y = -3x + (z + y)$, i.e., $-3x \in N_l$, $3x \in N_l$, element $3x$ is central for any $x \in Q$. □

A group $(G, \cdot)$ with identity $(xy)^3 = x^3y^3$ is called 3-abelian. In such groups the map $\delta(x) = x^3$ is the central endomorphism [171]. There exist non-abelian 3-abelian groups. See [565, 136].

Example 2.111. [407]. The group

$$(G, \cdot) = \langle a, b, c \,|\, a^3 = b^3 = c^3 = 1,\, ac = ca,\, bc = cb, ab = bac \rangle$$

has the order $3^3 = 27$ and it fulfills the identity $x^3 = 1$. It is clear that this group is non-abelian 3-abelian.

Example 2.112. If $(G, \cdot)$ is a 3-abelian group, then $(G, *)$, where $x * y = x^{-1}yx^2$, is a commutative Moufang loop.

More information on this construction is in [167, 72, 136]. Notice that the minimal order of proper commutative Moufang loop is equal to 81.

Example 2.113. [136]. Let $Z_3 = \{-1, 0, 1\}$, $(Z_3, +)$ be the cyclic group of order three, and group $(G, \cdot)$ be a 3-abelian group, for instance, from Example 2.111. Let $Z_3 \times G = \{(p, x) \,|\, p \in Z_3, x \in G\}$. Define operation $\circ$ on the set $Q = Z_3 \times G$ in the following way

$$(p, x) \circ (q, y) = (p + q, z_{q-p}(x, y)),$$

where $z_{-1}(x, y) = y \cdot x$, $z_0(x, y) = x^{-1} \cdot y \cdot x^2$, $z_1(x, y) = x \cdot y$. Then $(Q, \circ)$ is a commutative Moufang loop.

Example 2.114. [181]. Let H be a set of tetrads (a, b, c, d), $a, b, c, d \in (Z_3, +, \cdot)$, where $(Z_3, +, \cdot)$ is the ring of residues modulo 3, with the following operation

$$(a_1, b_1, c_1, d_1) \circ (a_2, b_2, c_2, d_2) = (a_1 + a_2, b_1 + b_2, c_1 + c_2, (c_1 - c_2) \cdot (a_1b_2 - b_1a_2).)$$

Then $(H, \circ)$ is a commutative Moufang loop.

Lemma 2.115. *In CML $(Q, +)$ the left, right, and middle nuclei are equal. The center $C(Q, +)$ of a CML $(Q, +)$ is a normal abelian subgroup of $(Q, +)$ and it coincides with the nucleus of $(Q, +)$ [173, 72, 685].*

R.H. Bruck proved the following:

Theorem 2.116. *If $(Q, +)$ is a finite CML, then $(Q, +)$ is isomorphic to the direct sum of an abelian group $(A, +)$ whose order is not divisible by 3 and a centrally nilpotent commutative Moufang loop $(B, +)$ whose order is a power of 3 [173, p. 101].*

Taking into consideration Lemma 2.110 we can prove Theorem 2.116 using an approach which is stated in Chapter 6.

Corollary 2.117. *If finite CML $(Q, +)$ is isomorphic to the direct sum $(A, +) \oplus (B, +)$, then $Aut(Q, +) \cong Aut(A, +) \oplus Aut(B, +)$.*

Proof. By Theorems 2.13 and 2.116 the loops $(A, +)$ and $(B, +)$ have elements of different orders.

We recall that any CML is diassociative. Let $\alpha \in Aut(Q, +)$. Then the order of an element x coincides with the order of element $\alpha(x)$. Indeed, if $nx = 0$, then $n(\alpha x) = \alpha(nx) = 0$.

Therefore the loops $(A, +)$ and $(B, +)$ are invariant relative to any automorphism of the loop $(Q, +)$. Then $Aut(Q, +) \cong Aut(A, +) \oplus Aut(B, +)$. □

From Theorem 2.116 we obtain the following:

Theorem 2.118. *If $(Q,+)$ is a finite commutative Moufang loop of order $p_1^{\alpha_1}\dots p_n^{\alpha_n}$, where p_i denotes prime, then $(Q,+)$ is isomorphic to the direct product of the loops $(Q_i,+)$ of order $p_i^{\alpha_i}$, $i=\overline{1,n}$, where $(Q_i,+)$ is a commutative group, if $p_i \neq 3$.*

Theorem 2.119. *Bruck-Slaby theorem [173, p. 157]. Let n be a positive integer, $n \geqslant 3$. Then every commutative Moufang loop G which can be generated by n elements is a centrally nilpotent loop of class at most $(n-1)$.*

From the Bruck-Slaby theorem it follows that any finite CML has a non-identity center and that any finite simple CML is a finite simple abelian group, i.e., it is the group Z_p for any prime p. Notice that there exists an infinite CML with an identity center [173, p. 132].

2.5.2 Distributive quasigroups

We recall that a quasigroup $(Q,\cdot)$ with identities

$$x\cdot yz = xy\cdot xz \tag{2.48}$$

(left distributive identity) and

$$xy\cdot z = xz\cdot yz \tag{2.49}$$

(right distributive identity) is called a distributive quasigroup.

Notice that C. Burstin and W. Mayer published an article devoted to distributive quasigroups in 1929 [185].

Example 2.120. [185]. We denote the ring of residues modulo p, p is prime, by $(R,+,\cdot)$. It is clear that the ring $(R,+,\cdot)$ coincides with the Galois Field $GF(p)$. We construct quasigroup $(R,\circ)$ of the form $x\circ y = a\cdot x + b\cdot y \mod p$, where a,b are fixed non-zero elements of R, $a+b\equiv 1 \mod p$. Quasigroup $(Q,\circ)$ is distributive, medial.

If we take $x=y=z$ in identity (2.48), then we obtain that any left distributive quasigroup is idempotent. Therefore any distributive quasigroup is an idempotent SM-quasigroup. Recall, any loop with left SM-identity is a CML (Lemma 2.20).

The class of distributive quasigroups is historically one of the first discovered quasigroup classes [185]. This class is one of the most researched, most known and most important quasigroup classes.

Left distributive and distributive groupoids are studied in [858, 463, 464, 230, 508, 839, 840, 456].

The fulfillment of left distributive identity in a quasigroup $(Q,\cdot)$ is equivalent to an affirmation that in this quasigroup any left translation L_x is an automorphism of this quasigroup. Indeed, we can re-write left distributive identity in a such manner $L_x yz = L_x y\cdot L_x z$.

Similarly, the fulfillment of the right distributive identity in a quasigroup $(Q,\cdot)$ is equivalent to an affirmation that in this quasigroup any right translation R_x is an automorphism of this quasigroup.

Notice, any CML is an SM-loop and any distributive quasigroup is an idempotent SM-quasigroup. Distributive quasigroups have the following fundamental property.

Theorem 2.121. *Belousov theorem. Any distributive quasigroup $(Q,\cdot)$ is an isotope of a CML $(Q,+)$ [72, 80, 685].*

In more details: If $(Q,\cdot)$ is a distributive quasigroup, $x+y=R_a^{-1}x\cdot L_a^{-1}y$, then $(Q,+)$ is a CML. If $(Q,+)$ is a CML and there exist commuting automorphisms $\psi,\varphi\in Aut(Q,+)$

such that $x + (y + z) = (\psi x + y) + (\varphi x + z)$ for all $x, y, z \in Q$, then $(Q, \cdot)$, $x \cdot y = \psi x + \varphi y$, is a distributive quasigroup.

Belousov Theorem 2.121 gives the possibility to transfer results about commutative Moufang loops (Theorem 2.118) into corresponding theorems about distributive quasigroups [339] and vice versa. We recall, mainly the theory of commutative Moufang loops was developed by R.H. Bruck.

Another important result on distributive quasigroups is proved by B. Fisher [323].

Theorem 2.122. *The left (the right) multiplication group of any finite distributive quasigroup is solvable.*

This theorem has opened a way to study properties (for example, solvability, simplicity) of finite distributive quasigroups using multiplication groups of these quasigroups and their subgroups. The following result is deduced from the Fisher theorem (Theorem 2.122) (more exactly, from the Smith version of the Fisher theorem [814]).

Theorem 2.123. *If $(Q, \cdot)$ is a finite distributive quasigroup of order $p_1^{\alpha_1} \dots p_n^{\alpha_n}$, where p_i denotes prime, then $(Q, \cdot)$ is isomorphic to the direct product of the quasigroups $(Q_i, \cdot)$ of order $p_i^{\alpha_i}$, $i = \overline{1, n}$, where $(Q_i, \cdot)$ is a medial idempotent quasigroup, if $p_i \neq 3$ [339].*

As a corollary of this theorem it is possible to deduce that any finite simple distributive quasigroup is medial.

We prove that any translation of a distributive quasigroup is its automorphism.

Lemma 2.124. *If $(Q, \circ)$ is an isotope of a quasigroup $(Q, \cdot)$ of the form $(R_a^{-1}, L_b^{-1}, \varepsilon)$, then $P_c^{\circ} = L_b P_c^{\cdot} R_a^{-1}$.*

Proof. If $x \circ y = R_a^{-1}x \cdot L_b^{-1}y = c$, then $P_c^{\cdot} R_a^{-1}x = L_b^{-1}y$, $L_b P_c^{\cdot} R_a^{-1}x = y$. From the other side, $P_c^{\circ}x = y$. Therefore $P_c^{\circ} = L_b P_c^{\cdot} R_a^{-1}$. □

Lemma 2.125. *In LIP-quasigroup $(Q, \cdot)$, $P_c x = R_c(\lambda x)$.*

Proof. Indeed, if $P_c x = y$, i.e., $x \cdot y = c$, then $\lambda x \cdot (x \cdot y) = \lambda x \cdot c$. By the LIP-inverse property we can re-write the last equality in the form $y = \lambda x \cdot c$, or, in the language of translations, in the form $y = R_c \lambda x$. Therefore in the LIP-quasigroup, $P_c x = R_c \lambda x$ (see also [758]). □

Lemma 2.126. *In a distributive quasigroup $(Q, \cdot)$, translations L_x, L_x^{-1}, R_x, R_x^{-1}, P_x, P_x^{-1} are automorphisms of the quasigroup $(Q, \cdot)$ for all $x \in Q$.*

Proof. From the left distributivity it follows that $L_x, L_x^{-1} \in Aut(Q, \cdot)$; from the right distributivity it follows that $R_x, R_x^{-1} \in Aut(Q, \cdot)$.

Let $(Q, +)$ be a CML that is an isotope of distributive quasigroup $(Q, \cdot)$ of the form $(R_a^{-1}, L_a^{-1}, \varepsilon)$, i.e., $x + y = R_a^{-1}x \cdot L_a^{-1}y$, or $x \cdot y = R_a x + L_a y$ for all $x, y \in Q$.

We recall that any Moufang loop has the LIP-inverse property. Then from Lemmas 2.124, 2.125 and the Belousov theorem (Theorem 2.121) it follows that $P_c^{\cdot} \in Aut(Q, \cdot)$, if permutations R_c^{+} and I ($Ix + x = 0$ for all $x \in Q$) are automorphisms of distributive quasigroup $(Q, \cdot)$.

From Theorem 1.202 it follows that for distributive quasigroup $(Q, \cdot)$ and CML $(Q, +)$ the following inclusions are true

$$M(Q, +) \subseteq M(Q, \cdot) \subseteq Aut(Q, \cdot).$$

Therefore $R_c^{+} \in Aut(Q, \cdot)$.

From the proof of the Belousov theorem it follows that translations R_a^{-1} and L_a^{-1} are automorphisms of CML $(Q, +)$, the permutation I is an automorphism of $(Q, +)$ and $IR_a^{-1} = R_a^{-1}I$, $IL_a^{-1} = L_a^{-1}I$. Using this information it is possible to prove that $I \in Aut(Q, \cdot)$. We have $I(x \cdot y) = I(R_a^{\cdot}x + L_a^{\cdot}y) = IR_a^{\cdot}x + IL_a^{\cdot}y = R_a^{\cdot}Ix + L_a^{\cdot}Iy = Ix \cdot Iy$. □

Corollary 2.127. *In a distributive quasigroup $FM \subseteq Aut$.*

Theorem 2.128. *Any parastrophe of a distributive quasigroup $(Q, \cdot)$ is a distributive quasigroup [72, 80, 685].*

Proof. The proof follows from Lemma 2.126 and Table 1.1.

We prove, if a quasigroup $(Q, \cdot)$ is distributive, then its (123)-parastrophe $(Q, *)$ is a distributive quasigroup too. It is clear that any parastrophe of a quasigroup is a quasigroup. From Table 1.1 it follows that $L_x^{123} = R_x^{-1}$, $R_x^{123} = P_x^{-1}$, $P_x^{123} = L_x$.

We recall (Theorem 1.153) that if α is an automorphism of a quasigroup $(Q, \cdot)$, then α is an automorphism of any parastrophe of this quasigroup.

Translations R_x^{-1}, P_x^{-1} and L_x are automorphisms of quasigroup $(Q, \cdot)$. Then they are automorphisms of the quasigroup $(Q, *)$. Therefore the quasigroup $(Q, *)$ is distributive.

Other cases are proved similarly. □

2.6 Left distributive quasigroups

In the class of left distributive quasigroups there exist interesting problems that are not solved till now.

The main progress in the theory of left distributive quasigroups is connected with the works of S. Stein and D.A. Norton [663, 845], V.D. Belousov [72], V.M. Galkin [343], V.I. Onoi [669, 670], A. Stein [844], and some other mathematicians. The functional equation of the left distributivity is researched in [73]. In this chapter we do not give information on left distributive groupoids that play an important role in geometry (symmetric spaces) [576, 840, 839].

2.6.1 Examples, constructions, orders

We start from constructions of left distributive quasigroups.

Definition 2.129. A groupoid (Q, A) is homogenous from the left over a loop $(Q, \cdot)$, if there exists a permutation φ of the set Q such that $A(x, y) = x \cdot \varphi(x\backslash y)$ for all $x, y \in Q$, where $x\backslash y = z \Leftrightarrow x \cdot z = y$.

A groupoid (Q, B) is homogenous from the right over a loop $(Q, \cdot)$, if there exists a permutation ψ of the set Q such that $B(x, y) = \psi(x/y) \cdot y$ for all $x, y \in Q$, where $x/y = z \Leftrightarrow z \cdot y = x$ [427], [72, p. 154].

Remark 2.130. If loop $(Q, \cdot)$ is an IP-loop, then the operation A takes the form $A(x, y) = x \cdot \varphi(x^{-1} \cdot y)$.

Theorem 2.131. *A distributive quasigroup is homogenous from the left (from the right) over a commutative Moufang loop [72, p. 154].*

Theorem 2.132. *Let $(Q, +)$ be a finite non-abelian group and φ be its fixed point free automorphism (regular), i.e., $\varphi(x) = x \Rightarrow x = 0$ for all $x \in Q$. Groupoid $(Q, \cdot)$, where $x \cdot y = x + \varphi(-x + y) = (x - \varphi x) + \varphi y$, $-x + x = 0$ for all $x \in Q$, is a left distributive quasigroup [845], [72, p. 155]. If $(Q, +)$ is commutative, then $(Q, \cdot)$ is distributive.*

Any left distributive quasigroup that is isotopic to a group is possible to obtain using this construction [72, p. 155], and vice versa; any left distributive quasigroup constructed in this way is a group isotope since the mapping $(x - \varphi x)$ is a permutation of the set Q [845]. Recall, any finite group that has regular automorphism is solvable [719].

Example 2.133. The non-abelian group from Example 2.111 has the following regular automorphism φ: $\varphi(a) = b$; $\varphi(b) = ab$; $\varphi(c) = c^2$ [848]. The corresponding left distributive quasigroup is isotopic to a non-abelian group, therefore it is not medial.

Example 2.134. Let $(Q, +)$ be a Moufang loop of odd order. Then $(Q, \cdot)$, $x \cdot y = x - y + x$, is a left distributive quasigroup [64, 72]. In Bruck terminology $(Q, \cdot)$ is a core of $(Q, +)$ [173]. It is also a right distributive if and only if $(Q, +)$ satisfies the identity

$$x + 2y + x = y + 2x + y. \tag{2.50}$$

The left distributive quasigroup constructed in this way is isotopic to a left Bol loop [64].

Example 2.135. Suppose $(Q, +)$ is a group of order $3^\alpha q^3$, where q is a prime of the form $3k + 2$ and this group is defined by the generators a, b, c, d and the following relations $3^\alpha a = 0, qb = qc = qd = 0, c + b = b + c + d, -a + b + a = c, -a + c + a = -b - c, d + a = a + d, d + b = b + d, d + c = c + d$ [812, p.225]. The identity (2.50) is not true in this group [64, 72]. Therefore the left distributive quasigroup constructed from the group $(Q, +)$ is isotopic to a left Bol loop [64].

Example 2.136. Take the group $(Q, +) = \langle a, b \mid a^3 = b^7 = 1, a^{-1}ba = b^2 \rangle$. Information on this group is in [471]. The identity (2.50) is not true in this group [80]. Indeed, suppose that this identity is true. Then $(aba)^2 = (a^2 \cdot a^{-1}ba)^2 = (a^2b^2)^2$ From the other side, $(aba)^2 = a(ba^2b)a \overset{(2.50)}{=} a(ab^2a)a = a^2b^2a^2$. Then $(a^2b^2)^2 = a^2b^2a^2$, $a^2b^2a^2b^2 = a^2b^2a^2$, $b^2 = 1$. But $b^7 = 1$. Then $b = 1$. This is impossible. Therefore the left distributive quasigroup constructed from the group $(Q, +)$ is isotopic to a left Bol loop [64].

Example 2.137. Left distributive quasigroups that are not isotopic to Bol loops are constructed in [670, 343]. We follow [343]. On the set $GF(5) \times GF(3)$ ($GF(5)$ and $GF(3)$ are the fields of orders five and three, respectively), define the following operation

$$(a, x) \circ (b, y) = (\mu(-x + y)a - b - x + y, -x - y), \tag{2.51}$$

where $a, b \in GF(5)$, $x, y \in GF(3)$. Notice, in expression $\mu(-x + y)a - b - x + y$ we associate elements of $GF(3)$ with the corresponding elements of $GF(5)$.

If we suppose that $\mu z = 2$, if $z = 0$, and $\mu z = -1$ otherwise, then we obtain a left distributive quasigroup of order 15 that is not isotopic to Bol loop.

Universal construction of any left distributive quasigroup and the theory of extensions of the left distribute quasigroups is given in [343].

Theorem 2.138. *S. Stein theorem. There exist left-distributive quasigroups of order n only for $n \equiv 0, 1, 3 \pmod 4$ [663, 845, 343].*

In the first variant of the proof of this theorem, topological ideas have been used [663, 845]. At present these ideas are developed in [509, 510]. More algebraic proof of this theorem is given in [343].

2.6.2 Properties, simple quasigroups, loop isotopes

We give some results from [340, 343] on left distributive quasigroups.

Theorem 2.139. *Let $(Q, \cdot)$ be a left distributive quasigroup. If $(S, \cdot)$ is a subquasigroup of $(Q, \cdot)$, then $|S| \leqslant |Q|/3$. If the equality holds, then $(S, \cdot)$ is a normal subquasigroup.*

Theorem 2.140. *Let $(Q, \cdot)$ be a finite left distributive quasigroup. If a subquasigroup $(S, \cdot)$ of $(Q, \cdot)$ is of minimal non-identity order, then the orders of Q and S cannot be relatively prime.*

A left distributive quasigroup is called minimal if it does not contain a proper non-identity subquasigroup [340].

Theorem 2.141. *Any finite minimal left distributive quasigroup is medial [340].*

Corollary 2.142. *Any finite minimal left distributive quasigroup is simple.*

Proof. This follows from the fact that in any medial quasigroup, its subquasigroup is normal. □

Recall, simple medial quasigroups are described by Jezek-Kepka Theorem (see Theorem 2.78).

Left-distributive quasigroups with identity $a(ab) = b$ are called symmetric [341]. Similar to finite distributive quasigroups, any finite symmetric quasigroup is solvable [341].

Example 2.143. An infinite non-solvable symmetric quasigroup is also constructed in [341]. Let

$$Q = \{x = (x_1, x_2, x_3) \mid x_1^2 + x_2^2 - x_3^2 = -1, x_3 > 0, x_1, x_2, x_3 \in \mathbb{R}\}$$

and $x \circ y = 2\langle x, y\rangle x - y$, $\langle x, y\rangle = -x_1y_1 - x_2y_2 + x_3y_3$. It is proved that groupoid $(Q, \circ)$ is a non-solvable symmetric quasigroup. By the proof, complexification (a passage into the field of complex numbers) and simple three-dimensional Lie algebra are used.

An isotope of the form $(R_a^{-1}, L_b^{-1}, \varepsilon)$ of a left distributive quasigroup $(Q, \cdot)$, where a, b are arbitrary fixed elements of the set Q, is an M-loop, i.e., in this loop the following equality $x \circ (y \circ z) = (x \circ (y \circ I\varphi x)) \circ (\varphi x \circ z)$ is fulfilled, where $Ix = x^{-1}$, $x \circ x^{-1} = 1$ (1 is the identity element of the loop $(Q, \circ)$), and φ is a map of the set Q [80, p. 107].

An isotope of the form $(R_a^{-1}, L_a^{-1}, \varepsilon)$ of a left distributive quasigroup $(Q, \cdot)$, where a is a fixed element of the set Q, is called an S-loop [90]. It is easy to see that any S-loop is an M-loop.

Definition 2.144. An automorphism ψ of a loop $(Q, \circ)$ is called *complete*, if there exists a permutation φ of the set Q such that $\varphi x \circ \psi x = x$ for all $x \in Q$. Permutation φ is called a complement of automorphism ψ.

Any complete automorphism is a complete mapping of the set Q (Definition 1.380). Indeed, we can re-write equality $\varphi x \circ \psi x = x$ in the following form $x \circ \psi\varphi^{-1}x = \varphi^{-1}x$.

The following theorem is proved in [90].

Theorem 2.145. *A loop $(Q, \circ)$ is an S-loop, if and only if there exists a full automorphism ψ of the loop $(Q, \circ)$ such that at least one of the following conditions is fulfilled:*

a) $\varphi(x \circ \varphi^{-1}y) \circ (\psi x \circ z) = x \circ (y \circ z)$

b) $L^{\circ}_{x,y}\psi = \psi L^{\circ}_{x,y}$ *and* $\varphi x \circ (\psi x \circ y) = x \circ y$ *for all* $x, y \in Q$, $x, y \in Q, L^{\circ}_{x,y} \in LI(Q, \circ)$.

Thus $(Q, \cdot)$, where $x \cdot y = \varphi x \circ \psi y$, is a left distributive quasigroup which corresponds to the loop $(Q, \circ)$.

Similar to the distributive case for left distributive quasigroups, the following result is true.

Theorem 2.146. *Let $(Q,\cdot)$ be a finite left distributive quasigroup. Then $LM(Q,\cdot)$ is solvable [844].*

It is possible to use this theorem in the study of finite simple and solvable left distributive quasigroups.

Exercise 2.147. Suppose there exist left distributive quasigroups $(Q,\circ)$ and $(Q,\cdot)$ with the forms $x\circ y = x+\varphi(-x+y)$, $x\cdot y = x+\psi(-x+y)$, respectively, where $(Q,+)$ is a group, and φ,ψ are regular automorphisms of $(Q,+)$. Quasigroups $(Q,\circ)$ and $(Q,\cdot)$ are isomorphic if and only if there exists an automorphism α of the group $(Q,+)$ such that $\psi = \alpha\varphi\alpha^{-1}$ [760].

2.7 TS-quasigroups

In this section we follow mainly [80, 91]. We start from the following:

Definition 2.148. A commutative groupoid $(Q,\cdot)$ with identity

$$x(xy) = y \text{ (left semi-symmetric identity)} \tag{2.52}$$

is called a TS-quasigroup (totally symmetric quasigroup).

The following quasigroup is a non-idempotent TS-quasigroup.

$\circ$	0	1	2
0	1	0	2
1	0	2	1
2	2	1	0

Definition 2.149. If TS-quasigroup $(Q,\cdot)$ is a loop, then $(Q,\cdot)$ is called a Steiner loop. The idempotent TS-quasigroup $(x^2 = x)$ is called a Steiner quasigroup.

Exercise 2.150. In any TS-loop $(Q,\cdot)$, $x\cdot x = 1$ for all $x \in Q$.

2.7.1 Constructions, loop isotopes

Example 2.151. A quasigroup $(Q.\cdot)$ of the form $x\cdot y = c - x - y$, where c is a fixed element of an abelian group $(Q,+)$, is a TS-quasigroup. It is clear that quasigroup $(Q,\cdot)$ is an isotope of the form (I,I,L_{-c}) of abelian group $(Q,+)$ [592].

In any TS-quasigroup $(Q,\cdot)$ the following identities are true $x\cdot yx = y$, $xy\cdot y = x$, $xy\cdot x = y$, i.e., any parastrophe of TS-quasigroup $(Q,\cdot)$ coincides with it. See Lemma 1.186. Any TS-quasigroup is an IP-quasigroup with $\lambda = \rho = \varepsilon$.

Theorem 2.152. *A quasigroup $(Q,\circ)$ of the form $x\circ y = \alpha x\cdot\beta y$, where $(Q,\cdot)$ is a TS-quasigroup, α,β are permutations of the set Q, is a TS-quasigroup if and only if $\beta(xy) = x\cdot\beta^{-1}y$, $\alpha(xy) = \alpha^{-1}x\cdot y$, $\alpha\beta = \beta\alpha$.*

Proof. From the identity $x \circ (x \circ y) = y$ we have $\alpha x \cdot \beta(\alpha x \cdot \beta y) = y$, $\alpha x(\alpha x \cdot \beta(\alpha x \cdot \beta y)) = \alpha x \cdot y$. Applying to the last equality the identity (2.52) we obtain that $\beta(\alpha x \cdot \beta y) = \alpha x \cdot y$, $\beta(xy) = x \cdot \beta^{-1} y$.

The condition $\alpha(xy) = x \cdot \alpha^{-1} y$ is obtained in a similar way from the identity $(x \circ y) \circ y = x$.

We prove that the permutations α and β commute. We have $\alpha\beta(xy) = \alpha(x \cdot \beta^{-1}y) = \alpha(\beta^{-1}y \cdot x) = \beta^{-1}y \cdot \alpha^{-1}x = \alpha^{-1}x \cdot \beta^{-1}y$, $\beta\alpha(xy) = \beta\alpha(yx) = \beta(y \cdot \alpha^{-1}x) = \beta(\alpha^{-1}x \cdot y) = \alpha^{-1}x \cdot \beta^{-1}y$. Therefore $\alpha\beta = \beta\alpha$.

Converse. We have $x \circ y = \alpha x \cdot \beta y = \beta^{-1}(\alpha x \cdot y) = \beta^{-1}(y \cdot \alpha x) = \beta^{-1}\alpha^{-1}(y \cdot x) = \alpha^{-1}\beta^{-1}(y \cdot x) = \alpha^{-1}(y \cdot \beta x) = \alpha^{-1}(\beta x \cdot y) = \beta x \cdot \alpha y = \alpha y \cdot \beta x = y \circ x$ and $x \circ (x \circ y) = \alpha x \cdot \beta(\alpha x \cdot \beta y) = \alpha x \cdot (\alpha x \cdot \beta^{-1}\beta y) = \alpha x \cdot (\alpha x \cdot y) = y$. □

Remark 2.153. The condition $\alpha(xy) = x \cdot \alpha^{-1} y$ means that triple $(\varepsilon, \alpha^{-1}, \alpha)$ lies in the right autotopy nucleus N_r^A of a TS-quasigroup.

Definition 2.154. A commutative loop with the following equality

$$x + J(y + x) = Jy, \tag{2.53}$$

where J is a permutation of the set Q, is called a commutative W-loop [91].

If we add the condition $x + Jx = 0$ for all $x \in Q$ to the definition of a commutative W-loop, then the commutative W-loop is transformed into a commutative WIP-loop.

Theorem 2.155. *A commutative W-loop $(Q, +)$ is a commutative WIP-loop if and only if $J0 \in N$, where N is the nucleus of loop $(Q, +)$ [91].*

Recall that a quasigroup with the equality (2.53) is called a WIP-quasigroup [488].

In the terminology of (r, s, t)-inverse quasigroups we can say that a quasigroup with the equality (2.53) is a $(-1, 0, -1)$-inverse quasigroup (see Lemma 3.30).

From Lemma 4.165 it follows that the left, right and middle A-nucleus in such quasigroups are isomorphic. From Corollary 4.170 it follows that the left, right and middle nucleus in such loops coincide [91].

Theorem 2.156. *Any TS-quasigroup $(Q, \cdot)$ is isotopic to a commutative W-loop $(Q, +)$ with the property $J^2 = \varepsilon$ [91].*

Proof. Let $0 \in Q$. We consider the following isotope of quasigroup $(Q, \cdot)$: $x + y = 0 \cdot (x \cdot y)$. In other words $x + y = L_0(x \cdot y), x \cdot y = L_0^{-1}(x + y)$. Quasigroup $(Q, +)$ is commutative. This follows from commutativity of $(Q, \cdot)$ and the form of isotopy.

Quasigroup $(Q, +)$ is a loop. Indeed, $0 + y = 0 \cdot (0 \cdot y) = y$. From the identity $x \cdot (x \cdot y) = y$ we have $L_0^{-1}(x + L_0^{-1}(x + y)) = y$,

$$x + L_0^{-1}(x + y) = L_0 y. \tag{2.54}$$

From the identity $x \cdot (x \cdot y) = y$ we have $L_x^2 = \varepsilon$, $L_x = L_x^{-1}$ for all $x \in Q$. If we denote translation L_0 by J, then we can rewrite the equality (2.54) in the form (2.53). □

Lemma 2.157. *If TS-quasigroup $(Q, \cdot)$ has at least one idempotent element, say 0, then its isotope $(Q, +)$ of the form $x + y = 0 \cdot (x \cdot y)$ is a commutative WIP-loop.*

Proof. We find the form of inverse permutation in the loop $(Q, +)$. We have $Ix + x = x + Ix = 0$, $0 \cdot (x \cdot Ix) = 0$, $x \cdot Ix = 0 \cdot 0$, $Ix = x \cdot (0 \cdot 0) = x \cdot 0^2$, $I = R_{0^2} = L_{0^2}$, since quasigroup $(Q, \cdot)$ is commutative. Since $0 \cdot 0 = 0$, then in the loop $(Q, +)$ $L_{0^2} = L_0$, $J = I$, loop $(Q, +)$ is a WIP-loop. □

Theorem 2.158. *TS-quasigroup $(Q, \cdot)$ is isotopic to a Steiner quasigroup if and only if quasigroup $(Q, \cdot)$ satisfies the identity $x^2y = xy^2$ [80, 91].*

Theorem 2.159. *Steiner quasigroup $(Q, \cdot)$ is isotopic to commutative WIP-loop $(Q, +)$ with condition $x + x + x = 0$ for all $x \in Q$ [80, 91].*

Proof. From Lemma 2.157 it follows that $(Q, \cdot)$ is isotopic to commutative WIP-loop $(Q, +)$. Since $(Q, +)$ is commutative, then $(x + x) + x = x + (x + x)$ and we can omit brackets.

Finally we prove that $x + x + x = 0$ for all $x \in Q$. We have $x + (x + x) = x + 0 \cdot (x \cdot x) = x + 0 \cdot x = x + Ix = 0$. □

Theorem 2.160. *TS-quasigroup $(Q, \cdot)$ is isotopic to a commutative Moufang loop if and only if quasigroup $(Q, \cdot)$ is an F-quasigroup [91, 332].*

We give the example of a TS-quasigroup that is not a Steiner quasigroup $(0 * 0 = 7 \neq 0)$ and the example of a W-loop that is not a WIP-loop (if $(Q, +)$ is a WIP-loop, then $1 + I(7 + 1) = I7, 1 + I4 = I7, 1 + 6 = 7$, but $1 + 6 = 2 \neq 7$) [91].

*	0	1	2	3	4	5	6	7
0	7	2	1	4	3	6	5	0
1	2	6	0	7	5	4	1	3
2	1	0	4	6	2	7	3	5
3	4	7	6	3	0	5	2	1
4	3	5	2	0	4	1	7	6
5	6	4	7	5	1	3	0	2
6	5	1	3	2	7	0	6	4
7	0	3	5	1	6	2	4	7

+	0	1	2	3	4	5	6	7
0	0	1	2	3	4	5	6	7
1	1	5	7	0	6	3	2	4
2	2	7	3	5	1	0	4	6
3	3	0	5	4	7	6	1	2
4	4	6	1	7	3	2	0	5
5	5	3	0	6	2	4	7	1
6	6	2	4	1	0	7	5	3
7	7	4	6	2	5	1	3	0

Here we give a standard Bruck prolongation construction (see Section 1.7.5 or [72, p. 129]) of obtaining the Steiner loop of order $(n + 1)$ using a TS-quasigroup of order n.

Example 2.161.

*	1	2	3
1	**1**	3	2
2	3	**2**	1
3	2	1	**3**

$\longrightarrow$

∘	1	2	3	4
1	□	3	2	**1**
2	3	□	1	**2**
3	2	1	□	**3**
4	**1**	**2**	**3**	□

$\longrightarrow$

+	1	2	3	4
1	4	3	2	**1**
2	3	4	1	**2**
3	2	1	4	**3**
4	**1**	**2**	**3**	4

In Example 2.161 the symbol 4 is an identity element.

We prove that the object constructed in Example 2.161 is a Steiner loop. Bruck prolongation construction gives us that $(Q', +)$ is a loop since $(Q, *)$ is an idempotent quasigroup.

It is easy to see that $(Q', +)$ is commutative. Prove that in $(Q', +)$ the identity $x + (x + y) = y$ is true. If $x \neq 4$, $y \neq 4$, then this identity is true because $(Q, *)$ is a TS-quasigroup.

Suppose that $x = 4$, $y \neq 4$. Then using a Cayley table of the loop $(Q', +)$ we have $4 + (4 + y) = 4 + y = y$.

Further suppose that $x \neq 4$, $y = 4$. Then using a Cayley table of the loop $(Q', +)$ we have $x + (x + 4) = x + x = 4$. Finally the equality $4 + (4 + 4) = 4$ follows from the Cayley table of the loop $(Q', +)$.

2.7.2 2-nilpotent TS-loops

Algebraic theory of centrally k-nilpotent (especially 2-nilpotent) TS-loops is developed in [377, 378, 379]. Notice, there are several different definitions of the associator.

Definition 2.162. Associator of degree 1 of the elements x_1, x_2, x_3 of a quasigroup $(Q, \cdot)$ is the following element:

$$[x_1, x_2, x_3] = x_1(x_2(x_3(x_1(x_2x_3)))). \tag{2.55}$$

Associator of degree 2 of the elements x_1, x_2, x_3, x_4, x_5 of a quasigroup $(Q, \cdot)$ is the following element: $[[x_1, x_2, x_3], x_4, x_5]$. And so on, i.e., an associator of degree $k > 2$ is defined inductively.

Definition 2.163. TS-loop $(Q, \cdot)$ is called centrally k-nilpotent, if in $(Q, \cdot)$ the following identity is true:

$$[[\dots[x_1, x_2, x_3], \dots, x_{2k-1}], x_{2k}, x_{2k+1}] = 1. \tag{2.56}$$

Therefore a TS-loop with identity $[[x_1, x_2, x_3], x_4, x_5] = 1$ is centrally 2-nilpotent. Notice, in other classes of quasigroups and loops, the definition of k-nilpotency is more complicated [740, 659, 220].

Any 2-nilpotent TS-loop $(Q, *)$ can be constructed in the following way [377, 378]. The set Q is the direct product of the sets B and A such that $Q = B \times A = \{(x, a) \mid x \in B, a \in A\}$, and the groups $(A, +)$ and $(B, +)$ satisfy the identity $x + x = 0$. It is clear that these groups are elementary abelian 2-groups [471]. Notice that on such group it is easy to define vector space structure [378, 920]. The operation $*$ is defined as follows:

$$(x, a) * (y, b) = (x + y, a + b + \xi(x, y)), \tag{2.57}$$

and the mapping $\xi : B \times B \to A$ has the following properties: $\xi(x, y) = \xi(y, x) = \xi(x, x + y), \xi(x, 0) = 0$.

2.7.3 Some properties of TS-quasigroups

Automorphism groups of Steiner triple systems and Steiner quasigroups are researched in [408, 577, 772], multiplication groups in [859], and coloring in [253, 295]. TS-quasigroups, which are associated with the composition of points of the set of points of a cubic hypersurface by means of drawing of straight lines, are studied in [592, 593]. In article [592] TS-quasigroups with the following identity

$$x \cdot y(xy \cdot z) = yy \cdot z \tag{2.58}$$

are studied. In any TS-quasigroup $(Q, \cdot)$ which is constructed in Example 2.151 the identity (2.58) is true. There exist TS-quasigroups which satisfy the identity (2.58) and which are non-isotopic to an abelian group [592].

2.8 Schröder quasigroups

In [750] Ernst Schröder (a German mathematician mainly known for his work on algebraic logic) introduced and studied the following identity of generalized associativity:

$$(yz)\backslash x = z \cdot xy. \tag{2.59}$$

See [436, 437, 552] for details.

Exercise 2.164. In the quasigroup case the identity (2.59) is equivalent to the following identity:

$$(yz)(z \cdot xy) = x. \tag{2.60}$$

It is convenient to call this identity the Schröder identity of generalized associativity. Often various variants of associative identity, which are true in a quasigroup, guarantee that this quasigroup is a loop. It is not so in the case with the identity (2.60). See the following example.

Example 2.165. We give an example of quasigroup with the identity (2.60).

*	0	1	2	3	4	5	6	7
0	1	4	7	0	6	5	2	3
1	5	2	3	6	0	1	4	7
2	0	7	4	1	5	6	3	2
3	6	3	2	5	1	0	7	4
4	4	1	0	7	3	2	5	6
5	3	6	5	2	4	7	0	1
6	7	0	1	4	2	3	6	5
7	2	5	6	3	7	4	1	0

The left cancellation (left division) groupoid with the identity (2.60) and with the identity (1.25) (in a quasigroup this identity guarantees existence of the left identity element) is a commutative group of exponent two [707]. The similar results are true for the right case [707].

If in idempotent quasigroup $(Q, \cdot)$ in identity (2.60) we put $x = y$, then we obtain the following "standard" Schröder identity:

$$xy \cdot yx = x. \tag{2.61}$$

Any quasigroup with the identity (2.61) is called a Schröder quasigroup.

The group $(Z_2 \oplus Z_2, +)$ is an example of a quasigroup in which the identity $(xz)(z \cdot xx) = x$ ($x = y$ in the identity (2.60)) is true and the identity (2.61) is not true.

Lemma 2.166. *In any Schröder quasigroup $(Q, \cdot)$ the equality $x \cdot x = y \cdot y$ implies $x = y$, and the equality $x \cdot y = y \cdot x$ implies $x = y$.*

Proof. Suppose $x \cdot x = y \cdot y$. Then from the identity (2.61) we have $x = (x \cdot x) \cdot (x \cdot x) = (y \cdot y) \cdot (y \cdot y) = y$.

Suppose $x \cdot y = y \cdot x$. Then we have $x = (x \cdot y) \cdot (y \cdot x) = (y \cdot x) \cdot (x \cdot y) = y$. □

Theorem 2.167. *A necessary condition for the existence of an idempotent Schröder quasigroup $(Q, \cdot)$ of order v is that $v \equiv 0$ or $1 \pmod 4$ [572].*

Proof. We follow [572]. For any pair x, y of distinct elements in the quasigroup, put

$$F(x, y) = \{(x, y), (y, x), (x \cdot y, y \cdot x), (y \cdot x, x \cdot y)\}.$$

Now, if $x = x \cdot y$, then using the idempotent identity $x \cdot x = x \cdot y$, $x = y$. Similarly $x = y \cdot x$ implies $x = y$. Hence $F(x, y)$ contains four ordered pairs of distinct elements.

Also, $F(x, y) = F(y, x) = F(x \cdot y, y \cdot x) = F(y \cdot x, x \cdot y)$ so that two sets $F(x, y)$, $F(u, v)$ are either identical or disjoint. Hence the number of ordered pairs of distinct elements is a multiple of four, i.e., $v(v - 1) \equiv 0 \pmod 4$. Hence, $v \equiv 0 \pmod 4$ or $v \equiv 1 \pmod 4$. □

Schröder quasigroups of order n exist for all $n \equiv 0, 1 \pmod 4$ except $n = 5$ and possibly $n = 21$ [572, 145]. These quasigroups are studied in [552, 145].

Example 2.168. Define groupoid $(GF(2^r), \circ)$ over the Galois field $(GF(2^r), +, \cdot, 0, 1)$, $r \geqslant 2$, in the following way: $x \circ y = a \cdot x + (a+1) \cdot y$, where $x, y \in GF(2^r)$, the element a is a fixed element of the set $GF(2^r)$, $a \neq 0$, $a \neq 1$, the operations $+$ and $\cdot$ are binary operations of this field. The groupoid $(GF(2^r), \circ)$ is an idempotent medial Schröder quasigroup [572].

There exist non-idempotent Schröder quasigroups [572, 145], Example 2.169. Notice, n-ary analogues of Schröder quasigroups are defined and studied in [852].

Example 2.169. The following quasigroups are non-idempotent and idempotent Schröder quasigroups, respectively.

$*$	0	1	2	3
0	1	3	2	0
1	2	0	1	3
2	0	2	3	1
3	3	1	0	2

$\circ$	0	1	2	3
0	0	2	3	1
1	3	1	0	2
2	1	3	2	0
3	2	0	1	3

Exercise 2.170. Construct a (12)-parastrophe of quasigroup $(Q, *)$ from Example 2.169.

The fulfillment Schröder identity in a finite quasigroup is the sufficient condition for orthogonality of this quasigroup and its (12)-parastrophe [85]. See also Theorem 9.72.

Problem 2.1. Give the algebraic classification of Schröder quasigroups. Which loops are isotopic to these quasigroups?

2.9 Incidence systems and block designs

2.9.1 Introduction

Definition 2.171. An incidence system is a triple $(S, \mathcal{B}, I)$, where S, $\mathcal{B}$ are some non-empty sets, I is an incidence relation between elements of these sets; the $(0, 1)$-incidence matrix of size $|S| \times |\mathcal{B}|$ corresponds to any incidence system [152, p. 98].

The fact that the elements $s \in S$ and $b \in \mathcal{B}$ are incident is usually denoted in the following way: xIb. Therefore the incidence relation is binary. The situation when $\mathcal{B}$ is a family (a multiset) of subsets of the set S, and relation I is the relation of membership is often studied in combinatorics. Note, if $S = \mathcal{B}$, then Definition 2.171 is transformed in the definition of a graph (Definition 1.409).

Below t-subset means a subset of a set S which consists of t elements. A block design in which all the blocks have the same size is called uniform.

Definition 2.172. An incidence system $(S, \mathcal{B}, I)$ is called a t-(v, k, λ)-design (t-design, tactical schema [918]), if the set S is a finite set of order v, and the family $\mathcal{B}$ consists of subsets of the set S, called blocks. The number of blocks is denoted by letter b, any block contains k elements of the set S, any t-subset of the set S is a subset of λ blocks, any element $x \in S$ appears in exactly r blocks, and the following inequalities are true: $0 < t < k < v$ [409, 152, 918].

It is clear that Definition 2.172 is close to Definition 1.418. The word "family" in the above definition can be replaced by the word "set" if repeated blocks are not allowed. Designs in which the repeated blocks are not allowed are called simple (see also Definition 1.420). Sometimes instead of the word "family" the word "multiset" is used.

The numbers t, v, k, λ determine b and r and the four numbers themselves cannot be chosen arbitrarily [918]. Two basic equations connecting block design parameters are

$$\begin{cases} bk = vr \\ \lambda(v-1) = r(k-1). \end{cases} \tag{2.62}$$

These conditions are necessary but they are not sufficient. There are no known examples of non-trivial t-$(v, k, 1)$-designs with $t > 5$ [918].

There exists the following equality which is true in any t-design:

$$\lambda_i = \frac{\lambda\binom{v-i}{t-i}}{\binom{k-i}{t-i}}, \tag{2.63}$$

where λ_i is the number of blocks that contain a fixed i-subset, $i \in \overline{0,t}$ [918]. Here

$$\binom{v-i}{t-i} = \frac{(v-i)!}{(t-i)!(v-t)!}, \binom{k-i}{t-i} = \frac{(k-i)!}{(t-i)!(k-t)!}.$$

Therefore

$$\lambda_i = \frac{\lambda(v-i)!(k-t)!}{(k-i)!(v-t)!}.$$

Theorem 2.173. *Any t-(v, k, λ)-design is also an i-(v, k, λ_i)-design for any i with $1 \leqslant i \leqslant t$ [850, p. 203], [918].*

A corollary of this theorem is that every t-design with $t \geqslant 2$ is also a 2-design.

Definition 2.174. Any 2-design is called a block design or BIBD, standing for balanced (the number λ is constant) incomplete ($v > k$) block design.

Theorem 2.175. *A necessary condition for the existence of balanced incomplete block designs 2-(v, k, λ) is that $\lambda(v-1) \equiv 0 \pmod{(k-1)}$ and $\lambda v(v-1) \equiv 0 \pmod{k(k-1)}$ [414].*

Proof. We follow [414]. We have that $r = \lambda(v-1)/(k-1)$ is the replication number of every point of this design and $b = \lambda v(v-1)/(k(k-1))$ is the total number of blocks. □

If we take all pair sets $\{x, y\} \subset S$, λ times each, then we construct a BIBD with $k = 2$ [414]. In [413] it is proved that conditions of Theorem 2.175 are sufficient for $k = 3$ and $k = 4$.

Tarry proved that there is no pair of orthogonal Latin squares of order six [882, 883]. A $(43, 7, 1)$-design does not exist, because the 2-design with such parameters is equivalent to the existence of five mutually orthogonal Latin squares of order six [414, 918].

The following theorem is named after the statistician Ronald Fisher.

Theorem 2.176. *In any 2-design the following inequality $b \geqslant v$ (Fisher's inequality) is true.*

Proof. The proof is taken from [929]. Let the incidence matrix $\mathbf{M}$ be a $v \times b$ matrix defined so that $M_{i,j}$ is 1 if element i is in block j and 0 otherwise. Then $\mathbf{B} = \mathbf{M}\mathbf{M}^\mathbf{T}$ is a $v \times v$ matrix such that $B_{i,i} = r$ and $B_{i,j} = \lambda$ for $i \neq j$. Since $r \neq \lambda$, $det(\mathbf{B}) \neq 0$, so $rank(\mathbf{B}) = v$; on the other hand, $rank(\mathbf{B}) = rank(\mathbf{M}) \leqslant b$, so $v \leqslant b$. □

The more detailed proof of Theorem 2.176 is given in [722, Chap. 8].

Theorem 2.177. *Balanced incomplete block designs* $2\text{-}(v, k, \lambda)$ *exist for all sufficiently large integers* v *satisfying the congruences* $\lambda(v-1) \equiv 0 \pmod{(k-1)}$ *and* $\lambda v(v-1) \equiv 0 \pmod{k(k-1)}$ *[945].*

See also Theorem 2.197.

2.9.2 3-nets and binary quasigroups

Quasigroups have not only combinatorial interpretation as Latin squares, but they have some geometrical interpretations, too. Here we give one of the most known interpretations, the so-called net interpretation.

Definition 2.178. A 3-net of order n is an incidence structure $\mathfrak{S} = (\mathfrak{P}, \mathfrak{L})$ which consists of an n^2-element set $\mathfrak{P}$ of points and a $3n$-element set $\mathfrak{L}$ of lines. The set $\mathfrak{L}$ is partitioned into 3 disjoint families L_1, L_2, L_3 of (parallel) lines, for which the following conditions are true:

(i) every point is incident with exactly one line of each family L_i $(i \in \{1, 2, 3\})$;
(ii) two lines of different families have exactly one point in common;
(iii) two lines in the same family do not have a common point [72, 76, 34, 419].

The families L_1, L_2, L_3 are sometimes called the directions or parallel classes of $\mathfrak{S}$.

Each Latin square M of order n naturally produces a 3-net. Points of this net are formed by the cells of M, while lines correspond to the left, right, and middle translations of the square M.

Therefore a 3-net $\mathfrak{S}$ is an incidence structure with the parameters $v = n^2$, $b = 3n$, $k = n$, $r = 3$. Note, using conditions (2.62) it is possible to check that 3-net $\mathfrak{S}$ is a t-(v, k, λ)-design only for $n = 2$.

The principle of duality is well known in projective geometry [948]. Using this principle it is possible to consider a dual structure to the structure $\mathfrak{S}$, namely $\mathfrak{S}^T = (\mathfrak{L}, \mathfrak{P})$, which has $\mathfrak{L}$ as points and $\mathfrak{P}$ as lines, and with the incidence relation transposed. In this case the incidence matrix is also transposed. Then $\mathfrak{S}^T$ has three families of points each of cardinality n, which are called groups, and n^2 blocks (lines). Parameters of $\mathfrak{S}^T$ are $v = 3n$, $b = n^2$, $k = 3$, $r = n$.

The structure $\mathfrak{S}^T$ is called a transversal design $TD(3, n)$ [213]. Note that a transversal design $TD(3, n)$ is a particular case of a partial linear space [164].

2.9.3 On orders of finite projective planes

"The case of equality in Fisher's inequality, that is, a 2-design with an equal number of points and blocks, is called a symmetric design. Symmetric designs have the smallest number of blocks amongst all the 2-designs with the same number of points.

"In a symmetric design $r = k$ holds as well as $b = v$, and, while it is generally not true in arbitrary 2-designs, in a symmetric design every two distinct blocks meet in λ points "[918]. The following theorem of Ryser provides the converse:

Theorem 2.179. *If* X *is a* v*-element set, and* $\mathcal{B}$ *is a* v*-element set of* k*-element subsets (the "blocks"), such that any two distinct blocks have exactly* λ *points in common, then* $(X, \mathcal{B})$ *is a symmetric block design [722].*

The parameters of a symmetric design satisfy $\lambda(v-1) = k(k-1)$. This imposes strong restrictions on v, so the number of points is far from arbitrary. The Bruck-Ryser-Chowla theorem gives necessary, but not sufficient, conditions for the existence of a symmetric design in terms of these parameters.

Theorem 2.180. *Bruck-Ryser-Chowla theorem. If a 2-(v,k,λ)-design exists with $v = b$ (a symmetric block design), then: if v is even, then $k - \lambda$ is a square; if v is odd, then the following Diophantine equation has a nontrivial solution:* $x^2-(k-\lambda)y^2-(-1)^{(v-1)/2}\lambda z^2 = 0$.

Definition 2.181. Finite projective planes are symmetric 2-designs with $\lambda = 1$.

Theorem 2.180 was proved in the case of projective planes in [175]. It was extended to symmetric designs in [210]. See also [722].

The number $n = k - 1$ is called the order of projective plane. For these designs the symmetric design equation becomes:

$$v - 1 = k(k - 1). \tag{2.64}$$

From equation (2.64) we obtain $v = (n + 1)n + 1 = n^2 + n + 1$ points in a projective plane of order n.

"As a projective plane is a symmetric design, we have $b = v$, meaning that $b = n^2+n+1$ also. The number b is the number of lines of the projective plane. There can be no repeated lines since $\lambda = 1$, so a projective plane is a simple 2-design in which the number of lines and the number of points are always the same. For a projective plane, k is the number of points on each line and it is equal to $n + 1$. Similarly, $r = n + 1$ is the number of lines with which any given point is incident.

For $n = 2$ we get a projective plane of order 2, also called the Fano plane, with $v = 4+2+1 = 7$ points and 7 lines. In the Fano plane, each line has $n+1 = 3$ points and each point belongs to $n + 1 = 3$ lines.

Projective planes are known to exist for all orders which are prime numbers or powers of primes. They form the only known infinite family (with respect to having a constant λ value) of symmetric block designs" [918].

The Bruck-Ryser theorem (a particular case of Theorem 2.180) gives necessary but not sufficient conditions for the existence of finite projective planes.

Theorem 2.182. *If the order n of a projective plane is congruent to* 1 *or* 2 (mod 4)*, then it must be the sum of two squares.*

"Theorem 2.182 rules out $n = 6$. The next case $n = 10$ has been ruled out by massive computer calculations. Nothing more is known; in particular, the question of whether there exists a finite projective plane of order $n = 12$ is still open. Another longstanding open problem is whether there exist finite projective planes of prime order which are not finite field planes (equivalently, whether there exists a non-Desarguesian projective plane of prime order)" [928].

2.9.4 Steiner systems

Definition 2.183. A Steiner system is a t-design with $\lambda = 1$ and $t \geqslant 2$.

Often a t-$(v,k,1)$-design (a Steiner system) is denoted by $S(t,k,n)$. Here n is the order of the set S. "As of 2012, an outstanding problem in design theory is if any nontrivial $(t < k < n)$ Steiner systems have $t \geqslant 6$. It is also unknown if infinitely many have $t = 4$ or 5" [937].

If $t = 2$, then we get the following classical definition:

Definition 2.184. A Steiner triple system $S(2,3,n)$ ($STS(n)$ for short) $(Q,3)$ defined on the set Q is a set of unordered triples $\{a,b,c\}$ such that

(i) a, b, c are distinct elements of Q;
(ii) to any $a, b \in Q$ such that $a \neq b$, there exists a unique triple $\{a,b,c\} \in (Q,3)$ [685].

Lemma 2.185. *In any Steiner triple system $S(n)$ the number of triples is $n(n-1)/6$.*

Proof. From Definition 2.184 it follows that any pair of elements defines a triple. Therefore we can have no more than $n(n-1)$ triples. We must divide the last expression by 6, because any ordered pair from the triple $\{a,b,c\}$ defines this triple in an unique way. □

If the triple of distinct elements a, b, c forms a Steiner triple, then on these elements we define the operation of Steiner quasigroup $(Q,\circ)$ as follows: $a \circ b = b \circ a = c$, $a \circ c = c \circ a = b$, $b \circ c = c \circ b = a$, $a \circ a = a$, $b \circ b = b$, $c \circ c = c$.

Theorem 2.186. *There is a one-to-one correspondence between all Steiner quasigroups which are defined on a set Q and all $STS(|Q|)$ defined on Q [685, V.1.11 Theorem].*

The following theorem, in fact, was proved in 1847 by Rev. T.P. Kirkman.

Theorem 2.187. *A necessary and sufficient condition for the existence of an $STS(n)$ S is that $n \equiv 1$ or $n \equiv 3 \pmod 6$.*

Proof. We follow [567, 204]. We find conditions when the expression $L = n(n-1)/6$ (see Lemma 2.185) is an integer. For any $x \in S$, the triples containing x partition $S\backslash\{x\}$ into pairs, thus $n-1$ is even, so n is odd.

Therefore, $n \equiv 1, 3$ or $5 \pmod 6$. However, if $n = 6k+5$, computing L gives: $L = (6k+5)(6k+4)/6 = (36k^2+54k+20)/6$ which is not an integer, so this case is eliminated.

In order to prove that conditions of this theorem are sufficient we must prove existence of Steiner triple systems of the announced orders.

Case $n \equiv 3 \pmod 6$. A quasigroup $(Q,\circ)$ of the form $x \circ y = (n+1)x + (n+1)y$, where $(Q,+,\cdot)$ is the ring of residues modulo $(2n+1)$, is a commutative idempotent quasigroup of order $(2n+1)$.

Further we give the Bose construction [158] of an $STS(6n+3)$ for any natural number n. We use the commutative idempotent quasigroup $(Q,\circ)$ of order $(2n+1)$ and the group $(Z_3,+)$ of residues modulo 3 that is defined on the set $\{0,1,2\}$.

The set $S = Q \times \{0,1,2\}$ consists of $6n+3$ ordered pairs and the set T of triples consists of triples of two types:

Type 1 : $\{(i,0),(i,1),(i,2)\}$ for each $i \in Q$.

Type 2 : $\{(i,k),(j,k),(i \circ j, k+1 \pmod 3)\}$ for $i \neq j$.

To show that this construction gives an STS we first count the number of triples: there are $2n+1$ triples of type 1 and $3(2n+1)(2n)/2 = 6n^2+3n$ triples of type 2. Thus, $T = 6n^2+5n+1 = (6n+3)(6n+2)/6 = v(v-1)/6$.

To prove that the set T is an STS we need only to show that each pair of distinct elements of S are contained in a triple (since the number of triples is correct this will force each pair to be in a unique triple). Let (a,b) and (c,d) be distinct elements of S. If $a = c$, then this pair is in a triple of type 1.

We now assume that $a \neq c$. If $b = d$, the pair is in a triple of type 2. We now also assume that $b \neq d$. We have two possibilities: either $d \equiv b+1 \pmod 3$ or $d \equiv b-1 \pmod 3$.

In the first case, let x be the unique solution of equation $a \circ x = c$ in quasigroup $(Q,\circ)$. The triple containing the pair is thus $\{(a,b),(x,b),(c,d)\}$. In the second case, let y be the unique solution of $y \circ c = a$ in $(Q,\circ)$. The triple is then $\{(y,d),(c,d),(a,b)\}$" [204].

Case $n \equiv 1 \pmod 6$.

Definition 2.188. A Latin square (quasigroup) L of order $2n$ is half-idempotent if the cells (i, i) and $(n+i, n+i)$ contain the symbol i, for every $1 \leqslant i \leqslant n$.

There exist commutative half-idempotent Latin squares for all even orders $2n (n \geqslant 1)$. We start from the elementary observation that the Cayley table of any cyclic group $(Z_{2n}, +)$ of order $2n$ contains, on the main diagonal, n distinct elements any of which appears twice. It is clear because in the group $(Z_{2n}, +)$ the mapping $x \mapsto 2x$ is an endomorphism with the kernel of order two.

Here we suppose that the group $(Z_{2n}, +)$ is defined on the set $\{0, 1, \dots, 2n-1\}$. The isotopic image of the group $(Z_{2n}, +)$ of the form $(\varepsilon, \varepsilon, \gamma^{-1})$, where $\gamma^{-1}(2i) = i$, $\gamma^{-1}(2i+1) = n+i$ for all $i \in \overline{\{0, n-1\}}$, is a half-idempotent quasigroup.

We give Skolem construction of $STS(6n+1)$ [808].

"This construction of an $STS(6n+1)$ starts with a set S consisting of the $6n$ ordered pairs of $Q \times \{0, 1, 2\}$, where $(Q, \circ)$ is a commutative half-idempotent quasigroup of order $2n$, together with a special symbol called ∞. To describe the triples we assume that the quasigroup $(Q, \circ)$ has symbols $\{1, 2, \dots, 2n\}$. The triples are then:

Type 1: $\{(i, 0), (i, 1), (i, 2)\}$ for $1 \leqslant i \leqslant n$.
Type 2: $\{\infty, (i, k), (n+i, k-1 \pmod 3)\}$ for $1 \leqslant i \leqslant n$.
Type 3: $\{(i, k), (j, k), (i \circ j, k+1 \pmod 3)\}$ for $1 \leqslant i < j \leqslant 2n$.

Note that the type 3 triples here are precisely the same as the type 2 triples in the Bose construction.

We again count the number of triples: There are n triples of type 1, $3n$ triples of type 2 and $3(2n)(2n-1)/2 = 6n^2 - 3n$ triples of type 3. This gives $|T| = 6n^2 + n = (6n+1)(6n)/6 = v(v-1)/6$.

Again, to show that we have constructed an STS, we need to show that each pair of elements is contained in a triple.

Any pair including the symbol ∞ is contained in a type 2 triple. Suppose (a, b) and (c, d) are a pair of elements of S. If $a = c$ and $a \leqslant n$, then the pair is contained in a triple of type 1.

Suppose (a, b) and (c, d) are a pair of elements of S. Now suppose that $a = c$ and $a > n$. Since $b \neq d$, either $d = b+1 \pmod 3$ or $d = b-1 \pmod 3$. In the first case, let x be the unique solution of $a \circ x = a$ in $(Q, \circ)$. Since $a > n$, $x \neq a$. The triple containing the pair is thus $\{(a, b), (x, b), (a, d)\}$. In the second case, let y be the unique solution of $y \circ a = a$ in $(Q, \circ)$. Again, $y \neq a$ and the triple is then $\{(y, d), (a, d), (a, b)\}$.

We can now assume that $a \neq c$. If $b = d$, then a triple of type 3 contains the pair, so we can also assume that $b \neq d$. Again, either d = b+1(mod 3) or d = b - 1 (mod 3). In the first case, let x be the unique solution of $a \circ x = c$ in $(Q, \circ)$. If $x \neq a$, then the type 3 triple $\{(a, b), (x, b), (c, d)\}$ contains the pair. If on the other hand $x = a$, then $a > n$, since $a \neq c$. In this case, $a = n + c$ and the pair is in the type 2 triple $\{\infty, (c, d), (n+c, b)\}$. The other possibility for d is treated in the similar way" [204]. □

Theorem 2.189. *Doyen-Wilson Theorem. There exists an $STS(v)$ which contains an $STS(w)$ as a proper subdesign if and only if $v > 2w+1$, $v \equiv 1$ or $3 \pmod 6$, and $w \equiv 1$ or $3 \pmod 6$ [266].*

Definition 2.190. A finite affine plane of order q, with the lines as blocks, is an $S(2, q, q^2)$, i.e., it is an 2-$(q^2, q, 1)$-design.

An affine plane of order q can be obtained from a projective plane of the same order by removing one block and all of the points in that block from the projective plane. Choosing different blocks to remove in this way can lead to non-isomorphic affine planes.

Non-formally speaking, STS(7) is the projective plane of order 2 (the Fano plane), an STS(9) is the affine plane of order 3.

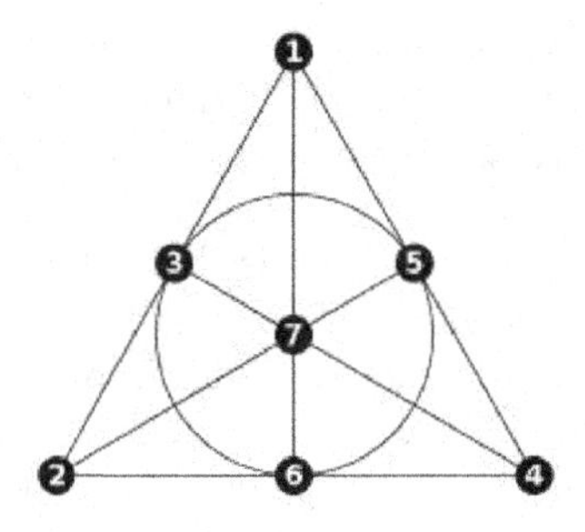

Figure 2.1: The Fano plane.

Using the following Steiner quasigroup of order 9 we can construct the corresponding STS(9) and affine plane of order 3.

*	0	1	2	3	4	5	6	7	8
0	0	2	1	4	3	8	7	6	5
1	2	1	0	5	7	3	8	4	6
2	1	0	2	6	8	7	3	5	4
3	4	5	6	3	0	1	2	8	7
4	3	7	8	0	4	6	5	1	2
5	8	3	7	1	6	5	4	2	0
6	7	8	3	2	5	4	6	0	1
7	6	4	5	8	1	2	0	7	3
8	5	6	4	7	2	0	1	3	8

Steiner quadruple systems $S(3,4,n)$ exist for all $n \equiv 2 \pmod 6$ or $n \equiv 4 \pmod 6$ [412, 415].

We give a sketch description of the Steiner system S(5, 6, 12) [937]. There is a unique Steiner system S(5, 6, 12); its automorphism group is the Mathieu group M_{12}. To construct it, take a 12-point set and think of it as the projective line over $GF(11)$, in other words, the integers mod 11 together with a point called infinity. Among the integers mod 11, the six ones are perfect squares (perfect square is an element which is equal to the square of another element):{0, 1, 3, 4, 5, 9}. Call this set a "block." From this, we may obtain other blocks by applying fractional linear transformations:

$$z \mapsto \frac{az+b}{cz+d}.$$

These blocks then form a (5, 6, 12) Steiner system.

Steiner systems $S(5,6,v)$ now are known to exist for $v = 12, 24, 48, 72, 84, 108,$ and 132. These are precisely the first seven admissible values of v which are 1 more than a prime $p \equiv 3 \pmod 4$ and this fact is utilized in constructing the systems from orbits of 6-sets under the group $PSL_2(v-1)$ [381]. Information about Mathieu groups and projective special linear (PSL) groups can be found in many books, for example in [407, 471].

2.9.5 Mendelsohn design

Definition 2.191. A (v, k, λ)-Mendelsohn design, or MD(v, k, λ), is a v-set V and a collection β of ordered k-tuples of distinct elements of V (called blocks), such that each ordered pair (x, y) with $x \neq y$ of elements of V is cyclically adjacent in λ blocks. The ordered pair (x, y) of distinct elements is cyclically adjacent in a block if the elements appear in the block as $(\ldots, x, y, \ldots)$ or $(y, \ldots, x)$. An $MD(v, 3, \lambda)$ is a Mendelsohn triple system, $MTS(v, \lambda)$ [213, 940].

In a Mendelsohn triple system the containment is cyclic: an ordered triple (x, y, z) contains the ordered pairs (x, y), (y, z) and (z, x) [382].

Example 2.192. An example of an $MTS(4, 1)$ on $V = \{0, 1, 2, 3\}$ is: $(0, 1, 2)$, $(1, 0, 3)$, $(2, 1, 3)$, $(0, 2, 3)$.

Example 2.193. If $(Q, *)$ is an idempotent semisymmetric quasigroup, that is, $x * x = x$ (idempotent) and $x * (y * x) = y$ (semisymmetric) for all $x, y \in Q$, let $\beta = \{(x, y, x * y) : x, y \in Q\}$. Then (Q, β) is a Mendelsohn triple system $MTS(|Q|, 3)$. This construction is reversible [213]. We give the example of an idempotent semisymmetric quasigroup of order 4.

*	0	1	2	3
0	0	2	3	1
1	3	1	0	2
2	1	3	2	0
3	2	0	1	3

This quasigroup defines $MTS(4, 3)$: $(0, 1, 2)$, $(0, 2, 3)$, $(0, 3, 1)$, $(1, 0, 3)$, $(1, 2, 0)$, $(1, 3, 2)$, $(2, 0, 1)$, $(2, 1, 3)$, $(2, 3, 0)$, $(3, 0, 2)$, $(3, 1, 0)$, $(3, 2, 1)$. For example, $(0, 1) \in \{(0, 1, 2), (1, 2, 0), (2, 0, 1)\}$.

Directed triple systems are defined similar to the Mendelsohn triple system with one exception, namely, the containment is transitive: an ordered triple (x, y, z) contains the ordered pairs (x, y), (y, z) and (x, z) [434, 284]. Notice that the Mendelsohn triple system $MTS(4, 3)$ constructed in Example 2.193 is also a directed triple system. For example, $(0, 1) \in \{(0, 1, 2), (0, 3, 1), (2, 0, 1)\}$.

Notice, $MTS(4, 1)$ from Example 2.192 is not a directed triple system. Indeed, the pairs $(2, 0), (3, 1), (3, 2)$ are not in the containment of the system $MTS(4, 1)$ which is considered as a directed triple system.

The book [214] is devoted to various triple systems.

2.9.6 Spectra of quasigroups with 2-variable identities

The spectrum of a quasigroup property T (for example, of an identity) refers to the possible orders of quasigroups in which the property T is fulfilled. Therefore the words "spectrum of Steiner quasigroups" mean the set of numbers (of orders) for which Steiner quasigroups exist.

Definition 2.194. A pairwise balanced design (or PBD) is a set S together with a family of subsets of S (which need not have the same size and may contain repeats) such that every pair of distinct elements of S is contained in exactly λ (a positive integer) subsets. The set S is allowed to be one of the subsets, and if all the subsets are copies of S, the PBD is called trivial. The size of S is v and the number of subsets in the family (counted with multiplicity) is b [918].

Fisher's inequality holds for PBDs. Pairwise balanced designs (PBDs) are examples of block designs that are not necessarily uniform. Pairwise balanced designs of order $v = 6n+4$ are constructed in [42], of order $v = 6n+5$ are constructed in [567, 204].

The following theorem establishes a connection between some idempotent quasigroups and block designs.

Theorem 2.195. *Let V be a variety (more generally universal class) of algebras that is idempotent and that is based on 2-variable identities. If there are models of V of orders $k_1, k_2, \dots$ and if there is a pairwise balanced block design with $\lambda = 1$, order v, and blocks of sizes $k_1, k_2, \dots$, then there is a model of V of order v [572].*

Earlier versions of this theorem are given in [846, Section 4], [680].

Example 2.196. This example of the application of Theorem 2.195 is taken from [572]. The following conditions $v \equiv 1$ or $4 \pmod{12}$ are necessary for the existence of BIBDs with $\lambda = 1$ and $k = 4$ [414, Lemma 5.11].

Taking into consideration Theorem 2.195 and the fact that there exists an idempotent Schröder quasigroup of order four (Examples 2.168 and 2.169), we can conclude that idempotent Schröder quasigroups exist for all values of the parameter v, for which the corresponding block designs are constructed.

In [570] the singular direct product for construction of idempotent quasigroups with some 2-variable identities is used.

Theorem 2.197 was proved by Richard M. Wilson. A generalization of the Wilson theorem for 3-designs is given in [415].

Theorem 2.197. *Let $K = \{k_1, k_2, \dots\}$ be a set of positive integers. Let $\alpha(K) = g.c.d.(k(k-1) \mid k \in K)$ and $\beta(K) = g.c.d.(k-1 \mid k \in K)$. Then there is an integer v_0 such that $v \geqslant v_0$ and if $v(v-1) \equiv 0 \pmod{\alpha(K)}$ and $v - 1 \equiv 0 \pmod{\beta(K)}$, then there is a pairwise balanced block design with $\lambda = 1$, order v and block sizes $k_1, k_2, \dots$ [945, 572].*

Using Example 2.168, Theorems 2.195 and 2.197, following [572], we can estimate the spectrum of the finite idempotent Schröder quasigroup. We recall, natural numbers q_i, such that there exist quasigroups $(Q, \cdot)$ of orders q_i, form a spectrum of $(Q, \cdot)$, i.e., the spectrum consists of natural numbers for which there exist quasigroups of this order.

From Example 2.168 it follows that in a particular case we can take $K = \{4, 8\}$. Then $\alpha(K) = 4$, $\beta(K) = 1$ implies that there is an idempotent Schröder quasigroup of order $v \equiv 0 \pmod 4$ or $v \equiv 1 \pmod 4$, provided $v \geqslant v_0$ (compare the last result with Theorem 2.167).

The following quasigroups are idempotent C_3-quasigroups of orders 4 and 7.

*	0	1	2	3
0	0	2	3	1
1	3	1	0	2
2	1	3	2	0
3	2	0	1	3

∘	0	1	2	3	4	5	6
0	0	2	3	1	5	6	4
1	2	1	4	6	0	3	5
2	3	4	2	5	6	0	1
3	1	6	5	3	2	4	0
4	5	0	6	2	4	1	3
5	6	3	0	4	1	5	2
6	4	5	1	0	3	2	6

Therefore in terms of the Wilson theorem (Theorem 2.197) $K = \{4, 7\}$, $\alpha K = 6$, $\beta K = 3$, $v(v-1) \equiv 0 \pmod 6$ and $v - 1 \equiv 0 \pmod 3$. The last system is equivalent to the following congruence: $v \equiv 1 \pmod 3$. Application of Theorem 2.195 yields the following result: there exist idempotent C_3-quasigroups for any finite "big" order v, $v \equiv 1 \pmod 3$.

It is clear that the groups Z_3 and $Z_3 \times Z_3$ of orders 3 and 9, respectively, are C_3-quasigroups. Application of Theorem 2.197 gives us the following result: there are pairwise balanced block designs with $\lambda = 1$ of order $v \equiv 0 \pmod 3$. Since the groups Z_3 and $Z_3 \times Z_3$ are not idempotent, we cannot apply Theorem 2.195. Therefore we must prove in another way, that using designs of this kind we can construct C_3-quasigroups of any order $v \equiv 0 \pmod 3$.

The final result is the following: spectrum of C_3-quasigroups is $q \equiv 0$ or $1 \pmod 3$ [139, 141]. In very short manner this result is reproved in [193] using the properties of orthogonal operations and binary selectors [74]. Information on some algebraic properties of C_3-quasigroups is given in [866].

Stein quasigroups satisfy the following identity $x \cdot xy = yx$. Any Stein quasigroup is idempotent. Indeed, by $x = y$ in the Stein identity we have $x \cdot xx = xx, xx = x$. The following quasigroups are Stein quasigroups of orders 4, 5, and 11.

$*$	0	1	2	3
0	0	2	3	1
1	3	1	0	2
2	1	3	2	0
3	2	0	1	3

$\circ$	0	1	2	3	4
0	0	2	3	4	1
1	3	1	4	2	0
2	4	0	2	1	3
3	1	4	0	3	2
4	2	3	1	0	4

$\star$	0	1	2	3	4	5	6	7	8	9	10
0	0	2	3	5	1	6	8	10	9	7	4
1	3	1	8	2	7	4	9	6	10	0	5
2	5	10	2	9	6	3	1	0	7	4	8
3	6	8	4	3	9	7	5	1	2	10	0
4	2	6	1	10	4	8	7	3	0	5	9
5	8	7	9	1	0	5	10	4	6	3	2
6	9	0	10	7	3	2	6	5	4	8	1
7	4	9	5	8	10	0	2	7	1	6	3
8	7	5	0	4	2	10	3	9	8	1	6
9	10	3	6	0	8	1	4	2	5	9	7
10	1	4	7	6	5	9	0	8	3	2	10

Therefore $K = \{4, 5, 11\}$, $\alpha(K) = 2, \beta(K) = 1$. We obtain the following system of congruences: $v(v-1) \equiv 0 \pmod 2$ and $v - 1 \equiv 0 \pmod 1$. From this system we have that any v, $v \geqslant 2$ satisfies this congruence. Using Theorem 2.197 and 2.195 we get the following result: there exist Stein quasigroups of any finite "big" order.

From the results of Pelling and Rogers [680] it follows that for Stein quasigroups $v_0 \leqslant 1042$. In this article many Stein quasigroups of small (< 1042) orders are also constructed.

Linear quasigroups (in partial case, medial and T-quasigroups) are a convenient tool for construction of various quasigroup examples and researching spectra of the corresponding quasigroups.

Lemma 2.198. *A T-quasigroup of the form $x \cdot y = \varphi x + \psi y$ is a Stein quasigroup if and only if $\varphi = \psi^2$, $\psi^2 + \psi - \varepsilon = 0$.*

Proof. It is easy to see. See also [680]. Notice, expression $\psi^2 + \psi - \varepsilon = 0$ can be written in the form $\psi^{-1}(\psi^2 + \psi - \varepsilon) = 0$, $\psi + \varepsilon = \psi^{-1}$. □

Ternary quasigroup (Q, f) is called 2-idempotent, if the following identity is true: $f(x, x, y) = f(x, y, x) = f(y, x, x) = y$. The spectrum of 2-idempotent 3-ary quasigroups with additional conditions is studied in [465].

2.10 Linear quasigroups

2.10.1 Introduction

As well as in other areas of mathematics, the idea of linearity plays an important role in quasigroup theory. This idea is used by definition of linear quasigroups.

We recall that nowadays it is usual to call a binary quasigroup $(Q, \cdot)$ of the form $x \cdot y = (\alpha x + \beta y) + c$, where $(Q, +)$ is a "good" loop, $\alpha, \beta \in Aut(Q, +)$, c is a fixed element of the set Q, *a linear quasigroup*. I.e., quasigroups that are linear relative to "good" loops (or relative to "good" quasigroups) are studied.

The majority of classical classes of quasigroups are linear or generalized linear quasigroups. For instance, medial quasigroups (Toyoda theorem, [890]), distributive quasigroups (Belousov theorem, [58]), distributive Steiner quasigroups, left distributive quasigroups (Belousov–Onoi theorem, [90]), CH-quasigroups (Manin theorem, [593]), T-quasigroups, n-ary groups (Gluskin-Hosszu theorem, [426, 364, 837]), n-ary medial quasigroups (Evans theorem [308] and Belousov theorem [77]), F-quasigroups (Kepka-Kinyon-Phillips theorem, [500]) are quasigroups of this kind.

There exist quasigroups that are also linear over other quasigroups. For example, from the proof of the fact that every commutative IP-quasigroup is isotopic to a TS-quasigroup [685, V.1.18 Theorem], it follows that some TS-quasigroups are linear over commutative IP-quasigroups.

It makes sense to call a quasigroup, in which the permutations α, β are "good" permutations of the set S_Q, *a generalized linear quasigroup*: for example, if α, β are antiautomorphisms of a "good" loop [875, 874, 877, 878].

V.D. Belousov introduced linear quasigroups (over groups) in the mid-1960s [69]. In the 1970s, linear quasigroups over abelian groups (T-quasigroups) and some other classes of generalized linear quasigroups were studied intensively by T. Kepka, P. Nemec, J. Jezek, V.D. Belousov, V.I. Onoi and other mathematicians [667, 506, 461, 497, 498, 507, 90]. It is easy to see that the concept of the generalized linear quasigroup can be extended to the n-ary case.

Generalized linear quasigroups can be stored in sufficiently compact form in computer memory. This gives rich possibilities for the use of such quasigroups in practice. We recall, in the general case, in order to define an 100-ary quasigroup of order 10 we should set 10^{100} equalities.

In the late 1980s and early 1990s, members of Belousov's quasigroup school (G.B. Belyavskaya and her pupils P.N. Syrbu and A.Kh. Tabarov, W.A. Dudek, V.I. Izbash, V.A. Shcherbacov, F.N. Sokhatsky and his pupils O.E. Kirnasovski and P. Syvakivskyi) and K.K. Shchukin started more active researches of generalized linear quasigroups. We recall that V.D. Belousov and K.K. Shchukin were pupils of A.G. Kurosh.

Researches of congruences, automorphisms, isomorphisms, endomorphisms, identities, nuclei, center, associators, commutators, multiplication groups and the property of orthogonality of generalized linear quasigroups (including the n-ary case) were carried out. Some numerical evaluations were obtained. Notice that research in this direction continues today.

Here we give some of the articles of the members of Belousov school devoted to these research areas: [104, 108, 110, 133, 134, 132, 290, 294, 441, 444, 443, 525, 523, 524, 760, 761, 763, 762, 764, 765, 769, 767, 824, 831, 832, 825, 826, 827, 828, 871, 874, 876, 878, 880, 881, 799, 801, 736].

Researches in this direction were carried out by L. Beneteau [201], T. Kepka [421], P. Němec [665, 666], J.D.H. Smith [201, 818, 819], and some other mathematicians. A good overview of results of a variety of group isotopes and new results on this variety are given in [362].

The idea of more or less complete research of generalized linear quasigroups (including n-ary case) was born during discussions of K.K. Shchukin and the author of this book. In the published form this idea (binary case) is reflected in [768, 767]. Linear quasigroups are used in code theory and cryptology. See [123, 790] and the following chapters.

The Toyoda and Belousov theorems were bases for the concept of a linear quasigroup. Notice, in his article Toyoda studied n-ary medial quasigroups, too. Probably other bases of the concept of a linear quasigroup are researches of the functional equation of associativity and balanced identities on quasigroups [63, 69, 5]. For the n-ary variant of the Toyoda theorem see in [308, 77].

The n-ary variant of the Belousov theorem on the properties of n-ary distributive quasigroups, as well as a suitable definition of n-ary distributive quasigroup, are not known.

2.10.2 Definitions

We start from definitions.

Definition 2.199. A quasigroup $(Q,\cdot)$ is called left linear, if $x \cdot y = \varphi x + a + \beta y$, where $(Q,+)$ is a group, the element a is a fixed element of the set Q, $\varphi \in Aut(Q,+)$, $\beta \in S_Q$.

Lemma 2.200. *If a left linear quasigroup $(Q,\cdot)$ has the form $x \cdot y = \varphi x + a + \beta y$ over a group $(Q,+)$, then it also has the form $x \cdot y = \varphi x + J_a \beta y + a$ over the group $(Q,+)$, where $J_a x = a + x - a$, and vice versa.*

Proof. Indeed, from equality $x \cdot y = \varphi x + a + \beta y$ we have $x \cdot y = \varphi x + a + \beta y - a + a = \varphi x + (a + \beta y - a) + a = \varphi x + J_a \beta y + a$. Notice, the map J_a is an inner automorphism of the group $(Q,\cdot)$ [471]. It is clear that the map J_a is a permutation of the set Q. □

Definition 2.201. A quasigroup $(Q,\cdot)$ is called left alinear if $x \cdot y = \overline{\varphi} x + a + \beta y$, where $(Q,+)$ is a group, the element a is a fixed element of the set Q, $\overline{\varphi}$ is an anti-automorphism, and $\beta \in S_Q$.

Lemma 2.202. *If a left alinear quasigroup $(Q,\cdot)$ has the form $x \cdot y = \overline{\varphi} x + a + \beta y$ over a group $(Q,+)$, then it also has the form $x \cdot y = \overline{\varphi} x + J_a \beta y + a$ over the group $(Q,+)$, where $J_a x = a + x - a$, and vice versa.*

Proof. The proof is similar to the proof of Lemma 2.200. □

In [818] left (right) linear quasigroups over an abelian group are called *semicentral.*

Definition 2.203. A quasigroup $(Q,\cdot)$ of the form $x \cdot y = \varphi x + a + \psi y$, where $(Q,+)$ is a group and the element a is a fixed element of the set Q, $\varphi, \psi \in Aut(Q,+)$, is called a linear quasigroup (over the group $(Q,+)$).

If the maps φ and ψ are antiautomorphisms of the group $(Q,+)$, then the quasigroup is called alinear.

Corollary 2.204.
1. *A linear quasigroup $(Q,\cdot)$ has the form $x \cdot y = \varphi x + a + \psi y$ if and only if it has the following form: $x \cdot y = \varphi x + J_a \psi y + a$.*
2. *An alinear quasigroup $(Q,\cdot)$ has the form $x \cdot y = \overline{\varphi} x + a + \overline{\psi} y$ if and only if it has the form $x \cdot y = \overline{\varphi} x + J_a \overline{\psi} y + a$.*

Proof. The proof follows from Lemmas 2.200 and 2.202. □

It is clear that any alinear quasigroup over an abelian group is linear since in any abelian group any antiautomorphism is an automorphism.

2.10.3 Group isotopes and identities

Information for this subsection is taken from [4, 63, 77, 69, 540, 875, 830]. We formulate the famous four quasigroups theorem [4, 63, 77, 830] as follows.

Theorem 2.205. *A quadruple (f_1, f_2, f_3, f_4) of binary quasigroup operation defined on a nonempty set Q is the general solution of the generalized associativity equation*

$$A_1(A_2(x, y), z) = A_3(x, A_4(y, z))$$

if and only if there exists a group $(Q, +)$ and permutations $\alpha, \beta, \gamma, \mu, \nu$ of the set Q such that $f_1(t, z) = \mu t + \gamma z$, $f_2(x, y) = \mu^{-1}(\alpha x + \beta y)$, $f_3(x, u) = \alpha x + \nu u$, $f_4(y, z) = \nu^{-1}(\beta y + \gamma z)$.

Lemma 2.206. *Belousov criteria. If in a group $(Q, +)$ the equality $\alpha x + \beta y = \gamma y + \delta x$ holds for all $x, y \in Q$, where $\alpha, \beta, \gamma, \delta$ are some fixed permutations of Q, then $(Q, +)$ is an abelian group [69].*

There exists also the following corollary adapted for our aims from the results of F.N. Sokhatskii (Sokhatskii criteria) [830, Theorem 6.7.2].

Corollary 2.207. *If in a principal group isotope $(Q, \cdot)$ of a group $(Q, +)$ the equality $\alpha x \cdot \beta y = \gamma y \cdot \delta x$ holds for all $x, y \in Q$, where $\alpha, \beta, \gamma, \delta$ are some fixed permutations of Q, then $(Q, +)$ is an abelian group.*

Proof. If $x \cdot y = \xi x + \chi y$, then we can re-write the equality $\alpha x \cdot \beta y = \gamma y \cdot \delta x$ in the form $\xi \alpha x + \chi \beta y = \xi \gamma y + \chi \delta x$. Now we can apply the Belousov criteria (Lemma 4.91). □

Lemma 2.208. *1. For any principal group isotope $(Q, \cdot)$ there exists its form $x \cdot y = \alpha x + \beta y$ such that $\alpha 0 = 0$.*

2. *For any principal group isotope $(Q, \cdot)$ there exists its form $x \cdot y = \alpha x + \beta y$ such that $\beta 0 = 0$.*

3. *For any right linear quasigroup $(Q, \cdot)$ there exists its form $x \cdot y = \alpha x + \psi y + c$ such that $\alpha 0 = 0$.*

4. *For any left linear quasigroup $(Q, \cdot)$ there exists its form $x \cdot y = \varphi x + \beta y + c$ such that $\beta 0 = 0$.*

5. *For any left linear quasigroup $(Q, \cdot)$ with idempotent element 0 there exists its form $x \cdot y = \varphi x + \beta y$ such that $\beta 0 = 0$.*

6. *For any right linear quasigroup $(Q, \cdot)$ with idempotent element 0 there exists its form $x \cdot y = \alpha x + \psi y$ such that $\alpha 0 = 0$ [825].*

Proof. 1). We have $x \cdot y = \alpha x + \beta y = R_{-\alpha 0}\alpha x + L_{\alpha 0}\beta y = \alpha' x + \beta' y$, $\alpha' 0 = 0$.

2). We have $x \cdot y = \alpha x + \beta y = R_{\beta 0}\alpha x + L_{-\beta 0}\beta y = \alpha' x + \beta' y$, $\beta' 0 = 0$.

3). We have $x \cdot y = \alpha x + \psi y + c = R_{-\alpha 0}\alpha x + I_{\alpha 0}\psi y + \alpha 0 + c = \alpha' x + \psi' y + c'$, where $I_{\alpha 0}\psi y = \alpha 0 + \psi y - \alpha 0$, $\alpha' 0 = 0$. Since $I_{\alpha 0}$ is an inner automorphism of the group $(Q, +)$, we obtain $I_{\alpha 0}\psi \in Aut(Q, +)$.

4). We have $x \cdot y = \varphi x + \beta y + c = \varphi x + \beta y - \beta 0 + \beta 0 + c = \varphi x + R_{-\beta 0}\beta y + \beta 0 + c = \varphi x + \beta' y + c'$, where $\beta' = R_{-\beta 0}\beta$, $c' = \beta 0 + c$.

5). If $x \cdot y = \varphi x + \beta y + c$, then $0 = 0 \cdot 0 = \varphi 0 + \beta 0 + c = \beta 0 + c$, $\beta 0 = -c$. Therefore $x \cdot y = \varphi x + R_c \beta y = \varphi x + \beta' y$ and $\beta' 0 = R_c \beta 0 = -c + c = 0$.

6). If $x \cdot y = \alpha x + \psi y + c$, then $0 = 0 \cdot 0 = \alpha 0 + \psi 0 + c = \alpha 0 + c$, $\alpha 0 = -c$. Therefore $x \cdot y = \alpha x + c - c + \psi y + c = R_c \alpha x + I_{-c}\psi y = R_c \alpha' x + \psi' y$ and $\alpha' 0 = R_c \alpha 0 = -c + c = 0$. Moreover, ψ' is an automorphism of $(Q, +)$ as the product of two automorphisms of the group $(Q, +)$. □

Lemma 2.209. *For any left linear quasigroup $(Q, \cdot)$ there exists its form such that $x \cdot y = \varphi x + \beta y$.*

For any right linear quasigroup $(Q, \cdot)$ there exists its form such that $x \cdot y = \alpha x + \psi y$.

Proof. We can re-write the form $x \cdot y = \varphi x + \beta y + c$ of a left linear quasigroup $(Q, \cdot)$ as follows: $x \cdot y = \varphi x + R_c \beta y = \varphi x + \beta' y$, where $\beta' = R_c \beta$.

We can re-write the form $x \cdot y = \alpha x + \psi y + c$ of a right linear quasigroup $(Q, \cdot)$ as follows: $x \cdot y = \alpha x + c - c + \psi y + c = R_c \alpha x + I_c \psi y = \alpha' x + \psi' y$, where $I_{-c} \psi y = -c + \psi y + c$. □

Classical criteria of a linearity of a quasigroup are given by V.D. Belousov in [69]. We give a partial case of the F.N. Sokhatskii result [830, Theorem 6.8.6], [828, Theorem 3], [826].

We recall, up to isomorphism, that any isotope is principal (Remark 1.118).

Theorem 2.210. *Let $(Q, \cdot)$ be a principal isotope of a group $(Q, +)$, $x \cdot y = \alpha x + \beta y$.*

If $(\alpha_1 x \cdot \alpha_2 y) \cdot a = \alpha_3 x \cdot \alpha_4 y$ is true for all $x, y \in Q$, where $\alpha_1, \alpha_2, \alpha_3, \alpha_4$ are permutations of the set Q, a is a fixed element of the set Q, then $(Q, \cdot)$ is a left linear quasigroup.

If $a \cdot (\alpha_1 x \cdot \alpha_2 y) = \alpha_3 x \cdot \alpha_4 y$ is true for all $x, y \in Q$, where $\alpha_1, \alpha_2, \alpha_3, \alpha_4$ are permutations of the set Q, a is a fixed element of the set Q, then $(Q, \cdot)$ is a right linear quasigroup.

Proof. We follow [830]. By Lemma 2.208, quasigroup $(Q, \cdot)$ can have the form $x \cdot y = \alpha x + \beta y$ over a group $(Q, +)$ such that $\alpha 0 = 0$. If we pass in the equality $(\alpha_1 x \cdot \alpha_2 y) \cdot a = \alpha_3 x \cdot \alpha_4 y$ to the operation "+ ", then we obtain $\alpha(\alpha \alpha_1 x + \beta \alpha_2 y) + \beta a = \alpha \alpha_3 x + \beta \alpha_4 y$, $\alpha(x + y) = \alpha \alpha_3 \alpha_1^{-1} \alpha^{-1} x + \beta \alpha_4 \alpha_2^{-1} \beta^{-1} y - \beta a$.

Then the permutation α is a group quasiautomorphism. It is known that any group quasiautomorphism has the form $L_a \varphi$, where $\varphi \in Aut(Q, +)$. See [80, 72] or Corollary 1.150. Therefore $\alpha \in Aut(Q, +)$, since $\alpha 0 = 0$.

By Lemma 2.208 there exists the form $x \cdot y = \alpha x + \beta y$ of quasigroup $(Q, \cdot)$ such that $\beta 0 = 0$. If we pass in the equality $a \cdot (\alpha_1 x \cdot \alpha_2 y) = \alpha_3 x \cdot \alpha_4 y$ to the operation "+ ", then we obtain $\alpha a + \beta(\alpha \alpha_1 x + \beta \alpha_2 y) = \alpha \alpha_3 x + \beta \alpha_4 y$, $\beta(x + y) = -\alpha a + \alpha \alpha_3 \alpha_1^{-1} \alpha^{-1} x + \beta \alpha_4 \alpha_2^{-1} \beta^{-1} y$.

Then the permutation β is a group quasiautomorphism. Therefore $\beta \in Aut(Q, +)$, since $\beta 0 = 0$. □

Corollary 2.211. *If a left F-quasigroup (E-quasigroup, SM-quasigroup) is a group isotope, then this quasigroup is right linear.*

If a right F-quasigroup (E-quasigroup, SM-quasigroup) is a group isotope, then this quasigroup is left linear [828].

Proof. It is possible to apply direct calculations and Theorem 2.210. □

Lemma 2.212. *1. If in a right linear quasigroup $(Q, \cdot)$ over a group $(Q, +)$ the equality $k \cdot yx = xy \cdot b$ holds for all $x, y \in Q$ and fixed $k, b \in Q$, then $(Q, +)$ is an abelian group.*

2. If in a left linear quasigroup $(Q, \cdot)$ over a group $(Q, +)$ the equality $k \cdot yx = xy \cdot b$ holds for all $x, y \in Q$ and fixed $k, b \in Q$, then $(Q, +)$ is an abelian group.

Proof. 1. By Lemma 2.209 we can take the following form of $(Q, \cdot)$: $x \cdot y = \alpha x + \psi y$. Thus we have $\alpha k + \psi(\alpha y + \psi x) = \alpha(\alpha x + \psi y) + \psi b$, $\alpha k + \psi \alpha y + \psi^2 x - \psi b = \alpha(\alpha x + \psi y)$, $\alpha(x + y) = \alpha k + \psi \alpha \psi^{-1} y + \psi^2 \alpha^{-1} x - \psi b$. Therefore α is a quasiautomorphism of the group $(Q, +)$. Let $\alpha = L_d \varphi$, where $\varphi \in Aut(Q, +)$.

Further we have $d + \varphi x + \varphi y = \alpha k + \psi \alpha \psi^{-1} y + \psi^2 \alpha^{-1} x - \psi b$, $L_d \varphi x + \varphi y = L_{\alpha k} \psi \alpha \psi^{-1} y + R_{-\psi b} \psi^2 \alpha^{-1} x$. Finally, we can apply Lemma 4.91.

Case 2 is proved in a similar way. □

Definition 2.213. Quasigroup $(Q, \cdot)$ with equality

$$xy \cdot z = x \cdot (y \circ z) \tag{2.65}$$

for all $x, y, z \in Q$ is called a *quasigroup which fulfills Sushkevich postulate A* [72]. Quasigroup $(Q, \cdot)$ with equality $x \cdot yz = (x \circ y) \cdot z$ for all $x, y, z \in Q$ will be called a *quasigroup which fulfills Sushkevich postulate A**.

Here in fact we have the so-called functional equation and in the following theorem we solve this functional equation [4, 69, 5].

Theorem 2.214. *1. If quasigroup $(Q, \cdot)$ fulfills Sushkevich postulate A, then $(Q, \cdot)$ is isotopic to the group $(Q, \circ)$, $(Q, \cdot) = (Q, \circ)(\varphi, \varepsilon, \varphi)$ ([80], Theorem 1.7).*

2. If quasigroup $(Q, \cdot)$ fulfills Sushkevich postulate A, then $(Q, \cdot)$ is isotopic to the group $(Q, \circ)$, $(Q, \cdot) = (Q, \circ)(\varepsilon, \psi, \psi)$.*

Proof. Case 1 is proved in [80].

The proof of Case 2 is similar to the proof of Case 1. It is easy to see that $(Q, \circ)$ is a quasigroup. Indeed, if $z = c$, then we have $x \cdot R^{\cdot}_c y = R^{\cdot}_c(x \circ y)$, $(Q, \circ)$ is an isotope of quasigroup $(Q, \cdot)$. Therefore $(Q, \circ)$ is a quasigroup. Moreover, $x \cdot y = R_c(x \circ R_c^{-1} y)$, $(Q, \cdot) = (Q, \circ)(\varepsilon, \psi, \psi)$, where $\psi = R_c^{-1}$.

Quasigroup $(Q, \circ)$ is a group. It is possible to use Theorem 2.205 but we give direct proof similar to the proof from [80]. We have $(x \circ (y \circ z)) \cdot w = x \cdot ((y \circ z) \cdot w) = x \cdot (y \cdot (z \cdot w)) = (x \circ y) \cdot (z \cdot w) = ((x \circ y) \circ z) \cdot w$, $x \circ (y \circ z) = (x \circ y) \circ z$. □

Quasigroup $(Q, \cdot)$ with the generalized identity $xy \cdot z = x \cdot y\delta(z)$, where δ is a fixed permutation of the set Q, is called a quasigroup which fulfills Sushkevich postulate B.

Quasigroup $(Q, \cdot)$ with the generalized identity $x \cdot yz = (\delta(x) \cdot y) \cdot z$, where δ is a fixed permutation of the set Q, will be called a quasigroup which fulfills Sushkevich postulate B*.

It is easy to see that any quasigroup with postulate B (B*) is a quasigroup with postulate A (A*).

Theorem 2.215. *1. If quasigroup $(Q, \cdot)$ fulfills Sushkevich postulate B, then $(Q, \cdot)$ is isotopic to the group $(Q, \circ)$, $(Q, \cdot) = (Q, \circ)(\varepsilon, \psi, \varepsilon)$, where $\psi \in Aut(Q, \circ)$, $\psi \in Aut(Q, \cdot)$ ([80], Theorem 1.8).*

2. If quasigroup $(Q, \cdot)$ fulfills Sushkevich postulate B, then $(Q, \cdot)$ is isotopic to the group $(Q, \circ)$, $(Q, \cdot) = (Q, \circ)(\varphi, \varepsilon, \varepsilon)$, where $\varphi \in Aut(Q, \circ)$, $\varphi \in Aut(Q, \cdot)$.*

Proof. Case 1 is proved in [80]. It is easy to see that quasigroup $(Q, \cdot)$ has the right identity element, i.e., $(Q, \cdot)$ is a right loop. Indeed, $x \cdot 0 = x \circ \psi 0 = x$ for all $x \in Q$, where 0 is zero of group $(Q, \circ)$.

Case 2. The proof of Case 2 is similar to the proof of Case 1. Here we give the direct proof. Since the quasigroup $(Q, \cdot)$ fulfills postulates A* and B*, then by Theorem 2.214, Case 2, groupoid (magma) $(Q, \circ)$, $x \circ y = \delta(x) \cdot y$, is a group and $(Q, \cdot) = (Q, \circ)(\delta^{-1}, \varepsilon, \varepsilon)$. By the same theorem $(Q, \cdot) = (Q, \circ)(\varepsilon, \psi, \psi)$. Therefore (δ, ψ, ψ) is an autotopy of the group $(Q, \circ)$. By Corollary 1.149 $\delta \in Aut(Q, \circ)$. Therefore $\varphi = \delta^{-1} \in Aut(Q, \circ)$. It is easy to see that $(Q, \cdot)$ is a left loop. □

2.10.4 Nuclei, identities

Below in this section by left h-nucleus $N_l(h)$ (Definition 1.236) we understand the Belyavskaya h-nucleus and so on.

Theorem 2.216. *The left h-nucleus of a quasigroup coincides with a quasigroup if and only if this quasigroup is left linear [875].*

Theorem 2.217. *The right h-nucleus of a quasigroup coincides with a quasigroup if and only if this quasigroup is right linear [875].*

Theorem 2.218. *The middle h-nucleus of a quasigroup $(Q,\cdot)$ coincides with a quasigroup if and only if this quasigroup has the form $x\cdot y = \overline{\varphi}\alpha x + c + \alpha y$, where $\overline{\varphi}$ is an antiautomorphism of the group $(Q,+)$, α is a permutation of the set Q [875].*

Identities that guarantee that a quasigroup $(Q,\cdot,/,\backslash)$ is isotopic to a group (to an abelian group) were found by V.D. Belousov in [69]. A quasigroup is isotopic to a group if and only if the following identity is true:

$$x(y\backslash((z/u)v)) = ((x(y\backslash z))/u)v; \tag{2.66}$$

it is isotopic to an abelian group if and only if the following identity is true:

$$x\backslash(y(u\backslash v)) = u\backslash(y(x\backslash v)). \tag{2.67}$$

We call these identities Belousov identity I and Belousov identity II, respectively. These Belousov identities are not unique identities which guarantee that a quasigroup is a group (abelian group) isotope. Belousov writes that the identities (2.66) and (2.67) are closely connected with Reidemeister ($x_1y_2 = x_2y_1, x_1y_4 = x_2y_3, x_3y_2 = x_4y_1 \Rightarrow x_3y_4 = x_4y_3$) and Thomsen ($x_1y_2 = x_2y_1, x_1y_3 = x_3y_1 \Rightarrow x_2y_3 = x_3y_2$) closure conditions, respectively [76]. F. N. Sokhatsky proved that in the identity (2.66) it is possible to put $y = u$ [826]. Many mathematicians worked in this direction and many various results were obtained. See [878] for details.

Definition 2.219. An identity that involves n variables, any of which appear once on both sides of the equation is called balanced of length n [69].

Balanced identities of length four that guarantee that a quasigroup is an abelian group isotope are also listed in [878].

G.B. Belyavskaya and A.Kh. Tabarov characterize some classes of linear quasigroups by identities [105, 106, 133, 131, 132, 134].

Theorem 2.220. *Tabarov theorem [875]. A quasigroup $(Q,\cdot,/,\backslash)$ is left linear if and only if it satisfies the identity*

$$(x(u\backslash y))z = (x(u\backslash u))(u\backslash yz). \tag{2.68}$$

A quasigroup $(Q,\cdot,/,\backslash)$ is right linear if and only if it satisfies the identity

$$x((y/u)z) = (xy/u)((u/u)z). \tag{2.69}$$

A quasigroup $(Q,\cdot,/,\backslash)$ has the form $x\cdot y = \overline{\varphi}\alpha x + c + \alpha y$, where $\overline{\varphi}$ is an antiautomorphism of the group $(Q,+)$, α is a permutation of the set Q, if and only if it satisfies the identity

$$(u\backslash((x/u)y))v = y(u\backslash((u\backslash x)v). \tag{2.70}$$

Theorem 2.221. *A quasigroup $(Q, \cdot)$ is linear if and only if it satisfies the following identity*

$$xy \cdot uv = xu \cdot (\alpha_u y \cdot v), \tag{2.71}$$

where $\alpha_u = R^{-1}_{e(u)}\sigma_u^{-1}L_{f(u)}$, $\sigma_u^{-1} = L_u^{-1}R_u$, i.e., $\alpha_u y = (u\backslash((u/u)y \cdot u))/(u\backslash u)$ [131].

Theorem 2.222. *A quasigroup $(Q, \cdot)$ is alinear if and only if it satisfies the following identity*

$$xy \cdot uv = (\beta_x v \cdot y) \cdot (ux), \tag{2.72}$$

where $\beta_x v = R^{-1}_{e(x)}\sigma_x L_{f(u)} = R^{-1}_{e_x}R^{-1}_x L_x L_{f_x} v = (x((x/x)v))/x))/(x\backslash x)$ [131].

Theorem 2.223. *A quasigroup $(Q, \cdot)$ is a T-quasigroup if and only if it satisfies the identities (2.71) and (2.72) [106, 131].*

Theorem 2.224. *$N_l^h = N_r^h = Q$ if and only if a quasigroup $(Q, \cdot)$ is linear over a group [133].*

Theorem 2.225. *In a quasigroup $(Q, \cdot)$, $N_l(h) = N_r(h) = N_m(h) = Q$ if and only if this quasigroup is a T-quasigroup [875].*

Theorem 2.226. *A quasigroup $(Q, \cdot)$ is a T-quasigroup if and only if it satisfies the identities (2.68), (2.69) and (2.70) [875].*

Theorem 2.227. *Let $(Q, \cdot)$ be a T-quasigroup. Then $N_h^l(Q, \cdot) = N_h^r(Q, \cdot)$ for any $h \in Q$; $Z_h(Q, \cdot) = (Q, \cdot)$.*

In any distributive quasigroup $N_l^h = N_r^h = N_m^h = Z_h = Q$. Identities with permutations that guarantee linearity (alinearity) of a quasigroup are given in [132].

2.10.5 Parastrophes of linear quasigroups

Lemma 2.228. *Suppose that a quasigroup $(Q, \cdot)$ is linear with the form $x \cdot y = \varphi x + \psi y + c$ over a group $(Q, +)$. Then its parastrophes have the following forms:*

1. $x \overset{(12)}{\cdot} y = \varphi y + \psi x + c$;

2. $x \overset{(13)}{\cdot} y = \varphi^{-1}x + IJ_{I\varphi^{-1}c}\varphi^{-1}\psi y + I\varphi^{-1}c$, *where* $J_{I\varphi^{-1}c}x = -\varphi^{-1}c + x + \varphi^{-1}c$;

3. $x \overset{(23)}{\cdot} y = I\psi^{-1}\varphi x + \psi^{-1}y + I\psi^{-1}c$;

4. $x \overset{(123)}{\cdot} y = \varphi^{-1}y + IJ_{I\varphi^{-1}c}\varphi^{-1}\psi x + I\varphi^{-1}c$;

5. $x \overset{(132)}{\cdot} y = I\psi^{-1}\varphi y + \psi^{-1}x + I\psi^{-1}c$.

Proof. Case 1 is clear.

Case 2. By definition of parastrophy, $x \cdot y = z \Leftrightarrow z \overset{(13)}{\cdot} y = x$. From the equality $x \cdot y = \varphi x + \psi y + c = z$ we have $\varphi x = z - c - \psi y$, $x = z \overset{(13)}{\cdot} y = \varphi^{-1}z - \varphi^{-1}c - \varphi^{-1}\psi y$. If we replace the letter z with the letter x, we obtain $x \overset{(13)}{\cdot} y = \varphi^{-1}x - \varphi^{-1}c - \varphi^{-1}\psi y = \varphi^{-1}x - \varphi^{-1}c - \varphi^{-1}\psi y + \varphi^{-1}c - \varphi^{-1}c = \varphi^{-1}x + J_{I\varphi^{-1}c}I\varphi^{-1}\psi y + I\varphi^{-1}c = \varphi^{-1}x + IJ_{I\varphi^{-1}c}\varphi^{-1}\psi y + I\varphi^{-1}c$.

Case 3. By definition of parastrophy, $x \cdot y = z \Leftrightarrow x \overset{(23)}{\cdot} z = y$. From the equality

$x \cdot y = \varphi x + \psi y + c = z$ we have $\psi y = -\varphi x + z - c$, $y = x \overset{(23)}{\cdot} z = \psi^{-1} I \varphi x + \psi^{-1} z + \psi^{-1} I c$. If we replace the letter z with the letter y, we obtain $x \overset{(23)}{\cdot} y = \psi^{-1} I \varphi x + \psi^{-1} y + \psi^{-1} I c = I\psi^{-1}\varphi x + \psi^{-1} y + I\psi^{-1} c$.

Cases 4 and 5 are "(12)-parastrophes" of Cases 2 and 3, respectively. □

Recall, by Lemma 1.302, $\overline{\varphi} = I\varphi$, $\overline{\psi} = I\psi$. Below we shall use these relations without any additional comments.

Lemma 2.229. *Suppose that a quasigroup $(Q, \cdot)$ is alinear with the form $x \cdot y = \overline{\varphi} x + \overline{\psi} y + c$ over a group $(Q, +)$. Then its parastrophes have the following forms:*

1. $x \overset{(12)}{\cdot} y = \overline{\varphi} y + \overline{\psi} x + c$;
2. $x \overset{(13)}{\cdot} y = I\varphi^{-1}\psi y + IJ_{\varphi^{-1}c}\varphi^{-1} x + \varphi^{-1} c$;
3. $x \overset{(23)}{\cdot} y = IJ_{\psi^{-1}c}\psi^{-1} y + IJ_{\psi^{-1}c}\psi^{-1}\varphi x + \psi^{-1} c$;
4. $x \overset{(123)}{\cdot} y = I\varphi^{-1}\psi x + IJ_{\varphi^{-1}c}\varphi^{-1} y + \varphi^{-1} c$;
5. $x \overset{(132)}{\cdot} y = IJ_{\psi^{-1}c}\psi^{-1} x + IJ_{\psi^{-1}c}\psi^{-1}\varphi y + \psi^{-1} c$.

Proof. Case 1 is clear.

Case 2. By definition of parastrophy, $x \cdot y = z \Leftrightarrow z \overset{(13)}{\cdot} y = x$. From the equality $x \cdot y = \overline{\varphi} x + \overline{\psi} y + c = z$ we have $\overline{\varphi} x = z - c - \overline{\psi} y$, $x = z \overset{(13)}{\cdot} y = -I\varphi^{-1}\overline{\psi} y - I\varphi^{-1} c + I\varphi^{-1} z = \varphi^{-1}\overline{\psi} y + \varphi^{-1} c - \varphi^{-1} z = I\varphi^{-1}\psi y + \varphi^{-1} c + I\varphi^{-1} z = I\varphi^{-1}\psi y + J_{\varphi^{-1}c} I\varphi^{-1} z + \varphi^{-1} c$.

If we replace the letter z with the letter x, we obtain $x \overset{(13)}{\cdot} y = I\varphi^{-1}\psi y + IJ_{\varphi^{-1}c}\varphi^{-1} x + \varphi^{-1} c$.

Case 3. By definition of parastrophy, $x \cdot y = z \Leftrightarrow x \overset{(23)}{\cdot} z = y$. From the equality $x \cdot y = I\varphi x + I\psi y + c = z$ we have $I\psi y = -I\varphi x + z - c$, $y = x \overset{(23)}{\cdot} z = I\psi^{-1}(-I\varphi x + z - c) = \psi^{-1} c + I\psi^{-1} z + I\psi^{-1}\varphi x$.

If we replace the letter z with the letter y, we obtain $x \overset{(23)}{\cdot} y = \psi^{-1} c + I\psi^{-1} y + I\psi^{-1}\varphi x = J_{\psi^{-1}c} I\psi^{-1} y + J_{\psi^{-1}c} I\psi^{-1}\varphi x + \psi^{-1} c = IJ_{\psi^{-1}c}\psi^{-1} y + IJ_{\psi^{-1}c}\psi^{-1}\varphi x + \psi^{-1} c$.

Cases 4 and 5 are "(12)-parastrophes" of Cases 2 and 3, respectively. □

Remark 2.230. From Lemma 2.229 it follows that any parastrophe of an alinear quasigroup is alinear.

Lemma 2.231. *Suppose that $(Q, \cdot)$ is a left linear right alinear quasigroup with the form $x \cdot y = \varphi x + I\psi y + c$ over a group $(Q, +)$. Then its parastrophes have the following forms:*

1. $x \overset{(12)}{\cdot} y = \varphi y + I\psi x + c$;
2. $x \overset{(13)}{\cdot} y = \varphi^{-1} x + J_{I\varphi^{-1}c}\varphi^{-1}\psi y + I\varphi^{-1} c$;
3. $x \overset{(23)}{\cdot} y = IJ_{\psi^{-1}c}\psi^{-1} y + J_{\psi^{-1}c}\psi^{-1}\varphi x + \psi^{-1} c$;
4. $x \overset{(123)}{\cdot} y = \varphi^{-1} y + J_{I\varphi^{-1}c}\varphi^{-1}\psi x + I\varphi^{-1} c$;
5. $x \overset{(132)}{\cdot} y = IJ_{\psi^{-1}c}\psi^{-1} x + J_{\psi^{-1}c}\psi^{-1}\varphi y + \psi^{-1} c$.

Proof. Case 1 is clear.

Case 2. From the equality $x\cdot y = \varphi x + \overline{\psi} y + c = z$ we have $\varphi x = z - c - \overline{\psi} y$, $x = z \overset{(13)}{\cdot} y = \varphi^{-1}z - \varphi^{-1}c + \varphi^{-1}\psi y = \varphi^{-1}z + J_{I\varphi^{-1}c}\varphi^{-1}\psi y + I\varphi^{-1}c$.

If we replace the letter z with the letter x, we obtain $x \overset{(13)}{\cdot} y = \varphi^{-1}x + J_{I\varphi^{-1}c}\varphi^{-1}\psi y + I\varphi^{-1}c$.

Case 3. From the equality $x \cdot y = \varphi x + I\psi y + c = z$ we have $I\psi y = I\varphi x + z + Ic$, $y = x \overset{(23)}{\cdot} z = IJ_{\psi^{-1}c}\psi^{-1}z + J_{\psi^{-1}c}\psi^{-1}\varphi x + \psi^{-1}c$.

If we replace the letter z with the letter y, we obtain $x \overset{(23)}{\cdot} y = IJ_{\psi^{-1}c}\psi^{-1}y + J_{\psi^{-1}c}\psi^{-1}\varphi x + \psi^{-1}c$.

Cases 4 and 5 are "(12)-parastrophes" of Cases 2 and 3, respectively. □

Lemma 2.232. *Suppose that $(Q,\cdot)$ is a left alinear right linear quasigroup with the form $x \cdot y = I\varphi x + \psi y + c$ over a group $(Q,+)$. Then its parastrophes have the following forms:*

1. $x \overset{(12)}{\cdot} y = I\varphi y + \psi x + c$;
2. $x \overset{(13)}{\cdot} y = \varphi^{-1}\psi y + IJ_{\varphi^{-1}c}\varphi^{-1}x + \varphi^{-1}c$;
3. $x \overset{(23)}{\cdot} y = \psi^{-1}\varphi x + \psi^{-1}y + I\psi^{-1}c$;
4. $x \overset{(123)}{\cdot} y = \varphi^{-1}\psi x + IJ_{\varphi^{-1}c}\varphi^{-1}y + \varphi^{-1}c$;
5. $x \overset{(132)}{\cdot} y = \psi^{-1}\varphi y + \psi^{-1}x + I\psi^{-1}c$.

Proof. Case 1 is clear.

Case 2. We have $x = z \overset{(13)}{\cdot} y = \varphi^{-1}\psi y + IJ_{\varphi^{-1}c}\varphi^{-1}z + \varphi^{-1}c$. If we replace the letter z with the letter x, we obtain $x \overset{(13)}{\cdot} y = \varphi^{-1}\psi y + IJ_{\varphi^{-1}c}\varphi^{-1}x + \varphi^{-1}c$.

Case 3. From the equality $x\cdot y = I\varphi x + \psi y + c = z$ we have $y = \psi^{-1}\varphi x + \psi^{-1}z + I\psi^{-1}c$.

If we replace the letter z with the letter y, we obtain $x \overset{(23)}{\cdot} y = \psi^{-1}\varphi x + \psi^{-1}y + I\psi^{-1}c$.

Cases 4 and 5 are "(12)-parastrophes" of Cases 2 and 3, respectively. □

2.10.6 On the forms of n-T-quasigroups

We give some propositions on the forms of n-T-quasigroups in order to demonstrate that the theories of binary and n-ary T-quasigroups are quite similar.

Proposition 2.233. *An n-T-quasigroup (Q,f) is an idempotent quasigroup if and only if (Q,f) has the form $f(x_1^n) = \varphi_1 x_1 + \dots + \varphi_n x_n$ over an abelian group $(Q,+)$ and $\varphi_1 + \dots + \varphi_n = \varepsilon$.*

Proof. From definition of n-T-quasigroup it follows that any n-T-quasigroup has the form $f(x_1^n) = \varphi_1 x_1 + \dots + \varphi_n x_n + a$ over an abelian group $(Q,+)$.

Let (Q,f) be an idempotent quasigroup. If we take $x_1 = x_2 = \dots = x_n = 0$, then $\varphi_1 0 + \dots + \varphi_n 0 + a = 0$, $a = 0$. Therefore $\varphi_1 x + \dots + \varphi_n x = x$ for all $x \in Q$, i.e., $\varphi_1 + \dots + \varphi_n = \varepsilon$.

It is easy to check that an n-T-quasigroup (Q,f) over an abelian group $(Q,+)$ with the form $f(x_1^n) = \varphi_1 x_1 + \dots + \varphi_n x_n$ and the property $\varphi_1 + \dots + \varphi_n = \varepsilon$ is an idempotent quasigroup. □

Exercise 2.234. An idempotent binary T-quasigroup is medial and distributive.

Proposition 2.235. *An n-T-quasigroup (Q, f) with the form $f(x_1^n) = \varphi_1 x_1 + \dots + \varphi_n x_n + a$ over an abelian group $(Q, +)$ has exactly one idempotent element if and only if endomorphism $\varphi_1 + \dots + \varphi_n - \varepsilon$ is a permutation of the set Q, i.e., it is an automorphism of the group $(Q, +)$. In these conditions the quasigroup (Q, f) is isomorphic to n-T-quasigroup (Q, g) with the form $g(x_1^n) = \varphi_1 x_1 + \dots + \varphi_n x_n$ over the same abelian group.*

Proof. Let u be an idempotent element of the n-T-quasigroup (Q, f). Then we have $\varphi_1 u + \dots + \varphi_n u + a = u$, i.e., $\varphi_1 u + \dots + \varphi_n u - \varepsilon u = -a$, $(\varphi_1 + \dots + \varphi_n - \varepsilon)u = -a$.

Therefore the element u is a unique idempotent of the quasigroup (Q, f) if and only if the endomorphism $\mu = (\varphi_1 + \dots + \varphi_n - \varepsilon)$ has an identity kernel, i.e., this endomorphism is an automorphism of the group $(Q, +)$.

Let $d = -\mu^{-1}a$, i.e., $\mu d = -a$, $\varphi_1 d + \varphi_2 d + \dots + \varphi_n d - d = -a$. Then we have

$$
\begin{aligned}
&-d + f(d + x_1, d + x_2, \dots, d + x_n) = \\
&-d + \varphi_1 d + \varphi_1 x_1 + \varphi_2 d + \varphi_2 x_2 + \dots + \varphi_n d + \varphi_n x_n + a = \\
&\varphi_1 x_1 + \varphi_2 x_2 + \dots + \varphi_n x_n + \varphi_1 d + \varphi_2 d + \dots + \varphi_n d - d + a = \\
&\varphi_1 x_1 + \varphi_2 x_2 + \dots + \varphi_n x_n - a + a = \\
&\varphi_1 x_1 + \varphi_2 x_2 + \dots + \varphi_n x_n = g(x_1^n),
\end{aligned}
$$

i.e., $(Q, f) \cong (Q, g)$. □

Some analogs of Proposition 7.89 are proved in [769, 600].

The following identity of n-ary quasigroup (Q, g)

$$
\begin{aligned}
&g(g(x_{11}, x_{12}, \dots, x_{1n}), g(x_{21}, x_{22}, \dots, x_{2n}), \dots, g(x_{n1}, x_{n2}, \dots, x_{nn})) = \\
&\quad g(g(x_{11}, x_{21}, \dots, x_{n1}), g(x_{12}, x_{22}, \dots, x_{n2}), \dots, g(x_{1n}, x_{2n}, \dots, x_{nn}))
\end{aligned}
\tag{2.73}
$$

is called medial identity [77].

An n-ary quasigroup with the identity (2.73) is called a *medial n-ary quasigroup.*

In the binary case from the identity (2.73) we obtain the usual medial identity: $xy \cdot uv = xu \cdot yv$ [646, 864].

2.10.7 (m, n)-Linear quasigroups

We define left, right, and middle order of an element a of a quasigroup $(Q, \cdot)$ using the orders of corresponding translations of the element a.

Definition 2.236. An element a of a quasigroup $(Q, \cdot)$ has the *left order* l (the number l is a natural number or the symbol of infinity ∞), if the equality $L_a^l = \varepsilon$ is true and a number k such that $1 \leqslant k < l$ and $L^k a = \varepsilon$ does not exist.

An element a of a quasigroup $(Q, \cdot)$ has the *right order* r (the number r is a natural number or ∞), if the equality $R_a^r = \varepsilon$ is true and a number k such that $1 \leqslant k < r$ and $R^k a = \varepsilon$ does not exist.

An element a of a quasigroup $(Q, \cdot)$ has the *middle order* m (the number m is a natural number or ∞), if the equality $P_a^m = \varepsilon$ is true and a number k such that $1 \leqslant k < m$ and $P^k a = \varepsilon$ does not exist.

Definition 2.237. An element a of a quasigroup $(Q, \cdot)$ has the *order* (m, n) (or element a is an (m, n)*-element*) if there exist natural numbers m, n such that $L_a^m = R_a^n = \varepsilon$ and the element a is not the (m_1, n_1)-element for any integers m_1, n_1 such that $1 \leqslant m_1 < m$, $1 \leqslant n_1 < n$.

Remark 2.238. It is obvious that m is the order of the element L_a in the group $Mlt(Q,\cdot)$ and n is the order of the element R_a in this group. Therefore it is possible to name the (m,n)-order of an element a as well as the (L,R)-order or the left-right order of an element a.

Remark 2.239. In the theory of non-associative rings [458], one often uses the so-called left and right order of brackets by multiplying elements of a ring $(R,+,\cdot)$, namely $(\dots(((a_1\cdot a_2)\cdot a_3)\cdot a_4)\dots)$ is called the left order of brackets and $(\dots(a_4\cdot(a_3\cdot(a_2\cdot a_1)))\dots)$ is called the right order of brackets.

So the (m,n)-order of an element a of a quasigroup $(Q,\cdot)$ is similar to the order of an element a of a non-associative ring $(R,+,\cdot)$ with the right and the left orders of brackets respectively.

Proposition 2.240. *In a diassociative loop $(Q,\cdot)$ there exist only (n,n)-elements.*

Proof. If we suppose that there exists an element $a\in Q$ of a diassociative loop of order (m,n), then in this case we have $L_a^m x = a\cdot(a\cdot\dots(a\cdot x)\dots) = a^m x = L_{a^m}x$.

Therefore $L_a^m = \varepsilon$ if and only if $a^m = 1$, where 1 is the identity element of the loop $(Q,\cdot)$. Similarly $R_a^n = \varepsilon$ if and only if $a^n = 1$.

From the last two equivalences and Definition 2.237 (from the minimality of numbers m,n) it follows that in a diassociative loop $m=n$, i.e., in diassociative loop there exist only (n,n)-elements. □

Remark 2.241. It is clear that Proposition 2.240 is true for Moufang loops and groups since these algebraic objects are diassociative.

From Definition 2.237 it follows that the $(1,1)$-element is the identity element of a quasigroup $(Q,\cdot)$, i.e., in this case the quasigroup $(Q,\cdot)$ is a loop.

Proposition 2.242. *Any $(1,n)$-element is a left identity element of a quasigroup $(Q,\cdot)$. In any quasigroup, this element is unique and in this case the quasigroup $(Q,\cdot)$ is a left loop i.e., $(Q,\cdot)$ is a quasigroup with the left identity element.*

Any $(m,1)$-element is the right identity element of a quasigroup $(Q,\cdot)$ and the quasigroup $(Q,\cdot)$ is a right loop.

Proof. If in a quasigroup $(Q,\cdot)$ an element a has the order $(1,n)$, then $a\cdot x = L_a^{\cdot}x = x$ for all $x\in Q$. If we suppose that in a quasigroup $(Q,\cdot)$ there exist left identity elements e and f, then we find that equality $x\cdot a = a$, where a is some fixed element of the set Q, will have two solutions, namely, e and f are such solutions. We obtain a contradiction. Therefore in a quasigroup there exists a unique left identity element. □

Using the language of quasigroup translations it is possible to re-write the definition of an (n,m)-identity element from [208, 207] in the following form:

Definition 2.243. An idempotent element e of a quasigroup $(Q,\cdot)$ is called an (m,n)-*identity element* if and only if there exist natural numbers m,n such that $(L_e^{\cdot})^m = (R_e^{\cdot})^n = \varepsilon$.

Hence any (m,n)-identity element of a quasigroup $(Q,\cdot)$ can be called an *idempotent element of order* (m,n) as well as an *idempotent* (m,n)*-element.*

Theorem 2.244. *A quasigroup $(Q,\cdot)$ has an (m,n)-identity element 0 if and only if there exists a loop $(Q,+)$ with the identity element 0 and permutations φ,ψ of the set Q such that $\varphi 0 = \psi 0 = 0$, $\varphi^n = \psi^m = \varepsilon$, $x\cdot y = \varphi x + \psi y$ for all $x,y\in Q$.*

Proof. Let a quasigroup $(Q,\cdot)$ have an idempotent element 0 of order (m,n). Then the isotope $(R_0^{-1}, L_0^{-1}, \varepsilon)$ of the quasigroup $(Q,\cdot)$ is a loop $(Q,+)$ with the identity element 0, i.e., $x+y = R_0^{-1}x \cdot L_0^{-1}y$ for all $x, y \in Q$ [72]. From the last equality we have $x \cdot y = R_0x + L_0y$, $R_00 = L_00 = 0$. Then $\varphi = R_0$, $\psi = L_0$, $L_0^m = \psi^m = R_0^n = \varphi^n = \varepsilon$.

Conversely, let $x \cdot y = \varphi x + \psi y$, where $(Q, +)$ is a loop with the identity element 0, $\varphi 0 = \psi 0 = 0$, $\varphi^m = \psi^n = \varepsilon$. Then the element 0 is an idempotent element of quasigroup $(Q, \cdot)$ of order (m, n) since $L_0^{\cdot}y = \psi y$, $R_0^{\cdot}x = \varphi x$ and $(L_0^{\cdot})^m = \psi^m = \varepsilon$, $(R_0^{\cdot})^n = \varphi^n = \varepsilon$. □

Definition 2.245. A quasigroup $(Q, \cdot)$ of the form $x \cdot y = \varphi x + \psi y$, where φ, ψ are automorphisms of a loop $(Q, +)$ such that $\varphi^n = \psi^m = \varepsilon$, is called an *$(m, n)$-linear quasigroup.*

Taking into consideration Theorem 2.244 we see that any (m, n)-linear quasigroup $(Q, \cdot)$ is a linear quasigroup over a loop $(Q, +)$ with at least one (m, n)-idempotent element.

Lemma 2.246. *In an (m, n)-linear quasigroup $(Q, \cdot)$ of the form $x \cdot y = \varphi x + \psi y$, where $(Q, +)$ is a group, we have*

$$L_a^{\cdot} = L_{\varphi a}^{+}\psi,\ (L_a^{\cdot})^k = L_c^{+}\psi^k,\ c = \varphi a + \psi\varphi a + \dots + \psi^{k-1}\varphi a,$$
$$R_a^{\cdot} = R_{\psi a}^{+}\varphi,\ (R_a^{\cdot})^r = R_d^{+}\varphi^r,\ d = \psi a + \varphi\psi a + \dots + \varphi^{r-1}\psi a.$$

Proof. It is well known that if $\varphi \in Aut(Q, \cdot)$, i.e., if $\varphi(x \cdot y) = \varphi x \cdot \varphi y$ for all $x, y \in Q$, then $\varphi L_x y = L_{\varphi x}\varphi y$, $\varphi R_y x = R_{\varphi y}\varphi x$. Indeed, we have $\varphi L_a x = \varphi(a \cdot x) = \varphi a \cdot \varphi x = L_{\varphi a}\varphi x$, $\varphi R_b x = \varphi(x \cdot b) = \varphi x \cdot \varphi b = R_{\varphi b}\varphi x$.

Using these last equalities we have

$$(L_x^{\cdot})^2 = L_{\varphi x}^{+}\psi L_{\varphi x}^{+}\psi = L_{\varphi x+\psi\varphi x}^{+}\psi^2,\quad (L_x^{\cdot})^3 = L_{(\varphi x+\psi\varphi x)+\psi^2\varphi x}^{+}\psi^3,$$

and so on. □

Proposition 2.247. *An element a of an (m, n)-linear quasigroup $(Q, \cdot)$ over a group $(Q, +)$ has order (k, r) if and only if $\varphi a + \psi\varphi a + \dots + \psi^{k-1}\varphi a = 0$, $\psi a + \varphi\psi a + \dots + \varphi^{r-1}\psi a = 0$, $k = m \cdot i$, $r = n \cdot j$, where i, j are some natural numbers.*

Proof. It is possible to use Lemma 2.246. If an element $a \in Q$ has an order $(k, \_)$, then the permutation $L_a^k = L_c^{+}\psi^k$, where $c = \varphi a + \psi\varphi a + \dots + \psi^{k-1}\varphi a$ is the identity permutation. This is possible only in two cases: (i) $L_c^{+} = \psi^{-k} \neq \varepsilon$; (ii) $L_c^{+} = \varepsilon$ and $\psi^k = \varepsilon$.

Case (i) is impossible. Indeed, if we suppose that $L_c^{+} = \psi^{-k}$, then we have $L_c^{+}0 = \psi^{-k}0$, where 0 is the identity element of the group $(Q, +)$. Further we have $\psi^{-k}0 = 0$, $L_c^{+}0 = 0$, $c = 0$, $L_c^{+} = \varepsilon$, $\psi^k = \varepsilon$. Therefore, if the element a has the order $(k, \_)$, then $L_c^{+} = \varepsilon$ and $\psi^k = \varepsilon$. Further, since $\psi^m = \varepsilon$, we have that $k = m \cdot i$ for some natural number $i \in N$.

Converse. If $\varphi a + \psi\varphi a + \dots + \psi^{k-1}\varphi a = 0$, $L_c^{+} = \varepsilon$ and $\psi^k = \varepsilon$ for some element a, then this element has the order $(k, \_)$.

Therefore an element a of an (m, n)-linear quasigroup $(Q, \cdot)$ over a group $(Q, +)$ will have the order $(k, \_)$ if and only if $L_c^{+} = \varepsilon$, i.e., $c = 0$, where $c = \varphi a + \psi\varphi a + \dots + \psi^{k-1}\varphi a$ and $\psi^k = \varepsilon$, i.e., $k = m \cdot i$ for some natural number $i \in N$.

Similarly, any element a of an (m, n)-linear quasigroup $(Q, \cdot)$ over a group $(Q, +)$ will have the order $(\_, r)$ if and only if $R_d^{+} = \varepsilon$, i.e., $d = 0$, where $d = \psi a + \varphi\psi a + \dots + \varphi^{r-1}\psi a$ and $\varphi^r = \varepsilon$. Further, since $\varphi^r = \varepsilon$, we have that $r = n \cdot j$ for some natural number $j \in N$.

Proposition 2.248. *The number M of elements of order (mi, nj) in an (m, n)-linear quasigroup $(Q, \cdot)$ over a group $(Q, +)$ is equal to $|K(\varphi) \cap K(\psi)|$, where $K(\varphi) = \{x \in Q \mid \psi x + \varphi\psi x + \dots + \varphi^{nj-1}\psi x = 0\}$, $K(\psi) = \{x \in Q \mid \varphi x + \psi\varphi x + \dots + \psi^{mi-1}\varphi x = 0\}$.*

Proof. From Proposition 2.247 it follows that an element a of an (m, n)-linear quasigroup $(Q, \cdot)$ over a group $(Q, +)$ has the order (mi, nj) if and only if $\varphi a + \psi\varphi a + \dots + \psi^{mi-1}\varphi a = 0$ and $\psi a + \varphi\psi a + \dots + \varphi^{nj-1}\psi a = 0$.

In other words an element a of (m, n)-linear quasigroup $(Q, \cdot)$ over a group $(Q, +)$ has the order (mi, nj) if and only if $a \in K(\varphi) \cap K(\psi)$.

Therefore $M = |K(\varphi) \cap K(\psi)|$. □

Corollary 2.249. *(i) Any $(m,1)$-linear quasigroup over a group has exactly one element of order $(m,1)$, namely* 0.

(ii) Any $(1,n)$-linear quasigroup over a group has exactly one element of order $(1,n)$, namely 0.

Proof. (i) In this case $K(\varphi) = \{x \in Q \mid \psi x + \varphi\psi x + \ldots + \varphi^{n-1}\psi x = 0\} = \{x \in Q \mid \psi x = 0\}$. Therefore $x = \psi^{-1}0$, $x = 0$, $|K(\varphi)| = 1$.

(ii) In this case $|K(\psi)| = 1$. □

Theorem 2.250. *Any $(2,2)$-linear quasigroup $(Q,\cdot)$ over a loop $(Q,+)$ such that all elements of $(Q,\cdot)$ have the order $(2,2)$ can be represented in the form $x \cdot y = Ix + Iy$, where $x + Ix = 0$ for all $x \in Q$.*

Proof. In this case we have $(L^{\cdot}_x)^2 = L^+_{\varphi x}L^+_{\psi\varphi x}\psi^2 = L^+_{\varphi x}L^+_{\psi\varphi x} = \varepsilon$ for any $x \in Q$. Then $\varphi x + (\psi\varphi x + 0) = \varepsilon 0 = 0$ for all $x \in Q$. Therefore $x + \psi x = 0$, $\psi x = -x = Ix$.

By analogy we have that $\varphi x = -x = Ix$ for all $x \in Q$. Indeed, $(R^{\cdot}_x)^2 = R^+_{\psi x}R^+_{\varphi\psi x}\varphi^2 = R^+_{\psi x}R^+_{\varphi\psi x} = \varepsilon$, $\psi x + \varphi\psi x = 0$, $x + \varphi x = 0$, $\varphi x = Ix$. □

Remark 2.251. From Theorem 2.250 it follows that any $(2,2)$-linear quasigroup $(Q,\cdot)$ such that all elements of $(Q,\cdot)$ have the order $(2,2)$ exists only over a loop with the property $I(x+y) = Ix + Iy$ for all $x, y \in Q$, where $x + Ix = 0$ for all $x \in Q$. A loop with this property is called an *automorphic-inverse property loop (AIP-loop)*.

We notice that the Bruck loops, the commutative Moufang loops, and the abelian groups are AIP-loops.

Theorem 2.252. *If in an (m,n)-linear T-quasigroup $(Q,\cdot)$ of the form $x \cdot y = \varphi x + \psi y$ over an abelian group $(Q,+)$ the maps $\varepsilon - \varphi, \varepsilon - \psi$ are permutations of the set Q, then all elements of the quasigroup $(Q,\cdot)$ have order (m,n).*

Proof. It is easy to see that if the maps $\varepsilon - \varphi, \varepsilon - \psi$ are permutations of the set Q, then $m > 1, n > 1$. From Proposition 2.248 it follows that the number M of elements of the order (m,n) is equal to the number $|K(\varphi) \cap K(\psi)|$, where

$$K(\varphi) = \{x \in Q | (\varepsilon + \varphi + \ldots + \varphi^{n-1})\psi x = 0\},$$
$$K(\psi) = \{x \in Q | (\varepsilon + \psi + \ldots + \psi^{m-1})\varphi x = 0\}.$$

Since the map $\varepsilon - \varphi$ is a permutation of the set Q, we have: $\varepsilon + \varphi + \ldots + \varphi^{n-1} = (\varepsilon + \varphi + \ldots + \varphi^{n-1})(\varepsilon - \varphi)(\varepsilon - \varphi)^{-1} = (\varepsilon - \varphi + \varphi - \varphi^2 + \varphi^2 - \ldots - \varphi^n)(\varepsilon - \varphi)^{-1} = (\varepsilon - \varphi^n)(\varepsilon - \varphi)^{-1}$. Since $\varphi^n = \varepsilon$ we obtain that $K(\varphi) = Q$.

By analogy it is proved that $K(\psi) = Q$. Therefore $K(\varphi) \cap K(\psi) = Q$. □

A quasigroup $(Q,\cdot)$ with the identities $x \cdot (y \cdot z) = (x \cdot y) \cdot (x \cdot z)$, $(x \cdot y) \cdot z = (x \cdot z) \cdot (y \cdot z)$ is called a *distributive quasigroup* [72].

Corollary 2.253. *In any medial distributive (m,n)-linear quasigroup all its elements have order (m,n).*

Proof. It is known that any medial distributive quasigroup $(Q,\cdot)$ can be presented in the form $x \cdot y = \varphi x + \psi y$, where $(Q,+)$ is an abelian group and $\varphi + \psi = \varepsilon$ [72, 769]. Therefore conditions of Theorem 2.252 are fulfilled in any medial distributive (m,n)-linear quasigroup. □

2.11 Miscellaneous

2.11.1 Groups with triality

G. Glauberman [350, 351] proved that in the group $M(Q,\cdot)$, where $(Q,\cdot)$ is a Moufang loop with identity nucleus, there exists automorphism ρ of order three such that: $\rho(L_a) = R_a$, $\rho(R_a) = R_a^{-1}L_a^{-1}$, $\rho(R_a^{-1}L_a^{-1}) = L_a$. In the group $M(Q,\cdot)$ it is possible to define the following automorphism σ: $\sigma(L_a) = R_a^{-1} = R_{a^{-1}}$, $\sigma(R_a) = L_a^{-1} = L_{a^{-1}}$, too. See [688] for details, too.

S. Doro, developing Glauberman's ideas, introduced the concept of an abstract group with triality [265]. This concept is useful, because some heavy problems about Moufang and Bol loops were reformulated in the language of group theory and after this were solved [566, 221].

In fact S. Doro, using the concept of triality, proved that if a finite Moufang loop is simple, then a subgroup of the group $M(Q,\cdot)$ is a simple group with triality. Later M. Liebeck [566], using classification of finite simple groups and Doro's results on groups with triality, proved that only Paige loops [676], any of which is constructed over the octonion algebra $\mathbb{O}(F)$ [653], are non-associative finite simple Moufang loops.

Moreover, some Moufang loops are used by construction of finite simple groups. See, for example, [389, 390, 391, 181].

Definition 2.254. A group G is a group with triality, if $S \subseteq Aut(G)$, $S = \langle \sigma, \rho \mid \sigma^2 = \rho^3 = (\sigma\rho)^2 = \varepsilon \rangle \cong S_3$, and for all $g \in G$ the triality identity

$$((\sigma g)g^{-1})\ \rho((\sigma g)g^{-1})\ \rho^2((\sigma g)g^{-1}) = 1 \tag{2.74}$$

holds [265, 653, 688].

We give some additional definitions in order to present other forms of the identity (2.74). Let G be a group. If $x, y \in G$, then $x^y = y^{-1}xy$ and $[x,y] = x^{-1}y^{-1}xy = x^{-1}x^y$. Let α be an automorphism of G. Then $\alpha(x)$ will be denoted by x^α, and $[x,\alpha] = x^{-1}x^\alpha$. The element $\alpha^y = y^{-1}\alpha y = I_y\alpha \in Aut(G)$ maps x to $x^{I_y\alpha}$. The identity (2.74) can be written in the following form $g^{-1}g^\sigma(g^{-1})^\rho g^{\sigma\rho}(g^{-1})^{\rho\rho}g^{\rho\sigma} = 1$ [688]. The identity (2.74) (see [406, 653]) also admits the following record

$$[g,\sigma][g,\sigma]^\rho[g,\sigma]^{\rho^2} = 1. \tag{2.75}$$

The following lemma demonstrates that there exist many groups connected with quasigroups and loops that have the group S_3 as automorphism group. Therefore Doro was forced to add the identity of triality to characterize namely Moufang loops enveloping groups.

Lemma 2.255. *If $(Q,\cdot)$ is an IP-loop, then $S_3 \cong H \subseteq Aut(Avt(Q,\cdot))$.*

Proof. Corollary 2.39 lets us define the following mappings on the set of all autotopies of loop $(Q,\cdot)$: $\sigma(\alpha,\beta,\gamma) = (I\alpha I, \gamma, \beta)$, $\rho(\alpha,\beta,\gamma) = (I\gamma I, \alpha, I\beta I)$, where $x \cdot Ix = 1$ for all $x \in Q$. It is clear that mappings σ and ρ are bijective.

Taking into consideration that $I^2 = \varepsilon$ (Theorem 2.34, Case 1), we conclude that $\sigma^2 = \rho^3 = \varepsilon$. Indeed, $\rho^3(\alpha,\beta,\gamma) = \rho^2(I\gamma I, \alpha, I\beta I) = \rho(\beta, I\gamma I, I\alpha I) = (\alpha,\beta,\gamma)$.

Prove that the maps σ and ρ are automorphisms of the group $Avt(Q,\cdot)$. We have $\sigma(\alpha_1\alpha_2, \beta_1\beta_2, \gamma_1\gamma_2) = (I\alpha_1\alpha_2 I, \gamma_1\gamma_2, \beta_1\beta_2)$, $\sigma(\alpha_1,\beta_1,\gamma_1)\sigma(\alpha_2,\beta_2,\gamma_2) = (I\alpha_1\alpha_2 I, \gamma_1\gamma_2, \beta_1\beta_2)$. For the map ρ the proof is similar.

It is easy to check that $\langle \sigma, \rho \rangle \cong S_3$. □

Note that Corollary 2.39 allows us to take other maps as automorphisms σ and ρ.

Lemma 2.256. *If $(Q,\cdot)$ is an IP-loop with identity $x^2 = 1$, then the group $Avt(Q,\cdot)$ is a group with triality.*

Proof. Taking into consideration Lemma 2.255 we must only prove that in conditions of this lemma, identity (2.75) is true. We have:

$$\begin{aligned}
&[(\alpha,\beta,\gamma),\sigma] = (\alpha^{-1},\beta^{-1},\gamma^{-1})(I\alpha I,\gamma,\beta) = (\alpha^{-1}I\alpha I,\beta^{-1}\gamma,\gamma^{-1}\beta),\\
&[(\alpha,\beta,\gamma),\sigma]^{\rho} = (\alpha^{-1}I\alpha I,\beta^{-1}\gamma,\gamma^{-1}\beta)^{\rho} = (I\gamma^{-1}\beta I,\alpha^{-1}I\alpha I, I\beta^{-1}\gamma I),\\
&[(\alpha,\beta,\gamma),\sigma]^{\rho^2} = (I\gamma^{-1}\beta I,\alpha^{-1}I\alpha I, I\beta^{-1}\gamma I)^{\rho} = (I\gamma^{-1}\beta I,\alpha^{-1}I\alpha I, I\beta^{-1}\gamma I) =\\
&(I\beta^{-1}\gamma I, I\gamma^{-1}\beta I, I\alpha^{-1}I\alpha).\\
&[(\alpha,\beta,\gamma),\sigma][(\alpha,\beta,\gamma),\sigma]^{\rho}[(\alpha,\beta,\gamma),\sigma]^{\rho^2} =\\
&(\alpha^{-1}I\alpha I,\beta^{-1}\gamma,\gamma^{-1}\beta)(I\gamma^{-1}\beta I,\alpha^{-1}I\alpha I, I\beta^{-1}\gamma I)(I\beta^{-1}\gamma I, I\gamma^{-1}\beta I, I\alpha^{-1}I\alpha) =\\
&(\alpha^{-1}I\alpha II\gamma^{-1}\beta II\beta^{-1}\gamma I,\beta^{-1}\gamma\alpha^{-1}I\alpha II\gamma^{-1}\beta I,\gamma^{-1}\beta I\beta^{-1}\gamma II\alpha^{-1}I\alpha) =\\
&(\alpha^{-1}I\alpha I,\beta^{-1}\gamma\alpha^{-1}I\alpha\gamma^{-1}\beta I,\gamma^{-1}\beta I\beta^{-1}\gamma\alpha^{-1}I\alpha).
\end{aligned} \tag{2.76}$$

If an IP-loop satisfies the identity $x^2 = 1$, then in this loop $I = \varepsilon$. It is easy to see, that in this case $\alpha I = I\alpha$, $\beta I = I\beta$, $\gamma I = I\gamma$, and the triality identity is true. □

Exercise 2.257. Any IP-loop with identity $x^2 = 1$ is a TS-loop, in other words it is a Steiner loop.

Here we give Pavel Mikheev's construction of groups with triality which are enveloping groups of Moufang loops [623]. We start from the following generalization of Doro's theorem [265] given in [623].

Theorem 2.258. *Let G be a group, $\sigma, \rho \in Aut(G)$, $\sigma^2 = \rho^3 = (\sigma\rho)^2 = \varepsilon$ and for any element $g \in G$ the identity (2.74) is true. Let $H = \{g \in G \mid \sigma g = g\}$ be a subgroup of the group G and let $M = \{\rho^2((\sigma g)g^{-1})\}_{g\in G}$ be a subset of the set G. Then M is a system of representatives (see Definition 1.97) of the left cosets aH, $a \in G$, in group G to subgroup H, i.e., M is a left transversal in group G to subgroup H. Relative to the following operation defined on the set M $a * b = ab \pmod H$, $a, b \in M$, groupoid $(M, *)$ is a Moufang loop.*

Sabinin's construction (see Section 1.6) lets us construct a group using a loop and a group.

Theorem 2.259. *Mikheev theorem. Let $(M,\cdot)$ be a Moufang loop, and $\Pi_l(M,\cdot)$ be the group of its left pseudoautomorphisms (see Remark 1.166). Then the set $G = M \times \Pi_l(M,\cdot)$ with the following composition law*

$$\begin{aligned}
&(x,(a,\theta))(y,(b,\delta)) =\\
&(x\cdot\theta y,(x\cdot\theta y\cdot x^{-1}\cdot\theta y^{-1}, l(x,\theta y))(\theta y\cdot a\cdot\theta y^{-1}, l(a,\theta y)^{-1})(a,\theta)(b,\delta)),
\end{aligned} \tag{2.77}$$

where $x, y \in M$, $(a,\theta),(b,\delta) \in \Pi_l(M,\cdot)$, is a group, the maps

$$\sigma(x,(a,\theta)) = (x^{-1},(x^{-3}, L_x R_x^{-1})(a,\theta)) \tag{2.78}$$

$$\rho(x,(a,\theta)) = (x^{-2},(x^{-3}, L_x R_x^{-1}))(a,(a,\theta)) \tag{2.79}$$

are automorphisms of the group G and satisfy conditions of Definition 2.254. Loop $(M',*)$ which is defined on the set of representatives $M' = \{\rho^2((\sigma g)g^{-1})\}_{g\in G}$ of the left cosets in group G to subgroup $H = \{g \in G \mid \sigma g = g\}$ (M' is a transversal) with composition law $a * b = ab \pmod H$, $a, b \in M'$, is a Moufang loop which is isomorphic to the loop $(M,\cdot)$.

In fact the equation (2.77) defines a composition law of autotopisms of a Moufang loop (see Theorem 2.50). Theorem 2.259 and Doro's results confirm the fact that there exists a close relationship between an autotopy group and a multiplication group of any Moufang loop.

Exercise 2.260. Check that Theorems 2.258 and 2.259 are true for the group $(Z_2 \oplus Z_2, +)$ and its autotopy group.

The TS-loop which satisfies conditions of Lemma 2.256 is given in Example 2.266.

The detailed historical survey of the concept of triality is given in A. Schleiermacher's review (MR2239539 (2007g:20062)), in the article of Jonathan I. Hall and Gabor P. Nagy [406] and in survey [653].

"The principle of triality had been introduced by Cartan [187] in 1938 as a property of orthogonal groups of dimension 8, and these examples are motivated by Tits [887]" [653]. Definition and some properties of groups with triality are given in [819].

Jonathan Hall [405] has analyzed Pavel Mikheev's construction of groups with triality [623] and discovered that autotopy groups of Moufang and Bol loops are groups with triality. Notice that the multiplication group and some of its subgroups of the Moufang loop, autotopy group, and some "net" groups "generated" by the Moufang loop are groups with triality. Groups with triality have clear geometrical sense in the language of 3-nets [406].

2.11.2 Universal properties of quasigroups

It is well known that properties of universal algebras that are invariant relative to isomorphisms (up to isomorphisms) are called algebraic [72, p. 186] or abstract [591, p. 50] properties of these algebras.

Definition 2.261. Let A be a universal algebra with a property P. Suppose that $\mathfrak{M}$ is a class of morphisms (of operators) that has a sense for algebra A. If the property P is true for any algebra AM, $M \in \mathfrak{M}$, then the property P is called universal (stable, invariant) in this class of algebras relative to the indicated morphisms (or operators).

In fact V.D. Belousov discovered the following direction in algebraic quasigroup theory: the study of quasigroup (loop) properties invariant (universal) relatively to quasigroup (loop) isotopy [72, p. 186]. It is clear that this problem has a geometrical origin [76]. The problem of the research of identities invariant (universal) relative to quasigroup (loop) isotopy was posed by V.D. Belousov [76].

From the definition of the G-loop it follows that any algebraic property of G-loop is universal. Any identity which guarantees that a quasigroup $(Q, \cdot)$ is isotopic to a G-loop is universal relative to any isotopy of $(Q, \cdot)$. In [69] it is proved that Belousov identities I and II are invariant relative to any quasigroup isotopy, because these identities guarantee that a quasigroup is isotopic to a group or to a commutative group, respectively.

Some closure conditions, i.e., quasi-identities of a certain kind, are invariant relative to loop isotopy [92].

Varieties, quasi-varieties, and pseudo-varieties [939, 591, 184, 400] of quasigroups that are stable relative to isotopy are researched in [400]. The general theory of construction of identities, quasi-identities, and pseudo-identities of a quasigroup Q, that are invariant (stable, universal) relative to any isotopy of a quasigroup Q is also built in [400].

Definition 2.262. If P is a property of a loop A, $\mathfrak{M}$ is the class of all loop isotopies, and any loop AT, $T \in \mathfrak{M}$, has the property P, then the property P is called universal [72, p. 186].

We give some typical examples of universal properties. From the Albert theorem (Theorem 1.123) it follows that associative identity is universal in the class of loops relative to any LP-isotopy. In other words, any LP-isotope of a group is also a group (we use the fact that any associative loop is a group).

1. If any LP-isotope of IP-loop $(Q, \cdot)$ is an IP-loop, then $(Q, \cdot)$ is a Moufang loop [173].
2. If any LP-isotope of LIP-loop $(Q, \cdot)$ is an LIP-loop, then $(Q, \cdot)$ is a left Bol loop [72].
3. If any LP-isotope of CI-loop $(Q, \cdot)$ is a CI-loop, then $(Q, \cdot)$ is a commutative group [72].
4. If any LP-isotope of WIP-loop $(Q, \cdot)$ is a WIP-loop, then $(Q, \cdot)$ satisfies the identity $xy \cdot \theta_x zx = (x \cdot yz)x$, where $\theta_x \in Aut(Q, \cdot)$ and θ_x depends on the element x [72].
5. If any LP-isotope of AAIP-loop $(Q, \cdot)$ is an AAIP-loop, then $(Q, \cdot)$ satisfies the identity $z((xy)\backslash z) = (z/y)(x\backslash z)$ [76].
6. If any LP-isotope of left alternative loop $(Q, \cdot)$ is a left alternative loop, then $(Q, \cdot)$ is a left Bol loop [76].

For the generalized Moufang loops the property WIP is universal [35]. Universal AI-loops were studied in [473, 41]. Universality of Osborn loops is researched in [40, 38, 453]. Universality of loops with the identity of flexibility (elasticity) $(x \cdot yx = xy \cdot x)$ is studied in [869, 870].

2.11.3 Alternative and various conjugate closed quasigroups and loops

In our opinion, left (right) alternative quasigroups are sufficiently important classes of quasigroups worthy of separate study. It is clear that Moufang loops are alternative. Some Moufang loops are multiplication parts of alternative rings and algebras [952, 746].

Lemma 2.263. *Any left alternative quasigroup $(Q, \cdot)$ has a left identity element.*

Proof. From identity $x \cdot xy = xx \cdot y$ by $x = f_y$ we have: $f_y \cdot f_y y = f_y f_y \cdot y$, $f_y y = f_y f_y \cdot y$, $f_y = f_y f_y$. Using the last equality further we have $f_y \cdot f_y x = f_y f_y \cdot x = f_y \cdot x$. After cancellation from the left in equality $f_y \cdot f_y x = f_y \cdot x$ we obtain $f_y x = x$ for all $x, y \in Q$. □

Corollary 2.264. *Any alternative quasigroup $(Q, \cdot)$ is a loop.*

Proof. It is possible to use Lemma 2.263 and its right analog. □

Any group, Mofang loop, and TS-loop is alternative and flexible. Information on left alternative (right alternative, flexible, alternative) T-quasigroups is given in [523].

Lemma 2.265. *A left alternative left linear quasigroup is a group.*

Proof. From Lemma 2.263 it follows that any left alternative quasigroup has an idempotent element, say 0.

From Lemma 2.208 Case 5 it follows that any left linear quasigroup $(Q, \cdot)$ with an idempotent element 0 over a group $(Q, +)$ has the following form: $x \cdot y = \varphi x + \beta y$, $\beta 0 = 0$.

We prove that $\varphi = \beta = \varepsilon$. Since element 0 is a left identity element of quasigroup $(Q, \cdot)$ (Lemma 2.263), we have $0 \cdot y = \varphi 0 + \beta y = 0 + \beta y = y$. Therefore $\beta y = y$, $\beta = \varepsilon$.

Then $x \cdot y = \varphi x + y$. Therefore we can write the left alternative identity in the following form: $\varphi(\varphi x + x) + y = \varphi x + \varphi x + y$. Thus $\varphi^2 x + \varphi x = \varphi x + \varphi x$, $\varphi^2 x = \varphi x$, $\varphi = \varepsilon$. □

Example 2.266. We give an example of a TS-loop of order ten.

$\blacksquare$	0	1	2	3	4	5	6	7	8	9
0	0	1	2	3	4	5	6	7	8	9
1	1	0	3	2	5	4	8	9	6	7
2	2	3	0	1	6	9	4	8	7	5
3	3	2	1	0	7	8	9	4	5	6
4	4	5	6	7	0	1	2	3	9	8
5	5	4	9	8	1	0	7	6	3	2
6	6	8	4	9	2	7	0	5	1	3
7	7	9	8	4	3	6	5	0	2	1
8	8	6	7	5	9	3	1	2	0	4
9	9	7	5	6	8	2	3	1	4	0

It is clear that in the loop $(Q, \blacksquare) \times (Z_3, +)$ the identity $x \blacksquare x = 0$ is not true. Therefore this loop is alternative flexible non-Moufang and non-TS-loop. Loop $(Q, \diamond)$ is also of this kind.

$\diamond$	0	1	2	3	4	5	6	7	8	9	10	11
0	0	1	2	3	4	5	6	7	8	9	10	11
1	1	0	3	2	5	4	8	9	6	7	11	10
2	2	3	0	1	6	7	4	5	10	11	8	9
3	3	2	1	0	7	6	5	4	11	10	9	8
4	4	10	6	7	0	11	2	3	9	8	1	5
5	5	11	8	9	1	10	3	2	7	6	0	4
6	6	8	4	10	2	9	0	11	1	5	3	7
7	7	9	10	4	3	8	11	0	5	1	2	6
8	8	6	5	11	9	2	1	10	0	4	7	3
9	9	7	11	5	8	3	10	1	4	0	6	2
10	10	4	7	6	11	0	9	8	2	3	5	1
11	11	5	9	8	10	1	7	6	3	2	4	0

In any alternative algebra, the following identities $x \cdot xy = xx \cdot y$, $x \cdot yy = xy \cdot y$, and $x \cdot yx = xy \cdot x$ are "two-equivalent," i.e., from any two identities the third one follows [952, 746]. The following loop is alternative and non-flexible.

$\circ$	0	1	2	3	4	5	6	7	8	9	10	11
0	0	1	2	3	4	5	6	7	8	9	10	11
1	1	0	3	2	6	7	4	5	9	8	11	10
2	2	4	0	8	1	9	11	10	3	5	7	6
3	3	5	1	9	0	8	10	11	6	7	2	4
4	4	2	10	0	11	1	8	9	5	3	6	7
5	5	3	9	10	8	11	0	1	7	6	4	2
6	6	7	11	1	10	0	9	8	4	2	3	5
7	7	6	8	11	9	10	1	0	2	4	5	3
8	8	9	7	5	2	4	3	6	11	10	0	1
9	9	8	5	7	3	6	2	4	10	11	1	0
10	10	11	4	6	5	3	7	2	0	1	9	8
11	11	10	6	4	7	2	5	3	1	0	8	9

Loop $(Q, \star)$ is left alternative, non-right alternative, and non-flexible. Loop $(Q, *)$ is flexible, non-right alternative, and non-left alternative.

⋆	0	1	2	3	4	5
0	0	1	2	3	4	5
1	1	0	3	2	5	4
2	2	3	5	4	1	0
3	3	5	4	0	2	1
4	4	2	1	5	0	3
5	5	4	0	1	3	2

*	0	1	2	3	4
0	0	1	2	3	4
1	1	0	3	4	2
2	2	4	0	1	3
3	3	2	4	0	1
4	4	3	1	2	0

In [374] E.G. Goodaire and D.A. Robinson defined and studied LCC-, RCC-, and CC-loops. In [547] Kenneth Kunen studied these loop classes and notices that there exists a possibility to study various corresponding conjugate closed quasigroups. Here we follow [222] and use the left record of maps.

Definition 2.267. A loop $(Q, \cdot)$ is left conjugacy closed (or an LCC-loop) if for all $a, b \in Q$ there exists an element $c \in Q$ such that $L_a^{-1} L_b L_a = L_c$.

A loop $(Q, \cdot)$ is right conjugacy closed (or an RCC-loop) if for all $a, b \in Q$ there exists an element $c \in Q$ such that $R_a^{-1} R_b R_a = R_c$.

A loop $(Q, \cdot)$ is middle conjugacy closed (or a PCC-loop) if for all $a, b \in Q$ there exists an element $c \in Q$ such that $P_a^{-1} P_b P_a = P_c$.

Any LCC- and RCC-loop is called a CC-loop [516, 514, 515].

Lemma 2.268. *A loop is an LCC-loop if and only if the following identity is true:*

$$y\backslash(z \cdot yx) = (y\backslash zy)x; \tag{2.80}$$

a loop is an RCC-loop if and only if the following identity is true:

$$(xy \cdot z)/y = x(yz/y); \tag{2.81}$$

a loop is a PCC-loop if and only if the following identity is true:

$$(z\backslash x)\backslash y = (z\backslash(x/(x\backslash y)))\backslash x. \tag{2.82}$$

The following loop is a non-associative PCC-loop.

*	0	1	2	3	4	5	6	7
0	0	1	2	3	4	5	6	7
1	1	0	3	5	2	4	7	6
2	2	3	0	1	6	7	4	5
3	3	5	1	0	7	6	2	4
4	4	2	6	7	0	1	5	3
5	5	4	7	6	1	0	3	2
6	6	7	4	2	5	3	0	1
7	7	6	5	4	3	2	1	0

Indeed, $1 * (1 * 2) \neq (1 * 1) * 2$. Any PCC-loop is commutative. Any commutative group is a PCC-loop.

Definition 2.269. A quasigroup with identity (2.80) is called an LCC-quasigroup;

A quasigroup with identity (2.81) is called an RCC-quasigroup;

A quasigroup with identity (2.82) is called a PCC-quasigroup.

"In a quasigroup, RCC implies that there is a left identity (apply RCC with $zy = z$ to show that $yx = x$ for all x), so that every CC quasigroup is a loop" [547]. Any PCC-quasigroup has a left identity element. Probably there exists a reason to research LCC-, RCC-, and PCC-quasigroups and loops.

Chapter 3

Binary inverse quasigroups

3.1 Definitions 183
3.1.1 Definitions of "general" inverse quasigroups 183
3.2 (r,s,t)-Inverse quasigroups 188
3.2.1 Elementary properties and examples 188
3.2.2 Left-linear quasigroups which are (r,s,t)-inverse 190
3.2.3 Main theorems 194
3.2.4 Direct product of (r,s,t)-quasigroups 197
3.2.5 The existence of (r,s,t)-inverse quasigroups 201
3.2.6 WIP-quasigroups 202
3.2.7 Examples of WIP-quasigroups 205
3.2.8 Generalized balanced parastrophic identities 206
3.2.9 Historical notes 207

Belousov defined λ-inverse and ρ-inverse quasigroups [82]. These quasigroups are a generalization of LIP- and RIP-quasigroups.

Most recently, (r,s,t)-inverse quasigroups were defined as a generalization of various kinds of "crossed-inverse" property quasigroups and loops: in particular, they generalize CI-, WIP- and m-inverse loops [472].

We demonstrate that all the above mentioned kinds of inverse property can be classified into three types which we call λ-inverse, ρ-inverse and (α,β,γ)-inverse. (The last of these three types was introduced in [487, 488].)

In some sense these inverse properties are equivalent to the following facts: a quasigroup has the property of "generalized commutativity"; in a quasigroup some nuclei or A-nuclei coincide or are isomorphic; in a quasigroup there exist autotopisms or autostrophisms of a special kind. See below.

In this chapter relatively much attention is given to the class of (r,s,t)-inverse quasigroups which is a proper subclass of the class of (α,β,γ)-inverse quasigroups.

At the end of the chapter we study some generalized balanced parastrophic identities.

3.1 Definitions

3.1.1 Definitions of "general" inverse quasigroups

We give the definitions of some well-known classes of quasigroups and loops which have an inverse property.

Definition 3.1. A quasigroup $(Q,\circ)$ has the λ-*inverse-property* if there exist permutations $\lambda_1,\lambda_2,\lambda_3$ of the set Q such that

$$\lambda_1 x\circ\lambda_2(x\circ y)=\lambda_3 y \tag{3.1}$$

for all $x, y \in Q$ [82].

Definition 3.2. A quasigroup $(Q, \circ)$ has the *ρ-inverse-property* if there exist permutations ρ_1, ρ_2, ρ_3 of the set Q such that

$$\rho_1(x \circ y) \circ \rho_2 y = \rho_3 x \tag{3.2}$$

for all $x, y \in Q$ [82].

Definition 3.3. A quasigroup $(Q, \circ)$ is an (α, β, γ)-inverse quasigroup if there exist permutations α, β, γ of the set Q such that

$$\alpha(x \circ y) \circ \beta x = \gamma y \tag{3.3}$$

for all $x, y \in Q$ [488, 487].

Definition 3.4. A quasigroup $(Q, \circ)$ has the *μ-inverse-property* if there exist permutations μ_1, μ_2, μ_3 of the set Q such that

$$\mu_1 y \circ \mu_2 x = \mu_3(x \circ y) \tag{3.4}$$

for all $x, y \in Q$ [787].

We give definitions of the main classes of inverse quasigroups using autostrophy [82, 489, 787].

Theorem 3.5. *A quasigroup $(Q, \cdot)$ is:*

1. *a λ-inverse quasigroup if and only if it has $[(2\ 3), (\lambda_1, \lambda_2, \lambda_3)]$ autostrophy;*
2. *a ρ-inverse quasigroup if and only if it has $[(1\ 3), (\rho_1, \rho_2, \rho_3)]$ autostrophy;*
3. *an (α, β, γ)-inverse quasigroup if and only if it has $[(1\ 2\ 3), (\alpha, \beta, \gamma)]$ autostrophy;*
4. *a μ-inverse quasigroup if and only if it has $[(12), (\mu_1, \mu_2, \mu_3)]$ autostrophy.*

Proof. If $(\sigma, (\alpha_1, \alpha_2, \alpha_3))$ is an autostrophism of a quasigroup (Q, A), then from formula (1.47) we obtain the following formula

$$A(x_1, x_2, x_3) = A(\alpha_1 x_{\sigma^{-1}1}, \alpha_2 x_{\sigma^{-1}2}, \alpha_3 x_{\sigma^{-1}3}) \tag{3.5}$$

Case 1. Using the equality (3.5) we find values of σ and α_i. Since $\alpha_1 x_{\sigma^{-1}1} = \lambda_1 x_1$, we have $\alpha_1 = \lambda_1$, $x_{\sigma^{-1}1} = x_1$, $\sigma 1 = 1$, and so on.

Cases 2–4 are proved in similar way with Case 1. □

Notice, $[(12), (\mu_1, \mu_2, \mu_3)]$ autostrophy is, in fact, an anti-autotopy of quasigroup $(Q, \cdot)$. See Definition 1.192.

Lemma 3.6. *1. A λ-inverse quasigroup has $((2\ 3), (\lambda_1^{-1}, \lambda_3^{-1}, \lambda_2^{-1}))$ autostrophy;*

2. *a ρ-inverse quasigroup has $((1\ 3), (\rho_3^{-1}, \rho_2^{-1}, \rho_1^{-1}))$ autostrophy;*
3. *an (α, β, γ)-inverse quasigroup has $((1\ 3\ 2), (\beta^{-1}, \gamma^{-1}, \alpha^{-1}))$ autostrophy, i.e., in any (α, β, γ)-inverse quasigroup $(Q, \circ)$ the following equality is true $\beta^{-1} y \circ \gamma^{-1}(x \circ y) = \alpha^{-1} x$;*
4. *a μ-inverse quasigroup has $((12), (\mu_2^{-1}, \mu_1^{-1}, \mu_3^{-1}))$ autostrophy.*

Proof. From Lemma 1.177 we have: if $(\sigma, T) = (\sigma, (\alpha_1, \alpha_2, \alpha_3))$ is an autostrophism, then

$$(\sigma, T)^{-1} = (\sigma^{-1}, (T^{-1})^{\sigma^{-1}}) = (\sigma^{-1}, (\alpha_{\sigma 1}^{-1}, \alpha_{\sigma 2}^{-1}, \alpha_{\sigma 3}^{-1})). \tag{3.6}$$

□

We shall show that these last three classes of quasigroup (λ-inverse, ρ-inverse and (α, β, γ)-inverse quasigroups) can all be treated in a similar way.

Example 3.7. Any abelian 2-group $(Q, +)$ (i.e., $(Q, +) \cong \oplus_{i=1}^{k}(Z_2)_i$) is an (α, β, γ)-inverse quasigroup with $\alpha = L_a$, $\beta = L_b$ and $\gamma = L_{a+b}$. This group is a λ-inverse quasigroup with $\lambda_1 = L_a$, $\lambda_2 = L_b$, $\lambda_3 = L_{a+b}$. Moreover, this group is a ρ-inverse quasigroup with $\rho_1 = L_a$, $\rho_2 = L_b$ and $\rho_3 = L_{a+b}$.

Example 3.8. We give an example of an (α, β, γ)-inverse loop (identity element $= 1$) with $\alpha = \varepsilon$, $\beta = (0\,1)(2\,3)(4\,5)$, $\gamma = (0\,2\,4\,1)(3\,5)$.

*	0	1	2	3	4	5
0	2	0	3	5	1	4
1	0	1	2	3	4	5
2	4	2	5	0	3	1
3	1	3	4	2	5	0
4	5	4	0	1	2	3
5	3	5	1	4	0	2

We have used MACE 4 in construction of this example [614].

Definition 3.9. A quasigroup $(Q, \circ)$ that has both λ-inverse-property and ρ-inverse-property is called an I-quasigroup [82].

Definition 3.10. An I-quasigroup in which there exist permutations λ_1, λ_2 of the set Q such that $\lambda_1 x \circ \lambda_2(x \circ y) = y$ and there exist permutations ρ_1, ρ_2 of the set Q such that $\rho_1(x \circ y) \circ \rho_2 y = x$ for all $x, y \in Q$ is called a *PI-quasigroup* [82].

Some properties of LIP-quasigroups (RIP-quasigroups) are studied in [363].

Corollary 3.11. *A λ- and (α, β, γ)-inverse quasigroup is an I-inverse quasigroup. A ρ- and (α, β, γ)-inverse quasigroup is an I-inverse quasigroup. A λ- and μ-inverse quasigroup is an I-inverse quasigroup. A ρ- and μ-inverse quasigroup is an I-inverse quasigroup. A μ- and (α, β, γ)-inverse quasigroup is an I-inverse quasigroup.*

Proof. Any of these sets of elements $\{(23), (123)\}$, $\{(13), (123)\}$, $\{(23), (12)\}$, $\{(13), (12)\}$, $\{(12), (123)\}$ generates the group S_3. □

Corollary 3.12. *A quasigroup which has the λ-inverse-property has a $(\lambda_1\lambda_1, \lambda_3\lambda_2, \lambda_2\lambda_3)$ autotopism. A quasigroup which has the ρ-inverse-property has a $(\rho_3\rho_1, \rho_2\rho_2, \rho_1\rho_3)$ autotopism.*

Proof. By Theorem 3.5,

$$((2\,3), (\lambda_1, \lambda_2, \lambda_3)) \cdot ((2\,3), (\lambda_1, \lambda_2, \lambda_3)) = (\varepsilon, (\lambda_1, \lambda_3, \lambda_2)(\lambda_1, \lambda_2, \lambda_3))$$

and so $(\lambda_1\lambda_1, \lambda_3\lambda_2, \lambda_2\lambda_3)$ is an autotopism. Similarly,

$$\begin{aligned} &((1\,3), (\rho_1, \rho_2, \rho_3)) \cdot ((1\,3), (\rho_1, \rho_2, \rho_3)) = \\ &(\varepsilon, (\rho_3, \rho_2, \rho_1)(\rho_1, \rho_2, \rho_3)) = \\ &= (\varepsilon, (\rho_3\rho_1, \rho_2\rho_2, \rho_1\rho_3)). \end{aligned}$$

□

Definition 3.13. A loop $(Q, +)$ with the property that

$$I_0(x + y) + y = I_0 x \tag{3.7}$$

for all $x, y \in Q$, where I_0 is the permutation of Q such that $x + I_0 x = 0$, is called a *weak commutative inverse property loop* ($WCIP$-loop) [467].

This loop class is researched in [467, 468, 386].
We can give the following:

Definition 3.14. A quasigroup $(Q, \circ)$ is said to be a *$WCIP$-quasigroup* with respect to the permutation J of Q if

$$J(x \circ y) \circ y = Jx \tag{3.8}$$

for all $x, y \in Q$.

It is clear that any $WCIP$-quasigroup is a ρ-inverse quasigroup such that $\rho_1 = J$, $\rho_2 = \varepsilon$, $\rho_3 = J$. Therefore by Corollary 3.12 the triple (J^2, ε, J^2) is an autotopy of $WCIP$-quasigroup.

Lemma 3.15. *In any $WCIP$-quasigroup, $J^2 = \varepsilon$.*

Proof. If we substitute the term $x\backslash y$ instead of variable y in the equality (3.8), then we obtain the following equalities

$$J(x \circ (x\backslash y)) \circ (x\backslash y) \overset{(1.4)}{=} Jy \circ (x\backslash y) = Jx. \tag{3.9}$$

From the equality (3.9) we have

$$Jx = Jy \circ (x\backslash y). \tag{3.10}$$

If we make in the equality (3.10) the following substitutions $x := Jx$, $y := x$, then we get

$$J^2 x = Jx \circ (Jx\backslash x) \overset{(1.4)}{=} x.$$

□

From the proof of Lemma 3.15 it follows that this lemma is true for left division groupoids (see, for example, Theorem 1.58.) with the $WCIP$-property.

Example 3.16. We give an example of a $WCIP$-quasigroup of order five in which $J_1 = (1\,2)(3\,4)$, $J_2 = (0\,1)(2\,3)$ and an example of a commutative $WCIP$-loop with $J = (0\,1)(3\,4)$ and the identity element 2.

*	0	1	2	3	4
0	0	1	2	4	3
1	1	3	0	2	4
2	2	0	4	3	1
3	3	4	1	0	2
4	4	2	3	1	0

∘	0	1	2	3	4	5
0	1	2	0	4	5	3
1	2	3	1	5	0	4
2	0	1	2	3	4	5
3	4	5	3	1	2	0
4	5	0	4	2	3	1
5	3	4	5	0	1	2

We have used MACE 4 [614] in construction of these examples.

Exercise 3.17. In any $WCIP$-loop in the sense of Definition 3.14, $J = I_0$.

Theorem 3.18. *A quasigroup which has the (α, β, γ)-inverse property has a $(\beta\alpha\gamma, \gamma\beta\alpha, \alpha\gamma\beta)$ autotopism.*

Proof. By Theorem 3.5 any (α, β, γ)-inverse quasigroup has $((1\ 2\ 3), (\alpha, \beta, \gamma))$ autostrophism. Further we have

$$((123), (\alpha, \beta, \gamma))((123), (\alpha, \beta, \gamma)) = ((132), (\beta\alpha, \gamma\beta, \alpha\gamma)),$$

$$((132), (\beta\alpha, \gamma\beta, \alpha\gamma))((123), (\alpha, \beta, \gamma)) = (\varepsilon, (\gamma\beta\alpha, \alpha\gamma\beta, \beta\alpha\gamma)).$$

□

Theorem 3.19. *If the equality $(a \cdot b)\alpha \cdot a\beta = b\gamma$ holds in a quasigroup $(Q, \cdot)$ for all $a, b \in Q$, where α, β, γ are fixed cyclically (or pairwise) permutable permutations of the set Q, then $\alpha\beta\gamma$ is an automorphism of the quasigroup $(Q, \cdot)$.*

Proof. The proof follows from Theorem 3.18. □

Theorem 3.20. *(i) An $(\alpha_1, \beta_1, \gamma_1)$-inverse quasigroup $(Q, \cdot)$ is an $(\alpha_2, \beta_2, \gamma_2)$-inverse quasigroup if and only if the triple $(\beta_2\beta_1^{-1}, \gamma_2\gamma_1^{-1}, \alpha_2\alpha_1^{-1})$ is an autotopism of the quasigroup $(Q, \cdot)$.*

(ii) A $(\lambda_1, \lambda_2, \lambda_3)$-inverse quasigroup $(Q, \cdot)$, i.e., a λ-inverse quasigroup, is a $(\lambda'_1, \lambda'_2, \lambda'_3)$-inverse quasigroup if and only if the triple $(\lambda_1^{-1}\lambda'_1, \lambda_2^{-1}\lambda'_2, \lambda_3^{-1}\lambda'_3)$ is an autotopism of the quasigroup $(Q, \cdot)$.

(iii) A (ρ_1, ρ_2, ρ_3)-inverse quasigroup $(Q, \cdot)$, i.e., a ρ-inverse quasigroup is a $(\rho'_1, \rho'_2, \rho'_3)$-inverse quasigroup if and only if the triple $(\rho_1^{-1}\rho'_1, \rho_2^{-1}\rho'_2, \rho_3^{-1}\rho'_3,)$ is an autotopism of the quasigroup $(Q, \cdot)$.

Proof. (i) From Theorem 3.5 it follows that a quasigroup $(Q, \cdot)$ is an $(\alpha_1, \beta_1, \gamma_1)$-inverse quasigroup if and only if this quasigroup has an autostrophy $S_1 = ((123), (\alpha_1, \beta_1, \gamma_1))$. Similarly, quasigroup $(Q, \cdot)$ is an $(\alpha_2, \beta_2, \gamma_2)$-inverse quasigroup if and only if the quasigroup $(Q, \cdot)$ has an autostrophy $S_2 = ((123), (\alpha_2, \beta_2, \gamma_2))$.

An $(\alpha_1, \beta_1, \gamma_1)$-inverse quasigroup $(Q, \cdot)$ has an autostrophy S_2 if and only if there exists an autotopy T_1 such that $S_2 = T_1 S_1$.

Indeed, if S_1, S_2 are autostrophies of quasigroup $(Q, \cdot)$, then

$$T_1 = S_2 S_1^{-1} = ((1\,2\,3), (\alpha_2, \beta_2, \gamma_2))((1\ 3\ 2), (\beta_1^{-1}, \gamma_1^{-1}, \alpha_1^{-1})) = (\varepsilon, (\beta_2\beta_1^{-1}, \gamma_2\gamma_1^{-1}, \alpha_2\alpha_1^{-1}))$$

is an autotopy.

Converse. If S_1 is a σ-autostrophy of a quasigroup $(Q, \cdot)$ and T_1 is an autotopy of $(Q, \cdot)$, then $S_2 = T_1 S_1$ is σ-autostrophy of the quasigroup $(Q, \cdot)$.

Cases (ii) and (iii) are proved similarly. □

Theorem 3.21. *If a groupoid (Q, A) has a $(1\ 3)$-autostrophy, it is a right quasigroup. If it has a $(2\ 3)$-autostrophy, it is a left quasigroup. If it has a $(1\ 2\ 3)$-autostrophy, it also has a $(1\ 3\ 2)$-autostrophy and is a quasigroup.*

Proof. Suppose that the groupoid (Q, A) has the autostrophy $[(1\ 3), (\alpha_1, \alpha_2, \alpha_3)]$. This is equivalent to saying that (Q, A) is isotopic to $(Q, A^{(1\ 3)})$ and so the latter is a groupoid. That is, for all $a, b \in Q$, $A^{(1\ 3)}(a, b) = x$ is uniquely soluble for x. Equivalently, $A(x, b) = a$ is uniquely soluble for x and so (Q, A) is a right quasigroup.

The proof of the second statement is similar.

For the third statement, we note that the product of a $(1\ 2\ 3)$-autostrophy with itself is a $(1\ 3\ 2)$-autostrophy. The remainder of the proof is similar to the foregoing. □

Corollary 3.22. *A right quasigroup which has a (2 3)-autostrophy is a quasigroup. Likewise, a left quasigroup which has a (1 3)-autostrophy is a quasigroup.*

Remark 3.23. Theorem 3.21 and Corollary 3.22 are also true in the case when autostrophy has the form $[(\sigma), (\varepsilon, \varepsilon, \varepsilon)]$.

3.2 (r, s, t)-Inverse quasigroups

In this section we shall use the right order of multiplication (of composition) of maps: $(x)(\alpha\beta) = ((x)\alpha)\beta$, where α, β are maps.

We show that the well-known properties of CI-, WIP- and m-inverse loops can be obtained as corollaries of more general propositions. We present some constructions and examples of (r, s, t)-inverse quasigroups.

3.2.1 Elementary properties and examples

We recall that definitions of m-inverse and (r, s, t)-inverse quasigroups are given in section 3.1.1.

Lemma 3.24. *If in (r, s, t)-inverse loop $(Q, \circ)$ $1J = 1$, then $J^r = J^t$, $J^{s-r} = I_r$.*

Proof. If we take $x = 1$ in the equality 2.12, then we obtain $yJ^r = yJ^t$, $J^r = J^t$. If we take $y = 1$ in equality 2.12, then we have $xJ^r \circ xJ^s = 1$, $J^s = J^r I_r$. Further we have $J^{s-r} = I_r$. □

Corollary 3.25. *Every (r, s, t)-inverse loop $(Q, \circ)$ with the property $x \circ xJ = 1$, where 1 is the identity element of loop $(Q, \circ)$, is an $(r, r+1, r)$-inverse loop: that is, it is an r-inverse loop.*

Proof. In this case we have $J = I_r$, $J^s = I_r^{r+1}$. □

Example 3.26. Any group of the form $\oplus_{i=1}^{n}(Z_2)_i$ (i.e., any 2-group) is a (1, 1, 0)-inverse, (1, 0, 1)-inverse and (0, 1, 1)-inverse loop relative to any permutation $J = L_a$, where L_a is a left translation of this group.

Proof. This group has the (1, 1, 0)-inverse property, since $(x+y)L_a + xL_a = y$ is equivalent to $a + x + y + a + x = y$, $y = y$. □

Example 3.27. The following loop $(Q, \cdot)$

$\cdot$	0	1	2	3	4	5	6	7	8	9
0	0	1	2	3	4	5	6	7	8	9
1	1	8	3	9	2	4	5	6	7	0
2	2	7	8	0	9	3	4	5	6	1
3	3	0	9	6	7	8	1	4	5	2
4	4	2	5	8	6	9	7	1	0	3
5	5	3	6	7	0	2	8	9	1	4
6	6	4	0	1	3	7	9	8	2	5
7	7	5	1	4	8	0	2	3	9	6
8	8	9	4	5	1	6	0	2	3	7
9	9	6	7	2	5	1	3	0	4	8

is a (2, 1, 7)-inverse loop with respect to permutation $J = (0\,1\,2\,8\,5\,6\,7\,9\,4)$ [795].

Proposition 3.28. *A groupoid $(Q, \circ)$ which has the (r, s, t)-inverse property relative to some permutation J of Q is an (r, s, t)-inverse quasigroup (relative to J).*

Proof. Proposition 3.28 follows from Theorem 3.21. □

Remark 3.29. A similar result for WIP-groupoids was obtained by Steinberger in [847].

Lemma 3.30. *In any (r, s, t)-inverse quasigroup the following equivalence is fulfilled:*

$$\begin{aligned} (x \circ y)J^r \circ xJ^s = yJ^t \text{ for all } x, y \in Q \Leftrightarrow \\ xJ^{-s} \circ (y \circ x)J^{-t} = yJ^{-r} \text{ for all } x, y \in Q. \end{aligned} \tag{3.11}$$

Proof. This lemma is a corollary of Lemma 3.6. □

From Lemma 3.30, in particular, it follows that a WIP-loop, i.e., a weak-inverse-property loop, (which satisfies the relation $x \circ (y \circ x)J = yJ$) is a $(-1, 0, -1)$-inverse loop. We discuss such loops (and quasigroups) in more detail below in Section 3.2.6.

Moreover, since $(x \circ y) \circ xJ = y \Longrightarrow xJ^{-1} \circ (y \circ x) = y \Longrightarrow z \circ (y \circ zJ) = y$, where $z = xJ^{-1}$, we have the well-known result that a crossed inverse loop may be defined by the latter relation instead of the former.

Remark 3.31. From the relation (3.11) we get $J^r R_z = L_x^{-1} J^r$ in the special case when $r = t$. Thus, for any (r, s, r)-inverse quasigroup $(Q, \circ)$, the mapping J^r lies in the normalizer of the subgroup of S_Q which is generated by the left and right translations: that is, $J^r \in N(M)$, where M is the multiplication group of the quasigroup $(Q, \circ)$.

Corollary 3.32. *(1) J^{r+s+t} is an automorphism of an (r, s, t)-inverse quasigroup.*

(2) J^2 is an automorphism of a weak inverse property quasigroup.

(3) J^{3r+1} is an automorphism of an r-inverse property quasigroup.

Proof. (1) The proof follows from Theorem 3.19.

(2) To see this, notice that, since a WIP-quasigroup is a $(-1, 0, -1)$-inverse quasigroup, J^{-2} is an automorphism and so its inverse is an automorphism too.

(3) Any r-inverse quasigroup is an $(r, r+1, r)$-inverse quasigroup. □

Remark 3.33. If the quasigroup $(Q, \circ)$ is an (r, s, t)-inverse quasigroup with respect to the permutation J and $J^h \in Aut(Q, \circ)$ for some integer h, then $(Q, \circ)$ is $(r + uh, s + uh, t + uh)$-inverse for any $u \in Z$.

To see this, we have only to note that, if $J^h \in Aut(Q, \circ)$, then $(x \circ y)J^{r+uh} \circ xJ^{s+uh} = yJ^{t+uh}$ follows from the relation (2.12). This leads us to make the following definitions:

Definition 3.34. An (r, s, t)-inverse quasigroup is said to be *properly described* if $r, s, t \geqslant 0$ and it is not (r_1, s_1, t_1)-inverse for any integers r_1, s_1, t_1 such that $0 \leqslant r_1 < r$, $0 \leqslant s_1 < s$ and $0 \leqslant t_1 < t$.

Remark 3.35. Sometimes, as for example when we said earlier that a weak-inverse-property loop is a $(-1, 0, -1)$-inverse loop, it is more usual to use a description which involves negative integers.

[Notice that, in fact, since J^2 is an automorphism of any *WIP*-loop, it is "properly described" as a $(1, 2, 1)$-loop.]

Remark 3.36. Since J is an automorphism of a $(0,1,0)$-inverse quasigroup, it follows from Remark 3.33 that such a quasigroup is also a $(1,2,1)$-inverse quasigroup. That is, every CI-quasigroup has the weak inverse property. Thus, we have a new proof of this fact: the already well-known fact for loops.

Definition 3.37. The relations $(x \circ y)J^{r_1} \circ xJ^{s_1} = yJ^{t_1}$ and $(x \circ y)J^{r_2} \circ xJ^{s_2} = yJ^{t_2}$ are *equivalent* for the quasigroup $(Q, \circ)$ if from fulfilment of the first relation it follows the fulfilment of the second one and vice versa.

3.2.2 Left-linear quasigroups which are (r, s, t)-inverse

Let $(Q, +)$ be a loop with left nucleus $N_l = \{c : c + (x + y) = (c + x) + y \text{ for all } x, y \in Q\}$ and let $Aut(Q, +)$ denote the automorphism group of this loop.

Definition 3.38. A *left linear quasigroup* over the loop $(Q, +)$ is a quasigroup $(Q, \cdot)$ such that $x \cdot y = c + x\varphi + y\psi$ for all $x, y \in Q$, where φ is in $Aut(Q, +)$, ψ is a permutation of the set Q such that $\psi 0 = 0$ (where the symbol 0 denotes the identity element of the loop) and c is in the left nucleus N_l of the loop. It becomes a *linear quasigroup* if φ and ψ are both automorphisms of $(Q, +)$.

As a special case of this, a quasigroup $(Q, \cdot)$ defined over an abelian group $(Q, +)$ by $x \cdot y = c + x\varphi + y\psi$, where c is a fixed element of Q and φ and ψ are both automorphisms of the group $(Q, +)$, is called a *T-quasigroup*.

(The latter concept was first introduced in [667] and [506].)

Theorem 3.39. *A left linear quasigroup $(Q, \cdot)$ over a loop $(Q, +)$ is an (r, s, t)-inverse quasigroup with respect to the permutation J of the set Q, where $J^r \in Aut(Q, +)$ and $0J = 0$ if and only if*

$$\begin{array}{ll} \text{(i)}\ c + cJ^r\varphi = 0, & \text{(iii)}\ x\varphi J^r\varphi + xJ^s\psi = 0 \text{ for all } x \in Q, \\ \text{(ii)}\ \psi = J^t\varphi^{-1}J^{-r}, & \text{(iv)}\ (Q, +) \text{ is a } CI\text{-loop}. \end{array}$$

Proof. Proof of necessity. Suppose that $(x \cdot y)J^r \cdot xJ^s = yJ^t$ for all $x, y \in Q$. Then $c + (x \cdot y)J^r\varphi + xJ^s\psi = yJ^t$. That is,

$$[c + (c + x\varphi + y\psi)J^r\varphi] + xJ^s\psi = yJ^t. \tag{3.12}$$

Put $x = y = 0$ in the equation (3.12). Then $c + cJ^r\varphi = 0$, so (i) is a necessary condition.

Since $J^r\varphi$ is an automorphism of $(Q, +)$, $cJ^r\varphi \in N_l$, the left nucleus of $(Q, +)$, and so the equation (3.12) can be re-written in the form $[(c + cJ^r\varphi) + (x\varphi J^r\varphi + y\psi J^r\varphi)] + xJ^s\psi = yJ^t$.

That is,

$$(x\varphi J^r\varphi + y\psi J^r\varphi) + xJ^s\psi = yJ^t. \tag{3.13}$$

Putting $x = 0$ in Equation 3.13, we see that $y\psi J^r\varphi = yJ^t$ is necessary. That is, $\psi J^r\varphi = J^t$ or $\psi = J^t\varphi^{-1}J^{-r}$, which is condition (ii).

Similarly, putting $y = 0$ in the equation (3.13), we see that condition (iii) is necessary.

Thus, $(x\varphi J^r\varphi + y\psi J^r\varphi) + xJ^s\psi = yJ^t$ for all $x, y \in Q$, where, since (iii) holds, $xJ^s\psi$ is the right inverse of $x\varphi J^r\varphi$ in $(Q, +)$. Thence, $(Q, +)$ must be a *CI*-loop, which is condition (iv).

Proof of sufficiency. Let us define a left linear quasigroup $(Q, \cdot)$ over the *CI*-loop $(Q, +)$ by $x \cdot y = c + x\varphi + y\psi$, where $\varphi \in Aut(Q, +)$ and $\psi = J^t\varphi^{-1}J^{-r}$. We wish to show that $(Q, \cdot)$ is an (r, s, t)-inverse quasigroup with respect to the permutation J of the set Q, where $J^r \in Aut(Q, +)$.

We have

$$(x \cdot y)J^r \cdot xJ^s = (c + x\varphi + yJ^t\varphi^{-1}J^{-r})J^r \cdot xJ^s$$
$$= c + (c + x\varphi + yJ^t\varphi^{-1}J^{-r})J^r\varphi + xJ^sJ^t\varphi^{-1}J^{-r}$$
$$= (c + [cJ^r\varphi + (x\varphi J^r\varphi + yJ^t)]) + xJ^s\psi.$$

[We use in succession the facts that $J^r\varphi$ is an automorphism of $(Q,+)$ and that, in consequence, $c \in N_l$ implies that $cJ^r\varphi \in N_l$.] Further we have:

$$= [(c + cJ^r\varphi) + (x\varphi J^r\varphi + yJ^t)] + xJ^s\psi = (x\varphi J^r\varphi + yJ^t) + xJ^s\psi = yJ^t$$

because $xJ^s\psi$ is the right inverse of $x\varphi J^r\varphi$ in $(Q,+)$ and because $(Q,+)$ is a CI-loop.

It follows that $(Q,\cdot)$ is an (r,s,t)-inverse quasigroup. □

Remark 3.40. When the conditions of Theorem 3.39 hold, $J^{r+s+t} = (J^r\varphi)^3 I_0$, where I_0 is defined by $x + xI_0 = 0$ for all $x \in Q$. Consequently, J^{r+s+t} is in $Aut(Q,+)$ as well as being in $Aut(Q,\cdot)$.

[To see this, we observe that $J^s\psi = \varphi J^r\varphi I_0$ from condition (iii). Substituting for ψ from condition (ii), we get $J^sJ^t\varphi^{-1}J^{-r} = \varphi J^r\varphi I_0$. So, left-multiplying by J^r, $J^{r+s+t}(J^r\varphi)^{-1} = J^r\varphi J^r\varphi I_0$. That is, $J^{r+s+t} = (J^r\varphi)^2 I_0\ (J^r\varphi)$. But, I_0 commutes with every automorphism of the loop $(Q,+)$ as we shall show later, so the stated result follows.]

Remark 3.41. Since a non-abelian group cannot have the crossed inverse property, it follows immediately from Theorem 3.39 that (r,s,t)-inverse left (or right) linear quasigroups over a non-abelian group do not exist.

On the other hand, since every abelian group is a CI-loop, a T-quasigroup $(Q,\cdot)$ is an (r,s,t)-inverse quasigroup with $J^r \in Aut(Q,+)$, where J is a permutation of Q such that $0J = 0$, if and only if conditions (i), (ii) and (iii) of Theorem 3.39 hold.

In particular, sufficient conditions are that $\varphi = J^{-r}I_0$, $\psi = J^tI_0$ and $J^{r+s+t} = I$, where I is the identity map and $xI_0 = -x$ in the group $(Q,+)$ for all $x \in Q$.

[To see this, we note that, if $\varphi = J^{-r}I_0$, the condition $c + cJ^r\varphi = 0$ becomes $c + cI_0 = 0$ and is satisfied identically. Also, if $\varphi = J^{-r}I_0$, $\psi = J^t\varphi^{-1}J^{-r} = J^tI_0^{-1} = J^tI_0$ and the condition $x\varphi J^r\varphi + xJ^s\psi = 0$ becomes $xJ^{-r} + xJ^{s+t}I_0 = 0$. Since J^r is in $Aut(Q,+)$, the last statement is equivalent to $x - xJ^{r+s+t} = 0$. Since $J^{r+s+t} = I$, this is automatically satisfied.]

The above conditions say in effect that the quasigroup $(Q,\cdot)$ defined by $x \cdot y = c + xJ^{-r}I_0 + yJ^tI_0$, where J is a permutation of Q such that $0J = 0$, $J^{r+s+t} = I$ and $J^r \in Aut(Q,+)$, where $(Q,+)$ is an abelian group, is an (r,s,t)-inverse quasigroup. This can easily be verified directly.

Example 3.42. $J : z \longrightarrow 2z \pmod{11}$ in the cyclic group $(Z_{11},+)$. Since $2^{10} \equiv 1 \pmod{11}$, we require $r+s+t = 10$. Let $r = 6, s = t = 2$. Then the quasigroup $(Z_{11},\cdot)$ defined by $x \cdot y = c + (2^{-6}x)I_0 + (2^2y)I_0$ is a $(6,2,2)$-inverse quasigroup.

[We have $x\cdot y = c - 2^4x - 2^2y \equiv c - 5x - 4y$ mod 11. So, $(x\cdot y)J^6 \cdot xJ^2 = 2^6(c-5x-4y)\cdot 2^2x = (-2c+10x+8y)\cdot 4x = c - 5(-2c+10x+8y) - 4(4x) = 11c - 66x - 40y \equiv 4y = yJ^2 mod 11$. That is, $(x\cdot y)J^6 \cdot xJ^2 = yJ^2$.]

Example 3.43. $J : z \longrightarrow 2z \pmod 9$ in the cyclic group $(Z_9,+)$. Since $2^6 \equiv 1 \pmod 9$, we require $r+s+t = 6$. Let $r = 2, s = 3, t = 1$. Then the quasigroup $(Z_9,\cdot)$ defined by $x \cdot y = c + (2^{-2}x)I_0 + (2y)I_0$ is a $(2,3,1)$-inverse quasigroup.

[We have $x\cdot y = c - 2^4x - 2y \equiv c - 7x - 2y$ mod 9. So, $(x\cdot y)J^2 \cdot xJ^3 = 2^2(c-7x-2y)\cdot 2^3x = (4c - x - 8y)\cdot 8x = c - 7(4c - x - 8y) - 2(8x) = c - 28c + 7x + 56y - 16x \equiv 2y = yJ$ mod 9. That is, $(x\cdot y)J^2 \cdot xJ^3 = yJ$.]

Remark 3.44. Note that, in the above two examples, the mapping J is an automorphism of the cyclic group but not of the quasigroup constructed from it.

Remark 3.45. Since an m-inverse quasigroup is an $(m, m+1, m)$-inverse quasigroup, we can construct an m-inverse quasigroup over $(Z_n, +)$ in the above manner only when $J : z \longrightarrow hz$ with h relatively prime to n and $h^{3m+1} \equiv 1 \bmod n$.

Another construction of an (r, s, t)-inverse quasigroup over the cyclic group $(Z_n, +)$ is obtained by taking $J : z \longrightarrow hz$ with h relatively prime to n as before and defining the quasigroup $(Z_n, \cdot)$ by the relation $x \cdot y = fx + gy$, where f and g are relatively prime to n.

In this case the quasigroup is linear over $(Z_n, +)$. It is no longer necessary that r, s, t are such that $J^{r+s+t} = I$ but J is always an automorphism of $(Z_n, \cdot)$ as well as of $(Z_n, +)$ because $(x \cdot y)J = h(fx + gy) = f(hx) + g(hy) = hx \cdot hy = xJ \cdot yJ$. So, when the quasigroup is properly described, one of r, s, t is always zero. If $x \cdot y = fx + gy$ for all $x, y \in Z_n$ and $J : z \longrightarrow hz$, we have $(x \cdot y)J^r \cdot xJ^s = yJ^t \iff h^r(fx + gy) \cdot h^s x = h^t y \iff fh^r(fx + gy) + gh^s x = h^t y \iff f^2h^r + gh^s = 0$ (Equ. 1) and $fgh^r = h^t$ (Equ. 2).

From Equ. 1, $f^3h^r + fgh^s = 0$. Using Equ. 2, $f^3h^r + h^{t-r}h^s = 0$. So, $f^3 = -h^{s+t-2r}$ and $g = h^{t-r}f^{-1}$.

Example 3.46. Let $n = 11, h = 2, r = 1, s = 5, t = 3$. Then $f^3 = -2^6$ so $f \equiv -4 \bmod 11$ and $g = 2^2(-1/4) \equiv -1 \bmod 11$. Thence, $x \cdot y = -4x - y$ defines a $(1, 5, 3)$-inverse quasigroup. Since J is an automorphism of $(Z_{11}, \cdot)$, it follows that this quasigroup is $(0, 4, 2)$-inverse.

We have $(x \cdot y) \cdot xJ^4 = (-4x - y) \cdot 2^4 x = -4(-4x - y) - 5x = 11x + 4y \equiv 4y = yJ^2$ as claimed.

Example 3.47. In the case when $n = p$ is prime, all three of the above constructions yield (r, s, t)-inverse quasigroups $(Z_p, \cdot)$ with an inverse cycle of length $p - 1$ since $J : z \longrightarrow hz$, where h is relatively prime to p, in each case.

Finally, we observe that a special case of Theorem 3.39 is the following:

Theorem 3.48. *A left linear quasigroup $(Q, \cdot)$ over a loop $(Q, +)$, where $x \cdot y = c + x\varphi + y\psi$, is a CI-quasigroup relative to the permutation J, where $0J = 0$, if and only if $c + c\varphi = 0$, $\psi = \varphi^{-1}$, $x\varphi^3 + xJ = 0$ for all $x \in Q$ and $(Q, +)$ is a CI-loop.*

Proof. We put $r = 0$, $s = 1$, $t = 0$ in Theorem 3.39. Then, $J^r \equiv I \equiv J^t$. □

Remark 3.49. Since $\psi = \varphi^{-1}$, it follows that, if φ is an automorphism, so is ψ. Therefore, a left linear *CI*-quasigroup over a loop must in fact be a linear *CI*-quasigroup.

The following example illustrates Theorem 3.48 for the case when $(Q, +)$ is a proper loop, not a group.

Example 3.50. We construct a CI-loop $(Q, \oplus)$ as follows:

Let $(Z_{19}, \cdot)$ be the linear quasigroup over the cyclic group $(Z_{19}, +)$ which is defined by $x \cdot y = x\alpha + y\beta$, where $\alpha : z \rightarrow 12z$ and $\beta : z \rightarrow 8z \bmod 19$. This quasigroup is *medial* because α and β are commuting automorphisms of $(Z_{19}, +)$. Also, it is *idempotent* because $x \cdot x = 12x + 8x \equiv x \bmod 19$. Consequently, it satisfies the *distributive laws* and we can easily check that it also satisfies the *left semi-symmetric law.* (See [240], page 58.)

We have $(u \cdot v) \cdot (x \cdot y) = (u \cdot x) \cdot (v \cdot y) \Rightarrow (u \cdot u) \cdot (x \cdot y) = (u \cdot x) \cdot (u \cdot y) \Rightarrow u \cdot (x \cdot y) = (u \cdot x) \cdot (u \cdot y)$. The right distributive law is similarly proved. Also, $(x \cdot y) \cdot x = (12x + 8y) \cdot x = 12(12x + 8y) + 8x = 152x + 96y \equiv y \bmod 19$ so $(x \cdot y) \cdot x = y$.

By a prolongation of this idempotent quasigroup, we obtain a CI-loop $(Q, \oplus)$, not a group, of order 20. We define this as follows:

$Q = Z_{19} \cup \{e\}$ and

(i) $x \oplus y = x \cdot y$ when $x \neq y$ for all $x, y \in Z_{19}$;
(ii)$x \oplus x = e$ for all $x \in Q$;
(iii) $x \oplus e = x = e \oplus x$ for all $x \in Q$. (For a history of this construction, see page 40 of [240].)

To see that $(Q, \oplus)$ is a CI-loop, we first observe that each element of $(Q, \oplus)$ is self-inverse by (ii) above and so the validity of the semi-symmetric law in $(Z_{19}, \cdot)$ implies validity of the CI-property when $x \neq y$ in $(Q, \oplus)$. We have $(x \oplus y) \oplus xI_0 = (x \oplus y) \oplus x = (x \cdot y) \cdot x = y$ when $y \neq x$ and $x, y \in Z_{19}$. Also, $(x \oplus x) \oplus xI_0 = (x \oplus x) \oplus x = e \oplus x = x$ when $y = x$ and $x \in Z_{19}$; $(x \oplus e) \oplus xI_0 = x \oplus x = e$ when $y = e$; and $(e \oplus y) \oplus eI_0 = y \oplus e = y$ when $x = e$. Thus, $(Q, \oplus)$ is a CI-loop.

We define a linear quasigroup $(Q, \circ)$ over the loop $(Q, \oplus)$ by the relation $x \circ y = x\varphi \oplus y\varphi^{-1}$, where φ is an automorphism of the loop $(Q, \oplus)$. Then, as stated in Theorem 3.48, $(Q, \circ)$ is a CI-quasigroup relative to the permutation J of Q such that $x\varphi^3 \oplus xJ = e$ for all $x \in Q$.

Since $(Q, \oplus)$ is unipotent by construction, the latter relation implies that $J = \varphi^3$, so we have

$$(x \circ y) \circ xJ = (x\varphi \oplus y\varphi^{-1}) \circ x\varphi^3$$
$$= (x\varphi \oplus y\varphi^{-1})\varphi \oplus x\varphi^3\varphi^{-1} = (x\varphi^2 \oplus y) \oplus x\varphi^2 = y$$

because $(Q, \oplus)$ is a CI-loop and because each element $x\varphi^2$ is self-inverse in $(Q, \oplus)$. This confirms Theorem 3.48.

It is of interest to determine the automorphism group of $(Q, \oplus)$ and hence the spectrum of the mappings φ. In order to do so, we need the following lemma.

Lemma 3.51. *Let $(T, \oplus)$ be a prolongation of the idempotent quasigroup $(S, \cdot)$ defined as follows: $T = S \cup \{e\}$ and*

(i) $x \oplus y = x \cdot y$ when $x \neq y$ for all $x, y \in S$;
(ii)$x \oplus x = e$ for all $x \in T$;
(iii) $x \oplus e = x = e \oplus x$ for all $x \in T$. Then $Aut(T, \oplus) \cong Aut(S, \cdot)$.

Proof. Let α be a permutation of the set S. We define a corresponding permutation β of the set T by the statements that $x\alpha = x\beta$ for all $x \in S$ and $e\beta = e$.

Conversely, each permutation β on the set T which keeps the element e fixed induces a corresponding permutation α on the set S. It is easy to see that, if β_1, β_2 correspond to α_1, α_2 respectively, then $\beta_1\beta_2$ corresponds to $\alpha_1\alpha_2$, so it will be sufficient to show that $\alpha \in Aut(S, \cdot)$ if and only if $\beta \in Aut(T, \oplus)$.

Firstly, we note that, for all $x \in T$, we have $x \oplus x = e$ and $x \oplus e = x = e \oplus x$. Therefore, if β is any permutation of T which keeps the element e fixed, $(x \oplus x)\beta = e\beta = e$. Also, $x\beta \oplus x\beta = e$.

Consequently, $(x \oplus x)\beta = x\beta \oplus x\beta$. Furthermore, $(x \oplus e)\beta = x\beta$ and $x\beta \oplus e\beta = x\beta \oplus e = x\beta$. So, $(x \oplus e)\beta = x\beta \oplus e\beta$. Similarly, $(e \oplus x)\beta = e\beta \oplus x\beta$.

Secondly, let $\alpha \in Aut(S, \cdot)$ and let β be the corresponding permutation of T. For all $x, y \in S$, $x \neq y$, we have $x \oplus y = x \cdot y \in S$.

Consequently, $(x \oplus y)\beta = (x \cdot y)\beta = (x \cdot y)\alpha$ since $x \cdot y \in S$ $= x\alpha \cdot y\alpha$ since $\alpha \in Aut(S, \cdot)$ $= x\alpha \oplus y\alpha$ since $x\alpha, y\alpha \in S$ and $x\alpha \neq y\alpha$ $= x\beta \oplus y\beta$, since $x, y \in S$.
Therefore,$\alpha \in Aut(S, \cdot) \Rightarrow \beta \in Aut(T, \oplus)$.

Thirdly, let $\beta \in Aut(T, \oplus)$ and let α be the corresponding permutation of S. For all $x, y \in T$, $x \neq y$, $x \neq e$ and $y \neq e$, we have $x \oplus y = x \cdot y \in S$.

Consequently, $(x \cdot y)\alpha = (x \oplus y)\alpha = (x \oplus y)\beta$ since $x \oplus y \in S = x\beta \oplus y\beta$ since $\beta \in Aut(T, \oplus)$ $= x\beta \cdot y\beta$ since $x\beta, y\beta \in S$ and $x\beta \neq y\beta = x\alpha \cdot y\alpha$, since $x, y \in S$.
Therefore, $\beta \in Aut(T, \oplus) \Rightarrow \alpha \in Aut(S, \cdot)$.

Thus, $\alpha \in Aut(S, \cdot) \Leftrightarrow \beta \in Aut(T, \oplus)$, as required. □

It follows from Lemma 3.51 that $Aut(Q, \oplus) \cong Aut(Z_{19}, \cdot)$ since $(Q, \oplus)$ is a prolongation of $(Z_{19}, \cdot)$.

Also, it follows from the paper (see [769]) that, if $(G, \cdot)$ is a linear quasigroup of the form $x \cdot y = x\alpha + y\beta$ over an abelian group $(G, +)$, where $\alpha, \beta \in Aut(G, +)$ and $z\alpha + z\beta = z$ for all $z \in G$, then $Aut(G, \cdot) \cong G \rtimes C$, the semi-direct product of G and C, where C is the group of all automorphisms of $(G, +)$ which commute with α and β.

[In fact, in Proposition 7, Corollary 2 of [769] it was proved: "Let $(G, \cdot)$ be a quasigroup isotopic to an abelian group $(G, +)$ given by $x \cdot y = x\alpha + y\beta$, where $\alpha, \beta \in Aut(G, +)$, then $Aut(G, \cdot)$ is the semi-direct product of K and C, where $K = \langle L_b^+ : b \in G \ \text{ and } \ b(\alpha + \beta - \varepsilon) = 0 \rangle$ and C is defined as above."]

In the present case, since every element of $(Z_{19}, +)$ satisfies $z(\alpha + \beta - \varepsilon) = 0$ (that is, $12z + 8z = z$), we find that K consists of all left translations of $(Z_{19}, +)$.

Also, $C \cong Aut(Z_{19}, +)$ which comprises all left multiplications of Z_{19} by non-zero elements of Z_{19} and has the order 18. Hence, the automorphism φ of $(Q, \circ)$ may be any combination of a multiplication by a non-zero element of Z_{19} and a left translation by an element of Z_{19}. For example, we might define $(Q, \circ)$ by $x \circ y = xL_{16}^+ \oplus yL_3^+$ or by $x \circ y = 5x \oplus 4y$.

[In the first case, $x \circ y = 12(16 + x) + 8(3 + y) \equiv 7 + 12x + 8y \bmod 19$ and $zJ = z(L_{16}^+)^3 = 48 + z \equiv 10 + x$. Thence, $(x \circ y) \circ xJ = (7 + 12x + 8y) \circ (10 + x) = 7 + 12(7 + 12x + 8y) + 8(10 + x) = 171 + 152x + 96y \equiv y \bmod 19$.

In the second case, $x \circ y = 12(5x) + 8(4y) \equiv 3x - 6y \bmod 19$ and $zJ = 5^3x \equiv 11x$. Thence, $(x \circ y) \circ xJ = (3x - 6y) \circ 11x = 3(3x - 6y) - 6(11x) = -57x - 18y \equiv y \bmod 19$.]

3.2.3 Main theorems

We give a more general set of necessary and sufficient conditions for a T-quasigroup to be an (r, s, t)-inverse quasigroup than given in the previous subsection.

Theorem 3.52. *A T-quasigroup $(Q, \circ)$ defined over an abelian group $(Q, +)$ by $x \circ y = c + x\phi + y\psi$ is an (r, s, t)-inverse quasigroup with respect to the permutation $J = L_b^+\beta$, where $\beta \in Aut(Q, +)$ and L_b^+ is the left translation of the group $(Q, +)$ defined by the element $b \in Q$, if and only if*

(i) $c + B_r\phi + c\beta^r\phi + B_s\psi = B_t$; (ii) $\beta^s\psi = \phi\beta^r\phi I_0$; (iii) $\beta^t = \psi\beta^r\phi$, where $B_u = b\beta + b\beta^2 + \dots + b\beta^u$ and I_0 is defined by $x + xI_0 = 0$ for all $x \in Q$.

Proof. Proof of necessity. We first show that $h(L_b^+\beta)^u = B_u + h\beta^u$. We have $h(L_b^+\beta)^u = (b + h)\beta(L_b^+\beta)^{u-1} = (b\beta + h\beta)(L_b^+\beta)^{u-1}$
$= (b + b\beta + h\beta)\beta(L_b^+\beta)^{u-2} = (b\beta + b\beta^2 + h\beta^2)(L_b^+\beta)^{u-2} = \dots = \dots$
$= (b\beta + b\beta^2 + b\beta^3 + \dots + b\beta^{u-1} + h\beta^{u-1})(L_b^+\beta)^{u-(u-1)}$
$= (b + b\beta + b\beta^2 + b\beta^3 + \dots + b\beta^{u-1} + h\beta^{u-1})\beta = B_u + h\beta^u$ as stated.

We require that $(x \circ y)J^r \circ xJ^s = yJ^t$ for all $x, y \in Q$, where $J = L_b^+\beta$. That is, we require $(c + x\phi + y\psi)J^r \circ xJ^s = yJ^t$ or
$c + [(c + x\phi + y\psi)(L_b^+\beta)^r]\phi + x(L_b^+\beta)^s\psi = y(L_b^+\beta)^t$. So, we require that
$c + [B_r + (c + x\phi + y\psi)\beta^r]\phi + [B_s + x\beta^s]\psi = B_t + y\beta^t$ or
$c + B_r\phi + c\beta^r\phi + x\phi\beta^r\phi + y\psi\beta^r\phi + B_s\psi + x\beta^s\psi = B_t + y\beta^t$ for all $x, y \in Q$. Therefore,

it is necessary that $c + B_r\phi + c\beta^r\phi + B_s\psi = B_t$ and $\phi\beta^r\phi + \beta^s\psi = 0$ and $\psi\beta^r\phi = \beta^t$. These are conditions (i), (ii) and (iii) of the theorem.

Proof of sufficiency. It is easy to see that, when conditions (i), (ii) and (iii) hold,

$$(x \circ y)(L_b^+\beta)^r \circ x(L_b^+\beta)^s = B_t + y\beta^t = y(L_b^+\beta)^t$$

for all $x, y \in Q$ and so $(Q, \circ)$ is an (r, s, t)-inverse quasigroup with respect to the permutation $J = L_b^+\beta$. □

Corollary 3.53. *A T-quasigroup $(Q, \circ)$ defined over an abelian group $(Q, +)$ by $x \circ y = c + x\phi + y\psi$ is an (r, s, t)-inverse quasigroup with respect to the permutation $J = \beta$, where $\beta \in Aut(Q, +)$, if and only if*

(i) $c + c\beta^r\phi = 0$; (ii) $\beta^s\psi = \phi\beta^r\phi I_0$; (iii) $\beta^t = \psi\beta^r\phi$.

Proof. This is a special case of Theorem 3.39. □

Corollary 3.54. *A T-quasigroup $(Q, \circ)$ defined over an abelian group $(Q, +)$ by $x \circ y = x\phi + y\psi$ is an (r, s, t)-inverse quasigroup with respect to the permutation $J = L_b^+$ if and only if (i) $\psi = \phi^{-1}$; (ii) $\phi^3 = I_0$; and (iii) $rb\phi + sb\psi = tb$.*

Proof. When $c = 0$ and $\beta = I$ in Theorem 3.52, we find that $B_u = ub$ and the stated result follows at once since $\psi = \phi^2 I_0$ and $\psi\phi = I$ imply that $\psi = \phi^{-1}$ and so $\phi^3 = I_0$. □

Since an automorphism maps generators to generators, every automorphism of $(Z_m, +)$ is of the form $z \to hz$, where h is relatively prime to m. Thus, in the T-quasigroup $(Z_m, \cdot)$ defined over $(Z_m, +)$ by $x \cdot y = x\phi + y\psi$, we have $\psi = \phi^{-1}$, where $\phi : z \to hz$ and $h^3 = -1$ (because $z\phi^3 = zI_0$ for all $z \in Z_m$).

It follows that $h = -1$ or, in the case when $\sqrt{-3} \in Z_m$, $h = h_1$ or h_2, where $h_1 = (1 + \sqrt{-3})/2$ and $h_2 = (1 - \sqrt{-3})/2$.

[In the latter case, h_1 and h_2 are the roots of the equation $h^2 - h + 1 = 0$ so $h_1h_2 = 1$. Consequently, both h_1 and h_2 are relatively prime to m and so they do define automorphisms of $(Z_m, +)$.]

Lemma 3.55. *We shall show that $\sqrt{-3}$ lies in Z_m if and only if $m = 2^k p_1^{k_1} p_2^{k_2} ... p_u^{k_u}$, where $k = 0, 1$ or 2 and each prime p_i has the form $p_i = 6v_i + 1$.*

Proof. We remark that the condition that -3 is a quadratic residue of the prime p is that the Legendre symbol $(\frac{-3}{p}) = 1$. We note that

$(\frac{-3}{p}) = (\frac{-1}{p})(\frac{3}{p})$. Also, by Gauss's Reciprocity Theorem,

$(\frac{3}{p})(\frac{p}{3}) = (-1)^{\frac{1}{2}(3-1)\cdot\frac{1}{2}(p-1)} = (-1)^{\frac{1}{2}(p-1)}$. Hence, we have

$(\frac{-3}{p}) = (\frac{-1}{p})(\frac{3}{p}) = (\frac{-1}{p})(\frac{p}{3})(-1)^{\frac{1}{2}(p-1)}$. From Wilson's Theorem, $(\frac{-1}{p}) = 1$ if and only if $p = 2$ or $p \equiv 1 \bmod 4$.

Thus, if $p \equiv 1 \bmod 4$, $(\frac{-1}{p}) = 1$ and $(-1)^{\frac{1}{2}(p-1)} = 1$. If $p \equiv 3 \bmod 4$, $(\frac{-1}{p}) = -1$ and $(-1)^{\frac{1}{2}(p-1)} = -1$. Hence, $(\frac{-3}{p}) = (\frac{p}{3})$ for all odd primes p.

We observe that, for an odd prime p, $(\frac{p}{3}) = 1$ if and only if $p \equiv 1 \bmod 3$ since $1^2 = 2^2 \equiv 1 \bmod 3$. Since $p \equiv 1 \bmod 3$ implies that $p \equiv 1 \bmod 6$ if p is odd, we conclude that -3 is a square modulo p if and only if $p \equiv 1 \bmod 6$ (or if $p = 2$). □

Remark 3.56. Conditions of Lemma 3.55 coincide with the conditions of the existence of roots of the equation $h^2 - h + 1 = 0$ with the exception of $m = 2$. In the last case the equation does not have any root.

We have the following theorem (see, for example, [178], Theorem 226, page 200):

Theorem 3.57. *If $m = 2^k p_1^{k_1} p_2^{k_2} ... p_u^{k_u}$, where $p_1, p_2, \dots , p_u$ are distinct prime numbers, then the equation $x^2 \equiv a \bmod m$, where a is relatively prime to m, has solutions for x if and only if (i) a is a quadratic residue for each of the primes $p_1, p_2, \dots , p_u$; and (ii) if and only if $a \equiv 1 \bmod 4$ in the case when $k = 2$ or $a \equiv 1 \bmod 8$ in the case when $k \geqslant 3$.*

Also, the number of solutions of the equation $x^2 \equiv a \bmod m$ when solutions exist is 2^u when $k = 0$ or 1, 2^{u+1} when $k = 2$ and 2^{u+2} when $k \geqslant 3$.

It follows from Theorem 3.57 that the equation $x^2 \equiv -3 \bmod m$ has 2^u solutions if $m = p_1^{k_1} p_2^{k_2} ... p_u^{k_u}$ or if $m = 2p_1^{k_1} p_2^{k_2} ... p_u^{k_u}$; 2^{u+1} solutions if $m = 4p_1^{k_1} p_2^{k_2} ... p_u^{k_u}$ where, in each case, $p^i \equiv 1 \bmod 6$ for $i = 1, 2, ..., u$; and otherwise has no solutions (since $-3 \not\equiv 1 \bmod 8$).

The following lemma is proved in [80].

Lemma 3.58. *Any autotopism of an abelian group $(Q, +)$ takes the form $(\theta R_a, \theta R_b, \theta R_{a+b})$, where R_a, R_b, R_{a+b} are right translations of $(Q, +)$ and θ is an automorphism of $(Q, +)$.*

This lemma is the corollary of Theorem 1.148 and we omit the proof.

Theorem 3.59. *A T-quasigroup $(Z_m, \cdot)$ defined over the cyclic group $(Z_m, +)$ by $x \cdot y = x\phi + y\psi$ which is an (r, s, t)-inverse quasigroup with respect to the permutation $J = L_b^+$ for some fixed element $b \in Z_m$ must be of one of the following two kinds:*

(a) a totally symmetric quasigroup of the form $x \cdot y = -x - y$; or

(b) a medial distributive quasigroup of the form $x \cdot y = x\phi + y\phi^{-1}$, where $\phi : z \to hz$ and h is a root of the equation $h^2 - h + 1 = 0$.

In each case, r, s and t are related by the equation $h^2 r + s = ht$.

Up to isomorphism, there is just one (r, s, t)-inverse T-quasigroup of the first kind for every positive integer m. But (r, s, t)-inverse T-quasigroups of the second kind exist only if m is a positive integer of one of the following three types: (i) $m = p_1^{k_1} p_2^{k_2} ... p_u^{k_u}$, (ii) $m = 2p_1^{k_1} p_2^{k_2} ... p_u^{k_u}$, (iii) $m = 4p_1^{k_1} p_2^{k_2} ... p_u^{k_u}$ where, in each case, $p_i \equiv 1 \bmod 6$ for $i = 1, 2, ..., u$.

When they exist, there are 2^u isomorphically distinct (r, s, t)-inverse T-quasigroups of types (b)(i) and (b)(ii) and 2^{u+1} isomorphically distinct (r, s, t)-inverse T-quasigroups of type (b)(iii).

Proof. We have already proved the first part of the theorem since $h = -1$ gives a quasigroup of the first kind. Also, the fact that r, s, t must be related by the equation $h^2 r + s = ht$ follows directly from condition (iii) of Corollary 3.54 above.

Now suppose that the T-quasigroups $x \circ y = x\phi + y\phi^{-1}$ and $x \cdot y = x\varphi + y\varphi^{-1}$ are isomorphic, where $\phi : z \to h_i z$ and $\varphi : z \to h_j z$. Then there is a permutation α of Z_m such that $x \circ y = z \Leftrightarrow x\alpha \cdot y\alpha = z\alpha$. That is, $x\alpha \cdot y\alpha = (x \circ y)\alpha$ or

$$(x\alpha)h_j + (y\alpha)h_j^{-1} = (xh_i + yh_i^{-1})\alpha \tag{3.14}$$

for all $x, y \in Z_m$.

Let us replace x by xh_i^{-1} and y by yh_i in this equality. Then we have

$$[(xh_i^{-1})\alpha]h_j + [(yh_i)\alpha]h_j^{-1} = (x + y)\alpha$$

for all $x, y \in Z_m$. So $(h_i^{-1}\alpha h_j, h_i\alpha h_j^{-1}, \alpha)$ is an autotopism of the abelian group $(Z_m, +)$.

By Lemma 3.58, every such autotopism takes the form $(\theta R_a, \theta R_b, \theta R_{a+b})$, where $\theta \in Aut(Z_m, +)$, so $\alpha = \theta R_t$ for some $t = a + b \in Z_m$.

Then, from the equality 3.14, $(x\theta + t)h_j + (y\theta + t)h_j^{-1} = (xh_i + yh_i^{-1})\theta + t$. Since this equality holds when $x = y = 0$, we have $th_j + th_j^{-1} = t$ and so $(x\theta)h_j + (y\theta)h_j^{-1} = (xh_i + yh_i^{-1})\theta$. The case when $y = 0$ gives $(x\theta)h_j = (xh_i)\theta$. But

$$(x\theta)h_j = x\theta + x\theta + ... + x\theta = (x + x + ... + x)\theta = xh_j\theta,$$

so $h_i = h_j$. Thus the number of non-isomorphic (r, s, t)-inverse T-quasigroups of type (b) is equal to the number of solutions of the congruence $x^2 \equiv -3 \bmod m$. We deduced this number from Theorem 3.57. This completes the proof. □

Example 3.60. Let $m = 14$. The congruence $x^2 \equiv -3 \bmod 14$ has two solutions: namely, $5(\equiv -9)$ and $9(\equiv -5)$. So the equation $h^3 = -1$ has solutions $h = -1$, $h = \frac{1+5}{2} = 3$ and $h = \frac{1+9}{2} = 5$ modulo 14. Also, $3^{-1} \equiv 5$ and $5^{-1} \equiv 3 \bmod 14$, so we get three isomorphically distinct T-quasigroups which are (r, s, t)-inverse quasigroups relative to the mapping $J = L_b^+$ of Z_{14}. These are as follows.

(a) $(Z_{14}, \circ)$ defined by $x \circ y = -x - y$, where $r + s + t = 0$ (since $h = -1$);
(b) $(Z_{14}, \cdot)$ defined by $x \cdot y = 3x + 5y$, where $s = 3t - 9r$ (since $h = 3$);
(c) $(Z_{14}, *)$ defined by $x * y = 5x + 3y$, where $s = 5t - 11r$ (since $h = 5$).

[Check: $(x \circ y)J^r \circ xJ^s = (rb + x \circ y) \circ (sb + x) = -(rb - x - y) - (sb + x) = (-r - s)b + y = tb + y = yJ^t$, since $r + s + t = 0$.

$(x \cdot y)J^r \cdot xJ^s = (rb + x \cdot y) \cdot (sb + x) = 3(rb + 3x + 5y) + 5(sb + x) = (3r + 5s)b + 14x + 15y = (3r + 15t - 45r)b + 14x + 15y$ since $s = 3t - 9r \equiv tb + y \bmod 14 = yJ^t$.

The quasigroup $(Z_{14}, *)$ is anti-isomorphic to the quasigroup $(Z_{14}, \cdot)$.]

Corollary 3.61. *A T-quasigroup $(Z_m, \cdot)$, defined over the cyclic group $(Z_m, +)$ by $x \cdot y = x\phi + y\psi$, which is an r-inverse quasigroup with respect to the permutation $J = L_b^+$ for some fixed element $b \in Z_m$, must be a totally symmetric r-inverse quasigroup, with $3r \equiv -1 \bmod m$, of the form $x \cdot y = -x - y$.*

Proof. Since, by Theorem 3.59, in an (r, s, t)-inverse quasigroup of the above kind, r, s and t are related by the equation $h^2 r + s = ht$, we find that the equality $h^2 r + (r + 1) \equiv hr \bmod m$ must hold if it is an r-inverse quasigroup. So, $(h^2 - h + 1)r \equiv -1 \bmod m$. This equation is not soluble for r when $h^2 - h + 1 = 0$ and so the quasigroup must be of type (a) with $h = -1$. Then $3r \equiv -1 \bmod m$. □

Example 3.62. Let $m = 14$. Then the T-quasigroup defined over the cyclic group $(Z_m, +)$ by $x \cdot y = -x - y$ is a 9-inverse quasigroup relative to the permutation $J = L_b^+$ of Z_{14}.

3.2.4 Direct product of (r, s, t)-quasigroups

Definition 3.63. Let $(Q_1, \cdot)$ and $(Q_2, \circ)$ be, respectively, an (r_1, s_1, t_1)-inverse quasigroup with respect to the permutation J_1 of Q_1 and an (r_2, s_2, t_2)-inverse quasigroup with respect to the permutation J_2 of Q_2. Define $(x_1, x_2)J = (x_1 J_1, x_2 J_2)$, where J is a permutation of $Q_1 \times Q_2$. Let the binary operation $(*)$ be defined on $Q = Q_1 \times Q_2$ by $(x_1, x_2) * (y_1, y_2) = (x_1 \cdot y_1, x_2 \circ y_2)$. Then $(Q, *)$ is the *direct product* of the quasigroups $(Q_1, \cdot)$ and $(Q_2, \circ)$.

Note 3.64. Throughout this section, we shall suppose that $|Q_1| = n_1$ and $|Q_2| = n_2$, that h_1, h_2 are the least positive integers for which $J^{h_1} \in Aut(Q_1, \cdot)$ and $J^{h_2} \in Aut(Q_2, \cdot)$, and that H_1, H_2 are the least positive integers for which $J_1^{H_1} = I$ and $J_2^{H_2} = I$.

Theorem 3.65. *Let $(Q_1, \cdot)$ and $(Q_2, \circ)$ be, respectively, an m_1-inverse quasigroup with respect to the permutation J_1 of Q_1 and an m_2-inverse quasigroup with respect to the permutation J_2 of Q_2. Then the direct product $(Q, *) = (Q_1, \cdot) \times (Q_2, \circ)$ will be an m-inverse* *quasigroup of order n_1n_2 relative to the permutation J if there exists a natural number t such that $m_1 - m_2 = (h_1, h_2)t$. In this case, m is a solution of the two congruences given below and $J^H = I$, where H is the least common multiple of H_1 and H_2.*

Proof. The binary operation $(*)$ is defined on $Q = Q_1 \times Q_2$ in the usual way by $(x_1, x_2) * (y_1, y_2) = (x_1 \cdot y_1, x_2 \circ y_2)$. Then, since $(x_1 \cdot y_1)J_1^{m_1} \cdot x_1 J_1^{m_1+1} = y_1 J_1^{m_1}$ and $(x_2 \circ y_2)J_2^{m_2} \circ x_2 J_2^{m_2+1} = y_2 J_2^{m_2}$, we have

$$[(x_1, x_2) * (y_1, y_2)]J^m * (x_1, x_2)J^{m+1} = (x_1 \cdot y_1, x_2 \circ y_2)J^m * (x_1, x_2)J^{m+1}$$
$$= [(x_1 \cdot y_1)J_1^m, (x_2 \circ y_2)J_2^m] * (x_1 J_1^{m+1}, x_2 J_2^{m+1})$$
$$= [(x_1 \cdot y_1)J_1^m \cdot x_1 J_1^{m+1}, (x_2 \circ y_2)J_2^m \circ x_2 J_2^{m+1}].$$

$(Q, *)$ will be an m-inverse quasigroup if the last expression is equal to $(y_1, y_2)J^m$: that is, equal to $(y_1 J_1^m, y_2 J_2^m)$.

So, we require that $(x_1 \cdot y_1)J_1^m \cdot x_1 J_1^{m+1} = y_1 J_1^m$ and $(x_2 \circ y_2)J_2^m \circ x_2 J_2^{m+1} = y_2 J_2^m$.

Since $(x_1 \cdot y_1)J_1^{m_1+u_1h_1} \cdot x_1 J_1^{m_1+1+u_1h_1} = y_1 J_1^{m_1+u_1h_1}$ and $(x_2 \cdot y_2)J_2^{m_2+u_2h_2} \cdot x_2 J_2^{m_2+1+u_2h_2} = y_2 J_2^{m_2+u_2h_2}$ for any integers u_1 and u_2 (by Remark 3.33), we require that $m_1 + u_1h_1 = m = m_2 + u_2h_2$ for some integers u_1 and u_2 for the above equalities to hold. That is, the direct product $(Q, *) = (Q_1, \cdot) \times (Q_2, \circ)$ will be an m-inverse quasigroup relative to the permutation J if there exists a solution of the system of equations

$$\begin{cases} m \equiv m_1 \pmod{h_1} \\ m \equiv m_2 \pmod{h_2}. \end{cases}$$

As is well known see for example [660, page 32, problem 14(c)], or [178, page 120], a solution of this system of equations exists if and only if the greatest common divisor (h_1, h_2) of h_1 and h_2 divides $m_1 - m_2$: that is, if and only if $m_1 - m_2 = (h_1, h_2)t$ for some $t \in N$. This completes the proof. □

Lemma 3.66. *Let $(Q, \cdot)$ be a quasigroup and J be a permutation of Q of order H (so that $J^H = I$) and suppose that $J^f \in Aut(Q, \cdot)$. Then also $J^h \in Aut(Q, \cdot)$, where $h = (f, H)$ is the greatest common divisor of f and H.*

Proof. Since h is the G.C.D. of f and H, there exist integers c and d such that $h = cf + dH$. Then, $(x \cdot y)J^f = xJ^f \cdot yJ^f$ and $(x \cdot y)J^H = xJ^H \cdot yJ^H$ together imply that $(x \cdot y)J^h = xJ^h \cdot yJ^h$. □

Remark 3.67. By making use of Corollary 3.32 and Lemma 3.66, we can deduce that, in Theorem 3.65, $h_1 \leqslant (3m + 1, H_1)$ and $h_2 \leqslant (3m + 1, H_2)$.

Example 3.68. Let $(Q_1, \cdot)$ and $(Q_2, \circ)$ be the 2-inverse quasigroup of order 7 [486, Figure 3d.3] and the 1-inverse quasigroup of order 8 [486, Figure 3e.2], respectively. Cayley tables of these quasigroups are given below. In these quasigroups $J_1 = (0\,1\,2\,3\,4\,5\,6)$, $J_2 = (0\,1\,2\,3\,4\,5\,6\,7)$, respectively. It is easy to see that $J_1^7 = I$ and $J_2^8 = I$.

$\cdot$	0	1	2	3	4	5	6
0	1	3	0	5	2	6	4
1	5	4	2	0	6	3	1
2	2	6	1	4	5	0	3
3	3	0	5	2	1	4	6
4	0	1	3	6	4	2	5
5	4	5	6	3	0	1	2
6	6	2	4	1	3	5	0

$\circ$	0	1	2	3	4	5	6	7
0	2	3	4	7	1	0	6	5
1	3	0	6	2	5	1	7	4
2	7	2	5	6	0	3	4	1
3	0	6	1	4	2	5	3	7
4	5	4	2	1	6	7	0	3
5	1	5	3	0	7	4	2	6
6	4	7	0	5	3	6	1	2
7	6	1	7	3	4	2	5	0

Also, J_2^4 is an automorphism of quasigroup $(Q_2, \circ)$. The direct product of these two quasigroups is a 9-inverse quasigroup of order 56 (since $m = 9$ is the least positive solution of the congruences $m \equiv 2 \bmod 7$ and $m \equiv 1 \bmod 4$).

Theorem 3.69. *Let $(Q_1, \cdot)$ and $(Q_2, \circ)$ be respectively an (r_1, s_1, t_1)-inverse quasigroup $(Q_1, \cdot)$ with respect to the permutation J_1 of Q_1 and an (r_2, s_2, t_2)-inverse quasigroup $(Q_2, \circ)$ with respect to the permutation J_2 of Q_2. Then the direct product $(Q, *) = (Q_1, \cdot) \times (Q_2, \circ)$ will be an (r, s, t)-inverse quasigroup relative to the permutation J of Q for the particular integers r, s, t if and only if*

$$(x_1 \cdot y_1)J_1^r \cdot x_1 J_1^s = y_1 J_1^t \quad \textit{and} \quad (x_2 \circ y_2)J_2^r \circ x_2 J_2^s = y_2 J_2^t. \tag{3.15}$$

Proof. By an argument exactly similar to that of Theorem 3.65, we easily see that

$$[(x_1, x_2) * (y_1, y_2)]J^r * (x_1, x_2)J^s = (y_1, y_2)J^t$$

if and only if the equations (3.15) hold. □

Also it follows from Corollary 3.32 and Lemma 3.66 that $h_1 \leqslant (r_1 + s_1 + t_1,\ h_1)$ and $h_2 \leqslant (r_2 + s_2 + t_2, h_2)$. We can state the following theorem:

Theorem 3.70. *The direct product $(Q, *) = (Q_1, \cdot) \times (Q_2, \circ)$ is an (r, s, t)-inverse quasigroup relative to the permutation J for the particular integers r, s, t if there exist integers u_1 and u_2 such that*

$$r - r_1 = s - s_1 = t - t_1 = u_1 h_1 \quad \textit{and} \quad r - r_2 = s - s_2 = t - t_2 = u_2 h_2, \tag{3.16}$$

where h_1 and h_2 are defined in the same way as before.

Proof. Since $(Q_1, \cdot)$ is an (r_1, s_1, t_1)-inverse quasigroup with respect to the permutation J_1 of Q_1 and $J_1^{h_1} \in Aut(Q_1, \cdot)$, we have $(x_1 \cdot y_1)J_1^{r_1 + u_1 h_1} \cdot x_1 J_1^{s_1 + u_1 h_1} = y_1 J_1^{t_1 + u_1 h_1}$ for all integers u_1 by Remark 3.33.

Similarly, we have $(x_2 \circ y_2)J_2^{r_2 + u_2 h_2} \circ x_2 J_2^{s_2 + u_2 h_2} = y_2 J_2^{t_2 + u_2 h_2}$ for all integers u_2. Therefore, when the equations (3.16) are satisfied, so are the equations (3.15). □

Remark 3.71. Theorem 3.70 can be stated as "if and only if" provided that the two equations which appear in the proof of the theorem are not satisfied for any indices except $r_i + u_i h_i$, etc. $(i = 1, 2)$. However, this is not always the case as the next Remark and Theorem show.

Remark 3.72. A quasigroup $(Q, \cdot)$ which is an (r, s, t)-inverse quasigroup relative to the permutation J is also an $(r + uh, s + uh, t + uh)$-inverse quasigroup for all $u \in \mathbf{Z}$, where $J^h \in Aut(Q, \cdot)$, but it may happen that $(Q, \cdot)$ is also an (R, S, T)-inverse quasigroup, where $(R, S, T) \notin \{(r + uh, s + uh, t + uh) : u \in \mathbf{Z}\}$ for any choice of h such that $J^h \in Aut(Q, \cdot)$.

Theorem 3.73. *Let $(Q, \cdot)$ be a quasigroup which is an (r_1, s_1, t_1)-quasigroup relative to the permutation J of Q. Then $(Q, \cdot)$ is also an (r_2, s_2, t_2)-quasigroup (relative to J), where $(r_2, s_2, t_2) \notin \{(r_1 + uh, s_1 + uh, t_1 + uh) : u \in \mathbf{Z}\}$ for any choice of h such that $J^h \in Aut(Q, \cdot)$ if and only if $(J^{s_2 - s_1}, J^{t_2 - t_1}, J^{r_2 - r_1})$ is an autotopism of the quasigroup $(Q, \cdot)$.*

Proof. Suppose that both of the identities

$$\begin{cases} (x \cdot y)J^{r_1} \cdot xJ^{s_1} = yJ^{t_1} \\ (x \cdot y)J^{r_2} \cdot xJ^{s_2} = yJ^{t_2} \end{cases} \tag{3.17}$$

are fulfilled in the quasigroup $(Q, \cdot)$. If we replace the variables x and y in the first identity by $xJ^{s_2-s_1}$ and $yJ^{t_2-t_1}$ respectively, then we shall have the following equalities

$$\begin{cases} (xJ^{s_2-s_1} \cdot yJ^{t_2-t_1})J^{r_1} \cdot xJ^{s_2} = yJ^{t_2} \\ (x \cdot y)J^{r_2} \cdot xJ^{s_2} = yJ^{t_2}. \end{cases}$$

Comparing the left sides of these equalities, we see that $(xJ^{s_2-s_1} \cdot yJ^{t_2-t_1})J^{r_1} \cdot xJ^{s_2} = (x \cdot y)J^{r_2} \cdot xJ^{s_2}$, whence $(xJ^{s_2-s_1} \cdot yJ^{t_2-t_1})J^{r_1} = (x \cdot y)J^{r_2}$ or $xJ^{s_2-s_1} \cdot yJ^{t_2-t_1} = (x \cdot y)J^{r_2-r_1}$.

Therefore the triple $(J^{s_2-s_1}, J^{t_2-t_1}, J^{r_2-r_1})$ is an autotopism of the quasigroup $(Q, \cdot)$.

Conversely, suppose that a quasigroup $(Q, \cdot)$ satisfying the identity $(x \cdot y)J^{r_1} \cdot xJ^{s_1} = yJ^{t_1}$ has an autotopism $(J^{s_2-s_1}, J^{t_2-t_1}, J^{r_2-r_1})$. Then from $xJ^{s_2-s_1} \cdot yJ^{t_2-t_1} = (x \cdot y)J^{r_2-r_1}$ we get $(xJ^{s_2-s_1} \cdot yJ^{t_2-t_1})J^{r_1} = (x \cdot y)J^{r_2}$.

Therefore $(xJ^{s_2-s_1} \cdot yJ^{t_2-t_1})J^{r_1} \cdot xJ^{s_2} = (x \cdot y)J^{r_2} \cdot xJ^{s_2}$. If, to the left side of the last identity, we apply our identity $(x \cdot y)J^{r_1} \cdot xJ^{s_1} = yJ^{t_1}$ in the form $(xJ^{s_2-s_1} \cdot yJ^{t_2-t_1})J^{r_1} \cdot xJ^{s_2} = yJ^{t_2}$, then we obtain $yJ^{t_2} = (x \cdot y)J^{r_2} \cdot xJ^{s_2}$: that is, we obtain $(x \cdot y)J^{r_2} \cdot xJ^{s_2} = yJ^{t_2}$. □

Remark 3.74. It is possible to use Theorem 3.20 by proving Theorem 3.73, but we preferred to give the direct proof.

Example 3.75. From Theorem 3.59 we see that the T-quasigroup $(Z_m, \cdot)$ defined by $x \cdot y = -x - y$ is (r, s, t)-inverse relative to the permutation $J = L_b^+$ of Z_m for all sets of integers r, s, t such that $r + s + t = 0$. Therefore any triple of the form $(J^u, J^v, J^{-(u+v)})$ is an autotopism of this quasigroup.

Remark 3.76. Let $J^h \in Aut(Q, \cdot)$. If $s_2 - s_1 = t_2 - t_1 = r_2 - r_1 = uh$ for some integer u, then $xJ^{s_2-s_1} \cdot yJ^{t_2-t_1} = (x \cdot y)J^{r_2-r_1}$ is trivially true, so the proof of Theorem 3.73 is another proof of Remark 3.33.

Taking into account both Theorem 3.70 and Theorem 3.73, we may state:

Theorem 3.77. *Suppose that $(Q_1, \cdot)$ and $(Q_2, \circ)$ are an (r_1, s_1, t_1)-inverse quasigroup $(Q_1, \cdot)$ with respect to the permutation J_1 of Q_1 and an (r_2, s_2, t_2)-inverse quasigroup $(Q_2, \circ)$ with respect to the permutation J_2 of Q_2, and that $(Q_1, \cdot)$ and $(Q_2, \circ)$ have no autotopisms of the forms $(J_1^{a_1}, J_1^{b_1}, J_1^{c_1})$ and $(J_2^{a_2}, J_2^{b_2}, J_2^{c_2})$ respectively, other than automorphisms.*

Then the direct product $(Q, *) = (Q_1, \cdot) \times (Q_2, \circ)$ will be an (r, s, t)-inverse quasigroup relative to the permutation J of Q for the particular integers r, s, t if and only if there exist integers u_1 and u_2 such that

$$r - r_1 = s - s_1 = t - t_1 = u_1 h_1 \quad \textit{and} \quad r - r_2 = s - s_2 = t - t_2 = u_2 h_2.$$

Remark 3.78. Clearly, it is not possible to meet the conditions of Theorem 3.77 unless $r_1 - r_2 = s_1 - s_2 = t_1 - t_2$ and unless the greatest common divisor of h_1 and h_2 divides each of these integers.

Lemma 3.79. *Let $(Q, \circ)$ be a quasigroup defined over the cyclic group $(Z_n, +)$ by $x \circ y = c + fx + gy$, where $c, f, g \in Z_n$, and let $J : z \to hz$, where h, n are relatively prime and m is the smallest positive integer such that $h^m \equiv 1 \bmod n$. Then $(Q, \circ)$ has no non-trivial autotopisms of type (J^u, J^v, J^w).*

Proof. $xJ^u \circ yJ^v = (h^u x) \circ (h^v y) = c + fh^u x + gh^v y$ and $(x \circ y)J^w = h^w(c + fx + gy)$. Therefore, $xJ^u \circ yJ^v = (x \circ y)J^w$ for all $x, y \in Z^n$ implies that $fh^u = fh^w$, $gh^v = gh^w$ and (if $c \neq 0$) $h^w = 1$. Thus, $w \equiv u \equiv v \bmod m$ (so the only autotopisms of the given type are automorphisms) and, if $c \neq 0$, $w \equiv 0 \bmod m$, so then the only automorphism is the identity map. □

Corollary 3.80. *The quasigroups constructed by the methods of Examples 3.42, 3.43 and 3.46 have no non-trivial autotopisms of type (J^u, J^v, J^w) and so they are properly defined (r,s,t)-inverse quasigroups relative to J for only one set of integers r, s, t.*

Lemma 3.81. *Let $(Q_1, \cdot)$ and $(Q_2, \circ)$ be, respectively, an (r_1, s_1, t_1)-inverse quasigroup $(Q_1, \cdot)$ with respect to the permutation J_1 of Q_1 and an (r_2, s_2, t_2)-inverse quasigroup $(Q_2, \circ)$ with respect to the permutation J_2 of Q_2. Then the direct product $(Q, *) = (Q_1, \cdot) \times (Q_2, \circ)$ has no non-trivial autotopisms of type (J^u, J^v, J^w) relative to the permutation J of Q unless (J_1^u, J_1^v, J_1^w) and (J_2^u, J_2^v, J_2^w) are, respectively, non-trivial autotopisms of $(Q_1, \cdot)$ and $(Q_2, \circ)$.*

Proof. Let (J^u, J^v, J^w) be an autotopism of $(Q, *) = (Q_1, \cdot) \times (Q_2, \circ)$. Then

$$[(x_1, x_2) * (y_1, y_2)]J^w = (x_1 \cdot y_1, x_2 \circ y_2)J^w = ((x_1 \cdot y_1)J_1^w, (x_2 \circ y_2)J_2^w)$$

and

$$\begin{aligned}&(x_1, x_2)J^u * (y_1, y_2)J^v = (x_1 J_1^u, x_2 J_2^u) * (y_1 J_1^v, y_2 J_2^v) = \\ &(x_1 J_1^u \cdot y_1 J_1^v, x_2 J_2^u \circ y_2 J_2^v).\end{aligned}$$

These are equal only if (J_1^u, J_1^v, J_1^w) and (J_2^u, J_2^v, J_2^w) are autotopisms of $(Q_1, \cdot)$ and $(Q_2, \circ)$ respectively. □

Corollary 3.82. *The direct product of quasigroups constructed by the methods of Examples 3.42, 3.43 and 3.46 is a properly defined (r,s,t)-inverse quasigroup relative to the permutation J defined as in Definition 2.22 for only one set of integers r, s, t.*

3.2.5 The existence of (r, s, t)-inverse quasigroups

We prove the existence of (r,s,t)-inverse quasigroups for every choice of positive integers r, s, t and we discuss the existence of properly defined direct products in more detail.

We can use Corollary 3.54 to give a construction for an (r,s,t)-inverse quasigroup of any order $3n$ such that n is not relatively prime to $r+s+t$ as follows:

Let $(G, +)$ be an abelian group which has an element c whose order divides $r+s+t$. Let $F = G \times G \times G$ so that $(F, +)$ is the group of ordered triples of elements from G. Let ϕ be the automorphism of $(F, +)$ which is defined by $\phi : (x_1\ x_2\ x_3) \to (-x_3\ \ -x_1\ \ -x_2)$. Then, ϕ^3 maps $(x_1\ x_2\ x_3)$ to $(-x_1\ \ -x_2\ \ -x_3)$, so $\phi^3 = I_0$. Also, $\phi^{-1} = \psi$ is the automorphism $(x_1\ x_2\ x_3) \to (-x_2\ \ -x_3\ \ -x_1)$.

We define the T-quasigroup $(F, \circ)$ by $\vec{x} \circ \vec{y} = \vec{x}\phi + \vec{y}\psi$, where $\vec{x} = (x_1\ x_2\ x_3)$, and let $\vec{b} = (c\ c\ c)$. Then, by Corollary 3.54, $(F, \circ)$ is an (r,s,t)-inverse quasigroup with respect to the permutation $J = L_{\vec{b}}^{+}$ if $r\vec{b}\phi + s\vec{b}\psi = t\vec{b}$. That is, if $r(-c\ -c\ -c) + s(-c\ -c\ -c) = t(c\ c\ c)$.

This condition is satisfied if the order of the element c in $(G, +)$ divides $r+s+t$. In that case, the permutation J has order equal to *ord* c in $(G, +)$ and so the inverse cycles of $(F, \circ)$ have length equal to *ord* c.

[CHECK:

$$\begin{aligned}&(\vec{x} \circ \vec{y})J^r \circ \vec{x}J^s = [(x_1\ x_2\ x_3)\phi + (y_1\ y_2\ y_3)\psi]J^r \circ \vec{x}J^s = \\ &[(-x_3\ \ -x_1\ \ -x_2) + (-y_2\ \ -y_3\ \ -y_1)]\ J^r \circ (x_1\ x_2\ x_3)J^s = \\ &(-x_3 - y_2 + rc\quad -x_1 - y_3 + rc\quad -x_2 - y_1 + rc)\circ \\ &(x_1 + sc\quad x_2 + sc\quad x_3 + sc) = \\ &(-x_3 - y_2 + rc\quad -x_1 - y_3 + rc\quad -x_2 - y_1 + rc)\phi + \\ &(x_1 + sc\quad x_2 + sc\quad x_3 + sc)\psi = \\ &(x_2 + y_1 - rc\quad x_3 + y_2 - rc\quad x_1 + y_3 - rc) + \\ &(-x_2 - sc\quad -x_3 - sc\quad -x_1 - sc) = \\ &[y_1 - (r+s)c\quad y_2 - (r+s)c\quad y_3 - (r+s)c]\,.\end{aligned}$$

Also, $\vec{y}J^t = (y_1 + tc\, y_2 + tc\, y_3 + tc)$. So $(\vec{x} \circ \vec{y})J^r \circ \vec{x}J^s = \vec{y}J^t$ if $(r+s+t)c = 0$: that is, if *ord* c in $(G,+)$ divides $r+s+t$.]

We can use this construction to prove:

Theorem 3.83. *(1) There exist (r,s,t)-inverse quasigroups of every order $3n$ such that n is not relatively prime to $r+s+t$ with inverse cycles of length equal to the G.C.D. of n and $r+s+t$.*

(2) There exist (r,s,t)-inverse quasigroups of order $3(r+s+t)$ with inverse cycles of length $r+s+t$.

Proof. Proof of (1). Let k be the G.C.D. of n and $r+s+t$. Let $(G,+)$ be an abelian group of order n which has an element of order k and choose the element c in the construction described above to be this element. (For example, we may take $(G,+)$ to be the direct sum of cyclic groups of order k.) The result follows.

Proof of (2). In the special case of the construction described above when $(G,+)$ is the cyclic group of order $n = r+s+t$ and c is a generator of that group, the quasigroup $(F,\circ)$ has order $3n$ and its inverse cycles are of length n. □

Corollary 3.84. *(r,s,t)-inverse quasigroups exist for every choice of positive integers r, s and t.*

3.2.6 WIP-quasigroups

In Section 3.2.1 we pointed out that a weak-inverse-property loop (WIP-loop) is a $(-1,0,-1)$-inverse loop. If we use additive notation for the loop and denote the identity by 0, we have the following:

Definition 3.85. A loop $(Q,+)$ with the property that

$$x + (y+x)I_0 = yI_0 \tag{3.18}$$

for all $x, y \in Q$, where I_0 is the permutation of Q such that $x + xI_0 = 0$, is called a *weak inverse property loop.*

Evidently, we can generalize this definition to that of a WIP-quasigroup and a WIP-loop as follows:

Definition 3.86. A quasigroup $(Q,\circ)$ is said to be a *WIP-quasigroup* with respect to the permutation J of Q if

$$x \circ (y \circ x)J = yJ \tag{3.19}$$

for all $x, y \in Q$.

In Examples 2.26 and 2.27 are constructed WIP-loops as in Definition 3.86 that are not weak inverse property loops as in Definition 3.85, i.e., the loops are constructed in which $J \neq I_0$.

In this section we obtain necessary and sufficient conditions for such a quasigroup to be a principal isotope of a WIP-loop as in Definition 3.85. (This then provides a method by which WIP-quasigroups may be constructed.)

Remark 3.87. A definition very similar to that of Definition 3.86 was earlier made by Steinberger [847] who studied what he called *T-WI-groupoids.*

Lemma 3.88. *Let $(Q, \circ)$ be a quasigroup defined over the loop $(Q, +)$ by $x \circ y = x\varphi + y\psi$, where φ and ψ are permutations of Q such that $0\varphi = 0$ and $0\psi = 0$. Then, the sufficient conditions for $(Q, \circ)$ to be a WIP-quasigroup with respect to the permutation J of Q are (i) $J\psi = \varphi^{-1}J \in Aut(Q, +)$; (ii) $x\varphi + x\psi J\psi = 0$ for all $x \in Q$; and (iii) $(Q, +)$ is a CI-loop.*

Proof. We have

$$x \circ (y \circ x)J = x \circ (y\varphi + x\psi)J = x\varphi + (y\varphi + x\psi)J\psi$$

$$= x\varphi + (y\varphi + x\psi)\varphi^{-1}J = x\varphi + (yJ + x\psi\varphi^{-1}J)$$

by condition (i),

$$= x\varphi + (yJ + x\psi J\psi) = x\varphi + (yJ + x\varphi I_o)$$

because $x\psi J\psi = x\varphi I_0$ by condition (ii).

Since $(Q, +)$ is a CI-loop, we have $u+(v+uI_0) = v$ for all $u, v \in Q$. Therefore $x\circ(y\circ x)J = yJ$. □

Lemma 3.89. *Let $(Q, \circ)$ be a quasigroup defined over the loop $(Q, +)$ by $x \circ y = x\varphi + y\psi$, where φ and ψ are permutations of Q such that $0\varphi = 0$ and $0\psi = 0$. Then, necessary conditions for $(Q, \circ)$ to be a WIP-quasigroup with respect to the permutation J of Q such that $0J = 0$ are (i) $J\psi = \varphi^{-1}J$ and (ii) $x\varphi + x\psi J\psi = 0$ for all $x \in Q$; or, equivalently, (i)* $J = \psi^{-1}\varphi I_0\psi^{-1}$ and (ii)* $[\varphi^{-1}, \psi] = I_0\psi^{-1}I_0^{-1}\varphi^{-1}$.*

Proof. Since, by definition of a WIP-quasigroup, $x \circ (y \circ x)J = yJ$, we have $x\varphi + (y\varphi + x\psi)J\psi = yJ$ for all $x, y \in Q$. Putting $x = 0$ in the last equality, we get $y\varphi J\psi = yJ$ for all $y \in Q$ and so $\varphi J\psi = J$ (Equ. 1). Similarly, putting $y = 0$, we get $x\varphi + x\psi J\psi = 0$ for all $x \in Q$. (These are conditions (i) and (ii).)

By definition of I_0, $(x\varphi)+(x\varphi)I_0 = 0$ for all $x \in Q$. But, from condition (ii), $x\varphi+x\psi J\psi = 0$ for all $x \in Q$. Since $(Q, +)$ is a loop, we deduce that $x\varphi I_0 = x\psi J\psi$ for all $x \in Q$. Therefore, $\varphi I_0 = \psi J\psi$ or, equivalently, $J = \psi^{-1}\varphi I_0\psi^{-1}$, which is condition (i)*.

Substituting in Equ. 1, we find that $\varphi(\psi^{-1}\varphi I_0\psi^{-1})\psi = \psi^{-1}\varphi I_0\psi^{-1}$. That is,

$$\varphi\psi^{-1}\varphi I_0 = \psi^{-1}\varphi I_0\psi^{-1} \quad \text{or} \quad \varphi^{-1}\psi(\varphi\psi^{-1}\varphi I_0) = I_0\psi^{-1}$$

or $[\varphi^{-1}, \psi] = I_0\psi^{-1}I_0^{-1}\varphi^{-1}$, where $[\varphi^{-1}, \psi] = \varphi^{-1}\psi\varphi\psi^{-1}$ is the commutator of φ^{-1} and ψ. This is condition (ii)*. □

From Lemmas 3.88 and 3.89, we easily obtain the following theorem:

Theorem 3.90. *Let $(Q, \circ)$ be a quasigroup defined over the loop $(Q, +)$ by $x \circ y = x\varphi + y\psi$, where φ and ψ are permutations of Q such that $0\,\varphi = 0$ and $0\,\psi = 0$. Then, if the permutation J of Q is such that $J\psi = \varphi^{-1}J \in Aut(Q, +)$, the necessary and sufficient conditions for $(Q, \circ)$ to be a WIP-quasigroup with respect to the permutation J of Q are (i) $x\varphi + x\psi J\psi = 0$ for all $x \in Q$; and (ii) $(Q, +)$ is a CI-loop.*

Proof. The conditions are sufficient by Lemma 3.88, so we have only to show that they are necessary.

Since $(Q, \circ)$ is a WIP-quasigroup with respect to the permutation J, we have $x \circ (y \circ x)J = yJ$ for all $x, y \in Q$ or, equivalently, $x\varphi + (y\varphi + x\psi)J\psi = yJ$ or

$$x\varphi + (y\varphi J\psi + x\psi J\psi) = yJ, \tag{3.20}$$

since $J\psi \in Aut(Q, +)$.

Putting $x = 0$ and $y = 0$ in turn in the equality (3.20), we get $\varphi J\psi = J$, or $J\psi = \varphi^{-1}J$, and $x\varphi + x\psi J\psi = 0$ for all $x \in Q$, or $\psi J\psi = \varphi I_0$ as in Lemma 3.89.

Therefore, substituting J instead of $\varphi J\psi$ and φI_0 instead of $\psi J\psi$ in the equality (3.20), we obtain $x\varphi + (yJ + x\varphi I_0) = yJ$ for all $x, y \in Q$. That is, $u + (v + uI_0) = u$ for all $u, v \in Q$. Thus, it is necessary that $(Q, +)$ is a CI-loop. □

Remark 3.91. The first sentence in the statements of Lemmas 3.88, 3.89 and Theorem 3.90 could alternatively be re-phrased as "Let $(Q, \circ)$ be any principal isotope $(\varphi, \psi, \epsilon)$ of the loop $(Q, +)$ such that $0\varphi = 0\psi = 0$, where 0 is the identity of $(Q, +)$."

The next theorem gives an alternative set of necessary and sufficient conditions for a quasigroup $(Q, \circ)$ of the above form to be a WIP-quasigroup.

Theorem 3.92. *Let $(Q, \circ)$ be a quasigroup defined over the loop $(Q, +)$ by $x \circ y = x\varphi + y\psi$, where φ, $\psi \in Aut(Q, +)$. Then $(Q, \circ)$ is a WIP-quasigroup with respect to the permutation J of Q if and only if (i) $J = \psi^{-1}\varphi I_0 \psi^{-1}$; (ii) $[\varphi^{-1}, \psi] = \psi^{-1}\varphi^{-1}$; and (iii) $(Q, +)$ is a WIP-loop.*

Proof. Proof that the conditions are necessary. The necessity of conditions (i) and (ii) follows immediately from Lemma 3.89 because the permutation I_0 defined by $z + zI_0 = 0$ for all $z \in Q$ commutes with the automorphisms φ and ψ.

[If $\alpha \in Aut(Q, +)$, then $z + zI_0 = 0 \Longrightarrow z\alpha + zI_0\alpha = 0\alpha = 0$. But, by definition of I_0, $(z\alpha) + (z\alpha)I_0 = 0$. Since the equation $z + u = 0$ is uniquely soluble for $u \in (Q, +)$, we find that $zI_0\alpha = z\alpha I_0$ for all $z \in Q$ and so $I_0\alpha = \alpha I_0$.]

The fact that $(Q, \circ)$ has the weak inverse property with respect to J implies that $x\varphi + (y\varphi + x\psi)J\psi = yJ$ for all $x, y \in Q$ as in Lemma 3.89.

Replacing $J\psi$ by $\psi^{-1}\varphi I_0$ using the condition (i), we get

$$x\varphi + (y\varphi + x\psi)\psi^{-1}\varphi I_0 = yJ.$$

That is, $x\varphi + (y\varphi\psi^{-1}\varphi + x\varphi)I_0 = yJ$ by using the facts that ψ^{-1} and φ are automorphisms. From condition (ii), $\varphi^{-1}\psi\varphi\psi^{-1} = \psi^{-1}\varphi^{-1}$ or $(\varphi^{-1}\psi)\varphi\ \psi^{-1}\varphi = \psi^{-1}$. So $\varphi\psi^{-1}\varphi = \psi^{-1}\varphi\psi^{-1} = JI_0^{-1}$ by condition (i). Hence, $x\varphi + (yJI_0^{-1} + x\varphi)I_0 = yJ$. That is, $x\varphi + (z + x\varphi)I_0 = zI_0$, where $z = yJI_0^{-1}$, which implies that $u + (v + u)I_0 = vI_0$ for all $u, v \in Q$. Therefore, $(Q, +)$ is a WIP-loop.

Proof that the conditions are sufficient. We have

$$x \circ (y \circ x)J = x\varphi + (y\varphi + x\psi)J\psi = x\varphi + (y\varphi + x\psi)(\psi^{-1}\varphi I_0 \psi^{-1})\psi$$

using condition (i),

$$= x\varphi + (y\varphi + x\psi)\psi^{-1}\varphi I_0 = x\varphi + (y\varphi\psi^{-1}\varphi + x\varphi)I_0 = y\varphi\psi^{-1}\varphi I_0$$

since $(Q, +)$ is a WIP-loop. But, we showed above that condition (ii) implies that $\varphi\psi^{-1}\varphi = JI_0^{-1}$. Therefore, $x \circ (y \circ x)J = yJ$ as required for a WIP-quasigroup. □

Remark 3.93. It follows from Theorem 3.48 that a linear quasigroup $(Q, \circ)$ of the form $x \circ y = x\varphi + y\varphi^{-1}$, defined over a loop $(Q, +)$, is a CI-quasigroup relative to the permutation $J = \varphi^3 I_0$ if and only if $(Q, +)$, is a CI-loop.

Similarly, it follows from Theorem 3.92 that a linear quasigroup of the above form is a WIP-quasigroup relative to the permutation $J = \varphi^3 I_0$ if and only if $(Q, +)$ is a WIP-loop.

[To see that the latter statement is true, we observe that, when $\psi = \varphi^{-1} \in Aut(Q, +)$, statement (i) of Theorem 3.92 reduces to $J = \varphi^3 I_0$ and that statement (ii) becomes $[\psi, \psi] = \varphi\varphi^{-1}$ which is vacuously true.]

3.2.7 Examples of WIP-quasigroups

We start from the following:

Proposition 3.94. *If $(Q, \circ)$ is a quasigroup of the form $x \circ y = x\varphi^{-1} \cdot y\varphi$, where $(Q, \cdot)$ is a TS-quasigroup and permutation φ is an automorphism of $(Q, \cdot)$ with the property that there exists an element $a \in Q$ such that $a\varphi = a$, then $(Q, \circ)$ is a WIP-quasigroup with $J \in Aut(Q, \cdot)$ if and only if $J = \varphi^{-3}$.*

Proof. The WIP-condition for quasigroup $(Q, \circ)$ we can write as $x \circ (y \circ x)J = yJ$ or, passing to the operation $\cdot$, as $x\varphi^{-1} \cdot (y\varphi^{-1} \cdot x\varphi)J\varphi = yJ$.

Then we have

$$x\varphi^{-1} \cdot (y\varphi^{-1}J\varphi \cdot x\varphi J\varphi) = yJ. \tag{3.21}$$

Using commutativity of $(Q, \cdot)$ further we obtain $x\varphi^{-1} \cdot (x\varphi J\varphi \cdot y\varphi^{-1}J\varphi) = yJ$. By $x = a$ we have $y\varphi^{-1}J\varphi = yJ$.

Then we can re-write the equality (3.21) in the form $x\varphi^{-1} \cdot (x\varphi J\varphi \cdot yJ) = yJ$. If we replace the element x with element $x\varphi$ and element y with element yJ^{-1} then we have $x \cdot (x\varphi^2 J\varphi \cdot y) = y$.

In a TS-quasigroup the identity $x \cdot (x \cdot y) = y$ holds. Therefore we obtain that $\varphi^2 J\varphi = \varepsilon$, i.e., $J = \varphi^{-3}$.

Converse. We can write WIP-property of quasigroup $(Q, \circ)$ as the equality (3.21). Taking into consideration that $J = \varphi^{-3}$ we can re-write the equality (3.21) as $x\varphi^{-1}\cdot(x\varphi^{-1}\cdot yJ) = yJ$. The last identity is fulfilled in TS-quasigroup $(Q, \cdot)$. □

Remark 3.95. In Proposition 3.94 in fact we use the concept of a quasigroup which is linear over another quasigroup.

Example 3.96. We take TS-quasigroup $(Q, \cdot)$ with the following Cayley table [72, p. 110]:

$\cdot$	1	2	3	4	5	6	7
1	1	3	2	5	4	7	6
2	3	2	1	6	7	4	5
3	2	1	3	7	6	5	4
4	5	6	7	4	1	2	3
5	4	7	6	1	5	3	2
6	7	4	5	2	3	6	1
7	6	5	4	3	2	1	7.

It is possible to check that the permutations $(1)(2534)(67)$, $(1)(265)(374)$ are automorphisms of quasigroup $(Q, \cdot)$.

Therefore, if we take $\varphi = (1)(2534)(67)$, then quasigroup $(Q, \circ)$, $x \circ y = x\varphi^{-1} \cdot y\varphi$, is a WIP-quasigroup.

Often WIP-loops are used for construction of G-loops. See Theorem 4.163 and the following:

Theorem 3.97. *If $(L, \cdot)$ is a loop such that*

$$x \cdot (xy)I_0 = (xz) \cdot (x \cdot yz)I_0 \tag{3.22}$$

holds for all $x, y, z \in L$, where $x \cdot xI_0 = 1$, then $(L, \cdot)$ is a G-WIP-loop [942, 685].

In any loop $(Q, \cdot)$, from the equality (3.22) equality follows which defines a WIP-loop (equality 3.18). Indeed, we can put $x = 0$ in (3.22).

The following loop satisfies the identity (3.22), i.e., it is a G-WIP-loop.

*	0	1	2	3	4	5
0	0	1	2	3	4	5
1	1	5	0	4	2	3
2	2	4	5	0	3	1
3	3	0	4	5	1	2
4	4	3	1	2	5	0
5	5	2	3	1	0	4

3.2.8 Generalized balanced parastrophic identities

In this subsection we follow [489]. If we examine the equations that define various inverse quasigroups carefully, we notice that they are all of the form

$$A^{\sigma}(x\nu_1, y\nu_2) = [A(x, y)]\nu_3, \tag{3.23}$$

where A^{σ} is some parastrophe of the operation A and ν_1, ν_2, ν_3 are permutations of the set Q.

For example, the equation (3.19) is $x \otimes (y \otimes x)J = yJ$ and is equivalent to $x \otimes^{(2\,3)} yJ = (y \otimes x)J$ or to $x \otimes^{(2\,3)} yJ = (x \otimes^{(1\,2)} y)J$. If we write $\otimes^{(1\,2)} = \oplus$, we get $\otimes^{(2\,3)} = \oplus^{(1\,2)(2\,3)} = \oplus^{(1\,3\,2)}$ and so $x \oplus^{(1\,3\,2)} yJ = (x \oplus y)J$, which is of the above form.

This fact suggests that we make the following generalization:

Definition 3.98. An identity of the form

$$[A^{\sigma}(x\nu_1, y\nu_2)]\nu_3 = [A^{\tau}(x\nu_4, y\nu_5)]\nu_6 \tag{3.24}$$

on a groupoid (Q, A), where A^{σ}, A^{τ} are some parastrophes of the operation A, $x, y \in Q$ and ν_i for $i = 1, 2, ..., 6$ are some permutations of the set Q, will be called a *generalized balanced parastrophic identity of length two* on the groupoid (Q, A).

We call this identity generalized, because in Definition 3.98, permutations are used. Therefore it is possible to call such identities, identities with permutations. Notice that identities with permutations are usual objects of quasigroup theory. See Section 2.1 or [72, 80, 685]. We call it as parastrophic, because in the definition, parastrophes of a fixed operation are used.

The identities (3.23) and (3.24) have length equal to two, because the length of balanced identity (Definition 2.219) is equal to the number of occurrences of variables in the left (or the right) part of this equality [69].

Theorem 3.99. *Any generalized balanced parastrophic identity of length two on a quasigroup (Q, A) is equivalent to an identity of type (3.23).*

Proof. It will be convenient to use (x_1, x_2, x_3) in place of (x, y, z) and to write $A(x_1, x_2, x_3)$ for denoting that $A(x_1, x_2) = x_3$ in the quasigroup (Q, A).

From the identity (3.24), we have $[A^{\sigma}(x_1\nu_1, x_2\nu_2)]\nu_3\nu_6^{-1} = A^{\tau}(x_1\nu_4, x_2\nu_5)$.
Put $y_1 = x_1\nu_4$, $y_2 = x_2\nu_5$. Then $[A^{\sigma}(y_1\nu_4^{-1}\nu_1, y_2\nu_5^{-1}\nu_2)]\nu_3\nu_6^{-1} = A^{\tau}(y_1, y_2)$.

That is, $[A^{\sigma}(y_1\theta_1, y_2\theta_2)]\theta_3^{-1} = A^{\tau}(y_1, y_2)$, where $\theta_1 = \nu_4^{-1}\nu_1$, $\theta_2 = \nu_5^{-1}\nu_2$ and $\theta_3 = \nu_3\nu_6^{-1}$. So there exists an isotopism $(\theta_1, \theta_2, \theta_3)$ from (Q, A^{σ}) to (Q, A^{τ}) and we have

$$A^{\sigma}(y_1\theta_1, y_2\theta_2, y_3\theta_3) \Leftrightarrow A^{\tau}(y_1, y_2, y_3) \Leftrightarrow A(y_{1\tau^{-1}}, y_{2\tau^{-1}}, y_{3\tau^{-1}}).$$

Therefore,

$$A^{\sigma}(z_{1_T}\theta_1, z_{2_T}\theta_2, z_{3_T}\theta_3) \Leftrightarrow A(z_1, z_2, z_3), \text{ where } z_i = y_{i_{T^{-1}}}.$$

So, $A^{\sigma T^{-1}}(z_1\theta_{1_{T^{-1}}}, z_2\theta_{2_{T^{-1}}}, z_3\theta_{3_{T^{-1}}}) \Leftrightarrow A(z_1, z_2, z_3)$.

That is, $A^{\sigma T^{-1}}(z_1\theta_{1_{T^{-1}}}, z_2\theta_{2_{T^{-1}}})\theta^{-1}_{3_{T^{-1}}} = A(z_1, z_2)$ or $A^{\sigma T^{-1}}(z_1\theta_{1_{T^{-1}}}, z_2\theta_{2_{T^{-1}}}) = [A(z_1, z_2)]\theta_{3_{T^{-1}}}$, which is of the form (3.23). □

Remark 3.100. It is easy to see that the identity $A^{(12)}(x, y) = A(x, y)$, which is a special form of the identity (3.23), is fulfilled in a quasigroup (Q, A) if and only if the quasigroup (Q, A) is a commutative quasigroup. In other words, a quasigroup (Q, A) is commutative if and only if this quasigroup has the autostrophy $[(12), (\varepsilon, \varepsilon, \varepsilon)]$.

This allows us to say that in a class of quasigroups the property of commutativity is a kind of inverse property and in some sense is true converse.

3.2.9 Historical notes

Among the classical objects of the theory of quasigroups there are weak-inverse property loops (WIP-loops) and crossed-inverse property loops (CI-loops). The first were defined by R. Baer [25] in one of the first articles devoted to quasigroup theory and the second by R. Artzy [17].

Important results on WIP-loops were obtained by M. Osborn [672] while a fairly detailed investigation of CI-loops was made by Artzy in a series of papers. Belousov and Tsurkan studied CI-quasigroups [95].

Later, a generalization of both these types of loop was introduced by B. B. Karklin´š and V. B. Karklin´ [472] which these authors called an m-inverse loop. (Some further observations about such loops were made in [671] and [43].)

In [486], the concept of an m-inverse quasigroup was introduced and the existence of m-inverse loops and quasigroups which have long inverse cycles was investigated with a view to the application of such structures to cryptography.

This concept was further generalized by defining (r, s, t)-inverse loops and quasigroups, which include, as special cases, WIP-, CI- and m-inverse loops (and quasigroups).

Chapter 4

A-nuclei of quasigroups

4.1	Preliminaries	210
4.1.1	Isotopism	210
4.1.2	Quasigroup derivatives	212
4.1.2.1	G-quasigroups	213
4.1.2.2	Garrison's nuclei in quasigroups	214
4.1.2.3	Mixed derivatives	214
4.1.3	Set of maps	215
4.2	Garrison's nuclei and A-nuclei	216
4.2.1	Definitions of nuclei and A-nuclei	216
4.2.2	Components of A-nuclei and identity elements	219
4.2.3	A-nuclei of loops by isostrophy	222
4.2.4	Isomorphisms of A-nuclei	222
4.2.5	A-nuclei by some isotopisms	225
4.2.6	Quasigroup bundle and nuclei	226
4.2.7	A-nuclei actions	228
4.2.8	A-nuclear quasigroups	230
4.2.9	Identities with permutation and group isotopes	232
4.3	A-centers of a quasigroup	235
4.3.1	Normality of A-nuclei and autotopy group	235
4.3.2	A-centers of a loop	239
4.3.3	A-centers of a quasigroup	241
4.4	A-nuclei and quasigroup congruences	243
4.4.1	Normality of equivalences in quasigroups	243
4.4.2	Additional conditions of normality of equivalences	244
4.4.3	A-nuclei and quasigroup congruences	248
4.4.4	A-nuclei and loop congruences	251
4.4.5	On loops with nucleus of index two	253
4.5	Coincidence of A-nuclei in inverse quasigroups	254
4.5.1	(α, β, γ)-inverse quasigroups	254
4.5.2	λ-, ρ-, and μ-inverse quasigroups	256
4.6	Relations between a loop and its inverses	257
4.6.1	Nuclei of inverse loops in Belousov sense	257
4.6.2	LIP- and AAIP-loops	258
4.6.3	Invariants of reciprocally inverse loops	259
4.6.3.1	Middle Bol loops	259
4.6.3.2	Some invariants	259
4.6.3.3	Term-equivalent loops	260
4.6.4	Nuclei of loops that are inverse to a fixed loop	261

4.1 Preliminaries

V.D. Belousov discovered connections between nuclei and the groups of regular mappings of quasigroups [59, 72, 80, 491, 488, 489, 500].

Belousov also studied autotopisms of the form $(\alpha, \varepsilon, \alpha)$ of a quasigroup $(Q, \circ)$. Here we study autotopisms of the form $(\alpha, \varepsilon, \gamma)$, i.e., we use Kepka's generalization [491, 493]. By this approach "usual" nuclei of a quasigroup are obtained as some orbits by the action of components of A-nuclei on the set Q.

The study of A-nuclei of (α, β, γ)- and $(r; s; t)$-inverse quasigroups was initiated in [488, 489]. Belousov and Gvaramia developed a point of view on quasigroup nuclei as on quasigroup, which satisfies some partial identities [89].

We develop Belousov's approach to parastrophes of quasigroup nuclei [65] using the concept of middle translation [75]. Also we make an attempt to extend some results of A. Drapal and P. Jedlička [274, 276] about loop nuclei.

We apply an autotopical approach to the concept of the center of a quasigroup and introduce the concept of the A-center (autotopical center) of a quasigroup. We obtain results about A-nuclei of some inverse quasigroups and loops.

4.1.1 Isotopism

We shall use the following:

Lemma 4.1. *If quasigroups $(Q, \circ)$ and $(Q, \cdot)$ are isotopic with isotopy T, i.e., $(Q, \circ) = (Q, \cdot)T$, then $Avt(Q, \circ) = T^{-1}Avt(Q, \cdot)T$ ([80], Lemma 1.4).*

See also Theorem 1.133.

Corollary 4.2. *If H is a subgroup of the group $Avt(Q, \cdot)$, then $T^{-1}HT$ is a subgroup of the group $Avt(Q, \circ)$.*

Proof. The proof follows from Theorem 1.133 and standard algebraic facts [471]. □

Lemma 4.3. *1. If $(Q, \circ) = (Q, \cdot)(\alpha, \varepsilon, \varepsilon)$, then $L^{\circ}_x = L^{\cdot}_{\alpha x}$, $R^{\circ}_x = R^{\cdot}_x\alpha$, $P^{\circ}_x = P^{\cdot}_x\alpha$ for all $x \in Q$.*

2. If $(Q, \circ) = (Q, \cdot)(\varepsilon, \beta, \varepsilon)$, then $L^{\circ}_x = L^{\cdot}_x\beta$, $R^{\circ}_x = R^{\cdot}_{\beta x}$, $(P^{\circ}_x)^{-1} = (P^{\cdot}_x)^{-1}\beta$ for all $x \in Q$.

3. If $(Q, \circ) = (Q, \cdot)(\varepsilon, \varepsilon, \gamma)$, then $L^{\circ}_x = \gamma^{-1}L^{\cdot}_x$, $R^{\circ}_x = \gamma^{-1}R^{\cdot}_x$, $P^{\circ}_x = P^{\cdot}_{\gamma x}$ for all $x \in Q$.

Proof. Case 1. We can re-write the equality $(Q, \circ) = (Q, \cdot)(\alpha, \varepsilon, \varepsilon)$ in the form $x \circ y = \alpha x \cdot y = z$ for all $x, y \in Q$. Therefore $L^{\circ}_x y = L^{\cdot}_{\alpha x} y$, $R^{\circ}_y x = R^{\cdot}_y \alpha x$, $P^{\circ}_z x = P^{\cdot}_z \alpha x$.

Case 2. We can re-write the equality $(Q, \circ) = (Q, \cdot)(\varepsilon, \beta, \varepsilon)$ in the form $x \circ y = x \cdot \beta y = z$ for all $x, y \in Q$. Therefore $L^{\circ}_x y = L^{\cdot}_x \beta y$, $R^{\circ}_y x = R^{\cdot}_{\beta y} x$, $(P^{\circ}_z)^{-1} y = (P^{\cdot}_z)^{-1} \beta y$.

Case 3. We can re-write the equality $(Q, \circ) = (Q, \cdot)(\varepsilon, \varepsilon, \gamma)$ in the form $x \circ y = \gamma^{-1}(x \cdot y)$ for all $x, y \in Q$. Therefore $L^{\circ}_x = \gamma^{-1}L^{\cdot}_x$, $R^{\circ}_y = \gamma^{-1}R^{\cdot}_y$. If $x \circ y = \gamma^{-1}(x \cdot y) = z$, then $P^{\circ}_z x = y$, $P^{\cdot}_{\gamma z} x = y$, $P^{\circ}_z = P^{\cdot}_{\gamma z}$. □

Corollary 4.4. *In conditions of Lemma 4.3:*

1. *(i) if $\alpha = (R^{\cdot}_a)^{-1}$, then $R^{\circ}_a = \varepsilon$, $(Q, \circ)$ is a right loop.*
   (ii) if $\alpha = (P^{\cdot}_a)^{-1}$, then $P^{\circ}_a = \varepsilon$, $(Q, \circ)$ is a unipotent quasigroup.
2. *(i) if $\beta = (L^{\cdot}_b)^{-1}$, then $L^{\circ}_b = \varepsilon$, $(Q, \circ)$ is a left loop.*

(ii) if $\beta = P_b^{\cdot}$, then $P_b^{\circ} = \varepsilon$, $(Q, \circ)$ is a unipotent quasigroup.

3. *(i) if $\gamma = L_c^{\cdot}$, then $L_c^{\circ} = \varepsilon$, $(Q, \circ)$ is a left loop.*

   (ii) if $\gamma = R_c^{\cdot}$, then $R_c^{\circ} = \varepsilon$, $(Q, \circ)$ is a right loop.

Proof. The proof follows from Lemma 4.3. □

We recall Definition 1.119. Isotopism of the form $(R_a^{-1}, L_b^{-1}, \varepsilon)$, where L_b, R_a are left and right translations of the quasigroup $(Q, \cdot)$, is called LP-isotopism.

Theorem 4.5. *Any LP-isotope of a quasigroup $(Q, \cdot)$ is a loop [72, 80].*

If $(Q, \circ) = (Q, \cdot)(\varepsilon, P_a, R_b)$, where $(Q, \cdot)$ is a quasigroup, then $(Q, \circ)$ is a unipotent right loop.

If $(Q, \circ) = (Q, \cdot)(P_a^{-1}, \varepsilon, L_b)$, where $(Q, \cdot)$ is a quasigroup, then $(Q, \circ)$ is a unipotent left loop.

Proof. Case 1. Theorem 1.120.

Case 2. Prove that the quasigroup $(Q, \circ)$, where $x \circ y = R_b^{-1}(x \cdot P_a y)$, is a right unipotent loop. Let $1 = R_b^{-1} a = a/b$. If we take $y = 1$, then $x \circ 1 = R_b^{-1}(x \cdot P_a R_b^{-1} a) = R_b^{-1}(x \cdot b) = x$, since $P_a R_b^{-1} a = b$. Indeed, if $P_a R_b^{-1} a = b$, then $R_b^{-1} a \cdot b = a$, $a = a$. Also we can use the identity (1.5). Therefore element 1 is the right identity element of the quasigroup $(Q, \circ)$.

If we take $x = y$, then we have $x \circ x = R_b^{-1}(x \cdot P_a x) = R_b^{-1} a = 1$ since by definition of middle translation $x \cdot P_a x = a$, $(Q, \circ)$ is a unipotent quasigroup.

Case 3. Prove that the quasigroup $(Q, \circ)$, where $x \circ y = L_b^{-1}(P_a^{-1} x \cdot y)$, is a left unipotent loop. Let $1 = L_b^{-1} a = b \backslash a$. If we take $x = 1$, then $1 \circ y = L_b^{-1}(P_a^{-1} L_b^{-1} a \cdot y) = L_b^{-1}(b \cdot y) = y$, since $P_a^{-1} L_b^{-1} a = b$. Indeed, if $P_a^{-1} L_b^{-1} a = b$, then $b \cdot L_b^{-1} a = a$, $a = a$. Therefore element 1 is the left identity element of the quasigroup $(Q, \circ)$.

If we take $x = y$, then we have $x \circ x = L_b^{-1}(P_a^{-1} x \cdot x) = L_b^{-1} a = 1$ since by definition of middle translation $P_a^{-1} x \cdot x = a$, $(Q, \circ)$ is a unipotent quasigroup. □

Remark 4.6. Instead of direct proof of Theorem 4.5 it is possible to use Corollary 4.4.

There exists the well-known connection between translations of a quasigroup and translations of its LP-isotopes [799, 438]. In the following lemma we extend this connection.

Lemma 4.7. *If $x \circ y = R_a^{-1} x \cdot L_b^{-1} y$, then $L_x^{\circ} = L_{R_a^{-1} x} L_b^{-1}$, $R_y^{\circ} = R_{L_b^{-1} y} R_a^{-1}$, $P_z^{\circ} = L_b P_z R_a^{-1}$.*

If $x \circ y = R_b^{-1}(x \cdot P_a y)$, then $L_x^{\circ} = R_b^{-1} L_x P_a$, $R_y^{\circ} = R_b^{-1} R_{P_a y}$, $P_z^{\circ} = P_a^{-1} P_{R_b z}$.

If $x \circ y = L_b^{-1}(P_a^{-1} x \cdot y)$, then $L_x^{\circ} = L_b^{-1} L_{P_a^{-1} x}$, $R_y^{\circ} = L_b^{-1} R_y P_a^{-1}$, $P_z^{\circ} = P_{b \cdot z} P_a^{-1}$.

Proof. Case 1. If $x \circ y = R_a^{-1} x \cdot L_b^{-1} y = z$, then $P_z^{\circ} x = y$, $P_z R_a^{-1} x = L_b^{-1} y$, $L_b P_z R_a^{-1} x = y$, $P_z^{\circ} x = L_b P_z R_a^{-1} x$.

Case 2. If $x \circ y = R_b^{-1}(x \cdot P_a y) = z$, then $R_y^{\circ} x = R_b^{-1} R_{P_a y} x$, $P_z^{\circ} x = y$, $R_b^{-1}(x \cdot P_a y) = z$, $P_{R_b z} x = P_a y$, $P_a^{-1} P_{R_b z} x = y$, $P_z^{\circ} x = P_a^{-1} P_{R_b z} x$, $L_x^{\circ} y = R_b^{-1} L_x P_a y$.

Case 3. If $x \circ y = L_b^{-1}(P_a^{-1} x \cdot y) = z$, then $P_z^{\circ} x = y$, $P_a^{-1} x \cdot y = L_b z$, $P_{L_b z} P_a^{-1} x = y$, $P_z^{\circ} = P_{b \cdot z} P_a^{-1}$, $L_x^{\circ} y = L_b^{-1} L_{P_a^{-1} x} y$, $R_y^{\circ} x = L_b^{-1} R_y P_a^{-1} x$. □

We shall use the following well-known fact.

Lemma 4.8. *If a quasigroup $(Q, \cdot)$ is a group isotope, i.e., $(Q, \cdot) \sim (Q, +)$, where $(Q, +)$ is a group, then any parastrophe of this quasigroup is also a group isotope [825].*

Proof. The proof is based on the fact that any parastrophe of a group is an isotope of this group, i.e., $(Q, +)^{\sigma} \sim (Q, +)$ [72, Lemma 5.1], [80, p. 53]. If $(Q, \cdot) \sim (Q, +)$, then $(Q, \cdot)^{\sigma} \sim (Q, +)^{\sigma} \sim (Q, +)$. □

4.1.2 Quasigroup derivatives

Quasigroup derivatives are used in the study of G-loops [72, 80]. We follow [59, 60, 72, 80, 685]. It is clear that identity of associativity is not true in any quasigroup. But we can replace it with the following equality which is true in any quasigroup (Q, A):

$$A(A(a,b),c) = A(a, A_a(b,c)), \tag{4.1}$$

where $a, b, c \in Q$, A_a is some binary operation which depends on the element a. The equality (4.1) can be obtained from the following equation $A(A(a,b),c) = A(a,x)$. In this case the solution of the equation can be denoted as $x = A_a(b,c)$.

Definition 4.9. The operation A_a is called a right derivative operation of the operation A relative to element a [72].

Definition 4.10. The operation ${}_aA$, which is defined from the equation $A(b, A(c,a)) = A(y,a)$, is called a left derivative operation of the operation A relative to element a, i.e., $y = {}_aA(b,c)$ [72].

Here we give an isotopical approach to the concept of quasigroup derivatives [80, 685]. For a quasigroup $(Q, \cdot)$ we can rewrite the equality (4.1) in the following form:

$$(a \cdot x) \cdot y = a \cdot (x \circ y), \tag{4.2}$$

where x, y are arbitrary elements of the set Q and element a is a fixed element of Q. The equality (4.2) defines a groupoid $(Q, \circ)$. Moreover, $(Q, \circ) = (Q, \cdot)(L_a, \varepsilon, L_a)$, i.e., the groupoid $(Q, \circ)$ is an isotope of quasigroup $(Q, \cdot)$ with isotopy (L_a, ε, L_a). Therefore, the groupoid $(Q, \circ)$ is a quasigroup and it is a *right* derivative of quasigroup $(Q, \cdot)$ relative to element a as in Definition 4.9.

Quasigroup $(Q, *) = (Q, \cdot)(\varepsilon, R_a, R_a)$ is a *left* derivative of quasigroup $(Q, \cdot)$ with respect to element a as in Definition 4.10 [80].

Definition 4.11. Quasigroup $(Q, \star) = (Q, \cdot)(R_a, L_a^{-1}, \varepsilon)$ is called a *middle* derivative of quasigroup $(Q, \cdot)$ with respect to element a.

Quasigroup $(Q, \diamond) = (Q, \cdot)(R_a^{-1}, L_a, \varepsilon)$ is called a *middle inverse* derivative of quasigroup $(Q, \cdot)$ with respect to element a.

Definition 4.12. Isotopisms of the form (L_a, ε, L_a), (ε, R_a, R_a), $(R_a, L_a^{-1}, \varepsilon)$, $(R_a^{-1}, L_a, \varepsilon)$ can be called nuclear isotopisms.

Lemma 4.13. *1. Any right derivative $(Q, \circ)$ of a quasigroup $(Q, \cdot)$ has the left identity element, i.e., $(Q, \circ)$ is a left loop [80].*

2. Any left derivative $(Q, *)$ of a quasigroup $(Q, \cdot)$ has the right identity element, i.e., $(Q, *)$ is a right loop [80].

3. Any middle derivative $(Q, \star)$ of a quasigroup $(Q, \cdot)$ is a left loop.

4. Any middle inverse derivative $(Q, \diamond)$ of a quasigroup $(Q, \cdot)$ is a right loop.

Proof. Case 1. As it was noticed above, the right derivative $(Q, \circ)$ of a quasigroup $(Q, \cdot)$ is also a quasigroup. In equality (4.2) we put $x = e_a$ and obtain $(a \cdot e_a) \cdot y = a \cdot (e_a \circ y)$, $a \cdot y = a \cdot (e_a \circ y)$, $y = e_a \circ y$ for all $y \in Q$. Then $(Q, \circ)$ is a left loop with left identity element e_a.

Case 2. In the equality $(x * y) \cdot a = x \cdot ya$ we put $y = f_a$ and obtain $(x * f_a) \cdot a = xa$, $x * f_a = x$ for all $x \in Q$. Then $(Q, *)$ is a right loop with right identity element f_a.

Case 3. In the equality $x \star y = xa \cdot L_a^{-1}y$ we put $x = f_a$ and obtain $f_a \star y = a \cdot L_a^{-1}y = L_a L_a^{-1}y = y$. Therefore, the quasigroup $(Q, \star)$ is a left loop with left identity element f_a.

Case 4. In the equality $x \diamond y = R_a^{-1}x \cdot ay$ we put $y = e_a$ and obtain $x \diamond e_a = R_a^{-1}x \cdot ae_a = R_a^{-1}x \cdot a = x$. Therefore, the quasigroup $(Q, \diamond)$ is a right loop with right identity element e_a. □

Remark 4.14. Additionally there exist at least 12 "translation nuclear" isotopisms which guarantee that an isotopic image has a left or right identity element:

$(P_a^{-1}, L_a^{-1}, \varepsilon)$ (left identity element); $(P_a^{-1}, \varepsilon, L_a)$ (left identity element); $(\varepsilon, P_a^{-1}, R_a)$ (right identity element);

$(R_a^{-1}, P_a^{-1}, \varepsilon)$ (right identity element); $(L_a, L_a^{-1}, \varepsilon)$ (left identity element); (ε, L_a, R_a) (right identity element);

$(R_a^{-1}, R_a, \varepsilon)$ (right identity element); (R_a, ε, L_a) (left identity element); $(P_a, L_a^{-1}, \varepsilon)$ (left identity element);

(P_a, ε, L_a) (left identity element); (ε, P_a, R_a) (right identity element); $(R_a^{-1}, P_a, \varepsilon)$ (right identity element).

Probably there is a reason to define twelve additional quasigroup derivatives using the defined above "translation nuclear" isotopisms.

It is easy to see that the 16 nuclear isotopisms obtained above can be used by the description of identities that guarantee the existence of identity element in a quasigroup. See Problem # 18 in [72, p. 217] and Section A.3.

For example, using the triple (R_a, ε, L_a) we can construct the following identity $z \cdot xy = xz \cdot y$ that guarantees the existence in quasigroup $(Q, \cdot)$ with this identity of the left identity element. Indeed, if we rewrite the last identity in the form $xy = L_z^{-1}(xz \cdot y)$ and put $x := f_z$, then we get $f_z y = L_z^{-1}(f_z z \cdot y)$, $f_z y = L_z^{-1}(z \cdot y) = L_z^{-1} L_z y = y$. Therefore $f_z y = y$ for all $y \in Q$. Notice, it is not complicated to prove that quasigroup $(Q, \cdot)$ with identity $z \cdot xy = xz \cdot y$ is an abelian group.

4.1.2.1 G-quasigroups

Definition 4.15. A quasigroup (Q, A) is called a right (a left) G-quasigroup if all its right (left) derivatives are isomorphic to (Q, A); a quasigroup (Q, A) is called a G-quasigroup if all its left and right derivatives are isomorphic to (Q, A) [72, p. 48].

From Lemma 4.13 it follows that any G-quasigroup is a loop. In [72, Theorem 3.8] it is proved that definitions of a G-loop (Definitions 1.167 and 4.15) are equivalent. See also Theorem 4.25.

In [72, Theorem 3.9] it is proved that an F-quasigroup (a quasigroup $(Q, *)$ with the identities $x * (y * z) = (x * y) * ((x \backslash x) * z)$ and $(x * y) * z = (x * (z/z)) * (y * z))$ with the left identity element is a right G-quasigroup.

Example 4.16. We give examples of F-quasigroups $(Q, *)$, $(Q, \circ)$, and $(Q, \bullet)$ with the left identity element. Therefore, the quasigroups $(Q, *)$, $(Q, \circ)$, and $(Q, \bullet)$ are right G-quasigroups.

$*$	0	1	2
0	0	1	2
1	2	0	1
2	1	2	0

$\circ$	0	1	2	3
0	0	1	2	3
1	2	0	3	1
2	1	3	0	2
3	3	2	1	0

$\bullet$	0	1	2	3	4
0	0	1	2	3	4
1	2	0	4	1	3
2	1	3	0	4	2
3	4	2	3	0	1
4	3	4	1	2	0

Various systems of axioms for a group are given in [549]. Here we prove the following:

Corollary 4.17. *A quasigroup $(Q,\cdot)$ with the identity of associativity is a loop.*

Proof. We can write the identity of associativity $a \cdot bc = ab \cdot c$ in the following form: $L_a(bc) = L_ab \cdot c$, i.e., the triple (L_a, ε, L_a) is an autotopy of quasigroup $(Q,\cdot)$. Therefore, by Lemma 4.13 Case 1 the quasigroup $(Q,\cdot)$ is a left loop. Similarly, the triple (ε, R_c, R_c) is an autotopy and by Lemma 4.13 Case 2 the left loop $(Q,\cdot)$ is a right loop. Therefore, it is a loop. □

4.1.2.2 Garrison's nuclei in quasigroups

Corollary 4.18. *1. If a quasigroup $(Q,\cdot)$ has a non-trivial left nucleus, then $(Q,\cdot)$ is a left loop.*

2. If a quasigroup $(Q,\cdot)$ has a non-trivial right nucleus, then $(Q,\cdot)$ is a right loop.

3. If a quasigroup $(Q,\cdot)$ has a non-trivial middle nucleus, then $(Q,\cdot)$ is a loop.

Proof. Case 1. If a quasigroup $(Q,\cdot)$ has a non-trivial left nucleus, then it has the autotopy (L_a, ε, L_a). Remember that autotopy is a partial case of isotopy. Therefore, we can apply Lemma 4.13 Case 1.

Similarly, the proof of Case 2 follows from Lemma 4.13, Case 2, the proof of Case 3 follows from Lemma 4.13, Cases 3 and 4. □

Theorem 4.19. *Let $(Q,\cdot)$ be a quasigroup.*

1. *If $N_l \neq \emptyset$, then $(N_l,\cdot)$ is a subgroup of $(Q,\cdot)$ and the identity element of $(N_l,\cdot)$ is a left identity element of $(Q,\cdot)$.*

2. *If $N_r \neq \emptyset$, then $(N_r,\cdot)$ is a subgroup of $(Q,\cdot)$ and the identity element of $(N_r,\cdot)$ is a right identity element of $(Q,\cdot)$.*

3. *If $N_m \neq \emptyset$, then $(N_m,\cdot)$ is a subgroup of $(Q,\cdot)$ and the identity element e of $(N_m,\cdot)$ is the identity element of $(Q,\cdot)$, i.e., $(Q,\cdot)$ is a loop [346] [685, p. 17].*

Proof. The fact that $(N_l,\cdot)$ is a subgroup of $(Q,\cdot)$ follows from Corollary 1.250 and Lemma 1.247. The existence of a left, right, two-sided identity element in quasigroup $(Q,\cdot)$ follows from Corollary 4.18. The fact that these identity elements are identity elements of the groups $(N_l,\cdot)$, $(N_r,\cdot)$, $(N_m,\cdot)$ can also be checked.

Case 1. From the fact that $e_ax = x$ for all $x \in Q$ and the equality $e_a \cdot xy = e_ax \cdot y$, it follows that $e_a \in N_l$. Suppose that $b \in N_l$. Then $be_a \cdot x = b \cdot e_ax = bx$, $be_a = b$ for all $b \in N_l$.

Case 2 is proved in the similar way. Case 3 is obvious. □

Notice, in [346], [685, p. 17] the direct proof of Cases 1, 2, 3 is given. The direct proof of Cases 1 and 2 is also given in [72, 80]. The proof of Case 3, which uses the corresponding group of regular permutations, is given in [72, 80].

4.1.2.3 Mixed derivatives

Lemma 4.20. *Any right (left) derivative $(Q,\circ)$ of a loop $(Q,\cdot)$ is a loop with the same identity element [80].*

Proof. From the equality $x \circ y = L_a^{-1}(L_ax \cdot y)$ we have: if we put $x = 1^{\cdot}$ in this equality, then we obtain $1^{\cdot} \circ y = L_a^{-1}(L_ay) = y$; if we put $y = 1^{\cdot}$, then $x \circ 1^{\cdot} = L_a^{-1}(L_ax) = x$.

From the equality $x \circ y = R_a^{-1}(x \cdot R_ay)$ we have: if we put $x = 1^{\cdot}$ in this equality, then we obtain $1^{\cdot} \circ y = R_a^{-1}(R_ay) = y$; if we put $y = 1^{\cdot}$, then $x \circ 1^{\cdot} = R_a^{-1}(x \cdot a) = R_a^{-1}R_ax = x$. □

Definition 4.21. Right-right derivative $(Q, \bullet)$ of a quasigroup $(Q, \cdot)$ is defined in the following way: $(Q, \bullet) = (Q, \circ)(L^{\circ}_a, \varepsilon, L^{\circ}_a)$, $(Q, \circ) = (Q, \cdot)(L_b, \varepsilon, L_b)$,

$$(Q, \bullet) = (Q, \cdot)(L_b, \varepsilon, L_b)(L^{\circ}_a, \varepsilon, L^{\circ}_a). \tag{4.3}$$

Left-left derivative $(Q, \bullet)$ of a quasigroup $(Q, \cdot)$ is defined in the following way: $(Q, \bullet) = (Q, \circ)(\varepsilon, R^{\circ}_a, R^{\circ}_a)$, $(Q, \circ) = (Q, \cdot)(\varepsilon, R_b, R_b)$,

$$(Q, \bullet) = (Q, \cdot)(\varepsilon, R_b, R_b)(\varepsilon, R^{\circ}_a, R^{\circ}_a). \tag{4.4}$$

Lemma 4.22. *Any right-right (left-left) derivative of a quasigroup $(Q, \cdot)$ is its right (left) derivative [80].*

Proof. Case 1. If we express translation L°_a using translations of quasigroup $(Q, \cdot)$, then we have $x \circ y = L_b^{-1}(L_b x \cdot y)$, $L^{\circ}_a y = L_b^{-1} L_{b \cdot a} y$, $L^{\circ}_a = L_b^{-1} L_{b \cdot a}$. After substitution of the last expression in equality (4.3) we have $(Q, \bullet) = (Q, \cdot)(L_{b \cdot a}, \varepsilon, L_{b \cdot a})$.

Case 2 is proved similarly to Case 1. □

Definition 4.23. The left-right (mixed) derivative $(Q, \bullet)$ of a quasigroup $(Q, \cdot)$ is defined as follows: $(Q, \bullet) = (Q, \circ)(L^{\circ}_a, \varepsilon, L^{\circ}_a)$, $(Q, \circ) = (Q, \cdot)(\varepsilon, R_b, R_b)$,

$$(Q, \bullet) = (Q, \cdot)(\varepsilon, R_b, R_b)(L^{\circ}_a, \varepsilon, L^{\circ}_a). \tag{4.5}$$

The light-left derivative is defined similarly.

Lemma 4.24. *Any left-right derivative of a quasigroup $(Q, \cdot)$ can be presented in the following form: $(Q, \bullet) = (Q, \cdot)(R_b^{-1}, L_a^{-1}, \varepsilon)(L_a R_a, L_a R_a, L_a R_a)$.*

Proof. From equality $x \circ y = R_b^{-1}(x \cdot R_b y)$ we obtain $L^{\circ}_x = R_b^{-1} L_x R_b$. Therefore

$$\begin{aligned}(Q, \bullet) = (Q, \cdot)(\varepsilon, R_b, R_b)(R_b^{-1} L_a R_b, \varepsilon, R_b^{-1} L_a R_b) = \\ (Q, \cdot)(R_b^{-1}, L_a^{-1}, \varepsilon)(L_a R_a, L_a R_a, L_a R_a).\end{aligned}$$

□

Theorem 4.25. *Any loop which is isotopic to a loop $(Q, \cdot)$ is isomorphic to a mixed derivative of the loop $(Q, \cdot)$ [72, Theorem 3.6].*

Proof. The proof follows from Lemmas 1.121 and 4.24. □

4.1.3 Set of maps

Denote the property of a set of permutations $\{p_1, p_2, \dots, p_m\}$ of an m-element set Q "$p_i p_j^{-1}$ $(i \neq j)$ leaves no variable unchanged" [594] as the τ-property. An m-tuple of permutations T can also have the τ-property. We shall call the m-tuple T a τ-m-tuple.

In [594], in fact, Mann proves the following:

Theorem 4.26. *A set $P = \{p_1, p_2, \dots, p_m\}$ of m permutations of a finite set Q of order m defines a Cayley table of a quasigroup if and only if P has the τ-property.*

A permutation α of a finite non-empty set Q which leaves no elements of the set Q unchanged will be called a *fixed point free permutation.*

Definition 4.27. A set M of maps of the set Q into itself is called *simply transitive* (more precisely, the set M acts on the set Q simply transitively) if for any pair of elements x, y of the set Q there exists a unique element μ_j of the set M such that $\mu_j(x) = y$.

In Definition 4.27 we do not suppose that the set M forms a group relative to usual composition of the maps.

Theorem 4.28. *A set $T = \{p_1, p_2, \dots, p_n\}$ of n permutations of a finite set Q of order n is simply transitive if and only if the set T has the τ-property [594].*

Lemma 4.29. *If $(Q, \cdot)$ is a quasigroup, then the sets $\mathbb{L}, \mathbb{R}, \mathbb{P}$ of all left, right, middle translations of quasigroup $(Q, \cdot)$ are simply transitive sets of permutations of the set Q.*

Proof. Indeed, if a, b are the fixed elements of the set Q, then there exists a unique element $x \in Q$ such that $x \cdot a = b$, there exists a unique element $y \in Q$ such that $a \cdot y = b$, and there exists a unique element $z \in Q$ such that $a \cdot b = z$. We establish the following bijections between elements of the sets Q and elements of the sets $\mathbb{L}(Q, \cdot), \mathbb{R}(Q, \cdot), \mathbb{P}(Q, \cdot)$: $\varphi_1 : x \leftrightarrow L_x$, $\varphi_2 : x \leftrightarrow R_x$, $\varphi_3 : x \leftrightarrow P_x$. Therefore, there exists a unique element $L_x \in \mathcal{L}(Q, \cdot)$ such that $L_x a = b$, there exists a unique element $R_y \in \mathcal{R}(Q, \cdot)$ such that $R_y a = b$, and there exists a unique element $P_z \in \mathcal{P}(Q, \cdot)$ such that $P_z a = b$. □

Definition 4.30. The centralizer of a set S of a group G (written as $C_G(S)$) is the set of elements of G which commute with any element of S; in other words, $C_G(S) = \{x \in G \mid xa = ax \text{ for all } a \in S\}$ [471, 407].

Definition 4.31. The normalizer of a set S in a group G, written as $N_G(S)$, is defined as $N_G(S) = \{x \in G \mid xS = Sx\}$.

The following theorem is called the NC-theorem [910].

Theorem 4.32. *$C(S)$ is always a normal subgroup of $N(S)$ [275, 910].*

Proof. The proof follows [910]. If c is in $C(S)$ and n is in $N(S)$, we have to show that $n^{-1}cn$ is in $C(S)$. To that end, pick s in S and let $t = nsn^{-1}$. Then t is in S, so therefore $ct = tc$. Then note that $ns = tn$ and $n^{-1}t = sn^{-1}$. So $(n^{-1}cn)s = (n^{-1}c)tn = n^{-1}(tc)n = (sn^{-1})cn = s(n^{-1}cn)$ which is what we needed. □

4.2 Garrison's nuclei and A-nuclei

4.2.1 Definitions of nuclei and A-nuclei

Classical definitions of quasigroup nuclei and center are given in Section 1.4.

We recall the standard Garrison [346] definition of quasigroup nuclei.

Definition 4.33. Let $(Q, \circ)$ be a quasigroup. Then $N_l = \{a \in Q \mid (a \circ x) \circ y = a \circ (x \circ y)\}$, $N_r = \{a \in Q \mid (x \circ y) \circ a = x \circ (y \circ a)\}$, and $N_m = \{a \in Q \mid (x \circ a) \circ y = x \circ (a \circ y)\}$ are respectively its left, right, and middle nuclei [346, 72, 685]. Nucleus is given by $N = N_l \cap N_r \cap N_m$ [685].

Garrison names an element of a middle quasigroup nucleus as a *center element* [346]. The importance of Garrison's quasigroup nuclei is in the fact that N_l, N_r and N_m all are subgroups of a quasigroup $(Q, \cdot)$ [346].

The weakness of Garrison's definition is in the fact that, if a quasigroup $(Q, \cdot)$ has a non-trivial left nucleus, then $(Q, \cdot)$ is a left loop, i.e., $(Q, \cdot)$ has a left identity element; if a quasigroup $(Q, \cdot)$ has a non-trivial right nucleus, then $(Q, \cdot)$ is a right loop, i.e., $(Q, \cdot)$ has a right identity element; if a quasigroup $(Q, \cdot)$ has a non-trivial middle nucleus, then $(Q, \cdot)$ is a loop, i.e., $(Q, \cdot)$ has an identity element (Corollary 4.18, Theorem 4.19). See also [346], [685, I.3.4 Theorem].

The connection between autotopies and nuclei is well known [59, 65, 72, 491]. Namely the set of autotopies of the form (L_a, ε, L_a) of a quasigroup $(Q, \circ)$ corresponds to the left nucleus of $(Q, \circ)$ and vice versa. Similarly, the set of autotopies of the form (ε, R_a, R_a) of a quasigroup $(Q, \circ)$ corresponds to the right nucleus of $(Q, \circ)$, the set of autotopies of the form $(R_a, L_a^{-1}, \varepsilon)$ of a quasigroup $(Q, \circ)$ corresponds to the middle nucleus of $(Q, \circ)$.

It is easy to see that from the Garrison definition of left nucleus of a loop $(Q, \cdot)$ it follows that $R_{xy}^{-1} R_y R_x a = a$ for all $x, y \in Q$ and all $a \in N_l$. Permutations of the form $R_{xy}^{-1} R_y R_x$ generate a right inner multiplication group of $(Q, \cdot)$. It is clear that any element of the left nucleus is invariant relative to any element of the group $\langle R_{xy}^{-1} R_y R_x \mid x, y \in Q \rangle$.

Similarly, from Garrison's definition of the right nucleus of a loop $(Q, \cdot)$ it follows that $L_{xy}^{-1} L_x L_y a = a$ for all $x, y \in Q$ and all $a \in N_l$. Permutations of the form $L_{xy}^{-1} L_x L_y$ generate a left inner multiplication group of $(Q, \cdot)$. It is clear that any element of the left nucleus is invariant relative to any element of the group $\langle L_{xy}^{-1} L_x L_y \mid x, y \in Q \rangle$.

For the middle nucleus the situation is slightly different and of course Garrison was right when calling elements of the middle loop nucleus central elements.

P.I. Gramma [380], M.D. Kitaroagă [529], and G.B. Belyavskaya [104, 108, 110] generalized on the quasigroup case the concepts of nuclei and center using namely this nuclear property. G.B. Belyavskaya obtained the most general results in this direction.

In [489] the following definition is given.

Definition 4.34. The set of all autotopisms of the form $(\alpha, \varepsilon, \gamma)$ of a quasigroup $(Q, \circ)$, where ε is the identity mapping, is called the *left autotopy nucleus* (left A-nucleus) of quasigroup $(Q, \circ)$.

Similarly, the sets of autotopisms of the forms $(\alpha, \beta, \varepsilon)$ and $(\varepsilon, \beta, \gamma)$ form the *middle and right A-nuclei* of $(Q, \circ)$. We shall denote these three sets of mappings as N_l^A, N_m^A and N_r^A respectively.

Using Definition 4.34 we can say that the elements of a left Kitaroagă nucleus of a quasigroup $(Q, \cdot)$ correspond to autotopies of the form $(L_a L_h^{-1}, \varepsilon, L_a L_h^{-1})$, where $a \in Q$ (in fact $a \in N_l$ in Kitaroagă sense), h is a fixed element of the set Q [529].

Autotopies of the form $(L_a L_h^{-1}, \varepsilon, L_{a \cdot e(h)} L_h^{-1})$, where $a \in Q$ (in fact $a \in N_l$ in Belyavskaya sense) correspond to the elements of a left Belyavskaya nucleus [104].

Remark 4.35. Often, by a generalization of some objects or concepts, we not only win in generality, but also lose some important properties of the generalized objects or concepts. The weakness of Definition 4.34 is in the fact that it is not easy to define an A-nucleus similarly to a quasigroup nucleus as an intersection of left, right and middle quasigroup A-nuclei.

Lemma 4.36.

1. *The first components of the autotopisms of any subgroup $K(N_l^A)$ of N_l^A themselves form a group $K_1(N_l^A)$.*
2. *The third components of the autotopisms of any subgroup $K(N_l^A)$ of N_l^A themselves form a group $K_3(N_l^A)$.*
3. *The first components of the autotopisms of any subgroup $K(N_m^A)$ of N_m^A themselves form a group $K_1(N_m^A)$.*
4. *The second components of the autotopisms of any subgroup $K(N_m^A)$ of N_m^A themselves form a group $K_2(N_l^A)$.*

5. *The second components of the autotopisms of any subgroup $K(N_r^A)$ of N_r^A themselves form a group $K_2(N_r^A)$.*

6. *The third components of the autotopisms of any subgroup $K(N_r^A)$ of N_r^A themselves form a group $K_3(N_r^A)$ [489].*

Proof. Case 1 is proved in Lemma 1.247. Other cases are proved in a similar way to Case 1. □

From Lemma 4.36 it follows that the first, second and third components of N_l^A, N_m^A and N_r^A each form groups. For brevity, we shall denote these nine groups by ${}_1N_l^A$, ${}_2N_l^A$, ${}_3N_l^A$, ${}_1N_m^A$, ${}_2N_m^A$, ${}_3N_m^A$, ${}_1N_r^A$, ${}_2N_r^A$ and ${}_3N_r^A$.

The next two lemmas demonstrate that A-nuclei have some advantages in comparison with Garrison's nuclei.

Lemma 4.37. *In any quasigroup Q its left, right, middle A-nucleus is a normal subgroup of the group $Avt(Q)$.*

Proof. Let $(\mu, \varepsilon, \nu) \in N_l^A$, $(\alpha, \beta, \gamma) \in Avt(Q)$. Then

$$(\alpha^{-1}, \beta^{-1}, \gamma^{-1})(\mu, \varepsilon, \nu)(\alpha, \beta, \gamma) = (\alpha^{-1}\mu\alpha, \varepsilon, \gamma^{-1}\nu\gamma) \in N_l^A.$$

The proofs for right and middle nucleus are similar. □

Lemma 4.38. *If quasigroups $(Q, \cdot)$ and $(Q, \circ)$ are isotopic, then these quasigroups have isomorphic autotopy nuclei and isomorphic components of the autotopy nuclei, i.e., $N_l^A(Q, \cdot) \cong N_l^A(Q, \circ)$, ${}_1N_l^A(Q, \cdot) \cong {}_1N_l^A(Q, \circ)$, ${}_3N_l^A(Q, \cdot) \cong {}_3N_l^A(Q, \circ)$, and so on.*

Proof. We can use Lemmas 4.1 and 4.36. □

Corollary 4.39. *If a quasigroup $(Q, \cdot)$ is an isotope of a group $(Q, +)$, then all its A-nuclei and components of A-nuclei are isomorphic with the group $(Q, +)$.*

Proof. It is clear that in any group all its A-nuclei and components of A-nuclei are isomorphic with the group $(Q, +)$. Further we can apply Lemma 4.38. □

Lemma 4.40. *Isomorphic components of A-nuclei of a quasigroup $(Q, \cdot)$ act on the set Q in such a manner that the numbers and lengths of orbits by these actions are equal.*

Proof. In Lemma 4.38 it is established that A-nuclei and the same components of A-nuclei are isomorphic in pairs. Notice that all A-nuclei of a quasigroup $(Q, \cdot)$ are subgroups of the group $S_Q \times S_Q \times S_Q$. Components of these A-nuclei are subgroups of the group S_Q.

If two components of A-nuclei, say B and C are isomorphic, then this isomorphism is an isomorphism of permutation groups which act on the set Q. Isomorphic permutation groups are called *similar* [471, p. 111].

Let $\psi B = C$, where B, C are isomorphic components of A-nuclei, ψ is an isomorphism. Then we can establish a bijection φ of the set Q such that $\psi(g)(\varphi(m)) = \varphi(g(m))$ for all $m \in Q, g \in B$ [471, p. 111]. □

4.2.2 Components of A-nuclei and identity elements

There exist some connections between components of A-nuclei and local identity elements in any quasigroup.

Lemma 4.41. *1. If $(Q,\cdot)$ is a quasigroup and $(\alpha,\varepsilon,\gamma)\in Avt(Q,\cdot)$, then $\alpha f(x)\cdot x=\gamma x$, $\alpha x\cdot e(x)=\gamma x$, $\alpha x\cdot x=\gamma s(x)$ for all $x\in Q$.*

2. If $(Q,\cdot)$ is a quasigroup and $(\varepsilon,\beta,\gamma)\in Avt(Q,\cdot)$, then $f(x)\cdot\beta x=\gamma x$, $x\cdot\beta e(x)=\gamma x$, $x\cdot\beta x=\gamma s(x)$ for all $x\in Q$.

3. If $(Q,\cdot)$ is a quasigroup and $(\alpha,\beta,\varepsilon)\in Avt(Q,\cdot)$, then $\alpha f(x)\cdot\beta x=x$, $\alpha x\cdot\beta e(x)=x$, $\alpha x\cdot\beta x=s(x)$ for all $x\in Q$.

Proof. Case 1. If we put $x=f(y)$, then we obtain $\alpha f(y)\cdot y=\gamma y$. If we put $y=e(x)$, then we obtain $\alpha(x)\cdot e(x)=\gamma x$. If we put $x=y$, then we obtain $\alpha x\cdot x=\gamma s(x)$.

Cases 2 and 3 are proved similarly. □

Corollary 4.42. *1. If $(Q,\cdot)$ is an idempotent quasigroup and $(\alpha,\varepsilon,\gamma)\in Avt(Q,\cdot)$, then $\alpha x\cdot x=\gamma x$, $\alpha x=P^{-1}_{\gamma x}$ for all $x\in Q$.*

2. If $(Q,\cdot)$ is an idempotent quasigroup and $(\varepsilon,\beta,\gamma)\in Avt(Q,\cdot)$, then $x\cdot\beta x=\gamma x$, $\beta x=P_{\gamma x}x$ for all $x\in Q$.

3. If $(Q,\cdot)$ is an idempotent quasigroup and $(\alpha,\beta,\varepsilon)\in Avt(Q,\cdot)$, then $\alpha x\cdot\beta x=x$, $P_x\alpha x=\beta x$ for all $x\in Q$.

Proof. The proof follows from Lemma 4.41 and the fact that in an idempotent quasigroup $(Q,\cdot)$ the maps f,e,s are all equal to identity permutation of the set Q. Further we have $P_{\gamma x}\alpha x=x$, $\alpha x=P^{-1}_{\gamma x}$ (Case 1). From $x\cdot\beta x=\gamma x$ we have $P_{\gamma x}x=\beta x$ in Case 2. From $\alpha x\cdot\beta x=x$ we have $P_x\alpha x=\beta x$ in Case 3. □

Remark 4.43. We can also rewrite:

equality $\alpha x\cdot x=\gamma x$ in the form $x\cdot\alpha^{-1}x=\gamma\alpha^{-1}x$, $\alpha\gamma^{-1}x\cdot\gamma^{-1}x=x$ (Case 1);
equality $x\cdot\beta x=\gamma x$ in the form $\beta^{-1}x\cdot x=\gamma\beta^{-1}x$, $\gamma^{-1}x\cdot\beta\gamma^{-1}x=x$ (Case 2);
equality $\alpha x\cdot\beta x=x$ in the form $x\cdot\beta\alpha^{-1}x=\alpha^{-1}x$, $\alpha\beta^{-1}x\cdot x=\beta^{-1}x$ (Case 3).

Corollary 4.44. *1. Let $(Q,\cdot)$ be a left loop.*

(a) If $(\alpha,\varepsilon,\gamma)\in Avt(Q,\cdot)$, then $\alpha 1\cdot x=\gamma x$ for all $x\in Q$, i.e., $\gamma=L_{\alpha 1}$.

(b) If $(\varepsilon,\beta,\gamma)\in Avt(Q,\cdot)$, then $\beta=\gamma$.

(c) If $(\alpha,\beta,\varepsilon)\in Avt(Q,\cdot)$, then $\alpha 1\cdot\beta x=x$ for all $x\in Q$, i.e., $\beta=L^{-1}_{\alpha 1}$.

2. Let $(Q,\cdot)$ be a right loop.

(a) If $(\alpha,\varepsilon,\gamma)\in Avt(Q,\cdot)$, then $\alpha=\gamma$.

(b) If $(\varepsilon,\beta,\gamma)\in Avt(Q,\cdot)$, then $x\cdot\beta 1=\gamma x$ for all $x\in Q$, i.e., $\gamma=R_{\beta 1}$.

(c) If $(\alpha,\beta,\varepsilon)\in Avt(Q,\cdot)$, then $\alpha x\cdot\beta 1=x$ for all $x\in Q$, i.e., $\alpha=R^{-1}_{\beta 1}$.

3. Let $(Q,\cdot)$ be a unipotent quasigroup.

(a) If $(\alpha,\varepsilon,\gamma)\in Avt(Q,\cdot)$, then $\alpha x\cdot x=\gamma 1$ for all $x\in Q$, i.e., $\alpha=P^{-1}_{\gamma 1}$.

(b) If $(\varepsilon,\beta,\gamma)\in Avt(Q,\cdot)$, then $x\cdot\beta x=\gamma 1$ for all $x\in Q$, i.e., $\beta=P_{\gamma 1}$.

(c) If $(\alpha,\beta,\varepsilon)\in Avt(Q,\cdot)$, then $\alpha=\beta$.

Proof. The proof follows from Lemma 4.41 and Remarks 1.79 and 1.81.

In Case 3(c) we have the following: if $(Q, \cdot)$ is a unipotent quasigroup and $(\alpha, \beta, \varepsilon) \in Avt(Q, \cdot)$, then $\alpha x \cdot \beta x = 1$ for all $x \in Q$, i.e., $\beta = P_1 \alpha$. But in the unipotent quasigroup $P_1 = \varepsilon$. □

Theorem 4.45. *1. Let $(Q, \cdot)$ be a loop.*

(a) If $(\alpha, \varepsilon, \gamma) \in Avt(Q, \cdot)$, then $\alpha = \gamma = L_{\alpha 1}$.

(b) If $(\varepsilon, \beta, \gamma) \in Avt(Q, \cdot)$, then $\beta = \gamma = R_{\beta 1}$.

(c) If $(\alpha, \beta, \varepsilon) \in Avt(Q, \cdot)$, then $\alpha = R_{\beta 1}$, $\beta = L^{-1}_{\beta 1}$.

2. Let $(Q, \cdot)$ be a unipotent left loop.

(a) If $(\alpha, \varepsilon, \gamma) \in Avt(Q, \cdot)$, then $\alpha = P_{\alpha 1}$, $\gamma = L_{\alpha 1}$.

(b) If $(\varepsilon, \beta, \gamma) \in Avt(Q, \cdot)$, then $\beta = \gamma = P_{\gamma 1}$.

(c) If $(\alpha, \beta, \varepsilon) \in Avt(Q, \cdot)$, then $\alpha = \beta = L^{-1}_{\alpha 1}$.

3. Let $(Q, \cdot)$ be a unipotent right loop.

(a) If $(\alpha, \varepsilon, \gamma) \in Avt(Q, \cdot)$, then $\alpha = \gamma = P^{-1}_{\gamma 1}$.

(b) If $(\varepsilon, \beta, \gamma) \in Avt(Q, \cdot)$, then $\beta = P^{-1}_{\beta 1}$, $\gamma = R_{\beta 1}$.

(c) If $(\alpha, \beta, \varepsilon) \in Avt(Q, \cdot)$, then $\alpha = \beta = R^{-1}_{\beta 1}$.

Proof. Case 1(c) is well known [489]. From Corollary 4.44 it follows that in the loop case, $\alpha = R^{-1}_{\beta 1}$ and $\beta = L^{-1}_{\alpha 1}$ for any autotopy of the form $(\alpha, \beta, \varepsilon)$. If $R^{-1}_a \in {}_1N^A_m(Q, \cdot)$, then $R_a \in {}_1N^A_m(Q, \cdot)$, since ${}_1N^A_m(Q, \cdot)$ is a group. Therefore, in the loop case any autotopy of the kind $(\alpha, \beta, \varepsilon)$ takes the following form:

$$(R_{\beta 1}, L_{\alpha 1}, \varepsilon). \tag{4.6}$$

Taking into consideration the equality (4.6) and the fact that ${}_1N^A_m(Q, \cdot)$ is a group we have: if $R_a, R_b \in {}_1N^A_m(Q, \cdot)$, then $R_a R_b = R_c$. Thus $R_a R_b 1 = R_c 1$, $c = b \cdot a$.

We find now the form of the inverse element to the element $R_a \in {}_1N^A_m(Q, \cdot)$. Notice that if $R^{-1}_a = R_b$ for some $b \in Q$, then $R_a R_b = R_{b \cdot a} = \varepsilon = R_1$. Then $b \cdot a = 1$ and since any right (left, middle) quasigroup translation is defined in a unique way by its index element, we have $b = {}^{-1}a$. From the other side, $R_b R_a = R_{a \cdot b} = \varepsilon = R_1$, $b = a^{-1}$. Therefore in this situation $a^{-1} = {}^{-1}a$ for any suitable element $a \in Q$, $R^{-1}_a = R_{({}^{-1}a)} = R_{a^{-1}}$.

Similarly we can obtain that $L^{-1}_a = L_{({}^{-1}a)} = L_{a^{-1}}$ for any $L_a \in {}_2N^A_m(Q, \cdot)$, where $(Q, \cdot)$ is a loop.

In loop $(Q, \cdot)$ from equality $\alpha x \cdot \beta y = x \cdot y$ by $x = y = 1$ we have $\alpha 1 \cdot \beta 1 = 1$. Then $\beta 1 = (\alpha 1)^{-1}$, $\alpha 1 = {}^{-1}(\beta 1)$. Moreover, we have that $\beta 1 = {}^{-1}(\alpha 1)$, $\alpha 1 = (\beta 1)^{-1}$, $R_{\beta 1} = R^{-1}_{\alpha 1}$.

Finally we obtain that any element of the middle A-nucleus of a loop has the form

$$(R_b, L^{-1}_b, \varepsilon). \tag{4.7}$$

It is easy to see that $(R_{\alpha 1}, L^{-1}_{\alpha 1}, \varepsilon) \in N^A_m \Leftrightarrow \alpha 1 \in N_m$.

Case 2(a). From Corollary 4.44 it follows that in a unipotent left loop $\alpha = P^{-1}_{\gamma 1} = P^{-1}_{\alpha 1 \cdot 1}$, $\gamma = L_{\alpha 1}$.

In a unipotent loop, any left A-nucleus autotopy $(\alpha, \varepsilon, \gamma)$ takes the following form:

$$(P^{-1}_{\gamma 1}, \varepsilon, L_{\alpha 1}). \tag{4.8}$$

It is clear that if $(P_{\gamma 1}^{-1}, \varepsilon, L_{\alpha 1}) \in N_l^A$, then any element of the group N_l^A has the form $(P_{\gamma 1}, \varepsilon, L_{\alpha 1}^{-1})$. We find now the form of the inverse element to the element $P_a \in {}_1N_l^A(Q, \cdot)$.

At first we have: if $P_a, P_b \in {}_1N_l^A(Q, \cdot)$, then $P_aP_b = P_c$ for some element $c \in Q$. Thus $P_aP_b1 = P_c1$. If $P_c1 = d$, then $1 \cdot d = c$, $c = d$. Therefore $P_ab = c$, $b \cdot c = a$, $c = b\backslash a$.

Notice, if $P_a^{-1} = P_b$ for some $b \in Q$, then $P_aP_b = P_{b\backslash a} = \varepsilon = P_1$ since $(Q, \cdot)$ is a unipotent quasigroup. Since any right (left, middle) quasigroup translation is defined in a unique way by its index element, then $b\backslash a = 1$, $a = b \cdot 1$, $b = a/1$. From the other side, $P_bP_a = P_{a\backslash b} = \varepsilon = P_1$, $a\backslash b = 1$, $b = a \cdot 1$. Thus, if $(Q, \cdot)$ is a left unipotent loop, then $a/1 = a \cdot 1$ for any element $a \in Q$ such that $P_a \in {}_1N_l^A(Q, \cdot)$. Then $P_a^{-1} = P_{a/1} = P_{a\cdot 1}$, $P_a = P_{a\cdot 1}^{-1}$, $\alpha = P_{\alpha 1\cdot 1}^{-1} = P_{\alpha 1}$. And we obtain that any element of the left A-nucleus of a unipotent left loop has the form

$$(P_{\alpha 1}, \varepsilon, L_{\alpha 1}). \tag{4.9}$$

Case 3(b). From Corollary 4.44 it follows that in a unipotent right loop, $\beta = P_{1\cdot\beta 1}$, $\gamma = R_{\beta 1}$. Further proof can be similar to Case 2(a), but we shall use parastrophic ideas.

It is easy to see that any unipotent right loop $(Q, \circ)$ is a (12)-parastrophic image of a unipotent left loop $(Q, \cdot)$ (Table 1.2). In this case ${}_1N_l^A(Q, \circ) = {}_2N_r^A(Q, \cdot)$, ${}_3N_l^A(Q, \circ) = {}_3N_r^A(Q, \cdot)$ (Table 1.4).

From formula (4.9) it follows that any element of the right A-nucleus of a unipotent right loop has the form

$$(\varepsilon, P^{\cdot}_{\beta 1}, L^{\cdot}_{\beta 1}). \tag{4.10}$$

Finally, the use of Table 1.1 gives us that in the language of translations of unipotent right loop $(Q, \cdot)$ we have that any element of the right A-nucleus of $(Q, \circ)$ has the form

$$(\varepsilon, P_{\beta 1}^{-1}, R_{\beta 1}). \tag{4.11}$$

□

Remark 4.46. In fact, from Theorem 4.45 it follows that in the loop case Belousov's concept of regular permutations coincides with the concepts of left, right, and middle A-nucleus [72, 80, 685].

Lemma 4.47. *Let $(Q, \cdot)$ be a quasigroup, $(Q, \circ) = (Q, \cdot)(R_a^{-1}, L_b^{-1}, \varepsilon)$. Then*

1. $N_l^A = (R_a^{-1}L_{R_a^{-1}l}L_b^{-1}R_a, \varepsilon, L_{R_a^{-1}l}L_b^{-1})$,
2. $N_r^A = (\varepsilon, L_b^{-1}R_{L_b^{-1}r}R_a^{-1}L_b, R_{L_b^{-1}r}R_a^{-1})$,
3. $N_m^A = (R_a^{-1}R_{L_b^{-1}m}, L_b^{-1}L_{R_a^{-1}m}, \varepsilon)$,

where a, b are the fixed elements of the set Q, $l \in N_l(Q, \circ), r \in N_r(Q, \circ), m \in N_m(Q, \circ)$.

Proof. Case 1. Let $(Q, \cdot)$ be a quasigroup, $(Q, \circ)$ be a loop of the form $x \circ y = R_a^{-1}x \cdot L_b^{-1}y$, i.e., $(Q, \circ) = (Q, \cdot)T$, where $T = (R_a^{-1}, L_b^{-1}, \varepsilon)$. Then

$$L_x^{\circ}y = L_{R_a^{-1}x}L_b^{-1}y; R_y^{\circ}x = R_{L_b^{-1}y}R_a^{-1}y. \tag{4.12}$$

The left A-nucleus of $(Q, \circ)$ has the form $(L_l^{\circ}, \varepsilon, L_l^{\circ})$. Using the equality (4.12) we can re-write it in the following form: $(L_{R_a^{-1}l}L_b^{-1}, \varepsilon, L_{R_a^{-1}l}L_b^{-1})$.

By Lemma 4.1 $Avt(Q, \cdot) = TAvt(Q, \circ)T^{-1}$. Thus

$$N_l^A(Q, \cdot) = (R_a^{-1}L_{R_a^{-1}l}L_b^{-1}R_a, \varepsilon, L_{R_a^{-1}l}L_b^{-1}).$$

Cases 2 and 3 are proved in a similar way. □

Corollary 4.48. *In any quasigroup $(Q,\cdot)$ any of the groups ${}_1N_l^A$, ${}_3N_l^A$, ${}_2N_r^A$, ${}_3N_r^A$, ${}_1N_m^A, {}_2N_m^A$ is a subgroup of the group $M(Q,\cdot)$.*

Proof. The proof follows from Lemma 4.47. □

Exercise 4.49. Isotope $(Q,\cdot)$ of a group $(Q,+)$ of the form (φ,ψ,R_{-a}), where $\varphi,\psi \in Aut(Q,+)$, is called a linear quasigroup (over the group $(Q,+)$). Research the groups N_r^A, N_l^A, N_m^A of quasigroup $(Q,\cdot)$ using Theorem 1.133 and Lemma 1.122.

4.2.3 A-nuclei of loops by isostrophy

Taking into consideration the importance of loops in the class of all quasigroups we give some information on A-nuclear components of isostrophic images of a loop. From Table 1.2 this lemma follows:

Lemma 4.50. *The (12)-parastrophe of a loop $(Q,\cdot)$ is a loop, the (13)-parastrophe of a loop $(Q,\cdot)$ is a unipotent right loop, the (23)-parastrophe of a loop $(Q,\cdot)$ is a unipotent left loop, the (123)-parastrophe of a loop $(Q,\cdot)$ is a unipotent left loop, and the (132)-parastrophe of a loop $(Q,\cdot)$ is a unipotent right loop.*

Taking into consideration Theorem 4.45, and Tables 1.1 and 1.4 we can give more detailed connections between components of an A-nuclei of a loop $(Q,\cdot)$ and its isostrophic images of the form $(Q,\circ) = (Q,\cdot)((\sigma)(\alpha,\beta,\gamma))$, where $\sigma \in S_3$, $\alpha,\beta,\gamma \in S_Q$.

Example 4.51. If $(Q,\circ) = (Q,\cdot)((123)(\alpha,\beta,\gamma))$, then from Table 1.4 we have ${}_2N_r^A(Q,\circ) = \beta^{-1}{}_1N_m^A\beta(Q,\cdot)$. By Theorem 4.45 any element of the group ${}_1N_m^A(Q,\cdot)$ is some right translation R_a of the loop $(Q,\cdot)$.

If, additionally, $\beta = \varepsilon$, then using Table 1.1 and Lemma 4.50 we can say that in the unipotent left loop $(Q,\circ)$ any element of the group ${}_2N_r^A(Q,\circ)$ is a middle translation P_a^{-1}.

In Table 4.1 $(Q,\circ) = (Q,\cdot)((\sigma)(\alpha,\beta,\gamma))$, where $(Q,\cdot)$ is a loop. We suppose that elements of the left, right and middle loop nuclei have the following forms (L_a,ε,L_a), (ε,R_b,R_b), and $(R_c,L_c^{-1},\varepsilon)$, respectively.

Table 4.1: Components of loop nuclei at isostrophy.

	(ε,T)	$((12),T)$	$((13),T)$	$((23),T)$	$((132),T)$	$((123),T)$
${}_1N_l^\circ$	$\alpha^{-1}L^{\cdot}_a\alpha$	$\alpha^{-1}R^{\cdot}_b\alpha$	$\alpha^{-1}L^{\cdot}_a\alpha$	$\alpha^{-1}R^{\cdot}_c\alpha$	$\alpha^{-1}(L^{\cdot}_c)^{-1}\alpha$	$\alpha^{-1}R^{\cdot}_b\alpha$
${}_3N_l^\circ$	$\gamma^{-1}L^{\cdot}_a\gamma$	$\gamma^{-1}R^{\cdot}_b\gamma$	$\gamma^{-1}L^{\cdot}_a\gamma$	$\gamma^{-1}L_c^{-1}\cdot\gamma$	$\gamma^{-1}R_c\cdot\gamma$	$\gamma^{-1}R_b\cdot\gamma$
${}_2N_r^\circ$	$\beta^{-1}R^{\cdot}_b\beta$	$\beta^{-1}L^{\cdot}_a\beta$	$\beta^{-1}(L_c^{-1})^{\cdot}\beta$	$\beta^{-1}R^{\cdot}_b\beta$	$\beta^{-1}L^{\cdot}_a\beta$	$\beta^{-1}R^{\cdot}_c\beta$
${}_3N_r^\circ$	$\gamma^{-1}R^{\cdot}_b\gamma$	$\gamma^{-1}L^{\cdot}_a\gamma$	$\gamma^{-1}R^{\cdot}_c\gamma$	$\gamma^{-1}R^{\cdot}_b\gamma$	$\gamma^{-1}L^{\cdot}_a\gamma$	$\gamma^{-1}(L_c^{-1})^{\cdot}\gamma$
${}_1N_m^\circ$	$\alpha^{-1}R^{\cdot}_c\alpha$	$\alpha^{-1}(L_c^{-1})^{\cdot}\alpha$	$\alpha^{-1}R^{\cdot}_b\alpha$	$\alpha^{-1}L^{\cdot}_a\alpha$	$\alpha^{-1}R^{\cdot}_b\alpha$	$\alpha^{-1}L^{\cdot}_a\alpha$
${}_2N_m^\circ$	$\beta^{-1}(L_c^{-1})^{\cdot}\beta$	$\beta^{-1}R^{\cdot}_c\beta$	$\beta^{-1}R^{\cdot}_b\beta$	$\beta^{-1}L^{\cdot}_a\beta$	$\beta^{-1}R^{\cdot}_b\beta$	$\beta^{-1}L^{\cdot}_a\beta$

4.2.4 Isomorphisms of A-nuclei

Lemma 4.52. *In any quasigroup $(Q,\circ)$ we have*

$$N_l^A = \{(\alpha,\varepsilon,R_c\alpha R_c^{-1}) \mid \textit{for all}\, c \in Q\} = \{(R_c^{-1}\gamma R_c,\varepsilon,\gamma) \mid \textit{for all}\, c \in Q\}, \tag{4.13}$$

$$N_r^A = \{(\varepsilon,\beta,L_c\beta L_c^{-1}) \mid \textit{for all}\, c \in Q\} = \{(\varepsilon,L_c^{-1}\gamma L_c,\gamma) \mid \textit{for all}\, c \in Q\}, \tag{4.14}$$

$$N_m^A = \{(\alpha,P_c\alpha P_c^{-1},\varepsilon) \mid \textit{for all}\, c \in Q\} = \{(P_c^{-1}\beta P_c,\beta,\varepsilon) \mid \textit{for all}\, c \in Q\}. \tag{4.15}$$

Proof. Re-write the left A-nuclear equality $\alpha x \circ y = \gamma(x \circ y)$ in the following form: $R_y \alpha x = \gamma R_y x$, $\gamma = R_y \alpha R_y^{-1}$ for all $y \in Q$.

Re-write the right A-nuclear equality $x \circ \beta y = \gamma(x \circ y)$ in the following form: $L_x \beta y = \gamma L_x y$, $\gamma = L_x \beta L_x^{-1}$ for all $x \in Q$.

In the middle A-nuclear equality $\alpha x \circ \beta y = x \circ y$ we put $y = x'$, where $x \circ x' = a$ and a is a fixed element of the set Q, i.e., $x' = P_a x$, where P_a is a permutation of the set Q.

Then $\alpha x \circ \beta x' = x \circ x' = a$ for all $x \in Q$, $P_a \alpha x = \beta x' = \beta P_a x$. Therefore $\beta = P_a \alpha P_a^{-1}$ for all $a \in Q$. □

Lemma 4.53. *If K is a subgroup of the group ${}_1N_l^A$, then $K \cong H \subseteq {}_3N_l^A$, moreover $H = R_x K R_x^{-1}$ for any $x \in Q$.*

If K is a subgroup of the group ${}_2N_r^A$, then $K \cong H \subseteq {}_3N_r^A$, moreover $H = L_x K L_x^{-1}$ for any $x \in Q$.

If K is a subgroup of the group ${}_1N_m^A$, then $K \cong H \subseteq {}_2N_m^A$, moreover $H = P_x K P_x^{-1}$ for any $x \in Q$.

Proof. We can use Lemma 4.52. □

Lemma 4.54. *In any quasigroup $(Q, \circ)$ the groups N_l^A, ${}_1N_l^A$ and ${}_3N_l^A$; N_m^A, ${}_1N_m^A$ and ${}_2N_m^A$; N_r^A, ${}_2N_r^A$ and ${}_3N_r^A$ are isomorphic in pairs.*

Proof. We can use Lemma 4.53. □

Corollary 4.55. *In any quasigroup $(Q, \circ)$ we have $|N_l^A| = |{}_1N_l^A| = |{}_3N_l^A|$; $|N_m^A| = |{}_1N_m^A| = |{}_2N_m^A|$; $|{}_2N_r^A| = |{}_2N_r^A| = |{}_3N_r^A|$.*

Proof. The proof follows from Lemma 4.54. □

Theorem 4.56. *In any quasigroup:*

1. ${}_1N_l^A = {}_3N_l^A$ *if and only if* $RM \subseteq N_{S_Q}({}_1N_l^A)$;
2. ${}_1N_l^A = {}_3N_l^A$ *if and only if* $RM \subseteq N_{S_Q}({}_3N_l^A)$;
3. ${}_2N_r^A = {}_3N_r^A$ *if and only if* $LM \subseteq N_{S_Q}({}_2N_r^A)$;
4. ${}_2N_r^A = {}_3N_r^A$ *if and only if* $LM \subseteq N_{S_Q}({}_3N_r^A)$;
5. ${}_1N_m^A = {}_2N_m^A$ *if and only if* $PM \subseteq N_{S_Q}({}_1N_m^A)$;
6. ${}_1N_m^A = {}_2N_m^A$ *if and only if* $PM \subseteq N_{S_Q}({}_2N_m^A)$.

Proof. Case 1. If ${}_1N_l^A = {}_3N_l^A$, then from Lemma 4.53 it follows that $R_c \in N_{S_Q}({}_1N_l^A)$ for any $c \in Q$. Since $N_{S_Q}({}_1N_l^A)$ is a group, further we have that $RM \subseteq N_{S_Q}({}_1N_l^A)$.

Converse. Let $\alpha \in {}_1N_l^A$. Then by Lemma 4.53 $R_c \alpha R_c^{-1} \in {}_3N_l^A$. But $RM \subseteq N_{S_Q}({}_1N_l^A)$. Then $R_c \alpha R_c^{-1} = \beta \in {}_1N_l^A$. Thus ${}_3N_l^A \subseteq {}_1N_l^A$.

Let $\gamma \in {}_3N_l^A$. Then by Lemma 4.53 $R_c^{-1} \gamma R_c \in {}_1N_l^A$. But $RM \subseteq N_{S_Q}({}_1N_l^A)$. Then $R_c R_c^{-1} \gamma R_c R_c^{-1} = \gamma \in {}_1N_l^A$. Thus ${}_3N_l^A \subseteq {}_1N_l^A$. Therefore ${}_1N_l^A = {}_3N_l^A$.

Cases 2–6 are proved in a similar way. □

Moreover we can specify Theorem 4.56.

Theorem 4.57.

1. *In the group N_l^A of a quasigroup $(Q, \circ)$, the first and the third components of any element $T \in N_l^A$ coincide if and only if $RM(Q, \circ) \subseteq C_{S_Q}({}_1N_l^A)$.*
2. *In the group N_l^A of a quasigroup $(Q, \circ)$, the first and the third components of any element $T \in N_l^A$ coincide if and only if $RM(Q, \circ) \subseteq C_{S_Q}({}_3N_l^A)$.*

3. *In the group N_r^A of a quasigroup $(Q, \circ)$, the second and the third components of any element $T \in N_r^A$ coincide if and only if $LM(Q, \circ) \subseteq C_{S_Q}({}_2N_r^A)$.*

4. *In the group N_r^A of a quasigroup $(Q, \circ)$, the second and the third components of any element $T \in N_r^A$ coincide if and only if $LM(Q, \circ) \subseteq C_{S_Q}({}_3N_r^A)$.*

5. *In the group N_m^A of a quasigroup $(Q, \circ)$, the first and the second components of any element $T \in N_m^A$ coincide if and only if $PM(Q, \circ) \subseteq C_{S_Q}({}_1N_m^A)$.*

6. *In the group N_m^A of a quasigroup $(Q, \circ)$, the first and the second components of any element $T \in N_m^A$ coincide if and only if $PM(Q, \circ) \subseteq C_{S_Q}({}_2N_m^A)$.*

Proof. The proof follows from Lemma 4.53. □

Corollary 4.58. 1. *If $(Q, \cdot)$ is a right loop, then the first and the third components of any element $T \in N_l^A$ coincide and ${}_1N_l^A = {}_3N_l^A = C_{S_Q}(\mathbb{R}) = C_{S_Q}(RM)$.*

2. *If $(Q, \cdot)$ is a left loop, then the second and the third components of any element $T \in N_r^A$ coincide and ${}_2N_r^A = {}_3N_r^A = C_{S_Q}(\mathbb{L}) = C_{S_Q}(LM)$.*

3. *If $(Q, \cdot)$ is a unipotent quasigroup, then the first and the second components of any element $T \in N_m^A$ coincide and ${}_1N_m^A = {}_2N_m^A = C_{S_Q}(\mathbb{P}) = C_{S_Q}(PM)$.*

4. *If $(Q, \cdot)$ is a loop, then ${}_1N_l^A = {}_3N_l^A = C_{S_Q}(\mathbb{R}) = C_{S_Q}(RM) = C_M(RM) = C_{FM}(RM)$; ${}_2N_r^A = {}_3N_r^A = C_{S_Q}(\mathbb{L}) = C_{S_Q}(LM) = C_M(LM) = C_{FM}(LM)$.*

5. *If $(Q, \cdot)$ is a unipotent left loop, then ${}_2N_r^A = {}_3N_r^A = C_{S_Q}(\mathbb{L}) = C_{S_Q}(LM) = C_{PLM}(LM) = C_{FM}(LM)$; ${}_1N_m^A = {}_2N_m^A = C_{S_Q}(\mathbb{P}) = C_{S_Q}(PM) = C_{PLM}(PM) = C_{FM}(PM)$.*

6. *If $(Q, \cdot)$ is a unipotent right loop, then ${}_1N_l^A = {}_3N_l^A = C_{S_Q}(\mathbb{R}) = C_{S_Q}(RM) = C_{PRM}(RM) = C_{FM}(RM)$; ${}_1N_m^A = {}_2N_m^A = C_{S_Q}(\mathbb{P}) = C_{S_Q}(PM) = C_{PRM}(PM) = C_{FM}(PM)$.*

Proof. Case 1. The proof follows from Lemma 4.53. In a right loop $(Q, \circ)$ there is a right identity element 1. Then $R_c = \varepsilon$, if $c = 1$. Thus $\gamma = R_c\alpha R_c^{-1} = \alpha$. See also Corollaries 4.44 and 4.45. From equality $R_c\alpha = \alpha R_c$ that is true for all $c \in Q$ and for all $\alpha \in {}_1N_l^A$, we conclude that ${}_1N_l^A \subseteq C_{S_Q}(\mathbb{R})$.

Prove that $C_{S_Q}(\mathbb{R}) \subseteq {}_1N_l^A$. We rewrite the proof from ([275], Lemma 2.6). Any element ψ of the group $C_{S_Q}(\mathbb{R})$ fulfills the equality $\psi R_y x = R_y \psi x$, i.e., $\psi(xy) = \psi(x) \cdot y$, i.e., $(\psi, \varepsilon, \psi) \in N_l^A$. Therefore $C_{S_Q}(\mathbb{R}) \subseteq {}_1N_l^A$, ${}_1N_l^A = C_{S_Q}(\mathbb{R})$. The equality $C_{S_Q}(\mathbb{R}) = C_{S_Q}(RM)$ is true since $\langle \mathbb{R} \rangle = RM$.

Case 2 is a mirror of Case 1 and we omit its proof.

Case 3. From the equality $P_c\alpha = \alpha P_c$ (Lemma 4.53) that is true for all $c \in Q$ and for all $\alpha \in {}_1N_m^A$, we conclude that ${}_1N_m^A \subseteq C_{S_Q}(\mathbb{P})$.

Any element ψ of the group $C_{S_Q}(\mathbb{P})$ satisfies the equality $\psi P_y x = P_y \psi x = z$. If $\psi P_y x = z$, then $x \cdot \psi^{-1} z = y$. If $P_y \psi x = z$, then $\psi x \cdot z = y$. Then $x \cdot \psi^{-1} z = \psi x \cdot z$, $(\psi, \psi, \varepsilon) \in N_m^A$. Therefore $C_{S_Q}(\mathbb{P}) \subseteq {}_1N_m^A$. Thus ${}_1N_m^A = C_{S_Q}(\mathbb{P})$. The equality $C_{S_Q}(\mathbb{P}) = C_{S_Q}(PM)$ is true since $\langle \mathbb{P} \rangle = PM$.

Case 4. The equality $C_{S_Q}(\mathbb{R}) = C_{S_Q}(RM)$ follows from the fact that $\langle \mathbb{R} \rangle = RM$.

The equality $C_{S_Q}(RM) = C_M(RM)$ follows from the fact that ${}_1N_l^A \subseteq M$ and $RM \subseteq M$ since in the loop case any element of the group N_l^A has the form (L_c, ε, L_c) (Corollary 4.45).

The equality $C_M(RM) = C_{FM}(RM)$ follows from the fact that $C_M(RM) = C_{S_Q}(RM)$, $M \subseteq FM \subseteq S_Q$, ${}_1N_l^A \subseteq FM$ and $RM \subseteq FM$.

The proofs of Cases 5 and 6 are similar to the proof of Case 4. □

Remark 4.59. 1. From the equality $R_c\alpha = \alpha R_c$ (Corollary 4.58, Case 1) that is true for all $c \in Q$ and for all $\alpha \in {}_1N_l^A$, we can make some additional conclusions: $\mathbb{R} \subseteq C_{S_Q}(\alpha)$; $\mathbb{R} \subseteq C_{S_Q}({}_1N_l^A)$; $RM(Q) \subseteq C_{S_Q}(\alpha)$; $RM(Q) \subseteq C_{S_Q}({}_1N_l^A)$.

2. From the equality $L_c\beta = \beta L_c$ (Corollary 4.58, Case 2) that is true for all $c \in Q$ and for all $\beta \in {}_2N_r^A$, we can make some additional conclusions, namely: $\mathbb{L} \subseteq C_{S_Q}(\beta)$; $\mathbb{L} \subseteq C_{S_Q}({}_2N_r^A)$; $LM(Q) \subseteq C_{S_Q}(\beta)$; $LM(Q) \subseteq C_{S_Q}({}_2N_r^A)$.

3. From the equality $P_c\alpha = \alpha P_c$ (Corollary 4.58, Case 3) that is true for all $c \in Q$ and for all $\alpha \in {}_1N_m^A$, we can make some additional conclusions, namely: $\mathbb{P} \subseteq C_{S_Q}(\alpha)$; $\mathbb{P} \subseteq C_{S_Q}({}_1N_m^A)$; $PM(Q) \subseteq C_{S_Q}(\alpha)$; $PM(Q) \subseteq C_{S_Q}({}_1N_m^A)$.

4.2.5 A-nuclei by some isotopisms

Mainly we shall use Lemma 4.1 [80, Lemma 1.4]: if quasigroups $(Q, \circ)$ and $(Q, \cdot)$ are isotopic with isotopy T, i.e., $(Q, \circ) = (Q, \cdot)T$, then $Avt(Q, \circ) = T^{-1}Avt(Q, \cdot)T$.

Lemma 4.60. *For any quasigroup $(Q, \cdot)$ there exists its isotopic image $(Q, \circ)$ such that:*

1. *any autotopy of the form $(\alpha, \varepsilon, \gamma)$ of quasigroup $(Q, \cdot)$ is transformed into an autotopy of the form $(\gamma, \varepsilon, \gamma)$ of quasigroup $(Q, \circ)$;*

2. *any autotopy of the form $(\alpha, \varepsilon, \gamma)$ of quasigroup $(Q, \cdot)$ is transformed into an autotopy of the form $(\alpha, \varepsilon, \alpha)$ of quasigroup $(Q, \circ)$;*

3. *any autotopy of the form $(\varepsilon, \beta, \gamma)$ of quasigroup $(Q, \cdot)$ is transformed into an autotopy of the form $(\varepsilon, \beta, \beta)$ of quasigroup $(Q, \circ)$;*

4. *any autotopy of the form $(\varepsilon, \beta, \gamma)$ of quasigroup $(Q, \cdot)$ is transformed into an autotopy of the form $(\varepsilon, \gamma, \gamma)$ of quasigroup $(Q, \circ)$;*

5. *any autotopy of the form $(\alpha, \beta, \varepsilon)$ of quasigroup $(Q, \cdot)$ is transformed into an autotopy of the form $(\alpha, \alpha, \varepsilon)$ of quasigroup $(Q, \circ)$;*

6. *any autotopy of the form $(\alpha, \beta, \varepsilon)$ of quasigroup $(Q, \cdot)$ is transformed into an autotopy of the form $(\beta, \beta, \varepsilon)$ of quasigroup $(Q, \circ)$.*

Proof. Case 1. If $(Q, \circ) = (Q, \cdot)(R_a^{-1}, \varepsilon, \varepsilon)$, where element a is a fixed element of the set Q, then $(Q, \circ)$ is a right loop (Corollary 4.4).

By Lemma 4.1, if quasigroups $(Q, \circ)$ and $(Q, \cdot)$ are isotopic with isotopy, then $Avt(Q, \circ) = T^{-1}Avt(Q, \cdot)T$. Then autotopy of the form $(\alpha, \varepsilon, \gamma)$ of $(Q, \cdot)$ passes to autotopy of the form $(R_a\alpha R_a^{-1}, \varepsilon, \gamma)$.

In any right loop, any element of the group $N_l^A(Q, \circ)$ has equal first and third components, i.e., it has the form $(\gamma, \varepsilon, \gamma)$ (Corollary 4.44).

Case 2. We can take the isotopy $(Q, \circ) = (Q, \cdot)(\varepsilon, \varepsilon, R_a)$. In this case $(Q, \circ)$ is a right loop.

Case 3. We can take the isotopy $(Q, \circ) = (Q, \cdot)(\varepsilon, \varepsilon, L_a)$. In this case $(Q, \circ)$ is a left loop.

Case 4. We can take the isotopy $(Q, \circ) = (Q, \cdot)(\varepsilon, L_a^{-1}, \varepsilon)$. In this case $(Q, \circ)$ is a left loop.

Case 5. We can take the isotopy $(Q, \circ) = (Q, \cdot)(\varepsilon, P_a, \varepsilon)$. In this case $(Q, \circ)$ is a unipotent quasigroup.

Case 6. We can take the isotopy $(Q, \circ) = (Q, \cdot)(P_a^{-1}, \varepsilon, \varepsilon)$. In this case $(Q, \circ)$ is a unipotent quasigroup. □

Corollary 4.61. *For any quasigroup $(Q,\cdot)$ there exists its isotopic image $(Q,\circ)$ such that:*

1. ${}_1N_l^A(Q,\circ) = {}_3N_l^A(Q,\circ) = {}_3N_l^A(Q,\cdot)$;
2. ${}_1N_l^A(Q,\circ) = {}_3N_l^A(Q,\circ) = {}_1N_l^A(Q,\cdot)$.
3. ${}_2N_r^A(Q,\circ) = {}_3N_r^A(Q,\circ) = {}_2N_r^A(Q,\cdot)$;
4. ${}_2N_r^A(Q,\circ) = {}_3N_r^A(Q,\circ) = {}_3N_r^A(Q,\cdot)$;
5. ${}_1N_m^A(Q,\circ) = {}_2N_m^A(Q,\circ) = {}_1N_m^A(Q,\cdot)$;
6. ${}_1N_m^A(Q,\circ) = {}_2N_m^A(Q,\circ) = {}_2N_m^A(Q,\cdot)$.

Proof. The proof follows from Lemma 4.60. □

Lemma 4.62. 1. *Let $(Q,\cdot)$ be a loop. If $(Q,\circ) = (Q,\cdot)(R_a^{-1}, L_b^{-1}, \varepsilon)$, then*

(i) $N_l^A(Q,\circ) = N_l^A(Q,\cdot)$,
(ii) $N_r^A(Q,\circ) = N_r^A(Q,\cdot)$.

2. *Let $(Q,\cdot)$ be a unipotent right loop. If $(Q,\circ) = (Q,\cdot)(\varepsilon, P_a, R_b)$, then*

(i) $N_l^A(Q,\circ) = N_l^A(Q,\cdot)$,
(ii) $N_m^A(Q,\circ) = N_m^A(Q,\cdot)$.

3. *Let $(Q,\cdot)$ be a unipotent left loop. If $(Q,\circ) = (Q,\cdot)(P_a^{-1}, \varepsilon, L_b)$, then*

(i) $N_r^A(Q,\circ) = N_r^A(Q,\cdot)$,
(ii) $N_m^A(Q,\circ) = N_m^A(Q,\cdot)$.

Proof. In this case $(Q,\circ)$ is a loop.

Case 1, (i). Let $(L_c, \varepsilon, L_c) \in N_l^A(Q,\cdot)$. By the isotopy, any element of the group $N_l^A(Q,\circ)$ has the form $(R_aL_cR_a^{-1}, \varepsilon, L_c)$. Since $(Q,\circ)$ (or $(Q,\cdot)$) is a loop, then $R_aL_cR_a^{-1} = L_c$, $N_l^A(Q,\cdot) \subseteq N_l^A(Q,\circ)$. Inverse inclusion is proved in a similar way. Therefore $N_l^A(Q,\cdot) = N_l^A(Q,\circ)$.

Case 1, (ii) is proved in a similar way to Case 1, (i).

Case 2 is a (13)-parastrophe image of Case 1. In this case $(Q,\circ)$ is a unipotent right loop (Theorem 4.5). The forms of the components of the left nucleus and middle nucleus of an unipotent right loop are given in Theorem 4.45.

Further proofs of Cases 2, (i) and 2, (ii) are similar to the proof of Case 1, (i) and we omit it.

Case 3 is a (23)-parastrophe image of Case 1. □

Notice, Case 3 can be considered as a (12)-parastrophe image of Case 2.

4.2.6 Quasigroup bundle and nuclei

Definition 4.63. Let $(Q,\circ)$ be a quasigroup.

Denote the set of all elements of the form $L_cL_d^{-1}$, where L_c, L_d are left translations of $(Q,\circ)$, by $\mathcal{M}_{\mathcal{L}}$.

Denote the set of all elements of the form $R_cR_d^{-1}$, where R_c, R_d are right translations of $(Q,\circ)$, by $\mathcal{M}_{\mathcal{R}}$.

Denote the set of all elements of the form $P_cP_d^{-1}$, where P_c, P_d are middle translations of $(Q,\circ)$, by $\mathcal{M}_{\mathcal{P}}$.

Further define the following sets: $\mathcal{M}^*_{\mathcal{L}} = \{L_c^{-1}L_d \mid a, b \in Q\}$, $\mathcal{M}^*_{\mathcal{R}} = \{R_c^{-1}R_d \mid a, b \in Q\}$, $\mathcal{M}^*_{\mathcal{P}} = \{P_c^{-1}P_d \mid a, b \in Q\}$.

It is clear that in any quasigroup $\mathcal{M}_{\mathcal{L}} \subseteq LM$, $\mathcal{M}^*_{\mathcal{L}} \subseteq LM$, $\mathcal{M}_{\mathcal{R}} \subseteq RM$, $\mathcal{M}^*_{\mathcal{R}} \subseteq RM$, $\mathcal{M}_{\mathcal{P}} \subseteq PM$, $\mathcal{M}^*_{\mathcal{P}} \subseteq PM$.

Notice that $(L_c L_d^{-1})^{-1} = L_d L_c^{-1}$. Therefore $(\mathcal{M}_{\mathcal{L}})^{-1} = \mathcal{M}_{\mathcal{L}}$, $(\mathcal{M}_{\mathcal{R}})^{-1} = \mathcal{M}_{\mathcal{R}}$ and so on.

Remark 4.64. It is clear that $\mathcal{M}_{\mathcal{L}} \subseteq LM(Q, \cdot)$. Then from Definition 4.30 it follows that $C_{S_Q}(\mathcal{M}_{\mathcal{L}}) \supseteq C_{S_Q} LM(Q, \cdot)$ and so on.

In [799] the set of all LP-isotopes of a fixed quasigroup $(Q, \cdot)$ is called a bundle of a quasigroup $(Q, \cdot)$.

Remark 4.65. If $|Q| = n$, then by Lemma 4.7 any of the sets $\mathcal{M}_{\mathcal{L}}$, $\mathcal{M}_{\mathcal{R}}$, $\mathcal{M}_{\mathcal{P}}$, $\mathcal{M}^*_{\mathcal{L}}$, $\mathcal{M}^*_{\mathcal{R}}$, and $\mathcal{M}^*_{\mathcal{P}}$ can contain n (not necessary different) simply transitive subsets of order n, which correspond to the sets of left, right and middle translations of corresponding loops, unipotent left loops, and unipotent right loops of a quasigroup $(Q, \cdot)$.

Theorem 4.66. *In any quasigroup* $(Q, \cdot)$*:*

$$\begin{array}{llllll} (1) & {}_1N_l^A(Q,\cdot) \subseteq C_{S_Q}(\mathcal{M}^*_{\mathcal{R}}), & (2) & {}_3N_l^A(Q,\cdot) \subseteq C_{S_Q}(\mathcal{M}_{\mathcal{R}}), \\ (3) & {}_2N_r^A(Q,\cdot) \subseteq C_{S_Q}(\mathcal{M}^*_{\mathcal{L}}), & (4) & {}_3N_r^A(Q,\cdot) \subseteq C_{S_Q}(\mathcal{M}_{\mathcal{L}}), \\ (5) & {}_1N_m^A(Q,\cdot) \subseteq C_{S_Q}(\mathcal{M}^*_{\mathcal{P}}), & (6) & {}_2N_m^A(Q,\cdot) \subseteq C_{S_Q}(\mathcal{M}_{\mathcal{P}}). \end{array}$$

Proof. Case 1. From Lemma 1.137 it follows that in any autotopy $(\alpha, \varepsilon, \gamma)$ of a quasigroup $(Q, \circ)$, permutations α and ε uniquely determine permutation γ. Therefore from Lemma 4.53 in this case $R_c \alpha R_c^{-1} = R_d \alpha R_d^{-1}$, $\alpha = R_c^{-1} R_d \alpha R_d^{-1} R_c = R_c^{-1} R_d \alpha (R_c^{-1} R_d)^{-1}$ for all $c, d \in Q$.

Case 2 is proved in a similar way.

Case 3. From Lemma 1.137 it follows that in any autotopy $(\varepsilon, \beta, \gamma)$ of a quasigroup $(Q, \circ)$, permutations ε and β uniquely determine permutation γ. Therefore from Lemma 4.53 in this case $L_c \beta L_c^{-1} = L_d \beta L_d^{-1}$, $\beta = L_c^{-1} L_d \beta L_d^{-1} L_c = L_c^{-1} L_d \beta (L_c^{-1} L_d)^{-1}$ for all $c, d \in Q$.

Case 4 is proved similarly.

Case 5. From Lemma 1.137 it follows that in any autotopy $(\alpha, \beta, \varepsilon)$ of a quasigroup $(Q, \circ)$, permutations β and ε uniquely determine permutation α. Therefore from Lemma 4.53 in this case $P_c \alpha P_c^{-1} = P_d \alpha P_d^{-1}$, $\alpha = P_c^{-1} P_d \alpha P_d^{-1} P_c = P_c^{-1} P_d \alpha (P_c^{-1} P_d)^{-1}$ for all $c, d \in Q$.

Case 6 is proved similarly. □

Corollary 4.67. *1. In any right loop* $(Q, \cdot)$

(a) ${}_1N_l^A(Q,\cdot) \subseteq C_{S_Q} RM(Q,\cdot)$,

(b) ${}_3N_l^A(Q,\cdot) \subseteq C_{S_Q} RM(Q,\cdot)$.

2. In any left loop $(Q, \cdot)$

(a) ${}_2N_r^A(Q,\cdot) \subseteq C_{S_Q} LM(Q,\cdot)$,

(b) ${}_3N_r^A(Q,\cdot) \subseteq C_{S_Q} LM(Q,\cdot)$.

3. In any unipotent quasigroup $(Q, \cdot)$

(a) ${}_1N_m^A(Q,\cdot) \subseteq C_{S_Q} PM(Q,\cdot)$,

(b) ${}_2N_m^A(Q,\cdot) \subseteq C_{S_Q} PM(Q,\cdot)$.

Proof. Case 1, (a). Since any right loop $(Q, \cdot)$ has right identity element 1, then R_1 is the identity permutation of the set Q. Therefore the set $\mathcal{M}^*_{\mathcal{R}}$ contains any right translation of this right loop $(Q, \cdot)$. It is clear that in this case $C_{S_Q}\mathbb{R} \supseteq C_{S_Q}(\mathcal{M}_{\mathcal{R}}) \supseteq C_{S_Q} RM$. But $C_{S_Q}\mathbb{R} = C_{S_Q} RM$ since $\langle \mathbb{R} \rangle = RM$. Then $C_{S_Q}(\mathcal{M}^*_{\mathcal{R}}) = C_{S_Q} RM$.

All other cases are proved in a similar way. □

Corollary 4.68. *Let $x \circ y = R_a^{-1}x \cdot y$ for all $x, y \in Q$. Then ${}_1N_l^A(Q,\cdot) \subseteq C_{S_Q}(\mathbb{R}(Q,\circ))$; ${}_3N_l^A(Q,\cdot) \subseteq C_{S_Q}(\mathbb{R}(Q,\circ))$.*

Let $x \circ y = x \cdot L_b^{-1}y$ for all $x, y \in Q$. Then ${}_2N_r^A(Q,\cdot) \subseteq C_{S_Q}(\mathbb{L}(Q,\circ))$; ${}_3N_r^A(Q,\cdot) \subseteq C_{S_Q}(\mathbb{L}(Q,\circ))$.

4.2.7 A-nuclei actions

Actions of subgroups of a multiplication group of a loop $(Q,\cdot)$ on the set Q are studied in [10, 11].

Theorem 4.69. *Any of the groups ${}_1N_l^A$, ${}_3N_l^A$, ${}_1N_m^A$, ${}_2N_m^A$, ${}_2N_r^A$ and ${}_3N_r^A$ of a quasigroup $(Q,\cdot)$ acts freely (semiregularly) on the set Q and in such a manner that the stabilizer of any element $x \in Q$ by this action is the identity group, i.e., $|\,{}_1N_l^A{}_x| = |\,{}_3N_l^A{}_x| = |\,{}_2N_r^A{}_x| = |\,{}_3N_r^A{}_x| = |\,{}_1N_{mx}^A| = |\,{}_3N_{mx}^A| = 1$.*

Proof. By Theorem 1.139 any autotopy (α,β,γ) $((\alpha,\beta,\gamma) \equiv (\alpha_1,\alpha_2,\alpha_3)\ )$ is uniquely defined by any autotopy component α_i, $i \in \{1,2,3\}$, and by element $b = \alpha_j a$, $i \neq j$. By definition of ${}_1N_l^A$ any element α of the group ${}_1N_l^A$ is the first component of an autotopy of the form $(\alpha, \varepsilon, R_c\alpha R_c^{-1})$ (Lemma 4.53).

In order to apply Theorem 1.139 to the group N_l^A we take $i = 2$ (in this case $\beta = \varepsilon$ for any $T \in N_l^A$). And, in fact, any left nuclear autotopy T is defined by the image αa of a fixed element $a \in Q$ by action of the permutation $\alpha \in {}_1N_l^A$.

Therefore, if $\alpha_1, \alpha_2 \in {}_1N_l^A$, then $\alpha_1 x \neq \alpha_2 x$ for any $x \in Q$, i.e., ${}_1N_l^A$ acts freely (fixed point free) on the set Q, or equivalently, if $\alpha(x) = x$ for some $x \in Q$ then $\alpha = \varepsilon$.

If $\alpha a = \beta a$ for some $\alpha, \beta \in {}_1N_l^A$, $a \in Q$, then $\beta^{-1}\alpha x = x$, $\beta^{-1}\alpha = \varepsilon$. Therefore $|\,{}_1N_l^A{}_x| = |\,{}_3N_l^A{}_x| = |\,{}_2N_r^A{}_x| = |\,{}_3N_r^A{}_x| = |\,{}_1N_{mx}^A| = |\,{}_3N_{mx}^A| = 1$. □

Corollary 4.70. *If H is a subgroup of the group ${}_1N_l^A$ (${}_3N_l^A$, ${}_1N_m^A$, ${}_2N_m^A$, ${}_2N_r^A$ and ${}_3N_r^A$), then the group H acts freely on the set Q and in such manner that a stabilizer of any element $x \in Q$ by this action is the identity group.*

Proof. The proof follows from Theorem 4.69. □

Corollary 4.71. *In finite quasigroup $(Q,\cdot)$ the order of any orbit of the group $H \subseteq {}_1N_l^A$ ($H \subseteq {}_3N_l^A$, $H \subseteq {}_1N_m^A$, $H \subseteq {}_2N_m^A$, $H \subseteq {}_2N_r^A$ and $H \subseteq {}_3N_r^A$) by its action on the set Q is equal to the order of the group H.*

Proof. The proof follows from Corollary 4.70. □

Lemma 4.72. *In any finite quasigroup $(Q,\cdot)$ the order of any subgroup of the group N_l^A (N_m^A, N_r^A, ${}_1N_l^A$, ${}_3N_l^A$, ${}_1N_m^A$, ${}_2N_m^A$, ${}_2N_r^A$ and ${}_3N_r^A$) divides the order of the set Q.*

Proof. We can use Corollary 4.71. □

Corollary 4.73. *In any finite quasigroup $(Q,\cdot)$ we have $|N_l^A| \leqslant |Q|$, $|N_m^A| \leqslant |Q|$, $|N_r^A| \leqslant |Q|$, $|{}_1N_l^A| \leqslant |Q|$, $|{}_3N_l^A| \leqslant |Q|$, $|{}_1N_m^A| \leqslant |Q|$, $|{}_2N_m^A| \leqslant |Q|$, $|{}_2N_r^A| \leqslant |Q|$, and $|{}_3N_r^A| \leqslant |Q|$.*

Proof. The proof follows from Lemma 4.72. □

Corollary 4.74. *In any finite loop $(Q,\cdot)$ we have: 1. $|\,{}_1N_l^A| = |\,{}_3N_l^A| = |N_l|$. 2. $|\,{}_2N_r^A| = |\,{}_3N_r^A| = |N_r|$. 3. $|\,{}_1N_m^A| = |\,{}_2N_m^A| = |N_m|$.*

Proof. Case 1. The proof follows from Corollary 4.71 and the fact that in any loop ${}_1N_l^A 1 = {}_3N_l^A 1 = N_l$ (Corollary 1.250). Cases 2 and 3 are proved similarly. □

Corollary 4.75. *In any finite loop $(Q, \cdot)$ the order of any subgroup of its left, right, middle nucleus divides the order of the set Q.*

Proof. The proof follows from Lemma 4.72 and Corollary 4.74. □

Definition 4.76. If Q is a loop, and $N_l(N_r, N_m)$ is its left (right, middle) nucleus, then the index of $N_l(N_r, N_m)$ in Q is equal to the quotient of the orders of two loops: $|Q : N_l| = |Q|/|N_l|$, $(|Q : N_r| = |Q|/|N_r|$, $|Q : N_m| = |Q|/|N_m|)$, respectively.

Remark 4.77. For the infinite case we can re-formulate Corollary 4.71 in the language of bijections. For example, there exists a bijection between any orbit of the group ${}_1N_l^A$ by its action on the set Q and the group ${}_1N_l^A$, and so on.

Lemma 4.78. *If the order of a quasigroup $(Q, \cdot)$ is equal to the order of the group N_l^A, then the orders of all groups N_m^A, N_r^A, ${}_1N_l^A$, ${}_3N_l^A$, ${}_1N_m^A$, ${}_2N_m^A$, ${}_2N_r^A$ and ${}_3N_r^A$ are equal to $|Q|$.*

Proof. It is clear that the order of a quasigroup $(Q, \cdot)$ is invariant relative to the parastrophy. I.e., $|(Q, \cdot)| = |(Q, \cdot)^\sigma|$, for any $\sigma \in S_3$. From Table 3 it follows that $\left(N_l^A\right)^{(12)} = N_r^A$, $\left(N_l^A\right)^{(23)} = N_m^A$. Therefore $|N_l^A| = |N_r^A| = |N_m^A| = |Q|$. Further we can apply Lemma 4.53. □

Remark 4.79. Analogues of Lemma 4.78 are true for any group from the following list: N_m^A, N_r^A, ${}_1N_l^A$, ${}_3N_l^A$, ${}_1N_m^A$, ${}_2N_m^A$, ${}_2N_r^A$ and ${}_3N_r^A$.

Corollary 4.80.

1. *If $(Q, \cdot)$ is a finite right loop, then $|C_{S_Q}(FM)| \leqslant |C_{S_Q}(M)| \leqslant |C_{S_Q}(RM)| \leqslant |Q|$.*
2. *If $(Q, \cdot)$ is a finite left loop, then $|C_{S_Q}(FM)| \leqslant |C_{S_Q}(M)| \leqslant |C_{S_Q}(LM)| \leqslant |Q|$.*
3. *If $(Q, \cdot)$ is a finite unipotent quasigroup, then $|C_{S_Q}(FM)| \leqslant |C_{S_Q}(PM)| \leqslant |Q|$.*

Proof. Case 1. Inclusions $FM \supseteq M \supseteq RM$ follow from definitions of these groups. Therefore $|C_{S_Q}(FM)| \leqslant |C_{S_Q}(M)| \leqslant |C_{S_Q}(RM)|$. Inequality $|C_{S_Q}(RM)| \leqslant |Q|$ follows from Corollaries 4.58 and 4.73.

Cases 2 and 3 are proved similarly. □

Example 4.81. We give an example of a loop with $N_m = \{0, 1\}$. Notice that this nucleus is a non-normal subloop of loop $(Q, *)$. Indeed, $5 * \{0, 1\} = \{5, 3\}$, $\{0, 1\} * 5 = \{5, 4\}$. Then ${}_1N_m^A = \{\varepsilon, R_1\}$, where $R_1 = (0\,1)(2\,4)(3\,5)$, ${}_2N_m^A = \{\varepsilon, L_1^{-1}\}$, where $L_1 = L_1^{-1} = (0\,1)(2\,3)(4\,5)$. The group ${}_1N_m^A$ by action on the set Q has the following set of orbits $\{\{0, 1\}, \{2, 4\}, \{3, 5\}\}$, and the group ${}_2N_m^A$ has the following set $\{\{0, 1\}, \{2, 3\}, \{4, 5\}\}$.

In the loop $(Q, *)$ we have: $N_l = \{0, 3\}$; $N_r = \{0, 4\}$. These nuclei are also non-normal: $\{0, 3\} * 4 = \{4, 2\}$, $4 * \{0, 3\} = \{4, 1\}$, $\{0, 4\} * 3 = \{3, 1\}$, $3 * \{0, 4\} = \{3, 2\}$. We have $L_3 = (0\,3)(1\,5)(2\,4)$, $R_4 = (0\,4)(1\,5)(2\,3)$.

*	0	1	2	3	4	5
0	0	1	2	3	4	5
1	1	0	3	2	5	4
2	2	4	1	5	3	0
3	3	5	4	0	2	1
4	4	2	5	1	0	3
5	5	3	0	4	1	2

4.2.8 A-nuclear quasigroups

We start from a theorem which is a little generalization of Belousov's regular theorem [72, Theorem 2.2]. In [491, 493] Kepka studied regular mappings of groupoids including n-ary case.

Theorem 4.82. *1. If an orbit K by the action of a subgroup H of the group ${}_1N_l^A$ on a quasigroup $(Q, \cdot)$ is a subquasigroup of $(Q, \cdot)$, then $(K, \cdot)$ is an isotope of the group H.*

2. If an orbit K by the action of a subgroup H of the group ${}_3N_l^A$ on a quasigroup $(Q, \cdot)$ is a subquasigroup of $(Q, \cdot)$, then $(K, \cdot)$ is an isotope of the group H.

3. If an orbit K by the action of a subgroup H of the group ${}_2N_r^A$ on a quasigroup $(Q, \cdot)$ is a subquasigroup of $(Q, \cdot)$, then $(K, \cdot)$ is an isotope of the group H.

4. If an orbit K by the action of a subgroup H of the group ${}_3N_r^A$ on a quasigroup $(Q, \cdot)$ is a subquasigroup of $(Q, \cdot)$, then $(K, \cdot)$ is an isotope of the group H.

5. If an orbit K by the action of a subgroup H of the group ${}_1N_m^A$ on a quasigroup $(Q, \cdot)$ is a subquasigroup of $(Q, \cdot)$, then $(K, \cdot)$ is an isotope of the group H.

6. If an orbit K by the action of a subgroup H of the group ${}_2N_m^A$ on a quasigroup $(Q, \cdot)$ is a subquasigroup of $(Q, \cdot)$, then $(K, \cdot)$ is an isotope of the group H.

Proof. Sometimes in the proof we shall denote by the symbol $*$ operation of composition of permutations (bijections) of the set Q.

Case 1. Let $k \in K$. By Lemma 4.60 for any quasigroup $(Q, \cdot)$ there exists its isotopic image $(Q, \star) = (Q, \cdot)(\varepsilon, \varepsilon, R_k)$ such that: ${}_1N_l^A(Q, \star) = {}_3N_l^A(Q, \star) = {}_1N_l^A(Q, \cdot)$. Notice that by isotopy of the form $(k \in K)$, subquasigroup $(K, \cdot)$ passes to subquasigroup $(K, \star)$ of quasigroup $(Q, \star)$.

Moreover, in the right loop $(Q, \star)$ any autotopy of the form $(\alpha, \varepsilon, \gamma)$ of quasigroup $(Q, \cdot)$ takes the form $(\alpha, \varepsilon, \alpha)$. Therefore $H \subseteq {}_1N_l^A(Q, \star)$.

We follow [72, Theorem 2.2]. In this part of the proof we "are" in a right loop $(Q, \star)$.

It is possible to present any element $l \in K$ in the form $l = \delta k$, where $\delta \in H \subseteq {}_1N_l^A$. Notice, $k = \varepsilon k$. If $l = \delta k$, $r = \lambda k$, then $\delta k \star \lambda k \in K$ since $(K, \star)$ is a subquasigroup of quasigroup $(Q, \star)$. Thus there exists $\mu \in H$ such that $\delta k \star \lambda k = \mu k$ since $Hk = K$.

On the set H we can define operation $\circ$ in the following way: $\delta \circ \lambda = \mu$ if and only if $\delta k \star \lambda k = \mu k$.

Prove that

$$(H, \circ) \cong (K, \star). \tag{4.16}$$

Define the map φ in the following way: $\varphi : \lambda \mapsto \lambda k$, $\varphi(\delta \circ \lambda) = \varphi(\delta) \star \varphi(\lambda) = \delta k \star \lambda k$. The map φ is bijective, since action of the group of permutations $H \subseteq {}_1N_l^A$ on the set K is regular (i.e., it is simply transitive). Therefore $(H, \circ) \cong (K, \star)$.

Let $\alpha, \lambda, \mu \in H$. Notice, the corresponding permutation to the permutation α from the set ${}_3N_l^A(Q, \star)$ is also a permutation α (Lemma 4.60). From the definition of the set K we have $\alpha K = K$. Indeed $\alpha K = \alpha(Hk) = (\alpha H)k = Hk = K$.

Then restriction of the action of the triple $(\alpha, \varepsilon, \alpha) \in N_l^A(Q, \star)$ on subquasigroup $(K, \star) \subseteq (Q, \star)$ is an autotopy of subquasigroup $(K, \star)$. We have $\alpha((\lambda \circ \mu)k) = \alpha(\lambda k \star \mu k) = (\alpha\lambda)k \star \mu k = ((\alpha\lambda) \circ \mu)k$.

Therefore we obtain

$$\alpha(\lambda \circ \mu) = (\alpha\lambda) \circ \mu. \tag{4.17}$$

If we put $\lambda = \varepsilon$ in the equality (4.17), then we obtain $\alpha(\varepsilon \circ \mu) = \alpha \circ \mu$.

Since $(H, \circ)$ is a quasigroup, $\varepsilon \in H$, then the mapping L°_{ε}, $L^{\circ}_{\varepsilon}\mu = \varepsilon \circ \mu$ for all $\mu \in H$ is a permutation of the set H. Then

$$\alpha * L^{\circ}_{\varepsilon}\mu = \alpha \circ \mu, \tag{4.18}$$

where $*$ is an operation of the group H, i.e., it is a composition of bijections of the set Q.

Then from the equality (4.18) we conclude that quasigroup $(H, \circ)$ is isotopic to the group $(H, *)$, i.e., $(H, \circ) \sim (H, *)$, $(H, \circ) = (H, *)(\varepsilon, L^{\circ}_{\varepsilon}, \varepsilon)$.

Further we have

$$(H, *) \sim (H, \circ) \cong (K, \star) \sim (K, \cdot). \tag{4.19}$$

Case 2. The proof of Case 2 is similar to the proof of Case 1 and we omit it.

Case 3. In the proof of Case 3 instead of the equality (4.17) we obtain the following equality

$$\beta * (\lambda \circ \mu) = \lambda \circ (\beta * \mu), \tag{4.20}$$

where $\beta, \lambda, \mu \in H \subseteq {}_2N^A_r$. If we put $\mu = \varepsilon$, then we have $\beta * \lambda' = \lambda \circ \beta$. Notice, since $(H, *)$ is a group, then its (12)-parastrophe $(H, \overset{(12)}{*})$ is also a group. Therefore quasigroup $(H, \circ)$ is isotopic to a group.

Case 4. The proof of this case is similar to the proof of Case 3.

Case 5. In this case the equality (4.17) is transformed to the following equality:

$$(\alpha * \lambda) \circ (\alpha * \mu) = \lambda \circ \mu. \tag{4.21}$$

In this case $H \subseteq {}_1N^A_m$. If we put $\mu = \varepsilon$ in the equality (4.21), then we have $(\alpha * \lambda) \circ \alpha = \lambda'$, and using the operation of left division of quasigroup $(H, \circ)$, further we obtain $\lambda' / \alpha = \alpha * \lambda$.

If we take the (12)-parastrophe of the operation $/$, then we obtain $\alpha \overset{(12)}{/} \lambda' = \alpha * \lambda$. In fact the operation $\overset{(12)}{/}$ is the (132)-parastrophe of the operation $\circ$.

Therefore the quasigroup $(H, \overset{(132)}{\circ})$ is a group isotope. Then the quasigroup $(H, \circ)$ is also a group isotope as a parastrophe of a group isotope (Lemma 4.8).

Case 6. The proof of this case is similar to the proof of Case 5. □

From Theorem 4.82 we obtain the following:

Theorem 4.83. *If an orbit K by the action of the group ${}_1N^A_l$ (${}_3N^A_l$, ${}_2N^A_r$, ${}_3N^A_r$, ${}_1N^A_m$, ${}_2N^A_m$) on a quasigroup $(Q, \cdot)$ is a subquasigroup of $(Q, \cdot)$, then $(K, \cdot)$ is an isotope of the group ${}_1N^A_l$ (${}_3N^A_l$, ${}_2N^A_r$, ${}_3N^A_r$, ${}_1N^A_m$, ${}_2N^A_m$), respectively.*

Proof. This theorem is the partial case of Theorem 4.82. In Case 1 $H = {}_1N^A_l$ and so on. □

Theorem 4.84. *The group ${}_1N^A_l$ (${}_3N^A_l$, ${}_2N^A_r$, ${}_3N^A_r$, ${}_1N^A_m$, ${}_2N^A_m$) of a quasigroup $(Q, \cdot)$ acts simply transitively on the set Q if and only if the quasigroup $(Q, \cdot)$ is a group isotope.*

Proof. If the group ${}_1N^A_l$ of a quasigroup $(Q, \cdot)$ acts on the set Q simply transitively, then by Theorem 4.83 $(Q, \cdot) \sim {}_1N^A_l$.

Let $(Q, \cdot) \sim (Q, +)$. Prove that in this case the group ${}_1N^A_l(Q, \cdot)$ acts simply transitively on the set Q. Any element of the left A-nucleus of the group $(Q, +)$ has the form $(L^+_a, \varepsilon, L^+_a)$ for all $a \in Q$. The group ${}_1N^A_l(Q, +)$ acts simply transitively on the set Q.

If $(Q, \cdot) = (Q, +)(\alpha, \beta, \gamma)$, then ${}_1N^A_l(Q, \cdot) = \alpha^{-1}\, {}_1N^A_l(Q, +)\alpha$ (Lemma 4.1).

Let $a, b \in Q$. Prove that there exists permutation $\psi \in {}_1N^A_l(Q, \cdot)$ such that $\psi a = b$. We can write permutation ψ in the form $\alpha^{-1}L^+_x\alpha$. Since the group ${}_1N^A_l(Q, +)$ is transitive on the set Q, we can take element x such that $L^+_x\alpha a = \alpha b$. Then $\psi a = \alpha^{-1}L^+_x\alpha a = \alpha^{-1}\alpha b = b$.

The fact that action of the group ${}_1N^A_l$ on the set Q is semiregular follows from Theorem 4.69.

Other cases are proved in a similar way. □

Definition 4.85. A quasigroup $(Q, \cdot)$ with transitive action on the set Q of at least one of its components of the A-nuclei will be called an *A-nuclear quasigroup.*

Definition 4.85 is a generalization of corresponding definition from [72].

Corollary 4.86. *If at least one component of a quasigroup A-nucleus is transitive, then all components of quasigroup A-nuclei are transitive.*

Proof. The proof follows from Theorem 4.84. □

We can reformulate Theorem 4.84 in the following form.

Theorem 4.87. *A quasigroup is A-nuclear if and only if it is a group isotope.*

We give a slightly different proof of Lemma 4.78.

Lemma 4.88. *If the order of a finite quasigroup $(Q, \cdot)$ is equal to the order of the group ${}_1N_l^A$, then the orders of groups N_m^A, N_r^A, N_l^A, ${}_3N_l^A$, ${}_1N_m^A$, ${}_2N_m^A$, ${}_2N_r^A$ and ${}_3N_r^A$ are equal to $|Q|$.*

Proof. By Theorem 4.69 the group N_l^A acts freely (semiregularly) on the set Q. From the condition of the lemma it follows that the group N_l^A acts regularly (simply transitively) on the set Q. Further we can apply Theorem 4.84. □

4.2.9 Identities with permutation and group isotopes

In this subsection we study identities with permutations [878, 119] using an A-nuclear approach. The author thanks Prof. A.A. Gvaramiya who drew his attention to these kinds of identities.

Definition 4.89. Let (Q, Ω) be an algebra. We shall name an identity of algebra (Q, Ω) with permutations of the set Q incorporated in this identity as an identity with permutations (permutation identity) [489].

Identities with permutations (with additional unary operations) are usual objects in quasigroup theory. See Section 3.2. We recall that in any LIP-quasigroup $(Q, \cdot)$ the following permutation identity $\lambda x(xy) = y$ holds true [72, 685].

Identities with permutations (or with the sets of permutations) can be obtained from "usual" quasigroup identities by rewriting these identities using quasigroup translations for some variables and fixation (or non-fixation) of these variables. Notice, any permutation of the set Q can be viewed as special kind of unary operation.

Permutation identities in explicit or implicit form have been researched in works of V.D. Belousov, G.B. Belyavskaya, A.A. Gvaramiya, A.D. Keedwell, A. Krapez, F.N. Sokhatsky, A.Kh. Tabarov, and many other mathematicians who study quasigroup identities.

Here we use the A-nuclei approach and identities with permutations to obtain conditions when a quasigroup is a group isotope.

The conditions when a quasigroup is a group isotope are studied in classical article of V.D. Belousov [69]. Functional equations on quasigroups are studied in [63, 540, 829, 539, 542]. Linearity, one-sided linearity, anti-linearity, and one-sided anti-linearity of group and abelian group isotopes is studied in the articles of V.D. Belousov [69], T. Kepka and P. Nemec [667, 506], G.B. Belyavskaya and A.Kh. Tabarov [133, 131, 134, 110, 874, 876], F.M. Sokhatskii [828], J.D.H. Smith [818], and many other mathematicians.

We give the following based on the Theorem 4.87 procedure to answer the following question: is a quasigroup with an identity a group isotope?

Procedure 4.1. 1. If we can write a permutation identity of a quasigroup $(Q, \cdot)$ in the form $Ax \cdot y = \Gamma(x \cdot y)$, where $A, \Gamma \subseteq S_Q$, and we can prove that the set A or the set Γ acts transitively on the set Q, then the quasigroup $(Q, \cdot)$ is a group isotope.

2. If we can write a permutation identity of a quasigroup $(Q, \cdot)$ in the form $x \cdot By = \Gamma(x \cdot y)$, where $B, \Gamma \subseteq S_Q$, and we can prove that the set B or the set Γ acts transitively on the set Q, then the quasigroup $(Q, \cdot)$ is a group isotope.

3. If we can write a permutation identity of a quasigroup $(Q, \cdot)$ in the form $Ax \cdot By = x \cdot y$, where $A, B \subseteq S_Q$ and we can prove that the set A or the set B acts transitively on the set Q, then the quasigroup $(Q, \cdot)$ is a group isotope.

A procedure similar to Procedure 4.1 is given in [825, 826, 827].

Remark 4.90. Procedure 4.1 shows that it is possible to generalize Definition 4.89 changing the words "fixed permutations of the set Q" to the words "fixed sets of permutations of the set Q". Notice, identities with sets of permutations are close to partial identities [399, 400].

Lemma 4.91. *Belousov criteria. If in a group $(Q, +)$ the equality $\alpha x + \beta y = \gamma y + \delta x$ holds for all $x, y \in Q$, where $\alpha, \beta, \gamma, \delta$ are some fixed permutations of Q, then $(Q, +)$ is an abelian group [69].*

Proof. From equality $\alpha x + \beta y = \gamma y + \delta x$ we have

$$\alpha\delta^{-1}x + \beta\gamma^{-1}y = y + x. \tag{4.22}$$

If we put $x = 0$ in the equality (4.22), then $\beta\gamma^{-1} = L_k^+$, where $k = -\alpha\delta^{-1}0$.

If we put $y = 0$ in the equality (4.22), then $\alpha\delta^{-1} = R_d^+$, where $d = -\beta\gamma^{-1}0$.

We can rewrite the equality (4.22) in the form

$$R_d^+ x + L_k^+ y = y + x. \tag{4.23}$$

If we put $x = y = 0$ in the equality (4.23), then $d + k = 0$ and the equality (4.23) takes the form $x + y = y + x$. □

There also exists the following corollary from results of F.N. Sokhatskii [830, Theorem 6.7.2]. Recall, up to isomorphism any isotope is principal.

Corollary 4.92. *If in a principal group isotope $(Q, \cdot)$ of a group $(Q, +)$ the equality $\alpha x \cdot \beta y = \gamma y \cdot \delta x$ holds for all $x, y \in Q$, where $\alpha, \beta, \gamma, \delta$ are some fixed permutations of Q, then $(Q, +)$ is an abelian group.*

Proof. If $x \cdot y = \xi x + \chi y$, then we can re-write the equality $\alpha x \cdot \beta y = \gamma y \cdot \delta x$ in the form $\xi\alpha x + \chi\beta y = \xi\gamma y + \chi\delta x$. Now we can apply the Belousov criteria (Lemma 4.91). □

Lemma 4.93. *A quasigroup $(Q, \cdot)$ with identity*

$$\alpha_1 x \cdot \alpha_2(y \cdot z) = y \cdot \alpha_3(x \cdot \alpha_4 z), \tag{4.24}$$

where $\alpha_1, \alpha_2, \alpha_3, \alpha_4$ are some fixed permutations of the set Q, is an abelian group isotope.

Proof. We can re-write the identity (4.24) in the form $L_{\alpha_1 x}\alpha_2(y \cdot z) = y \cdot \alpha_3 L_x \alpha_4 z$. Since $(Q, \cdot)$ is a quasigroup, then for any fixed elements $a, b \in Q$ there exists a unique element t such that $ta = b$, i.e., $L_t a = b$. Then the set of translations $\{L_x \mid x \in Q\}$ acts on the set Q transitively. Therefore the set of permutations of the form $\alpha_3 L_x \alpha_4 \subseteq {}_2N_r^A$ also acts transitively on Q. By Procedure 4.1 the quasigroup $(Q, \cdot)$ is a group isotope.

We can re-write the identity (4.24) in the form $\alpha_1 x \cdot \alpha_2 R_z y = y \cdot \alpha_3 R_{\alpha_4 z} x$. From Corollary 4.92 it follows that the quasigroup $(Q, \cdot)$ is an abelian group isotope. □

Definition 4.94. Let $(Q, \cdot)$ be a groupoid. The identity of the form

$$\alpha_1(\alpha_2 x \cdot \alpha_3 y) \cdot \alpha_4(\alpha_5 u \cdot \alpha_6 v) = \alpha_7(x \cdot u) \cdot \alpha_8(y \cdot v), \tag{4.25}$$

where $\alpha_1, \dots, \alpha_8$ are permutations of the set Q, we shall call a *permutation medial identity.*

The identity of the form

$$\alpha_1(\alpha_2 x \cdot \alpha_3 y) \cdot \alpha_4(\alpha_5 u \cdot \alpha_6 v) = \alpha_7(v \cdot y) \cdot \alpha_8(u \cdot x), \tag{4.26}$$

where $\alpha_1, \dots, \alpha_8$ are permutations of the set Q, we shall call a *permutation paramedial identity.*

Permutation medial identity and permutation paramedial identity are partial cases of a functional equation of mediality (generalized bisymmetry). For binary quasigroups, the functional equation of generalized bisymmetry is solved in [4]; for quasigroups of arity $n > 2$ this equation is solved in [78].

Theorem 4.95. *1. Permutation medial quasigroup $(Q, \cdot)$ is an abelian group isotope.*

2. Permutation paramedial quasigroup $(Q, \cdot)$ is an abelian group isotope.

Proof. Case 1. Using the language of translations we can rewrite the identity (4.25) in the form $\beta_1 x \cdot \beta_2 v = \beta_3 x \cdot \beta_4 v$, where $\beta_1 = \alpha_1 R_{\alpha_3 y}\alpha_2$, $\beta_2 = \alpha_4 L_{\alpha_5 u}\alpha_6$, $\beta_3 = \alpha_7 R_u$, $\beta_4 = \alpha_8 L_y$.

Then $\beta_1\beta_3^{-1} \subseteq {}_1N_m^A$ and the set of permutations of the form $\beta_1\beta_3^{-1}$ acts on the set Q transitively (we can take $u = a$, where the element a is a fixed element of the set Q). Therefore, by Procedure 4.1, a medial quasigroup is a group isotope.

We can write the identity (4.25) in the form $\gamma_1 y \cdot \gamma_2 u = \gamma_3 u \cdot \gamma_4 y$, where $\gamma_1 = \alpha_1 L_{\alpha_2 x}\alpha_3$, $\gamma_2 = \alpha_4 R_{\alpha_6 v}\alpha_5$, $\gamma_3 = \alpha_7 L_x$, $\gamma_4 = \alpha_8 R_v$. From Corollary 4.92 it follows that any permutation medial quasigroup is an isotope of an abelian group.

Case 2 is proved similar to Case 1. □

Corollary 4.96. *Medial quasigroup $(Q, \cdot)$ is an abelian group isotope [72, p. 33]. Paramedial quasigroup $(Q, \circ)$ is an abelian group isotope [667].*

Proof. The proof follows from Theorem 4.95. □

Lemma 4.97. *If a quasigroup $(Q, \cdot)$ satisfies the identity of associativity with permutations*

$$\alpha_1(\alpha_2(\alpha_3 x \cdot \alpha_4 y) \cdot \alpha_5 z) = \alpha_6 x \cdot \alpha_7(\alpha_8 y \cdot \alpha_9 z) \tag{4.27}$$

and $\alpha_3 = \alpha_6$, then this quasigroup is a group isotope.

Proof. The identity (4.27) can be rewritten in the form:

$$\alpha_1(\alpha_2(x \cdot y) \cdot z) = x \cdot \alpha_3(\alpha_4 y \cdot \alpha_5 z), \tag{4.28}$$

and further in the following form:

$$\alpha_1\alpha_2 R_z(x \cdot y) = x \cdot \alpha_3 R_{\alpha_5 z}\alpha_4 y. \tag{4.29}$$

Then by Procedure 4.1 Case 2, quasigroup $(Q, \cdot)$ with equality $\alpha_1\alpha_2 R_z(x \cdot y) = x \cdot \alpha_3 R_{\alpha_5 z}\alpha_4 y$ is a group isotope. □

In [115], using the famous four quasigroups theorem [4, 77], the more general result is proved:

Theorem 4.98. *If a quasigroup $(Q, \cdot)$ satisfies the identity (4.27), then this quasigroup is a group isotope.*

In [119] the following results are proved.

Theorem 4.99. *A quasigroup $(Q, \cdot)$ is isotopic to a group if and only if for all $x, y, z \in Q$ the following identity with permutations is true: $R_a^{-1}(x \cdot L_b^{-1}y) \cdot z = x \cdot L_b^{-1}(R_a^{-1}y \cdot z)$, where $L_a x = a \cdot x$, $R_b y = y \cdot b$, elements a, b are some fixed elements of the set Q.*

A quasigroup $(Q, \cdot)$ is isotopic to an abelian group if and only if $R_a^{-1}(y{\cdot}x){\cdot}z = R_a^{-1}(y{\cdot}z){\cdot}x$ for all $x, y, z \in Q$ and some fixed element $a \in Q$ [119].

We give the proof of the following well-known theorem [173], [72, p.175].

Theorem 4.100. *If a G-loop is commutative, then it is an abelian group.*

Proof. Recall that a G-loop is a loop that is isomorphic to all its LP-isotopes (Definition 1.167). Let $(Q, \cdot)$ be a commutative G-loop. Then its LP-isotope also is commutative. Therefore we obtain the following equality:

$$R_a^{-1}x \cdot L_b^{-1}y = R_a^{-1}y \cdot L_b^{-1}x, \tag{4.30}$$

which is true for any fixed $a, b \in Q$ and all $x, y \in Q$. From the equality (4.30) we have

$$R_a^{-1}x \cdot L_b^{-1}y = L_b^{-1}x \cdot R_a^{-1}y, R_a^{-1}L_b x \cdot L_b^{-1}R_a y = x \cdot y. \tag{4.31}$$

It is clear that the set of permutations $\{R_a^{-1}L_b \mid a, b \in Q\}$ acts on the set Q transitively. Notice that it is possible to put $a = 1$. Further, using Procedure 4.1, Case 3, we conclude that the loop $(Q, \cdot)$ is a group isotope, i.e., it is a group. Moreover, it is an abelian group. $\square$

4.3 A-centers of a quasigroup

4.3.1 Normality of A-nuclei and autotopy group

We recall the definition of center in a loop $(Q, \cdot)$ (Definition 1.213). Let $(Q, \cdot)$ be a loop. Then center Z of the loop $(Q, \cdot)$ is the following set $Z = N \cap C$, where $C = \{a \in Q \mid a \cdot x = x \cdot a \; \forall x \in Q\}$ [10, 173, 685]. In the loop case, $(Z, \cdot)$ is a normal abelian subgroup of the loop $(Q, \cdot)$ [173].

Isotopic quasigroups have isomorphic autotopy groups. See Lemma 4.1 or [72]. It is well known that any quasigroup $(Q, \cdot)$ that is an isotope of a group Q has the following structure of its autotopy group [580, 769]:

$$Avt(Q, \cdot) \cong (Q \times Q) \rtimes Aut(Q).$$

Lemma 4.101. *1. If $\mu \in {}_3N_l^A$, $\nu \in {}_3N_r^A$, then $\mu\nu = \nu\mu$.*

2. If $\mu \in {}_1N_l^A$, $\nu \in {}_1N_m^A$, then $\mu\nu = \nu\mu$.

3. If $\mu \in {}_2N_r^A$, $\nu \in {}_2N_m^A$, then $\mu\nu = \nu\mu$.

Proof. Case 1. Let $(\delta, \varepsilon, \lambda) \in N_l^A$, $(\varepsilon, \mu, \psi) \in N_r^A$. Here the sign "$\cdot$" denotes the operation of multiplication of triplets of the group $Avt(Q)$. Any element of the set $N_l^A \cdot N_r^A$ has the form $(\delta, \mu, \lambda\psi)$ and any element of the set $N_r^A \cdot N_l^A$ has the form $(\delta, \mu, \psi\lambda)$. By Lemma 1.137 any two autotopy components define the third component in a unique way, therefore $\psi\lambda = \lambda\psi$ in our case.

Cases 2 and 3 are proved in the similar way. □

Corollary 4.102. *In any quasigroup:* ${}_1N_l^A \cdot {}_1N_m^A = {}_1N_m^A \cdot {}_1N_l^A$; ${}_2N_r^A \cdot {}_2N_m^A = {}_2N_m^A \cdot {}_2N_r^A$; ${}_3N_r^A \cdot {}_3N_l^A = {}_3N_l^A \cdot {}_3N_r^A$.

Proof. The proof follows from Lemma 4.101. □

Lemma 4.103. *In any quasigroup Q we have:*

$$ {}_1N_l^A \subseteq C_{S_Q}({}_1N_m^A) \cap M(Q);\ {}_1N_m^A \subseteq C_{S_Q}({}_1N_l^A) \cap M(Q); $$

$$ {}_2N_r^A \subseteq C_{S_Q}({}_2N_m^A) \cap M(Q);\ {}_2N_m^A \subseteq C_{S_Q}({}_2N_l^A) \cap M(Q); $$

$$ {}_3N_r^A \subseteq C_{S_Q}({}_3N_l^A) \cap M(Q);\ {}_3N_l^A \subseteq C_{S_Q}({}_3N_r^A) \cap M(Q). $$

Proof. The fact that ${}_1N_l^A \subseteq C_{S_Q}({}_1N_m^A)$ follows from Lemma 4.101. But the group ${}_1N_l^A$ of a quasigroup Q is a subgroup of the group $M(Q)$ (Corollary 4.48). Other cases are proved in a similar way. □

Corollary 4.104. *In any quasigroup Q we have:*

$$ {}_1N_l^A \subseteq C_{M(Q)}({}_1N_m^A);\ {}_1N_m^A \subseteq C_{M(Q)}({}_1N_l^A);\ {}_2N_r^A \subseteq C_{M(Q)}({}_2N_m^A); $$

$$ {}_2N_m^A \subseteq C_{M(Q)}({}_2N_r^A);\ {}_3N_r^A \subseteq C_{M(Q)}({}_3N_l^A);\ {}_3N_l^A \subseteq C_{M(Q)}({}_3N_r^A). $$

Proof. The proof follows from Lemma 4.103. □

Theorem 4.105. *1. If in a quasigroup Q ${}_1N_m^A \trianglelefteq M(Q)$ and ${}_1N_l^A = C_{M(Q)}({}_1N_m^A)$, then ${}_1N_l^A \trianglelefteq M(Q)$.*

2. If in a quasigroup Q ${}_1N_l^A \trianglelefteq M(Q)$ and ${}_1N_m^A = C_{M(Q)}({}_1N_l^A)$, then ${}_1N_m^A \trianglelefteq M(Q)$.

3. If in a quasigroup Q ${}_2N_r^A \trianglelefteq M(Q)$ and ${}_2N_m^A = C_{M(Q)}({}_2N_r^A)$, then ${}_2N_m^A \trianglelefteq M(Q)$.

4. If in a quasigroup Q ${}_2N_m^A \trianglelefteq M(Q)$ and ${}_2N_r^A = C_{M(Q)}({}_2N_m^A)$, then ${}_2N_r^A \trianglelefteq M(Q)$.

5. If in a quasigroup Q ${}_3N_r^A \trianglelefteq M(Q)$ and ${}_3N_l^A = C_{M(Q)}({}_3N_r^A)$, then ${}_3N_l^A \trianglelefteq M(Q)$.

6. If in a quasigroup Q ${}_3N_l^A \trianglelefteq M(Q)$ and ${}_3N_r^A = C_{M(Q)}({}_3N_l^A)$, then ${}_3N_r^A \trianglelefteq M(Q)$.

Proof. Case 1. The proof follows from the NC-theorem (Theorem 4.32), since in this case the normalizer of the group ${}_1N_m^A$ in the group $M(Q)$ is equal to $M(Q)$.

Cases 2–6 are proved in a similar way. □

Lemma 4.106. *In any quasigroup Q:*

1. $N_l^A \cdot N_r^A \cong N_l^A \times N_r^A$;
2. $N_l^A \cdot N_m^A \cong N_l^A \times N_m^A$;
3. $N_r^A \cdot N_m^A \cong N_r^A \times N_m^A$.

Proof. Case 1. Notice, that N_l^A, N_r^A are subgroups of the group $Avt(Q)$. We demonstrate that $N_l^A \cap N_r^A = (\varepsilon, \varepsilon, \varepsilon)$. Any element of the group $N_l^A \cap N_r^A$ has the form $(\varepsilon, \varepsilon, \gamma)$. By Corollary 1.138 $\gamma = \varepsilon$ in this case.

The fact that $N_l^A \cdot N_r^A = N_r^A \cdot N_l^A$ follows from Lemma 4.101. Thus $N_l^A \cdot N_r^A = N_l^A \times N_r^A$. Cases 2 and 3 are proved in a similar way. □

Theorem 4.107. *For any loop Q we have:*

1. $(N_l^A \times N_r^A) \rightthreetimes Aut(Q) \cong H \subseteq Avt(Q)$;
2. $(N_l^A \times N_m^A) \rightthreetimes Aut(Q) \cong H \subseteq Avt(Q)$;
3. $(N_m^A \times N_r^A) \rightthreetimes Aut(Q) \cong H \subseteq Avt(Q)$.

Proof. Case 1. It is clear that the groups N_l^A, N_r^A and $Aut(Q)$ are the subgroup of the group $Avt(Q)$.

Further proof is sufficiently standard for group theory [471]. Let $H = N_l^A \cdot N_r^A \cdot Aut(Q)$. By Lemma 4.106 $N_l^A \cdot N_r^A \cong N_l^A \times N_r^A$.

Prove that $N_l^A \times N_r^A \trianglelefteq Avt(Q)$. It is easy to see that $N_l^A \trianglelefteq Avt(Q)$, $N_r^A \trianglelefteq Avt(Q)$. Indeed, we have

$$(\alpha^{-1}, \beta^{-1}, \gamma^{-1})(\delta_1, \varepsilon, \lambda_1)(\alpha, \beta, \gamma) = (\delta_2, \varepsilon, \lambda_2),$$

where $(\alpha, \beta, \gamma) \in Avt(Q)$, $(\delta_1, \varepsilon, \lambda_1), (\delta_2, \varepsilon, \lambda_2) \in N_l^A$.

Similarly

$$(\alpha^{-1}, \beta^{-1}, \gamma^{-1})(\varepsilon, \mu_1, \psi_1)(\alpha, \beta, \gamma) = (\varepsilon, \mu_2, \psi_2).$$

Then $N_l^A \trianglelefteq H$, $N_r^A \trianglelefteq H$.

Further we have

$$\begin{aligned}
&(\alpha^{-1}, \beta^{-1}, \gamma^{-1})(\delta, \mu, \lambda\psi)(\alpha, \beta, \gamma) = \\
&= (\alpha^{-1}, \beta^{-1}, \gamma^{-1})(\delta, \varepsilon, \lambda)(\varepsilon, \mu, \psi)(\alpha, \beta, \gamma) = \\
&= (\alpha^{-1}, \beta^{-1}, \gamma^{-1})(\delta, \varepsilon, \lambda)(\alpha, \beta, \gamma)(\alpha^{-1}, \beta^{-1}, \gamma^{-1})(\varepsilon, \mu, \psi)(\alpha, \beta, \gamma) = \\
&= (\delta_1, \varepsilon, \lambda_1)(\varepsilon, \mu_1, \psi_1) = (\delta_1, \mu_1, \lambda_1\psi_1) \in N_l^A \times N_r^A.
\end{aligned} \tag{4.32}$$

Therefore $N_l^A \times N_r^A \trianglelefteq Avt(Q)$ in any quasigroup Q.

Prove that $(N_l^A \times N_r^A) \cap Aut(Q) = (\varepsilon, \varepsilon, \varepsilon)$. In this place we shall use the fact that Q is a loop. In a loop, any element of the group $(N_l^A \times N_r^A)$ has the form (L_a, R_b, R_bL_a). Since any automorphism is an autotopy with equal components, we have $L_a = R_bL_a$, $R_b = \varepsilon$, $R_b = R_bL_a$, $L_a = \varepsilon$.

Therefore we have that $H \cong (N_l^A \times N_r^A) \rightthreetimes Aut(Q)$.

Cases 2 and 3 are proved similar to Case 1. □

Remark 4.108. The fact that $N_l^A \cap N_r^A = (\varepsilon, \varepsilon, \varepsilon)$ demonstrates that there is a difference between Garrison left nucleus N_l and left A-nucleus N_l^A, between Garrison right nucleus N_r and right A-nucleus N_r^A, and so on.

Remark 4.109. Theorem 4.107 is true for any quasigroup $(Q, \cdot)$ but by isotopy $((Q, \cdot) \sim (Q, +))$, in general, the group $Aut(Q, +)$ passes to a subgroup of the group $Avt(Q, \cdot)$ [769].

Corollary 4.110. *For any quasigroup Q we have*

$$(N_l^A \times N_r^A) \trianglelefteq Avt(Q),$$

$$(N_l^A \times N_m^A) \trianglelefteq Avt(Q),$$

$$(N_m^A \times N_r^A) \trianglelefteq Avt(Q).$$

Proof. The proof follows from the proof of Theorem 4.107 (equality (4.32)). □

Theorem 4.111. *In any quasigroup, an autotopy group is isomorphic to the following semidirect products*

$$Avt \cong N_r^A \rightthreetimes Avt_1 \cong N_l^A \rightthreetimes Avt_2 \cong N_m^A \rightthreetimes Avt_3, \tag{4.33}$$

where Avt_i is the group of the i-th autotopy components.

Proof. Indeed any autotopy (α, β, γ) can be decomposed in the following way:

$$\begin{array}{l}(\varepsilon, \beta\alpha^{-1}, \gamma\alpha^{-1})(\alpha, \alpha, \alpha) = \\ (\alpha\beta^{-1}, \varepsilon, \gamma\beta^{-1})(\beta, \beta, \beta) = \\ (\alpha\gamma^{-1}, \beta\gamma^{-1}, \varepsilon)(\gamma, \gamma, \gamma).\end{array}$$

Further proof is sufficiently standard and we omit it. □

The analog of Theorem 4.111 for loops and the usual middle nucleus is proved in [905]. In order to prove Wedel's result it is also possible to use Lemma 1.140, Theorem 1.249 and Lemma 1.247.

Example 4.112. Loop $(Q, \cdot)$ has an identity automorphism group $Aut(Q)$ [802].

$\cdot$	1	2	3	4	5
1	1	2	3	4	5
2	2	3	1	5	4
3	3	4	5	1	2
4	4	5	2	3	1
5	5	1	4	2	3

This loop is not a group and it is of prime order. Therefore by Corollary 4.72, all A-nuclei (and "usual" nuclei, too) of loop $(Q, \cdot)$ are identity groups. Then the groups $(N_l^A \times N_r^A) \rightthreetimes Aut(Q)$, $(N_l^A \times N_m^A) \rightthreetimes Aut(Q)$, and $(N_r^A \times N_r^A) \rightthreetimes Aut(Q)$ are also identity groups. From Theorem 4.111 it follows that in loop $(Q, \cdot)$ $Avt \cong Avt_1 \cong Avt_2 \cong Avt_3$.

The autotopy group $Avt(Q, \cdot)$ consists of the following triplets:

$$\begin{array}{ll}(\varepsilon, \varepsilon, \varepsilon), & ((3\,5\,4), (1\,2\,3), (1\,2\,3)), \\ ((1\,3\,4), (2\,5\,3), (1\,3\,4)), & ((3\,4\,5), (1\,3\,2), (1\,3\,2)), \\ ((1\,5\,4), (1\,2\,5), (2\,3\,4)), & ((1\,5)(3\,4), (1\,5)(2\,3), (1\,3)(2\,4)), \\ ((1\,3)(4\,5), (1\,2)(3\,5), (1\,4)(2\,3)), & ((1\,5\,3), (1\,3\,5), (1\,4\,2)), \\ ((1\,4\,5), (1\,5\,2), (2\,4\,3)), & ((1\,4\,3), (2\,3\,5), (1\,4\,3)), \\ ((1\,3\,5), (1\,5\,3), (1\,2\,4)), & ((1\,4)(3\,5), (1\,3)(2\,5), (1\,2)(3\,4)),\end{array}$$

and is isomorphic to the alternating group A_4 [407].

Notice, triples $(\varepsilon, \varepsilon, \varepsilon)$, $((3\,5\,4), (1\,2\,3), (1\,2\,3))$, and $((3\,4\,5), (1\,3\,2), (1\,3\,2))$ form the group of right pseudo-automorphisms of loop $(Q, \cdot)$, triples $(\varepsilon, \varepsilon, \varepsilon)$, $((1\,3\,4), (2\,5\,3), (1\,3\,4))$, and $((1\,4\,3), (2\,3\,5), (1\,4\,3))$ form the group of left pseudo-automorphisms. We recall that $|A_4| = 12$ and this number is the divisor of the number $5! \cdot 5 = 600$.

4.3.2 A-centers of a loop

Definition 4.113. Let Q be a loop.

An autotopy of the form $\{(L_a, \varepsilon, L_a) \mid a \in Z(Q)\}$ we shall name the left central autotopy. A group of all left central autotopies we shall denote by Z_l^A.

An autotopy of the form $\{(\varepsilon, L_a, L_a) \mid a \in Z(Q)\}$ we shall name the right central autotopy. A group of all right central autotopies we shall denote by Z_r^A.

An autotopy of the form $\{(L_a, L_a^{-1}, \varepsilon) \mid a \in Z(Q)\}$ we shall name the middle central autotopy. A group of all middle central autotopies we shall denote by Z_m^A.

Lemma 4.114. *In any loop the groups Z_l^A, Z_r^A, Z_m^A, ${}_1Z_l^A$, ${}_3Z_l^A$, ${}_2Z_r^A$, ${}_3Z_r^A$, ${}_1Z_m^A$, ${}_2Z_m^A$, and Z are isomorphic. In more detail $Z \cong Z_l^A \cong Z_r^A \cong Z_m^A \cong {}_1Z_l^A = {}_3Z_l^A = {}_2Z_r^A = {}_3Z_r^A = {}_1Z_m^A = {}_2Z_m^A$.*

Proof. The proof follows from the definition of the center of a loop (Definition 1.213) and Definition 4.113. The map $\xi : a \mapsto L_a$, where $a \in Z$, gives necessary isomorphism of the group Z and ${}_1Z_l^A$, and so on. □

Lemma 4.115. *If Q is a loop, then ${}_1Z_l^A1 = {}_3Z_l^A1 = {}_2Z_r^A1 = {}_3Z_r^A1 = {}_1Z_m^A1 = {}_2Z_m^A1 = Z$.*

Proof. It is possible to use Definition 4.113. □

Corollary 4.116. *In any loop Q the orbit ${}_1Z_l^A1$ (${}_3Z_l^A1$, ${}_2Z_r^A1$, ${}_3Z_r^A1$, ${}_1Z_m^A1$, ${}_2Z_m^A1$) coincides with the set Q if and only if Q is an abelian group.*

Theorem 4.117. *Let Q be a loop. Then*

1. $(N_l^A \times N_r^A) \cap N_m^A = \{(L_a, L_a^{-1}, \varepsilon) \mid a \in Z(Q)\} = Z_m^A$;
2. $(N_l^A \times N_m^A) \cap N_r^A = \{(\varepsilon, L_a, L_a) \mid a \in Z(Q)\} = Z_r^A$;
3. $(N_r^A \times N_m^A) \cap N_l^A = \{(L_a, \varepsilon, L_a) \mid a \in Z(Q)\} = Z_l^A$.
4. *The groups Z_l^A, Z_r^A and Z_m^A are the abelian subgroups of the group $Avt(Q)$.*

Proof. Case 1. Recall that the groups N_l^A, N_r^A are subgroups of the group $Avt(Q)$. Then any element of the group $N_l^A \times N_r^A$ has the form

$$(L_a, R_b, L_aR_b), \tag{4.34}$$

any element of the group N_m^A has the form

$$(R_c, L_c^{-1}, \varepsilon) \tag{4.35}$$

(Theorem 4.45). Then $R_bL_a = \varepsilon$, $R_b = L_a^{-1}$. Expression (4.34) takes the form $(L_a, L_a^{-1}, L_aL_a^{-1}) = (L_a, L_a^{-1}, \varepsilon)$. Comparing it with the expression (4.35) we have that $L_a = R_a$, i.e., $ax = xa$ for any element $x \in Q$. In addition, taking into consideration that $(L_a, L_a^{-1}, \varepsilon) \in N_m^A, (L_a, \varepsilon, L_a) \in N_l^A, (\varepsilon, L_a^{-1}, L_a^{-1}) \in N_r^A$, we obtain that the element a is central, i.e., $a \in Z(Q)$.

Therefore $(N_l^A \times N_r^A) \cap N_m^A \subseteq \{(L_a, L_a^{-1}, \varepsilon) \mid a \in Z(Q)\}$.

Converse. It is easy to see that any element from the center $Z(Q)$ of a loop Q "generates" autotopies of the left, right, and middle A-nuclei of the loop Q, i.e., $\{(L_a, L_a^{-1}, \varepsilon) \mid a \in Z(Q)\} \subseteq (N_l^A \times N_r^A) \cap N_m^A$.

And finally we have $(N_l^A \times N_r^A) \cap N_m^A = \{(L_a, L_a^{-1}, \varepsilon) \mid a \in Z(Q)\}$.

Case 2. The proof of Case 2 is similar to the proof of Case 1 and we omit some details. Any element of the group $N_l^A \times N_m^A$ has the form

$$(L_aR_b, L_b^{-1}, L_a), \tag{4.36}$$

and any element of the group N_r^A has the form

$$(\varepsilon, R_c, R_c) \tag{4.37}$$

(Theorem 4.45). Since the expressions (4.36) and (4.37) in the intersection should be componentwise equal, we have the following system of equations:

$$\begin{cases} L_a R_b = \varepsilon \\ L_b^{-1} = R_c \\ L_a = R_c. \end{cases}$$

We remember that in any loop for any nuclear translation we have $L_a^{-1} = L_{a^{-1}}$, $R_a^{-1} = R_{a^{-1}}$ (Theorem 4.45).

From the equality $L_b^{-1} = L_a$ we have that $a = b^{-1}$ and $b = a^{-1}$. Then from the equality $L_a R_b = \varepsilon$ it follows that $L_b = R_b$ for all elements of the group $(N_l^A \times N_m^A) \cap N_r^A$.

Expression (4.36) takes the form $(L_a L_a^{-1}, L_a, L_a) = (\varepsilon, L_a, L_a)$. Expression (4.35) takes the same form.

Taking into consideration that $L_a = R_a$, $(L_a^{-1}, L_a, \varepsilon) \in N_m^A, (L_a, \varepsilon, L_a) \in N_l^A, (\varepsilon, L_a, L_a) \in N_r^A$, we obtain that the element a is central, i.e., $a \in Z(Q)$.

Therefore $(N_l^A \times N_r^A) \cap N_m^A \subseteq \{(L_a, L_a^{-1}, \varepsilon) \mid a \in Z(Q)\}$.

Converse inclusion is proved similar to the proof in Case 1.

Finally we have $(N_l^A \times N_m^A) \cap N_r^A = \{(\varepsilon, L_a, L_a) \mid a \in Z(Q)\}$.

Case 3. The proof of Case 3 is similar to the proof of Cases 1 and 2. Any element of the group $N_r^A \times N_m^A$ has the form

$$(R_b, R_a L_b^{-1}, R_a), \tag{4.38}$$

and any element of the group N_l^A has the form

$$(L_c, \varepsilon, L_c) \tag{4.39}$$

(Theorem 4.45). Since the expressions (4.38) and (4.39) in the intersection should be equal componentwise, we have the following system of equations:

$$\begin{cases} R_b = L_c \\ R_a L_b^{-1} = \varepsilon \\ R_a = L_c. \end{cases}$$

From the equality $R_b = R_a$ we have that $a = b$. Then from the equality $R_a L_b^{-1} = \varepsilon$ it follows that $L_b = R_b$ for all elements of the group $(N_r^A \times N_m^A) \cap N_l^A$.

Expression (4.38) takes the form $(L_a, L_a L_a^{-1}, L_a) = (L_a, \varepsilon, L_a)$.

Taking into consideration that $L_a = R_a$, $(L_a, L_a^{-1}, \varepsilon) \in N_m^A, (L_a, \varepsilon, L_a) \in N_l^A, (\varepsilon, L_a, L_a) \in N_r^A$, we obtain that the element a is central, i.e., $a \in Z(Q)$.

Therefore $(N_r^A \times N_m^A) \cap N_l^A = \{(L_a, \varepsilon, L_a) \mid a \in Z(Q)\}$.

Case 4. The proof follows from Lemma 4.114. Indeed, in this case, if we denote the loop operation by $+$, then we have $L_a L_b = L_{a+b} = L_{b+a} = L_b L_a$. □

Theorem 4.118. *Let Q be a loop, $K = N_l^A \cdot N_r^A \cdot N_m^A$. Then*

(i) $K/Z_l^A \cong (N_r^A \times N_m^A)/Z_l^A \times N_l^A/Z_l^A$;

(ii) $K/Z_r^A \cong (N_l^A \times N_m^A)/Z_r^A \times N_r^A/Z_r^A$;

(iii) $K/Z_m^A \cong (N_l^A \times N_r^A)/Z_m^A \times N_m^A/Z_m^A$.

Proof. Case (i). The groups N_l^A, N_r^A and N_m^A are normal subgroups of the group $Avt(Q)$. From Theorem 4.107 we have that the groups N_l^A, N_r^A and N_m^A are intersected in pairs by identity group.

From Lemma 4.37 and Corollary 4.110 we have that $N_l^A \trianglelefteq Avt(Q)$, $(N_r^A \times N_m^A) \trianglelefteq Avt(Q)$. Then $N_l^A \trianglelefteq K$, $(N_r^A \times N_m^A) \trianglelefteq K$.

From Theorem 4.117 and Theorem 4.2.1 [471, p. 47] we have that groups $(N_r^A \times N_m^A)/Z_l^A$ and N_l^A/Z_l^A are normal subgroups of the group K/Z_l^A that are intersected by the identity subgroup. Therefore $K/Z_l^A \cong (N_r^A \times N_m^A)/Z_l^A \times N_l^A/Z_l^A$.

Cases (ii) and (iii) are proved in a similar way. □

4.3.3 A-centers of a quasigroup

Definition 4.119. Let Q be a quasigroup.

(i) The group $Z_l^A = (N_r^A \cdot N_m^A) \cap N_l^A$ is the left A-center of Q.

(ii) The group $Z_r^A = (N_l^A \cdot N_m^A) \cap N_r^A$ is the right A-center of Q.

(iii) The group $Z_m^A = (N_l^A \cdot N_r^A) \cap N_m^A$ is the middle A-center of Q.

Lemma 4.120. *(i)* $Z_l^A \cong (N_r^A \times N_m^A) \cap N_l^A$;

(ii) $Z_r^A \cong (N_l^A \times N_m^A) \cap N_r^A$;

(iii) $Z_m^A \cong (N_l^A \times N_r^A) \cap N_m^A$.

Proof. The proof follows from Lemma 4.106. □

Theorem 4.121. *In any quasigroup* $Z_l^A \trianglelefteq Avt(Q), Z_r^A \trianglelefteq Avt(Q), Z_m^A \trianglelefteq Avt(Q)$.

Proof. The proof follows from Corollary 4.110, Lemma 4.37, and the fact that the intersection of normal subgroups of a group is a normal subgroup. □

Lemma 4.122. *Left, right, and middle A-centers and their components of isotopic quasigroups $(Q, \circ)$ and $(Q, \cdot)$ are isomorphic, i.e., if $(Q, \circ) \sim (Q, \cdot)$, then*

1. $Z_l^A(Q,\cdot) \cong Z_l^A(Q,\circ)$; 2. ${}_1Z_l^A(Q,\cdot) \cong {}_1Z_l^A(Q,\circ)$; 3. ${}_3Z_l^A(Q,\cdot) \cong {}_3Z_l^A(Q,\circ)$;
4. $Z_r^A(Q,\cdot) \cong Z_r^A(Q,\circ)$; 5. ${}_2Z_r^A(Q,\cdot) \cong {}_2Z_r^A(Q,\circ)$; 6. ${}_3Z_r^A(Q,\cdot) \cong {}_3Z_r^A(Q,\circ)$;
7. $Z_l^A(Q,\cdot) \cong Z_l^A(Q,\circ)$; 8. ${}_1Z_m^A(Q,\cdot) \cong {}_1Z_m^A(Q,\circ)$; 9. ${}_2Z_m^A(Q,\cdot) \cong {}_2Z_m^A(Q,\circ)$.

Proof. Case 1. Suppose that $(Q, \circ) = (Q, \cdot)T$, where T is an isotopy. By Lemma 4.1 $Avt(Q, \circ) = T^{-1}Avt(Q, \cdot)T$.

Notice, $T^{-1}Z_l^A(Q,\cdot)T = T^{-1}((N_r^A(Q,\cdot) \times N_m^A(Q,\cdot)) \cap N_l^A(Q,\cdot))T$. Briefly, below we shall omit denotation of quasigroups $(Q, \cdot)$ and $(Q, \circ)$.

Denote $(N_r^A \times N_m^A)$ by A, N_l^A by B, and the conjugation isomorphism by the letter φ. Prove that $\varphi(A \cap B) = \varphi A \cap \varphi B$.

Let $\varphi x \in \varphi(A \cap B)$. Then $x \in A \cap B$ since φ is a bijective map. If $x \in A \cap B$, then $x \in A$, $x \in B$, $\varphi(x) \in \varphi(A)$, $\varphi(x) \in \varphi(B)$, $\varphi(x) \in \varphi(A) \cap \varphi(B)$, $\varphi(A \cap B) \subseteq \varphi A \cap \varphi B$.

Let $\varphi x \in \varphi A \cap \varphi B$. Then $\varphi x \in \varphi A$, $\varphi x \in \varphi B$. Since φ is a bijective map, then $x \in A$, $x \in B$, $x \in A \cap B$, $\varphi x \in \varphi(A \cap B)$, $\varphi A \cap \varphi B \subseteq \varphi(A \cap B)$.

Finally, $\varphi A \cap \varphi B = \varphi(A \cap B)$, i.e., $T^{-1}((N_r^A \times N_m^A) \cap N_l^A)T = T^{-1}(N_r^A \times N_m^A)T \cap T^{-1}N_l^AT$.

The fact that $T^{-1}(N_r^A \times N_m^A)T = T^{-1}N_r^AT \times T^{-1}N_m^AT$ is well known. Indeed, in the group of all triplets $S_Q \times S_Q \times S_Q$ an isomorphic image of the direct product of two subgroups is the direct product of their isomorphic images. Also it is possible to prove this fact using the equality (4.32).

We have the following chain of equalities:

$$T^{-1}Z_l^A(Q,\cdot)T = T^{-1}((N_r^A \times N_m^A) \cap N_l^A)T =$$
$$T^{-1}(N_r^A \times N_m^A)T \cap T^{-1}N_l^AT = ((T^{-1}N_r^AT) \times (T^{-1}N_m^AT)) \cap T^{-1}N_l^AT =$$
$$(N_r^A(Q,\circ) \times N_m^A(Q,\circ)) \cap N_l^A(Q,\circ) = Z_l^A(Q,\circ).$$

Therefore $Z_l^A(Q,\cdot) \cong Z_l^A(Q,\circ)$.

Cases 2 and 3 follow from Case 1. Cases 4 and 7 are proved in a similar way to Case 1. Cases 5 and 6, and 8 and 9 follow from Cases 2 and 3, respectively. □

Corollary 4.123. *In any quasigroup $(Q,\cdot)$ the groups Z_l^A, Z_r^A, Z_m^A, ${}_1Z_l^A$, ${}_3Z_l^A$, ${}_2Z_r^A$, ${}_3Z_r^A$, ${}_1Z_m^A$, ${}_2Z_m^A$ are the isomorphic abelian groups, i.e.,*

$$Z_l^A(Q,\cdot) \cong Z_r^A(Q,\cdot) \cong Z_m^A(Q,\cdot) \cong$$
$${}_1Z_l^A(Q,\cdot) \cong {}_3Z_l^A(Q,\cdot) \cong {}_2Z_r^A(Q,\cdot) \cong$$
$${}_3Z_r^A(Q,\cdot) \cong {}_1Z_m^A(Q,\cdot) \cong {}_2Z_m^A(Q,\cdot).$$

Proof. Suppose that quasigroup $(Q,\cdot)$ is isomorphic to a loop $(Q,\circ)$. Then by Lemma 4.122 $Z_l^A(Q,\cdot) \cong Z_l^A(Q,\circ)$; $Z_r^A(Q,\cdot) \cong Z_r^A(Q,\circ)$; $Z_m^A(Q,\cdot) \cong Z_m^A(Q,\circ)$, and so on.

By Lemma 4.114 the groups $Z_l^A(Q,\circ)$, ${}_1Z_l^A(Q,\circ)$, ${}_3Z_l^A(Q,\circ)$, $Z_r^A(Q,\circ)$, ${}_2Z_r^A(Q,\circ)$, ${}_3Z_r^A(Q,\circ)$, $Z_m^A(Q,\circ)$, ${}_1Z_m^A(Q,\circ)$, ${}_2Z_m^A(Q,\circ)$ are the isomorphic abelian groups. Therefore the same is true for A-centers of any quasigroup. □

Theorem 4.124. *If an orbit K by the action of a subgroup H of the group ${}_1Z_l^A(Q,\cdot)$ (${}_3Z_l^A(Q,\cdot)$, ${}_2Z_r^A(Q,\cdot)$, ${}_3Z_r^A(Q,\cdot)$, ${}_1Z_m^A(Q,\cdot)$, ${}_2Z_m^A(Q,\cdot)$) on the set Q is a subquasigroup of quasigroup $(Q,\cdot)$, then $(K,\cdot)$ is an isotope of abelian group H.*

Proof. From Corollary 4.123 it follows that the group ${}_1Z_l^A$ is abelian. Then any subgroup H of the group ${}_1Z_l^A$ is also abelian. The fact that $(K,\cdot)$ is a group isotope follows from the definition of left A-center Z_l^A and Theorem 4.82.

Other cases are proved in a similar way. □

Corollary 4.125. *If an orbit K by the action of group ${}_1Z_l^A(Q,\cdot)$ (${}_3Z_l^A(Q,\cdot)$, ${}_2Z_r^A(Q,\cdot)$, ${}_3Z_r^A(Q,\cdot)$, ${}_1Z_m^A(Q,\cdot)$, ${}_2Z_m^A(Q,\cdot)$) on the set Q is a subquasigroup of quasigroup $(Q,\cdot)$, then $(K,\cdot)$ is an isotope of abelian group ${}_1Z_l^A(Q,\cdot)$ (${}_3Z_l^A(Q,\cdot)$, ${}_2Z_r^A(Q,\cdot)$, ${}_3Z_r^A(Q,\cdot)$, ${}_1Z_m^A(Q,\cdot)$, ${}_2Z_m^A(Q,\cdot)$, respectively).*

Proof. This corollary is the partial case of Theorem 4.124. For example, in Case 1, $H = {}_1Z_l^A(Q,\cdot)$. □

Theorem 4.126. *The group ${}_1Z_l^A$ (${}_3Z_l^A$, ${}_2Z_r^A$, ${}_3Z_r^A$, ${}_1Z_m^A$, ${}_2Z_m^A$) of a quasigroup $(Q,\cdot)$ acts on the set Q simply transitively if and only if quasigroup $(Q,\cdot)$ is the abelian group isotope.*

Proof. If the group ${}_1Z_l^A$ of a quasigroup $(Q,\cdot)$ acts on the set Q transitively, then the orbit by the action of group ${}_1Z_l^A$ coincides with the set Q, and by Corollary 4.125, $(Q,\cdot) \sim {}_1Z_l^A$.

Let $(Q,\cdot) \sim (Q,+)$, where $(Q,+)$ is an abelian group. Prove that in this case the group ${}_1Z_l^A(Q,\cdot)$ acts on the set Q simply transitively. Any element of the left A-center of the group $(Q,+)$ has the form $(L_a^+, \varepsilon, L_a^+)$ for all $a \in Q$.

It is clear that group ${}_1Z_l^A(Q,+)$ acts on the set Q simply transitively.

If $(Q,\cdot) = (Q,+)(\alpha,\beta,\gamma)$, then ${}_1Z_l^A(Q,\cdot) = \alpha^{-1}\,{}_1Z_l^A(Q,+)\alpha$ (Corollary 4.2).

Prove that action of the group ${}_1Z_l^A(Q,\cdot)$ on the set Q is transitive. Let $a,b \in Q$. Prove that there exists permutation $\psi \in {}_1Z_l^A(Q,\cdot)$ such that $\psi a = b$. We can write permutation ψ in the form $\alpha^{-1}L_x^+\alpha$. Since the group ${}_1Z_l^A(Q,+)$ is transitive on the set Q, we can take element x such that $L_x^+\alpha a = \alpha b$. Then $\psi a = \alpha^{-1}L_x^+\alpha a = \alpha^{-1}\alpha b = b$.

The fact that action of the group ${}_1Z_l^A$ on the set Q has identity stabilizers for any element $x \in Q$ follows from Theorem 4.69 since in any abelian group ${}_1N_l^A = {}_1Z_l^A$.

Other cases are proved in a similar way. □

Corollary 4.127. *If at least one component of a quasigroup A-center is transitive, then all components of A-centers are transitive.*

Proof. The proof follows from Theorem 4.126. □

Definition 4.128. *An A-central quasigroup* is a quasigroup Q with transitive action of at least one of its components of A-centers on Q.

Using this definition we formulate the main theorem of this section:

Theorem 4.129. *A quasigroup is A-central if and only if it is an abelian group isotope.*

Proof. We can use Theorem 4.126. □

Corollary 4.130. *Any central quasigroup in the Belyavskaya-Smith sense is an A-central quasigroup.*

Proof. The proof follows from Theorems 1.242 and 4.87. Any T-quasigroup is an isotope of abelian group. □

It is easy to see that the converse is not true.

4.4 A-nuclei and quasigroup congruences

4.4.1 Normality of equivalences in quasigroups

Basic information on quasigroup congruences is given in Section 1.5. Here we give additional information on this topic.

Definition 4.131. An equivalence θ is a left congruence of a groupoid $(Q,\circ)$, if the following implication is true for all $x, y, z \in Q$: $x\,\theta\,y \Longrightarrow (z\circ x)\,\theta\,(z\circ y)$. In other words, equivalence θ is stable relative to any left translation of $(Q,\circ)$.

An equivalence θ is a right congruence of a groupoid $(Q,\circ)$ if the following implication is true for all $x, y, z \in Q$: $x\,\theta\,y \Longrightarrow (x\circ z)\,\theta\,(y\circ z)$. In other words, equivalence θ is stable relative to any right translation of $(Q,\circ)$ [212].

Definition 4.132. An equivalence θ of a groupoid $(Q,\circ)$ is called cancellative from the left, if the following implication is true for all $x, y, z \in Q$: $(z\circ x)\,\theta\,(z\circ y) \Longrightarrow x\,\theta\,y$.

An equivalence θ of a groupoid $(Q,\circ)$ is called cancellative from the right, if the following implication is true for all $x, y, z \in Q$: $(z\circ x)\,\theta\,(z\circ y) \Longrightarrow x\,\theta\,y, (x\circ z)\,\theta\,(y\circ z) \Longrightarrow x\,\theta\,y$ [72, 80].

An equivalence θ of a groupoid $(Q,\circ)$ is called normal from the right (right normal), if it is stable and cancellative from the right.

An equivalence θ of a groupoid $(Q,\circ)$ is called normal from the left (left normal), if it is stable and cancellative from the left.

An equivalence θ of a groupoid $(Q,\circ)$ is called normal, if it is left and right normal.

Remark 4.133. In [462, 463, 786] a somewhat different approach to normality of groupoid congruences is presented.

In the following lemma we give information on the behavior of a left quasigroup congruence relative to quasigroup parastrophy.

Lemma 4.134. *The following propositions are equivalent:*

1. *An equivalence θ is stable relative to translation L_c of a quasigroup $(Q, \circ)$.*
2. *An equivalence θ is stable relative to translation R_c of a quasigroup $(Q, \overset{(12)}{\circ})$.*
3. *An equivalence θ is stable relative to translation P_c^{-1} of a quasigroup $(Q, \overset{(13)}{\circ})$.*
4. *An equivalence θ is stable relative to translation L_c^{-1} of a quasigroup $(Q, \overset{(23)}{\circ})$.*
5. *An equivalence θ is stable relative to translation R_c^{-1} of a quasigroup $(Q, \overset{(123)}{\circ})$.*
6. *An equivalence θ is stable relative to translation P_c of a quasigroup $(Q, \overset{(132)}{\circ})$.*

Proof. From Table 1.1 it follows that $L_c^{(12)} = R_c$, $L_c^{(13)} = P_c^{-1}$, and so on. □

Remark 4.135. It is easy to see that similar equivalences (Lemma 4.134) are true for the other five kinds of translations and its combinations of a quasigroup $(Q, \cdot)$.

Lemma 4.136. 1. *If an equivalence θ of a quasigroup $(Q, \cdot)$ is admissible relative to any left and right translation of this quasigroup, then θ is a normal congruence.*

2. *If an equivalence θ of a quasigroup $(Q, \cdot)$ is admissible relative to any left and middle translation of this quasigroup, then θ is a normal congruence.*
3. *If an equivalence θ of a quasigroup $(Q, \cdot)$ is admissible relative to any right and middle translation of this quasigroup, then θ is a normal congruence.*

Proof. Case 1. Definition 1.262.

Case 2. Equivalence θ is a normal congruence of quasigroup $(Q, \backslash)$ (Table 1.1, Definition 1.262) and we apply Corollary 1.268.

Case 3. Equivalence θ is a normal congruence of quasigroup $(Q, /)$ (Table 1.1, Definition 1.262) and we apply Corollary 1.268. □

4.4.2 Additional conditions of normality of equivalences

We give additional conditions when an equivalence is a left, right or "middle" quasigroup congruence. We use the H. Thurston [885, 884] and A.I. Maltsev [589] approaches. Results similar to the results given in this section are in [782, 798].

The set of all left and right translations of a quasigroup $(Q, \cdot)$ will be denoted by $\mathbb{T}(Q, \cdot)$. If φ and ψ are binary relations on Q, then their product is defined in the following way: $(a, b) \in \varphi \circ \psi$ if there is an element $c \in Q$ such that $(a, c) \in \varphi$ and $(c, b) \in \psi$. If φ is a binary relation on Q, then $\varphi^{-1} = \{(y, x) \mid (x, y) \in \varphi\}$. The operation of the product of binary relations is associative [211, 678, 813, 711].

Remark 4.137. Translations of a quasigroup can be considered as binary relations: $(x, y) \in L_a$, if and only if $y = a \cdot x$; $(x, y) \in R_b$, if and only if $y = x \cdot b$; and $(x, y) \in P_c$, if and only if $y = x \backslash c$ [766, 767, 782].

Remark 4.138. To coordinate the multiplication of translations with their multiplication as binary relations, we use the following multiplication of translations: if α, β are translations, x is an element of the set Q, then $(\alpha\beta)(x) = \beta(\alpha(x))$.

Proposition 4.139. *1. An equivalence θ is a left congruence of a quasigroup $(Q, \cdot)$ if and only if $\theta\omega \subseteq \omega\theta$ for all $\omega \in \mathbb{L}$.*

2. An equivalence θ is a right congruence of a quasigroup $(Q, \cdot)$ if and only if $\theta\omega \subseteq \omega\theta$ for all $\omega \in \mathbb{R}$.

3. An equivalence θ is a "middle" congruence of a quasigroup $(Q, \cdot)$ if and only if $\theta\omega \subseteq \omega\theta$ for all $\omega \in \mathbb{P}$.

4. An equivalence θ is a congruence of a quasigroup $(Q, \cdot)$ if and only if $\theta\omega \subseteq \omega\theta$ for all $\omega \in \mathbb{T}$ [766, 767, 782].

Proof. Case 1. Let θ be an equivalence, $\omega = L_a$. It is clear that $(x, z) \in \theta L_a$ is equivalent to the following statement "there exists an element $y \in Q$ such that $(x, y) \in \theta$ and $(y, z) \in L_a$." But if $(y, z) \in L_a$, $z = ay$, then $y = L_a^{-1}z$. Therefore, from the relation $(x, z) \in \theta L_a$ it follows that $(x, L_a^{-1}z) \in \theta$.

Let us prove that from $(x, L_a^{-1}z) \in \theta$ it follows $(x, z) \in \theta L_a$. We have $(x, L_a^{-1}z) \in \theta$ and $(L_a^{-1}z, z) \in L_a$, $(x, z) \in \theta L_a$. Thus $(x, z) \in \theta L_a$ is equivalent to $(x, L_a^{-1}z) \in \theta$.

Similarly, $(x, z) \in L_a\theta$ is equivalent to $(ax, z) \in \theta$. Now we can say that the inclusion $\theta\omega \subseteq \omega\theta$ by $\omega = L_a$ is equivalent to the following implication:

$$(x, L_a^{-1}z) \in \theta \Longrightarrow (ax, z) \in \theta$$

for all suitable $a, x, z \in Q$.

If in the last implication we replace z with $L_a z$, we shall obtain the implication:

$$(x, z) \in \theta \Longrightarrow (ax, az) \in \theta$$

for all $a \in Q$.

Thus, the inclusion $\theta L_a \subseteq L_a\theta$ is equivalent to the stability of the relation θ from the left relative to an element a. Since the element a is an arbitrary element of the set Q, we have that the inclusion $\theta\omega \subseteq \omega\theta$ by $\omega \in \mathbb{L}$ is equivalent to the stability of the relation θ from the left.

Case 2. Similarly, the inclusion $\theta\omega \subseteq \omega\theta$ for any $\omega \in \mathbb{R}$ is equivalent to the stability from the right of relation θ.

Case 3. Applying Case 1 to the quasigroup $(Q, //) = (Q, \cdot)^{(132)}$ we obtain that an equivalence θ is a left congruence of a quasigroup $(Q, \cdot)^{(132)}$ if and only if $\theta\omega \subseteq \omega\theta$ for all $\omega \in \mathbb{L}^{(132)}$.

Using parastrophic equivalence of translations (Table 1.1 or Lemma 4.134) we conclude that an equivalence θ is stable relative to any middle translation of a quasigroup $(Q, \cdot)$ if and only if $\theta\omega \subseteq \omega\theta$ for all $\omega \in \mathbb{P}$.

In other words, an equivalence θ is a "middle" congruence of a quasigroup $(Q, \cdot)$ if and only if $\theta\omega \subseteq \omega\theta$ for all $\omega \in \mathbb{P}$.

Case 4. Uniting Cases 1 and 2 we obtain the required equivalence. □

Let us remark that Proposition 4.139 can be deduced from the results of H. Thurston [884].

Proposition 4.140. *1. An equivalence θ is a left-cancellative congruence of a quasigroup $(Q, \cdot)$ if and only if $\omega\theta \subseteq \theta\omega$ for all $\omega \in \mathbb{L}$.*

2. *An equivalence θ is a right-cancellative congruence of a quasigroup $(Q, \cdot)$ if and only if $\omega\theta \subseteq \theta\omega$ for all $\omega \in \mathbb{R}$.*

3. *An equivalence θ is a middle-cancellative congruence of a quasigroup $(Q, \cdot)$ if and only if $\omega\theta \subseteq \theta\omega$ for all $\omega \in \mathbb{P}$.*

Proof. As it is proved in Proposition 4.139, the inclusion $\theta L_a \subseteq L_a\theta$ is equivalent to the implication $x\theta y \Longrightarrow ax\theta ay$.

Let us check up that the inclusion $L_a\theta \subseteq \theta L_a$ is equivalent to the implication

$$ax\theta ay \Rightarrow x\theta y.$$

Indeed, as it is proved in Proposition 4.139, $(x, z) \in \theta L_a$ is equivalent to $(x, L_a^{-1}z) \in \theta$. Similarly, $(x, z) \in L_a\theta$ is equivalent to $(ax, z) \in \theta$. The inclusion $\omega\theta \subseteq \theta\omega$ by $\omega = L_a$ has the form $L_a\theta \subseteq \theta L_a$ and it is equivalent to the following implication:

$$(ax, z) \in \theta \Longrightarrow (x, L_a^{-1}z) \in \theta$$

for all $a, x, z \in Q$. If in the last implication we change the element z to the element $L_a z$, we shall obtain that the inclusion $\theta L_a \supseteq L_a\theta$ is equivalent to the implication $ax\theta ay \Rightarrow x\theta y$. Therefore, the equivalence θ is cancellative from the left.

Similarly, the inclusion $R_b\theta \subseteq \theta R_b$ is equivalent to the implication:

$$(xa, za) \in \theta \Longrightarrow (x, z) \in \theta.$$

Case 3 is proved similarly. It is possible to use the identity (1.8). □

Below we shall denote by the symbol $\langle \mathbb{L}, \mathbb{P} \rangle$ the group generated by all left and middle translations of a quasigroup.

Corollary 4.141. 1. *An equivalence θ is a left normal congruence of a quasigroup $(Q, \cdot)$ if and only if $\omega\theta = \theta\omega$ for all $\omega \in \mathbb{L}$.*

2. *An equivalence θ is a right normal congruence of a quasigroup $(Q, \cdot)$ if and only if $\omega\theta = \theta\omega$ for all $\omega \in \mathbb{R}$.*

3. *An equivalence θ is a middle normal congruence of a quasigroup $(Q, \cdot)$ if and only if $\omega\theta = \theta\omega$ for all $\omega \in \mathbb{P}$.*

4. *An equivalence θ is a normal congruence of a quasigroup $(Q, \cdot)$ if and only if $\omega\theta = \theta\omega$ for all $\omega \in \mathbb{L} \cup \mathbb{P}$.*

5. *An equivalence θ is a normal congruence of a quasigroup $(Q, \cdot)$ if and only if $\omega\theta = \theta\omega$ for all $\omega \in \mathbb{R} \cup \mathbb{P}$.*

6. *An equivalence θ is a normal congruence of a quasigroup $(Q, \cdot)$ if and only if $\omega\theta = \theta\omega$ for all $\omega \in \langle \mathbb{L}, \mathbb{R} \rangle = M(Q, \cdot)$.*

7. *An equivalence θ is a normal congruence of a quasigroup $(Q, \cdot)$ if and only if $\omega\theta = \theta\omega$ for all $\omega \in \langle \mathbb{L}, \mathbb{P} \rangle$.*

8. *An equivalence θ is a normal congruence of a quasigroup $(Q, \cdot)$ if and only if $\omega\theta = \theta\omega$ for all $\omega \in \langle \mathbb{R}, \mathbb{P} \rangle$.*

Proof. The proof follows from Propositions 4.139, 4.140 and Lemma 4.136.

In the proving of Cases 6–8 we can use the following fact: if $\omega\theta = \theta\omega$, then $\omega^{-1}\theta = \theta\omega^{-1}$. □

The following proposition is an almost obvious corollary of Theorem 5 from [589].

Proposition 4.142. *1. An equivalence θ is a left congruence of a quasigroup Q if and only if $\omega\theta(x) \subseteq \theta(\omega x)$ for all $x \in Q$, $\omega \in \mathbb{L}$.*

2. An equivalence θ is a right congruence of a quasigroup Q if and only if $\omega\theta(x) \subseteq \theta(\omega x)$ for all $x \in Q$, $\omega \in \mathbb{R}$.

3. An equivalence θ is a "middle" congruence of a quasigroup Q if and only if $\omega\theta(x) \subseteq \theta(\omega x)$ for all $x \in Q$, $\omega \in \mathbb{P}$.

4. An equivalence θ is a left-cancellative congruence of a quasigroup Q if and only if $\theta(\omega x) \subseteq \omega\theta(x)$ for all $x \in Q$, $\omega \in \mathbb{L}$.

5. An equivalence θ is a right-cancellative congruence of a quasigroup Q if and only if $\theta(\omega x) \subseteq \omega\theta(x)$ for all $x \in Q$, $\omega \in \mathbb{R}$.

6. An equivalence θ is a middle-cancellative congruence (i.e., $P_x a\, \theta\, P_x b \Rightarrow a\theta b$) of a quasigroup Q if and only if $\theta(\omega x) \subseteq \omega\theta(x)$ for all $x \in Q$, $\omega \in \mathbb{P}$.

Proof. Case 1. Let θ be an equivalence relation and for all $\omega \in \mathbb{L}$, $\omega\theta(x) \subseteq \theta(\omega x)$. We shall prove that from $a\theta b$, $ca\theta cb$ follows for all $c \in Q$.

By definition of the equivalence θ, $a\theta b$ is equivalent to $a \in \theta(b)$. Then $ca \in c\theta(b) \subseteq \theta(cb)$, $ca\theta cb$.

Converse. Let θ be a left congruence. We shall prove that $c\,\theta(a) \subseteq \theta(ca)$ for all $c, a \in Q$. Let $x \in c\,\theta(a)$. Then $x = cy$, where $y \in \theta(a)$, that is, $y\theta a$. Then, since θ is a left congruence, we obtain $cy\theta ca$. Therefore $x = cy \in \theta(ca)$. Thus, $L_c\theta \subseteq \theta(ca)$.

Case 2 is proved similarly.

Case 3. We can use an approach similar to the approach used in the proof of Case 4 of Proposition 4.139.

Cases 4–6 are proved in a similar way to Cases 1–3. □

Proposition 4.143. *1. An equivalence θ of a quasigroup Q is a left normal if and only if $\theta(\omega x) = \omega\theta(x)$ for all $\omega \in \mathbb{L}$, $x \in Q$;*

2. A equivalence θ of a quasigroup Q is a right normal if and only if $\theta(\omega x) = \omega\theta(x)$ for all $\omega \in \mathbb{R}$, $x \in Q$.

3. An equivalence θ of a quasigroup Q is a middle normal if and only if $\theta(\omega x) = \omega\theta(x)$ for all $\omega \in \mathbb{P}$, $x \in Q$.

Proof. The proof follows from Proposition 4.142. □

Proposition 4.144. *1. An equivalence θ is a normal congruence of a quasigroup Q if and only if $\theta(\omega x) = \omega\theta(x)$ for all $\omega \in \mathbb{L} \cup \mathbb{R}$, $x \in Q$.*

2. An equivalence θ is a normal congruence of a quasigroup Q if and only if $\theta(\omega x) = \omega\theta(x)$ for all $\omega \in \mathbb{L} \cup \mathbb{P}$, $x \in Q$.

3. An equivalence θ is a normal congruence of a quasigroup Q if and only if $\theta(\omega x) = \omega\theta(x)$ for all $\omega \in \mathbb{R} \cup \mathbb{P}$, $x \in Q$.

4. An equivalence θ is a normal congruence of a quasigroup Q if and only if $\theta(\omega x) = \omega\theta(x)$ for all $\omega \in \langle \mathbb{L}, \mathbb{R} \rangle$, $x \in Q$.

5. An equivalence θ is a normal congruence of a quasigroup Q if and only if $\theta(\omega x) = \omega\theta(x)$ for all $\omega \in \langle \mathbb{L}, \mathbb{P} \rangle$, $x \in Q$.

6. *An equivalence θ is a normal congruence of a quasigroup Q if and only if $\theta(\omega x) = \omega\theta(x)$ for all $\omega \in \langle \mathbb{R}, \mathbb{P}\rangle$, $x \in Q$.*

Proof. The proof follows from Proposition 4.143 and Lemma 4.136.

In proving Cases 4–6 we can use the following fact. If $\theta(\omega x) = \omega\theta(x)$, then $\theta(\omega\omega^{-1}x) = \omega\theta(\omega^{-1}x) = \omega\omega^{-1}\theta(x)$. Thus $\omega\theta(\omega^{-1}x) = \omega\omega^{-1}\theta(x)$, $\theta(\omega^{-1}x) = \omega^{-1}\theta(x)$. □

4.4.3 A-nuclei and quasigroup congruences

Definition 4.145. We define the following binary relation on a quasigroup $(Q, \cdot)$ that corresponds to a subgroup H of the group ${}_1N_l^A$: $a\,({}_1\theta_l^A)\,b$ if and only if there exists a permutation $\alpha \in H \subseteq {}_1N_l^A$ such that $b = \alpha a$.

For a subgroup H of the group ${}_3N_l^A$ (${}_1N_m^A$, ${}_2N_m^A$, ${}_2N_r^A$ and ${}_3N_r^A$) binary relation ${}_3\theta_l^A$ (${}_1\theta_m^A$, ${}_2\theta_m^A$, ${}_2\theta_r^A$ and ${}_3\theta_r^A$, respectively) is defined in a similar way.

Definition 4.145 is similar to the Albert definition [10] which was given for a loop and subgroups of its multiplication group.

Lemma 4.146. *Binary relation ${}_1\theta_l^A$ is an equivalence of the set Q.*

Proof. Binary relation ${}_1\theta_l^A$ is defined using orbits by the action of a subgroup of the group ${}_1N_l^A$ on the set Q. It is clear that any two orbits either do not intersect, or coincide, therefore we obtain a partition of the set Q. Then we conclude that this binary relation is an equivalence (Theorem 1.3). Notice, all these equivalence classes are of equal order (Corollary 4.70). □

Analogs of Lemma 4.146 are true for any subgroup of the groups ${}_3N_l^A$, ${}_1N_m^A$, ${}_2N_m^A$, ${}_2N_r^A$ and ${}_3N_r^A$.

These equivalences have additional properties.

Lemma 4.147. *Let $(Q, \cdot)$ be a quasigroup.*

1. *Let ${}_1\theta_l^A$ be an equivalence that is defined by a subgroup K of the group ${}_1N_l^A$ and equivalence ${}_3\theta_l^A$ is defined by isomorphic to K subgroup $R_xKR_x^{-1}$ of the group ${}_3N_l^A$ (Lemma 4.53). Then we have the following implications:*
   (a) *if $a({}_1\theta_l^A)b$, then $(a \cdot c)({}_3\theta_l^A)(b \cdot c)$ for all $a, b, c \in Q$;*
   (b) *if $(a \cdot c)({}_1\theta_l^A)(b \cdot c)$, then $a({}_3\theta_l^A)b$ for all $a, b, c \in Q$.*
2. *Let ${}_3\theta_l^A$ be an equivalence that is defined by a subgroup K of the group ${}_3N_l^A$ and equivalence ${}_1\theta_l^A$ is defined by isomorphic to K subgroup $R_x^{-1}KR_x$ of the group ${}_1N_l^A$ (Lemma 4.53). Then we have the following implications:*
   (a) *if $a({}_3\theta_l^A)b$, then $(a \cdot c)({}_1\theta_l^A)(b \cdot c)$ for all $a, b, c \in Q$;*
   (b) *if $(a \cdot c)({}_3\theta_l^A)(b \cdot c)$, then $a({}_1\theta_l^A)b$ for all $a, b, c \in Q$.*
3. *Let ${}_2\theta_r^A$ be an equivalence that is defined by a subgroup K of the group ${}_2N_r^A$ and equivalence ${}_3\theta_r^A$ is defined by isomorphic to K subgroup $L_xKL_x^{-1}$ of the group ${}_3N_r^A$ (Lemma 4.53). Then we have the following implications:*
   (a) *if $a({}_2\theta_r^A)b$, then $(c \cdot a)({}_3\theta_r^A)(c \cdot b)$ for all $a, b, c \in Q$;*
   (b) *if $(c \cdot a)({}_2\theta_r^A)(c \cdot b)$, then $a({}_3\theta_r^A)b$ for all $a, b, c \in Q$.*

4. *Let ${}_3\theta_r^A$ be an equivalence that is defined by a subgroup K of the group ${}_3N_r^A$ and equivalence ${}_2\theta_r^A$ is defined by isomorphic to K subgroup $L_x^{-1}KL_x$ of the group ${}_3N_r^A$ (Lemma 4.53). Then we have the following implications:*

   (a) if $a({}_3\theta_r^A)b$, then $(c \cdot a)({}_2\theta_r^A)(c \cdot b)$ for all $a, b, c \in Q$;

   (b) if $(c \cdot a)({}_3\theta_r^A)(c \cdot b)$, then $a({}_2\theta_r^A)b$ for all $a, b, c \in Q$.

5. *Let ${}_1\theta_m^A$ be an equivalence that is defined by a subgroup K of the group ${}_1N_m^A$ and equivalence ${}_2\theta_m^A$ is defined by isomorphic to K subgroup $P_xKP_x^{-1}$ of the group ${}_2N_m^A$ (Lemma 4.53). Then we have the following implications:*

   (a) if $a({}_1\theta_m^A)b$, then $P_ca({}_2\theta_m^A)P_cb$ for all $a, b, c \in Q$;

   (b) if $P_ca({}_1\theta_m^A)P_cb$, then $a({}_2\theta_m^A)b$ for all $a, b, c \in Q$.

6. *Let ${}_2\theta_m^A$ be an equivalence that is defined by a subgroup K of the group ${}_2N_m^A$ and equivalence ${}_1\theta_m^A$ is defined by isomorphic to K subgroup $P_x^{-1}KP_x$ of the group ${}_1N_m^A$ (Lemma 4.53). Then we have the following implications:*

   (a) if $a({}_2\theta_m^A)b$, then $P_c^{-1}a({}_1\theta_m^A)P_c^{-1}b$ for all $a, b, c \in Q$;

   (b) if $P_c^{-1}a({}_2\theta_m^A)P_c^{-1}b$, then $a({}_1\theta_m^A)b$ for all $a, b, c \in Q$.

Proof. Case 1, (a). Expression $a({}_1\theta_l^A)b$ means that there exists a permutation $\alpha \in K \subseteq {}_1N_l^A$ such that $b = \alpha a$. Expression $(a \cdot c)({}_3\theta_l^A)(b \cdot c)$ means that there exists a permutation $\gamma \in R_xKR_x^{-1} \subseteq {}_3N_l^A$ such that $b \cdot c = \gamma(a \cdot c)$. We can take $\gamma = R_c\alpha R_c^{-1}$, i.e., we can take $x = c$. Then $b \cdot c = \alpha a \cdot c$.

Then we obtain the following implication: $b = \alpha a \Longrightarrow b \cdot c = \alpha a \cdot c$ for all $a, b, c \in Q$. It is clear that this implication is true for all $a, b, c \in Q$, since $(Q, \cdot)$ is a quasigroup. Therefore, implication

$$\text{if } a({}_1\theta_l^A)b, \text{ then } (a \cdot c)({}_3\theta_l^A)(b \cdot c) \text{ for all } a, b, c \in Q$$

is also true.

Case 1, (b). Expression $(a \cdot c)({}_1\theta_l^A)(b \cdot c)$ means that there exists a permutation $\alpha \in K \subseteq {}_1N_l^A$ such that $b \cdot c = \alpha(a \cdot c)$, i.e., $b = R_c^{-1}\alpha R_c a$. Expression $a({}_3\theta_l^A)b$ means that there exists a permutation $\gamma \in H = R_xKR_x^{-1} \subseteq {}_3N_l^A$ such that $b = \gamma a$. By Lemma 4.53 we can take $\gamma = R_c^{-1}\alpha R_c$.

Then we obtain the following implication:

$$b = R_c^{-1}\alpha R_c a \Longrightarrow b = R_c^{-1}\alpha R_c a \text{ for all } a, b, c \in Q.$$

It is clear that this implication is true.

Cases 2–4 are proved in a similar way to Case 1.

Case 5, (a). Expression $a({}_1\theta_m^A)b$ means that there exists a permutation $\alpha \in K \subseteq {}_1N_m^A$ such that $b = \alpha a$. Expression $P_ca({}_2\theta_m^A)P_cb$ means that there exists a permutation $\beta \in H \subseteq {}_2N_m^A$ such that $P_cb = \beta P_ca$. By Lemma 4.53 we can take $\beta = P_c\alpha P_c^{-1}$. Then $P_cb = \beta P_ca = P_c\alpha P_c^{-1}P_ca = P_c\alpha a$.

Then we obtain the following implication $b = \alpha a \Longrightarrow P_cb = P_c\alpha a$ for all $a, b, c \in Q$. It is clear that this implication is true for all $a, b, c \in Q$.

Case 5, (b). Expression $P_ca({}_1\theta_m^A)P_cb$ means that there exists a permutation $\alpha \in H \subseteq {}_1N_m^A$ such that $P_ca = \alpha P_cb$. By Lemma 4.53 we can take $\alpha = P_c\beta P_c^{-1}$, where $\beta \in {}_2N_m^A$. Thus $P_ca = P_c\beta P_c^{-1}P_cb$, $a = \beta b$, $a({}_2\theta_m^A)b$.

Case 6 is proved in a similar way to Case 5. □

Cases 1 and 2 show that the pair of equivalences ${}_1\theta_l^A$ and ${}_3\theta_l^A$ is normal from the right reciprocally.

Cases 3 and 4 show that the pair of equivalences ${}_2\theta_r^A$ and ${}_3\theta_r^A$ is normal from the left reciprocally.

Cases 5 and 6 show that the pair of equivalences ${}_1\theta_m^A$ and ${}_2\theta_m^A$ is stable relative to the middle quasigroup translations and their inverses.

Corollary 4.148. *If in a quasigroup $(Q, \cdot)$:*

${}_1\theta_l^A = {}_3\theta_l^A = \theta_l^A$, *then the equivalence* ${}_1\theta_l^A$ $({}_3\theta_l^A)$ *is normal from the right;*

${}_2\theta_r^A = {}_3\theta_r^A = \theta_r^A$, *then the equivalence* ${}_2\theta_r^A$ $({}_3\theta_r^A)$ *is normal from the left;*

${}_1\theta_m^A = {}_2\theta_m^A = \theta_m^A$, *then the equivalence* ${}_1\theta_m^A$ $({}_2\theta_m^A)$ *is stable relative to middle quasigroup translations and its inverse;*

${}_1\theta_l^A = {}_3\theta_l^A = \theta_l^A = {}_2\theta_r^A = {}_3\theta_r^A = \theta_r^A$, *then equivalences* ${}_1\theta_l^A$, ${}_3\theta_l^A$, ${}_2\theta_r^A$, ${}_3\theta_r^A$ *are normal congruences;*

${}_1\theta_l^A = {}_3\theta_l^A = \theta_l^A = \theta_m^A = {}_1\theta_m^A = {}_2\theta_m^A$, *then equivalences* ${}_1\theta_l^A$, ${}_3\theta_l^A$, ${}_1\theta_m^A$, ${}_2\theta_m^A$ *are normal congruences;*

${}_2\theta_r^A = {}_3\theta_r^A = \theta_r^A = \theta_m^A = {}_1\theta_m^A = {}_2\theta_m^A$, *then equivalences* ${}_2\theta_r^A$, ${}_3\theta_r^A$, ${}_1\theta_m^A$, ${}_2\theta_m^A$ *are normal congruences.*

Proof. We use Lemmas 4.147 and 4.136. □

Theorem 4.149. *1. If in a quasigroup $(Q, \cdot)$ $H = {}_1N_l^A \cap {}_3N_l^A \cap {}_2N_r^A \cap {}_3N_r^A$, then H induces a normal congruence of $(Q, \cdot)$.*

2. If in a quasigroup $(Q, \cdot)$ $H = {}_1N_l^A \cap {}_3N_l^A \cap {}_1N_m^A \cap {}_2N_m^A$, then H induces a normal congruence of $(Q, \cdot)$.

3. If in a quasigroup $(Q, \cdot)$ $H = {}_2N_r^A \cap {}_3N_r^A \cap {}_1N_m^A \cap {}_2N_m^A$, then H induces a normal congruence of $(Q, \cdot)$.

Proof. The proof follows from Lemma 4.147 and Corollary 4.148. □

Corollary 4.150. *1. If $(Q, \circ)$ is a right loop, then ${}_1\theta_l^A = {}_3\theta_l^A = \theta_l^A$ and equivalence ${}_1\theta_l^A$ is normal from the right. Any subgroup of the left nucleus N_l (coset class $\theta_l^A(1)$) is normal from the right.*

2. *If $(Q, \circ)$ is a left loop, then ${}_2\theta_r^A = {}_3\theta_r^A = \theta_r^A$ and equivalence ${}_2\theta_r^A$ is normal from the left. Any subgroup of the right nucleus N_r (coset class $\theta_r^A(1)$) is normal from the left.*

3. *If $(Q, \circ)$ is a unipotent quasigroup, then ${}_1\theta_m^A = {}_2\theta_m^A = \theta_m^A$ and equivalence ${}_1\theta_m^A$ is stable relative to middle quasigroup translations and its inverse. Any subgroup of the middle nucleus N_m (coset class $\theta_m^A(1)$) is stable relative to middle quasigroup translations and its inverse.*

Proof. The proof follows from Corollary 4.58 and Corollary 4.148. □

If in conditions of Lemma 4.147 $K = {}_1N_l^A$, then we shall denote by ${}_1\Theta_l^A$ the corresponding equivalence and so on.

Corollary 4.151. *1. If $(Q, \circ)$ is a commutative loop, then any of equivalences θ_l^A and θ_r^A is a normal congruence. Left nucleus N_l (coset class $\Theta_l^A(1)$) and right nucleus N_r (coset class $\Theta_r^A(1)$) are equal and N_l is a normal subgroup of $(Q, \circ)$.*

2. *If $(Q, \circ)$ is a unipotent right loop with identity $x \circ (x \circ y) = y$, then any of equivalences θ_l^A and θ_m^A is a normal congruence. Left nucleus N_l (coset class $\Theta_l^A(1)$) and middle nucleus N_m (coset class $\Theta_m^A(1)$) are equal and N_l is a normal subgroup of $(Q, \circ)$.*

3. *If $(Q, \circ)$ is a unipotent left loop with identity $(x \circ y) \circ y = x$, then any of equivalences θ_r^A and θ_m^A is a normal congruence. Right nucleus N_r (coset class $\Theta_r^A(1)$) and middle nucleus N_m (coset class $\Theta_m^A(1)$) are equal and N_r is a normal subgroup of $(Q, \circ)$.*

Proof. Case 1. In any right loop, the first and the third components of any element of the group N_l^A coincide, and in any left loop the second and the third components of the group N_r^A coincide (Corollary 4.45). In a commutative loop, in addition, $L_a = R_a$ for any element $a \in Q$. From Corollary 4.148 it follows that in this case any equivalence θ_l^A is a normal congruence and therefore any subgroup of left nucleus N_l is normal as coset class $\theta_l^A(1)$ of normal congruence θ_l^A.

Cases 2 and 3 are proved using the passage to parastrophy images ((23) and (13), respectively) of the loop from Case 1. □

4.4.4 A-nuclei and loop congruences

Lemma 4.152. *In a loop $(Q, \cdot)$ any subgroup of the group Z (Z_l^A, Z_r^A, Z_m^A, ${}_1Z_l^A$, ${}_3Z_l^A$, ${}_2Z_r^A$, ${}_3Z_r^A$, ${}_1Z_m^A$, ${}_2Z_m^A$) defines a normal congruence.*

Proof. We use Lemma 4.114 and Corollary 4.148. If H is a subgroup of the group Z, then subgroups of the groups ${}_1Z_l^A$, ${}_3Z_l^A$ the corresponding to H are equal, since $(Q, \cdot)$ is a loop. Therefore the corresponding equivalences ${}_1\zeta_l^A = {}_3\zeta_l^A$ are equal. Since ${}_1Z_l^A = {}_2Z_r^A = {}_3Z_r^A$ (Lemma 4.114) we obtain that ${}_1\zeta_l^A = {}_2\zeta_r^A = {}_3\zeta_r^A$ and equivalence ${}_1\zeta_l^A$ is a normal congruence (Corollary 4.148) of loop $(Q, \cdot)$.

Other cases are proved in a similar way. □

Corollary 4.153. *In a loop $(Q, \cdot)$ an A-center defines a normal congruence.*

Proof. We use Lemma 4.152. □

For loops we give conditions of normality of congruences which are induced by A-nuclei. We also give conditions of normality of Garrison's nuclei, since Garrison's nuclei are coset classes by action of corresponding A-nuclei.

Theorem 4.154. *Let $(Q, \cdot)$ be a loop.*

1. *Let H be a subgroup of the group ${}_1N_l^A$. Denote by ${}_1\theta_l^A$ the corresponding equivalence to the group H.*

   (a) *If equivalence ${}_1\theta_l^A$ is admissible relative to any translation L_x of loop $(Q, \cdot)$, then ${}_1\theta_l^A$ is a normal congruence of $(Q, \cdot)$.*

   (b) *If equivalence ${}_1\theta_l^A$ is admissible relative to any translation P_x of loop $(Q, \cdot)$, then ${}_1\theta_l^A$ is a normal congruence of $(Q, \cdot)$.*

2. *Let H be a subgroup of the group ${}_2N_r^A$. Denote by ${}_2\theta_r^A$ the corresponding equivalence to the group H.*

   (a) *If equivalence ${}_2\theta_r^A$ is admissible relative to any translation R_x of loop $(Q, \cdot)$, then ${}_2\theta_r^A$ is a normal congruence of $(Q, \cdot)$.*

   (b) *If equivalence ${}_2\theta_r^A$ is admissible relative to any translation P_x of loop $(Q, \cdot)$, then ${}_2\theta_r^A$ is a normal congruence of $(Q, \cdot)$.*

3. Let H be a subgroup of the group ${}_1N_m^A \cap {}_2N_m^A$. Denote by ${}_1\theta_m^A$ the corresponding equivalence to the group H.

(a) If equivalence ${}_1\theta_m^A$ is admissible relative to any translation R_x of loop $(Q, \cdot)$, then ${}_1\theta_m^A$ is a normal congruence of $(Q, \cdot)$.

(b) If equivalence ${}_1\theta_m^A$ is admissible relative to any translation L_x of loop $(Q, \cdot)$, then ${}_1\theta_m^A$ is a normal congruence of $(Q, \cdot)$.

Proof. The proof follows from Lemma 4.136 and Corollary 4.150. □

Lemma 4.155. *In any loop $(Q, \cdot)$ the subgroup $N_l \cap N_r$ is normal.*

Proof. We use Theorem 4.149. In the loop case we have ${}_1N_l^A \cap {}_3N_l^A \cap {}_2N_r^A \cap {}_3N_r^A = {}_1N_l^A \cap {}_2N_r^A$, since ${}_1N_l^A = {}_3N_l^A$, ${}_2N_r^A = {}_3N_r^A$.

Further we have $({}_1N_l^A \cap {}_2N_r^A)1 = {}_1N_l^A 1 \cap {}_2N_r^A 1 = N_l \cap N_r$ (the last equality follows from Theorem 1.249). By Theorem 4.149 the group ${}_1N_l^A \cap {}_2N_r^A$ induces normal congruence such that the group $N_l \cap N_r$ is a coset class of this congruence.

Notice, from Proposition 5.66, it follows that any normal congruence of a quasigroup $(Q, \cdot)$ is regular, i.e., it is defined by a unique coset class. Therefore the coset class $N_l \cap N_r$ defines normal congruence that is induced by the group ${}_1N_l^A \cap {}_2N_r^A$. □

Corollary 4.156. *If in a loop $(Q, \cdot)$ $N_m \trianglelefteq Q$, then $N \trianglelefteq Q$.*

Proof. The proof follows from Lemma 4.155 and the fact that intersection of any two normal subloops of the loop $(Q, \cdot)$ is a normal subloop of this loop. □

For a group G, Lemma 4.155 is transformed into the fact that $G \trianglelefteq G$.

Lemma 4.157. *If in a loop $(Q, \cdot)$ $N_l = N_m = N_r = N$, then nucleus N is a normal subgroup of loop $(Q, \cdot)$, i.e., $N \trianglelefteq Q$.*

Proof. From equalities $N_l = N_m = N_r = N$ we have that $N_l \cap N_r = N$, and we can apply Lemma 4.155. □

Remark 4.158. From Lemma 4.155 it follows that "non-standard behavior" of the middle A-nucleus in a loop is the reason why in the general case, the loop nucleus is a non-normal subgroup of the loop.

Lemma 4.159. *If a loop Q has the order $p_1^{n_1} p_2^{n_2} \dots p_t^{n_t}$, where $p_1, p_2, \dots, p_t$ are prime numbers, $1 < p_1 < p_2 < \dots < p_t$, nucleus N is of index p_1, i.e., $|Q : N| = p_1$, then $N \trianglelefteq Q$.*

Proof. The case that N_l, N_m, or N_r is equal to Q is impossible in the conditions of this lemma, because in this case $Q = N$, $|Q : N| = 1$.

Therefore $N_l \subsetneq Q$, $N_r \subsetneq Q$, $N_m \subsetneq Q$. Then $|Q : N_l| \geqslant p_1$, $|Q : N_m| \geqslant p_1$, $|Q : N_r| \geqslant p_1$. From the other side $|Q : N_l| \leqslant |Q : N| = p_1$, $|Q : N_m| \leqslant |Q : N| = p_1$, $|Q : N_r| \leqslant |Q : N| = p_1$. Therefore $|Q : N_l| = p_1$, $|Q : N_m| = p_1$, $|Q : N_r| = p_1$.

If we suppose that $N_l \neq N_m$ (either $N_l \neq N_r$, or $N_r \neq N_m$, or $N_r \neq N_m \neq N_l$), then $|Q : (N_l \cap N_m \cap N_r = N)| > p_1$ and we obtain a contradiction. Thus $N_l = N_m = N_r = N$ and by Lemma 4.157 $N \trianglelefteq Q$. □

Lemma 4.160. *If in loop Q, nucleus N has the lowest prime index p_1, then $N \trianglelefteq Q$.*

Proof. The proof is similar to the proof of Lemma 4.159 and we omit it. □

Remark 4.161. Case $p_1 = 2$ in Lemma 4.160 is proved in [72, Lemma 10.1]. See also [685]. The result, that any subgroup of index lowest prime p_1 is normal, is proved in [930, 557] for finite and infinite groups.

4.4.5 On loops with nucleus of index two

Lemma 4.162. *A loop $(Q,\cdot)$ with nucleus N of index two is a WIP-loop [72, 685].*

Proof. We shall use the following implicative definition of a WIP-loop: if $x \cdot yz = 1$, then $xy \cdot z = 1$ (Lemma 2.61). It is clear that $Q = \{N, aN\}$, where $a \notin N$. If at least one from the elements x, y, z lies in the subgroup N of loop Q, then this implication is true.

By Lemma 4.160, $N \trianglelefteq Q$. Then $Q/N \cong (Z_2, +)$, since cyclic group of order two is a unique loop of this order. Here $Z_2 = \{0, 1\}$. In this situation coset class N plays the role of element 0 and coset class aN plays the role of element 1. Then $aN \cdot aN = a(N \cdot aN) = a(Na \cdot N) = a(aN \cdot N) = a(a \cdot NN) = a(aN) = aa \cdot N = N$. Thus $a \cdot a \in N$. Therefore, if $x, y \in aN$, then $xy \in N$.

Suppose that $x, y, z \notin N$. Then the expression $x \cdot yz \in aN$, $x \cdot yz \neq 1$ for any $x, y, z \in aN$ holds. Implication, which defines a WIP-loop, is true, since the premise of this implication is false. Notice that the conclusion of this implication is false too. □

Theorem 4.163. *A loop $(Q,\cdot)$ with nucleus N of index two is a G-loop [67], [72, Theorem 10.3], [685, p. 84].*

Proof. We prove this theorem, if we demonstrate that any left and right derivative of loop $(Q,\cdot)$ is isomorphic to $(Q,\cdot)$ (see Definition 4.15).

We start from the right derivatives. Let

$$x \circ y = L_b^{-1}(bx \cdot y). \tag{4.40}$$

If at least one element of the elements b, x, y lies in the nucleus N, then the equality (4.40) takes the form: $x \circ y = L_b^{-1}(b \cdot xy) = x \cdot y$.

Suppose that $b, x, y \notin N$. Then $b = ac_1, x = ac_2, y = ac_3$, where $a \notin N$, $c_1, c_2, c_3 \in N$. From Lemma 4.159 it follows that we can put $b = c_4a$ $(c_4 \in N)$ and from Lemma 4.22 it follows that we can put $b = a$. Therefore the equality (4.40) takes the form:

$$ac_2 \circ ac_3 = L_a^{-1}((a \cdot ac_2) \cdot ac_3). \tag{4.41}$$

Further, we have

$$\begin{aligned} L_a^{-1}((a \cdot ac_2) \cdot ac_3) = L_a^{-1}((a^2 \cdot c_2) \cdot ac_3) = L_a^{-1}(a^2(c_2 \cdot ac_3)) \overset{a^2 \in N}{=} \\ (L_a^{-1}a^2)(c_2 \cdot ac_3) = a(c_2 \cdot ac_3) = ac_2 \cdot ac_3. \end{aligned}$$

The proof for the left derivative is similar and we omit it. □

Lemma 4.164. *In a loop $(Q,\cdot)$ with the identity $x^2 = 1$ the following subgroups are normal: 1. $N_l \cap N_m$; 2. $N_m \cap N_r$; 3. $N = N_l \cap N_r \cap N_m$.*

Proof. Case 1. In a loop $(Q,\cdot)$ with the identity $x^2 = 1$ we have ${}_1N_m^A = {}_2N_m^A$ (Corollary 4.58, Case 3). Further, from Theorem 4.149 Case 2, we have that the group ${}_1N_l^A \cap {}_3N_l^A \cap {}_1N_m^A \cap {}_2N_m^A = {}_1N_l^A \cap {}_1N_m^A$ induces a normal congruence of $(Q,\cdot)$ such that the group $N_l \cap N_r$ is a coset class of this congruence.

Cases 2 and 3 are proved in a similar way. □

Notice, a group with the identity $x^2 = 1$ is commutative. Indeed, $abab = 1$, $bab = a$, $ab = ba$. For loops it is not so. There exist non-commutative unipotent loops of order five, six (Example 1.402) and so on.

4.5 Coincidence of A-nuclei in inverse quasigroups

Information on the coincidence of A-nuclei and nuclei of inverse quasigroups and loops is mainly taken from [489]. Also we give results on normality of nuclei of some inverse quasigroups and loops. We use some definitions from Chapter 3.

4.5.1 (α, β, γ)-inverse quasigroups

Lemma 4.165. *In (α, β, γ)-inverse quasigroup we have the following relations between components of A-nuclei:*

$$ {}_1N_l^A = \alpha^{-1}{}_3N_r^A\alpha = \beta\,{}_2N_m^A\beta^{-1}, \tag{4.42}$$
$$ {}_3N_l^A = \gamma^{-1}{}_2N_r^A\gamma = \alpha\,{}_1N_m^A\alpha^{-1}, \tag{4.43}$$
$$ {}_2N_r^A = \beta^{-1}{}_1N_m^A\beta = \gamma\,{}_3N_l^A\gamma^{-1}, \tag{4.44}$$
$$ {}_3N_r^A = \gamma^{-1}{}_2N_m^A\gamma = \alpha\,{}_1N_l^A\alpha^{-1}, \tag{4.45}$$
$$ {}_1N_m^A = \alpha^{-1}{}_3N_l^A\alpha = \beta\,{}_2N_r^A\beta^{-1}, \tag{4.46}$$
$$ {}_2N_m^A = \beta^{-1}{}_1N_l^A\beta = \gamma\,{}_3N_r^A\gamma^{-1}. \tag{4.47}$$

Proof. The proof follows from Theorem 3.5, Lemma 3.6 and Table 1.4. □

Theorem 4.166. *1. In $(\varepsilon; \beta; \gamma)$-inverse loop $(Q, \circ)$, ${}_1N_l^A = {}_3N_l^A = {}_2N_r^A = {}_3N_r^A = {}_1N_m^A$, $N_l = N_r = N_m \trianglelefteq Q$.*

2. In $(\alpha; \varepsilon; \gamma)$-inverse loop $(Q, \circ)$, ${}_1N_l^A = {}_3N_l^A = {}_2N_m^A$, ${}_2N_r^A = {}_3N_r^A = {}_1N_m^A$, $N_l = N_r = N_m \trianglelefteq Q$.

3. In $(\alpha; \beta; \varepsilon)$-inverse loop $(Q, \circ)$, ${}_1N_l^A = {}_3N_l^A = {}_2N_r^A = {}_3N_r^A = {}_2N_m^A$, $N_l = N_r = N_m \trianglelefteq Q$.

4. In $(\alpha; \alpha^{-1}; \gamma)$-inverse loop $(Q, \circ)$, ${}_1N_l^A = {}_3N_l^A = {}_2N_r^A = {}_3N_r^A = {}_2N_m^A$, $N_l = N_r = N_m \trianglelefteq Q$.

5. In $(\alpha; \beta; \alpha^{-1})$-inverse loop $(Q, \circ)$, ${}_1N_l^A = {}_3N_l^A = {}_2N_m^A$, ${}_2N_r^A = {}_3N_r^A = {}_1N_m^A$, $N_l = N_r = N_m \trianglelefteq Q$.

6. In $(\alpha; \beta; \beta^{-1})$-inverse loop $(Q, \circ)$, ${}_1N_l^A = {}_3N_l^A = {}_2N_r^A = {}_3N_r^A = {}_1N_m^A$, $N_l = N_r = N_m \trianglelefteq Q$.

Proof. The proof follows from Lemma 4.165 and the fact that in any loop ${}_1N_l^A = {}_3N_l^A$, ${}_2N_r^A = {}_3N_r^A$. Normality of Garrison's nuclei follows from Corollary 4.157. □

Theorem 4.167. *If in $(\alpha; \beta; \gamma)$-inverse loop $(Q, \cdot)$, $\alpha\,{}_3N_r^A\alpha^{-1} \subseteq {}_3N_r^A$, $\beta\,{}_1N_m^A\beta^{-1} \subseteq {}_1N_m^A$, $\alpha^{-1}{}_3N_l^A\alpha \subseteq {}_3N_l^A$, then in this loop $N_l = N_m = N_r$.*

Proof. By proving this theorem we use the standard way [472]. Let $L_a \in {}_1N_l^A$, i.e., $a \in N_l$. Then using (4.42) we have $\alpha^{-1}L_a\alpha \in {}_3N_r^A$. Therefore $L_a \in \alpha\,{}_3N_r^A\alpha^{-1} \subseteq {}_3N_r^A$. Thus $L_a1 = a \in {}_3N_r^A1 = N_r$.

If $a \in N_r$, then $(\varepsilon, R_a, R_a) \in N_r^A$ and using (4.44) we see that $\beta^{-1}R_a\beta \in {}_1N_m^A$. Then $R_a \in \beta\,{}_1N_m^A\beta^{-1} \subseteq {}_1N_m^A$. Therefore $a = R_a1 \in {}_1N_m^A1 = N_m$.

If $a \in N_m$, then $(R_a, L_a^{-1}, \varepsilon) \in N_m^A$ and using (4.46) we see that $\alpha R_a\alpha^{-1} \in {}_3N_l^A$. Then $R_a \in \alpha^{-1}{}_3N_l^A\alpha \subseteq {}_3N_l^A$. Therefore $a = R_a1 \in {}_3N_l^A1 = N_l$.

We have obtained $N_l \subseteq N_r \subseteq N_m \subseteq N_l$, therefore $N_l = N_r = N_m$. □

We repeat the following:

Definition 4.168. A quasigroup $(Q, \circ)$ is

1. an (r, s, t)-inverse quasigroup if there exists a permutation J of the set Q such that $J^r(x \circ y) \circ J^s x = J^t y$ for all $x, y \in Q$ [488, 487];
2. an m-inverse quasigroup if there exists a permutation J of the set Q such that $J^m(x \circ y) \circ J^{m+1}x = J^m y$ for all $x, y \in Q$ [472, 488, 487];
3. a WIP-inverse quasigroup if there exists a permutation J of the set Q such that $J(x \circ y) \circ x = Jy$ for all $x, y \in Q$ [72, 25, 847, 487];
4. a CI-inverse quasigroup if there exists a permutation J of the set Q such that $(x \circ y) \circ Jx = y$ for all $x, y \in Q$ [16, 17, 95, 72].

It is easy to see that classes of (r, s, t)-inverse, m-inverse, WIP-inverse, and CI-inverse quasigroups are included in the class of (α, β, γ)-inverse quasigroups.

Lemma 4.169. *An (r, s, t)-inverse quasigroup has $((1\,2\,3), (J^r, J^s, J^t))$ autostrophy.*
An m-inverse quasigroup has $((1\,2\,3), (J^m, J^{m+1}, J^m))$ autostrophy.
A WIP-inverse quasigroup has $((1\,2\,3), (J, \varepsilon, J))$ autostrophy.
A CI-inverse quasigroup has $((1\,2\,3), (\varepsilon, J, \varepsilon))$ autostrophy.

Proof. The proof follows from Theorem 3.5. □

Corollary 4.170. *1. If in m-inverse loop $(Q, \circ)$ $J = I_r$, where $x \circ I_r x = 1$, then $N_l = N_r = N_m = N \trianglelefteq Q$.*

2. In a WIP-loop, ${}_1N_l^A = {}_3N_l^A = {}_2N_m^A$, ${}_2N_r^A = {}_3N_r^A = {}_1N_m^A$, $N_l = N_r = N_m = N \trianglelefteq Q$.

3. In a commutative WIP-loop, ${}_1N_l^A = {}_3N_l^A = {}_2N_m^A = {}_2N_r^A = {}_3N_r^A = {}_1N_m^A$, $N_l = N_r = N_m = N \trianglelefteq Q$.

4. In a W-loop, ${}_1N_l^A = {}_3N_l^A = {}_2N_m^A = {}_2N_r^A = {}_3N_r^A = {}_1N_m^A$, $N_l = N_r = N_m = N \trianglelefteq Q$.

5. In a CI-loop, ${}_1N_l^A = {}_3N_l^A = {}_2N_m^A = {}_2N_r^A = {}_3N_r^A = {}_1N_m^A$, $N_l = N_r = N_m = N \trianglelefteq Q$.

Proof. Case 1. The proof follows from Theorem 4.167 and Corollary 4.157. Cases 2–5. The proof follows from Theorem 4.166 and Corollary 4.150. □

The coincidence of nuclei in CI-loops and WIP-loops is proved in [18].

Example 4.171. This example shows WIP-loop in which $N_l = N_r = N_m = N = \{0\}$.

*	0	1	2	3	4	5	6	7
0	0	1	2	3	4	5	6	7
1	1	0	3	4	2	7	5	6
2	2	3	4	6	5	1	7	0
3	3	4	6	2	7	0	1	5
4	4	2	5	7	3	6	0	1
5	5	7	1	0	6	2	4	3
6	6	5	7	1	0	4	3	2
7	7	6	0	5	1	3	2	4

WIP-loop $(Q, \circ)$ has the following nucleus $N = \{0, 5\}$; WIP-loop $(Q, *)$ has the following nucleus $N = \{0, 1, 2\}$. Notice, $N \lhd Q$ in both cases. Loop $(Q, *)$ is a G-loop (Theorem 4.163).

∘	0	1	2	3	4	5
0	0	1	2	3	4	5
1	1	2	0	4	5	3
2	2	0	3	5	1	4
3	3	4	5	2	0	1
4	4	5	1	0	3	2
5	5	3	4	1	2	0

*	0	1	2	3	4	5
0	0	1	2	3	4	5
1	1	2	0	4	5	3
2	2	0	1	5	3	4
3	3	5	4	1	0	2
4	4	3	5	2	1	0
5	5	4	3	0	2	1

4.5.2 λ-, ρ-, and μ-inverse quasigroups

Lemma 4.172. *1. In a λ-inverse quasigroup*

(a) ${}_1N_l^A = \lambda_1^{-1}{}_1N_m^A\lambda_1 = \lambda_1\, {}_1N_m^A\lambda_1^{-1}$;

(b) ${}_3N_l^A = \lambda_3^{-1}{}_2N_m^A\lambda_3 = \lambda_2\, {}_2N_m^A\lambda_2^{-1}$;

(c) ${}_2N_r^A = \lambda_2^{-1}{}_3N_r^A\lambda_2 = \lambda_3\, {}_3N_r^A\lambda_3^{-1}$.

2. In a ρ-inverse quasigroup

(a) ${}_1N_l^A = \rho_1^{-1}{}_3N_l^A\rho_1 = \rho_3\, {}_3N_l^A\rho_3^{-1}$;

(b) ${}_2N_r^A = \rho_2^{-1}{}_2N_m^A\rho_2 = \rho_2\, {}_2N_m^A\rho_2^{-1}$;

(c) ${}_3N_r^A = \rho_3^{-1}{}_1N_m^A\rho_3 = \rho_1\, {}_1N_m^A\rho_1^{-1}$.

3. In a μ-inverse quasigroup

(a) ${}_1N_l^A = \mu_1^{-1}{}_2N_r^A\mu_1 = \mu_2\, {}_2N_r^A\mu_2^{-1}$;

(b) ${}_3N_l^A = \mu_3^{-1}{}_3N_r^A\mu_3 = \mu_3\, {}_3N_r^A\mu_3^{-1}$;

(c) ${}_1N_m^A = \mu_1^{-1}{}_2N_m^A\mu_1 = \mu_2\, {}_2N_m^A\mu_2^{-1}$.

Proof. We can use Theorem 3.5, Lemma 3.6 and Table 1.4. □

We recall that a quasigroup $(Q, \cdot)$ with the identities $xy = yx$, $x \cdot xy = y$ is called a TS-quasigroup. In a TS-quasigroup any parastrophy is an autostrophy [72], i.e., $(Q, \cdot)^\sigma = (Q, \cdot)$ for all parastrophies $\sigma \in S_3$.

Corollary 4.173. *1. In a LIP-quasigroup, ${}_1N_l^A = \lambda\ ({}_1N_m^A)\lambda$, $\lambda^2 = \varepsilon$, ${}_3N_l^A = {}_2N_m^A$, ${}_2N_r^A = {}_3N_r^A$.*

2. In a RIP-quasigroup, ${}_2N_r^A = \rho\ ({}_2N_m^A)\rho$, $\rho^2 = \varepsilon$, ${}_1N_l^A = {}_3N_l^A$, ${}_1N_m^A = {}_3N_r^A$.

3. In an IP-quasigroup, $\lambda^2 = \varepsilon$, $\rho^2 = \varepsilon$, ${}_1N_l^A = {}_3N_l^A = {}_2N_m^A \cong {}_1N_m^A = {}_2N_r^A = {}_3N_r^A$.

4. In a $WCIP$-quasigroup, $J^2 = \varepsilon$, ${}_1N_l^A = J({}_3N_l^A)J$, ${}_2N_r^A = {}_2N_m^A$, ${}_3N_r^A = J_1N_m^AJ$.

5. In a TS-quasigroup, ${}_1N_l^A = {}_3N_l^A = {}_1N_m^A = {}_2N_m^A = {}_2N_r^A = {}_3N_r^A$.

Proof. The proof follows from Lemma 4.172. In a LIP-quasigroup, $\lambda_2 = \lambda_3 = \varepsilon$. In a RIP-quasigroup, $\rho_1 = \rho_3 = \varepsilon$. In a $WCIP$-quasigroup, $\rho_1 = \rho_3 = J$, $J^2 = \varepsilon$ (Lemma 3.15). □

Theorem 4.174. *1. In a LIP-quasigroup, ${}_2\Theta_r^A = {}_3\Theta_r^A$ and this equivalence is normal from the left.*

2. *In a RIP-quasigroup,* ${}_1\Theta_l^A = {}_3\Theta_l^A$*, this equivalence is normal from the right.*

3. *In a TS-quasigroup we have* ${}_1\Theta_l^A = {}_3\Theta_l^A = {}_2\Theta_r^A = {}_3\Theta_r^A = {}_1\Theta_m^A = {}_2\Theta_m^A$ *and all these equivalences are normal congruences.*

Proof. Cases 1 and 2. The proof follows from Lemma 4.172, Corollaries 4.148 and 4.151.
By proving Case 3 it is possible to use Table 1.4 and Corollary 4.148. □

Corollary 4.175. 1. *In a LIP-loop* $N_l = N_m$*; in a RIP-loop* $N_r = N_m$.

2. *In a Moufang loop* $N_l = N_r = N_m \trianglelefteq Q$.

3. *In an IP-loop* $N_l = N_r = N_m \trianglelefteq Q$.

4. *In a commutative LIP-loop, and a commutative RIP-loop* $N_l = N_r = N_m \trianglelefteq Q$.

5. *In* $(\varepsilon; \mu_2; \mu_3)$*-,* $(\mu_1; \varepsilon; \mu_3)$*-, and* $(\mu_1; \mu_2; \varepsilon)$*-inverse loop,* $N_l = N_r \trianglelefteq Q$.

6. *In a* $WCIP$*-loop* $N_r = N_m$*, and in a commutative* $WCIP$*-loop* $N_l = N_r = N_m \trianglelefteq Q$.

Proof. The proof follows from Corollary 4.173 and Lemma 4.172. To prove normality it is possible to use Corollaries 4.148 and 4.157. □

Cases 1, 2 are well known [18, 72, 80, 173, 685].
Loop $(Q, \cdot)$ with the identity $xx \cdot yz = xy \cdot xz$ is a commutative Moufang loop (CML).

Corollary 4.176. *In any CML* ${}_1N_l^A = {}_3N_l^A = {}_2N_m^A = {}_1N_m^A = {}_2N_r^A = {}_3N_r^A$, $N_l = N_r = N_m = Z \trianglelefteq Q$.

Proof. Any CML is a commutative IP-loop [72]. We can apply Corollary 4.175. □

4.6 Relations between a loop and its inverses

4.6.1 Nuclei of inverse loops in Belousov sense

The definition of the inverse of a fixed loop Q is in [72, 65]. See Lemma 1.187. We give relations between nuclei of inverse loops in the Belousov sense.

Theorem 4.177. 1. *Between A-nuclei of loops* (Q, λ) *and* $(Q, \cdot)$ *there exist the following relations:* ${}_1N_l^{\lambda} = {}_3N_l^{\cdot}$, ${}_3N_l^{\lambda} = {}_1N_l^{\cdot}$, ${}_2N_r^{\lambda} = I^{-1}{}_2N_m^{\cdot}I$, ${}_3N_r^{\lambda} = {}_1N_m^{\cdot}$, ${}_1N_m^{\lambda} = {}_3N_r^{\cdot}$, ${}_2N_m^{\lambda} = I^{-1}{}_2N_r^{\cdot}I$.

2. *Between nuclei of loops* (Q, λ) *and* $(Q, \cdot)$ *there exist the following relations:* $N_l^{\lambda} = N_l^{\cdot}$, $N_r^{\lambda} = N_m^{\cdot}$, $N_m^{\lambda} = N_r^{\cdot}$, $N^{\lambda} = N^{\cdot}$.

3. *Between A-nuclei of loops* (Q, ρ) *and* $(Q, \cdot)$ *there exist the following relations:* ${}_1N_l^{\rho} = I{}_3N_l^{\cdot}I^{-1}$, ${}_3N_l^{\rho} = {}_2N_m^{\cdot}$, ${}_2N_r^{\rho} = {}_3N_r^{\cdot}$, ${}_3N_r^{\rho} = {}_2N_r^{\cdot}$, ${}_1N_m^{\rho} = I{}_1N_l^{\cdot}I^{-1}$, ${}_2N_m^{\rho} = {}_3N_l^{\cdot}$.

4. *Between nuclei of loops* (Q, ρ) *and* $(Q, \cdot)$ *there exist the following relations:* $N_l^{\rho} = N_m^{\cdot}$, $N_r^{\rho} = N_r^{\cdot}$, $N_m^{\rho} = N_l^{\cdot}$, $N^{\rho} = N^{\cdot}$.

5. *Between A-nuclei of loops* $(Q, *)$ *and* $(Q, \cdot)$ *there exist the following relations:* ${}_1N_l^{*} = I^{-1}{}_2N_r^{\cdot}I$, ${}_3N_l^{*} = I^{-1}{}_3N_r^{\cdot}I$, ${}_2N_r^{*} = I^{-1}{}_1N_l^{\cdot}I$, ${}_3N_r^{*} = I^{-1}{}_3N_l^{\cdot}I$, ${}_1N_m^{*} = I^{-1}{}_2N_m^{\cdot}I$, ${}_2N_m^{*} = I^{-1}{}_1N_m^{\cdot}I$.

6. *Between nuclei of loops* $(Q, *)$ *and* $(Q, \cdot)$ *there exist the following relations:* $N_l^* = N_r^{\cdot}$, $N_r^* = N_l^{\cdot}$, $N_m^* = N_m^{\cdot}$, $N^* = N^{\cdot}$.

7. *Between A-nuclei of loops* (Q, ρ^*) *and* $(Q, \cdot)$ *there exist the following relations:* ${}_1N_l^{\rho^*} = I^{-1}{}_3N_r^{\cdot}I$, ${}_3N_l^{\rho^*} = I^{-1}{}_2N_r^{\cdot}I$, ${}_2N_r^{\rho^*} = {}_1N_m^{\cdot}$, ${}_3N_r^{\rho^*} = I^{-1}{}_2N_m^{\cdot}I$, ${}_1N_m^{\rho^*} = I^{-1}{}_3N_l^{\cdot}I$, ${}_2N_m^{\rho^*} = {}_1N_l^{\cdot}$.

8. *Between nuclei of loops* (Q, ρ^*) *and* $(Q, \cdot)$ *there exist the following relations:* $N_l^{\rho^*} = N_r^{\cdot}$, $N_r^{\rho^*} = N_m^{\cdot}$, $N_m^{\rho^*} = N_l^{\cdot}$, $N^{\rho^*} = N^{\cdot}$.

9. *Between A-nuclei of loops* (Q, λ^*) *and* $(Q, \cdot)$ *there exist the following relations:* ${}_1N_l^{\lambda^*} = {}_2N_m^{\cdot}$, ${}_3N_l^{\lambda^*} = I^{-1}{}_1N_m^{\cdot}I$, ${}_2N_r^{\lambda^*} = I^{-1}{}_1N_m^{\cdot}I$, ${}_3N_r^{\lambda^*} = I^{-1}{}_1N_l^{\cdot}I$, ${}_1N_m^{\lambda^*} = {}_2N_r^{\cdot}$, ${}_2N_m^{\lambda^*} = I^{-1}{}_3N_r^{\cdot}I$.

10. *Between nuclei of loops* (Q, λ^*) *and* $(Q, \cdot)$ *there exist the following relations:* $N_l^{\lambda^*} = N_m^{\cdot}$, $N_r^{\lambda^*} = N_l^{\cdot}$, $N_m^{\lambda^*} = N_r^{\cdot}$, $N^{\lambda^*} = N^{\cdot}$.

Proof. The proof follows from Table 1.4 and the fact that in a loop Q, the left nucleus is an orbit by the action of group ${}_1N_l^A$ or group ${}_3N_l^A$ on the identity element of Q, and so on. □

4.6.2 LIP- and AAIP-loops

Definition 4.178. A loop $(Q, \circ)$ with the property $Ix \circ Iy = I(y \circ x)$ we shall call a *MIP*-loop (middle inverse property loop).

Say that a loop $(Q, \cdot)$ satisfies the *antiautomorphic inverse property* (AAIP) if $I^2 = \varepsilon$ and $I(xy) = IyIx$ for all $x, y \in Q$.

The following example demonstrates that there exists MIP-loop that is not an AAIP-loop.

$\circ$	0	1	2	3	4
0	0	1	2	3	4
1	1	0	3	4	2
2	2	4	0	1	3
3	3	2	4	0	1
4	4	3	1	2	0

In the loop $(Q, \circ)$ permutation I has the following form $I = (1\,2\,4\,3)$.

From Definition 4.178 it follows that any *MIP*-loop has autostrophy of the form $((1\,2), (I, I, I))$ and that $((1\,2), (I, I, I))^2 = ((\varepsilon), (I^2, I^2, I^2))$. The last means that permutation I^2 is an automorphism of *MIP*-loop $(Q, \circ)$. See also Lemma 1.187.

Lemma 4.179. *A left inverse loop* $(Q, \circ)$ *to an LIP-loop* $(Q, \cdot)$ *is an AAIP-loop and vice versa, a left inverse loop to an AAIP-loop is an LIP-loop.*

Proof. Since loop $(Q, \circ)$ is a left inverse loop to the loop $(Q, \cdot)$, then $x \circ y = x/Iy$.

From *LIP*-equality $Ix \cdot (xy) = y$ we obtain $y/(xy) = Ix$. If we denote xy by z, then x passes to z/y. Then $y/z = I(z/y)$, $y \circ I^{-1}z = I(z \circ I^{-1}y)$, $Iy \circ Iz = I(z \circ y)$, since in any *LIP*-loop $I^2 = \varepsilon$, i.e., $I^{-1} = I$ [72].

The converse is proved using the inverse order of steps in the previous proof.

By proving this lemma we can also use isotrophy-autostrophy arguments. A *LIP*-property means that a loop $(Q, \cdot)$ has autostrophy of the form $((23), (I, \varepsilon, \varepsilon))$, i.e., $(Q, \cdot) =$

$(Q,\cdot)((23),(I,\varepsilon,\varepsilon))$. A left inverse loop to a loop $(Q,\cdot)$ is defined by the following equality $(Q,\lambda) = (Q,\cdot)((1\,3),(\varepsilon,I,\varepsilon))$. Therefore

$$\begin{aligned}
&(Q,\lambda) = \\
&(Q,\lambda)((1\,3)(\varepsilon,I,\varepsilon))^{-1}((2\,3)(I,\varepsilon,\varepsilon))((1\,3)(\varepsilon,I,\varepsilon)) = \\
&(Q,\lambda)((1\,3)(\varepsilon,I,\varepsilon))((2\,3)(I,\varepsilon,\varepsilon))((1\,3)(\varepsilon,I,\varepsilon)) = \\
&(Q,\lambda)((1\,2\,3)(I,\varepsilon,I))((1\,3)(\varepsilon,I,\varepsilon)) = \\
&(Q,\lambda)((1\,2)(I,I,I)).
\end{aligned}$$

Converse

$$\begin{aligned}
&(Q,\lambda) = \\
&(Q,\lambda)((1\,3)(\varepsilon,I,\varepsilon))^{-1}((1\,2)(I,I,I))((1\,3)(\varepsilon,I,\varepsilon)) = \\
&(Q,\lambda)((1\,3)(\varepsilon,I,\varepsilon))((1\,2)(I,I,I))((1\,3)(\varepsilon,I,\varepsilon)) = \\
&(Q,\lambda)((1\,3\,2)(I^2,I,I))((1\,3)(\varepsilon,I,\varepsilon)) = \\
&(Q,\lambda)((2\,3)(I,I^2,I^2)) = (Q,\lambda)((2\,3)(I,\varepsilon,\varepsilon)).
\end{aligned}$$

□

Corollary 4.180. *In an AAIP-loop we have* $I1 = 1$.

Lemma 4.181. *In any AAIP-loop,* $N_l = N_r$ *[18].*

Proof. The proof follows from Lemma 4.179, Corollary 4.175 and Theorem 4.177 (Case 2). □

4.6.3 Invariants of reciprocally inverse loops

In this subsection we use information from [285].

4.6.3.1 Middle Bol loops

A loop $(Q,\cdot)$ is called the middle Bol if in the corresponding primitive loop $(Q,\cdot,/,\backslash)$ the following identity $x(yz\backslash x) = (x/z)(y\backslash x)$ holds true. Middle Bol loops were defined by Belousov [72].

If $(Q,\cdot)$ is a left Bol loop and $(Q,\circ)$ is the corresponding middle Bol loop, then $x\circ y = x/y^{-1}$, where "/" is the left division in $(Q,\cdot)$ [385, 873]. I.e., the middle Bol loop is a left inverse loop to the left Bol loop; see Lemma 1.187, Case 1.

Proposition 4.182. *In any middle Bol loop,* $N_l = N_r$ *[398, 873].*

Proof. The proof follows from Lemma 4.181 since any middle Bol loop is an $AAIP$-loop. □

4.6.3.2 Some invariants

We recall, if θ is a normal quasigroup congruence of a quasigroup Q, then θ is a normal congruence of any parastrophe of Q. See Corollary 1.268 or [75]. If θ is a normal congruence of a loop Q, then it is a normal congruence of any inverse loop to loop Q (Corollary 1.269).

In Corollary 1.269, in fact, it is proved that lattices of normal congruences of a loop Q and any inverse loop to loop Q are equal.

It is easy to see that any subloop of a loop $(Q,\cdot)$ is invariant relative to mapping I.

Therefore groupoid $(H, \cdot)$ is a subloop of a loop $(Q, \cdot)$ if and only if (H, λ) is a subloop of the loop (Q, λ). And this is true for any pair of loop operations from Lemma 1.187.

In [65, Theorem 3] it is proved that any inverse loop to a G-loop [72, 80, 685] also is a G-loop.

Therefore, results about the lattice of normal congruences of a loop $(Q, \cdot)$, about subloops of loop $(Q, \cdot)$ and about the G-property of the loop $(Q, \cdot)$ can be extended to similar results in the inverse loops to the loop $(Q, \cdot)$.

For example, it is possible to spread results about left Bol loops (simplicity, direct decompositions, central series, G-property [650, 45]) on middle Bol loops. See [384] for more details.

The commutant of a loop is the set of elements which commute with all elements of the loop.

In [518] it is proved that the commutant of a left Bol loop does not form a subloop. For groups, a similar example is constructed in [471].

In a middle Bol loop Q, the commutant $C(Q)$ is always a subloop [385]. These two results do not contradict one another, since the commutant of the middle Bol loop does not pass into the commutant of the left Bol loop by the isostrophy $((1\,3), (\varepsilon, I, \varepsilon))$. Some properties of middle Bol loops are given in [452].

Definition 4.183. An element a of a loop $(Q, \cdot)$ is called a CI-element if $x \cdot (a \cdot Ix) = a$ for all $x \in Q$.

Notice that elements of the middle Bol loop commutant are transformed into CI-elements of the corresponding left Bol loop. From the Grecu-Syrbu result [385] it follows that the set of all CI-elements of a left Bol loop Q forms a subloop of the loop Q.

4.6.3.3 Term-equivalent loops

Definition 4.184. If operations of an algebra (Q, A) can be expressed using operations of an algebra (Q, B) and vice versa, then such algebras are called *term-equivalent* [705, 620, 912].

From Lemma 1.187 it is possible to deduce the following fact. The reciprocally-inverse loops in signature with three binary operations "$\cdot$", "/", "\", one nul-ary operation "1" and with the identities $x \cdot (x\backslash y) = y$, $(y/x) \cdot x = y$, $x\backslash(x \cdot y) = y$, $(y \cdot x)/x = y$, $1 \cdot x = x$, $x \cdot 1 = x$ are term-equivalent.

Such loops share subalgebras and congruences. A subset S is a subloop of $(Q, \cdot, /, \backslash, 1)$ if and only if it is a subloop of $(Q, *, //, \backslash\backslash, 1)$, where loop operation $*$ is an inverse loop operation to the loop operation $\cdot$. Furthermore it is a normal subloop in one of the loops if and only if it is a normal subloop in the other one.

Note also that a mapping $f : (Q_1, \cdot) \to (Q_2, \bar{\cdot})$ is a loop homomorphism if and only if it is a loop homomorphism of an inverse loop $(Q_1, *) \to (Q_2, \bar{*})$.

The mentioned functorial properties imply that $(Q, \cdot)$ and $(Q, *)$ coincide in those structural features that can be obtained by homomorphisms onto loops listed in Corollary 1.189, in general, and onto groups, in particular.

For example, they must coincide in the *associator subloop* $A(Q)$ (the least normal subloop A such that Q/A is a group) and in the *commutant* Q' (the least normal subloop S such that Q/S is an abelian group). The higher derived subloops are also the same, of course. In particular, $(Q, \cdot, /, \backslash, 1)$ is soluble if and only if $(Q, *, //, \backslash\backslash, 1)$ is soluble.

Proposition 4.185. *Let Q be a loop. Both the lower and the upper central series of Q coincide with those of any inverse of Q.*

Proof. In any loop, central congruence (congruence in which the loop center is a coset class) is normal and we can apply Corollary 1.269. □

4.6.4 Nuclei of loops that are inverse to a fixed loop

We give relations between nuclei of inverse loops in the sense of Definition 1.190.

Theorem 4.186. *1. Between A-nuclei of loops (Q, λ) and $(Q, \cdot)$ there exist the following relations: ${}_1N_l^{\lambda} = {}_3N_l^{\cdot}$, ${}_3N_l^{\lambda} = {}_1N_l^{\cdot}$, ${}_2N_r^{\lambda} = I^{-1}{}_2N_m^{\cdot}I$, ${}_3N_r^{\lambda} = {}_1N_m^{\cdot}$, ${}_1N_m^{\lambda} = {}_3N_r^{\cdot}$, ${}_2N_m^{\lambda} = I^{-1}{}_2N_r^{\cdot}I$.*

2. Between nuclei of loops (Q, λ) and $(Q, \cdot)$ there exist the following relations: $N_l^{\lambda} = N_l^{\cdot}$, $N_r^{\lambda} = N_m^{\cdot}$, $N_m^{\lambda} = N_r^{\cdot}$, $N^{\lambda} = N^{\cdot}$.

3. Between A-nuclei of loops (Q, ρ) and $(Q, \cdot)$ there exist the following relations: ${}_1N_l^{\rho} = I{}_3N_l^{\cdot}I^{-1}$, ${}_3N_l^{\rho} = {}_2N_m^{\cdot}$, ${}_2N_r^{\rho} = {}_3N_r^{\cdot}$, ${}_3N_r^{\rho} = {}_2N_r^{\cdot}$, ${}_1N_m^{\rho} = I{}_1N_l^{\cdot}I^{-1}$, ${}_2N_m^{\rho} = {}_3N_l^{\cdot}$.

4. Between nuclei of loops (Q, ρ) and $(Q, \cdot)$ there exist the following relations: $N_l^{\rho} = N_m^{\cdot}$, $N_r^{\rho} = N_r^{\cdot}$, $N_m^{\rho} = N_l^{\cdot}$, $N^{\rho} = N^{\cdot}$.

5. Between A-nuclei of loops $(Q, *)$ and $(Q, \cdot)$ there exist the following relations: ${}_1N_l^{*} = {}_2N_r^{\cdot}$, ${}_3N_l^{*} = {}_3N_r^{\cdot}$, ${}_2N_r^{*} = {}_1N_l^{\cdot}$, ${}_3N_r^{*} = {}_3N_l^{\cdot}$, ${}_1N_m^{*} = {}_2N_m^{\cdot}$, ${}_2N_m^{*} = {}_1N_m^{\cdot}$.

6. Between nuclei of loops $(Q, *)$ and $(Q, \cdot)$ there exist the following relations: $N_l^{*} = N_r^{\cdot}$, $N_r^{*} = N_l^{\cdot}$, $N_m^{*} = N_m^{\cdot}$, $N^{*} = N^{\cdot}$.

7. Between A-nuclei of loops (Q, ρ^*) and $(Q, \cdot)$ there exist the following relations: ${}_1N_l^{\rho^*} = {}_3N_r^{\cdot}$, ${}_3N_l^{\rho^*} = {}_2N_r^{\cdot}$, ${}_2N_r^{\rho^*} = I{}_1N_l^{\cdot}I^{-1}$, ${}_3N_r^{\rho^*} = {}_2N_m^{\cdot}$, ${}_1N_m^{\rho^*} = {}_3N_l^{\cdot}$, ${}_2N_m^{\rho^*} = I{}_1N_l^{\cdot}I^{-1}$.

8. Between nuclei of loops (Q, ρ^*) and $(Q, \cdot)$ there exist the following relations: $N_l^{\rho^*} = N_r^{\cdot}$, $N_r^{\rho^*} = N_m^{\cdot}$, $N_m^{\rho^*} = N_l^{\cdot}$, $N^{\rho^*} = N^{\cdot}$.

9. Between A-nuclei of loops (Q, λ^*) and $(Q, \cdot)$ there exist the following relations: ${}_1N_l^{\lambda^*} = I^{-1}{}_2N_m^{\cdot}I$, ${}_3N_l^{\lambda^*} = {}_1N_m^{\cdot}$, ${}_2N_r^{\lambda^*} = {}_1N_m^{\cdot}$, ${}_3N_r^{\lambda^*} = {}_1N_l^{\cdot}$, ${}_1N_m^{\lambda^*} = I^{-1}{}_2N_r^{\cdot}I$, ${}_2N_m^{\lambda^*} = {}_3N_r^{\cdot}$.

10. Between nuclei of loops (Q, λ^*) and $(Q, \cdot)$ there exist the following relations: $N_l^{\lambda^*} = N_m^{\cdot}$, $N_r^{\lambda^*} = N_l^{\cdot}$, $N_m^{\lambda^*} = N_r^{\cdot}$, $N^{\lambda^*} = N^{\cdot}$.

Proof. The proof follows from Table 1.4 and the fact that in a loop Q, the left nucleus is an orbit by the action of group ${}_1N_l^A$ or group ${}_3N_l^A$ on the identity element of Q, and so on. □

Part II

Theory

Chapter 5

On two Belousov problems

5.1 The existence of identity elements in quasigroups 265
5.1.1 On quasigroups with Moufang identities 265
5.1.2 Identities that define a CML 272
5.2 Bruck-Belousov problem 274
5.2.1 Introduction 274
5.2.2 Congruences of quasigroups 276
5.2.3 Congruences of inverse quasigroups 282
5.2.4 Behavior of congruences by an isotopy 283
5.2.5 Regularity of quasigroup congruences 284

In this chapter some quasigroup properties that relate to Burmistrovich-Belousov and Bruck-Belousov problems (Problems no. 18 and no. 20 from Belousov's book [72]) are studied.

5.1 The existence of identity elements in quasigroups

In [72, p. 217] the following Burmistrovich-Belousov problem is posed:

Problem 18. From what identities, that are true in a quasigroup $Q(\cdot)$, does it follow that the quasigroup $Q(\cdot)$ is a loop? (An example of such identity is the identity of associativity).

The identity of associativity, the Buchsteiner identity [482], is not a unique identity, that guarantees existence of an identity element in a quasigroup. In the next sections we shall mainly deal with Bol-Moufang-type identities.

Taking into consideration Lemma 1.92 we can generalize Problem 18 in the following way:

Problem 18*. From what identities, that are true in a quasigroup $Q(\cdot)$, does it follow that the quasigroup $Q(\cdot)$ is a left (right) loop?

5.1.1 On quasigroups with Moufang identities

In this section it is proved that a quasigroup with any of the Moufang identities is a loop. The similar result was obtained by K. Kunen using automatic reasoning tools (a version of Otter) [545].

Here we give a proof of this result in which no automatic reasoning tools were used. This result was announced in [445] and was published in [794].

The question about the existence of an identity element in a quasigroup with Moufang identity (5.3) was posed by V.M. Galkin in his private letter to V.D. Belousov in late autumn of 1987.

The identities

$$(x \cdot yz)x = xy \cdot zx, \tag{5.1}$$

$$x(yz \cdot x) = xy \cdot zx, \tag{5.2}$$

$$x(y \cdot xz) = (xy \cdot x)z, \tag{5.3}$$

$$(zx \cdot y)x = z(x \cdot yx) \tag{5.4}$$

are called, respectively, middle left, middle right, left, and right Moufang identities (laws). It is known that in a loop, i.e., in a quasigroup with an identity element, all these identities are equivalent, i.e., if one of the laws (5.1)–(5.4) holds in a loop, then the remaining Moufang identities also hold in the loop [72, 173, 685].

As V.M. Galkin has noticed, if the identity (5.1) or the identity (5.2) is true in a quasigroup, then the quasigroup is a loop [342]. Below we give a proof of this fact.

Proposition 5.1. *If the identity (5.1) or (5.2) holds in a quasigroup $(Q, \cdot)$, then this quasigroup is a loop [342].*

Proof. Suppose that in a quasigroup $(Q, \cdot)$ the identity (5.1) is fulfilled. Substitute $z = f_x$ in (5.1), where, as usual, f_x is the left local unit of the element x, i.e., $x = f_x \cdot x$. We obtain $(x \cdot yf_x)x = xy \cdot f_x x = xy \cdot x$. Thus $(x \cdot yf_x)x = xy \cdot x$. First reducing by x from the right and then from the left, we have $yf_x = y$ for all $x, y \in Q$, i.e., f_x is a right unit of the quasigroup $(Q, \cdot)$. The uniqueness of the right unit is obvious.

We show that left local units of the quasigroup $(Q, \cdot)$ coincide. If f_x and f_y are left local units, then $xf_x = xf_y = x$, and reducing by x, we have $f_x = f_y$ for all $x, y \in Q$. Thus all left and right units of any quasigroup $(Q, \cdot)$ with the identity (5.1) coincide, i.e., $(Q, \cdot)$ is a loop.

Proof for the identity (5.2) is similar. □

The identities (5.1) and (5.2) are usually not distinguished in loops because in a loop from the identity (5.1) or (5.2) the identity of elasticity $x \cdot yx = xy \cdot x$ follows (if $z = 1$).

To prove that if the identity (5.3) holds in a quasigroup, then this quasigroup is a loop, seems to be much harder. Our proof extends over several pages. For brevity, we should call a quasigroup which satisfies the identity (5.3) an LM-quasigroup (from the words "left Moufang quasigroup").

Definition 5.2. We recall that a quasigroup $(Q, \cdot)$ is called an LIP-quasigroup if there exists a mapping λ on Q such that $\lambda x(xy) = y$ for all $x, y \in Q$ [173, 72, 80].

It is easy to check that the mapping λ is a permutation on the set Q [173].

Lemma 5.3. *Any LM-quasigroup is an LIP-quasigroup.*

Proof. It is easy to see that in a quasigroup $(Q, \cdot)$, for each a there exists an element x such that the relation $ax \cdot a = a$ is fulfilled. Indeed, the element x is a solution of the equation $ax = a/a$, where symbol / denotes the operation of the left division, i.e., if $l \cdot m = n$, then $l = n/m$.

Further $x = a\backslash(a/a)$, where symbol \ denotes the operation of the right division in the quasigroup $(Q, \cdot)$, i.e., if $l \cdot m = n$, then $m = l\backslash n$ [72, 80].

Then, using the identity (5.3), we have $ay = (ax \cdot a)y = a(x \cdot ay)$, thus $ay = a(x \cdot ay)$. After the cancelation in the last equality by the element a we obtain $y = x \cdot ay$ for all $y \in Q$, where $x = a\backslash(a/a)$.

Thus for arbitrary element a there is an element x such that for all $y \in Q$ $y = x \cdot ay$, where $x = \lambda a$. From Definition 5.2 it follows that the LM-quasigroup $(Q, \cdot)$ is an LIP-quasigroup. □

The symbol e_x denotes the right local unit of the element x, i.e., the relation $x \cdot e_x = x$ is fulfilled.

The following lemma was proved by V.I. Izbash.

Lemma 5.4. *In an LM-quasigroup $(Q, \cdot)$ the element $f_z f_z$ is an idempotent for any $z \in Q$.*

Proof. In the identity (5.3) let us assume $x = y = f_z$, where f_z is a left local unit of the element z. Then we receive

$$f_z(f_z \cdot f_z z) = (f_z f_z \cdot f_z)z. \tag{5.5}$$

As $f_z(f_z \cdot f_z z) = f_z z$, the following equality is obtained from (5.5)

$$f_z z = (f_z f_z \cdot f_z)z. \tag{5.6}$$

Reducing in the equality (5.6) by z, we obtain

$$f_z = f_z f_z \cdot f_z. \tag{5.7}$$

This equality is true for any element z from the set Q. The equality (5.7) can be written also as

$$f_z f_z = f_{f_z}. \tag{5.8}$$

From the definition of the left local unit we have $f_z \cdot f_z z = z$ for any element z from the set Q.

On the other hand, in view of the LIP-property of an LM-quasigroup $(Q, \cdot)$, we obtain $\lambda f_z \cdot f_z z = z$, therefore, comparing the last two equalities, we get $\lambda f_z = f_z$. Therefore the equalities $\lambda f_z \cdot f_z f_z = f_z \cdot f_z f_z = f_z$ are true, i.e.,

$$f_z f_z = e_{f_z}. \tag{5.9}$$

In the identity (5.3) put $z = y, x = f_y$. Then $f_y(y \cdot f_y y) = (f_y y \cdot f_y)y$, whence we obtain

$$f_y \cdot yy = y f_y \cdot y. \tag{5.10}$$

Now we are ready to prove the existence of an idempotent.

$$f_z f_z \cdot f_z f_z \overset{(5.8)}{=} f_{f_z} \cdot f_z f_z \overset{(5.10)}{=} (f_z f_{f_z}) \cdot f_z \overset{(5.9)}{=} (f_z \cdot e_{f_z}) f_z = f_z f_z.$$

Thus we have received $f_z f_z \cdot f_z f_z = f_z f_z$. The lemma is proved. □

For some fixed element $z \in Q$ we shall denote the element $f_z f_z$ by 1. Note that $1 \cdot 1 = 1$. Consider an LP-isotope of an LM-quasigroup $(Q, \cdot)$ of the form $(R_1^{-1}, L_1^{-1}, \varepsilon)$, where ε is the identity permutation.

It is known [72, 173, 685] that any isotope of a quasigroup of such a form is a loop with the unit $1 \cdot 1 = 1$. Denote this loop by $(Q, \circ)$, where $x \circ y = R_1^{-1}x \cdot L_1^{-1}y$ for all $x, y \in Q$. By replacing in the last expression x by $R_1 x$ and y by $L_1 y$, we obtain $x \cdot y = R_1 x \circ L_1 y$.

For convenience denote R_1 by α and L_1 by β, thus $x \cdot y = \alpha x \circ \beta y$ for all $x, y \in Q$. Note that $\alpha 1 = R_1 1 = 1 \cdot 1 = 1$, $\beta 1 = L_1 1 = 1 \cdot 1 = 1$.

Now let us study properties of the permutations α, β defined above and the loop $(Q, \circ)$.

Lemma 5.5. *a) $\beta^2 = \varepsilon$;*
b) $\beta\alpha\beta = \alpha^2$;
c) $\alpha^3 = \varepsilon$;
d) the permutation β is an automorphism of the loop $(Q, \circ)$.

Proof. a) As it follows from Lemma 5.3, the LM-quasigroup $(Q, \cdot)$ is an LIP-quasigroup. Then $\lambda 1(1 \cdot 1) = 1$, whence we obtain the equality $\lambda 1 \cdot 1 = 1 \cdot 1$, and reducing by 1, we obtain $\lambda 1 = 1$.

Change over in the equality $\lambda x(xy) = y$ to the operation $\circ$:

$$\alpha\lambda x \circ \beta(\alpha x \circ \beta y) = y.$$

If we put here $x = 1$, then we obtain $\alpha\lambda 1 \circ \beta(\alpha 1 \circ \beta y) = 1 \circ \beta(1 \circ \beta y) = \beta^2 y = y$, whence $\beta^2 = \varepsilon$.

b) Change over in the identity (5.3) to the operation $\circ$:

$$\alpha x \circ \beta(\alpha y \circ \beta(\alpha x \circ \beta z)) = \alpha(\alpha(\alpha x \circ \beta y) \circ \beta x) \circ \beta z. \tag{5.11}$$

By substituting $x = z = 1$ in the equality (5.11) we obtain $\beta(\alpha y) = \alpha^2 \beta y$, i.e., $\beta\alpha = \alpha^2\beta$. Taking into account that $\beta^2 = \varepsilon$, we have $\beta\alpha\beta = \alpha^2$.

c) From the last relation we obtain $\alpha = \beta\alpha^2\beta$. As $\beta^2 = \varepsilon$, then

$$\alpha = \beta\alpha^2\beta = \beta\alpha\beta\beta\alpha\beta = \alpha^2\alpha^2 = \alpha^4,$$

whence $\alpha^3 = \varepsilon$.

d) If we put $x = 1$ in the equality (5.11), then we receive $\beta(\alpha y \circ \beta^2 z) = \alpha^2\beta y \circ \beta z$. Taking in consideration that $\beta^2 = \varepsilon$, $\alpha^2\beta = \beta\alpha$, we have $\beta(\alpha y \circ z) = \beta\alpha y \circ \beta z$. Replacing αy by y, we finally obtain $\beta(y \circ z) = \beta y \circ \beta z$ for all $y, z \in Q$, i.e., β is an automorphism of the loop $(Q, \circ)$. □

We recall that the identity $x \circ (y \circ (x \circ z)) = (x \circ (y \circ x)) \circ z$ is called the left Bol identity. We shall name a loop satisfying the left Bol identity a Bol loop.

Lemma 5.6. *The loop $(Q, \circ)$ is a Bol loop for which the relation*

$$x \circ (y \circ x) = \alpha(\alpha(x \circ \alpha y) \circ \alpha\beta x) \tag{5.12}$$

is fulfilled for all $x, y \in Q$.

Proof. If $z = 1$, then from the equality (5.11) we obtain

$$\alpha x \circ \beta(\alpha y \circ \beta(\alpha x)) = \alpha(\alpha(\alpha x \circ \beta y) \circ \beta x). \tag{5.13}$$

If we substitute the left-hand side of the equality (5.13) in the right-hand side of the equality (5.11), then we obtain $\alpha x \circ \beta(\alpha y \circ \beta(\alpha x \circ \beta z)) = (\alpha x \circ \beta(\alpha y \circ \beta\alpha x)) \circ \beta z$.

Taking into account that β is an automorphism of the loop $(Q, \circ)$ and $\beta^2 = \varepsilon$, we obtain from the last equality the following relation:

$$\alpha x \circ (\beta\alpha y \circ (\alpha x \circ \beta z)) =$$
$$(\alpha x \circ (\beta\alpha y \circ \alpha x)) \circ \beta z.$$

Replacing αx by x, $\beta\alpha y$ by y, and βz by z in this relation we obtain that in the loop $(Q, \circ)$ the identity $x \circ (y \circ (x \circ z)) = (x \circ (y \circ x)) \circ z$ is true.

If we replace αx with x and $\beta\alpha y$ with y in the equality (5.13), then y will be changed with $(\beta\alpha)^{-1}y = \alpha^{-1}\beta^{-1}y = \alpha^2\beta y$. Thus we have $x \circ (y \circ x) = \alpha(\alpha(x \circ \beta\alpha^2\beta y) \circ \beta\alpha^{-1}x)$.

Using relations of Lemma 5.5 we shall transform the last equality to the form: $x \circ (y \circ x) = \alpha(\alpha(x \circ \alpha y) \circ \alpha\beta x)$. The lemma is proved. □

Corollary 5.7. *The relation*

$$x \circ x = \alpha(\alpha x \circ \alpha\beta x) \tag{5.14}$$

is fulfilled in the Bol loop $(Q, \circ)$ for all $x \in Q$.

Proof. To receive the relation (5.14) we shall put $y = 1$ in the relation (5.12).

If $\alpha = \beta = \varepsilon$, then $(Q, \cdot) = (Q, \circ)$, the LM-quasigroup $(Q, \cdot)$ becomes a Moufang loop, and the condition (5.12) takes the form $x \circ (y \circ x) = (x \circ y) \circ x$. □

Corollary 5.8. *a) If $\beta = \varepsilon$, then $\alpha = \varepsilon$, and the LM-quasigroup $(Q, \cdot)$ is a loop;*
b) if $\alpha = \varepsilon$, then $\beta = \varepsilon$, and the LM-quasigroup $(Q, \cdot)$ is a loop.

Proof. a) If $\beta = \varepsilon$, then by Lemma 5.5 b) $\alpha = \alpha^2$, thus $\alpha = \varepsilon$.

b) If $\alpha = \varepsilon$, then using the relation (5.14) we obtain $x \circ x = x \circ \beta x$ for any $x \in Q$, therefore $\beta x = x$, i.e., $\beta = \varepsilon$.

If $\alpha^3 = \varepsilon$, then in a disjoint decomposition of the permutation α in cycles, the length of any cycle is either 1 or 3. □

Lemma 5.9. *There is no element $a \in Q$ of the Bol loop $(Q, \circ)$ such that the relations $\alpha a \neq a$, $\beta a = a$ are true.*

Proof. Suppose the contrary, namely, that there is an element a such that the elements $a, \alpha a, \alpha^2 a$ are pairwise different and $\beta a = a$. Consider the element $a \circ a$. As the permutation β is an automorphism of the loop $(Q, \circ)$, we have $\beta(a \circ a) = \beta a \circ \beta a = a \circ a$.

Then we have the following chain of equalities

$$\begin{aligned}&\alpha\beta(a \circ a) = \alpha(\beta(a \circ a)) = \alpha(a \circ a) = \alpha^{-2}(a \circ a) = \\ &\alpha^{-1}(\alpha^{-1}(a \circ a)) \overset{(5.14)}{=} \alpha^{-1}(\alpha a \circ \alpha\beta a) = \alpha^{-1}(\alpha a \circ \alpha a) \overset{(5.14)}{=} \\ &\alpha^2 a \circ \alpha\beta\alpha a = \alpha^2 a \circ \beta a = \alpha^2 a \circ a.\end{aligned}$$

On the other hand

$$\begin{aligned}&\alpha\beta(a \circ a) = \beta\alpha^{-1}(a \circ a) \overset{(5.14)}{=} \beta(\alpha a \circ \alpha\beta a) = \\ &\beta\alpha a \circ \beta\alpha\beta a = \alpha^2\beta a \circ \alpha^2 a = \alpha^2 a \circ \alpha^2 a.\end{aligned}$$

We used the relation (5.14) and the relations of Lemma 5.5.

Comparing the right-hand sides of the last chains of equalities, we obtain that $\alpha^2 a \circ a = \alpha^2 a \circ \alpha^2 a$ whence $\alpha^2 a = a, \alpha a = a$. We obtain a contradiction with a condition of the lemma. The lemma is proved. □

Corollary 5.10. *If $\beta a = a$ for some element $a \in (Q, \circ)$, then $\alpha a = a$.*

Lemma 5.11. *If $\alpha a = a$, then $\beta a = a$.*

Proof. Suppose the contrary, namely, that there is an element b, $b \neq a$, such that $\beta a = b$ and $\alpha a = a$. Then $\beta b = a$ by Lemma 5.5 (b). Note that $\alpha b = b$. Indeed, if $\beta a = b$, then $\alpha\beta b = \alpha a = a$. By Lemma 5.5 (b) $\alpha\beta = \beta\alpha^2$. Therefore $\beta\alpha^2 b = a$, whence we obtain $\beta(\beta\alpha^2)b = \beta a, \alpha^2 b = b, \alpha b = b$.

Then $a \circ a \overset{(5.14)}{=} \alpha(\alpha a \circ \alpha\beta a) = \alpha(a \circ b)$, $\beta(a \circ a) = b \circ b = \beta\alpha(a \circ b) = \alpha^{-1}(b \circ a)$. Thus

$$b \circ b = \alpha^{-1}(b \circ a). \tag{5.15}$$

But $b \circ b \overset{(5.14)}{=} \alpha(\alpha b \circ \alpha\beta b) = \alpha(b \circ a)$, i.e., $b \circ a = \alpha^{-1}(b \circ b)$. If we substitute the right-hand side of the last relation for the element $b \circ a$ in the equality (5.15), we obtain $b \circ b = \alpha^{-1}\alpha^{-1}(b \circ b) = \alpha(b \circ b)$, i.e., $b \circ b = \alpha(b \circ b)$.

But from the relation (5.15) we have $b \circ a = \alpha(b \circ b)$. From the last two expressions it follows that $b \circ a = b \circ b, b = a$. We obtain a contradiction. Thus our assumption is not true. If $\alpha a = a$, then $\beta a = a$. □

Corollary 5.12. *The equivalence $\alpha x = x \iff \beta x = x$ is true in the loop $(Q, \circ)$ for any $x \in Q$.*

Proof. The proof follows from Lemma 5.11 and Corollary 5.10. □

Let I be the permutation on the set Q such that $Ix = x^{-1}$. It is known that for any automorphism β of loop $(Q, \circ)$ the relation $\beta I = I\beta$ or, in other words, $(\beta x)^{-1} = \beta x^{-1}$, is true for any $x \in Q$.

Indeed, $\beta x \circ \beta x^{-1} = \beta(x \circ x^{-1}) = 1 = \beta x \circ (\beta x)^{-1}$.

Lemma 5.13. *In the loop $(Q, \circ)$ the equivalence $\beta x = x \iff \beta x^{-1} = x^{-1}$ is true for any $x \in Q$.*

Proof. Let $\beta x = x$ for some element x. Consider the product $x \circ x^{-1} = 1$. Then $\beta x \circ \beta x^{-1} = 1$. Using the condition of the lemma we have $x \circ \beta x^{-1} = 1$, and by the LIP-property of the loop $(Q, \circ)$, we obtain $\beta x^{-1} = x^{-1}$. Proof of the inverse implication is similar. □

Remember that a Bol loop $(Q, \circ)$ satisfies the left alternative law $x \circ (x \circ y) = (x \circ x) \circ y$ ([72, 173]).

Lemma 5.14. *If $\beta x \neq x$, then $\beta x \neq x^{-1}$.*

Proof. If $\beta x \neq x$, then $\alpha x \neq x$ by Corollary 5.12. Suppose that there is an element x such that $\beta x \neq x$ and $\beta x = x^{-1}$.

Then we have

$$\alpha x \circ (\beta x \circ \alpha x) \overset{(5.12)}{=} \alpha(\alpha(\alpha x \circ \alpha\beta x) \circ \alpha\beta\alpha x) \overset{(5.14)}{=} \alpha((x \circ x) \circ \alpha\beta\alpha x) = \alpha((x \circ x) \circ \beta x).$$

Applying the left alternative law we obtain $\alpha((x \circ x) \circ \beta x) = \alpha(x \circ (x \circ \beta x))$.

Further, using our assumption, we obtain $\alpha(x \circ (x \circ \beta x)) = \alpha x$.

Thus $\alpha x \circ (\beta x \circ \alpha x) = \alpha x$. Reducing by αx, we obtain $\beta x \circ \alpha x = 1$. Since $\beta x = x^{-1}$, we get $x^{-1} \circ \alpha x = 1$, whence we have $\alpha x = x$. We have received a contradiction with the relation $\alpha x \neq x$. The lemma is proved. □

Hereinafter we shall also use the equivalent form of Lemma 5.14: if $\beta x = x^{-1}$, then $\beta x = x$.

Lemma 5.15. *If there exists an element a such that $\beta a \neq a$, then there is an element k such that $\beta a = k \circ a$, where $\beta k = k$ and $k^2 = 1$.*

Proof. If $\beta a \neq a$, then there exists an element $k \neq 1, k \in Q$, such that $\beta a = k \circ a$, because $(Q, \circ)$ is a quasigroup. As the loop $(Q, \circ)$ is an *LIP*-loop, $k^{-1} \circ \beta a = a$. Therefore $\beta(k^{-1} \circ \beta a) = \beta k^{-1} \circ \beta\beta a = \beta a$.

Thus $\beta k^{-1} \circ a = \beta a$. Since $\beta a = k \circ a$, we obtain $\beta k^{-1} \circ a = k \circ a$. Reducing in the last equality by the element a we have $\beta k^{-1} = k$, $k^{-1} = \beta k$. Then it follows by Lemma 5.14 that $\beta k = k$.

If $\beta a = k \circ a$, then $\beta\beta a = a = \beta(k \circ a) = \beta k \circ \beta a = k \circ \beta a$ and therefore $a = k \circ \beta a$. Substituting the right-hand side of the last equality for the element a in the equality $\beta a = k \circ a$, we have $\beta a = k \circ (k \circ \beta a)$. Using the left alternative law we obtain $\beta a = k^2 \circ \beta a, k^2 = 1$. The lemma is proved. □

Lemma 5.16. *If the inequality $\alpha a \neq a$ holds for some element a, then there exists an element k such that $\alpha a = a \circ k$, where $\alpha k = k$, $\beta k = k$ and $k^2 = 1$.*

Proof. Consider the expression $a^{-1}\circ(a\circ a^{-1}) = a^{-1}$. Using (5.12), we have $a^{-1} = \alpha(\alpha(a^{-1}\circ \alpha a)\circ\alpha\beta a^{-1})$ or $\alpha^{-1}a^{-1} = \alpha^2 a^{-1} = \alpha(a^{-1}\circ\alpha a)\circ\alpha\beta a^{-1}$.

From the last relation it follows that $\alpha(a^{-1}\circ\alpha a)\circ\beta\alpha^2 a^{-1} = \alpha^2 a^{-1}$. Then we receive $\beta\alpha(a^{-1}\circ\alpha a)\circ\alpha^2 a^{-1} = \beta\alpha^2 a^{-1}$.

Show that the elements $\beta\alpha^2 a^{-1}$ and $\alpha^2 a^{-1}$ are different and that we can apply Lemma 5.15 to the last equality.

Suppose $\beta\alpha^2 a^{-1} = \alpha^2 a^{-1}$. Then by Corollary 5.12 $\alpha\alpha^2 a^{-1} = \alpha^2 a^{-1}$, by Lemma 5.5 $a^{-1} = \alpha^2 a^{-1}$ and $\alpha a^{-1} = a^{-1}$, then again by Corollary 5.12 $\beta a^{-1} = a^{-1}$. By Lemma 5.13 we obtain $\beta a = a$.

Therefore the elements $\beta\alpha^2 a^{-1}$ and $\alpha^2 a^{-1}$ are different and, finally, by applying Lemma 5.15, we obtain $\alpha(a^{-1}\circ\alpha a) = k$, where $k^2 = 1, \beta k = k$.

From the last relation it follows that $a^{-1}\circ\alpha a = \alpha^{-1}k$. The equality $\alpha k = k$ follows from Corollary 5.12, therefore we receive $a^{-1}\circ\alpha a = k$. As the loop $(Q,\circ)$ is an LIP-loop, from the last relation it follows that $a\circ(a^{-1}\circ\alpha a) = a\circ k$, $\alpha a = a\circ k$. The lemma is proved. □

To distinguish between the elements k, corresponding to different elements βa, we shall write the relation $\beta a = k\circ a$ as

$$\beta a = k_{\beta a}\circ a. \tag{5.16}$$

Remark 5.17. If we put $\beta = \varepsilon$ in the equality (5.16), then we have $a = k_a\circ a$. Since $(Q,\circ)$ is a loop, then in this case $k_a = 1$.

Lemma 5.18. *In the quasigroup $(Q,\cdot)$ the relation $a = k\circ\beta a$ takes the form $f_a\cdot a = a$, i.e., $k = f_a$.*

Proof. Taking into consideration Lemma 5.15 it is easy to see that relations $k\circ a = \beta a$ and $k\circ\beta a = a$ are equivalent. Passing to the operation $\cdot$ in the last equality we have $\alpha^{-1}k\cdot a = a$, $k\cdot a = a$, $k = f_a$. □

Lemma 5.19. *There is no element a of the loop $(Q,\circ)$ such that $\alpha a \neq a$ and $\beta a\neq a$.*

Proof. Suppose the contrary, that such an element a exists. Let's say that then the elements $a, \alpha a, \alpha^2 a$ are pairwise different, and we can apply Lemmas 5.15 and 5.16 to the element a.

Show that $k_{\beta\alpha a} = k_{\alpha a}$. Replace the element a by the element $\beta\alpha a$ in the equality (5.16). It is possible, as $\beta(\beta\alpha a)\neq\beta\alpha a$.

Indeed, if we suppose that $\beta(\beta\alpha a) = \beta\alpha a$, then $\beta\alpha a = \alpha a$, and by Corollary 5.12, we get $\alpha(\alpha a) = \alpha a, \alpha a = a$. We obtain a contradiction with the assumption.

After the replacement we have the relation $\alpha a = k_{\alpha a}\circ\beta\alpha a$. Then we receive

$$\beta\alpha a = k_{\alpha a}\circ\alpha a, \tag{5.17}$$

since $\beta k_{\alpha a} = k_{\alpha a}$ by Lemma 5.15.

We may substitute the element αa in (5.16) for the element a. Indeed, if we suppose that $\beta\alpha a = \alpha a$, then by Corollary 5.12 $\alpha(\alpha a) = \alpha a$, whence $\alpha a = a$. We obtain a contradiction with the assumption. After the substitution we get

$$\beta\alpha a = k_{\beta\alpha a}\circ\alpha a. \tag{5.18}$$

Comparing the equality (5.17) with the equality (5.18), we obtain

$$k_{\beta\alpha a} = k_{\alpha a}. \tag{5.19}$$

If we change the element αa by the element a in the equality (5.19), then the equality (5.19) takes the form

$$k_{\beta a} = k_a. \tag{5.20}$$

Thus from the equality (5.20), taking into consideration Remark 5.17, it follows that $k_{\beta a} = 1$. Therefore the equality (5.16) takes the form $\beta a = a$. We have received a contradiction with the assumption that $\beta a \neq a$. The lemma is proved. □

Now we can prove the following:

Theorem 5.20. *If the identity $x(y \cdot xz) = (xy \cdot x)z$ holds in a quasigroup $(Q, \cdot)$, then the quasigroup is a loop.*

Proof. It follows from Lemma 5.9 that the case $\alpha a \neq a$, $\beta a = a$ is impossible. By Lemma 5.19 it follows that the case $\alpha a \neq a$, $\beta a \neq a$ is impossible, too, for any element a of the Bol loop $(Q, \circ)$.

Then the only possibility for the permutation α is $\alpha = \varepsilon$. Then by Corollary 5.12 it follows that $\beta = \varepsilon$. Since $\alpha = \beta = \varepsilon$, the quasigroup $(Q, \cdot)$ coincides with the loop $(Q, \circ)$, thus the LM-quasigroup $(Q, \cdot)$ is a loop. □

Corollary 5.21. *If the identity $(zx \cdot y)x = z(x \cdot yx)$ is fulfilled in a quasigroup, then this quasigroup is a loop.*

Proof. It is known that if a quasigroup $(Q, \circ)$ is a loop, then its parastrophe $(Q, \star)$, $(x \star y = y \circ x)$ is a loop [72]. It is easy to check that if in a quasigroup $(Q, \cdot)$ the identity (5.3) is fulfilled, then in its parastrophe of this form, i.e., in a quasigroup $(Q, \circledast)$, the identity (5.4) is true. Further, if the quasigroup $(Q, \cdot)$ with the identity (5.3) is a loop by Theorem 5.20, then in the quasigroup $(Q, \circledast)$ the identity (5.4) is fulfilled and it is a loop. □

Theorem 5.22 follows from Proposition 5.1, Theorem 5.20 and Corollary 5.21.

Theorem 5.22. *If in a quasigroup at least one of the Moufang identities is fulfilled, then this quasigroup is a loop.*

5.1.2 Identities that define a CML

Recall, the following identity

$$xx \cdot yz = xy \cdot xz \tag{5.21}$$

is called left semi-medial identity, and the identity

$$xy \cdot zz = xz \cdot yz \tag{5.22}$$

is called right semi-medial. See Chapter 6 for more detailed information about left (right) SM-quasigroups. These identities are the standard identities defining CML [72].

The quasigroup given in Example 5.23 is left semi-medial.

Example 5.23.

$*$	0	1	2
0	1	2	0
1	0	1	2
2	2	0	1

Theorem 5.24. *The following identity*

$$x(xy \cdot z) = (z \cdot xx)y \tag{5.23}$$

([219, identity A2I4]) defines the class of commutative Moufang loops [219, Theorem 4.5].

Proof. Letting $x = 1$ in (5.23), $x(xy \cdot z) = (z \cdot xx)y$, gives commutativity, $yz = zy$. Similarly, by setting $y = 1$, we have $x \cdot xz = z \cdot xx$. Using these,

$$x(z \cdot xy) = x(xy \cdot z) \text{ (by commutativity)} = (z \cdot xx)y \text{ (assumption)} =$$
$$(x \cdot xz)y = (xz \cdot x)y \text{ (by commutativity)}.$$

Thus in the class of loops, the identity (5.23) defines commutative Moufang loops. □

Example 5.25. The following quasigroup satisfies identity (5.23) and does not have any left and right identity element.

*	0	1	2
0	0	2	1
1	2	1	0
2	1	0	2

Using Prover 9 [615] it is possible to see that in quasigroup case, left semi-medial identity follows from the identity (5.23). The inverse is not true, i.e., there exist quasigroups in which left semi-medial identity is true and the identity (5.23) is not true. The quasigroup from Example 5.23 is just such a quasigroup.

One more identity, which defines a CML in loop case, is given in [593]

$$x(y \cdot xz) = x^2y \cdot z. \tag{5.24}$$

A quasigroup which satisfies the identity (5.24) is given in Example 5.25.

After a little modernization of Moufang identities we can obtain the following identities:

$$(xy \cdot x)z = (y \cdot xz)x \tag{5.25}$$

and

$$x(y \cdot zy) = y(xy \cdot z). \tag{5.26}$$

Notice, the identities (5.25) and (5.26) are Bol-Mofang-type identities. We can call these identities left and right CML identities, respectively.

Theorem 5.26. *If in a quasigroup the identity (5.25) or identity (5.26) is true, then this quasigroup is a loop.*

Proof. It is possible to use Prover 9 [615]. □

Theorem 5.27. *In the loop case the identities (5.21)–(5.26) are equivalent. Any of these identities defines a commutative Moufang loop.*

Proof. It is possible to use Prover 9 [615]. □

It is clear that in the quasigroup case these identities are not equivalent.

In [321, 322] the following result is proved: "Let x, y, z denote variables and let e be a distinguished element in a groupoid G. With the aid of an automatic theorem prover the author has obtained a proof that the identity $e(x((x((x(yy))z))y)) = z$ is a shortest single identity axiom for the variety of commutative Moufang loops of exponent 3 with neutral element e." MR2381424 (2008m:20112), A. Schleiermacher.

5.2 Bruck-Belousov problem

In this section, in the language of subgroups of the multiplication group of a quasigroup (of the associated group of a quasigroup), necessary and sufficient conditions of normality of congruences of a left (right) loop are given. These conditions can be considered as a partial answer to the problem posed in the books of R. H. Bruck and V.D. Belousov about conditions of normality of all congruences of quasigroups.

Results on the behavior of quasigroup congruences by isotopy are given.

5.2.1 Introduction

It is proved that the class of quasigroups with one binary operation in signature (see Definition 1.21) is not closed relative to homomorphic images, i.e., the homomorphic image of a quasigroup in the general case is only a division groupoid [43].

Various questions of homomorphic theory of quasigroups with one binary operation in signature were studied in [418, 346, 10, 11, 167, 522, 43, 72, 460, 52, 895, 799, 105].

The main purpose of this section is an attempt at solving the Bruck-Belousov problem: *What loops G have the property that every image of G under a multiplicative homomorphism is also a loop ([173, p. 92])? What are the quasigroups or loops in which all congruences are normal ([72, Problem 20, p. 217])?*

We notice, it is well known, that if a homomorphic image of a multiplicative homomorphism φ of a loop is also a loop, then congruence θ which corresponds to φ, is a normal congruence.

This section is based on results which were published in [774, 782], see, also, [766, 767, 768]. We shall use standard quasigroup notations and definitions from [72, 77, 173, 685]. Information on lattices and universal algebras can be found in [151, 548, 813, 358], on groups in [549, 471], and on semigroups in [211].

By $\Pi(Q)$ or by Π for short, we shall designate a semigroup generated by the left and right translations of a quasigroup Q, i.e., elements of a semigroup $\Pi(Q)$ are words of the form $T_1^{\alpha_1}T_2^{\alpha_2}\dots T_n^{\alpha_n}$, where $T_i \in \mathbb{T}$, $\alpha_i \in \mathbb{N}$.

The group generated by all left and right translations of a quasigroup Q will be denoted by $M(Q)$, or by M for short. Elements of the group M are words of the form $T_1^{\alpha_1}T_2^{\alpha_2}\dots T_n^{\alpha_n}$, where $T_i \in \mathbb{T}$, $\alpha_i \in \mathbb{Z}$.

A binary relation φ on a set Q is a subset of the Cartesian product $Q \times Q$ [151, 678].

If φ and ψ are binary relations on Q, then their product is defined in the following way: $(a, b) \in \varphi \circ \psi$ if there is an element $c \in Q$ such that $(a, c) \in \varphi$ and $(c, b) \in \psi$. If φ is a binary relation on Q, then $\varphi^{-1} = \{(y, x) \mid (x, y) \in \varphi\}$. The operation of the product of binary relations is associative [211, 678, 813, 711].

For convenience of readers we recall the following definitions.

Definition 5.28. An equivalence θ of a quasigroup $(Q, \cdot)$ such that from $a\theta b$ there follows $(c \cdot a)\theta(c \cdot b)$ for all $a, b, c \in Q$ is called a *left congruence* of a quasigroup $(Q, \cdot)$.

An equivalence θ of a quasigroup $(Q, \cdot)$ such that from $a\theta b$ there follows $(a \cdot c)\theta(b \cdot c)$ for all $a, b, c \in Q$ is called a *right congruence* of a quasigroup $(Q, \cdot)$.

A left and right congruence θ of a quasigroup $(Q, \cdot)$ is called a *congruence* of a quasigroup $(Q, \cdot)$ [72, 77].

A congruence θ of a quasigroup $(Q, \cdot)$ is called *left normal,* if from $(c \cdot a)\theta(c \cdot b)$ there follows $a\theta b$ for all $a, b, c \in Q$.

A congruence θ of a quasigroup $(Q, \cdot)$ is called *right normal,* if from $(a \cdot c)\theta(b \cdot c)$ there follows $a\theta b$ for all $a, b, c \in Q$.

A congruence θ of a quasigroup $(Q, \cdot)$ is called *normal,* if it is left normal and right normal.

We shall call a binary relation θ of a groupoid $(Q, \cdot)$ *stable from the left* (accordingly from the right) if from $x\theta y$ there follows $(a \cdot x)\theta(a \cdot y)$, (accordingly $(x \cdot a)\theta(y \cdot a)$) for all $a \in Q$.

It is easy to see that a stable from the left (from the right) equivalence of a quasigroup $(Q, \cdot)$ is called a *left (right) congruence.* A congruence is an equivalence relation which is stable from the left and from the right.

If θ is a binary relation on a set Q, α is a permutation of the set Q and from $x\theta y$ there follows $\alpha x\theta\alpha y$ and $\alpha^{-1}x\theta\alpha^{-1}y$ for all $(x, y) \in \theta$, then we shall say that the permutation α is an *admissible* permutation relative to the binary relation θ.

Below we shall designate the product of binary relations and quasigroup operation by a point, by letter Q we shall designate a quasigroup $(Q, \cdot)$ and a set on which this quasigroup is defined.

Lemma 5.29. *For all binary relations $\varphi, \psi, \theta \subseteq Q^2$ from $\varphi \subseteq \psi$ there follows $\varphi\,\theta \subseteq \psi\,\theta$, $\theta\,\varphi \subseteq \theta\,\psi$, i.e., it is possible to say that a binary relation of set-theoretic inclusion of binary relations is stable from the left and from the right relative to the multiplication of binary relations.*

Proof. If $(x, z) \in \varphi\,\theta$, then there exists an element $y \in Q$, such that $(x, y) \in \varphi$ and $(y, z) \in \theta$. Since $\varphi \subseteq \psi$, then we have $(x, y) \in \psi$, $(x, z) \in \psi\,\theta$. □

Remark 5.30. Translations of a quasigroup can be considered as binary relations: $(x, y) \in L_a$, if and only if $y = a \cdot x$, $(x, y) \in R_b$, if and only if $y = x \cdot b$.

Remark 5.31. To coordinate the multiplication of translations with their multiplication as binary relations, we use the following multiplication of translations: if α, β are translations, x is an element of the set Q, then $(\alpha\beta)(x) = \beta(\alpha(x))$, i.e., $(\alpha\beta)x = \beta\alpha\, x$.

Therefore in this section we use the following order of multiplication (of composition) of maps: $(\alpha\beta)(x) = \beta(\alpha(x))$, where α, β are maps.

A partially ordered set $(L, \subseteq)$ is called a lower (an upper) semilattice if its two-element subset has an exact lower (upper) bound, i.e., in a set L there exists $\inf(a, b)$ $(\sup(a, b))$ for all $a, b \in L$ [151, 548].

If a partially ordered set is simultaneously the lower and upper semilattice, then it is called a *lattice.*

We can define a lattice as algebra $(L, \vee, \wedge)$ satisfying the following axioms [151]:

$$\begin{array}{ll} (a \vee b) \vee c = a \vee (b \vee c); & a \vee b = b \vee a; \\ a \vee a = a; & (a \vee b) \wedge a = a; \\ (a \wedge b) \wedge c = a \wedge (b \wedge c); & a \wedge b = b \wedge a; \\ a \wedge a = a; & (a \wedge b) \vee a = a. \end{array}$$

We notice that similar to quasigroups, which are defined in a signature with one and three binary operations, for the lattices which are defined in a signature with one binary operation $\leqslant$ and with two binary operations $\vee$ and $\wedge$, the concepts of a sublattice do not coincide. Namely, the sublattice of a lattice $(L, \vee, \wedge)$ always is a sublattice of a lattice $(L, \leqslant)$, but an inverse is not always correct [548].

5.2.2 Congruences of quasigroups

Connections between normal subloops of a loop Q and normal subgroups of the group $M(Q)$ were studied by A. Albert [10, 11]. Generalizations of Albert's results on some classes of quasigroups can be found in works of V.A. Beglaryan [52] and K.K. Shchukin [799], on some classes of groupoids in [895]. In these works, questions on the lattice embedding the lattices of some normal congruences of a quasigroup Q into the lattice of normal subgroups of the group $M(Q)$ are studied. Information on lattice isomorphic medial distributive quasigroups is in [770].

Here we give results on normality of congruences, which can be found in more detailed form in Section 4.4.2.

Corollary 5.32. *An equivalence θ of a quasigroup $(Q,\cdot)$ is a congruence if and only if $\theta\omega \subseteq \omega\theta$ for any $\omega \in \Pi$.*

Proof. The multiplication of binary relations is associative, therefore, if $\theta\omega_1 \subseteq \omega_1\theta$, $\theta\omega_2 \subseteq \omega_2\theta$, where $\omega_1, \omega_2 \in \Pi$, then $\theta(\omega_1\omega_2) = (\theta\omega_1)\omega_2 \subseteq (\omega_1\theta)\omega_2 = \omega_1(\theta\omega_2) \subseteq \omega_1(\omega_2\theta) = (\omega_1\omega_2)\theta$. □

Corollary 5.33. *An equivalence θ is a congruence of a quasigroup Q if and only if $\omega\theta(x) \subseteq \theta(\omega x)$ for all $x \in Q$, $\omega \in \Pi$.*

Proof. The proof is similar to the previous one. □

Corollary 5.34. *A congruence θ of a quasigroup Q is normal if and only if $\omega\theta = \theta\omega$ for all $\omega \in \Pi$.*

A congruence θ of a quasigroup Q is normal if and only if $\omega\theta = \theta\omega$ for all $\omega \in M$.

Proof. The proof is obvious. □

It is easy to see that an equivalence q of a set M is a congruence of the group M if and only if q is admissible (even semi-admissible) relative to all elements of the set $\mathbb{T} \cup \mathbb{T}^{-1}$, where $\mathbb{T}^{-1} = \{L_x^{-1}, R_x^{-1} \mid \forall x \in Q\}$.

Theorem 5.35. *The lattice of congruences $(L(Q), \leqslant_1)$ of a one-sided loop (in particular, of a loop) Q is isomorphically embedded in the lattice $(L(M(Q)), \leqslant_2)$ of the left congruences of group M, which are semi-admissible from the right relative to all permutations of the semigroup Π.*

Proof. The proof of this theorem in some parts repeats the proof of the theorem on an isomorphic embedding of normal congruences of a quasigroup Q in the lattice of congruences of the group $M(Q)$ [885, 884].

By a quasigroup Q during the proof of this theorem we shall understand a quasigroup with the right unit, i.e., right loop.

Let q be a congruence of a quasigroup Q. We shall define the relation $q^{\top}$ in group M as follows: $\theta q^{\top}\varphi \Longleftrightarrow \theta^{-1}\varphi \subseteq q$ for all $\theta, \varphi \in M$.

We prove that $q^{\top}$ is a left congruence of the group M which is admissible from the right relative to all permutations α, $\alpha \in \Pi$.

Reflexivity of $q^{\top}$. Since $\varepsilon \subseteq q$, $\alpha q^{\top}\alpha$ for all $\alpha \in M$.

Symmetry of $q^{\top}$. The equivalence $\theta q^{\top}\varphi \leftrightarrow \varphi q^{\top}\theta$ is equivalent to the equivalence $\theta^{-1}\varphi \subseteq q \leftrightarrow \varphi^{-1}\theta \subseteq q$. The last equivalence is true since, if $\theta^{-1}\varphi \subseteq q$, then $(\theta^{-1}\varphi)^{-1} \subseteq q^{-1}$, $\varphi^{-1}\theta \subseteq q^{-1} = q$. It is clear that in the same way it is possible to also receive an inverse implication: $(\varphi^{-1}\theta \subseteq q) \rightarrow (\theta^{-1}\varphi \subseteq q)$.

Transitivity of $q^{\top}$. An implication $\theta q^{\top}\varphi \wedge \varphi q^{\top}\psi \to \theta q^{\top}\psi$ is equivalent to the implications $\theta^{-1}\varphi \subseteq q \wedge \varphi^{-1}\psi \subseteq q \to \theta^{-1}\psi \subseteq q$. We shall show that the last implication is fulfilled. Indeed, if $\theta^{-1}\varphi \subseteq q \wedge \varphi^{-1}\psi \subseteq q$, then $\theta^{-1}\varphi\varphi^{-1}\psi = \theta^{-1}\psi \subseteq q^2 = q$.

Let us show that $q^{\top}$ is semi-admissible from the left relative to any permutation $\alpha \in M$. Indeed, the condition "if $\theta q^{\top}\varphi$, then $\alpha\theta q^{\top}\alpha\varphi$" is equivalent to the following condition: if $\theta^{-1}\varphi \subseteq q$, then $\theta^{-1}\alpha^{-1}\alpha\varphi = \theta^{-1}\varphi \subseteq q$.

Let us show that the binary relation $q^{\top}$ (we have already proved that $q^{\top}$ is a left congruence of M) is semi-admissible from the right relative to any permutation $\alpha \in \Pi$. For this purpose we shall show that $\theta\alpha q^{\top}\varphi\alpha$ for all $\alpha \in \Pi$. We shall pass, using Proposition 4.139, to the needed inclusions.

Then we have $\theta q^{\top}\varphi \leftrightarrow \theta^{-1}\varphi \subseteq q$, $\theta\alpha q^{\top}\varphi\alpha \leftrightarrow \alpha^{-1}\theta^{-1}\varphi\alpha \subseteq q$. Since q is a congruence, then by Proposition 4.139 we have $\alpha^{-1}q\alpha \subseteq q$ for all $\alpha \in \Pi$. Therefore, if $\theta^{-1}\varphi \subseteq q$, then $\alpha^{-1}\theta^{-1}\varphi\alpha \subseteq \alpha^{-1}q\alpha \subseteq q$.

Thus, we have proved that an arbitrary congruence of a quasigroup Q corresponds the left congruence $q^{\top}$ of the group M which is semi-admissible from the right relative to all permutations of the semigroup Π.

Let p be a left congruence of the group M, that is semi-admissible from the right relative to all $\alpha \in \Pi$. We shall define a binary relation on a quasigroup Q in the following way: $p^{\perp} = \cup\, \theta^{-1}\varphi$ for all $\theta, \varphi \in M$, such that $\theta\, p\, \varphi$.

We demonstrate that $p^{\perp}$ is a congruence of a quasigroup Q.

Reflexivity of $p^{\perp}$. Since $\theta\, p\, \theta$ for all $\theta \in M$, then $\theta^{-1}\theta = \varepsilon \subseteq p^{\perp}$.

Symmetry of $p^{\perp}$. $(p^{\perp})^{-1} = p^{\perp}$, since $p^{-1} = p$ and $p^{\perp} = \cup\,\theta^{-1}\varphi$ for all $\theta, \varphi \in M$, such that $\theta\, p\, \varphi$.

Transitivity of $p^{\perp}$. Let $(a, b) \in (p^{\perp})^2$, i.e., there exists element c such that $ap^{\perp}c$ and $cp^{\perp}b$. Hence, there exist $\theta, \varphi, \psi, \xi \in M$, $\theta p\varphi$, $\psi p\xi$, such that $a\theta^{-1}\varphi c$ and $c\psi^{-1}\xi b$. Then $c = (\theta^{-1}\varphi)\, a, b = (\psi^{-1}\xi)\, c$, and $b = (\theta^{-1}\varphi\psi^{-1}\xi)\, a$, i.e., $(a, b) \in (\varphi^{-1}\theta)^{-1}\psi^{-1}\xi$.

We need to prove that $\varphi^{-1}\theta\, p\, \psi^{-1}\xi$. If $\theta\, p\, \varphi$, then taking into account that the binary relation p is stable from the left relative to any permutation $\alpha \in M$, we obtain, $\varphi^{-1}\theta\, p\, \varphi^{-1}\varphi$, $\varphi^{-1}\theta\, p\, \varepsilon$.

Similarly, $\varepsilon\, p\psi^{-1}\xi$, and by transitivity of the relation p we have: $\varphi^{-1}\theta\, p\, \psi^{-1}\xi$. Thus, we have proved that p is an equivalence on Q.

Let us show that $p^{\perp}$ is a congruence of a quasigroup Q. For this purpose it is sufficient, taking into account Corollary 5.32, to prove that for all $\omega \in \Pi$, $\omega^{-1}\, p^{\perp}\, \omega \subseteq p^{\perp}$.

Let $(a, b) \in \omega^{-1}p^{\perp}\omega$. Then there exist $\varphi, \theta \in M$, $\theta p\varphi$ such that $(a, b) \in \omega^{-1}\theta^{-1}\varphi\omega = (\theta\omega)^{-1}\varphi\omega$.

Since $\theta\, p\, \varphi$, then for all $\omega \in \Pi$, $\theta\omega\, p\, \varphi\omega$, and then $(\theta\omega)^{-1}\varphi\omega \subseteq p^{\perp}$.

Thus $(a, b) \in p^{\perp}$, $\omega^{-1}\, p^{\perp}\, \omega \subseteq p^{\perp}$ for all $\omega \in \Pi$, i.e., $p^{\perp}$ is a congruence of a quasigroup Q.

We prove that if q is a congruence of a quasigroup Q, then $q^{\top\perp} = q$, i.e., we establish that the map $\top$ is a bijective map and that $(\top)^{-1} = \perp$.

It is easy to understand that $q^{\top\perp} \subseteq q$. Indeed, if $(a, b) \in (q^{\perp})^{\top}$, there is a pair of permutations $\varphi, \theta \in M$ such that $\theta q^{\top}\varphi$, $(a, b) \in \theta^{-1}\varphi$.

By definition of the relation $q^{\top}$, $\varphi q^{\top}\theta$ if and only if $\varphi^{-1}\theta \subseteq q$, and then $(a, b) \in q$.

Let us prove a converse inclusion. Now we use the property that the quasigroup Q has the right identity element.

Let $(a, b) \in q$. Then for all x from Q the relation $ax\, q\, bx$ is equivalent with $L_a x\, q\, L_b x$. Having replaced x by $L_a^{-1}x$, we obtain $x\, q\, (L_a^{-1}L_b)x$, i.e., $L_a^{-1}L_b \subseteq q$, and then $L_a q^{\top} L_b$,

$L_a^{-1}L_b \subseteq (q^{\top})^{\perp}$. From the last relation we have $(a, (L_a^{-1}L_b)a) = (a, L_b e_a) = (a, b) \in (q^{\top})^{\perp}$, since $e_a = e_b$. Therefore $q \subseteq (q^{\top})^{\perp}$.

If we have a quasigroup with the left unit, then instead of translations L_a, L_b we take translations R_a, R_b. Thus $(q^{\top})^{\perp} \supseteq q$, the map $\top$ is a bijective map, $(\top)^{-1} = \perp$.

Let us recall that lattices $\mathcal{L}_1 = (L_1, \leqslant_1)$ and $\mathcal{L}_2 = (L_2, \leqslant_2)$ are called isomorphic if there is a bijective map σ such that $a \leqslant_1 b$ in $\mathcal{L}_1$ if and only if $\sigma(a) \leqslant_2 \sigma(b)$ in $\mathcal{L}_2$ [151].

In order to prove that $\top$ is a lattice isomorphism, we need to prove: if $q_1 \subseteq q_2$, then $q_1^{\top} \subseteq q_2^{\top}$; if $p_1 \subseteq p_2$, then $p_1^{\perp} \subseteq p_2^{\perp}$, where q_1, q_2 are congruences of a quasigroup Q, p_1, p_2 are the left congruences of group M that are semi-admissible relative to multiplication from the right on permutations α from the semigroup Π. These two implications, taking into account the definition of maps $\perp, \top$, are obvious. □

Proposition 5.36. *If the lattice of congruences is considered as an algebra of the form $(L, \vee, \wedge)$, i.e., in a signature with two binary operations, then $(q_1 \wedge q_2)^{\top} = q_1^{\top} \wedge q_2^{\top}$.*

Proof. Indeed, the operation $\wedge$ both in a lattice of congruences of a quasigroup and in a lattice of the left congruences of group M coincides with the set-theoretic intersection of congruences. Therefore, if $(\alpha, \beta) \in (q_1 \wedge q_2)^{\top}$, then $\alpha^{-1}\beta \subseteq q_1 \wedge q_2$, $\alpha^{-1}\beta \subseteq q_1 \cap q_2$, $\alpha^{-1}\beta \subseteq q_1$, $\alpha^{-1}\beta \subseteq q_2$, $(\alpha, \beta) \in q_1^{\top}$, $(\alpha, \beta) \in q_2^{\top}$, $(\alpha, \beta) \in q_1^{\top} \cap q_2^{\top} = q_1^{\top} \wedge q_2^{\top}$.

Conversely, let $(\alpha, \beta) \in q_1^{\top} \wedge q_2^{\top}$. Then $\alpha^{-1}\beta \subseteq q_1$, $\alpha^{-1}\beta \subseteq q_2$, $\alpha^{-1}\beta \subseteq q_1 \cap q_2 = q_1 \wedge q_2$, $(\alpha, \beta) \in (q_1 \wedge q_2)^{\top}$. Thus, $(q_1 \wedge q_2)^{\top} = q_1^{\top} \wedge q_2^{\top}$. □

Remark 5.37. It is easy to see that $q_1^{\top} \vee q_2^{\top} \subseteq (q_1 \vee q_2)^{\top}$. Indeed, $q_1^{\top} \subseteq (q_1 \vee q_2)^{\top}$, $q_2^{\top} \subseteq (q_1 \vee q_2)^{\top}$, $q_1^{\top} \vee q_2^{\top} \subseteq (q_1 \vee q_2)^{\top} \vee (q_1 \vee q_2)^{\top} = (q_1 \vee q_2)^{\top}$.

Probably, in general, there exist examples such that $q_1^{\top} \vee q_2^{\top} \subsetneqq (q_1 \vee q_2)^{\top}$. Results from [10, 11, 52, 799] strengthen our guess.

Since in Theorem 5.35 it is proved that the map $\top$ is bijective, we can formulate the following theorem.

Theorem 5.38. *The lower semilattice of congruences of a one-sided loop Q is isomorphically embedded in the lower semilattice of the left congruences of the group $M(Q)$ that are semi-admissible relative to all elements of the semigroup $\Pi(Q)$.*

Corollary 5.39. *The lower semilattice of congruences of a one-sided loop Q is isomorphically embedded in the lower semilattices of congruences: of the semigroup $L\Pi$, of the semigroup Π and of the left congruences of the group LM.*

Proof. By the intersection of the left congruences of group M with the set $L\Pi \times L\Pi$, for example, we obtain some binary relations of semigroup $L\Pi$.

It is easy to understand that these binary relations are equivalences which are semi-admissible relative to multiplication from the left and from the right by elements of the semigroup $L\Pi$, i.e., these equivalences are congruences of semigroup $L\Pi$.

Now we should prove: if $p_1 \subset p_2$ and $p_1^{\perp} \subset p_2^{\perp}$, then p_1, p_2 are elements of lattice of the left congruences of the group M, $p_1 \cap (L\Pi)^2 \subset p_2 \cap (L\Pi)^2$.

If $p_1 \subset p_2$ and $p_1^{\perp} \subset p_2^{\perp}$, then there is a pair (a, b), such that $(a, b) \in p_2^{\perp}$ and $(a, b) \notin p_1^{\perp}$. Then $(L_a, L_b) \in p_2 = ((p_2)^{\perp})^{\top}$ and $(L_a, L_b) \notin p_1$.

If we suppose that $(L_a, L_b) \in p_1$, then $L_a^{-1}L_b \subseteq p_1^{\perp}$, $(a, L_a^{-1}L_b a) = (a, b) \in p_1^{\perp}$. We have received a contradiction. Thus $p_1 \cap (L\Pi)^2 \subset p_2 \cap (L\Pi)^2$.

The remaining inclusion maps are proved similarly. □

Corollary 5.40. *A lower semilattice of congruences of a loop is isomorphically embedded in the lower semilattices of congruences of semigroups $L\Pi$, $R\Pi$, Π, the left congruences of groups LM, RM.*

Theorem 5.41. *In a one-sided loop Q all congruences are normal if and only if in the group M, all left congruences, which are semi-admissible from the right relative to all elements of the semigroup Π, are congruences.*

Proof. We suppose that in the group M all left congruences, which are semi-admissible from the right relative to permutations from the semigroup Π, are congruences. We shall show that then they induce in Q only normal congruences. Indeed, let p be a congruence of the group M. We demonstrate that then $p^{\perp}$ is a normal congruence of a quasigroup Q.

For this purpose it is enough to prove, taking into account Theorem 5.35, Proposition 4.139, that $\omega p^{\perp}\omega^{-1} \subseteq p^{\perp}$ for all $\omega \in \Pi$.

Let $(a, b) \in \omega p^{\perp}\omega^{-1}$. Then there exist $\theta, \varphi \in M$, $\theta\, p\, \varphi$ such that $(a, b) \in \omega\theta^{-1}\varphi\omega^{-1} = (\theta\omega^{-1})^{-1}\varphi\omega^{-1}$. Since p is a congruence of the group M, then from $\theta\, p\, \varphi$ there follows $\theta\omega^{-1}\, p\, \varphi\omega^{-1}$ for all $\omega \in M$. Thus, $(a, b) \in p^{\perp}$, $\omega p^{\perp}\omega^{-1} \subseteq p^{\perp}$.

Converse. Let all congruences in a one-sided loop Q be normal. We shall then prove that in the group M all left congruences, which are semi-admissible from the right relative to permutations from Π, are congruences.

We suppose the converse, that in the group M there exists a left congruence p which is not semi-admissible relative to multiplication on the right by at least one element from the set $\mathbb{T}^{-1}$. We denote such element by R_c^{-1}. In other words there exist elements α, β such that $\alpha\, p\, \beta$, but αR_c^{-1} is not congruent with the element βR_c^{-1}.

Passing to the congruence $p^{\perp}$, we obtain $\alpha^{-1}\beta \subseteq p^{\perp}$, but $R_c\alpha^{-1}\beta R_c^{-1} \nsubseteq p^{\perp}$, i.e. there is an element $x \in Q$ such that $(x, (R_c\alpha^{-1}\beta R_c^{-1})x) \notin p^{\perp}$.

Since $p^{\perp}$ is a normal congruence of a one-sided loop Q, then, if $(a, b) \notin p^{\perp}$, then for all $x \in Q$ we obtain $(a\,x, b\,x) \notin p^{\perp}$.

Thus, if $(x, (R_c\alpha^{-1}\beta R_c^{-1})x) \notin p^{\perp}$, $(R_c x, (R_c\alpha^{-1}\beta)x) \notin p^{\perp}$ or $(xc, (\alpha^{-1}\beta)xc) \notin p^{\perp}$, i.e., $\alpha^{-1}\beta \nsubseteq p^{\perp}$. We have a contradiction.

Therefore left but not right congruence θ of the group M, which is semi-admissible from the right relative to permutations of semigroup Π defines a non-normal congruence of a one-sided loop. □

Definition 5.42. A subgroup H of a group M will be called *A-invariant* relative to a set A of elements of the group M, if $a^{-1}Ha \subseteq H$ for all $a \in A$.

In the language of Definition 5.42 any normal subgroup H of a group G is G-invariant subgroup of the group G [471].

We reformulate Theorems 5.38 and 5.41 as follows.

Theorem 5.43. *The lower semilattice of congruences of a one-sided loop is isomorphically embedded in the lower semilattice of Π-invariant subgroups of the group M.*

Proof. We shall show that the kernel of a left congruence θ of the group M is some of its subgroup H, but the left congruence θ is a partition of the group M in left coset classes by this subgroup. Indeed, if $\alpha\,\theta\,\varepsilon$ and $\beta\,\theta\,\varepsilon$, then $\alpha\beta\,\theta\,\alpha$, whence, $\alpha\beta\,\theta\,\varepsilon$.

If $\alpha\,\theta\,\varepsilon$, then $\alpha^{-1}\alpha\,\theta\,\alpha^{-1}\varepsilon$, $\alpha^{-1}\,\theta\,\varepsilon$. Thus, the kernel of left congruence θ is a subgroup of a group M.

We notice that various left congruences of the group M define various kernels. Indeed, if we suppose the converse, that $\alpha\,\theta_1\,\beta$ and it is not true that $\alpha\,\theta_2\,\beta$, but $\beta^{-1}\alpha\,\theta_1\,\varepsilon$ and $\beta^{-1}\alpha\,\theta_2\,\varepsilon$, then $\beta(\beta^{-1}\alpha)\,\theta_2\,\beta\varepsilon$, $\alpha\,\theta_2\,\beta$. We have received a contradiction.

Since any subgroup H of a group M defines the left congruence ($\alpha \sim \beta \Longleftrightarrow \alpha H = \beta H$), we proved that there is a bijection between the left congruences of the group M and its subgroups.

We shall show that the left congruence θ of the group M is semi-admissible from the right relative to all permutations of the semigroup Π if and only if its kernel H fulfills the relation $H\gamma \subseteq \gamma H$ for all elements $\gamma \in \Pi$, or, equivalently, $\gamma^{-1}H\gamma \subseteq H$.

Indeed, if the left congruence θ of the group M is semi-admissible from the right relative to permutations of the semigroup Π, then for the kernel H of the congruence θ we have: let $\alpha \in H$, i.e., $\alpha\,\theta\,\varepsilon$.

Then, taking into consideration the semi-admissibility from the right of the congruence θ, we obtain $\alpha\gamma\,\theta\,\gamma$ for all $\gamma \in \Pi$. Since θ is a left congruence, then $\gamma^{-1}\alpha\gamma\,\theta\,\gamma^{-1}\gamma$, $\gamma^{-1}\alpha\gamma\,\theta\,\varepsilon$. Therefore, for all $\gamma \in \Pi$ we have $\gamma^{-1}H\gamma \subseteq H$.

Converse. Let the kernel H of a congruence θ satisfy the relation $\gamma^{-1}H\gamma \subseteq H$ for all $\gamma \in \Pi$. If $\alpha\,\theta\,\beta$, then $\beta^{-1}\alpha\,\theta\,\varepsilon$, whence $\gamma^{-1}\beta^{-1}\alpha\gamma\,\theta\,\varepsilon$, $\alpha\gamma\,\theta\,\beta\gamma$ for all $\gamma \in \Pi$. □

Theorem 5.44. *Congruences of a one-sided loop are normal if and only if Π-invariant subgroups of the group M are normal in M.*

We can give sufficient conditions of normality of all congruences of a quasigroup.

Proposition 5.45. *If a quasigroup Q satisfies the condition $\mathbb{T}^{-1} \subseteq \Pi$, then in Q all congruences are normal.*

Proof. If θ is a congruence of a quasigroup Q, then, obviously, from $a\theta b$ there follows $\alpha a\,\theta\,\alpha b$ for all $\alpha \in \Pi$.

Since $\mathbb{T}^{-1} \subseteq \Pi$, then from $ab\,\theta\,ac$ there follows $L_a^{-1}(ab)\theta L_a^{-1}(ac)$, $b\theta c$, from $ca\,\theta\,ba$ there follows $R_a^{-1}(ca)\theta R_a^{-1}(ba)$, $c\theta b$. □

Corollary 5.46. *If in a quasigroup Q the condition $M = \Pi$ is fulfilled, then in the quasigroup Q all congruences are normal.*

Proof. It is easy to see that conditions $\mathbb{T}^{-1} \subseteq \Pi$ and $M = \Pi$ are equivalent. □

Conditions of Proposition 5.45 and Corollary 5.46 can be used for concrete classes of quasigroups. See some examples below.

But, in general, these conditions are only sufficient, since there exists an example of a quasigroup, in which all congruences are normal, but $\mathbb{T} \subsetneq \Pi$, or, equivalently, $M \neq \Pi$.

Example 5.47. Let $A = \{\frac{a}{2^n} \mid a \in \mathbb{Z}, n \in \mathbb{N}\}$, where $\mathbb{Z}$ is the set of integers, and $\mathbb{N}$ is the set of natural numbers.

The set A forms a torsion-free abelian group of rank 1 relative to the operation of addition of elements of the set A [549].

Using the group $(A, +)$ we define on the set A a new quasigroup operation $\circ$. Let φ be a map of the set A into itself such that $\varphi x = \frac{1}{2}x$ for all $x \in A$.

It is easy to check that φ is an automorphism of the group $(A, +)$. Then $(A, \cdot)$ with the form $x \cdot y = \varphi x + y$ for all $x, y \in A$ is a left loop with the left identity 0. Indeed, $0 \cdot x = \varphi 0 + x = x$.

We prove that in the quasigroup $(A, \cdot)$ $M(A, \cdot) \neq \Pi(A, \cdot)$, and all congruences are normal.

For this purpose in the beginning we calculate the form of translations of a quasigroup $(A,\cdot)$. We have $R_a x = x \cdot a = \varphi x + a = (\varphi R_a^+)x$, $L_a x = a \cdot x = \varphi a + x = L_{\varphi a}^+ x$. Using the results from [471, 799] further, it is possible to deduce the following relations:

$$\begin{aligned}
&LM(A,\cdot) = LM(A,+) \cong (A,+),\\
&RM(A,\cdot) \cong RM(A,+)\rtimes\langle\varphi\rangle \cong (A,+)\rtimes(\mathbb{Z},+),\\
&L\Pi(A,\cdot) = L\Pi(A,+),\\
&R\Pi(A,\cdot) = \{(\varphi^n R_a^+) \mid a \in A, n \in \mathbb{N}\}.
\end{aligned}$$

It is easy to see that $M(A,\cdot) = RM(A,\cdot) = \{(\varphi^n R_a^+) \mid a \in A, n \in \mathbb{Z}\}$, $\Pi(A,\cdot) = \{(\varphi^n R_a^+) \mid a \in A, n \in \mathbb{N} \cup \{0\}\}$. Thus, $\Pi(A,\cdot) \subsetneqq M(A,\cdot)$. Moreover, if we denote by $\Pi^{-1}(A)$ the set $\{(\varphi^n R_a^+) \mid a \in A, n \in -\mathbb{N}\}$, then $M(A) = \Pi(A) \cup \Pi^{-1}(A)$.

Since $(A,\cdot)$ is a left loop, we can use Theorem 5.43. As it follows from Theorem 5.43, the subgroups of the group $M(A,\cdot)$ that are invariant relative to all permutations of the semigroup $\Pi(A,\cdot)$ correspond to congruences of the quasigroup $(A,\cdot)$.

We demonstrate that any Π-invariant subgroup of the group $M(A,\cdot)$ is a normal subgroup of the group $M(A,\cdot)$.

We notice that following our agreements we have $(R_a^+\varphi)(x) = \varphi(x+a) = \varphi x + \varphi a = (\varphi R_{\varphi a}^+)(x)$. Below in this example we shall write R_x instead of R_x^+. We have $(\varphi^k R_a)(\varphi^l R_b) = \varphi^{k+l} R_{\varphi^l a + b}$, $(\varphi^n R_a)^{-1} = \varphi^{-n} R_{-\varphi^{-n} a}$.

It is clear that any element of a subgroup H of the group M has the form $\varphi^k R_b$. If H is a Π-invariant subgroup of the group M, then we have: if $\varphi^k R_b \in H$, then $\varphi^{-n} R_{-\varphi^{-n}a} \varphi^k R_b \varphi^n R_a = \varphi^k R_c \in H$ for all $\varphi^k R_b \in H$, $\varphi^n R_a \in \Pi$, where $c = -\varphi^k a + \varphi^n b + a$.

In other words, If H is a Π-invariant subgroup of the group M, then: if $\varphi^k R_b \in H$, then $\varphi^k R_{\varphi^n b} R_{-\varphi^k a + a} \in H$ for all $\varphi^k R_b \in H$, $n \in \mathbb{N} \cup \{0\}$, $a \in A$.

If we change a by $-a$ in the last implication, then we obtain the following implication:

$$\text{if } \varphi^k R_b \in H, \text{ then } \varphi^k R_{\varphi^n b} R_{\varphi^k a - a} \in H \text{ for all } \varphi^k R_b \in H,\ n \in \mathbb{N} \cup \{0\},\ a \in A. \tag{$*$}$$

From the implication $(*)$ by $a = 0$ it follows:

$$\text{if } \varphi^k R_b \in H, \text{ then } \varphi^k R_{\varphi^n b} \in H \text{ for all } \varphi^k R_b \in H,\ n \in \mathbb{N} \cup \{0\}. \tag{$**$}$$

We can write the condition that the Π-invariant subgroup H of group M is a normal subgroup of group M, in the form: if $\varphi^k R_b \in H$, then $\varphi^n R_a \varphi^k R_b \varphi^{-n} R_{-\varphi^{-n} a} = \varphi^k R_d \in H$, where $d = -\varphi^{-n}(-\varphi^k a - b + a)$, for all $\varphi^k R_b \in H$, $\varphi^n R_a \in \Pi$.

Applying the condition $(**)$ to the last implication, we obtain the following equivalent condition of normality of Π-invariant group H: if $\varphi^k R_b \in H$, then $\varphi^k R_h \in H$, where $h = \varphi^k a + b - a$ for all $\varphi^k R_b \in H$, $a \in A$.

The last implication we can re-write in the form: if $\varphi^k R_b \in H$, then $\varphi^k R_b R_{\varphi^k a - a} \in H$ for all $\varphi^k R_b \in H$, $a \in A$.

It is easy to see that the last implication follows from the implication $(*)$ by $n = 0$.

Example 5.48. Using the group $(A,+)$ from Example 5.47 we define a binary operation $*$ on the set A in the following way: $x * y = 2 \cdot x + y$ for all $x, y \in A$. The operation $*$ is a quasigroup operation, since the map $2 : x \mapsto 2 \cdot x$ for all $x \in A$ is an automorphism of the group $(A,+)$, moreover, a left loop operation, see Example 5.47.

We denote by H the following subgroup of the group $M(A,*)$: $H = \langle R_1^+ \mid 1 \in A\rangle = \{\dots R_{-2}^+, R_{-1}^+, R_0^+, R_1^+, R_2^+, \dots\}$. It is easy to see that $H \cong (\mathbb{Z},+)$.

We check that the group H is a Π-invariant non-normal subgroup of the group M.

We use results of Example 5.47 by $\varphi = 2$. Thus, if H is a Π-invariant subgroup of the group M, then we have: $2^{-n}R_{-2^{-n}a}R_1 2^n R_a = R_{2^n} \in H$ for all $2^n R_a \in \Pi$.

We prove that the group H is a non-normal subgroup of the group M. We have $2^n R_a R_1 2^{-n} R_{-2^{-n}a} = R_{2^{-n}} \notin H$ for all $2^n R_a \in M$ such that $n > 1$.

As it follows from Theorems 5.38 and 5.43, the subgroup H of the group $M(A, *)$ induces a non-normal congruence of the quasigroup $(A, *)$.

Remark 5.49. The fact that the group H induces a congruence of the quasigroup $(A, *)$ can be deduced from results of the article of T. Kepka and P. Nemec [16, Theorem 42], since the quasigroup $(A, *)$ is a T-quasigroup, moreover, it is a medial quasigroup.

5.2.3 Congruences of inverse quasigroups

We recall definitions of some inverse quasigroups from Section 3.1.

A quasigroup $(Q, \circ)$ is called an (α, β, γ)-inverse quasigroup, if there exist permutations α, β, γ of the set Q such that in quasigroup $(Q, \circ)$ for all $x, y \in Q$ the relation $\alpha(x \circ y) \circ \beta x = \gamma y$ is fulfilled.

A quasigroup $(Q, \circ)$ is called an (r, s, t)-inverse quasigroup, if there exists a permutation J of the set Q and some fixed integers r, s, t such that in $(Q, \circ)$ the relation $J^r(x \circ y) \circ J^s x = J^t y$ is fulfilled for all $x, y \in Q$.

A $(0, 1, 0)$-inverse quasigroup is called a CI-quasigroup, a $(-1, 0, -1)$-inverse quasigroup is called a WIP-quasigroup, and an $(m, m+1, m)$-inverse quasigroup is called an m-inverse quasigroup [72, 472].

Proposition 5.50. *In an (α, β, γ)-quasigroup $(Q, \cdot)$ all congruences are normal if permutations α and γ^{-1} are semi-admissible relative to any congruence of $(Q, \cdot)$.*

Proof. In the language of translations we can re-write the definition of an (α, β, γ)-inverse quasigroup in the form $L_x \alpha R_{\beta x} = \gamma$. Then $L_x^{-1} = \alpha R_{\beta x} \gamma^{-1}$, $R_{\beta x}^{-1} = \gamma^{-1} L_x \alpha$.

Using Proposition 5.45 we obtain the required. □

Proposition 5.51. *In an (r, s, t)-quasigroup $(Q, \cdot)$ all congruences are normal if permutations J^r and J^{-t} are semi-admissible relative to any congruence of $(Q, \cdot)$.*

Proof. The proof repeats the proof of Proposition 5.50. □

Corollary 5.52. *In a CI-quasigroup all congruences are normal.*

Corollary 5.53. *In a WIP-quasigroup $(Q, \cdot)$ all congruences are normal if the permutation J is admissible relative to any congruence of $(Q, \cdot)$.*

Corollary 5.54. *In an m-inverse quasigroup $(Q, \cdot)$ all congruences are normal if the permutation J^m is admissible relative to any congruence of $(Q, \cdot)$.*

We recall, a quasigroup $(Q, \circ)$ has the λ-inverse property if there exist permutations $\lambda_1, \lambda_2, \lambda_3$ of the set Q such that $\lambda_1 x \circ \lambda_2(x \circ y) = \lambda_3 y$ for all $x, y \in Q$.

Lemma 5.55. *In a λ-inverse quasigroup $(Q, \cdot)$ all congruences are left normal if permutations λ_2, λ_3^{-1} are semi-admissible relative to any congruence of $(Q, \cdot)$.*

Proof. From the last equality we have $L_x \lambda_2 L_{\lambda_1 x} = \lambda_3$. Then $L_x^{-1} = \lambda_2 L_{\lambda_1 x} \lambda_3^{-1}$. □

A quasigroup $(Q, \circ)$ has the ρ-inverse property if there exist permutations ρ_1, ρ_2, ρ_3 of the set Q such that $\rho_1(x \circ y) \circ \rho_2 y = \rho_3 x$ for all $x, y \in Q$ ([82] or Definition 2.22).

Lemma 5.56. *In a ρ-inverse quasigroup $(Q,\cdot)$ all congruences are right normal if permutations ρ_1 and ρ_3^{-1} are semi-admissible relative to any congruence of $(Q,\cdot)$.*

Proof. From the last equality we have $R_y\rho_1 R_{\rho_2 y} = \rho_3$. Therefore $R_y^{-1} = \rho_1 R_{\rho_2 y}\rho_3^{-1}$. □

A quasigroup $(Q,\circ)$ that has the λ-inverse property and ρ-inverse property is called an I-quasigroup ([82] or Section 3.1).

Proposition 5.57. *In an I-quasigroup $(Q,\cdot)$ all congruences are normal if permutations λ_2, λ_3^{-1}, ρ_1 and ρ_3^{-1} are semi-admissible relative to any congruence of $(Q,\cdot)$.*

Proof. The proof follows from the definition of an I-inverse quasigroup, Lemmas 5.55 and 5.56. □

If in I-quasigroup $(Q,\circ)$ $\lambda_2 = \lambda_3 = \rho_1 = \rho_3 = \varepsilon$, then $(Q,\circ)$ is called an IP-quasigroup.

Corollary 5.58. *In an IP-quasigroup all congruences are normal [72].*

Proof. The proof follows from the definition of an IP-quasigroup and Proposition 5.57. □

5.2.4 Behavior of congruences by an isotopy

Proposition 5.59. *The lattice $(L,\vee,\wedge)$ of normal congruences of a quasigroup $(Q,\cdot)$ is isomorphic to a sublattice of the lattice $(L_1,\vee,\wedge)$ of normal congruences of isotope loop $(Q,\circ)$ [103].*

Proof. By an LP-isotopy T ($T = (R_a^{-1}, L_b^{-1}, \varepsilon)$) a normal congruence θ of quasigroup $(Q,\cdot)$ is also a normal congruence of a loop $(Q,\star)$, $(Q,\star) = (Q,\cdot)T$ (Corollary 1.307).

Since the operation $\wedge$ in sets of congruences of a quasigroup $(Q,\cdot)$ and loops $(Q,\star)$ coincides with the set-theoretic intersection, and the operation $\vee$ coincides, in view of the permutability of normal congruences, with their product as binary relations [813], we can state that the lattice of normal congruences of a quasigroup $(Q,\cdot)$ is a sublattice of the lattice of normal congruences of the loop $(Q,\star)$. This proposition is proved, since any isotopy between a loop and a quasigroup has the form $(R_a^{-1}, L_b^{-1}, \varepsilon)(\varphi,\varphi,\varphi)$. □

Obviously, any permutation of the semigroup $\Pi(Q,\cdot)$ is semi-admissible relative to any congruence of a quasigroup $(Q,\cdot)$. An isotopy is semi-admissible, if all permutations included in it are semi-admissible.

Proposition 5.60. *Let θ be a congruence of a quasigroup $(Q,\cdot)$. If a quasigroup $(Q,\circ)$ is isotopic to $(Q,\cdot)$, and the isotopy T is semi-admissible relative to θ, then θ is also a congruence in $(Q,\circ)$.*

Proof. We suppose that the isotopy T has the form $T = (\alpha,\beta,\gamma)$. If $a\theta b$, then $\beta a\theta\beta b$, $\alpha c\cdot\beta a\theta\alpha c\cdot\beta b$, $\gamma^{-1}(\alpha c\cdot\beta a)\theta\gamma^{-1}(\alpha c\cdot\beta b)$. Finally, we obtain $(c\circ a)\theta(c\circ b)$.

Similarly, if $a\theta b$, then $a\circ c\theta b\circ c$. □

Proposition 5.61. *If in a quasigroup $(Q,\cdot)$ there exist elements a,b such that R_a^{-1}, $L_b^{-1}\in\Pi$, then the lower semilattice $(L_1,\wedge)$ of congruences of a quasigroup $(Q,\cdot)$ is a subsemilattice of the semilattice $(L_2,\wedge)$ of congruences of the loop $(Q,\circ)$ which is an isotope of a quasigroup $(Q,\cdot)$ of the form $(R_a^{-1}, L_b^{-1}, \varepsilon)$.*

Proof. If $R_a^{-1}, L_b^{-1}\in\Pi$, then the isotopy $(R_a^{-1}, L_b^{-1}, \varepsilon)$ is admissible relative to any congruence of quasigroup $(Q,\cdot)$. The corollary is true, since the operations $\wedge$ in $(L_1,\wedge)$ and $(L_2,\wedge)$ coincide with the set-theoretic intersection of congruences. □

In any IP-loop $(Q, \circ)$ with the identity element 1, the map $J : a \mapsto a^{-1}$ for all $a \in Q$, where $a \circ a^{-1} = 1$, is a permutation of the set Q, $J^2 = \varepsilon$ [173].

Proposition 5.62. *If $(Q, \circ)$ is an IP-loop, $(Q, \cdot)$ is its isotope of the form $(\alpha J^{\tau}, \beta J^{\kappa}, \varepsilon)$, where $\alpha, \beta \in M(Q, \circ)$, $\tau, \kappa \in \{0, 1\}$, i.e., $x \cdot y = \alpha J^{\tau} x \circ \beta J^{\kappa} y$ for all $x, y \in Q$, then $Con(Q, \circ) = nCon(Q, \cdot)$.*

Proof. The permutation J is an antiautomorphism in $(Q, \circ)$ and any normal congruence in $(Q, \circ)$ is admissible relative to this permutation. Indeed, if $x\theta y$, then $1\theta x^{-1} \circ y$, $y^{-1}\theta(x^{-1} \circ y) \circ y^{-1}$, $y^{-1}\theta x^{-1}$, $x^{-1}\theta y^{-1}$ and in a similar way, $x\theta y$ follows from $x^{-1}\theta y^{-1}$.

By Corollary 5.58, in any IP-loop all congruences are normal, i.e., $Con(Q, \circ) = nCon(Q, \circ)$. Then permutations α, β and J are admissible relative to any congruence of the loop $(Q, \circ)$, by Lemma 1.306 $Con(Q, \circ) \subseteq nCon(Q, \cdot)$.

Since $(Q, \cdot) = (Q, \circ)(\alpha J^{\tau}, \beta J^{\kappa}, \varepsilon)$, then

$$(Q, \circ) = (Q, \cdot)((\alpha J^{\tau})^{-1}, (\beta J^{\kappa})^{-1}, \varepsilon).$$

It is known that every principal isotopy (the third component of such isotopy is an identity mapping) of a quasigroup $(Q, \cdot)$ to a loop $(Q, \circ)$ has the form $(R_a^{-1}, L_b^{-1}, \varepsilon)$, where $R_a x = x \cdot a$, $L_b x = b \cdot x$ [72, 77].

Thus, taking into consideration Corollary 1.307, we have: $nCon(Q, \cdot) \subseteq Con(Q, \circ)$. Therefore $nCon(Q, \cdot) = Con(Q, \circ)$. □

Proposition 5.63. *If $(Q, \circ)$ is a CI-loop, $(Q, \cdot)$ is its isotope of the form $x \cdot y = \alpha J^{\tau} x \circ \beta J^{\kappa} y$ for all $x, y \in Q$, where $\alpha, \beta \in M(Q, \circ)$, $\tau, \kappa \in \{0, 1\}$, then $Con(Q, \circ) = nCon(Q, \cdot)$.*

Proof. The permutation J is an automorphism in $(Q, \circ)$ ([72] or Theorem 3.19) and any normal congruence in $(Q, \circ)$ is admissible relative to this permutation. Indeed, if $x\theta y$, then $1\theta y \circ Jx$, $Jy\theta(y \circ Jx) \circ Jy$, $Jy\theta Jx$, $Jx\theta Jy$.

In any CI-quasigroup $(Q, \circ)$ the following equality is true $x \circ (y \circ Jx) = y$ for all $x, y \in Q$ ([487] or Lemma 3.30). If $Jx\theta Jy$, then $y \circ Jx \,\theta\, y \circ Jy$, $y \circ Jx \,\theta\, 1$, $x \circ (y \circ Jx) \,\theta\, x$, $y\theta x$.

By Corollary 5.52 in the loop $(Q, \circ)$ all congruences are normal. Therefore, permutations α, β are admissible relative to any congruence of the loop $(Q, \circ)$. □

5.2.5 Regularity of quasigroup congruences

Definition 5.64. A congruence is called *regular* if it is uniquely defined by its coset, and the coset of a congruence is called regular if it is a coset of only one congruence.

In [589] A.I. Mal'tsev has given necessary and sufficient conditions that a normal complex K of an algebraic systems A is a coset of only one congruence, i.e., K is a coset of only one congruence of a system A.

For this purpose for any set $S \subseteq A$ the congruence mod S is constructed. Elements a, b are equivalent $a \sim b \pmod{S}$ if either $a = b$, or $a, b \in S$, or $a = \alpha u$, $b = \alpha v$, where $u, v \in S$, α is a translation of the algebraic system A.

A.I. Mal'tsev calls elements a and b *comparable* if there exists a sequence $x_1, \dots, x_n$ of elements from A such that: $a \sim x_1, x_1 \sim x_2, \dots, x_n \sim b \pmod{S}$.

The binary relation (mod S) is a congruence on an algebraic system A, and the congruence (mod S) is minimal among all congruences for which elements of the set S are comparable with each other [589].

Theorem 5.65. *The normal complex K is a coset of only one congruence of an algebraic system A if and only if elements $a, b \in A$, for which by any translation α the statements $\alpha a \in K$ and $\alpha b \in K$ are equivalent, are comparable* $(\mathrm{mod\,K})$ *[589].*

We notice that if in Theorem 5.65 A is a binary quasigroup, then α is an element of $\Pi(A)$.

If in the Mal'tsev theorem we pass from a quasigroup A to its homomorphic image $\bar{A} = A/\mathrm{mod\,K}$, then we shall have the following conditions of regularity of a normal complex K of a quasigroup A.

Proposition 5.66. *The normal complex K is a coset of only one congruence of a quasigroup A if and only if for each pair of elements $\bar{a}, \bar{b} \in \bar{A}$ for which by any translation $\bar{\alpha} \in \bar{A}$ the statements $\bar{\alpha}\bar{a} = \bar{k}$ and $\bar{\alpha}\bar{b} = \bar{k}$ are equivalent, the equality $\bar{a} = \bar{b}$ is fulfilled.*

Remark 5.67. Let's note that if $\bar{A}$ is a binary quasigroup, then the conditions of Proposition 5.66 are fulfilled. Indeed, if we take translation $\bar{\alpha}$ such that $\bar{\alpha} = \bar{L}_c$ and $\bar{c} \cdot \bar{a} = \bar{k}$, then we have $\bar{c} \cdot \bar{b} = \bar{k}$ by conditions of the proposition. Then $\bar{a} = \bar{c} \backslash \bar{k} = \bar{b}$.

Example 5.68. It is possible to construct a division groupoid in which the conditions of Proposition 5.66 are satisfied. We denote by $(\mathbb{Q}, +)$ the group of rational numbers relative to the operation of addition, and by $(\mathbb{Z}, +)$ the group of integers relative to the operation of addition. On the factor group $\bar{A} = (\mathbb{Q}/\mathbb{Z}, +)$ we define operation $x \circ y = 2x + y$ for all $x, y \in \bar{A}$.

It is easy to check that $(\bar{A}, \circ)$ is a division groupoid. We shall demonstrate that this groupoid satisfies the conditions of Proposition 5.66. Since $(\bar{A}, \circ)$ is a division groupoid, then for any $\bar{k} \in \bar{A}$ there exists $\bar{c} \in \bar{A}$ such that $\bar{c} \circ \bar{a} = \bar{k}$, and then by the conditions of this proposition also $\bar{c} \circ \bar{b} = \bar{k}$. Therefore $2\bar{c} + \bar{a} = \bar{k}$, $2\bar{c} + \bar{b} = \bar{k}$, $\bar{a} = \bar{b} = \bar{k} - 2\bar{c}$.

Proposition 5.69. *There exist a quasigroup Q and its subset K such that K is a coset of more than one congruence.*

Proof. It is known (see [173], p. 10]) that any division groupoid is a homomorphic image of some quasigroups.

From the Mal'tsev theorem it follows that to give an example of a quasigroup in which not all congruences are regular it is necessary to find a pair of elements $a, b \in Q$ such that $a \not\sim b\,(\mathrm{mod\,K})$, where K is a coset of some congruence, but for which by any translation, α statements $\alpha a \in K$ and $\alpha b \in K$ are equivalent.

We pass to homomorphic image $P = Q/\mathrm{modK}$ of quasigroups Q. Then conditions that the coset K is not regular are the following: $\bar{a} \neq \bar{b}$, but for any $c \in Q$ the equality $\bar{c} \cdot \bar{a} = \bar{k}$ is equivalent to the equality $\bar{c} \cdot \bar{b} = \bar{k}$, the equality $\bar{a} \cdot \bar{c} = \bar{k}$ is equivalent to the equality $\bar{b} \cdot \bar{c} = \bar{k}$, where $\bar{k}$ is an image of the set K in the groupoid P.

We construct the following division groupoid. Let $\mathbb{C}$ be a set of complex numbers, $x \circ y = (xy)^2$ for all $x, y \in \mathbb{C}$. It is easy to check that $(\mathbb{C}, \circ)$ is a commutative division groupoid.

Let $\bar{k} = 4$. Then the equation $a \circ y = 4 \iff (ay)^2 = 4$, $ay = \pm 2$, $y = \pm\frac{2}{a}$ has two solutions. And, if one of the radicals is a solution of the equations $a \circ y = 4$, so is the other, for any $a \in \mathbb{C}$. If we take in a quasigroup Q pre-images of elements of 2 and -2, then we find the necessary pair. □

Chapter 6

Quasigroups which have an endomorphism

6.1 Introduction 287
6.1.1 Parastrophe invariants and isostrophisms 291
6.2 Left and right F-, E-, SM-quasigroups 293
6.2.1 Direct decompositions 297
6.2.2 F-quasigroups 301
6.2.3 E-quasigroups 307
6.2.4 SM-quasigroups 310
6.2.5 Finite simple quasigroups 311
6.2.6 Left FESM-quasigroups 312
6.2.7 CML as an SM-quasigroup 314
6.3 Loop isotopes 315
6.3.1 Left F-quasigroups 315
6.3.2 F-quasigroups 321
6.3.3 Left SM-quasigroups 321
6.3.4 Left E-quasigroups 322

The cause that finite medial and paramedial quasigroups are directly decomposable lies in the fact that these quasigroups have an endomorphism or an antiendomorphism.

In this chapter we give an overview of properties of quasigroups which have an endomorphism [789, 788]. It is easy to see that if a quasigroup is isomorphic to the direct product of its subquasigroups, then it has some endomorphisms.

6.1 Introduction

We list some quasigroups and loops that have an endomorphism. This fact implies (entails) that finite quasigroups from these quasigroup classes are decomposable into a direct product.

D.C. Murdoch introduced F-quasigroups in [645]. At that time A.K. Sushkevich studied quasigroups with weak associative properties [863, 864]. F-quasigroups got their name in an article by V.D. Belousov [61]. Later Belousov and his pupils I.A. Golovko and I.A. Florja, M.I. Ursul, T. Kepka, M. Kinyon, J.D. Phillips, L.V. Sabinin, L.V. Sbitneva, L.L. Sabinina and many other mathematicians studied F-quasigroups and left F-quasigroups [72, 80, 87, 369, 370, 332, 329, 330, 498, 725, 726, 196, 500]. In [500, 502, 503] it is proved that any F-quasigroup is linear over a Moufang loop. The structure of F-quasigroups is described in [500, 502, 503].

Left and right SM-quasigroups (semimedial quasigroups) are defined by T. Kepka. In [492] Kepka has called these quasigroups LWA-quasigroups and RWA-quasigroups, respectively. SM-quasigroups are connected with trimedial quasigroups. These quasigroup classes are studied in [492, 494, 497, 799, 51, 800, 197, 519, 520]. In is clear that any idempotent left (right) SM-quasigroup is a left (right) distributive quasigroup.

M. Kinyon and J.D. Phillips have defined and studied left and right E-quasigroups [520].

In fact all above mentioned quasigroup classes are defined with the help of some generalized distributive identities. All these quasigroups have an endomorphism which we shall use to study their structure.

The idea to apply an endomorphism is used in the study of many loop and quasigroup classes, for example, in the study of commutative Moufang loops, commutative diassociative loops, CC-loops (LK-loops), F-quasigroups, SM-quasigroups, trimedial quasigroups and so on [173, 685, 72, 174, 170, 403, 52, 37, 514, 515, 521, 689]. This idea is expressed especially clearly in K.K. Shchukin's book [799].

Using the language of identities of quasigroups with three operations in signature, i.e., of quasigroups of the form $(Q, \cdot, /, \backslash)$, we can say that we study some quasigroups from the following quasigroup classes: (i) $(xy)\backslash(xy) = (x\backslash x)\cdot(y\backslash y)$; (ii) $(xy)/(xy) = (x/x)\cdot(y/y)$; (iii) $(xy)\cdot(xy) = (xx)\cdot(yy)$.

We recall here definitions of main quasigroup classes that are studied in this chapter. See Definition 2.2.

Definition 6.1. A quasigroup $(Q, \cdot)$ is called:

1. *left F-quasigroup,* if $x \cdot yz = xy \cdot e(x)z$ for all $x, y, z \in Q$, where $x \cdot e(x) = x$ for all x;
2. *right F-quasigroup,* if $xy \cdot z = xf(z) \cdot yz$ for all $x, y, z \in Q$, where $f(x) \cdot x = x$ for all x;
3. *F-quasigroup,* if it is left and right F-quasigroup;
4. *left semi-medial quasigroup (left SM-quasigroup) or LWA-quasigroup,* if $s(x) \cdot yz = xx \cdot yz = xy \cdot xz$ for all $x, y, z \in Q$, where $s(x) = x \cdot x$ for all x;
5. *right semi-medial quasigroup (right SM-quasigroup) or RWA-quasigroup,* if $zy \cdot s(x) = zx \cdot yx$ for all $x, y, z \in Q$;
6. *semi-medial quasigroup, either SM-quasigroup, or WA-quasigroup,* if it is left and right SM-quasigroup;
7. *left E-quasigroup,* if $x \cdot yz = f(x)y \cdot xz$ for all $x, y, z \in Q$, where $f(x) \cdot x = x$ for all x;
8. *right E-quasigroup,* if $zy \cdot x = zx \cdot ye(x)$ for all $x, y, z \in Q$, where $x \cdot e(x) = x$ for all x;
9. *E-quasigroup,* if it is left and right E-quasigroup [520].

Definition 6.2. A loop $(Q, \cdot)$ is:

1. *left M-loop,* if $x \cdot (y \cdot z) = (x \cdot (y \cdot I\varphi x)) \cdot (\varphi x \cdot z)$ for all $x, y, z \in Q$, where φ is a mapping of the set Q, $x \cdot Ix = 1$ for all $x \in Q$;
2. *right M-loop,* if $(y \cdot z) \cdot x = (y \cdot \psi x) \cdot ((I^{-1}\psi x \cdot z) \cdot x)$ for all $x, y, z \in Q$, where ψ is a mapping of the set Q;
3. *M-loop,* if it is left M- and right M-loop;
4. *left special,* if $S_{a,b} = L_b^{-1}L_a^{-1}L_{ab}$ is an automorphism of $(Q, \cdot)$ for any pair $a, b \in Q$ [823];
5. *right special,* if $T_{a,b} = R_b^{-1}R_a^{-1}R_{ba}$ is an automorphism of $(Q, \cdot)$ for any pair $a, b \in Q$ [823].

In [72] the left special loop is called *special.* Left semimedial quasigroups are studied in [494, 799, 520].

A quasigroup is trimedial if any three of its elements generate a medial sub-quasigroup [494]. A quasigroup is trimedial if and only if it satisfies left and right E-quasigroup equality [520]. Information on properties of trimedial quasigroups is given in [519].

Every semimedial quasigroup is isotopic to a commutative Moufang loop [494]. For trimedial quasigroups the isotopy has a more restrictive form [494].

In a quasigroup $(Q,\cdot,\backslash,/)$ the equalities $x\cdot yz = xy\cdot e(x)z$, $xy\cdot z = xf(z)\cdot yz$, $x\cdot yz = f(x)y\cdot xz$ and $zy\cdot x = zx\cdot ye(x)$ take the form $x\cdot yz = xy\cdot(x\backslash x)z$, $xy\cdot z = x(z/z)\cdot yz$, $x\cdot yz = (x/x)y\cdot xz$ and $zy\cdot x = zx\cdot y(x\backslash x)$, respectively, and they are identities in $(Q,\cdot,\backslash,/)$.

Therefore any subquasigroup of a left F-quasigroup $(Q,\cdot,\backslash,/)$ is a left F-quasigroup, and any homomorphic image of a left F-quasigroup $(Q,\cdot,\backslash,/)$ is a left F-quasigroup [184, 591]. It is clear that the same situation is true for right F-quasigroups, and left and right E- and SM-quasigroups.

Lemma 6.3. *Any medial quasigroup $(Q,\cdot)$ is both a left and right F-, SM- and E-quasigroup.*

Proof. Equality $x\cdot uv = xu\cdot e(x)v$ follows from medial identity $xy\cdot uv = xu\cdot yv$ by $y = e(x)$. Respectively by $u = e(x)$ we have $xy\cdot e(x)v = x\cdot yv$, i.e., $(Q,\cdot)$ is a left F-quasigroup in these cases, and so on. □

Lemma 6.4. *1. Any left distributive quasigroup $(Q,\cdot)$ is a left F-, SM- and E-quasigroup.*
2. Any right distributive quasigroup $(Q,\cdot)$ is a right F-, SM- and E-quasigroup.

Proof. 1. It is easy to see that $(Q,\cdot)$ is an idempotent quasigroup. Therefore $x\cdot x = x/x = x\backslash x = x$. Then $x\cdot yz = xy\cdot xz = ((x/x)\cdot y)\cdot xz = ((x\backslash x)\cdot y)\cdot xz = xy\cdot(x/x)z = xy\cdot(x\backslash x)z = xx\cdot yz$.

2. The proof of this case is similar to the proof of Case 1. □

Lemma 6.5. *A quasigroup $(Q,\cdot)$ in which:*

1. *the equality $x\cdot yz = xy\cdot\delta(x)z$ is true for all $x,y,z\in Q$, where δ is a map of the set Q, is a left F-quasigroup [80];*

2. *the equality $xy\cdot z = x\delta(z)\cdot yz$ is true for all $x,y,z\in Q$, where δ is a map of the set Q, is a right F-quasigroup;*

3. *the equality $\delta(x)\cdot yz = xy\cdot xz$ is true for all $x,y,z\in Q$, where δ is a map of the set Q, is a left semi-medial quasigroup;*

4. *the equality $zy\cdot\delta(x) = zx\cdot yx$ is true for all $x,y,z\in Q$, where δ is a map of the set Q, is a right semi-medial quasigroup;*

5. *the equality $x\cdot yz = \delta(x)y\cdot xz$ is true for all $x,y,z\in Q$, where δ is a map of the set Q, is a left E-quasigroup;*

6. *the equality $zy\cdot x = zx\cdot y\delta(x)$ is true for all $x,y,z\in Q$, where δ is a map of the set Q, is a right E-quasigroup.*

Proof. If we take $y = e(x)$, then we have $x\cdot e(x)z = x\cdot\delta(x)z$, $e(x) = \delta(x)$. Cases 2–6 are proved similarly. □

A left (right) F-quasigroup is isotopic to a left (right) M-loop [370, 80]. A left (right) F-quasigroup is isotopic to a left (right) special loop [61, 87, 72, 369]. An F-quasigroup is isotopic to a Moufang loop [500].

If a loop $(Q, \circ)$ is isotopic to a left distributive quasigroup $(Q, \cdot)$ with isotopy of the form $x \circ y = R_a^{-1}x \cdot L_a^{-1}y$, then $(Q, \circ)$ will be called a *left S-loop*. Loop $(Q, \circ)$ and quasigroup $(Q, \cdot)$ are said to be *related*.

If a loop $(Q, \circ)$ is isotopic to a right distributive quasigroup $(Q, \cdot)$ with isotopy of the form $x \circ y = R_a^{-1}x \cdot L_a^{-1}y$, then $(Q, \circ)$ will be called a *right S-loop*.

Definition 6.6. An automorphism ψ of a loop $(Q, \circ)$ is called *complete*, if there exists a permutation φ of the set Q such that $\varphi x \circ \psi x = x$ for all $x \in Q$. Permutation φ is called a complement of automorphism ψ.

The following theorem is proved in [90].

Theorem 6.7. *A loop $(Q, \circ)$ is a left S-loop, if and only if there exists a complete automorphism ψ of the loop $(Q, \circ)$ such that at least one of the following conditions is fulfilled:*

a) $\varphi(x \circ \varphi^{-1}y) \circ (\psi x \circ z) = x \circ (y \circ z)$;
b) $L^{\circ}_{x,y}\psi = \psi L^{\circ}_{x,y}$ and $\varphi x \circ (\psi x \circ y) = x \circ y$ for all $x, y \in Q$, $x, y \in Q$, $L^{\circ}_{x,y} \in LI(Q, \circ)$.

Thus $(Q, \cdot)$, where $x \cdot y = \varphi x \circ \psi y$, is a left distributive quasigroup which corresponds to the loop $(Q, \circ)$.

Remark 6.8. In [90, 670] a left S-loop is called an *S-loop*.

A left distributive quasigroup $(Q, \cdot)$ with identity $x \cdot xy = y$ is isotopic to a left Bol loop [87, 72, 80]. The last results of G. Nagy [650] allow us to hope that researches of left distributive quasigroups will progress. Some properties of distributive and left distribute quasigroups are described in [339, 340, 343, 844].

Theorem 6.9. *Any loop which is isotopic to a left F-quasigroup is a left M-loop [80, Theorem 3.17, p. 109].*

Theorem 6.10. *If a simple distributive quasigroup $(Q, \circ)$ is isotopic to a finitely generated commutative Moufang loop $(Q, +)$, then $(Q, \circ)$ is a finite medial distributive quasigroup [769, 767, 768].*

Proof. It is known [173] that any finitely generated CML $(Q, +)$ has a non-identity center $C(Q, +)$ (for short C).

We verify if the center of CML $(Q, +)$ is invariant (is a characteristic subloop) relative to any automorphism of the loop $(Q, +)$ and the quasigroup $(Q, \circ)$.

Indeed, if $\varphi \in Aut(Q, +)$, $a \in C(Q, +)$ (see Remark 2.115), then we have $\varphi(a+(x+y)) = \varphi((a+x)+y) = \varphi a+(\varphi x+\varphi y) = (\varphi a+\varphi x)+\varphi y$. Thus $\varphi a \in C(Q, +)$, $\varphi C(Q, +) \subseteq C(Q, +)$.

For any distributive quasigroup $(Q, \circ)$ of the form $x \circ y = \varphi x + \psi y$ we have

$$Aut(Q, \circ) \cong M(Q, +) \rtimes (\mathbb{C}/I),$$

where I is the group of inner permutations of commutative Moufang loop $(Q, +)$,

$$\mathbb{C} = \{\omega \in Aut(Q, +) \,|\, \omega\varphi = \varphi\omega\}.$$

Therefore any automorphism of $(Q, \circ)$ has the form $L_a^+\alpha$, where $\alpha \in Aut(Q, +)$ [772].

The center C defines normal congruence θ of the loop $(Q, +)$ in the following way $x\theta y \iff x + C = y + C$. We give a little part of this standard proof: $(x + a) + C = (y + a) + C \iff (x + C) + a = (y + C) + a \iff x + C = y + C$. In fact C is coset class of θ containing zero element of $(Q, +)$.

The congruence θ is admissible relative to any permutation of the form $L_a^+\alpha$, where

$\alpha \in Aut(Q,+)$, since θ is a central congruence. Therefore, θ is a congruence in the quasigroup $(Q,\circ)$.

Since $(Q,\circ)$ is a simple quasigroup and θ cannot be a diagonal congruence, then $\theta = Q \times Q$, $C(Q,+) = (Q,+)$, $(Q,\circ)$ is medial. From Jezek-Kepka result [461] it follows that $(Q,\circ)$ is finite. □

Lemma 6.11. *If simple quasigroup $(Q,\cdot)$ is an isotope either of the form $(f,\varepsilon,\ \varepsilon)$, or of the form $(\varepsilon,e,\varepsilon)$, or of the form $(\varepsilon,\varepsilon,s)$ of a distributive quasigroup $(Q,\circ)$, where $f,e,s \in Aut(Q,\circ)$ and $(Q,\circ)$ is isotopic to finitely generated commutative Moufang loop $(Q,+)$, then $(Q,\cdot)$ is a finite medial quasigroup.*

Proof. Since $(Q,+)$ is finitely generated, then $|\,C(Q,+)\,| > 1$ [173] and $C(Q,+)$ is invariant relative to any automorphism of $(Q,\circ)$ (Theorem 6.10).

Therefore, a necessary condition of simplicity of $(Q,\cdot)$ is the fact that $C(Q,+) = (Q,+)$. Then $(Q,\circ)$ is medial.

Prove that $(Q,\cdot)$ is medial, if $x\cdot y = fx\circ y$. We have $xy\cdot uv = f(fx\circ y)\circ(fu\circ v) = (f^2x\circ fu)\circ(fy\circ v) = (xu)\cdot(yv)$ [646].

Prove that $(Q,\cdot)$ is medial, if $x\cdot y = x\circ ey$. We have $xy\cdot uv = (x\circ ey)\circ(eu\circ e^2v) = (x\circ eu)\circ(ey\circ e^2v) = (xu)\cdot(yv)$ [646].

Prove that $(Q,\cdot)$ is medial, if $x\cdot y = s(x\circ y)$. We have $xy\cdot uv = (s^2x\circ s^2y)\circ(s^2u\circ s^2v) = (s^2x\circ s^2u)\circ(s^2y\circ s^2v) = (xu)\cdot(yv)$. □

6.1.1 Parastrophe invariants and isostrophisms

Parastrophe invariants and isostrophisms are studied in [75].

Lemma 6.12. *If a quasigroup Q is the direct product of a quasigroup A and a quasigroup B, then $Q^\sigma = A^\sigma \times B^\sigma$, where σ is a parastrophy.*

Proof. From Theorem 7.36 it follows that the direct product $A\times B$ defines two quasigroup congruences. From Theorem 1.279 it follows that these congruences are normal. By Corollary 1.268 these congruences are invariant relative to any parastrophy of the quasigroup Q. □

Lemma 6.13. *If Q is a quasigroup and $\alpha \in Aut(Q)$, then $\alpha \in Aut(Q^\sigma)$, where σ is a parastrophy.*

Proof. It is easy to check [764, 767]. □

Lemma 6.14. *(1) A quasigroup $(Q,\cdot)$ is a left F-quasigroup if and only if its (12)-parastrophe is a right F-quasigroup.*

(2) A quasigroup $(Q,\cdot)$ is a left E-quasigroup if and only if its (12)-parastrophe is a right E-quasigroup.

(3) A quasigroup $(Q,\cdot)$ is a left SM-quasigroup if and only if its (12)-parastrophe is a right SM-quasigroup.

(4) A quasigroup $(Q,\cdot)$ is a left distributive quasigroup if and only if its (12)-parastrophe is a right distributive quasigroup.

(5) A quasigroup $(Q,\cdot)$ is a left distributive quasigroup if and only if its (23)-parastrophe is a left distributive quasigroup.

(6) A quasigroup $(Q,\cdot)$ is a left SM-quasigroup if and only if $(Q,\backslash)$ is a left F-quasigroup.

(7) A quasigroup $(Q,\cdot)$ is a right SM-quasigroup if and only if $(Q,/)$ is a right F-quasigroup.

(8) A quasigroup $(Q,\cdot)$ is a left E-quasigroup if and only if $(Q,\backslash)$ is a left E-quasigroup ([520], Lemma 2.2).

(9) A quasigroup $(Q, \cdot)$ is a right E-quasigroup if and only if $(Q, /)$ is a right E-quasigroup ([520], Lemma 2.2).

Proof. It is easy to check Cases 1–4.

Case 5. The fulfilment of the left distributive identity in a quasigroup $(Q, \cdot)$ is equivalent to the fact that in this quasigroup any left translation L_x is an automorphism of this quasigroup. Indeed, we can re-write the left distributive identity in this manner: $L_x yz = L_x y \cdot L_x z$. Using Table 1.1 we have that $L_x^{\backslash} = L_x^{-1}$. Thus by Lemma 6.13 $L_x^{\backslash} \in Aut(Q, \backslash)$. Therefore, if $(Q, \cdot)$ is a left distributive quasigroup, then $(Q, \backslash)$ is also a left distributive quasigroup and vice versa.

Case 6. Let $(Q, \backslash)$ be a left F-quasigroup. Then

$$x\backslash(y\backslash z) = (x\backslash y)\backslash(e^{(\backslash)}(x)\backslash z) = v.$$

If $x\backslash(y\backslash z) = v$, then $x \cdot v = (y\backslash z)$, $y \cdot (x \cdot v) = z$. We notice, if $x\backslash e^{(\backslash)}(x) = x$, then $e^{(\backslash)}(x) = x \cdot x \overset{def}{=} s(x)$. See Table 1.4.

We can re-write the equality $(x\backslash y)\backslash(e^{(\backslash)}(x)\backslash z) = v$ in the following form $(x\backslash y)\cdot v = s(x)\backslash z$, $s(x) \cdot ((x\backslash y) \cdot v) = z$. Now we have the equality $s(x) \cdot ((x\backslash y) \cdot v) = y \cdot (x \cdot v)$. If we denote $(x\backslash y)$ by u, then $x \cdot u = y$.

Therefore we can re-write the equality $s(x)\cdot((x\backslash y)\cdot v) = y\cdot(x\cdot v)$ in the form $s(x)\cdot(u\cdot v) = (x \cdot u) \cdot (x \cdot v)$, i.e., in the form $(x \cdot x) \cdot (u \cdot v) = (x \cdot u) \cdot (x \cdot v)$.

In a similar way it is possible to check the converse: if $(Q, \backslash)$ is a left SM-quasigroup, then $(Q, \cdot)$ is a left F-quasigroup.

Cases 7–9 are proved in a similar way. □

Corollary 6.15. *If $(Q, \cdot)$ is a group, then: 1. $(Q, \backslash)$ is a left SM-quasigroup; 2. $(Q, /)$ is a right SM-quasigroup.*

Proof. 1. Any group is a left F-quasigroup since in this case $e(x) = 1$ for all $x \in Q$. Therefore we can use Lemma 6.14, Case 6.

2. We can use Lemma 6.14, Case 7. □

Definition 6.16. A quasigroup (Q, B) is an isostrophic image of a quasigroup (Q, A) if there exists a collection of permutations $(\sigma, (\alpha_1, \alpha_2, \alpha_3)) = (\sigma, T)$, where $\sigma \in S_3$, $T = (\alpha_1, \alpha_2, \alpha_3)$ and $\alpha_1, \alpha_2, \alpha_3$ are permutations of the set Q such that

$$B(x_1, x_2) = A(x_1, x_2)(\sigma, T) = A^{\sigma}(x_1, x_2)T = \alpha_3^{-1}A(\alpha_1 x_{\sigma^{-1}1}, \alpha_2 x_{\sigma^{-1}2})$$

for all $x_1, x_2 \in Q$ [77].

A collection of permutations $(\sigma, (\alpha_1, \alpha_2, \alpha_3)) = (\sigma, T)$ will be called an *isostrophism* or an *isostrophy* of a quasigroup (Q, A). We can re-write the equality from Definition 6.16 in the form $(A^{\sigma})T = B$.

Lemma 6.17. *An isostrophic image of a quasigroup is a quasigroup [77].*

Proof. The proof follows from the fact that any parastrophic image of a quasigroup is a quasigroup and any isotopic image of a quasigroup is a quasigroup. □

From Lemma 6.17 it follows that it is possible to define the multiplication of isostrophies of a quasigroup operation defined on a set Q.

Definition 6.18. If (σ, S) and (τ, T) are isostrophisms of a quasigroup (Q, A), then

$$(\sigma, S)(\tau, T) = (\sigma\tau, S^{\tau}T),$$

where $A^{\sigma\tau} = (A^{\sigma})^{\tau}$ and $(x_1, x_2, x_3)(S^{\tau}T) = ((x_1, x_2, x_3)\ S^{\tau})T$ for any quasigroup triplet (x_1, x_2, x_3) [787].

A somewhat different operation on the set of all isostrophies (multiplication of quasigroup isostrophies) is defined in [77]. The definition from [580] is very close to Definition 6.18. See, also, [75, 488].

Corollary 6.19. $(\varepsilon, S)(\tau, \varepsilon) = (\tau, S^{\tau}) = (\tau, \varepsilon)(\varepsilon, S^{\tau})$.

Lemma 6.20. $(\sigma, S)^{-1} = (\sigma^{-1}, (S^{-1})^{\sigma^{-1}})$.

Proof. Let $S = (\alpha_1, \alpha_2, \alpha_3)$ be an isotopy of a quasigroup A, $S^{-1} = (\alpha_1^{-1}, \alpha_2^{-1}, \alpha_3^{-1})$, $S^{\sigma} = (\alpha_{\sigma^{-1}1}, \alpha_{\sigma^{-1}2}, \alpha_{\sigma^{-1}3})$. Then

$$(\sigma, S)(\sigma^{-1}, (S^{-1})^{\sigma^{-1}}) = (\varepsilon', S^{\sigma^{-1}}(S^{-1})^{\sigma^{-1}}) \overset{(Lemma\ 1.175)}{=} (\varepsilon', (SS^{-1})^{\sigma^{-1}}) = (\varepsilon', (\varepsilon, \varepsilon, \varepsilon)).$$

□

6.2 Left and right F-, E-, SM-quasigroups

In order to study the structure of left F-quasigroups we shall use an approach from [646, 784]. As usual $e(e(x)) = e^2(x)$ and so on.

Lemma 6.21. *1. In a left F-quasigroup $(Q, \cdot)$ the map e^i is an endomorphism of $(Q, \cdot)$, and $e^i(Q, \cdot)$ is a subquasigroup of quasigroup $(Q, \cdot)$ for all suitable values of the index i [645, 80].*

2. In a right F-quasigroup $(Q, \cdot)$ the map f^i is an endomorphism of $(Q, \cdot)$, and $f^i(Q, \cdot)$ is a subquasigroup of quasigroup $(Q, \cdot)$ for all suitable values of the index i [645, 80].

Proof. 1. From the identity $x \cdot yz = xy \cdot e(x)z$ by $z = e(y)$ we have $xy = xy \cdot e(x)e(y)$, i.e., $e(x \cdot y) = e(x) \cdot e(y)$. Further we have $e^2(x \cdot y) = e(e(x \cdot y)) = e(e(x) \cdot e(y)) = e^2(x) \cdot e^2(y)$ and so on. Therefore e^m is an endomorphism of the quasigroup $(Q, \cdot)$.

The fact that $e^m(Q, \cdot)$ is a subquasigroup of quasigroup $(Q, \cdot)$ follows from Lemma 1.298.

2. The proof is similar. □

The proof of the following lemma was taken from [80, p. 33].

Lemma 6.22. *1. Endomorphism e of a left F-quasigroup $(Q, \cdot)$ is a zero endomorphism, i.e., $e(x) = k$ for all $x \in Q$, if and only if a left F-quasigroup $(Q, \cdot)$ is a right loop. The quasigroup $(Q, \cdot)$ is an isotope of a group $(Q, +)$ of the form $(Q, \cdot) = (Q, +)(\varepsilon, \psi, \varepsilon)$, where $\psi \in Aut(Q, +)$, $k = 0$.*

2. Endomorphism f of a right F-quasigroup $(Q, \cdot)$ is a zero endomorphism, i.e., $f(x) = k$ for all $x \in Q$, if and only if a right F-quasigroup $(Q, \cdot)$ is a left loop. The quasigroup $(Q, \cdot)$ is an isotope of a group $(Q, +)$ of the form $(Q, \cdot) = (Q, +)(\varphi, \varepsilon, \varepsilon)$, where $\varphi \in Aut(Q, +)$, $k = 0$.

Proof. 1. We can rewrite equality $x \cdot yz = xy \cdot L_k z$ in the form $xy \cdot z = x(yL_k^{-1}z) = x(y \cdot \delta z)$, where $\delta = L_k^{-1}$. Therefore Sushkevich postulate B is fulfilled in $(Q, \cdot)$ and we can apply Theorem 2.215. Further we have $x \cdot 0 = x + \psi 0 = x$. From the other side $x \cdot k = x$. Therefore, $k = 0$. It is easy to see that the converse is also true.

2. We can use "mirror" principles. □

Lemma 6.23. *1. The endomorphism e of a left F-quasigroup $(Q, \cdot)$ is a permutation of the set Q if and only if quasigroup $(Q, \circ)$ of the form $x \circ y = x \cdot e(y)$ is a left distributive quasigroup and $e \in Aut(Q, \circ)$ [80].*

2. The endomorphism f of a right F-quasigroup $(Q, \cdot)$ is a permutation of the set Q if and only if quasigroup $(Q, \circ)$ of the form $x \circ y = f(x) \cdot y$ is a right distributive quasigroup and $f \in Aut(Q, \circ)$ [80].

Proof. 1. Prove that $(Q, \circ)$ is a left distributive quasigroup. We have

$$\begin{aligned} & x \circ (y \circ z) = x \cdot e(y \cdot e(z)) = x \cdot (e(y) \cdot e^2(z)) = \\ & (x \cdot e(y)) \cdot (e(x) \cdot e^2(z)) = \\ & (x \cdot e(y)) \cdot e(x \cdot e(z)) = (x \circ y) \circ (x \circ z). \end{aligned} \tag{6.1}$$

Prove that $e \in Aut(Q, \circ)$. We have $e(x \circ y) = e(x \cdot e(y)) = e(x) \cdot e^2(y) = e(x) \circ e(y)$ [600].

Conversely, let $(Q, \cdot)$ be an isotope of the form $x \cdot y = x \circ \psi(y)$, where $\psi \in Aut(Q, \circ)$ of a left distributive quasigroup $(Q, \circ)$. The fact that $\psi \in Aut(Q, \cdot)$ follows from Lemma 1.154.

We can use the equalities (6.1) by proving that $(Q, \cdot)$ is a left F-quasigroup. The fact that $\psi = e^{-1}$ follows from Lemma 6.5.

2. The proof is similar. □

In a finite left F-quasigroup $(Q, \cdot)$ define the following chain

$$Q \supset e(Q) \supset e^2(Q) \supset \dots \supset e^m(Q) \supset \dots \tag{6.2}$$

Notice, in chain (6.2) we suppose that $|Ker\, e^i| > 0$ for all suitable values of index i.

Definition 6.24. Chain (6.2) becomes stable means that there exists a number m such that $e^m(Q) = e^{m+1}(Q) = e^{m+2}(Q) \dots$. In this case we shall say that endomorphism e has the order m.

Lemma 6.25. *In any left F-quasigroup Q, chain (6.2) becomes stable, i.e., the map $e|_{e^m(Q)}$ is an automorphism of quasigroup $e^m(Q)$.*

Proof. Chain (6.2) becomes stable on a finite step m. It is clear that in this case $e|_{e^m(Q)}$ is an automorphism of $e^m(Q, \cdot)$. □

Example 6.26. Quasigroup $(Z, \cdot)$, where $x \cdot y = -x + y$, $(Z, +)$ is infinite cyclic group, is medial, unipotent, left F-quasigroup such that $e(x) = x + x = 2x$. Notice, in this case $Ker\, e = \{0\}$, the length of the chain (6.2) is equal to 1. In [591, p. 59] a mapping similar to the mapping e is called an isomorphism and the embedding of an algebra in its subalgebra.

Theorem 6.27. *1. Any finite left F-quasigroup $(Q, \cdot)$ has the following structure*

$$(Q, \cdot) \cong (A, \circ) \times (B, \cdot),$$

where $(A, \circ)$ is a quasigroup with a unique idempotent element; $(B, \cdot)$ is isotope of a left distributive quasigroup $(B, \star)$, $x \cdot y = x \star \psi y$ for all $x, y \in B$, $\psi \in Aut(B, \cdot)$, $\psi \in Aut(B, \star)$.

2. Any finite right F-quasigroup $(Q, \cdot)$ has the following structure

$$(Q, \cdot) \cong (A, \circ) \times (B, \cdot),$$

where $(A, \circ)$ is a quasigroup with a unique idempotent element; $(B, \cdot)$ is isotope of a right distributive quasigroup $(B, \star)$, $x \cdot y = \varphi x \star y$ for all $x, y \in B$, $\varphi \in Aut(B, \cdot)$, $\varphi \in Aut(B, \star)$.

Proof. Case 1. The proof of this theorem mainly repeats the proof of Theorem 6 from [784].

If the map e is a permutation of the set Q, then by Lemma 6.23, $(Q, \cdot)$ is an isotope of a left distributive quasigroup.

If $e(Q) = k$, where k is a fixed element of the set Q, then the quasigroup $(Q, \cdot)$ is a quasigroup with right identity element k, i.e., it is a right loop, which is isotopic to a group $(Q, +)$ (Lemma 6.22).

Let us suppose that $e^m = e^{m+1}$, where $m > 1$.

From Lemma 7.52 it follows that $e^m(Q, \cdot) = (B, \cdot)$ is a subquasigroup of quasigroup $(Q, \cdot)$. It is clear that $(B, \cdot)$ is a left F-quasigroup in which the map $\overline{e} = e|_{e^m(Q)}$ is a permutation of the set $B \subset Q$. In other words $e(B, \cdot) = (B, \cdot)$.

Define binary relation δ on quasigroup $(Q, \cdot)$ by the following rule: $x\delta y$ if and only if $e^m(x) = e^m(y)$.

Define binary relation ρ on quasigroup $(Q, \cdot)$ by the rule: $x\rho y$ if and only if $B \cdot x = B \cdot y$, i.e., for any $b_1 \in B$ there exists exactly one element $b_2 \in B$ such that $b_1 \cdot x = b_2 \cdot y$ and vice versa, for any $b_2 \in B$ there exists exactly one element $b_1 \in B$ such that $b_1 \cdot x = b_2 \cdot y$.

From Theorem 1.279 and Lemma 1.298 it follows that δ is a normal congruence.

It is easy to check that binary relation ρ is an equivalence relation (see Theorem 1.3).

We prove that binary relation ρ is a congruence, i.e., that the following implication is true: $x_1\rho y_1,\ x_2\rho y_2, \Longrightarrow (x_1 \cdot x_2)\rho(y_1 \cdot y_2)$.

Using the definition of relation ρ we can re-write the last implication in the following equivalent form: if

$$B \cdot x_1 = B \cdot y_1, B \cdot x_2 = B \cdot y_2, \tag{6.3}$$

then $B \cdot (x_1 \cdot x_2) = B \cdot (y_1 \cdot y_2)$.

If we multiply both sides of the equalities (6.3), respectively, then we obtain the following equality

$$(\overset{x}{B} \cdot \overset{y}{x_1}) \cdot (\overset{e(x)}{B} \cdot \overset{z}{x_2}) = (B \cdot y_1) \cdot (B \cdot y_2).$$

Using the left F-quasigroup equality $(x \cdot yz = xy \cdot e(x)z)$ from the right to the left and, taking into consideration that if $x \in B$, then $e(x) \in B$, i.e., $eB = B$, and we can re-write the last equality in the following form

$$B \cdot (x_1 \cdot x_2) = B \cdot (y_1 \cdot y_2),$$

since $(B, \cdot)$ is a subquasigroup and, therefore, $B \cdot B = B$. Thus the binary relation ρ is a congruence.

Prove that $\delta \cap \rho = \hat{Q} = \{(x, x) | \, \forall \, x \in Q\}$. From reflexivity of relations δ, ρ it follows that $\delta \cap \rho \supseteq \hat{Q}$.

Let $(x, y) \in \delta \cap \rho$, i.e., let $x\ \delta\ y$ and $x\ \rho\ y$, where $x, y \in Q$. Using the definitions of relations δ, ρ we have $e^m(x) = e^m(y)$ and $(B, \cdot) \cdot x = (B, \cdot) \cdot y$. Then there exist $a, b \in B$ such that $a \cdot x = b \cdot y$. Applying to the both sides of last equality the map e^m, we obtain $e^m(a) \cdot e^m(x) = e^m(b) \cdot e^m(y)$, $e^m(a) = e^m(b)$, $a = b$, since the map $e^m|_B$ is a permutation of the set B. If $a = b$, then from the equality $a \cdot x = b \cdot y$ we obtain $x = y$.

Prove that $\delta \circ \rho = Q \times Q$. Let a, c be any fixed elements of the set Q. We prove the equality if it is shown that there exists the element $y \in Q$ such that $a\delta y$ and $y\rho c$.

From the definition of congruence δ we have that the condition $a\delta y$ is equivalent to the equality $e^m(a) = e^m(y)$. From the definition of congruence ρ it follows that the condition $y\rho c$ is equivalent to the following condition: $y \in \rho(c) = B \cdot c$.

We prove the equality if it is shown that there exists element $y \in B \cdot c$ such that $e^m(a) = e^m(y)$. Such element y exists since $e^m(B \cdot c) = e^m(B) \cdot e^m(c) = B = e^m(Q)$.

Prove that $\rho \circ \delta = Q \times Q$. Let a, c be any fixed elements of the set Q. We prove the equality if it will be shown that there exists element $y \in Q$ such that $a\rho y$ and $y\delta c$.

From definition of congruence δ we have that the condition $y\delta c$ is equivalent to the equality $e^m(c) = e^m(y)$. From definition of congruence ρ it follows that the condition $a\rho y$ is equivalent to the following condition: $y \in \rho(a) = B \cdot a$.

We prove the equality if it will be shown that there exists the element $y \in B \cdot a$ such that $e^m(c) = e^m(y)$. Such element y exists since $e^m(B \cdot a) = e^m(B) \cdot e^m(a) = B = e^m(Q)$.

Therefore $\rho \circ \delta = Q \times Q = \delta \circ \rho$, $\delta \cap \rho = \hat{Q}$ and we can use Theorem 7.36. Now we can say that quasigroup $(Q, \cdot)$ is isomorphic to the direct product of a quasigroup $(Q, \cdot)/\delta \cong (B, \cdot)$ (Theorem 1.279) and a division groupoid $(Q, \cdot)/\rho \cong (A, \circ)$ [43, 173].

From Definition 1.313 it follows, if $(Q, \cdot) \cong (B, \cdot) \times (A, \circ)$, where $(Q, \cdot)$, $(B, \cdot)$ are quasigroups, then $(A, \circ)$ is also a quasigroup. Then by Theorem 1.279 the congruence ρ is normal, $(B, \cdot) \trianglelefteq (Q, \cdot)$.

Left F-quasigroup equality holds in the quasigroup $(B, \cdot)$ since $(B, \cdot) \subseteq (Q, \cdot)$.

If the quasigroups $(Q, \cdot)$ and $(B, \cdot)$ are left F-quasigroups, $(Q, \cdot) \cong (A, \circ) \times (B, \cdot)$, then $(A, \circ)$ is also a left F-quasigroup (Lemma 1.317).

Prove that the quasigroup $(A, \circ) \cong (Q, \cdot)/(B, \cdot)$, where $e^m(Q, \cdot) = (B, \cdot)$, has a unique idempotent element.

We can identify elements of quasigroup $(Q, \cdot)/(B, \cdot)$ with cosets of the form $B \cdot c$, where $c \in Q$.

From the properties of the quasigroup $(A, \circ)$ we have that $e^m(A) = a$, where the element a is a fixed element of the set A that corresponds to the coset class B. Further, taking into consideration the properties of endomorphism e of the quasigroup $(A, \circ)$, we obtain $e^{m+1}A = e(e^m A) = e(a) = a$. Therefore $e(a) = a$, i.e., the element a is an idempotent element of quasigroup $(A, \circ)$.

Prove that there exists exactly one idempotent element in the quasigroup $(A, \circ)$. Suppose that there exists an element c of the set A such that $c \circ c = c$, i.e., such that $e(c) = c$. Then we have $e^m(c) = c = a$, since $e^m(A) = a$.

The fact that $(B, \cdot)$ is an isotope of a left distributive quasigroup $(B, \star)$ follows from Lemma 6.23.

Case 2. Properties of right F-quasigroups coincide with the "mirror" properties of left F-quasigroups. □

We add some details on the structure of left F-quasigroup $(Q, \cdot)$. By $e^j(Q, \cdot)$ we denote an endomorphic image of the quasigroup $(Q, \cdot)$ relative to the endomorphism e^j.

Corollary 6.28. *If $(Q, \cdot)$ is a left F-quasigroup, then $e^m(Q, \cdot) \trianglelefteq (Q, \cdot)$.*

Proof. This follows from the fact that the binary relation ρ from Theorem 6.27 is a normal congruence in $(Q, \cdot)$ and subquasigroup $e^m(Q, \cdot) = (B, \cdot)$ is an equivalence class of ρ. □

Remark 6.29. For brevity we shall denote the endomorphism $e|_{e^j(Q,\cdot)}$ such that

$$e|_{e^j(Q,\cdot)} : e^j(Q, \cdot) \rightarrow e^{j+1}(Q, \cdot)$$

by e_j, the endomorphism $f|_{f^j(Q,\cdot)}$ by f_j, the endomorphism $s|_{s^j(Q,\cdot)}$ by s_j.

Corollary 6.30. *If $(Q, \cdot)$ is a left F-quasigroup with an idempotent element, then equivalence class (cell) $\bar{a}$ of the normal congruence $Ker\, e_j$ containing an idempotent element $a \in Q$ forms a linear right loop $(\bar{a}, \cdot)$ for all suitable values of j.*

Proof. By Proposition 7.96 $(\bar{a}, \cdot)$ is a quasigroup. From the properties of the endomorphism e we have that in $(\bar{a}, \cdot)$, endomorphism e is a zero endomorphism. Therefore in this case we can apply Lemma 6.22. Then $(\bar{a}, \cdot)$ is isotopic to a group with isotopy of the form $(\varepsilon, \psi, \varepsilon)$, where $\psi \in Aut(\bar{a}, \cdot)$. □

Corollary 6.31. *If $(Q,\cdot)$ is a right F-quasigroup, then $f^m(Q,\cdot) \trianglelefteq (Q,\cdot)$.*

Proof. The proof is similar to the proof of Corollary 6.28. □

Corollary 6.32. *If $(Q,\cdot)$ is a right F-quasigroup with an idempotent element, then equivalence class $\bar{a}$ of the normal congruence $Ker\, f_j$ containing an idempotent element $a \in Q$ forms linear left loop $(\bar{a},\cdot)$ for all suitable values of j.*

Proof. The proof is similar to the proof of Corollary 6.30. □

6.2.1 Direct decompositions

Theorem 6.33. 1. *If endomorphism s of a left semimedial quasigroup $(Q,\cdot)$ is a zero endomorphism, i.e., $s(x) = 0$ for all $x \in Q$, then $(Q,\cdot)$ is a unipotent quasigroup, $(Q,\cdot) \cong (Q,\circ)$, where $x \circ y = -\varphi x + \varphi y$, $(Q,+)$ is a group, $\varphi \in Aut(Q,+)$.*

2. *If endomorphism s of a right semimedial quasigroup $(Q,\cdot)$ is a zero endomorphism, i.e., $s(x) = 0$ for all $x \in Q$, then $(Q,\cdot)$ is a unipotent quasigroup, $(Q,\cdot) \cong (Q,\circ)$, where $x \circ y = \varphi x - \varphi y$, $(Q,+)$ is a group, $\varphi \in Aut(Q,+)$.*

3. *If endomorphism f of a left E-quasigroup $(Q,\cdot)$ is a zero endomorphism, i.e., $f(x) = 0$ for all $x \in Q$, then up to isomorphism $(Q,\cdot)$ is a left loop, $x \cdot y = \alpha x + y$, $(Q,+)$ is an abelian group, $\alpha 0 = 0$.*

4. *If endomorphism e of a right E-quasigroup $(Q,\cdot)$ is a zero endomorphism, i.e., $e(x) = 0$ for all $x \in Q$, then up to isomorphism $(Q,\cdot)$ is a right loop, $x \cdot y = x + \beta y$, $(Q,+)$ is an abelian group, $\beta 0 = 0$.*

Proof. **Case 1.** We can rewrite the equality $xx\cdot yz = xy\cdot xz$ in the form $k\cdot yz = xy\cdot xz$, where $s(x) = k$ for all $x \in Q$. If we denote xz by v, then $z = x\backslash v$ and the equality $k \cdot yz = xy \cdot xz$ takes the form $k \cdot (y \cdot (x\backslash v)) = xy \cdot v$, $k \cdot (y \cdot (x\backslash v)) = (y * x) \cdot v$.

Then the last equality has the form $A_1(y, A_2(x, v)) = A_3(A_4(y, x), v)$, where A_1, A_2, A_3, A_4 are quasigroup operations, namely, $A_1(y,t) = L_k(y \cdot t)$, $t = A_2(x, v) = x\backslash v$, $A_3(u, v) = u \cdot v$, $u = A_4(y, x) = x \cdot y$.

From the four quasigroups theorem (Theorem 2.205) it follows that quasigroup $(Q,\cdot)$ is an isotope of a group $(Q,+)$.

If in the equality $k \cdot yz = xy \cdot xz$ we fix the variable $x = b$, then we obtain the following equality $k \cdot yz = by \cdot bz$, $k \cdot yz = L_b y \cdot L_b z$. From Theorem 2.210 it follows that $(Q,\cdot)$ is a right linear quasigroup.

If in $k \cdot yz = xy \cdot xz$ we put $x = z$, then we obtain $k \cdot yx = xy \cdot k$. From Lemma 2.212 it follows that $(Q,+)$ is a commutative group.

From Lemma 2.208 we have that there exists a group $(Q,+)$ such that $x\cdot y = \alpha x+\psi y+c$, where α is a permutation of the set Q, $\alpha 0 = 0$, $\psi \in Aut(Q,+)$.

Further we have $s(0) = k = 0 \cdot 0 = c$, $k = c$. Then $s(x) = k = x \cdot x = \alpha x + \psi x + k$. Therefore, $\alpha x + \psi x = 0$ for all $x \in Q$. Then $\alpha = I\psi$, where $x + I(x) = 0$ for all $x \in Q$. Therefore α is an antiautomorphism of the group $(Q,+)$, $x \cdot y = I\psi x + \psi y + k$.

Finally $L_k^{-1}(L_k x \cdot L_k y) = L_k^{-1}(I\psi x + I\psi k + \psi k + \psi y + k) = L_k^{-1}(I\psi x + \psi y + k) = -k+I\psi x+k-k+\psi y+k = I_k I\psi x+I_k\psi y = II_k\psi x+I_k\psi y = x\circ y$, where $I_k x = -k+x+k$ is an inner automorphism of $(Q,+)$. It is easy to see that $s^{\circ}(x) = 0$ for all $x \in Q$.

Below we shall suppose that any left semimedial quasigroup $(Q,\cdot)$ with zero endomorphism s is a unipotent quasigroup with the form $x\cdot y = -\varphi x+\varphi y$, where $(Q,+)$ is a group, $\varphi \in Aut(Q,+)$.

Case 2. We can rewrite the equality $zy \cdot s(x) = zx \cdot yx$ in the form $zy \cdot k = zx \cdot yx$, where $s(x) = k$. If we denote zx by v, then $z = v/x$ and the equality $zy \cdot k = zx \cdot yx$ takes the form $((v/x)y)k = v \cdot yx = v \cdot (x * y)$.

We re-write the last equality in the form $A_1(A_2(v, x), y) = A_3(v, A_4(x, y))$, where A_1, A_2, A_3, A_4 are quasigroup operations, namely, $A_1(t, y) = R_k(t \cdot y)$, $t = A_2(v, x) = v/x$, $A_3(v, u) = v \cdot u$, $u = A_4(x, y) = x * y$.

From the four quasigroups theorem it follows that quasigroup $(Q, \cdot)$ is an isotope of a group $(Q, +)$.

If in the equality $zy \cdot k = zx \cdot yx$ we fix the variable $x = b$, then we obtain the following equality $zy \cdot k = zb \cdot yb$, $zy \cdot k = R_b z \cdot R_b y$. From Theorem 2.210 it follows that $(Q, \cdot)$ is a left linear quasigroup.

If in the equality $zy \cdot k = zx \cdot yx$ we put $x = z$, then we obtain $xy \cdot k = k \cdot yx$. Thus from Lemma 2.212 it follows that $(Q, +)$ is a commutative group.

From Lemma 2.208 we have that there exists a group $(Q, +)$ such that $x \cdot y = \varphi x + \beta y + c$, where β is a permutation of the set Q, $\beta 0 = 0$, $\varphi \in Aut(Q, +)$.

Further we have $s(0) = k = 0 \cdot 0 = c$, $k = c$. Then $s(x) = k = x \cdot x = \varphi x + \beta x + k$. Therefore, $\varphi x + \beta x = 0$ for all $x \in Q$. Then $\beta = I\varphi$, where $Ix + x = 0$ for all $x \in Q$.

Therefore $\beta = I\varphi \in Aut(Q, +)$, $x \cdot y = \varphi x - \varphi y + k$.

We have $R_k^{-1}(R_k x \cdot R_k y) = R_k^{-1}(\varphi x + \varphi k - \varphi k - \varphi y + k) = \varphi x - \varphi y + k - k = \varphi x - \varphi y = x \circ y$. It is easy to see that $s^{\circ}(x) = 0$ for all $x \in Q$.

Below we shall suppose that any right semimedial quasigroup $(Q, \cdot)$ with zero endomorphism s is a unipotent quasigroup with the form $x \cdot y = \varphi x - \varphi y$, where $(Q, +)$ is a group, $\varphi \in Aut(Q, +)$.

Case 3. We can rewrite the equality $x \cdot yz = f(x)y \cdot xz$ in the form $x \cdot yz = ky \cdot xz = y \cdot xz$, $x \cdot (z * y) = xz * y$, where $f(x) = k$ for all $x \in Q$.

Then $A_1(x, A_2(z, y)) = A_3(A_4(x, z), y)$, where A_1, A_2, A_3, A_4 are quasigroup operations, namely, $A_1(x, t) = x \cdot t$, $t = A_2(z, y) = z * y$, $A_3(u, y) = u * y$, $u = A_4(x, z) = x \cdot z$. From the four quasigroups theorem it follows that quasigroup $(Q, \cdot)$ is a group isotope.

If in the equality $x \cdot yz = y \cdot xz$ we fix the variable z, i.e., if we take $z = a$, then we have $x \cdot R_a y = y \cdot R_a x$. From Corollary 4.92 it follows that the group $(Q, +)$ is commutative.

If in the equality $x \cdot yz = y \cdot xz$ we fix the variable x, i.e., if we take $x = a$, then we have $a \cdot yz = y \cdot az$, $a \cdot (yz) = y \cdot L_a z$. The application of Theorem 2.210 to the last equality gives us that $(Q, \cdot)$ is a right linear quasigroup, i.e., $x \cdot y = \alpha x + \psi y + c$.

Then $f(x) \cdot x = k \cdot x = \alpha k + \psi x + c = x$. By $x = 0$ we have $\alpha k + \psi 0 + c = 0$, $\alpha k = -c$. Therefore, $k \cdot x = x = \psi x$ for all $x \in Q$. Then $\psi = \varepsilon$, $x \cdot y = \alpha x + y + c = L_c \alpha x + y$ for all $x, y \in Q$. In other words $x \cdot y = \alpha x + y$ for all $x, y \in Q$.

Further let $a + \alpha 0 = 0$. Then $L_a^{-1}(L_a \alpha x + L_a y) = -a + a + \alpha x + a + y = a + \alpha x + y = \alpha' x + y = x \circ y$, where $\alpha' = L_a \alpha$ and $\alpha' 0 = 0$.

Case 4. Case 4 is a "mirror" case of Case 3, but we give the direct proof. We can rewrite the equality $zy \cdot x = zx \cdot ye(x)$ in the form $zy \cdot x = zx \cdot yk = zx \cdot y$, $(y * z) \cdot x = y * zx$, where $e(x) = k$.

Then $A_1(A_2(y, z), x) = A_3(y, A_4(z, x))$, where A_1, A_2, A_3, A_4 are quasigroup operations, namely, $A_1(t, x) = t \cdot x$, $t = A_2(y, z) = y * z$, $A_3(y, v) = y * v$, $v = A_4(z, x) = z \cdot x$.

From the four quasigroups theorem it follows that quasigroup $(Q, \cdot)$ is an isotope of a group $(Q, +)$.

If in the equality $zy \cdot x = zx \cdot y$ we fix the variable z, i.e., if we take $z = a$, then we have $L_a y \cdot x = L_a x \cdot y$. From Corollary 4.92 it follows that the group $(Q, +)$ is commutative.

If in the equality $zy \cdot x = zx \cdot y$ we fix the variable x, i.e., if we take $x = a$, then we have $zy \cdot a = za \cdot y$, $zy \cdot a = R_a z \cdot y$. The application of Theorem 2.210 to the last equality gives us that $(Q, \cdot)$ is a left linear quasigroup, i.e., $x \cdot y = \varphi x + \beta y + c$.

Then $x \cdot e(x) = x \cdot k = \varphi x + \beta k + c = x$. By $x = 0$ we have $\varphi 0 + \beta k + c = 0$, $\beta k = -c$. Therefore, $x \cdot k = x = \varphi x$ for all $x \in Q$. Then $\varphi = \varepsilon$, $x \cdot y = x + \beta y + c = x + R_c \beta y$ for all $x, y \in Q$. In other words $x \cdot y = x + \beta y$ for all $x, y \in Q$.

Further let $a + \beta 0 = 0$. Then $L_a^{-1}(L_a x + L_a \beta y) = -a + a + x + a + \beta y = x + a + \beta y = x + \beta' y = x \circ y$, where $\beta' = L_a \beta$ and $\beta' 0 = 0$. □

In proof of the following lemma we use ideas from [80].

Lemma 6.34. *1. If endomorphism s of a left semimedial quasigroup $(Q, \cdot)$ is a permutation of the set Q, then the quasigroup $(Q, \circ)$ of the form $x \circ y = s^{-1}(x \cdot y)$ is a left distributive quasigroup and $s \in Aut(Q, \circ)$.*

2. If endomorphism s of a right semimedial quasigroup $(Q, \cdot)$ is a permutation of the set Q, then the quasigroup $(Q, \circ)$ of the form $x \circ y = s^{-1}(x \cdot y)$ is a right distributive quasigroup and $s \in Aut(Q, \circ)$.

3. If endomorphism f of a left E-quasigroup $(Q, \cdot)$ is a permutation of the set Q, then the quasigroup $(Q, \circ)$ of the form $x \circ y = f(x) \cdot y$ is a left distributive quasigroup and $f \in Aut(Q, \circ)$.

4. If endomorphism e of a right E-quasigroup $(Q, \cdot)$ is a permutation of the set Q, then the quasigroup $(Q, \circ)$ of the form $x \circ y = x \cdot e(y)$ is a right distributive quasigroup and $e \in Aut(Q, \circ)$.

Proof. Case 1. We prove that $(Q, \circ)$ is left distributive. It is clear that $s^{-1} \in Aut(Q, \cdot)$. We have

$$\begin{array}{l} x \circ (y \circ z) = s^{-1}(x \cdot s^{-1}(y \cdot z)), \\ (x \circ y) \circ (x \circ z) = s^{-2}((x \cdot y) \cdot (x \cdot z)) = s^{-2}(s(x) \cdot (y \cdot z)) = \\ s^{-1}(x \cdot s^{-1}(y \cdot z)). \end{array}$$

Prove that $s \in Aut(Q, \circ)$. We have $s(x \circ y) = x \cdot y$, $s(x) \circ s(y) = s^{-1}(s(x) \cdot s(y)) = x \cdot y$. See, also [600].

Case 2. We prove that $(Q, \circ)$ is right distributive. It is clear that $s^{-1} \in Aut(Q, \cdot)$. We have

$$\begin{array}{l} (x \circ y) \circ z = s^{-1}(s^{-1}(x \cdot y) \cdot z), \\ (x \circ z) \circ (y \circ z) = s^{-2}((x \cdot z) \cdot (y \cdot z)) = \\ s^{-2}((x \cdot y) \cdot s(z)) = s^{-1}(s^{-1}(x \cdot y) \cdot z). \end{array}$$

Prove that $s \in Aut(Q, \circ)$. We have $s(x \circ y) = x \cdot y$, $s(x) \circ s(y) = s^{-1}(s(x) \cdot s(y)) = x \cdot y$.

Case 3. If endomorphism f is a permutation of the set Q, then $f, f^{-1} \in Aut(Q, \cdot)$. We have

$$\begin{array}{l} x \circ (y \circ z) = f(x) \cdot (f(y) \cdot z), \\ (x \circ y) \circ (x \circ z) = f(f(x) \cdot y) \cdot (f(x) \cdot z) = \\ (f^2(x) \cdot f(y)) \cdot (f(x) \cdot z) = f(x) \cdot (f(y) \cdot z). \end{array}$$

Prove that $f \in Aut(Q, \circ)$. We have $f(x \circ y) = f(f(x) \cdot y) = f^2(x) \cdot f(y) = f(x) \circ f(y)$.

Case 4. If endomorphism e is a permutation of the set Q, then $e, e^{-1} \in Aut(Q, \cdot)$. We have

$$\begin{array}{l} (x \circ y) \circ z = (x \cdot e(y)) \cdot e(z), \\ (x \circ z) \circ (y \circ z) = (x \cdot e(z)) \cdot e(y \cdot e(z)) = \\ (x \cdot e(z)) \cdot (e(y) \cdot e^2(z)) = (x \cdot e(y)) \cdot e(z). \end{array}$$

Prove that $e \in Aut(Q, \circ)$. We have $e(x \circ y) = e(x \cdot e(y)) = e(x) \cdot e^2(y) = e(x) \circ e(y)$. □

In this subsection we give an overview of results from [788, 789] on direct decompositions of some quasigroups.

Theorem 6.35. *1. Any finite left SM-quasigroup $(Q,\cdot)$ has the following structure*

$$(Q,\cdot) \cong (A,\circ) \times (B,\cdot),$$

where $(A,\circ)$ is a quasigroup with a unique idempotent element and there exists a number m such that $|s^m(A,\circ)| = 1$; $(B,\cdot)$ is an isotope of a left distributive quasigroup $(B,\star)$, $x \cdot y = s(x \star y)$ for all $x, y \in B$, $s \in Aut(B,\cdot)$, $s \in Aut(B,\star)$.

2. Any finite right SM-quasigroup $(Q,\cdot)$ has the following structure

$$(Q,\cdot) \cong (A,\circ) \times (B,\cdot),$$

where $(A,\circ)$ is a quasigroup with a unique idempotent element and there exists an ordinal number m such that $|s^m(A,\circ)| = 1$; $(B,\cdot)$ is an isotope of a right distributive quasigroup $(B,\star)$, $x \cdot y = s(x \star y)$ for all $x, y \in B$, $s \in Aut(B,\cdot)$, $s \in Aut(B,\star)$.

3. Any finite left E-quasigroup $(Q,\cdot)$ has the following structure

$$(Q,\cdot) \cong (A,\circ) \times (B,\cdot),$$

where $(A,\circ)$ is a quasigroup with a unique idempotent element and there exists a number m such that $|f^m(A,\circ)| = 1$; $(B,\cdot)$ is an isotope of a left distributive quasigroup $(B,\star)$, $x \cdot y = f^{-1}(x) \star y$ for all $x, y \in B$, $f \in Aut(B,\cdot)$, $f \in Aut(B,\star)$.

4. Any finite right E-quasigroup $(Q,\cdot)$ has the following structure

$$(Q,\cdot) \cong (A,\circ) \times (B,\cdot),$$

where $(A,\circ)$ is a quasigroup with a unique idempotent element and there exists a number m such that $|e^m(A,\circ)| = 1$; $(B,\cdot)$ is an isotope of a right distributive quasigroup $(B,\star)$, $x \cdot y = x \star e^{-1}(y)$ for all $x, y \in B$, $e \in Aut(B,\cdot)$, $e \in Aut(B,\star)$.

Proof. The proof of this theorem is similar to the proof of Theorem 2.105. See [788, 789] for details. □

Corollary 6.36. *If $(Q,\cdot)$ is a left SM-quasigroup with an idempotent element, then equivalence class $\bar{a}$ of the normal congruence $Ker\, s_j$ containing an idempotent element $a \in Q$ forms a unipotent quasigroup $(\bar{a},\cdot)$ isotopic to a group with isotopy of the form $(-\psi, \psi, \varepsilon)$, where $\psi \in Aut(\bar{a},\cdot)$ for all suitable values of j.*

If $(Q,\cdot)$ is a right SM-quasigroup with an idempotent element, then equivalence class $\bar{a}$ of the normal congruence $Ker\, s_j$ containing an idempotent element $a \in Q$ forms a unipotent quasigroup $(\bar{a},\cdot)$ isotopic to a group with isotopy of the form $(\varphi, -\varphi, \varepsilon)$, where $\varphi \in Aut(\bar{a},\cdot)$ for all suitable values of j.

If $(Q,\cdot)$ is a left E-quasigroup with an idempotent element, then equivalence class $\bar{a}$ of the normal congruence $Ker\, f_j$ containing an idempotent element $a \in Q$ forms a left loop isotopic to an abelian group with isotopy of the form $(\alpha, \varepsilon, \varepsilon)$ for all suitable values of j.

If $(Q,\cdot)$ is a right E-quasigroup with an idempotent element, then equivalence class $\bar{a}$ of the normal congruence $Ker\, e_j$ containing an idempotent element $a \in Q$ forms a right loop isotopic to an abelian group with isotopy of the form $(\varepsilon, \beta, \varepsilon)$ for all suitable values of j.

Proof. Mainly the proof repeats the proof of Corollary 6.30. It is possible to use Theorem 6.33. □

6.2.2 F-quasigroups

Simple F-quasigroups isotopic to groups (FG-quasigroups) are described in [501]. The authors prove that any simple FG-quasigroup is a simple group or a simple medial quasigroup. We notice that simple medial quasigroups are described in [461]. See also [783, 785]. The conditions when a group isotope is a left (right) F-quasigroup are given in [523, 828].

The following examples demonstrate that in an F-, E-, or an SM-quasigroup, the order of map e does not coincide with the order of map f, i.e., there exists some independence of the orders of maps e, f and s.

Example 6.37. By $(Z_3, +)$ we denote the cyclic group of order 3 and we take $Z_3 = \{0, 1, 2\}$. Groupoid $(Z_3, \cdot)$, where $x \cdot y = x - y$, is a medial E-, F-, SM-quasigroup and $e^{\cdot}(Z_3) = s^{\cdot}(Z_3) = \{0\}$, $f^{\cdot}(Z_3) = Z_3$.

Example 6.38. By $(Z_6, +)$ we denote the cyclic group of order 6 and we take $Z_6 = \{0, 1, 2, 3, 4, 5\}$. Groupoid $(Z_6, \cdot)$, where $x \cdot y = x - y$, is a medial E-, F-, SM-quasigroup and $e^{\cdot}(Z_6) = s^{\cdot}(Z_6) = \{0\}$, $f^{\cdot}(Z_6) = \{0, 2, 4\}$.

The following lemmas give connections between the maps e and f in F-quasigroups.

Lemma 6.39. *1. Endomorphism e of an F-quasigroup $(Q, \cdot)$ is a zero endomorphism, i.e., $e(x) = 0$ for all $x \in Q$ if and only if $x \cdot y = x + \psi y$, $(Q, +)$ is a group, $\psi \in Aut(Q, +)$, $(Q, \cdot)$ contains a unique idempotent element 0, $x + fy = fy + x$ for all $x, y \in Q$.*

2. Endomorphism f of an F-quasigroup $(Q, \cdot)$ is a zero endomorphism, i.e., $f(x) = 0$ for all $x \in Q$ if and only if $x \cdot y = \varphi x + y$, $(Q, +)$ is a group, $\varphi \in Aut(Q, +)$, $(Q, \cdot)$ contains a unique idempotent element 0, $x + ey = ey + x$ for all $x, y \in Q$.

Proof. 1. From Lemma 6.22 Case 1 it follows that $(Q, \cdot)$ is a right loop, and an isotope of a group $(Q, +)$ of the form $x \cdot y = x + \psi y$, where $\psi \in Aut(Q, +)$.

If $a \cdot a = a$, then $a + \psi a = a$, $\psi a = 0$, $a = 0$.

If we rewrite the right F-quasigroup equality in terms of the operation $+$, then we obtain $x + \psi y + \psi z = x + \psi f(z) + \psi y + \psi^2 z$, $\psi y + \psi z = \psi f(z) + \psi y + \psi^2 z$. If we take $y = 0$ in the last equality, then $\psi z = \psi f(z) + \psi^2 z$. Therefore $\psi y + \psi f(z) + \psi^2 z = \psi f(z) + \psi y + \psi^2 z$, $\psi y + \psi f(z) = \psi f(z) + \psi y$, $y + f(z) = f(z) + y$.

Conversely, from $x \cdot y = x + \psi y$ we have $x \cdot e(x) = x + \psi e(x) = x$, $e(x) = 0$ for all $x \in Q$.

2. This case is proved in a similar way with Case 1. □

Lemma 6.40. *1. If endomorphism e of an F-quasigroup $(Q, \cdot)$ is a zero endomorphism, i.e., $e(x) = 0$ for all $x \in Q$, then*

(i) $f(x) = x - \psi x$, $f \in End(Q, +)$;

(ii) $f(Q, +) \subseteq C(Q, +)$;

(iii) $(H, +) \trianglelefteq (Q, +)$, $f(Q, +) \trianglelefteq (Q, +)$, $(Q, +)/(H, +) \cong f(Q, +)$, where $(H, +)$ is an equivalence class of the congruence $Ker\, f$ containing identity element of $(Q, +)$;

(iv) $f(Q, \cdot)$ is a medial F-quasigroup; $(H, \cdot) = (H, +)$ is a group;

(v) $(\bar{a}, \cdot)$, where $\bar{a}$ is an equivalence class of the normal congruence $Ker\, f_j$ containing an idempotent element $a \in Q$, $i \geqslant 1$, is an abelian group.

2. If endomorphism f of an F-quasigroup $(Q, \cdot)$ is a zero endomorphism, i.e., $f(x) = 0$ for all $x \in Q$, then

(i) $e(x) = -\varphi x + x$, $e \in End(Q, +)$;

(ii) $e(Q, +) \subseteq C(Q, +)$.

(iii) $(H, +) \trianglelefteq (Q, +)$, $e(Q, +) \trianglelefteq (Q, +)$, $(Q, +)/(H, +) \cong e(Q, +)$, where $(H, +)$ is an equivalence class of the congruence $Ker\, e$ containing identity element of $(Q, +)$;

(iv) $e(Q,\cdot)$ *is a medial F-quasigroup;* $(H,\cdot) = (H,+)$ *is a group;*

(v) $(\bar{a},\cdot)$, *where* $\bar{a}$ *is an equivalence class of the normal congruence* $Ker\, e_j$ *containing an idempotent element* $a \in Q$, $i \geqslant 1$, *is an abelian group.*

Proof. 1. (i) From Lemma 6.39, Case 1 we have $f(x) \cdot x = f(x) + \psi x = x$, $f(x) = x - \psi x$. We can rewrite the equality $f(x \cdot y) = f(x) \cdot f(y)$ in the form $f(x + \psi y) = f(x) + \psi f(y)$. If $x = y = 0$, then we have $f(0) = 0$. If $x = 0$, then $f\psi(y) = \psi f(y)$. Therefore

$$f(x + \psi y) = f(x) + f\psi(y). \tag{6.4}$$

(ii) If we apply the equality $f(z) = z - \psi z$ to the equality (6.4), then we obtain $x + \psi y - \psi(x + \psi y) = x - \psi x + \psi y - \psi^2 y$, $x + \psi y - \psi^2 y - \psi x = x - \psi x + \psi y - \psi^2 y$, $\psi y - \psi^2 y - \psi x = -\psi x + \psi y - \psi^2 y$, $y - \psi y - x = -x + y - \psi y$, $fy - x = -x + fy$, $x + fy = fy + x$, i.e., $f(Q,+) \subseteq C(Q,+)$.

(iii) From definitions and Case (ii) it follows that $(H,+) \trianglelefteq (Q,+)$, $f(Q,+) \trianglelefteq (Q,+)$. The last follows from definition of $(H,+)$.

(iv) $f(Q,\cdot)$ is a medial F-quasigroup since from Case (ii) it follows that $f(Q,+)$ is an abelian group. Quasigroup $(H,\cdot)$ is a group since in this quasigroup the maps e and f are zero endomorphisms and we can use Case (i).

(v) $(\bar{a},\cdot) \cong f^i(Q,\cdot)/f^{i+1}(Q,\cdot)$ is an abelian group since in this quasigroup the maps e and f are zero endomorphisms and $f^i(Q,\cdot)$ is a medial quasigroup for any suitable value of the index i. Moreover, it is well known that in a medial quasigroup, its subquasigroup is normal [506]. Then $f^{i+1}(Q,\cdot) \trianglelefteq f^i(Q,\cdot)$.

2. This case is proved in a similar way with Case 1. □

Corollary 6.41. *Both endomorphisms* e *and* f *of an F-quasigroup* $(Q,\cdot)$ *are zero endomorphisms if and only if* $(Q,\cdot)$ *is a group.*

Proof. By Lemma 6.39, Case 1, $x \cdot y = x + \psi y$. By Lemma 6.40, Case 1, (i), $f(x) = x - \psi x$. Since $f(x) = 0$ for all $x \in Q$, further we have $\psi = \varepsilon$.

Conversely, it is clear that in any group $e(x) = f(x) = 0$ for all $x \in Q$. □

Example 6.42. By $(Z_4,+)$ we denote the cyclic group of order 4 and we take $Z_4 = \{0, 1, 2, 3\}$. Groupoid $(Z_4,\cdot)$, where $x \cdot y = x + 3y$, is a medial E-, F-, SM-quasigroup, $e^{\cdot}(Z_4) = s^{\cdot}(Z_4) = \{0\}$ and $f^{\cdot}(Z_4) = \{0, 2\} = H$.

Corollary 6.43. *1. If in F-quasigroup* $(Q,\cdot)$ *endomorphism* e *is a zero endomorphism and the group* $(Q,+)$ *has identity center, then* $(Q,\cdot) = (Q,+)$.

2. If in F-quasigroup $(Q,\cdot)$ *endomorphism* f *is a zero endomorphism and the group* $(Q,+)$ *has identity center, then* $(Q,\cdot) = (Q,+)$.

Proof. The proof follows from Cases (iii) and (ii) of Lemma 6.40. □

Corollary 6.44. *1. If endomorphism* e *of an F-quasigroup* $(Q,\cdot)$ *is a zero endomorphism, i.e.,* $e(x) = 0$ *for all* $x \in Q$, *endomorphism* f *is a permutation of the set* Q, *then* $x \cdot y = x + \psi y$, $(Q,+)$ *is an abelian group,* $\psi \in Aut(Q,+)$ *and* $(Q,\circ)$, $x \circ y = fx + \psi y$, *is a medial distributive quasigroup.*

2. If endomorphism f *of an F-quasigroup* $(Q,\cdot)$ *is a zero endomorphism, i.e.,* $f(x) = 0$ *for all* $x \in Q$, *endomorphism* e *is a permutation of the set* Q, *then* $x \cdot y = \varphi x + y$, $(Q,+)$ *is an abelian group,* $\varphi \in Aut(Q,+)$ *and* $(Q,\circ)$, $x \circ y = \varphi x + ey$, *is a medial distributive quasigroup.*

Proof. The proof follows from Lemma 6.40. It is a quasigroup folklore that an idempotent medial quasigroup is distributive [760, 759]. □

Remark 6.45. It is easy to see that condition "$(D, \cdot)$ is a medial F-quasigroup of the form $x \cdot y = x + \psi y$ such that $(D, \circ)$, $x \circ y = fx + \psi y$, is a medial distributive quasigroup" in Corollary 6.44 is equivalent to the condition that the automorphism ψ of the group $(D, +)$ is complete (Definition 6.6).

Lemma 6.46. *1. If endomorphism e of an F-quasigroup $(Q, \cdot)$ is a permutation of the set Q, i.e., e is an automorphism of $(Q, \cdot)$, then $(Q, \circ)$, $x \circ y = x \cdot e(y)$, is a left distributive quasigroup which satisfies the equality $(x \circ y) \circ z = (x \circ fz) \circ (y \circ e^{-1}z)$, for all $x, y, z \in Q$.*

2. If endomorphism f of an F-quasigroup $(Q, \cdot)$ is a permutation of the set Q, i.e., f is an automorphism of $(Q, \cdot)$, then $(Q, \circ)$, $x \circ y = f(x) \cdot y$, is a right distributive quasigroup which satisfies the equality $x \circ (y \circ z) = (f^{-1}x \circ y) \circ (ex \circ z)$, for all $x, y, z \in Q$.

Proof. 1. The fact that $(Q, \circ)$, $x \circ y = x \cdot e(y)$, is a left distributive quasigroup, follows from Lemma 6.23. If we rewrite right F-quasigroup equality in terms of the operation $\circ$, then $(x \circ e^{-1}y) \circ e^{-1}z = (x \circ e^{-1}fz) \circ (e^{-1}y \circ e^{-2}z)$. If we replace $e^{-1}y$ by y, $e^{-1}z$ by z and take into consideration that $e^{-1}f = fe^{-1}$, then we obtain the equality $(x \circ y) \circ z = (x \circ fz) \circ (y \circ e^{-1}z)$.

2. The proof is similar to Case 1. □

Corollary 6.47. *1. If endomorphism e of an F-quasigroup $(Q, \cdot)$ is an identity permutation of the set Q, then $(Q, \cdot)$ is a distributive quasigroup.*

2. If endomorphism f of an F-quasigroup $(Q, \cdot)$ is an identity permutation of the set Q, then $(Q, \cdot)$ is a distributive quasigroup.

Proof. 1. If $fx \cdot x = x$, then $fx \circ e^{-1}x = x$. Further proof follows from Lemma 6.46. Indeed from $fx \circ e^{-1}x = x$ it follows $fx \circ x = x$, $fx = x$, since $(Q, \circ)$ is an idempotent quasigroup. Then $f = \varepsilon$.

2. The proof is similar to Case 1. □

The following proof belongs to OTTER 3.3 [613]. The author of this program is Professor W. McCune. We have also used a lot J.D. Phillips' article [690]. Here we give the adopted (humanized) form of this proof.

Theorem 6.48. *If in a left distributive quasigroup $(Q, \circ)$ the equality*

$$(x \circ y) \circ z = (x \circ fz) \circ (y \circ ez) \tag{6.5}$$

is fulfilled for all $x, y, z \in Q$, where f, e are the maps of Q, then the following equality is fulfilled in $(Q, \circ)$: $(x \circ y) \circ fz = (x \circ fz) \circ (y \circ fz)$.

Proof. If we pass to operation / in the equality (6.5), then we obtain

$$((x \circ y) \circ z)/(y \circ e(z)) = x \circ fz. \tag{6.6}$$

From the equality (6.5) by $x = y$ we obtain $x \circ z = (x \circ fz) \circ (x \circ ez)$ and using left distributivity we have $x \circ z = x \circ (fz \circ e(z))$,

$$z = fz \circ e(z), \qquad e(z) = fz \backslash z. \tag{6.7}$$

If we change the expression $e(z)$ in the equality (6.6) using equality (6.7), then we obtain

$$((x \circ y) \circ z)/(y \circ (fz \backslash z)) = x \circ fz. \tag{6.8}$$

We make the following replacements in (6.8): $x \to x/z$, $y \to z$, $z \to y$. Then we obtain $(x \circ y) \circ z \to ((x/z) \circ z) \circ y = x \circ y$ and the following equality is fulfilled

$$(x \circ y)/(z \circ (f(y) \backslash y)) = (x/z) \circ f(y). \tag{6.9}$$

Using the operation / we can rewrite left distributive identity in the following form

$$(x \circ (y \circ z))/(x \circ z) = x \circ y. \tag{6.10}$$

If we change $(y \circ z)$ by y in the identity (6.10), then the variable y passes in y/z. Indeed, if $y \circ z = t$, then $y = t/z$. Therefore, we have

$$(x \circ y)/(x \circ z) = x \circ (y/z). \tag{6.11}$$

From the equality (6.5) using left distributivity to the right side of this equality we obtain $(x \circ y) \circ z = ((x \circ fz) \circ y) \circ ((x \circ fz) \circ ez)$. After applying the operation / to the last equality we obtain

$$((x \circ y) \circ z)/((x \circ f(z)) \circ e(z)) = (x \circ f(z)) \circ y. \tag{6.12}$$

After substitution of (6.7) in (6.12) we obtain

$$((x \circ y) \circ z)/((x \circ f(z)) \circ (fz \backslash z)) = (x \circ f(z)) \circ y. \tag{6.13}$$

Now we show the most unexpected OTTER step. We apply the left side of the equality (6.9) to the left side of the equality (6.13). In this case the expression $((x \circ y) \circ z)$ from (6.13) plays the role of $(x \circ y)$, $(x \circ f(z))$ – the role of z and $(fz \backslash z)$ plays the role of $(f(y)\backslash y)$.

Therefore we obtain

$$((x \circ y)/(x \circ f(z))) \circ f(z) = (x \circ fz) \circ y. \tag{6.14}$$

After application of the equality (6.11) to the left side of the equality (6.14) we have

$$((x \circ (y/fz)) \circ f(z) = (x \circ fz) \circ y. \tag{6.15}$$

If we change (y/fz) by y in the equality (6.15), then the variable y passes in $y \circ fz$. Therefore $(x \circ y) \circ fz = (x \circ fz) \circ (y \circ fz)$. □

Corollary 6.49. *If in a left distributive quasigroup $(Q, \circ)$ the equality*

$$(x \circ y) \circ z = (x \circ fz) \circ (y \circ ez)$$

is fulfilled for all $x, y, z \in Q$, where e is a map, f is a permutation of the set Q, then $(Q, \circ)$ is a distributive quasigroup.

Proof. The proof follows from Theorem 6.48. □

Theorem 6.50. *If in F-quasigroup $(Q, \cdot)$ endomorphisms e and f are permutations of the set Q, then $(Q, \cdot)$ is an isotope of the form $x \cdot y = x \circ e^{-1}y$ of a distributive quasigroup $(Q, \circ)$.*

Proof. Quasigroup $(Q, \circ)$ of the form $x \circ y = x \cdot e(y)$ is a left distributive quasigroup (Lemma 6.23) in which the equality $(x \circ y) \circ z = (x \circ fz) \circ (y \circ e^{-1}z)$ is true (Lemma 6.46). By Corollary 6.49 $(Q, \circ)$ is distributive. □

Theorem 6.51. *A finite F-quasigroup $(Q, \cdot)$ is simple if and only if $(Q, \cdot)$ lies in one of the following quasigroup classes:*

(i) $(Q, \cdot)$ is a simple group in the case when the maps e and f are zero endomorphisms;
(ii) $(Q, \cdot)$ has the form $x \cdot y = x + \psi y$, where $(Q, +)$ is a ψ-simple abelian group,

$\psi \in Aut(Q,+)$, *in the case when the map* e *is a zero endomorphism and the map* f *is a permutation; in this case* $e = -\psi, fx + \psi x = x$ *for all* $x \in Q$;

(iii) $(Q,\cdot)$ *has the form* $x \cdot y = \varphi x + y$, *where* $(Q,+)$ *is a* φ*-simple abelian group,* $\varphi \in Aut(Q,+)$, *in the case when the map* f *is a zero endomorphism and the map* e *is a permutation; in this case* $f = -\varphi, \varphi x + ex = x$ *for all* $x \in Q$;

(iv) $(Q,\cdot)$ *has the form* $x \cdot y = x \circ \psi y$, *where* $(Q,\circ)$ *is a* ψ*-simple distributive quasigroup* $\psi \in Aut(Q,\circ)$, *in the case when the maps* e *and* f *are permutations; in this case* $e = \psi^{-1}$, $fx \circ \psi x = x$ *for all* $x \in Q$.

Proof. ($\Longrightarrow$) (i) It is clear that in this case left and right F-quasigroup equalities are transformed in the identity of associativity.

(ii) From Lemma 6.40 (iii) and the fact that the map f is a permutation of the set Q it follows that $(Q,+)$ is an abelian group.

(iii) This case is similar to Case (ii).

(iv) By the Belousov result [80] (see Lemma 6.23 of this book) if the endomorphism e of a left F-quasigroup $(Q,\cdot)$ is a permutation of the set Q, then quasigroup $(Q,\cdot)$ has the form $x\cdot y = x\circ\psi y$, where $(Q,\circ)$ is a left distributive quasigroup and $\psi \in Aut(Q,\circ)$, $\psi \in Aut(Q,\cdot)$. The right distributivity of $(Q,\circ)$ follows from Theorem 6.50.

($\Longleftarrow$) Using Corollary 1.309 we can say that F-quasigroups from these quasigroup classes are simple. □

Remark 6.52. There exists a possibility to formulate Case (iv) of Theorem 6.51 in the following way:

*(iv)** $(Q,\cdot)$ has the form $x\cdot y = \varphi x\circ y$, where $(Q,\circ)$ is a φ-simple distributive quasigroup, in the case when the maps e and f are permutations; in this case $f = \varphi^{-1}, \varphi x \circ ex = x$ for all $x \in Q$.

Corollary 6.53. *Finite simple F-quasigroup* $(Q,\cdot)$ *is a simple group or a simple medial quasigroup.*

Proof. Case (i) of Theorem 6.51 demonstrates that a simple F-quasigroup can be a simple group.

Taking into consideration the Toyoda theorem (Theorem 2.65) we see that Cases (ii) and (iii) of Theorem 6.51 tell us that simple F-quasigroups can be simple medial quasigroups.

We shall prove that in Case (iv) of Theorem 6.51 we also obtain medial quasigroups.

The quasigroup $(Q,\cdot)$ is isotopic to distributive quasigroup $(Q,\circ)$, quasigroup $(Q,\circ)$ is isotopic to CML $(Q,+)$. Therefore $(Q,\cdot)$ is isotopic to the $(Q,+)$ and we can apply Lemma 6.11. □

Taking into consideration Lemma 6.3 we can say that some properties of finite simple medial F-quasigroups are described in Theorem 2.78.

Using the results obtained in this section we can add information on the structure of F-quasigroups [501].

Theorem 6.54. *Any finite F-quasigroup* $(Q,\cdot)$ *has the following structure*

$$(Q,\cdot) \cong (A,\circ) \times (B,\cdot),$$

where $(A,\circ)$ *is a quasigroup with a unique idempotent element;* $(B,\cdot)$ *is an isotope of a left distributive quasigroup* $(B,\star)$, $x\cdot y = x \star \psi y$, $\psi \in Aut(B,\cdot)$, $\psi \in Aut(B,\star)$. *In the quasigroups* $(A,\circ)$ *and* $(B,\cdot)$ *there exist the following chains*

$$A \supset e(A) \supset \cdots \supset e^{m-1}(A) \supset e^m(A) = 0, B \supset f(B) \supset \cdots \supset f^r(B) = f^{r+1}(B)$$

where:

1. *Let D_i be an equivalence class of the normal congruence $Ker\, e_i$ containing an idempotent element $a \in A$, $i \geqslant 0$. Then:*

   (a) *$(D_i, \circ)$ is linear right loop of the form $x \circ y = x + \psi y$, where $\psi \in Aut(D_i, +)$;*

   (b) *$Ker\,(f|_{(D_i,\circ)})$ is a group;*

   (c) *if $j \geqslant 1$, then $Ker\,(f_j|_{(D_i,\circ)})$ is an abelian group;*

   (d) *if f is a permutation of $f^l(D_i, \circ)$, then $f^l(D_i, \circ)$ is a medial right loop of the form $x \circ y = x + \psi y$, where ψ is a complete automorphism of the group $f^l(D_i, +)$;*

   (e) *$(D_i, \circ) \cong (E_i, +) \times f^l(D_i, \circ)$, where $(E_i, +)$ is a linear right loop, an extension of an abelian group by abelian groups and by a group.*

2. *Let H_j be an equivalence class of the normal congruence $Ker\, f_j$ containing an idempotent element $b \in B$, $j \geqslant 0$. Then:*

   (a) *$(H_0, \cdot)$ is a linear left loop of the form $x \cdot y = \varphi x + y$;*

   (b) *$f(B, \cdot)$ is an isotope of a distributive quasigroup $f(B, \star)$ of the form $x \cdot y = x \star e^{-1}y$;*

   (c) *if $0 < j < r$, then $(H_j, \cdot)$ is a medial left loop of the form $x \cdot y = \varphi x + y$, where $(H_j, +)$ is an abelian group, $\varphi \in Aut(H_j, +)$ and $(H_j, \star)$, $x \star y = \varphi x + ey$, is a medial distributive quasigroup;*

   (d) *$(B, \cdot) \cong (G, +) \times f^r(B, \cdot)$, where $(G, +)$ has a unique idempotent element, is an extension of an abelian group by abelian groups and by a linear left loop $(H_0, \cdot)$, $f^r(B, \cdot)$ is a distributive quasigroup.*

Proof. From Theorem 6.27, Case 1 it follows that F-quasigroup $(Q, \cdot)$ is isomorphic to the direct product of quasigroups $(A, \circ)$ and $(B, \cdot)$.

In F-quasigroup $(A, \circ)$ the chain

$$A \supset e(A) \supset e^2(A) \supset \cdots \supset e^{m-1}(A) \supset e^m(A) = e^{m+1}(A) = 0$$

becomes stable on a number m, where 0 is an idempotent element.

Case 1, (a). If $0 \leqslant j < m$, then by Lemma 6.39 any quasigroup $(D_j, \circ)$ is a right loop, isotope of a group $(D_j, +)$ of the form $(D_j, \circ) = (D_j, +)(\varepsilon, \psi, \varepsilon)$, where $\psi \in Aut(D_j, +)$.

Case 1, (b). The "behavior" of the map f in the right loop $(D_j, \circ)$ is described by Lemma 6.40. If f is a zero endomorphism, then $(D_j, \circ)$ is a group in case $j = 0$ (Lemma 6.40, Case (i)) and it is an abelian group in the case $j > 0$ (Lemma 6.40, Case (ii)).

If f is a non-zero endomorphism of $(D_j, \circ)$, then information on the structure of $(D_j, \circ)$ follows from Lemma 6.40 and Corollary 6.44.

Case 1, (c). The proof follows from Lemma 6.40, (ii), (iv) and the fact that in the quasigroup $Ker\,(f_j|_{(D_j,\circ)})$ the maps e and f are zero endomorphisms.

Case 1, (d). The proof follows from Corollary 6.44, Case 1.

Case 1, (e). The proof follows from results of the previous cases of this theorem and Theorem 6.27, Case 2.

Using Lemma 6.46 we can state that F-quasigroup $(B, \cdot)$ is isotopic to left distributive quasigroup $(B, \star)$, where $x \star y = x \cdot e(y)$.

In order to have more detailed information on the structure of the quasigroup $e^m(Q, \cdot)$ we study the following chain

$$B \supset f(B) \supset \cdots \supset f^r(B) = f^{r+1}(B),$$

which becomes stable on a number r.

Case 2, (a). The proof follows from Corollary 6.44, Case 2.

Case 2, (b). The proof follows from Theorem 6.48.

Case 2, (c). Since f is a zero endomorphism of quasigroup $(H_j, \cdot)$, $e_j|_{H_j}$ is a permutation of the set H_j, then by Corollary 6.44 quasigroup $(H_j, \cdot)$ has the form $x \cdot y = \varphi x + y$, where $(H_j, +)$ is an abelian group, $\varphi \in Aut(H_j, +)$ and $(H_j, \circ)$, $x \circ y = \varphi x + ey$, is a medial distributive quasigroup.

Case 2, (d). Existence of direct decomposition follows from Theorem 6.27, Case 2. □

We note that information on the structure of finite medial quasigroups is given in [785].

6.2.3 E-quasigroups

We recall, a quasigroup $(Q, \cdot)$ is trimedial if and only if $(Q, \cdot)$ is an E-quasigroup [520]. Any trimedial quasigroup is isotopic to CML [494]. The structure of trimedial quasigroups have been studied in [137, 499, 800, 505].

Notice, from the formulated results and Remark 2.19, it follows that CH-quasigroup can be defined as E- and TS-quasigroup (for short ETS-quasigroup).

Here a slightly different point of view on the structure of trimedial quasigroups is presented.

Lemma 6.55. *1. If endomorphism f of an E-quasigroup $(Q, \cdot)$ is a zero endomorphism, i.e., $f(x) = 0$ for all $x \in Q$, then $x \cdot y = \varphi x + y$, $(Q, +)$ is an abelian group, $\varphi \in Aut(Q, +)$.*

2. If endomorphism e of an E-quasigroup $(Q, \cdot)$ is a zero endomorphism, i.e., $e(x) = 0$ for all $x \in Q$, then $x \cdot y = x + \psi y$, $(Q, +)$ is an abelian group, $\psi \in Aut(Q, +)$.

Proof. 1. From Theorem 6.33 Case 3 it follows that $(Q, \cdot)$ is a left loop, $x \cdot y = \alpha x + y$, $(Q, +)$ is an abelian group, $\alpha \in S_Q$, $\alpha 0 = 0$.

Further we have $x \cdot e(x) = \alpha x + e(x) = x$, $\alpha x = x - e(x) = (\varepsilon - e)x$. Therefore α is an endomorphism of $(Q, +)$, and moreover, it is an automorphism of $(Q, +)$, since α is a permutation of the set Q.

2. The proof of Case 2 is similar to the proof of Case 1. □

Corollary 6.56. *If endomorphisms f and e of an E-quasigroup $(Q, \cdot)$ are zero endomorphisms, i.e., $f(x) = e(x) = 0$ for all $x \in Q$, then $x \cdot y = x + y$, $(Q, +)$ is an abelian group.*

Proof. From the equality $\alpha x + e(x) = x$ of Lemma 6.55 we have $\alpha x = x$, $\alpha = \varepsilon$. □

Corollary 6.57. *1. If endomorphism f of an E-quasigroup $(Q, \cdot)$ is a zero endomorphism and endomorphism e is a permutation of the set Q, then $x \cdot y = \varphi x + y$, $(Q, +)$ is an abelian group, $\varphi \in Aut(Q, +)$ and $(Q, \circ)$, $x \circ y = \varphi x + ey$, is a medial distributive quasigroup.*

2. If endomorphism e of an E-quasigroup $(Q, \cdot)$ is a zero endomorphism and endomorphism f is a permutation of the set Q, then $x \cdot y = x + \psi y$, $(Q, +)$ is an abelian group, $\psi \in Aut(Q, +)$ and $(Q, \circ)$, $x \circ y = fx + \psi y$, is a medial distributive quasigroup.

Proof. Case 1. From Lemma 6.55 it follows that in this case $(Q, \cdot)$ has the form $x \cdot y = \varphi x + y$ over abelian group $(Q, +)$. Then $x \cdot e(x) = \varphi x + e(x) = x$, $e(x) = x - \varphi x$, $e(0) = 0$. We can rewrite the equality $e(x \cdot y) = e(x) \cdot e(y)$ in the form $e(\varphi x + y) = \varphi e(x) + e(y)$. By $y = 0$ we have $e\varphi(x) = \varphi e(x)$. Then $e(\varphi x + y) = e\varphi x + ey$, the map e is an endomorphism of $(Q, +)$. Moreover, the map e is an automorphism of $(Q, +)$.

From the Toyoda theorem and the equality $e(x) = x - \varphi x$ it follows that quasigroup $(Q, \circ)$ is medial idempotent. It is well known that a medial idempotent quasigroup is distributive.

Case 2 is proved similar to Case 1. □

Theorem 6.58. *If the endomorphisms f and e of an E-quasigroup $(Q, \cdot)$ are permutations of the set Q, then quasigroup $(Q, \circ)$ of the form $x \circ y = f(x) \cdot y$ is a distributive quasigroup and $f, e \in Aut(Q, \circ)$.*

Proof. The proof of this theorem is similar to the proof of Theorem 6.50.

By Lemma 6.34 $(Q, \cdot)$ is an isotope of the form $x \cdot y = f^{-1}x \circ y$ of a left distributive quasigroup $(Q, \circ)$ and $f \in Aut(Q, \circ)$.

Moreover, by Lemma 6.34 $(Q, \cdot)$ is an isotope of the form $x \cdot y = x \diamond e^{-1}y$ of a right distributive quasigroup and $e \in Aut(Q, \diamond)$. Therefore $f^{-1}x \circ y = x \diamond e^{-1}y$, $x \circ y = fx \diamond e^{-1}y$.

Automorphisms e, f of the quasigroup $(Q, \cdot)$ lie in $Aut(Q, \circ)$ (Lemma 1.154 or [600], Corollary 12). We recall, $ef = fe$ (Lemma 1.301).

Now we need to rewrite the right distributive identity in terms of operation $\circ$. We have

$$f(fx \circ e^{-1}y) \circ e^{-1}z = f(fx \circ e^{-1}z) \circ e^{-1}(fy \circ e^{-1}z),$$
$$(f^2x \circ fe^{-1}y) \circ e^{-1}z = (f^2x \circ fe^{-1}z) \circ (e^{-1}fy \circ e^{-2}z).$$

If in the last equality we change element f^2x by element x, element $fe^{-1}y = e^{-1}fy$ by element y, element $e^{-1}z$ by element z, then we obtain

$$(x \circ y) \circ z = (x \circ fz) \circ (y \circ e^{-1}z).$$

In order to finish this proof we shall apply Corollary 6.49. □

Corollary 6.59. *Finite E-quasigroup $(Q, \cdot)$ is simple if and only if this quasigroup lies in one of the following quasigroup classes:*

(i) $(Q, \cdot)$ is a simple abelian group in the case when the maps e and f are zero endomorphisms;

(ii) $(Q, \cdot)$ is a simple medial quasigroup of the form $x \cdot y = \varphi x + y$ in the case when the map f is a zero endomorphism and the map e is a permutation;

(iii) $(Q, \cdot)$ is a simple medial quasigroup of the form $x \cdot y = x + \psi y$ in the case when the map e is a zero endomorphism and the map f is a permutation;

(iv) $(Q, \cdot)$ has the form $x \cdot y = x \circ \psi y$, where $(Q, \circ)$ is a ψ-simple distributive quasigroup, $\psi \in Aut(Q, \circ)$, in the case when the maps e and f are permutations.

Proof. ($\Longrightarrow$) (i) The proof follows from Corollary 6.56. (ii) The proof follows from Lemma 6.55 Case 1. (iii) The proof follows from Lemma 6.55 Case 2. (iv) The proof is similar to the proof of Case (iv) of Theorem 6.51.

($\Longleftarrow$) It is clear that any quasigroup of these quasigroup classes is a simple E-quasigroup. □

Corollary 6.60. *Finite simple E-quasigroup $(Q, \cdot)$ is a simple medial quasigroup.*

Proof. The proof follows from Corollary 6.59 and is similar to the proof of Corollary 6.53. We can use Lemma 6.11. □

Taking into consideration Corollary 6.60 we can say that properties of finite simple E-quasigroups are described by Theorem 2.78.

Lemma 6.61. *1. If endomorphism f of an E-quasigroup $(Q, \cdot)$ is a zero endomorphism, then $(Q, \cdot) \cong (A, \circ) \times (B, \cdot)$, where $(A, \circ)$ a medial E-quasigroup of the form $x \cdot y = \varphi x + y$ and there exists a number m such that $|e^m(A, \circ)| = 1$, $(B, \cdot)$ is a medial E-quasigroup of the form $x \cdot y = \varphi x + y$ such that $(B, \star)$, $x \star y = \varphi x + ey$, is a medial distributive quasigroup.*

2. If endomorphism e of an E-quasigroup $(Q, \cdot)$ is a zero endomorphism, then $(Q, \cdot) \cong (A, +) \times (B, \cdot)$, where $(A, \circ)$ is a medial E-quasigroup of the form $x \cdot y = x + \psi y$ and there exists a number m such that $|f^m(A, \circ)| = 1$, $(B, \cdot)$ is a medial E-quasigroup of the form $x \cdot y = x + \psi y$ such that $(B, \star)$, $x \star y = fx + \psi y$, is a medial distributive quasigroup.

Proof. 1. By Theorem 6.35 Case 4 any right E-quasigroup $(Q, \cdot)$ has the structure $(Q, \cdot) \cong (A, \circ) \times (B, \cdot)$, where $(A, \circ)$ is a quasigroup with a unique idempotent element and there exists a number m such that $|e^m(A, \circ)| = 1$; $(B, \cdot)$ is an isotope of a right distributive quasigroup $(B, \star)$, $x \cdot y = x \star e^{-1}(y)$ for all $x, y \in B$, $e \in Aut(B, \cdot)$, $e \in Aut(B, \star)$.

From Lemma 6.55 it follows that $(Q, \cdot)$ has the form $x \cdot y = \varphi x + y$ over an abelian group $(Q, +)$.

We recall that $e = \varepsilon - \varphi$, $e\varphi = \varphi e$ (Corollary 6.57). From equalities $x \cdot y = \varphi x + y$ and $x \cdot y = x \star e^{-1}(y)$ we have $x \star y = \varphi x + ey$. Then $(B, \star)$ is medial, idempotent, therefore it is distributive.

2. The proof is similar to Case 1. □

Remark 6.62. If $m = 1$, then $(A, \circ)$ is an abelian group (Corollary 6.56).

If $m = 2$, then $(A, \circ)$ is an extension of an abelian group by an abelian group. If, in addition, the conditions of Lemma 1.321 are fulfilled, then $(A, \circ)$ is an abelian group.

If the number m is finite and the conditions of Lemma 1.321 are fulfilled, then after application of Lemma 1.321 $(m - 1)$ times we obtain that $(A, \circ)$ is an abelian group.

Now we have a possibility to give more detailed information on the structure of finite E-quasigroups. The proof of the following theorem, in many ways, is similar to the proof of Theorem 6.54.

Let D_i be an equivalence class of the normal congruence $Ker\, e_i$ containing an idempotent element $a \in A$, $i \geqslant 0$. Let H_j be an equivalence class of the normal congruence $Ker\, f_j$ containing an idempotent element, $j \geqslant 0$.

Theorem 6.63. *In any finite E-quasigroup $(Q, \cdot)$ there exists the following finite chain*

$$Q \supset e(Q) \supset \cdots \supset e^{m-1}(Q) \supset e^m(Q) = e^{m+1}(Q),$$
$$e^m(Q) \supset fe^m(Q) \supset \cdots \supset f^r e^m(Q) = f^{r+1}e^m(Q),$$

where the following holds

1. *If $i < m$, then $(D_i, \cdot) \cong (H_i, +) \times (G_i, \cdot)$, where right loop $(H_i, +)$ is an extension of an abelian group by abelian groups, $(G_i, \cdot)$ is a medial E-quasigroup of the form $x \cdot y = x + \psi y$ such that ψ is a complete automorphism of the group $(G_i, +)$.*

2. *If $i = m$, then $(e^m Q, \cdot)$ is an isotope of right distributive quasigroup $(e^m Q, \circ)$, where $x \circ y = x \cdot ey$.*

   (a) *If $j < r$, then $(H_j, \cdot)$ is medial left loop, $(H_j, \cdot)$ has the form $x \cdot y = \varphi x + y$, where $(H_j, +)$ is an abelian group, $\varphi \in Aut(H_j, +)$ and $(H_j, \circ)$, $x \circ y = \varphi x + ey$, is a medial distributive quasigroup.*

   (b) *If $j = r$, then $(f^r e^m Q, \cdot)$ is an isotope of the form $x \circ y = f(x) \cdot y$ of a distributive quasigroup $(f^r e^m Q, \circ)$.*

Proof. It is clear that in finite E-quasigroup $(Q, \cdot)$ the chain (6.2)

$$Q \supset e(Q) \supset e^2(Q) \supset \cdots \supset e^{m-1}(Q) \supset e^m(Q) = e^{m+1}(Q)$$

becomes stable.

Case 1, $i < m$. By Lemma 6.55, Case 2, any quasigroup $(D_i, \cdot)$ is a medial right loop, and an isotope of an abelian group $(D_i, +)$ of the form $(D_i, \cdot) = (D_i, +)(\varepsilon, \psi, \varepsilon)$, where $\psi \in Aut(D_i, +)$, for all suitable values of index i, since in the quasigroup $(D_i, \cdot)$ endomorphism e is a zero endomorphism.

If f is a zero endomorphism, then in this case $(D_i, \cdot)$ is an abelian group (Corollary 6.56).

If f is a non-zero endomorphism of $(D_i, \cdot)$, then we can use Lemma 6.61 Case 2.

Case 1, $i = m$. From Lemma 6.34 Case 4 it follows that E-quasigroup $(e^m Q, \cdot)$ is isotopic to right distributive quasigroup $(e^m Q, \circ)$, where $x \circ y = x \cdot e(y)$.

In order to have more detailed information on the structure of the quasigroup $e^m(Q, \cdot)$ we study the following chain

$$e^m(Q) \supset fe^m(Q) \supset f^2e^m(Q) \cdots \supset f^re^m(Q) = f^{r+1}e^m(Q).$$

Case 2, $j < r$. From Lemma 6.34 Case 4 it follows that E-quasigroup $(H_j, \cdot)$ is isotopic to right distributive quasigroup $(H_j, \circ)$, $x \circ y = x \cdot e(y)$.

From Lemma 6.55 Case 1 it follows that $(H_j, \cdot)$ has the form $x \cdot y = \varphi x + y$, where $(H_j, +)$ is an abelian group, $\varphi \in Aut(H_j, +)$.

From equalities $x \circ e^{-1}y = x \cdot y$ and $x \cdot y = \varphi x + y$, we have $x \circ e^{-1}y = \varphi x + y$, $x \circ y = \varphi x + ey$. Then right distributive quasigroup $(H_j, \circ)$ is isotopic to abelian group $(H_j, +)$.

If we rewrite identity $(x \circ y) \circ z = (x \circ z) \circ (y \circ z)$ in terms of the operation $+$, then $\varphi^2 x + \varphi e y + ez = \varphi^2 x + \varphi e z + e\varphi y + e^2 z$, $\varphi e y + ez = \varphi e z + e\varphi y + e^2 z$. By $z = 0$ from the last equality it follows that $\varphi e = e\varphi$. Then $(H_j, \circ)$ is a medial quasigroup. Moreover, $(H_j, \circ)$ is a medial distributive quasigroup, since any medial right distributive quasigroup is distributive.

Case 2, $j = r$. If e and f are permutations of the set $f^r e^m Q$, then by Theorem 6.58 $(f^r e^m Q, \cdot)$ is an isotope of the form $x \circ y = f(x) \cdot y$ of a distributive quasigroup $(f^r e^m Q, \circ)$.

□

6.2.4 SM-quasigroups

We recall, left and right SM-quasigroup is called an SM-quasigroup. The structure theory of SM-quasigroups has been mainly developed by T. Kepka and K. K. Shchukin [497, 496, 799, 51].

If finite SM-quasigroup $(Q, \cdot)$ is simple, then the endomorphism s is a zero endomorphism or a permutation of the set Q.

If $s(x) = 0$, then from Theorem 6.33 it follows:

Corollary 6.64. *If the endomorphism s of a semimedial quasigroup $(Q, \cdot)$ is a zero endomorphism, i.e., $s(x) = 0$ for all $x \in Q$, then $(Q, \cdot)$ is a medial unipotent quasigroup, $(Q, \cdot) \cong (Q, \circ)$, where $x \circ y = \varphi x - \varphi y$, $(Q, +)$ is an abelian group, $\varphi \in Aut(Q, +)$.*

Remark 6.65. By Corollary 6.64, equivalence class D_i of the congruence $Ker\, s_i$ containing an idempotent element is a medial unipotent quasigroup $(D_i, \cdot)$ of the form $x \circ y = \varphi x - \varphi y$, where $(D_i, +)$ is an abelian group, and $\varphi \in Aut(D_i, +)$ for all suitable values of index i.

Information on the structure of medial unipotent quasigroups is in [785].

If $s(x)$ is a permutation of the set Q, then from Lemma 6.34 it follows:

Lemma 6.66. *If the endomorphism s of a semimedial quasigroup $(Q, \cdot)$ is a permutation of the set Q, then quasigroup $(Q, \circ)$ of the form $x \circ y = s^{-1}(x \cdot y)$ is a distributive quasigroup and $s \in Aut(Q, \circ)$.*

Corollary 6.67. *Any finite semimedial quasigroup $(Q, \cdot)$ has the following structure: $(Q, \cdot) \cong (A, \circ) \times (B, \cdot)$, where $(A, \circ)$ is a quasigroup with a unique idempotent element and there exists a number m such that $|s^m(A, \circ)| = 1$; $(B, \cdot)$ is an isotope of a distributive quasigroup $(B, \star)$, $x \cdot y = s(x \star y)$ for all $x, y \in B$, $s \in Aut(B, \cdot)$, $s \in Aut(B, \star)$.*

Proof. The proof follows from Cases 3 and 4 of Theorem 6.35. □

Corollary 6.68. *Finite SM-quasigroup $(Q, \cdot)$ is simple if and only if it lies in one of the following quasigroup classes:*

(i) $(Q, \cdot)$ is a medial unipotent quasigroup of the form $x \circ y = \varphi x - \varphi y$, $(Q, +)$ is an abelian group, $\varphi \in Aut(Q, +)$ and the group $(Q, +)$ is φ-simple;
(ii) $(Q, \cdot)$ has the form $x \cdot y = \varphi(x \circ y)$, where $\varphi \in Aut(Q, \circ)$ and $(Q, \circ)$ is a φ-simple distributive quasigroup.

Proof. The proof follows from Cases 3 and 4 of Theorem 6.71. □

A similar result on properties of simple SM-quasigroups is given in [799, Corollary 4.13].

Corollary 6.69. *Any finite simple semimedial quasigroup $(Q, \cdot)$ is medial [799].*

Proof. Conditions of Lemma 6.11 are fulfilled and we can apply them. □

6.2.5 Finite simple quasigroups

Definition 6.70. A quasigroup $(Q, \cdot)$ is α-simple, if this quasigroup does not contain a non-trivial congruence that is admissible relative to a permutation α of the set Q.

Theorem 6.71. *1. Finite left F-quasigroup $(Q, \cdot)$ is simple if and only if it lies in one of the following quasigroup classes:*

(i) $(Q, \cdot)$ is a right loop of the form $x \cdot y = x + \psi y$, where $\psi \in Aut(Q, +)$ and the group $(Q, +)$ is ψ-simple;

(ii) $(Q, \cdot)$ has the form $x \cdot y = x \circ \psi y$, where $\psi \in Aut(Q, \circ)$ and $(Q, \circ)$ is a ψ-simple left distributive quasigroup.

2. Finite right F-quasigroup $(Q, \cdot)$ is simple if and only if it lies in one of the following quasigroup classes:

(i) $(Q, \cdot)$ is a left loop of the form $x \cdot y = \varphi x + y$, where $\varphi \in Aut(Q, +)$ and the group $(Q, +)$ is φ-simple;

(ii) $(Q, \cdot)$ has the form $x \cdot y = \varphi x \circ y$, where $\varphi \in Aut(Q, \circ)$ and $(Q, \circ)$ is a φ-simple left distributive quasigroup.

3. Finite left SM-quasigroup $(Q, \cdot)$ is simple if and only if it lies in one of the following quasigroup classes:

(i) $(Q, \cdot)$ is a unipotent quasigroup of the form $x \circ y = -\varphi x + \varphi y$, $(Q, +)$ is a group, $\varphi \in Aut(Q, +)$ and the group $(Q, +)$ is φ-simple;

(ii) $(Q, \cdot)$ has the form $x \cdot y = \varphi(x \circ y)$, where $\varphi \in Aut(Q, \circ)$ and $(Q, \circ)$ is a φ-simple left distributive quasigroup.

4. Finite right SM-quasigroup $(Q, \cdot)$ is simple if and only if it lies in one of the following quasigroup classes:

(i) $(Q, \cdot)$ is a unipotent quasigroup of the form $x \circ y = \varphi x - \varphi y$, $(Q, +)$ is a group, $\varphi \in Aut(Q, +)$ and the group $(Q, +)$ is φ-simple;

(ii) $(Q, \cdot)$ has the form $x \cdot y = \varphi(x \circ y)$, where $\varphi \in Aut(Q, \circ)$ and $(Q, \circ)$ is a φ-simple right distributive quasigroup.

5. *Finite left E-quasigroup* $(Q, \cdot)$ *is simple if and only if it lies in one of the following quasigroup classes:*

   (i) $(Q, \cdot)$ *is a left loop of the form* $x \cdot y = \alpha x + y$, $\alpha 0 = 0$, *and* $(Q, +)$ *is an* α*-simple abelian group;*

   (ii) $(Q, \cdot)$ *has the form* $x \cdot y = \varphi x \circ y$, *where* $\varphi \in Aut(Q, \circ)$ *and* $(Q, \circ)$ *is a* φ*-simple left distributive quasigroup.*

6. *Finite right E-quasigroup* $(Q, \cdot)$ *is simple if and only if it lies in one of the following quasigroup classes:*

   (i) $(Q, \cdot)$ *is a right loop of the form* $x \cdot y = x + \beta y$, $\beta 0 = 0$, *and* $(Q, +)$ *is a* β*-simple abelian group;*

   (ii) $(Q, \cdot)$ *has the form* $x \cdot y = x \circ \psi y$, *where* $\psi \in Aut(Q, \circ)$ *and* $(Q, \circ)$ *is a* ψ*-simple right distributive quasigroup.*

Proof. Case 1. Suppose that $(Q, \cdot)$ is a simple left F-quasigroup. From Theorem 6.27 it follows that $(Q, \cdot)$ can be a quasigroup with a unique idempotent element or an isotope of a left distributive quasigroup.

By Theorem 1.279 the endomorphism e defines the corresponding normal congruence $Ker\, e$. Since $(Q, \cdot)$ is simple, then this congruence is the diagonal $\hat{Q} = \{(q, q) \,|\, q \in Q\}$ or the universal congruence $Q \times Q$.

From Theorem 6.27 it follows that in a simple left F-quasigroup the map e is a zero endomorphism or a permutation.

The structure of left F-quasigroups in the case when e is a zero endomorphism follows from Lemma 6.22.

The structure of left F-quasigroups in the case when e is an automorphism follows from Lemma 6.23.

Conversely, using Corollary 1.309 we can say that left F-quasigroups from these quasigroup classes are simple.

Cases 2–6 are proved in a similar way. □

6.2.6 Left FESM-quasigroups

M. Kinyon and J.D. Phillips defined left FESM-quasigroups in [520].

Definition 6.72. A quasigroup $(Q, \cdot)$ which simultaneously is a left F-, E- and SM-quasigroup we shall call *left FESM-quasigroup.*

Remark 6.73. Other quasigroup classes are also possible to define from these six quasigroup classes (from left and right F-, E-, and SM-quasigroups).

From Definition 6.72 it follows that in FESM-quasigroup the maps e, f, s are its endomorphisms.

Lemma 6.74. *1. If endomorphism* e *of a left FESM-quasigroup* $(Q, \cdot)$ *is a zero endomorphism, then* $(Q, \cdot)$ *is a medial right loop,* $x \cdot y = x + \psi y$, $(Q, +)$ *is an abelian group,* $\psi \in Aut(Q, +)$, $\psi^2 = \varepsilon$, $\psi s = s, \psi f = f\psi = -f$.

2. If endomorphism f *of an FESM-quasigroup* $(Q, \cdot)$ *is a zero endomorphism, then* $(Q, \cdot)$ *is a medial left loop, i.e.,* $x \cdot y = \varphi x + y$, $(Q, +)$ *is an abelian group,* $\varphi \in Aut(Q, +)$, $\varphi^2 = \varepsilon$, $\varphi s = s, \varphi e = e\varphi = -e$.

3. If endomorphism s *of a left FESM-quasigroup* $(Q, \cdot)$ *is a zero endomorphism, then* $(Q, \cdot)$ *is a medial unipotent quasigroup of the form* $x \cdot y = \varphi x - \varphi y$, *where* $(Q, +)$ *is an abelian group,* $\varphi \in Aut(Q, +)$, $\varphi f = f\varphi$, $\varphi e = e\varphi$.

Proof. 1. From Lemma 6.22 it follows that $(Q,\cdot)$ has the form $x \cdot y = x + \psi y$, where $(Q,+)$ is a group, $\psi \in Aut(Q,+)$.

Then $s(x) = x \cdot x = x + \psi x$. Since s is an endomorphism of $(Q,\cdot)$, further we have $s(x \cdot y) = x + y + \psi x + \psi y$, $sx \cdot sy = sx + \psi sy = x + \psi x + \psi y + \psi^2 y$. Then $x + y + \psi x + \psi y = x + \psi x + \psi y + \psi^2 y$, $y + \psi x + \psi y = \psi x + \psi y + \psi^2 y$. By $x = 0$ we have $y + \psi y = \psi(y + \psi y)$, $sy = \psi sy$. Then $y + \psi x + \psi y = \psi x + \psi y + \psi^2 y = \psi x + \psi sy = \psi x + sy = \psi x + y + \psi y$. Therefore $y + \psi x + \psi y = \psi x + y + \psi y$, $y + \psi x = \psi x + y$, and the group $(Q,+)$ is commutative. From the equality $y + \psi x + \psi y = \psi x + \psi y + \psi^2 y$ we obtain $y = \psi^2 y$, $\psi^2 = \varepsilon$.

Further we have $f(x) \cdot x = fx + \psi x = x$, $fx = x - \psi x$, $\psi fx = \psi x - x = -fx$, $f\psi x = \psi x - x$.

2. From Theorem 6.33 Case 3 it follows that $(Q,\cdot)$ is a left loop, $x \cdot y = \varphi x + y$, $(Q,+)$ is an abelian group, $\varphi \in S_Q$, $\varphi 0 = 0$.

Further we have $x \cdot e(x) = \varphi x + e(x) = x$, $\varphi x = x - e(x) = (\varepsilon - e)x$. Therefore φ is an endomorphism of $(Q,+)$, moreover, it is an automorphism of $(Q,+)$, since φ is a permutation of the set Q.

Then $sx = x \cdot x = \varphi x + x$, $s(x \cdot y) = \varphi x + \varphi y + x + y = sx \cdot sy = \varphi sx + sy = \varphi^2 x + \varphi x + \varphi y + y$. From the equality $\varphi x + \varphi y + x + y = \varphi^2 x + \varphi x + \varphi y + y$ we obtain $\varphi^2 = \varepsilon$. Then $\varphi sx = \varphi(\varphi x + x) = sx$.

Further, $x \cdot ex = \varphi x + ex = x$. Then $ex = x - \varphi x$, $\varphi ex = \varphi x - x = -ex$, $e\varphi x = \varphi x - x$.

3. From Theorem 6.33 Case 1 it follows that $(Q,\cdot)$ is a unipotent quasigroup of the form $x \cdot y = -\varphi x + \varphi y$, where $(Q,+)$ is a group, $\varphi \in Aut(Q,+)$.

Since f is an endomorphism of quasigroup $(Q,\cdot)$ we have $f(x \cdot y) = e(x) \cdot f(y)$, $f(-\varphi x + \varphi y) = -\varphi f(x) + \varphi f(y)$. If $y = 0$, then $f(-\varphi) = -\varphi f$. If $x = 0$, then $f\varphi = \varphi f$. Then f is an endomorphism of the group $(Q,+)$. Similarly, $e(-\varphi) = -\varphi e$, $e\varphi = \varphi e$, e is an endomorphism of the group $(Q,+)$.

From $x \cdot ex = x$ we have $-\varphi x + e\varphi x = x$, $e\varphi x = \varphi x + x$, $ex = x + \varphi^{-1}x$. Then

$$\begin{aligned} e(x \cdot y) &= x \cdot y + \varphi^{-1}(x \cdot y) = -\varphi x + \varphi y - x + y, \\ ex \cdot ey &= -\varphi(x + \varphi^{-1}x) + \varphi(y + \varphi^{-1}y) = -\varphi x - x + \varphi y + y. \end{aligned} \tag{6.16}$$

Comparing the right sides of the equalities (6.16) we obtain that $(Q,+)$ is a commutative group. □

Lemma 6.75. *If endomorphisms e, f and s of a left FESM-quasigroup $(Q,\cdot)$ are permutations of the set Q, then quasigroup $(Q,\circ)$ of the form $x \circ y = x \cdot e(y)$ is a left distributive quasigroup and $e, f, s \in Aut(Q,\circ)$.*

Proof. By Lemma 6.23 endomorphism e of a left F-quasigroup $(Q,\cdot)$ is a permutation of the set Q if and only if quasigroup $(Q,\circ)$ of the form $x \circ y = x \cdot e(y)$ is a left distributive quasigroup and $e \in Aut(Q,\circ)$ [80]. Then $x \cdot y = x \circ e^{-1}y$, $s(x) = x \circ e^{-1}x$, $f(x) \cdot x = fx \circ e^{-1}x = x$.

The fact that $e, f, s \in Aut(Q,\circ)$ follows from Lemma 1.154 Case 2. □

Theorem 6.76. *If $(Q,\cdot)$ is a finite simple left FESM-quasigroup, then*

(i) $(Q,\cdot)$ is a simple medial quasigroup in the case when at least one of the maps e, f and s is a zero endomorphism;

(ii) $(Q,\cdot)$ has the form $x \cdot y = x \circ \psi y$, where $(Q,\circ)$ is a ψ-simple left distributive quasigroup, $\psi \in Aut(Q,\circ)$, in the case when the maps e, f and s are permutations; in this case $e = \psi^{-1}$, $fx \circ \psi x = x$, $s(x) = x \circ \psi x$ for all $x \in Q$.

Proof. It is possible to use Lemma 6.74 for the proof of Case (i) and Lemma 6.75 for the proof of Case (ii). □

Example 6.77. By $(Z_7, +)$ we denote a cyclic group of order 7 and we take $Z_7 = \{0, 1, 2, 3, 4, 5, 6\}$.

Quasigroup $(Z_7, \circ)$, where $x \circ y = x + 6y = x - y$, is a simple medial FESM-quasigroup in which the maps e and s are zero endomorphisms, the map f is a permutation of the set Z_7 ($f(x) = 2x$ for all $x \in Z_7$).

Quasigroup $(Z_7, \cdot)$, where $x \cdot y = 2x + 3y$, is a simple medial FESM-quasigroup in which endomorphisms e, f, s are permutations of the set Z_7.

Remark 6.78. There exists a possibility to research combinations of left and right F-, E- and SM-property both on quasigroup class, and on some generalizations of this class. See for example [519, 520].

6.2.7 CML as an SM-quasigroup

In this subsection we give some (mainly well-known) information about CML using the quasigroup approach.

Recall, any CML is an SM-loop. A quasigroup $(Q, \cdot)$ with identities $xy = yx$, $x \cdot xy = y$, $x \cdot yz = xy \cdot xz$ is called a distributive Steiner quasigroup [72, 80].

Theorem 6.79. *1. Any finite commutative Moufang loop $(Q, +)$ has the following structure*

$$(Q, +) \cong (A, \oplus) \times (B, +),$$

where $(A, \oplus)$ is an abelian group of the order 2^k, $k \in \mathbb{N}$, and there exists a number m such that $|s^m(A, \oplus)| = 1$; $(B, +)$ is an isotope of a distributive quasigroup $(B, \star)$, $x + y = s(x \star y)$ for all $x, y \in B$, $s \in Aut(B, +)$, $s \in Aut(B, \star)$.

2. $C(Q, +) \cong (A, \oplus) \times C(B, +)$.

3. $(Q, +)/C(Q, +) \cong (B, +)/C(B, +) \cong (D, +)$ is 3-CML in which the endomorphism s is a permutation I such that $Ix = -x$.

4. Quasigroup $(D, \star)$, $x \star y = -x - y$, $x, y \in (D, +)$, is a distributive Steiner quasigroup.

Proof. Case 1. The existence of the decomposition of $(Q, +)$ into two factors follows from Theorem 6.35.

From Corollary 6.36 it follows that any equivalence class $\bar{a} \equiv H_j$ of the normal congruence $Ker\, s_j$ containing an idempotent element $0 \in Q$ is an unipotent loop $(H_j, \cdot)$ isotopic to an abelian group with isotopy of the form $(\varphi, -\varphi, \varepsilon)$, where $\varphi \in Aut(H_j, \cdot)$ for all suitable values of j.

Since $(H_j, \cdot)$ is a commutative loop, we have that $x \cdot 0 = \varphi x - \varphi 0 = \varphi x = x$, $\varphi = \varepsilon$, $0 \cdot x = \varphi 0 - \varphi x = -\varphi x = x$, $-\varphi = \varepsilon$. Thus $x \cdot y = x + y$ for all $x, y \in (H_j, \cdot)$.

We notice, in a commutative Moufang loop $(Q, +)$ the map s^i takes the form 2^i, i.e., $s^i(x) = 2^i(x)$. Then in the loop $(A, \oplus)$ any non-zero element has the order 2^i.

If an element x of the loop $(A, \oplus)$ has a finite order, then $x \in C(A, \oplus)$, where $C(A, \oplus)$ is a center of $(A, \oplus)$ since $G.C.D.(2^i, 3) = 1$.

Therefore $(A, \oplus) \cong 3(A, \oplus) \subseteq C(A, \oplus)$, $(A, \oplus)$ is an abelian group.

From Cases 1, 2 of Theorem 6.35 it follows that $(B, +)$ is an isotope of a left and right distributive quasigroup. Therefore, $(B, +)$ is an isotope of a distributive quasigroup.

Case 2. From Lemma 1.316 it follows that $C(Q, +) \cong C(A, \oplus) \times C(B, +)$. Therefore $C(Q, +) \cong (A, \oplus) \times C(B, +)$ since $C(A, \oplus) = (A, \oplus)$.

Case 3. The fact that $(Q, +)/C(Q, +)$ is 3-CML is well known and it follows from Lemma 2.110. Isomorphism $(Q, +)/C(Q, +) \cong ((A, \oplus) \times (B, +))/((A, \oplus) \times C(B, +))$ follows from Cases 1, 2.

Isomorphism

$$((A, \oplus) \times (B, +))/((A, \oplus) \times C(B, +)) \cong (B, +)/C(B, +)$$

follows from the second isomorphism theorem ([184, p. 51], for group case see [471]) and the fact that $(A, \oplus) \cap C(B, +) = \{(0, 0)\}$, i.e., $|(A, \oplus) \cap C(B, +)| = 1$.

Case 4. It is clear that in 3-CML $(D, +)$ the map s takes the form $s(x) = 2x = -x = Ix$. Moreover, $I^{-1} = I$. It is easy to see that the quasigroup $(D, \star)$ is a distributive Steiner quasigroup. □

Remark 6.80. Theorem 6.79 is a specification of Theorem 2.116. It is easy to see that Theorem 6.79 is also true without significant modification of the proof for CML $(Q, +)$ in which the endomorphism s has a finite order.

Corollary 6.81. *If in CML $(Q, +)$ the endomorphism s has a finite order m, then: (i) any non-zero element of the group $(A, \oplus)$ has the order 2^i, $1 \leqslant i \leqslant m$; (ii) $Aut(Q, +) \cong Aut(A, \oplus) \times Aut(B, +)$.*

Proof. Case (i). It is easy to see. Case (ii). The proof is similar to the proof of Corollary 2.117 and we omit it. □

6.3 Loop isotopes

Often the answer to the question "to which loop a quasigroup isotopic?" helps very much in the study of corresponding isotopic quasigroups and loops.

We know that medial and paramedial quasigroups are isotopic to abelian groups, and that quasigroups that are linear over groups are isotopic to groups (by definition). Belousov proved an unexpected and very fine and deep result that any distributive quasigroup is isotopic to a commutative Moufang loop. Moreover in collaboration with Onoi and Florea he studied isotopes of left distributive quasigroups. For example, it was proved that any left distributive quasigroup with the identity $x \cdot xy = y$ is isotopic to a left Bol loop.

Further Kepka and coauthors proved a series of outstanding results on isotopes of some quasigroups. The first in this series is the result that any F-quasigroup is isotopic to a Moufang loop [500, 502]. It was proved that any SM-quasigroup is isotopic to a commutative Moufang loop [496]. Since any E-quasigroup is an SM-quasigroup [497, 520], then any E-quasigroup is also isotopic to a commutative Moufang loop.

Manin proved that CH-quasigroups are isotopic to commutative Moufang loops [593].

We give information on loops and left loops which are isotopic to finite left F-, SM-, E- and $FESM$-quasigroups [789].

6.3.1 Left F-quasigroups

Taking into consideration Theorems 6.27, 6.35, Lemma 1.319 and Corollary 1.320 we can study loop isotopes of the factors of direct decompositions of left and right F- and E-quasigroups.

Theorem 6.82. *1. Finite left F-quasigroup $(Q, \cdot)$ is isotopic to the direct product of a group $(A, +)$ and a left S-loop $(B, \diamond)$, i.e., $(Q, \cdot) \sim (A, +) \times (B, \diamond)$.*

2. Finite right F-quasigroup $(Q, \cdot)$ is isotopic to the direct product of a group $(A, +)$ and right S-loop $(B, \diamond)$, i.e., $(Q, \cdot) \sim (A, +) \times (B, \diamond)$.

Proof. 1. By Theorem 6.27, Case 1, any left F-quasigroup $(Q, \cdot)$ has the following structure $(Q, \cdot) \cong (A, \circ) \times (B, \cdot)$, where $(A, \circ)$ is a quasigroup with a unique idempotent element; $(B, \cdot)$ is an isotope of a left distributive quasigroup $(B, \star)$, $x \cdot y = x \star \psi y$ for all $x, y \in B$, $\psi \in Aut(B, \cdot)$, $\psi \in Aut(B, \star)$.

By Corollary 1.320, if a quasigroup Q is the direct product of quasigroups A and B, then there exists an isotopy $T = (T_1, T_2)$ of Q such that $QT \cong AT_1 \times BT_2$ is a loop.

Therefore we have a possibility to divide our proof in two steps.

Step 1. Denote a unique idempotent element of $(A, \circ)$ by 0. We notice that $e^{\circ}0 = 0$. Indeed, from $(e^{\circ})^m A = 0$ we have $(e^{\circ})^{m+1} A = e^{\circ} 0 = 0$.

From left F-equality $x \circ (y \circ z) = (x \circ y) \circ (e^{\circ}(x) \circ z)$ by $x = 0$ we have $0 \circ (y \circ z) = (0 \circ y) \circ (0 \circ z)$. Then $L_0 \in Aut(A, \circ)$.

Consider isotope $(A, \oplus)$ of the quasigroup $(A, \circ)$: $x \oplus y = x \circ L_0^{-1} y$. We notice that $(A, \oplus)$ is a left loop. Indeed, $0 \oplus y = 0 \circ L_0^{-1} y = y$. Further we have $x \circ y = x \oplus L_0 y$, $x \oplus e^{\oplus} x = x = x \circ L_0^{-1} e^{\oplus} x = x \circ e^{\circ} x$, $e^{\oplus}(x) = L_0 e^{\circ}(x)$, $e^{\oplus}(0) = L_0 e^{\circ}(0) = 0 \circ 0 = 0$.

Prove that $L_0 \in Aut(A, \oplus)$. From the equality $L_0(x \circ y) = L_0 x \circ L_0 y$ we have $L_0(x \circ y) = L_0(x \oplus L_0 y)$, $L_0 x \circ L_0 y = L_0 x \oplus L_0^2 y$, $L_0(x \oplus L_0 y) = L_0 x \oplus L_0^2 y$.

If we pass in the left F-equality to the operation $\oplus$, then we obtain $x \oplus (L_0 y \oplus L_0^2 z) = (x \oplus L_0 y) \oplus (L_0 e^{\circ}(x) \oplus L_0^2 z)$. If we change $L_0 y$ by y, $L_0^2 z$ by z, then we obtain

$$x \oplus (y \oplus z) = (x \oplus y) \oplus (L_0 e^{\circ}(x) \oplus z) = (x \oplus y) \oplus (e^{\oplus}(x) \oplus z). \tag{6.17}$$

Then $(A, \oplus)$ is a left F-quasigroup with the left identity element. Briefly, below in this theorem we shall use denotation e instead of $e^{\oplus}$.

Further we pass from the operation $\oplus$ to the operation $+$: $x + y = R_0^{-1} x \oplus y$, $x \oplus y = (x \oplus 0) + y$. Then $x + y = (x/0) \oplus y$, where $x/y = z$ if and only if $z \oplus y = x$. We notice, $R_0^{-1} 0 = 0$, since $R_0 0 = 0$, $0 \oplus 0 = 0$, $0 = 0$.

It is well known [685, 72, 80], that $(A, +)$ is a loop. Indeed, $0 + y = R_0^{-1} 0 \oplus y = 0 \oplus y = y$; $x + 0 = R_0^{-1} x \oplus 0 = R_0 R_0^{-1} x = x$.

We express the map $e(x)$ in terms of the operation $+$. We have $x \oplus e(x) = x$. Then $(x \oplus 0) + e(x) = x$, $e(x) = (x \oplus 0)\backslash\backslash x$, where $x \backslash\backslash y = z$ if and only if $x + z = y$.

If we denote the map $R_0^{\oplus}$ by α, then $x \oplus y = \alpha x + y$, $e(x) = \alpha x \backslash\backslash x$. We can rewrite (6.17) in terms of the loop operation $+$ as follows

$$\alpha x + (\alpha y + z) = \alpha(\alpha x + y) + (\alpha e(x) + z). \tag{6.18}$$

From $e(x \oplus y) = ex \oplus ey$ we have $e(\alpha x + y) = \alpha e(x) + e(y)$. By $y = 0$ from the last equality we have

$$e\alpha = \alpha e. \tag{6.19}$$

Therefore $e(\alpha x + y) = e\alpha(x) + e(y)$, e is a normal endomorphism of $(A, +)$. Changing αx by x and taking into consideration (6.19) we obtain from the equality (6.18) the following equality

$$x + (\alpha y + z) = \alpha(x + y) + (ex + z). \tag{6.20}$$

Next part of the proof was obtained using Prover 9 which was developed by Prof. W. McCune [615].

If we put a quasigroup in the equality (6.20), then $\alpha x + ex = x$, or, equivalently,

$$\alpha x = x // ex. \tag{6.21}$$

If we put $y = 0$ in the equality (6.20), then

$$\alpha x + (ex + z) = x + z. \tag{6.22}$$

If we apply the equality (6.21) to the equality (6.22), then

$$(x//ex) + (ex + z) = x + z. \tag{6.23}$$

If we apply the equality (6.21) to the equality (6.20), then

$$x + ((y//ey) + z) = ((x + y)//e(x + y)) + (ex + z). \tag{6.24}$$

If we change x by $x + y$ in the equality (6.23), then

$$((x + y)//e(x + y)) + (e(x + y) + z) = (x + y) + z. \tag{6.25}$$

Taking into consideration Lemma 7.52 and Theorem 6.27 we can say that there exists a minimal number n (finite or infinite) such that $e^n(a) = 0$ for any $a \in A$.

If we change x by $e^{n-1}x$ in the equality (6.20), then

$$e^{n-1}x + (\alpha y + z) = \alpha(e^{n-1}x + y) + z. \tag{6.26}$$

If we change αx by $x//ex$ (equality 6.21) in (6.26), then

$$e^{n-1}x + ((y//ey) + z) = ((e^{n-1}x + y)//ey) + z. \tag{6.27}$$

We change in the equality (6.25) x by $e^{n-1}x$. Then

$$((e^{n-1}x + y)//ey) + (ey + z) = (e^{n-1}x + y) + z. \tag{6.28}$$

We rewrite the left side of the equality (6.28) as follows

$$((e^{n-1}x + y)//ey) + (ey + z) \overset{6.27}{=} e^{n-1}x + ((y//ey) + (ey + z)) \overset{6.23}{=} e^{n-1}x + (y + z). \tag{6.29}$$

From (6.28) and (6.29) we have

$$e^{n-1}x + (y + z) = (e^{n-1}x + y) + z. \tag{6.30}$$

We change x by $e^{n-2}x$ in the equality (6.25), then

$$((e^{n-2}x + y)//e(e^{n-2}x + y)) + (e(e^{n-2}x + y) + z) = (e^{n-2}x + y) + z. \tag{6.31}$$

We rewrite the left side of the equality (6.31) as follows

$$\begin{gathered} ((e^{n-2}x + y)//e(e^{n-2}x + y)) + (e(e^{n-2}x + y) + z) \overset{6.30}{=} \\ ((e^{n-2}x + y)//e(e^{n-2}x + y)) + (e^{n-1}x + (ey + z)) \overset{6.24}{=} \\ e^{n-2}x + ((y//ey) + (ey + z)) \overset{6.23}{=} e^{n-2}x + (y + z). \end{gathered} \tag{6.32}$$

From (6.31) and (6.32) we have

$$e^{n-2}x + (y + z) = (e^{n-2}x + y) + z. \tag{6.33}$$

Begin Cycle

We change x by $e^{n-3}x$ in the equality (6.25). Then

$$((e^{n-3}x + y)//e(e^{n-3}x + y)) + (e(e^{n-3}x + y) + z) = (e^{n-3}x + y) + z. \tag{6.34}$$

We rewrite the left side of the equality (6.34) as follows

$$\begin{aligned}
&((e^{n-3}x + y)//e(e^{n-3}x + y)) + (e(e^{n-3}x + y) + z) \overset{6.33}{=} \\
&((e^{n-3}x + y)//e(e^{n-3}x + y)) + (e^{n-2}x + (ey + z)) \overset{6.24}{=} \\
&e^{n-3}x + ((y//ey) + (ey + z)) \overset{6.23}{=} e^{n-3}x + (y + z).
\end{aligned} \tag{6.35}$$

From (6.34) and (6.35) we have

$$e^{n-3}x + (y + z) = (e^{n-3}x + y) + z. \tag{6.36}$$

End Cycle

Therefore

$$e^{n-i}x + (y + z) = (e^{n-i}x + y) + z \tag{6.37}$$

for any natural number i. If the number n is finite, then repeating the **Cycle** a necessary number of times we shall obtain that $x + (y + z) = (x + y) + z$ for all $x, y, z \in A$.

Since n is a fixed finite number, then we can apply the **Cycle** the necessary number of times to obtain associativity.

Step 2. From Theorems 6.27 and 6.7 it follows that

$$(B, \diamond) = (B, \cdot)(\varepsilon, \psi, \varepsilon)((R^{\star}_{a})^{-1}, (L^{\star}_{a})^{-1}, \varepsilon) = (B, \cdot)((R^{\star}_{a})^{-1}, \psi(L^{\star}_{a})^{-1}, \varepsilon)$$

is a left S-loop.

2. This case is proved similar to Case 1. □

Corollary 6.83. *A loop $(Q, *)$, which is the direct product of a group $(A, +)$ and a left S-loop $(B, \diamond)$, is a left special loop.*

Proof. Indeed, any group is left special. Any left S-loop is also a special loop ([670], p. 61). Therefore $(Q, *)$ is a left special loop. □

Lemma 6.84. *The fulfilment of the equality (6.20) in the group $(A, +)$ is equivalent to the fact that the triple $T_x = (\alpha L_x \alpha^{-1}, \varepsilon, L_{\alpha x})(\varepsilon, L_{e(x)}, L_{e(x)})$ is an autotopy of $(A, +)$ for all $x \in A$.*

Proof. From (6.20) by $y = 0$ we have

$$x + z = \alpha x + (ex + z), \tag{6.38}$$

i.e., $L_x = L_{\alpha x} L_{e(x)}$.

If we change y by $\alpha^{-1}y$ in (6.20), then

$$x + (y + z) = \alpha(x + \alpha^{-1}y) + (ex + z). \tag{6.39}$$

Equality (6.39) means that the group $(A, +)$ has an autotopy of the form

$$T_x = (\alpha L_x \alpha^{-1}, L_{e(x)}, L_x)$$

for all $x \in A$. Taking into consideration that $L_x = L_{\alpha x} L_{e(x)}$, we can rewrite T_x in the form

$$T_x = (\alpha L_x \alpha^{-1}, L_{e(x)}, L_{\alpha x} L_{e(x)}) = (\alpha L_x \alpha^{-1}, \varepsilon, L_{\alpha x})(\varepsilon, L_{e(x)}, L_{e(x)}).$$

□

Corollary 6.85. *If the group $(A,+)$ has the property $[L_d, \alpha^{-1}] \in LM(A,+)$ for all $d \in A$, then*

(i) $e(A,+) \trianglelefteq C(A,+) \trianglelefteq (A,+)$;
(ii) $\alpha \in Aut(A,+)$;
(iii) $\alpha|_{(Ker\, e,+)} = \varepsilon$.

Proof. (i) It is well known that any autotopy of a group $(A,+)$ has the form $(L_a\delta, R_b\delta, L_aR_b\delta)$, where L_a is a left translation of the group $(A,+)$, R_b is a right translation of this group, and δ is an automorphism of this group [80].

Therefore if the triple T_d is an autotopy of the loop $(A,+)$, then we have

$$\alpha L_d\alpha^{-1} = L_a\delta, L_{e(d)} = R_b\delta, L_d = L_aR_b\delta. \tag{6.40}$$

Then $L_{e(d)}0 = R_b\delta 0$, $e(d) = b$. From $L_d0 = L_aR_b\delta 0$ we have $d = a + b$, $d = a + e(d)$. But $d = \alpha d + e(d)$. Therefore, $a = \alpha d$.

We can rewrite the equalities (6.40) in the form

$$\alpha L_d\alpha^{-1} = L_{\alpha d}\delta, L_{e(d)} = R_{e(d)}\delta, L_d = L_{\alpha d}R_{e(d)}\delta. \tag{6.41}$$

Then

$$\begin{gathered}\delta = R_{-e(d)}L_{e(d)} = L_{e(d)}R_{-e(d)}, \alpha L_d\alpha^{-1} = L_{\alpha d}L_{e(d)}R_{-e(d)} = L_dR_{-e(d)},\\ L_{-d}\alpha L_d\alpha^{-1} = R_{-e(d)}, [L_d, \alpha^{-1}] = R_{-e(d)}.\end{gathered}$$

We notice, all permutations of the form $\{R_{-e(d)} \mid d \in A\}$ form a subgroup H' of the group $RM(A,+)$, since e is an endomorphism of the group $(A,+)$.

By our supposition $H' \subseteq LM(A,+)$. Then

$$H' \subseteq RM(A,+) \cap LM(A,+).$$

But $LM\langle A,\alpha\rangle \cap RM\langle A,\alpha\rangle \subseteq C\langle A,\alpha\rangle$ [407, 758]. Therefore $R_{-e(d)} = L_{-e(d)}$ for all $d \in A$, $e(A) \subseteq C(A)$.

(ii) From (i) it follows that the triple $(\varepsilon, L_{e(b)}, L_{e(b)})$ is an autotopy of $(A,+)$. Indeed, the equality $y + (e(b) + z) = e(b) + (y + z)$ is true for all $b, y, z \in A$ since $e(b) \in C(A)$.

Then the triple $(\alpha L_b\alpha^{-1}, \varepsilon, L_{\alpha b})$ is an autotopy of $(A,+)$, i.e., $\alpha L_b\alpha^{-1}y + z = L_{\alpha b}(y+z)$. By $z = 0$ we have $\alpha L_b\alpha^{-1}y = L_{\alpha b}y$. Then the triple $(L_{\alpha b}, \varepsilon, L_{\alpha b})$ is a loop autotopy.

The equality $\alpha L_b\alpha^{-1} = L_{\alpha b}$ means that $\alpha b + y = \alpha(b + \alpha^{-1}y)$ for all $b, y \in A$. If we change y by αy, then $\alpha b + \alpha y = \alpha(b + y)$ for all $b, y \in A$, $\alpha \in Aut(A,+)$.

(iii) From the equality $\alpha x + e(x) = x$ by $e(x) = 0$ we have $\alpha x = x$. □

Remark 6.86. Conditions $[L_d, \alpha^{-1}] \in LM(A,+)$ for all $d \in A$ and $\alpha \in Aut(A,+)$ are equivalent.

Corollary 6.87. *If $e(x) = 0$ for all $x \in A$, then $\alpha \in Aut(A,+)$.*

Proof. In this case the equality (6.39) takes the form $(\alpha L_x\alpha^{-1}, \varepsilon, L_{\alpha x})$. If the autotopy of such form is true in a loop, then $\alpha L_x\alpha^{-1} = L_{\alpha x}$, $\alpha L_x = L_{\alpha x}\alpha$. □

F.N. Sokhatsky has proved the following theorem [828, Theorem 17].

Theorem 6.88. *A group isotope $(Q,\cdot)$ with the form $x \cdot y = \alpha x + a + \beta y$ is a left F-quasigroup if and only if β is an automorphism of the group $(Q,+)$, β commutes with α and α satisfies the identity $\alpha(x + y) = x + \alpha y - x + \alpha x$.*

Example 6.89. Dihedral group $(D_8, +)$ with the following Cayley table

+	0	1	2	3	4	5	6	7
0	0	1	2	3	4	5	6	7
1	1	0	3	2	6	7	4	5
2	2	6	7	1	0	3	5	4
3	3	4	5	0	1	2	7	6
4	4	3	0	5	7	6	1	2
5	5	7	6	4	3	0	2	1
6	6	2	1	7	5	4	0	3
7	7	5	4	6	2	1	3	0

has endomorphism

$$e = \begin{pmatrix} 0 & 1 & 2 & 3 & 4 & 5 & 6 & 7 \\ 0 & 3 & 3 & 0 & 3 & 3 & 0 & 0 \end{pmatrix}, \qquad e^2 = 0,$$

and permutation $\alpha = (1\,2)(4\,5)$ such that $\alpha \notin Aut(D_8)$. Using this permutation and taking into consideration Theorem 6.88 we may construct left F-quasigroups $(D_8, \cdot)$ and $(D_8, *)$ with the forms $x \cdot y = \alpha x + a + y$ and $x * y = \alpha x + a + \beta y$, where $\beta = (15)(24)$. These quasigroups are right-linear group isotopes but they are not left linear quasigroups ($\alpha \notin Aut(D_8, +)$). This example was constructed using Mace 4 [614].

Corollary 6.90. *A left special loop $(Q, \oplus)$ is an isotope of a finite left F-quasigroup $(Q, \cdot)$ if and only if $(Q, \oplus)$ is isotopic to the direct product of a group $(A, +)$ and a left S-loop $(B, \diamond)$.*

Proof. If a left special loop $(Q, \oplus)$ is an isotope of a left F-quasigroup $(Q, \cdot)$, then from Theorem 6.82 it follows that $(Q, \oplus)$ is isotopic to a loop $(Q, *)$ which is the direct product of a group $(A, +)$ and a left S-loop $(B, \diamond)$.

Conversely, suppose that a left special loop $(Q, \oplus)$ is an isotope of a loop $(Q, *)$ which is the direct product of a group $(A, +)$ and a left S-loop $(B, \diamond)$. It is easy to see that an isotopic image of group $(A, +)$ of the form $(\varepsilon, \psi, \varepsilon)$, where $\psi \in Aut(A, +)$, is a left F-quasigroup.

From Theorem 6.7 we have that an isotopic image of the loop $(B, \diamond)$ of the form $(\alpha, \psi^{\diamond}, \varepsilon)$, where $\psi^{\diamond}$ is a complete automorphism of $(B, \diamond)$, is a left distributive quasigroup $(B, \circ)$. By Lemma 6.23 (see also [80]) isotope of the form $x \cdot y = x \cdot \psi^{\circ} y$, where $\psi^{\circ} \in Aut(B, \circ)$, is a left F-quasigroup.

Therefore, among isotopic images of the left special loop $(Q, \oplus)$ there exists a left F-quasigroup. □

Corollary 6.90 gives an answer to the Belousov 1a Problem [72] in the finite case.

Corollary 6.91. *If $(Q, *)$ is a left M-loop which is isotopic to finite left F-quasigroup $(Q, \cdot)$, then $(Q, *)$ is isotopic to the direct product of a group and LP-isotope of a left S-loop.*

Proof. By Theorem 6.82 any left F-quasigroup $(Q, \cdot)$ is LP-isotopic to a loop $(Q, \oplus)$ which is the direct product of a group $(A, \oplus)$ and left S-loop $(B, \diamond)$.

By Theorem 6.9 any loop which is isotopic to a left F-quasigroup is a left M-loop.

Up to isomorphism $(Q, *)$ is an LP-isotope of $(Q, \cdot)$. Then the loops $(Q, *)$ and $(Q, \oplus)$ are isotopic with an isotopy $(\alpha, \beta, \varepsilon)$. Moreover, they are LP-isotopic ([80], Lemma 1.1).

From the proof of Lemma 1.319 it follows that an LP-isotopic image of a loop that is a direct product of two subloops is also isomorphic to the direct product of some subloops.

By the Albert theorem (Theorem 1.123) an LP-isotopic image of a group is a group. □

6.3.2 F-quasigroups

Theorem 6.92. *Any finite F-quasigroup $(Q,\cdot)$ is isotopic to the direct product of a group $(A,\oplus)\times(G,+)$ and a commutative Moufang loop $(K,\diamond)$, i.e., $(Q,\cdot)\sim(A,\oplus)\times(G,+)\times(K,\diamond)$.*

Proof. By Theorem 6.27, Case 1 any left F-quasigroup $(Q,\cdot)$ has the following structure $(Q,\cdot)\cong(A,\circ)\times(B,\cdot)$, where $(A,\circ)$ is a quasigroup with a unique idempotent element; $(B,\cdot)$ is isotope of a left distributive quasigroup $(B,\star)$, $x\cdot y = x\star\psi y$ for all $x,y\in B$, $\psi\in Aut(B,\cdot)$, $\psi\in Aut(B,\star)$.

By Theorem 6.27, Case 2 the quasigroup $(B,\cdot)$ has the following structure $(B,\cdot)\cong(G,\circ)\times(K,\cdot)$, where $(G,\circ)$ is a quasigroup with a unique idempotent element; $(K,\cdot)$ is an isotope of a right distributive quasigroup $(K,\star)$, $x\cdot y=\varphi x\star y$ for all $x,y\in K$, $\varphi\in Aut(K,\cdot)$, $\varphi\in Aut(K,\star)$.

By Theorem 6.82, Case 1, the quasigroup $(A,\circ)$ is a group isotope. By Theorem 6.82, Case 2, the quasigroup $(G,\circ)$ is a group isotope.

In the quasigroup $(K,\cdot)$ the endomorphisms e and f are permutations of the set K and by Theorem 6.50, $(K,\cdot)$ is an isotope of a distributive quasigroup. Then by the Belousov theorem (Theorem 2.121) quasigroup $(K,\cdot)$ is an isotope of a CML $(K,\diamond)$. Therefore

$$(Q,\cdot)\sim(A,\oplus)\times(G,+)\times(K,\diamond).$$

□

Theorem 6.93. *(Kepka-Kinyon-Phillips theorem). Any loop $(Q,*)$ that is isotopic to a finite F-quasigroup $(Q,\cdot)$ is isomorphic to the direct product of a group and a Moufang loop [500, 502].*

Proof. By Theorem 6.92 an F-quasigroup $(Q,\cdot)$ is isotopic to a loop $(Q,+)\cong(A,+)\times(B,+)$ which is the direct product of a group and a commutative Moufang loop. Then any left translation L of $(Q,+)$ can be present as a pair (L_1,L_2), where L_1 is a left translation of the loop $(A,+)$, and L_2 is a left translation of the loop $(B,+)$.

From Lemma 1.306 it follows that any LP-isotope of the loop $(Q,+)$ is the direct product of its subloops.

By generalized Albert theorem, an LP-isotope of a group is a group. Any LP-isotope of a commutative Moufang loop is a Moufang loop [72]. □

Corollary 6.94. *If $(Q,*)$ is an M-loop which is isotopic to a finite F-quasigroup, then $(Q,*)$ is a Moufang loop.*

Proof. The proof follows from Theorem 6.93. It is well known that any group is a Moufang loop. □

6.3.3 Left SM-quasigroups

Theorem 6.95. *Any finite left SM-quasigroup $(Q,\cdot)$ is isotopic to the direct product of a group $(A,\oplus)$ and a left S-loop $(B,\diamond)$, i.e., $(Q,\cdot)\sim(A,\oplus)\times(B,\diamond)$.*

Proof. In many ways the proof of this theorem repeats the proof of Theorem 6.82.

By Theorem 6.35 any left SM-quasigroup $(Q,\cdot)$ has the structure $(Q,\cdot)\cong(A,\circ)\times(B,\cdot)$, where $(A,\circ)$ is a quasigroup with a unique idempotent element and there exists a number m such that $|s^m(A,\circ)|=1$; $(B,\cdot)$ is an isotope of a left distributive quasigroup $(B,\star)$, $x\cdot y=s(x\star y)$ for all $x,y\in B$, $s\in Aut(B,\cdot)$, $s\in Aut(B,\star)$.

By Corollary 1.320, if a quasigroup Q is the direct product of quasigroups A and B, then there exists an isotopy $T = (T_1, T_2)$ of Q such that $QT \cong AT_1 \times BT_2$ is a loop.

Therefore we can divide our proof in two steps.

Step 1. Denote a unique idempotent element of $(A, \circ)$ by 0. It is easy to check that $s^\circ 0 = 0$. Indeed, from $(s^\circ)^m A = 0$ we have $(s^\circ)^{m+1} A = s^\circ 0 = 0$.

The (23)-parastrophe of $(A, \circ)$ is a left F-quasigroup $(A, \cdot)$ (Lemma 6.14, (5)) such that $|e^m(A, \cdot)| = 1$. Then $(A, \cdot)$ also has a unique idempotent element. By Theorem 6.82 the principal isotope of $(A, \cdot)$ is a group $(A, \oplus)$.

We shall use multiplication of isostrophies (Definition 6.18, Corollary 6.19 and Lemma 6.20). The (23)-parastrophe image of group $(A, \oplus)$ coincides with its isotope of the form $(I, \varepsilon, \varepsilon)$, where $x \oplus Ix = 0$ for all $x \in A$. Indeed, if $x \oplus y = z$, then $x \oplus^{23} z = y$. But $y = Ix \oplus z$. Therefore $x \oplus^{23} z = Ix \oplus z$, i.e., $(\oplus)((23), \varepsilon) = (\oplus)(\varepsilon, (I, \varepsilon, \varepsilon))$. Then $(\oplus) = (\oplus)((23), (I, \varepsilon, \varepsilon))$, since $I^2 = \varepsilon$.

We have

$$\begin{aligned}&(\oplus) = (\circ)((23), \varepsilon)(\varepsilon, (\alpha, \beta, \varepsilon)) = (\circ)((23), (\alpha, \beta, \varepsilon)),\\ &(\oplus) = (\oplus)((23), (I, \varepsilon, \varepsilon)) = (\circ)((23), (\alpha, \beta, \varepsilon))((23), (I, \varepsilon, \varepsilon)) =\\ &(\circ)(\varepsilon, (\alpha I, \varepsilon, \beta)).\end{aligned}$$

Step 2. The proof of this step is similar to the proof of Step 2 from Theorem 6.82 and we omit them. □

6.3.4 Left E-quasigroups

Lemma 6.96. *Any finite left E-quasigroup $(Q, \cdot)$ is isotopic to the direct product of a left loop $(A, \oplus)$ with equality $(\delta x \oplus x) \oplus (y \oplus z) = (\delta x \oplus y) \oplus (x \oplus z)$, where δ is an endomorphism of the loop $(A, \oplus)$, and a left S-loop $(B, \diamond)$, i.e., $(Q, \cdot) \sim (A, \oplus) \times (B, \diamond)$.*

Proof. In some ways the proof of Lemma 6.96 repeats the proof of Theorem 6.82. By Theorem 6.35 any left E-quasigroup $(Q, \cdot)$ has the following structure $(Q, \cdot) \cong (A, \circ) \times (B, \cdot)$, where $(A, \circ)$ is a quasigroup with a unique idempotent element and there exists a number m such that $|f^m(A, \circ)| = 1$; $(B, \cdot)$ is an isotope of a left distributive quasigroup $(B, \star)$, $x \cdot y = f^{-1}(x) \star y$ for all $x, y \in B$, $f \in Aut(B, \cdot)$, $f \in Aut(B, \star)$.

By Corollary 1.320, if a quasigroup Q is the direct product of quasigroups A and B, then there exists an isotopy $T = (T_1, T_2)$ of Q such that $QT \cong AT_1 \times BT_2$ is a loop.

Therefore we can divide our proof in two steps.

Step 1. We shall prove that $(A, \oplus)$ is a left loop. Denote a unique idempotent element of $(A, \circ)$ by 0. It is easy to check that $f^\circ 0 = 0$. Indeed, from $(f^\circ)^m A = 0$ we have $(f^\circ)^{m+1} A = f^\circ 0 = 0$.

From left E-equality $x \circ (y \circ z) = (f^\circ(x) \circ y) \circ (x \circ z)$ by $x = 0$ we have $0 \circ (y \circ z) = (0 \circ y) \circ (0 \circ z)$. Then $L_0 \in Aut(A, \circ)$.

Consider isotope $(A, \oplus)$ of the quasigroup $(A, \circ)$: $x \oplus y = x \circ L_0^{-1} y$. We notice that $(A, \oplus)$ is a left loop. Indeed, $0 \oplus y = 0 \circ L_0^{-1} y = y$.

Prove that $L_0 \in Aut(A, \oplus)$. From the equality $L_0(x \circ y) = L_0 x \circ L_0 y$ we have $L_0(x \circ y) = L_0(x \oplus L_0 y)$, $L_0 x \circ L_0 y = L_0 x \oplus L_0^2 y$, $L_0(x \oplus L_0 y) = L_0 x \oplus L_0^2 y$.

If we pass in left E-equality to the operation $\oplus$, then we obtain $x \oplus (L_0 y \oplus L_0^2 z) = (f^\circ x \oplus L_0 y) \oplus (L_0 x \oplus L_0^2 z)$. If we change $L_0 y$ by y, $L_0^2 z$ by z, then we obtain

$$x \oplus (y \oplus z) = (f^\circ x \oplus y) \oplus (L_0 x \oplus z). \tag{6.42}$$

We notice, $f^\circ x \circ x = x$. Then $f^\circ x \oplus L_0 x = x$. Moreover, from $f^\circ(x \circ y) = f^\circ x \circ f^\circ y$

we have $f^\circ(x \oplus L_0 y) = f^\circ(x) \oplus L_0 f^\circ(y)$. If $x = 0$, then $f^\circ L_0(y) = L_0 f^\circ(y)$, $f^\circ(x \oplus L_0 y) = f^\circ(x) \oplus f^\circ L_0(y)$, f° is an endomorphism of the left loop $(A, \oplus)$.

We can rewrite the equality (6.42) in the following form

$$(f^\circ x \oplus L_0 x) \oplus (y \oplus z) = (f^\circ x \oplus y) \oplus (L_0 x \oplus z). \tag{6.43}$$

If we change in (6.43) x by L_x^{-1}, then we obtain

$$(f^\circ L_0^{-1} x \oplus x) \oplus (y \oplus z) = (f^\circ L_0^{-1} x \oplus y) \oplus (x \oplus z). \tag{6.44}$$

If we denote the map $f^\circ L_0^{-1}$ of the set Q by δ, then from (6.44) we have $(\delta x \oplus x) \oplus (y \oplus z) = (\delta x \oplus y) \oplus (x \oplus z)$. The map $\delta = f^\circ L_0^{-1}$ is an endomorphism of the left loop $(A, \oplus)$ since f° is an endomorphism and L_0^{-1} is an automorphism of $(A, \oplus)$. We notice, $f^\circ L_0^{-1} 0 = 0$.

Step 2. From Theorems 6.35 and 6.7 it follows that

$$(B, \diamond) = (B, \cdot)(f, \varepsilon, \varepsilon)((R_a^\star)^{-1}, (L_a^\star)^{-1}, \varepsilon) = (B, \cdot)(f(R_a^\star)^{-1}, (L_a^\star)^{-1}, \varepsilon)$$

is a left S-loop. □

Remark 6.97. If we take $f^\circ a = 0$, then from $f^\circ a \oplus L_0 a = a$ we have $L_0 a = a$. Thus from (6.43) we have $a \oplus (y \oplus z) = y \oplus (a \oplus z)$.

Lemma 6.98. *A finite left E-quasigroup $(Q, \cdot)$ is isotopic to the direct product of a loop $(A, +)$ with the equality $(\delta x + x) + (y + z) = (\delta x + y) + (x + z)$, where δ is an endomorphism of the loop $(A, +)$, and a left S-loop $(B, \diamond)$, i.e., $(Q, \cdot) \sim (A, +) \times (B, \diamond)$.*

Proof. We pass from the operation $\oplus$ to operation $+$: $x + y = R_0^{-1} x \oplus y$, $x \oplus y = (x \oplus 0) + y$. Then $x + y = (x/0) \oplus y$, where $x/y = z$ if and only if $z \oplus y = x$. We notice, $R_0^{-1} 0 = 0$, since $R_0 0 = 0$, $0 \oplus 0 = 0$.

If we denote the map $R_0^\oplus$ by α, then $x \oplus y = \alpha x + y$. We can rewrite (6.44) in terms of the loop operation $+$ as follows

$$\alpha(\delta\alpha x + x) + (\alpha y + z) = \alpha(\delta\alpha x + y) + (\alpha x + z). \tag{6.45}$$

Prove that $\alpha\delta = \delta\alpha$. Notice that $R_y^\oplus x = R^\circ_{L_0^{-1} y} x$. Then $R_0^\oplus = R_0^\circ$. Thus

$$\begin{aligned} L_0 R_0^\oplus x &= L_0 R_0 x = 0 \circ (x \circ 0) = (0 \circ x) \circ 0 = R_0^\oplus L_0 x, \\ f^\circ R_0^\oplus x &= f^\circ(x \circ 0) = f^\circ x \circ 0 = R_0^\oplus f^\circ x. \end{aligned}$$

Then δ is an endomorphism of the loop $(A, +)$. Indeed, $\delta(x + y) = \delta(\alpha^{-1} x \oplus y) = \delta\alpha^{-1} x \oplus \delta y = \alpha^{-1}\delta x \oplus \delta y = \delta x + \delta y$.

The equality (6.45) takes the form

$$\alpha(\delta x + x) + (\alpha y + z) = \alpha(\delta x + y) + (x + z). \tag{6.46}$$

If we put $x = y$ in the equality (6.46), then $\alpha x = x$, $\alpha = \varepsilon$ and equality (6.46) takes the form

$$(\delta x + x) + (y + z) = (\delta x + y) + (x + z). \tag{6.47}$$

□

Lemma 6.99. *If $\delta x = 0$ for all $x \in A$, then $(A, +)$ is a commutative group.*

Proof. If we put $z = 0$ in the equality (6.47), then $x + y = y + x$. Therefore, from $x + (y + z) = y + (x + z)$ we have $(y + z) + x = y + (z + x)$. □

Lemma 6.100. *There exists a number m such that in the loop $(A,+)$ the chain*

$$(A,+) \supset \delta(A,+) \supset \delta^2(A,+) \supset \dots \supset \delta^m(A,+) = (0,+) \tag{6.48}$$

is stabilized on the element 0.

Proof. From Theorem 6.35 it follows that $(A, \circ)$ is a left E-quasigroup with a unique idempotent element 0 such that the chain

$$(A,\circ) \supset f^{\circ}(A,\circ) \supset (f^{\circ})^2(A,\circ) \supset \dots \supset (f^{\circ})^m(A,\circ) = (0,\circ) \tag{6.49}$$

is stabilized on the element 0.

From Lemmas 6.96 and 6.98 it follows that $(A,+) = (A,\circ)T$, where isotopy T has the form $(R_0^{-1}, L_0^{-1}, \varepsilon)$. Since $0 \in (f^{\circ})^i(A,\circ)$, then $((f^{\circ})^i(A,\circ))T = (f^{\circ})^i(A,+)$ is a subloop of the loop $(A,+)$ (Lemma 1.207) for all suitable values of i.

Thus we obtain that the isotopic image of the chain (6.49) is the following chain

$$(A,+) \supset f^{\circ}(A,+) \supset (f^{\circ})^2(A,+) \supset \dots \supset (f^{\circ})^m(A,+) = (0,+). \tag{6.50}$$

We recall that $\delta = f^{\circ}L_0^{-1}$ and $f^{\circ}L_0^{-1} = L_0^{-1}f^{\circ}$ (Lemma 6.96). Then $\delta^i = (f^{\circ})^i L_0^{-i}$ and $\delta^i(A,+) = (f^{\circ})^i L_0^{-i}(A,+)$. It is clear that L_0^{-i} is a bijection of the set A for all suitable values of i.

Thus we can establish the following bijection: $(f^{\circ})^i(A,+) \leftrightarrow \delta^i(A,+)$. Then $\delta^i(A,+) \supset \delta^{i+1}(A,+)$, since $(f^{\circ})^i(A,+) \supset (f^{\circ})^{i+1}(A,+)$. Therefore $(f^{\circ})^m(A,+) \leftrightarrow \delta^m(A,+)$, $\delta^m(A,+) = (0,+)$. □

Lemma 6.101. *The loop $(A,+)$ is a commutative group.*

Proof. From Lemma 6.100 it follows that in $(A,+)$ there exists a number m such that $\delta^m x = 0$ for all $x \in A$. We have used Prover's 9 help [615]. From (6.47) by $y = 0$ we obtain

$$(\delta x + x) + y = \delta x + (x + y). \tag{6.51}$$

If we change y by $y + z$ in the equality (6.51), then we obtain

$$(\delta x + x) + (y + z) = \delta x + (x + (y + z)). \tag{6.52}$$

From (6.47) by $z = 0$ using (6.51) we have

$$(\delta x + y) + x = \delta x + (x + y). \tag{6.53}$$

If we change y by $\delta x \backslash y$ in (6.53), then

$$(\delta x + (\delta x \backslash y)) + x = \delta x + (x + (\delta x \backslash y)). \tag{6.54}$$

But $(\delta x + (\delta x \backslash y)) = y$ (Definition 1.54, identity (1.4)). Therefore

$$\delta x + (x + (\delta x \backslash y)) = y + x. \tag{6.55}$$

If we change x by $\delta^{m-1}x$ in (6.47), then, using condition $\delta^m x = 0$, we have

$$\delta^{m-1}x + (y + z) = y + (\delta^{m-1}x + z). \tag{6.56}$$

Begin Cycle

If we change the element x by the element $\delta^{m-2}x$ in the equality (6.55), then we have

$$\delta^{m-1}x + (\delta^{m-2}x + (\delta^{m-1}x \backslash y)) = y + \delta^{m-2}x. \tag{6.57}$$

If we change in (6.56) z by $\delta^{m-1}x\backslash z$, then, using Definition 1.54, identity (1.4), we obtain

$$\delta^{m-1}x + (y + (\delta^{m-1}x\backslash z)) = y + z. \tag{6.58}$$

If we change y by $\delta^{m-2}x$, z by y in (6.58) and compare (6.58) with (6.57), then we obtain

$$\delta^{m-2}x + y = y + \delta^{m-2}x. \tag{6.59}$$

We have $\delta^{m-1}(A) \subseteq \delta^{m-2}(A)$ since δ is an endomorphism of the loop $(A, +)$. Notice, from the equalities (6.56) and (6.59) it follows that $\delta^{m-1}(A) \subseteq N_l(A)$.

From the equality $(y/\delta^{m-2}x) + \delta^{m-2}x = y$ (Definition 1.54, identity (1.5)) using the commutativity (6.59) we obtain

$$\delta^{m-2}x + (y/\delta^{m-2}x) = y. \tag{6.60}$$

From the equality (6.60) and definition of the operation $\backslash$ we have

$$\delta^{m-2}x\backslash y = y/\delta^{m-2}x. \tag{6.61}$$

If we change $y + z$ by y in (6.56), then y passes in y/z and we have

$$\delta^{m-1}x + y = (y/z) + (\delta^{m-1}x + z). \tag{6.62}$$

Applying the operation $/$ to (6.62) we have

$$(\delta^{m-1}x + y)/(\delta^{m-1}x + z) = (y/z). \tag{6.63}$$

Write the equality (6.47) in the form

$$(\delta x + y)\backslash((\delta x + x) + (y + z)) = x + z. \tag{6.64}$$

From (6.64) using (6.52) we obtain

$$(\delta x + y)\backslash(\delta x + (x + (y + z))) = x + z. \tag{6.65}$$

From the equality (6.65) using (6.61) we have

$$(\delta x + (x + (y + z)))/(\delta x + y) = x + z. \tag{6.66}$$

If we change x by δ^{m-2} in the equality (6.66), then we obtain

$$(\delta^{m-1}x + (\delta^{m-2}x + (y + z)))/(\delta^{m-1}x + y) = \delta^{m-2}x + z. \tag{6.67}$$

Using the equality (6.63) in the equality (6.67) we have

$$(\delta^{m-2}x + (y + z))/y = \delta^{m-2}x + z. \tag{6.68}$$

Therefore

$$\delta^{m-2}x + (y + z) = (\delta^{m-2}x + z) + y, \tag{6.69}$$

and

$$\delta^{m-2}x + (y + z) = y + (\delta^{m-2}x + z). \tag{6.70}$$

End Cycle

Therefore we can change the equality (6.56) by the equality (6.70) and start a new step of the cycle.

After m steps we obtain that in the loop $(A, +)$ the equality $x + (y + z) = y + (x + z)$ is fulfilled, i.e., $(A, +)$ is an abelian group. $\square$

Theorem 6.102. *1. A finite left E-quasigroup* $(Q, \cdot)$ *is isotopic to the direct product of an abelian group* $(A, +)$ *and a left S-loop* $(B, \diamond)$*, i.e.,* $(Q, \cdot) \sim (A, +) \times (B, \diamond)$.

2. A finite right E-quasigroup $(Q, \cdot)$ *is isotopic to the direct product of an abelian group* $(A, +)$ *and a right S-loop* $(B, \diamond)$*, i.e.,* $(Q, \cdot) \sim (A, +) \times (B, \diamond)$.

Proof. 1. The proof follows from Lemmas 6.98 and 6.99. □

Theorem 6.102 gives an answer to the Kinyon-Phillips problems ([520], Problem 2.8, (1)) in the finite case.

Corollary 6.103. *A finite left FESM-quasigroup* $(Q, \cdot)$ *is isotopic to the direct product of an abelian group* $(A, \oplus)$ *and a left S-loop* $(B, \diamond)$.

Proof. We can use Theorem 6.102. □

Corollary 6.103 gives an answer to the Kinyon-Phillips problem ([520], Problem 2.8, (2)) in the finite case.

Chapter 7

Structure of n-ary medial quasigroups

7.1	On n-ary medial quasigroups	327
7.1.1	n-ary quasigroups: Isotopy and translations	327
7.1.2	Linear n-ary quasigroups	329
7.1.3	n-Ary medial quasigroups	330
7.1.4	Homomorphisms of n-ary quasigroups	331
7.1.5	Direct product of n-ary quasigroups	334
7.1.6	Multiplication group of n-ary T-quasigroup	335
7.1.7	Homomorphisms of n-ary linear quasigroups	336
7.1.8	n-Ary analog of Murdoch theorem	339
7.2	Properties of n-ary simple T-quasigroups	344
7.2.1	Simple n-ary quasigroups	344
7.2.2	Congruences of linear n-ary quasigroups	345
7.2.3	Simple n-T-quasigroups	348
7.2.4	Simple n-ary medial quasigroups	349
7.3	Solvability of finite n-ary medial quasigroups	354

In this chapter we prove the n-ary analog of the Murdoch theorem about direct decomposition of finite n-ary medial quasigroups and describe the structure of simple n-ary medial quasigroups.

7.1 On n-ary medial quasigroups

We shall use basic terms and concepts from books and articles [72, 80, 77, 685, 345, 871, 832, 524].

7.1.1 n-ary quasigroups: Isotopy and translations

We recall some known facts. Let Q be a nonempty set, let n be natural number, $n \geqslant 2$. A map f that maps all n-tuples over Q into elements of the set Q is called an *n-ary operation*, i.e., $f(x_1, x_2, \dots, x_n) = x_{n+1}$ for all $(x_1, x_2, \dots, x_n) \in Q^n$ and $x_{n+1} \in Q$. .

We can define the n-ary operation f as a set $\mathfrak{F}$ of $(n+1)$-tuples of the following form $(x_1, x_2, \dots, x_n, f(x_1, x_2, \dots, x_n))$, where $x_1, x_2, \dots, x_n$, $f(x_1, x_2, \dots, x_n) \in Q$. The n-ary operations f and g defined on a set Q are equal if and only if $\mathfrak{F} = \mathfrak{G}$.

A sequence $x_m, x_{m+1}, \dots, x_n$ will be denoted by x_m^n. Of course m, n are natural numbers with $m \leqslant n$. As usual in the study of n-quasigroups, $\overline{1, n} = \{1, 2, \dots, n\}$ [77].

Definition 7.1. A nonempty set Q with n-ary operation f such that in the equation $f(x_1, x_2, \dots, x_n) = x_{n+1}$ the knowledge of any n elements of $x_1, x_2, \dots, x_n$, x_{n+1} uniquely specifies the remaining one is called an *n-ary quasigroup* [93, 77].

This definition of an n-ary quasigroup is in the V.D. Belousov and M.D. Sandik article [93]. Notice, Definition 7.1 is a part of definition of an n-ary group [264, 701].

Another definition of an n-ary quasigroup is in [739]. A definition of 3-ary quasigroup is in [708]. See also article of E.L. Post about more early definitions of 3-ary objects [701].

It is possible to give an equational definition of an n-ary quasigroup as a generalization of Definition 1.54. We follow [77, 683].

Definition 7.2. An n-ary groupoid (Q, A) is called an n-ary quasigroup if on the set Q there exist operations ${}^{(1,n+1)}A$, ${}^{(2,n+1)}A$, $\dots$, ${}^{(n,n+1)}A$ such that in the algebra $(Q, A, {}^{(1,n+1)}A, \dots, {}^{(n,n+1)}A)$ the following identities are fulfilled for all $i \in \overline{1,n}$:

$$A(x_1^{i-1}, {}^{(i,n+1)}A(x_1^n), x_{i+1}^n) = x_i, \tag{7.1}$$

$${}^{(i,n+1)}A(x_1^{i-1}, A(x_1^n), x_{i+1}^n) = x_i. \tag{7.2}$$

In [357] it is proved that any n-ary quasigroup of order $k \geqslant 7$ is a special composition of binary quasigroups isotopic to a fixed quasigroup.

We say that n-ary quasigroup (Q, f) is an *isotope of n-ary quasigroup* (Q, g) if there exist permutations $\mu_1, \mu_2, \dots, \mu_n, \mu$ of the set Q such that

$$f(x_1, x_2, \dots, x_n) = \mu^{-1}g(\mu_1 x_1, \dots, \mu_n x_n) \tag{7.3}$$

for all $x_1, \dots, x_n \in Q$. We can also write this fact in the form $(Q, f) = (Q, g)T$, where $T = (\mu_1, \mu_2, \dots, \mu_n, \mu)$.

If in (7.3) $f = g$, then an $(n+1)$-tuple $(\mu_1, \mu_2, \dots, \mu_n, \mu)$ of permutations of the set Q is called an *autotopy of n-quasigroup* (Q, f). The last component of an autotopy of an n-quasigroup is called a *quasiautomorphism* (by analogy with binary case).

If in (7.3) $\mu_1 = \mu_2 = \dots = \mu_n = \mu$, then quasigroups (Q, f) and (Q, g) are isomorphic.

At last, if in (7.3) the n-ary operations f and g are equal and $\mu_1 = \mu_2 = \dots = \mu_n = \mu$, then we obtain an *automorphism of quasigroup* (Q, f), i.e., a permutation μ of the set Q is called an automorphism of an n-quasigroup (Q, f) if for all $x_1, \dots, x_n \in Q$ the following relation is fulfilled: $\mu f(x_1, \dots, x_n) = f(\mu x_1, \dots, \mu x_n)$. We denote by $Aut(Q, f)$ the automorphism group of an n-ary quasigroup (Q, f).

A translation of n-ary quasigroup (Q, f) $(n > 2)$ will be denoted as $T(a_1, \dots, a_{i-1}, -, a_{i+1}, \dots, a_n)$, where $a_i \in Q$ for all $i \in \overline{1,n}$ and

$$T(a_1, \dots, a_{i-1}, -, a_{i+1}, \dots, a_n)x = f(a_1, \dots, a_{i-1}, x, a_{i+1}, \dots, a_n)$$

for all $x \in Q$. From the definition of an n-ary quasigroup it follows that any translation of n-ary quasigroup (Q, f) is a permutation of the set Q.

Let $\mathbf{T}(Q, f)$ be a set of all translations of n-ary quasigroup (Q, f), $\mathbf{T}^{-1}(Q, f) = \{T^{-1} \mid T \in \mathbf{T}(Q, f)\}$.

The semigroup generated by the set $\mathbf{T}$ will be denoted by $\Pi\mathbf{T}(Q, f)$ (often for short $\Pi(Q, f)$), the group generated by the set $\mathbf{T}$ will be denoted by $M\mathbf{T}(Q, f)$ or as $M(Q, f)$ for short. $M(Q, f)$ is called the *multiplication group of a quasigroup* (Q, f).

An element d of an n-ary quasigroup (Q, f) such that $f(d, \dots, d) = d$ is called an *idempotent element* of quasigroup (Q, f) (in brackets we have taken the element d exactly n times). In the binary case, an element $d \in Q$, such that $d \cdot d = d$, is called an idempotent element of quasigroup $(Q, \cdot)$.

7.1.2 Linear n-ary quasigroups

An n-ary quasigroup (Q,g) of the form

$$\gamma g(x_1,x_2,\dots,x_n)=\gamma_1x_1+\gamma_2x_2+\dots+\gamma_nx_n, \tag{7.4}$$

where $(Q,+)$ is a group, $\gamma,\gamma_1,\dots,\gamma_n$ are some permutations of the set Q, we shall call an *n-ary group isotope*. This equality (as well as analogous equalities that will appear later in this article) is true for all $x_1,x_2,\dots,x_n\in Q$.

Remark 7.3. It is easy to see that $(Q,g)=(Q,f)T$, where $f(x_1^n)=x_1+x_2+\dots+x_n$ and the isotopy T has the form $(\gamma_1,\gamma_2,\dots,\gamma_n,\gamma)$.

Following articles [506, 667] we shall call the equality (7.4) *the form of quasigroup* (Q,g).

An n-ary quasigroup (Q,f) with the form $f(x_1^n)=x_1+x_2+\dots+x_n$, where $(Q,+)$ is a binary group, will be called an *n-ary derivative group of a binary group* $(Q,+)$ [264, 77, 720]. Sometimes we shall denote this quasigroup as $(Q,\overset{n}{+})$.

Remark 7.4. The form of a quasigroup (Q,g) is some analytical definition of the n-ary operation g over a group $(Q,+)$ similar to the definition of a function with the help of a formula over a field of real numbers or over a field of some other kind of numbers.

Remark 7.5. In algebra quasigroups are usually studied up to isomorphism. Therefore in some cases, without loss of generality, we will be able to choose an isotopy T of an n-ary quasigroup (Q,g) so that its last component is the identity map.

Theorem 7.6. *An n-ary quasigroup (Q,f) of the form $f(x_1^n)=\alpha_1x_1+\alpha_2x_2+\dots+\alpha_nx_n$ is isomorphic to an n-ary quasigroup (Q,g) of the form $g(x_1^n)=\beta_1x_1+\beta_2x_2+\dots+\beta_nx_n$, where $(Q,+)$ is a group, $\alpha_i0=\beta_i0=0$ for all $i\in\overline{1,n}$, if and only if there exists an automorphism ψ of the group $(Q,+)$ such that $\psi\alpha_i=\beta_i\psi$ for all $i\in\overline{1,n}$.*

Proof. If we take $x_1=x_2=\dots=x_{i-1}=x_{i+1}=\dots=x_n=0$, then $\psi\alpha_i=\beta_i\psi$ for all suitable $i\in\overline{1,n}$. If we put $x_3=x_4=\dots=x_n=0$, then $\psi(\alpha_1x_1+\alpha_2x_2)=\beta_1\psi x_1+\beta_2\psi x_2=\psi\alpha_1x_1+\psi\alpha_2x_2$, i.e., $\psi\in Aut(Q,+)$.

It is easy to see that the converse is also true. □

We notice, there are many "isomorphism" theorems of this kind. See, for example, [506, 760, 871].

Definition 7.7. An n-quasigroup (Q,g) of the form

$$g(x_1^n)=\alpha_1x_1+\alpha_2x_2+\dots+\alpha_nx_n+a=\sum_{i=1}^{n}\alpha_ix_i+a, \tag{7.5}$$

where $(Q,+)$ is a group, $\alpha_1,\dots,\alpha_n$ are some automorphisms of the group $(Q,+)$, and the element a is some fixed element of the set Q, will be called a *linear n-ary quasigroup* (over group $(Q,+)$).

Example 7.8. The Gluskin-Hosszu theorem [426, 364, 77, 837, 264, 720, 345, 291, 292]. Any n-ary group (Q,g) has the following form

$$g(x_1^n)=x_1+\theta x_2+\theta^2x_3+\dots+\theta^{n-2}x_{n-1}+\theta^{n-1}x_n+c, \tag{7.6}$$

where $(Q,+)$ is a binary group, $\theta\in Aut(Q,+)$, $c\in Q$, $\theta^{n-1}x=c+x-c$ and $\theta c=c$ [77]. Information on n-ary groups is given, for example, in [701, 344].

From this theorem it follows that any n-ary group is a linear n-ary quasigroup. In some sense this theorem can be considered as a definition of n-ary groups.

Lemma 7.9. *The form of a linear n-ary quasigroup (Q, g) over a fixed group $(Q, +)$ defines the quasigroup (Q, g) in a unique way.*

Proof. If we suppose that

$$g(x_1^n) = \alpha_1 x_1 + \alpha_2 x_2 + \cdots + \alpha_n x_n + a = \beta_1 x_1 + \beta_2 x_2 + \cdots + \beta_n x_n + b, \tag{7.7}$$

then, if we put $x_1 = \cdots = x_n = 0$ in the equality (7.7), we obtain $a = b$. Further, if we suppose that in the equality (7.7) $x_2 = \cdots = x_n = 0$, then $\alpha_1 = \beta_1$, since $a = b$. If we take in (7.7) $x_1 = x_3 = x_4 = \cdots = x_n = 0$, then $\alpha_2 = \beta_2$ and so on. □

Definition 7.10. An n-ary linear quasigroup (Q, g) over an abelian group $(Q, +)$ is called an *n-T-quasigroup* [871]. If $n = 2$, then a quasigroup from this quasigroup class is called a *T-quasigroup* [667, 506].

7.1.3 n-Ary medial quasigroups

The following identity of n-ary quasigroup (Q, g)

$$\begin{aligned} &g(g(x_{11}, x_{12}, \dots, x_{1n}), g(x_{21}, x_{22}, \dots, x_{2n}), \dots, g(x_{n1}, x_{n2}, \dots, x_{nn})) = \\ &g(g(x_{11}, x_{21}, \dots, x_{n1}), g(x_{12}, x_{22}, \dots, x_{n2}), \dots, g(x_{1n}, x_{2n}, \dots, x_{nn})) \end{aligned} \tag{7.8}$$

is called a medial identity [77]. An n-ary quasigroup with identity (7.8) is called a *medial n-ary quasigroup.*

In the binary case from the identity (7.8) we obtain the usual *medial identity*: $xy \cdot uv = xu \cdot yv$.

Medial quasigroups, as well as the other classes of quasigroups isotopic to groups, let us construct quasigroups with preassigned properties. Often it is possible to express these properties in the language of properties of groups and components of isotopy. Systematically this approach was used in the study of T-quasigroups in [462, 461, 506, 667].

In [646] D.C. Murdoch proved that any finite binary medial quasigroup $(Q, \cdot)$ is either a quasigroup with a unique idempotent element, or it is a quasigroup in which the map e ($e : x \mapsto e(x)$, where $x \cdot e(x) = x$), is a permutation, or it is isomorphic to the direct product of a quasigroup $(A, \cdot)$ with a unique idempotent element and a quasigroup $(B, \cdot)$ in which the map e ($e : x \mapsto e(x)$ is a permutation.

It is easy to see that in the Murdoch theorem it is possible to use the map f, $f(x) \cdot x = x$. A little bit different proof of this theorem is for the map $s : x \mapsto s(x), s(x) = x \cdot x$. For us the map s is more suitable in the n-ary ($n \geqslant 3$) case to prove the n-ary analog of the Murdoch theorem. See below.

In [77] V.D. Belousov proved the following theorem. This theorem follows from the results of T. Evans ([308], Theorem 6.2.), too.

Theorem 7.11. *Let (Q, f) be a medial n-quasigroup. Then there exists an abelian group $(Q, +)$, its commuting in pairs automorphisms $\alpha_1, \dots, \alpha_n$, and a fixed element a of the set Q such that $f(x_1, x_2, \dots, x_n) = \alpha_1 x_1 + \alpha_2 x_2 + \cdots + \alpha_n x_n + a = \sum_{i=1}^{n} \alpha_i x_i + a$ for all $x_i \in Q$, $i \in \overline{1, n}$.*

In the binary case from Theorem 7.11 the classical Toyoda theorem (T-theorem) follows [890, 646, 166, 72, 80].

Remark 7.12. From Theorem 7.11 it follows that any n-ary medial quasigroup is an n-T-quasigroup.

Remark 7.13. From the Gluskin-Hossu theorem and Theorem 7.11 it follows that any n-group over an abelian group $(Q, +)$ is an n-ary medial quasigroup with the form (7.6).

Remark 7.14. We notice that n-ary quasigroups arise in different areas of mathematics in the study of various objects. Properties of n-ary medial quasigroups are described in many articles from various points of view, see, for example [308, 2, 890, 206, 94, 289, 352, 699, 698, 655, 78].

Remark 7.15. From Theorem 7.11 it follows that any n-ary medial quasigroup (Q, f) can be represented as an isotope of an n-ary derivative group $(Q, \overset{n}{+})$ of an abelian group $(Q, +)$, namely $(Q, f) = (Q, \overset{n}{+})T$, where the isotopy T has the form $(\alpha_1, \dots, \alpha_{n-1}, L_a^+\alpha_n, \varepsilon)$, $\alpha_1, \dots, \alpha_n \in Aut(Q, +)$, $L_a^+x = a + x$.

7.1.4 Homomorphisms of n-ary quasigroups

We need the n-quasigroup homomorphic theory and some facts from universal algebra [72, 80, 685, 212, 813, 871]. Most of the results are true for any Ω-algebra [212], but we shall formulate them for n-ary quasigroups. We do not give the definition of Ω-algebra here but we would like to note that an n-ary quasigroup is an Ω-algebra. Almost all lemmas in this paragraph are well known in the binary case [72, 685, 460].

Definition 7.16. Let (Q, f) and (H, g) be n-ary quasigroups, and let φ be a single-valued mapping of Q into H such that $\varphi f(x_1, \dots, x_n) = g(\varphi x_1, \dots, \varphi x_n)$, then φ is called a *homomorphism* of (Q, f) into (H, g) and the set $\{\varphi x | x \in Q\}$ is called a *homomorphic image* of (Q, f) under φ [685].

Even in the case $n = 2$ a homomorphic image of an n-ary quasigroup (Q, f) does not have to be a quasigroup [43], i.e., the class of n-ary quasigroups with the signature that consists of one n-ary operation is not closed in regard to homomorphic images.

To avoid the appearance of non-quasigroup homomorphic images, the notion of normal congruence is defined. Any normal congruence induces the homomorphism that its (homomorphic) image is a quasigroup. To prove this fact we give some definitions.

If (B, f) and (C, g) are n-ary quasigroups (Ω-algebras), C is a subset of B (written $C \subseteq B$) and C is closed under the action f, then (C, f) is said to be a *subalgebra of* (B, f), written $(C, f) \leqslant (B, f)$ [813].

As usual, a binary relation θ is an equivalence relation on Q if and only if θ is a reflexive, symmetric and transitive subset of Q^2 [212, 383]. Let V be an equivalence relation on Q. The notation xVy will often be used instead of $(x, y) \in V$. The equivalence class $V(x)$ is $\{y \in Q | xVy\}$.

Definition 7.17. Equivalence θ is a *congruence* of n-ary quasigroup (Q, f) if and only if this implication is true: $(a_1\theta b_1, a_2\theta b_2, \dots, a_n\theta b_n) \Longrightarrow f(a_1^n)\theta f(b_1^n)$.

In other words equivalence θ is a *congruence* of (Q, f) if and only if θ is a subalgebra of $(Q \times Q, (f, f))$.

Definition 7.18. If θ is an equivalence on the set Q, α is a permutation of a set Q and from $x\theta y$ there follows $\alpha x\theta\alpha y$ for all $(x, y) \in \theta$, then we shall say that permutation α is *semi-admissible relative to equivalence* θ.

Remark 7.19. We notice, in [212], that a permutation with such property is called an admissible permutation.

Definition 7.20. If θ is an equivalence on the set Q, α is a permutation of a set Q and from $x\theta y$ it follows $\alpha x\theta\alpha y$ and $\alpha^{-1}x\theta\alpha^{-1}y$ for all $(x, y) \in \theta$, then we shall say that permutation α is an *admissible permutation relative to equivalence* θ [72].

In the language of admissibility of permutations it is possible to re-write Definition 7.17 in the following form.

Definition 7.21. Equivalence θ is a congruence of n-ary quasigroup (Q, f) if and only if θ semi-admits any element of the set $\mathbf{T}(Q, f)$.

The equivalence of Definitions 7.17 and 7.21 is well known for the binary case, but we recall this equivalence here.

Definition 7.17 $\Longrightarrow$ Definition 7.21. If $a_1\theta b_1$, $a_2\theta a_2$, ..., $a_n\theta a_n$, then from Definition 7.17 it follows that $f(a_1, a_2, \dots, a_n)\theta f(b_1, a_2, \dots, a_n)$. Therefore the equivalence θ semi-admits translation $T(-, a_2, \dots, a_n)$ and so on.

Definition 7.21 $\Longrightarrow$ Definition 7.17. If $a_1\theta b_1, a_2\theta b_2, \dots, a_n\theta b_n$ and equivalence θ semi-admits any translation of the set $\mathbf{T}(Q, f)$, we have

$$\begin{array}{l} f(a_1, a_2, \dots, a_n)\theta f(b_1, a_2, \dots, a_n) \\ f(b_1, a_2, \dots, a_n)\theta f(b_1, b_2, a_3 \dots, a_n) \\ \dots \\ f(b_1, b_2, b_3 \dots, b_{n-1}, a_n)\theta f(b_1, b_2, \dots, b_{n-1}, b_n). \end{array}$$

Binary relation θ is transitive, then from the last expressions we conclude that $f(a_1^n)\theta f(b_1^n)$.

Moreover, it is easy to deduce that every congruence θ semi-admits any element of the semigroup $\Pi\mathbf{T}(Q, f)$ ([212], proposition II.6.1).

Definition 7.22. The congruence θ on n-quasigroup (Q, f) is called *normal* if for every $i = \overline{1, n}$ and for every $(c_1^n) \in Q^n$ the following implication is true:

$$f(c_1^{i-1}, a, c_{i+1}^n)\theta f(c_1^{i-1}, b, c_{i+1}^n) \Longrightarrow a\theta b$$

where $a, b \in Q$ [871].

We can give the following definition of normal congruence of an n-ary quasigroup (Q, f) as well:

Definition 7.23. An equivalence θ is a *normal congruence* of n-ary quasigroup (Q, f) if and only if θ admits every element of the set $\mathbf{T}(Q, f)$.

Lemma 7.24. *If a congruence θ semi-admits any element of the set $\mathbf{T}^{-1}$, then θ is a normal congruence.*

Proof. Let $f(c_1^{i-1}, a, c_{i+1}^n)\theta f(c_1^{i-1}, b, c_{i+1}^n)$, i.e., in the language of translations

$$T(c_1^{i-1}, -, c_{i+1}^n)a\theta T(c_1^{i-1}, -, c_{i+1}^n)b.$$

Since congruence θ admits permutations from the set $\mathbf{T}^{-1}$, then we have

$$T^{-1}(c_1^{i-1}, -, c_{i+1}^n)T(c_1^{i-1}, -, c_{i+1}^n)a\theta T^{-1}(c_1^{i-1}, -, c_{i+1}^n)T(c_1^{i-1}, -, c_{i+1}^n)b,$$

therefore $a\theta b$. □

Corollary 7.25. *A normal congruence θ of n-quasigroup (Q, f) admits any element of the group $M\,\mathbf{T}(Q, f)$.*

Proof. It is possible to use Lemma 7.24 and induction. □

Proposition 7.26. *An equivalence θ is a normal congruence of an n-ary quasigroup (Q, f) if and only if θ admits any element of the group $\mathbf{M}(Q, f)$.*

Proof. If equivalence θ admits any element of the group $\mathbf{M}(Q, f)$, then it admits any element of the set $\mathbf{T}(Q, f)$. Then by Definition 7.23 this equivalence is a normal congruence of the quasigroup (Q, f).

If θ is a normal congruence of quasigroup (Q, f), then this congruence admits any element of the set $\mathbf{T}(Q, f)$ and any finite composition of elements of this set, therefore θ admits any element of the group $\mathbf{M}(Q, f)$. □

Lemma 7.27. *An equivalence θ of a set Q is a normal congruence of an n-ary quasigroup (Q, f) if and only if this equivalence has the property $T\theta(x) = \theta(Tx)$ for all $x \in Q$, $T \in M(Q, f)$.*

Proof. Let θ be a normal congruence of the quasigroup (Q, f). Let $y \in T\theta(x)$, i.e., there exists $z \in \theta(x)$ such that $y = Tz$. Since $z\theta x$ and θ is a normal congruence of the quasigroup (Q, f), further we have $Tz\theta Tx$, $y = Tz \in \theta(Tx)$, $T\theta(x) \subseteq \theta(Tx)$.

Let $y \in \theta\,(Tx)$, i.e., $y\,\theta\,Tx$. Since θ is a normal congruence of the quasigroup (Q, f), further we obtain $T^{-1}y\,\theta\,T^{-1}Tx$, $T^{-1}y\,\theta\,x$, $T^{-1}y \in \theta(x)$, $y \in T\theta(x)$, $\theta(Tx) \subseteq T\theta(x)$, $\theta(Tx) = T\theta(x)$.

Let θ be an equivalence of a set Q with the property $T\theta(x) = \theta(Tx)$ for all $T \in M(Q, f), x \in Q$. Let $a\theta b$, i.e., $a \in \theta(b)$. Then $Ta \in T\theta(b) = \theta(Tb)$. Therefore $Ta\,\theta Tb$ for all $T \in M(Q, f)$. Thus by Proposition 7.26, θ is a normal congruence of the quasigroup (Q, f). □

Lemma 7.28. *If φ is a homomorphism of an n-quasigroup (Q, f), then φ induces a congruence.*

Proof. Let φ be a homomorphism of an n-quasigroup (Q, f) onto an n-quasigroup (H, g). Then φ induces an equivalence relation $\eta : a\eta b$ if $\varphi a = \varphi b$. This equivalence is a congruence. Really, for the equivalence η implication $a_i\eta b_i, i = \overline{1, n} \Rightarrow f(a_1^n)\eta f(b_1^n)$ we can rewrite as $\varphi a_i = \varphi b_i, i = \overline{1, n} \Rightarrow \varphi(f(a_1^n)) = \varphi(f(b_1^n))$.

Since φ is a homomorphism we have $\varphi(f(a_1^n)) = \varphi(f(b_1^n)) \Leftrightarrow g(\varphi a_1, \dots, \varphi a_n) = g(\varphi b_1, \dots, \varphi b_n)$. Therefore we have the following true implication $\varphi a_i = \varphi b_i, i = \overline{1, n} \Rightarrow g(\varphi a_1, \dots, \varphi a_n) = g(\varphi b_1, \dots, \varphi b_n)$. □

Corollary 7.29. *If φ is a homomorphism of an n-quasigroup (Q, f) onto an n-quasigroup (H, g), then φ induces a normal congruence η.*

Proof. Prove that η is normal congruence. Let $f(c_1^{i-1}, a, c_{i+1}^n)\eta f(c_1^{i-1}, b, c_{i+1}^n)$. Then

$$\varphi(f(c_1^{i-1}, a, c_{i+1}^n)) = \varphi(f(c_1^{i-1}, b, c_{i+1}^n)),$$
$$g(\varphi c_1^{i-1}, \varphi a, \varphi c_{i+1}^n) = g(\varphi c_1^{i-1}, \varphi b, \varphi c_{i+1}^n).$$

Since (H, g) is a quasigroup we have $\varphi a = \varphi b$, $a\eta b$. □

Lemma 7.30. *If θ is a normal congruence on an n-quasigroup (Q, f), then θ determines a homomorphism of quasigroup (Q, f) onto n-quasigroup (Q^θ, f).*

Proof. Let θ be a normal congruence, i.e., θ admits any element of group $\mathbf{MT}$. The set Q^θ with n-ary operation $f(a_1^\theta, \dots, a_n^\theta) = (f(a_1, \dots, a_n))^\theta$ forms a quasigroup.

It is easy to see that the element $f(a_1^\theta, \dots, a_n^\theta)$ is determined in a unique way. The equation $f(a_1^\theta, \dots, a_{i-1}^\theta, x^\theta, a_{i+1}^\theta, \dots, a_n^\theta) = a_{n+1}^\theta$ has the solution b^θ, where $f(a_1, \dots, a_{i-1}, b, a_{i+1}, \dots, a_n) = a_{n+1}$.

We prove that this solution is unique. Let

$$f(a_1^\theta, \dots, a_{i-1}^\theta, b_1^\theta, a_{i+1}^\theta, \dots, a_n^\theta) = \\ f(a_1^\theta, \dots, a_{i-1}^\theta, b_2^\theta, a_{i+1}^\theta, \dots, a_n^\theta).$$

Then

$$f(a_1, \dots, a_{i-1}, b_1, a_{i+1}, \dots, a_n)\theta f(a_1, \dots, a_{i-1}, b_2, a_{i+1}, \dots, a_n) \Rightarrow \\ b_1 \theta b_2 \Rightarrow b_1^\theta = b_2^\theta.$$

□

Lemma 7.31. *If (Q, f) is a finite n-ary quasigroup, then its congruence is normal and its homomorphic image is an n-ary quasigroup (H, g).*

Proof. In a finite n-ary quasigroup $\Pi\mathbf{T} = \mathrm{M}\mathbf{T}$. Then from Lemma 7.24 it follows that in this case any congruence is normal.

Since a homomorphism φ induces a congruence $ker\ (\varphi)$, and any congruence of a finite n-quasigroup is normal, then any homomorphism induces normal congruence. Thus $nat\ ker\ (\varphi) = \varphi(Q, f)$ is an n-ary quasigroup. Really, if

$$g(\varphi a_1, \dots, \varphi a_{i-1}, \varphi x, \varphi a_{i+1}, \dots, \varphi a_n) = \\ g(\varphi a_1, \dots, \varphi a_{i-1}, \varphi y, \varphi a_{i+1}, \dots, \varphi a_n),$$

then

$$\varphi f(a_1, \dots, a_{i-1}, x, a_{i+1}, \dots, a_n) = \varphi f(a_1, \dots, a_{i-1}, y, a_{i+1}, \dots, a_n), \\ (f(a_1, \dots, a_{i-1}, x, a_{i+1}, \dots, a_n))(ker\ \varphi)(f(a_1, \dots, a_{i-1}, y, a_{i+1}, \dots, a_n)).$$

Since $ker\ (\varphi)$ is a normal congruence, from the last expression we have $x(ker\ (\varphi))y$, $\varphi x = \varphi y$. Therefore (H, g) is an n-ary quasigroup. □

Remark 7.32. It is possible to prove that homomorphic image $(H, \circ)$ of a homomorphism η of a quasigroup $(Q, \cdot)$ is a quasigroup if and only if its corresponding congruence θ is normal.

7.1.5 Direct product of n-ary quasigroups

The direct product of quasigroups is studied in many articles and books, see, for example, [212, 813, 667, 460, 103, 107]. The concept of a direct product of quasigroups was used already in [646]. In the group case it is possible to find these definitions, for example, in [422].

Definition 7.33. If (B, f) and (C, g) are n-ary quasigroups (Ω-algebras), then $B \times C$ with the action f on the first component and g on the second component is called a *direct product* of (B, f) and (C, g) and denoted by $(B \times C, (f, g))$.

We also give the more usual definition of a direct product of n-ary quasigroups.

Definition 7.34. If (Q_1, f_1), (Q_2, f_2) are n-ary quasigroups, then their *(external) direct product* $(Q, f) = (Q_1, f_1) \times (Q_2, f_2)$ is a set of all ordered pairs (a', a''), where $a' \in Q_1$, $a'' \in Q_2$, and where the operation in (Q, f) is defined component-wise, that is, $f(a_1^n) = (f_1((a')_1^n), f_2((a'')_1^n))$.

It is easy to see that $(Q, f) = (Q_1, f_1) \times (Q_2, f_2)$ is an n-ary quasigroup.

If we have an additive form of the group operation, then we can speak instead of direct products about the direct sums, and instead of factors we can speak about items and write $G = G_1 \oplus G_2$ [471].

In [813] there is a definition of the (internal) direct product of Ω-algebras. For our aims, this approach is more preferable. We notice that internal direct products of quasigroups and Ω-algebras were studied in many articles, and some of these articles are listed above.

If U and W are equivalence relations on a set A, let $U \circ W = \{(x, y) \in A^2 \mid \exists\, t \in A, xUtWy\}$ and $U \vee W = \{(x, y) \in A^2 \mid \exists\, n \in N,\ \exists\, t_0, t_1, \dots, t_{2n} \in A, x = t_0 U t_1\, W\, t_2\, U \dots U t_{2n-1} W t_{2n} = y\}$. $U \vee W$ is an equivalence relation on A called the join of U and W. If U and W are equivalence relations on A for which $U \circ W = W \circ U$, then $U \circ W = U \vee W$, U and W are said to commute [813].

If A is an Ω-algebra and U, W are congruences on A, then $U \vee W$, and $U \cap W$ are also congruences on A.

Definition 7.35. If U and W are congruences on the algebra A which commute and for which $U \cap W = \hat{A} = \{(a, a) \mid \forall\, a \in A\}$, then the join $U \circ W = U \vee W$ of U and W is called a *direct product $U \sqcap W$ of U and W* [813].

The following theorem establishes the connection between the concepts of internal and external direct products of Ω-algebras [813, p.16].

Theorem 7.36. *An Ω-algebra A is isomorphic to a direct product of Ω-algebras B and C with isomorphism e, i.e., $e : A \to B \times C$, if and only if there exist congruences U and W of A such that $A^2 = U \sqcap W$.*

Proof. If $e : A \to B \times C$ is an isomorphism, and $\varphi : B \times C \to B, \psi : B \times C \to C$ are projections, let $U = ker\ \varphi e, W = ker\ \psi e$. Then $A^2 = U \sqcap W$.

Conversely, if $A^2 = U \sqcap W$, then every element of A is uniquely specified by its U-class and its V-class. Thus $A^2 = A^U \times A^W$. □

Lemma 7.37. *If (Q, f) is a finite n-ary quasigroup, then each pair of congruences on (Q, f) commute.*

Proof. From [589] it follows that congruences commute on all algebras with transitive groups of invertible translations. A translation is invertible if it is a product of translations of this algebra.

Since in a finite n-quasigroup for any translation T there exists a finite natural number k such that $T^k = \varepsilon$, then $T^{k-1} = T^{-1}$ and every translation is invertible.

For any pair of elements a and b of a quasigroup (Q, f) there exists translation $T(-, c_2, \dots, c_n)$ such that $T(-, c_2, c_3, \dots, c_n)\ a = b$. For example, elements $c_3, \dots, c_n$ can be any fixed elements of the set Q and element c_2 must be a solution of equation $f(a, c_2, c_3, \dots, c_n) = b$. □

Lemmas 7.31 and 7.37 let us use Theorem 7.36 for finite n-ary quasigroups.

7.1.6 Multiplication group of n-ary T-quasigroup

Theorem 7.38. *The multiplication group $M\mathbf{T}(Q, f)$ of an n-ary T-quasigroup (Q, f) with the form $f(x_1, x_2, \dots, x_n) = \sum_{j=1}^{n} \alpha_j x_j + a$ over an abelian group $(Q, +)$ has the following structure*

$$M\mathbf{T}(Q, f) \cong (Q, +) \rtimes \langle \alpha_1, \dots, \alpha_n \rangle .$$

Proof. Any translation T_i of the quasigroup (Q,f) has the form

$$\begin{aligned} &T_i(a_1,\dots,a_{i-1},-,a_{i+1},\dots,a_n)x = \\ &\alpha_1 a_1 + \dots + \alpha_{i-1}a_{i-1} + \alpha_i x + \alpha_{i+1}a_{i+1} + \dots + \alpha_n a_n + a = \\ &\alpha_1 a_1 + \dots + \alpha_{i-1}a_{i-1} + \alpha_{i+1}a_{i+1} + \dots + \alpha_n a_n + a + \alpha_i x = \\ &L^+_{\alpha_1 a_1} L^+_{\alpha_2 a_2} \dots L^+_{\alpha_{i-1}a_{i-1}} L^+_{\alpha_{i+1}a_{i+1}} \dots L^+_{\alpha_n a_n + a}\alpha_i x = L_b \alpha_i x, \end{aligned}$$

where $b = \alpha_1 a_1 + \alpha_2 a_2 + \dots + \alpha_{i-1}a_{i-1} + \alpha_{i+1}a_{i+1} + \dots + \alpha_n a_n + a$.

From standard group theory knowledge [471] it follows that

$$M_i\mathbf{T}(Q,f) \cong (Q,+) \rtimes \langle \alpha_i \rangle .$$

Use of Lemma 1.122 helps us prove that any element of the group $M\mathbf{T}(Q,f)$ can be represented in the following form: $L^+_a \beta$, where $a \in Q$, $\beta \in \langle \alpha_1,\dots,\alpha_n\rangle$.

Further, the proof is also standard [600], and, we hope this part of the proof can be easily re-established. □

Corollary 7.39. *For an n-ary medial quasigroup (Q,g) the group $M(Q,g)$ is solvable.*

Proof. This follows from Theorem 7.38, the definition of solvability of a group [334, 471], and the fact that for an n-ary medial quasigroup the group $\langle \alpha_1,\dots,\alpha_n\rangle$ is an abelian group. □

7.1.7 Homomorphisms of n-ary linear quasigroups

At the Algebraic Seminar at Charles University (Prague, October, 2003) Prof. Jan Trlifaj raised the following question. *Let an n-ary medial quasigroup (Q,f) be a direct product of n-ary medial quasigroups (Q_1,f_1) and (Q_2,f_2), i.e., $(Q,f) \cong (Q_1,f_1) \times (Q_2,f_2)$. What connections exist there between the form f of the quasigroup (Q,f) and the forms f_1 and f_2 of the quasigroups (Q_1,f_1) and (Q_2,f_2)?*

In this subsection we attempt to answer this question. See [506, 667] for the binary case.

From the definition and properties of the direct product of quasigroups we obtain the following property. If (Q_1,f_1) is a linear quasigroup with the form $f_1((x')_1^n) = \beta_1 x'_1 +_1 \beta_2 x'_2 +_1 \dots +_1 \beta_n x'_n +_1 b$ over a group $(Q_1,+_1)$ and (Q_2,f_2) is a linear quasigroup with the form

$$f_2((x'')_1^n) = \gamma_1 x''_1 +_2 \gamma_2 x''_2 +_2 \dots +_2 \gamma_n x''_n +_2 c$$

over a group $(Q_2,+_2)$, then $(Q,f) = (Q_1,f_1)\times(Q_2,f_2)$ is a linear quasigroup with the form $f(x_1^n) = \alpha_1 x_1 + \alpha_2 x_2 + \dots + \alpha_n x_n + a$ over a group $(Q,+) = (Q_1,+_1)\times(Q_2,+_2)$, where $x_i = (x'_i, x''_i)$, $\alpha_i = (\beta_i,\gamma_i)$ for all $i \in \overline{1,n}$, $a = (b,c)$.

If $(Q,f) = (Q_1,f_1)\times(Q_2,f_2)$, T_1 is an isotopy of the quasigroup (Q_1,f_1), and T_2 is an isotopy of the quasigroup (Q_2,f_2), then $T = (T_1,T_2)$ is an isotopy of the quasigroup (Q,f) and vice versa.

Proposition 7.40. *Let (Q,f) be an n-ary medial quasigroup with the property $(Q,f) = (Q_1,f_1)\times(Q_2,f_2)$ and the forms of quasigroups (Q,f), (Q_1,f_1), and (Q_2,f_2) are defined over groups $(Q,+)$, $(Q_1,+_1)$, $(Q_2,+_2)$ respectively. Then*

$$(Q,+) \cong (Q_1,+_1)\times(Q_2,+_2).$$

Proof. If (Q,f) is a medial n-ary quasigroup and $(Q,f) \cong (Q_1,f_1)\times(Q_2,f_2)$, then (Q_1,f_1) and (Q_2,f_2) are medial quasigroups too. Thus from the Belousov theorem (Theorem 7.11) it follows (see Remarks 7.3 and 7.15) that there exist derivative n-groups $(Q,\overset{n}{+})$, $(Q_1,\overset{n}{+}_1)$,

$(Q_2, \overset{n}{+}_2)$ of abelian groups $(Q,+)$, $(Q_1,+_1)$, $(Q_2,+_2)$ respectively, isotopies T, T_1 and T_2 such that $(Q,f) = (Q,\overset{n}{+})T$, $(Q_1,f_1) = (Q_1,\overset{n}{+}_1)T_1$, $(Q_2,f_2) = (Q_2,\overset{n}{+}_2)T_2$ and

$$(Q,\overset{n}{+})T = (Q_1,\overset{n}{+}_1)T_1 \times (Q_2,\overset{n}{+}_2)T_2.$$

Taking into consideration properties of the direct product of universal algebras ([383, 212]), we conclude that from the last relation, it follows that $(Q,\overset{n}{+}) = (Q_1,\overset{n}{+}_1) \times (Q_2,\overset{n}{+}_2)$ and $T = (T_1, T_2)$.

If $(Q,\overset{n}{+}) = (Q_1,\overset{n}{+}_1) \times (Q_2,\overset{n}{+}_2)$, then $(Q,+) = (Q_1,+_1) \times (Q_2,+_2)$. Indeed, if $x_1 + x_2 + \dots + x_n = (x'_1 +_1 x'_2 +_1 \dots +_1 x'_n) \times (x''_1 +_2 x''_2 +_2 \dots +_2 x''_n)$ for all $x_1^n \in Q$, $(x')_1^n \in Q_1$, $(x'')_1^n \in Q_2$, then in the case when $x_3 = x_4 = \dots = x_n = 0$, $x'_3 = x'_4 = \dots = x'_n = 0'$, $x''_3 = x''_4 = \dots = x''_n = 0''$ we obtain $x_1 + x_2 = (x'_1 +_1 x'_2) \times (x''_1 +_2 x''_2)$, i.e., $(Q,+) = (Q_1,+_1) \times (Q_2,+_2)$. □

Remark 7.41. It is necessary to notice that an analog of Proposition 7.40 is true for the direct product of n-ary linear quasigroups.

Unfortunately, in the general case, decomposition of a quasigroup that is a direct product of two (or more) quasigroups, cannot be carried out in a unique way.

Lemma 7.42. *If $\gamma_1 x_1 + \gamma_2 x_2 + \dots + \gamma_n x_n + g$ is a form of n-T-quasigroup (Q, A) over a finite abelian group $(Q,+)$, then every congruence on (Q,A) is a congruence on $(Q,+)$.*

Proof. We repeat the proof of Proposition 5 from [871] since Lemma 7.42 is a direct corollary of this proposition. We have

$$a\Theta b \Longleftrightarrow A(\gamma_1^{-1}(a), \overset{n-2}{0}, \gamma_n^{-1}(-g))\Theta A(\gamma_1^{-1}(b), \overset{n-2}{0}, \gamma_n^{-1}(-g)) \Longleftrightarrow$$
$$\gamma_1^{-1}(a)\Theta\gamma_1^{-1}(b).$$

Therefore

$$A(\gamma_1^{-1}(a), \gamma_2^{-1}(c), \overset{n-3}{0}, \gamma_n^{-1}(-g))\Theta A(\gamma_1^{-1}(b), \gamma_2^{-1}(c), \overset{n-3}{0}, \gamma_n^{-1}(-g)),$$

or $(a+c)\Theta(b+c)$, i.e., Θ is a congruence on the abelian group $(Q,+)$. We notice that the first equivalence is true since any congruence Θ of an n-ary quasigroup (Q,A) is admissible relative to any element of the group $Mlt(Q,A)$. □

Remark 7.43. Results similar to Lemma 7.42 are also in [525].

Proposition 7.44. *If a map $\xi : Q \longrightarrow Q_1$ is a homomorphism of a finite n-ary medial quasigroup (Q,f) with the form $f(x_1^n) = \alpha_1 x_1 + \alpha_2 x_2 + \dots + \alpha_n x_n + a$ over an abelian group $(Q,+)$ into a finite medial n-ary quasigroup (Q_1, f_1), then there exists an abelian group $(Q_1,+_1)$ such that $\xi(Q,+) = (Q_1,+_1)$ and the quasigroup (Q_1,f_1) has the form $f_1(x'_1,\dots,x'_n) = \beta_1 x'_1 +_1 \beta_2 x'_2 +_1 \dots +_1 \beta x'_n +_1 b$, where $\xi\alpha_i = \beta_i\xi$, $x'_i \in Q_1$ for all $i \in \overline{1,n}$, $\xi a = b$.*

Proof. It is known that using the homomorphism ξ it is possible to define a congruence θ on the set Q in the following way: $a\theta b$ if and only if $\xi a = \xi b$. Since the quasigroup (Q,f) is finite, from Lemma 7.42 it follows that the congruence θ of quasigroup (Q,f) is a congruence of the group $(Q,+)$.

Using the congruence θ of the group $(Q,+)$ we can define a binary operation on the set Q_1 in such a manner: if $x' = \xi x$, $y' = \xi y$, then $x' +_1 y' = \xi(x+y)$ for all $x, y \in Q$. It is a standard check that $(Q,+_1)$ is a finite abelian group.

Since the map ξ is a homomorphism of the group $(Q, +)$, then we obtain

$$\xi(f(x_1^n)) = \xi(\alpha_1 x_1 + \alpha_2 x_2 + \dots + \alpha_n x_n + a) =$$
$$\xi(\alpha_1 x_1) +_1 \xi(\alpha_2 x_2) +_1 \dots +_1 \xi(\alpha_n x_n) +_1 \xi a.$$

Therefore the quasigroup (Q_1, f_1) has a form over the group $(Q_1, +_1)$, i.e., it is an isotope of the group $(Q_1, \overset{n}{+_1})$.

The quasigroup (Q_1, f_1) is a finite medial n-ary quasigroup as a homomorphic image of a finite medial n-ary quasigroup. Then by Theorem 7.11 the quasigroup (Q_1, f_1) over the group $(Q_1, +_1)$ has the form $f_1(x_1^n) = \beta_1 x_1 +_1 \beta_2 x_2 +_1 \dots +_1 \beta_n x_n +_1 b$, where $\beta_i \in Aut(Q, +_1)$, $\beta_i\beta_j = \beta_j\beta_i$, $x_i \in Q_1$ for all $i, j \in \overline{1, n}$.

Since the map ξ is a homomorphism of the quasigroup (Q, f) into the quasigroup (Q_1, f_1), then we have $\xi f(x_1^n) = f_1(\xi x_1, \xi x_2, \dots, \xi x_n)$, i.e., $\xi f(x_1^n) = \beta_1 \xi x_1 +_1 \beta_2 \xi x_2 +_1 \dots +_1 \beta \xi x_n +_1 b$.

Comparing the right sides of the last equalities, we obtain the following equality

$$\begin{aligned} &\xi(\alpha_1 x_1) +_1 \xi(\alpha_2 x_2) +_1 \dots +_1 \xi(\alpha_n x_n) +_1 \xi a = \\ &\beta_1 \xi x_1 +_1 \beta_2 \xi x_2 +_1 \dots +_1 \beta \xi x_n +_1 b. \end{aligned} \tag{7.9}$$

We notice that $\xi 0 = 0_1$, since ξ is a homomorphism of the group $(Q, +)$ into the group $(Q_1, +_1)$. Indeed, $\xi x = \xi x +_1 0_1 = \xi(x + 0) = \xi x +_1 \xi 0$. Thus $\xi 0 = 0_1$.

If we put $x_1 = x_2 = \dots = x_n = 0$ in the equality (7.9), where 0 is the identity element of the group $(Q, +)$, then $\xi a = b$. Thus from the equality (7.9) the equality follows:

$$\xi \alpha_1 x_1 +_1 \xi \alpha_2 x_2 +_1 \dots +_1 \xi \alpha_n x_n = \beta_1 \xi x_1 +_1 \beta_2 \xi x_2 +_1 \dots +_1 \beta \xi x_n. \tag{7.10}$$

If we put $x_2 = x_3 = \dots = x_n = 0$ in (7.10), then we obtain that $\xi \alpha_1 x = \beta_1 \xi x$ for all $x \in Q$.

For other values of the index i the equality $\xi \alpha_i = \beta_i \xi$ is proved similarly. □

Remark 7.45. Proposition 7.44 is n-ary analog of Lemma 28 from [667] and it is very close to Proposition 9 from [871].

Corollary 7.46. *If (Q, f) is an n-ary finite medial quasigroup with the form $f(x_1^n) = \alpha_1 x_1 + \alpha_2 x_2 + \dots + \alpha_n x_n + a$, $(Q, f) \cong (Q_1, f_1) \times (Q_2, f_2)$, homomorphisms ξ and χ are such that $\xi : \xi(Q, f) = (Q_1, f_1)$ and $\chi : \chi(Q, f) = (Q_2, f_2)$, then there exist abelian groups $(Q_1, +_1)$ and $(Q_2, +_2)$ such that $(Q, +) \cong (Q_1, +_1) \times (Q_2, +_2)$ and*

$$f_1(x')_1^n = \beta_1 x'_1 +_1 \beta_2 x'_2 +_1 \dots +_1 \beta_n x'_n +_1 \xi(a),$$
$$f_2(x'')_1^n = \gamma_1 x''_1 +_2 \gamma_2 x''_2 +_2 \dots +_2 \gamma_n x''_n +_2 \chi(a),$$

where $\beta_i \in Aut(Q_1, +_1)$, $\beta_i\beta_j = \beta_j\beta_i$, $\gamma_i \in Aut(Q_2, +_2)$, $\gamma_i\gamma_j = \gamma_j\gamma_i$, $x'_i \in Q_1$ and $x''_i \in Q_2$ for all $i, j \in \overline{1, n}$.

Proof. We can use Propositions 7.44 and 7.40. □

There exists a possibility to formulate an analog of Proposition 7.44 for isomorphisms of n-ary T-quasigroups.

Corollary 7.47. *If a map $\xi : Q \longrightarrow Q_1$ is an isomorphism of an n-ary T-quasigroup (Q, f) with the form $f(x_1^n) = \alpha_1 x_1 + \alpha_2 x_2 + \dots + \alpha_n x_n + a$ over an abelian group $(Q, +)$ and an n-ary T-quasigroup (Q_1, f_1), then there exists an abelian group $(Q_1, +_1)$ such that $\xi(Q, +) = (Q_1, +_1)$ and the quasigroup (Q_1, f_1) has the form $f_1(x'_1, \dots, x'_n) = \beta_1 x'_1 +_1 \beta_2 x'_2 +_1 \dots +_1 \beta x'_n +_1 b$, where $\xi \alpha_i = \beta_i \xi$, $x'_i \in Q_1$ for all $i \in \overline{1, n}$, $\xi a = b$.*

Proof. The proof is similar to the proof of Proposition 7.44. □

Remark 7.48. Corollary 7.47 is a variation of Proposition 9 from [871].

7.1.8 n-Ary analog of Murdoch theorem

For finite n-ary medial quasigroups we prove a theorem that any such quasigroup is either an isotope of the special form of an idempotent medial quasigroup, or it is a quasigroup with a unique idempotent element, or it is isomorphic to the direct product of the isotope of an idempotent medial quasigroup and a medial quasigroup with a unique idempotent element. In the binary case the similar result was proved by D.C. Murdoch [646]. Some preliminary results and ideas in this direction are also in [645, 418].

Use of this theorem lets us obtain information on an automorphism group of any finite n-ary medial quasigroup.

Definition 7.49. For n-ary quasigroup (Q, f) we define map s_f in this manner: $s_f(x) = f(\overline{x}^n)$ for any element $x \in Q$. Below, often by using the map s_f, we shall omit denotation of the n-ary operation and shall write s instead of s_f.

Remark 7.50. In a binary quasigroup $(Q, \cdot)$ the map s has the form $s(x) = x \cdot x$ for all $x \in Q$. In an abelian group $(Q, +)$ the map s has the form $s(x) = x + x = 2x$. D.C. Murdoch formulated and proved his famous results using the map e, where $x \cdot e(x) = x$ for all $x \in Q$.

Lemma 7.51. *In a medial n-ary quasigroup (Q, f) the map s is an endomorphism of the quasigroup (Q, f).*

Proof. We have $s(f(x_1^n)) = f(\overline{f(x_1^n)}^n)$. Usage of the medial identity (see identity 7.8) further yields us

$$f(f(\overline{x_1}^n), f(\overline{x_2}^n), \dots, f(\overline{x_n}^n)) = f(s(x_1), s(x_2), \dots, s(x_n)).$$

□

Lemma 7.52. *If an n-ary medial quasigroup (Q, f) has a finite order, then there exists a natural number m such that the map $s|_{s^m(Q)}$ is an automorphism of n-quasigroup $s^m(Q, f)$.*

Proof. By $s^j(Q)$ we denote the endomorphic image of the quasigroup (Q, f) relative to the endomorphism s^j.

Then, since medial n-ary quasigroup (Q, f) has a finite order, we obtain that there exists natural number m such that the following chain

$$Q \supset s(Q) \supset s^2(Q) \supset \dots \supset s^{m-1}(Q) \supset s^m(Q)$$

becomes stable, i.e., $s^m(Q) = s^{m+1}(Q) = s^{m+2}(Q)$ and so on. □

In this case we shall say that the endomorphism s has order m.

We denote a set $\{\tau \in Aut(Q, f) | \tau\alpha = \alpha\tau\}$, where (Q, f) is an n-ary quasigroup, α is a permutation of the set Q, by $C_{Aut(Q,f)}(\alpha)$. As it is well known, the set $C_{Aut(Q,f)}(\alpha)$ forms a group with respect to the usual multiplication of permutations ([471]).

By proving the next lemma we shall use the following theorem (Theorem 8.33). *If $(Q, f) = (Q, g)T_0$ is an isotope of n-ary idempotent quasigroup (Q, g) such that isotopy T_0 has the form $(\varepsilon, \dots, \varepsilon, \beta_{i+1}, \varepsilon, \dots, \varepsilon)$ (there are $(n+1)$ members in this sequence) and $i \in \overline{0, n}$, then $Aut(Q, f) = C_{Aut(Q,g)}(\beta_{i+1})$.*

Lemma 7.53. *Let (H, f) be an n-ary medial quasigroup and the map s_f be a bijection on the set H. Then an n-ary quasigroup (H, g) where $g(x_1^n) = s_f^{-1}(f(x_1^n))$ is a medial idempotent n-ary quasigroup and $s_f \in Aut(Q, g)$.*

Proof. It is obvious that $s_f \in Aut(H, f)$. Let us prove that the quasigroup (H, f) is an isotope of a medial idempotent n-ary quasigroup (H, g).

At first check that n-ary quasigroup (H, g) is an idempotent quasigroup:

$$s_g(x) = g(\overline{x}^n) = s_f^{-1}(f(\overline{x}^n)) = s_f^{-1}(s_f(x)) = x.$$

We prove that the quasigroup (H, g) is a medial quasigroup. We note that by Lemma 7.52 the map $s^{-1} = s_f^{-1}$ is an automorphism of the quasigroup (H, f). We have

$$\begin{aligned}
&g(g(x_{11}, x_{12}, \dots, x_{1n}), \dots, g(x_{n1}, x_{n2}, \dots, x_{nn})) = \\
&s^{-1}f(s^{-1}f(x_{11}, x_{12}, \dots, x_{1n}), \dots, s^{-1}f(x_{n1}, x_{n2}, \dots, x_{nn})) = \\
&s^{-1}f(f(s^{-1}x_{11}, s^{-1}x_{12}, \dots, s^{-1}x_{1n}), \dots, \\
&f(s^{-1}x_{n1}, s^{-1}x_{n2}, \dots, s^{-1}x_{nn})) = \\
&s^{-1}f(f(s^{-1}x_{11}, s^{-1}x_{21}, \dots, s^{-1}x_{n1}), \dots, \\
&f(s^{-1}x_{1n}, s^{-1}x_{2n}, \dots, s^{-1}x_{nn})) = \\
&s^{-1}f(s^{-1}f(x_{11}, x_{21}, \dots, x_{n1}), \dots, s^{-1}f(x_{1n}, x_{2n}, \dots, x_{nn})) = \\
&g(g(x_{11}, x_{21}, \dots, x_{n1}), \dots, g(x_{1n}, x_{2n}, \dots, x_{nn})).
\end{aligned}$$

From Theorem 8.33 it follows that $Aut(Q, f) \cong C_{Aut(Q,g)}(s_f)$.

Thus $Aut(Q, f) \subseteq Aut(Q, g)$ and $s_f \in Aut(Q, g)$ since $s_f \in Aut(Q, f)$. □

We recall that n-ary quasigroup (Q, f) is called *unipotent* if there exists an element $e \in Q$ such that for all $x \in Q$ $f(\overline{x}^n) = e$.

Lemma 7.54. *An n-ary T-quasigroup (Q, f) of the form $f(x_1^n) = \alpha_1 x_1 + \alpha_2 x_2 + \dots + \alpha_n x_n + a$ is unipotent if and only if $\alpha_1 + \alpha_2 + \dots + \alpha_n = 0$.*

Proof. The proof repeats the proof of Lemma 18 from [506]. If $\alpha_1 + \alpha_2 + \dots + \alpha_n = 0$, then $f(\overline{x}^n) = a = f(\overline{y}^n)$ for all $x, y \in Q$.

We notice that $f(\overline{0}^n) = a$. Since (Q, f) is unipotent, then $f(\overline{x}^n) = \alpha_1 x + \alpha_2 x + \dots + \alpha_n x + a = a$ for any $x \in Q$, i.e., $\alpha_1 + \alpha_2 + \dots + \alpha_n = 0$. □

Theorem 7.55. *Let (Q, f) be an n-ary finite medial quasigroup. Then (Q, f) is either an isotope of a special form of an idempotent medial quasigroup, or (Q, f) is a quasigroup with a unique idempotent element, or $(Q, f) \cong (A, f_1) \times (B, f_2)$, where the quasigroup (A, f_1) is a medial n-ary quasigroup with exactly one idempotent element and the quasigroup (B, f_2) is an isotope of an n-ary medial idempotent quasigroup.*

Proof. If the map s_f is a permutation of the set Q, then by Lemma 7.53 (Q, f) is an isotope of special form of an idempotent medial quasigroup.

If $s_f(Q) = a$, where a is a fixed element of the set Q, then the quasigroup (Q, f) is a quasigroup with unique idempotent element a.

Let us suppose that $s^m = s^{m+1}$, where $m > 1$.

We define binary relation δ on the n-ary quasigroup (Q, f) by the rule $x\delta y$ if and only if $s^m(x) = s^m(y)$, where $s(x) = f(\overline{x}^n)$ for any element $x \in Q$, m is order of the endomorphism s.

From Lemma 7.31 it follows that $s^m(Q, f) = (H, f)$ is a normal n-ary subquasigroup of the n-ary quasigroup (Q, f).

We define binary relation ρ on the n-ary quasigroup (Q, f) by the rule $x\rho y$ if and only if there exist elements $h_2, h_3, \dots, h_{n-1}$ of the n-ary subquasigroup (H, f) such that

$$f(x, h_2, h_3, \dots, h_{n-1}, H) = f(y, h_2, h_3, \dots, h_{n-1}, H),$$

where $f(x, h_2, h_3, \dots, h_{n-1}, H) = \{f(x, h_2, h_3, \dots, h_{n-1}, h') \mid \text{for all } h' \in H\}$.

Below we shall also use the following shorter form of the last equality: $f(x, (h_i)_2^{n-1}, H) = f(y, (h_i)_2^{n-1}, H)$ for all $i \in \overline{2, n-1}$.

It is easy to check that so defined binary relations δ and ρ are equivalence relations.

To prove that these equivalence relations δ and ρ are normal congruences, it is sufficient to check that these relations are congruences because in n-ary finite quasigroup (Q, f) all congruences are normal (Lemma 7.31).

To prove that equivalence relation δ is a congruence we must show that the following implication is true: $x_i \delta y_i$ for all $i \in \overline{1,n} \Rightarrow f(x_1^n) \delta f(y_1^n)$. Using the definition of the binary relation δ we re-write this implication in the following equivalent form: $s^m(x_i) = s^m(y_i)$ for all $i \in \overline{1,n} \Longrightarrow s^m(f(x_1^n)) = s^m(f(y_1^n))$. Since map s^m is an endomorphism of the quasigroup (Q, f) (Lemma 7.51), further we have

$$\begin{aligned}
&s^m(x_i) = s^m(y_i) \text{ for all } i \in \overline{1,n} \Longrightarrow \\
&f(s^m(x_1), s^m(x_2), \dots, s^m(x_n)) = f(s^m(y_1), s^m(y_2), \dots, s^m(y_n)).
\end{aligned}$$

The last implication is true. Therefore equivalence relation δ is a normal congruence of the quasigroup (Q, f).

We prove that binary relation ρ is a normal congruence, i.e., that the following implication is true: $x_i \rho y_i$ for all $i \in \overline{1,n} \Longrightarrow f(x_1^n) \rho f(y_1^n)$.

Using the definition of relation ρ we can re-write the last implication in the following equivalent form: if

$$\begin{aligned}
&f(x_1, h_{1,2}, h_{1,3}, \dots, h_{1,n-1}, H) = f(y_1, h_{1,2}, h_{1,3}, \dots, h_{1,n-1}, H), \\
&f(x_2, h_{2,2}, h_{2,3}, \dots, h_{2,n-1}, H) = f(y_2, h_{2,2}, h_{2,3}, \dots, h_{2,n-1}, H), \\
&\dots \\
&f(x_n, h_{n,2}, h_{n,3}, \dots, h_{n,n-1}, H) = f(y_n, h_{n,2}, h_{n,3}, \dots, h_{n,n-1}, H),
\end{aligned} \tag{7.11}$$

then there exist $h'_2, h'_3, \dots, h'_{n-1} \in H$ such that the following equality is true

$$f(f(x_1^n), h'_2, h'_3, \dots, h'_{n-1}, H) = f(f(y_1^n), h'_2, h'_3, \dots, h'_{n-1}, H).$$

If we apply the operation f to both sides of equalities (7.11), then we obtain the following equality

$$\begin{aligned}
&f(f(x_1, h_{1,2}, h_{1,3}, \dots, h_{1,n-1}, H), f(x_2, h_{2,2}, h_{2,3}, \dots, h_{2,n-1}, H), \dots, \\
&f(x_n, h_{n,2}, h_{n,3}, \dots, h_{n,n-1}, H)) = \\
&f(f(y_1, h_{1,2}, h_{1,3}, \dots, h_{1,n-1}, H), f(y_2, h_{2,2}, h_{2,3}, \dots, h_{2,n-1}, H), \dots, \\
&f(y_n, h_{n,2}, h_{n,3}, \dots, h_{n,n-1}, H)).
\end{aligned}$$

Using medial identity we can re-write the last equality in the following form

$$\begin{aligned}
&f(f(x_1, x_2, \dots, x_n), f(h_{1,2}, h_{2,2}, \dots, h_{n,2}), \dots, \\
&f(h_{1,n-1}, h_{2,n-1}, \dots, h_{n,n-1}), f(\overline{H}^n)) = \\
&f(f(y_1, y_2, \dots, y_n), f(h_{1,2}, h_{2,2}, \dots, h_{n,2}), \dots, \\
&f(h_{1,n-1}, h_{2,n-1}, \dots, h_{n,n-1}), f(\overline{H}^n)).
\end{aligned} \tag{7.12}$$

Since (H, f) is a subquasigroup of the quasigroup (Q, f), we have $f(h_{1,2}, h_{2,2}, \dots, h_{n,2}), \dots, f(h_{1,n-1}, h_{2,n-1}, \dots, h_{n,n-1}) \in H$, $f\overline{(H)}^n = H$. Then from the equality (7.12) we obtain

$$f(f(x_1^n), h'_2, h'_3, \dots, h'_{n-1}, H) = f(f(y_1^n), h'_2, h'_3, \dots, h'_{n-1}, H),$$

where $h'_2 = f(h_{1,2}, h_{2,2}, \dots, h_{n,2})$, $h'_3 = f(h_{1,3}, h_{2,3}, \dots, h_{n,3}), \dots, h'_{n-1} = f(h_{1,n-1}, h_{2,n-1}, \dots, h_{n,n-1})$. Therefore the binary relation ρ is a normal congruence.

We prove that $\delta \cap \rho = \hat{Q} = \{(x,x) | \forall x \in Q\}$. From reflexivity of relations δ, ρ it follows that $\delta \cap \rho \supseteq \hat{Q}$.

Let $(x,y) \in \delta \cap \rho$, i.e., let $x \; \delta \; y$ and $x \; \rho \; y$, where $x, y \in Q$. Using the definitions of relations δ, ρ we have $s^m(x) = s^m(y)$ and there exist elements $h_2^{n-1} \in H$ such that $f(x, h_2^{n-1}, H) = f(y, h_2^{n-1}, H)$. Then there exist elements $h', h'' \in H$ such that $f(x, h_2^{n-1}, h') = f(y, h_2^{n-1}, h'')$. Thus we have

$$s^m f(x, h_2^{n-1}, h') = s^m f(y, h_2^{n-1}, h''),$$
$$f(s^m(x), s^m(h_2^{n-1}), s^m(h')) = f(s^m(y), s^m(h_2^{n-1}), s^m(h'')).$$

Since in the last equality all elements lie in the subquasigroup (H, f), we can conclude that $s^m(h') = s^m(h'')$. Therefore $h' = h''$ since $s \mid_H$ is a permutation of the set H. Then $f(x, h_2^{n-1}, h') = f(y, h_2^{n-1}, h')$ and we obtain $x = y$. Therefore $\delta \cap \rho \subseteq \hat{Q}$, and, finally, we have $\delta \cap \rho = \hat{Q}$.

We prove that $\delta \vee \rho = Q \times Q$. Let a, b be any fixed elements of the set Q. We prove the equality if it is shown that there exists an element $y \in Q$ such that $a\delta y$ and $y\rho b$.

From the definition of congruence δ it follows that condition $a\delta y$ is equivalent to the equality $s^m(a) = s^m(y)$.

From the definition of congruence ρ it follows that condition $y\rho b$ is equivalent to the following conditions: there exist elements $c \in Q$, $h_2^{n-1}, h', h'' \in H$ such that $y = f(c, h_2^{n-1}, h')$ and $b = f(c, h_2^{n-1}, h'')$.

Then in our new denotations the condition "there exists an element $y \in Q$ such that $a\delta y$ and $y\rho b$" takes the following equivalent form: "there exists an element $h' \in H$ such that $s^m(a) = s^m f(c, h_2^{n-1}, h')$ and $b = f(c, h_2^{n-1}, h'')$, where $c \in Q$ and $h_2^{n-1}, h'' \in H$."

Passing to images of endomorphism s^m further we have

$$s^m(a) = f(s^m(c), s^m(h_2^{n-1}), s^m(h')).$$

In the last equality we have that all elements, possibly with the exception of the last element, are in the set H. Since (H, f) is a subquasigroup we have that $s^m(h') \in H$. But the map $s \mid_H$ is a permutation of the set H and we obtain that $h' \in H$ too.

Then there exists an element h' such that $y = f(c, h_2^{n-1}, h')$, i.e., such that $a\delta y$ and $y\rho b$ for any pair $(a,b) \in Q \times Q$. Thus $(a,b) \in \delta \vee \rho$ for any pair $(a,b) \in Q \times Q$, i.e., $Q \times Q \subseteq \delta \vee \rho$. Therefore $\delta \vee \rho = Q \times Q$.

Taking into consideration Theorem 7.36 now we can say that the n-ary quasigroup (Q, f) is isomorphic to a direct product of quasigroups (H, f) and $(Q, f)/(H, f)$.

It is easy to see that medial identity holds in quasigroups (H, f) and $(Q, f)/(H, f)$. The quasigroup (H, f) is a medial quasigroup as an endomorphic image of medial quasigroup (Q, f). It is possible to check that the quasigroup $(Q, f)/(H, f)$ is a medial quasigroup too.

[Check. Any element of the quasigroup $(Q, f)/(H, f)$ we can present as a^ρ. Let $x_{11}^\rho, \dots, x_{nn}^\rho \in (Q, f)/(H, f)$. Medial identity in quasigroup $(Q, f)/(H, f)$ takes the form

$$f(f(x_{11}^\rho, \dots, x_{1n}^\rho), \dots, f(x_{n1}^\rho, \dots, x_{nn}^\rho)) =$$
$$f(f(x_{11}^\rho, \dots, x_{n1}^\rho), \dots, f(x_{1n}^\rho, \dots, x_{nn}^\rho)).$$

If we suppose that there exist elements $b_{11}^\rho, \dots, b_{nn}^\rho \in (Q, f)/(H, f)$ such that medial identity is not true for these elements, then the medial identity will not be true for some elements of the quasigroup (Q, f). We received a contradiction that shows that our supposition was not true, and, really, medial identity holds in the quasigroup $(Q, f)/(H, f)$].

By Lemma 7.53 the quasigroup (H, f) is an isotope of a medial idempotent n-ary quasigroup (H, g), where $g(x_1^n) = s_f^{-1}(f(x_1^n))$.

Prove that the quasigroup $s^j(Q, f)/s^{j+1}(Q, f)$ is a unipotent quasigroup for all suitable values j.

Denote the quasigroup $s^{j+1}(Q, f)$ as K. Any element of the quasigroup $s^j(Q, f)/s^{j+1}(Q, f)$ we can write in this form: $f(a, h_2^{n-2}, K)$, where $a \in s^j(Q, f)$, $h_2^{n-2} \in K$. Further we have $s(f(a, h_2^{n-2}, K)) = f(s(a), s(h_2^{n-2}), s(K)) \subseteq K$ since $s(a) \in K$. Therefore we obtain that the quasigroup

$$s^j(Q, f)/s^{j+1}(Q, f)$$

is a unipotent quasigroup for all suitable values of j.

Prove that the quasigroup $(A, f) \cong (Q, f)/(H, f)$, where $s^m(Q, f) = (H, f)$, is an n-ary medial quasigroup with exactly one idempotent element over an abelian group $(A, +)$.

From the properties of quasigroup (A, f) it follows that $s^m(A) = a$, where the element a is a fixed element of the set A that corresponds to the coset class H. Further, taking into consideration the properties of endomorphism s of the quasigroup (A, f), we have $s^{m+1}A = s(s^m A) = s(a) = a$. Therefore $s(a) = a$, i.e., the element a is an idempotent element of n-ary quasigroup (A, f).

Prove that there exists exactly one idempotent element in the quasigroup (A, f). Suppose that there is an element b of the quasigroup (A, f) such that $f(\overset{n}{\overline{b}}) = b$, i.e., that $s(b) = b$. Then we have $s^m(b) = b = a$. □

Remark 7.56. In some cases Theorem 7.55 and results from Subsection 7.1.7 help us to find more detailed information on forms of n-ary medial quasigroups that are components in a decomposition of a finite n-ary medial quasigroup into a direct product. See the following examples.

Definition 7.57. An n-ary quasigroup (Q, f) is called a *unipotently solvable* quasigroup of degree m if there exists the following finite chain of unipotent quasigroups:

$$Q/s(Q), s(Q)/s^2(Q), \dots, s^{m-1}(Q)/s^m(Q),$$

where the number m is the minimal number with the property $|s^m(Q)| = 1$.

Definition 7.57 helps us to re-formulate Theorem 7.55 in the following form.

Theorem 7.58. *Any finite medial n-ary quasigroup (Q, f) is isomorphic to the direct product of a medial unipotently solvable quasigroup (Q_1, f_1) and an isotope (Q_2, f_2) of the form $(\varepsilon, \dots, \varepsilon, \gamma)$ of a medial idempotent quasigroup (Q_2, f_3), where $\gamma \in Aut(Q_2, f_3)$.*

It is clear that Theorem 7.58 reduces the study of finite n-ary medial quasigroup structure to the study of finite n-ary medial unipotent and idempotent quasigroup structure.

We notice, for any unipotent n-ary quasigroup (Q, f) with an idempotent element e, we have $s(Q) = e$, and for any idempotent n-ary quasigroup (Q, f) we have $s = \varepsilon$.

Therefore, in these cases we cannot say anything on the structure of n-ary medial unipotent and medial idempotent quasigroups using the endomorphism s. Some information on the structure of unipotent n-ary quasigroups and idempotent n-ary quasigroups is in Section 7.3.

7.2 Properties of n-ary simple T-quasigroups

Classification of binary simple medial quasigroups up to isomorphism was made by Jezek and Kepka in [461]. They proved that any simple binary medial quasigroup is finite, has order p^n, where p is a prime number, and that such quasigroups are isotopes of finite elementary abelian p-groups. Moreover, they indicated the form of any (up to isomorphism) simple medial binary quasigroup.

In this section we prove that any simple n-ary medial quasigroup is finite and give some description of these quasigroups.

We shall use basic terms and concepts from books and articles [72, 77, 80, 685, 345, 871, 832, 524].

7.2.1 Simple n-ary quasigroups

The following definitions are similar to the definitions of simple and strongly simple binary quasigroups [201, 799].

Definition 7.59. An n-ary quasigroup (Q, f) is *simple* if its only normal congruences are the diagonal $\hat{Q} = \{(q,q) \,|q \in Q\}$ and $Q \times Q$.

Definition 7.60. An n-ary quasigroup (Q, f) is *strongly simple* if its only congruences are the diagonal $\hat{Q} = \{(q,q) \,|q \in Q\}$ and $Q \times Q$.

In A.A. Albert's articles [10, 11] a point of view on multiplication group $M(Q, \cdot)$ of a quasigroup $(Q, \cdot)$ as on a set of permutations of the set Q that act on set Q is developed.

The important part of the Albert articles is the theory of connection between normal subloops of a loop and normal subgroups of a multiplication group of this loop. Later Albert's results were generalized on the quasigroup case by K.K. Shchukin, V.A. Beglaryan, J.D.H. Smith, T. Ihringer and some other mathematicians [52, 799, 438].

This connection is especially good and clear in case of quasigroups that are linear over "good" loops and, in particular, in case of n-T-quasigroups [667, 506, 831, 759, 871], since these quasigroups are defined over abelian groups with the help of its automorphisms and have "good" multiplication groups.

We recall some definitions from [334, 471].

Definition 7.61. A group G acts on a set M if for any pair of elements (g, m), $g \in G, m \in M$, an element $(gm) \in M$ is defined. Moreover, $g_1(g_2(m)) = (g_1g_2)m$ and $em = m$ for all $m \in M$, $g_1, g_2 \in G$. Here e is the identity element of the group G.

The set $Gm = \{gm \,|\, g \in G\}$ is called an orbit of element m. The orbits of any two elements of the set M coincide or are not intersected. Then the set M is divided into a set of non-intersected orbits. In other words, if we define on the set M a binary relation $\sim$ as:

$$m_1 \sim m_2 \text{ if and only if there exists } g \in G \text{ such that } m_2 = gm_1,$$

then $\sim$ is an equivalence relation on the set M.

A partition of the set M on disjoint subsets M_α is called a partition on blocks relative to the group G, if for any M_α and any $g \in G$ there exists a subset M_β such that $gM_\alpha = M_\beta$. It is obvious that there exist trivial partitions of the set M, namely, partition into one-element blocks and partition into a unique block.

If there does not exist a partition of the set M into non-trivial blocks, then the group G is called primitive.

It is proved [438, 799] that a quasigroup $(Q, \cdot)$ is simple if and only if the group $Mlt(Q, \cdot)$ acts on the set Q as a primitive group of permutations.

Repeating a part of the proof of Shchukin Theorem D from [801] we can prove the following theorem.

Theorem 7.62. *An n-ary quasigroup (Q, f) is simple if and only if the multiplication group $M(Q, f)$ is primitive.*

Proof. Let us suppose that the group $M(Q, f)$ is imprimitive, i.e., we assume that on the set Q there exists some non-trivial partition $\mathcal{P}$, $Q = \cup\, Q_b$, $b \in Q$ that is an imprimitivity system of group $M(Q, f)$. We define the equivalence θ on the set Q as $\theta(b) = Q_b$ if and only if $b \in Q_b$.

From the definition of action of the group $M(Q, f)$ on the set Q and the definition of an imprimitivity system we have that $T(Q_b) = Q_{T(b)}$ for all $T \in M(Q, f)$, $Q_b \in \mathcal{P}$. Passing to equivalence classes of the equivalence θ we obtain the following property of this equivalence: $T\theta(b) = \theta(Tb)$ for all $b \in Q, T \in M(Q, f)$. From Lemma 7.27 it follows that θ is a non-trivial normal congruence.

If we suppose that the quasigroup (Q, f) is non-simple, i.e., on the set Q there exists a non-trivial normal congruence θ, then equivalence classes $\theta(a)$, $a \in Q$, give us a partition of the set Q that is an imprimitivity system of the group $M(Q, f)$. □

7.2.2 Congruences of linear n-ary quasigroups

In this section we study connections between congruences of isotopes of n-ary linear quasigroups. We start from generalization of Lemma 1.306 ([72, p. 59]). Note that this lemma was generalized in various directions [667, 506, 107, 759, 784]. Here we give one more generalization of this lemma.

Lemma 7.63. *Let θ be a normal congruence of an n-ary quasigroup (Q, f). If (Q, g) is an isotope of (Q, f) with an isotopy $T = (\alpha_1, \alpha_2, \dots, \alpha_n, \alpha_{n+1})$, i.e., $(Q, g) = (Q, f)T$, and any component of isotopy T is admissible relative to θ, then θ is a normal congruence of the quasigroup (Q, g).*

Proof. If $a_1\theta b_1$, $a_2\theta b_2$, $\dots$, $a_n\theta b_n$, then $\alpha_1 a_1\theta\alpha_1 b_1$, $\alpha_2 a_2\theta\alpha_2 b_2$, $\dots$,$\alpha_n a_n\theta\alpha_n b_n$, for all $(a_i, b_i) \in \theta$, $i \in \overline{1, n}$. Since θ is a congruence of quasigroup (Q, f), we have

$$f(\alpha_1 a_1, \alpha_2 a_2, \dots, \alpha_n a_n)\, \theta f(\alpha_1 b_1, \alpha_2 b_2, \dots, \alpha_n b_n).$$

Since permutation α_{n+1} is an admissible permutation relative to congruence θ, further we obtain the following equality:

$$\alpha_{n+1}^{-1} f(\alpha_1 a_1, \alpha_2 a_2, \dots, \alpha_n a_n)\ \theta\ \alpha_{n+1}^{-1} f(\alpha_1 b_1, \alpha_2 b_2, \dots, \alpha_n b_n),$$

i.e., $g(a_1, \dots, a_n)\theta g(b_1, \dots, b_n)$. □

Definition 7.64. Let (Q, f) be an n-ary quasigroup, $f(x_1^n) = x_{n+1}$ for all $x_1, \dots, x_{n+1} \in Q$. Let m be a natural number, with $m \leqslant n$. If in the last expression we change elements $x_{k_1}, \dots, x_{k_m}$ respectively by some fixed elements $a_1, \dots, a_m \in Q$, then this expression takes the form

$$f(x_1^{k_1-1}, a_1, x_{k_1+1}^{k_2-1}, a_2, \dots, x_{k_m}^{n}),$$

i.e., we obtain a new operation $g(x_1^{k_1-1}, x_{k_1+1}^{k_2-1}, \dots, x_{k_m}^{n})$. An operation g obtained in such manner is called a *retract* of operation f [77].

The operation g is an $(n-m)$-ary quasigroup operation [77].

Lemma 7.65. *Let (Q,f) be an n-ary quasigroup, (Q,g) be an $(n-m)$-ary retract of quasigroup (Q,f), $(n-m)>1$. If θ is a congruence of the quasigroup (Q,f), then θ is a congruence of the quasigroup (Q,g). If θ is a normal congruence of the quasigroup (Q,f), then θ is a normal congruence of the quasigroup (Q,g).*

Proof. If an equivalence θ is semi-admissible (see Definition 1.264) (respectively, admissible (see Definition 1.265)) relative to any translation

$$T(x_1, x_2, \dots, x_{i-1}, -, x_{i+1}, \dots, x_n)$$

of quasigroup (Q,f), where $x_1, x_2, \dots, x_{i-1}, x_{i+1}, \dots, x_n$ are any elements of the set Q, then this equivalence is semi-admissible (respectively admissible) relative to any translation of the form $T(x_1^{k_1-1}, a_1, x_{k_1+1}^{k_2-1}, a_2, \dots, x_{k_m}^n)$, where $a_1, a_2, \dots, a_m$ are some fixed elements of the set Q. □

Lemma 7.66. *Let (Q,g) be an n-ary quasigroup of the form $g(x_1^n) = x_1+x_2+\dots+x_n+d$, where $(Q,+)$ is a group, d is a fixed element of the set Q. A congruence θ is a congruence of the group $(Q,+)$ if and only if θ is a congruence of the quasigroup (Q,g).*

Proof. If θ is a congruence of the quasigroup (Q,g), then by Lemma 7.65 θ is a congruence of the group $(Q,+)$. We can obtain the group operation $+$, for example, as a retract of the quasigroup (Q,f) of the form $x_1+x_2 = x_1+x_2+0\dots+0-d+d$. In the last equality we took the item 0 $(n-3)$ times, where 0 is the identity element of the group $(Q,+)$.

Let θ be a congruence of a group $(Q,+)$. Then from $a_1\theta b_1$, $a_2\theta b_2$ and definition of congruence it follows that $a_1+a_2\theta b_1+b_2$. Further from $a_3\theta b_3$ and $a_1+a_2\theta b_1+b_2$ it follows that $a_1+a_2+a_3\theta b_1+b_2+b_3$ and so on. Finally, from $a_1+a_2+a_3+\dots+a_{n-1}\theta b_1+b_2+b_3+\dots+b_{n-1}$ and $a_n\theta b_n$ it follows that $a_1+a_2+a_3+\dots+a_{n-1}+a_n\theta b_1+b_2+b_3+\dots+b_{n-1}+b_n$, i.e., from $a_i\theta b_i$ for all $i\in\{1,2,\dots,n\}$ it follows that

$$(a_1+a_2+a_3+\dots+a_{n-1}+a_n)\,\theta\,(b_1+b_2+b_3+\dots+b_{n-1}+b_n),$$
$$(a_1+a_2+a_3+\dots+a_{n-1}+a_n+d)\,\theta\,(b_1+b_2+b_3+\dots+b_{n-1}+b_n+d).$$

Therefore, θ is a congruence of the quasigroup (Q,g). □

Corollary 7.67. *Let (Q,g) be an n-ary quasigroup of the form $g(x_1^n) = x_1+x_2+\dots+x_n$, where $(Q,+)$ is a group. Then the lattice of congruences of the group $(Q,+)$ coincides with the lattice of congruences of the quasigroup (Q,g).*

Lemma 7.68. *If θ is a normal congruence of n-ary T-quasigroup (Q,f) with the form $f(x_1, x_2, \dots, x_n) = \alpha_1x_1+\alpha_2x_2+\dots+\alpha_nx_n+a$ over abelian group $(Q,+)$, then θ is a congruence of the group $(Q,+)$.*

Proof. We recall that any congruence of any group is a normal congruence ([72]) (this follows also from Lemma 1.122). If θ is a normal congruence of an n-ary T-quasigroup (Q,f), then θ admits any element of the group $\mathbf{M}(Q,f)$. Any element of the group $\mathbf{M}(Q,f)$ (Theorem 7.38) has the form $L^+_{a_1}L^+_{a_2}\dots L^+_{a_{n-1}}\beta$, where $a_1,\dots,a_{n-1}\in Q$, $\beta\in\langle\alpha_1,\dots,\alpha_n\rangle$.

Then the element $L^+_{a_1}L^+_0\dots L^+_0\varepsilon = L^+_{a_1}$ lies in this group. Therefore, if $x\theta y$, then $L^+_{a_1}x\theta L^+_{a_1}y$, i.e., the following implication is true: $x\theta y \Longrightarrow (a_1+x)\theta(a_1+y)$. Since the group $(Q,+)$ is commutative, we obtain that θ is a congruence of the group $(Q,+)$. □

Corollary 7.69. *If $\mathcal{L}(Q,f)$ is the set of all normal congruences of an n-ary T-quasigroup (Q,f) with the form $f(x_1,x_2,\dots,x_n) = \alpha_1x_1+\alpha_2x_2+\dots+\alpha_nx_n+a$ over an abelian group $(Q,+)$ and $\mathcal{L}(Q,+)$ is the set of all congruences of the group $(Q,+)$, then $\mathcal{L}(Q,f)\subseteq\mathcal{L}(Q,+)$.*

Proposition 7.70. *Let (Q, f) be an n-ary T-quasigroup with the form $f(x_1, x_2, \dots, x_n) = \alpha_1 x_1 + \alpha_2 x_2 + \dots + \alpha_n x_n + a$ over abelian group $(Q, +)$. A congruence θ of the group $(Q, +)$ is a congruence of the quasigroup (Q, f) if and only if θ admits any permutation from the set $\{\alpha_1, \alpha_2, \dots, \alpha_n\}$.*

Proof. If θ is a congruence of the group $(Q, +)$, then it admits any translation $L^+_{a_1}$ of the group $(Q, +)$. Then θ admits such permutation $L^+_{a_1} L^+_{a_2} \dots L^+_{a_{n-1}}$. Since this congruence admits any element from the set $\{\alpha_1, \alpha_2, \dots, \alpha_n\}$, it admits any element of the group $\langle \alpha_1, \alpha_2, \dots, \alpha_n \rangle$. Finally, we can say that this congruence admits any permutation of the form $L^+_{a_1} L^+_{a_2} \dots L^+_{a_{n-1}} \beta$, where $a_1, \dots, a_{n-1} \in Q$, $\beta \in \langle \alpha_1, \dots, \alpha_n \rangle$. The last means that θ is a normal congruence of the quasigroup (Q, f).

If θ is a normal congruence of the quasigroup (Q, f), then by Corollary 7.69 this congruence is the congruence of the group $(Q, +)$. From Proposition 7.26 it follows that θ admits any element from the set $\{\alpha_1, \alpha_2, \dots, \alpha_n\}$. □

Corollary 7.71. *If $\mathcal{L}(Q, f)$ is the set of all normal congruences of an n-ary T-quasigroup (Q, f) with the form $f(x_1, x_2, \dots, x_n) = \alpha_1 x_1 + \alpha_2 x_2 + \dots + \alpha_n x_n +$ a over abelian group $(Q, +)$ and $\mathcal{AL}(Q, +)$ is the set of all congruences of the group $(Q, +)$ admissible relative to any permutation of the set $\{\alpha_1, \alpha_2, \dots, \alpha_n\}$, then $\mathcal{L}(Q, f) = \mathcal{AL}(Q, +)$.*

Remark 7.72. If we define on the sets of congruences $\mathcal{L}(Q, f)$ and $\mathcal{AL}(Q, +)$ lattice operation $\vee$ and $\cap$, then we obtain, that these lattices coincide.

It is known [72, 471] that any congruence θ of a group $(Q, +)$ corresponds to a normal subgroup $(H, +)$ of the group $(Q, +)$. Moreover this map is bijective. On the language of properties of normal subgroups the property of admissibility of a congruence θ takes the following form.

Let $(H, +)$ be a normal subgroup of a group $(Q, +)$ that corresponds to the congruence θ, i.e., $H = \{h \in Q \mid h\theta 1\}$, where 1 is an identity element of the group $(Q, +)$. Therefore implications $a\theta b \Longrightarrow \alpha a \theta \alpha b$ and $a\theta b \Longrightarrow \alpha^{-1} a \theta \alpha^{-1} b$ for all $(a, b) \in \theta$ take the forms $h\theta 1 \Longrightarrow \alpha h \theta \alpha 1$ and $h\theta 1 \Longrightarrow \alpha^{-1} h \theta \alpha^{-1} 1$ for all $h \in H$ respectively.

If permutation α has the property $\alpha 1 = 1$, then the last implications, taking into consideration the equivalence $h\theta 1 \Longleftrightarrow h \in H$, have the following form: $h \in H \Longrightarrow \alpha h \in H$ and $h \in H \Longrightarrow \alpha^{-1} h \in H$. From these implications we can conclude that $H \subseteq \alpha H$ and $H \subseteq \alpha^{-1} H$. From the last inclusion we obtain $\alpha H \subseteq \alpha\alpha^{-1} H = H$. And finally, from inclusions $H \subseteq \alpha H$ and $\alpha H \subseteq H$ we obtain equality $\alpha H = H$.

If the permutation α is an automorphism of the group $(Q, +)$, then the property $\alpha H = H$ means that restriction of automorphism α on subgroup $(H, +)$ is an automorphism of this subgroup. In general such restriction of an automorphism α on a subgroup $(H, +)$ of the group $(G, +)$ is only an isomorphism of this subgroup ([549]).

If we denote the subgroup $(H, +)$ of a group $(Q, +)$ that corresponds to congruence θ as $Ker\theta$ and denote the restriction of automorphism α on subgroup $Ker\theta$ as $\alpha|_{Ker\theta}$, then we can re-formulate Proposition 7.70 in the following manner.

Proposition 7.73. *Let (Q, f) be an n-ary T-quasigroup with the form $f(x_1, x_2, \dots, x_n) = \alpha_1 x_1 + \alpha_2 x_2 + \dots + \alpha_n x_n + a$ over an abelian group $(Q, +)$. A congruence θ of the group $(Q, +)$ is a congruence of the quasigroup (Q, f) if and only if $\alpha_i|_{Ker\theta}$ is an automorphism of the subgroup $Ker\theta$ for all $i \in \overline{1, n}$.*

Proposition 7.73 was proved in [871]. The binary case of this proposition was proved by Kepka and Nemec in [667, 506]. Some generalizations of this result of Kepka and Nemec are in [759, 769].

Definition 7.74. Let Φ be a subset of the set of all endomorphisms of a group G. A subgroup H of a group G is *admissible relative to the subset* Φ (for short Φ-admissible) if $\varphi H \subseteq H$ for all $\varphi \in \Phi$ [471].

Definition 7.75. A group G is called Φ-*simple*, if only the group G and the identity subgroup of the group G are admissible relative to all endomorphisms of the set Φ.

Proposition 7.76. *There exists a bijection* $\mu : \theta \longrightarrow Ker\,\theta$ *between the set of all normal congruences of n-ary T-quasigroup* (Q, f) *with the form* $f(x_1, x_2, \dots, x_n) = \alpha_1 x_1 + \alpha_2 x_2 + \dots + \alpha_n x_n + a$ *over abelian group* $(Q, +)$ *and the set of* $\langle \alpha_1, \alpha_2, \dots, \alpha_n \rangle$*-admissible subgroups of the group* $(Q, +)$.

Proof. From Definition 7.23 we have that in our case a subgroup $Ker\theta$ is admissible relative to any finite product of permutations from the set $\{\alpha_1, \dots, \alpha_n, \alpha_1^{-1}, \dots, \alpha_n^{-1}\}$. Then any subgroup $Ker\theta$, $\theta \in \mathcal{L}(Q, f)$, is admissible relative to any element of the group $\langle \alpha_1, \dots, \alpha_n \rangle$. □

7.2.3 Simple n-T-quasigroups

From Proposition 7.76 we have the following:

Corollary 7.77. *An n-ary T-quasigroup* (Q, f) *with the form*

$$f(x_1, x_2, \dots, x_n) = \alpha_1 x_1 + \alpha_2 x_2 + \dots + \alpha_n x_n + a$$

over an abelian group $(Q, +)$ *is simple if and only if the group* $(Q, +)$ *is* $\langle \alpha_1, \dots, \alpha_n \rangle$*-simple, i.e., the group* $(Q, +)$ *does not contain any non-trivial* $\langle \alpha_1, \dots, \alpha_n \rangle$*-admissible subgroup.*

Remark 7.78. If the group $(Q, +)$ is of prime order, i.e., the group $(Q, +)$ is the simple cyclic group $(Z_p, +)$ of order p, then in this case the group $(Q, +)$ is simple independent of the properties of the group $\langle \alpha_1, \dots, \alpha_n \rangle$. Or, in other words, in this case the group $(Q, +)$ is $\langle \alpha_1, \dots, \alpha_n \rangle$-simple for any group $\langle \alpha_1, \dots, \alpha_n \rangle$.

Theorem 7.79. *If a finite n-ary T-quasigroup* (Q, f) *with the form* $f(x_1, x_2, \dots, x_n) = \alpha_1 x_1 + \alpha_2 x_2 + \dots + \alpha_n x_n + a$ *over an abelian group* $(Q, +)$ *is simple, then the group* $(Q, +)$ *is a finite elementary abelian p-group, i.e.,*

$$(Q, +) \cong \bigoplus_{i=1}^{k} (Z_p, +)_i,$$

where p is a prime number, $|Q| = p^k$.

Proof. From Corollary 7.77 it follows that the group $(Q, +)$ is an $\langle \alpha_1, \dots, \alpha_n \rangle$-simple group. A necessary condition for an $\langle \alpha_1, \dots, \alpha_n \rangle$-simple abelian group is the condition that the group $(Q, +)$ is a characteristically simple group, i.e., this group is $Aut(Q, +)$-simple [549, p. 85].

As it is proved in [861, Lemma 4, p. 35] only additive groups of vector spaces over finite simple fields have such property in the finite case. In other words $(Q, +) \cong \bigoplus_{i=1}^{k} (Z_p, +)_i$, where Z_p is the cyclic group of order p and p is a prime number. It is obvious that $|Q| = p^k$. □

We recall some definitions; see, for example, [538]. Let $V(k, p)$ be a k-dimensional vector space over a field $GF(p)$. We denote the group of all invertible linear operators on $V(k, p)$ as $GL(k, p)$, i.e., an automorphism group of the vector space $V(k, p)$.

Let G be a group. Any homomorphism $\Phi : G \longrightarrow GL(k,p)$ is called a *linear representation of the group G in the vector space $V(k,p)$*. This representation of the group G is denoted as $(\Phi, V(k,p))$.

Let $(\Phi, V(k,p))$ be a linear representation of a group G. A subspace U of the space $V(k,p)$ is called an invariant subspace relative to G if $\Phi(g)u \in U$ for all $u \in U$ and all $g \in G$. Zero subspace and the space $V(k,p)$ are called trivial invariant subspaces.

Definition 7.80. A representation $(\Phi, V(k,p))$ of a group G that has only trivial invariant subspaces is called an *irreducible representation.*

Summarizing Corollary 7.77, Remark 7.78, and Theorem 7.79 we can formulate the following:

Theorem 7.81. *A finite n-ary T-quasigroup (Q,f) with the form $f(x_1, x_2, \dots, x_n) = \alpha_1 x_1 + \alpha_2 x_2 + \dots + \alpha_n x_n + a$ over an abelian group $(Q,+)$ is simple if and only if:*

1. *the group $(Q,+)$ is a finite group of order p; or*

2. *the group $(Q,+)$ is a finite elementary abelian p-group, i.e., $(Q,+) \cong \bigoplus_{i=1}^{k}(Z_p,+)_i$, where p is a prime number, $|Q| = p^k$, $k > 1$, the group $\langle \alpha_1, \dots, \alpha_n \rangle$ is an irreducible subgroup of the group $GL(k,p)$.*

Proof. Suppose that the quasigroup (Q,f) is simple. Then by Theorem 7.79 only the following cases are possible: the group $(Q,+)$ is a finite group of order p or the group $(Q,+)$ is a finite elementary abelian p-group of order p^k, $k > 1$.

In the Case 2 $(Q,+) \cong \bigoplus_{i=1}^{k}(Z_p,+)_i$, $Aut(Q,+) \cong GL(k,p)$ [471, p. 63]. It is clear that the group $\langle \alpha_1, \dots, \alpha_n \rangle$ is a subgroup of the group $GL(k,p)$, the group $(Q,+)$ is $\langle \alpha_1, \dots, \alpha_n \rangle$-simple, and by Definition 7.80 the group $\langle \alpha_1, \dots, \alpha_n \rangle$ is an irreducible subgroup of the group $GL(k,p)$.

Converse. If Case 1 or Case 2 is true, then quasigroup (Q,f) is simple. □

7.2.4 Simple n-ary medial quasigroups

In this section we prove some results on the structure of simple n-ary medial quasigroups. Binary simple medial quasigroups are described in [461].

Definition 7.82. Let Φ be a set of operators of an abelian group $(Q,+)$, i.e., $\Phi \subseteq End(Q,+)$, where $End(Q,+)$ is the set of all endomorphisms of the group $(Q,+)$. An operator homomorphism h of an abelian group $(Q,+)$ into itself is a mapping $Q \longrightarrow Q$ such that $h(a+b) = h(a) + h(b)$, $h(\omega x) = \omega h(x)$ for all $\omega \in \Phi$, $a, b \in Q$.

The set C of all operator homomorphisms (Φ-homomorphisms) of abelian group $(Q,+)$ forms a ring C relative to the following operations of addition and multiplication:

- $h = h_1 + h_2$ means that $h(x) = h_1(x) + h_2(x)$ for all $x \in Q$,
- $h = h_1 \cdot h_2$ means that $h(x) = h_1(h_2(x))$ for all $x \in Q$,

where $h_1, h_2 \in C$ [549, 471].

Theorem 7.83. *If n-ary medial quasigroup (Q,f) with the form $f(x_1, x_2, \dots, x_n) = \alpha_1 x_1 + \alpha_2 x_2 + \dots + \alpha_n x_n + a$ over an abelian group $(Q,+)$ is simple, then the group $\langle \alpha_1, \dots, \alpha_n \rangle$ is finite.*

Proof. For short we denote the group $\langle\alpha_1, \ldots, \alpha_n\rangle$ in this proof by letter Ψ. From Theorem 7.11 it follows that the group Ψ in medial quasigroup (Q, f) is commutative.

Since the group $(Q, +)$ is Ψ-simple (Corollary 7.77), then from Corollary 1 [861, p. 121], it follows that the ring C of all Ψ-homomorphisms of the group $(Q, +)$ into itself is a skew field. The same result follows from the famous Schur lemma [538], too.

It is easy to see that $\{\alpha_1, \ldots, \alpha_n\} \subseteq C$. If we generate by the set $\{\alpha_1, \ldots, \alpha_n\}$ a commutative ring R, then, since the ring R is a sub-ring of the skew field C, the ring R is a field. Indeed, in the skew field C there do not exist zero divisors, i.e., there do not exist elements $a \neq 0$, $b \neq 0$ of the field C such that $a \cdot b = 0$.

By Lemma 2.7 from [461] if a commutative field K which is (as a ring) finitely generated, then K is finite.

Therefore we obtain that the commutative field R is finite. The group $\langle\alpha_1, \ldots, \alpha_n\rangle$ is a subgroup of the multiplication group of the field R, i.e., $\Psi \subseteq (R, \cdot)$. Therefore the group $\langle\alpha_1, \ldots, \alpha_n\rangle$ is finite. □

Theorem 7.84. *If n-ary medial quasigroup (Q, f) with the form $f(x_1, x_2, \ldots, x_n) = \alpha_1 x_1 + \alpha_2 x_2 + \cdots + \alpha_n x_n + a$ over an abelian group $(Q, +)$ is simple, then the group $(Q, +)$ is finite.*

Proof. We denote the group $\langle\alpha_1, \alpha_2, \ldots, \alpha_n\rangle$ by letter Ψ.

Recall, that we can consider the non-zero elements of the field R as permutations of the set Q. In other words (especially in infinite case) we can consider the non-zero elements of the field R as bijective maps of the set Q into itself. We look at the zero element of the field R as the map δ such that $\delta a 0$ for all $a \in Q$. From the proof of Theorem 7.83 it follows that $\Psi \subseteq R$.

If Ψ is the identity group, then the quasigroup (Q, f) takes the form $f(x_1, x_2, \ldots, x_n) = x_1 + x_2 + \cdots + x_n + a$. If (Q, f) is simple, then by Lemma 7.66 the abelian group $(Q, +)$ is simple, therefore [471] this group is the finite cyclic group of order p, where p is a prime number.

Let us suppose that $|\Psi| \geqslant 2$ and the set Q is infinite. Let us consider an action of elements of the field R on any fixed non-zero element a of the set Q. Since the field R is finite, the set Ra is finite, too.

Since $\delta \in R$ we have that $0 \in Ra$. If $x \in Ra$, then there exists an element $\varphi \in R$ such that $x = \varphi a$. Since R is a field, then $-\varphi \in R$. Then $-\varphi a = -x \in Ra$. If $x, y \in Ra$, then $x = \varphi a, y = \psi a$, where $\varphi, \psi \in R$. Thus $x + y \in Ra$ since $\varphi a + \psi a = (\varphi + \psi)a \in Ra$. Moreover, $x + y = y + x$ because $x + y = (\varphi + \psi)a = (\psi + \varphi)a = y + x$. Therefore $(Ra, +)$ is a commutative subgroup of the group $(Q, +)$.

Since $\Psi \subseteq R$ and R is closed relative to multiplication of elements, we have $\Psi R = R$, $\Psi Ra = Ra$.

Therefore Ra is the finite non-identity (since $|Ra| > 1$) Ψ-admissible normal (the group $(Q, +)$ is abelian) subgroup of the group $(Q, +)$. The group $(Q, +)$ is not a Ψ-simple group. We obtain a contradiction. Then the group $(Q, +)$ is finite in this case too. □

Remark 7.85. By proving the Theorems 7.83 and 7.84 we followed the ideas from the Shchukin article [801].

Theorem 7.86. *If an n-ary medial quasigroup (Q, f) with the form $f(x_1, x_2, \ldots, x_n) = \alpha_1 x_1 + \alpha_2 x_2 + \cdots + \alpha_n x_n + a$ over an abelian group $(Q, +)$ is simple, then the group $(Q, +)$ is a finite elementary abelian p-group, i.e.,*

$$(Q, +) \cong \bigoplus_{i=1}^{k} (Z_p, +)_i,$$

where p is a prime number, $|Q| = p^k$.

Proof. From Theorem 7.84 it follows that the group $(Q,+)$ is a finite Ψ-simple group. A necessary condition to be a Ψ-simple abelian group is the condition that the group $(Q,+)$ is characteristically simple group, i.e., this group is $Aut(Q,+)$-simple ([549], p. 85).

As it is proved in [861, Lemma 4, p. 35] only additive groups of vector spaces over finite simple fields have such property in the finite case. In other words, $(Q,+) \cong \bigoplus_{i=1}^{k} (Z_p,+)_i$, where Z_p is a cyclic group of order p and p is a prime number. It is obvious that $|Q| = p^k$. □

Theorem 7.87. *If an n-ary medial quasigroup (Q,f) with the form $f(x_1, x_2, \dots, x_n) = \alpha_1 x_1 + \alpha_2 x_2 + \dots + \alpha_n x_n + a$ over an abelian group $(Q,+)$ is simple, then $|Q| = p^k$, the group $\langle \alpha_1 + \dots + \alpha_n \rangle = \Psi$ is isomorphic to the multiplicative group of the field $GF(p^k)$, the group Ψ is the cyclic group of order $p^k - 1$ in case $k > 1$ and the group Ψ is isomorphic to any subgroup of the multiplicative group of the field $GF(p)$ in the case when $k = 1$.*

Proof. Let (Q,f) be an n-ary simple medial quasigroup with the form $f(x_1, x_2, \dots, x_n) = \alpha_1 x_1 + \alpha_2 x_2 + \dots + \alpha_n x_n + a$ over an abelian group $(Q,+)$.

From Theorem 7.86 it follows that we can consider the group $(Q,+)$ as the set of k-tuples over a field $GF(p)$ with the usual component-wise addition of these k-tuples (vectors).

Using this point of view it is easy to see that any subgroup $(P,+)$ of the group $(Q,+)$ is a subspace of the vector space $V(k,p)$.

Indeed, if $(a_1, \dots, a_k) \in P$, where $a_i, r \in GF(p)$, $i = \overline{1,k}$, then $(r \cdot a_1, \dots, r \cdot a_k) = (a_1, \dots, a_k) + \dots + (a_1, \dots, a_k)$. We take r items in the right side of the last equality.

Then any finite elementary abelian p-group $(Q,+)$ is Ψ-simple, where $\Psi \subseteq Aut(Q,+)$ if and only if the group Ψ is isomorphic to an irreducible subgroup of the group $GL(k,p)$.

We recall that for the medial quasigroup (Q,f), the group Ψ is abelian. As it is proved in [861], any irreducible abelian subgroup of the group $GL(k,p)$, where p is a prime number, $k > 1$, is isomorphic to a multiplicative group of the field $GF(p^k)$. Therefore the group Ψ is the cyclic group of order $p^k - 1$ in this case.

If $k = 1$, then the group $(Q,+)$ is the simple cyclic group $(Z_p,+)$ of order p. Therefore in this case the group $(Z_p,+)$ is the Ψ-simple group relative to any subgroup of the group $Aut(Z_p,+)$. Thus in this case any n-ary medial quasigroup (Z_p,f) is the simple quasigroup independently of the form of the group Ψ. □

Proposition 7.88. *An n-T-quasigroup (Q,f) is an idempotent quasigroup if and only if (Q,f) has the form $f(x_1^n) = \varphi_1 x_1 + \dots + \varphi_n x_n$ over an abelian group $(Q,+)$ and $\varphi_1 + \dots + \varphi_n = \varepsilon$.*

Proof. From the definition of n-T-quasigroup it follows that any n-T-quasigroup has the form $f(x_1^n) = \varphi_1 x_1 + \dots + \varphi_n x_n + a$ over an abelian group $(Q,+)$.

Let (Q,f) be an idempotent quasigroup. If we take $x_1 = x_2 = \dots = x_n = 0$, then $\varphi_1 0 + \dots + \varphi_n 0 + a = 0$, $a = 0$. Therefore $\varphi_1 x + \dots + \varphi_n x = x$ for all $x \in Q$, i.e., $\varphi_1 + \dots + \varphi_n = \varepsilon$.

It is easy to check that n-T-quasigroup (Q,f) over an abelian group $(Q,+)$ with the form $f(x_1^n) = \varphi_1 x_1 + \dots + \varphi_n x_n$ and the property $\varphi_1 + \dots + \varphi_n = \varepsilon$ is an idempotent quasigroup. □

Proposition 7.89. *An n-T-quasigroup (Q,f) with the form $f(x_1^n) = \varphi_1 x_1 + \dots + \varphi_n x_n + a$ over an abelian group $(Q,+)$ has exactly one idempotent element if and only if endomorphism $\varphi_1 + \dots + \varphi_n - \varepsilon$ is a permutation of the set Q, i.e., it is an automorphism of the group $(Q,+)$. In these conditions the quasigroup (Q,f) is isomorphic to n-T-quasigroup (Q,g) with the form $g(x_1^n) = \varphi_1 x_1 + \dots + \varphi_n x_n$ over the same abelian group.*

Proof. Let u be an idempotent element of the n-T-quasigroup (Q, f). Then we have $\varphi_1 u + \dots + \varphi_n u + a = u$, i.e., $\varphi_1 u + \dots + \varphi_n u - \varepsilon u = -a$, $(\varphi_1 + \dots + \varphi_n - \varepsilon)u = -a$.

Therefore the element u is a unique idempotent of the quasigroup (Q, f) if and only if the endomorphism $\mu = (\varphi_1 + \dots + \varphi_n - \varepsilon)$ has an identity kernel, i.e., this endomorphism is an automorphism of the group $(Q, +)$.

Let $d = -\mu^{-1}a$, i.e., $\mu d = -a$, $\varphi_1 d + \varphi_2 d + \dots + \varphi_n d - d = -a$. Then we have

$$\begin{aligned}
&-d + f(d + x_1, d + x_2, \dots, d + x_n) = \\
&-d + \varphi_1 d + \varphi_1 x_1 + \varphi_2 d + \varphi_2 x_2 + \dots + \varphi_n d + \varphi_n x_n + a = \\
&\varphi_1 x_1 + \varphi_2 x_2 + \dots + \varphi_n x_n + \varphi_1 d + \varphi_2 d + \dots + \varphi_n d - d + a = \\
&\varphi_1 x_1 + \varphi_2 x_2 + \dots + \varphi_n x_n - a + a = \\
&\varphi_1 x_1 + \varphi_2 x_2 + \dots + \varphi_n x_n = g(x_1^n),
\end{aligned}$$

i.e., $(Q, f) \cong (Q, g)$. □

Some analogs of Proposition 7.89 are proved in [769, 600].

Theorem 7.90. *An n-ary medial quasigroup (Q, f) with the form $f(x_1, x_2, \dots, x_n) = \alpha_1 x_1 + \alpha_2 x_2 + \dots + \alpha_n x_n + a$ over an abelian group $(Q, +)$ is simple if and only if*

1. *the group $(Q, +)$ is the additive group of a finite Galois field $GF(p^k)$;*
2. *the group $\langle \alpha_1, \dots, \alpha_n \rangle$ is the multiplicative group of the field $GF(p^k)$ in case $k > 1$, the group $\langle \alpha_1, \dots, \alpha_n \rangle$ is any subgroup of the group $Aut(Z_p, +)$ in case $k = 1$;*
3. *the quasigroup (Q, f) in case $|Q| > 1$ can be a quasigroup from one of the following disjoint quasigroup classes:*
   - *(a) $\alpha_1 + \alpha_2 + \dots + \alpha_n = \varepsilon, a = 0$; in this case the quasigroup (Q, f) is an idempotent quasigroup;*
   - *(b) $\alpha_1 + \alpha_2 + \dots + \alpha_n = \varepsilon$ and $a \neq 0$; in this case the quasigroup (Q, f) does not have any idempotent element, the quasigroup (Q, f) is isomorphic to quasigroup (Q, g) with the form $g(x_1^n) = \alpha_1 x_1 + \alpha_2 x_2 + \dots + \alpha_n x_n + 1$ over the same group $(Q, +)$;*
   - *(c) $\alpha_1 + \alpha_2 + \dots + \alpha_n \neq \varepsilon$; in this case the quasigroup (Q, f) has exactly one idempotent element, the quasigroup (Q, f) is isomorphic to quasigroup (Q, g) with the form $g(x_1^n) = \alpha_1 x_1 + \alpha_2 x_2 + \dots + \alpha_n x_n$ over the group $(Q, +)$.*

Proof. From Theorems 7.86 and 7.87 it follows that the group $(Q, +)$ is the additive group of a finite Galois field $GF(p^k)$, the group $\langle \alpha_1, \dots, \alpha_n \rangle$ is the multiplicative group of this field in case $k > 1$, and this group is any subgroup of the group $Aut(Z_p, +)$ in case $k = 1$.

(a) From Proposition 7.88 it follows the form of a simple n-ary medial quasigroup in this case.

(b) If we suppose that simple n-ary medial quasigroup (Q, f) with the form $f(x_1^n) = \alpha_1 x_1 + \alpha_2 x_2 + \dots + \alpha_n x_n + a$ and the properties $\alpha_1 + \alpha_2 + \dots + \alpha_n = \varepsilon$ and $a \neq 0$ has an idempotent element u, then we have $\alpha_1 u + \alpha_2 u + \dots + \alpha_n u + a = u$, $u + a = u$, $a = 0$. We obtain a contradiction. Therefore, this quasigroup does not contain any idempotent element.

Since in conditions of this theorem the group $(Q, +)$ is the additive group of a field $GF(p^k)$, the group $\langle \alpha_1, \dots, \alpha_n \rangle$ is the multiplicative group of this field or it is some subgroup of this multiplicative group, we can identify automorphisms $\alpha_1, \dots, \alpha_n$ with some non-zero elements $a_1, \dots, a_n$ of the field $GF(p^k)$. We can identify action of any automorphism

$\alpha_1, \dots, \alpha_n$ on the set Q with multiplication of the elements $a_1, \dots, a_n$ by all elements of the field $GF(p^k)$. In this situation the identity automorphism ε corresponds to the identity element 1 of the field $GF(p^k)$.

Let (Q, f) be a simple n-ary medial quasigroup with the form $f(x_1^n) = a_1x_1 + a_2x_2 + \dots + a_nx_n + a$ where $a_1 + \dots + a_n = 1$, $a \neq 0$. Then $(Q, f) \cong (Q, g)$, where $g(x_1^n) = a_1x_1 + a_2x_2 + \dots + a_nx_n + 1$. Indeed, if we take a multiplication by the element a as an isomorphism of the quasigroup (Q, f) to quasigroup (Q, g), then we have

$$\begin{aligned}
&a^{-1} \cdot f(a \cdot x_1, a \cdot x_2, \dots, a \cdot x_n) = \\
&a^{-1}(a_1 \cdot a \cdot x_1 + a_2 \cdot a \cdot x_2 + \dots + a_n \cdot a \cdot x_n + a) = \\
&a^{-1}(a \cdot a_1 \cdot x_1 + a \cdot a_2 \cdot x_2 + \dots + a \cdot a_n \cdot x_n + a) = \\
&a^{-1} \cdot a \cdot (a_1 \cdot x_1 + a_2 \cdot x_2 + \dots + a_n \cdot x_n + 1) = \\
&\alpha_1x_1 + \alpha_2x_2 + \dots + \alpha_nx_n + 1 = g(x_1^n).
\end{aligned}$$

(c) We can identify the endomorphism $\alpha_1 + \dots + \alpha_n - \varepsilon$ of the group $(Q, +)$ with some non-zero element d of the field $GF(p^k)$. But in any field any non-zero element is the invertible element, thus multiplication by the element d is an automorphism of the field $GF(p^k)$, $\alpha_1 + \dots + \alpha_n - \varepsilon$ is an automorphism of the group $(Q, +)$. Further we can apply Proposition 7.89 in this case.

It is easy to see that for any finite order $|Q| > 1$ classes (a), (b) and (c) of finite simple n-ary medial quasigroups are pairwise not intersected classes, i.e., they are disjoint classes.

Converse. It is easy to see, if a medial n-ary quasigroup satisfies conditions 1, 2, and one of the disjoint conditions 3(a), 3(b), 3(c), then this quasigroup is simple. □

Remark 7.91. Let $Q = \{1\}$. In this case there exist only two n-ary quasigroups, namely, (Q, f) with the form $f(x_1^n) = \varepsilon x_1 + \dots + \varepsilon x_n$ and the quasigroup (Q, g) with the form $g(x_1^n) = \varepsilon x_1 + \dots + \varepsilon x_n + 1$, $x_1, \dots, x_n \in Q$. It is easy to see that both (Q, f) and (Q, g) are idempotent quasigroups and it is clear that these quasigroups have a unique idempotent element.

Corollary 7.92. *The multiplication group $M(Q, f)$ of any simple n-ary medial quasigroup (Q, f) is a primitive finite solvable group.*

Proof. The proof follows from Theorems 7.62, 7.86 and Corollary 7.39. □

Example 7.93. Let $(Q, +) = (Z_3 \oplus Z_3, +)$ be a direct product of two cyclic groups of order 3. Let

$$\alpha_1 = \begin{pmatrix} 1 & 1 \\ 2 & 1 \end{pmatrix}, \quad \alpha_2 = \begin{pmatrix} 2 & 2 \\ 1 & 2 \end{pmatrix}, \quad \alpha_3 = \begin{pmatrix} 2 & 1 \\ 2 & 2 \end{pmatrix},$$

where, for example,

$$\begin{pmatrix} 2 & 2 \\ 1 & 2 \end{pmatrix} \begin{pmatrix} y_1 \\ y_2 \end{pmatrix} = \begin{pmatrix} 2 \cdot y_1 + 2 \cdot y_2 \\ 1 \cdot y_1 + 2 \cdot y_2 \end{pmatrix}$$

for all $(y_1, y_2) \in Z_3 \oplus Z_3$.

Notice that $\langle \alpha_1 \rangle \cong Z_8$, $\alpha_2 = \alpha_1^5, \alpha_3 = \alpha_1^{-1}$, $\langle \alpha_1 \rangle$ is an irreducible group.

A quasigroup (Q, f) of the form $f(x_1, x_2, x_3) = \alpha_1x_1 + \alpha_2x_2 + \alpha_3x_3$ for all $x_1, x_2, x_3 \in Q$ is a simple medial 3-ary quasigroup with a unique idempotent element $(0; 0)$, since

$$\alpha_1 + \alpha_2 + \alpha_3 = \begin{pmatrix} 2 & 1 \\ 2 & 2 \end{pmatrix} \neq \begin{pmatrix} 1 & 0 \\ 0 & 1 \end{pmatrix}.$$

7.3 Solvability of finite n-ary medial quasigroups

In this section we summarize information on the structure of finite n-ary medial quasigroups using the results of previous sections. Some results of this section were published in [776, 785].

Taking into consideration Theorem 7.58 in this section we concentrate on the structure of n-ary medial unipotent and medial idempotent quasigroups. We notice that any n-ary unipotent or idempotent quasigroup, similar to the binary case, has an idempotent element.

Proposition 7.94. *Any subquasigroup (H, f) of an n-ary medial quasigroup (Q, f) is normal, i.e., the set H coincides with an equivalence class of a normal congruence θ of the quasigroup (Q, f) ([871], Proposition 8).*

From Proposition 7.94 it follows that in a simple medial quasigroup $(Q, \cdot)$ its only subquasigroups are one-element subquasigroups and the quasigroup $(Q, \cdot)$.

Remark 7.95. We notice that in general there exist non-simple medial quasigroups with only trivial subquasigroups. For example, the quasigroup $(Z_9, \diamond)$ with the form $x \diamond y = 2 \cdot x + 8 \cdot y + 1$, where $(Z_9, +)$ is the additive group of residues modulo 9, is a non-simple quasigroup without proper subquasigroups.

Proposition 7.96. *An equivalence class $\theta(h)$ of a normal congruence θ of an n-ary quasigroup (Q, f) is a subquasigroup of (Q, f) if and only if $f(h, \dots, h)\, \theta\, h$.*

Proof. The proof is similar to the proof of the binary variant of Proposition 7.96 [72, 80]. Let $(\theta(h), f)$ be a subquasigroup of an n-ary quasigroup (Q, f). We notice that $a \in \theta(h)$ means that $a\, \theta\, h$. From Definition 7.17 it follows that the following implication is true $(a_1\, \theta\, h, a_2\, \theta\, h, \dots, a_n\, \theta\, h) \Longrightarrow f(a_1^n)\, \theta\, f(\overline{h})$. Since $(\theta(h), f)$ is a subquasigroup, then $f(a_1^n) \in \theta(h)$, i.e., $f(a_1^n)\, \theta\, h$. Therefore $f(\overline{h})\, \theta\, h$.

Let $f(h, \dots, h)\, \theta\, h$. We shall prove that in this case $(\theta(h), f)$ is a subquasigroup. If $a_1, a_2, \dots, a_n \in \theta(h)$, then $a_1\, \theta\, h, a_2\, \theta\, h, \dots, a_n\, \theta\, h$ and thus by Definition 7.17 $f(a_1^n)\, \theta\, f(\overline{h})$. From the condition $f(h, \dots, h)\, \theta\, h$ further we have that $f(a_1^n)\, \theta\, h$, i.e., $f(a_1^n) \in \theta(h)$, i.e., the set $\theta(h)$ is closed with respect to the operation f.

We prove that if $a_1, a_2, \dots, a_{i-1}, a_{i+1}, \dots, a_n, b \in \theta(h)$, then the solution of the equation $f(a_1, a_2, \dots, a_{i-1}, x, a_{i+1}, \dots, a_n) = b$ also lies in $\theta(h)$, where $i \in \overline{(1, n)}$.

We recall, the existence and uniqueness of this solution follow from the fact that $(\theta(h), f)$ is a subset of n-ary quasigroup (Q, f). Since $b \in \theta(h)$, then by Definition 7.17 we have

$$\begin{aligned} h\, \theta\, & f(a_1, a_2, \dots, a_{i-1}, x, a_{i+1}, \dots, a_n) \\ \theta\, & f(h, a_2, \dots, a_{i-1}, x, a_{i+1}, \dots, a_n) \\ \theta\, & f(h, h, \dots, a_{i-1}, x, a_{i+1}, \dots, a_n) \\ & \dots \\ \theta\, & f(h, h, \dots, h, x, h, \dots, h). \end{aligned}$$

Therefore $h\, \theta\, f(h, h, \dots, h, x, h, \dots, h)$. From the condition $f(h, \dots, h)\, \theta\, h$ further it follows that $f(h, \dots, h)\, \theta\, f(h, h, \dots, h, x, h, \dots, h)$. Since θ is a normal congruence, then from the last expression we have $x\, \theta\, h$. □

Remark 7.97. It is clear that $\theta(h)$ is a normal subquasigroup of the quasigroup (Q, f) since a subquasigroup (H, f) of a quasigroup (Q, f) is normal if and only if (H, f) is an equivalence class of a normal congruence θ.

Corollary 7.98. *If an equivalence class $\theta(h)$ of a normal congruence θ of an n-ary quasigroup (Q, f) contains an idempotent element, then $\theta(h)$ is a normal subquasigroup of (Q, f).*

Corollary 7.99. *If $\theta(h)$ is a normal subquasigroup of a quasigroup (Q, f), then $f(\overline{a})\,\theta\,a$ for any $a \in \theta(h)$.*

Proposition 7.100. *In an idempotent n-ary medial quasigroup or in a unipotent n-ary medial quasigroup (Q, f), any normal congruence θ contains at least one equivalence class $\theta(e)$ such that $(\theta(e), f)$ is a normal subquasigroup of the quasigroup $(Q, \cdot)$.*

Proof. Idempotent quasigroups and unipotent quasigroups contain idempotent elements. □

To concretize the structure of n-ary medial quasigroups we give the following:

Definition 7.101. We shall say that an n-ary quasigroup (Q, f) is *solvable* if there exists the following finite chain of n-ary quasigroups

$$Q/Q_1, Q_1/Q_2, \dots, Q_m/Q_{m+1},$$

where the quasigroup Q_{i+1} is a maximal normal subquasigroup of the quasigroup Q_i and m is the minimal number such that $|Q_m/Q_{m+1}| = 1$.

It is clear that Definition 7.101 differs from the definition of the solvability of groups [471].

Propositions 7.102 and 7.103 give additional information on the structure of n-ary medial quasigroups.

Proposition 7.102. *Any finite n-ary medial idempotent quasigroup (Q, f) is solvable and any quasigroup Q_i/Q_{i+1} is a finite simple n-ary medial idempotent quasigroup.*

Proof. The proof follows from Proposition 7.100 and the fact that the quasigroup (Q, f) is finite. □

Proposition 7.103. *Any finite n-ary medial unipotent quasigroup (Q, f) is solvable and any quasigroup Q_i/Q_{i+1} is a finite simple n-ary medial unipotent quasigroup.*

Proof. The proof is similar to the proof of Proposition 7.102. □

Information on simple n-ary medial idempotent quasigroups and simple n-ary medial unipotent quasigroups is given in Theorem 7.90. We only note that any n-ary simple unipotent quasigroup is isomorphic to an n-ary quasigroup with the form $g(x_1^n) = \alpha_1 x_1 + \alpha_2 x_2 + \dots + \alpha_n x_n$ such that $\alpha_1 + \alpha_2 + \dots + \alpha_n = 0$. This follows from Lemma 7.54 and Theorem 7.90.

Chapter 8

Automorphisms of some quasigroups

8.1	On autotopies of n-ary linear quasigroups	357
8.1.1	Autotopies of derivative groups	358
8.1.2	Automorphisms of n-T-quasigroups	363
8.1.3	Automorphisms of some quasigroup isotopes	368
8.1.4	Automorphisms of medial n-quasigroups	370
8.1.5	Examples	372
8.2	Automorphism groups of some binary quasigroups	374
8.2.1	Isomorphisms of IP-loop isotopes	374
8.2.2	Automorphisms of loop isotopes	377
8.2.3	Automorphisms of LD-quasigroups	379
8.2.4	Automorphisms of isotopes of LD-quasigroups	381
8.2.5	Quasigroups with transitive automorphism group	383
8.3	Non-isomorphic isotopic quasigroups	383

We describe the structure of automorphism groups of n-T-quasigroups and of some binary quasigroups including left distributive quasigroups.

8.1 On autotopies of n-ary linear quasigroups

In this section we study the structure of autotopies, automorphisms, autotopy groups and automorphism groups of n-ary linear quasigroups.

We find a connection between automorphism groups of some special kinds of n-ary quasigroups (idempotent quasigroups, loops) and some isotopes of these quasigroups. In the binary case we find the more detailed connections between an automorphism group of a loop and an automorphism group of its isotope. We prove that every finite medial n-ary quasigroup of order greater than 2 has a non-identity automorphism group.

We apply the obtained results to give some information on automorphism groups of n-ary quasigroups that correspond to the ISSN code, the EAN code and the UPC code. Some results of this section are announced in [780] and published in [600].

We shall use basic terms and concepts from books [72, 80, 77, 685, 345]. To give some n-ary definitions we take into consideration articles [871, 832, 524].

Automorphisms and automorphism groups of some binary and n-ary quasigroups were studied in many articles; see, for example, [646, 461, 462, 736, 760, 769, 290, 772, 832, 444, 524, 525, 600].

Definition 8.1. An $(n+1)$-tuple $(\mu_1, \mu_2, \dots, \mu_n, \mu)$ of permutations of a set Q is called an *autotopy of an n-quasigroup* (Q, f), if and only if

$$f(x_1, x_2, \dots, x_n) = \mu^{-1} f(\mu_1 x_1, \dots, \mu_n x_n) \tag{8.1}$$

for all $x_1, \dots, x_n \in Q$.

We can write this fact also in the form $(Q, f) = (Q, g)T$ where $T = (\mu_1, \mu_2, \dots, \mu_n, \mu)$.

A set of all autotopies of a quasigroup (Q, f) forms the group of autotopies relative to the usually defined operation on this set: if $T_1 = (\mu_1, \mu_2, \dots, \mu_n, \mu)$ and $T_2 = (\nu_1, \nu_2, \dots, \nu_n, \nu)$ are autotopies of quasigroup (Q, f), then $T_1 \circ T_2 = (\mu_1\nu_1, \mu_2\nu_2, \dots, \mu_n\nu_n, \mu\nu)$ is an autotopy of quasigroup (Q, f). The autotopy group of a quasigroup (Q, f) will be denoted as $\mathfrak{T}(Q, f)$.

If in (8.1) the n-ary operations f and g are equal and $\mu_1 = \mu_2 = \dots = \mu_n = \mu$, then we obtain an *automorphism of quasigroup* (Q, f), i.e., a permutation μ of the set Q is called an automorphism of an n-quasigroup (Q, f) if for all $x_1, \dots, x_n \in Q$ the following relation is fulfilled: $\mu f(x_1, \dots, x_n) = f(\mu x_1, \dots, \mu x_n)$. We denote by $Aut(Q, f)$ the automorphism group of an n-ary quasigroup (Q, f).

We shall use the following theorem [77].

Theorem 8.2. *If n-ary quasigroups (Q, f) and (Q, g) are isotopic with isotopy T, i.e., $(Q, f) = (Q, g)T$, then $\mathfrak{T}(Q, f) = T^{-1}\mathfrak{T}(Q, g)T$.*

It is clear that any automorphism is an autotopy with equal components. So, if we know the structure of autotopies of a "good" n-ary quasigroup (Q, f) and the form of isotopy T, then we can obtain information on autotopies and automorphisms of n-ary quasigroup $(Q, g) = (Q, f)T$.

Probably, for the first time this observation was used in the study of automorphism groups of quasigroups isotopic to groups in [762].

8.1.1 Autotopies of derivative groups

In this subsection we study structure of autotopies, automorphisms, autotopy groups and automorphism groups of some classes of n-ary derivative groups and some of their isotopes.

We recall that an n-ary quasigroup (Q, f) with the form $f(x_1^n) = x_1 + x_2 + \dots + x_n$, where $(Q, +)$ is a binary group and is called an *n-ary derivative group of a binary group* $(Q, +)$ ([264, 77, 720]).

We denote by $Z(Q, +)$ the center of a group $(Q, +)$, i.e., $Z(Q, +) = \{a \in Q \mid a + x = x + a \quad \forall x \in Q\}$.

Lemma 8.3. *Let (Q, g) be an n-ary quasigroup of the form $g(x_1^n) = x_1 + x_2 + \dots + x_n$, where $(Q, +)$ is a group. A permutation θ of the set Q is an automorphism of the quasigroup (Q, g) if and only if $\theta = L_k\varphi$, where $\varphi \in Aut(Q, +)$, $k \in Z(Q, +)$, $(n-1)k = 0$.*

Proof. Let $\theta \in Aut(Q, g)$. Then

$$\theta(x_1 + x_2 + \dots + x_n) = \theta x_1 + \theta x_2 + \dots + \theta x_n \tag{8.2}$$

for all $x_i \in Q, i \in \{1, 2, \dots, n\}$. If we take $x_1 = x_2 = \dots = x_n = 0$ in (8.2), then we have $n\theta 0 = \theta 0$, $(n-1)\theta 0 = 0$.

If we take $x_2 = x_3 = \dots = x_n = 0$ in (8.2), then $\theta x_1 = \theta x_1 + \theta 0 + \dots + \theta 0 = R_{(n-1)\theta 0}\theta x_1$.

If we substitute $x_1 = x_3 = \dots = x_n = 0$ in (8.2), then $\theta x_2 = \theta 0 + \theta x_2 + \theta 0 + \dots + \theta 0 = L_{\theta 0}R_{(n-2)\theta 0}\theta x_2$.

If in the last two equalities we rename x_1 and x_2 by x and compare the right sides of these equalities, then we obtain $\theta x + \theta 0 + \dots + \theta 0 = \theta 0 + \theta x + \theta 0 + \dots + \theta 0$, $\theta x + \theta 0 = \theta 0 + \theta x$. Thus $\theta 0 \in Z(Q, +)$.

If we take in (8.2) $x_3 = \dots = x_n = 0$, then $\theta(x_1 + x_2) = \theta x_1 + R_{(n-2)\theta 0}\theta x_2$.

Therefore the permutation θ is a quasiautomorphism of the group $(Q, +)$. Moreover,

$\theta = L_{\theta 0}\varphi$, where $\varphi \in Aut(Q,+)$. Indeed, any group quasiautomorphism has the form $L_a\varphi$. But $a = \theta 0$ because $L_a\varphi 0 = L_a 0 = a = \theta 0$.

Therefore we obtain that $\theta = L_{\theta 0}\varphi$, where $(n-1)\theta 0 = 0$, $\theta 0 \in Z(Q,+)$ and $\varphi \in Aut(Q,+)$.

Converse. Let $\theta = L_k\varphi$, where $\varphi \in Aut(Q,+)$, $k \in Z(Q,+)$, $(n-1)k = 0$. Let us prove that $\theta \in Aut(Q,g)$. We have $\theta(x_1 + x_2 + \cdots + x_n) = L_k\varphi(x_1 + x_2 + \cdots + x_n) = k + \varphi x_1 + \varphi x_2 + \cdots + \varphi x_n = nk + \varphi x_1 + \varphi x_2 + \cdots + \varphi x_n = k + \varphi x_1 + k + \varphi x_2 + \ldots + k + \varphi x_n = \theta x_1 + \theta x_2 + \cdots + \theta x_n$. □

Lemma 8.4. *Let (Q,g) be an n-ary quasigroup of the form $g(x_1^n) = x_1 + x_2 + \cdots + x_n$, where $(Q,+)$ is a group. An $(n+1)$-tuple $T = (\alpha_1, \ldots, \alpha_n, \gamma)$ of permutations of the set Q is an autotopy of the quasigroup (Q,g) if and only if*

$$T = (L_{a_1}I_{a_1}, L_{a_2}I_{a_1+a_2}, L_{a_3}I_{a_1+a_2+a_3}, \ldots, L_{a_n}I_t, R_t)\circ (\varphi, \varphi, \varphi, \ldots, \varphi), \qquad (\star)$$

where $t = a_1 + a_2 + a_3 + \cdots + a_n$, $\varphi \in Aut(Q,+)$, $a_1^n \in Q$.

Proof. Let $(\alpha_1, \ldots, \alpha_n, \gamma)$ be an autotopy of a quasigroup (Q,g). Then

$$\gamma(x_1 + \cdots + x_n) = \alpha_1 x_1 + \cdots + \alpha_n x_n. \qquad (8.3)$$

If in (8.3) all $x_i = 0$, then $\gamma 0 = \sum_{i=1}^n \alpha_i 0$. If in (8.3) only $x_i \neq 0$ for any fixed value i, then $\gamma x_i = L_{\alpha_1 0+\cdots+\alpha_{i-1}0}R_{\alpha_{i+1}0+\cdots+\alpha_n 0}\alpha_i x_i$. Further, if we take into consideration Lemma 1.122 (ii), we have

$$\alpha_i = R_{-(\alpha_{i+1}0+\cdots+\alpha_n 0)}L_{-(\alpha_1 0+\cdots+\alpha_{i-1}0)}\gamma = L_{-(\alpha_1 0+\cdots+\alpha_{i-1}0)}R_{-(\alpha_{i+1}0+\cdots+\alpha_n 0)}\gamma.$$

If in the equality (8.3) we re-write all permutations α_i in their last form, then we have

$$\gamma(x_1 + \cdots + x_n) = \gamma x_1 - \alpha_n 0 - \cdots - \alpha_1 0 + \gamma x_2 - \alpha_n 0 - \cdots - \alpha_1 0 + \gamma x_3 - \cdots + \gamma x_n. \qquad (8.4)$$

Let $d = (\alpha_1 0 + \cdots + \alpha_n 0)$. Then from the equality (8.4) we have $\gamma(x_1 + \cdots + x_n) - d = \gamma x_1 - d + \gamma x_2 - d + \cdots + \gamma x_n - d$. Therefore

$$R_{-(\alpha_1 0+\cdots+\alpha_n 0)}\gamma = R_{-d}\gamma \in Aut(Q,g).$$

By Lemma 8.3 any automorphism of the quasigroup (Q,g) has the form $L_k\varphi$, where $k \in Z(Q,+)$, $\varphi \in Aut(Q,+)$. Then $R_{-d}\gamma = L_k\varphi = R_k\varphi$, $\gamma = R_d R_k\varphi = R_{k+d}\varphi$.

We remark that, if $I_b x = -b + x + b$ is the inner automorphism of the group $(Q,+)$, then $I_b = L_{-b}R_b$ or $R_b = L_b I_b$. Further we have

$$\begin{aligned}\alpha_i &= L_{-(\alpha_1 0+\cdots+\alpha_{i-1}0)}R_{-(\alpha_{i+1}0+\cdots+\alpha_n 0)}\gamma = \\ & L_{-(\alpha_1 0+\cdots+\alpha_{i-1}0)}R_{-\alpha_n 0-\cdots-\alpha_{i+1}0}R_{\alpha_1 0+\cdots+\alpha_n 0}R_k\varphi = \\ & L_{-(\alpha_1 0+\cdots+\alpha_{i-1}0)}R_{\alpha_1 0+\cdots+\alpha_i 0}R_k\varphi = \\ & L_{\alpha_i 0}L_{-\alpha_i 0}L_{-(\alpha_1 0+\cdots+\alpha_{i-1}0)}R_{\alpha_1 0+\cdots+\alpha_i 0}R_k\varphi = \\ & L_{\alpha_i 0}L_{-(\alpha_1 0+\cdots+\alpha_i 0)}R_{\alpha_1 0+\cdots+\alpha_i 0}R_k\varphi = \\ & L_{\alpha_i 0}I_{\alpha_1 0+\cdots+\alpha_i 0}R_k\varphi = L_{\alpha_i 0}R_k I_{\alpha_1 0+\cdots+\alpha_i 0}\varphi.\end{aligned}$$

We have used that if $k \in Z(Q,+)$, then $R_k I_b = I_b R_k$. Indeed, $-b + x + b + k = -b + x + k + b = I_b R_k x$. Further we obtain $\alpha_i = L_{\alpha_i 0+k}I_{\alpha_1 0+\cdots+\alpha_i 0}\varphi$, since for every element k from the center of the group $(Q,+)$, $R_k = L_k$.

We denote $\alpha_i 0 + k$ as a_i and $d + k$ as t. Let us prove that $\sum_{i=1}^n a_i = t$. Indeed, we have

$a_1+a_2+\cdots+a_n = \alpha_1 0+k+\alpha_2 0+k+\cdots+\alpha_n 0+k = \alpha_1 0+\alpha_2 0+\cdots+\alpha_n 0+k = d+k = t$. Therefore we obtain that $(n+1)$-tuple

$$T = (L_{a_1}I_{a_1}\varphi, L_{a_2}I_{a_1+a_2}\varphi, L_{a_3}I_{a_1+a_2+a_3}\varphi, \dots, L_{a_n}I_t\varphi, R_t\varphi)$$

is an autotopy of the quasigroup (Q, g).

Converse. We shall prove that any $(n+1)$-tuple of such form is an autotopy of the quasigroup (Q, g). Let $n = 3$. We have

$$a_1 - a_1 + \varphi x_1 + a_1 + a_2 - a_2 - a_1 + \varphi x_2 + a_1 + a_2 + a_3 - a_3 - a_2 - a_1 + \\ \varphi x_3 + a_1 + a_2 + a_3 = R_t\varphi(x_1 + x_2 + x_3).$$

Further after cancellation in the left-hand side of the last relation we obtain $\varphi x_1 + \varphi x_3 + \varphi x_3 + a_1 + a_2 + a_3 = R_t\varphi(x_1 + x_2 + x_3)$. For other values of arity n the proof is similar. □

Corollary 8.5. *Any autotopy of a quasigroup (Q, g) with the form $g(x_1^n) = \Sigma_{i=1}^n(x_i)$ over a group $(Q, +)$ has a unique representation in the form $(\star)$.*

Proof. If we suppose that

$$(L_{a_1}I_{a_1}\varphi, L_{a_2}I_{a_1+a_2}\varphi, L_{a_3}I_{a_1+a_2+a_3}\varphi, \dots, L_{a_n}I_t\varphi, R_t\varphi) = \\ (L_{b_1}I_{b_1}\psi, L_{b_2}I_{b_1+b_2}\psi, L_{b_3}I_{b_1+b_2+b_3}\psi, \dots, L_{b_n}I_d\psi, R_d\psi),$$

then we have $L_{a_1}I_{a_1}\varphi = L_{b_1}I_{b_1}\psi$, $L_{a_1}I_{a_1}\varphi 0 = L_{b_1}I_{b_1}\psi 0$, $a_1 = b_1$, $\varphi = \psi$. Further we obtain $L_{a_2}I_{a_1+a_2} = L_{b_2}I_{b_1+b_2}$, $L_{a_2}I_{a_1+a_2}0 = L_{b_2}I_{b_1+b_2}0$, $a_2 = b_2$ and so on. Thus $a_i = b_i$ for all $i \in \overline{1,n}$, $t = d$, $\varphi = \psi$. □

Remark 8.6. The change of distribution of brackets in the equality (8.4) permits us to obtain from one autotopy of the quasigroup (Q, g) other autotopies of this quasigroup.

Proposition 8.7. *Let (Q, g) be a finite n-ary quasigroup of order $|Q|$ with the form $g(x_1^n) = x_1 + x_2 + \cdots + x_n$, where $(Q, +)$ is a group. Then*

$$|\mathfrak{T}(Q, g)| = |Q|^n \cdot |Aut(Q, +)|.$$

Proof. From Lemma 8.4 it follows that any autotopy T of the quasigroup (Q, g) has the form $T = T_1 \circ T_2$, where

$$T_1 = (L_{a_1}I_{a_1}, L_{a_2}I_{a_1+a_2}, L_{a_3}I_{a_1+a_2+a_3}, \dots, L_{a_n}I_t, R_t), \\ T_2 = (\varphi, \varphi, \varphi, \dots, \varphi), \\ t = a_1 + a_2 + a_3 + \cdots + a_n, \varphi \in Aut(Q, +), a_1^n \in Q.$$

Let

$$\mathfrak{T}_1 = \{(L_{a_1}I_{a_1}, L_{a_2}I_{a_1+a_2}, L_{a_3}I_{a_1+a_2+a_3}, \dots, L_{a_n}I_t, R_t) \mid \\ \forall\, a_1^n \in Q\}, \mathfrak{T}_2 = \{(\varphi, \varphi, \varphi, \dots, \varphi) \mid \varphi \in Aut(Q, +)\}.$$

Taking into consideration Corollary 8.5 it is easy to see that $|\mathfrak{T}_1| = |Q|^n$. It is clear that $|\mathfrak{T}_2| = |Aut(Q, +)|$.

We prove that $\mathfrak{T}_1 \cap \mathfrak{T}_2 = (\varepsilon, \varepsilon, \dots, \varepsilon)$. Indeed, if $L_{a_1}I_{a_1} = \varphi$, then $L_{a_1}I_{a_1}0 = \varphi 0$, $a_1 = 0$, $\varphi = \varepsilon$.

From Lemma 8.4 it follows that any n-tuple $T_1 \in \mathfrak{T}_1$ and any n-tuple $T_2 \in \mathfrak{T}_2$ is an autotopy of the quasigroup (Q, g). Therefore $|\mathfrak{T}(Q, g)| = |Q|^n \cdot |Aut(Q, +)|$. □

Corollary 8.8. *Let (Q, g) be a finite n-ary quasigroup of order $|Q|$ with the form $g(x_1^n) = \alpha_1 x_1 + \alpha_2 x_2 + \cdots + \alpha_n x_n$, where $(Q, +)$ is a group, α_1^n are permutations of the set Q. Then*

$$|\mathfrak{T}(Q, g)| = |Q|^n \cdot |Aut(Q, +)|.$$

Proof. This follows from Proposition 8.7 and Theorem 8.2 since finite isomorphic groups have equal orders. □

Corollary 8.9. *Let (Q, g) be an n-ary quasigroup of the form $g(x_1^n) = x_1 + x_2 + \cdots + x_n$, where $(Q,+)$ is an abelian group. An $(n+1)$-tuple $T = (\alpha_1, \dots, \alpha_n, \gamma)$ of permutations of the set Q is an autotopy of the quasigroup (Q, g) if and only if $T = (L_{a_1}\varphi, L_{a_2}\varphi, L_{a_3}\varphi, \dots, L_{a_n}\varphi, L_t\varphi)$, where $t = a_1 + a_2 + a_3 + \cdots + a_n$, $\varphi \in Aut(Q,+)$, $a_1^n \in Q$.*

Proof. In abelian group $I_a = \varepsilon$, $L_a = R_a$ for all $a \in Q$. □

Theorem 8.10. *If an n-ary quasigroup (Q, g) has the form $g(x_1^n) = \alpha_1 x_1 + \alpha_2 x_2 + \cdots + \alpha_n x_n$, where $(Q,+)$ is an abelian group, α_1^n are permutations of the set Q, then*

$$\mathfrak{T}(Q,g) \cong \bigoplus_{i=1}^{n} (Q,+)_i \leftthreetimes Aut(Q,+),$$

where $\bigoplus_{i=1}^{n}(Q,+)_i$ is the direct sum of n copies of the group $(Q,+)$.

Proof. By Theorem 8.2, autotopy groups of isotopic n-ary quasigroups are isomorphic (moreover, autotopy groups of isostrophic n-ary quasigroups are isomorphic [77]). Therefore it is sufficient to prove this theorem for an n-ary quasigroup (Q, f) with the form $f(x_1^n) = x_1 + x_2 + \cdots + x_n$, where $(Q,+)$ is an abelian group.

From Corollary 8.9 it follows that any autotopy T of the quasigroup (Q, g) has the form $T = T_1 \circ T_2$, where

$$\begin{aligned} T_1 &= (L_{a_1}, L_{a_2}, L_{a_3}, \dots, L_{a_n}, L_t), \\ T_2 &= (\varphi, \varphi, \varphi, \dots, \varphi), \\ t &= a_1 + a_2 + a_3 + \cdots + a_n, \varphi \in Aut(Q,+), a_1^n \in Q. \end{aligned}$$

Let

$$\begin{aligned} \mathfrak{T}_1 &= \{(L_{a_1}, L_{a_2}, L_{a_3}, \dots, L_{a_n}, L_t) \,|\, t = a_1 + \dots a_n, \text{ for all } a_1^n \in Q\}, \\ \mathfrak{T}_2 &= \{(\varphi, \varphi, \varphi, \dots, \varphi) \,|\, \text{for all } \varphi \in \text{Aut}(Q,+)\}. \end{aligned}$$

From Corollary 8.9 it follows that $\mathfrak{T}(Q, g) = \mathfrak{T}_1 \circ \mathfrak{T}_2$.

We shall prove that $\mathfrak{T}_1 \cap \mathfrak{T}_2 = (\varepsilon, \varepsilon, \dots, \varepsilon)$. Indeed, if $L_{a_1} = \varphi$, then $L_{a_1}0 = \varphi 0$, $a_1 = 0$, $\varphi = \varepsilon$.

From Lemma 8.4 it follows that any n-tuple $T_1 \in \mathfrak{T}_1$ and any n-tuple $T_2 \in \mathfrak{T}_2$ is an autotopy of the quasigroup (Q, g).

The set $\mathfrak{T}_1$ forms a group with respect to the term-by-term multiplication of $(n+1)$-tuples of the set $\mathfrak{T}_1$. Indeed, let

$$\begin{aligned} T_1 &= (L_{a_1}, L_{a_2}, \dots, L_{a_n}, L_d), \\ T_2 &= (L_{b_1}, L_{b_2}, \dots, L_{b_n}, L_t), \end{aligned}$$

$T_1, T_2 \in \mathfrak{T}_1$. We prove that $T_1 \circ T_2 \in \mathfrak{T}_1$.

We have $L_{a_1}L_{b_1} = L_{a_1+b_1}, \dots, L_{a_n}L_{b_n} = L_{a_n+b_n}, L_d L_t = L_{d+t}, a_1 + b_1 + \dots + a_n + b_n = d + t$. Then $T_1 \circ T_2 \in \mathfrak{T}_1$. From Lemma 1.122 (ii) it follows that $T_1^{-1} = (L_{-a_1}, L_{-a_2}, \dots, L_{-a_n}, L_{-d})$. Thus $T_1^{-1} \in \mathfrak{T}_1$.

Therefore $(\mathfrak{T}_1, \circ)$ is a group and it is easy to see that $(\mathfrak{T}_1, \circ) \cong \bigoplus_{i=1}^{n}(Q,+)_i$.

It is clear that $(\mathfrak{T}_2, \circ) \cong Aut(Q,+)$, $(\mathfrak{T}_1, \circ) \subseteq \mathfrak{T}(Q, g)$, $(\mathfrak{T}_2, \circ) \subseteq \mathfrak{T}(Q, g)$.

We shall prove that $(\mathfrak{T}_1, \circ) \trianglelefteq \mathfrak{T}(Q, g)$. Let $T = (L_{a_1}\varphi, L_{a_2}\varphi, L_{a_3}\varphi, \dots, L_{a_n}\varphi, L_t\varphi) \in \mathfrak{T}$, $T_1 = (L_{b_1}, L_{b_2}, L_{b_3}, \dots, L_{b_n}, L_d) \in \mathfrak{T}_1$.

From Lemma 1.122 it follows that

$$\begin{aligned} T^{-1} = \;& (L_{-\varphi^{-1}a_1}\varphi^{-1}, L_{-\varphi^{-1}a_2}\varphi^{-1}, L_{-\varphi^{-1}a_3}\varphi^{-1}, \dots, \\ & L_{-\varphi^{-1}a_n}\varphi^{-1}, L_{-\varphi^{-1}t}\varphi^{-1}). \end{aligned}$$

Further we have

$$\begin{aligned} T^{-1} \circ T_1 \circ T = \;& (L_{-\varphi^{-1}a_1}\varphi^{-1}L_{b_1}L_{a_1}\varphi, L_{-\varphi^{-1}a_2}\varphi^{-1}L_{b_2}L_{a_2}\varphi, \dots, \\ & L_{-\varphi^{-1}a_n}\varphi^{-1}L_{b_n}L_{a_n}\varphi, L_{-\varphi^{-1}t}\varphi^{-1}L_dL_t\varphi) = \\ & (L_{-\varphi^{-1}a_1+\varphi^{-1}b_1+\varphi^{-1}a_1}, L_{-\varphi^{-1}a_2+\varphi^{-1}b_2+\varphi^{-1}a_2}, \dots, \\ & L_{-\varphi^{-1}a_n+\varphi^{-1}b_n+\varphi^{-1}a_n}, L_{-\varphi^{-1}t+\varphi^{-1}d+\varphi^{-1}t}) \in \mathfrak{T}_1. \end{aligned}$$

By proving the last equality we have used Lemma 1.122.

Therefore $\mathfrak{T}(Q, g) \cong \bigoplus_{i=1}^{n}(Q, +)_i \rightthreetimes Aut(Q, +)$. □

Let $Z^{(n-1)}(Q, +) = \{a \in Z(Q, +) \mid (n-1)a = 0\}$, where $(Q, +)$ is a group and $Z(Q, +)$ is the center of this group. It is easy to see that $Z^{(n-1)}(Q, +)$ is a subgroup of the group $(Q, +)$. Let $\overline{Z}(Q, +)$ be a group that consists of all left translations of the group $(Q, +)$ such that $x \in Z^{(n-1)}(Q, +)$, i.e., $\overline{Z}(Q, +) \subseteq M(Q, +)$.

Proposition 8.11. *In a quasigroup (Q, g) with the form $g(x_1^n) = x_1 + x_2 + \dots + x_n$, where $(Q, +)$ is a binary group,*

$$Aut(Q, g) \cong \overline{Z}(Q, +) \rightthreetimes Aut(Q, +).$$

Proof. Really, $\overline{Z}(Q, +) \bigcap Aut(Q, +) = \varepsilon$, $\overline{Z}(Q, +) \lhd Aut(Q, g)$, and, further, we have

$$\begin{aligned} &\overline{Z}(Q, +) \cdot Aut(Q, +) = Aut(Q, g), \\ &Aut(Q, g)/\overline{Z}(Q, +) \cong Aut(Q, +), \\ &Aut(Q, g) \cong \overline{Z}(Q, +) \rightthreetimes Aut(Q, +). \end{aligned}$$

□

Corollary 8.12. *In a quasigroup (Q, g) with the form $g(x_1^n) = \alpha_1 x_1 + \alpha_2 x_2 + \dots + \alpha_n x_n$, where $(Q, +)$ is a binary group, α_1^n are permutations of the set Q,*

$$Aut(Q, g) \cong H \subseteq (Q, +) \rightthreetimes Aut(Q, +).$$

Proof. From the definition of isotopy it follows that $(Q, g) = (Q, f)(\alpha_1, \alpha_2, \dots, \alpha_n, \varepsilon)$, where the n-ary quasigroup (Q, f) has the form $f(x_1^n) = x_1 + x_2 + \dots + x_n$ over the group $(Q, +)$.

Then from Theorem 8.2 it follows that any quasiautomorphism of the n-ary quasigroup (Q, g) has the same form as the form of any quasiautomorphism of the n-ary quasigroup (Q, f).

Thus from Lemma 8.4 and Lemma 1.122 (iii) it follows that any quasiautomorphism of the n-ary quasigroup (Q, g) has the form $L_a^+\varphi$, where $\varphi \in Aut(Q, +)$. Therefore any automorphism of the quasigroup (Q, g) has the same form.

From Lemma 8.4 it follows that any permutation of the form $L_a^+\varphi$ is a quasiautomorphism of the n-ary quasigroups (Q, g) and (Q, f).

It is easy to check that a set of all quasiautomorphisms of the quasigroup (Q, g) forms a group with respect to the usual operation multiplication of permutations of the set Q. It is well known ([471]) that this group is isomorphic to the group $(Q, +) \rightthreetimes Aut(Q, +)$.

Therefore $Aut(Q, g) \cong H \subseteq (Q, +) \rightthreetimes Aut(Q, +)$. □

Remark 8.13. The binary analog of Corollary 8.12 is proved in [769].

8.1.2 Automorphisms of n-T-quasigroups

In this section we study automorphisms and automorphism groups of some classes of n-ary T-quasigroups. We prove that every finite medial n-ary quasigroup of order greater than two has a non-identity automorphism group. We recall an example of an infinite medial quasigroup with an identity automorphism group [832].

The binary analog of Proposition 8.14 was proved in [769], and an n-ary analog of Proposition 8.14 was claimed in [824].

It is known that the centralizer of elements $\varphi_1, \dots, \varphi_n$ in group $Aut(Q,+)$ is the set $\{\omega \in Aut(Q,+) \mid \omega\varphi_i = \varphi_i\omega \, \forall\, i \in \overline{1,n}\}$ [471]. Denote this set as $C_{Aut(Q,+)}(\varphi_1, \dots, \varphi_n)$. Sometimes we shall denote this set only by letter C. The set C forms a group with respect to usual operation of multiplication of permutations of the set Q [471].

Proposition 8.14. *A permutation γ of a set Q is an automorphism of an n-T-quasigroup (Q,f) of the form $f(x_1, x_2, \dots, x_n) = \varphi_1 x_1 + \varphi_2 x_2 + \dots + \varphi_n x_n + a$ if and only if $\gamma = L_b^+\beta$, where $\beta a - a = \delta b$, $\beta \in C_{Aut(Q,+)}(\varphi_1, \dots, \varphi_n)$, $\delta = \sum_{i=1}^n \varphi_i - \varepsilon$, $b \in Q$.*

Proof. From Corollary 8.9 and Theorem 8.2 it follows that any autotopy T of the n-T-quasigroup (Q,f) has the following form:

$$\begin{aligned} T = \;& (L_{\varphi_1^{-1}(a_1)}\varphi_1^{-1}\psi\varphi_1, L_{\varphi_2^{-1}(a_2)}\varphi_2^{-1}\psi\varphi_2, \dots, \\ & L_{\varphi_n^{-1}(a_n)}\varphi_n^{-1}\psi\varphi_n, L_t\psi), \end{aligned}$$

where ψ is an automorphism of abelian group $(Q,+)$, and all translations in the last equality are translations of the abelian group $(Q,+)$. It is so since $\varphi^{-1}L_x\psi\varphi = L_{\varphi^{-1}x}\varphi^{-1}\psi\varphi$ for all $x \in Q$.

Since any automorphism of n-T-quasigroup (Q,f) is an autotopy with equal components, then we can write any automorphism of quasigroup (Q,f) in the form $L_b\beta$, where $\beta \in Aut(Q,+)$ and L_b is a left translation of the group $(Q,+)$.

This presentation of this quasigroup automorphism is unique. Indeed, if $L_d\rho = L_b\beta$, then $L_d\rho 0 = L_b\beta 0$, $d = b$, $\rho = \beta$.

Let $L_b\beta \in Aut(Q,f)$, i.e., $\varphi_1 L_b\beta x_1 + \varphi_2 L_b\beta x_2 + \dots + \varphi_n L_b\beta x_n + a = L_b\beta(\varphi_1 x_1 + \varphi_2 x_2 + \dots + \varphi_n x_n)$. Since $\varphi_i \in Aut(Q,+)$ for any $i \in \overline{1,n}$, we have

$$\begin{aligned} & \varphi_1 b + \varphi_1\beta x_1 + \varphi_2 b + \varphi_2\beta x_2 + \dots + \varphi_n b + \varphi_n\beta x_n + a = \\ & b + \beta\varphi_1 x_1 + \beta\varphi_2 x_2 + \dots + \beta\varphi_n x_n + \beta a. \end{aligned} \tag{8.5}$$

If we take $x_i = 0$ in (8.5) for all i, then we have

$$\varphi_1 b + \varphi_2 b + \dots + \varphi_n b + a = b + \beta a. \tag{8.6}$$

If we denote $\sum_{i=1}^n \varphi_i - \varepsilon$ by δ, then we have $\beta a - a = \delta b$.

If we take into consideration equality (8.6), then from (8.5) it follows

$$\varphi_1\beta x_1 + \varphi_2\beta x_2 + \dots + \varphi_n\beta x_n = \beta\varphi_1 x_1 + \beta\varphi_2 x_2 + \dots + \beta\varphi_n x_n. \tag{8.7}$$

If we take $x_2 = x_3 = \dots = x_n = 0$ in (8.7), then we have $\varphi_1\beta = \beta\varphi_1$. Similarly we obtain that $\varphi_2\beta = \beta\varphi_2, \dots, \varphi_n\beta = \beta\varphi_n$, i.e., $\beta \in C_{Aut(Q,+)}(\varphi_1, \dots, \varphi_n) = C$.

It is easy to check that the converse is correct too. Indeed,

$$\begin{aligned} & \gamma(f(x_1^n)) = \gamma(\textstyle\sum_{i=1}^n(\varphi_i x_i) + a) = L_b\beta(\sum_{i=1}^n(\varphi_i x_i) + a) = \\ & b + \textstyle\sum_{i=1}^n \beta\varphi_i x_i + \beta a = \\ & \textstyle\sum_{i=1}^n(\varphi_i b) + \sum_{i=1}^n(\varphi_i\beta x_i) + a = \\ & \textstyle\sum_{i=1}^n(\varphi_i(L_b\beta x_i)) + a = f(\gamma x_1, \dots, \gamma x_n). \end{aligned}$$

□

Corollary 8.15. *If $L_b\beta \in Aut(Q, f)$, where (Q, f) is an n-T-quasigroup with the form $f(x_1^n) = \varphi_1 x_1 + \varphi_2 x_2 + \dots + \varphi_n x_n$, then $L_b \in Aut(Q, f)$ and $\beta \in Aut(Q, f)$.*

Proof. If $L_b\beta \in Aut(Q, f)$, then the equality (8.6) is fulfilled. The equality (8.6) in the case $a = 0$ has the form $b = \sum_{i=1}^{n}(\varphi_i b)$. Further we have

$$\begin{array}{l} L_b f(x_1^n) = L_b(\sum_{i=1}^{n}(\varphi_i x_i)) = \\ b + \sum_{i=1}^{n}\varphi_i x_i = \sum_{i=1}^{n}(\varphi_i b) + \sum_{i=1}^{n}\varphi_i x_i = \\ \sum_{i=1}^{n}\varphi_i(b + x_i) = f(L_b x_1, \dots, L_b x_n). \end{array}$$

Thus $L_b \in Aut(Q, f)$. If $L_b\beta \in Aut(Q, f)$, $L_b \in Aut(Q, f)$, then $\beta \in Aut(Q, f)$. □

Proposition 8.16. *If an n-T-quasigroup (Q, g) has the form $g(x_1^n) = \varphi_1 x_1 + \varphi_2 x_2 + \dots + \varphi_n x_n$, then*

$$Aut(Q, g) \cong K \rtimes C,$$

where $K = \{L_b^+ \mid b \in Q, \varphi_1 b + \varphi_2 b + \dots + \varphi_n b = b\}$, $C = \{\omega \in Aut(Q, +) \mid \omega\varphi_i = \varphi_i\omega \ \forall\, i \in \overline{1, n}\}$.

Proof. From Proposition 8.14 and Corollary 8.15 it follows that sets K and C are subgroups of group $Aut(Q, g)$ and that $Aut(Q, g) = K \cdot C$. Since $K \cap C = \varepsilon$, $K \trianglelefteq Aut(Q, g)$ we have $Aut(Q, g) \cong K \rtimes C$. □

An element d of an n-ary quasigroup (Q, f) such that $f(d, \dots, d) = d$ is called an *idempotent element* of quasigroup (Q, f) (in brackets we have taken the element d exactly n times). In the binary case an element $d \in Q$ such that $d \cdot d = d$ is called an idempotent element of quasigroup $(Q, \cdot)$.

Corollary 8.17. *If n-T-quasigroup (Q, g) with the form $g(x_1, \dots, x_n) = \sum_{i=1}^{n}\varphi_i x_i$ has exactly one idempotent element, then*

$$Aut(Q, g) \cong C,$$

where $C = \{\omega \in Aut(Q, +) \mid \omega\varphi_i = \varphi_i\omega \ \forall\, i \in \overline{1, n}\}$.

Proof. In this case $K = \{\varepsilon\}$. □

Corollary 8.18. *If n-T-quasigroup (Q, g) is an idempotent quasigroup with the form $g(x_1, \dots, x_n) = \sum_{i=1}^{n}\varphi_i x_i$ over an abelian group $(Q, +)$, then*

$$Aut(Q, g) \cong (Q, +) \rtimes C,$$

$C = \{\omega \in Aut(Q, +) \mid \omega\varphi_i = \varphi_i\omega \ \forall\, i \in \overline{1, n}\}$.

Proof. We have $K \cong (Q, +)$ in the case when (Q, g) is an idempotent quasigroup. □

Lemma 8.19. *An n-T-quasigroup (Q, f) $f(x_1, \dots, x_n) = \sum_{i=1}^{n}\varphi_i x_i + a$ has at least one idempotent element if and only if it is isomorphic to an n-T-quasigroup (Q, g) with the form $g(x_1^n) = \sum_{i=1}^{n}\varphi_i x_i$ over the same abelian group $(Q, +)$.*

Proof. Let element u be an idempotent element of n-ary T-quasigroup (Q, f) with the form $f(x_1, x_2, \dots, x_n) = \sum_{i=1}^{n}\varphi_i x_i + a$. If we take in the last equality $x_1 = x_2 = \dots = x_n = u$, then we have $\sum_{i=1}^{n}\varphi_i u + a = u$, i.e.,

$$\sum_{i=1}^{n}\varphi_i u + a - u = 0. \tag{8.8}$$

Then an isomorphic image of the n-T-quasigroup (Q, f) with an isomorphism T of the form $(L_u, \cdots, L_u)$ ((n+1) times), where L_u is a left translation of the group $(Q, +)$, will be an n-T-quasigroup (Q, g) with the form $g(x_1, x_2, \dots, x_n) = \sum_{i=1}^{n} \varphi_i x_i$. Really, we have

$$\begin{aligned}&L_{-u}f(L_u x_1, \cdots, L_u x_n) = \\ &-u + \varphi_1 u + \varphi_1 x_1 + \varphi_2 u + \varphi_2 x_2 + \cdots + \varphi_n u + \varphi x_n + a = \\ &-u + \varphi_1 u + \varphi_2 u + \cdots + \varphi_n u + a + \textstyle\sum_{i=1}^{n} \varphi_i x_i = \\ &\textstyle\sum_{i=1}^{n} \varphi_i u + a - u + \sum_{i=1}^{n} \varphi_i x_i = \\ &0 + \textstyle\sum_{i=1}^{n} \varphi_i x_i = g(x_1, \cdots, x_n).\end{aligned}$$

Any n-T-quasigroup (Q, g) with the form $\sum_{i=1}^{n} \varphi_i x_i$ has at least one idempotent element. Indeed, $\sum_{i=1}^{n} \varphi_i 0 = 0$. Then the n-T-quasigroup (Q, f), that is an isomorphic copy of n-T-quasigroup (Q, g), has an idempotent element. □

Condition (8.8) that the n-T-quasigroup (Q, f) with the form $f(x_1^n) = \sum_{i=1}^{n} \varphi_i x_i + a$ has an idempotent element u we can re-write in the form $\delta u = -a$, where δ is an endomorphism of abelian group $(Q, +)$ such that $\delta = \sum_{i=1}^{n} \varphi_i - \varepsilon$.

Hence we can formulate the following conditions when the n-T-quasigroup (Q, f) has an idempotent element.

Lemma 8.20. *An n-T-quasigroup (Q, f), $f(x_1, \dots, x_n) = \sum_{i=1}^{n} \varphi_i x_i + a$, has at least one idempotent element if and only if there exists an element $d \in Q$ such that $\delta d = -a$.*

Remark 8.21. An n-T-quasigroup (Q, f), $f(x_1, \dots, x_n) = \sum_{i=1}^{n} \varphi_i x_i + a$, has exactly one idempotent element if and only if the endomorphism δ is a permutation of the set Q, i.e., if and only if the endomorphism δ is an automorphism of the quasigroup (Q, f).

For the binary case, Lemmas 8.19 and 8.20 are easily received from the remark on page 109 in [769].

Lemma 8.22. *A T-quasigroup $(Q, \cdot)$, $x \cdot y = \varphi_1 x + \varphi_2 y + a$, has at least one idempotent element if and only if it is isomorphic to T-quasigroup $(Q, \circ)$ with the form $x \circ y = \varphi_1 x + \varphi_2 y$ over the same abelian group $(Q, +)$.*

Lemma 8.23. *A T-quasigroup $(Q, \cdot)$ with the form $x \cdot y = \varphi_1 x_1 + \varphi_2 y + a$ has at least one idempotent element if and only if there exists element $d \in Q$ such that $\delta d = -a$.*

Theorem 8.24. *If an n-T-quasigroup (Q, g) with the form $g(x_1, \dots, x_n) = \sum_{i=1}^{n} \varphi_i x_i + a$ has at least one idempotent element, then*

$$Aut(Q, g) \cong K \rtimes C,$$

where $K = \{L_b^+ \mid b \in Q, \varphi_1 b + \varphi_2 b + \cdots + \varphi_n b = b\}$, $C = \{\omega \in Aut(Q, +) \mid \omega\varphi_i = \varphi_i \omega \, \forall i \in \overline{1, n}\}$.

Proof. It is sufficient to take into consideration Lemma 8.19 and Proposition 8.16. □

Remark 8.25. In the binary case from idempotency of T-quasigroup $(Q, \cdot)$ it follows that this quasigroup is medial and distributive. Indeed, if $x \cdot x = \varphi x + \psi x = x$ for any $x \in Q$, then $\varphi + \psi = \varepsilon$, $\varphi = -\psi + \varepsilon$ and then $\varphi\psi = (-\psi + \varepsilon)\psi = -\psi^2 + \psi = \psi(-\psi + \varepsilon) = \psi\varphi$.

We notice that even in the ternary case there exist non-medial idempotent T-quasigroups.

Example 8.26. Let $(Q, +) = (Z_5 \oplus Z_5, +)$ be the direct product of two cyclic groups of order five. Let

$$\alpha_1 = \begin{pmatrix} 2 & 2 \\ 2 & 1 \end{pmatrix}, \quad \alpha_2 = \begin{pmatrix} 1 & 3 \\ 0 & 2 \end{pmatrix}, \quad \alpha_3 = \begin{pmatrix} 3 & 0 \\ 3 & 3 \end{pmatrix},$$

where, for example,

$$\begin{pmatrix} 2 & 2 \\ 2 & 1 \end{pmatrix} \begin{pmatrix} y_1 \\ y_2 \end{pmatrix} = \begin{pmatrix} 2 \cdot y_1 + 2 \cdot y_2 \\ 2 \cdot y_1 + 1 \cdot y_2 \end{pmatrix}$$

for all $(y_1, y_2) \in Z_5 \oplus Z_5$.

A quasigroup (Q, f) of the form $f(x_1, x_2, x_3) = \alpha_1 x_1 + \alpha_2 x_2 + \alpha_3 x_3$ for all $x_1, x_2, x_3 \in Q$ is an idempotent non-medial ($\alpha_1\alpha_2 \neq \alpha_2\alpha_1$) 3-ary T-quasigroup.

Remark 8.27. It is clear that with growing of a quasigroup operation arity the influence of idempotent elements on properties of a quasigroup will be weakened. Indeed, in binary idempotent quasigroup (Q, f_2) we have $|Q|^2$ quasigroup "words" of the form $(x, y, f_2(x, y))$ and $|Q|$ idempotent elements, in ternary quasigroup (Q, f_3) we have $|Q|^3$ quasigroup words and $|Q|$ idempotent elements, and so on.

Example 8.28. Medial quasigroup $(Z, \circ)$ over infinite cyclic group $(Z, +)$ of the form $x \circ y = -x - y + 1$ has identity automorphism group [832].

Proof. From Proposition 8.14 it follows that any automorphism of the quasigroup $(Z, \star)$, $x \star y = -x - y + a$ has the form $L_b^+\psi$, where $\psi \in Aut(Z, +) \cong (Z_2, \cdot)$ [471]. We suppose that $Z_2 = \{1, -1\}$.

Then we have $L_b^+\psi(x \circ y) = L_b^+\psi x \circ L_b^+\psi y$, $b - \psi x - \psi y + \psi a = -b - \psi x - b - \psi y + a$, $\psi a - a = -3b$. If $\psi = 1$, then $0 = -3b$, $b = 0$ and we obtain that $L_b^+\psi = \varepsilon$ is an automorphism of the quasigroup $(Z, \star)$. If $\psi = -1$, then we have $-2a = -3b$, $b = (2/3)a$.

Therefore if $a = 3k$, then $Aut(Z, \star) \cong Z_2$. If $a = 3k + 1$ or $a = 3k + 2$, then $Aut(Z, \star)$ consists only of the identity mapping. Thus $|Aut(Z, \circ)| = 1$. □

We remark that any other binary medial quasigroup over the group $(Z, +)$ can have either the form $x * y = x + y + a$, or $x \diamond y = x - y + a$, or $x \circledast y = -x + y + a$. Every one of these quasigroups has an idempotent element: $(-a) * (-a) = -a, a \diamond a = a, a \circledast a = a$.

Using Lemma 8.22 and Proposition 8.16 we can obtain that the automorphism group of any of these quasigroups is isomorphic to the group Z_2. See also [832].

Theorem 8.29 is similar to Proposition 8 [769] that was proved for finite binary T-quasigroups.

Let (Q, f) be an n-T-quasigroup of the form $f(x_1^n) = \sum_{i=1}^n (\varphi_i x_i) + a$ over an abelian group $(Q, +)$. Let $P = \{C(a) - a\} \cap \delta Q$, $C(a)$ be an orbit of the element a in the set Q under action of the group C and C be a centralizer of the elements $\varphi_1, \dots, \varphi_n$ in the group $Aut(Q, +)$, N be a kernel of the endomorphism δ, $\delta = \sum_{i=1}^n \varphi_i - \varepsilon$, S be a stabilizer of the element a under action of the group C on the set Q.

Theorem 8.29. *If (Q, f) is an n-T-quasigroup of the form $f(x_1^n) = \sum_{i=1}^n (\varphi_i x_i) + a$ over an abelian group $(Q, +)$, and the sets P, N, S have finite order, then*

$$|Aut(Q, f)| = |P| \cdot |N| \cdot |S|.$$

Proof. This proof is a re-written form of the proof of Proposition 8 from [769]. We should analyze the equality $\alpha a - a = \delta b$.

If there exists a pair of elements (b, α), $b \in Q$, $\alpha \in C$ such that the equality $\alpha a - a = \delta b$ is fulfilled, then $L_b^+\alpha \in Aut(Q, f)$.

Every element $p \in P$ corresponds to $|N|$ elements b_p of the set Q such that $\delta(b_p) = p$ and this element p corresponds to $|S|$ elements ξ_p of the set C such that $\xi_p(a) - a = p$. If $p, r \in P$, $p \neq r$, then $\delta(b_p) \neq \delta(b_r)$ and $\xi_p \neq \xi_r$. Therefore $|Aut(Q, f)| = |P| \cdot |N| \cdot |S|$. □

Corollary 8.30. *A finite n-ary T-quasigroup (Q, f) of the form $f(x_1^n) = \sum_{i=1}^n(\varphi_i x_i) + a$ over an abelian group $(Q, +)$ has the identity automorphism group if and only if*

$$(Q, +) \cong \bigoplus_{i=1}^{m} (Z_2)_i,$$

$(Q, f) \cong (Q, g)$, *where* $g(x_1^n) = \sum_{i=1}^n(\varphi_i x_i)$, *the endomorphism* δ *is a permutation of the set* Q, $|C| = 1$, *and* m *is a natural number.*

Proof. Let $|Aut(Q, f)| = 1$. From Theorem 8.29 it follows that in this case $|N| = 1$. Since the order of the set Q is finite, the endomorphism δ is a permutation of the set Q and it is an automorphism of the group $(Q, +)$.

Further from Lemma 8.20 it follows that the quasigroup (Q, f) has an idempotent element, $(Q, f) \cong (Q, g)$, where $g(x_1^n) = \sum_{i=1}^n(\varphi_i x_i)$. Moreover, since the endomorphism δ is an automorphism, we obtain that the quasigroup (Q, f) has exactly one idempotent element, and from Corollary 8.17 it follows that $Aut(Q, f) \cong C$. Therefore in this case we have that $|C| = 1$.

In any abelian group $(Q, +)$ the permutation I, $I(x) = -x$, is an automorphism of this group and $I\psi = \psi I$ for any $\psi \in Aut(Q, +)$, i.e., $I \in C$. Therefore, for a fulfillment of the condition $|C| = 1$ it is necessary that $I = \varepsilon$.

It is well known [471] that in the finite case, only elementary abelian group of order 2^m, where m is a natural number, has the property that $I = \varepsilon$. Therefore $(Q, +) \cong \bigoplus_{i=1}^m (Z_2)_i$.

Conversely, let (Q, f) be a finite n-ary T-quasigroup of the form $f(x_1^n) \cong \sum_{i=1}^n(\varphi_i x_i)$ over an abelian group $(Q, +) \cong \bigoplus_{i=1}^m (Z_2)_i$, endomorphism δ is a permutation of the set Q and $|C| = 1$.

Then $|N| = 1$, the quasigroup (Q, f) has exactly one idempotent element, by Corollary 8.17 $Aut(Q, f) \cong C$ and, further, $|Aut(Q, f)| = 1$, since $|C| = 1$. □

Corollary 8.31. *Any finite medial n-ary quasigroup (Q, f) such that $|Q| \geqslant 3$ has a non-identity automorphism group.*

Proof. From Theorem 8.29 it follows that if $|N| > 1$, then $Aut(Q, f) > 1$. If we suppose that $|N| = 1$, then, taking into consideration that quasigroup (Q, f) is finite, we have that endomorphism δ is a permutation of the set Q, quasigroup (Q, f) has exactly one idempotent element and $Aut(Q, f) \cong C$.

Since quasigroup Q is a medial quasigroup, $\varphi_i \varphi_j = \varphi_j \varphi_i$ for all suitable values i, j. Thus $\langle \varphi_1, \dots, \varphi_n \rangle \subseteq C$.

Then $|C| = 1$ only in the case when $\varphi_1 = \dots = \varphi_n = \varepsilon$. In this case, quasigroup (Q, f) will have the form $f(x_1^n) \cong \sum_{i=1}^n(x_i)$ and by Proposition 8.11, $Aut(Q, f) \cong Z^{n-1}(Q, +) \rightthreetimes Aut(Q, +)$. It is well known ([471]) that $|Aut(Q, +)| > 1$ for any group $(Q, +)$ such that $|Q| \geqslant 3$. Therefore $|Aut(Q, f)| > 1$ in this case too. □

Example 8.32. This example of a T-quasigroup with the identity automorphism group is given in [769]. Let $(Q, +) = Z_2 \oplus Z_2 \oplus Z_2$, $x \cdot y = \varphi x + \psi y$, where

$$\varphi = \begin{pmatrix} 0 & 1 & 0 \\ 1 & 0 & 0 \\ 0 & 1 & 1 \end{pmatrix}, \quad \psi = \begin{pmatrix} 1 & 0 & 1 \\ 1 & 0 & 0 \\ 1 & 1 & 1 \end{pmatrix}.$$

Then $|Aut(Q, \cdot)| = 1$.

8.1.3 Automorphisms of some quasigroup isotopes

In this section we study some connections between automorphism groups of an idempotent n-quasigroup and some of its isotopes.

We denote the set $\{\tau \in Aut(Q, f) | \tau\alpha = \alpha\tau\}$, where (Q, f) is an n-ary quasigroup and α is a permutation of the set Q, by $C_{Aut(Q,f)}(\alpha)$. As it was noted in the previous section, the set $C_{Aut(Q,f)}(\alpha)$ forms a group with respect to the usual multiplication of permutations ([471]).

Theorem 8.33. *If $(Q, f) = (Q, g)T_0$ is an isotope of an n-ary idempotent quasigroup (Q, g) such that isotopy T_0 has the form $(\varepsilon, \dots, \varepsilon, \beta_{i+1}, \varepsilon, \dots, \varepsilon)$ (there are $(n + 1)$ members in this sequence) and $i \in \overline{0, n}$, then $Aut(Q, f) = C_{Aut(Q,g)}(\beta_{i+1})$.*

Proof. (a) Let $i \in \overline{0, n-1}$. Let $\varphi \in Aut(Q, f)$, i.e., $\varphi f(x_1, x_2, \dots, x_n) = f(\varphi x_1, \varphi x_2, \dots, \varphi x_n)$ for all $x_1, x_2, \dots, x_n \in Q$. Passing to operation g we obtain

$$\varphi g(x_1, \dots, x_i, \beta_{i+1}x_{i+1}, x_{i+2}, \dots, x_n) = \\ g(\varphi x_1, \dots, \varphi x_i, \beta_{i+1}\varphi x_{i+1}, \varphi x_{i+2}, \dots, \varphi x_n).$$

If we change x_{i+1} by $\beta_{i+1}^{-1}x_{i+1}$, then further we have

$$\varphi g(x_1, \dots, x_i, x_{i+1}, x_{i+2}, \dots, x_n) = \\ g(\varphi x_1, \dots, \varphi x_i, \beta_{i+1}\varphi\beta_{i+1}^{-1}x_{i+1}, \varphi x_{i+2}, \dots, \varphi x_n). \tag{8.9}$$

If in (8.9) $x_1 = x_2 = \dots = x_n = x$, then taking into consideration that operation g is idempotent, we have $\varphi x = g(\varphi x, \dots, \varphi x, \beta_{i+1}\varphi\beta_{i+1}^{-1}x, \varphi x, \dots, \varphi x)$. But $g(\overline{\varphi x}^n) = g(\varphi x, \dots, \varphi x) = \varphi x$.

Therefore $g(\varphi x, \dots, \varphi x, \beta_{i+1}\varphi\beta_{i+1}^{-1}x, \varphi x, \dots, \varphi x) = g(\varphi x, \dots, \varphi x) = g(\overline{\varphi x}^n)$.

Since (Q, g) is a quasigroup, from the last equality we have $\beta_{i+1}\varphi\beta_{i+1}^{-1} = \varphi$. Then $\beta_{i+1}\varphi = \varphi\beta_{i+1}$.

Therefore we can re-write (8.9) in the following form: $\varphi g(x_1^n) = g(\varphi x_1, \dots, \varphi x_n)$. Then $\varphi \in Aut(Q, g)$, $Aut(Q, f) \subseteq C_{Aut(Q,g)}(\beta_{i+1})$.

Converse. Let $\varphi \in C_{Aut(Q,g)}(\beta_{i+1})$. Then

$$\varphi f(x_1^n) = \varphi g(x_1, \dots, x_i, \beta_{i+1}x_{i+1}, x_{i+2}, \dots, x_n) = \\ g(\varphi x_1, \dots, \varphi x_i, \varphi\beta_{i+1}x_{i+1}, \varphi x_{i+2}, \dots, \varphi x_n) = \\ g(\varphi x_1, \dots, \varphi x_i, \beta_{i+1}\varphi x_{i+1}, \varphi x_{i+2}, \dots, \varphi x_n) = f(\varphi x_1, \varphi x_2, \dots, \varphi x_n).$$

Therefore $C_{Aut(Q,g)}(\beta_{i+1}) \subseteq Aut(Q, f)$ and, finally, we obtain

$$Aut(Q, f) = C_{Aut(Q,g)}(\beta_{i+1}).$$

(b) Let $i = n$. Let $\varphi \in Aut(Q, f)$, i.e., $\varphi f(x_1, x_2, \dots, x_n) = f(\varphi x_1, \varphi x_2, \dots, \varphi x_n)$ for all $x_1, x_2, \dots, x_n \in Q$. Passing to operation g we obtain $\varphi\beta_{n+1}g(x_1, \dots, x_n) = \beta_{n+1}g(\varphi x_1, \dots, \varphi x_n)$.

If in the last equality we put $x_1 = x_2 = \dots = x_n = x$, then, taking into consideration that operation g is idempotent, we have $\varphi\beta_{n+1}x = \beta_{n+1}\varphi x$. Thus

$$\beta_{n+1}\varphi g(x_1, \dots, x_n) = \beta_{n+1}g(\varphi x_1, \dots, \varphi x_n), \\ \varphi g(x_1, \dots, x_n) = g(\varphi x_1, \dots, \varphi x_n), \\ \varphi \in Aut(Q, f), Aut(Q, f) \subseteq C_{Aut(Q,g)}(\beta_{i+1}).$$

Converse. Let $\varphi \in C_{Aut(Q,g)}(\beta_{i+1})$. Then

$$\begin{aligned}&\varphi f(x_1, \dots, x_n) = \varphi\beta_{n+1}g(x_1, \dots, x_n) = \\ &\beta_{n+1}\varphi g(x_1, \dots, x_n) = \beta_{n+1}g(\varphi x_1, \dots, \varphi x_n) = \\ &= f(\varphi x_1, \varphi x_2, \dots, \varphi x_n).\end{aligned}$$

Therefore $C_{Aut(Q,g)}(\beta_{n+1}) \subseteq Aut(Q, f)$.

Inclusions $Aut(Q, f) \subseteq C_{Aut(Q,g)}(\beta_{i+1})$ and $C_{Aut(Q,g)}(\beta_{n+1}) \subseteq Aut(Q, f)$ give us the following equality $Aut(Q, f) = C_{Aut(Q,g)}(\beta_{i+1})$. In case (b) this theorem is proved too. □

Corollary 8.34. *If n-ary quasigroup (Q, f) is an isotope of an n-ary idempotent T-quasigroup (Q, g), $g(x_1^n) = \sum_{i=1}^{n} \alpha_i x_i$, and the isotopy has the form $(\varepsilon, \dots, \varepsilon, \beta_{i+1}, \varepsilon, \dots, \varepsilon)$, $i \in \overline{0, n}$, $\beta_{i+1} = L_d^+$, then*

$$Aut(Q, f) \cong (Q, +)\leftthreetimes S,$$

where $S = \{\theta \in C | \theta d = d\}$, $C = \{\omega \in Aut(Q, +) \,|\, \omega\alpha_i = \alpha_i\omega \,\forall\, i \in \overline{1, n}\}$.

Proof. From Theorem 8.33 it follows that we need to find the condition when $L_b\theta L_d = L_d L_b\theta$, where $L_b\theta \in Aut(Q, g)$ (Corollary 8.18). We have $L_{b+\theta d}\theta = L_{b+d}\theta$, $L_{b+\theta d}\theta 0 = L_{b+d}\theta 0$, $b + \theta d = b + d$, $\theta d = d$. □

Corollary 8.35. *If $(Q, f) = (Q, g)T_0$ is an isotope of an n-ary idempotent T-quasigroup (Q, g) such that isotopy T_0 has the form $(\varepsilon, \dots, \varepsilon, \beta_{i+1}, \varepsilon, \dots, \varepsilon)$, $i \in \overline{0, n}$ and $\beta_{i+1} = \varphi \in Aut(Q, +)$, then*

$$Aut(Q, f) \cong B\leftthreetimes N,$$

where $B = \{L_b^+ | b \in Q, \varphi b = b\}$, $N = \{\sigma \in C | \sigma\varphi = \varphi\sigma\}$.

Proof. From Corollary 8.18 it follows that any automorphism of n-T-quasigroup (Q, g) has the form $L_b\theta$, where $b \in Q$, L_b is a left translation of an abelian group $(Q, +)$, $\theta \in C$. Taking into consideration Theorem 8.33 we find the condition when the component φ of the isotopy T_0 and an automorphism $L_b\theta$ of quasigroup (Q, g) commute: $\varphi L_b\theta = L_b\theta\varphi$. Further we have $L_{\varphi b}\varphi\theta = L_b\theta\varphi$, $L_{\varphi b}\varphi\theta 0 = L_b\theta\varphi 0$, $\varphi b = b$, $\varphi\theta = \theta\varphi$. □

Remark 8.36. It is easy to see that in Corollaries 8.34 and 8.35 the n-ary quasigroup (Q, f) is an n-ary T-quasigroup.

Corollary 8.37. *If $x \circ y = \alpha x \cdot y$, where $(Q, \cdot)$ is an idempotent quasigroup, α is a permutation of the set Q, then $Aut(Q, \circ) = C_{Aut(Q,\cdot)}(\alpha)$.*

Proof. If in the conditions of Theorem 8.33 we suppose that $n = 2$, $i = 1$, then we have the conditions of this corollary. □

Corollary 8.38. *If $x \circ y = \gamma(x \cdot y)$, where $(Q, \cdot)$ is an idempotent quasigroup, γ is a permutation of the set Q, then $Aut(Q, \circ) = C_{Aut(Q,\cdot)}(\gamma)$.*

Proof. The proof is analogous to the proof of Corollary 8.37. □

Example 8.39. Let $(\mathbb{Q}, +)$ be a group of rational numbers. It is known [471] that $Aut(\mathbb{Q}, +) \cong (\mathbb{Q}^*, \cdot)$, i.e., this group is isomorphic to a group of non-zero rational numbers with respect to the operation of multiplication of these numbers. Let $(\mathbb{Q}, f)$ be an n-ary medial quasigroup of the form $f(x_1^n) = (\sum_{i=1}^{n} \varphi_i x_i) + a$, $\varphi_1^n \in \mathbb{Q}^*$, $a \in \mathbb{Q}$.

a). If $\delta = 0$ and $a = 0$, then $Aut(\mathbb{Q}, f) \cong (\mathbb{Q}, +)\leftthreetimes(\mathbb{Q}^*, \cdot)$.
b). If $\delta = 0$ and $a \neq 0$, then $Aut(\mathbb{Q}, f) \cong (\mathbb{Q}, +)$.
c). If $\delta \neq 0$, then $Aut(\mathbb{Q}, f) \cong (\mathbb{Q}^*, \cdot)$.

Proof. a). In this case the quasigroup $(\mathbb{Q}, f)$ is an idempotent quasigroup and by Corollary 8.18 $Aut(\mathbb{Q}, f) \cong (\mathbb{Q}, +)\leftthreetimes(\mathbb{Q}^*, \cdot)$.

b). By Corollary 8.34 $Aut(\mathbb{Q}, f) \cong (\mathbb{Q}, +)\leftthreetimes S$. In our case $S = \varepsilon$.

c). In this case the endomorphism δ is a permutation of the set $\mathbb{Q}$, quasigroup $(\mathbb{Q}, f)$ has exactly one idempotent element, by Corollary 8.17 $Aut(\mathbb{Q}, f) \cong C$. In our case $C = (\mathbb{Q}^*, \cdot)$. Therefore $Aut(\mathbb{Q}, f) \cong (\mathbb{Q}^*, \cdot)$. □

8.1.4 Automorphisms of medial n-quasigroups

Since any medial quasigroup is a T-quasigroup, then the structure of the automorphism group of any medial quasigroup with at least one idempotent element is described in Theorem 8.24 and Corollaries 8.17 and 8.18. Information on automorphism groups of some isotopes of n-ary T-quasigroups is given in Corollaries 8.34 and 8.35.

In this section we apply an n-ary analog of the Murdoch theorem (Theorem 7.55) to obtain information about the structure of an automorphism group of any finite n-ary medial quasigroup.

Remark 8.40. It is easy to see that there exist four classes of n-ary medial quasigroup over the group Z_2, namely:

1) a $(2k+1)$-ary medial quasigroup (Z_2, f_1) of the form $f_1(x_1^{2k+1}) = \sum_{i=1}^{2k+1} x_i$;
2) a $(2k)$-ary medial quasigroup (Z_2, f_2) of the form $f_2(x_1^{2k}) = \sum_{i=1}^{2k} x_i$;
3) a $(2k+1)$-ary medial quasigroup (Z_2, f_3) of the form $f_3(x_1^{2k+1}) = \sum_{i=1}^{2k+1} x_i + 1$;
4) a $(2k)$-ary medial quasigroup (Z_2, f_4) of the form $f_4(x_1^{2k}) = \sum_{i=1}^{2k} x_i + 1$.

Lemma 8.41. *For the indicated cases we have:*

$$\begin{array}{lll} 1) & s_{f_1} = \varepsilon, & Aut(Z_2, f_1) \cong Z_2; \\ 2) & s_{f_2}0 = 0, s_{f_2}1 = 0, & Aut(Z_2, f_2) \cong \varepsilon; \\ 3) & s_{f_3} = (01), & Aut(Z_2, f_3) \cong Z_2; \\ 4) & s_{f_4}0 = 1, s_{f_4}1 = 1, & Aut(Z_2, f_4) \cong \varepsilon. \end{array}$$

Proof. 1). We have $s_{f_1} = \varepsilon$, since $s_{f_1}(0) = 0$, $s_{f_1}(1) = 1$. This quasigroup is a medial $(\alpha_i\alpha_j = \alpha_j\alpha_i)$ idempotent $((2k+1)1 = 1)$ $(2k+1)$-ary quasigroup. From Corollary 8.18 it follows that $Aut(Z_2, f_1) \cong Z_2\leftthreetimes C \cong Z_2$ since $|C| = 1$.

2). We have $s_{f_2}0 = 0, s_{f_2}1 = 0$. This quasigroup is a unipotent quasigroup and by Corollary 8.17 we obtain $Aut(Z_2, f_2) \cong Aut(Z_2) = \varepsilon$.

3). In this case a map s_{f_3} has the form $s_{f_3} = (01)$, and since $s_{f_3}(0) = 1$, $s_{f_3}(1) = 0$, the map s_{f_3} is a permutation of the set Z_2.

This quasigroup is an isotope of an idempotent T-quasigroup. From Corollary 8.34 it follows that $Aut(Z_2, f_3) \cong Z_2\leftthreetimes C \cong S \cong Z_2$ since $S = \{\theta \in C \mid \theta 1 = 1\} = \varepsilon$.

4). We have $s_{f_4}0 = 1, s_{f_4}1 = 1$. This quasigroup is a unipotent quasigroup, from Corollary 8.17 it follows that $Aut(Z_2, f_4) \cong Aut(Z_2) = \varepsilon$. □

Theorem 8.42. *If (Q, f) is a finite medial n-ary quasigroup of the form $f(x_1^n) = \sum_{i=1}^{n} \alpha_i x_i + a$, $s(x) = f(\overline{x}^n)$ for any element $x \in Q$ and m is the smallest natural number such that $s^m Q = s^{m+1}Q$, then*

$$Aut(Q, f) \cong Aut(Q_1, f_1) \times Aut(Q_2, f_2),$$

where

$$\begin{array}{l} (Q_1, f_1) = s^m(Q, f), Aut(Q_1, f_1) \cong C_{Aut(Q_1,g)}(s_{f_1}), g(x_1^n) = s_{f_1}^{-1}(f_1(x_1^n)), \\ Q_2 \cong Q/Q_1, f_2(x_1^n) = \sum_{i=1}^{n}(\overline{\alpha_i} x_i), \\ Aut(Q_2, f_2) \cong C, C = \{\omega \in Aut(Q_2, +) \mid \omega\overline{\alpha}_i = \overline{\alpha}_i\omega \; \forall\, i \in \overline{1,n}\}. \end{array}$$

Proof. From Theorem 7.55 it follows that any finite medial n-ary quasigroup (Q, f) can be presented as a direct product of two n-ary medial quasigroups, namely $(Q, f) \cong (Q_1, f_1) \times (Q_2, f_2)$, where $Q_1 = s^m Q$, $Q_2 \cong Q/Q_1$, $s(x) = f(\overline{x}^n)$ for any element $x \in Q$ and m is the smallest natural number such that $s^m Q = s^{m+1} Q$, quasigroup (Q, f_1) is an isotope of special form of a medial idempotent quasigroup, and quasigroup (Q, f_2) is a medial quasigroup with exactly one idempotent element.

Therefore we have $Aut(Q, f) \supseteq Aut(Q_1, f_1) \times Aut(Q_2, f_2)$. Now we shall prove that $Aut(Q, f) \subseteq Aut(Q_1, f_1) \times Aut(Q_2, f_2)$, i.e., that $Aut(Q, f) \cong Aut(Q_1, f_1) \times Aut(Q_2, f_2)$.

For $|Q| \geqslant 3$, quasigroups (Q_1, f_1) and (Q_2, f_2) are non-isomorphic since in the quasigroup (Q_1, f_1) the map s_{f_1} is a permutation of the set Q_1 and in the quasigroup (Q_2, f_2) we have $s_{f_2}(x) = e$ for any $x \in Q_2$ and some fixed element e of the set Q_2. Moreover, any pair of subquasigroups (S, f_1) and (S_2, f_2) of quasigroups (Q_1, f_1) and (Q_2, f_2), respectively, with $|S_1| \geqslant 2$ or $|S_2| \geqslant 2$ are non-isomorphic too.

Indeed, in any subquasigroup (S_1, f_1) $(|S_1| > 1)$ of quasigroup (Q_1, f_1) the endomorphism s is a permutation of the set S_1 but in any subquasigroup (S_2, f_2) $(|S_2| > 1)$ of quasigroup (Q_2, f_2) the endomorphism s is not a permutation of the set S_2.

Therefore in this case there does not exist an automorphism φ of the quasigroup (Q, f) such that φ is an isomorphism of the quasigroups (Q_1, f_1) and (Q_2, f_2) or the automorphism φ is an isomorphism of their subquasigroups, i.e., $\varphi(S_1, f_1) \cong (S_2, f_2)$, where quasigroup (S_1, f_1) $(|S_1| \geqslant 2)$ is a subquasigroup of quasigroup (Q_1, f_1) and quasigroup (S_2, f_2) is a subquasigroup of quasigroup (Q_2, f_2).

Let $|S_1| = |S_2| = 1$. We have to prove that in this case there is no automorphism φ of finite medial n-ary quasigroup (Q, f) such that $\varphi S_1 \cong S_2$ as well.

It is clear that $(S_1, f_1) \times (S_2, f_2)$ is an n-ary medial quasigroup of order 2 and in this quasigroup a map s cannot be a permutation.

It is well known that up to isomorphism there exists the unique abelian group of order 2, namely, Z_2. In Lemma 8.41 some properties of all up to isomorphism possible classes of n-ary medial quasigroups over the group Z_2 are described.

We see that only n-ary medial quasigroups from classes 2) and 4) satisfy our condition that the map s is not a permutation. From Lemma 8.41 it follows that in cases 2), 4) there exists only the identity automorphism. Therefore $\varphi Q_1 = Q_1$ and $\varphi Q_2 = Q_2$ for any automorphism of the quasigroup (Q, f).

The structure of the group $Aut(Q_1, f_1)$ follows from Theorem 8.24 and the structure of the group $Aut(Q_2, f_2)$ follows from Corollary 8.17. □

Corollary 8.43. *If (Q, f) is a finite medial n-ary quasigroup and $s_{f_1} \in Aut(Q_1, +)$, then*

$$Aut(Q, f) \cong (B \rtimes N) \times C_2,$$

where $B = \{L_b^+ \mid b \in Q_1, s_{f_1}(b) = b\}$, N is a centralizer of the map s_{f_1} in the group $C_1 = C_{Aut(Q_1,+)}(\alpha_1^n)$, $C_2 = \{\omega \in Aut(Q_2, +) \mid \omega\alpha_i = \alpha_i\omega \, \forall \, i \in \overline{1, n}\}$.

Proof. This follows from Theorem 8.42 and Corollary 8.35. □

Corollary 8.44. *If (Q, f) is a finite medial n-ary quasigroup and $s_{f_1} = L_d^+, d \in Q_1$, then*

$$Aut(Q, f) \cong ((Q_1, +) \rtimes S) \times C_2,$$

where $S = \{\theta \in C_1 | \theta d = d\}$, $C_1 = \{\omega \in Aut(Q_1, +) \mid \omega\alpha_i = \alpha_i\omega \, \forall \, i \in \overline{1, n}\}$, $C_2 = \{\omega \in Aut(Q_2, +) \mid \omega\alpha_i = \alpha_i\omega \, \forall \, i \in \overline{1, n}\}$.

Proof. This follows from Theorem 8.42 and Corollary 8.34. □

8.1.5 Examples

In this section we apply obtained results to describe automorphism groups of n-ary quasigroups that correspond to the ISSN code, the EAN code, and the UPC code.

Example 8.45. The International Standard Serial Number code (the ISSN code) which is used now, consists of eight digits. These are the Arabic numerals from 0 to 9 on places from the 1-st to the 7-th. On the 8-th place the Arabic numerals 0 to 9 and an upper case X can occur. Denote this code as $\mathfrak{C}$.

The first seven digits a_1^7 are the so-called information symbols and the 8-th digit a_8 is a check digit. Any eight right (without any error) digits of the ISSN code satisfy the following check equation:

$$8 \cdot a_1 + 7 \cdot a_2 + 6 \cdot a_3 + 5 \cdot a_4 + 4 \cdot a_5 + 3 \cdot a_6 + 2 \cdot a_7 + 1 \cdot a_8 \equiv 0 \pmod{11},$$

i.e., $a_1^8 \in \mathfrak{C}$ if and only if this code word satisfies the above-stated check equation.

We can associate the 7-ary medial quasigroup (Z_{11}, f) with the ISSN code in this manner: $8 \cdot y_1 + 7 \cdot y_2 + 6 \cdot y_3 + 5 \cdot y_4 + 4 \cdot y_5 + 3 \cdot y_6 + 2 \cdot y_7 \equiv -1 \cdot y_8 \pmod{11}$ for all $y_1^7 \in Z_{11}$, $y_8 \equiv 3 \cdot y_1 + 4 \cdot y_2 + 5 \cdot y_3 + 6 \cdot y_4 + 7 \cdot y_5 + 8 \cdot y_6 + 9 \cdot y_7 \pmod{11}$.

Therefore we have the 7-ary quasigroup (Z_{11}, f) of the form

$$f(y_1^7) = 3 \cdot y_1 + 4 \cdot y_2 + 5 \cdot y_3 + 6 \cdot y_4 + 7 \cdot y_5 + 8 \cdot y_6 + 9 \cdot y_7$$

over the group $(Z_{11}, +)$. It is easy to see that $f(y_1^7) = y_8$ if and only if $y_1^8 \in \mathfrak{C}$.

Prove that $Aut(Z_{11}, f) \cong Z_{10}$. It is easy to check that the quasigroup (Z_{11}, f) has exactly one idempotent element, namely the element 0. In conditions of this example we can apply Corollary 8.17. Since $Aut(Z_{11}) \cong Z_{10}$ is a commutative group, we obtain $C \cong Z_{10}$. Therefore

$$Aut(Z_{11}, f) \cong Z_{10}.$$

Example 8.46. The Universal Product Code (the UPC code) is the code with the check equation $1 \cdot 0 + 3 \cdot x_2 + 1 \cdot x_3 + 3 \cdot x_4 + 1 \cdot x_5 + 3 \cdot x_6 + 1 \cdot x_7 + 3 \cdot x_8 + 1 \cdot x_9 + 3 \cdot x_{10} + 1 \cdot x_{11} + 3 \cdot x_{12} + 1 \cdot x_{13} \equiv 0 \pmod{10}$, where $x_i \in Z_{10}$, $i \in \overline{1, 13}$, elements x_1^{12} are the information digits and element x_{13} is the check digit. In other words the UPC code is in fact a sub-code of the more general EAN code.

We can associate an 11-ary medial quasigroup (Z_{10}, f) of the form $f(x_2^{12}) = 7 \cdot x_2 + 9 \cdot x_3 + 7 \cdot x_4 + 9 \cdot x_5 + 7 \cdot x_6 + 9 \cdot x_7 + 7 \cdot x_8 + 9 \cdot x_9 + 7 \cdot x_{10} + 9 \cdot x_{11} + 7 \cdot x_{12}$ with this code.

We can also use Proposition 8.16 in this example. We have $K = \{L_0, L_5\}$, $K \cong Z_2$. Since $Aut(Z_{10}) \cong Z_4$, we obtain $C \cong Z_4$. Therefore $Aut(Z_{10}, f) \cong Z_2 \leftthreetimes Z_4$.

Since any element β of the group Z_4 acts on the group Z_2, $Z_2 = \{0, 1\}$, as an inner automorphism ([471]), then $\beta 0 = 0$ and, therefore, $\beta 1 = 1$.

Then any element of the group Z_4 acts on the group Z_2 as the identity automorphism, and, finally, we have

$$Aut(Z_{10}, f) \cong Z_2 \times Z_4.$$

Example 8.47. The European Article Number code (EAN) is the code with the check equation

$$\begin{aligned} &1 \cdot x_1 + 3 \cdot x_2 + 1 \cdot x_3 + 3 \cdot x_4 + 1 \cdot x_5 + 3 \cdot x_6 + 1 \cdot x_7 + \\ &+3 \cdot x_8 + 1 \cdot x_9 + 3 \cdot x_{10} + 1 \cdot x_{11} + 3 \cdot x_{12} + 1 \cdot x_{13} \equiv 0 \pmod{10}, \end{aligned}$$

where $x_i \in Z_{10}$, $i \in \overline{1, 13}$, elements x_1^{12} are the information digits and element x_{13} is a check digit ([752]).

We can associate a 12-ary medial quasigroup (Z_{10}, f) with this code in the following way. From the last check equation we have

$$\begin{aligned} -x_{13} \equiv\ & 1 \cdot x_1 + 3 \cdot x_2 + 1 \cdot x_3 + 3 \cdot x_4 + 1 \cdot x_5 + 3 \cdot x_6 + 1 \cdot x_7 + \\ & 3 \cdot x_8 + 1 \cdot x_9 + 3 \cdot x_{10} + 1 \cdot x_{11} + 3 \cdot x_{12} \pmod{10}, \end{aligned}$$

$$\begin{aligned} x_{13} \equiv\ & 9 \cdot x_1 + 7 \cdot x_2 + 9 \cdot x_3 + 7 \cdot x_4 + 9 \cdot x_5 + 7 \cdot x_6 + 9 \cdot x_7 + \\ & 7 \cdot x_8 + 9 \cdot x_9 + 7 \cdot x_{10} + 9 \cdot x_{11} + 7 \cdot x_{12} \pmod{10}. \end{aligned}$$

Therefore we obtain a 12-ary medial quasigroup (Z_{10}, f) with the form

$$\begin{aligned} f(x_1^{12}) \equiv\ & 9 \cdot x_1 + 7 \cdot x_2 + 9 \cdot x_3 + 7 \cdot x_4 + 9 \cdot x_5 + 7 \cdot x_6 + \\ & 9 \cdot x_7 + 7 \cdot x_8 + 9 \cdot x_9 + 7 \cdot x_{10} + 9 \cdot x_{11} + 7 \cdot x_{12} \pmod{10}. \end{aligned} \tag{8.10}$$

For this quasigroup we have $s_f(x) = 6 \cdot (9+7)x = 96x = 6x$ for any $x \in Z_{10}$, $s_f(Z_{10}) = \{0, 6, 2, 8, 4\} = A$, $s_f(A) = A$. Therefore $m = 2$ in this case. We notice that in this example the endomorphism s_f of the quasigroup (Z_{10}, f) is also an endomorphism of the group $(Z_{10}, +)$. It is easy to see that $s_f(Z_{10}, +) \cong (Z_5, +)$.

From Theorem 7.55 it follows that $(Z_{10}, f) \cong (Z_2, f_1) \times (Z_5, f_2)$.

Information from Section 7.1 lets us find the forms f_1 and f_2, if we define the group $(Z_{10}, +)$ as a direct sum of groups $(Z_2, +)$ and $(Z_5, +)$ and define an isomorphism between these groups.

Let $Z_{10} = \{0, 1, 2, 3, 4, 5, 6, 7, 8, 9\}$, $Z_5 = \{0, 1, 2, 3, 4\}$, $Z_2 = \{0, 1\}$, $Z_2 \oplus Z_5 = \{(0;0), (0;1), (0;2), (0;3), (0;4), (1;0), (1;1), (1;2), (1;3), (1;4)\}$.

Define isomorphism ξ between the group $(Z_{10}, +)$ and the group $(Z_2 \oplus Z_5, +)$ as follows: $\xi(0) = (0;0), \xi(1) = (1;1), \xi(2) = (0;2), \xi(3) = (1;3), \xi(4) = (0;4), \xi(5) = (1;0), \xi(6) = (0;1), \xi(7) = (1;2), \xi(8) = (0;3)$, $\xi(9) = (1;4)$.

Multiplication of elements of the group $(Z_{10}, +)$ by the element 7 or by the element 9 is an automorphism of this group. Since $\xi(7) = (1, 2)$, then the following ordered pair of automorphisms: $1 : x \mapsto 1 \cdot x \pmod 2$ and $2 : x \mapsto 2 \cdot x \pmod 5$ corresponds to the automorphism $7 : x \mapsto 7 \cdot x \pmod{10}$.

Similarly, since $\xi(9) = (1;4)$, we have, that the following ordered pair of automorphisms $1 : x \mapsto 1 \cdot x \pmod 2$ and $4 : x \mapsto 4 \cdot x \pmod 5$ corresponds to the automorphism $9x \mapsto 9 \cdot x \pmod{10}$.

Now we can say that a 12-ary medial quasigroup (Z_{10}, f) with the form (8.10) is isomorphic to the 12-ary medial quasigroup $(Z_2, f_1) \times (Z_5, f_2)$, where

$$\begin{aligned} f_1(x_1^{12}) \equiv\ & 1 \cdot x_1 + 1 \cdot x_2 + 1 \cdot x_3 + 1 \cdot x_4 + 1 \cdot x_5 + 1 \cdot x_6 + \\ & 1 \cdot x_7 + 1 \cdot x_8 + 1 \cdot x_9 + 1 \cdot x_{10} + 1 \cdot x_{11} + 1 \cdot x_{12} \pmod 2 \end{aligned}$$

and

$$\begin{aligned} f_2(x_1^{12}) \equiv\ & 4 \cdot x_1 + 2 \cdot x_2 + 4 \cdot x_3 + 2 \cdot x_4 + 4 \cdot x_5 + 2 \cdot x_6 + \\ & 4 \cdot x_7 + 2 \cdot x_8 + 4 \cdot x_9 + 2 \cdot x_{10} + 4 \cdot x_{11} + 2 \cdot x_{12} \pmod 5. \end{aligned}$$

From Theorem 8.42 it follows that $Aut(Z_{10}, f) \cong Aut(Z_2, f_1) \times Aut(Z_5, f_2)$.

From Lemma 8.41 case 2) it follows that $Aut(Z_2, f_1) = \langle \varepsilon \rangle$.

Further we have $s_{f_2}(x) = 6 \cdot (4+2)x = 1 \cdot x$ for every $x \in Z_5$. The quasigroup (Z_5, f_2) is a 12-ary medial idempotent quasigroup. We can use Corollary 8.18 in order to find $Aut(Z_5, f_2)$.

Since the group $Aut(Z_5, +)$ is a commutative group and $Aut(Z_5, +) \cong Z_4$, we have $Aut(Z_5, f_2) \cong Z_5 \leftthreetimes Z_4$.

Finally we obtain

$$Aut(Z_{10}, f) \cong (Z_5 \leftthreetimes Z_4) \times \langle \varepsilon \rangle \cong Z_5 \leftthreetimes Z_4.$$

The last group is named as the general affine group $GA(1, 5)$ [921].

Remark 8.48. We can use Proposition 8.16 or Corollary 8.43 in order to find $Aut(Z_{10}, f)$.

Indeed, by Proposition 8.16, $Aut(Z_{10}, f) \cong K \leftthreetimes C$. In conditions of Example 8.47 we have $K = \{L_0, L_2, L_4, L_6, L_8\}$, $K \cong Z_5$. Since $Aut(Z_{10}) \cong Z_4$ and Z_4 is a commutative group, we obtain $C \cong Z_4$. Therefore

$$Aut(Z_{10}, f) \cong Z_5 \leftthreetimes Z_4.$$

Remark 8.49. In [600] the automorphism group of the quasigroup (Z_{10}, f) was found without use of Theorem 7.55, since this quasigroup has an idempotent element (for example, the element 0 is such element) and we can use Theorem 8.24.

Example 8.50. We find the structure and automorphism group of ternary medial quasigroup (Z_{12}, f) over the group $(Z_{12}, +)$ with the form $f(x_1^3) = 1 \cdot x_1 + 7 \cdot x_2 + 1 \cdot x_3 + 7$.

We have: $s(x) = 9 \cdot x + 7$. It is easy to see that this quasigroup does not contain any idempotent element. Indeed, if $9 \cdot x + 7 = x \pmod{12}$, then $8 \cdot x = -7 = 5 \pmod{12}$. It is clear that the last equation does not have a solution and in this case we cannot directly apply Theorem 8.24 or its corollaries.

Further we have $s(Z_{12}) = \{7, 4, 1, 10\} = A$, $s(A) = A$. Therefore $m = 2$ in this case, $s(Z_{12}, +) \cong (Z_4, +)$.

As in the previous example, first of all we fix an isomorphism ξ between the group $(Z_{12}, +)$ defined on the set $\{0, 1, 2, 3, 4, 5, 6, 7, 8, 9, 10, 11\}$ and the group $(Z_3 \oplus Z_4, +)$ defined on the set $\{(0;0), (0;1), (0;2), (0;3), (1;0), (1;1), (1;2), (1;3), (2;0), (2;1), (2;2), (2;3)\}$. Let $\xi(1) = (1, 1)$. Then $\xi(7) = (1;3)$.

Thus from Theorem 7.55 it follows that $(Z_{12}, f) \cong (Z_3, f_1) \times (Z_4, f_2)$, where $f_1(x_1^3) = 1 \cdot x_1 + 1 \cdot x_2 + 1 \cdot x_3 + 1$ and $f_1(x_1^3) = 1 \cdot x_1 + 3 \cdot x_2 + 1 \cdot x_3 + 3$ are the forms of quasigroups (Z_3, f_1) and (Z_4, f_2) respectively over groups $(Z_3, +)$ and $(Z_4, +)$.

From Theorem 8.42 it follows that $Aut(Z_{12}, f) \cong Aut(Z_3, f_1) \times Aut(Z_4, f_2)$. The quasigroup $Aut(Z_3, f_1)$ has exactly one idempotent element (namely, the element 1 is an idempotent element). By Corollary 8.17 $Aut(Z_3, f_1) \cong C$. Therefore $Aut(Z_3, f_1) \cong Z_2$.

The quasigroup (Z_4, f_2) is an isotope of the form $(\varepsilon, \varepsilon, L_3^+, \varepsilon)$ of idempotent 3-ary medial quasigroup (Z_4, g) with the form $g(x_1^3) = 1 \cdot x_1 + 3 \cdot x_2 + 1 \cdot x_3$. Use of Corollary 8.34 gives us that $Aut(Z_4, f_2) \cong Z_4 \leftthreetimes < \varepsilon > \cong Z_4$ since out of two automorphisms of the group $(Z_4, +)$ only the identity automorphism fixes element 3.

Finally we obtain

$$Aut(Z_{12}, f) \cong Z_2 \times Z_4.$$

8.2 Automorphism groups of some binary quasigroups

8.2.1 Isomorphisms of IP-loop isotopes

Isomorphisms of group isotopes are studied in [441, 444], isomorphisms of quasigroups which are linear over a commutative Moufang loop are studied in [665, 666], isomorphisms of CH-quasigroups are studied in [593]. Some results presented in this section are published in [760, 764, 769, 767].

We recall that by Theorem 2.50 any autotopy of an IP-loop can be written in the form

$$(L_a\theta, R_aR_c\theta, L_aR_aR_c\theta), \tag{8.11}$$

where θ is a right pseudo-automorphism with companion c, element a is a middle Moufang element.

Lemma 8.51. *Let $(Q,+)$ be an IP-loop. Suppose that $(Q,\circ) = (Q,+)(\alpha_1,\beta_1,\gamma_1)$, $(Q,*) = (Q,+)(\alpha_2,\beta_2,\gamma_2)$, $(Q,*) = (Q,\circ)(\sigma,\sigma,\sigma)$. Then*

$$\sigma = \alpha_1^{-1}L_a\theta\alpha_2 = \beta_1^{-1}R_aR_c\theta\beta_2 = \gamma_1^{-1}L_aR_aR_c\theta\gamma_2, \tag{8.12}$$

where $(L_a\theta, R_aR_c\theta, L_aR_aR_c\theta)$ is an autotopy of the loop $(Q,+)$.

Proof. Since $(\circ) = (+)(\alpha_1,\beta_1,\gamma_1)$, $(*) = (+)(\alpha_2,\beta_2,\gamma_2)$, and $(*) = (\circ)(\sigma,\sigma,\sigma)$, then $(+)(\alpha_2,\beta_2,\gamma_2) = (+)(\alpha_1,\beta_1,\gamma_1)(\sigma,\sigma,\sigma)$,

$$(+) = (+)(\alpha_1,\beta_1,\gamma_1)(\sigma,\sigma,\sigma)(\alpha_2,\beta_2,\gamma_2)^{-1} = \\ (+)(\alpha_1\sigma\alpha_2^{-1}, \beta_1\sigma\beta_2^{-1}, \gamma_1\sigma\gamma_2^{-1})$$

is an autotopy of the loop $(Q,+)$. Then using the form (8.11) of an autotopy of an IP-loop we have $\alpha_1\sigma\alpha_2^{-1} = L_a\theta$ for some middle Moufang element a and pseudo-automorphism θ. Therefore we obtain that $\sigma = \alpha_1^{-1}L_a\theta\alpha_2$ and so on. □

Corollary 8.52. *If $(Q,\circ) = (Q,+)(\alpha,\beta,\gamma)$, where $(Q,+)$ is an IP-loop, then any automorphism φ of the quasigroup $(Q,\circ)$ has the form*

$$\varphi = \alpha^{-1}L_a\theta\alpha = \beta^{-1}R_aR_c\theta\beta = \gamma^{-1}L_aR_aR_c\theta\gamma, \tag{8.13}$$

where θ is a right pseudo-automorphism with companion c and element a is a middle Moufang element of loop $(Q,\cdot)$.

Proof. We use Lemma 8.51. In the case when $(Q,\circ) = (Q,*)$ isomorphism σ passes in automorphism φ. □

Lemma 8.53. *Let $(Q,+)$ be an IP-loop. Suppose that quasigroup $(Q,\circ)$ has the form $x\circ y = \alpha_1 x + \beta_1 y + c$, quasigroup $(Q,*)$ has the form $x * y = \alpha_2 x + \beta_2 y + d$, where $\alpha_1,\alpha_2,\beta_1,\beta_2$ are permutations of the set Q, $c,d \in N(Q,+)$. Then isomorphism σ has the form $\sigma = L_b\theta$, where b is a middle Moufang element, θ is a left pseudo-automorphism, L_b is some left translation of loop $(Q,+)$.*

Proof. Quasigroup $(Q,\circ)$ is a special form of isotope of IP-loop $(Q,+)$: $(\circ) = (+)(\alpha_1, R_c\beta_1, \varepsilon)$. Similarly, $(*) = (+)(\alpha_2, R_d\beta_2, \varepsilon)$.

Therefore from Lemma 8.51 it follows that the isomorphism σ has the following form: $L_aR_aR_c\theta$.

From Corollary 2.39 it follows that the third component of any autotopy of an IP-loop can be presented in the form $L_b\theta$, where b is a middle Moufang element, θ is a left pseudo-automorphism. Therefore $\sigma = L_b\theta$. □

Definition 8.54. The holomorph of a group is a group which simultaneously contains (copies of) the group and its automorphism group. The holomorph of group G denoted by $Hol(G)$ can be described as a semidirect product of a group and its automorphism group [471, 922].

Corollary 8.55. *Let $(Q,+)$ be a group. Suppose that quasigroup $(Q,\circ)$ has the form $x\circ y = \alpha_1 x + \beta_1 y + c$, quasigroup $(Q,*)$ has the form $x * y = \alpha_2 x + \beta_2 y + d$, where $\alpha_1,\alpha_2,\beta_1,\beta_2$ are permutations of the set Q. Then the isomorphism σ has the form $\sigma = L_b\theta$, i.e., any isomorphism σ lies in holomorph of the group $(Q,+)$ [441].*

Proof. Any group is an IP-loop. □

Definition 8.56. Quasigroup $(Q,\cdot)$ with the form $x \cdot y = \varphi x + \psi y + c$, where $(Q,+)$ is an IP-loop, $c \in N$ is called a linear quasigroup over IP-loop $(Q,+)$.

Quasigroup $(Q,\cdot)$ with the form $x \cdot y = \varphi x + \psi y + c$, where $(Q,+)$ is a commutative Moufang loop, $c \in Z$ is called linear quasigroup over CML-loop $(Q,+)$ [201, 665].

In fact quasigroup $(Q,\cdot)$ is a special form of isotope of IP-loop $(Q,+)$: $(\cdot) = (+)(\varphi, R_c\psi, \varepsilon)$.

We recall that in any commutative Moufang loop a center coincides with its nucleus (Lemma 2.115). Any distributive quasigroup is isotopic to a commutative Moufang loop (Theorem 2.121).

Theorem 8.57. *Let $(Q,+)$ be a commutative Moufang loop. Suppose that quasigroup $(Q,\circ)$ has the form $x \circ y = \varphi_1 x + \psi_1 y + c$, quasigroup $(Q,*)$ has the form $x * y = \varphi_2 x + \psi_2 y + d$, where $\varphi_1, \varphi_2, \psi_1, \psi_2$ are automorphisms of loop $(Q,+)$, $c, d \in Z(Q,+)$, $x \bullet y = \varphi_1 x + \psi_1 y$, $x \star y = \varphi_2 x + \psi_2 y$ are distributive quasigroups. Quasigroups $(Q,\circ)$ and $(Q,*)$ are isomorphic if and only if there exists an automorphism θ of loop $(Q,+)$ such that $\theta\varphi_1 = \varphi_2\theta$, $\theta c = d$.*

Proof. Suppose that quasigroups $(Q,\circ)$ and $(Q,*)$ are isomorphic. From Lemma 8.53 it follows that isomorphism has the form $L_a\theta$. Further we have

$$\begin{aligned} &L_a\theta(x \circ y) = L_a\theta x * L_a\theta y, \\ &a + \theta(x \circ y) = (a + \theta x) * (a + \theta y), \\ &a + \theta(\varphi_1 x + \psi_1 y + c) = (\varphi_2 a + \varphi_2\theta x) + (\psi_2 a + \psi_2 \theta y) + d. \end{aligned} \tag{8.14}$$

If we put $x = y = 0$ in the equality (8.14), then $a + \theta c = \varphi_2 a + \psi_2 a + d$. Notice that $\varphi_2 a + \psi_2 a = a$ since the quasigroup $(Q,*)$ is distributive, therefore it is an idempotent quasigroup. Thus we have $a + \theta c = a + d$, $\theta c = d$.

From Theorem 1.201 Case 1 and the fact that in a distributive quasigroup its left translation is its automorphism, we obtain that $L_a \in Aut(Q,\star)$, i.e.,

$$a + (\varphi_2 x + \psi y) = (\varphi_2 a + \varphi_2 x) + (\psi_2 a + \psi_2 y). \tag{8.15}$$

Applying the equality (8.15) to the right side of the equality (8.14) we have

$$\begin{aligned} &a + \theta(\varphi_1 x + \psi_1 y + c) = a + (\varphi_2\theta x + \psi_2\theta y) + d, \\ &\theta(\varphi_1 x + \psi_1 y + c) = (\varphi_2\theta x + \psi_2\theta y) + d. \end{aligned} \tag{8.16}$$

Taking into consideration that in a commutative loop any left pseudo-automorphism θ with a companion h is also a right pseudo-automorphism (Lemma 1.256) further we have

$$\begin{aligned} &\theta((\varphi_1 x + \psi_1 y) + c) = \\ &(\theta(\varphi_1 x + \psi_1 y) + (\theta(c) + h)) - h = \\ &(\theta(\varphi_1 x + \psi_1 y) + (d + h)) - h = \\ &((\theta(\varphi_1 x + \psi_1 y) + d) + h) - h \overset{RIP}{=} \\ &\theta(\varphi_1 x + \psi_1 y) + d. \end{aligned} \tag{8.17}$$

Comparing the right sides of the equalities (8.16) and (8.17) we have that $\theta(\varphi_1 x + \psi_1 y) = \varphi_2\theta x + \psi_2\theta y$. If we put $y = 0$ in the last equality, then we have $\theta\varphi_1 = \varphi_2\theta$, by $x = 0$ we have $\theta\psi_1 = \psi_2\theta$. Thus $\theta(\varphi_1 x + \psi_1 y) = \varphi_1 x + \theta\psi_1 y$, $\theta \in Aut(Q,+)$.

Converse. If $\theta\varphi_1 = \varphi_2\theta$, then $\theta\psi_1 = \psi_2\theta$. Indeed, since θ is an automorphism of loop $(Q,+)$, then $\theta x = \theta(\varphi_1 x + \psi_1 x) = \theta\varphi_1 x + \theta\psi_1 x = \varphi_2\theta x + \psi_2\theta x$. Since $\theta\varphi_1 = \varphi_2\theta$, further we have $\theta\psi_1 = \psi_2\theta$. Then $\theta(x \circ y) = \theta\varphi_1 x + \theta\psi_1 y + \theta c = \varphi_2\theta x + \psi_2\theta y + d = \theta x * \theta y$. □

Corollary 8.58. *The number of non-isomorphic quasigroups of the form $x \cdot y = \varphi x + \psi y + d$, where $x \circ y = \varphi x + \psi y$ is a fixed distributive quasigroup which is isotopic to a CML $(Q,+)$, $d \in Z(Q,+)$, is equal to the number of orbits by action of the group $C = \{\theta \in Aut(Q,+) \mid \theta\varphi = \varphi\theta\}$ on the group $Z(Q,+)$.*

Proof. The proof follows from Theorem 8.57 in the case $\varphi_1 = \varphi_2$. □

Corollary 8.59. *Let $(Q,+)$ be a commutative Moufang loop. Suppose that distributive quasigroup $(Q,\circ)$ has the form $x \circ y = \varphi_1 x + \psi_1 y$, distributive quasigroup $(Q,*)$ has the form $x * y = \varphi_2 x + \psi_2 y$, where $\varphi_1, \varphi_2, \psi_1, \psi_2$ are automorphisms of loop $(Q,+)$. Quasigroups $(Q,\circ)$ and $(Q,*)$ are isomorphic if and only if there exists automorphism θ of loop $(Q,+)$ such that $\theta\varphi_1 = \varphi_2\theta$.*

Proof. It is sufficient to put $c = d = 0$ in conditions of Theorem 8.57. □

Definition 8.60. An automorphism ψ of a CML $(Q,+)$ is called completely full, if there exists an automorphism φ of CML $(Q,+)$ such that $\varphi x + \psi y = x + \psi(-x + y)$.

Also see Definition 2.144.

Corollary 8.61. *The number of non-isomorphic in pairs distributive quasigroups which are isotopic to a fixed CML $(Q,+)$ is equal to the number of non-conjugated in the group $Aut(Q,+)$ completely full automorphisms of the group $Aut(Q,+)$.*

Proof. Recall that any distributive quasigroup is isotopic to a CML $(Q,+)$ (Belousov Theorem). We take into consideration that any distributive quasigroup is defined in a unique way by the completely full automorphism ψ (Theorem 6.7) and Corollary 8.59. □

8.2.2 Automorphisms of loop isotopes

In this section we study a connection between an automorphism group of a loop and an automorphism group of a loop isotope of a special form. We denote the identity element of a loop $(Q,+)$ as 0.

Proposition 8.62. *If $(Q,\circ)$ is a quasigroup with the form $x \circ y = \alpha x + y$, where $(Q,+)$ is a loop, α is a permutation of the set Q such that $\alpha 0 = 0$, then $Aut(Q,\circ) = C_{Aut(Q,+)}(\alpha)$.*

Proof. Let $\varphi \in Aut(Q,\circ)$, i.e., $\varphi(x \circ y) = \varphi x \circ \varphi y$ for all $x, y \in Q$. Passing to the operation $+$ we have

$$\varphi(\alpha x + y) = \alpha\varphi x + \varphi y. \tag{8.18}$$

If we take $x = y = 0$ in the equality (8.18), then $\varphi 0 = \alpha\varphi 0 + \varphi 0$, $\alpha\varphi 0 = 0$, $\varphi 0 = \alpha^{-1}0 = 0$, therefore $\varphi 0 = 0$.

If we assume that $y = 0$ in (8.18), then $\varphi\alpha x = \alpha\varphi x + \varphi 0 = \alpha\varphi x + 0 = \alpha\varphi x$, i.e., $\varphi\alpha x = \alpha\varphi x$. Therefore, $\varphi\alpha = \alpha\varphi$ and $\varphi \in Aut(Q,+)$, since $\varphi(\alpha x + y) = \alpha\varphi x + \varphi y = \varphi\alpha x + \varphi y$. Thus $\varphi \in Aut(Q,+)$, $Aut(Q,\circ) \subseteq C_{Aut(Q,+)}(\alpha)$.

Let $\varphi \in C_{Aut(Q,+)}(\alpha)$. Then $\varphi(x \circ y) = \varphi(\alpha x + y) = \varphi\alpha x + \varphi y = \alpha\varphi x + \varphi y = \varphi x \circ \varphi y$. Therefore $C_{Aut(Q,+)}(\alpha) \subseteq Aut(Q,\circ)$, $C_{Aut(Q,+)}(\alpha) = Aut(Q,\circ)$. □

Proposition 8.63. *If $(Q,\circ)$ is a quasigroup with the form $x \circ y = x + \beta y$, where $(Q,+)$ is a loop, β is a permutation of the set Q such that $\beta 0 = 0$, then $Aut(Q,\circ) = C_{Aut(Q,+)}(\beta)$.*

Proof. The proof is analogous to the proof of Proposition 8.62. □

Proposition 8.64. *If $(Q, \circ)$ is a quasigroup with the form $x \circ y = \gamma(x + y)$, where $(Q, +)$ is a loop, γ is a permutation of the set Q such that $\gamma 0 = 0$, then $\varphi \in Aut(Q, \circ)$ if and only if: (i) $(\varphi, \varphi, L_{\varphi 0}\varphi)$ is an autotopy of the loop $(Q, +)$; (ii) $\gamma^{-1}\varphi\gamma = L_{\varphi 0}\varphi$; (iii) $\varphi 0 + x = x + \varphi 0$ for all $x \in Q$; (iv) $\varphi 0 \circ \varphi 0 = \varphi 0$.*

Proof. Let $\varphi \in Aut(Q, \circ)$. Then

$$\varphi\gamma(x + y) = \gamma(\varphi x + \varphi y). \tag{8.19}$$

If in (8.19) $x = y = 0$, then $\varphi\gamma 0 = \gamma(\varphi 0 + \varphi 0)$, $\varphi 0 = \gamma(\varphi 0 + \varphi 0)$, $\varphi 0 = \varphi 0 \circ \varphi 0$, i.e., $\varphi 0$ is an idempotent element of the quasigroup $(Q, \circ)$. If we take $x = 0$ in (8.19), then we have $\varphi\gamma y = \gamma L_{\varphi 0}\varphi y$, $\gamma^{-1}\varphi\gamma = L_{\varphi 0}\varphi$.

By $y = 0$ in (8.19) we have $\varphi\gamma x = \gamma(\varphi x + \varphi 0)$, $\gamma^{-1}\varphi\gamma = R_{\varphi 0}\varphi$. Then we have

$$\gamma^{-1}\varphi\gamma = R_{\varphi 0}\varphi = L_{\varphi 0}\varphi. \tag{8.20}$$

We can re-write (8.19) in the following way: $\gamma^{-1}\varphi\gamma(x + y) = \varphi x + \varphi y$, and, taking into consideration the equality (8.20), as $L_{\varphi 0}\varphi(x+y) = \varphi x+\varphi y$, i.e., $(\varphi, \varphi, L_{\varphi 0}\varphi)$ is an autotopy of the loop $(Q, +)$.

From the equality (8.20) it follows that $R_{\varphi 0} = L_{\varphi 0}$, i.e., $x + \varphi 0 = \varphi 0 + x$ for any $x \in Q$.

Conversely, if $(\varphi, \varphi, L_{\varphi 0}\varphi)$ is an autotopy of the loop $(Q, +)$, then we have $L_{\varphi 0}\varphi(x+y) = \varphi x+\varphi y$, $\gamma L_{\varphi 0}\varphi(x+y) = \gamma(\varphi x+\varphi y)$. From (ii) we have $\gamma L_{\varphi 0}\varphi = \varphi\gamma$. $\varphi\gamma(x+y) = \gamma(\varphi x+\varphi y)$, $\varphi(x \circ y) = \varphi x \circ \varphi y$. Therefore $\varphi \in Aut(Q, \circ)$. □

We shall denote by $Z(Q, +)$ the subset of a loop $(Q, +)$ such that

$$Z(Q, +) = \{a \in Q | a + x = x + a \ \forall\ x \in Q\}.$$

Corollary 8.65. *Let the quasigroup $(Q, \circ)$ be an isotope of a loop $(Q, +)$ with $Z(Q, +) = 0$ of the form $x \circ y = \gamma(x + y)$, where γ is a permutation of the set Q such that $\gamma 0 = 0$. Then $Aut(Q, \circ) = C_{Aut(Q,+)}(\gamma)$.*

Proof. If $\varphi \in Aut(Q, \circ)$, then $(\varphi, \varphi, L_{\varphi 0}\varphi)$ is an autotopy of the loop $(Q, +)$. Since $Z(Q, +) = 0$, then $L_{\varphi 0} = \varepsilon$, the permutation φ is an automorphism of the loop $(Q, +)$. From condition (ii) of Proposition 8.64 it follows that $\gamma\varphi = \varphi\gamma$. Therefore $Aut(Q, \circ) \subseteq C_{Aut(Q,+)}(\gamma)$.

Conversely, let $\varphi \in C_{Aut(Q,+)}(\gamma)$. Then $\varphi(x \circ y) = \varphi\gamma(x+y) = \gamma\varphi(x+y) = \gamma(\varphi x+\varphi y) = \varphi x \circ \varphi y$. Therefore $Aut(Q, \circ) = C_{Aut(Q,+)}(\gamma)$. □

Remark 8.66. For groups the condition $Z(Q, +) = 0$ is equivalent to the condition that the center of the group $(Q, +)$ coincides with 0. In the condition of Corollary 8.65 the quasigroup $(Q, \circ)$ has exactly one idempotent element.

Proposition 8.67. *If a quasigroup $(Q, \circ)$ of the form $x \circ y = \alpha x + \beta y$ has a unique idempotent element, where $(Q, +)$ is a loop, α, β are the permutations of the set Q such that $\alpha 0 = \beta 0 = 0$, then $Aut(Q, \circ) = C_{Aut(Q,+)}(\alpha, \beta)$.*

Proof. If the quasigroup $(Q, \circ)$ has a unique idempotent element and has the form $x \circ y = \alpha x + \beta y$, where $\alpha 0 = \beta 0 = 0$, then $0 \circ 0 = 0$ is this idempotent element. If φ is an automorphism of the quasigroup $(Q, \circ)$, then $\varphi 0 = 0$. Indeed, $\varphi(0 \circ 0) = \varphi 0 \circ \varphi 0$, $\varphi 0 = \varphi 0 \circ \varphi 0$, $\varphi 0 = 0$ because there is only one idempotent element, namely 0.

From $\varphi(x \circ y) = \varphi x \circ \varphi y$ we have $\varphi(\alpha x + \beta y) = \alpha\varphi x + \beta\varphi y$. If we take $x = 0$ in the last equality, then $\varphi\beta y = \beta\varphi y$. By analogy $\varphi\alpha = \alpha\varphi$. Then $Aut(Q, \circ) \subseteq C_{Aut(Q,+)}(\alpha, \beta)$.

Further $\varphi(x \circ y) = \varphi(\alpha x + \beta y) = \varphi\alpha x + \varphi\beta y = \alpha\varphi x + \beta\varphi y = \varphi x \circ \varphi y$. Then $Aut(Q, \circ) = C_{Aut(Q,+)}(\alpha, \beta)$. □

Remark 8.68. It is possible to re-write the condition "the quasigroup $(Q, \circ)$ has a unique idempotent element" as $\alpha x + \beta x \neq x$ for all $x \in Q \backslash \{0\}$.

8.2.3 Automorphisms of LD-quasigroups

In this section some information about the structure of an automorphism group of the left distributive quasigroup Q and some of its isotopes is given in terms of subgroups associated with an S-loop, i.e., with an isotope of a special kind of quasigroup Q. The results of this section are published in [771, 772].

The problem of the study of automorphism groups of distributive quasigroups was posed in Galkin's Doctor of Sciences thesis [343].

A quasigroup $(Q, \cdot)$ is called *left distributive*, if it satisfies the identity

$$x \cdot yz = xy \cdot xz. \tag{8.21}$$

For short, we shall call a left distributive quasigroup an *LD-quasigroup*.

Recall, the group which is generated by all left translations of a quasigroup $(Q, \cdot)$ is called a *left multiplication group* of a quasigroup $(Q, \cdot)$. We shall denote this group by $LM(Q, \cdot)$. We denote by $M(Q, \cdot)$ the group which is generated by all the left and right translations of a quasigroup $(Q, \cdot)$.

We notice that any left translation of a left distributive quasigroup $(Q, \cdot)$ is its automorphism. Indeed, $L_a(xy) = a \cdot xy = ax \cdot ay = L_a x \cdot L_a y$.

Therefore, in any left distributive quasigroup $(Q, \cdot)$ we have $LM(Q, \cdot) \subseteq Aut(Q, \cdot)$, where $Aut(Q, \cdot)$ is the automorphism group of the quasigroup $(Q, \cdot)$.

Let $(Q, +)$ be a loop, i.e., a quasigroup with the identity element 0. The group $LI(Q, +) = \{\alpha \in LM(Q, +) \mid \alpha 0 = 0\}$ is called a *group of left inner permutations of a loop* $(Q, +)$, the group $I(Q, +) = \{\alpha \in M(Q, +) \mid \alpha 0 = 0\}$ is called *a group of inner permutations* of a loop $(Q, +)$.

It is known that the group $LI(Q, +)$ is generated by permutations of the form $L_{a,b} = L^{-1}_{a+b} L_a L_b$, the group $I(Q, +)$ is generated by permutations of the form $L_{a,b}$, $R_{a,b} = R^{-1}_{a+b} R_b R_a$, $T_a = L_a^{-1} R_a$, where a, b run the set Q ([80], p. 61). See, also, Subsection 1.2.6.

A loop in which the group LI is a subgroup of the group Aut is called *special*.

Below we shall use the fact that any translation L_a of a left distributive quasigroup $(Q, \cdot)$ commutes with any element of the group $LI(Q, \circ)$, i.e., $L_a = \psi$ lies in the centralizer of this group.

We start from the following theorem.

Theorem 8.69. *Any autotopy of a left M-loop has the form $T_a^{-1} T_1$, where $T_a = (L_a R_{I\varphi a}, L_{\varphi a}, L_a)$, $T_1 = (L_k\theta, \theta, L_k\theta)$, i.e., θ is a left pseudoautomorphism of this M-loop ([80, p. 111]).*

We recall that any S-loop is an M-loop.

Lemma 8.70. *Any automorphism of a left distributive quasigroup $(Q, \cdot)$ with the form $x \cdot y = \varphi x \circ \psi y$, where $(Q, \circ)$ is an S-loop, can be presented in the form $L_a^{-1} L_k \theta$, where L_k, L_a are translations of the loop $(Q, \circ)$, θ is an automorphism of $(Q, \circ)$, $k \in N_l^\circ$, $\theta\varphi = \varphi\theta$, $\theta\psi = \psi\theta$.*

Proof. From Theorems 8.69 and 1.133 and the form of isotopy between loop $(Q, \circ)$ and quasigroup $(Q, \cdot)$ it follows that any autotopy of quasigroup $(Q, \cdot)$ has the form

$$(\varphi^{-1}, \psi^{-1}, \varepsilon) T_a^{-1} T_1 (\varphi, \psi, \varepsilon),$$

where $T_a^{-1} T_1$ is an autotopy of the loop $(Q, \circ)$.

Since any automorphism of quasigroup $(Q, \cdot)$ is an autotopy with equal components, then any automorphism of quasigroup $(Q, \cdot)$ has the form $L_a^{-1} L_k \theta$, where L_a, L_k are some translations of loop $(Q, \circ)$, θ is some its pseudoautomorphism, i.e., any automorphism has the form of the third component of an autotopy of S-loop $(Q, \circ)$.

We prove that left translations of the loop $(Q, \circ)$ are automorphisms of the quasigroup $(Q, \cdot)$, and θ is an automorphism of the loop $(Q, \circ)$.

Indeed, $L^{\circ}_x y = x \circ y = \varphi^{-1}x \cdot \psi^{-1}y = L^{\cdot}_{\varphi_x^{-1}}\psi^{-1}y$. Since $\psi = L^{\cdot}_a$, then $L^{\circ}_x = L^{\cdot}_{\varphi_x^{-1}}L^{\cdot -1}_a$. Therefore, $LM(Q, \circ) \subseteq Aut(Q, \cdot)$.

If $L^{\circ -1}_a L^{\circ}_k$ is an automorphism of quasigroup $(Q, \cdot)$, and θ is an automorphism of quasigroup $(Q, \cdot)$, then $\theta(xy) = \theta(\varphi x \circ \psi y) = \theta x \cdot \theta y = \varphi\theta x \circ \psi\theta y$.

By $x = 1$ we obtain $\theta\psi y = \varphi\theta 1 \circ \psi\theta y$, since from $\varphi 1 \circ \psi 1 = 1$ it follows $\varphi 1 = 1$.

We notice, for any pseudoautomorphism θ of a loop $(Q, \circ)$ the following equality is true $\theta 1 = 1$. Indeed, $L^{\circ}_k\theta x \circ \theta y = L^{\circ}_k\theta(x \circ y)$ by definition of pseudoautomorphism [72, p. 45]. If we take $x = y = 1$, we have $(k \circ \theta 1) \circ \theta 1 = k \circ \theta 1$.

In the last equality we cancel from the right on $\theta 1$, after this we cancel from the left on k and we obtain: $\theta 1 = 1$. Therefore, $\theta\psi y = \psi\theta y$, i.e., $\theta\psi = \psi\theta$.

Similarly it is proved that $\theta\varphi = \varphi\theta$. Thus we have $\theta(\varphi x \circ \psi y) = \theta\varphi x \circ \theta\psi y$, i.e., θ is an automorphism of the loop $(Q, \circ)$. If $\theta \in Aut(Q, \circ)$, then $k \in N^{\circ}_l$, since $(k \circ \theta x) \circ \theta y = k \circ \theta(x \circ y) = k \circ (\theta x \circ \theta y)$. □

Let $C(\varphi, \psi) = \{\alpha \in Aut(Q, \circ) \mid \alpha\varphi = \varphi\alpha, \alpha\psi = \psi\alpha\}$, $C = \{\alpha \in Aut(Q, \circ) \mid \alpha\psi = \psi\alpha\}$.

Corollary 8.71. *In a left distributive quasigroup $(Q, \cdot)$*

$$Aut(Q, \cdot) = LM(Q, \circ) \cdot C.$$

Proof. From Lemma 8.70 it follows that $Aut(Q, \cdot) \subseteq LM(Q, \circ) \cdot C(\varphi, \psi)$. We prove that $C(\varphi, \psi) = C$. It is easy to see $C(\varphi, \psi) \subseteq C$.

Let us prove inverse inclusion. Let $\theta \in Aut(Q, \circ)$ and $\theta\psi = \psi\theta$. Prove that $\theta\varphi = \varphi\theta$. Since a left distributive quasigroup is an idempotent quasigroup, we have $\theta x = \theta x \cdot \theta x = \varphi\theta x \circ \psi\theta x$, $\theta x = \theta(x \cdot x) = \theta(\varphi x \circ \psi x) = \theta\varphi x \circ \psi\theta x$. Comparing the right sides of the last two equalities we have $\theta\varphi = \varphi\theta$. Therefore $C(\varphi, \psi) = C$.

We demonstrate that $LM(Q, \circ) \cdot C \subseteq Aut(Q, \cdot)$. Let $\alpha = \gamma\beta$, where $\gamma \in LM(Q, \circ)$, $\beta \in C$. From the proof of Lemma 8.70 it follows that $\gamma \in Aut(Q, \cdot)$.

We prove that β is an automorphism in the quasigroup $(Q, \cdot)$ too. Indeed, $\beta(xy) = \beta(\varphi x \circ \psi y) = \beta\varphi x \circ \beta\psi y = \varphi\beta x \circ \psi\beta y = \beta x \cdot \beta y$. □

Lemma 8.72. *In a special loop Q the group of left inner permutations is a normal subgroup of automorphism group of Q.*

Proof. In any loop Q we have $Aut(Q) \subseteq N(LI)$, where $N(LI) = \{\beta \in S_Q \mid \beta^{-1}(LI)\beta = LI\}$ is a normalizer in the symmetric group S_Q of the group LI. Indeed, the group LI is generated by the set $\{L^{-1}_{a \cdot b}L_a L_b \mid a, b \in Q\}$ (Lemma 1.109). Let $\beta \in Aut(Q)$.

We shall use the following equalities: $\beta L_a = L_{\beta a}\beta$, $\beta L^{-1}_a = L^{-1}_{\beta a}\beta$. Indeed, $\beta L_a x = \beta(ax) = \beta a \cdot \beta x = L_{\beta a}\beta x$, i.e., $\beta L_a = L_{\beta a}\beta$. If we multiply from the right by L^{-1}_a, further from the left by $L^{-1}_{\beta a}$, we obtain $\beta L^{-1}_a = L^{-1}_{\beta a}\beta$. Then $\beta L^{-1}_{a \cdot b}L_a L_b \beta^{-1} = L^{-1}_{\beta(a \cdot b)}L_{\beta a}L_{\beta b}$.

Since in a special loop $LI(Q) \subseteq Aut(Q)$ and LI is a normal subgroup of the group $N(LI)$, then $LI(Q) \trianglelefteq Aut(Q)$. □

We notice that any S-loop is a special loop [670, p. 61].

Theorem 8.73. *If $(Q, \cdot)$ is a left distributive quasigroup, $(Q, \circ)$ is an S-loop which is isotopic to $(Q, \cdot)$, then*

$$Aut(Q, \cdot) \cong LM(Q, \circ) \rtimes (C \,/\, LI(Q, \circ)).$$

Proof. From the properties of S-loop $(Q, \circ)$ which are pointed out in Theorem 6.7 and from the speciality of $(Q, \circ)$ it follows that $LI(Q, \circ) \subseteq C$. Since $LI(Q, \circ) \trianglelefteq Aut(Q, \circ)$ (Lemma 8.72) and $C \subseteq Aut(Q, \circ)$, then $LI(Q, \circ) \trianglelefteq C$.

We examine the intersection of the groups $LM(Q, \circ)$ and C. If a permutation β of the group $LM(Q, \circ)$ is its automorphism, then $\beta 1 = 1$. Therefore $\beta \in LI(Q, \circ)$, $LM(Q, \circ) \cap C = LI(Q, \circ)$.

Demonstrate that $LM(Q, \circ) \trianglelefteq Aut(Q, \cdot)$. From Lemma 8.70 it follows $L^{\circ}_x = L^{\cdot}_{\varphi^{-1}x} L^{\cdot -1}_a$.

Suppose $\alpha \in Aut(Q, \cdot)$. Then $\alpha L^{\circ}_x \alpha^{-1} = \alpha L_{\varphi^{-1}x} L^{-1}_a \alpha^{-1} = L_{\alpha\varphi^{-1}x} L^{-1}_{\alpha a} \in LM(Q, \circ)$, since it is known ([799], p. 3), that $LM(Q, \circ) = < L_x L^{-1}_y \mid x, y \in Q >$. Therefore $LM(Q, \circ) \trianglelefteq Aut(Q, \cdot)$.

Using the second homomorphism theorem [471, 422] further we obtain

$$Aut(Q, \cdot) \,/\, LM(Q, \circ) \cong (LM(Q, \circ) \cdot C) \,/\, LM(Q, \circ) \cong$$

$$C \,/\, (LM(Q, \circ) \cap C) \cong C \,/\, LI(Q, \circ).$$

Finally, by definition of the semi-direct product [471] we have

$$Aut(Q, \cdot) \cong LM(Q, \circ) \rtimes (C \,/\, LI(Q, \circ)).$$

□

Corollary 8.74. *In a left distributive quasigroup $(Q, \cdot)$ with the form $x \cdot y = \varphi x \circ \psi y$, where $(Q, \circ)$ is an S-loop, $\psi \in Aut(Q, \circ)$,*

$$Aut(Q, \cdot) \,/\, LM(Q, \cdot) \cong B \,/\, D,$$

where $B = C \,/\, LI(Q, \circ)$ and D is a factor-group of the group $\langle \psi \rangle$.

Proof. It is well known that $LM(Q, \cdot) \trianglelefteq Aut(Q, \cdot)$. Indeed, if $\alpha \in Aut(Q, \cdot)$, then $\alpha L_a = L_{\alpha a} \alpha$ (see Lemma 8.72).

From [799, p. 3] it follows that $LM(Q, \circ) \trianglelefteq LM(Q, \cdot)$ and $LM(Q, \cdot) = LM(Q, \circ) \langle \psi \rangle$, where $(Q, \circ)$ is an S-loop.

Then

$$Aut(Q, \cdot) \,/\, LM(Q, \cdot) \cong$$
$$(Aut(Q, \cdot) \,/\, LM(Q, \circ)) \,/\, (LM(Q, \cdot) \,/\, LM(Q, \circ)) \cong B \,/\, D.$$

□

8.2.4 Automorphisms of isotopes of LD-quasigroups

As usual, N_r denotes the right nucleus of a loop $(Q, \circ)$, i. e., $N_r = \{a \in Q \mid (x \circ y) \circ a = x \circ (y \circ a)$ for all $x, y \in Q\}$.

Theorem 8.75. *If $(Q, \circ)$ is a quasigroup of the form $x \circ y = (\varphi x + \psi y) + d$, where $x \cdot y = \varphi x + \psi y$ is a left distributive quasigroup, $d \in N_r(Q, +)$, $(Q, +)$ is an S-loop, then $Aut(Q, \circ) \cong LM(Q, +) \rtimes (H \,/\, LI(Q, +))$, where $H = \{\beta \in C \mid \beta d = d\}$.*

Proof. We prove that $Aut(Q, \circ) \subseteq Aut(Q, \cdot)$. Let $\beta \in Aut(Q, \circ)$, i.e., $\beta(x \circ y) = \beta x \circ \beta y$ for all $x, y \in Q$. If we express the operation $\circ$ through operations of the loop $(Q, +)$ and left distributive quasigroup $(Q, \cdot)$, then we have $x \circ y = x \cdot y + d$.

Then $\beta(x \circ y) = \beta(x \cdot y + d)$, $\beta x \circ \beta y = \beta x \cdot \beta y + d$. Comparing the right sides of the last two equalities we obtain $\beta(x \cdot y + d) = \beta x \cdot \beta y + d$.

If we suppose $x = y$ in the last equality, then, using the identity of the left distributivity, we have $\beta(x + d) = \beta x + d$.

Therefore $\beta(x \cdot y + d) = \beta(x \cdot y) + d = \beta x \cdot \beta y + d$. Then $\beta(x \cdot y) + d = \beta x \cdot \beta y + d$, $\beta(x \cdot y) = \beta x \cdot \beta y$. Thus $Aut(Q, \circ) \subseteq Aut(Q, \cdot)$.

Therefore any automorphism of a quasigroup $(Q, \circ)$ can be presented in the form $\alpha\beta$, where $\alpha \in LM(Q, +)$, $\beta \in C$.

Moreover, since group $LM(Q, +)$ acts transitively on the set Q, and the stabilizer of the element 0 is the group $LI(Q, +)$, then any element of the group $LM(Q, +)$ can be presented in the form $L_a\gamma$, where $\gamma \in LI(Q, +)$.

Since $LI(Q, +) \subseteq C$, then any automorphism of a quasigroup $(Q, \cdot)$ can be written in the form $L_a\gamma \cdot \beta = L_a(\gamma\beta) = L_a\delta$, where $\delta \in C$, L_a is a translation of the loop $(Q, +)$.

Let $L_a\beta \in Aut(Q, \cdot)$ and $L_a\beta$ be an automorphism of the quasigroup $(Q, \circ)$. Then $L_a\beta(x \circ y) = L_a\beta(x \cdot y + d) = a + (\beta x \cdot \beta y + \beta d)$,$L_a\beta x \circ L_a\beta y = (a + \beta x) \cdot (a + \beta y) + d$.

Comparing the right sides of the last equalities we have $a + (\beta x \cdot \beta y + \beta d) = (a + \beta x)(a + \beta y) + d$. If $x = y = 0$, then from the last equality it follows $a + \beta d = a + d$, $\beta d = d$.

Taking into consideration the last equality and the fact that $d \in N_r$, further we have $a + (\beta x \cdot \beta y + \beta d) = (a + \beta x \cdot \beta y) + d = (a + \beta x)(a + \beta y) + d$. Cancellation of d gives us the equality $a + \beta x \cdot \beta y = (a + \beta x)(a + \beta y)$. The received equality is true for all $x, y, a \in Q$, since $L_a^+ \in Aut(Q, \cdot)$.

Therefore, if $L_a\beta \in Aut(Q, \cdot)$ and $L_a\beta \in Aut(Q, \circ)$, then $\beta d = d$.

It is easy to check the converse: if $L_a\beta \in Aut(Q, \cdot)$ and $\beta d = d$, then $L_a\beta \in Aut(Q, \circ)$. Indeed, $L_a\beta(x \circ y) = a + (\beta x \cdot \beta y + \beta d) = a + (\beta x \cdot \beta y + d) = (a + \beta x \cdot \beta y) + d = (a + \beta x) \cdot (a + \beta y) + d = L_a\beta x \circ L_a\beta y$.

Therefore, $Aut(Q, \circ) = \{\beta \in Aut(Q, \cdot) \mid \beta d = d\}$.

We notice that any left translation of the loop $(Q, +)$ is an automorphism of the quasigroup $(Q, \circ)$, since $L_a(x \circ y) = L_a(x \cdot y + d) = L_a(x \cdot y) + d = L_a x \circ L_a y$.

Repeating the proof of Theorem 8.73, we obtain $LM(Q, +) \trianglelefteq Aut(Q, \circ)$.

Let $H = \{\beta \in C \mid \beta d = d\}$. Then, taking into consideration the proof of Theorem 8.73, we have $Aut(Q, \circ) \cong LM(Q, +) \rightthreetimes (H \,/\, LI(Q, +))$. □

We give some corollaries of Theorems 8.73 and 8.75.

Corollary 8.76. *In a distributive quasigroup* $(Q, \cdot)$ *we have* $Aut(Q, \cdot) \cong M(Q, +) \rightthreetimes (C \,/\, I)$, *where* I *is the group of inner permutations of a respective commutative Moufang loop.*

Proof. By the Belousov theorem [72], any S-loop $(Q, +)$, which is isotopic to a distributive quasigroup $(Q, \cdot)$, is a commutative Moufang loop (CML). The rest follows from the property of commutativity of CML $(Q, +)$. □

Definition 8.77. A quasigroup $(Q, \cdot)$ with the identities $xy = yx$, $x(xy) = y$, any three elements of which generate a medial subquasigroup, is called a CH-quasigroup.

The center of a $CLM(Q, +)$ is a set Z such that $Z = \{a \in Q \mid a + (x + y) = (a + x) + y$ for all $x, y \in Q\}$. In any $CLM(Q, +)$ center $Z(Q, +)$ is its normal abelian subgroup [173].

It is proved [593, p. 31], that any CH-quasigroup can be constructed in the following way: $x \cdot y = (-x - y) + d$, where the element d lies in the center of $CML(Q, +)$. A $CML(Q, +)$ with identity $3x = 0$ is called a 3-CML. Information about CH-quasigroups is in [740].

Corollary 8.78. *If* $(Q, \circ)$ *is a* CH*-quasigroup that is isotopic to a 3-*$CML(Q, +)$, *then* $Aut(Q, \circ) \cong M(Q, +) \rightthreetimes G$, *where*

$$G = \{\beta \in (Aut(Q, +) \,/\, I) \mid \beta d = d\}.$$

Proof. Since for a 3-$CML(Q,+)$ the quasigroup $(Q,\cdot)$ with the form $x\cdot y = (-x-y)$ is a distributive quasigroup, then we can apply Theorem 8.75.

If $(Q,+)$ is a CML, then $\beta(-x) = -\beta x$ for all $\beta \in Aut(Q,+)$, $x \in Q$. Thus $C = Aut(Q,+)$.

From commutativity of $(Q,+)$ we also have $LI = I$. We can apply Theorem 8.75. □

We recall, a quasigroup $(Q,\cdot)$ with identities $x\cdot y = y\cdot x, x\cdot(x\cdot y) = y, x\cdot(y\cdot z) = (x\cdot y)\cdot(x\cdot z)$ is called a distributive Steiner quasigroup. It is known that a distributive Steiner quasigroup is isotopic to a 3-$CML(Q,+)$.

The converse is true: any distributive Steiner quasigroup $(Q,\cdot)$ can be constructed from a $CML(Q,+)$ in the following way $x\cdot y = -x-y$, where $-x+x=0$ for all $x \in Q$.

It is easy to see that any distributive Steiner quasigroup is a CH-quasigroup. Therefore from Corollary 8.78 by $d=0$ we have the following result.

Corollary 8.79. *If $(Q,\circ)$ is a distributive Steiner quasigroup, then*

$$Aut(Q,\circ) \cong M(Q,+)\rtimes Aut(Q,+).$$

Example 8.80. The quasigroup from Example 3.96 is a non-distributive Steiner quasigroup. Indeed, $1\cdot 5\,7 = 3$, $1\,5\cdot 1\,7 = 2$.

8.2.5 Quasigroups with transitive automorphism group

It is clear that an automorphism group of any left distributive quasigroup $(Q,\cdot)$ acts on the set Q transitively. Quasigroups with transitive automorphism groups (especially n-ary linear quasigroups) are researched in [524].

Loops that have transitive automorphism groups also have been researched. We follow the well-written review of D.A. Robinson [714] in some places using the direct citation. "Let L be a loop with more than one element, and let L^* be the set of all elements in L which are not the identity element. If L has a transitive automorphism group (i.e., the automorphism group of L acts transitively on L^*), then it is convenient to employ an obvious acronym and to call such loops tag-loops" [714].

R.H. Bruck proved that a tag-loop with the ascending chain condition for normal subloops is either a simple loop or an abelian group of type $(p,p,\dots,p)$. However, a simple loop does not need to have a transitive automorphism group.

Geometrical classification of tag-loops is given by A. Barlotti and K. Strambach [34], and the algebraic classification by A. Drisko [287]. Classification of finite groupoids with 2-transitive group of automorphisms is given in [439]. Groups of automorphisms of some finitely defined quasigroups are studied in [404].

The spectrum for quasigroups with cyclic automorphisms and additional symmetries is researched in [177].

8.3 Non-isomorphic isotopic quasigroups

In this section we try to evaluate the number of non-isomorphic quasigroups that are isotopic to a fixed quasigroup of order n. Notice, here components of any isotopy are only permutations of the set Q. We shall use concepts which are defined in

Section 1.3. These results can also be found in [764, 767]. More complete information on the number of quasigroups, especially medial and T-quasigroups, is given in [760, 832, 290, 825, 826, 827, 523, 525, 282, 843]. The number of non-isomorphic left loops of a given order is estimated in [447].

Lemma 8.81. *The number B of isotopisms that map an m-ary quasigroup (Q, f) of order n in an isomorphic quasigroup is equal to*

$$B = \frac{|Avt| \cdot n!}{|Aut|}, \tag{8.22}$$

where $|Avt|$ is the order of the autotopism group of quasigroup (Q, f), $|Aut|$ is the order of the automorphism group of quasigroup (Q, f).

Proof. Any autotopism and any isomorphism maps m-ary quasigroup (Q, f) in an isomorphic quasigroup. Since in this case automorphisms are counted twice, we obtain formulae (8.22). □

Theorem 8.82. *The number k of non-isomorphic quasigroups isotopic to a finite fixed m-ary quasigroup (Q, f) of order n is included in the following formula*

$$\sum_{i=1}^{k} \frac{1}{|Aut_i|} = \frac{(n!)^{m-1}}{|Avt|}, \tag{8.23}$$

where $|Aut_i|$ are the orders of the automorphism groups of quasigroups which are isotopic to quasigroup (Q, f).

Proof. From Lemma 8.81 it follows that the number of isotopisms that move a fixed quasigroup (Q, f) in isomorphic ones is equal to B. It is clear that the number of all possible isotopisms over the set Q is equal to $(n!)^m$.

Therefore we obtain the following equality

$$\sum_{i=1}^{k} \frac{|Avt_i| \cdot n!}{|Aut_i|} = (n!)^m. \tag{8.24}$$

But isotopic quasigroups have isomorphic autotopy groups. Therefore $|Avt_i| = |Avt_j|$ for all possible values of indexes i and j. Thus we obtain the equality (8.23). □

We apply arguments used by proving of Theorem 8.82 to the case of G-loops. We recall that a G-loop is a loop such that all its loop isotopes are isomorphic to it. It is clear that isomorphic loops have isomorphic automorphism groups.

Lemma 8.83. *In any finite G-loop of order n the following equality $|Avt| = |Aut| \cdot n^2$ is true.*

Proof. We use the equality (8.22). There exist $n! \cdot n^2$ of isotopisms that map a loop in a loop. Therefore we obtain the following equality

$$\frac{|Avt| \cdot n!}{|Aut|} = n! \cdot n^2. \tag{8.25}$$

From the last equality we have $|Avt| = |Aut| \cdot n^2$. □

Corollary 8.84. *The number k of non-isomorphic quasigroups isotopic to finite binary group $(Q,+)$ of order n is included in the following formula*

$$\sum_{i=1}^{k} \frac{1}{|Aut_i|} = \frac{((n-1)!)^2}{|Aut|}, \tag{8.26}$$

where $|Aut|$ is the order of the automorphism group of group $(Q,+)$.

Proof. We can use formula (8.23) and the fact that $|Avt(Q,+)| = n^2 \cdot |Aut|$ (Theorem 1.151). □

It is clear that the above presented formulas are very general and cannot be very useful, but sometimes it is possible to obtain relatively interesting information.

Example 8.85. If $(Q,+) = Z_3$, then $|Aut(Q,+)| = 2$ and the formula (8.26) takes the form

$$\sum_{i=1}^{k} \frac{1}{|Aut_i|} = \frac{(2!)^2}{2} = 2. \tag{8.27}$$

It is not very complicated to find five non-isomorphic paired quasigroups of order three. We give Cayley tables of these quasigroups.

$\cdot$	1	2	3
1	1	2	3
2	2	3	1
3	3	1	2

$*$	1	2	3
1	1	2	3
2	3	1	2
3	2	3	1

$\star$	1	2	3
1	1	3	2
2	2	1	3
3	3	2	1

$\circ$	1	2	3
1	2	1	3
2	1	3	2
3	3	2	1

$\diamond$	1	2	3
1	1	3	2
2	3	2	1
3	2	1	3

The first quasigroup is a loop (unique loop of order three), the second one is the left loop, the third is the right loop, $|Aut_1| = |Aut_2| = |Aut_3| = 2$, $|Aut_4| = 3$, $|Aut_5| = 6$. Therefore these quasigroups are not isomorphic in pairs. Formula (8.27) helps us to stop our search, because $3 \cdot 1/2 + 1/3 + 1/6 = 2$. Thus $k = 5$. Taking into consideration that up to isomorphism there exist five linear quasigroups of order three [523], we conclude that any quasigroup of order three is simple medial.

Lemma 8.86. *If G is finite abelian group of order n, then the number k of non-isomorphic T-quasigroups which are isotopic to group G satisfies the following equality*

$$\sum_{i=1}^{k} \frac{1}{|Aut_i|} = |Aut|, \tag{8.28}$$

where $|Aut|$ is the order of automorphism group of the group G, $|Aut_i|$ are the orders of automorphism groups of T-quasigroups.

Proof. Using Corollary 8.55 we have that the number of isomorphisms that map a fixed quasigroup $(G,\cdot)$ in some other quasigroup is counted by formula

$$\frac{|H(G)|}{Aut(G,\cdot)} = \frac{|Aut| \cdot n}{Aut(G,\cdot)}.$$

The number of possible triplets, i.e., the number of all T-quasigroups, that we can obtain from the group G, is equal to $|Aut|^2 \cdot n$. Therefore

$$\sum_{i=1}^{k} \frac{|Aut| \cdot n}{|Aut_i|} = |Aut|^2 \cdot n, \; \sum_{i=1}^{k} \frac{1}{|Aut_i|} = |Aut|.$$

□

Chapter 9

Orthogonality of quasigroups

9.1 Orthogonality: Introduction 388
9.1.1 Squares and Latin squares 388
9.1.2 m-Tuples of maps and its product 389
9.1.3 m-Tuples of maps and groupoids 390
9.1.4 τ-Property 392
9.1.5 Definitions of orthogonality 395
9.1.6 Orthogonality in works of V.D. Belousov 397
9.1.7 Product of squares 398
9.2 Orthogonality and parastroph orthogonality 398
9.2.1 Orthogonality of left quasigroups 399
9.2.2 Orthogonality of quasigroup parastrophes 400
9.2.3 Orthogonality in the language of quasi-identities 402
9.2.4 Orthogonality of parastrophes in the language of identities 403
9.2.5 Spectra of some parastroph orthogonal quasigroups 405
9.3 Orthogonality of linear and alinear quasigroups 407
9.3.1 Orthogonality of one-sided linear quasigroups 408
9.3.2 Orthogonality of linear and alinear quasigroups 412
9.3.3 Orthogonality of parastrophes 416
9.3.4 Parastrophe orthogonality of T-quasigroups 420
9.3.5 (12)-parastrophe orthogonality 422
9.3.6 totCO-quasigroups 425
9.4 Nets and orthogonality of the systems of quasigroups 426
9.4.1 k-nets and systems of orthogonal binary quasigroups 426
9.4.2 Algebraic (k, n)-nets and systems of orthogonal n-ary quasigroups 427
9.4.3 Orthogonality of n-ary quasigroups and identities 428
9.5 Transformations which preserve orthogonality 429
9.5.1 Isotopy and (12)-isostrophy 430
9.5.2 Generalized isotopy 431
9.5.3 Gisotopy and orthogonality 433
9.5.4 Mann's operations 434

Orthogonality of a pair of binary groupoids, left quasigroups and quasigroups is studied from some points of view. Necessary and sufficient conditions of orthogonality of a finite quasigroup and its parastrophe (conjugate quasigroup in other terminology), including ones in the language of quasi-identities, are given. A new concept of gisotopy, which generalizes the concept of isotopy, is defined. There is information on quasigroups with self-orthogonal conjugates. In general we follow the article [643].

Orthogonal quasigroups are used by construction of codes. See [240, 241, 641, 642] and the next chapter.

In some places this chapter we recall in detail some known facts, many of which can be found in [72, 80, 77, 559, 685, 113, 594, 81, 86, 466, 74], maybe, in some other form, or, more or less easy, can be proved independently.

Notice that some results on orthogonality, especially, on orthogonality of quasigroups, which are included in this chapter, are also true in the infinite case.

9.1 Orthogonality: Introduction

9.1.1 Squares and Latin squares

We give some definitions of squares and Latin squares [594, 240, 559, 661].

Definition 9.1. For a positive integer n, *square* $S(Q)$ of order n is a $n \times n$ matrix, the entries (or values) of which belong to a set Q of k elements, $k \leqslant n$, and such that every element of Q has at least one occurrence in at least one row.

In combinatorics, squares without the condition $k \leqslant n$ are also studied, for example, magic squares [240].

We say that the square $S(Q)$ is defined over the set Q. Sometimes we also write this fact in the form $D(S) = Q$.

We write the fact that in a square $S(Q)$ in a cell with co-ordinates (i, j), where $i, j \in \{1, \dots, m\}$, an element $a \in Q$ is arranged, as $(i, j, a) \in S$.

The squares $S_1(Q)$ and $S_2(Q)$ are equal if and only if $(i, j, a) = (i, j, b)$ for all $(i, j, a) \in S_1$ and $(i, j, b) \in S_2$.

Definition 9.2. For a positive integer n, *row-Latin square* L of order n is an $n \times n$, matrix the entries (or values) of which belong to a set X of n elements, and such that every element of X has exactly one occurrence in each row.

Definition 9.3. For a positive integer n, *column-Latin square* L of order n is an $n \times n$ matrix, the entries (or values) of which belong to a set X of n elements, and such that every element of X has exactly one occurrence in each column.

In [240] there is the following definition of Latin square.

Definition 9.4. For a positive integer n, Latin square L of order n is a $n \times n$ matrix, the entries (or values) of which belong to a set X of n elements, and such that every element of X has exactly one occurrence in each row and each column.

We also give the earlier definition of a Latin square.

Definition 9.5. A *Latin square* is an arrangement of m variables $x_1, \dots, x_m$ into m rows and m columns such that no row and no column contains variables twice [594].

It is easy to see that the body of a Cayley table (i.e., a Cayley table without the bordering row and the bordering column) of a groupoid (G, A) is a *square* $S(G)$ and any square $S(G)$ can be a body of a Cayley table of a groupoid (G, A).

The body of Cayley table of a left quasigroup (Q, A) is a *permutation square* $L(Q, A)$ in which no row contains variables twice and the body of a Cayley table of a right quasigroup (Q, A) is a *permutation square* $R(Q, A)$ in which no column contains variables twice.

Any permutation square in which no row contains any of the variables twice can be a body of a Cayley table of a left quasigroup (Q, A). Any permutation square in which no column contains any of the variables twice can be a body of a Cayley table of a right quasigroup (Q, A).

The body of Cayley table of a quasigroup (Q, A) is a *Latin square* $L(Q, A)$ and any Latin square can be a body of a Cayley table of a quasigroup (Q, A).

9.1.2 m-Tuples of maps and its product

In [594] Mann defined the so-called set of permutations of a Latin square S. Mann wrote: "The rows of a Latin square are permutations of the row $x_1, x_2, \dots, x_m$. Let p_i be the permutation which transforms $x_1, x_2, \dots, x_m$ into the i-th row of the Latin square. Then $p_i p_j^{-1}$ leaves no variables unchanged for $i \neq j$. Otherwise one column contains a variable twice. On the other hand each set of m permutations $(p_1, p_2, \dots, p_m)$ such that $p_i p_j^{-1}$ $(i \neq j)$ leaves no variable unchanged generates a Latin square. We may therefore identify every Latin square with a set of m permutations $(p_1, p_2, \dots, p_m)$ such that $p_i p_j^{-1}$ $(i \neq j)$ leaves no variable unchanged."

Therefore Mann obtained a new presentation of a Latin square $S(Q)$, $|Q| = m$, as a set of m permutations of the set Q. As Denes and Keedwell wrote [240], the similar presentation of a Latin square as a set of permutations can be found in the earlier article of Schönhardt [749].

From the article [594] it follows that by permutations in a permutation square H.B. Mann understood rows of this square, i.e., left translations in the terminology of this section. Moreover, by the set of permutations $(s_1, s_2, \dots, s_m)$ he, in fact, understood an ordered set of permutations, i.e., an m-tuple of permutations of a set Q. Later many authors used this presentation in their articles [661, 559, 113].

Below we shall often consider a non-empty set Q of order m as the set of natural numbers $Q = \{1, 2, \dots, m\}$ with their natural order, i.e., $1 < 2 < 3 < \dots < m$. We do not lose the generality since there exists a bijective map between the set Q and any other set of finite order m.

Definition 9.6. Let $Q = \{1, 2, \dots, m\}$. By *m-tuple M of maps defined on the set Q* we shall understand any set of maps of Q indexed by elements of the set Q, i.e., $M = (\mu_1, \mu_2, \dots, \mu_m)$.

Definition 9.7. An m-tuple of maps $T = (\mu_1, \mu_2, \dots, \mu_m)$ of a set Q, such that any map μ_i is a permutation of the set Q, will be called an *m-tuple of permutations of the set Q*.

In other words, the tuple M is a vector, the co-ordinates of which are m fixed maps of the set Q, the permutation tuple T consists of m permutations of the group S_Q.

Example 9.8. Let $Q = \{1, 2, 3\}$. The following ordered sets of permutations are 3-tuples of permutations: $T_1 = (\varepsilon, (123), (132))$, $T_2 = ((12), (13), (132))$.

Definition 9.9. If $M_1 = (\mu_1, \mu_2, \dots, \mu_m)$ and $M_2 = (\nu_1, \nu_2, \dots, \nu_m)$ are m-tuples of maps defined on a set Q, $Q = \{1, 2, \dots, m\}$, then the product $M_1 * M_2$ is an m-tuple of the form $(\mu_1\nu_1, \mu_2\nu_2, \dots, \mu_m\nu_m)$.

Below we shall usually omit the symbol of operation of m-tuples product.

Proposition 9.10. *The set $\mathcal{M}$ of all m-tuples of maps, defined on a non-empty set Q of order m, forms a semigroup $(\mathcal{M}, *)$ relative to the operation $*$. The semigroup $(\mathcal{M}, *)$ is isomorphic to the direct product of m copies of the symmetric semigroup $\mathfrak{S}_Q$, i.e., $(\mathcal{M}, *) \simeq \mathfrak{S}_Q \times \mathfrak{S}_Q \times \dots \times \mathfrak{S}_Q = \bigotimes_{i=1}^{m} (\mathfrak{S}_Q)_i$, $|(\mathcal{M}, *)| = (m^m)^m$.*

Proof. We omit the proof, since it is easy and standard. □

Corollary 9.11. *The set $\mathcal{P}$ of all m-tuples of permutations, defined on a non-empty set Q of order m, forms a group $(\mathcal{P}, *)$ relative to the operation $*$. The group $(\mathcal{P}, *)$ is isomorphic to the direct product of m copies of the symmetric group S_Q, i.e., $(\mathcal{P}, *) \simeq S_Q \times S_Q \times \cdots \times S_Q = \bigotimes_{i=1}^{m} (S_Q)_i$, $|(\mathcal{P}, *)| = (m!)^m$ [661].*

Proof. We also omit the proof, since it is easy and standard. □

9.1.3 m-Tuples of maps and groupoids

Let (Q, A) be a finite groupoid of order m which is defined on the well-ordered set $Q = \{1, 2, \ldots, m\}$. This groupoid defines two sets of translations, namely the set of all left translations and the set of all right translations. It is clear that the last statement is true for any groupoid, not only for a finite groupoid (Q, A).

Moreover, the groupoid (Q, A) uniquely defines two m-tuples of maps, namely $T_1 = (L_1, L_2, \ldots, L_m)$ and $T_2 = (R_1, R_2, \ldots, R_m)$.

Any of the m-tuples T_1 and T_2 specifies the groupoid (Q, A) uniquely if we indicate a method (rows or columns) of filling a Cayley table of groupoid (Q, A).

It is easy to see that any m-tuple of maps $T = \{\mu_1, \mu_2, \ldots, \mu_m\}$ of a well-ordered set Q, $|Q| = m$, uniquely defines the following two groupoids: groupoid (Q, A), in which the maps μ_i are left translations and groupoid (Q, B), in which the maps μ_i are right translations.

Definition 9.12. We shall denote an m-tuple T of maps that consists of all left (respectively, right, middle, inverse left, inverse right, inverse middle) translations of a groupoid (Q, A) by T^l (respectively, by T^r, T^p, T^{Il}, T^{Ir}, T^{Ip}). In this case we shall say that the tuple T^l has the *kind* l, or we shall say that the tuple T^l is of the *kind* l.

Remark 9.13. A kind of maps (of permutations) defines the way of writing the maps (the permutations) in a square (in a permutation square) S. Left translations correspond to rows of the square S, right translations correspond to columns of the square S, and middle translations correspond to cells of the square S (see Example 9.34).

We denote by $\mathcal{T}^l(Q)$ the class of all m-tuples of maps of the kind l, which are defined on a well-ordered set Q, $|Q| = m$, and denote by $\mathcal{G}(Q)$ the class of all groupoids defined on the set Q.

By analogy with m-tuples of maps, we shall denote a square S that consists of all left (respectively, right, middle, inverse left, inverse right, inverse middle) translations of a groupoid (Q, A) by S^l (respectively, by S^r, S^p, S^{Il}, S^{Ir}, S^{Ip}).

We denote by $\mathbb{S}^l(Q)$ the class of all squares of the kind l that are defined on a set Q.

Proposition 9.14. *There exist bijections between the classes $\mathcal{T}^l(Q)$, $\mathbb{S}^l(Q)$, $\mathbb{S}^r(Q)$ and $\mathcal{G}(Q)$.*

It is well known that a Latin square L defines m-tuples of permutations of all six kinds. If $\mathcal{Q}(Q)$ denotes the class of all quasigroups that are defined on a well-ordered set Q, then Proposition 9.14 is also true for classes of tuples and classes of squares of the kinds $\{Il, Ir, p, Ip\}$.

Definition 9.15. A square which defines at least one m-tuple of permutations will be called a *permutation square.*

Proposition 9.16. *Any permutation square defines m-tuples of at least three kinds.*

Proof. From the definition of a permutation square it follows that a permutation square defines at least one m-tuple of permutations T. Since the tuple T is a permutation tuple, then T^{-1} is a permutation tuple, too. □

Example 9.17. The permutation square

$$S = \begin{matrix} 1 & 2 \\ 1 & 2 \end{matrix}$$

defines the following 2-tuples: $T_1 = (L_1, L_2) = (\varepsilon, \varepsilon)$ of the kind l, $T_2 = (L_1^{-1}, L_2^{-1}) = (\varepsilon, \varepsilon)$ of the kind Il and $T_3 = (R_1, R_2)$ of the kind r, where $R_1(1) = R_1(2) = 1$, $R_2(1) = R_2(2) = 2$.

Proposition 9.18. *If a square $S(Q)$ defines m-tuples of permutations of the kind l and r, then this square is a Latin square.*

Proof. Let $Q = \{a_1, \dots, a_m\}$. We denote by T_1 the m-tuple of permutations of kind l and by T_2 the m-tuple of permutations of the kind r which generate the square $S(Q)$.

If we suppose that there exist permutations p_1 and p_2 of the m-tuple T_1 such that $p_1(a_i) = p_2(a_i) = a_j$, then the column number a_i contains twice the element a_j, therefore the square $S(Q)$ does not define an m-tuple of the kind r.

If we suppose that there exist permutations p_1 and p_2 of the m-tuple T_2 such that $p_1(a_i) = p_2(a_i) = a_j$, then the row number a_i contains twice the element a_j, therefore the square $S(Q)$ does not define an m-tuple of the kind l.

Therefore, if a permutation square $S(Q)$ defines m-tuples of the kind l and r, then this square is a Latin square. □

Proposition 9.19. *A square $S(Q)$ defines an m-tuple of permutations of the kind p if and only if this square is a Latin square.*

Proof. Let $Q = \{a_1, \dots, a_m\}$. We suppose that the square $S(Q)$ defines the m-tuple T of permutations of the kind p. We recall that $p_x(y) = z$, where p_x is a middle translation of the groupoid (Q, A) which corresponds to the square $S(Q)$, means that in the square $S(Q)$ in the position (y, z) the element x is situated.

If we suppose that there exist permutations p_{a_x} and p_{a_y} $(a_x \neq a_y)$ of the m-tuple T such that $p_{a_x}(a_i) = p_{a_y}(a_i) = a_j$, then we have that in the position (a_i, a_j) the elements a_x and a_y are situated simultaneously.

Therefore we can conclude that the m-tuple T contains pairwise different permutations of the set Q such that "$p_i p_j^{-1}$ leaves no variables unchanged for $i \neq j$," i.e., the square $S(Q)$ is a Latin square.

It is easy to see that any Latin square defines an m-tuple of the kind p. □

Corollary 9.20. *Any permutation square S defines one or three m-tuples of permutations from the following set of the kinds of m-tuples of permutations $\{l, r, p\}$.*

Proof. The proof follows from Example 9.17, Propositions 9.18 and 9.19. □

We notice that it is possible to define tuples of maps of an infinite groupoid.

9.1.4 τ-Property

Denote the property of a set of permutations $\{p_1, p_2, \dots, p_m\}$ of an m-element set Q "$p_i p_j^{-1}$ $(i \neq j)$ leaves no variable unchanged" [594] as the τ-property. An m-tuple of permutations T can also have the τ-property. We shall call the m-tuple T a τ-m-tuple.

In [594], in fact, Mann proves the following:

Theorem 9.21. *A set $T = \{p_1, p_2, \dots, p_m\}$ of m permutations of a finite set Q of order m of a kind α, where $\alpha \in \{l, Il, r, Ir\}$, defines a Cayley table of a quasigroup if and only if T has the τ-property.*

A permutation α of a finite non-empty set Q which leaves no elements of the set Q unchanged will be called a *fixed point free permutation.*

Definition 9.22. Let Q be a non-empty finite set of an order m. A set $M = \{\mu_1, \mu_2, \dots, \mu_m\}$ of m maps of the set Q is called *strictly transitive* (more precisely, the set M acts on the set Q strictly transitively) if for any pair of elements x, y of the set Q there exists a unique map μ_j of the set Q such that $\mu_j(x) = y$.

Lemma 9.23. *A set $M = \{\mu_1, \mu_2, \dots, \mu_m\}$ of maps of a finite set Q of order m is a strictly transitive set if and only if M is a set of permutations of the set Q.*

Proof. Let $Q = \{1, 2, \dots, m\}$. We construct the map θ_M of the set Q^2 in the following way $\theta_M : (j; x) \mapsto (j; \mu_j(x))$, i.e.,

$$\begin{array}{lll}
(1;1) & \longrightarrow & (1;\mu_1(1)) \\
(1;2) & \longrightarrow & (1;\mu_1(2)) \\
\dots & \dots & \dots \\
(1;m) & \longrightarrow & (1;\mu_1(m)) \\
(2;1) & \longrightarrow & (2;\mu_2(1)) \\
(2;2) & \longrightarrow & (2;\mu_2(2)) \\
\dots & \dots & \dots \\
(m;m) & \longrightarrow & (m;\mu_m(m)).
\end{array}$$

The set M is a strictly transitive set of maps if and only if the map θ_M is a permutation of the set Q.

If θ_M is a permutation of the set Q^2, then $|Im\,\mu_i| = m$ for any map μ_i. Indeed, if we suppose that there exists a map μ_j such that $|Im\,\mu_i| < m$, then we obtain that θ_M is not a permutation of the set Q^2. □

An m-tuple of permutations can also have the property of strict transitivity.

Theorem 9.24. *A set $T = \{p_1, p_2, \dots, p_m\}$ of m permutations of a finite set Q of order m is strictly transitive if and only if the set T has the τ-property.*

Proof. Mann (Theorem 9.21) proved that the set T of m permutations of an m-element set Q defines a Latin square if and only if the set T has the τ-property. He also proved ([594]) that the set T has the property of strict transitivity if and only if the set T defines a Latin square.

Therefore we can conclude that for the set T the τ-property and the property of strict transitivity are equivalent. □

Proposition 9.25. *An m-tuple of permutations T is a tuple of the kind p or of the kind Ip if and only if the tuple T is a τ-m-tuple of permutations.*

Proof. The proof follows from Proposition 9.19. □

Any m-tuple $T = (p_1, p_2, \dots, p_m)$ defines m-tuple T^{-1} such that $T^{-1} = (p_1^{-1}, p_2^{-1}, \dots, p_m^{-1})$.

Proposition 9.26. *An m-tuple $T = (p_1, p_2, \dots, p_m)$ has the τ-property if and only if the m-tuple T^{-1} has the τ-property.*

Proof. From Theorem 9.24 it follows that this proposition will be proved if we prove the following equivalence: an m-tuple $T = (p_1, p_2, \dots, p_m)$ is a strictly transitive m-tuple if and only if the m-tuple T^{-1} is strictly transitive.

But it is easy to see that the following statements are equivalent: $(\forall\, a, b \in Q)$ $(\exists!\, p_i \in T)\, p_i(a) = b$ and $(\forall\, a, b \in Q)$ $(\exists!\, p_i^{-1} \in T^{-1})\, p_i^{-1}(b) = a$. □

Proposition 9.27. *An m-tuple of permutations $T = (p_1, p_2, \dots, p_m)$ has the τ-property if and only if the m-tuple $pTq = (pp_1q, pp_2q, \dots, pp_mq)$, where p, q are some fixed permutations of the set Q, has the τ-property.*

Proof. It is easy to see that the following statements are equivalent: "for any fixed elements $a, b \in Q$ there exists a unique permutation $p_i \in T$ such that $p_i(a) = b$" and "for any fixed elements $a, b \in Q$ there exists a unique permutation $p_iq \in Tq$ such that $p_iq(a) = q(b)$."

Since elements a, b are arbitrary fixed elements of the set Q, we can denote the element $q(b)$ by b_1. Therefore, we can re-write the last statement in the following equivalent form "for any fixed elements $a, b_1 \in Q$ there exists a unique permutation $p_iq \in Tq$ such that $p_iq(a) = b_1$."

The last statement is equivalent to the following one: "for any fixed elements $a, b_1 \in Q$ there exists a unique permutation $pp_iq \in pTq$ such that $pp_iq(a) = p(b_1)$."

Similarly, as it was pointed out above, further we can re-write the last statement in the following equivalent form "for any fixed elements $a, b_2 \in Q$ there exists a unique permutation $pp_iq \in pTq$ such that $pp_iq(a) = b_2$," where $b_2 = p(b_1)$. □

Remark 9.28. In fact, Proposition 9.27 in some other form can be found in the article by Mann [594].

Corollary 9.29. *An m-tuple of permutations $pT = (pp_1, pp_2, \dots, pp_m)$ has the τ-property if and only if the m-tuple $Tp = (p_1p, p_2p, \dots, p_mp)$ has the τ-property, where p is a permutation of the set Q.*

Proof. By Proposition 9.26 we have that an m-tuple pT has the τ-property if and only if the m-tuple $p^{-1}pTp = Tp$ has the τ-property. □

Corollary 9.30. *An m-tuple of permutations $T = (p_1, p_2, \dots, p_m)$ has the τ-property if and only if the m-tuple $p^{-1}Tp = (p^{-1}p_1p, p^{-1}p_2p, \dots, p^{-1}p_mp)$, where p is a permutation of a set Q, has the τ-property.*

Proof. It is evident. □

Lemma 9.31. *In a quasigroup $(Q, \cdot)$ any of the sets* $\mathbf{L}$, $\mathbf{R}$, $\mathbf{I}$, $\mathbf{L}^{-1}$, $\mathbf{R}^{-1}$ *and* $\mathbf{I}^{-1}$ *has the τ-property.*

Proof. Let us suppose the contrary, that there exist translations L_a, L_b, $a \neq b$, of a quasigroup $(Q, \cdot)$ and an element $x \in Q$ such that $L_aL_b^{-1}x = x$. If in the last equality we change the element x by the element L_bx, then we obtain $L_ax = L_bx$, $a \cdot x = b \cdot x$, $a = b$. We receive a contradiction. Therefore the set $\mathbf{L}(Q, \cdot)$ has the τ-property.

The remaining cases can be proved in a similar way. □

Remark 9.32. Often we shall suppose that the set Q is a well-ordered set. We shall fix an order of any bordering row and bordering column of any groupoid; moreover, we shall suppose that orders in bordering rows and bordering columns of any groupoid coincide with the order of the set Q.

Remark 9.33. Any of the sets $\mathbf{L}$, $\mathbf{R}$, $\mathbf{I}$, $\mathbf{L}^{-1}$, $\mathbf{R}^{-1}$ and $\mathbf{I}^{-1}$ of a quasigroup $(Q, \cdot)$ defines this quasigroup in a unique way. Indeed, we can take into consideration the agreements of Remark 9.32 and the fact that all these quasigroup translations are indexed by the elements of set Q.

Example 9.34. If $P_1^{-1} = (23)$, $P_2^{-1} = (13)$, $P_3^{-1} = (12)$ are inverse permutations for middle translations of a quasigroup $(Q, \circ)$, then we can construct $(Q, \circ)$ in the following way:

	1	2	3
1	1		
2			1
3		1	

	1	2	3
1	1		2
2		2	1
3	2	1	

$\circ$	1	2	3
1	1	3	2
2	3	2	1
3	2	1	3

We can supplement Proposition 9.14 in the following way. Let Q be a finite well-ordered set.

We denote:

by $\mathcal{LQ}(Q)$–the class of all left quasigroups, which are defined on the set Q;
by $\mathcal{Q}(Q)$–the class of all quasigroups, which are defined on the set Q;
by $\mathbb{PS}(\mathcal{LQ})$ ($\mathbb{PS}(\mathcal{Q})$)–the class of all permutation squares, which are bodies of Cayley tables of left quasigroups (quasigroups) from $\mathcal{LQ}(Q)$ ($\mathcal{Q}(Q)$);
by $\mathfrak{T}^{\alpha}(Q)$–the class of m-tuples of permutations of a kind α, $\alpha \in \{l, Il, r, Ir, p.Ip\}$ that are defined on the set Q.

Proposition 9.35. *(i) There exist bijections between the classes $\mathcal{LQ}(Q)$, $\mathbb{PS}(\mathcal{LQ})$ and $\mathfrak{T}^{\alpha}(Q)$, $\alpha \in \{l, Il\}$;*

(ii) there exist bijections between the classes $\mathcal{Q}(Q)$, $\mathbb{PS}(\mathcal{Q})$ and $\mathfrak{T}^{\alpha}(Q)$, $\alpha \in \{l, Il, r, Ir, p.Ip\}$.

Proof. The proof follows from the results of Subsection 9.1.3 and this subsection. □

Proposition 9.36. *A τ-m-tuple T of a finite set Q of order m "defines" six Latin squares, namely: L^l, L^r, L^p, L^{Il}, L^{Ir}, L^{Ip}, which correspond to six quasigroups (Q, A), $(Q, A^{(12)})$, $(Q, A^{(132)})$, $(Q, A^{(23)})$, $(Q, A^{(123)})$ and $(Q, A^{(13)})$, respectively.*

Proof. The tuple T can have the following kinds $\{l, r, p, Il, Ir, Ip\}$. Thus this tuple defines six Latin squares. If we denote the Latin square that corresponds to the tuple T^l by L^l, then we can denote other Latin squares by L^r, L^p, L^{Il}, L^{Ir}, L^{Ip}. If we denote the quasigroup that corresponds to the square L^l by (Q, A), then the other five quasigroups are $(Q, A^{(12)})$, $(Q, A^{(132)})$, $(Q, A^{(23)})$, $(Q, A^{(123)})$ and $(Q, A^{(13)})$, respectively. □

Corollary 9.37. *Latin squares L^l, L^r, L^p, L^{Il}, L^{Ir} and L^{Ip} which are constructed from a τ-m-tuple T define at most six τ-m-tuples, namely six tuples that are left, right, middle translations and their inverse ones of the quasigroup (Q, A) which corresponds to the square L^l.*

Proof. Any of Latin squares L^l, L^r, L^p, L^{Il}, L^{Ir} and L^{Ip} from Proposition 9.36 define six, in general various, τ-m-tuples of permutations. But, as it follows from Table 1.1, any of these 36 τ-m-tuples of permutations coincides with the τ-m-tuples which are left (=T), right, middle translations and their inverse ones of the quasigroup (Q, A). □

It is not very difficult to understand that the number n of various τ-m-tuples which can be constructed from a τ-m-tuple T, using Proposition 9.36, is equal to 1, 2, 3 or 6.

For example, it is easy to see, if the tuple T defines a TS-quasigroup, then $n = 1$, since in any TS-quasigroup $(Q, \cdot)$ all its parastrophes coincide with $(Q, \cdot)$ [72]. We notice that TS-quasigroup of order 3 is given in Example 9.34.

9.1.5 Definitions of orthogonality

One of the most frequently applied and historically one of the first studied properties of Latin squares is the property of orthogonality. Orthogonality of quasigroups and Latin squares is used in application of quasigroups in coding theory and cryptology [240]. The famous Euler problem on Latin squares is devoted to the question of the existence of a pair of orthogonal Latin squares of order $4k + 2$, $k \in N$ [240, 559].

Definition 9.38. Two $m \times m$ squares S_1 and S_2, defined on the sets Q_1 and Q_2 respectively, $|Q_1| = |Q_2| = m$, are called *orthogonal* if when one is superimposed upon the other, every ordered pair of variables occurs once in the resulting square, i.e., the resulting square S_{12} is defined on the set $Q_1 \times Q_2$ [594].

Example 9.39. Let a square S_1 be defined on the set $Q_1 = \{1, 2, 3, 4\}$ and a square S_2 be defined on the set $Q_2 = \{a, b, c, d\}$. Let

$$S_1 = \begin{matrix} 1 & 2 & 3 & 3 \\ 1 & 1 & 2 & 2 \\ 2 & 1 & 3 & 3 \\ 4 & 4 & 4 & 4, \end{matrix} \qquad S_2 = \begin{matrix} a & a & a & b \\ b & c & b & c \\ d & d & c & d \\ a & b & c & d. \end{matrix} \qquad \textit{Then } S_{12} = \begin{matrix} 1a & 2a & 3a & 3b \\ 1b & 1c & 2b & 2c \\ 2d & 1d & 3c & 3d \\ 4a & 4b & 4c & 4d. \end{matrix}$$

Proof. The squares S_1 and S_2 are orthogonal since the square S_{12} is defined on the set $Q_1 \times Q_2$. See Remark 9.45 below. □

We can give the following definition.

Definition 9.40. We suppose that $m \times m$ squares S_1 and S_2 are defined on the sets Q_1 and Q_2, respectively. We define the operation $\oplus$ of *superimposition of the squares* S_1 and S_2 in the following way: $S_1 \oplus S_2 = S_{12}$ is an $m \times m$ square such that in any position (i, j), $i, j \in \{1, 2, \dots, m\}$, in S_{12} there is arranged an ordered pair of elements (a, b), where the element a is arranged in position (i, j) in the square S_1 and the element b is arranged in position (i, j) in the square S_2.

It is easy to see that the square S_{12} is defined on the set Q_{12} such that $Q_{12} \subseteq Q_1 \times Q_2$. In language of notions of Definition 9.40 we can re-write Definition 9.38 in the following form.

Definition 9.41. Two $m \times m$ squares S_1 and S_2, defined on the sets Q_1 and Q_2 respectively, $|Q_1| = |Q_2| = m$, are called orthogonal if and only if $D(S_1 \oplus S_2) = Q_1 \times Q_2$.

Definition 9.42. If a square $S_2(Q_2)$ is a Cayley table of a groupoid (Q_2, B) and a square $S_1(Q_1)$ is a Cayley table of a groupoid (Q_1, A), then the square $S_2(Q_2)$ is an isotopic image of the square $S_1(Q_1)$ with an isotopy T if and only if $(Q_2, B) = (Q_1, A)T$.

We formulate the well-known lemma which is mathematical folklore.

Lemma 9.43. *Squares $S_1(Q_1)$ and $S_2(Q_2)$ are orthogonal if and only if their isotopic images are orthogonal with the isotopies of the form $T_1 = (\varepsilon, \varepsilon, \varphi)$ and $T_2 = (\varepsilon, \varepsilon, \psi)$, respectively.*

Proof. We recall, the isotopy T_1 changes an element $b \in S_1(Q_1)$ with co-ordinates (i,j) by the element $\varphi b \in S_1(Q_1)T_1$ with co-ordinates (i,j).

If $S_1(Q_1)T_1$ is an isotopic image of a square $S_1(Q_1)$, $S_2(Q_2)T_2$ is an isotopic image of a square $S_2(Q_2)$ and $D(S_1(Q_1) \oplus S_2(Q_2)) = Q_1 \times Q_2$, then $D(S_1(Q_1)T_1 \oplus S_2(Q_2)T_2) = \varphi(Q_1) \times \psi(Q_2) = Q_1 \times Q_2$. □

Corollary 9.44. *Squares $S_1(Q_1)$ and $S_2(Q_2)$ are orthogonal if and only if there are orthogonal squares $S_1(Q_1)$ and $S_2'(Q_1) = S_2(Q_2)T_2$, where ψ is a third component of isotopy T_2 such that $\psi(Q_2) = Q_1$.*

Proof. In conditions of Lemma 9.43 it is sufficient to suppose that $\varphi = \varepsilon$ and to choose the component ψ of the isotopy T_2 such that $\psi(Q_2) = Q_1$. □

Remark 9.45. Below, taking into consideration Corollary 9.44, we shall study only orthogonality of squares, which are defined on the same set Q.

Definition 9.46. Groupoids $(Q,\cdot)$ and $(Q,*)$ defined on the same set Q are said to be *orthogonal* if the system of equations $x \cdot y = a$ and $x * y = b$ (where a and b are two given arbitrary elements of Q) has a unique solution [240].

We shall denote a fact that groupoids $(Q,\cdot)$ and $(Q,*)$ are orthogonal by $(Q,\cdot) \perp (Q,*)$. There exist various generalizations of Definitions 9.38 and 9.46 on the n-ary case, i.e., on hypercubes and n-ary groupoids [527, 57].

It is well known that on a finite set G of an order m there exist $(m)^{m^2}$ binary groupoids and $(m)^{m^2}$ squares.

Following [240] we shall call any square S_2, which is orthogonal to a square S_1, an *orthogonal mate of a square* S_1. Similarly, we shall call any groupoid (G,B), which is orthogonal to a groupoid (G,A), orthogonal mate, too.

Lemma 9.47. *An $m \times m$ square S defined on the set $Q = \{1,2,\dots,m\}$ has an orthogonal mate if and only if in this square there are m entries of any element of the set Q.*

Proof. This proof is a version of the proof of Theorem 5.1.1 from [240]. Let $S(Q)$ be an $m \times m$ square with m entries of any element of the set Q. We are able to construct an orthogonal mate to the square S if and only if we are able to change all entries of the element 1 in the square S by all elements of the set Q in any order, all entries of the element 2 in the square S by all elements of the set Q in any order, and so on. □

In [240] (p. 155) there are similar conditions for a Latin square to have an orthogonal mate.

The proof of Lemma 9.47 provides, probably, one of the most universal and the simplest methods of construction of an orthogonal mate to any binary groupoid which has such a mate.

D.A. Norton in [661] has called the squares which correspond to groupoids from Lemma 9.47 *pseudo-Latin squares.*

In [62] V.D. Belousov called operations with the property that is similar to the property of squares from Lemma 9.47, *full operations.*

Corollary 9.48. *If a square S which is defined on n-element set Q, $Q = \{1,2,\dots,n\}$, has an orthogonal mate, then there exist at least $(n!)^n$ squares which are orthogonal to the square S.*

Proof. The proof follows from Lemma 9.47. We can fill all entries of the element 1 by all elements of the square Q in any order, we can fill all entries of the element 2 by all elements of the square Q in any order, and so on. □

9.1.6 Orthogonality in works of V.D. Belousov

In a series of articles, see, for example, [62, 74, 76, 81, 85, 86], V.D. Belousov studied the property of orthogonality of binary and n-ary operations and systems of operations from algebraic and geometric points of view. Later his researches were continued by his pupils and by many other mathematicians.

In this subsection we suppose that all binary operations are defined on the same non-empty set $Q = \{1, 2, \dots, m\}$.

V.D. Belousov, in his study of the property of orthogonality, used the idea that a pair of binary operations $A(x, y)$ and $B(x, y)$ defines a map θ of the set Q^2 such that $\theta(x, y) = (A(x, y), B(x, y))$. Notice that S. Stein also used a similar idea [845].

Remark 9.49. It is easy to see that the operations A and B are orthogonal if and only if θ is a permutation of the set $Q \times Q$.

Corollary 9.50. *There exist $(n^2)!/(n!)^n$ squares defined on a set Q of the order n that have an orthogonal mate.*

Proof. From Remark 9.49 it follows that on the set Q there exist $(n^2)!$ ordered pairs of orthogonal operations. From Corollary 9.48 it follows that any square has $(n!)^n$ orthogonal mates. □

Following [62, 74], we have:

Definition 9.51. The binary operation $F(x, y) = x$ for all $x, y \in Q$ is called *the left identity operation*, the operation $E(x, y) = y$ is called *the right identity operation.* It is easy to see that the squares F and E that correspond to the groupoids (Q, F) and (Q, E), respectively, have the forms

$$F = \begin{array}{cccc} 1 & 1 & \dots & 1 \\ 2 & 2 & \dots & 2 \\ \dots & \dots & \dots & \dots \\ m & m & \dots & m, \end{array} \qquad E = \begin{array}{cccc} 1 & 2 & \dots & m \\ 1 & 2 & \dots & m \\ \dots & \dots & \dots & \dots \\ 1 & 2 & \dots & m. \end{array}$$

The operation F (E) is called a left (right) binary selector [74].

It is easy to see that the square F corresponds to the m-tuple $T^r_\varepsilon = (\varepsilon, \varepsilon, \dots, \varepsilon)$ of the kind r and the square E corresponds to the m-tuple $T^l_\varepsilon = (\varepsilon, \varepsilon, \dots, \varepsilon)$ of the kind l.

We re-formulate the well-known [661, 74, 240, 559] results on orthogonality of left quasigroups, right quasigroups, and quasigroups with the identity permutation squares of the kind l and r in the following manner:

Lemma 9.52. *(i) A square S is a permutation square of the kind l or the kind Il if and only if $S \perp F$;*

(ii) a square S is a permutation square of the kind r or the kind Ir if and only if $S \perp E$;

(iii) a permutation square S of the kind l is a Latin square if and only if $S \perp E$;

(iv) a permutation square S of the kind r is a Latin square if and only if $S \perp F$;

(v) a square S is a Latin square if and only if $S \perp F$ and $S \perp E$.

We denote by F^r_p the square, which is determined by the following tuple of permutations $T^r = (p, p, \dots, p)$, and we denote by E^l_g the square, which is determined by the following tuple of permutations $T^l = (g, g, \dots, g)$, where p and g are permutations of the set $Q = \{1, 2, \dots, m\}$.

We can re-write Lemma 9.52 in the following form:

Lemma 9.53. *(i) A square S is a permutation square of the kind l or the kind Il if and only if $S \perp F_p^r$;*

(ii) a square S is a permutation square of the kind r or the kind Ir if and only if $S \perp E_g^l$;

(iii) a permutation square S of the kind l is a Latin square if and only if $S \perp E_g^l$;

(iv) a permutation square S of the kind r is a Latin square if and only if $S \perp F_p^r$;

(v) a square S is a Latin square if and only if $S \perp F_p^r$ and $S \perp E_g^l$.

9.1.7 Product of squares

In [594] H.B. Mann defined the product of two permutation squares in this way: "Denote now by an m sided square S any set of m permutations $(s_1, s_2, \dots, s_m)$ and by the product SS' of two squares S and S' the square $(s_1s_1', s_2s_2', \dots, s_ms_m')$."

Using Definition 9.7 we can give Mann's definition of product of the squares and corresponding groupoids in the following form.

Definition 9.54. If $(p_1, p_2, \dots, p_m)$ and $(q_1, q_2, \dots, q_m)$ are m-tuples of the kind l of the permutation squares L_1 and L_2 respectively, then the product L_1L_2 is the permutation square $(p_1q_1, p_2q_2, \dots, p_mq_m)$ of the same kind.

Using Definition 9.54 of the product of squares it is possible to define the concept of the power of a square and, in particular, the concept of the power of a Latin square.

We can give the following generalization of Definition 9.54.

Definition 9.55. If $(p_1, p_2, \dots, p_m)$ and $(q_1, q_2, \dots, q_m)$ are m-tuples of maps of some fixed kinds of squares L_1 and L_2, respectively, then the product L_1L_2 is the square $(p_1q_1, p_2q_2, \dots, p_mq_m)$ of an admissible kind.

We notice that in general for Latin squares L_1 and L_2 we have $L_1^lL_2^l \neq L_1^rL_2^r$.

In conditions of Definition 9.55 the multiplication of a pair of Latin squares L_1 and L_2 defines at least $6^2 \cdot 4 = 144$ squares. Indeed, $L_1^{\alpha}L_2^{\beta} = L^{\gamma}$, where $\alpha, \beta \in \{l, r, p, Il, Ir, Ip\}$, $\gamma \in \{l, r, Il, Ir\}$. The multiplication of a pair of Latin squares of equal kinds defines at least 24 squares.

Example 9.56. If T_1 and T_2 are some m-tuples of maps, then m-tuple $T = T_1T_2$ defines a square of kind α, where $\alpha \in \{l, r\}$.

If T_1 and T_2 are some m-tuples of permutations, then m-tuple $T = T_1T_2$ defines a permutation square of kind α, where $\alpha \in \{l, r, Il, Ir\}$.

If, in addition, the m-tuple T has the τ-property, then T defines six Latin squares of any kind from the set of kinds $\{l, r, p, Il, Ir, Ip\}$.

9.2 Orthogonality and parastroph orthogonality

We give necessary and sufficient conditions of orthogonality of permutation squares, Latin squares, quasigroups, and their parastrophes.

9.2.1 Orthogonality of left quasigroups

In this subsection we give necessary and sufficient conditions of orthogonality of permutation squares, Latin squares, quasigroups and left (right) quasigroups.

Below in this article we suppose that we multiply squares only of equal kinds α and that the resulting square has also the kind α.

For Latin squares H.B. Mann proved the following basic theorem [594], which we give in a little bit more general form. See also [661, 240, 559].

Theorem 9.57. *Permutation squares L_1 and L_2 of kind α, $\alpha \in \{l, Il, r, Ir\}$, are orthogonal if and only if there exists a Latin square L_3 such that $L_3L_1 = L_2$.*

Proof. Let $L_1 \perp L_2$, $\alpha = l$. We suppose that m-tuple $T_1 = (p_1, p_2, \dots, p_m)$ corresponds to the permutation square L_1 and m-tuple $T_2 = (q_1, q_2, \dots, q_m)$ corresponds to the permutation square L_2.

If we superimpose the square L_1 on the square L_2, then in any cell (i, j) of the square of pairs P we shall have the pair $(p_i(j), q_i(j))$. The fact that the squares L_1 and L_2 are orthogonal means that in the square P any pair of elements $(x, y) \in Q \times Q$ appears exactly one time.

In other words, for any pair of elements (a, b), where a, b are some fixed elements of the set Q, there exists a unique pair of elements $i, j \in Q$ such that $p_i(j) = a$ and $q_i(j) = b$.

Since $p_i^{-1}(a) = j$, further we have $q_ip_i^{-1}(a) = b$. The last equality means that the tuple $T_2T_1^{-1} = (q_1p_1^{-1}, q_2p_2^{-1}, \dots, q_mp_m^{-1})$ is a strictly transitive set of permutations that acts on the set Q.

From Theorem 9.24 it follows that the tuple $T_2T_1^{-1}$ has the τ-property. Then the tuple $T_2T_1^{-1}$ defines a Latin square. It is easy to see that if the tuple $T_2T_1^{-1}$ defines the Latin square L_3 of the kind l, then $L_3 = L_2L_1^{-1}$.

Converse. Let the τ-m-tuple $T_1 = (p_1, p_2, \dots, p_m)$ (respectively, $T_2 = (q_1, q_2, \dots, q_m)$, $T_3 = (s_1, s_2, \dots, s_m)$) of the kind l correspond to the permutation square L_1 (L_2, L_3, respectively).

The equality $L_3L_1 = L_2$ means that $s_i(p_i(j)) = q_i(j)$ for any fixed element $p_i(j) \in Q$. Since the squares L_1 and L_2 are permutation squares, then in every row of the square of pairs P, the set of the first components of pairs is equal to the set Q and the set of the second components of any row of the square P coincides with the set Q, too.

Since the tuple T_3 is a strictly transitive set of permutations that acts on the set Q, we obtain that for any pair of elements $(a, b) \in Q^2$ there exists a unique element $s_i \in T_3$ such that $s_i(a) = b$, i.e., $s_i(p_i(j)) = q_i(j)$, for an element j of the set Q.

Therefore, in the square P any ordered pair (a, b) appears exactly one time, i.e., $L_1 \perp L_2$.

For permutation squares of the kind Il, r, Ir the proof is similar. □

Remark 9.58. Theorem 9.57 describes all orthogonal mates of a permutation square L_1 which are permutation squares, but, in general, there exist orthogonal mates of the square L_1 which are not permutation squares. For example, $L_1 \perp L_2$, but L_2 is not a permutation square:

$$L_1 = \begin{matrix} 1 & 2 & 3 \\ 3 & 2 & 1 \\ 2 & 1 & 3, \end{matrix} \qquad L_2 = \begin{matrix} 1 & 1 & 1 \\ 3 & 3 & 2 \\ 2 & 3 & 2. \end{matrix}$$

We denote the number of all Latin squares that are defined on a set Q of an order m by $\mathfrak{L}(m)$.

Corollary 9.59. *The number of permutation squares of the kind l that are defined on a set Q of order m and that are orthogonal to a fixed permutation square $S(Q)$ of the kind l is equal to the number $\mathfrak{L}(m)$.*

Proof. This is a direct consequence of Theorem 9.57. □

Corollary 9.60. *Latin squares L_1 and L_2 of the kind α are orthogonal if and only if the square $L_2 L_1^{-1}$ is a Latin square of the kind α, where $\alpha \in \{l, r, Il, Ir\}$.*

Proof. It is easy to see. □

The problem of construction of a Latin square L_j which is orthogonal to a fixed Latin square L is reduced to the problem when Mann's product of two Latin squares (of a square L_j and the square L^{-1}) is a Latin square. We believe it is possible to use a computer to check if Mann's product of Latin squares is a Latin square.

Corollary 9.61. *Latin squares L_1 and L_2 are orthogonal if and only if $L_1 L_2^{-1}$ is a Latin square.*

Proof. The proof follows from Corollary 9.60 and the following notice: $L_1 \perp L_2$ if and only if $L_2 \perp L_1$. Then $L_1 \perp L_2$ if and only if $L_1 L_2^{-1}$ is a Latin square. □

Theorem 9.57 allows us to give the following definition of orthogonality of m-tuples of permutations.

Definition 9.62. m-Tuples of permutations T_1 and T_2 are called *orthogonal* if $T_1 T_2^{-1}$ has the τ-property.

Since the product of squares (Definition 9.54) is defined with the help of the notion of the product of permutation tuples, we can re-formulate Theorem 9.57 in the language of m-tuples of permutations.

Theorem 9.63. *Permutation squares L_1 and L_2 of the kind α, $\alpha \in \{l, Il, r, Ir\}$, are orthogonal if and only if $T_1 T_2^{-1}$ is a τ-m-tuple, where T_1, T_2 are m-tuples of permutations of the kind α that correspond to the squares L_1 and L_2, respectively.*

Taking into consideration bijections which we have formulated in Proposition 9.35 we can formulate Theorem 9.57 for finite right quasigroups, left quasigroups and for quasigroups.

Theorem 9.64. *Left (right) quasigroups (Q, A) and (Q, B) are orthogonal if and only if there exists a τ-m-tuple T_3 such that $T_3 T_A = T_B$, where T_A is an m-tuple of the kind α that corresponds to left (right) quasigroup (Q, A), T_B is an m-tuple of the kind α that corresponds to left (right) quasigroup (Q, B), $\alpha \in \{l, Il\}$ ($\alpha \in \{r, Ir\}$).*

Theorem 9.65. *Quasigroups (Q, A) and (Q, B) are orthogonal if and only if $T_3 = T_A^{\alpha}(T_B^{\alpha})^{-1}$ is a τ-m-tuple, where T_A^{α} is a τ-m-tuple of the kind α that corresponds to the quasigroup (Q, A), T_B^{α} is a τ-m-tuple of the kind α that corresponds to the quasigroup (Q, B), and $\alpha \in \{l, r, Il, Ir\}$.*

We notice, it is possible to formulate conditions of orthogonality of quasigroups on the language of τ-m-tuples of the kind p. But these conditions differ sufficiently sharply from conditions of orthogonality of quasigroups given in language of τ-m-tuples of the kinds l and r.

9.2.2 Orthogonality of quasigroup parastrophes

In this subsection we give some conditions of orthogonality of a quasigroup and its parastrophes.

We suppose that $Q = \{1, 2, \dots, m\}$. For convenience we denote a finite quasigroup (Q, A) by the letter A. We denote m-tuples of translations of the quasigroup A in the following way $\overline{L} = (L_1, L_2, \dots, L_m)$, $\overline{L^{-1}} = (L_1^{-1}, L_2^{-1}, \dots, L_m^{-1})$, $\overline{R} = (R_1, R_2, \dots, R_m)$, $\overline{R^{-1}} = (R_1^{-1}, R_2^{-1}, \dots, R_m^{-1})$.

Theorem 9.66. *For a finite quasigroup A, the following equivalences are fulfilled:*

(i) $A \perp A^{(12)} \Longleftrightarrow \overline{R}\,\overline{L^{-1}}$ *is a τ-m-tuple;*
(ii) $A \perp A^{(13)} \Longleftrightarrow \overline{R}\,\overline{R}$ *is a τ-m-tuple;*
(iii) $A \perp A^{(23)} \Longleftrightarrow \overline{L}\,\overline{L}$ *is a τ-m-tuple;*
(iv) $A \perp A^{(123)} \Longleftrightarrow \overline{L}\,\overline{R}$ *is a τ-m-tuple;*
(v) $A \perp A^{(132)} \Longleftrightarrow \overline{R}\,\overline{L}$ *is a τ-m-tuple.*

Proof. (i) We can identify the Cayley table of a quasigroup A with a τ-m-tuple T_1 of the kind r, i.e., T_1 is a vector, the components of which are all right translations R_a, $a \in Q$, of the quasigroup A.

It is possible to identify a Cayley table of the quasigroup $A^{(12)}$ with a τ-m-tuple T_2 of the kind r, too, where T_2 is composed of the permutations $R_a^{(12)}$, $a \in Q$. From Table 1.1 it follows that $R_a^{(12)} = L_a$.

In order to obtain a criterion of orthogonality of the quasigroups A and $A^{(12)}$ we can apply Theorem 9.65, since the τ-tuples T_1 and T_2 are of the same kind. Therefore we have that $A \perp A^{(12)}$ if and only if $T_1 T_2^{-1} = (R_1 L_1^{-1}, R_2 L_2^{-1}, \dots, R_m L_m^{-1}) = \overline{R}\,\overline{L^{-1}}$ is a τ-tuple.

(ii) In this case we identify Cayley tables of quasigroups A and $A^{(13)}$ with τ-m-tuples T_1 and T_2 of the kind r, too. From Table 1.1 it follows that $R_a^{(13)} = R_a^{-1}$, i.e., $T_2 = (R_1^{-1}, \dots, R_m^{-1})$, where R_a is a right translation of the quasigroup A, $a \in Q$. Application of Theorem 9.65 gives us that $A \perp A^{(13)}$ if and only if $T_1 T_2^{-1} = (R_1 R_1, R_2 R_2, \dots, R_m R_m) = \overline{R}\,\overline{R} = \overline{R^2}$ is a τ-tuple.

(iii) We identify Cayley tables of quasigroups A and $A^{(23)}$ with τ-tuples T_1 and T_2 of the kind l. From Table 1.1 it follows that $L_a^{(23)} = L_a^{-1}$, i.e., $T_2 = (L_1^{-1}, \dots, L_m^{-1})$. Then $T_1 (T_2)^{-1} = \overline{L}\,\overline{L} = \overline{L^2}$. From Theorem 9.65 it follows that $A \perp A^{(23)}$ if and only if the tuple $\overline{L^2}$ is a τ-m-tuple.

(iv) In this case we identify Cayley tables of quasigroups A and $A^{(123)}$ with τ-tuples T_1 and T_2 of the kind l. From Table 1.1 it follows that $L_a^{(123)} = R_a^{-1}$, i.e., $T_2 = (R_1^{-1}, \dots, R_m^{-1})$. Then $T_1 (T_2)^{-1} = \overline{L}\,\overline{R}$. From Theorem 9.65 it follows that $A \perp A^{(123)}$ if and only if the tuple $\overline{L}\,\overline{R}$ is a τ-m-tuple.

(v) In this case we identify Cayley tables of quasigroups A and $A^{(132)}$ with τ-tuples T_1 and T_2 of the kind r. From Table 1.1 it follows that $R_a^{(132)} = L_a^{-1}$, i.e., $T_2 = (L_1^{-1}, \dots, L_m^{-1})$ in this case. Then $T_1 (T_2)^{-1} = \overline{R}\,\overline{L}$. From Theorem 9.65 it follows that $A \perp A^{(132)}$ if and only if the tuple $\overline{R}\,\overline{L}$ is a τ-m-tuple. □

Let

$$\begin{array}{ll} \mathbf{L}^2(Q,\cdot) = \{L_a L_a \mid a \in Q\}; & \mathbf{L}^{-1}(Q,\cdot) = \{L_a^{-1} \mid a \in Q\}; \\ \mathbf{R}^2(Q,\cdot) = \{R_a R_a \mid a \in Q\}; & \mathbf{R}^{-1}(Q,\cdot) = \{R_a^{-1} \mid a \in Q\}; \\ \mathbf{LR}(Q,\cdot) = \{L_a R_a \mid a \in Q\}; & \mathbf{P}^{-1}(Q,\cdot) = \{P_a^{-1} \mid a \in Q\}; \\ \mathbf{RL}(Q,\cdot) = \{R_a L_a \mid a \in Q\}; & \mathbf{RL}^{-1}(Q,\cdot) = \{R_a L_a^{-1} \mid a \in Q\}. \end{array}$$

Since by proving Theorem 9.66 we do not use the property that the sets of permutations of a quasigroup $(Q,\cdot)$ are well-ordered sets, then it is possible to re-formulate Theorem 9.66 in the following form.

Theorem 9.67. *For a finite quasigroup A the following equivalences are fulfilled:*

(i) $A \perp A^{(12)} \Longleftrightarrow \mathbf{RL}^{-1}$ *has the τ-property;*
(ii) $A \perp A^{(13)} \Longleftrightarrow \mathbf{R}^2$ *has the τ-property;*
(iii) $A \perp A^{(23)} \Longleftrightarrow \mathbf{L}^2$ *has the τ-property;*
(iv) $A \perp A^{(123)} \Longleftrightarrow \mathbf{LR}$ *has the τ-property;*
(v) $A \perp A^{(132)} \Longleftrightarrow \mathbf{RL}$ *has the τ-property.*

We recall that if the set of permutations $\mathbf{RL}^{-1}$ ($\mathbf{R}^2$, $\mathbf{L}^2$, $\mathbf{LR}$, $\mathbf{RL}$) has the τ-property, then this set defines a Latin square of the kind α, where $\alpha \in \{l, r, p, Il, Ir, Ip\}$.

Corollary 9.68. *If L is a Latin square that corresponds to the body of the Cayley table of a quasigroup (Q, A), then:*

(i) the square $L^r L^{Il}$ is a Latin square if and only if $A \bot A^{(12)}$;
(ii) the square $L^r L^r$ is a Latin square if and only if $A \bot A^{(13)}$;
(iii) the square $L^l L^l$ is a Latin square if and only if $A \bot A^{(23)}$;
(iv) the square $L^l L^r$ is a Latin square if and only if $A \bot A^{(123)}$;
(v) the square $L^r L^l$ is a Latin square if and only if $A \bot A^{(132)}$.

Proof. The proof follows from Theorem 9.66. □

9.2.3 Orthogonality in the language of quasi-identities

Conditions of orthogonality of a quasigroup and its parastrophe in the language of identities have a long and rich history [845, 240, 241, 74, 81, 141, 85]. The profound results in this direction belong to T. Evans [310], [241, Chapter 7], [184, Chapter 3], [146].

We re-formulate Theorem 9.66 in the language of quasi-identities. A quasi-identity is an identity or a formula of the form $(p_1 \approx q_1 \& \ldots \& p_n \approx q_n) \to p \approx q$ which is to be true for all choices of the variables. It is a bit fuzzy, but is made precise in any textbook on universal algebra; see, for example, [184, 591]. The left cancellation law $xy \approx xz \to y \approx x$ is a quasi-identity.

Theorem 9.69. *For a finite quasigroup $(Q, \cdot)$ the following equivalences are fulfilled:*

(i) $(Q,\cdot)\bot(Q,\cdot)^{(12)} \Longleftrightarrow ((x\backslash z)\cdot x = (y\backslash z)\cdot y \Longrightarrow x = y) \Longleftrightarrow (P_{zy}P_{yz}x = x \Longrightarrow x = y)$;
(ii) $(Q,\cdot)\bot(Q,\cdot)^{(13)} \Longleftrightarrow (zx\cdot x = zy\cdot y \Longrightarrow x = y) \Longleftrightarrow (R_x^2 z = R_y^2 z \Longrightarrow x = y)$;
(iii) $(Q,\cdot)\bot(Q,\cdot)^{(23)} \Longleftrightarrow (x\cdot xz = y\cdot yz \Longrightarrow x = y) \Longleftrightarrow (L_x^2 z = L_y^2 z \Longrightarrow x = y)$;
(iv) $(Q,\cdot)\bot(Q,\cdot)^{(123)} \Leftrightarrow (x\cdot zx = y\cdot zy \Longrightarrow x = y) \Leftrightarrow (L_x R_x z = L_y R_y z \Longrightarrow x = y)$;
(v) $(Q,\cdot)\bot(Q,\cdot)^{(132)} \Leftrightarrow (xz\cdot x = yz\cdot y \Longrightarrow x = y) \Leftrightarrow (R_x L_x z = R_y L_y z \Longrightarrow x = y)$.

Proof. (i) From Theorem 9.66 it follows that the tuple $\overline{R}\ \overline{L^{-1}}$ has the τ-property. The τ-property means: if $x, y, z \in Q$, $x \neq y$, then the following inequality is fulfilled $(R_x L_x^{-1})(R_y L_y^{-1})^{-1} z \neq z$ for all $x, y, z \in Q$.

Further proof is only a simplification of the last inequality. We can write the last implication in an equivalent form: if $(R_x L_x^{-1})(R_y L_y^{-1})^{-1} z = z$, then $x = y$. In an equivalent form further we have: if $(R_x L_x^{-1}) z = (R_y L_y^{-1}) z$, then $x = y$. Therefore, if $(x\backslash z)\cdot x = (y\backslash z)\cdot y$, then $x = y$.

We can simplify the equality $(R_x L_x^{-1})(R_y L_y^{-1})^{-1} z = z$ in the following way:

$$R_x L_x^{-1} L_y R_y^{-1} z = z, (z \to R_y z), R_x L_x^{-1} L_y z = R_y z, R_x L_x^{-1}(yz) = zy,$$
$$(x\backslash yz)x = zy, (x\backslash yz)\backslash zy = x, R^{\backslash}_{zy} R^{\backslash}_{yz} x = x, (\text{Table1.1}), P_{zy}P_{yz}x = x,$$
$$P_{yz}x = P_{zy}^{-1}x, P_{yz}x = P_{y*x}x, \text{ where } y * z = t \text{ if and only if } z\cdot y = t.$$

(ii) From Theorem 9.66 it follows that the tuple $\overline{R}\,\overline{R}$ has the τ-property. Then we have: if $x, y, z \in Q$, $x \neq y$, then the following inequality is fulfilled:

$$(R_x R_x)(R_y R_y)^{-1} z \neq z$$

for all $x, y, z \in Q$. We can write the last implication in an equivalent form: if

$$(R_x R_x)(R_y R_y)^{-1} z = z,$$

then $x = y$. We simplify equality $(R_x R_x)(R_y R_y)^{-1} z = z$ in the following way: $R_x R_x R_y^{-1} R_y^{-1} z = z$, $(z \rightarrow R_y R_y z)$, $R_x R_x z = R_y R_y z$, $zx \cdot x = zy \cdot y$.

Cases (iii), (iv) and (v) are proved similar to Case (i) and we omit the proofs of these cases. □

We notice that it is possible to deduce Theorem 9.69 from the following Belousov criteria [74, Lemma 2] of orthogonality of two binary quasigroups which is also true for infinite quasigroups:

Theorem 9.70. *Belousov criteria. Quasigroups (Q, A) and (Q, B) are orthogonal if and only if the following binary operation $C(x, y) = A(x, B^{(23)}(x, y))$ is a right quasigroup [74].*

Proof. From Definition 9.46 it follows that $A \perp B$ if and only if the system

$$\begin{cases} A(x, y) = a \\ B(x, y) = b \end{cases}$$

has a unique solution $(x, y) \in Q^2$ for any fixed elements $a, b \in Q$. The last system is equivalent to the following

$$\begin{cases} A(x, y) = a \\ B^{(23)}(x, b) = y \end{cases} \iff \begin{cases} A(x, B^{(23)}(x, b)) = a \\ B^{(23)}(x, b) = y. \end{cases}$$

We denote the binary operation $A(x, B^{(23)}(x, y))$ by $C(x, y)$. It is easy to see that the operation C is a left quasigroup. Therefore, quasigroups (Q, A) and (Q, B) are orthogonal if and only if the operation C is a right quasigroup. Thus $A \perp B$ if and only if the operation C is a quasigroup. □

Suppose that any quasigroup is defined as an algebra with three binary operations; see Definition 1.54. We shall denote the quasigroup class: with the quasi-identity $(x \backslash yz)x = zy \Longrightarrow x = y$ by $\mathfrak{C}^{(12)}$; with the quasi-identity $zx \cdot x = zy \cdot y \Longrightarrow x = y$ by $\mathfrak{C}^{(13)}$; with the quasi-identity $x \cdot xz = y \cdot yz \Longrightarrow x = y$ by $\mathfrak{C}^{(23)}$; with the quasi-identity $x \cdot zx = y \cdot zy \Longrightarrow x = y$ by $\mathfrak{C}^{(123)}$; with the quasi-identity $xz \cdot x = yz \cdot y \Longrightarrow x = y$ by $\mathfrak{C}^{(132)}$.

Proposition 9.71. *Any of classes $\mathfrak{C}^{\sigma}$, where $\sigma \in S_3 \backslash \{\varepsilon\}$, forms a quasi-variety and this class is closed under the formation of subalgebras, products, ultraproducts, isomorphic algebras and it contains a trivial algebra.*

Proof. This follows from the definition of classes $\mathfrak{C}^{\sigma}$ and standard information on quasivarieties [591, 212, 809, 894]. □

A quasivariety $\mathfrak{Q}$ is a variety if and only if it is closed under the formation of homomorphic images [212, 809].

9.2.4 Orthogonality of parastrophes in the language of identities

Theorem 9.66 allows us to give some sufficient conditions of orthogonality of a quasigroup and its parastrophe in the language of identities. In the following theorem we list some identities which provide orthogonality of a quasigroup $(Q, \cdot)$ and its concrete parastrophe.

Theorem 9.72. *1. In a finite quasigroup $(Q, \cdot)$ the fulfillment of any of the identities $x \cdot xy = yx$ (I), $xy \cdot y = yx$ (II), $xy \cdot yx = x$ (III), $yx \cdot xy = x$ (IV) is a sufficient condition for orthogonality of $(Q, \cdot)$ and $(Q, \cdot)^{(12)}$;*

2. *In a finite quasigroup $(Q, \cdot)$ the fulfillment of any of the identities $(yx \cdot x)x = y$ (V), $yx \cdot x = xy$ (VI), $(xy \cdot x)x = y$ (VII), $y(yx \cdot x) = x$ (VIII), $(yx \cdot x)y = x$ (IX) is a sufficient condition for orthogonality of $(Q, \cdot)$ and $(Q, \cdot)^{(13)}$;*

3. *In a finite quasigroup $(Q, \cdot)$ the fulfillment of any of the identities $x(x \cdot xy) = y$ (X), $x \cdot xy = yx$ (XI), $x(x \cdot yx) = y$ (XII), $y(x \cdot xy) = x$ (XIII), $(x \cdot xy)y = x$ (XIV) is a sufficient condition for orthogonality of $(Q, \cdot)$ and $(Q, \cdot)^{(23)}$;*

4. *In a finite quasigroup $(Q, \cdot)$ the fulfillment of any of the identities $x(x \cdot yx) = y$ (XV), $x(yx \cdot x) = y$ (XVI), $y(x \cdot yx) = x$ (XVII), $(x \cdot yx)y = x$ (XVIII) is a sufficient condition for orthogonality of $(Q, \cdot)$ and $(Q, \cdot)^{(123)}$;*

5. *In a finite quasigroup $(Q, \cdot)$ the fulfillment of any of the identities $(xy \cdot x)x = y$ (XIX), $(x \cdot xy)x = y$ (XX), $y(xy \cdot x) = x$ (XXI), $(xy \cdot x)y = x$ (XXII) is a sufficient condition for orthogonality of $(Q, \cdot)$ and $(Q, \cdot)^{(132)}$.*

Proof. The truth of Case 1 follows easily from Belousov's results [74]. Also we can apply Theorem 9.66. From this theorem it follows that, if are able to find a τ-tuple T of permutations such that $\overline{R}\,\overline{L^{-1}} = T$, then $(Q, \cdot) \bot (Q, \cdot)^{(12)}$. The first candidates on the role of tuple T can be tuples of the left, right, and middle translations of the quasigroup $(Q, \cdot)$ and tuples of inverse translations.

It is easy to see that for our purpose, the equalities $R_x L_x^{-1} y = R_x y$ and $R_x L_x^{-1} y = L_x^{-1} y$ for all $x, y \in Q$ are not suitable, since in these cases we obtain $L_x^{-1} y = y$ and $R_x z = z$ for all $x, y, z \in Q$. The fulfillment of any of the last two equalities in a quasigroup $(Q, \cdot)$ means that $|Q| = 1$.

Another 4 possibilities give us the first 4 identities. Namely,
$R_x L_x^{-1} y = L_x y$, $y \to L_x y$, $R_x y = L_x^2 y$, $x \cdot xy = yx$ (I);
$R_x L_x^{-1} y = R_x^{-1} y$, $R_x^2 L_x^{-1} y = y$, $R_x^2 y = L_x y$, $yx \cdot x = xy$, $x \leftrightarrow y$, $xy \cdot y = yx$ (II);
$R_x L_x^{-1} y = P_x y$, $R_x y = P_x L_x y$, $yx = P_x(xy)$, $xy \cdot yx = x$ (III);
$R_x L_x^{-1} y = P_x^{-1} y$, $R_x y = P_x^{-1} L_x y$, $yx = P_x^{-1}(xy)$, $yx \cdot xy = x$ (IV).
Identities (V)–(XXII) are obtained similarly. □

Corollary 9.73. *If in a finite quasigroup $(Q, \cdot)$ the identity $x \cdot xy = yx$ (I) holds, then $(Q, \cdot) \bot (Q, \cdot)^{(12)}$, $(Q, \cdot) \bot (Q, \cdot)^{(23)}$;*

if in a quasigroup $(Q, \cdot)$ the identity $xy \cdot y = yx$ (II) holds, then $(Q, \cdot) \bot (Q, \cdot)^{(12)}$, $(Q, \cdot) \bot (Q, \cdot)^{(13)}$;

if in a quasigroup $(Q, \cdot)$ the identity $x(x \cdot yx) = y$ (XII) holds, then $(Q, \cdot) \bot (Q, \cdot)^{(23)}$, $(Q, \cdot) \bot (Q, \cdot)^{(123)}$;

if in a quasigroup $(Q, \cdot)$ the identity $(xy \cdot x)x = y$ (VII) holds, then $(Q, \cdot) \bot (Q, \cdot)^{(13)}$, $(Q, \cdot) \bot (Q, \cdot)^{(132)}$.

Proof. In Theorem 9.72 there are the following equalities or equivalences of identities: (XI)=(I), (VI)$\leftrightarrow$ (II) ($x \leftrightarrow y$), (XV)=(XII), (XIX) = (VII). □

It is easy to see that the identities from Theorem 9.72 have two variables and five occurrences of these variables in any identity.

In [81, 85] all identities of the form $A^{\alpha}(x, A^{\beta}(x, A^{\gamma}(x, y))) = y$, where A^{α}, A^{β}, A^{γ} are some parastrophes of a quasigroup operation A which provide orthogonality of operation A and its parastrophe (or parastrophes) are classified up to parastrophical equivalence.

V.D. Belousov proved that up to parastrophical equivalence, there exist exactly seven types of such identities:

$$x(x \cdot xy) = y, \tag{9.1}$$
$$x(x(x/y)) = y, \tag{9.2}$$
$$x(x(y/x)) = y, \tag{9.3}$$
$$x(x/(y/x)) = y, \tag{9.4}$$
$$(x/xy)/x = y, \tag{9.5}$$
$$x((y/x)\backslash x) = y, \tag{9.6}$$
$$x((y\backslash x)/x) = y. \tag{9.7}$$

He also proved that these identities are minimal, i.e., these identities have minimal number of variables (two), and a minimal number of occurrences of these variables in both sides of an identity (five). Using only the operation "·" the identities (9.1)–(9.7) can be rewritten in the following form:

$$x(x \cdot xy) = y \quad (C_3 \text{ law}), \tag{9.8}$$
$$x(y \cdot yx) = y \quad \text{of type } T_2 \text{ [81]}, \tag{9.9}$$
$$x \cdot xy = yx \quad \text{(Stein's 1st law)}, \tag{9.10}$$
$$xy \cdot x = y \cdot xy \quad \text{(Stein's 2nd law)}, \tag{9.11}$$
$$xy \cdot yx = y \quad \text{(Stein's 3rd law)}, \tag{9.12}$$
$$xy \cdot y = x \cdot xy \quad \text{(Schroder's 1st law)}, \tag{9.13}$$
$$yx \cdot xy = y \quad \text{(Schroder's 2nd law)}. \tag{9.14}$$

The name C_3 law (C_3 identity) is used in [141]. The names of identities (9.10)–(9.14) originate from Sade's paper [730]. V.D. Belousov researched quasigroups with these identities in [81, 85]. See also [872], [866].

All identities from Belousov's list (or their parastrophically equivalent forms) can also be found in [310, 141, 146]. Spectra of quasigroups with these identities are researched in [141, 140, 146]. See also [680, 681, 142, 866].

9.2.5 Spectra of some parastroph orthogonal quasigroups

Finite quasigroups that are orthogonal to its (12)-parastrophe exist for any order n with the exception of $n = 1, 2, 3, 6$ [163].

In [687] the author proves, using constructive methods, that for every finite natural number n ($n \neq 1, 2, 6$) there exists an $n \times n$ Latin square orthogonal to its (123)-parastrophe.

The similar result is true for a quasigroup and its (132)-, (13)-, and (23)-parastrophes [687, 143].

Notice that in [143] the following exceptional cases $\sigma = (13)$, $\sigma = (23)$ and $n = 14, 26$ are researched.

We give the easily obtained typical partial results about spectrum of T_2-quasigroups, i.e., quasigroups with the identity (9.9) [745]. See also [140, 193].

Theorem 9.74. *A T-quasigroup $(Q, \cdot)$ of the form*

$$x \cdot y = \varphi x + \psi y + b \tag{9.15}$$

satisfies the T_2-identity if and only if $\varphi = I\psi^3$, $\psi^5 + \psi^4 + 1 = (\psi^2 + \psi + 1)(\psi^3 - \psi + 1) = 0$, where 1 is identity automorphism of the group $(Q, +)$ and 0 is the zero endomorphism of this group, $\psi^2 b + \psi b + b = 0$.

Proof. We rewrite the T_2-identity using the right part of the form (9.15) as follows:

$$\varphi x + \psi(\varphi y + \psi(\varphi y + \psi x + b) + b) + b = y \tag{9.16}$$

or, taking into consideration that $(Q, +)$ is an abelian group, φ, ψ are its automorphisms, after simplification of the equality (9.16) we have

$$\varphi x + \psi\varphi y + \psi^2\varphi y + \psi^3 x + \psi^2 b + \psi b + b = y. \tag{9.17}$$

If we put $x = y = 0$ in the equality (9.17), then we obtain

$$\psi^2 b + \psi b + b = 0, \tag{9.18}$$

where 0 is the identity (neutral) element of the group $(Q, +)$.

Therefore we can rewrite the equality (9.17) in the following form

$$\varphi x + \psi\varphi y + \psi^2\varphi y + \psi^3 x = y. \tag{9.19}$$

If we put $y = 0$ in the equality (9.19), then we obtain that $\varphi x + \psi^3 x = 0$. Therefore $\varphi = I\psi^3$, where $x + Ix = 0$ for all $x \in Q$ as above.

Notice, in any abelian group $(Q, +)$ the map I is an automorphism of this group. Really, $I(x + y) = Iy + Ix = Ix + Iy$.

Moreover, $I\alpha = \alpha I$ for any automorphism of the group $(Q, +)$. Indeed, $\alpha x + I\alpha x = 0$. From the other side $\alpha x + \alpha I x = \alpha(x + Ix) = \alpha 0 = 0$. Comparing the left sides we have $\alpha x + I\alpha x = \alpha x + \alpha I x$, $I\alpha x = \alpha I x$, i.e., $\alpha I = I\alpha$.

It is well known that $I^2 = \varepsilon$, i.e., $-(-x) = x$. Indeed, from equality $x + Ix = 0$ using commutativity we have $Ix + x = 0$. From the other side $I(x + Ix) = 0$, $Ix + I^2 x = 0$. Then $Ix + x = Ix + I^2 x$, $x = I^2 x$ for all $x \in Q$.

If we put $x = 0$ in the equality (9.19), then we obtain that

$$\psi\varphi y + \psi^2\varphi y = y. \tag{9.20}$$

If in the equality (9.20) we substitute the expression $I\psi^3$ instead of φ, then we have $I\psi^5 y + I\psi^4 y = y$, $\psi^5 y + \psi^4 y = Iy$, $\psi^5 y + \psi^4 y + y = 0$. The last condition can be written in the form $\psi^5 + \psi^4 + 1 = 0$, where 1 is the identity automorphism of the group $(Q, +)$ and 0 is the zero endomorphism of this group.

It is easy to check that $\psi^5 + \psi^4 + 1 = (\psi^2 + \psi + 1)(\psi^3 - \psi + 1)$.

Converse. If we take into consideration that $\psi^2 b + \psi b + b = 0$, then from the equality (9.17) we obtain the equality (9.19). If in the equality (9.19) we apply the following equality $\varphi = I\psi^3$, then we obtain $\psi I\psi^3 y + \psi^2 I\psi^3 y = y$, $\psi^4 Iy + \psi^5 Iy = y$ which is equivalent to the equality $\psi^5 y + \psi^4 y + y = 0$. Therefore T-quasigroup $(Q, \cdot)$ is a T_2-quasigroup. □

Corollary 9.75. *Any T-T_2-quasigroup is medial.*

Proof. The proof follows from the equality $\varphi = I\psi^3$ (see Theorem 9.74). □

Corollary 9.76. *A T-quasigroup $(Q, \cdot)$ of the form $x \cdot y = \varphi x + \psi y + b$ satisfies the T_2-identity if $\varphi = I\psi^3$, $\psi^2 + \psi + 1 = 0$.*

Proof. The proof follows from Theorem 9.74 and the following fact: if $\psi^2 + \psi + 1 = 0$, then $\psi^5 + \psi^4 + 1 = 0$. □

Theorem 9.77. *There exist medial T_2-quasigroups of any prime order p such that $p = 6t+1$, where $t \in \mathbb{N}$.*

Proof. We use Corollary 9.76. Let $(Z_p, +, \cdot, 1)$ be a ring (a Galois field) of residues modulo p, where p is a prime of the form $6t+1$, $t \in \mathbb{N}$. Quadratic equation $\psi^2 + \psi + 1 = 0$ has two roots $h_1 = (-1-\sqrt{-3})/2$ and $h_2 = (-1+\sqrt{-3})/2$. Since p is prime, then $g.c.d(h_1, p) = g.c.d(h_2, p) = 1$.

It is known (Lemma 3.55) that the number -3 is a quadratic residue modulo any prime p of the form $6t+1$, $m \in \mathbb{N}$. Finally, if the number $(-1-\sqrt{-3})$ is odd, then the number $(-1-\sqrt{-3})+p$ is even. □

Using the computer program Mace4 [614] we construct the following examples of T_2-quasigroups.

⊠	0	1	2	3
0	0	2	3	1
1	1	3	2	0
2	2	0	1	3
3	3	1	0	2

∘	0	1	2	3	4
0	0	2	4	1	3
1	2	1	3	4	0
2	4	3	2	0	1
3	1	4	0	3	2
4	3	0	1	2	4

⋄	0	1	2	3	4	5	6	7
0	0	2	7	1	5	6	3	4
1	4	1	6	2	3	7	5	0
2	5	7	2	6	0	1	4	3
3	7	5	0	3	2	4	1	6
4	3	6	1	7	4	2	0	5
5	1	4	3	0	6	5	7	2
6	2	0	5	4	7	3	6	1
7	6	3	4	5	1	0	2	7

Lemma 9.78. *There exist medial T_2-quasigroups of order 2^k for any $k \geqslant 2$.*

Proof. Since the medial T_2-quasigroup with the operation ⊠ has the order 2^2 and the medial T_2-quasigroup with the operation ⋄ has the order 2^3, $g.c.d.(2,3) = 1$, it is clear that the direct product of a finite number of T_2-quasigroups is a T_2–quasigroup. □

Combining Lemma 9.78 and Theorem 9.77 we formulate the following:

Theorem 9.79. *There exist medial T_2-quasigroups of any order of the form $2^k p_1^{\alpha_1} p_2^{\alpha_2} \dots p_m^{\alpha_m}$, where p_i are prime numbers of the form $6t+1$, $\alpha_i \in \mathbb{N}$, $i \in \overline{1,m}$, $k = 0$ or $k \geqslant 2$.*

Proof. The direct product of a finite number of T_2-quasigroups is a T_2-quasigroup. □

9.3 Orthogonality of linear and alinear quasigroups

In this section we give the conditions of orthogonality of a pair of left (right) linear quasigroups over a group $(Q,+)$. See Section 2.10 for definitions.

9.3.1 Orthogonality of one-sided linear quasigroups

Lemma 9.80. *Suppose that finite left (right) linear (alinear) quasigroup $(Q, \cdot)$ and finite left (right) linear (alinear) quasigroup $(Q, \circ)$ have the forms $x \cdot y = \alpha x + \beta y + c$ and $x \circ y = \gamma x + \delta y + d$ over a group $(Q, +)$. Then without loss of generality for the study of orthogonality of these quasigroups, we can take $c = d = 0$.*

Proof. The inner part of the Cayley table of any quasigroup is a square in the sense of Lemma 9.43. The quasigroup $(Q, \cdot)$ is an isotope of the form $(\varepsilon, \varepsilon, R_c^{-1})$ of a quasigroup $(Q, \diamond)$ with the form $x \diamond y = \alpha x + \beta y$.

The quasigroup $(Q, \circ)$ is an isotope of the form $(\varepsilon, \varepsilon, R_d^{-1})$ of a quasigroup $(Q, *)$ with the form $x * y = \gamma x + \delta y$.

Then by Lemma 9.43 for the study of orthogonality of left (right) linear (alinear) quasigroups we can take $c = d = 0$ without loss of generality. □

By Lemma 2.209 any left linear quasigroup $(Q, \cdot)$ over a group $(Q, +)$ has the form $x \cdot y = \varphi x + \beta y$, where $\varphi \in Aut(Q, +)$, $\beta \in S_Q$.

Theorem 9.81. *Left linear quasigroups $(Q, \cdot)$ and $(Q, \circ)$ of the form $x \cdot y = \varphi x + \beta y$ and $x \circ y = \psi x + \delta y$, respectively, which are defined over a group $(Q, +)$, are orthogonal if and only if the mapping $(-\varphi^{-1}\beta + \psi^{-1}\delta)$ is a permutation of the set Q.*

Proof. In [833] the analogue of Theorem 9.81 (Theorem 7) is proved for linear quasigroups. Quasigroups $(Q, \cdot)$ and $(Q, \circ)$ are orthogonal if and only if the system of equations

$$\begin{cases} \varphi x + \beta y = a \\ \psi x + \delta y = b \end{cases} \tag{9.21}$$

has a unique solution for any fixed elements $a, b \in Q$. We solve this system of equations in the usual way.

$$\begin{cases} x + \varphi^{-1}\beta y = \varphi^{-1}a \\ x + \psi^{-1}\delta y = \psi^{-1}b \end{cases} \iff \begin{cases} -\varphi^{-1}\beta y - x = -\varphi^{-1}a \\ x + \psi^{-1}\delta y = \psi^{-1}b. \end{cases}$$

We do the following transformation: (I row + II row → I row) and obtain the system:

$$\begin{cases} -\varphi^{-1}\beta y + \psi^{-1}\delta y = -\varphi^{-1}a + \psi^{-1}b \\ x + \psi^{-1}\delta y = \psi^{-1}b. \end{cases}$$

Write the expression $-\varphi^{-1}\beta y + \psi^{-1}\delta y$ as follows: $(-\varphi^{-1}\beta + \psi^{-1}\delta)y$. Then the system (9.21) is equivalent to the following system:

$$\begin{cases} (-\varphi^{-1}\beta + \psi^{-1}\delta)y = -\varphi^{-1}a + \psi^{-1}b \\ x + \psi^{-1}\delta y = \psi^{-1}b. \end{cases}$$

It is clear that the system (9.21) has a unique solution if and only if the mapping $(-\varphi^{-1}\beta + \psi^{-1}\delta)$ is a permutation of the set Q. □

By Lemma 2.209 any right linear quasigroup $(Q, \cdot)$ over a group $(Q, +)$ has the form $x \cdot y = \alpha x + \varphi y$, where $\alpha \in S_Q$, $\varphi \in Aut(Q, +)$.

Theorem 9.82. *Right linear quasigroups $(Q, \cdot)$ and $(Q, \circ)$ of the form $x \cdot y = \alpha x + \varphi y$ and $x \circ y = \gamma x + \psi y$, respectively, which are defined over a group $(Q, +)$, are orthogonal if and only if the mapping $(\varphi^{-1}\alpha - \psi^{-1}\gamma)$ is a permutation of the set Q.*

The proof of Theorem 9.82 is similar to the proof of Theorem 9.81 and we omit it.

Theorem 9.83. *Left linear quasigroup* $(Q,\cdot)$ *and left alinear quasigroup* $(Q,\circ)$ *of the form* $x\cdot y = \varphi x+\beta y$ *and* $x\circ y = I\psi x+\delta y$, *respectively, defined over a group* $(Q,+)$, *are orthogonal if and only if the mapping* $(\psi^{-1}\delta + J_{I\psi^{-1}b}\varphi^{-1}\beta)$ *is a permutation of the set* Q *for any* $b\in Q$.

Proof. The quasigroups $(Q,\cdot)$ and $(Q,\circ)$ are orthogonal if and only if the system of equations

$$\begin{cases} \varphi x + \beta y = a \\ \overline{\psi} x + \delta y = b \end{cases} \tag{9.22}$$

has a unique solution for any fixed elements $a, b \in Q$. We solve this system of equations in the following way:

$$\begin{cases} x + \varphi^{-1}\beta y = \varphi^{-1}a \\ (\overline{\psi})^{-1}\delta y + x = (\overline{\psi})^{-1}b \end{cases} \iff \begin{cases} J_x\varphi^{-1}\beta y + x = \varphi^{-1}a \\ Ix + \psi^{-1}\delta y = \psi^{-1}b. \end{cases}$$

We make the transformation (I row + II row → I row), and obtain the system:

$$\begin{cases} J_x\varphi^{-1}\beta y + \psi^{-1}\delta y = \varphi^{-1}a + \psi^{-1}b \\ Ix = \psi^{-1}b + I\psi^{-1}\delta y. \end{cases} \tag{9.23}$$

In the system (9.23) we simplify the left part of the first equation using the second equation:

$$\begin{aligned} &J_x\varphi^{-1}\beta y + \psi^{-1}\delta y = \\ &x + \varphi^{-1}\beta y - x + \psi^{-1}\delta y = \\ &\psi^{-1}\delta y - \psi^{-1}b + \varphi^{-1}\beta y + \psi^{-1}b + I\psi^{-1}\delta y + \psi^{-1}\delta y = \\ &\psi^{-1}\delta y - \psi^{-1}b + \varphi^{-1}\beta y + \psi^{-1}b = \\ &\psi^{-1}\delta y + J_{I\psi^{-1}b}\varphi^{-1}\beta y. \end{aligned}$$

Write the expression $(\psi^{-1}\delta y + J_{I\psi^{-1}b}\varphi^{-1}\beta y)$ as follows: $(\psi^{-1}\delta + J_{I\psi^{-1}b}\varphi^{-1}\beta)y$. Then the system (9.23) is equivalent to the following system:

$$\begin{cases} (\psi^{-1}\delta + J_{I\psi^{-1}b}\varphi^{-1}\beta)y = \varphi^{-1}a + \psi^{-1}b \\ Ix = \psi^{-1}b + I\psi^{-1}\delta y. \end{cases}$$

It is clear that the system (9.22) has a unique solution if and only if the mapping $(\psi^{-1}\delta + J_{I\psi^{-1}b}\varphi^{-1}\beta)$ is a permutation of the set Q for any $b \in Q$. □

Theorem 9.84. *Left alinear quasigroup* $(Q,\cdot)$ *and right linear quasigroup* $(Q,\circ)$ *of the form* $x\cdot y = \overline{\varphi}x + \beta y$ *and* $x\circ y = \gamma y + \psi x$, *respectively, defined over a group* $(Q,+)$, *are orthogonal if and only if the mapping* $(\psi^{-1}\gamma + \varphi^{-1}\beta)$ *is a permutation of the set* Q.

Proof. The quasigroups $(Q,\cdot)$ and $(Q,\circ)$ are orthogonal if and only if the system of equations

$$\begin{cases} \overline{\varphi} x + \beta y = a \\ \gamma y + \psi x = b \end{cases} \tag{9.24}$$

has a unique solution for any fixed elements $a, b \in Q$. We solve this system of equations as follows:

$$\begin{cases} (\overline{\varphi})^{-1}\beta y + x = (\overline{\varphi})^{-1}a \\ \psi^{-1}\gamma y + x = \psi^{-1}b \end{cases} \iff \begin{cases} (\overline{\varphi})^{-1}\beta y + x = (\overline{\varphi})^{-1}a \\ -x - \psi^{-1}\gamma y = -\psi^{-1}b. \end{cases}$$

We do the transformation (I row + II row → I row), and obtain the system:

$$\begin{cases} (\overline{\varphi})^{-1}\beta y - \psi^{-1}\gamma y = (\overline{\varphi})^{-1}a - \psi^{-1}b \\ -x - \psi^{-1}\gamma y = -\psi^{-1}b. \end{cases}$$

Write expression $(\overline{\varphi})^{-1}\beta y - \psi^{-1}\gamma y$ as follows: $((\overline{\varphi})^{-1}\beta - \psi^{-1}\gamma)y$. Then the system (9.24) is equivalent to the following system:

$$\begin{cases} ((\overline{\varphi})^{-1}\beta - \psi^{-1}\gamma)y = (\overline{\varphi})^{-1}a - \psi^{-1}b \\ -x - \psi^{-1}\gamma y = -\psi^{-1}b. \end{cases}$$

It is clear that the system (9.24) has a unique solution if and only if the mapping $((\overline{\varphi})^{-1}\beta - \psi^{-1}\gamma) = (I\varphi^{-1}\beta + I\psi^{-1}\gamma) = I(\psi^{-1}\gamma + \varphi^{-1}\beta)$ is a permutation of the set Q. □

Theorem 9.85. *Left linear quasigroup $(Q, \cdot)$ and right alinear quasigroup $(Q, \circ)$ of the form $x \cdot y = \varphi x + \beta y$ and $x \circ y = \gamma y + \overline{\psi} x$, respectively, defined over a group $(Q, +)$, are orthogonal if and only if the mapping $(\psi^{-1}\gamma + \varphi^{-1}\beta)$ is a permutation of the set Q.*

Proof. The quasigroups $(Q, \cdot)$ and $(Q, \circ)$ are orthogonal if and only if the system of equations

$$\begin{cases} \varphi x + \beta y = a \\ \gamma y + \overline{\psi} x = b \end{cases} \tag{9.25}$$

has a unique solution for any fixed elements $a, b \in Q$. We solve this system of equations in the usual way:

$$\begin{cases} x + \varphi^{-1}\beta y = \varphi^{-1}a \\ x + (\overline{\psi})^{-1}\gamma y = (\overline{\psi})^{-1}b \end{cases} \iff \begin{cases} I\varphi^{-1}\beta y + Ix = I\varphi^{-1}a \\ x + (\overline{\psi})^{-1}\gamma y = (\overline{\psi})^{-1}b. \end{cases}$$

We make the transformation (I row + II row → I row), and obtain the system:

$$\begin{cases} I\varphi^{-1}\beta y + (\overline{\psi})^{-1}\gamma y = I\varphi^{-1}a + (\overline{\psi})^{-1}b \\ x + (\overline{\psi})^{-1}\gamma y = (\overline{\psi})^{-1}b. \end{cases}$$

Write the expression $I\varphi^{-1}\beta y + (\overline{\psi})^{-1}\gamma y$ as follows: $(-\varphi^{-1}\beta + (\overline{\psi})^{-1}\gamma)y$. Then the system (9.25) is equivalent to the following system:

$$\begin{cases} (-\varphi^{-1}\beta + (\overline{\psi})^{-1}\gamma)y = I\varphi^{-1}a + (\overline{\psi})^{-1}b \\ x + (\overline{\psi})^{-1}\gamma y = (\overline{\psi})^{-1}b. \end{cases}$$

It is clear that the system (9.25) has a unique solution if and only if the mapping $-\varphi^{-1}\beta + (\overline{\psi})^{-1}\gamma$ is a permutation of the set Q. We simplify the last expression:

$$-\varphi^{-1}\beta + (\overline{\psi})^{-1}\gamma = I\varphi^{-1}\beta + I\psi^{-1}\gamma = I(\psi^{-1}\gamma + \varphi^{-1}\beta).$$

Therefore, the system (9.25) has a unique solution if and only if the mapping $(\psi^{-1}\gamma + \varphi^{-1}\beta)$ is a permutation of the set Q. □

Theorem 9.86. *Left linear quasigroup $(Q, \cdot)$ and right alinear quasigroup $(Q, \circ)$ of the form $x \cdot y = \varphi y + \beta x$ and $x \circ y = \gamma x + \overline{\psi} y$, respectively, defined over a group $(Q, +)$, are orthogonal if and only if the mapping $(\psi^{-1}\gamma + \varphi^{-1}\beta)$ is a permutation of the set Q.*

Proof. The proof is similar to the proof of Theorem 9.85 and we omit it. □

Theorem 9.87. *Left alinear quasigroups $(Q,\cdot)$ and $(Q,\circ)$ of the form $x\cdot y=\overline{\varphi}x+\beta y$ and $x\circ y=\overline{\psi}x+\delta y$, respectively, defined over a group $(Q,+)$, are orthogonal if and only if the mapping $((\overline{\varphi})^{-1}\beta-(\overline{\psi})^{-1}\delta)$ is a permutation of the set Q.*

Proof. The quasigroups $(Q,\cdot)$ and $(Q,\circ)$ are orthogonal if and only if the system of equations

$$\begin{cases}\overline{\varphi}x+\beta y=a\\ \overline{\psi}x+\delta y=b\end{cases} \tag{9.26}$$

has a unique solution for any fixed elements $a,b\in Q$. We solve this system of equations in the usual way:

$$\begin{cases}(\overline{\varphi})^{-1}\beta y+x=(\overline{\varphi})^{-1}a\\ (\overline{\psi})^{-1}\delta y+x=(\overline{\psi})^{-1}b\end{cases}\iff\begin{cases}(\overline{\varphi})^{-1}\beta y+x=(\overline{\varphi})^{-1}a\\ Ix+I(\overline{\psi})^{-1}\delta y=I(\overline{\psi})^{-1}b.\end{cases}$$

We do the transformation (I row + II row $\to$ II row). and obtain the system:

$$\begin{cases}(\overline{\varphi})^{-1}\beta y+x=(\overline{\varphi})^{-1}a\\ (\overline{\varphi})^{-1}\beta y+I(\overline{\psi})^{-1}\delta y=(\overline{\varphi})^{-1}a-(\overline{\psi})^{-1}b.\end{cases}$$

Write the expression $(\overline{\varphi})^{-1}\beta y+I(\overline{\psi})^{-1}\delta y$ as follows: $((\overline{\varphi})^{-1}\beta-(\overline{\psi})^{-1}\delta)y$. Then the system (9.26) is equivalent to the following system:

$$\begin{cases}(\overline{\varphi})^{-1}\beta y+x=(\overline{\varphi})^{-1}a\\ ((\overline{\varphi})^{-1}\beta-(\overline{\psi})^{-1}\delta)y=(\overline{\varphi})^{-1}a-(\overline{\psi})^{-1}b.\end{cases}$$

It is clear that the system (9.26) has a unique solution if and only if the mapping $((\overline{\varphi})^{-1}\beta-(\overline{\psi})^{-1}\delta)$ is a permutation of the set Q. □

Remark 9.88. The mapping $((\overline{\varphi})^{-1}\beta-(\overline{\psi})^{-1}\delta)$ from Theorem 9.87 can also be written in the form $((\overline{\varphi})^{-1}\beta-(\overline{\psi})^{-1}\delta)=I\varphi^{-1}\beta-I\psi^{-1}\delta=I(-\psi^{-1}\delta+\varphi^{-1}\beta)$.

Theorem 9.89. *Left linear quasigroup $(Q,\cdot)$ of the form $x\cdot y=\varphi x+\beta y$ and right linear quasigroup $(Q,\circ)$ of the form $x\circ y=\gamma y+\psi x$, both defined over a group $(Q,+)$, where $\varphi,\psi\in Aut(Q,+)$, are orthogonal if and only if the mapping $(J_t\psi^{-1}\gamma-\varphi^{-1}\beta)$ is a permutation of the set Q for any element $t\in Q$.*

Proof. The quasigroups $(Q,\cdot)$ and $(Q,\circ)$ are orthogonal if and only if the system of equations

$$\begin{cases}\varphi x+\beta y=a\\ \gamma y+\psi x=b\end{cases}$$

has a unique solution for any fixed elements $a,b\in Q$.

We solve this system of equations as follows:

$$\begin{cases}I\varphi^{-1}\beta y+Ix=I\varphi^{-1}a\\ \psi^{-1}\gamma y+x=\psi^{-1}b\end{cases}\iff\begin{cases}I\varphi^{-1}\beta y+Ix=I\varphi^{-1}a\\ x+J_{-x}\psi^{-1}\gamma y=\psi^{-1}b,\end{cases}$$

where $J_{-x}\gamma y=-x+\gamma y+x$.

If in the last system we add the first and the second equation, and write the sum instead of the second equation (I row + II row $\to$ II row), then we obtain the following system:

$$\begin{cases}I\varphi^{-1}\beta y+Ix=I\varphi^{-1}a\\ I\varphi^{-1}\beta y+J_{-x}\psi^{-1}\gamma y=I\varphi^{-1}a+\psi^{-1}b.\end{cases} \tag{9.27}$$

Therefore we can rewrite the system (9.27) in the following form:

$$\begin{cases} x = \varphi^{-1}a + I\varphi^{-1}\beta y \\ I\varphi^{-1}\beta y + J_{-x}\psi^{-1}\gamma y = I\varphi^{-1}a + \psi^{-1}b. \end{cases} \quad (9.28)$$

Rewrite the left part of the second equation of the system (9.28) in the following form:

$$\begin{aligned} & I\varphi^{-1}\beta y + J_{-x}\psi^{-1}\gamma y = \\ & I\varphi^{-1}\beta y - x + \psi^{-1}\gamma y + x = \\ & I\varphi^{-1}\beta y + \varphi^{-1}\beta y - \varphi^{-1}a + \psi^{-1}\gamma y + \varphi^{-1}a + I\varphi^{-1}\beta y = \\ & -\varphi^{-1}a + \psi^{-1}\gamma y + \varphi^{-1}a + I\varphi^{-1}\beta y = \\ & J_{I\varphi^{-1}a}\psi^{-1}\gamma y - \varphi^{-1}\beta y. \end{aligned}$$

We write expression $J_{I\varphi^{-1}a}\psi^{-1}\gamma y - \varphi^{-1}\beta y$ in the following form: $(J_{I\varphi^{-1}a}\psi^{-1}\gamma - \varphi^{-1}\beta)y$. The system (9.28) takes the form

$$\begin{cases} x = \varphi^{-1}a + I\varphi^{-1}\beta y \\ (J_{I\varphi^{-1}a}\psi^{-1}\gamma - \varphi^{-1}\beta)y = I\varphi^{-1}a + \psi^{-1}b. \end{cases} \quad (9.29)$$

From the system (9.29) it follows that the quasigroups $(Q, \cdot)$ and $(Q, \circ)$ are orthogonal if and only if the mapping $(J_{I\varphi^{-1}a}\psi^{-1}\gamma - \varphi^{-1}\beta)$ is a permutation of the set Q for any element $a \in Q$.

Denote the expression $I\varphi^{-1}a$ by the letter t. We can reformulate the last condition as follows: the quasigroups $(Q, \cdot)$ and $(Q, \circ)$ are orthogonal if and only if the mapping $(J_t\psi^{-1}\gamma - \varphi^{-1}\beta)$ is a permutation of the set Q for any element $t \in Q$. □

9.3.2 Orthogonality of linear and alinear quasigroups

The more detailed information on orthogonality of various linear and alinear quasigroups is in [793]. As usual, if $(Q, +)$ is a group, then $Ix = -x$ for all $x \in Q$.

Theorem 7 [833] on conditions of orthogonality of linear quasigroups follows from Theorems 9.82 and 9.81.

Taking into consideration that we have not proved an analogue of Lemma 2.209 for linear quasigroup, we give proof for the following theorem, which is independent of Theorems 9.82 and 9.81.

Theorem 9.90. *Linear quasigroup $(Q, \cdot)$ of the form $x \cdot y = \alpha x + \beta y + c$ and linear quasigroup $(Q, \circ)$ of the form $x \circ y = \gamma x + \delta y + d$, both defined over a group $(Q, +)$, are orthogonal if and only if the map $(-\gamma^{-1}\delta + \alpha^{-1}\beta)$ is a permutation of the set Q [833].*

Proof. The quasigroups $(Q, \cdot)$ and $(Q, \circ)$ are orthogonal if and only if the system of equations

$$\begin{cases} \alpha x + \beta y + c = a \\ \gamma x + \delta y + d = b \end{cases}$$

has a unique solution for any fixed elements $a, b \in Q$.

We solve this system of equations as follows:

$$\begin{cases} \alpha x + \beta y = a - c \\ \gamma x + \delta y = b - d \end{cases} \iff \begin{cases} x + \alpha^{-1}\beta y = \alpha^{-1}(a - c) \\ -\gamma^{-1}\delta y - x = -\gamma^{-1}(b - d). \end{cases}$$

In the last system we add the second and the first row and write the sum instead of the second row (II row + I row → II row). We obtain the following system:

$$\begin{cases} x + \alpha^{-1}\beta y = \alpha^{-1}(a - c) \\ -\gamma^{-1}\delta y + \alpha^{-1}\beta y = -\gamma^{-1}(b - d) + \alpha^{-1}(a - c). \end{cases} \tag{9.30}$$

Similar to Theorem 9.81 we write the expression $-\gamma^{-1}\delta y + \alpha^{-1}\beta y$ in the form $(-\gamma^{-1}\delta + \alpha^{-1}\beta)y$. From the system (9.30) it follows that quasigroups $(Q, \cdot)$ and $(Q, \circ)$ are orthogonal if and only if the map $(-\gamma^{-1}\delta + \alpha^{-1}\beta)$ is a permutation of the set Q. □

Corollary 9.91. *T-quasigroup $(Q, \cdot)$ of the form $x \cdot y = \alpha x + \beta y + c$ and T-quasigroup $(Q, \circ)$ of the form $x \circ y = \gamma x + \delta y + d$, both defined over a group $(Q, +)$, are orthogonal if and only if the map $\alpha^{-1}\beta - \gamma^{-1}\delta$ is an automorphism of the group $(Q, +)$.*

Proof. The proof follows from Theorem 9.90 and the fact that in abelian group $-\gamma^{-1}\delta + \alpha^{-1}\beta = \alpha^{-1}\beta - \gamma^{-1}\delta$ and that the map $\alpha^{-1}\beta - \gamma^{-1}\delta$ is an endomorphism of the group $(Q, +)$. □

If the group $(Q, +)$ is a cyclic group $(\mathbb{Z}_n.+)$ of order n, then Theorem 2.4 from [742] follows from Corollary 9.91. Notice, for the case $(Q, +) = (\mathbb{Z}_n, +)$, Corollary 9.95 is also true.

Lemma 9.92. *If $(Q, +)$ is an abelian group, $\varphi, \psi \in Aut(Q, +)$, then $\varphi - \psi$ is an automorphism of the group $(Q, +)$ if and only if $\psi - \varphi$ is an automorphism of this group.*

Proof. Taking into consideration that the map $I(x) = -x$ is an automorphism of an abelian group $(Q, +)$ and a permutation of the set Q of order two, we have $-(\varphi - \psi) = -\varphi + \psi = \psi - \varphi$. □

Lemma 9.93. *In conditions of Theorem 9.91 the following statements are equivalent: "the endomorphism $(\alpha^{-1}\beta - \gamma^{-1}\delta)$ is an automorphism of $(Q, +)$" and "the endomorphism $(\beta^{-1}\alpha - \delta^{-1}\gamma)$ is an automorphism of $(Q, +)$."*

Proof. The map $(\alpha^{-1}\beta - \gamma^{-1}\delta)$ is a permutation of the set Q if and only if the map $\varepsilon - \alpha\gamma^{-1}\delta\beta^{-1}$ is a permutation of the set Q. Indeed, $\alpha(\alpha^{-1}\beta - \gamma^{-1}\delta)\beta = \varepsilon - \alpha\gamma^{-1}\delta\beta^{-1}$.

Similarly, $(\beta^{-1}\alpha - \delta^{-1}\gamma)$ is a permutation of the set Q if and only if the map

$$\varepsilon - \beta\delta^{-1}\gamma\alpha^{-1} \tag{9.31}$$

is a permutation of the set Q.

If we denote the map $\alpha\gamma^{-1}\delta\beta^{-1}$ by ψ, then $\beta\delta^{-1}\gamma\alpha^{-1} = \psi^{-1}$.

Further we have the following equivalence: the map $\varepsilon - \psi$ is a permutation if and only if the map $\varepsilon - \psi^{-1}$ is a permutation of the set Q.

Indeed, $\varepsilon - \psi$ is a permutation if and only if the map $\psi - \varepsilon$ is a permutation (Lemma 9.92), further $\psi - \varepsilon$ is a permutation if and only if $\psi^{-1}(\psi - \varepsilon) = \varepsilon - \psi^{-1}$ is a permutation. □

Corollary 9.94. *A T-quasigroup $(Q, \cdot)$ of the form $x \cdot y = \varphi x + \psi y + c$ over a group $(Q, +)$ and its (12)-parastrophe $(Q, \star)$ of the form $x \star y = \psi x + \varphi y + c$ are orthogonal if and only if the map $\varphi^{-1}\psi - \psi^{-1}\varphi$ is an automorphism of the group $(Q, +)$.*

Corollary 9.95. *A T-quasigroup $(Q, \cdot)$ of the form $x \cdot y = \alpha x + \beta y + c$ and a medial quasigroup $(Q, \circ)$ of the form $x \circ y = \gamma x + \delta y + d$, both defined over a group $(Q, +)$, are orthogonal if and only if the map $\alpha\delta - \gamma\beta$ is an automorphism of the group $(Q, +)$.*

Proof. From Lemma 9.93 and equality (9.31), it follows that quasigroups $(Q, \cdot)$ and $(Q, \circ)$ are orthogonal if and only if the map $\varepsilon - \beta\delta^{-1}\gamma\alpha^{-1}$ is a permutation of the set Q. Further, since $\delta\gamma = \gamma\delta$, we have $\beta\delta^{-1}\gamma\alpha^{-1} = \beta\gamma\delta^{-1}\alpha^{-1}$ and the map $\varepsilon - \beta\delta^{-1}\gamma\alpha^{-1}$ is a permutation of the set Q if and only if the map $(\varepsilon - \beta\gamma\delta^{-1}\alpha^{-1})\alpha\delta = \alpha\delta - \beta\gamma$ is a permutation of the set Q. □

Theorem 9.96. *Linear quasigroup $(Q, \cdot)$ of the form $x \cdot y = \alpha x + \beta y + c$ and linear quasigroup $(Q, \circ)$ of the form $x \circ y = \gamma y + \delta x + d$, both defined over a group $(Q, +)$, are orthogonal if and only if the mapping $(-J_t\gamma^{-1}\delta + \beta^{-1}\alpha)$ is a permutation of the set Q for any element $t \in Q$.*

Proof. The quasigroups $(Q, \cdot)$ and $(Q, \circ)$ are orthogonal if and only if the system of equations

$$\begin{cases} \alpha x + \beta y + c = a \\ \gamma y + \delta x + d = b \end{cases}$$

has a unique solution for any fixed elements $a, b \in Q$.

We solve this system of equations as follows:

$$\begin{cases} \alpha x + \beta y = a - c \\ \gamma y + \delta x = b - d \end{cases} \iff \begin{cases} \beta^{-1}\alpha x + y = \beta^{-1}(a - c) \\ J_{\gamma y}\delta x + \gamma y = (b - d), \end{cases}$$

where $J_{\gamma y}\delta x = \gamma y + \delta x - \gamma y$. Notice $\gamma^{-1}J_{\gamma y}\delta x = J_y\gamma^{-1}\delta x$.

Further we have:

$$\begin{cases} \beta^{-1}\alpha x + y = \beta^{-1}(a - c) \\ -y - \gamma^{-1}J_{\gamma y}\delta x = -\gamma^{-1}(b - d). \end{cases} \tag{9.32}$$

If in the system (9.32) we add the first and the second row, and write the sum instead of the second row (I row + II row → II row), then we obtain the following system:

$$\begin{cases} \beta^{-1}\alpha x + y = \beta^{-1}(a - c) \\ \beta^{-1}\alpha x - \gamma^{-1}J_{\gamma y}\delta x = \beta^{-1}(a - c) - \gamma^{-1}(b - d). \end{cases} \tag{9.33}$$

Then we can rewrite the system (9.33) in the following form:

$$\begin{cases} y = -\beta^{-1}\alpha x + \beta^{-1}(a - c) \\ \beta^{-1}\alpha x - J_y\gamma^{-1}\delta x = \beta^{-1}(a - c) - \gamma^{-1}(b - d). \end{cases} \tag{9.34}$$

Rewrite the left part of the second equation of the system (9.34) in the following form:

$$\beta^{-1}\alpha x + IJ_y\gamma^{-1}\delta x = \beta^{-1}\alpha x + J_yI\gamma^{-1}\delta x = \beta^{-1}\alpha x + y - \gamma^{-1}\delta x - y.$$

Further, taking into consideration the first equation of the system (9.34), we have:

$$\begin{aligned} &\beta^{-1}\alpha x + y - \gamma^{-1}\delta x - y = \\ &\beta^{-1}\alpha x - \beta^{-1}\alpha x + \beta^{-1}(a - c) - \gamma^{-1}\delta x - \beta^{-1}(a - c) + \beta^{-1}\alpha x = \\ &\beta^{-1}(a - c) - \gamma^{-1}\delta x - \beta^{-1}(a - c) + \beta^{-1}\alpha x = \\ &J_{\beta^{-1}(a-c)}I\gamma^{-1}\delta x + \beta^{-1}\alpha x = -J_{\beta^{-1}(a-c)}\gamma^{-1}\delta x + \beta^{-1}\alpha x. \end{aligned}$$

Similar to Theorem 9.81, we write the expression $-J_{\beta^{-1}(a-c)}\gamma^{-1}\delta x + \beta^{-1}\alpha x$ in the following form: $(-J_{\beta^{-1}(a-c)}\gamma^{-1}\delta + \beta^{-1}\alpha)x$. The system (9.34) takes the form

$$\begin{cases} y = -\beta^{-1}\alpha x + \beta^{-1}(a - c) \\ (-J_{\beta^{-1}(a-c)}\gamma^{-1}\delta + \beta^{-1}\alpha)x = \beta^{-1}(a - c) - \gamma^{-1}(b - d). \end{cases} \tag{9.35}$$

From the system (9.35) it follows that the quasigroups $(Q, \cdot)$ and $(Q, \circ)$ are orthogonal if and only if the mapping $(-J_{\beta^{-1}(a-c)}\gamma^{-1}\delta + \beta^{-1}\alpha)$ is a permutation of the set Q for any element $a \in Q$.

Denote the expression $\beta^{-1}(a-c)$ by the letter t. We can reformulate the last condition as follows: the quasigroups $(Q, \cdot)$ and $(Q, \circ)$ are orthogonal if and only if the mapping $(-J_t\gamma^{-1}\delta + \beta^{-1}\alpha)$ is a permutation of the set Q for any element $t \in Q$. □

Theorem 9.97. *An alinear quasigroup $(Q, \cdot)$ of the form $x \cdot y = I\alpha x + I\beta y + c$ and an alinear quasigroup $(Q, \circ)$ of the form $x \circ y = I\gamma y + I\delta x + d$, both defined over a group $(Q, +)$, where $\alpha, \beta, \gamma, \delta \in Aut(Q, +)$, are orthogonal if and only if the mapping $(\beta^{-1}\alpha - J_t\gamma^{-1}\delta)$ is a permutation of the set Q for any element $t \in Q$.*

Proof. The quasigroups $(Q, \cdot)$ and $(Q, \circ)$ are orthogonal if and only if the system of equations

$$\begin{cases} I\alpha x + I\beta y + c = a \\ I\gamma y + I\delta x + d = b \end{cases}$$

has a unique solution for any fixed elements $a, b \in Q$.

We solve this system of equations as follows:

$$\begin{cases} I\alpha x + I\beta y = a - c \\ I\gamma y + I\delta x = b - d \end{cases} \iff \begin{cases} y + \beta^{-1}\alpha x = I\beta^{-1}(a-c) \\ J_{I\gamma y}I\delta x + I\gamma y = (b-d), \end{cases}$$

where $J_{I\gamma y}I\delta x = I\gamma y + I\delta x - I\gamma y$. Notice $\gamma^{-1}J_{I\gamma y}I\delta x = J_{Iy}\gamma^{-1}I\delta x$.

Further we have:

$$\begin{cases} y + \beta^{-1}\alpha x = I\beta^{-1}(a-c) \\ J_{Iy}\gamma^{-1}I\delta x + Iy = \gamma^{-1}(b-d). \end{cases} \tag{9.36}$$

If in the system (9.36) we add the second row and the first row, and write the sum instead of the second row (II row + I row → II row), then we obtain the following system:

$$\begin{cases} y + \beta^{-1}\alpha x = I\beta^{-1}(a-c) \\ J_{Iy}\gamma^{-1}I\delta x + \beta^{-1}\alpha x = \gamma^{-1}(b-d) - \beta^{-1}(a-c). \end{cases} \tag{9.37}$$

Therefore, we can rewrite the system (9.37) in the following form:

$$\begin{cases} y = I\beta^{-1}(a-c) + I\beta^{-1}\alpha x \\ J_{Iy}\gamma^{-1}I\delta x + \beta^{-1}\alpha x = \gamma^{-1}(b-d) - \beta^{-1}(a-c). \end{cases} \tag{9.38}$$

Rewrite the left part of the second equation of the system (9.38) in the following form:

$$\begin{aligned} & J_{Iy}\gamma^{-1}I\delta x + \beta^{-1}\alpha x = -(J_{Iy}\gamma^{-1}\delta x) + \beta^{-1}\alpha x = \\ & -(-y + \gamma^{-1}\delta x + y) + \beta^{-1}\alpha x = \\ & -y - \gamma^{-1}\delta x + y + \beta^{-1}\alpha x \overset{(9.38)}{=} \\ & -(I\beta^{-1}(a-c) + I\beta^{-1}\alpha x) - \gamma^{-1}\delta x + I\beta^{-1}(a-c) + I\beta^{-1}\alpha x + \beta^{-1}\alpha x = \\ & \beta^{-1}\alpha x + \beta^{-1}(a-c) - \gamma^{-1}\delta x - \beta^{-1}(a-c) = \\ & \beta^{-1}\alpha x - J_{\beta^{-1}(a-c)}\gamma^{-1}\delta x. \end{aligned}$$

Similar to Theorem 9.81, we write the expression $\beta^{-1}\alpha x - J_{\beta^{-1}(a-c)}\gamma^{-1}\delta x$ in the following form: $(\beta^{-1}\alpha - J_{\beta^{-1}(a-c)}\gamma^{-1}\delta)x$. The system (9.38) takes the form

$$\begin{cases} y = I\beta^{-1}(a-c) + I\beta^{-1}\alpha x \\ (\beta^{-1}\alpha - J_{\beta^{-1}(a-c)}\gamma^{-1}\delta)x = \gamma^{-1}(b-d) - \beta^{-1}(a-c). \end{cases} \tag{9.39}$$

From the system (9.39) it follows that the quasigroups $(Q, \cdot)$ and $(Q, \circ)$ are orthogonal if and only if the mapping $(\beta^{-1}\alpha - J_{\beta^{-1}(a-c)}\gamma^{-1}\delta)$ is a permutation of the set Q for any element $a \in Q$.

Denote the expression $\beta^{-1}(a-c)$ by the letter t. We can reformulate the last condition as follows: the quasigroups $(Q, \cdot)$ and $(Q, \circ)$ are orthogonal if and only if the mapping $(\beta^{-1}\alpha - J_t\gamma^{-1}\delta)$ is a permutation of the set Q for any element $t \in Q$. □

Corollary 9.98. *If in conditions of Theorem 9.97 the group $Inn(Q, +)$ of inner automorphisms of the group $(Q, +)$ acts transitively on the group $Aut(Q, +)$, then orthogonal quasigroups $(Q, \cdot)$ and $(Q, \circ)$ do not exist.*

Proof. Since the group $Inn(Q, +)$ acts transitively, then in conditions of Theorem 9.97 there exists an element $d \in Q$ such that $(\beta^{-1}\alpha - J_d\gamma^{-1}\delta)x = 0$ for any $x \in Q$. □

Corollary 9.99. *If in conditions of Theorem 9.97 the group $(Q, +)$ is a symmetric group S_n $(n \neq 6)$, then there exist no orthogonal quasigroups $(S_n, \cdot)$ and $(S_n, \circ)$.*

Proof. By the Gölder theorem, $Aut(S_n) = Inn(S_n)$ for any natural number n, $n \neq 2; 6$ [471, p. 67]. It is well known that orthogonal quasigroups of order 2 do not exist [163]. □

It is known that there exist Latin squares which are orthogonal to the group S_n for all $n > 3$ [410, Theorem 2].

9.3.3 Orthogonality of parastrophes

Theorem 9.100. *For a linear quasigroup (Q, A) of the form $A(x, y) = \varphi x + \psi y + c$ over a group $(Q, +)$ the following equivalences are true:*

1. $A \bot A^{12} \Longleftrightarrow$ *the mapping $(-J_t\varphi^{-1}\psi + \psi^{-1}\varphi)$ is a permutation of the set Q for any $t \in Q$;*
2. $A \bot A^{13} \Longleftrightarrow$ *the mapping $(\varphi J_{I\varphi^{-1}c} + \varepsilon)$ is a permutation of the set Q;*
3. $A \bot A^{23} \Longleftrightarrow$ *the mapping $(\varepsilon + \psi)$ is a permutation of the set Q;*
4. $A \bot A^{123} \Longleftrightarrow$ *the mapping $(\varphi J^{-1}_{\psi^{-1}c} + \psi^2)$ is a permutation of the set Q;*
5. $A \bot A^{132} \Longleftrightarrow$ *the mapping $(\varphi^2 + \psi)$ is a permutation of the set Q.*

Proof. The forms of parastrophes of quasigroup (Q, A) are given in Lemma 2.228.

Case 1. The proof follows from Theorem 9.96.

Case 2. Using Theorem 9.81 we have: $A \bot A^{13}$ if and only if the mapping $I\varphi^{-1}\psi + \varphi I J_{I\varphi^{-1}c}\varphi^{-1}\psi$ is a permutation of the set Q.

We make the following transformations: $I\varphi^{-1}\psi + \varphi I J_{I\varphi^{-1}c}\varphi^{-1}\psi = (I + \varphi I J_{I\varphi^{-1}c})\varphi^{-1}\psi = (I + \varphi J_{I\varphi^{-1}c} I)\varphi^{-1}\psi = (\varphi J_{I\varphi^{-1}c} + \varepsilon) I \varphi^{-1}\psi$. The last mapping is a permutation if and only if the mapping $(\varphi J_{I\varphi^{-1}c} + \varepsilon)$ is a permutation of the set Q.

Case 3. Using Theorem 9.82 we have: $A \bot A^{23}$ if and only if the mapping $\psi^{-1}\varphi - \psi I \psi^{-1}\varphi$ is a permutation of the set Q. We simplify the last equality in the following way:

$$\psi^{-1}\varphi - \psi I \psi^{-1}\varphi = \psi^{-1}\varphi + \psi\psi^{-1}\varphi = (\varepsilon + \psi)\psi^{-1}\varphi.$$

Therefore $A \bot A^{23}$ if and only if the mapping $(\varepsilon + \psi)$ is a permutation of the set Q.

Case 4. From Theorem 9.85 it follows that $A \bot A^{123}$ if and only if the mapping

$$I\varphi^{-1}\psi + (I J_{\varphi^{-1}c}\varphi^{-1}\psi)^{-1}\varphi^{-1} \tag{9.40}$$

is a permutation of the set Q. We make the following transformation of the expression (9.40):

$$\begin{aligned}&I\varphi^{-1}\psi + (IJ_{\varphi^{-1}c}\varphi^{-1}\psi)^{-1}\varphi^{-1} =\\&I\varphi^{-1}\psi + \psi^{-1}\varphi J^{-1}_{\varphi^{-1}c}I\varphi^{-1} = I(\psi^{-1}\varphi J^{-1}_{\varphi^{-1}c}\varphi^{-1} + \varphi^{-1}\psi) =\\&I(\psi^{-1}J_c^{-1}\varphi\varphi^{-1} + \varphi^{-1}\psi) = I(\psi^{-1}J_c^{-1} + \varphi^{-1}\psi) =\\&I\varphi^{-1}(\varphi\psi^{-1}J_c^{-1} + \psi) = I\varphi^{-1}(\varphi J^{-1}_{\psi^{-1}c}\psi^{-1} + \psi) =\\&I\varphi^{-1}(\varphi J^{-1}_{\psi^{-1}c} + \psi^2)\psi^{-1}.\end{aligned} \tag{9.41}$$

We obtain: $A\bot A^{123}$ if and only if the mapping $(\varphi J^{-1}_{\psi^{-1}c} + \psi^2)$ is a permutation of the set Q.

Case 5. From Theorem 9.84 we have: $A\bot A^{132}$ if and only if the mapping $(I\psi^{-1}\varphi)^{-1}\psi^{-1} - \psi^{-1}\varphi$ is a permutation of the set Q. We simplify the last equality in the following way:

$$\begin{aligned}&(I\psi^{-1}\varphi)^{-1}\psi^{-1} - \psi^{-1}\varphi =\\&\varphi^{-1}\psi I\psi^{-1} + I\psi^{-1}\varphi = I\varphi^{-1} + I\psi^{-1}\varphi =\\&I(\psi^{-1}\varphi + \varphi^{-1}) = I\psi^{-1}(\varphi^2 + \psi)\varphi^{-1}.\end{aligned}$$

Therefore $A\bot A^{132}$ if and only if the mapping $(\varphi^2 + \psi)$ is a permutation of the set Q. □

Taking into consideration Lemma 9.80 we can, without loss of generality, take $c = 0$ in the finite case in formulation of Theorem 9.100. Therefore we can reformulate Theorem 9.100 in the following form:

Theorem 9.101. *For a linear quasigroup (Q, A) of the form $A(x, y) = \varphi x + \psi y + c$ over a finite group $(Q, +)$ the following equivalences are fulfilled:*

1. $A\bot A^{12} \Longleftrightarrow$ *the mapping $(-J_t\varphi^{-1}\psi + \psi^{-1}\varphi)$ is a permutation of the set Q for any $t \in Q$;*
2. $A\bot A^{13} \Longleftrightarrow$ *the mapping $(\varphi + \varepsilon)$ is a permutation of the set Q;*
3. $A\bot A^{23} \Longleftrightarrow$ *the mapping $(\varepsilon + \psi)$ is a permutation of the set Q;*
4. $A\bot A^{123} \Longleftrightarrow$ *the mapping $(\varphi + \psi^2)$ is a permutation of the set Q;*
5. $A\bot A^{132} \Longleftrightarrow$ *the mapping $(\varphi^2 + \psi)$ is a permutation of the set Q.*

Corollary 9.102. *Any linear quasigroup over the group S_n $(n \neq 6)$ is not orthogonal to its (12)-parastrophe.*

Proof. The proof follows from Theorem 9.100 and Lemma 9.99. □

Theorem 9.103. *For an alinear quasigroup (Q, A) of the form $A(x, y) = I\varphi x + I\psi y + c$ over a group $(Q, +)$ the following equivalences are true:*

1. $A\bot A^{12} \Longleftrightarrow$ *the mapping $(\psi^{-1}\varphi - J_t\varphi^{-1}\psi)$ is a permutation of the set Q for any $t \in Q$;*
2. $A\bot A^{13} \Longleftrightarrow$ *the mapping $(\varphi - J_{\psi t}J_c)$ is a permutation of the set Q for any $t \in Q$;*
3. $A\bot A^{23} \Longleftrightarrow$ *the mapping $(\varepsilon + I\psi J_t)$ is a permutation of the set Q for any $t \in Q$;*
4. $A\bot A^{123} \Longleftrightarrow$ *the mapping $(\psi^2 - \varphi J_{\psi^{-1}c})$ is a permutation of the set Q;*

5. $A \perp A^{132} \Longleftrightarrow$ *the mapping* $(\psi - \varphi^2)$ *is a permutation of the set* Q.

Proof. The forms of parastrophes of the quasigroup (Q, A) are given in Lemma 2.229.

Case 1. The proof follows from Theorem 9.97.

Case 2. Using Theorem 9.97 we have $A \perp A^{13}$ if and only if the mapping $\psi^{-1}\varphi - J_t\psi^{-1}\varphi J_{\varphi^{-1}c}\varphi^{-1}$ is a permutation of the set Q for any $t \in Q$.

We make the following transformations: $\psi^{-1}\varphi - J_t\psi^{-1}\varphi J_{\varphi^{-1}c}\varphi^{-1} = \psi^{-1}\varphi - \psi^{-1}J_{\psi t}J_c\varphi\varphi^{-1} = \psi^{-1}\varphi - \psi^{-1}J_{\psi t}J_c = \psi^{-1}(\varphi - J_{\psi t}J_c)$. The last mapping is a permutation if and only if the mapping $(\varphi - J_{\psi t}J_c)$ is a permutation of the set Q for any $t \in Q$.

Case 3. Using Theorem 9.97 we have $A \perp A^{23}$ if and only if the mapping $\psi^{-1}\varphi - J_t\psi J^{-1}_{\psi^{-1}c}J_{\psi^{-1}c}I\psi^{-1}\varphi$ is a bijection (a permutation) of the set Q for any $t \in Q$.

We simplify the last expression in the following way:

$$\psi^{-1}\varphi - J_t\psi J^{-1}_{\psi^{-1}c}J_{\psi^{-1}c}I\psi^{-1}\varphi = \psi^{-1}\varphi - J_t\varphi = \psi^{-1}(\varepsilon - \psi J_t)\varphi.$$

Therefore, $A \perp A^{23}$ if and only if the mapping $(\varepsilon + I\psi J_t)$ is a bijection of the set Q for any $t \in Q$.

Case 4. From Theorem 9.87 it follows that $A \perp A^{123}$ if and only if the mapping

$$I\varphi^{-1}I\psi + \psi^{-1}\varphi IJ_{\varphi^{-1}c}\varphi^{-1} \tag{9.42}$$

is a permutation of the set Q. We make the following transformation of the expression (9.42):

$$\begin{aligned} &I\varphi^{-1}I\psi + \psi^{-1}\varphi IJ_{\varphi^{-1}c}\varphi^{-1} = \\ &\varphi^{-1}\psi - \psi^{-1}J_c\varphi\varphi^{-1} = \varphi^{-1}\psi - \psi^{-1}J_c. \end{aligned} \tag{9.43}$$

We obtain $A \perp A^{123}$ if and only if the mapping $(\varphi^{-1}\psi - \psi^{-1}J_c)$ is a permutation of the set Q.

Further we have: the mapping $(\varphi^{-1}\psi - \psi^{-1}J_c)$ is a permutation of the set Q if and only if the mapping $\varphi(\varphi^{-1}\psi - \psi^{-1}J_c)\psi = (\psi^2 - \varphi J_{\psi^{-1}c})$ is a permutation of the set Q.

Therefore, $A \perp A^{123}$ if and only if the mapping $(\psi^2 - \varphi J_{\psi^{-1}c})$ is a permutation of the set Q.

Case 5. From Theorem 9.87 we have $A \perp A^{132}$ if and only if the mapping $(\varphi^{-1}\psi - \varphi)$ is a permutation of the set Q.

Therefore, $A \perp A^{132}$ if and only if the mapping $(\varphi^{-1}\psi - \varphi) = \varphi^{-1}(\psi - \varphi^2)$ is a permutation of the set Q. □

Corollary 9.104. *Any alinear quasigroup over the group* S_n $(n \neq 6)$ *is not orthogonal to its*

(i) *(12)–parastrophe*; (ii) *(13)–parastrophe*; (iii) *(23)–parastrophe.*

Proof. It is possible to use Theorem 9.103 and Lemma 9.99 but we give the direct proof.

(i). From Case 1 of Theorem 9.103 and properties of the group S_n it follows that there exists an element $w \in S_n$ such that $(\psi^{-1}\varphi - J_w\varphi^{-1}\psi)x = 0$ for any $x \in S_n$.

Cases (ii) and (iii) are proved in a similar way. □

Theorem 9.105. *For a left linear right alinear quasigroup* (Q, A) *of the form* $A(x, y) = \varphi x + I\psi y + c$ *over a group* $(Q, +)$ *the following equivalences are true:*

1. $A \perp A^{12} \Longleftrightarrow$ *the mapping* $(\varphi^{-1}\psi - \psi^{-1}\varphi)$ *is a permutation of the set* Q;

2. $A \perp A^{13} \Longleftrightarrow$ *the mapping* $(\varepsilon + \varphi J_{Ic})$ *is a permutation of the set* Q;

3. $A \perp A^{23} \Longleftrightarrow$ *the mapping* $(J_t + \psi)$ *is a permutation of the set* Q *for any* $t \in Q$;

4. $A \perp A^{123} \Longleftrightarrow$ *the mapping* $(\varphi + J_{\psi Ic}\psi^2)$ *is a permutation of the set* Q;

5. $A \perp A^{132} \Longleftrightarrow$ *the mapping* $(\varphi^2 + IJ_k\psi)$ *is a permutation of the set* Q *for any* $k \in Q$.

Proof. The forms of parastrophes of the quasigroup (Q, A) are given in Lemma 2.231.

Case 1. The proof follows from Theorem 9.85.

Case 2. Using Theorem 9.81 we have $A \perp A^{13}$ if and only if the mapping $I\varphi^{-1}I\psi + \varphi J_{I\varphi^1 c}\varphi^{-1}\psi$ is a permutation of the set Q.

We make the following transformations: $I\varphi^{-1}I\psi + \varphi J_{I\varphi^{-1}c}\varphi^{-1}\psi = \varphi^{-1}\psi + J_{Ic}\varphi\varphi^{-1}\psi = (\varphi^{-1} + J_{Ic})\psi$. The last mapping is a permutation if and only if the mapping $(\varphi^{-1} + J_{Ic}) = \varphi^{-1}(\varepsilon + \varphi J_{Ic})$ is a permutation of the set Q.

Case 3. Using Theorem 9.89 we have $A \perp A^{23}$ if and only if the mappings $J_t\varphi^{-1}\psi J^{-1}_{\psi^{-1}c}J_{\psi^{-1}c}\psi^{-1} - \varphi^{-1}I\psi$ are permutations of the set Q.

We simplify the last expression in the following way:

$$J_t\varphi^{-1}\psi J^{-1}_{\psi^{-1}c}J_{\psi^{-1}c}\psi^{-1} - \varphi^{-1}I\psi = \varphi^{-1}(IJ_{\varphi t} + \psi) = \varphi^{-1}(J_{I\varphi t} + \psi).$$

Therefore, $A \perp A^{23}$ if and only if the mapping $(J_t + \psi)$ is a permutation of the set Q for any $t \in Q$.

Case 4. From Theorem 9.86 it follows that $A \perp A^{123}$ if and only if the mapping

$$\psi^{-1}\varphi + \varphi J_{I\varphi^{-1}c}\varphi^{-1}\psi = \psi^{-1}\varphi + J_{Ic}\psi \tag{9.44}$$

is a permutation of the set Q. We obtain $A \perp A^{123}$ if and only if the mapping $(\psi^{-1}\varphi + J_{Ic}\psi) = \psi^{-1}(\varphi + J_{\psi Ic}\psi^2)$ is a permutation of the set Q, i.e., $A \perp A^{123}$ if and only if the mapping $(\varphi + J_{\psi Ic}\psi^2)$ is a permutation of the set Q.

Case 5. From Theorem 9.83 we have $A \perp A^{132}$ if and only if the mapping $\psi J^{-1}_{\psi^{-1}c}J_{\psi^{-1}c}\psi^{-1}\varphi + J_{I\psi^{-1}b}\varphi^{-1}I\psi = \varphi + IJ_{I\psi^{-1}b}\varphi^{-1}\psi$ is a permutation of the set Q for any $b \in Q$. Denote the expression $I\psi^{-1}b$ by the letter t.

Then $A \perp A^{132}$ if and only if the mappings $\varphi + IJ_t\varphi^{-1}\psi$ are permutations of the set Q for any $t \in Q$. But $\varphi + IJ_t\varphi^{-1}\psi = \varphi^{-1}(\varphi^2 + IJ_{\varphi t}\psi)$. Therefore $A \perp A^{132}$ if and only if the mappings $(\varphi^2 + IJ_{\varphi t}\psi)$ are permutations of the set Q for any $t \in Q$. Denote the expression φt by the letter k.

Then $A \perp A^{132}$ if and only if the mappings $(\varphi^2 + IJ_k\psi)$ are permutations of the set Q for any $k \in Q$. □

Corollary 9.106. *Any left linear right alinear quasigroup over the group* S_n $(n \neq 2; 6)$ *is not orthogonal to its* (132)*–parastrophe.*

Proof. The proof follows from Theorem 9.105 and Lemma 9.99. In this case we can find element d such that $J_d\psi = \varphi^2$. □

Theorem 9.107. *For a left alinear right linear quasigroup* (Q, A) *of the form* $A(x, y) = I\varphi x + \psi y + c$ *over a group* $(Q, +)$ *the following equivalences are true:*

1. $A \perp A^{12} \Longleftrightarrow$ *the mapping* $(-\varphi^{-1}\psi + \psi^{-1}\varphi)$ *is a permutation of the set* Q;

2. $A \perp A^{13} \Longleftrightarrow$ *the mapping* $(\varphi + IJ_{(Ib+c)})$ *is a permutation of the set* Q *for any* $b \in Q$;

3. $A \perp A^{23} \Longleftrightarrow$ *the mapping* $(\psi + \varepsilon)$ *is a permutation of the set* Q;

4. $A \perp A^{123} \Longleftrightarrow$ *the mapping* $(\psi^2 + \varphi IJ_t)$ *is a permutation of the set* Q *for any* $t \in Q$;

5. $A \perp A^{132} \Longleftrightarrow$ *the mapping* $(\varphi^2 + \psi)$ *is a permutation of the set* Q.

Proof. The proof is similar to the proof of Theorem 9.105 and we omit it. □

9.3.4 Parastrophe orthogonality of T-quasigroups

We start from the following:

Lemma 9.108. *In any T-quasigroup $(Q,\cdot)$ with the form $x\cdot y=\varphi x+\psi y+c$ the following conditions are equivalent:*

(the maps $\varphi-\psi$ and $\varphi+\psi$ are bijections of the set Q) and
(the map $\varphi^{-1}\psi-\psi^{-1}\varphi$ is a bijection of the set Q).

Proof. We notice that the map $\varphi-\psi$ is a bijection if and only if the map $\varphi^{-1}-\psi^{-1}$ is a bijection of the set Q since we have the following equality $\psi^{-1}(\varphi-\psi)\varphi^{-1}=\psi^{-1}-\varphi^{-1}$.

By Lemma 9.92 the map $\psi^{-1}-\varphi^{-1}$ is a bijection if and only if the map $\varphi^{-1}-\psi^{-1}$ is a bijection.

Then we have the following equivalence:

(the maps $\varphi-\psi$ and $\varphi+\psi$ are bijections of the set Q) $\Longleftrightarrow$
(the maps $\varphi^{-1}-\psi^{-1}$ and $\varphi+\psi$ are bijections of the set Q).

Since $(\varphi^{-1}-\psi^{-1})(\varphi+\psi)=\varepsilon+\varphi^{-1}\psi-\psi^{-1}\varphi-\varepsilon=\varphi^{-1}\psi-\psi^{-1}\varphi$ we can say that the following conditions:

(the maps $\varphi-\psi$ and $\varphi+\psi$ are bijections of the set Q) and
(the map $\varphi^{-1}\psi-\psi^{-1}\varphi$ is a bijection of the set Q)

are equivalent too. □

Theorem 9.109. *For a T-quasigroup (Q,A) of the form $A(x,y)=\varphi x+\psi y+a$ over an abelian group $(Q,+)$ the following equivalences are fulfilled:*

(i) $A\perp A^{12}\Longleftrightarrow(\varphi-\psi),(\varphi+\psi)$ are bijections of the set Q;
(ii) $A\perp A^{13}\Longleftrightarrow(\varepsilon+\varphi)$ is a bijection of the set Q;
(iii) $A\perp A^{23}\Longleftrightarrow(\varepsilon+\psi)$ is a bijection of the set Q;
(iv) $A\perp A^{123}\Longleftrightarrow(\varphi+\psi^2)$ is a bijection of the set Q;
(v) $A\perp A^{132}\Longleftrightarrow(\varphi^2+\psi)$ is a bijection of the set Q.

Proof. (i) From Theorem 9.100 (or Theorem 9.91) it follows that the T-quasigroup $(Q,\cdot)$ of the form $x\cdot y=\varphi x+\psi y+c$ over a commutative group $(Q,+)$ and its (12)-parastrophe $(Q,\star)$ of the form $x\cdot y=\psi x+\varphi y+c$ are orthogonal if and only if the map $\varphi^{-1}\psi-\psi^{-1}\varphi$ is a bijection of the set Q (i.e., this map is an automorphism of the group $(Q,+)$).

The rest follows from Lemma 9.108.

Cases (ii)–(v) coincide with the corresponding cases of Theorem 9.100.

□

Remark 9.110. It is possible to use Theorem 9.66 by proving Theorem 9.109 at least for finite quasigroups.

From the form of quasigroup $(Q,\cdot)$ it follows that $L^{\cdot}_x y=L_{\varphi x+a}\psi y$, $R^{\cdot}_y x=L_{\psi y+a}\varphi y$. For instance, using Theorem 9.66 we can prove Case (ii) in the following way.

Any map $R^{\cdot}_y R^{\cdot}_y x$ has the following form:

$$\begin{aligned}&R^{\cdot}_y R^{\cdot}_y x=L_{\psi y+a}\varphi L_{\psi y+a}\varphi x=\\&\psi y+a+\varphi(\psi y+a+\varphi x)=\varphi^2 x+(\varphi+\varepsilon)\psi y+\varphi a+a=\\&L_{(\varphi+\varepsilon)\psi y+\varphi a+a}\varphi^2 x.\end{aligned}$$

It is easy to see that the m-tuple $(R^{\cdot}_y R^{\cdot}_y)$, where variable y runs over all the set Q, will have the τ-property if and only if the map $\varphi+\varepsilon$ is a permutation of the set Q.

Corollary 9.111. *If L is a Latin square that is a Cayley table of a finite T-quasigroup $(Q, \cdot)$ of the form $x \cdot y = \varphi x + \psi y + a$, then:*

(i) the square $L^r L^{l l}$ is a Latin square if and only if $(\varphi - \psi), (\varphi + \psi)$ are permutations of the set Q;

(ii) the square $L^r L^r$ is a Latin square if and only if $(\varepsilon + \varphi)$ is a permutation of the set Q;

(iii) the square $L^l L^l$ is a Latin square if and only if $(\varepsilon + \psi)$ is a permutation of the set Q;

(iv) the square $L^l L^r$ is a Latin square if and only if $(\varphi + \psi^2)$ is a permutation of the set Q;

(v) the square $L^r L^l$ is a Latin square if and only if $(\varphi^2 + \psi)$ is a permutation of the set Q.

Proof. The proof follows from Corollary 9.68 and Theorem 9.109. □

Corollary 9.112. *T-quasigroup $(Z_p, \circ)$ of the form $x \circ y = k \cdot x + m \cdot y + c$, where $(Z_p, +)$ is the cyclic group of a prime order p,*

$$k, m, c \in Z_p, k, m, k + m, k - m, k + 1, m + 1, k^2 + m, k + m^2 \neq 0 \pmod{p},$$

where the operation $\cdot$ is multiplication modulo p, is orthogonal to any of its parastrophes.

Example 9.113. The quasigroup $(Z_p, \circ)$ of the form $x \circ y = 1 \cdot x + 2 \cdot y$, where $(Z_p, +)$ is the additive group of residues modulo p, p is a prime number, $p \geqslant 7$, is orthogonal to any of its parastrophes.

Example 9.114. The quasigroup $(Z_{11}, \circ)$ of the form $x \circ y = 3 \cdot x + 9 \cdot y$, where $(Z_{11}, +)$ is the additive group of residues modulo 11, is an idempotent quasigroup, which is orthogonal to any of its parastrophes.

Example 9.115. Denote elements of the group $(Z_2 \oplus Z_2, +)$ as follows: $\{(0;0), (1;0), (0;1), (1;1)\}$. The group $Aut(Z_2 \oplus Z_2, +)$ consists of the following automorphisms:

$$\begin{pmatrix} 1 & 0 \\ 0 & 1 \end{pmatrix}, \begin{pmatrix} 1 & 0 \\ 1 & 1 \end{pmatrix}, \begin{pmatrix} 1 & 1 \\ 0 & 1 \end{pmatrix}, \begin{pmatrix} 0 & 1 \\ 1 & 0 \end{pmatrix}, \begin{pmatrix} 1 & 1 \\ 1 & 0 \end{pmatrix}, \begin{pmatrix} 0 & 1 \\ 1 & 1 \end{pmatrix}.$$

Denote these automorphisms by the letters $\varepsilon, \varphi_2, \varphi_3, \varphi_4, \varphi_5, \varphi_6$, respectively.

Notice that $\varphi_2^2 = \varphi_3^2 = \varphi_4^2 = \varepsilon, \varphi_5^2 = \varphi_6, \varphi_6^2 = \varphi_5$. It is known that $Aut(Z_2 \oplus Z_2, +) \cong S_3$ [407, 471].

For convenience we give the Cayley table of the group $Aut(Z_2 \oplus Z_2, +)$.

$\cdot$	ε	φ_2	φ_3	φ_4	φ_5	φ_6
ε	ε	φ_2	φ_3	φ_4	φ_5	φ_6
φ_2	φ_2	ε	φ_5	φ_6	φ_3	φ_4
φ_3	φ_3	φ_6	ε	φ_5	φ_4	φ_2
φ_4	φ_4	φ_5	φ_6	ε	φ_2	φ_3
φ_5	φ_5	φ_4	φ_2	φ_3	φ_6	ε
φ_6	φ_6	φ_3	φ_4	φ_2	ε	φ_5

Further we construct three T-quasigroups over the group $(Z_2 \oplus Z_2, +)$:

$(Z_2 \oplus Z_2, D)$ with the form $D(x, y) = \varphi_3 x + \varphi_6 y + a_1$;
$(Z_2 \oplus Z_2, E)$ with the form $E(x, y) = \varphi_2 x + \varphi_5 y + a_2$;
$(Z_2 \oplus Z_2, F)$ with the form $F(x, y) = \varphi_3 x + \varphi_5 y + a_3$.

Lemma 9.116. *The quasigroups $(Z_2 \oplus Z_2, D)$, $(Z_2 \oplus Z_2, E)$, and $(Z_2 \oplus Z_2, F)$ are orthogonal in pairs.*

Proof. We can use Theorem 9.91 and the Cayley table of the group $Aut(Z_2 \oplus Z_2, +)$. □

9.3.5 (12)-parastrophe orthogonality

Orthogonality of a quasigroup and its (12)-parastrophe is clearer from the intuitive point of view and this orthogonality was studied in many articles [74, 732, 240, 144]. See also Theorem 9.69 and Lemma 9.70.

In this section we study connections between properties of anti-commutativity and (12)-parastrophe orthogonality of T-quasigroups.

A. Sade [732, 240] called a quasigroup $(Q, \cdot)$ anti-abelian if it is orthogonal to its (12)-parastrophe $(Q, \star)$: that is, if $x \cdot y = z \cdot t$ and $y \cdot x = t \cdot z$ $(x \star y = z \star t)$ imply $x = z$ and $y = t$.

Definition 9.117. A binary quasigroup $(Q, \cdot)$ is called *anti-commutative* (sometimes this quasigroup is called an *anti-symmetric quasigroup* [227]) if and only if the following implication is true: $x \cdot y = y \cdot x \Rightarrow x = y$ for all $x, y \in Q$ [72].

We shall call a binary anti-commutative quasigroup $(Q, \cdot)$ *totally anti-commutative* (sometimes TAC-quasigroup for short) if the following implication is true: $x \cdot x = y \cdot y \Rightarrow x = y$ for all $x, y \in Q$.

Remark 9.118. We notice, that the definition of a TAC-quasigroup is close to the definition of a totally anti-symmetric quasigroup [227]: an anti-symmetric quasigroup $(Q, \cdot)$ is *totally anti-symmetric* if the following implication is true $(c \cdot x) \cdot y = (c \cdot y) \cdot x \Rightarrow x = y$ for all $x, y \in Q$.

Theorem 9.119. *A binary T-quasigroup $(Q, \cdot)$ of the form $x \cdot y = \alpha x + \beta y + a$ is a totally anti-commutative quasigroup if and only if the mappings $\alpha - \beta$ and $\alpha + \beta$ are automorphisms of the group $(Q, +)$ (i.e., they are permutations of the set Q).*

Proof. For a T-quasigroup $(Q, \cdot)$ the property of anti-commutativity $x \cdot y = y \cdot x \Rightarrow x = y$ for all $x, y \in Q$ can be rewritten in the form:

$$\begin{gathered}(\alpha x + \beta y = \alpha y + \beta x \Rightarrow x = y) \Leftrightarrow \\ ((\alpha - \beta)x = (\alpha - \beta)y \Rightarrow x = y) \Leftrightarrow \\ ((\alpha - \beta)(x - y) = 0 \Rightarrow x = y).\end{gathered}$$

The last implication is true only if $\alpha - \beta$ is an automorphism of group $(Q, +)$ (in the general case the mapping $\alpha - \beta$ is an endomorphism of the group $(Q, +)$).

The implication $x \cdot x = y \cdot y \Rightarrow x = y$ for all $x, y \in Q$ can be rewritten in the form

$$\begin{gathered}(\alpha x + \beta x = \alpha y + \beta y \Rightarrow x = y) \Leftrightarrow \\ ((\alpha + \beta)(x - y) = 0 \Rightarrow x = y).\end{gathered}$$

The last implication is true only if $\alpha + \beta$ is an automorphism.

Conversely, if the map $\alpha - \beta$ is an automorphism (a permutation on the set Q), then the implication $(\alpha - \beta)(x - y) = 0 \Rightarrow x = y$ is true since the automorphism $\alpha - \beta$ has the identity as its kernel.

If the map $\alpha + \beta$ is an automorphism, then the implication $x \cdot x = y \cdot y \Rightarrow x = y$ holds in the T-quasigroup $(Q, \cdot)$. □

M. Damm [227] proved that any anti-abelian quasigroup is a totally anti-commutative quasigroup. We may prove the following:

Theorem 9.120. *A T-quasigroup $(Q,\cdot)$ of the form $x\cdot y=\varphi x+\psi y+c$ is a totally anti-commutative quasigroup if and only if it is an anti-abelian quasigroup.*

Proof. From Theorem 9.119 it follows that a T-quasigroup $(Q,\cdot)$ of the form $x\cdot y=\varphi x+\psi y+c$ is totally anti-commutative if and only if the maps $\varphi-\psi$ and $\varphi+\psi$ are permutations of the set Q. From Corollary 9.94 it follows that a T-quasigroup of the form $x\cdot y=\varphi x+\psi y+c$ is anti-abelian if and only if the map $\varphi^{-1}\psi-\psi^{-1}\varphi$ is a permutation of the set Q. Further we can apply Lemma 9.108. □

Let us remark that the implication "$\Leftarrow$" in Theorem 9.120 follows from the above mentioned result of M. Damm [227].

Example 9.121. The following quasigroup is a totally anti-commutative quasigroup but it is not an anti-abelian quasigroup [228].

$\cdot$	0	1	2	3	4	5
0	0	1	2	3	4	5
1	2	5	0	4	1	3
2	3	4	1	0	5	2
3	4	3	5	2	0	1
4	5	2	4	1	3	0
5	1	0	3	5	2	4

Example 9.122. Using Mace [614] the following example of a totally anti-commutative quasigroup which is not an anti-abelian quasigroup is constructed [795].

$\cdot$	0	1	2	3	4	5	6	7	8	9
0	1	5	2	4	3	7	8	6	9	0
1	2	0	6	1	7	8	3	4	5	9
2	8	2	7	9	5	6	0	3	4	1
3	5	6	1	3	4	9	7	8	0	2
4	6	8	4	7	9	5	2	0	1	3
5	0	7	5	8	1	2	6	9	3	4
6	9	1	8	6	0	3	4	2	7	5
7	7	3	9	0	2	4	1	5	8	6
8	4	9	3	2	8	0	5	1	6	7
9	3	4	0	5	6	1	9	7	2	8

Theorem 9.123. *For a T-quasigroup $(Q,\cdot)$ of the form $x\cdot y=\varphi x+\psi y+c$ over a commutative group $(Q,+)$ the following conditions are equivalent:*

- $(x\cdot y=y\cdot x)\Rightarrow(x=y)$, $(x\cdot x=y\cdot y)\Rightarrow(x=y)$ *for all* $x,y\in Q$;
- $(x\cdot y=z\cdot t$ *and* $y\cdot x=t\cdot z)\Rightarrow(x=z$ *and* $y=t)$ *for all* $x,y,z,t\in Q$;
- *the maps* $\varphi-\psi$ *and* $\varphi+\psi$ *are permutations of the set* Q;
- *the maps* $\varphi^{-1}-\psi^{-1}$ *and* $\varphi+\psi$ *are permutations of the set* Q;
- *the map* $\varphi^{-1}\psi-\psi^{-1}\varphi$ *is a permutation of the set* Q;
- *the T-quasigroup* $(Q,\cdot)$ *and its* (12)*-parastrophe* $(Q,\star)$ *are orthogonal.*

Proof. The proof follows from Theorem 9.94 and Theorem 9.120. □

Remark 9.124. For a medial quasigroup $(Q, \cdot)$ of the form $x \cdot y = \varphi x + \psi y + c$ over a commutative group $(Q, +)$ the following conditions are equivalent: "the maps $\varphi - \psi$ and $\varphi + \psi$ are permutations of the set Q" $\iff$ "the map $\varphi^2 - \psi^2$ is a permutation of the set Q."

Proof. From the definition of a medial quasigroup we have that $\varphi\psi = \psi\varphi$. Then $(\varphi-\psi)(\varphi+\psi) = \varphi^2 + \varphi\psi - \psi\varphi - \psi^2 = \varphi^2 - \psi^2$. □

In [144] Bennett and Zhang study Latin squares with self-orthogonal conjugates. In the language of this work, Latin squares with self-orthogonal conjugates correspond to quasigroups with the property: $(Q, A^{\sigma}) \perp (Q, A^{\sigma})^{(12)}$ for any $\sigma \in S_3$. For short, we shall call quasigroups with this property SOC-quasigroups.

For *SOC-T*-quasigroups we can prove the following:

Theorem 9.125. *A T-quasigroup $(Q, \cdot)$ of the form $x \cdot y = \varphi x + \psi y + c$ over a group $(Q, +)$ is a SOC-quasigroup if and only if the maps $\varphi - \psi$, $\varphi + \psi$, $\varepsilon - \psi$, $\varepsilon + \psi$, $\varepsilon - \varphi$ and $\varepsilon + \varphi$ are permutations of the set Q.*

Proof. If $(Q, \cdot)$ is a T-quasigroup of the form $x \cdot y = \varphi x + \psi y + c$, then its parastrophes have the following forms, respectively:

$$\begin{aligned} x \overset{(12)}{\cdot} y &= \psi x + \varphi y + c, \\ x \overset{(13)}{\cdot} y &= \varphi^{-1} x - \varphi^{-1}\psi y - \varphi^{-1} c, \\ x \overset{(23)}{\cdot} y &= -\psi^{-1}\varphi x + \psi^{-1} y - \psi^{-1} c, \\ x \overset{(123)}{\cdot} y &= -\varphi^{-1}\psi x + \varphi^{-1} y - \varphi^{-1} c, \\ x \overset{(132)}{\cdot} y &= \psi^{-1} x - \psi^{-1}\varphi y - \psi^{-1} c. \end{aligned} \tag{9.45}$$

From Theorem 9.109 Case (i) it follows that $(Q, \cdot) \perp (Q, \overset{(12)}{\cdot})$ if and only if $\varphi - \psi$, $\varphi + \psi$ are permutations of the set Q (see, also, Theorem 9.123).

By Theorem 9.109 quasigroup $(Q, \overset{(13)}{\cdot})$ is orthogonal to its (12)-parastrophe if and only if $\varphi^{-1} - \varphi^{-1}\psi = \varphi^{-1}(\varepsilon - \psi)$ and $\varphi^{-1} + \varphi^{-1}\psi = \varphi^{-1}(\varepsilon + \psi)$ are permutations of the set Q. The last two statements are equivalent to the following: the maps $(\varepsilon - \psi)$ and $(\varepsilon + \psi)$ are permutations of the set Q.

Similarly, by Theorem 9.109, quasigroup $(Q, \overset{(23)}{\cdot})$ is orthogonal to its (12)-parastrophe if and only if $-\psi^{-1}\varphi + \psi^{-1} = \psi^{-1}(-\varphi + \varepsilon)$ and $-\psi^{-1}\varphi - \psi^{-1} = \psi^{-1}(-\varphi - \varepsilon)$ are permutations of the set Q. The last two equalities are equivalent to the following: $(\varepsilon - \varphi)$ and $(\varepsilon + \varphi)$ are permutations of the set Q. □

Example 9.126. The quasigroup $(Z_7, \circ)$ of the form $x \circ y = 3 \cdot x + 5 \cdot y$, where $(Z_7, +)$ is the additive group of residues modulo 7, is a *SOC*-quasigroup of order 7.

Proof. The proof follows from Theorem 9.125, since $3 + 5 \equiv 1 \pmod 7$, $3 - 5 \equiv 5 \pmod 7$, $1 - 3 \equiv 5 \pmod 7$, $1 + 3 \equiv 4 \pmod 7$, $1 - 5 \equiv 3 \pmod 7$, $1 + 5 \equiv 6 \pmod 7$. □

Example 9.127. The quasigroup $(Z_{11}, \circ)$ of the form $x \circ y = 3 \cdot x + 9 \cdot y$, where $(Z_{11}, +)$ is the additive group of residues modulo 11, is a SOC-quasigroup of order 11.

Proof. The proof follows from Theorem 9.125, since $3+9 \equiv 1 \pmod{11}$, $3-9 \equiv 5 \pmod{11}$, $1-3 \equiv 9 \pmod{11}$, $1+3 \equiv 4 \pmod{11}$, $1-9 \equiv 3 \pmod{11}$, $1+9 \equiv 10 \pmod{11}$. □

Therefore, from Examples 9.126 and 9.127 it follows that there exist Latin squares with self-orthogonal conjugates of order 7 and 11. These examples supplement the results of Bennett and Zhang [144].

Proposition 9.128. *A quasigroup* $(\mathbb{Q}, \circ)$ *of the form* $x \circ y = a \cdot x + b \cdot y + c$, *where* $(\mathbb{Q}, +)$ *is the additive group of rational numbers,* $a \neq b$, $a \neq 1$, $a \neq 0$, $b \neq 1$,$b \neq 0$, *is an infinite SOC-quasigroup.*

Proof. The proof follows from Theorem 9.125. □

Remark 9.129. It is easy to see that classes of SOC-quasigroups and quasigroups which are orthogonal to all its parastrophes, intersect (Example 9.127 = Example 9.114), but neither of these two classes is included in the other class.

The quasigroups constructed in Example 9.113 are orthogonal to all its parastrophes and are not SOC-quasigroups ($1-1 \equiv 0 \pmod{p}$).

In Example 9.126 a SOC-quasigroup is constructed which is not orthogonal to all its parastrophes ($3^2+5 \equiv 0 \pmod{7}$).

9.3.6 totCO-quasigroups

Definition 9.130. A quasigroup (Q, A) is called a totally parastroph (or conjugate) orthogonal quasigroup (for short totCO-quasigroup), if all its parastrophes are orthogonal in pairs [127].

It is easy to see that there exist fifteen unordered pairs of parastrophes. These pairs are listed in Lemma 1.172.

Theorem 9.131. *A T-quasigroup* $(Q, \cdot)$ *of the form* $x \cdot y = \varphi x + \psi y + c$ *over a group* $(Q, +)$ *is a totCO-quasigroup if and only if the maps* $\varphi - \psi$, $\varphi + \psi$, $\varepsilon - \psi$, $\varepsilon + \psi$, $\varepsilon - \varphi$, $\varepsilon + \varphi$, $\varphi + \psi^2$, $\varphi^2 + \psi$, *and* $\varphi\psi - \varepsilon$ *are permutations of the set* Q *[127].*

Proof. From Proposition 9.140 it follows that orthogonality is invariant relative to (1 2)-parastrophy (s-parastrophy in Belousov's designations). From Lemma 1.172 it follows that we can split the proof of this theorem into 9 cases. In numeration of the cases we follow Lemma 1.172.

Cases I, II, III, IV, and V are considered in Theorem 9.109, cases VI and VII in Theorem 9.125.

Therefore to finish the proof we must consider Cases VIII and IX. The forms of parastrophes are given in equations (9.45) and criteria of orthogonality are taken from Corollary 9.91.

Case VIII. We have $(Q, \overset{(13)}{\cdot}) \perp (Q, \overset{(23)}{\cdot})$ if and only if $\varepsilon - \varphi\psi$ is a permutation of the set Q.

Case IX. We have $(Q, \overset{(23)}{\cdot}) \perp (Q, \overset{(123)}{\cdot})$ if and only if $\varphi\psi - \varepsilon$ is a permutation of the set Q.

It is easy to see that the conditions "the endomorphism $\varepsilon - \varphi\psi$ is a permutation of the set Q" and "the endomorphism $\varphi\psi - \varepsilon$ is a permutation of the set Q" are equivalent. Indeed, $I(\varepsilon - \varphi\psi) = \varphi\psi - \varepsilon$. □

Example 9.132. Quasigroup $(Z_p, \circ)$ of the form $x \circ y = 2 \cdot x + 3 \cdot y$, where $(Z_p, +)$ is the additive group of residues modulo p, p is a prime number, $p \geqslant 13$, is orthogonal to any of its parastrophes, moreover, it is a totCO-quasigroup.

More details about the information presented in this section is given in [120, 127, 836, 835, 793].

9.4 Nets and orthogonality of the systems of quasigroups

9.4.1 k-nets and systems of orthogonal binary quasigroups

Here we present the next generalization step of 3-nets. We give a definition of a geometrical object (k-net) that corresponds to a pair ($k = 4$), triple ($k = 5$), s ($k = s+2$) mutually orthogonal binary quasigroups (the system of orthogonal quasigroups).

Since the letter k is used in denotation of t-(v, k, λ)-design we shall use letter m instead of letter k in the denotation of k-net.

Definition 9.133. An m-net ($m \geqslant 3$) of order n is an incidence structure $\mathfrak{S} = (\mathfrak{P}, \mathfrak{L})$ which consists of an n^2-element set $\mathfrak{P}$ of points and an mn-element set $\mathfrak{L}$ of lines. The set $\mathfrak{L}$ is partitioned into m disjoint families $L_1, \dots, L_m$ of (parallel) lines, for which the following conditions are true:

(i) every point is incident with exactly one line of each family L_i ($i \in \{1, 2, \dots, m\}$);
(ii) two lines of different families have exactly one point in common;
(iii) two lines in the same family do not have a common point [76, 580].

The families $L_1, \dots, L_m$ sometimes are called directions or parallel classes of $\mathfrak{S}$. An m-net $\mathfrak{S}$ is an incidence structure with the parameters $v = n^2$, $b = mn$, $k = n$, $r = m$.

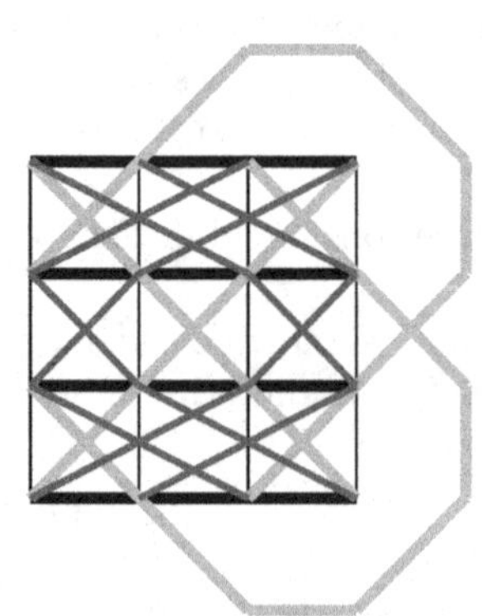

Figure 9.1: 4-net of order four.

Example 9.134. Using Example 1.419 we construct a 4-net of order 3. We denote points of this net in the following way $\{(0,0), (0,1), (0,2), (1,0), (1,1), (1,2), (2,0), (2,1), (2,2)\}$. We list all 12 lines (they are triplets of points) of this net:

$$\begin{array}{lll} \{(0,0),(0,1),(0,2)\}, & \{(1,0),(1,1),(1,2)\}, & \{(2,0),(2,1),(2,2)\}, \\ \{(0,0),(1,0),(2,0)\}, & \{(0,1),(1,1),(2,1)\}, & \{(0,2),(1,2),(2,2)\}, \\ \{(0,0),(1,2),(2,1)\}, & \{(0,1),(1,0),(2,2)\}, & \{(0,2),(1,1),(2,0)\}, \\ \{(0,0),(1,1),(2,2)\}, & \{(0,1),(1,2),(2,0)\}, & \{(0,2),(1,0),(2,1)\}. \end{array}$$

Dual structure $\mathfrak{S}^T$ has m families of points each of cardinality n, which are called groups, and n^2 blocks (lines). Algebraic properties of k-nets and corresponding systems of orthogonal quasigroups (loops) are studied in [233, 234, 236, 235, 579, 580, 581, 582]. Often these properties are similar to corresponding properties of "usual" binary quasigroups.

Any finite k-net is coordinatized by the system of orthogonal binary operations, by the system of orthogonal binary quasigroups [76, 580].

9.4.2 Algebraic (k, n)-nets and systems of orthogonal n-ary quasigroups

We see that k-nets correspond to the systems of mutually orthogonal binary quasigroups. Further we define geometrical objects which correspond to the systems of mutually orthogonal n-ary quasigroups. We recall concept of orthogonality of n-ary operations.

Definition 9.135. A system Σ ($|\Sigma| \geqslant n$) of n-ary operations defined on a set Q is said to be orthogonal if for every subset $M = \{A_1, \dots, A_n\}$ of the set Σ the mapping $\sigma(A_1(x_1^n), \dots, A_n(x_1^n)) \mapsto (x_1^n)$ is a permutation of Q^n [57].

Definition 9.136. The system of objects $S = (N, \mathfrak{L})$ ($|N| = n, |\mathfrak{L}| = k \geqslant n+1$) of two kinds ($N$ is the set of points, $\mathfrak{L}$ is the set of classes of hyperplanes) is called a (k, n)-net, if the following conditions are true:

(i) every class $L_i \in \mathfrak{L}$ ($i \in \overline{1,k}$) contains at least two objects;

(ii) for every point $P \in N$ there exists exactly one hyperplane $H_i \in L_i$ ($i \in \overline{1,k}$) such that $P \in H_i$;

(iii) if $H = \{H_i \mid H_i \in L_i, i \in \overline{1,n}\}$ is a set of n hyperplanes that lie in n different classes L_i, then there exists a unique point P such that $P = \cap_{i=1}^n H_i$.

If $k = n+1$, then Definition 9.136 is transformed in a definition of an n-dimensional net which corresponds to an n-ary quasigroup.

The number k is called a genus of the (k, n)-net and number n is called dimension of this net. A finite (k, n)-net contains $|N| = q^n$ points; any hyperplane consists of q^{n-1} points and it is an $(n-1)$-dimension net [44, 53].

The following inequality is true:

$$k \leqslant (n-1)q + 1. \tag{9.46}$$

See [432, 53] for details. Compare inequality (9.46) with the results presented in Theorem 1.335.

The following results are similar to the results about coordinatization of m-nets [580]. Any finite (k, n)-net $S = (N, L_1, \dots, L_k)$ can be coordinatized using a system Σ of orthogonal n-ary quasigroups (OSQ).

It is proved: if OSQ Σ_1 and Σ_2 coordinatize a finite (k, n)-net S in which numbering of classes is fixed, then these systems are isostrophic [53].

Example 9.137. Let N be the set of all points of n-dimensional Euclidean space $\mathbb{R}_n$. We can associate to any point $c \in N$ its coordinate, namely, n-tuple $(c_1, \dots, c_n)$, where $c_i \in \mathbb{R}$. Let L_i $(i \in \overline{1,n})$ be a class of hyperplanes that satisfies the following equation:

$$x_i = b_i, \tag{9.47}$$

where b_i are fixed real numbers. Let L_j $(j \in \overline{n+1,k})$ be a class of hyperplanes that satisfy the following equation:

$$\sum_{i=1}^{n} a_{i,j} x_i = b_j, \tag{9.48}$$

where $a_{i,j}, b_j$ are some fixed real numbers. If any system of n equations that consists of equations of the form (9.47) and (9.48) has a unique solution, then the system of objects $S = (N, L_1, \dots, L_k)$ forms a (k,n)-net [53].

An example of a (k,n)-net constructed using affine n-dimensional space over the field $GF(p^k)$ is given in [297].

Finite nets of dimension d are defined in [558, 297]. The authors find connections of these nets with association schemes [232, 26, 33, 419]. Algebraic (k,n)-nets and some similar objects are studied in [708, 44, 54, 55, 56].

9.4.3 Orthogonality of n-ary quasigroups and identities

In [309] Evans defined the following identity on ternary quasigroup $(Q, \langle , , \rangle)$:

$$\langle x, \langle x, y, z\rangle, \langle z, x, y\rangle\rangle = \langle y, z, x\rangle . \tag{9.49}$$

This identity is a generalization of the following Stein identity $x(xy) = yx$ (Definition 2.2). Any binary quasigroup $(Q, \cdot)$ which satisfies this Stein identity is orthogonal to its (12)- and (23)-parastrophe (see Corollary 9.73). The situation with the ternary analog of Stein identity is similar. The following quasigroups are orthogonal: $(Q, \langle , , \rangle)$, $(Q, \langle , , \rangle_1)$, and $(Q, \langle , , \rangle_2)$, where $\langle x, y, z\rangle_1 = \langle z, x, y\rangle$, $\langle x, y, z\rangle_2 = \langle y, z, x\rangle$.

Indeed, suppose that $\langle x, y, z\rangle = a$, $\langle z, x, y\rangle = b$, $\langle y, z, x\rangle = c$. From the identity (9.49) we have $\langle x, a, b\rangle = c$. We can also rewrite the identity (9.49) in the following forms: $\langle y, \langle y, z, x\rangle, \langle x, y, z\rangle\rangle = \langle z, x, y\rangle$ and $\langle z, \langle z, x, y\rangle, \langle y, z, x\rangle\rangle = \langle x, y, z\rangle$. From the last forms of the identity (9.49) we obtain these equalities: $\langle y, c, a\rangle = b$, $\langle z, b, c\rangle = a$. Therefore the system of equations

$$\begin{cases} \langle x, y, z\rangle = a \\ \langle x, y, z\rangle_1 = \langle z, x, y\rangle = b \\ \langle x, y, z\rangle_2 = \langle y, x, z\rangle = c \end{cases} \tag{9.50}$$

has a unique solution for any fixed elements $a, b, c \in Q$, i.e., quasigroups $(Q, \langle , , \rangle)$, $(Q, \langle , , \rangle_1)$, and $(Q, \langle , , \rangle_2)$ are orthogonal.

Further steps in the study of orthogonality of n-ary quasigroups, also including parastroph orthogonality, were realized in the articles [855, 854, 852, 851, 130, 867, 868, 114, 125, 116, 126, 118, 294].

A quasigroup (Q, f) satisfying the identity $x(yx) = y$ (or the equivalent identity $(xy)x = y$) is called semi-symmetric. An n-quasigroup (Q, f) satisfying the identity $f(f(x_1, \dots, x_n), x_1, ..., x_{n-1}) = x_n$ is called cyclic. So, cyclic n-quasigroups are a generalization of semi-symmetric quasigroups.

The n-ary operation f on Q gives rise to n-ary operations $f_1, f_2 \dots, f_{n-1}$ on Q as follows:

$$\begin{aligned} f_1(x_1, x_2, \dots, x_n) &= f(x_2, x_3, \dots, x_n, x_1), \\ f_2(x_1, x_2, \dots, x_n) &= f(x_3, x_4, \dots, x_n, x_1, x_2), \\ \dots, \\ f_{n-1}(x_1, x_2, \dots, x_n) &= f(x_n, x_1, x_2, \dots, x_{n-2}, x_{n-1}). \end{aligned} \tag{9.51}$$

It is clear that the equalities (9.51) define cyclic parastrophes of operation f.

An n-ary quasigroup (Q, f) is said to be self-orthogonal if the following system of equations

$$\begin{cases} f(x_1, x_2, \dots, x_n) = a_1 \\ f_1(x_1, x_2, \dots, x_n) = a_2 \\ \dots \\ f_{n-1}(x_1, x_2, \dots, x_n) = a_n \end{cases} \tag{9.52}$$

has a unique solution for any fixed elements $a_1, a_2, \dots, a_{n-1}, a_n \in Q$, where operations $f_1, \dots, f_n$ are defined by the equalities (9.51) (i.e., are some parastrophes of operation f).

Example 9.138. Let $(Q, +)$ be an abelian group, $\varphi \in Aut(Q, +)$, such that $\varphi a = -a$ for a fixed element $a \in Q$, $\varphi^{n+1}x = -x$ for any element $x \in Q$, if n is even, and $\varphi^{n+1} = \varepsilon$, if n is odd. Then by the equality

$$f(x_1, x_2, \dots, x_n) = \varphi x_1 - \varphi^2 x_2 + \varphi^3 x_3 - \dots + (-1)^{n+1}\varphi^n x_n + a$$

a cyclic n-quasigroup (Q, f) is defined [853]. It is easy to see that the automorphism $Ix = -x$ of the group $(Q, +)$ satisfies the conditions of this example.

We give some results from article [854]. Generally, we follow MR843666 (87j:20124).

Denote by n the degree (arity) of operation f and by q, the order of the set Q. If n and q are odd integers with $q \geqslant 3$, then there exists an SOCQ (self-orthogonal cyclic quasigroup) of degree n and order q; if k, r, and s are positive integers with $k \geqslant 2$, $s \geqslant 2$, and $2^k - 1 = rs$, then for every positive integer t there exists an SOCQ of degree $ts - 1$ and order $2k$; if k, r, s, and p are positive integers such that p is an odd prime, s is odd with $s \geqslant 3$, and $p^k - 1 = 2rs$, then for every positive odd integer t there exists an SOCQ of degree $ts - 1$ and order p^k. Additional results are obtained by reformulating these results and by taking direct products.

In [130] self-orthogonal cyclic n-groupoids and their generalizations are considered. Let A be an n-ary operation on the set Q, let S_{n+1} be the symmetric group of degree $n + 1$, and let β be a permutation in S_{n+1} such that there exists a parastrophe ${}^{\beta}A$ of A. Notice, conditions that an n-ary groupoid (G, A) has a parastrophe are relatively strong. See, for example, Theorem 1.65 for binary case.

Let $H = \{\alpha_1, \dots, \alpha_n\}$ be a subgroup of S_{n+1} with $\alpha_i(n+1) = n+1$ for $i = \overline{1, n}$. Authors study n-groupoids satisfying the following identity:

$${}^{\beta}A({}^{\alpha_1}A(x_1, \dots, x_n), \dots, {}^{\alpha_n}A(x_1, \dots, x_n)) = x_1. \tag{9.53}$$

They prove that every finite n-groupoid which satisfies the identity (9.53) is self-orthogonal (see MR1037948 (91g:20105)). Notice that the identity (9.53) is a generalization of Evans' identity (9.49).

9.5 Transformations which preserve orthogonality

9.5.1 Isotopy and (12)-isostrophy

From definitions of orthogonality of squares, groupoids and m-tuples of permutations and properties of these objects, it follows that the property of orthogonality is symmetric, i.e., $A \perp B$ if and only if $B \perp A$, where A and B are squares, groupoids or m-tuples of permutations.

Let (Q, A) be a groupoid and $S_1(Q)$ be a square that coincides with the body of the Cayley table of this groupoid and let (Q, B) be a groupoid and $S_2(Q)$ be a square that coincides with the body of the Cayley table of this groupoid. Then $(Q, A) \perp (Q, B)$ if and only if $S_1(Q) \perp S_2(Q)$.

Proposition 9.139. *Groupoids (Q, A) and (Q, B) are orthogonal if and only if groupoids $(Q, A)T$ and $(Q, B)T$ are orthogonal, where T is an isotopy.*

Proof. Let $T = (\alpha, \beta, \gamma)$. We can decompose this isotopy in the product of isotopies $T_1T_2T_3$, where $T_1 = (\alpha, \varepsilon, \varepsilon)$, $T_2 = (\varepsilon, \beta, \varepsilon)$ and $T_3 = (\varepsilon, \varepsilon, \gamma)$ [72].

It is well known that the isotopy T_1 changes the order of rows in the Cayley tables of the groupoids (Q, A) and (Q, B), the isotopy T_2 changes the order of columns in these Cayley tables, and the isotopy T_3 changes elements in these Cayley tables.

It is easy to see, if $(Q, A) \perp (Q, B)$, then

$$
\begin{aligned}
&(Q, A)T_1 \perp (Q, B)T_1, \\
&(Q, A)T_1T_2 \perp (Q, B)T_1T_2, \\
&(Q, A)T_1T_2T_3 \perp (Q, B)T_1T_2T_3.
\end{aligned}
$$

Therefore, if $(Q, A) \perp (Q, B)$, then $(Q, A)T \perp (Q, B)T$.

It is clear that the implication $(Q, A)T \perp (Q, B)T \Rightarrow (Q, A) \perp (Q, B)$ is fulfilled, too. □

We notice that in [86] Proposition 9.139 is proved for quasigroups. See, also, Lemma 9.43.

Proposition 9.140. *Groupoids (Q, A) and (Q, B) are orthogonal if and only if groupoids $(Q, A^{(12)})$ and $(Q, B^{(12)})$ are orthogonal.*

Proof. From the properties of (12)-parastrophy of groupoids it follows that the i-th row of the body of the Cayley table of groupoid $(Q, A^{(12)})$ coincides with the i-th column of the body of the Cayley table of groupoid (Q, A). A similar situation exists with groupoids (Q, B) and $(Q, B^{(12)})$.

We denote by S_1 and S_2 the bodies of the Cayley tables of groupoids (Q, A) and (Q, B) respectively. By Definition 9.41 if $S_1 \perp S_2$, then $D\,(S_1 \oplus S_2) = Q \times Q$. Since the i-th row in the square $S_1^{(12)} \oplus S_2^{(12)}$ coincides with the i-th column of the square $S_1 \oplus S_2$, we conclude that $D\,(S_1^{(12)} \oplus S_2^{(12)}) = Q \times Q$, i.e., $S_1^{(12)} \perp S_2^{(12)}$.

Therefore, if $(Q, A) \perp (Q, B)$, then $(Q, A^{(12)}) \perp (Q, B^{(12)})$. Using the same arguments we can prove that if $(Q, A^{(12)}) \perp (Q, B^{(12)})$, then $(Q, A) \perp (Q, B)$. □

In [81, 85] Proposition 9.140 is proved for quasigroups.

Corollary 9.141. *Groupoids (Q, A) and (Q, B) are orthogonal if and only if $(Q, A)\beta \perp (Q, B)\beta$, where $\beta \in ISOS_{(12)}(Q)$.*

Proof. The proof follows from Propositions 9.139 and 9.140. □

Unfortunately, in general, it is impossible to extend the result of Proposition 9.140 to other types of parastrophy of quasigroups.

Example 9.142. There exists a pair of orthogonal quasigroups (Q, A) and (Q, B) such that quasigroups $(Q, A^{(23)})$ and $(Q, B^{(23)})$ are not orthogonal.

A, B	0	1	2	3	4
0	00	23	41	14	32
1	11	34	02	20	43
2	22	40	13	31	04
3	33	01	24	42	10
4	44	12	30	03	21

$A^{(23)}, B^{(23)}$	0	1	2	3	4
0	00	32	14	41	23
1	23	00	32	14	41
2	41	23	00	32	14
3	14	41	23	00	32
4	32	14	41	23	00

Proposition 9.143. *The permutation squares S_1 and S_2 of a kind α, $\alpha \in \{l, Il, r, Ir\}$ are orthogonal if and only if $S_1S_3 \perp S_2S_3$, where S_3 is a permutation square of the kind α [661].*

Proof. From Theorem 9.57 it follows that the permutation squares S_1 and S_2 are orthogonal if and only if the tuple $T_2T_1^{-1}$ of a kind α, $\alpha \in \{l, Il, r, Ir\}$ is strictly transitive, where T_1 and T_2 are m-tuples of a kind α of the squares S_1 and S_2 respectively.

The permutation squares S_1S_3 and S_2S_3 are orthogonal if and only if the tuple $T_2T_3T_3^{-1}T_1^{-1}$ of the kind α is a strictly transitive set of permutations. But we have $T_2T_3T_3^{-1}T_1^{-1} = T_2T_1^{-1}$. □

Proposition 9.143 leads us to the following generalization of the concept of isotopy.

9.5.2 Generalized isotopy

It is possible to give a concept which is a generalization of the concept of isotopy in many cases. As we saw, the concept of isotopy makes sense for any square and any groupoid. The concept of generalized isotopy (probably, gisotopy for convenience) also makes sense for squares and groupoids. As we shall see, for permutation squares and left (right) quasigroups the concept of generalized isotopy is more general than the concept of "usual" isotopy.

Definition 9.144. Any m-tuple of permutations P of a kind α, $\alpha \in \{l, Il, r, Ir, p, Ip\}$, will be called a *generalized isotopy of the kind α* or a *gisotopy of the kind α*.

Definition 9.145. A groupoid (Q, A) is a *gisotope (a gisotopic image) of a kind α of a groupoid* (Q, B), where $\alpha \in \{l, r\}$, if there exists an m-tuple of permutations P of the set Q of the kind α such that $T_A^{\alpha} = T_B^{\alpha} P$, i.e., $(t_A^{\alpha})_i = (t_B^{\alpha})_i p_i$ for all suitable values of the index i, where $T_A^{\alpha}, T_B^{\alpha}$ are m-tuples of maps of the kind α that correspond to the groupoids (Q, A), (Q, B), respectively.

Remark 9.146. We write a gisotopy P from the right from a groupoid (Q, A), as we write an isotopy T from the right from a groupoid (Q, A).

Remark 9.147. The concept of gisotopy is closely related to the concepts of derivation [512, p. 107] and Belousov's crossed isotopy [84]. We thank Dr. E.A. Kuznetsov for this remark.

It is easy to see that the concept of gisotopy makes sense for m-tuples of maps and for squares, since gisotopy is defined using the concept of multiplication of m-tuples.

For left (and right) quasigroups the kind α from Definition 9.145 can be any element of the set $\{l, Il, r, Ir\}$. Using Mann's product of permutation squares, we can give the following:

Definition 9.148. A permutation square S_1 of a kind α, $\alpha \in \{l, Il, r, Ir\}$, is a *gisotopic image of a permutation square* S_2 of the kind α if and only if $S_1 = S_2P$, where P is a gisotopy of the kind α.

Example 9.149. It is well known that the groups $Z_2 \oplus Z_2$ and Z_4

$Z_2 \oplus Z_2$	0	1	2	3
0	0	1	2	3
1	1	0	3	2
2	2	3	0	1
3	3	2	1	0

Z_4	0	1	2	3
0	0	1	2	3
1	1	2	3	0
2	2	3	0	1
3	3	0	1	2

are non-isotopic (for example, this follows from the Albert theorem (Theorem 1.123) but these groups are gisotopic with the left gisotopy $P = (\varepsilon, (13), \varepsilon, (13))$.

Theorem 9.150. *If (Q, A) is a left quasigroup and T is an isotopy, then there exists a gisotopy GT of the kind l such that $(Q, A)T = (Q, A)GT$, i.e., any isotopy of a left quasigroup is a gisotopy.*

Proof. If (Q, A) is a left quasigroup and T is an isotopy, then by Corollary 1.131 $(Q, A)T$ is a left quasigroup.

If S_1 is a permutation square of the kind l which corresponds to the left quasigroup (Q, A), S_2 is a permutation square of the kind l which corresponds to the left quasigroup $(Q, A)T$, then $S_2 = S_1(S_1^{-1}S_2)$. Thus an m-tuple of permutation GT of the kind l that corresponds to the square $S_1^{-1}S_2$ is a gisotopy such that $(Q, A)T = (Q, A)GT$.

Therefore any isotopy of a left quasigroup is a gisotopy. □

Remark 9.151. It is easy to see that the similar theorem is true for right quasigroups.

Corollary 9.152. *Any isotopy of a quasigroup is a generalized isotopy.*

Proof. The proof is a direct consequence of Theorem 9.150 and Remark 9.151. □

It is easy to see that, generally speaking, a gisotopic image of a square is a square, a gisotopic image of a permutation square is a permutation square, and a gisotopic image of a Latin square is a permutation square.

Proposition 9.153. *The action of gisotopy $P = (p_1, p_2, \dots, p_n, \dots)$ of the kind l on a groupoid $(Q, \cdot)$ coincides with the action of the tuple T of isotopies of the form $T = ((\varepsilon, p_1, \varepsilon), (\varepsilon, p_2, \varepsilon), \dots, (\varepsilon, p_i, \varepsilon), \dots)$, where the isotopy $(\varepsilon, p_i, \varepsilon)$ acts only on the i-th row of Cayley table of the groupoid $(Q, \cdot)$.*

Proof. If L_{a_i} is the i-th left translation of the groupoid $(Q, \cdot)$, then in groupoid $(Q, \cdot)P$ the i-th row has the form $L^{\cdot}_{a_i} p_i(x) = a_i \cdot (p_i(x))$.

If we apply the isotopy $(\varepsilon, p_i, \varepsilon)$ to the groupoid $(Q, \cdot)$, then we have $x \circ y = x \cdot p_i(y)$. The i-th left translation of the groupoid $(Q, \circ)$ has the form $L^{\circ}_{a_i} y = L^{\cdot}_{a_i} p_i(y)$. □

Proposition 9.154. *The action of gisotopy $P = (p_1, p_2, \dots, p_n, \dots)$ of the kind r on a groupoid $(Q, \cdot)$ coincides with the action of the tuple T of isotopies of the form $T = ((p_1, \varepsilon, \varepsilon), (p_2, \varepsilon, \varepsilon), \dots, (p_i, \varepsilon, \varepsilon), \dots)$, where the isotopy $(p_i, \varepsilon, \varepsilon)$ acts only on the i-th column of the Cayley table of the groupoid $(Q, \cdot)$.*

Proof. If R_{a_i} is the i-th right translation of the groupoid $(Q, \cdot)$, then in groupoid $(Q, \cdot)P$ the i-th column has the form $R^{\cdot}_{a_i} p_i(x) = (p_i(x)) \cdot a_i$.

If we apply the isotopy $(p_i, \varepsilon, \varepsilon)$ to the groupoid $(Q, \cdot)$, then we have $x \circ y = p_i(x) \cdot y$. The i-th right translation of the groupoid $(Q, \circ)$ has the form $R^{\circ}_{a_i} x = R^{\cdot}_{a_i} p_i(x)$. □

Corollary 9.155. *If a gisotopy T has the form $T = (p, \dots, p)$ and the kind l or r, where $p \in S_Q$, then a gisotopic image LT of a Latin square L, which is defined on the set Q, is a Latin square.*

Proof. This follows from Propositions 9.153 and 9.154 and from the well-known fact that an isotopic image of a Latin square is a Latin square. □

The class of all permutation squares defined on a set Q will be denoted by $\mathfrak{S}(Q)$.

Proposition 9.156. *$\mathfrak{S}(Q)P \subseteq \mathfrak{S}(Q)$ for any $P \in \mathfrak{S}(Q)$, i.e., $\mathfrak{S}(Q)\mathfrak{S}(Q) \subseteq \mathfrak{S}(Q)$.*

Proof. The product of two m-tuples of permutations is an m-tuple of permutations. □

Proposition 9.157. *If S_1, S_2 are permutation squares of a kind α, $\alpha \in \{l, Il, r, Ir\}$, then there exists a generalized isotopy P of the kind α such that $S_1 P = S_2$.*

Proof. Indeed, $P = S_1^{-1}S_2$. □

Corollary 9.158. *If S_1 and S_2 are Latin squares of a kind α, $\alpha \in \{l, Il, r, Ir, p, Ip\}$, then there exists a generalized isotopy P of the kind α such that $S_1 P = S_2$.*

Proof. Indeed, $P = S_1^{-1}S_2$. □

Remark 9.159. Corollary 9.158 gives us a way of starting from a fixed quasigroup of order n (for example, from the cyclic group Z_n of order n) to obtain all quasigroups of order n using generalized isotopy.

We notice that Proposition 9.157 and Corollary 9.158 are true for a pair of left (right) quasigroups and for a pair of quasigroups, respectively.

Proposition 9.160. *If L_1 is a Latin square of a kind α, $\alpha \in \{l, Il, r, Ir\}$, P is a permutation square of the kind α, and $P = L_1^{-1}L_2$, where L_2 is a Latin square, then $L_1 P$ is a Latin square.*

Proof. It is easy to see that $L_1 P = L_1 L_1^{-1} L_2 = L_2$. □

9.5.3 Gisotopy and orthogonality

Gisotopy is a transformation which preserves the property of orthogonality of squares, groupoids and m-tuples of maps. We formulate the following proposition for squares.

Proposition 9.161. *Squares S_1 and S_2, both of a kind α, $\alpha \in \{l, r\}$, are orthogonal if and only if any of their gisotopic images $S_1 P$ and $S_2 P$ are orthogonal, where P is a gisotopy of the kind α.*

Proof. We denote by T_1 an m-tuple of the kind l that corresponds to the square S_1, i.e., $T_1 = (L_1, L_2, \dots, L_m)$, and $T_2 = (L'_1, L'_2, \dots, L'_m)$ is an m-tuple of the kind l which corresponds to the square S_2. Thus in the square of pairs E in the position (i, j) the pair $(L_i j, L'_i j)$ is situated.

If $P = (p_1, p_2, \dots, p_m)$ is a left gisotopy, then in the square of pairs EP, the pair $(L_i p_i(j), L'_i p_i(j))$ is situated in cell (i, j) and the pair $(L_i j, L'_i j)$ is situated in the cell $(i, p_i^{-1}(j))$.

Thus, any gisotopy P of kind l changes the order of pairs in any row of the square of pairs E.

Similarly, any gisotopy P of kind r changes the order of pairs in any column of the square E.

Therefore, if $S_1 \perp S_2$, then $S_1 P \perp S_2 P$ for any gisotopy P of kind l or kind r. □

In article [594] H.B. Mann, in fact, proved the following:

Theorem 9.162. *If Latin squares L_1 and L_2 are orthogonal, then the Latin squares L_1P_1 and L_2P_2 are also orthogonal, where P_1 and P_2 are gisotopies of the form $P_1 = (p_1, p_1, \dots, p_1)$ and $P_2 = (p_2, p_2, \dots, p_2)$, respectively.*

Proof. We suppose that Latin squares L_1 and L_2 have the kind l. If $L_1 \perp L_2$, then by Theorem 9.57 $L_3 = L_2L_1^{-1}$ is a Latin square.

This theorem is proved if we prove that the permutation square $L_4 = (L_2P_2)\,(P_1^{-1}L_1^{-1})$ is a Latin square.

From Corollary 9.29 it follows that $(P_2P_1^{-1})L_1^{-1}$ is a Latin square if and only if $L_1^{-1}(P_2P_1^{-1})$ is a Latin square. Therefore L_4 is a Latin square if and only if $L_2L_1^{-1}(P_2P_1^{-1})$ is a Latin square.

It is easy to see that $L_2L_1^{-1}(P_2P_1^{-1})$ is a Latin square. This follows from the forms of gisotopies P_1, P_2 and Corollary 9.155. □

A theorem that is a generalization of Theorem 9.162 can be found in [240].

Corollary 9.163. *If Latin squares L_1 and L_2 are orthogonal, then the Latin squares L_1 and L_2P are also orthogonal, where P is a gisotopy of the form $P = (p, p, \dots, p)$.*

Proof. It is easy to see. □

9.5.4 Mann's operations

In [70] V.D. Belousov has defined the following Mann's operations.

Definition 9.164. In a set Σ of binary operations that are defined on a set Q it is possible to define the following operation of composition of binary operations: the right Mann multiplication [596]

$$AB(x, y) = A(x, B(x, y)) \tag{9.54}$$

and the left Mann multiplication

$$A \circ B(x, y) = A(B(x, y), y). \tag{9.55}$$

Operation $C(x, y) = A(x, B^{(23)}(x, y))$ in the proof of Theorem 9.70 is an example of the right Mann multiplication of binary operations.

Multiplication of m-tuples of maps of the kind l corresponds to the right Mann multiplication of operations and multiplication of m-tuples of maps of the kind r corresponds to the left Mann multiplication of operations.

Definition 9.165. [70]. Suppose that system Σ contains the identity operations E and F (Definition 9.51). The system $Q(\Sigma)$ is called the S-system, if the following conditions are true:

the system $\Sigma' = \Sigma \backslash F$ is a group relative to the right Mann multiplication;

the system $\Sigma'' = \Sigma \backslash E$ is a group relative to the left Mann multiplication.

In [62] it is proved that any operation $A \neq E, F$ of an S-system Σ is a quasigroup operation. All operations of an S-system Σ are orthogonal in pairs. Notice that some S-systems are closely connected with Grätzer algebras [704, 70].

Definition 9.166. A groupoid $(Q, \circ)$ is called an *orthogonal isotope* of a quasigroup $(Q, \cdot)$ if there exist orthogonal binary operations A, B and a permutation γ on Q such that $\gamma(x \circ y) = A[x, y] \cdot B[x, y]$ for all $x, y \in Q$. The triple (A, B, γ) is said to be an orthogonal isotopy [84].

If $A[x, y] = \alpha y$, where α is a permutation of Q, then (α, B, γ) is called a right crossed isotopy of a quasigroup $(Q, \cdot)$. If $B[x, y] = \beta y$, where β is a permutation of Q, then (A, β, γ) is called a left crossed isotopy.

In [84] the author gives necessary and sufficient conditions for a right crossed isotopy of a quasigroup be a quasigroup.

In some sense the concept of orthogonal isotopy develops the concepts of left (right) Mann operations. The fact that a pair of orthogonal operations defined on a set Q determines a permutation of the set Q^2 (see Section 9.1.6 of this chapter) is also used.

Part III

Applications

Chapter 10

Quasigroups and codes

10.1 One check symbol codes and quasigroups 439
10.1.1 Introduction 439
10.1.2 On possibilities of quasigroup codes 443
10.1.3 TAC-quasigroups and n-quasigroup codes 445
10.1.4 5-n-quasigroup codes 450
10.1.5 Phonetic errors 451
10.1.6 Examples of codes 452
10.2 Recursive MDS-codes 458
10.2.1 Some definitions 459
10.2.2 Singleton bound 459
10.2.3 MDS-codes 460
10.2.4 Recursive codes 460
10.2.5 Gonsales-Couselo-Markov-Nechaev construction 462
10.2.6 Orthogonal quasigroups of order ten 464
10.2.7 Additional information 465
10.3 On signs of Bol loop translations 466

In this chapter we study codes with one check symbol and based on quasigroups recursive MDS-codes. Results on signs of left, right, middle translations in left Bol loops are also presented.

10.1 One check symbol codes and quasigroups

We study properties of check character systems with one check symbol using an n-ary quasigroup approach. Check character systems detecting all five types of the most frequent errors made by human operators during transmission of data are constructed. Results of this section were published in [641, 642]. A good introduction to code theory is in [559].

Codes based on Moufang loops (including Parker loop) are studied in [389, 391, 710, 215, 430, 648, 286]. These codes and loops were discovered by construction of a Monster group [923]. Codes based on loop transversals are researched in [817].

10.1.1 Introduction

Statistical investigations of J. Verhoeff [896] and D.F. Beckley [49] have shown that the most frequent errors made by human operators during data transmission are single errors (i.e., errors in one component exactly), adjacent transpositions (in other words errors made by interchanging adjacent digits, i.e., errors of the form $ab \longrightarrow ba$), and insertion or deletion errors. We note, if all codewords are of equal length, insertion and deletion errors can be detected easily.

Table 10.1: Error types and their frequencies [752].

error type		relative frequency in %	
		Verhoeff	Beckley
single error	$a \to b$	79.0 (60-95)	86
adjacent transposition	$ab \to ba$	10.2	8
jump transposition	$abc \to cba$	0.8	
twin error	$aa \to bb$	0.6	6
phonetic error ($a \geqslant 2$)	$a0 \to 1a$	0.5	
jump twin error	$aca \to bcb$	0.3	
other error		8.6	

The numbers in Table 10.1 can vary from sample to sample and may depend on the location of the affected digits; e.g., the rightmost two digits may be affected by single errors more than the other digits together ([896] p. 14, [752, 751]).

It is well known that there exists a possibility to detect some types of the most frequent errors made by human operators during transmission of data using a code with one check symbol [49, 896, 123, 227, 301, 397, 732, 752, 751, 122].

In this chapter we construct systematic error detecting codes that allow us to detect practically all errors from Table 10.1, the appearance of which arise with some regularity: namely, single errors, adjacent transposition errors, jump transposition errors, twin errors, jump twin errors and phonetic errors, i.e., all errors from Table 10.1 with the exception of "other errors."

To detect single errors and adjacent transpositions, one often uses check digit systems; these usually consist of codewords $a_1 \dots a_{n+1}$ containing, besides the information digits $a_1 \dots a_n$, one control (check) character a_{n+1}.

Definition 10.1. [752, 751]. A check digit system with one check character is a systematic error detecting code over an alphabet Q which arises by appending *a check digit* a_{n+1} to every word $a_1 a_2 \dots a_n \in Q^n$:

$$\mathfrak{C} : \begin{cases} Q^n & \longrightarrow & Q^{n+1} \\ a_1 a_2 \dots a_n & \longmapsto & a_1 a_2 \dots a_n a_{n+1}. \end{cases}$$

Here the word "systematic" means that the check character is the last symbol of any codeword of the code $\mathfrak{C}$.

We shall use basic quasigroup terms and concepts from the books [72, 80, 559, 685]. Check character systems over quasigroups have been studied in [123, 227, 301, 778].

Let Q be a non-empty set, let n be a natural number, $n \geqslant 2$. A map f that maps all n-tuples over Q into elements of the set Q is called an *n-ary operation*, i.e., $f(x_1, x_2, \dots, x_n) = x_{n+1}$ for all $(x_1, x_2, \dots, x_n) \in Q^n$ and $x_{n+1} \in Q$.

We recall that a non-empty set Q with an n-ary operation f such that in the equation $f(x_1, x_2, \dots, x_n) = x_{n+1}$ knowledge of any n elements of $x_1, x_2, \dots, x_n, x_{n+1}$ uniquely specifies the remaining one is called *n-ary quasigroup*.

We can view the code $\mathfrak{C}$ as a mapping over an alphabet Q such that the check symbol a_{n+1} is obtained from information symbols $a_1, a_2, \dots, a_n$ in the following manner: $g(a_1, a_2, \dots, a_n) = a_{n+1}$, where g is an n-ary operation on the set Q.

Definition 10.2. We shall call the code $\mathfrak{C}$ with one check character a_{n+1} over an alphabet Q an *n-ary code* (Q, g). If in an n-ary code (Q, g) the operation g is an n-ary quasigroup operation, then this code will be called an *n-quasigroup code* (Q, g).

We shall say that codewords $a_1 \dots a_{n+1}$ and $b_1 \dots b_{n+1}$ are equal if and only if $a_i = b_i$ for all $i \in \{1, \dots, n+1\}$. Sometimes a codeword $a_1 \dots a_{n+1}$ will be denoted as a_1^{n+1}.

By an error in a codeword a_1^{n+1} of a code $\mathfrak{C}$ over an alphabet Q we mean any word $b_1^{n+1} \in Q^{n+1}$ such that there exists at least one index $j \in \overline{1, n+1}$ such that $a_j \neq b_j$.

As usual, an n-ary code (Q, g) detects an error in a received transmission word $a_1 \dots a_n a_{n+1}$ if and only if $g(a_1^n) \neq a_{n+1}$.

Theorem 10.3. *Any n-ary code (Q, g) detects all single errors if and only if it is an n-quasigroup code, i.e., an n-ary operation g is an n-ary quasigroup operation.*

Proof. This fact follows from properties of an n-ary quasigroup and of n-ary quasigroup code (Q, g) since any n elements uniquely specify the remaining one in both cases. □

Some variants of Theorem 10.3 are given in [397, 227].

With any n-ary quasigroup (Q, f) it is possible to associate $((n+1)! - 1)$ n-ary quasigroups, so-called *parastrophes of the quasigroup* (Q, f) [77].

Let σ be a permutation of the set $\overline{1, n+1}$. Operation f^σ is called a σ-parastrophe of the operation f if and only if the following equalities are equivalent: $f^\sigma(x_{\sigma 1}, x_{\sigma 2}, \dots, x_{\sigma n}) = x_{\sigma(n+1)}$ and $f(x_1, x_2, \dots, x_n) = x_{n+1}$ for all $x_1^n \in Q$.

For example, $f^{(132)}(x_{(132)1}, x_{(132)2}) = x_{(132)3}$ if and only if $f(x_1, x_2) = x_3$ and we have $f^{(132)}(x_3, x_1) = x_2$ if and only if $f(x_1, x_2) = x_3$.

Let (Q, f) be an n-ary quasigroup, $f(x_1^n) = x_{n+1}$ for all $x_1, \dots, x_{n+1} \in Q$. Let m be a natural number, with $m \leqslant n$. If in the last expression we replace elements $x_{k_1}, \dots, x_{k_m}$ by some fixed elements $a_1, \dots, a_m \in Q$ respectively, then this expression takes the form

$$f(x_1^{k_1-1}, a_1, x_{k_1+1}^{k_2-1}, a_2, \dots, x_{k_m}^n),$$

i.e., we obtain a new operation $g(x_1^{k_1-1}, x_{k_1+1}^{k_2-1}, \dots, x_{k_m}^n)$. The operation g is an $(n-m)$-ary quasigroup operation. An operation g obtained in this manner is called a *retract* of operation f [77].

Lemma 10.4. *By using an n-ary quasigroup retract we can fix in the equality $f(x_1, x_2, \dots, x_n) = x_{n+1}$ the last element x_{n+1}.*

Proof. Indeed, from the definition of a parastrophy we have $f(x_1^n) = x_{n+1}$ if and only if $f^\sigma(x_1^{n-1}, x_{n+1}) = x_n$, where the operation f^σ is a σ-parastrophe of the quasigroup operation f and $\sigma = (n, n+1)$. Then $f(x_1^n) = a_{n+1}$ if and only if $f^\sigma(x_1^{n-1}, a_{n+1}) = x_n$. Since in this case any $(n-1)$ elements uniquely specify the remaining one, we also obtain an $(n-1)$-ary quasigroup operation $g(x_1^{n-1}) = x_n$. We shall call the $(n-1)$-ary operation g an $(n+1)$-*retract of the n-ary quasigroup operation* f. □

Let (Q, f) and (Q, g) be n-ary and m-ary quasigroups respectively. Let

$$h(x_1^{m+n-1}) = g(x_1^{i-1}, f(x_i^{i+n-1}), x_{i+n}^{m+n-1}).$$

Then (Q, h) is an $(m+n-1)$-ary quasigroup. Quasigroup (Q, h) is obtained by the *superposition* of the quasigroup (Q, f) with the quasigroup (Q, g) on the i-th place [77].

In order to define a systematic n-ary code $\mathfrak{C}$ one often uses a check equation of the following form: $f(x_1^{n+1}) = e$, where elements $x_1, \dots, x_n$ are information symbols, element

x_{n+1} is a check symbol, the element e is a fixed element of the set Q, and the operation f is an $(n+1)$-ary operation.

It is easy to see that an n-ary code (Q, g) is defined with the help of the check equation $f(x_1^{n+1}) = e$ if and only if the equality $f(x_1^n, g(x_1^n)) = e$ is true for all elements $x_1, \dots, x_n \in Q$.

If in an n-ary code (Q, g) with check equation $f(x_1^{n+1}) = e$ the operation f is a quasigroup operation, then $x_{n+1} = f^{\sigma}(x_1^n, e)$, where the operation f^{σ} is a σ-parastrophe of the quasigroup operation f and $\sigma = (n+1, n+2)$. Therefore in this case we have $x_{n+1} = g(x_1^n) = f^{\sigma}(x_1^n, e)$.

Proposition 10.5. *Any n-ary quasigroup code (Q, g) is possible to define with the help of a check equation $f(x_1^{n+1}) = e$ such that this equation is an $(n+2)$-th retract (we fix the $(n+2)$-th place) of an $(n+1)$-ary quasigroup operation.*

Proof. Let $A(x, y) = z$ be a binary group operation on the set Q with the identity element e. We construct the following $(n+1)$-ary quasigroup operation $A(g(x_1^n), y) = z$. Then $A((g(x_1^n), e) = x_{n+1}$, where the element x_{n+1} is a check digit of the information symbols x_1^n of the code (Q, g).

Thus the equation $f(x_1^{n+1}) = A^{(23)}(g(x_1^n), x_{n+1}) = e$ is a check equation of the code (Q, g) such that this equation is an $(n+2)$-th retract of an $(n+1)$-ary quasigroup operation. □

Remark 10.6. We suppose that the check equation $f(x_1^{n+1}) = e$ of an n-ary quasigroup code (Q, g) is obtained as the $(n+2)$-th retract of an $(n+1)$-ary quasigroup operation $f(x_1^{n+1}) = x_{n+2}$.

The systems most commonly in use are defined over alphabets endowed with a group structure. For a group G = (A, $\cdot$) one can determine the check digit a_n such that the following (check) equation holds (for fixed permutations δ_i of G, $i = 1, ..., n$, and an element e of G, for instance the identity element)

$$\delta_1(a_1)\delta_2(a_2)\dots\delta_n(a_n) = e. \tag{10.1}$$

Such a system detects all single errors; and it detects all adjacent transpositions if and only if for all $x, y \in G$ with $x \neq y$

$$x \cdot \delta_{i+1}\delta_i^{-1}(y) \neq y \cdot \delta_{i+1}\delta_i^{-1}(x).$$

The proofs are straightforward, see [752]. We denote this code as $\mathfrak{C}_1$.

We give one more definition from [752]: Let $(Q, \star_i)$ be quasigroups; then one uses as the check equation the following formula

$$(\dots(x_n \star_n x_{n-1}) \star_{n-1} x_{n-2}) \dots) \star_1 x_0 = e. \tag{10.2}$$

In this definition the element e is any fixed element of the set Q. If elements x_0^{n-1} are information symbols, then element x_n is some check symbol. We denote this code as $\mathfrak{C}_2$.

Proposition 10.7. *The code $\mathfrak{C}_1$ is an $(n-1)$-ary quasigroup code and the code $\mathfrak{C}_2$ is an n-ary quasigroup code.*

Proof. The left-hand side of the check equation (10.1) defines an n-ary quasigroup operation f [77]. Check equation (10.1) is obtained as the $(n+1)$-th retract (we fix the $(n+1)$-th place) of the n-ary quasigroup operation f. Therefore the code $\mathfrak{C}_1$ is an $(n-1)$-ary quasigroup code.

The code $\mathfrak{C}_2$ is an n-ary quasigroup code by the same arguments. We see that the check equation (10.2) is an n-ary operation $h : h(x_0^n) = e$. This operation is the $(n+2)$-th retract of the operation $h^\star : h^\star(x_0^n) = x_{n+1}$ for all $x_0, \dots, x_{n+1} \in Q$. Since the operation $h^\star$ is a superposition of n binary quasigroup operations $\star_1, \star_2, \dots, \star_n$, then by [77] the operation $h^\star$ is an $(n+1)$-ary quasigroup operation. Therefore the code $\mathfrak{C}_2$ is an n-ary quasigroup code. □

10.1.2 On possibilities of quasigroup codes

Now we would like to show that all n-ary quasigroup codes (Q, d) over the same alphabet Q and with different quasigroup operations d (arity n is fixed) have equal possibilities to detect errors.

As usual

$$\binom{n}{k} = \frac{n!}{k!(n-k)!}$$

denotes a binomial coefficient. We shall call an error on k places $(k > 1)$ in a codeword a *k-error*.

Theorem 10.8. *Any n-quasigroup code (Q, d) over a fixed alphabet Q, $(|Q| = q)$, with a fixed finite number n of information symbols and with one check digit can detect:*

a) any one of $q^{n+1} - q^n$ possible erroneous words that could be received by mistake for a particular codeword (a_1^{n+1}), where the error (or errors) occur either among the n information digits and/or in the single check digit;

b) any one of $q^n - q^{n-1}$ possible erroneous words that could be received by mistake for a particular codeword (a_1^{n+1}), where the error (or errors) occur only among the n information digits;

c) any one of $\binom{n+1}{k}(q-1)^{k-1}(q-2)$ words among the erroneous words which differ from the given transmitted quasigroup codeword in exactly k places;

d) any one of $\binom{n}{k}(q-1)^{k-1}(q-2)$ words among the erroneous words which differ from the given transmitted quasigroup codeword in exactly k places when the error (or errors) occurs only among the information digits.

Proof. a) If we fix a codeword (a_1^{n+1}) of an n-ary quasigroup code (Q, d) (i.e., for this word the equality $a_{n+1} = d(a_1, \dots, a_n)$ holds), then all other possible words (x_1^{n+1}), where $x_1, \dots, x_{n+1} \in Q$ will be erroneous. Thus, there exist $q^{n+1} - 1$ erroneous words. However, $q^n - 1$ of these are codewords because they satisfy the check equation $x_{n+1} = d(x_1, \cdots x_n)$ and so these words will not be recognized as erroneous. Therefore, an n-ary quasigroup code detects $(q^{n+1} - 1) - (q^n - 1) = q^{n+1} - q^n$ of the $q^{n+1} - 1$ erroneous words which may be received.

b) In this case we suppose that a check symbol a_{n+1} was transmitted without error. The case b) is proved by analogy with the case a). There exist $q^n - 1$ possible errors and there exist $q^{n-1} - 1$ quasigroup codewords for which the check equation $a_{n+1} = d(x_1^n)$ does not detect any error.

Therefore an n-ary quasigroup code (Q, d) detects $q^n - q^{n-1}$ errors in the first n information symbols of any quasigroup code word (a_1^{n+1}).

c) There are $(q-1)^k$ words which differ from the given transmitted codeword (a_1^{n+1}) in k fixed places and there are $\binom{n+1}{k}$ choices for the k fixed places, so there are $\binom{n+1}{k}(q-1)^k$ words which differ from the given transmitted codeword (a_1^{n+1}) in exactly k places. However, among the $(q-1)^k$ words which differ from the given transmitted codeword (a_1^{n+1}) in k fixed places, there are $(q-1)^{k-1}$ ones which satisfy the check equation $x_{n+1} = d(x_1, \cdots x_n)$ and so these words will not be recognized as erroneous. (To see this, observe that there are $(q-1)^{k-1}$ erroneous ways of filling $k-1$ of the k fixed places and that, for each of these, there is exactly one way of filling the remaining fixed place in such a way that the whole word satisfies the check equation.) It follows that $\binom{n+1}{k}(q-1)^k - \binom{n+1}{k}(q-1)^{k-1} = \binom{n+1}{k}(q-1)^{k-1}(q-2)$ of the possible erroneous words will be detected.

d) In this case we suppose that the check symbol a_{n+1} was transmitted without error. The case d) is proved by analogy with the case c). □

Definition 10.9. We say that n-ary quasigroup codes (Q,d) and (Q,g) are *isotopic* if their n-ary quasigroup operations (Q,d) and (Q,g) are isotopic.

Remark 10.10. From Theorem 10.8 it follows that isotopic n-ary quasigroup codes detect equal numbers of errors of all types and an equal number of k-errors for any suitable k.

If we assume that the probability of receiving a word with multiple errors in place of a transmitted codeword (a_1^{n+1}) is the same as that of receiving an erroneous word which contains only a single error, then we obtain the following:

Corollary 10.11. *a) The relative frequency of errors detected by an n-ary quasigroup code (Q,d) in n information digits and in one check symbol of any fixed quasigroup codeword (a_1^{n+1}) is equal to*

$$\frac{q^n}{q^n + q^{n-1} + \cdots + 1} > \frac{q-1}{q}.$$

b) The relative frequency of errors detected by n-ary quasigroup code (Q,d) of all possible types in the first n information symbols of any fixed quasigroup codeword (a_1^{n+1}) is equal to

$$\frac{q^{n-1}}{q^{n-1} + q^{n-2} + \cdots + 1} > \frac{q-1}{q}.$$

c) The relative frequency k-errors of detected by n-ary quasigroup code (Q,d) in n information digits and in one check symbol of any fixed quasigroup codeword (a_1^{n+1}) is equal to

$$\frac{q-2}{q-1}.$$

d) The relative frequency of k-errors detected by n-ary quasigroup code (Q,d) in the first n information symbols of any fixed quasigroup codeword (a_1^{n+1}) is equal to

$$\frac{q-2}{q-1}.$$

Proof. a) From Theorem 10.8 we have

$$\frac{q^{n+1} - q^n}{q^{n+1} - 1} = \frac{q^n(q-1)}{(q-1)(q^n + q^{n-1} + \cdots + 1)} = \frac{q^n}{(q^n + q^{n-1} + \cdots + 1)}.$$

Further

$$\frac{q^{n+1} - q^n}{q^{n+1} - 1} > \frac{q^{n+1} - q^n}{q^{n+1}} = \frac{q-1}{q}.$$

b) This case is proved by analogy with the case a).

c) We have

$$\frac{\binom{n+1}{k}(q-1)^{k-1}(q-2)}{\binom{n+1}{k}(q-1)^k} = \frac{q-2}{q-1}.$$

d) Case d) is proved by analogy with the case c). □

Remark 10.12. We can consider transposition, jump transposition, twin and jump twin errors as two types of errors in two types of positions (places), namely, transpositions $ab \to ba$ and twin errors $aa \to bb$ on places $(i, i+1)$ and $(i, i+2)$ for all suitable $i \in \overline{1, n+1}$.

In fact, from Table 10.1 it follows that human operators often make single errors and three types of errors on two types of places, namely: transposition, twin and phonetic errors $(a0 \to 1a, a \neq 0, a \neq 1)$ on places of the form $(i, i+1)$, $(i, i+2)$ for all suitable $i \in \overline{1, n+1}$.

We notice, in a fixed code word (a_1^n) in any fixed position $(i, i+1)$ transposition, twin and phonetic errors eliminate each other, i.e., they cannot appear simultaneously.

Moreover, in any fixed codeword in a fixed position $(i, i+k)$, "generalized" transposition $((\underset{i}{a}, \dots, \underset{i+k}{b}) \to (\underset{i}{b}, \dots, \underset{i+k}{a}))$ and twin $((\underset{i}{a}, \dots, \underset{i+k}{a}) \to (\underset{i}{b}, \dots, \underset{i+k}{b}))$ errors cannot be in the same time.

10.1.3 TAC-quasigroups and n-quasigroup codes

We recall that a binary quasigroup $(Q, \cdot)$ is called *anti-commutative* (sometimes such quasigroup is called as *anti-symmetric quasigroup* [227]) if and only if the following implication is true: $x \cdot y = y \cdot x \Rightarrow x = y$ for all $x, y \in Q$ [72].

A binary anti-commutative quasigroup $(Q, \cdot)$ is called *totally anti-commutative* (sometimes TAC-quasigroup for short) if the following implication is true: $x \cdot x = y \cdot y \Rightarrow x = y$ for all $x, y \in Q$ (Definition 9.117).

Definition 10.13. A retract of the form $f(a_1^{i-1}, x_i, a_{i+1}^{i+k-1}, x_{i+k}, a_{i+k+1}^n)$ of an n-ary quasigroup (Q, f), where $a_1^{i-1}, a_{i+1}^{i+k-1}, a_{i+k+1}^n$ are some fixed elements of the set Q, $i \in \overline{1, n-k}$, $k \in \overline{1, n}$ is called an $(i, i+k)$ *binary retract of an n-ary quasigroup* (Q, f).

Theorem 10.14. *An n-ary quasigroup code (Q, d) detects any transposition and twin error on places of the form $(i, i+k)$ $(i \in \overline{1, n-k}, k \in \overline{1, n-1}, i+k \leqslant n)$ if and only if all $(i, i+k)$ binary retracts of n-ary quasigroup (Q, d) are totally anti-commutative quasigroups.*

Proof. If we suppose that all $(i, i+k)$ binary retracts of an n-ary quasigroup (Q, d) are totally anti-commutative quasigroups, then from the definition of totally anti-commutative binary quasigroups it follows that the code (Q, g) detects any transposition and twin error in the place $(i, i+k)$.

Conversely, if we suppose that there is a place $(i, i+k)$ and there are elements $a_1^{i-1}, b, a_{i+1}^{i+k-1}$, c, a_{i+k+1}^n $(b \neq c)$ such that

$$d(a_1^{i-1}, b, a_{i+1}^{i+k-1}, c, a_{i+k+1}^n) = d(a_1^{i-1}, c, a_{i+1}^{i+k-1}, b, a_{i+k+1}^n),$$

then the binary retract $d(a_1^{i-1}, x, a_{i+1}^{i+k-1}, y, a_{i+k+1}^n)$ is not an anti-commutative quasigroup, and we have a contradiction.

If we suppose that there is a place $(i, i+k)$ and there are elements a_1^{i-1}, b, a_{i+1}^{i+k-1}, c, a_{i+k+1}^n $(b \neq c)$ such that

$$d(a_1^{i-1}, b, a_{i+1}^{i+k-1}, b, a_{i+k+1}^n) = d(a_1^{i-1}, c, a_{i+1}^{i+k-1}, c, a_{i+k+1}^n),$$

then the binary retract $d(a_1^{i-1}, x, a_{i+1}^{i+k-1}, y, a_{i+k+1}^n)$ is not an anti-commutative quasigroup, and again we have a contradiction. □

Theorem 10.15. *Let (Q, d) be a finite n-ary quasigroup of order q. The $(n-1)$-ary quasigroup codes (Q, g_i) $(i \in \overline{1, q})$ with check equations $d_i(x_1^n) = e_i$, where the elements e_i are fixed different elements of the set Q, detect any transposition and twin error on places of the form $(i, i+k)$ $(i \in \overline{1, n-k}$, $k \in \overline{1, n-1}$, $i+k \leqslant n)$ if and only if all $(i, i+k)$ binary retracts of n-ary quasigroup (Q, d) are totally anti-commutative quasigroups.*

Proof. Proof of this theorem is similar to the proof of Theorem 10.14. □

Theorem 10.16. *An $(n-1)$-ary quasigroup code (Q, g) with check equation $d(x_1^n) = \gamma_1 x_1 + \gamma_2 x_2 + \cdots + \gamma_n x_n = e$, where the element e is a fixed element of the set Q, $(Q, +)$ is a group, detects any transposition and twin error on places of the form $(i, i+k)$ $(i \in \overline{1, n-k},\ k \in \overline{1, n-1},\ i+k \leqslant n)$ with the exception of errors on place $(1, n)$ if and only if all $(i, i+k)$ $((i, i+k) \neq (1, n))$ binary retracts of n-ary quasigroup (Q, d) are totally anti-commutative quasigroups.*

Proof. If we suppose that all $(i, i+k)$ binary retracts of an n-ary quasigroup (Q, d) are totally anti-commutative quasigroups, then from the definition of totally anti-commutative binary quasigroups it follows that the code (Q, g) detects any transposition and twin error in the place $(i, i+k)$.

Conversely, if we suppose that there is a place $(i, i+k)$ and there are elements $a_1^{i-1}, b, a_{i+1}^{i+k-1}$, c, a_{i+k+1}^n $(b \neq c)$ such that

$$d(a_1^{i-1}, b, a_{i+1}^{i+k-1}, c, a_{i+k+1}^n) = d(a_1^{i-1}, c, a_{i+1}^{i+k-1}, b, a_{i+k+1}^n) = l,$$

then, since $i \neq 1$ or $i+k \neq n$, we can change the element a_1 or the element a_n so that

$$d(a_1^{i-1}, b, a_{i+1}^{i+k-1}, c, a_{i+k+1}^n) = d(a_1^{i-1}, c, a_{i+1}^{i+k-1}, b, a_{i+k+1}^n) = e.$$

Therefore the binary retract $d(a_1^{i-1}, x, a_{i+1}^{i+k-1}, y, a_{i+k+1}^n)$ is not an anti-commutative quasigroup, and we have a contradiction.

If we suppose that there is a place $(i, i+k)$ and there are elements a_1^{i-1}, b, a_{i+1}^{i+k-1}, c, a_{i+k+1}^n $(b \neq c)$ such that

$$d(a_1^{i-1}, b, a_{i+1}^{i+k-1}, b, a_{i+k+1}^n) = d(a_1^{i-1}, c, a_{i+1}^{i+k-1}, c, a_{i+k+1}^n) = l,$$

then, since $i \neq 1$, or $i+k \neq n$, we can change the element a_1 or the element a_n so that

$$d(a_1^{i-1}, b, a_{i+1}^{i+k-1}, b, a_{i+k+1}^n) = d(a_1^{i-1}, c, a_{i+1}^{i+k-1}, c, a_{i+k+1}^n) = e,$$

then the binary retract $d(a_1^{i-1}, x, a_{i+1}^{i+k-1}, y, a_{i+k+1}^n)$ is not an anti-commutative quasigroup, and again we have a contradiction. □

Corollary 10.17. *An $(n-1)$-ary quasigroup code (Q, g) with check equation $d(x_1^n) = \gamma_1 x_1 + \gamma_2 x_2 + \cdots + \gamma_n x_n = e$, where the element e is a fixed element of the set Q, $(Q, +)$ is an abelian group, detects any transposition and twin error on places of the form $(i, i+k)$ $(i \in \overline{1, n-k},\ k \in \overline{1, n-1},\ i+k \leqslant n)$ if and only if all $(i, i+k)$ binary retracts of n-ary quasigroup (Q, d) are totally anti-commutative quasigroups.*

Proof. We only need to prove that anti-commutativity of binary retracts is a necessary condition to detect any transposition and twin error on the place $(1, n)$.

If we suppose that there are elements a_2^{n-1}, b, c, $(b \neq c)$ such that

$$d(b, a_2^{n-1}, c) = d(c, a_2^{n-1}, b) = k,$$

then, since the group $(Q, +)$ is an abelian group, we can change the element a_2 so that

$$d(b, a_2^{n-1}, c) = d(c, a_2^{n-1}, b) = e.$$

Therefore the binary retract $d(x, a_2^{n-1}, y)$ is not an anti-commutative quasigroup, the code (Q, g) cannot detect an error on the place $(1, n)$, and we have a contradiction.

If we suppose that there are elements a_2^{n-1}, b, c $(b \neq c)$ such that

$$d(b, a_2^{n-1}, b) = d(c, a_2^{n-1}, c) = k,$$

then, since the group $(Q, +)$ is an abelian group, we can change the element a_2 so that

$$d(b, a_2^{n-1}, b) = d(c, a_2^{n-1}, c) = e.$$

Therefore the binary retract $d(x, a_2^{n-1}, y)$ is not an anti-commutative quasigroup, and again we have a contradiction. □

Corollary 10.18. *In an n-ary group isotope (Q, g) of the form $g(x_1^n) = \gamma_1 x_1 + \gamma_2 x_2 + \dots + \gamma_n x_n$:*

a) all its $(i, i+1)$ $(i \in \overline{1, n-1})$ binary retracts are totally anti-commutative quasigroups if and only if all its binary retracts of the form $\gamma_i x_i + \gamma_{i+1} x_{i+1}$ are totally anti-commutative quasigroups;

b) all its $(i, i+k)$ $(i \in \overline{1, n-k}, k \in \overline{1, n-1})$ binary retracts are totally anti-commutative quasigroups if and only if all its binary retracts of the form $\gamma_i x_i + a_{i+1} + \dots + a_{i+k-1} + \gamma_{i+k} x_{i+k}$ for any fixed elements $a_{i+1}, \dots, a_{i+k-1} \in Q$ are totally anti-commutative quasigroups.

Proof. a) Assume all binary retracts of the form $\gamma_i x_i + \gamma_{i+1} x_{i+1}$ of n-ary group isotope (Q, g) are totally anti-commutative quasigroups.

If we suppose that there is a place $(i, i+1)$ and there are elements $a_1^{i-1}, b, c, a_{i+2}^n (b \neq c)$ such that $g(a_1^{i-1}, b, c, a_{i+2}^n) = g(a_1^{i-1}, c, b, a_{i+2}^n)$, i.e., that

$$\begin{aligned}\gamma_1 a_1 + \dots + \gamma_{i-1} a_{i-1} + \gamma_i b + \gamma_{i+1} c + \gamma_{i+1} a_{i+1} + \dots + \gamma_n a_n = \\ \gamma_1 a_1 + \dots + \gamma_{i-1} a_{i-1} + \gamma_i c + \gamma_{i+1} b + \gamma_{i+1} a_{i+1} + \dots + \gamma_n a_n\end{aligned}$$

or, upon cancellation, that $\gamma_i b + \gamma_{i+1} c = \gamma_i c + \gamma_{i+1} b$, then we obtain a retract of the form $\gamma_i x_i + \gamma_{i+1} x_{i+1}$ which is not a totally anti-commutative quasigroup. We have a contradiction with conditions of this corollary.

If we suppose that there is a place $(i, i+1)$ and there are elements $a_1^{i-1}, b, c, a_{i+2}^n$ $(b \neq c)$ such that $g(a_1^{i-1}, b, b, a_{i+2}^n) = g(a_1^{i-1}, c, c, a_{i+2}^n)$, i.e., that

$$\begin{aligned}\gamma_1 a_1 + \dots + \gamma_{i-1} a_{i-1} + \gamma_i b + \gamma_{i+1} b + \gamma_{i+1} a_{i+1} + \dots + \gamma_n a_n = \\ \gamma_1 a_1 + \dots + \gamma_{i-1} a_{i-1} + \gamma_i c + \gamma_{i+1} c + \gamma_{i+1} a_{i+1} + \dots + \gamma_n a_n\end{aligned}$$

or, after cancellation, that $\gamma_i b + \gamma_{i+1} b = \gamma_i c + \gamma_{i+1} c$, then we obtain a retract of the form $\gamma_i x_i + \gamma_{i+1} x_{i+1}$ which is not a totally anti-commutative quasigroup. We have again a contradiction with conditions of this corollary.

Therefore, if all binary retracts of the form $\gamma_i x_i + \gamma_{i+1} x_{i+1}$ of the n-ary group isotope (Q, g) are totally anti-commutative quasigroups, then all $(i, i+1)$ $(i \in \overline{1, n-1})$ binary retracts of this n-ary group isotope are totally anti-commutative quasigroups. The converse assertion is obvious.

b) This case is proved by analogy with case a). □

There is a possibility to re-formulate Corollary 10.18 in the language of binary quasigroups which are retracts of the n-ary quasigroup (Q, g).

Corollary 10.19. *In an n-ary group isotope (Q, g) of the form*

$$g(x_1, x_2, \dots, x_n) = \gamma_1 x_1 + \gamma_2 x_2 + \dots + \gamma_n x_n :$$

a) all of the $(i, i+1)$ $(i \in \overline{1, n-1})$ binary retracts are totally anti-commutative quasigroups if and only if all binary quasigroups of the form $\gamma_i x_i + \gamma_{i+1} x_{i+1}$ are totally anti-commutative quasigroups;

b) all of the $(i, i+k)$ $(i \in \overline{1, n-k}, k \in \overline{1, n-1})$ binary retracts are totally anti-commutative quasigroups if and only if all binary quasigroups of the form $\gamma_i x_i + t + \gamma_{i+k} x_{i+k}$, for any fixed element t, are totally anti-commutative quasigroups.

Proof. It is sufficient to denote the element $a_{i+1} + \dots + a_{i+k-1}$ by the letter t. □

Corollary 10.20. *An $(n-1)$-ary group isotope code (Q, g) with the check equation $\sum\limits_{i=1}^{n} \gamma_i x_i = 0$, where the element 0 is the identity element of the group $(Q, +)$, detects any transposition and twin error on places $(i, i+1)$ $(i \in \overline{1, n-1})$, $(i, i+2)$ $(i \in \overline{1, n-2})$ if and only if all quasigroups of the form $\gamma_i x_i + \gamma_{i+1} x_{i+1}$ and of the form $\gamma_i x_i + a_i + \gamma_{i+2} x_{i+2}$ for any fixed $a_i \in Q$, are totally anti-commutative quasigroups.*

Proof. This follows from Corollary 10.17 and Corollary 10.18. □

Corollary 10.21. *In an n-ary group isotope (Q, g) of the form $g(x_1^n) = \gamma_1 x_1 + \gamma_2 x_2 + \dots + \gamma_n x_n$, where the group $(Q, +)$ is abelian, all of its $(i, i+k)$ $(i \in \overline{1, n-k}, k \in \overline{1, n-1})$ binary retracts are totally anti-commutative quasigroups if and only if all binary retracts of the form $\gamma_i x_i + \gamma_{i+k} x_{i+k}$ are totally anti-commutative quasigroups.*

Proof. If we suppose that there is a place $(i, i+k)$ and there are elements a_1^{i-1}, b, a_{i+1}^{i+k-1}, c, a_{i+k+1}^n $(b \neq c)$ such that

$$g(a_1^{i-1}, b, a_{i+1}^{i+k-1}, c, a_{i+k+1}^n) = g(a_1^{i-1}, c, a_{i+1}^{i+k-1}, b, a_{i+k+1}^n),$$

i.e., that

$$\begin{aligned} &\gamma_1 a_1 + \dots + \gamma_{i-1} a_{i-1} + \gamma_i b + \gamma_{i+1} a_{i+1} + \dots + \\ &\gamma_{i+k-1} a_{i+k-1} + \gamma_{i+k} c + \gamma_{i+k+1} a_{i+k+1} + \dots + \gamma_n a_n = \\ &\gamma_1 a_1 + \dots + \gamma_{i-1} a_{i-1} + \gamma_i c + \gamma_{i+1} a_{i+1} + \dots + \gamma_{i+k-1} a_{i+k-1} + \\ &\gamma_{i+k} b + \gamma_{i+k+1} a_{i+k+1} + \dots + \gamma_n a_n, \end{aligned}$$

then upon cancellation we have $\gamma_i b + \gamma_{i+1} a_{i+1} + \dots + \gamma_{i+k-1} a_{i+k-1} + \gamma_{i+k} c = \gamma_i c + \gamma_{i+1} a_{i+1} + \dots + \gamma_{i+k-1} a_{i+k-1} + \gamma_{i+k} b$.

Since the group $(Q, +)$ is commutative we further obtain that $\gamma_i b + \gamma_{i+k} c = \gamma_i c + \gamma_{i+k} b$.

Thus we see that the retract of the form $\gamma_i x_i + \gamma_{i+k} x_{i+k}$ is not a totally anti-commutative quasigroup. We have a contradiction with conditions of this corollary.

If we suppose that there is a place $(i, i+k)$ and there are elements a_1^{i-1}, b, a_{i+1}^{i+k-1}, c, a_{i+k+1}^n $(b \neq c)$ such that

$$g(a_1^{i-1}, b, a_{i+1}^{i+k-1}, b, a_{i+k+1}^n) = g(a_1^{i-1}, c, a_{i+1}^{i+k-1}, c, a_{i+k+1}^n),$$

i.e., that

$$\begin{aligned} &\gamma_1 a_1 + \dots + \gamma_{i-1} a_{i-1} + \gamma_i b + \gamma_{i+1} a_{i+1} + \dots + \gamma_{i+k-1} a_{i+k-1} + \gamma_{i+k} b + \\ &\gamma_{i+k+1} a_{i+k+1} + \dots + \gamma_n a_n = \\ &\gamma_1 a_1 + \dots + \gamma_{i-1} a_{i-1} + \gamma_i c + \gamma_{i+1} a_{i+1} + \dots + \gamma_{i+k-1} a_{i+k-1} + \\ &\gamma_{i+k} c + \gamma_{i+k+1} a_{i+k+1} + \dots + \gamma_n a_n, \end{aligned}$$

then after cancellation we have

$$\begin{aligned} &\gamma_i b + \gamma_{i+1} a_{i+1} + \dots + \gamma_{i+k-1} a_{i+k-1} + \gamma_{i+k} b = \\ &\gamma_i c + \gamma_{i+1} a_{i+1} + \dots + \gamma_{i+k-1} a_{i+k-1} + \gamma_{i+k} c. \end{aligned}$$

Since the group $(Q, +)$ is commutative we further obtain that $\gamma_i b + \gamma_{i+k} b = \gamma_i c + \gamma_{i+k} c$.

Thus we have that a retract of the form $\gamma_i x_i + \gamma_{i+k} x_{i+k}$ is not totally anti-commutative quasigroup, and we have a contradiction with conditions of this corollary.

Therefore, if all binary retracts of the form $\gamma_i x_i + \gamma_{i+k} x_{i+k}$ of the n-ary group isotope (Q, g) over an abelian group $(Q, +)$ are totally anti-commutative quasigroups, then all $(i, i+k)$ $(i \in \overline{1, n-k}, k \in \overline{1, n-1})$ binary retracts of this n-ary group isotope are totally anti-commutative quasigroups.

The converse is obvious. □

Corollary 10.22. *An $(n-1)$-ary abelian group isotope code (Q, g) with check equation $\sum_{i=1}^{n} \gamma_i x_i = 0$, where the element 0 is the identity element of the abelian group $(Q, +)$, detects any transposition and twin error on places $(i, i+k)$ $(i \in \overline{1, n-k}$, $k \in \overline{1, n-1}$, $i+k \leqslant n)$ if and only if all quasigroups of the form $\gamma_i x_i + \gamma_{i+k} x_{i+k}$ are totally anti-commutative quasigroups.*

Proof. This follows from Corollaries 10.17 and 10.21. □

Definition 10.23. If in an n-ary quasigroup code (Q, g) the operation g or the operation d from the check equation $d(x_1^{n+1}) = e$ of this code is an n-T-quasigroup operation, then the code (Q, g) is called an *n-T-quasigroup code* (Q, g).

Theorem 10.24. *Any $(n-1)$-T-quasigroup code (Q, g) with check equation*

$$d(x_1^n) = \alpha_1 x_1 + \alpha_2 x_2 + \cdots + \alpha_n x_n = 0$$

detects:

- *any transposition error on the place $(i, i+k)$, $(i \in \overline{1, n-k}$, $k \in \overline{1, n-1}$, $i+k \leqslant n)$ if and only if the mapping $\alpha_i - \alpha_{i+k}$ is an automorphism of the group $(Q, +)$;*
- *any twin error on the place $(i, i+k)$, $(i \in \overline{1, n-k}$, $k \in \overline{1, n-1}$, $i+k \leqslant n)$ if and only if the mapping $\alpha_i + \alpha_{i+k}$ is an automorphism of the group $(Q, +)$.*

Proof. This follows from Corollary 10.22 and Theorem 9.119. □

Theorem 10.25. *Any $(n-1)$-T-quasigroup code (Q, g) with check equation $d(x_1^n) = \alpha_1 x_1 + \alpha_2 x_2 + \cdots + \alpha_n x_n = 0$ detects:*

a) any transposition error on the place $(i, i+1)$, $i \in \overline{1, n-1}$, if and only if the mapping $\alpha_i - \alpha_{i+1}$ is an automorphism of the group $(Q, +)$;

b) any transposition error on the place $(i, i+2)$ (i.e., jump transposition error), $i \in \overline{1, n-2}$, if and only if the mapping $\alpha_i - \alpha_{i+2}$ is an automorphism of the group $(Q, +)$;

c) any twin error on the place $(i, i+1)$, $i \in \overline{1, n-1}$, if and only if the mapping $\alpha_i + \alpha_{i+1}$ is an automorphism of the group $(Q, +)$;

d) any twin error on the place $(i, i+2)$ (i.e., jump twin error), $i \in \overline{1, n-2}$, if and only if the mapping $\alpha_i + \alpha_{i+2}$ is an automorphism of the group $(Q, +)$.

Proof. This follows from Theorem 10.24. □

10.1.4 5-n-quasigroup codes

Definition 10.26. We shall call an n-quasigroup code (Q,d) that detects any transposition and twin error in places $(i,i+1)$, where $i \in \overline{1,n-1}$, and in places $(i,i+2)$ where $i \in \overline{1,n-2}$ a *5-n-quasigroup code* (Q,d) (since such code detects five types of errors).

The following theorem helps to construct 5-n-quasigroup codes.

Theorem 10.27. *The direct product of a 5-n-quasigroup code* (Q_1,d) *and 5-n-quasigroup code* (Q_2,g) *is a 5-n-quasigroup code* $(Q_1 \times Q_2, f)$, *where* $f = d \circ g$.

Proof. This follows from the standard definition of the direct product and the statement that the direct product of anti-commutative quasigroups is an anti-commutative quasigroup. The last statement follows from the well-known fact that a class of universal algebras of fixed signature defined by identities and quasi-identities is closed with respect to the direct product [212]. □

Remark 10.28. Theorem 10.27 is true for n-quasigroup codes that detect any transposition and twin error on the same set of places of the form $(i,i+k)$.

Theorem 10.29. *Existence of at least three different automorphisms* α,β,γ *of an abelian group* $(Q,+)$ *such that endomorphisms* $\alpha+\beta$, $\alpha+\gamma$, $\beta+\gamma$, $\alpha-\beta$, $\alpha-\gamma$, $\beta-\gamma$ *are automorphisms of this group is a necessary and sufficient condition for existence of a 5-n-T-quasigroup code* (Q,d) *with check equation* $d(x_1^n) = \alpha_1 x_1 + \alpha_2 x_2 + \cdots + \alpha_n x_n = 0$.

Proof. Sufficiency. Suppose that we have three different automorphisms of the group $(Q,+)$. If we take $\alpha_{3l+1} = \alpha$, $\alpha_{3l+2} = \beta$, $\alpha_{3l+3} = \gamma$, then we have a code (Q,d) with the following check equation $d(x_1^n) = \alpha x_1 + \beta x_2 + \gamma x_3 + \alpha x_4 + \cdots + \delta x_n = 0$, where ($\delta = \alpha$, if $n = 3k+1$, $\delta = \beta$, if $n = 3k+2$, $\delta = \gamma$, if $n = 3k$).

From Theorem 10.25 it follows that this code is a 5-n-T-quasigroup code, since it is easy to see that, if the endomorphism $\alpha-\beta$ is an automorphism of the group $(Q,+)$, then the endomorphism $\beta-\alpha$ is an automorphism of the group $(Q,+)$ too, and so on.

Necessity. If we suppose that we have only two different automorphisms α and β of the group $(Q,+)$ (or one automorphism), then it easy to see that it is impossible to construct a 5-n-T-quasigroup code. In particular, if we take $\alpha_1 = \alpha$ and $\alpha_2 = \beta$, then: if $\alpha_3 = \alpha$, then this code cannot detect jump transposition errors; if $\alpha_3 = \beta$, then this code cannot detect transposition errors in the place $(2,3)$. □

In [896, 397, 301, 751, 227] and some other articles, the systems that detect all single and transposition errors are studied.

Corollary 10.30. *Existence of at least two different automorphisms* α,β *of an abelian group* $(Q,+)$ *such that the endomorphism* $\alpha-\beta$ *is an automorphism of this group, is the necessary and sufficient condition for existence of an n-T-quasigroup code* (Q,d) *with check equation* $d(x_1^n) = \alpha_1 x_1 + \alpha_2 x_2 + \cdots + \alpha_n x_n = 0$ *that detects all single and transposition errors.*

Proof. The proof of this corollary is easy to obtain from Theorem 10.29. □

Remark 10.31. It is easy to see that in general in the conditions of Corollary 10.30 it is impossible that $\alpha-\beta = \alpha$, but it is possible that $\alpha-\beta = \beta$.

The following theorem is in the spirit of the work [227].

Theorem 10.32. *There does not exist a 5-n-T-quasigroup code over cyclic groups* Z_{2k} *and* Z_{3k}, *where k is an odd number.*

Proof. We use Theorem 10.25. In the first case it is easy to see that all automorphisms of the group Z_{2m} multiply the elements of this group by some odd number. But a sum of two odd numbers is an even number.

In the second case, any automorphism has the form $3h+1$ or $3l+2$. Then either a sum of any two automorphisms is not an automorphism (if one of the automorphisms is of the form $3h+1$ and other is of the form $3l+2$), or their difference is not an automorphism (if both automorphisms are of the same form, i.e., both have the form $3h+1$ or $3l+2$). $\square$

Corollary 10.33. *There does not exist an n-T-quasigroup code over a cyclic group* Z_{2k} *that detects all single and transposition errors.*

10.1.5 Phonetic errors

In this section we assume that all quasigroups are defined on a set Q such that $Q = \{0, 1, \dots, m\}$. We can see a phonetic error $a0 \to 1a, a \neq 0, a \neq 1$, as on a special kind of double error, in a codeword in the place of the form $(i, i+1)$.

Theorem 10.34. *A binary T-quasigroup* $(Q, \cdot)$ *of the form* $x \cdot y = \alpha x + \beta y + b$ *detects all phonetic errors if and only if* $(\alpha - \beta)a \neq \alpha 1$ *for all* $a \in Q$ *such that* $a \neq 0, a \neq 1$.

Proof. We find conditions when $a \cdot 0 = 1 \cdot a$. We have $a \cdot 0 = \alpha a + b$, $1 \cdot a = \alpha 1 + \beta a + b$. Then the quasigroup $(Q, \cdot)$ cannot detect a phonetic error if and only if $\alpha a + b = \alpha 1 + \beta a + b$, i.e., if and only if $(\alpha - \beta)a = \alpha 1$.

Therefore the T-quasigroup $(Q, \cdot)$ detects all phonetic errors if and only if $(\alpha - \beta)a \neq \alpha 1$ for all $a \in Q, a \neq 0, a \neq 1$. $\square$

Theorem 10.35. *There does not exist an* $(n-1)$*-T-quasigroup code* (Q, g) *with check equation* $d(x_1^n) = \alpha_1 x_1 + \alpha_2 x_2 + \dots + \alpha_n x_n = 0$ *that simultaneously detects all transposition errors and all phonetic errors in all places of the form* $(i, i+1)$.

Proof. From Theorem 10.25 it follows that to detect all transposition errors on a fixed place of the form $(i, i+1)$ the following condition must be fulfilled: the map $\alpha_i - \alpha_{i+1}$ is a permutation of the set Q.

From Theorem 10.34 it follows that the code (Q, g) detects all phonetic errors on a place $(i, i+1)$ if and only if $(\alpha_i - \alpha_{i+1})a \neq \alpha_i 1$. But if the map $(\alpha_i - \alpha_{i+1})$ is a permutation of the set Q, then $a \neq (\alpha_i - \alpha_{i+1})^{-1}\alpha_i 1$.

We prove that $a \neq 0, 1$. If $a = 0$, then $\alpha_i^{-1}(\alpha_i - \alpha_{i+1})0 = 0$, since $\alpha_i, (\alpha_i - \alpha_{i+1})$ are automorphisms of the abelian group $(Q, +)$. Thus $0 = 1$ and we have a contradiction. If $a = 1$, then $(\alpha_i - \alpha_{i+1})1 = \alpha_i 1$, $\alpha_i 1 - \alpha_{i+1} 1 = \alpha_i 1$, $\alpha_{i+1} 1 = 0$, $0 = 1$ and we again have a contradiction.

Therefore if the map $(\alpha_i - \alpha_{i+1})$ is a permutation of the set Q, then there exists an element a of the set Q, $a \neq 0, a \neq 1$, such that $(\alpha_i - \alpha_{i+1})a = \alpha_i 1$. In this case the code (Q, g) cannot exactly detect one phonetic error on a fixed place $(i, i+1)$. $\square$

Corollary 10.36. *If in a binary T-quasigroup* $(Q, \cdot)$ *of the form* $x \cdot y = \alpha x + \beta y + b$ *the map* $(\alpha - \beta)$ *is a permutation of the set* Q*, then this quasigroup detects all phonetic errors with the exception of the following error* $c0 \to 1c$*, where* $c = (\alpha - \beta)^{-1}\alpha 1$.

Proof. This follows from Theorem 10.35. $\square$

Theorem 10.37. *Any 5-n-T-quasigroup code* (Q, g) *with check equation*

$$d(x_1^{n+1}) = \alpha_1 x_1 + \alpha_2 x_2 + \dots + \alpha_n x_n + \alpha_{n+1} x_{n+1} = 0$$

detects all phonetic errors in all possible places of the form $(i, i+1)$ with the exception of one phonetic error in every place of the form $(i, i+1)$.

Proof. It follows from Theorem 10.25 that in the code (Q, g) the maps $\alpha_i - \alpha_{i+1}$ are permutations of the set Q for all $i \in \overline{1, n}$.

Then from Corollary 10.36 it follows that there exists exactly one element $c, c \neq 0, c \neq 1$ in the set Q such that $(\alpha_i - \alpha_{i+1})c = \alpha_i 1$. Therefore in the place $(i, i+1)$, the code (Q, g) cannot detect only one phonetic error, namely, the error $c0 \to 1c$. □

10.1.6 Examples of codes

Example 10.38. The International Standard Book Number (ISBN) code with $(Z_{11}, +)$, $n = 10$, and the check equation

$$\begin{aligned} &1 \cdot x_1 + 2 \cdot x_2 + 3 \cdot x_3 + 4 \cdot x_4 + 5 \cdot x_5 + 6 \cdot x_6 + \\ &7 \cdot x_7 + 8 \cdot x_8 + 9 \cdot x_9 + 10 \cdot x_{10} \equiv 0 \pmod{11} \end{aligned}$$

is used presently.

"... this system detects all adjacent transpositions but needs an element $X \notin \{0, ..., 9\}$" [752].

Proof. Using Theorem 10.25 we can say that this system detects all single errors, transposition, and twin errors in places $(i, i+1)$, $(i, i+2)$ for any possible value of index i with the exception of twin error in the place $(5, 6)$.

From Corollary 10.36 it follows that this code cannot detect the following phonetic errors:

$10\,0 \to 1\,10$	in the place $(1, 2)$;	$9\,0 \to 1\,9$	in the place $(2, 3)$;
$8\,0 \to 1\,8$	in the place $(3, 4)$;	$7\,0 \to 1\,7$	in the place $(4, 5)$;
$6\,0 \to 1\,6$	in the place $(5, 6)$;	$5\,0 \to 1\,5$	in the place $(6, 7)$;
$4\,0 \to 1\,4$	in the place $(7, 8)$;	$3\,0 \to 1\,3$	in the place $(8, 9)$;
$2\,0 \to 1\,2$	in the place $(9, 10)$.		

In particular, in the place $(1, 2)$ we have

$$c = (\alpha - \beta)^{-1}\alpha 1 = (1-2)^{-1} \cdot 1 \cdot 1 = (-1)^{-1}1 = -1 \cdot 1 = -1 = 10,$$

in the place $(2, 3)$ we have $c = (2-3)^{-1} \cdot 2 \cdot 1 = (-1)^{-1}2 = -1 \cdot 2 = -2 = 9$, and so on.

We notice that in the ISBN code the number 10 is not used in places $1, 2, \dots, 9$.

Example 10.39. If we take the check equation

$$\begin{aligned} &1 \cdot x_1 + 2 \cdot x_2 + 3 \cdot x_3 + 4 \cdot x_4 + 5 \cdot x_5 + 10 \cdot x_6 + 9 \cdot x_7 + \\ &8 \cdot x_8 + 7 \cdot x_9 + 6 \cdot x_{10} \equiv 0 \pmod{11}, \end{aligned}$$

then these 9-ary-T-quasigroup codes over the group $(Z_{11}, +)$ detect all single errors, transposition, and twin errors in places $(i, i+1)$, $(i, i+2)$ for any permissible value of index i.

We enumerate phonetic errors which the (Z_{11}, g)-code cannot detect.

$10\,0 \to 1\,10$	in the place $(1, 2)$;	$9\,0 \to 1\,9$	in the place $(2, 3)$;
$8\,0 \to 1\,8$	in the place $(3, 4)$;	$7\,0 \to 1\,7$	in the place $(4, 5)$;
$10\,0 \to 1\,10$	in the place $(5, 6)$;	$10\,0 \to 1\,10$	in the place $(6, 7)$;
$9\,0 \to 1\,9$	in the place $(7, 8)$;	$8\,0 \to 1\,8$	in the place $(8, 9)$;
$7\,0 \to 1\,7$	in the place $(9, 10)$.		

Code with the following check equation

$$1 \cdot x_1 + 3 \cdot x_2 + 5 \cdot x_3 + 7 \cdot x_4 + 9 \cdot x_5 + 10 \cdot x_6 +$$
$$8 \cdot x_7 + 6 \cdot x_8 + 4 \cdot x_9 + 2 \cdot x_{10} \equiv 0 \pmod{11}$$

has similar properties.

Of course, in the last check equations we may change the group $(Z_{11}, +)$ by any group $(Z_p, +)$, where p is a prime number, $p \geqslant 7$ and we can take any finite number $n \geqslant 4$ of items (4 because we must have the check symbol and a possibility to receive jump errors).

For $p = 7$ we have the following systematic code with the check equation

$$x_1 + 2x_2 + 3x_3 + 6x_4 + 5x_5 + 4x_6 + x_7 + 2x_8 + \cdots + ax_n \equiv 0 \pmod{7},$$

where $a = 1$, if $n \equiv 1 \pmod 6$, $a = 2$, if $n \equiv 2 \pmod 6$, $a = 3$, if $n \equiv 3 \pmod 6$, $a = 6$, if $n \equiv 4 \pmod 6$, $a = 5$, if $n \equiv 5 \pmod 6$, $a = 4$, if $n \equiv 0 \pmod 6$ or

$$x_1 + 3x_2 + 5x_3 + 6x_4 + 4x_5 + 2x_6 + x_7 + 3x_8 + \cdots + ax_n \equiv 0 \pmod{7},$$

where $a = 1$, if $n \equiv 1 \pmod 6$, $a = 3$, if $n \equiv 2 \pmod 6$, $a = 5$, if $n \equiv 3 \pmod 6$, $a = 6$, if $n \equiv 4 \pmod 6$, $a = 4$, if $n \equiv 5 \pmod 6$, $a = 2$, if $n \equiv 0 \pmod 6$, $x_i \in Z_7$ for any $i \in \overline{1, n}$. □

Example 10.40. "The European Article Number code (EAN) and (after adding 0 as the first digit) the Universal Product Code (UPC) with $G = (Z_{10}, +)$, $n = 13$, $e = 0$, $\delta_{2i-1}(a) = a = L_1(a)$ and $\delta_{2i}(a) = 3a = L_3(a)$ ... " [752].

In other words, the EAN code is the code with the check equation

$$x_1 + 3x_2 + x_3 + 3x_4 + x_5 + 3x_6 + x_7 + 3x_8 + x_9 + 3x_{10} + x_{11} + 3x_{12} + x_{13} = 0,$$

where $x_i \in Z_{10}$, $i \in \overline{1, 13}$.

Notice that "... this system does not detect adjacent transpositions $\dots ab \dots \to \dots ba \dots$ for $|a - b| = 5$ "[752].

Proof. Moreover, this system does not detect errors of the form $\dots\ acb\ \dots \to \dots\ bca\ \dots$, i.e., the so-called jump transposition errors for any pair of elements $a, b \in Z_{10}$.

Indeed, let us have jump transposition $\dots acb \dots \to \dots bca$. Passing to group operation we have the following expressions $\dots 3a + c + 3b \dots$, $\dots 3b + c + 3a \dots$ or $\dots a + 3c + b \dots$, $\dots b + 3c + a \dots$. Since the group Z_{10} is commutative, we obtain $\dots 3a + c + 3b \dots = \dots 3b + c + 3a \dots$ or $\dots a + 3c + b \dots = \dots b + 3c + a \dots$. Therefore EAN code does not detect "jump transpositions errors."

The EAN code does not detect twin errors $(\dots aa \dots \to \dots bb \dots)$ for $|a - b| = 5$. Really, passing to the group operation, we obtain the following expressions $\dots + 3a + a + \dots = \dots + 4a \dots$, $\dots + 3b + b + \dots = \dots + 4b + \dots$ or $\dots + a + 3a + \dots = \dots + 4a + \dots$, $\dots + b + 3b + \dots = \dots + 4b + \dots$.

In the case when $4a = 4b$, the EAN code will not detect this twin error. We can rewrite the last equality in this form: $4(a - b) \equiv 0 \pmod{10}$. Therefore $(a - b) \equiv 0 \pmod 5$, $|a - b| = 5$ since $a, b \in \{0, 1, 2, 3, 4, 5, 6, 7, 8, 9\}$.

By analogy it is possible to prove that the EAN code cannot also detect jump twin errors $(\dots aca \dots \to \dots bcb \dots)$ for any pair of elements $a, b \in Z_{10}$ such that $|a - b| = 5$.

Therefore the EAN code does not detect adjacent transpositions, twin errors, or jump twin errors for any pair of elements $a, b \in Z_{10}$ such that $|a - b| = 5$ and it does not detect jump transposition for any pair of elements $a, b \in Z_{10}$.

Example 10.41. We propose the following code with the check equation

$$x_1 + 3x_2 + 9x_3 + 7x_4 + x_5 + 3x_6 + 9x_7 + 7x_8 + x_9 + 3x_{10} + 9x_{11} + 7x_{12} + x_{13} = 0,$$

where $x_i \in Z_{10}, i \in \overline{1, 13}$ as a small modification of EAN code. For convenience we shall call it *the EAN-1 code*.

As it follows from Theorem 10.25, the EAN-1 code detects all single errors since multiplication of elements of the group Z_{10} by elements 1, 3, 7, 9 is automorphism of this group.

The EAN-1 code does not detect transposition errors, jump transposition errors, or twin errors for any pair of elements $a, b \in Z_{10}$ such that $|a - b| = 5$ and it does not detect jump twin errors for any pair of elements $a, b \in Z_{10}$. □

Our computer investigations show that there is no anti-commutative quasigroup $(Z_{10}, \circ)$ of the form $x \circ y = \alpha x + y$ over the group Z_{10} with $\alpha \in S_{10}$, but there exist more than 140,000 totally anti-commutative quasigroups of the form $x \circ y = \alpha x + \beta y$ over the group Z_7 with $\alpha, \beta \in S_7$.

Example 10.42. We can propose a code $\mathfrak{C}$ over Z_{10} with the following check equation:

$$x_1 + \alpha x_2 + \beta x_3 + x_4 + \alpha x_5 + \beta x_6 + \cdots \equiv 0 \pmod{10},$$

where $\alpha = (0\,8\,7\,6\,3\,9\,1\,2\,5)(4)$, $\beta = (0\,4\,5\,7\,8\,1\,6\,3\,2)(9)$.

This code does not detect two transposition errors, two jump transposition errors, not more than eight twin errors and jump twin errors in any place of form $(i, i+1), (i, i+2)$.

Proof. The Cayley tables of quasigroups $x \cdot y = x + \alpha y$, $x \circ y = \alpha x + \beta y$ and $x * y = \beta x + y$ are the following:

$\cdot$	0	1	2	3	4	5	6	7	8	9
0	8	2	5	9	4	0	3	6	7	1
1	9	3	6	0	5	1	4	7	8	2
2	0	4	7	1	6	2	5	8	9	3
3	1	5	8	2	7	3	6	9	0	4
4	2	6	9	3	8	4	7	0	1	5
5	3	7	0	4	9	5	8	1	2	6
6	4	8	1	5	0	6	9	2	3	7
7	5	9	2	6	1	7	0	3	**4**	8
8	6	0	3	7	2	8	1	**4**	5	9
9	7	1	4	8	3	9	2	5	6	0

$\circ$	0	1	2	3	4	5	6	7	8	9
0	2	4	8	0	3	5	1	6	9	7
1	6	8	2	4	7	9	5	0	**3**	1
2	9	1	5	7	0	2	8	3	6	4
3	3	5	9	1	4	6	2	7	0	8
4	8	0	4	6	9	1	7	2	5	3
5	4	6	0	2	5	7	3	8	1	9
6	7	9	3	5	8	0	6	1	4	2
7	0	2	6	8	1	3	9	4	7	5
8	1	**3**	7	9	2	4	0	5	8	6
9	5	7	1	3	6	8	4	9	2	0

*	0	1	2	3	4	5	6	7	8	9
0	4	5	6	7	8	9	0	1	2	3
1	6	7	8	9	0	1	2	3	4	5
2	0	1	2	3	4	5	6	7	8	9
3	2	3	4	5	6	7	8	9	0	1
4	5	6	7	8	9	0	1	**2**	3	4
5	7	8	9	0	1	2	3	4	5	6
6	3	4	5	6	7	8	9	0	1	2
7	8	9	0	1	**2**	3	4	5	6	7
8	1	2	3	4	5	6	7	8	9	0
9	9	0	1	2	3	4	5	6	7	8

In the quasigroup $(Q, \cdot)$ only elements 7 and 8 are permutable: $8 \cdot 7 = 7 \cdot 8 = 4$. Therefore this code does not detect only transposition errors $78 \to 87$ and $87 \to 78$ in places of the form $(1+3j; 2+3j)$ for any suitable j. In these places this code does not detect the following twin errors $00 \leftrightarrow 44$ (i.e., $00 \to 44$, $44 \to 00$), $11 \leftrightarrow 77$, $55 \leftrightarrow 88$.

In the quasigroup $(Q, \circ)$ only the elements 1 and 8 are permutable. Therefore the code $\mathfrak{C}$ does not detect transposition errors $18 \leftrightarrow 81$ in places of the form $(2+3j; 3+3j)$ for any suitable j. In these places the code does not detect twin errors $11 \leftrightarrow 88$, either.

Only the elements 4 and 7 commute in the quasigroup $(Q, *)$. Therefore the code $\mathfrak{C}$ does not detect transposition errors $47 \leftrightarrow 74$ in places of the form $(3+3j; 4+3j)$ for any suitable j. In these places the code does not detect the following eight twin errors: $33 \leftrightarrow 77$, $44 \leftrightarrow 66 \leftrightarrow 88$.

In places of the form $(1+3j, 3+3j)$ the code $\mathfrak{C}$ cannot detect the same set of errors as in the places of the form $(3+3j; 4+3j)$, in places of the form $(2+3j, 4+3j)$ the code $\mathfrak{C}$ cannot detect the same set errors as in the places of the form $(1+3j; 2+3j)$, and in places of the form $(3+3j, 5+3j)$ the code $\mathfrak{C}$ cannot detect the same set of errors as in the places of the form $(2+3j; 3+3j)$. □

Example 10.43. Let $(Z_{2n+1}, +)$ be a cyclic group of order $(2n+1) \geqslant 7$ and the number $2n+1$ is a prime. An n-ary quasigroup code (Z_{2n+1}, d) with check equation

$$\begin{aligned} &1 \cdot x_1 + 2 \cdot x_2 + 3 \cdot x_3 + 4 \cdot x_4 + \cdots + n \cdot x_n + \\ &1 \cdot x_{n+1} + 2 \cdot x_{n+2} + \cdots \equiv 0 \pmod{2n+1}, \end{aligned}$$

where element 0 is the identity element of the group $(Z_{2n+1}, +)$, detects single errors, and any transposition and twin errors in places $(i, i+1)$, $(i, i+2)$ for all suitable values of natural number i.

Proof. It is known that multiplying of elements of the group $(Z_{2n+1}, +)$ ($2n+1$ is a prime number) on element k, $k \in \{1, 2, 3, \dots, 2n\}$, is an automorphism of the group Z_{2n+1}.

Taking into consideration Theorem 10.25 we only have to show that the following sums of automorphisms of the group $(Z_{2n+1}, +)$ $1+2=3, 1+3=4, 2+3=5, \dots, n-1+n = 2n-1, \dots, 1-2=-1, 1-3=-2, \dots, n-1-1=n-2, n-1, n-2$ are automorphisms of the group $(Z_{2n+1}, +)$. It is easy to see that it is so.

Therefore our code can detect all single errors, transposition, and twin errors in places $(i, i+1)$, $(i, i+2)$ for any suitable value of i. □

Example 10.44. Let $(Z_p, +)$ be a cyclic group of prime order $p \geqslant 7$. An $(n-1)$-ary quasigroup code (Z_p, d) with the check equation

$$1 \cdot x_1 + 2 \cdot x_2 + 3 \cdot x_3 + 1 \cdot x_4 + 2 \cdot x_5 + 3 \cdot x_6 + \cdots + \alpha x_n \equiv 0 \pmod{p},$$

where elements x_1^{n-1} are information symbols and element x_n is a check character, ($\alpha = 1$, if $n = 3k+1$, $\alpha = 2$, if $n = 3k+2$, $\alpha = 3$, if $n = 3k$), detects any transposition and twin errors in places $(i, i+1)$, where $i \in \overline{1, n-1}$, and $(i, i+2)$, where $i \in \overline{1, n-2}$.

Proof. Taking into consideration Theorem 10.25 we only have to show that the following sums of automorphisms of the group $(Z_p, +)$ $1+2 = 3, 1+3 = 4, 2+3 = 5, 1-2 = -1 = p-1, 1-3 = -2 = p-2, 2-3 = -1 = p-1, 2-1 = 1, 3-1 = 2, 3-2 = 1$ are automorphisms of the group $(Z_p, +)$. Since multiplication of the elements of the group Z_p on the numbers $1, 2, 3, 4, 5, p-2, p-1$ are automorphisms of the group $(Z_p, +)$, our code can detect all single errors, transposition, and twin errors on places $(i, i+1)$, where $i \in \overline{1, n-1}$, and $(i, i+2)$, where $i \in \overline{1, n-2}$.

This code cannot detect the following phonetic errors: $(-1; 0) \to (1; -1)$ in places of the form $(1+3k; 2+3k)$; $(-2; 0) \to (1; -2)$ in places of the form $(2+3k; 3+3k)$; or $(2^{-1} \cdot 3; 0) \to (1; 2^{-1} \cdot 3)$ on places of the form $(3+3k; 4+3k)$. □

Example 10.45. Let $(Q, +) = (Z_m \times Z_m, +)$, where m is a natural number, $m \geqslant 2$, G.C.D.$(m, 3) = 1$. For instance, let $m = 2$ (minimal possible value of m) or $m = 5$. Let

$$\alpha = \begin{pmatrix} 1 & 0 \\ 0 & 1 \end{pmatrix}, \quad \beta = \begin{pmatrix} 1 & 1 \\ 1 & 0 \end{pmatrix}, \quad \gamma = \begin{pmatrix} 0 & 1 \\ 1 & 1 \end{pmatrix}.$$

We define $(n-1)$-ary quasigroup code (Q, d) with check equation

$$\alpha x_1 + \beta x_2 + \gamma x_3 + \alpha x_4 + \beta x_5 + \cdots + \delta x_n = 0,$$

where elements x_1^{n-1} are information symbols and the element x_n is a check character, $x_1^n \in Q$, ($\delta = \alpha$, if $n = 3k+1$, $\delta = \beta$, if $n = 3k+2$, $\delta = \gamma$, if $n = 3k$). This code detects any transposition and twin errors in places $(i, i+1)$, where $i \in \overline{1, n-1}$ and in places $(i, i+2)$, where $i \in \overline{1, n-2}$.

Proof. Taking into consideration Theorem 10.25 we only have to show that the following sums of automorphisms $\alpha+\beta$, $\alpha-\beta$, $\beta-\alpha$, $\alpha+\gamma$, $\alpha-\gamma$, $\gamma-\alpha$, $\beta+\gamma$, $\beta-\gamma$, $\gamma-\beta$ are automorphisms of the group $(Z_m \times Z_m, +)$.

As usual $det(\alpha)$ denotes a determinant of the matrix α. We have $det(\alpha+\beta) = 1$, $det(\alpha-\beta) = -1$, $det(\beta-\alpha) = -1$, $det(\alpha+\gamma) = 1$, $det(\alpha-\gamma) = -1$, $det(\gamma-\alpha) = -1$, $det(\beta+\gamma) = -3$, $det(\beta-\gamma) = -1$, $det(\gamma-\beta) = -1$. Therefore all these sums of automorphisms are also automorphisms of the group $(Z_m \times Z_m, +)$.

Thus our code can detect all single errors, transposition, and twin errors in places $(i, i+1)$, where $i \in \overline{1, n-1}$ and in places $(i, i+2)$, where $i \in \overline{1, n-2}$.

We calculate phonetic errors that this code cannot detect. Denote the element $(1; 0)$ as 1 and the element $(0; 0)$ as 0. The code cannot detect the error $(-1; -1)0 \to 1\,(-1; -1)$ in places of the form $(1+3k; 2+3k)$, the error $(1; -1)0 \to 1\,(1; -1)$ in places of the form $(2+3k; 3+3k)$, or the error $(1; 1)0 \to 1\,(1; 1)$ in places of the form $(3+3k; 4+3k)$, where k is a natural number. □

Remark 10.46. The last example shows that existence of three different automorphisms of an abelian group $(Q, +)$ is a sufficient condition for existence of a 5-n-T-quasigroup code (Q, d) over this group with the check equation $d(x_1^n) = \alpha_1 x_1 + \alpha_2 x_2 + \cdots + \alpha_n x_n = 0$.

Proof. If $m = 2$, then we have $\alpha = -\alpha$, $\beta = -\beta$, $\gamma = -\gamma$, $\alpha+\beta = \gamma$, $\alpha+\gamma = \beta$ and $\beta+\gamma = \alpha$. □

Example 10.47. Let $(Q, +) = (Z_m \times Z_m, +)$, where m is a natural number, $m \geqslant 3$, G.C.D.$(m, 5) = 1$. For example, let $m = 3$ (minimal possible value of m) or $m = 7$. Let

$$\alpha = \begin{pmatrix} 1 & 0 \\ 0 & 1 \end{pmatrix}, \ \beta = \begin{pmatrix} -1 & 1 \\ 1 & 0 \end{pmatrix}, \ \gamma = \begin{pmatrix} 0 & 1 \\ 1 & 1 \end{pmatrix}.$$

We define $(n-1)$-ary quasigroup code (Q, d) with check equation

$$\alpha x_1 + \beta x_2 + \gamma x_3 + \alpha x_4 + \beta x_5 + \cdots + \delta x_n = 0,$$

where elements x_1^{n-1} are information symbols and the element x_n is a check character, $x_1^n \in Q$, ($\delta = \alpha$, if $n = 3k+1$, $\delta = \beta$, if $n = 3k+2$, $\delta = \gamma$, if $n = 3k$), $0 = (0, 0)$. This code detects any transposition and twin errors in places $(i, i+1)$, where $i \in \overline{1, n-1}$ and in places $(i, i+2)$, where $i \in \overline{1, n-2}$.

Proof. The proof is similar to the proof of Example 10.45. □

Using Example 10.47 it is possible to construct a 5-n-T-quasigroup code over the group $(Z_3 \times Z_3, +)$.

Example 10.48. We suppose that the Cayley table of the group $(Z_3 \times Z_3, +)$ has the form

+	1	2	3	4	5	6	7	8	9
1	1	2	3	4	5	6	7	8	9
2	2	3	1	5	6	4	8	9	7
3	3	1	2	6	4	5	9	7	8
4	4	5	6	7	8	9	1	2	3
5	5	6	4	8	9	7	2	3	1
6	6	4	5	9	7	8	3	1	2
7	7	8	9	1	2	3	4	5	6
8	8	9	7	2	3	1	5	6	4
9	9	7	8	3	1	2	6	4	5

and we take $\rho = \varepsilon$, $\sigma = (1)\,(2\,4\,8\,9\,3\,7\,6\,5)$, $\tau = (1)\,(2\,5\,6\,7\,3\,9\,8\,4)$.

We define $(n-1)$-ary quasigroup code $(Z_3 \times Z_3, f)$ with check equation

$$\rho x_1 + \sigma x_2 + \tau x_3 + \rho x_4 + \sigma x_5 + \cdots + \delta x_n = 1,$$

where elements x_1^{n-1} are information symbols and the element x_n is a check character, $x_1^n \in Z_3 \times Z_3$, ($\delta = \rho$, if $n = 3k+1$; $\delta = \sigma$, if $n = 3k+2$; $\delta = \tau$, if $n = 3k$). This code detects any transposition and twin errors in places $(i, i+1)$, where $i \in \overline{1, n-1}$ and in places $(i, i+2)$, where $i \in \overline{1, n-2}$. In this code there cannot be phonetic errors of the form $a0 \to 1a, a \neq 0, a \neq 1$.

Proof. We take only a re-written form of automorphisms α, β, γ from Example 10.47 for the group $(Z_3 \times Z_3, +)$. It is clear that $\alpha = \rho$, $\beta = \sigma$, $\gamma = \tau$.

It follows from Example 10.47, that this code detects all five types of errors. Since in the alphabet of this code there is no zero element, this code allows us to avoid phonetic errors of the form $a0 \to 1a, a \neq 0, a \neq 1$. Therefore this code detects all errors from Table 10.1. □

Example 10.49. Let $(Z_p, +)$, $(Z_q, +)$ be cyclic groups of prime order $p, q \geqslant 7$. An $(n-1)$-ary quasigroup code $(Q, d) = (Z_p \times Z_q, d)$ with the check equation

$$1 \cdot x_1 + 2 \cdot x_2 + 3 \cdot x_3 + 1 \cdot x_4 + 2 \cdot x_5 + 3 \cdot x_6 + \cdots + \alpha x_n \equiv 0 \pmod{pq},$$

where binary operation + is the operation of the group $(Z_p \times Z_q, +)$, elements x_1^{n-1} are information symbols and element x_n is a check character, $x_1^n \in Q$, $\alpha = 1$, if $n = 3k+1$, $\alpha = 2$, if $n = 3k+2$, $\alpha = 3$, if $n = 3k$, detects any transposition and twin error in places $(i, i+1)$, where $i \in \overline{1, n-1}$, and $(i, i+2)$, where $i \in \overline{1, n-2}$.

Proof. We can take into consideration Theorem 10.25 and Example 10.45. □

Example 10.50. An $(n-1)$-ary quasigroup code $(Q, d) = (Z_p \times Z_p \times Z_q, d)$ with the check equation

$$\alpha_1 \cdot x_1 + \alpha_2 \cdot x_2 + \alpha_3 \cdot x_3 + \alpha_1 \cdot x_4 + \alpha_2 \cdot x_5 + \alpha_3 \cdot x_6 + \cdots + \delta x_n = 0,$$

where elements x_1^{n-1} are information symbols and element x_n is a check character, $x_1^n \in Q$, $\delta = \alpha_1$, if $n = 3k+1$, $\delta = \alpha_2$, if $n = 3k+2$, $\delta = \alpha_3$, if $n = 3k$, detects any transposition and twin errors in places $(i, i+1)$, where $i \in \overline{1, n-1}$ and in places of the form $(i, i+2)$, where $i \in \overline{1, n-2}$, $\alpha_1 x_i = (\alpha x_i'; 1 \cdot y_i')$, $\alpha_2 x_i = (\beta x_i'; 2 \cdot y_i')$, $\alpha_3 x_i = (\gamma x_i'; 3 \cdot y_i')$, where α, β, γ are defined as in Example 10.49, $x_i' \in Z_p \times Z_p$, $y_i' \in Z_q$.

Proof. We can take into consideration Theorem 10.25, and Examples 10.45 and 10.49. In other words, we construct this code taking the direct product of codes defined in Examples 10.45 and 10.49. □

Example 10.51. Let $(Z_{2n+1}, +)$ be a cyclic group of order $(2n+1) \geqslant 7$ and the number $2n+1$ is a prime. An n-ary quasigroup code (Z_{2n+1}, g) with check equation

$$1 \cdot x_1 + 2 \cdot x_2 + 3 \cdot x_3 + 4 \cdot x_4 + \cdots + n \cdot x_n \equiv 0 \pmod{2n+1},$$

where element 0 is zero of the group $(Z_{2n+1}, +)$, detects single errors, any transposition, and twin errors on all places of the form $(i, i+k)$ for all suitable values of natural numbers i, k.

Proof. It is known that multiplying the elements of group $(Z_{2n+1}, +)$ ($2n+1$ is a prime number) by element k, $k \in \{1, 2, 3, \dots, 2n\}$, is an automorphism of the group Z_{2n+1}.

Taking into consideration Theorem 10.25 we only have to show that all sums and differences of different automorphisms of the set of automorphisms $\{1, 2, \dots, n\}$ are automorphisms of the group $(Z_{2n+1}, +)$. It is easy to see it is true.

Therefore our code can detect all single errors, transposition, and twin errors in all places $(i, i+k)$, for all suitable values of i, k. □

Theorem 10.52. *There exist 5-n-T-quasigroup codes: of any prime order $p \geqslant 7$; of any order m^2, where $m > 1$; of any composite order d such that $d = m^2 p_1 p_2 \dots p_s$, where $m \geqslant 1$, $p_i \geqslant 7$.*

Proof. The proof follows from Examples 10.44, 10.45, 10.47, 10.49 and Theorem 10.27. □

Remark 10.53. It is possible to check that there does not exist a 5-T-quasigroup code (i.e., a code that detects 5 types of the errors) over an alphabet of order 3, 5, 6. Some other results of this kind are available in [227].

10.2 Recursive MDS-codes

10.2.1 Some definitions

Let Q be an alphabet with $q \geqslant 2$ elements. Any subset of $K \subseteq Q^n$, $|Q^n| = q^n$, is called a code of length n, or n-code. In fact, there is a defined block-code because all words are of equal length.

A code is called $[n,k]_Q$-code, if $|K| = q^k$, $k \in \mathbb{R}$. Here $\mathbb{R}$ means the set of real numbers. It is easy to see that $k = \log_q |K|$.

A linear code of length n and rank k is a linear subspace K with dimension k of the vector space $\mathbb{F}_q^n$, where $\mathbb{F}_q$ is the finite field with q elements [914].

A code K of length n over the alphabet Q is called generalized linear, if there exists a binary operation $+$ on Q such that $(Q,+)$ is an Abelian group and K is a subgroup of $(Q^n,+)$ [371]. Any linear code can be considered as a generalized linear code over the group $(\mathbb{F}_q,+)$.

It is known that t mutually orthogonal Latin squares of order n generate a $\lfloor \frac{t}{2} \rfloor$ -error correcting code with n^2 codewords [240].

10.2.2 Singleton bound

Definition 10.54. The Hamming distance between two strings of equal length is the number of positions at which the corresponding symbols of these strings are different [913]. Similarly the Hamming distance $d(u,v)$ between two codewords $u, v \in K$ is defined. The Hamming distance $d(K)$ of a code K is a minimum of distances between different codewords.

Theorem 10.55. *The expression $A_q(n,d)$ represents the maximum number of possible codewords in a q-ary block code of length n and minimum distance d. The Singleton bound states that $A_q(n,d) \leqslant q^{n-d+1}$.*

Proof. We follow [915]. First, observe that there are q^n q-ary words of length n. Now let K be an arbitrary q-ary block code of minimum distance d. Clearly, all codewords are distinct.

If we delete the first $(d-1)$ letters of each codeword, then all resulting codewords must still be pairwise different, since all original codewords in K are at a Hamming distance of at least d from each other. Thus the size of the code remains unchanged.

The newly obtained codewords each have the length $n-(d-1) = n-d+1$ and thus there can be at most q^{n-d+1} of them. Hence the original code shares the same bound on its size:

$$|K| \leqslant A_q(n,d) \leqslant q^{n-d+1}.$$

□

In some articles, for example in [446], the Singleton bound is called the Joshi bound.

We further follow [915]. Block codes that achieve equality in Singleton bound are called MDS (maximum distance separable) codes.

Examples of such codes include codes that have only one codeword (minimum distance n), codes that use the whole of Ω^n (minimum distance 1), codes with a single parity symbol (minimum distance 2) and their dual codes. These are often called trivial MDS codes. In the case of binary alphabets, only trivial MDS codes exist [897].

The following theorem demonstrates the importance of codes with big Hamming distance.

Theorem 10.56. *A code K can detect up to s errors in any codeword if $d(K) > s+1$. A code K can correct up to t errors if $d(K) > 2t+1$ [559].*

In coding theory, a systematic code is any error-correcting code in which the input data is embedded in the encoded output [916]. It is also possible to say that the code in which the first symbols are information and the last symbols are check digits is called systematic.

10.2.3 MDS-codes

Let $K \subseteq Q^n$ be a systematic code $[n, k]$-code with the first k information (input data) positions. Then there exists a system of functions $f_s : Q^k \to Q$, $s \in \overline{1, n-k}$ such that

$$K = \{\mathbf{u} = (x_1^k, f_1(x_1^k), f_2(x_1^k), \dots, f_{n-k}(x_1^k)) \mid x_1^k \in Q^k\}. \tag{10.3}$$

We recall the following:

Definition 10.57. A system $\Sigma = \{f_1, f_2, \dots, f_t\}_{t \geqslant k}$ of k-ary operations defined on a set Q is called orthogonal if all k operations from Σ are orthogonal [446].

Definition 10.58. A system $\mathbf{\Sigma} = \{f_1, f_2, \dots, f_s\}_{s \geqslant 1}$ of k-ary operations defined on a set Q is called *strong orthogonal* if the system $\{E_1, \dots, E_k, f_1, f_2, \dots, f_s\}$ is orthogonal, where $E_i(x_1^k) = x_i$, for all $(x_1, \dots, x_k) \in Q^k$ and for all $i = 1, 2, \dots, k$ (the k-ary selectors) [446].

Remark 10.59. From Definition 10.58 it follows that any operation f_i of the system Σ, which is not a selector, is a quasigroup operation. Indeed, any system of equations consisting of $(k-1)$ selectors and one operation from Σ has exactly one solution for any suitable values of its right side [371].

Any orthogonal system of binary quasigroups is strong orthogonal [446].
We call the following theorem the Couselo-Gonsales-Markov-Nechaev (CKMN) criteria.

Theorem 10.60. *A code defined by formula (10.3) is an MDS-code if and only if the system of functions $\{f_1(x_1^k), f_2(x_1^k), \dots, f_{n-k}(x_1^k)\}$ is an orthogonal system of k-ary quasigroups [371].*

For construction of the following example we use Example 9.115.

Example 10.61. Any code $K \subseteq (Z_2 \oplus Z_2)^5$ of the form

$$K = \{\mathbf{u} = (x, y, D(x, y), E(x, y), F(x, y)) \mid x, y \in (Z_2 \oplus Z_2)\}, \tag{10.4}$$

is an MDS-code. This fact follows from Lemma 9.116 (orthogonality of operations D, E, F) and GKMN criteria.

If Hamming distance d of the code K defined by equation (10.4) is equal to five, then, by Theorem 10.55, $|K| \leqslant |Z_2 \oplus Z_2| = 4$; if $d = 4$, then $|K| \leqslant |Z_2 \oplus Z_2|^2 = 4^2 = 16$; if $d = 3$, then, $|K| \leqslant 4^3 = 64$.

10.2.4 Recursive codes

Definition 10.62. [371]. Let Q be an alphabet with $q \geqslant 2$ elements and $f : Q^k \to Q$ be a function of k $(k \leqslant n)$ variables.

A code $K(n, f)$ of length n over an alphabet Q is called a *complete k-recursive code*, if $K(n, f)$ consists of all words $(u_0, u_1, , u_{n-1})$, satisfying the condition

$$u_{i+k} = f(u_i, \dots, u_{i+k-1}), i = 0, 1, \dots, n-k. \tag{10.5}$$

In other words, we say that k is a k-recursive code if it consists of all words of length $n \geqslant k$, the coordinates of which are obtained from the first k coordinates using some fixed recursion law [1]. Any sub-code of $K(n,f)$ is called recursive.

We can rewrite the equation (10.5) in more detail [371, 446] in the following way:

$$K(n,f) = \{(x_1,\dots,x_k, f^{(0)}(x_1^k),\dots,f^{(n-k-1)}(x_1^k)) \mid x_1,\dots,x_k \in Q\}. \tag{10.6}$$

The functions $f^{(0)}, f^{(1)}, \dots, f^{(n-k-1)}$ are called k-recursive derivatives of f and are defined as follows:

$$\begin{aligned}
f^{(0)}(x_1^k) &= f(x_1^k),\\
f^{(1)}(x_1^k) &= f(x_2^k, f^{(0)}(x_1^k)),\\
f^{(2)}(x_1^k) &= f(x_3^k, f^{(0)}(x_1^k), f^{(1)}(x_1^k)),\\
&\dots,\\
f^{(t)}(x_1^k) &= f(x_{t+1}^k, f^{(0)}(x_1^k),\dots,f^{(t-1)}(x_1^k)), \quad \text{for} \quad t<k,\\
f^{(t)}(x_1^k) &= f(f^{(t-k-1)}(x_1^k),\dots,f^{(t-1)}(x_1^k)), \quad \text{for} \quad t \geqslant k.
\end{aligned} \tag{10.7}$$

In the binary case the first derivative of a function f has the form $f^{(1)}(x,y) = f(y, f(x,y))$.

Lemma 10.63. *If (Q,f) and $(Q,f^{(1)})$ are operations of binary quasigroups, then $(Q,f) \perp (Q,f^{(1)})$.*

Proof. Indeed, in this case the system of equations

$$\begin{cases} f(x,y) = a \\ f(y, f(x,y)) = b \end{cases}$$

has a unique solution for any fixed elements $a, b \in Q$. □

A quasigroup (Q,f) such that $(Q,f^{(1)})$ is also a quasigroup is called *differentiable (1-differentiable)*.

Lemma 10.64. *A binary group $(G,+)$ is differentiable if and only if the mapping $x \to (x+x)$ is a bijection [446].*

Corollary 10.65. *Any code $K \subseteq Q^4$ of the form*

$$K = \{\mathbf{u} = (x, y, f(x,y), f^{(1)}(x,y)) \mid x, y \in Q\}, \tag{10.8}$$

where (Q,f) is a binary quasigroup such that its derivative $(Q,f^{(1)})$ is also a binary quasigroup, is an MDS-code.

Proof. The proof follows from Lemma 10.63 and Theorem 10.60. □

If as defined by the equation (10.8) code K, $d = 4$ (the number d means the minimum distance between codewords,) then, by Theorem 10.55, $|K| \leqslant |Q| = q$; if $d = 3$, then $|K| \leqslant |Q|^2 = q^2$; if $d = 2$, then, $|K| \leqslant |Q|^3 = q^3$.

As corollary of Theorem 10.60 we formulate the following:

Theorem 10.66. *A complete k-recursive code $K(n,f)$ defined by the formula (10.6) is an MDS-code if and only if the system of k-recursive derivatives $f^{(0)}, f^{(1)}, \dots, f^{(n-k-1)}$ is a system of orthogonal k-ary quasigroups [371, 446].*

G.B. Belyavskaya proves the following:

Theorem 10.67. *If (Q, f) is a binary quasigroup, then $f^{(i)} = f\theta^i$ for all $i \in \mathbb{N}$, where θ is the following mapping: $\theta : Q^2 \to Q^2$, $\theta(x, y) = (y, f(x, y))$, for all $(x, y) \in Q^2$ [112].*

Theorem 10.67 is generalized by V. Izbash and P. Syrbu as follows:

Theorem 10.68. *If f is a k-ary operation ($k \geqslant 2$), then $f^{(i)} = f\theta^i$ for all $i \in \mathbb{N}$, where θ is the following mapping: $\theta : Q^k \to Q^k$, $\theta(x_1^k) = (x_2, \dots, x_k, f(x_1^k))$, for all $x_1^k \in Q^k$ [446].*

Lemma 10.69. *Let (Q, f) be a k-ary quasigroup, $k \geqslant 2$. Every $(k + 1)$ consecutive k-recursive derivatives $\{f^{(i)}, f^{(i+1)}, \dots, f^{(i+k)}\}$ of f are orthogonal [446].*

10.2.5 Gonsales-Couselo-Markov-Nechaev construction

We describe Gonsales-Couselo-Markov-Nechaev prolongation construction of quasigroups [371]. In fact, this is a formalized description of Bruck-Belousov quasigroup prolongation with more than one transversal from Subsection 1.7.5.

Below, any transversal of a quasigroup $(Q, \circ)$ consists of triplets of the form $(a; b; a \circ b)$, where a denotes the row and b denotes the column of a Cayley table of quasigroup $(Q, \circ)$, in which the element $a \circ b$ is situated.

Suppose that $(Q, \circ)$ is a finite quasigroup of order q, $\{T^{(1)}, \dots, T^{(m)}\}$ is the m-element set of transversals of quasigroup $(Q, \circ)$, such that $T^{(i)} \cap T^{(j)} = \varnothing$ for $i \neq j$. It is clear that $m \leqslant q$.

For any $s \in \overline{1, m}$ we define the following set

$$A_s = \{(a, b) \in Q^2 : \exists c \in Q : (a, b, c) \in T^{(s)}\} = \{(a, T_{12}^{(s)}(a)) : a \in Q\}.$$

Here $T_{12}^{(s)}(a)$ denotes the second element of a unique triplet of the transversal $T^{(s)}$ such that its first element is equal to a;

$T_{23}^{(s)}(b)$ denotes the third element of a unique triplet of the transversal $T^{(s)}$ such that its second element is equal to b;

$T_{13}^{(s)}(a)$ denotes the third element of a unique triplet of the transversal $T^{(s)}$ such that its first element is equal to a.

Notice, from the fact that $T^{(i)} \cap T^{(j)} = \varnothing$, it follows that $A_s \cap A_t = \varnothing$. Let

$$A = \bigcup_{s=1}^{m} A_s.$$

Further take a quasigroup $(H, \bullet)$, $H = \{\xi_1, \dots, \xi_m\}$, of order m such that $Q \cap H = \varnothing$.

On the set $\Omega = Q \cup H$ define operation $*$ for all $a, b \in Q$, $s, t \in \overline{1, m}$, as follows.

$$\begin{aligned} a * b &= a \circ b, \quad (a, b) \notin A, \\ a * b &= \xi_s, \quad (a, b) \in A_s, \\ a * \xi_t &= T_{13}^{(t)}(a), \\ \xi_s * b &= T_{23}^{(s)}(b), \\ \xi_s * \xi_t &= \xi_s \bullet \xi_t. \end{aligned} \tag{10.9}$$

Theorem 10.70. *Groupoid $(\Omega, *)$ is a quasigroup.*

Example 10.71. In the capacity of quasigroup $(Q, \circ)$ we shall use quasigroup from Example 1.362 and its two transversals $T^{(1)}$–"italic" main diagonal transversal and $T^{(2)}$–"bold" transversal transversal. Write these transversals as the sets of ordered triplets. We have $T^{(1)} = \{(1;1;1), (2;2;3), (3;3;2)\}$, $T^{(2)} = \{(1;3;3), (2;1;2), (3;2;1)\}$.

$\circ$	1	2	3
1	*1*	2	**3**
2	**2**	*3*	1
3	3	**1**	*2*

Here $m = 2$ and we construct the sets A_1, A_2, A.

$$A_1 = \{(1;1), (2;2), (3;3)\},$$
$$A_2 = \{(1;3), (2;1), (3;2)\},$$
$$A = \{(1;1), (2;2), (3;3), (1;3), (2;1), (3;2)\}.$$

In the capacity of quasigroup $(H, \bullet)$ we take the cyclic group of order two Z_2 defined on the elements 4 and 5, namely:

$\bullet$	4	5
4	4	5
5	5	4

We put $\xi_1 = 4$, $\xi_2 = 5$.

On the set $\Omega = \{1, 2, 3, 4, 5\}$ we construct quasigroup $(\Omega, *)$ using formulas (10.9). After applying equalities (1), (2), and (5) from (10.9) we obtain the following table:

$*$	1	2	3	4	5
1	4	2	5	□	□
2	5	4	1	□	□
3	3	5	4	□	□
4	□	□	□	4	5
5	□	□	□	5	4

Further, using equalities (3) and (4) from (10.9), we have; $1 * 4 = 1 * \xi_1 = T^{(1)}_{13}(1) = 1$; $1 * 5 = 1 * \xi_2 = T^{(2)}_{13}(1) = 3$; $2 * 4 = 2 * \xi_1 = T^{(1)}_{13}(2) = 1$; $4 * 1 = \xi_1 * 1 = T^{(1)}_{23}(1) = 1$; $4 * 2 = \xi_1 * 2 = T^{(1)}_{23}(2) = 3$, and so on. The following table is our final result.

$*$	1	2	3	4	5
1	4	2	5	*1*	**3**
2	5	4	1	*3*	**2**
3	3	5	4	*2*	**1**
4	*1*	*3*	*2*	4	5
5	**2**	**1**	**3**	5	4

Quasigroup which is constructed in Example 10.71 differs from the quasigroup that is constructed in Example 1.372 only in the bottom right 2×2 square which can be replaced by the square

$$\begin{matrix} 5 & 4 \\ 4 & 5. \end{matrix}$$

Reference [371] gives conditions when quasigroup $(\Omega, *)$ is differentiable, and conditions when $(\Omega, *)$ is differentiable and $(Q, \circ)$ is also a differentiable T-quasigroup.

The last conditions we give in the following theorem in an adapted form.

Theorem 10.72. *Let $(Q,+)$ be a finite commutative group of order q, $\varphi, \psi, \alpha \in Aut(Q,+)$, and $C = \{b_1, \dots, b_m\} \subseteq Q$, $(H, \bullet)$ be a differentiable quasigroup of order m, $m \leqslant q$.*

Suppose that the following conditions are true:

$$\begin{aligned}
&\varphi + \psi^2, \varphi\alpha^{-1} - \alpha\psi, \psi + \alpha^2 \in Aut(Q,+);\\
&\psi\alpha = \varphi - \alpha^2;\\
&\varphi\alpha^{-1}C \cap C = (\psi + \alpha^2)C \cap (\varphi\alpha^{-1} - \alpha\psi)C = \varnothing;\\
&(\varphi\alpha^{-1} - \alpha\psi)C \cup (\psi + \alpha^2)C = (\varphi + \psi^2)(C \cup \alpha\varphi^{-1}C).
\end{aligned} \tag{10.10}$$

Then quasigroup $(Q, \circ)$ with the form $x \circ y = \varphi x + \psi y$ for all $x, y \in Q$ is a differentiable quasigroup, the family of sets

$$T^{(s)} = \{(x, y, \varphi x + \psi y) \mid x, y \in Q, \alpha x + y = b_s\}, s = \overline{1,m},$$

is a family of disjoint transversals, and quasigroup $(\Omega, *)$ is differentiable.

Corollary 10.73. *If g.c.d.$(l, 6) = 1$, then there exists differentiable quasigroup of order $7 \cdot l + 3$ [371].*

Proof. In conditions of Theorem 10.72 it is sufficient to take $(Q,+) \cong Z_7 \oplus Z_l$, $\varphi = (2,2)$, $\psi = (1,1)$, $\alpha = (-2,1)$, $m = 3$, $C = \{(1,0),(2,0),(4,0)\}$, $(H, \bullet) \cong Z_3$. □

In [371] it is proved that finite binary differentiable quasigroups $(Q, \cdot)$ exist for all values $|Q| = q$ such that $q \neq 2, 6, 14, 18, 26, 42$.

Therefore for these values of q there exist recursive MDS codes with codewords of length n such that $n \geqslant 4$ (see Definition 10.62, Corollary 10.65).

10.2.6 Orthogonal quasigroups of order ten

Example 10.74. [371]. Using Corollary 10.73 it is possible to construct a differentiable quasigroup of order ten, i.e., using this construction it is possible to construct a pair of orthogonal quasigroups (Latin squares) of order ten.

We start from quasigroup $(Z_7, \circ)$, with the form $x \circ y = 2 \cdot x + y \pmod 7$ for all $x, y \in Z_7$, where $(Z_7, +)$ is the cyclic group of order seven and $(H, \bullet) \cong (Z_3, +)$, $\alpha = -2 = 5$, $\alpha^{-1} = 3$, $m = 3$, $C = \{1, 2, 4\}$. Transversal $T^{(1)}$ with $b_1 = 1$ is written using the "normal" font, transversal $T^{(2)}$ with $b_2 = 2$ is written using "italic" font, and transversal $T^{(3)}$ with $b_3 = 4$ is written using "bold" font. See also Section 1.7.5.

0	1	2	**3**	4	*5*	6	□	□	□
2	3	4	5	6	**0**	1	□	□	□
4	5	*6*	0	1	2	3	□	□	□
6	0	**1**	2	*3*	4	5	□	□	□
1	2	3	4	**5**	6	*0*	□	□	□
3	*4*	5	6	0	1	**2**	□	□	□
5	**6**	0	*1*	2	3	4	□	□	□
□	□	□	□	□	□	□	7	8	9
□	□	□	□	□	□	□	8	9	7
□	□	□	□	□	□	□	9	7	8

In empty boxes of transversal $T^{(1)}$ we write number 7, in empty boxes of transversal $T^{(2)}$ we write number 8, and in empty boxes of transversal $T^{(3)}$ we write number 9. Below the right-hand quasigroup $(Q, \circ)$ is obtained from the left-hand quasigroup $(Q, \cdot)$ using the following formula $x \circ y = y \cdot (x \cdot y)$. See Lemma 10.63.

$\cdot$	0	1	2	3	4	5	6	7	8	9
0	0	1	2	9	4	8	7	6	5	3
1	8	7	4	5	6	9	1	3	2	0
2	9	5	8	7	1	2	3	0	6	4
3	6	0	9	2	8	7	5	4	3	1
4	7	2	3	4	9	6	8	1	0	5
5	3	8	7	6	0	1	9	5	4	2
6	5	9	0	8	7	3	4	2	1	6
7	1	3	5	0	2	4	6	7	8	9
8	2	4	6	1	3	5	0	8	9	7
9	4	6	1	3	5	0	2	9	7	8

$\circ$	0	1	2	3	4	5	6	7	8	9
0	0	7	8	1	9	4	2	6	5	3
1	5	3	1	7	8	2	9	0	6	4
2	3	9	6	4	2	7	8	1	0	5
3	7	8	4	9	0	5	3	2	1	6
4	6	4	7	8	5	9	1	3	2	0
5	9	2	0	5	7	8	6	4	3	1
6	8	0	9	3	1	6	7	5	4	2
7	1	5	2	6	3	0	4	7	9	8
8	2	6	3	0	4	1	5	8	7	9
9	4	1	5	2	6	3	0	9	8	7

From the results of Section 10.2.5 it follows that the above Latin squares (quasigroups) of order ten are orthogonal.

00	17	28	91	49	84	72	66	55	33
85	73	41	57	68	92	19	30	26	04
93	59	86	74	12	27	38	01	60	45
67	08	94	29	80	75	53	42	31	16
76	24	37	48	95	69	81	13	02	50
39	82	70	65	07	18	96	54	43	21
58	90	09	83	71	36	47	25	14	62
11	35	52	06	23	40	64	77	89	98
22	46	63	10	34	51	05	88	97	79
44	61	15	32	56	03	20	99	78	87

10.2.7 Additional information

In [1], "The maximal possible length of a 3-dimensional linear recursive MDS-code and a 2-dimensional idempotent linear recursive MDS-code (i.e., a code containing all vectors of equal coordinates) are determined in the case when the size of the alphabet is a prime power." S. Dodunekov, MR1783081 (2001m:94052).

In [742] Latin squares of order n made up of elements taken from the cyclic group $(\mathbb{Z}_n, +)$ are considered. It is proved that Latin square generates a linear code if and only if the quasigroup $(\mathbb{Z}_n, \cdot)$ corresponding to this Latin square is medial. A method of constructing a maximal family of mutually orthogonal Latin squares that form linear codes is obtained. It is shown that no pair of mutually orthogonal Latin squares of even order generates a linear code.

Further development of the theory of MDS-codes is given in [703]. Loop codes are studied in [372]. From the abstract: "We give a complete description (with the use of computation) of the best parameters of linear codes that correspond to the left ideals in the loop algebras F_qL for $q \in \{2, 3, 4, 5\}$ and $|L| \leqslant 7$, and also in the group algebras F_qG for groups G of order $|G| \leqslant 12$."

10.3 On signs of Bol loop translations

Signs of left, right, middle translations and signs of the product of all left (right, middle) translations of a finite Bol loop are studied in this section. The results of this section are published in [773] and [599]. By construction of some codes information on signs of code words can be usable. Information on signs of group translations and group elements is used in [227].

By the sign function we mean a homomorphism of the symmetric group S_n onto the group Z_2 of order 2, $Z_2 = \{1, -1\}$.

If $\alpha \in S_n$ is a product of an even number of cycles of length 2 (2-cycles), then $\operatorname{sgn}\alpha = 1$. If α is a product of an odd number of 2-cycles, then $\operatorname{sgn}\alpha = -1$.

We shall mention some properties of the sign function. Let $\alpha, \beta, \gamma \in S_n$. Then $\operatorname{sgn}(\alpha\beta) = \operatorname{sgn}\alpha \cdot \operatorname{sgn}\beta = \operatorname{sgn}\beta \cdot \operatorname{sgn}\alpha = \operatorname{sgn}(\beta\alpha)$, $\operatorname{sgn}(\alpha(\beta\gamma)) = \operatorname{sgn}((\alpha\beta)\gamma)$, because associative and commutative identities hold in the group Z_2. Let $(Q, \cdot)$ be a finite quasigroup of order n.

We use the known notions

$$\operatorname{sgn}\mathbf{L} = \prod_{i=1}^{n} \operatorname{sgn}(L_{a_i}), \quad \operatorname{sgn}\mathbf{R} = \prod_{i=1}^{n} \operatorname{sgn}(R_{a_i}), \quad \operatorname{sgn}\mathbf{P} = \prod_{i=1}^{n} \operatorname{sgn}(P_{a_i}),$$

where $a_i \in Q$; moreover let's define $\operatorname{tsgn} Q = \langle \operatorname{sgn}\mathbf{L}, \operatorname{sgn}\mathbf{R}, \operatorname{sgn}\mathbf{P}\rangle$.

A loop $(Q, \cdot)$ with identity $x(y \cdot xz) = (xy \cdot x)z$ is called a Moufang loop; a loop with identity $x(y \cdot xz) = (x \cdot yx)z$ is called a left Bol loop.

We shall consider only left Bol loops and shall call them Bol loops omitting the word "left" for short. A Moufang loop is a diassociative loop, i.e., every two elements generate a subgroup; a Bol loop is a monoassociative loop, i.e., every element generates a subgroup [72, 685].

Let Q be a finite Bol loop, $a \in Q$. A natural number n is called the order of element a if $a^n = 1$ and it is the minimum integer with such a property. We denote by ε the identity permutation on the set Q.

Lemma 10.75. *Let $a \in Q$, where Q is a Bol loop. Then $a^n = 1$ if and only if ${L_a}^n = \varepsilon$.*

Proof. Taking into account the following property of a Bol loop: $a^n x = a^{n-1}(ax)$ for all $a, x \in Q, n \in Z$ [2], we have:

$$x = L_a^n x = \overbrace{a \cdot (a \cdot (a \cdot \dots \cdot (a}^{n} \cdot x) \dots)) = a^n x \ .$$

□

We recall some properties of Bol loop translations.

Proposition 10.76. *If α is a left translation of a finite Bol loop, then α is a product of disjoint cycles of equal length. The order of a left translation of a finite Bol loop divides the order of the loop [182].*

In general this proposition is not true for right translations of a Bol loop. But Proposition 10.76 and Lemma 10.75 are true for all left and right translations in every Moufang loop, because left and right $((yx \cdot z)x = y(xz \cdot x))$ Bol identities hold in every Moufang loop [72, 685].

Let Q be a Moufang loop, and P be the set of left and right translations of loop Q. By Proposition 10.76, if $\alpha, \beta \in P$ and α, β have the same order, then $\operatorname{sgn}\alpha = \operatorname{sgn}\beta$.

Corollary 10.77. *If Q is a Moufang loop, then* $\operatorname{sgn} L_a = \operatorname{sgn} R_a$ *for every* $a \in Q$, $\operatorname{sgn} \mathbf{L} = \operatorname{sgn} \mathbf{R}$.

We recall that a loop $(Q, \cdot)$ is called a LIP-loop if the equality $x^{-1} \cdot (x \cdot y) = y$ holds for all $x, y \in Q$, where $Ix = x^{-1}$ is a permutation on the set Q [72, 685].

Every middle translation of an LIP-loop can be represented in the form $P_a = R_a I$. Indeed, from the definition of middle translation and the definition of LIP-loop we have $x \cdot P_a x = a$, $P_a x = x^{-1} a$, $P_a x = R_a x^{-1} = R_a I x$. Therefore $P_a = R_a I$. We recall that $I = I_1, I^2 = \varepsilon$, and that every Bol loop is a LIP-loop [72, 685].

Lemma 10.78. *If Q is a Bol loop of order n, then* $\operatorname{sgn} \mathbf{P} = (\operatorname{sgn} I)^n \operatorname{sgn} \mathbf{R}$.

Proof. We may use properties of the sign function.

$$\operatorname{sgn} \mathbf{P} = \prod_{i=1}^{n} \operatorname{sgn}(P_{a_i}) = \prod_{i=1}^{n} \operatorname{sgn}(R_{a_i} I) = (\operatorname{sgn} I)^n \prod_{i=1}^{n} \operatorname{sgn}(R_{a_i}) = (\operatorname{sgn} I)^n \operatorname{sgn} \mathbf{R}.$$

□

Remark 10.79. Lemma 10.78 holds for *LIP*-loops.

Corollary 10.80. *If Q is a Bol loop of even order, then* $\operatorname{sgn} \mathbf{P} = \operatorname{sgn} \mathbf{R}$.

Proof. In this case $(\operatorname{sgn} I)^n = (\operatorname{sgn} I)^{2k} = ((\operatorname{sgn} I)^2)^k = 1$. □

Lemma 10.81. *If Q is a Bol loop, then* $\operatorname{sgn} \mathbf{R} = \operatorname{sgn} R_p$, *where p is a product of all elements of loop Q in any fixed order at any fixed bracket arrangement.*

Proof. From the left Bol identity $x(y \cdot xz) = (x \cdot yx)z$ and the definition of translations we have $L_x R_{xz} y = R_z L_x R_x y$.

Then for all x, z from Q we obtain the translation equality $L_x R_{xz} = R_z L_x R_x$. Using properties of the sign function we get $\operatorname{sgn}(L_x R_{xz}) = \operatorname{sgn}(R_z L_x R_x)$, $\operatorname{sgn} R_{xz} = \operatorname{sgn} L_x \operatorname{sgn} R_z \operatorname{sgn} L_x \operatorname{sgn} R_x$, $\operatorname{sgn} R_{xz} = \operatorname{sgn} R_x \operatorname{sgn} R_z$.

If we apply the last relation to the equality $\operatorname{sgn} \mathbf{R} = \prod_{i=1}^{n} \operatorname{sgn}(R_{a_i})$ the necessary number of times, then we receive that $\operatorname{sgn} \mathbf{R} = \operatorname{sgn} R_p$ for some p. The lemma is proved. □

We note that the element p can be equal to the product of all elements of order 2. If

$$p = (\dots(((a_1 a_1^{-1}) a_2) a_2^{-1}) \cdot \dots) \cdot a_s) a_s^{-1}) b_1) b_2) \dots b_r),$$

where $a_i (i = 1, 2, \dots, s)$ are elements of order $m > 2$ and $b_i (i = 1, 2, \dots, r)$ are elements of order 2, then the element p is a product of all elements of order 2.

Lemma 10.82. *If Q is a Bol loop, then* $\operatorname{sgn} \mathbf{L}$ *and* $\operatorname{sgn} \mathbf{R}$ *are determined by the elements of order two in Q.*

Proof. The number of elements of order $m, m > 2$, is even in a finite Bol loop, because we can separate them in pairs $(a, a^{-1}s)$, where $a \neq a^{-1}$. Hence, the sign of a product of all left translations of order m $(m > 2)$ is equal to 1.

It is sufficient to show that $\operatorname{sgn} \mathbf{R}$ is determined by right translations of elements of order two in Q too. From Lemma 10.81 the equality $\operatorname{sgn} R_{xz} = \operatorname{sgn} R_x \operatorname{sgn} R_z$ follows. If $z = x^{-1}$, then $\operatorname{sgn} R_x \operatorname{sgn} R_{x^{-1}} = \operatorname{sgn} R_1 = \operatorname{sgn} \varepsilon = 1$ for all elements x of order greater than 2. Hence the sign of a product of all right translations of elements of order more than 2 is equal to 1. □

Corollary 10.83. *If Q is a Bol loop of odd order, then* $\operatorname{sgn} L_x = 1$ *for all* $x \in Q$, $\operatorname{sgn} \mathbf{L} = \operatorname{sgn} \mathbf{R} = 1$.

Proof. Let $| Q |= 2n + 1$. Then by Proposition 10.76 every Bol loop has no elements of order 2. Using Lemma 10.82 in this case we get $\operatorname{sgn} \mathbf{L} = \operatorname{sgn} \mathbf{R} = 1$. □

Corollary 10.84. *a) If there exists an element x of a finite Bol loop such that $x^2 = 1$, then translations L_x, R_x generate a dihedral group;*

b) For any element x of a finite Bol loop the following equality is true: $\operatorname{sgn} R_x = \operatorname{sgn} R_{x^{-1}}$.

Proof. a) In the translation equality $L_x R_{xz} = R_z L_x R_x$ let us assume $z = x, x^2 = 1$. Then we receive the equality $L_x = R_x L_x R_x$. If $x^2 = 1$, then by Lemma 10.75 it follows that ${L^2}_x = 1$. Therefore $1 = L_x R_x L_x R_x$, $(L_x R_x)^2 = 1$.

b) From the last equality of Lemma 10.82 it follows that $\operatorname{sgn} R_x \operatorname{sgn} R_{x^{-1}} = 1$. Then $\operatorname{sgn} R_{x^{-1}} = (\operatorname{sgn} R_x)^{-1} = \operatorname{sgn} R_x$. □

Proposition 10.85. *Cycle types of translations R_x and $R_{x^{-1}}$ coincide for any element x of a finite Bol loop.*

Proof. From the translation equality $L_x R_{xz} = R_z L_x R_x$ (Lemma 10.81) we have $L_x R_{xz} R_x^{-1} L_x^{-1} = R_z$. If $z = x^{-1}$, then we obtain the relation $L_x R_x^{-1} L_x^{-1} = R_{x^{-1}}$. Then cycle types of the translations R_x and $R_{x^{-1}}$ are equal, because cycle types of the translations R_x^{-1} and R_x, R_x^{-1} and $R_{x^{-1}}$ coincide. □

Lemma 10.86. *If there exists an element a of a finite Bol loop of order n such that* $\operatorname{sgn} R_a = -1$, *then there exist $n/2$ elements of the loop such that* $\operatorname{sgn} R_x = -1$.

Proof. Let $\operatorname{sgn} R_a = -1$ for some $a \in Q$. If $a = x \cdot z$, then from the translations equality $\operatorname{sgn} R_a = \operatorname{sgn} R_{xz} = \operatorname{sgn} R_x \operatorname{sgn} R_z$ we obtain either $\operatorname{sgn} R_x = -1$ and $\operatorname{sgn} R_z = 1$, or $\operatorname{sgn} R_x = 1$ and $\operatorname{sgn} R_z = -1$. From the definition of a middle translation we have $x \cdot I_a x = a$, i.e., $x \cdot z = a$. Hence there exist n pairs $(x; z)$ of elements of loop Q such that $a = x \cdot z$ in a Bol loop of order n.

If $\operatorname{sgn} R_a = -1$ and $a = x \cdot z$, then $x \neq z$. Indeed, if $x = z$, then $R_x = R_z$, $\operatorname{sgn} R_x = \operatorname{sgn} R_z$, $\operatorname{sgn} R_a = \operatorname{sgn} R_{xz} = \operatorname{sgn} R_x \operatorname{sgn} R_z = 1$.

Let us assume that in a Bol loop of order n there exists an element a such that $\operatorname{sgn} R_a = -1$. Let $M^+ = \{x \in Q \mid \operatorname{sgn} R_x = 1\}$, $| M^+ |= m$, $M^- = \{y \in Q \mid y = I_a x, \forall x \in M^+\}$. We can notice that $\operatorname{sgn} R_y = -1, \forall y \in M^-$. Taking into account that I_a is a permutation of the finite set Q, we obtain $| M^+ |=| M^- |= m$. It is easy to see that $M^+ \cap M^- = \varnothing$, $M^+ \cup M^- = Q$. Hence $m + m = n$, $m = n/2$.

Let $M(Q) = \langle L_x, R_x \mid \forall x \in Q \rangle$ be a group generated by all left and right translations of a loop Q, i.e., a multiplication group of a loop Q. □

Corollary 10.87. *Let Q be a Bol loop of odd order n. Then*

a) $\operatorname{sgn} R_x = 1$ *for all* $x \in Q$*;*

b) If $\alpha \in M(Q)$, *then* $\operatorname{sgn} \alpha = 1$.

Proof. a) By Lemma 10.86 the number of elements $x \in Q$ with $\operatorname{sgn} R_x = -1$ is equal to $n/2$ and must be an integer. This is impossible. Hence elements of this finite Bol loop such that $\operatorname{sgn} R_x = -1$ for any $x \in Q$ don't exist.

b) By Corollary 10.87 a) $\operatorname{sgn} R_x = 1$ for all $x \in Q$ and by Corollary 10.83 $\operatorname{sgn} L_a = 1$ for all $a \in Q$. □

Lemma 10.88. *Let Q be a finite Bol loop of odd order. If* $| Q |= 4k + 1$, *then* $\operatorname{sgn} P_x = 1$ *for all* $x \in Q$, $\operatorname{tsgn} Q = \langle 1, 1, 1 \rangle$*; if* $| Q |= 4k + 3$, *then* $\operatorname{sgn} P_x = -1$ *for all* $x \in Q$, $\operatorname{tsgn} Q = \langle 1, 1, -1 \rangle$.

Proof. Let $| Q |= 2n + 1$, where $n \in N$. Then by Proposition 10.76 every left Bol loop translation (i.e., every row of a Bol loop Latin square) is a permutation of odd order. By Proposition 10.76 the cycles in disjoint decomposition of left translation have the same odd order. Therefore $\operatorname{sgn} L_a = 1$ for all $a \in Q$. Hence in this case we have $\operatorname{sgn} \mathbf{L} = \operatorname{sgn} \mathbf{R} = 1$. We used Corollary 10.83.

The cycles in a disjoint decomposition of the permutation I have the length 1 or 2. If $| Q |= 4k + 1$, then the Bol loop Q has no elements of even order (see Lemma 10.75 and Proposition 10.76).

Therefore the permutation I has exactly 1 cycle of length 1, in which the identity element of the loop Q lies. Thus the permutation I has $4k : 2 = 2k$ 2-cycles. Hence $\operatorname{sgn} I = 1$.

Taking into account that $P_a = R_a I$ and properties of the sign function, we have $\operatorname{sgn} P_a = \operatorname{sgn}(R_a I) = \operatorname{sgn} R_a \cdot \operatorname{sgn} I = \operatorname{sgn} R_a = 1$. We used Corollary 10.87.

Therefore, if $| Q |= 4k + 1$, then $\operatorname{sgn} \mathbf{P} = 1, \operatorname{tsgn} Q = \langle 1, 1, 1 \rangle$.

If $| Q |= 4k + 3$, then the Bol loop Q also has no elements of even order. Hence the number of 2-cycles in the permutation I is $(4k+2) : 2 = 2k+1$. Therefore $\operatorname{sgn} I = -1$. Thus $\operatorname{sgn} P_a = \operatorname{sgn}(R_a I) = sgn(R_a) \cdot (-1) = -sgn(R_a) = -1$ for all $a \in Q$, $\operatorname{sgn} \mathbf{P} = (-1)^{4k+3} = -1$. Hence, if $| Q |= 4k + 3$, then $\operatorname{tsgn} Q = \langle 1, 1, -1 \rangle$. □

Lemma 10.89. *Let Q be a finite Bol loop of even order. If $| Q |= 4k+2$, then $\operatorname{sgn} \mathbf{L} = -1$; if $| Q |= 4k$, then $\operatorname{tsgn} Q = \langle 1, 1, 1 \rangle$.*

Proof. Let $| Q |= 4k + 2$. By Lemma 10.82 $\operatorname{sgn} \mathbf{L}$ and $\operatorname{sgn} \mathbf{R}$ are determined by the set of elements of order two.

Let t be the number of elements of order 2 and $2d$ be the number of elements of order m greater than 2. Then $t = 4k + 2 - 2d - 1 = 2(2k - d) + 1$. Every element of order 2 of loop Q ($| Q |= 4k + 2$) has $(4k + 2) : 2 = 2k + 1$ 2-cycles in its disjoint decomposition. Hence, if $a^2 = 1$, then $\operatorname{sgn} L_a = -1$ in Bol loop of order $4k + 2$.

Therefore $\operatorname{sgn} \mathbf{L} = (-1)^{2(2k-d)+1} = -1$. Using Corollary 10.80 we get $\operatorname{sgn} \mathbf{P} = \operatorname{sgn} \mathbf{R}$.

Let $| Q |= 4k$. Then $\operatorname{sgn} \alpha = (-1)^{4k/2} = (-1)^{2k} = 1$ for every left translation α of order 2 of a Bol loop. Then $\operatorname{sgn} \mathbf{L} = 1$.

Let us assume that in a Bol loop of order 4k there exists an element a such that $\operatorname{sgn} R_a = -1$. Therefore by Lemma 10.86 we have $2k$ right translations of the Bol loop with $\operatorname{sgn} R_a = -1$ and $2k$ right translations with $\operatorname{sgn} R_a = 1$. Then $\operatorname{sgn} \mathbf{R} = 1$ and by Corollary 10.80 we have $\operatorname{sgn} \mathbf{P} = 1$.

Obviously, if $\operatorname{sgn} R_x = 1$ for all $a \in Q$, then $\operatorname{sgn} \mathbf{R} = \operatorname{sgn} \mathbf{I} = 1$. The lemma is proved. □

Corollary 10.90. *Let Q be a Bol loop of order $4k$. Then there exists an element x of the loop Q such that $x^2 = 1$, $\operatorname{sgn} R_x = 1$.*

Proof. There is an odd number of elements of order two in a Bol loop of even order. By Lemma 10.89, $\operatorname{sgn} \mathbf{R} = 1$ for a Bol loop of order $4k$. The sign of a product of right translations of all elements of order more than two is equal to 1, because we can separate them in pairs (x, x^{-1}) and by Corollary 10.83 $\operatorname{sgn} R_x = \operatorname{sgn} R_{x^{-1}}$.

Hence the sign of a product of right translations of all elements of order two is also equal to 1. Indeed, if we assume that for every element x of order two $\operatorname{sgn} R_x = -1$, then we obtain a contradiction with the condition $\operatorname{sgn} \mathbf{R} = 1$. □

As we can notice, there exist elements $x \in Q$ of a Bol loop of order 8 such that $\operatorname{sgn} R_x = -1$ [182].

Analogously as for finite groups [598], for finite Moufang loops the following theorem is true.

Theorem 10.91. *Let Q be a finite Moufang loop. If* $\mid Q \mid= 4k$, *then* $\operatorname{tsgn} Q = \langle 1, 1, 1\rangle$; *if* $\mid Q \mid= 4k+1$, *then* $\operatorname{tsgn} Q = \langle 1, 1, 1\rangle$; *if* $\mid Q \mid= 4k+2$, *then* $\operatorname{tsgn} Q = \langle -1, -1, -1\rangle$; *if* $\mid Q \mid= 4k+3$, *then* $\operatorname{tsgn} Q = \langle 1, 1, -1\rangle$ *[773].*

Proof. If $\mid Q \mid= 4k+1$, then $\operatorname{tsgn} Q = \langle 1, 1, 1\rangle$; if $\mid Q \mid= 4k+3$, then $\operatorname{tsgn} Q = \langle 1, 1, -1\rangle$. It follows from Lemma 6, because every Moufang loop is a Bol loop.

Let $\mid Q \mid= 4k+2$. Using Lemma 10.89 we have $\operatorname{sgn} \mathbf{L} = -1$. From Corollary 10.77 it follows that $\operatorname{sgn} \mathbf{L} = \operatorname{sgn} \mathbf{R}$ and from Corollary 10.80 it follows that $\operatorname{sgn} \mathbf{P} = \operatorname{sgn} \mathbf{R}$. Therefore, if $\mid Q \mid= 4k+2$, then $\operatorname{tsgn} Q = \langle -1, -1, -1\rangle$.

Analogously, if $\mid Q \mid= 4k$, then $\operatorname{sgn} \mathbf{L} = \operatorname{sgn} \mathbf{R} = \operatorname{sgn} \mathbf{P}$; this follows from Corollaries 10.77 and 10.80. Using Lemma 10.89, we obtain $\operatorname{tsgn} Q = \langle 1, 1, 1\rangle$. The theorem is proved. □

Chapter 11

Quasigroups in cryptology

11.1	Introduction	472
	11.1.1 Quasigroups in "classical" cryptology	473
11.2	Quasigroup-based stream ciphers	474
	11.2.1 Introduction	474
	11.2.2 Modifications and generalizations	475
	11.2.3 Further development	476
	11.2.4 Some applications	478
	11.2.5 Additional modifications of Algorithm 1.69	478
	11.2.6 n-Ary analogs of binary algorithms	479
	11.2.7 Further development of Algorithm 11.10	481
11.3	Cryptanalysis of some stream ciphers	482
	11.3.1 Chosen ciphertext attack	482
	11.3.2 Chosen plaintext attack	482
11.4	Combined algorithms	483
	11.4.1 Ciphers based on the systems of orthogonal n-ary operation	483
	11.4.2 Modifications of Algorithm 11.14	483
	11.4.3 Stream cipher based on orthogonal system of quasigroups	485
	11.4.4 T-quasigroup-based stream cipher	485
	11.4.5 Generalization of functions of Algorithm 11.16	487
	11.4.6 On quasigroup-based cryptcode	488
	11.4.6.1 Code part	488
	11.4.6.2 Cryptographical part	489
	11.4.6.3 Decoding	490
	11.4.6.4 Resistance	490
	11.4.6.5 A code-crypt algorithm	491
	11.4.7 Comparison of the power of the proposed algorithms	491
11.5	One-way and hash functions	492
	11.5.1 One-way function	492
	11.5.2 Hash function	493
11.6	Secret-sharing schemes	494
	11.6.1 Critical sets	494
	11.6.2 Youden squares	495
	11.6.3 Reed-Solomon codes	496
	11.6.4 Orthogonality and secret-sharing schemes	496
11.7	Some algebraic systems in cryptology	497
	11.7.1 Inverse quasigroups in cryptology	498
	11.7.2 Some groups in cryptology	498
	11.7.2.1 El Gamal cryptosystem	499
	11.7.2.2 De-symmetrization of Algorithm 1.69	499
	11.7.2.3 RSA and GM cryptosystems	500

11.7.2.4 Homomorphic encryption 501
11.7.2.5 MOR cryptosystem 501
11.7.3 El Gamal signature scheme 502
11.7.4 Polynomially complete quasigroups in cryptology 503
11.7.5 Cryptosystems which are based on row-Latin squares .. 504
11.7.6 Non-binary pseudo-random sequences over Galois fields 505
11.7.7 Authentication of a message 505
11.7.8 Zero-knowledge protocol 506
11.7.9 Hamming distance between quasigroups 507
11.7.10 Generation of quasigroups for cryptographical needs ... 507

The theory of quasigroup applications in cryptology has gone through a period of rapid growth. Therefore any review of results in the given area of research quite quickly becomes outdated. Here we give a re-written and supplemented form of earlier versions [779, 781, 790] of such reviews. See also [359, 949, 607, 395, 624].

11.1 Introduction

Almost all results obtained in the domain of quasigroup applications in cryptology and coding theory through the end of the 1980s are described in [240, 241, 243]. In the present survey the main attention is devoted to the later articles in this direction.

Basic facts on quasigroup theory can be found in [72, 80, 77, 685, 559, 779]. Some information is in the previous chapters. Information on basic facts in cryptology can be found in many books; see, for example, [29, 148, 632, 633].

Cryptology is a science that consists of two parts: cryptography and cryptanalysis. Cryptography is a science of methods of transformation (ciphering) of information to protect it from an unlawful user. Cryptography can also be defined as the science of transmitting and managing information in the presence of an adversary [532]. Cryptanalysis is a science of methods and ways of breaking down the ciphers [263].

In some sense cryptography is a *defense*, i.e., this is a science of construction of new ciphers, but cryptanalysis is an *attack*, i.e., this is a science and some kind of "art," a set of methods for the breaking the ciphers. This situation is similar to the situation with intelligence and counterintelligence.

These two sciences (cryptography and cryptanalysis) are very close and there does not exist a good cryptographer who does not know methods of cryptanalysis.

It is clear that cryptology depends on the level of development of the society, science and technology.

We recall that a cipher is a way (a method, an algorithm) of information transformation with the purpose of its defense. A key is some hidden part (usually, a little one) or parameter of a cipher.

Steganography is a set of means and methods of hiding the fact of sending (or passing) information, for example, a communication or a letter. Now there exist methods of hiddenness of the fact of information sending by usual post, by e-mail and so on.

In this survey a science of defense of information from accidental errors caused by transformation and sending (passing) this information will be called Coding Theory (Code Theory).

When sending the important and confidential information, as it seems to us, there exists a sense to use methods of Code Theory, Cryptology, and Steganography all together [536].

In cryptology one often uses Kerkhoff's (1835–1903) rule: an opponent (an unlawful user) knows all ciphering procedures (sometimes a part of plaintext or ciphertext) with the exception of the key.

Many authors of books devoted to cryptology divide this science (sometimes not paying attention to this fact) in two parts: before the article of Diffie and Hellman [250] (the so-called cryptology with a non-public (symmetric) key) and after this work (a cryptology with a public or non-symmetric key). Practically, Diffie and Hellman opened a new era in cryptology. Moreover, it is possible to apply these new approaches in practice.

Especially fast development of the second part of cryptology is connected with very fast development of personal computers and nets of personal computers, and other electronic technical devices in the end of the 20th century. Many new mathematical and cryptographical problems appeared in this direction and some of them are not solved. Solving these problems has great importance for practice.

Almost all known constructions of error detecting and error correcting codes, cryptographic algorithms and enciphering systems have made use of associative algebraic structures such as groups and fields; see, for example, [584, 231].

We can use such non-associative structures as quasigroups and neo-fields in almost all branches of coding theory, and especially in cryptology.

Often the codes and ciphers based on non-associative systems show better possibilities than the known codes and ciphers based on associative systems [243, 534]. Many results of non-associative public-key cryptography are reflected in [470].

Notice that in recent years, quantum code theory and quantum cryptology [805, 302, 951, 162, 682] have been developed intensively. Quantum cryptology also uses theoretical achievements of "usual" cryptology [138].

Efficacy of applications of quasigroups in cryptology is based on the fact that quasigroups are "generalized permutations" of some kind and the number of quasigroups of order n is larger than $n! \cdot (n-1)! \cdot \ldots \cdot 2! \cdot 1!$ [240]. What does the use of quasigroups in cryptography give us? It gives the same permutations and substitutions but easily generated and not requiring a lot of device memory, acting "locally" only on one block of a plaintext.

It is worth noting that several of the early professional cryptographers, in particular, A.A. Albert, A. Drisko, M.M. Glukhov, J.B. Rosser, E. Schönhardt, C.I. Mendelson, and R. Schaufler were connected with the development of Quasigroup Theory. The main known users of quasigroups in cryptology were (and are) J. Denes and A.D. Keedwell [237, 240, 241, 243, 238].

Of course, one of the most effective cipher methods is to use an unknown, non-standard or very rare language. Probably the best enciphering method was (and is) to have a good agent.

11.1.1 Quasigroups in "classical" cryptology

There exist two main elementary methods when ciphering information:

(i). Symbols in plaintext (or in its piece (its bit)) are permuted by some law. The first known cipher of this kind is Scital (Sparta, 2500 years ago).

(ii). All symbols in a fixed alphabet are changed by a law on other letters of this alphabet. One of the first ciphers of this kind is Caesar's cipher ($x \to x+3$ for any letter of the Latin alphabet, for example $a \to d, b \to e$ and so on).

In many contemporary ciphers (DES, Russian GOST, Blowfish [632, 255]), methods (i) and (ii) are used with some modifications.

The Trithemius cipher makes use of a 26×26 square array containing 26 letters of the alphabet (assuming that the language is English) arranged in a Latin square. Different rows of this square array are used for enciphering various letters of the plaintext in a manner prescribed by the keyword or key-phrase [29, 469]. Since a Latin square is the multiplication table of a quasigroup, this may be regarded as the earliest use of a non-associative algebraic structure in cryptology. There exists a possibility to develop this direction using the quasigroup approach, in particular, using orthogonal systems of binary or n-ary quasigroups.

R. Schaufler in his Ph.D. dissertation discussed the minimum amount of plaintext and the corresponding ciphertext which would be required to break the Vigenere cipher (a modification of Trithemius cipher) [747]. That is, he considered the minimum number of entries of a partial Latin square which would determine the square completely.

Recently this problem has re-arisen as the problem of determining the so-called critical sets in Latin squares, see [476, 256, 257, 262, 260, 480]. See, also, articles devoted to Latin trades, for example, [48].

More recent enciphering systems which may be regarded as extensions of Vigenere's idea are mechanical machines such as Jefferson's wheel and the M-209 Converter (used by the U.S. Army until the early 1950s) and the electronically produced stream ciphers of the present day [533, 632].

During the second World War, R. Shauffler, while working for the German Cryptography service, developed a method of error detection based on the use of generalized identities (as they were later called by V.D. Belousov) in which the check digits are calculated by means of an associative system of quasigroups (see also [227]). He pointed out that the resulting message would be more difficult to decode by an unauthorized receiver than in the case when a single associative operation is used for calculation [748].

Therefore it is possible to assume that information on systems of quasigroups with generalized identities (see, for example, works of Yu. Movsisyan [639]) may be applied in cryptography of the present day.

A direct application of orthomorphisms to cryptography is described in [628, 627].

11.2 Quasigroup-based stream ciphers

11.2.1 Introduction

"Stream ciphers are an important class of encryption algorithms. They encrypt individual characters (usually binary digits) of a plaintext message one at a time, using an encryption transformation which varies with time.

By contrast, block ciphers tend to simultaneously encrypt groups of characters of a plaintext message using a fixed encryption transformation. Stream ciphers are generally faster than block ciphers in hardware, and have less complex hardware circuitry.

They are also more appropriate, and in some cases mandatory (e.g., in some telecommunications applications), when buffering is limited or when characters must be individually processed as they are received. Because they have limited or no error propagation, stream ciphers may also be advantageous in situations where transmission errors are highly probable" [621].

Often for ciphering a block (a letter) B_i of a plaintext, the previous ciphered block C_{i-1} is used. Notice that Horst Feistel was one of the first who proposed this method of encryption (Feistel net) [316].

In [555] quasigroup concepts (isotopy, quasigroup operation) are used in construction of a block encryption standard. In [533] (see also [534, 535]) C. Koscielny has shown how quasigroups/neofields-based stream ciphers may be produced which are both more efficient and more secure than those based on groups/fields.

In [668, 606] Smile Markovsky and coauthors proposed to use quasigroups for secure encoding.

A quasigroup $(Q, \cdot)$ and its (23)-parastrophe $(Q, \backslash)$ satisfy the following identities $x\backslash(x \cdot y) = y$, $x \cdot (x\backslash y) = y$. The authors (S. Markovsky and his coauthors) propose to use this property of the quasigroups to construct a stream cipher (Algorithm 1.69).

Probably (Algorithm 1.69) and its generalizations are now the most known and the most used quasigroup-based stream ciphers.

Below we concentrate on various generalizations of Algorithm 1.69.

11.2.2 Modifications and generalizations

The improvements and researches of Algorithm 1.69 were carried out intensively. Some information on this process is given in [779]. We thank our colleagues A. Krapez, V. Bakeva, V. Dimitrova and A. Popovska-Mitrovikj for the following new information.

Remark 11.1. In article [30], the authors find the distribution of k-tuples of letters after n applications of quasigroup transformation ($k > n$) (i.e., Algorithm 1.69) and give an algorithm for statistical attack in order to discover the original message. Also, they give some conclusions on how to protect the original messages.

In work [542], Krapez defines parastrophic quasigroup transformation. In [31], the authors propose a modification of this transformation and give a new classification of quasigroups of order 4. Finally, in [17] the authors presented this transformation and gave the relationship between the new classification and the symmetries of quasigroups.

Notice that parastrophic transformations from [542, 251] are promising for further applications and research.

We recall the classical equational definition (Definition 1.54) of a binary quasigroup.

Definition 11.2. A binary groupoid (Q, A) is called a binary quasigroup if on the set Q there exist operations ${}^{(13)}A$ and ${}^{(23)}A$ such that in the algebra $(Q, A, {}^{(13)}A, {}^{(23)}A)$ the following identities are fulfilled:

$$A({}^{(13)}A(x, y), y) = x, \tag{11.1}$$

$${}^{(13)}A(A(x, y), y) = x, \tag{11.2}$$

$$A(x, {}^{(23)}A(x, y)) = y, \tag{11.3}$$

$${}^{(23)}A(x, A(x, y)) = y. \tag{11.4}$$

By tradition the operation A is denoted by $\cdot$, ${}^{(23)}A$ by $\backslash$ and ${}^{(13)}A$ by $/$.

In Algorithm 1.69 it is also possible to use a quasigroup (Q, A) and its (13)-, (123)-, and (132)-parastrophe since quasigroup (Q, A) and these parastrophes fulfill the following identities, namely, identities (11.2), (11.5), and (11.6), respectively [786, 542, 251].

$${}^{(123)}A(A(x, y), x) = y \tag{11.5}$$

$${}^{(132)}A(y, A(x, y)) = x \tag{11.6}$$

More details in this direction are given in [542].

In [602], the authors claimed that this cipher is resistant to the brute force attack (exhaustive search) and to the statistical attack (in many languages some letters are used more frequently than others). Later, similar results were presented in [668].

In the dissertation of Milan Vojvoda [901] it is proved that this cipher is not resistant to chosen ciphertext attack and chosen plaintext attack. It is claimed that this cipher is not resistant to a special kind of statistical attack (Slovak language) [901].

There exist a few other ways to generalize Algorithm 1.69. The most obvious way is to increase arity of a quasigroup, i.e., instead of binary, apply n-ary ($n \geqslant 3$) quasigroups. This way was proposed in [779, 781] and was realized in [684, 683]. Notice that Prof. A. Petrescu writes that he found this n-ary generalization independently.

Some modifications to make Algorithm 1.69 more resistant against known attacks can be found in [542, 251]. One of these attempts, taking into consideration Vojvoda's results [901], was proposed in [790]. Namely, additionally a system of n n-ary orthogonal operations (groupoids) was proposed to use.

11.2.3 Further development

Further development of Algorithm 1.69 is presented in [354].

Definition 11.3. Let r be a positive integer. Let $(Q, *)$ be a quasigroup and $a_j, b_j \in Q$. For each fixed $m \in Q$, define first the transformation $Q_m : Q^r \longrightarrow Q^r$ by

$$Q_m(a_0, a_1, \ldots, a_{r-1}) = (b_0, b_1, \ldots, b_{r-1}) \Longleftrightarrow$$

$$b_i = \begin{cases} m * a_0; & i = 0 \\ b_{i-1} * a_i; & 1 \leqslant i \leqslant (r-1). \end{cases}$$

Then define $\mathcal{R}_1$ as a composition of transformations of kind Q_m, for suitable choices of the indexes m, as follows

$$\mathcal{R}_1(a_0, a_1, \ldots, a_{r-1}) = Q_{a_0}(Q_{a_1} \ldots (Q_{a_{r-1}}(a_0, a_1, \ldots, a_{r-1}))).$$

Definition 11.4. [354] (Shapeless quasigroup) A quasigroup $(Q, *)$ of order n is said to be shapeless if it is non-commutative and non-associative, it does not have a left or right unit, it does not contain proper subquasigroups, and there is no $k < 2n$ for which the identities of the following kinds are satisfied:

$$\underbrace{x * (x \ldots x * (x(x}_{k} * y)) = y; \; y = ((y * \underbrace{x) * \ldots) * x) * x}_{k}. \qquad (11.7)$$

Remark 11.5. Condition $k < 2n$ for the identities (11.7) means that any left and right translation of quasigroup $(Q, *)$ should have the order $k \geqslant (2n + 1)$.

In [354] it is proposed to construct shapeless quasigroups using the transversal approach [409]. Shapeless quasigroups are also constructed in [625]. Simple quasigroups without subquasigroups and with an identity automorphism group are studied in [551, 495, 443, 769]. Such quasigroups are functionally complete [551]. Notice that simple infinite groupoids are constructed in [561, 562].

A block cipher based on Algorithm 1.69 is proposed in [353]. Let $(Q, *)$ be a quasigroup of finite order 2^d. Using the operation $*$, the authors define the following vector-valued Boolean function (v.v.b.f.) $a * b = c \Leftrightarrow *_{vv}(x_1, x_2, ..., x_d, y_1, y_2, ..., y_d) = (z_1, z_2, ..., z_d)$, where $x_1 ... x_d, y_1 ... y_d, z_1 ... z_d$ are binary representations of a, b, c respectively.

Each element z_i depends on the bits $x_1, x_2, ..., x_d, y_1, y_2, ..., y_d$ and is uniquely determined by them. So, each z_i can be seen as a $2d$-ary Boolean function $z_i =$

$f_i(x_1, x_2, ..., x_d, y_1, y_2, ..., y_d)$, where $f_i : \{0,1\}^{2d} \to \{0,1\}$ strictly depends on, and is uniquely determined by $*$.

The authors state that for every quasigroup $(Q, *)$ of order 2^d and for each bijection $Q \to \{0, 1..., 2^d - 1\}$ there are uniquely determined v.v.b.f. $*_{vv}$ and d uniquely determined $2d$-ary Boolean functions $f_1, f_2, ..., f_d$ such that for each $a, b, c \in Q$

$$a * b = c \Leftrightarrow *_{vv}(x_1, ..., x_d, y_1, ..., y_d) = (f_1(x_1, ..., x_d, y_1, ..., y_d), ..., f_d(x_1, ..., x_d, y_1, ..., y_d)).$$

Each k-ary Boolean function $f(x_1, ..., x_k)$ can be represented in a unique way by its algebraic normal form (ANF), i.e., as a sum of products

$$ANF(f) = \alpha_0 + \sum_{i=1}^{k} \alpha_i x_i + \sum_{1 \leqslant i \leqslant j \leqslant k}^{k} \alpha_{i,j} x_i x_j + \sum_{1 \leqslant i \leqslant j \leqslant s \leqslant k}^{k} \alpha_{i,j,s} x_i x_j x_s + ...,$$

where the coefficients $\alpha_0, \alpha_i, \alpha_{i,j}, ...$ are in the set $\{0, 1\}$ and the addition and multiplication are in the field $GF(2)$.

The ANFs of the functions f_i give information about the complexity of the quasigroup (Q, .) via the degrees of the Boolean functions f_i. The degrees of the polynomials $ANF(f_i)$ rise with the order of the quasigroup. In general, for a randomly generated quasigroup of order 2^d, $d \geqslant 4$, the degrees are higher than 2.

Definition 11.6. A quasigroup $(Q, *)$ of order 2^d is called a Multivariate Quadratic Quasigroup (MQQ) of type $Quad_{d-k}Lin_k$ if exactly $d - k$ of the polynomials f_i are of degree 2 (i.e., are quadratic) and k of them are of degree 1 (i.e., are linear), where $0 \leqslant k < d$ [353].

Authors prove the following:

Theorem 11.7. *Let $A1 = [f_{ij}]$ and $A2 = [g_{ij}]$ be two $d \times d$ matrices of linear Boolean expressions, and let $b_1 = [u_i]$ and $b_2 = [v_i]$ be two $d \times 1$ vectors of linear or quadratic Boolean expressions. Let the functions f_{ij} and u_i depend only on variables $x_1, ..., x_d$, and let the functions g_{ij} and v_i depend only on variables $x_{d+1}, ..., x_{2d}$. If $Det(A_1) = Det(A_2) = 1$ in $GF(2)$ and if*

$$A_1 \cdot (x_{d+1}, ..., x_{2d})^T + b_1 \equiv A_2 \cdot (x_1, ..., x_d)^T + b_2,$$

then the vector-valued operation $*_{vv}(x_1, ..., x_{2d}) = A_1 \cdot (x_{d+1}, ..., x_{2d})^T + b_1$ defines a quasigroup $(Q, *)$ of order 2^d that is MQQ [353].

The authors researched the existence of MQQs of order 8, 16 and 32.

Problem 11.1. Finding of MQQs of orders 2^d, $d \geqslant 6$, the authors consider as an open research problem.

The authors show that the proposed cipher is resistant relative to the chosen plaintext attack, attacks with differential cryptanalysis, XL attacks, Gröbner basis attacks and some other kinds of attacks.

Algebraic cryptanalysis of a MQQ public key cryptosystem is given in [629]: " ... we present an efficient attack of the multivariate Quadratic Quasigroups (MQQ) cryptosystem. Our cryptanalysis breaks MQQ cryptosystems by solving systems of multivariate quadratic polynomial equations using a modified version of the MutantXL algorithm."

11.2.4 Some applications

In [668] (see also [606]) it is proposed to use Algorithm 1.69 for secure encoding of a file system. A survey of security mechanisms in mobile communication systems is in [899].

SMS (Short Message Service) messages are sometimes used for the interchange of confidential data such as social security number, bank account number, password, etc. A typing error in selecting a number when sending such a message can have severe consequences if the message is readable to any receiver.

Most mobile operators encrypt all mobile communication data, including SMS messages. But sometimes, when encrypted, the data is readable for the operator.

Among others, these needs give rise for the necessity to develop additional encryption for SMS messages, so that only accredited parties are able to be engaged in a communication. In [416] an approach to this problem using Algorithm 1.69 is described. In [417] differential cryptanalysis of the quasigroup cipher is given. Definition of the encryption method is presented.

In [606] the authors introduce a stream cipher with an almost public key, based on quasigroups for defining suitable encryption and decryption. They consider the security of this method. It is shown that the key (quasigroups) can be public and still has sufficient security. A software implementation is also given.

In [537] a public-key cryptosystem that uses generalized quasigroup-based stream ciphers is presented. It is shown that such a cryptosystem allows one to transmit securely both a cryptogram and a secret portion of the enciphering key using the same insecure channel. The system is illustrated by means of a simple, but nontrivial, example.

11.2.5 Additional modifications of Algorithm 1.69

Sometimes only the use of other records of a mathematical fact leads to a generalization. For example, we can re-write Algorithm 1.69 using the concept of translation in the following way:

Algorithm 11.8. Let Q be a non-empty finite alphabet. Define a quasigroup $(Q, \cdot)$. It is clear that the quasigroup $(Q, \overset{(23)}{\cdot})$ is defined in a unique way.

Take a fixed element l $(l \in Q)$, which is called a leader.
Let $u_1 u_2 ... u_k$ be a k-tuple of letters from Q.
It is proposed the following ciphering procedure
$v_1 = l \cdot u_1 = L_l u_1$,
$v_2 = v_1 \cdot u_2 = L_{v_1} u_2$.
$v_i = v_{i-1} \cdot u_i = L_{v_{i-1}} u_i$, $i = 3, ..., k$.
Therefore we obtain the following ciphertext $v_1 v_2 \dots v_k$.

The deciphering algorithm is constructed in the following way. We have the following ciphertext: $v_1 v_2 \dots v_k$. Recall that $L_a^{\overset{(23)}{\cdot}} = (L_a^{\cdot})^{-1}$ for any $a \in Q$ [779]. Below we shall denote translation $L_a^{\overset{(23)}{\cdot}}$ as L_a^*, translation $L_a^{\cdot}$ as L_a for any $a \in Q$. Then

$$\begin{aligned}
&u_1 = l \overset{(23)}{\cdot} v_1 = L_l^* (v_1) = L_l^* (L_l u_1) = \\
&L_l^{-1} (L_l u_1) = u_1; \\
&u_i = v_{i-1} \overset{(23)}{\cdot} v_i = L_{v_{i-1}}^* (v_i) = L_{v_{i-1}}^* \left(L_{v_{i-1}} u_i\right) = \\
&L_{v_{i-1}}^{-1} \left(L_{v_{i-1}} u_i\right) = u_i
\end{aligned} \tag{11.8}$$

for all $i \in \overline{2, k}$.

From this form of Algorithm 1.69 we can obtain the following generalization. Instead of translations L_x, $x \in Q$, we propose to use powers of these translations in the enciphering part of this algorithm, i.e., to use permutations of the form L_x^k, $k \in \mathbb{Z}$, instead of permutations of the form L_x.

The proposed modification also forces us to use permutations of the form L_x^k, $k \in \mathbb{Z}$, also in the decryption procedure.

Algorithm 11.9. Let Q be a non-empty finite alphabet. Define a quasigroup $(Q, \cdot)$. It is clear that the quasigroup $(Q, \overset{(23)}{\cdot})$ is defined in a unique way. Take a fixed element l ($l \in Q$), which is called a leader.

Let $u_1 u_2 ... u_k$ be a k-tuple of letters from Q.

The following ciphering procedure is proposed:

$$\begin{aligned} v_1 &= L_l^{c_1} u_1, c_1 \in \mathbb{Z}, \\ v_2 &= L_{v_1}^{c_2} u_2, c_2 \in \mathbb{Z}, \\ &\dots, \\ v_i &= L_{v_{i-1}}^{c_i} u_i, c_i \in \mathbb{Z}, i \in \overline{3,k}. \end{aligned} \tag{11.9}$$

Therefore we obtain the following ciphertext $v_1 v_2 \dots v_k$.

The elements c_i, $i \in \overline{1,k}$, in the equalities (11.9) should vary from step to step in order to protect this algorithm against chosen plaintext and ciphertext attack. It is clear that the rule of generating the numbers c_i must be known to the receiver.

For example, it is possible to use the following rule: $c_1 = |u_1|/2$, where $|u_1|$ is the left (right, middle) order of the element u_1, $c_2 = |u_2|/2$, $c_3 = |u_3|/2$ and so on.

The deciphering algorithm is constructed in the following way. We use notations of Algorithm 11.8. Recall that $(L_x^*)^a = L_x^{-a}$ for all $x \in Q$. Then

$$\begin{aligned} (L_l^*)^{c_1} (v_1) &= (L_l^*)^{c_1} (L_l^{c_1} u_1) = u_1, \\ (L_{v_1}^*)^{c_2} (v_2) &= (L_{v_1}^*)^{c_2} (L_{v_1}^{c_2} u_2) = u_2, \\ (L_{v_{i-1}}^*)^{c_i} (v_i) &= (L_{v_{i-1}}^*)^{c_i} (L_{v_{i-1}}^{c_i} u_i) = u_i, i \in \overline{3,k}. \end{aligned} \tag{11.10}$$

It is clear that the right and middle [779] translations are also possible to use in Algorithm 11.9 instead of the left translations. See below.

11.2.6 n-Ary analogs of binary algorithms

We construct an n-ary analog of Algorithm 1.69. See also [683].

Algorithm 11.10. Let Q be a non-empty finite alphabet and k be a natural number, $u_i, v_i \in Q$, $i \in \{1, ..., k\}$. Define an n-ary quasigroup (Q, f). It is clear that any quasigroup $(Q, {}^{(i,n+1)}f)$ for any fixed value i is defined in a unique way. Below for simplicity we put $i = n$.

Take the fixed elements $l_1^{(n-1)(n-1)}$ ($l_i \in Q$), which are called leaders.

Let $u_1 u_2 ... u_k$ be a k-tuple of letters from Q.

The following ciphering (encryption) procedure is proposed:

$$\begin{aligned}
v_1 &= f(l_1^{n-1}, u_1),\\
v_2 &= f(l_n^{2n-2}, u_2),\\
&\dots,\\
v_{n-1} &= f(l_{n^2-3n+3}^{(n-1)(n-1)}, u_{n-1}),\\
v_n &= f(v_1^{n-1}, u_n),\\
v_{n+1} &= f(v_2^{n}, u_{n+1}),\\
v_{n+2} &= f(v_3^{n+1}, u_{n+2}),\\
&\dots
\end{aligned} \tag{11.11}$$

Therefore we obtain the following ciphertext: $v_1 v_2 \dots, v_{n-1}, v_n, v_{n+1}, \dots$.

The deciphering algorithm is constructed similarly with binary case:

$$\begin{aligned}
u_1 &= {}^{(n,n+1)}f(l_1^{n-1}, v_1),\\
u_2 &= {}^{(n,n+1)}f(l_n^{2n-2}, v_2),\\
&\dots,\\
u_{n-1} &= {}^{(n,n+1)}f(l_{n^2-3n+3}^{(n-1)(n-1)}, v_{n-1})\\
u_n &= {}^{(n,n+1)}f(v_1^{n-1}, v_n),\\
u_{n+1} &= {}^{(n,n+1)}f(v_2^{n}, v_{n+1}),\\
u_{n+2} &= {}^{(n,n+1)}f(v_3^{n+1}, v_{n+2}),\\
&\dots
\end{aligned} \tag{11.12}$$

Indeed, for example, ${}^{(n,n+1)}f(v_1^{n-1}, v_n) = {}^{(n,n+1)}f(v_1^{n-1}, f(v_1^{n-1}, u_n)) \overset{(7.2)}{=} u_n$.

Below we shall denote this encryption algorithm as $G(u)$, because at any step only one element of a plaintext is enciphered. Probably it makes sense to use an irreducible 3-ary or 4-ary finite quasigroup [77, 157, 8, 9] in Algorithm 11.10.

Remark 11.11. In the encryption procedure (equalities (11.11)) and, therefore, in the decryption procedure (equalities (11.12)) it is possible to use more than one n-quasigroup operation.

A translation of n-ary quasigroup (Q, f) $(n > 2)$ will be denoted as $T(a_1, \dots, a_{i-1}, -, a_{i+1}, \dots, a_n)$, where $a_i \in Q$ for all $i \in \overline{1,n}$ and

$$T(a_1, \dots, a_{i-1}, -, a_{i+1}, \dots, a_n)x = f(a_1, \dots, a_{i-1}, x, a_{i+1}, \dots, a_n)$$

for all $x \in Q$.

From the definition of an n-ary quasigroup it follows that any translation of n-ary quasigroup (Q, f) is a permutation of the set Q.

Lemma 11.12. *If* ${}_fT(a_1, \dots, a_{n-1}, -)$ *is a translation of a quasigroup* (Q, f)*, then*

$${}_fT^{-1}(a_1, \dots, a_{n-1}, -) = {}_{(n,n+1)f}T(a_1, \dots, a_{n-1}, -).$$

Proof. In the proof we omit the symbol f in the notation of translations of quasigroup

(Q, f). We have

$$\begin{aligned} &T^{-1}(a_1, \dots, a_{n-1}, -)(T(a_1, \dots, a_{n-1}, -)x) = \\ &T^{-1}(a_1, \dots, a_{n-1}, -)f(a_1, \dots, a_{n-1}, x) = \\ &{}^{(n,n+1)}f(a_1, \dots, a_{n-1}, f(a_1, \dots, a_{n-1}, x)) \overset{(7.2)}{=} x. \end{aligned} \tag{11.13}$$

□

We propose an n-ary analogue of Algorithm 11.9.

Algorithm 11.13. Let Q be a non-empty finite alphabet. Define an n-ary quasigroup (Q, f). It is clear that the quasigroup $(Q, {}^{(n,n+1)}f)$ is defined in a unique way.

Take the fixed elements $l_1^{(n-1)(n-1)}$ ($l_i \in Q$), which are called leaders.
Let $u_1 u_2 ... u_k$ be a k-tuple of letters from Q.
The following ciphering (encryption) procedure is proposed:

$$\begin{aligned} &v_1 = T^a(l_1, l_2, \dots, l_{n-1}, u_1), \\ &v_2 = T^b(l_n, l_{n+1}, \dots, l_{2n-2}, u_2), \\ &\dots, \\ &v_{n-1} = T^c(l_{n^2-3n+3}, \dots, l_{(n-1)(n-1)}, u_{n-1}), \\ &v_n = T^d(v_1, \dots, v_{n-1}, u_n), \\ &v_{n+1} = T^e(v_2, \dots, v_n, u_{n+1}), \\ &v_{n+2} = T^t(v_3, \dots, v_{n+1}, u_{n+2}), \\ &\dots \end{aligned} \tag{11.14}$$

Therefore we obtain the following ciphertext $v_1 v_2 \dots v_k$.

Taking into consideration Lemma 11.12 we can say that the deciphering algorithm can be constructed similar to the deciphering one in Algorithm 11.9.

11.2.7 Further development of Algorithm 11.10

In the article [194] the following generalizations of Algorithm 11.10 are proposed:

(i) It is proposed to use composition of binary quasigroups in order to construct some n-ary ($n \geqslant 2$) quasigroups. For example, if (Q, A) and (Q, B) are some binary quasigroups, then $C(x, y, z) = A(x, B(y, z))$ is a ternary quasigroup.

(ii) It is proposed to use isotopy of n-ary quasigroups by generation of a key quasigroup (in Algorithm 11.10, quasigroup (Q, f) is a key quasigroup).

In [194] it is written: "In this paper we have proposed a modified 3-quasigroup based stream cipher. The scheme is designed to increase the key complexity exponentially so that it may be used for present day practical applications. The first part of the key randomly generates the initial quasigroup. It selects different parameters for generation of the seed 3-quasigroup and in a way acts as a different algorithm for encryption. Since the choice of 3-quasigroups is large even for small order and it also increases exponentially, we include this structure and use it to customize the algorithm based on the key used for encryption. In this paper, we have considered only reducible 3-quasigroups. Extension of this scheme for selecting suitable randomly generated 3-quasigroup based on the key is in progress. This would help to improve the cryptographic strength of our scheme."

11.3 Cryptanalysis of some stream ciphers

Information for this section is taken from the paper [224].

11.3.1 Chosen ciphertext attack

We describe the chosen ciphertext attack on a cipher defined in Algorithm 11.10. A binary analog of this attack is described in [901]. Let $Q = \{q_1, q_2, \dots, q_k\}$, $|Q| = q$ and assume the cryptanalyst has access to the decryption device loaded with an unknown key. Then one can construct the following ciphertext:

$v_1 = q_1, v_2 = q_1, \dots, v_{n-1} = q_1, v_n = q_1, v_{n+1} = q_2$. Then $u_n = {}^{(n,n+1)}A(q_1, \dots, q_1)$. Since $v_{n+1} = q_2$, then $u_{n+1} = {}^{(n,n+1)}A(q_1, \dots, q_1, q_2)$. If $v_{n+2} = q_3$, then $u_{n+2} = {}^{(n,n+1)}A(q_1, \dots, q_1, q_2, q_3)$ and so on. Continuing in such manner we can find a multiplication table of quasigroup $(Q, {}^{(n,n+1)}A)$, and therefore multiplication table of quasigroup (Q, A) too. Notice, $|(Q, A)| = q^n$.

Having the multiplication table of quasigroup $(Q, {}^{(n,n+1)}A)$, we can easy encipher any ciphertext starting from the symbol v_n.

In order to decrypt the elements $v_1, \dots, v_{n-1}$ we should know the action of $(n-1)$-tuples of the leader elements on any element of the set Q. In other words we should know the action of translation $T_1(l_1, l_2, \dots, l_{n-1}, -)$, $T_2(l_n, l_{n+1}, \dots, l_{2n-2}, -)$, $\dots$, T_{n-1} on the set Q.

It is not difficult to find the element-leader using quasigroup $(Q, {}^{(n,n+1)}A)$ in the binary case. It is sufficient to solve the equation ${}^{(23)}A(l, a) = b$ for the fixed elements $a, b \in Q$.

For simplicity now we are using the ternary case. Notice, for cryptographical purposes it is not necessary to find element-leaders l_1, l_2. It is sufficient to find a pair of elements c, d such that ${}^{(34)}A(l_1, l_2, x) = {}^{(34)}A(c, d, x)$ for all $x \in Q$. For these aims there exists a possibility to decrypt q letters of a ciphertext $q_1, q_2, \dots, q_n$ in any order.

In order to establish the action of element-leaders l_3, l_4 on the set Q (action of translation $T(l_3, l_4, -)$) we can decrypt q pairs of elements of the form (a, q_1), $(a, q_2, \dots, (a, q_q)$, where a is a fixed element of the set Q.

We can unite calculation of the action of translations $T(l_1, l_2, -)$ and $T(l_3, l_4, -)$ in one procedure using by decryption q pairs of elements (q_i, q_j), where $q_i \neq q_j$, $\cup_{i=1}^{q} q_i = Q$, $\cup_{j=1}^{q} q_j = Q$.

In a similar way it is possible to operate in n-ary case ($n \geqslant 4$).

11.3.2 Chosen plaintext attack

The chosen plaintext attack is similar to the chosen ciphertext attack.

Let us assume the cryptanalyst has access to the encryption device loaded with an unknown key. Then one can construct the following plaintexts:

$u_1 = q_1, u_2 = q_1, \dots, u_{n-1} = q_1, u_n = q_1, u_{n+1} = q_2$. Then $v_n = A(q_1, \dots, q_1)$. If $u_{n+1} = q_2$, then $v_{n+1} = A(q_1, \dots, q_1, q_2)$. If $u_{n+2} = q_3$, then $v_{n+2} = A(q_1, \dots, q_1, q_2, q_3)$ and so on. Continuing it in such manner we can find the multiplication table of quasigroup (Q, A). Notice $|(Q, A)| = q^n$.

The chosen plaintext attack on leader elements is similar to the chosen ciphertext attack on leader elements and we omit it.

11.4 Combined algorithms

11.4.1 Ciphers based on the systems of orthogonal n-ary operation

Here we propose to use a system of orthogonal n-ary groupoids as an additional procedure in order to construct an almost-stream cipher [790].

Orthogonal systems of n-ary quasigroups were studied in [309, 854, 867, 294, 852]. Such systems have more uniform distribution of elements of the base set and therefore such systems may be more preferable in protection against statistical cryptanalytic attacks.

Algorithm 11.14. Let A be a non-empty finite alphabet, k be a natural number, and x_1^t be a plaintext. Take a system of n n-ary orthogonal operations (A, f_i), $i = 1, 2, \dots, n$. This system defines a permutation F of the set A^n. We propose the following enciphering procedure.

- Step 1: $y_1^n = F^l(x_1^n)$, where $l \geqslant 1$, l is a natural number, values of l vary from one enciphering round to another. If $t < n$, then we can add some "neutral" symbols to plaintext.
- At the Steps $\geqslant 2$ it is possible to use the Feistel schema [316, 621]. For example, we can do the following enciphering procedure $z_1^n = F^s(y_2, y_3, \dots, y_n, x_{n+1})$, if arity $n \geqslant 2$, or $z_1^n = F^s(y_3, y_4, \dots, y_n, x_{n+1}, x_{n+2})$, if $n \geqslant 3$, and so on.

The deciphering algorithm is based on the fact that an orthogonal system of n n-ary operations (1.96) has a unique solution for any tuple of elements $a_1, \dots, a_n$.

Algorithm 11.14 is sufficiently safe relative to the chosen ciphertext and plaintext attack since the key is a non-periodic sequence of applications of permutation F, i.e., sequence of powers of permutation F. Therefore any permutation of the group $\langle F \rangle$ can be used by ciphering information using Algorithm 11.14.

Recall that application of only one step of Algorithm 11.14 is not very safe since this procedure is not resistant relative to the chosen ciphertext attack and chosen plaintext attack.

11.4.2 Modifications of Algorithm 11.14

In our opinion some modifications of this algorithm are desirable. Following "vector ideas" [634] we propose as the first step to write any letter u_i of a plaintext as an n-tuple (n-vector) and after that to apply Procedure 11.14. For example it is possible to use a binary representation of characters of the alphabet A.

It is possible to divide plaintext $u_1, \dots, u_n$ in parts and to apply Algorithm 11.14 to some parts of plaintext, to a text, a part of which has been ciphered by Algorithm 11.14 at a previous ciphering round.

It is possible to change variables $x_1, \dots, x_k$ $(1 \leqslant k \leqslant (n-1))$ in Algorithm 11.14 by some fixed elements of the set Q and name these elements as leaders. Notice, if $k = n-1$, then we obtain n ciphering images from any plaintext letter u.

If in a system of orthogonal n-ary operations there is at least one n-ary quasigroup, then we can apply Algorithm 11.10 and Algorithm 11.14 together with some non-periodic frequency by ciphering of information, i.e., for example, we can apply Algorithm 11.10 four times and then apply Algorithm 11.14 five times, and so on.

It is possible to use for construction of this non-periodical sequence a part of a decimal representation of an irrational or transcendent number. In this case we can take as a key the sequence of applications of Algorithm 11.10 and Algorithm 11.14.

The proposed modifications make realization of the chosen plaintext attack and chosen ciphertext attack more complicated.

Taking into consideration that in the binary case one application of Algorithm 11.14 generates two cipher symbols, say v_1, v_2, from one plaintext symbol u, we may propose to apply Procedure 11.14 for two plaintext symbols (either to one cipher symbol and one plain symbol, or to two cipher symbols) simultaneously.

We propose to use Algorithm 11.10 and Algorithm 11.14 simultaneously.

Algorithm 11.15. Suppose that we have a plaintext x_1^t, $t \geqslant n$.

Divide plaintext into n-tuples.

We apply n-ary permutation $F^l(x_1^n) = y_1^n$ to any n-tuple of plaintext.

We apply Algorithm 11.10 (its binary or k-ary variant) to n-tuple y_1^n: $G(y_1^n) = z_1^n$. Probably it will be better if $k < n$.

We apply n-ary permutation F^s to n-tuple z_1^n: $F^s(z_1^n) = t_1^n$.

The deciphering algorithm is clear.

Below we denote by the symbol ${}_{g_1}T_{l_1}^a(u_1)$ the action of the left (right, middle) translation T_{l_1} of a binary quasigroup (Q, g_1) in the power a on the element u_1, and so on.

Algorithm 11.16. Enciphering. Initially we have plaintext $u_1, u_2, \dots, u_6$.

$$\begin{aligned}
&\textit{Step } 1.\\
&{}_{g_1}T_{l_1}^a(u_1) = v_1\\
&{}_{g_2}T_{l_2}^b(u_2) = v_2\\
&F_1^c(v_1, v_2) = (v_1', v_2')\\
&\textit{Step } 2.\\
&{}_{g_3}T_{v_1'}^d(u_3) = v_3\\
&{}_{g_4}T_{v_2'}^e(u_4) = v_4\\
&F_2^f(v_3, v_4) = (v_3', v_4')\\
&\textit{Step } 3.\\
&{}_{g_5}T_{v_3'}^g(u_5) = v_5\\
&{}_{g_6}T_{v_4'}^h(u_6) = v_6\\
&F_3^i(v_5, v_6) = (v_5', v_6')
\end{aligned} \tag{11.15}$$

And so on. We obtain ciphertext $v_1', v_2', \dots, v_6'$.

Deciphering. Initially we have ciphertext $v'_1, v'_2, \dots, v'_6$.

$$\begin{aligned}
&\textit{Step } 1.\\
&F_1^{-c}(v'_1, v'_2) = (v_1, v_2)\\
&{}_{g_1}T_{l_1}^{-a}(v_1) = u_1\\
&{}_{g_2}T_{l_2}^{-b}(v_2) = u_2\\
&\textit{Step } 2.\\
&F_2^{-f}(v'_3, v'_4) = (v_3, v_4)\\
&{}_{g_3}T_{v'_1}^{-d}(v_3) = u_3\\
&{}_{g_4}T_{v'_2}^{-e}(v_4) = u_4\\
&\textit{Step } 3.\\
&F_3^{-i}(v'_5, v'_6) = (v_5, v_6)\\
&{}_{g_5}T_{v'_3}^{-g}(v_5) = u_5\\
&{}_{g_6}T_{v'_4}^{-h}(v_6) = u_6
\end{aligned} \tag{11.16}$$

We obtain plaintext $u_1, u_2, \dots, u_6$.

It is clear that Algorithm 11.9 is a partial case of Algorithm 11.16.

As in Algorithm 11.9, in Algorithm 11.16 the elements $a, b, c, \dots, h$ should vary in order to protect this algorithm against chosen plaintext and chosen ciphertext attacks.

Algorithm 11.16 allows us to obtain an almost "natural" stream cipher, i.e., a stream cipher that encodes a pair of elements of a plaintext at any step. It is easy to see that Algorithm 11.16 can be generalized for an n-ary ($n \geqslant 3$) case. One of the possible generalizations is realized in Algorithm 11.22.

Additional research is necessary for the modifications proposed in this section.

11.4.3 Stream cipher based on orthogonal system of quasigroups

For construction of Algorithms 11.10 and 11.14 we propose the use of orthogonal systems of binary parastrophic quasigroups, in particular, orthogonal systems of binary parastrophic T-quasigroups.

For this purpose we can use Theorem 9.69. Notice, in order to construct quasigroups mentioned in Theorem 9.69 a computer search is probably preferable. It is possible to use GAP and Prover [615].

In order to construct a T-quasigroup (Q, A) that is orthogonal with its parastrophe in a more theoretical way it is possible to use Theorem 9.109 and the equalities (9.45).

Quasigroups from Corollary 9.112 are suitable objects to construct the above-mentioned algorithms (binary case).

11.4.4 T-quasigroup-based stream cipher

We give a numerical example of encryption Algorithm 11.16 based on T-quasigroups. Notice, the number 257 is prime.

Example 11.17. Take the cyclic group $(Z_{257}, +) = (A, +)$.

1. Define T-quasigroup $(A, *)$ with the form $x * y = 2 \cdot x + 131 \cdot y + 3$ with a leader element

l, say, $l = 17$. Denote the mapping $x \mapsto x * l$ by the letter g_1, i.e., $g_1(x) = x * l$ for all $x \in A$.

In order to find the mapping g_1^{-1} we find the form of operation $\overset{(13)}{*}$ using formula (9.45). We have $x \overset{(13)}{*} y = 129 \cdot x + 63 \cdot y + 127$, $f^{-1}x = x \overset{(13)}{*} l$. Then $g_1^{-1}(g_1(x)) = g_1^{-1}(x * l) = (x * l) \overset{(13)}{*} l \overset{(11.2)}{=} x$.

In some sense, quasigroup $(A, \overset{(13)}{*})$ is the "right inverse quasigroup" to quasigroup $(A, *)$. From the identity (11.6) it follows that quasigroup $(A, \overset{(132)}{*})$ is the "left inverse" quasigroup to quasigroup $(A, *)$.

Notice, from Corollary 9.112 it follows that $(A, *) \perp (A, \overset{(13)}{*})$.

2. Define T-quasigroup $(A, \circ)$ with the form $x \circ y = 10 \cdot x + 81 \cdot y + 53$ with a leader element l, say, $l = 71$. Denote the mapping $x \mapsto l * x$ by the letter g_2, i.e., $g_2(x) = l \circ x$ for all $x \in A$.

   In order to find the mapping g_2^{-1} we find the form of operation $\overset{(23)}{\circ}$ using formula (9.45). We have $x \overset{(23)}{\circ} y = 149 \cdot x + 165 \cdot y + 250$.

3. Define a system of two parastroph orthogonal T-quasigroups $(A, \cdot)$ and $(A, \overset{(23)}{\cdot})$ in the following way

$$\begin{cases} x \cdot y = 3 \cdot x + 5 \cdot y + 6 \\ x \overset{(23)}{\cdot} y = 205 \cdot x + 103 \cdot y + 153. \end{cases}$$

   Denote quasigroup system $(A, \cdot, \overset{(23)}{\cdot})$ by $F(x, y)$, since this system is a function of two variables.

   In order to find the mapping $F^{-1}(x, y)$ we solve the system of linear equations

$$\begin{cases} 3 \cdot x + 5 \cdot y + 6 = a \\ 205 \cdot x + 103 \cdot y + 153 = b. \end{cases}$$

   We have $\Delta = 55$, $1/\Delta = 243$, $x = 100 \cdot a + 70 \cdot b + 255$, $y = 43 \cdot a + 215 \cdot b$. Therefore we have, if $F(x, y) = (a, b)$, then $F^{-1}(a, b) = (100 \cdot a + 70 \cdot b + 255, 43 \cdot a + 215 \cdot b)$, i.e.,

$$\begin{cases} x = 100 \cdot a + 70 \cdot b + 255 \\ y = 43 \cdot a + 215 \cdot b. \end{cases}$$

We have defined the mappings g_1, g_2, F and now we can use them in Algorithm 11.16.

Let 212; 17; 65; 117 be a plaintext. We take the following values in formula (11.15): $a = b = d = e = f = 1; c = 2$. Below we use Gothic font to distinguish leader elements, i.e., the numbers 17 and 71 are leader elements. Then

Step 1.
$g_1(212) = 212 * 17 = 2 \cdot 212 + 131 \cdot 17 + 3 = 84$
$g_2(17) = 71 \circ 17 = 10 \cdot 71 + 81 \cdot 17 + 53 = 84$
$F(84; 84) = (3 \cdot 84 + 5 \cdot 84 + 6; 205 \cdot 84 + 103 \cdot 84 + 153) = (164; 68)$
$F(164; 68) = (3 \cdot 164 + 5 \cdot 68 + 6; 205 \cdot 164 + 103 \cdot 68 + 153) = \mathbf{(67; 171)}$

Step 2.

$g_1(65) = 65 * 67 = 2 \cdot 65 + 131 \cdot 67 + 3 = 172$
$g_2(117) = 171 \circ 117 = 10 \cdot 171 + 81 \cdot 117 + 53 = 189$
$F(172; 189) = (3 \cdot 172 + 5 \cdot 189 + 6; 205 \cdot 172 + 103 \cdot 189 + 153) = \mathbf{(182;\ 139)}$
We obtain the following ciphertext 67; 171; 182; 139.

For deciphering we use formula (11.16).

Step 1.
$F^{-1}(67; 171) = (100 \cdot 67 + 70 \cdot 171 + 255, 43 \cdot 67 + 215 \cdot 171) = (164; 68)$
$F^{-1}(164; 68) = (100 \cdot 164 + 70 \cdot 68 + 255, 43 \cdot 164 + 215 \cdot 68) = (84; 84)$
$g_1^{-1}(84) = 84 \overset{(13)}{*} 17 = 129 \cdot 84 + 63 \cdot 17 + 127 = \mathbf{212}$
$g_2^{-1}(84) = 71 \overset{(23)}{\circ} 84 = 149 \cdot 71 + 165 \cdot 84 + 250 = \mathbf{17}$

Step 2.
$F^{-1}(182; 139) = (100 \cdot 182 + 70 \cdot 139 + 255, 43 \cdot 182 + 215 \cdot 139) = (172; 189)$
$g_1^{-1}(172) = 172 \overset{(13)}{*} 67 = 129 \cdot 172 + 63 \cdot 67 + 127 = \mathbf{65}$
$g_2^{-1}(189) = 171 \overset{(23)}{\circ} 189 = 149 \cdot 171 + 165 \cdot 189 + 250 = \mathbf{117}$

Experiments demonstrate that encoding-decoding is executed sufficiently fast.

Remark 11.18. Proper binary groupoids are more preferable than linear quasigroups in construction of the mapping $F(x, y)$ in order to make encryption safer, but in this case decryption may be slower than in the linear quasigroup case since definition of such groupoids needs more computer (or some other device) memory. The same remark is true for the choice of the function g. Maybe a golden mean in this choice problem is to use linear quasigroups over non-abelian, especially simple, groups.

Remark 11.19. In this cipher there exists a possibility of protection against standard statistical attack. For this scope it is possible to denote the more-often used letters or pairs of letters by more than one integer or by more than one pair of integers.

11.4.5 Generalization of functions of Algorithm 11.16

We give a method for constructing the functions that it is possible to use in cryptographical procedures. Suppose that all functions are defined on a set Q. Functions $F(x_1^n)$ and $g(x_1^n)$ are functions of n variables.

Function F (n orthogonal binary groupoids that define a permutation of the set Q^n) has an inverse function of n variables $F^{-1}(x_1^n)$ such that $F(F^{-1}(x_1^n)) = F^{-1}(F(x_1^n)) = x_1^n$.

Recall that a function g of two variables (binary quasigroup) has four "local" inverse functions

$$\begin{aligned} g_1^{-1}(g(x, y), y) &= x \quad \text{(identity (11.2))} \\ g_2^{-1}(x, g(x, y)) &= y \quad \text{(identity (11.4))} \\ g_3^{-1}(g(x, y), x) &= y \quad \text{(identity (11.5))} \\ g_4^{-1}(y, g(x, y)) &= x \quad \text{(identity (11.6)).} \end{aligned}$$

Notice that if g is an n-ary quasigroup operation, then, in general, we cannot decode values x, y, for example, from equality $g(\overline{a}^{n-2}, x, y) = b$, but we can easily solve equation $g(\overline{a}^{n-1}, x) = b$ of one variable, i.e., we can decode value of variable x.

Taking into consideration this quasigroup feature, we describe the set (clone) of functions that it is possible to use in cryptology based on these two kinds of functions, namely, functions F and g. We shall use the concept of term [917] to define cryptographical terms (cryptographical functions) inductively.

The cryptographical function (cryptographical term) below in Case 3 means that encoding and decoding of a text using this function (this term) is performed uniquely.

Definition 11.20. 1. Any individual constant is a cryptographical term.

2. Any individual variable is a cryptographical term.

3. (a) If g is an n-ary quasigroup functional constant ((Q, g) is an n-ary quasigroup) and t is a term, b_1^n are individual constants, then $g^a(b_1^{i-1}, t, b_{i+1}^n)$, $i \in \overline{1,n}$, where $a \in \mathbb{Z}$, is a cryptographical term.

   (b) If F is a permutation of a set Q^n which is constructed using n orthogonal n-ary groupoids and $t_1, t_2, \dots, t_n$ are quasigroup cryptographical terms, then $F^a(t_1, \dots, t_n)$, where $a \in \mathbb{Z}$ is a cryptographical term.

Example 11.21. Let $Q = B \times B$ be a non-empty set, F be a pair of orthogonal groupoids each of which is defined on the set B, and (Q, g) be a ternary quasigroup. Then $g(q_1, q_2, F)$, where q_1, q_2 are fixed elements of Q, is a cryptographical term constructed following Rule 3 (a), Definition 11.20.

In Example 11.17 cryptographical term $F^a(g_1, g_2)$ is constructed following Rule 3, (b), Algorithm 11.20. Indeed, the function F is a pair of parastrophic orthogonal T-quasigroups that are defined on the set Z_{257}, i.e., F is a permutation of the set $Z_{257} \times Z_{257}$; (Z_{257}, g_1), (Z_{257}, g_2) are binary T-quasigroups; and $a = 1; 2$.

Algorithm 11.22. Suppose that we have an n-ary permutation F, n procedures G_j (they can be of various arity and it is supposed that the leader elements are used) and plaintext x_1^t.

By the letter y with an index we denote an element of enciphered text or a leader element. We propose the following enciphering procedure.

The i-th step of this procedure can have the following form

$$_iF^k(G_1(y_1^m, x_i), \dots, G_n(y_1^r, x_i)) = {}_iy_1^n. \tag{11.17}$$

The deciphering algorithm is executed "from the top to the bottom" in general and "from the bottom to the top" at any step. See more details in Algorithm 11.16.

11.4.6 On quasigroup-based cryptcode

Using possibilities that give us Algorithms 11.20 and 11.22 we construct an example of a quasigroup-based hybrid of a code and a cipher. The hybrid idea is sufficiently known see, for example, [631, 779, 781]. Following Markovski, Gligoroski, and Kocarev [605, 604], we call this hybrid a cryptcode.

11.4.6.1 Code part

Information on codes is in [153]. We shall use Example 10.45. Suppose that the symbols x, y are informational and the symbol z is a check symbol. Remember, $x, y, z \in (Z_2 \oplus Z_2)$. We propose the following check equation $x + \varphi_5 y + \varphi_6 z = 0$, i.e., we propose the formula to find the element

$$z = \varphi_5 x + \varphi_6 y. \tag{11.18}$$

As it is proved in Chapter 10 that the proposed code detects any single, transposition, and twin errors.

Suppose we have a word of the form ab, $a, b \in Z_2 \oplus Z_2$. There exist $3 \cdot 3 = 9$ double errors that can be occur in this word. It is easy to see that the given code detects 6 errors and it cannot detect 3 double errors.

Thus this code detects 12 from a theoretically possible 15 errors in any word of the form ab, $a, b \in Z_2 \oplus Z_2$, i.e., it detects 80% of errors in information symbols by supposition that the check symbol was transmitted without error.

11.4.6.2 Cryptographical part

We shall use three orthogonal T-quasigroups from Example 9.115 defined over the group $(Z_2 \oplus Z_2, +)$ as follows:

$(Z_2 \oplus Z_2, D)$ with the form $D(x, y) = \varphi_3 x + \varphi_6 y + a_1$;
$(Z_2 \oplus Z_2, E)$ with the form $E(x, y) = \varphi_2 x + \varphi_5 y + a_2$;
$(Z_2 \oplus Z_2, F)$ with the form $F(x, y) = \varphi_3 x + \varphi_5 y + a_3$.

Define three ternary operations in the following way: $K_1(D(x, y), z) = D(x, y) + z$, $K_2(E(x, y), z) = E(x, y) + z$, $K_3(F(x, y), z) = F(x, y) + z$. It is clear that these operations can be replaced by a more complex system of operations.

Lemma 11.23. *The triple of ternary operations $K_1(x, y, z), K_2(x, y, z), K_3(x, y, z)$ forms an orthogonal system of operations.*

Proof. We solve the following system of equations

$$\begin{cases} \varphi_3 x + \varphi_6 y + a_1 + z = b_1 \\ \varphi_2 x + \varphi_5 y + a_2 + z = b_2 \\ \varphi_3 x + \varphi_5 y + a_3 + z = b_3, \end{cases} \tag{11.19}$$

where b_1, b_2, b_3 are fixed elements of the set $Z_2 \oplus Z_2$.

We use properties of the groups $(Z_2 \oplus Z_2)$ and $Aut(Z_2 \oplus Z_2)$.

$$\begin{cases} \varphi_3 x + \varphi_6 y + z = b_1 + a_1 \\ \varphi_2 x + \varphi_5 y + z = b_2 + a_2 \\ \varphi_3 x + \varphi_5 y + z = b_3 + a_3. \end{cases} \tag{11.20}$$

We do the following transformations of the system (11.20): (first row + third row) $\rightarrow$ first row; (second row + third row) $\rightarrow$ second row; and obtain the system:

$$\begin{cases} y = b_1 + a_1 + b_3 + a_3 \\ x = \varphi_4(b_2 + a_2 + b_3 + b_4) \\ \varphi_3 x + \varphi_5 y + z = b_3 + a_3. \end{cases} \tag{11.21}$$

In the third equation of the system (11.21) we replace x by $\varphi_4(b_2 + a_2 + b_3 + b_4)$ and y by $b_1 + a_1 + b_3 + a_3$, obtaining:

$$\begin{cases} x = \varphi_4(b_2 + a_2 + b_3 + a_3) \\ y = b_1 + a_1 + b_3 + a_3 \\ z = b_3 + a_3 + \varphi_5(b_1 + a_1 + b_2 + a_2). \end{cases} \tag{11.22}$$

Therefore the system (11.19) has a unique solution for any fixed elements $b_1, b_2, b_3 \in (Z_2 \oplus Z_2)$, and operations $K_1(x, y, z), K_2(x, y, z), K_3(x, y, z)$ are orthogonal. $\square$

The triple of orthogonal operations $K_1(x, y, z), K_2(x, y, z), K_3(x, y, z)$ defines a permutation on the set Q^3. Denote this permutation by the letter K.

By $a_1 = a_2 = a_3 = (0; 0)$ permutation K has the following cycle type: $1^2 2^1 4^1 7^2 14^1 28^1$, i.e., this permutation contains two cycles of order 1, one cycle of order 2, and so on.

The order of permutation K is equal to 28. Notice that using isotopy or generalized isotopy it is possible to change the order of permutation K.

We shall use the system of three ternary orthogonal groupoids (Q, A), (Q, B), (Q, C) of order 4 from Example 1.342. See also [224]. Denote the permutation that defines this system of three ternary orthogonal groupoids by the letter M, $M = M(A(x, y, z), B(x, y, z), C(x, y, z))$. This permutation has the following cycle type: $1^1 17^1 20^1 26^1$. The order of this permutation is equal to $17 \cdot 20 \cdot 13 = 4420$.

In order to use the system of orthogonal groupoids and the system of orthogonal T-quasigroups simultaneously, we redefine the basic set of the T-quasigroups in the following (non-unique) way $(0; 0) \to 0$, $(1; 0) \to 1$, $(0; 1) \to 2$, $(1; 1) \to 3$.

We propose the following cryptographical term (a cryptographical primitive):

$$H(x, y, z) = M^k(K^l(x, y, z)), k, l \in Z. \tag{11.23}$$

Transformation H is a permutation of the set Q^3. Indeed, this transformation is a composition of two permutations: K^l and M^k.

Remark 11.24. It is possible to use the following cryptographical procedure:

$$H_1(x, y, z) = K^t(M^k(K^l(x, y, z))), t, k, l \in \mathbb{Z},$$

and so on.

Therefore we can propose the following:

Algorithm 11.25.

1. Take a pair of information symbols $a, b \in (Z_2 \oplus Z_2)$.
2. Using the formula (11.18) (or its analogue), find the value of check symbol c.
3. Apply the cryptographical term H to the triple (a, b, c).
4. Therefore, we obtain the first three elements of the ciphertext.
5. Take a pair of information symbols $d, e \in (Z_2 \oplus Z_2)$.
6. Using the formula (11.18), find the value of check symbol f.
7. Change the values of the numbers k, l in the cryptographical term H; also it is possible to change the term H to another term of this or other type.
8. Apply the cryptographical term H to the triple (d, e, f).
9. We obtain the next three elements of the ciphertext.
10. And so on.

Remark 11.26. At Step 7 of Algorithm 11.25 it is possible to use ideas of the Feistel schema. Namely, it is possible to calculate the numbers k, l using some bijective functions, where the numbers of triplet $H(a, b, c)$ and previous values of k and l are used as arguments.

11.4.6.3 Decoding

Using permutations K^{-1} and M^{-1}, we can construct the corresponding triplets of orthogonal 3-ary groupoids and so on.

11.4.6.4 Resistance

Taking into consideration Remark 11.24, we can estimate the number of possible keys in the presented cryptcode. This number is equal to (64!). The length of any key is equal to $64 \cdot 3 \cdot 2 = 384$ bits.

At each step of the proposed algorithm only three symbols (six bits) are ciphered. Moreover, after any step this key can be changed. Therefore, brute-force attack is difficult.

A statistical attack also seems to be difficult. It is possible to present the following argument: the symmetric group S_{64} acts on the set, which consists of 64 triplets 64-transitively [471].

11.4.6.5 A code-crypt algorithm

Denote the coding procedure from Algorithm 11.25 as $C(x, y)$ since this procedure is a function of two variables. Therefore, we can describe procedures of coding and enciphering in Algorithm 11.25 by the following formula:

$$H(x, y, C(x, y)), \tag{11.24}$$

where H is taken from the equation (11.23).

It is possible to construct a code-crypt algorithm by the formula $C_1(H(x, y))$ since there exists a possibility to use an analogue of the code C for three information symbols [642, Example 19], i.e., we can transpose the procedures C and H.

Moreover, this approach is more suitable in practice, if we use block ciphers. See [631] for details.

Recall, the number $N(n)$ of mutually (in pairs) orthogonal Latin squares of order n fulfills the following inequality $N(n) \leqslant (n-1)$ [559]. Then for $n = 4$ we have $N(4) \leqslant 3$.

Almost all constructions in this paper are performed over the field $GF(2^2)$. An analog of Algorithm 11.25 can be constructed over a field of the order more than four. Also we can use an alternating more powerful code. See Section 10.2, [32], [371].

11.4.7 Comparison of the power of the proposed algorithms

We shall compare how many permutations and of what length can be generated and used by working some of the above-mentioned algorithms.

Algorithm 1.69. If we shall use only one quasigroup $(Q, \cdot)$, $|Q| = n$, then we can obtain by encoding not more than n permutations of the group S_n.

Algorithm 11.9. If we shall use only one quasigroup $(Q, \cdot)$, $|Q| = n$, then we shall use by encoding the set $S = \cup_{i=1}^{n} \langle L_{a_i} \rangle$ of permutations which is a subset of the left multiplication group LM of quasigroup $(Q, \cdot)$. We recall $LM(Q, \cdot) = \langle L_x \mid x \in Q \rangle$ [72, 685, 779].

It is possible to construct a quasigroup $(Q, \cdot)$ such that $LM(Q, \cdot) = S_Q$. Notice that it is proved [279] that there exist quasigroups with the property $LM(Q, \cdot) = A_Q$, where A_O is the alternating group defined on the set Q [471, 407].

Therefore by encoding using Algorithm 11.9 we can obtain not more than $|S_n| = n!$ permutations.

The situation with Algorithm 11.10 is similar to the situation with Algorithm 1.69. Since by encoding the translations of an m-ary quasigroup (Q, f) are used, we can obtain not more than $|S_n| = n!$ permutations. The properties of the multiplication group (more exactly, multiplication groups) of n-ary quasigroups are not researched well.

Information on the multiplication groups of linear n-ary quasigroups is in [600]. These quasigroups are used in [684, 683] in construction of some ciphers (see above).

Algorithm 11.13 is a synthesis of Algorithms 11.9 and 1.69. Here by the symbol T_i

we denote translations of an n-ary quasigroup (Q, f). It is clear that the order of the set $S = \cup \langle T_i \rangle$ can be large but cannot be more than $|S_n| = n!$.

In Algorithm 11.14 elements of the cyclic group $\langle F \rangle \subset S_{n^m}$, where $|Q| = n$, m is the arity of orthogonal groupoids, can appear. Notice, in the above-mentioned inclusion, the equality cannot be even theoretically, since the minimum number of generators of the symmetric group is equal to two [471, 407].

It is well known that a cycle of order n and a cycle of order two generate the symmetric group S_n [471, 407].

The group S_{n^m} is an upper bound of the sets of permutations that can be generated during the work of Algorithms 11.15, 11.22. For Algorithm 11.16 the group S_{n^2} is such an upper bound. It is clear that in Algorithm 11.16 by the encryption any permutation of the group S_{n^2} may be realized. But it is also clear that this is not necessary from the cryptographical point of view.

The possible number of permutations generated during the work of the algorithm from Example 11.17 is bounded by the number $(257^2)! = 66049!$ and during the work of Algorithm 11.25 is bounded by the number $(64)!$.

11.5 One-way and hash functions

11.5.1 One-way function

A function $F : X \to Y$ is called a one-way function if the following conditions are fulfilled:

- there exists a polynomial algorithm of calculation of $F(x)$ for any $x \in X$;
- there does not exist a polynomial algorithm of inverting of the function F, i.e., there does not exist any polynomial time algorithm for solving the equation $F(x) = y$ relatively variable x.

It is proved that the problem of the existence of a one-way function is equivalent to the well-known problem of coincidence of classes P and NP.

A better candidate for a one-way function is the so-called function of discrete logarithms [559].

We recall that a neofield $(N, +, \cdot)$ of order n consists of a set N of n symbols on which two binary operations "+" and "·" are defined such that $(N, +)$ is a loop with identity element, say 0, $(N\backslash\{0\}, \cdot)$ is a group and the operation "·" distributes from the left and right over "+" [243].

Let $(N, +, \cdot)$ be a finite Galois field or a cyclic ($(N\backslash\{0\}, \cdot)$ is a cyclic group) neofield. Then each non-zero element u of the additive group or loop $(N, +)$ can be represented in the form $u = a^{\nu}$, where a is a generator of the multiplication group $(N\backslash\{0\}, \cdot)$ and ν is called the discrete logarithm of u with base a, or, sometimes, the exponent or index of u.

Given ν and a, it is easy to compute u in a finite field, but if the order of the finite field is a sufficiently large prime p and is also appropriately chosen, it is believed to be difficult to compute ν when u (as a residue modulo p) and a are given.

In [243] discrete logarithms are studied over a cyclic neofield whose addition is a CI-loop.

In [559] the discrete logarithm problem for the group RL_n of all row-Latin squares of order n is defined (p.103) and, on pages 138 and 139, some illustrations of applications to cryptography are given.

11.5.2 Hash function

Definition 11.27. A function $H()$ that maps an arbitrary-length message M to a fixed-length hash value $H(M)$ is a OneWay Hash Function (OWHF), if it satisfies the following properties:

1. The description of $H()$ is publicly known and should not require any secret information for its operation.
2. Given M, it is easy to compute $H(M)$.
3. Given $H(M)$ in the range of $H()$, it is hard to find a message M for given $H(M)$, and given M and $H(M)$, it is hard to find a message $M_0(\neq M)$ such that $H(M_0) = H(M)$.

Definition 11.28. A OneWay Hash Function $H()$ is called a Collision-Free Hash Function (CFHF), if it is hard to find two distinct messages M and M_0 that hash to the same result $(H(M) = H(M_0))$[300, 299].

In [300, 299] an approach for construction of a hash function using quasigroups is described. We give the construction of a hashing function based on a quasigroup [300].

Definition 11.29. Let $H_Q() : Q \longrightarrow Q$ be a projection defined as

$$H_Q(q_1 q_2 \dots q_n) = ((\dots(a \star q_1) \star q_2 \star \dots) \star q_n. \tag{11.25}$$

Then $H_Q()$ is said to be a hash function over quasigroup $(Q; \star)$. The element a is a fixed element from Q.

Example 11.30. We take the ring $(Z_4, +, \cdot)$ of residues modulo 4 and define quasigroup $(Z_4, \star)$ in the following way: $a \star b = (a - b) \pmod 4$. This quasigroup has the following multiplication table:

$\star$	0	1	2	3
0	0	3	2	1
1	1	0	3	2
2	2	1	0	3
3	3	2	1	0

The value of the hash function is $H_2(0013) = (((2 \star 0) \star 0) \star 1) \star 3 = 2$.

There exists a possibility to apply an n-ary quasigroup approach to study hash functions of this kind, since, in fact, the equality (11.25) defines an n-ary operation. This approach is realized in [946].

Remark 11.31. We notice that a safe hash function must have at least a 128-bit image, i.e., $H_Q(q_1 q_2 \dots q_n)$ must consist of at least a 128-digit number [633].

In [900, 901, 810] hash functions, proposed in [300, 299], are discussed. The author shows that for some types of quasigroups these hash functions are not secure.

From [603] we give the following summary: "In this paper we consider two quasigroup transformations $QM1$: $A^{2m} \rightarrow A^{2m}$ and $QM2$: $A^m \rightarrow A^{2m}$, where A is the carrier of a quasigroup. Based on these transformations we show that different kinds of hash functions can be designed with suitable security."

Further development of quasigroup-based hash functions is reflected in [355, 356, 821]. Some attacks on quasigroup-based systems (on Edon-R) are described in [564].

The binary additive stream cipher Edon-80 is described in many articles. See for example [335]. Notice, the keystream in Edon-80 is generated as a row of a certain matrix whose elements are defined iteratively using quasigroup operations.

In [743], based on Algorithm 1.69, a encrypter that has good scrambling properties is proposed.

11.6 Secret-sharing schemes

Definition 11.32. Let $1 < k \leqslant n$. A secret-sharing scheme between n users is called an (n, k)-*threshold* if any group of k from n users can restore a secret, but none of the groups with the smaller number of users can obtain information about the secret key [12].

In some cases the (n, k)-threshold scheme is called a (k, n)-secret sharing scheme. Therefore we can give the following definition.

Definition 11.33. If the scheme has n participants, a (k, n)-secret sharing scheme is a system where n pieces of information, called shares or shadows of a secret key K, are distributed so that each participant has a share such that

(1) the key K can be reconstructed from knowledge of any k or more shares;
(2) the key K cannot be reconstructed from knowledge of fewer than k shares.

Secret-sharing schemes were invented independently [934] by George Blakley and Adi Shamir in 1979 [154, 754]. Shamir used the idea that any polynomial $f(x) = a_0 + a_1x + a_2x^2 + \dots a_{k-1}x^{k-1}$ is determined by any k of its points. Lagrange basis polynomials are used for construction of polynomial f using these known k points. The secret is the number a_0 [936]. Later this idea was generalized.

Blakley uses the following idea. Suppose that a system

$$\begin{cases} a_{11}x_1 + a_{12}x_2 + \dots a_{1k}x_k = b_1 \\ a_{21}x_1 + a_{22}x_2 + \dots a_{2k}x_k = b_2 \\ \dots \\ a_{k1}x_1 + a_{k2}x_2 + \dots a_{kk}x_k = b_k \end{cases} \tag{11.26}$$

of k linear equations in k variables over a field F has exactly one solution $(b_1, b_2, \dots, b_k)$ (this solution is the shared secret).

It is easy to see and it is well known (see, for example, [550]) that the solution of the system (11.26) remains the same, if an equation, say the i-th equation, $1 \leqslant i \leqslant k$, of the system (11.26) will be changed by the following equation $\gamma_1(a_{11}x_1 + a_{12}x_2 + \dots a_{1k}x_k - b_1) + \gamma_2(a_{21}x_1 + a_{22}x_2 + \dots a_{2k}x_k - b_2) + \dots + \gamma_n(a_{k1}x_1 + a_{k2}x_2 + \dots a_{kk}x_k - b_k) = 0$, where γ_j, $j \in \overline{1,k}$ $\gamma_i \neq 0$, are some elements of the field F.

There are secret-sharing schemes that make use of the Chinese Remainder Theorem (Mignotte's and Asmuth-Bloom's Schemes) and orthogonal arrays (see below). Some secret-sharing schemes are proposed in [135]. In [807] Simmons surveyed various secret-sharing schemes known at that time.

11.6.1 Critical sets

Secret-sharing schemes based on critical sets in Latin squares (see Definition 1.354) are studied in [216]. We note that critical sets of Latin squares give rise to the possibilities to construct secret-sharing systems.

Critical sets of Latin squares were studied in many articles. We survey results from some of these articles. In [259] the spectrum of critical sets in Latin squares of order 2^n is studied. The paper [250] gives constructive proofs that critical sets exist for all sizes between $[n^2/4]$ and $[(n^2 - n)/2]$, with the exception of size $n^2/4 + 1$ for even values of n.

For Latin squares of order n, the size of a smallest critical set is denoted by $\mathrm{scs}(n)$ in [189]. The main result of [189] is that $\mathrm{scs}(n) \geqslant n\lfloor \frac{1}{2}(\log n)^{1/3} \rfloor$ for all positive integers n.

In [425] the authors show that any critical set in a Latin square of order $n \geqslant 7$ must have at least $\lfloor \frac{7n-\sqrt{n}-20}{2} \rfloor$ empty cells. See also [424].

The paper [257] contains lists of (a) theorems on possible sizes of critical sets in Latin squares of order less than 11, (b) publications, where these theorems are proved, and (c) concrete examples of such type of critical sets. In [262] an algorithm for writing any Latin interchange as a sum of intercalates is corrected.

In [411] the author proposes a greedy algorithm to find critical sets in Latin squares. He applies this algorithm to Latin squares which are abelian 2-groups to find new critical sets in these Latin squares. The critical sets have the nice property that they all intersect some 2×2 Latin subsquare in a unique element so that it is easy to show the criticality.

In [47] the author gives an example of a critical set of size 121 in the elementary abelian 2-group of order 16.

In [630] critical sets of symmetric Latin squares are studied. Therefore the authors require all elements in their critical sets and uniquely completable partial Latin squares to lie on or above the main diagonal. For $n > 2$, a general procedure is given for writing down a uniquely completable partial symmetric $2n \times 2n$ Latin square L'_{2n} containing $n^2 - n + 2$ entries, of which $2n - 2$ are identical and lie on the main diagonal.

Paper [256] presents a solution to the interesting combinatorial problem of finding a minimal number of elements in a given Latin square of odd order n by which one may restore the initial form of this square. In particular, it is proved that in every cyclic Latin square of odd order n the minimal number of elements equals $n(n-1)/2$.

The study of flaws within certain theoretical cryptographic schemes based on critical sets in Latin squares is continued in [261].

Surveys of critical sets of Latin squares are given in [476, 480]. See also [481].

The concept of Latin trades is closely connected with the concept of critical sets in Latin squares [191]. Information on Latin trades is given in [190]. Some definitions are in Subsection 1.7.2 of Chapter 1.

Other applications of critical sets of Latin squares in cryptology is mentioned in the Introduction to this chapter.

"For a given triple of permutations $T = (\alpha, \beta, \gamma)$ the set of all Latin squares L such that T is its autotopy is denoted by $LS(T)$. The cardinality of $LS(T)$ is denoted by $\Delta(T)$. Specifically, the computation of $\Delta(T)$ for any triple T is at the moment an open problem having relevance in secret sharing schemes related to Latin squares" [313, 314, 857].

11.6.2 Youden squares

Definition 11.34. A Youden square is a $k \times v$ rectangular array ($k < v$) of v symbols such that each symbol appears exactly once in each row and the symbols appearing in any column form a block of a symmetric (v, k, λ) design, all the blocks of which occur in this manner [940].

A Youden square is a Latin rectangle. The term "square" in the name comes from an older definition which did use a square array [213].

An example of a 4×7 Youden square is given by

$$\begin{array}{ccccccc} 1 & 2 & 3 & 4 & 5 & 6 & 7 \\ 2 & 3 & 4 & 5 & 6 & 7 & 1 \\ 3 & 4 & 5 & 6 & 7 & 1 & 2 \\ 5 & 6 & 7 & 1 & 2 & 3 & 4 \end{array}$$

The seven blocks (columns) form a symmetric (7,4,2)-design [940].

We give the summary from [325]:

"We investigate subsets of critical sets of some Youden squares in the context of secret-sharing schemes. A subset $\mathcal{C}$ of a Youden square is called a critical set if $\mathcal{C}$ can be uniquely completed to a Youden square but no proper subset of $\mathcal{C}$ has a unique completion to a Youden square."

"That part of a Youden square Y which is inaccessible to subsets of a critical set $\mathcal{C}$ of Y, called the strongbox of $\mathcal{C}$, may be thought to contain secret information. We study the size of the secret. J. R. Seberry and A. P. Street [753] have shown how strongboxes may be used in hierarchical and compartmentalized secret-sharing schemes."

11.6.3 Reed-Solomon codes

Some secret-sharing systems are described in [241]. One such system is the Reed-Solomon code over a Galois field $GF[q]$ with generating matrix $C(a_{ij})$ of size $k \times (q-1)$, $k \leqslant q-1$. Reed-Solomon code over a finite field $GF(q)$ with words of length $q-1$ is defined by the following matrix G:

$$G = \begin{pmatrix} 1 & 1 & \cdots & 1 \\ a_1 & a_2 & \cdots & a_{q-1} \\ a_1^2 & a_2^2 & \cdots & a_{q-1}^2 \\ & \cdots & & \\ a_1^{k-1} & a_2^{k-1} & \cdots & a_{q-1}^{k-1} \end{pmatrix}$$

where $a_0 = 0, a_1 = 1, a_2, ..., a_{q-1}$ are the different elements of $GF(q)$ with $q = p^m$ (p is prime). The determinant formed by any k columns of G is a non-zero element of $GF[q]$. The Hamming distance d of this code is maximal ($d = q - k$) and any k from $q-1$ keys unlock the secret.

Some modifications of this schema are in an article by R.J. McEliece and D.V. Sarwarte [616].

Every k-tuple $s = (s_0, s_1, ..., s_{k-1})$ in coded form is the $(q-1)$-tuple $b = (b_1, b_2, ..., b_{q-1})$, where $b = sG$. In this case $b_i = q(a_i)$, where $q(x) = s_0 + s_1 x + ... + s_{k-1}x^{k-1}$. The subset of $n \leqslant (q-1)$ pairs of the set $\{(i, b_i) \mid i = 1, 2, ..., q-1\}$, any k of which unlock the secret, can be the secret key.

11.6.4 Orthogonality and secret-sharing schemes

Here we give results of G.B. Belyavskaya [117], which suggests a general secret-sharing scheme based on orthogonal systems of partial (in particular, everywhere determined) k-ary operations which generalizes some of the known schemes. Also the orthogonal systems of k-ary operations concerning these known schemes are found.

Let Q be a nonempty set and $D \subseteq Q^k$, $D \neq \varnothing$. If A is a mapping of D into Q, then (Q, A) is said to be a partial k-ary operation (or a partial k-ary groupoid). If $D = Q^k$, we have a usual k-ary operation given on the set Q.

Let $(Q, A_1), (Q, A_2), ..., (Q, A_k)$ be a k-tuple of a partial k-ary operation with the same domain $D = D(A_1) = D(A_2) = ... = D(A_k)$, $D \subseteq Q^k$. The k-tuple A_1^k is said to be orthogonal if for every $(a_1, a_2, ..., a_k) \in Q^k$ for which the system $\{A_i^k(x_1^k) = a_i\}_1^k$ has a solution, this solution is unique.

The set of different partial k-operations of the same domain is said to be an orthogonal system of partial k-operations (k-OSPO) if each k-tuple of the k-operations of this set is orthogonal [4].

It is proved that in an (n, k)-threshold secret-sharing system between n users, any k of them can unlock the secret can be constructed by any orthogonal system $\sum = \{A_1, A_2, ..., A_t\}$ of partial (or everywhere defined) k-ary operations, $1 < k \leqslant n, n \leqslant t$.

The following method of construction of such schemes is proposed.

1) Choose the sufficiently big order q of a set Q and k-OSPO $\sum = \{A_1, A_2, ..., A_t\}$ with a big domain D, which is given on Q.

2) Choose a k-tuple $a = (a_0, a_1, ..., a_{k-1})$ in D.

3) Choose an n-tuple $i_1, i_2, ..., i_n$ of $\{1, 2, ..., t\}$, $k < n \leqslant t$.

4) Calculate the n-tuple

$$b = (b_1, b_2, ..., b_n) = (A_{i_1}(a_0^{k-1}), A_{i_2}(a_0^{k-1}), ..., A_{i_n}(a_0^{k-1})).$$

5) The pairs $(i_1, b_{i_1}), (i_2, b_{i_2}), ..., (i_n, b_{i_n})$ are separated among n keepers of the secret.

We obtain a (n, k) threshold scheme. In this scheme it is possible to increase the number of keepers of the secret up to the number t.

Example 11.35. In order to construct a $(5, 2)$-threshold we shall use the system of five orthogonal in pairs operations (medial quasigroups) defined over the field $(\mathbb{Q}, +, \cdot, 0, 1, ^{-1})$ of rational numbers: $(\mathbb{Q}, \bullet)$, $x \bullet y = 1 \cdot x + 2 \cdot y$; $(\mathbb{Q}, \circ)$, $x \circ y = 3 \cdot x + 4 \cdot y$; $(\mathbb{Q}, *)$, $x * y = 5 \cdot x + 6 \cdot y$; $(\mathbb{Q}, \star)$, $x \star y = 7 \cdot x + 8 \cdot y$; $(\mathbb{Q}, \diamond)$, $x \diamond y = 9 \cdot x + 10 \cdot y$.

Suppose that the secret is the pair of numbers $(11; 12)$. Then between five keepers of the secret we can distribute the following numbers $35 = 1 \cdot 11 + 2 \cdot 12$; $81 = 3 \cdot 11 + 4 \cdot 12$; $127 = 5 \cdot 11 + 6 \cdot 12$; $173 = 7 \cdot 11 + 8 \cdot 12$; $219 = 9 \cdot 11 + 10 \cdot 12$. Also we should indicate the used operation (or the number of used operations), if this is necessary. Therefore, finally we have the following five triples: $(1; 2; 35)$; $(3; 4; 81)$; $(5; 6; 127)$; $(7; 8; 173)$; $(9; 10; 219)$.

The procedure of "opening of the door" is the following. Suppose that the keepers number I and number V would like to open a door. For this purpose it is necessary to solve the system of equations:

$$\begin{cases} x + 2 \cdot y = 35 \\ 9 \cdot x + 10 \cdot y = 219. \end{cases}$$

It is easy to check that $x = 11$ and $y = 12$.

Remark 11.36. Notice, the list of orthogonal quasigroups given in Example 11.35 ($l \neq k$) can be extended up to any finite natural number m. Indeed,

$$\begin{vmatrix} n+l & n+l+1 \\ n+k & n+k+1 \end{vmatrix} = n^2 + nk + n + ln + lk + \\ +l - n^2 - nl - n - kn - kl - k = l - k.$$

Therefore we have constructed an infinite number of orthogonal in pairs binary quasigroups that are defined over the field of rational or real numbers.

In a similar way it is possible to construct an infinite number of n-ary ($n \geqslant 3$) quasigroups, any n of which form a system of orthogonal operations. A bunch of hyperplanes in n-dimensional Euclidian space over the field $\mathbb{R}$ can be regarded as a geometrical interpretation of this construction.

The number of linearly independent binary vectors is studied in many articles. See for example [225].

11.7 Some algebraic systems in cryptology

11.7.1 Inverse quasigroups in cryptology

In [243, 367] some applications of CI-quasigroups in cryptology with non-symmetric keys are described.

Definition 11.37. Suppose that there exists a permutation J of the elements of a quasigroup $(Q, \circ)$ such that, for all $x, y \in Q$

$$J^r(x \circ y) \circ J^s x = J^t y,$$

where r, s, t are integers. Then $(Q, \circ)$ is called an (r, s, t)-inverse quasigroup ([487]).

In the special case when $r = t = 0$, $s = 1$, we have a definition of a CI-quasigroup.

Example 11.38. A CI-quasigroup can be used to provide a one-time pad for key exchange (without the intervention of a key distributing center) [243, 477].

The sender S, using a physical random number generator (see [534] on a random number generator based on quasigroups), selects an arbitrary element $c^{(u)}$ of the CI-quasigroup $(Q, \circ)$ and sends both $c^{(u)}$ and the enciphered key (message) $c^{(u)} \circ m$. The receiver R uses this knowledge of the algorithm for obtaining $Jc^{(u)} = c^{(u+1)}$ from $c^{(u)}$ and hence he computes $(c^{(u)} \circ m) \circ c^{(u+1)} = m$.

Example 11.39. We can propose the following application of rst-inverse quasigroups in a situation similar to the situation described in Example 11.38. It is possible to re-write definitive equality of a rst-inverse quasigroup in the following manner $J^r(J^k u \circ m) \circ J^{s+k} u = J^t m$.

Then the schema of the previous example can be re-written in the following manner. The sender S selects an arbitrary element $J^k u$ of the rst-quasigroup $(Q, \circ)$ and sends both $J^k u$ and the enciphered key (message) $J^r(J^k u \circ m)$. The receiver R uses this knowledge of the algorithm for obtaining $J^{k+s}(u)$ from $J^k(u)$ and hence he computes $J^r(J^k u \circ m) \circ J^{s+k} u = J^t m$ and after this he computes the message m. Of course this example can be modified.

Example 11.40. [243]. Take a CI-quasigroup with a long inverse cycle $(c\, c'\, c'' \dots\, c^{t-1})$ of length t. Suppose that all the users U_i $(i = 1, 2, \dots)$ are provided with an apparatus (for example, a chip card) which will compute $a \circ b$ for any given $a, b \in Q$. We assume that only the key distributing center has knowledge of the long inverse cycle which serves as a look-up table for keys.

Each user U_i has a public key $u_i \in Q$ and a private key Ju_i, both supplied in advance by the key distributing center. User U_s wishes to send a message m to user U_t. He uses U_t's public key u_t to compute $u_t \circ m$ and sends that to the user U_t. The user U_t computes $(u_t \circ m) \circ Ju_t = m$.

Remark 11.41. It is not very difficult to understand that an opponent which knows the permutation J may decipher a message encrypted by this method.

Remark 11.42. There exists a possibility to generalize Example 11.40 using some m-inverse quasigroups [486], or (r, s, t)-inverse quasigroups [487, 488], or (α, β, γ)-inverse quasigroups [489].

11.7.2 Some groups in cryptology

11.7.2.1 El Gamal cryptosystem

Usually the classical Taher El Gamal (Egyptian cryptographer) encryption system is formulated in the language of number theory using a multiplication modulo of a prime number [303].

It is easy to see that this system can also be formulated in terms of a ring of residues modulo p, or, equivalently, using the language of Galois field $GF(p)$.

Moreover, it is possible to use the concept of action of a group of automorphisms of the cyclic group $(Z_p, +)$ on this group. It is clear that different points of view on the same mathematical idea (phenomenon) led to various generalizations. Taking into consideration further generalization presented in the next subsection we prefer here the group approach.

Let $(Z_p, +)$ be a cyclic group of residues of big (say 200 to 300 digits) prime order relative to addition of residues and a be a generator of the group $(Z_{p-1}, \cdot) \cong Aut(Z_p, +)$ $(gcd(a, p-1) = 1)$.

Alice's keys are as follows:

Public Key p, a, and a^m, $m \in \mathbb{N}$.

Private Key m.

Encryption

To send a message $b \in (Z_{p-1}, \cdot)$, Bob computes a^r and a^{mr} for a random $r \in \mathbb{N}$ (sometimes the number r is called an ephemeral key [925]).

The ciphertext is $(a^r, a^{mr} \cdot b)$.

Decryption

Alice knows m, so if she receives the ciphertext $(a^r, a^{mr} \cdot b)$, she computes a^{rm} from a^r and then a^{-mr} and then from $a^{mr} \cdot b$ computes b.

Example 11.43. Alice chooses $p = 107$, $a = 2$, $m = 67$, and she computes $a^m = 2^{67} \equiv 94 \mod 107$. Her public key is $(p, a^m) = (107, 94)$, and her private key is $m = 67$. Notice, in practice, this choice is carried out in a key distribution center.

Bob wants to send the message "B" (66 in ASCII) to Alice. He chooses a random integer $r = 45$ and encrypts $B = 66$ as $(a^m)^r \cdot B$. Bob receives: $(2^{45}, 94^{45} \cdot 66) \equiv (28, 5 \cdot 66) \equiv (28, 9) \mod 107$.

He sends the encrypted message $(28, 9)$ to Alice. Alice receives this message, and using her private key $m = 67$ she decrypts in the following way: $28^{-67} \cdot 9 = 28^{106-67} \cdot 9 = 28^{39} \cdot 9 \equiv 43 \cdot 9 \equiv 66 \mod 107$.

The complexity of this system is based on the complexity of the discrete logarithm problem. The El Gamal encryption system is not secure under the chosen ciphertext attack [925]. The El Gamal cryptosystem is usually used in a hybrid cryptosystem, i.e., the message itself is encrypted using a symmetric cryptosystem and El Gamal is then used to encrypt the key used for the symmetric cryptosystem.

11.7.2.2 De-symmetrization of Algorithm 1.69

We give an analogue of the El Gamal encryption system based on the Markovski algorithm [635, 757].

Let (Q, f) be a binary quasigroup and $T = (\alpha, \beta, \gamma)$ be its isotopy.

Alice's keys are as follows:

Public Key is (Q, f), T, $T^{(m,n,k)} = (\alpha^m, \beta^n, \gamma^k)$, $m, n, k \in \mathbb{N}$, and Markovski algorithm.

Private Key m, n, k.

Encryption

To send a message $b \in (Q, f)$, Bob computes $T^{(r,s,t)}$, $T^{(mr,ns,kt)}$ for a random $r, s, t \in \mathbb{N}$ and $(T^{(mr,ns,kt)}(Q, f))$.

The ciphertext is $(T^{(r,s,t)}, (T^{(mr,ns,kt)}(Q, f))b)$.

To obtain $(T^{(mr,ns,kt)}(Q, f))b$, Bob uses the Markovski algorithm which is known to Alice.

Decryption

Alice knows m, n, k, so if she receives the ciphertext

$$(T^{(r,s,t)}, (T^{(mr,ns,kt)}(Q, f))b),$$

she computes $(T^{(mr,ns,kt)}(Q, f))^{-1}$ using $T^{(r,s,t)}$ and, finally, she computes b.

In this algorithm the isostrophy [789] can also be used instead of isotopy, Algorithm 11.16 instead of the Markovski algorithm, and n-ary ($n > 2$) quasigroups [77, 790] instead of binary quasigroups.

A generalization of the Diffie-Hellman schema of open key distribution is given in [474]. The generalization is based on the concepts of the left and right powers of elements of some non-associative groupoids. For medial quasigroups this approach is realized in [387].

A protocol of the elaboration of a common secret key based on Moufang loops is given in [387]. This protocol is a generalization of the results from [849]. Generalizations of the El Gamal schema on Moufang loops is given in [387]. In [611] a discrete logarithm problem in Moufang loops is reduced to the same problem over finite simple fields. One more generalization of the El Gamal schema based on quasiautomorphisms of quasigroups is presented in [387].

11.7.2.3 RSA and GM cryptosystems

Taking into consideration the theoretical importance of the RSA (Rivest-Shamir-Adleman) [712] and GM (Goldwasser-Micaly) [365, 366] cryptosystems, we give their sketch description. In the description of the RSA system we use [927]. Proofs of the facts used from the number theory can be found in many books; see, for example, [178].

Setting up an RSA system involves:

1. Choosing large prime numbers p and q.
2. Computing $n = pq$ and $k = \varphi(n)$; we recall, if $u = p_1^{k_1} p_2^{k_2} \dots p_r^{k_r}$, then

$$\varphi(u) = u(1 - \frac{1}{p_1})(1 - \frac{1}{p_2}) \dots (1 - \frac{1}{p_r}).$$

In our case $\varphi(n) = k = (p - 1)(q - 1)$.

3. Finding two numbers e and d such that $ed \equiv 1 \bmod k$.

Public Key is the numbers n and e. Private Key is d.

Encryption

To send a message (an integer) m, where $0 < m < n$, Bob computes $S \equiv m^e \bmod n$.

Decryption

Alice knows d and S. Alice decrypts the message by computing $t = S^d = (m^e)^d = m^{ed} \bmod n$.

Euler's theorem ($a^{\varphi(n)} \equiv 1 \bmod n$, if $g.c.d.(a, n) = 1$, [178]) can be used to show that if $0 < t < n$, then $t = m$. Indeed, since $k = \varphi(n)$ we have: $ed \equiv 1 \bmod \varphi(n)$, i.e., there exists a number r such that $ed = r\varphi(n) + 1$. Therefore we can rewrite equality $t = m^{ed} \bmod n$ in the following form: $t = (m^{\varphi(n)})^r m \bmod n$. By Euler's theorem further we have $t = 1^r m = m \bmod n$.

The security of an RSA system would be compromised if the number n could be factored or if $\varphi(n)$ could be computed without factoring n [927].

In the description of the GM cryptosystem we use [941].

Key generation

1. Alice generates two distinct large prime numbers p and q, randomly and independently of each other.

2. Alice computes $N = pq$.

3. She finds some non-residue x such that the Legendre symbols satisfy $\left(\frac{x}{p}\right) = \left(\frac{x}{q}\right) = -1$ and hence the Jacobi symbol $\left(\frac{x}{N}\right)$ is equal to $+1$.

The public key consists of (x, N). The secret key is the factorization (p, q).

Message encryption

Suppose Bob wishes to send a message m to Alice:

1. Bob first encodes m as a string of bits $(m_1, \dots, m_n)$.

2. For every bit m_i, Bob generates a random value y_i from the group of units modulo N, i.e., $g.c.d.(y_i, N) = 1$. He outputs the value $c_i = y_i^2 x^{m_i} \bmod N$. Bob sends the ciphertext $(c_1, \dots, c_n)$.

Message decryption

Alice receives $(c_1, \dots, c_n)$. She can recover m using the following procedure:

1. For each i, using the prime factorization (p, q), Alice determines whether the value c_i is a quadratic residue; if so, $m_i = 0$, otherwise $m_i = 1$. Alice outputs the message $m = (m_1, \dots, m_n)$.

A cryptosystem is called probabilistic if in the encryption process some random parameters are used [633]. It is clear that the GM cryptosystem and Blum-Goldwasser cryptosystem [155] are probabilistic.

In cryptography, a cryptosystem is semantically secure if any probabilistic, polynomial-time algorithm (PPTA [365, 366]) that is given, the ciphertext of a certain message (taken from any distribution of messages), and the message's length, cannot determine any partial information of the message with probability non-negligibly higher than all other PPTAs that only have access to the message length (and not the ciphertext).

In other words, knowledge of the ciphertext (and length) of some unknown message does not reveal any additional information on the message that can be feasibly extracted [935]. The Goldwasser-Micaly cryptosystem is semantically secure [365, 366]. See also [532].

Recall, there exist polynomial time algorithms for prime factorization and discrete logarithms on a quantum computer [804, 805]. Therefore the RSA and El Gamal schemes, their analogs and many generalizations cannot be safe, if an illegal user can use a quantum computer.

11.7.2.4 Homomorphic encryption

"Homomorphic" is an adjective which describes a property of an encryption scheme. That property, in simple terms, is the ability to perform computations on the ciphertext without decrypting it first [347, 348].

The El Gamal cryptosystem is homomorphic relative to multiplication. The notation $\epsilon(b)$ is used to denote the encryption of the message b. We have $\epsilon(b_1) \cdot \epsilon(b_2) = (a^{r_1}, b_1 \cdot a^{mr_1}) \cdot (a^{r_2}, b_2 \cdot a^{mr_2}) = (a^{r_1+r_2}, b_1 b_2 \cdot a^{mr_1} a^{mr_2}) = (a^{r_1+r_2}, b_1 b_2 \cdot a^{m(r_1+r_2)}) = \epsilon(b_1 \cdot b_2)$. GM cryptosystem also has the homomorphic property.

11.7.2.5 MOR cryptosystem

In [674, 673] the following MOR cryptosystem is proposed. We follow [586].

Let G be a group and $\psi : G \to G$ be an automorphism. Usually the map ψ is an inner automorphism of the group G.

Alice's keys are as follows:
Public Key ψ and ψ^m, $m \in \mathbb{N}$.
Private Key m.

Encryption

To send a message $a \in G$, Bob computes ψ^r and ψ^{mr} for a random $r \in \mathbb{N}$.
The ciphertext is $(\psi^r, \psi^{mr}(a))$.

Decryption

Alice knows m, so if she receives the ciphertext $(\psi^r, \psi^{mr}(a))$, she computes ψ^{rm} from ψ^r, and then ψ^{-mr} and then from $\psi^{mr}(a)$ computes a.

The complexity of this system is comparable to the complexity of the discrete logarithm problem. Cryptanalyses of the MOR cryptosystem is given in [888]. The semantic security of the MOR cryptosystem depends on the group used [888].

In [359] cryptanalysis of some modification of the MOR system ([634]) is given. It is demonstrated that the complexity of the proposed cryptoalgorithm is comparable to the complexity of the discrete logarithms problem.

This system and its modifications are researched actively [560, 585, 587, 588]. It is easy to see that in the generalized MOR system in fact, any algebraic system (including n-ary groupoids) and its automorphism group can be used.

Probably, many other encryption systems based on arithmetics of residues (more generally, on modular arithmetics) can be generalized on the case of non-abelian groups.

In [584] a cryptosystem, in which certain epimorphism f (epimorphism is an analogue of surjective function) from a group G onto a group H is a secret key, is described.

A cryptosystem based on the complexity of random covers in finite groups is researched in [563]. An encryption scheme based on the coverings of a loop is presented in [387].

11.7.3 El Gamal signature scheme

Here we follow [303] and [926]. Let H be a collision-resistant hash function (see Definition 11.28). Let p be a large prime such that computing discrete logarithms modulo p is difficult. Let $g < p$ be a randomly chosen generator of the multiplicative group of integers modulo p, i.e., of the group $(Z_{p-1}, \cdot)$. These system parameters may be shared between users.

Key generation.

Randomly choose a secret key x with $1 < x < p-1$.
Compute $y \equiv g^x \mod p$.
The public key is (p, g, y).
The secret key is x.
These steps are performed once by the signer.

Signature generation

To sign a message m the signer performs the following steps.
Choose a random k such that $1 < k < p-1$ and $gcd(k, p-1) = 1$.
Compute $r \equiv g^k \mod p$.
Compute $s \equiv (H(m) - xr)k^{-1} \mod p-1$.
If $s \equiv 0 \mod p-1$ start over again.

Then the pair (r, s) is the digital signature of m. The signer repeats these steps for every signature.

Verification

A signature (r, s) of a message m is verified as follows.
$0 < r < p$ and $0 < s < p-1$;
$g^{H(m)} \equiv y^r r^s \mod p$.

The verifier accepts a signature if all conditions are satisfied and rejects it otherwise.

Check

From construction of a signature we have

$g^{H(m)} \equiv g^{xr} g^{ks} \equiv (g^x)^r (g^k)^s \equiv y^r r^s \mod p.$

The original paper [303] did not include a hash function as a system parameter. Message m was used directly in the algorithm instead of $H(m)$. It enables an attack called an existential forgery [303, 697, 926]. An improved version (with a hash) is known as the Pointcheval-Stern signature algorithm [697].

11.7.4 Polynomially complete quasigroups in cryptology

We give the abstract from article [388]. "Nowadays the most popular public key cryptosystems are RSA, the El Gamal cryptosystem and encryption schemes based on the Diffie-Hellman problem. We construct a similar cryptosystem by means of a non-associative structure, namely, quasigroup ring. Some modifications increasing the security of this scheme against possible attacks are described. Several concrete non-associative algebraic structures acceptable for cryptosystem constructions were considered and analyzed also."

Notice that quasigroup rings are constructed in a way that is similar to construction of group rings.

Public key cryptography based on (mainly commutative and with an identity element) semigroup actions is studied in [612]. A practical example which is described in Section 5 of article [612] consists of a two-sided action. Notice that a two-sided group action is used for cryptological needs in [806].

Public key cryptography based on semimodules over quotient semirings is studied in [23].

Cryptographically suitable quasigroups via functional equations are studied in [541].

Definition 11.44. A polynomial quasigroup is said to be a quasigroup that can be defined by a polynomial over a ring [738, 737].

Example 11.45. Quasigroup $(Z_{2^3}, *)$ is defined over the ring Z_{2^3} using the following polynomial: $x * y = 3 + 5x + 7y + 4xy + 2xy^2$ [737].

Quasigroup $(Z_{2^3}, *)$ has the following Cayley table [737]:

*	0	1	2	3	4	5	6	7
0	3	2	1	0	7	6	5	4
1	0	5	6	3	4	1	2	7
2	5	0	3	6	1	4	7	2
3	2	3	0	1	6	7	4	5
4	7	6	5	4	3	2	1	0
5	4	1	2	7	0	5	6	3
6	1	4	7	2	5	0	3	6
7	6	7	4	5	2	3	0	1

Definition 11.46. A universal algebra $\mathfrak{A}$ is called polynomially complete if every function on $\mathfrak{A}$ with values in $\mathfrak{A}$ is a polynomial function [440].

Theorem 11.47. *Let $\mathfrak{V}$ be the variety of loops. $E \in \mathfrak{V}$ is polynomially complete if and only if E is a finite simple non-abelian loop or a finite simple non-associative loop [440, p. 108].*

Polynomially complete quasigroups are studied in [252, 15].

A finite quasigroup $Q = \{x_1, \dots, x_n\}$ is generated by a right shift [240] if the following property is satisfied: for all $1 \leqslant i, k, j \leqslant n$ we have

$$x_{i+k \pmod n} x_j = x_i x_{j-k \bmod n}.$$

It means that $a_{pq} = a_{rs}$, provided $pr \equiv sq \bmod n$.

Similarly a quasigroup Q is generated by a left shift if

$$x_{i+k \bmod n} x_j = x_i x_{j+k \bmod n}$$

for any indices $1 \leqslant i, j, k \leqslant n$. Properties of quasigroups generated by right or left shift are researched in [240, 15].

11.7.5 Cryptosystems which are based on row-Latin squares

A possible application of Latin power sets in cryptology is proposed in [244].

In [238] an encrypting device is described, based on row-Latin squares with a maximal period equal to the von Mangoldt function [933].

In our opinion, big perspectives has an application of row-Latin squares in various branches of contemporary cryptology ("neo-cryptology").

In [559] it is proposed to use: 1) row-Latin squares to generate an open key; 2) a conventional system for transmission of a message that is the form of a Latin square; 3) a row-Latin square analogue of the RSA system; and 4) a procedure of digital signature based on row-Latin squares.

Example 11.48. Let

$$L = \begin{matrix} 2 & 3 & 4 & 1 \\ 4 & 1 & 3 & 2 \\ 3 & 2 & 4 & 1 \\ 4 & 3 & 1 & 2 \end{matrix}$$

Then

$$L^7 = \begin{matrix} 4 & 1 & 2 & 3 \\ 4 & 1 & 2 & 3 \\ 3 & 2 & 4 & 1 \\ 3 & 4 & 2 & 1 \end{matrix}$$

$$L^3 = \begin{matrix} 4 & 1 & 2 & 3 \\ 1 & 2 & 3 & 4 \\ 1 & 2 & 3 & 4 \\ 3 & 4 & 2 & 1 \end{matrix}$$

Then

$$L^{21} = \begin{matrix} 2 & 3 & 4 & 1 \\ 1 & 2 & 3 & 4 \\ 1 & 2 & 3 & 4 \\ 4 & 3 & 1 & 2 \end{matrix}$$

is a common key for a user A with the key L^3 and a user B with the key L^7.

A public-key cryptosystem, using generalized quasigroup-based stream ciphers, as it noted earlier, is presented in [537].

11.7.6 Non-binary pseudo-random sequences over Galois fields

Non-binary pseudo-random sequences over GF[q] of length $q^m - 1$, called PN sequences, have been known for a long time [368]. PN sequences over a finite field GF[q] are unsuitable directly for cryptology because of their strong linear structure [534]. Usually PN sequences are defined over a finite field and often an irreducible polynomial is used for their generation.

In article [534] definition of the PN sequence was generalized with the purpose to use these sequences in cryptology.

We notice that in some sense ciphering is making a "pseudo-random sequence" from a plaintext, and cryptanalysis is a science how to reduce a check of all possible variants (cases) by deciphering some ciphertext.

These new sequences were called NLPN-sequences (non-linear pseudo-noise sequences). C. Koscielny proposed the following method for construction of NLPN-sequences.

Let $\overrightarrow{a}$ be a PN sequence of length $q^m - 1$ over GF[q], $q > 2$, i.e.,

$$\overrightarrow{a} = a_0 a_1 \dots a_{q^m-2}.$$

Let $\overrightarrow{a}^i$ be its cyclic i places shifted to the right. For example

$$\overrightarrow{a}^1 = a_1 \dots a_{q^m-2} a_0.$$

Let $Q = (SQ, \cdot)$ be a quasigroup of order q defined on the set of elements of the field GF[q].

Then $\overrightarrow{b} = \overrightarrow{a} \cdot \overrightarrow{a}^i$, $\overrightarrow{c} = \overrightarrow{a}^i \cdot \overrightarrow{a}$, where $b_j = a_j \cdot a_j^i$, $c_j = a_j^i \cdot a_j$ for any suitable value of index j ($j \in \{1, 2, \dots, q^m - 1\}$) are called NLPN sequences [534].

NLPN sequences have much more randomness than PN sequences. As C. Koscielny noticed, the method of construction of NLPN sequences is especially convenient for fast software encryption. It is proposed to use NLPN sequences by generation of keys. See also [530].

11.7.7 Authentication of a message

By authentication of a message we mean that it is possible for a receiver of a message to verify that the message has not been modified in transit, so that it is not possible for an interceptor to substitute a false message for a legitimate one.

By identification of a message we mean that it is possible for the receiver of a message to ascertain its origin, so that it is not possible for an intruder to masquerade as someone else.

By non-repudiation we mean that a sender should not be later able to falsely deny that he sent a message.

In [243] some quasigroup approaches to problems of identification of a message, non-repudiation of a message, production of a dynamic password, and digital fingerprinting are discussed. See also [217].

In [242] the authors suggested a new authentication scheme based on quasigroups (Latin squares). See also [241, 243, 229].

In [741] several cryptosystems based on quasigroups upon various combinatorial objects such as orthogonal Latin squares and frequency squares, block designs, and room squares are considered.

Definition 11.49. Let $2 \leqslant t < k < v$. A generalized $S(t, k, v)$ Steiner system is a finite block design $(T, \mathcal{B})$ such that (1) $|T| = v$; (2) $\mathcal{B} = \mathcal{B}' \cup \mathcal{B}''$, where any $B' \in \mathcal{B}'$, called a maximal block, has k points and $2 \leqslant |B''| < k$ for any $B'' \in \mathcal{B}''$, called a small block; (3) for any $B'' \in \mathcal{B}''$ there exists a $B' \in \mathcal{B}'$ such that $B'' \subseteq B'$; (4) every subset of T with t elements not belonging to the same $B'' \in \mathcal{B}''$ is contained in exactly one maximal block.

In [610] (see also [305]) an application of generalized $S(t, k, v)$ Steiner systems in cryptology is proposed, namely, there is introduced a new authentication scheme based on the generalized Steiner systems, and the properties of such scheme are studied in the generalized affine planes.

Authentication using matrix conjugation in groups is studied in [392]. In 2009 in her thesis [636, 637], Natalia Mosina presented a new probabilistic approach to prove the vulnerability of the authentication protocols on the braid groups without solving the underlying problem.

In [622] a new message authentication code called QMAC, whose security is based on non-associativity of quasigroups, is created.

11.7.8 Zero-knowledge protocol

In [713] Rivest introduced the All-or-Nothing (AON) encryption mode in order to devise a means to make brute-force search more difficult, by appropriately pre-processing a message before encrypting it. The method is general, but it was initially discussed for block-cipher encryption, using fixed-length blocks.

It is an unkeyed transformation, mapping a sequence of input blocks $(x_1, x_2, \dots, x_s)$ to a sequence of output blocks $(y_1, y_2, \dots, y_t)$ having the following properties:

Having all blocks $(y_1, y_2, \dots, y_t)$ it is easy to compute $(x_1, x_2, \dots, x_s)$.

If any output block y_j is missing, then it is computationally infeasible to obtain any information about any input block x_J.

The main idea is to preserve a small-length key (e.g., 64-bit) for the main encryption that can be handled by special hardware without enough processing power or memory. This gives a strong advantage to the method, since we can have strong encryption for devices that have minimum performance.

Several transformation methods have been proposed in the literature for AON. In the article [609] a special transform is proposed which is based on the use of a quasigroup (it is used in algorithm 1.69).

In [239] it is proposed to use isotopy of quasigroups in a zero-knowledge protocol.

Assume that the users $(u_1, u_2, ..., u_k)$ form a network. The user u_i has the public-key L_{u_i}, L'_{u_i} (denotes two isotopic Latin squares of order n) and the secret-key I_{u_i} (denotes the isotopism of L_{u_i} upon L'_{u_i}). The user u_i wants to prove the identity for u_j but he doesn't want to reveal the secret-key (zero-knowledge proof).

1. u_i randomly permutes L_{u_i} to produce another Latin square H.
2. u_i sends H to u_j.
3. u_j asks u_i to:
   a. prove that H and L'_{u_i} are isotopic, or
   b. prove that H and L_{u_i} are isotopic.
4. u_i complies. He either
   a. proves that H and L'_{u_i} are isotopic, or
   b. proves that H and L_{u_i} are isotopic.
5. u_i and u_j repeat steps 1 through 4 n times.

Remark 11.50. In the last procedure it is possible to use isotopy of n-ary groupoids.

11.7.9 Hamming distance between quasigroups

The following question is very important in construction of quasigroup-based cryptosystems: how big is the distance between different binary or n-ary quasigroups? Information on the Hamming distance between quasigroup operations is in the articles [272, 273, 270, 267, 271, 281, 898].

We recall, if α and β are two n-ary operations on a finite set Ω, then the Hamming distance of α and β is defined by

$$\text{dist}(\alpha, \beta) = |\{(u_1, \ldots, u_n) \in \Omega^n : \alpha(u_1, \ldots, u_n) \neq \beta(u_1, \ldots, u_n)\}|.$$

The author in [272] discusses Hamming distances of algebraic objects with binary operations. He also explains how the distance set of two quasigroups yields a 2-complex, and points out a connection with dissections of equilateral triangles.

For a fixed group $(G, \circ)$, $\delta(G, \circ)$ is defined to be the minimum of all such distances for $(G, \star)$ not equal to $(G, \circ)$, and $\nu(G, \circ)$ is the minimum for $(G, \star)$ not isomorphic to $(G, \circ)$.

In [267] it is proved that $\delta(G, \circ)$ is $6n - 18$ if n is odd, $6n - 20$ if $(G, \circ)$ is dihedral of twice odd order and $6n - 24$ otherwise for any group $(G, \circ)$ of order greater than 50. In [898] it is shown that $\delta(G, \circ) = 6p - 18$ for $n = p$, a prime, and $p > 7$.

In the article [270] there are listed a number of group orders for which the distance is less than the value suggested by the above theorems. The results obtained in this direction are also given [281].

11.7.10 Generation of quasigroups for cryptographical needs

An important cryptographical problem is the generation of "big" quasigroups which it is possible to keep easily in a compact form in computer memory. It is clear that for these aims the most suitable is a way to keep a little base and some procedures of obtaining a necessary element.

Therefore, we should have easily generated objects (cyclic group, abelian group, group), fast and complicated methods for their transformation (parastrophy, isotopy, isostrophy, crossed isotopy [756], homotopy, generalized isotopy), and their glue and blowing (direct product, semi-direct product, wreath product [471], crossed product, generalized crossed product). For these aims various linear quasigroups (especially n-ary quasigrous) are quite suitable [77, 600, 787].

In [664] the Boolean function is proposed for use in construction of n-ary and binary quasigroups.

A method of generating a practically unlimited number of quasigroups of an arbitrary (theoretically) order using the computer algebra system Maple 7 is presented in [535].

In this article [535] it is proposed to use isotopy of quasigroups and direct products of quasigroups. If we start from a class of finite groups, then, using these ways, it is possible to obtain only a class of quasigroups that are isotopic to groups.

We notice that there are many quasigroups (especially of large order) that are not isotopic to a group. Therefore for construction of quasigroups that are not isotopic to groups it is probably better to use the concept of n-ary isostrophy [57] and gisotopy [643, 787].

In [608], ']'the authors give an efficient algorithm for constructing huge quasigroups of orders $(2^s)^{2^t}$ by using extended Feistel networks. ... I see this work as a significant contribution to the application of quasigroups in cryptography."(MR2536710, D. Gligoroski).

Definition 11.51. [608] Let $(G, +)$ be an Abelian group, let $f : G \to G$ be a mapping, and let $a, b, c \in G$ be constants. The extended Feistel network $F_{a,b,c} : G^2 \to G^2$ created by f is defined for every $l, r \in G$ by

$$F_{a,b,c}(l, r) = (r + a, l + b + f(r + c)). \tag{11.27}$$

The extended Feistel network $F_{a,b,c}$ is a bijection with inverse

$$F^{-1}_{a,b,c}(l, r) = (r - b - f(l + c - a), l - a).$$

Notice, it is important from a cryptographical point of view that by construction of an extended Feistel network $F_{a,b,c}$ the non-linear transformations are used.

In many cases in cryptography it is possible to change associative systems by non-associative ones and in practically any case this change gives in some sense better results than use of associative systems. Quasigroups in spite of their simplicity, have various applications in cryptology. Many new cryptographical algorithms can be formed on the basis of quasigroups.

Appendix A

Appendix

A.1 The system of German banknotes 509
A.2 Outline of the history of quasigroup theory 510
A.3 On 20 Belousov problems 512

A.1 The system of German banknotes

The system of the serial numbers of German banknotes is one of the oldest and the most famous check digit systems with one check symbol.

This system was constructed over the dihedral group $(D_{10}, +)$ of order 10 with the check equation $\delta^1 a_1 + \delta^2 a_2 + \cdots + \delta^{10} a_{10} + a_{11} = 0$, where numbers a_1^{10} are information symbols, and number a_{11} is a check digit, and $\delta = (0\,1\,5\,8\,9\,4\,2\,7)(3\,6)$ is an anti-symmetric mapping [752, 751]. This mapping was found by J. Verhoeff [896]. It is well known that this code detects all single and all transposition errors [227].

We enumerate twin, spring twin, spring transposition and phonetic errors that this code cannot detect. We used the following Cayley table of the group $(D_{10}, +)$.

+	0	1	2	3	4	5	6	7	8	9
0	0	1	2	3	4	5	6	7	8	9
1	1	2	3	4	0	6	7	8	9	5
2	2	3	4	0	1	7	8	9	5	6
3	3	4	0	1	2	8	9	5	6	7
4	4	0	1	2	3	9	5	6	7	8
5	5	9	8	7	6	0	4	3	2	1
6	6	5	9	8	7	1	0	4	3	2
7	7	6	5	9	8	2	1	0	4	3
8	8	7	6	5	9	3	2	1	0	4
9	9	8	7	6	5	4	3	2	1	0

The system of the serial numbers of German banknotes cannot detect the following twin errors:

$66 \leftrightarrow 44 \quad 77 \leftrightarrow 99$ in place $(1,2)$;
$33 \leftrightarrow 99 \quad 22 \leftrightarrow 88$ in place $(2,3)$;
$44 \leftrightarrow 55 \quad 66 \leftrightarrow 88$ in place $(3,4)$;
$33 \leftrightarrow 55 \quad 99 \leftrightarrow 11$ in place $(4,5)$;
$11 \leftrightarrow 66 \quad 00 \leftrightarrow 88$ in place $(5,6)$;
$00 \leftrightarrow 33 \quad 55 \leftrightarrow 77$ in place $(6,7)$;
$11 \leftrightarrow 22 \quad 66 \leftrightarrow 77$ in place $(7,8)$;
$22 \leftrightarrow 33 \quad 00 \leftrightarrow 44$ in place $(8,9)$;
$66 \leftrightarrow 44 \quad 77 \leftrightarrow 99$ in place $(9,10)$;
$11 \leftrightarrow 88 \quad 44 \leftrightarrow 77$ in place $(10,11)$.

We notice, in place $(10, 11)$, this code cannot detect the following transposition errors $2\,9 \leftrightarrow 9\,2$, $5\,8 \leftrightarrow 8\,5$.

In place $(1, 3)$ the code cannot detect the following errors:

$$\begin{array}{l} 000 \leftrightarrow 202, 303 \leftrightarrow 404, 505 \leftrightarrow 909, 002 \leftrightarrow 200, 304 \leftrightarrow 403, 509 \leftrightarrow 905, \\ 212 \leftrightarrow 919, 717 \leftrightarrow 818, 219 \leftrightarrow 912, 718 \leftrightarrow 817, 020 \leftrightarrow 525, 222 \leftrightarrow 929, \\ 121 \leftrightarrow 626, 025 \leftrightarrow 520, 126 \leftrightarrow 621, 229 \leftrightarrow 922, 232 \leftrightarrow 737, 838 \leftrightarrow 939, \\ 237 \leftrightarrow 732, 839 \leftrightarrow 938, 040 \leftrightarrow 545, 141 \leftrightarrow 646, 848 \leftrightarrow 949, 045 \leftrightarrow 540, \\ 146 \leftrightarrow 641, 849 \leftrightarrow 948, 050 \leftrightarrow 858, 252 \leftrightarrow 757, 058 \leftrightarrow 850, 257 \leftrightarrow 752, \\ 161 \leftrightarrow 464, 363 \leftrightarrow 666, 565 \leftrightarrow 767, 164 \leftrightarrow 461, 161 \leftrightarrow 464, 366 \leftrightarrow 663, \\ 070 \leftrightarrow 272, 777 \leftrightarrow 878, 072 \leftrightarrow 270, 778 \leftrightarrow 877, 080 \leftrightarrow 888, 383 \leftrightarrow 484, \\ 585 \leftrightarrow 989, 088 \leftrightarrow 880, 384 \leftrightarrow 483, 589 \leftrightarrow 985, 191 \leftrightarrow 494, 393 \leftrightarrow 696, \\ 595 \leftrightarrow 797, 194 \leftrightarrow 491, 396 \leftrightarrow 693, 597 \leftrightarrow 795. \end{array}$$

As in place $(1, 3)$, in places $(2, 4)$, $(3, 5)$, $(4, 6)$, $(5, 7)$, $(6, 8)$, $(7, 9)$, $(8, 10)$ this code cannot detect 104 transposition and twin errors. In place $(9, 11)$ this code cannot detect 144 transposition and twin errors, namely:

$$\begin{array}{l} 202 \leftrightarrow 303, 404 \leftrightarrow 505, 004 \leftrightarrow 400, 405 \leftrightarrow 504, 212 \leftrightarrow 313 \leftrightarrow 717, \\ 010 \leftrightarrow 818, 111 \leftrightarrow 919, 014 \leftrightarrow 410, 018 \leftrightarrow 810, 119 \leftrightarrow 911, 217 \leftrightarrow 712, \\ 317 \leftrightarrow 713, 020 \leftrightarrow 424 \leftrightarrow 828, 121 \leftrightarrow 616, 323 \leftrightarrow 929, 028 \leftrightarrow 820, \\ 126 \leftrightarrow 621, 223 \leftrightarrow 322, 329 \leftrightarrow 923, 428 \leftrightarrow 824, 030 \leftrightarrow 434, 131 \leftrightarrow 939, \\ 232 \leftrightarrow 636, 333 \leftrightarrow 737, 139 \leftrightarrow 931, 233 \leftrightarrow 332, 236 \leftrightarrow 632, 337 \leftrightarrow 733, \\ 242 \leftrightarrow 343 \leftrightarrow 949, 141 \leftrightarrow 646, 444 \leftrightarrow 848, 044 \leftrightarrow 440, 146 \leftrightarrow 641, \\ 249 \leftrightarrow 942, 349 \leftrightarrow 943, 448 \leftrightarrow 844, 151 \leftrightarrow 757, 252 \leftrightarrow 353, 054 \leftrightarrow 450, \\ 157 \leftrightarrow 751, 262 \leftrightarrow 363 \leftrightarrow 666, 060 \leftrightarrow 565, 064 \leftrightarrow 460, 065 \leftrightarrow 560, \\ 266 \leftrightarrow 662, 366 \leftrightarrow 663, 070 \leftrightarrow 474 \leftrightarrow 575, 171 \leftrightarrow 777, 272 \leftrightarrow 979, \\ 075 \leftrightarrow 570, 177 \leftrightarrow 771, 273 \leftrightarrow 372, 475 \leftrightarrow 574, 279 \leftrightarrow 972, 080 \leftrightarrow 484, \\ 282 \leftrightarrow 787, 283 \leftrightarrow 382, 287 \leftrightarrow 782, 090 \leftrightarrow 494, 393 \leftrightarrow 696, 293 \leftrightarrow 392, \\ 396 \leftrightarrow 693. \end{array}$$

The system of the serial numbers of German banknotes cannot detect the following phonetic errors: $20 \leftrightarrow 12$ in place $(5, 6)$, $50 \leftrightarrow 15$ in place $(8, 9)$ and $70 \leftrightarrow 17$ in place $(10, 11)$.

By writing computer programs in order to research possibilities of the system of serial numbers of German banknotes, we have used algorithms of generation of all elements of the group S_n which are described in [574].

A.2 Outline of the history of quasigroup theory

A good historical survey of the development of loop and quasigroup theory is in the article of Hala Orlyk Pflügfelder [686] and the Habilitation Dissertation of Hubert Kiechle [512].

Quasigroup theory began its development in the 1920s and 1930s. At the end of 19th century David Hilbert published fundamental works devoted to the axiomatization of mathematics, in particular, devoted to the axiomatization of geometry. Subsequent to these publications, quite intensive researches which were devoted to the basics of geometry were started.

Mainly, various systems of axioms of finite and infinite geometries, including Euclidian geometry, projective geometries, and hyperbolic geometries in dimensions 2 and 3 were studied.

Since systems of coordinates of these geometries are connected with various algebraic systems (with fields, skew field, near-fields, groups, semigroups), then various systems of axioms of these algebraic systems were also studied. For example, Dickson studied semigroups in 1905 [249].

Therefore, at the end of 19th century the study of objects, which appear by some changing of a system of axioms for a group or a geometry, was started. Partly this has been caused by the aspiration to research questions of independence of these systems of axioms.

Notice that Wilhelm Dörnte, following the advice of Emmy Noether, studied n-ary quasigroups as some generalizations of binary groups [264], Anton Suschkewitsch (Sushkevich in more modern English transliteration) [863, 864] studied quasigroups with some additional conditions (postulates), and Burstin and Mayer studied distributive quasigroups [185].

Clifford and Preston in their book [211] have written, that the "theory (i.e., semigroup theory) really began in 1928 with the publication of a paper of fundamental importance by A.K. Suschkewitch" [862]. Therefore, Suschkewitsch is one of the founders of both quasigroup and semigroup theories.

The term "quasigroup" appeared in Ruth Moufang's article [638], which was devoted to problems of coordinatization of some projective planes [22, 407]. It is possible also to say that the term "quasigroup" appeared in the study of independence of axioms of a projective plane.

By quasigroup, Ruth Moufang has called an object which is now called a Moufang loop.

She defined it as an IP-loop $(Q, \cdot)$ which satisfies an identity of a weak associativity. R. Moufang studied IP-loops with the identity $\alpha(\gamma \cdot \alpha\beta) = (\alpha\gamma \cdot \alpha)\beta$ ("Quasi-gruppe Q^*") and the identity $(\alpha\beta)(\gamma\alpha) = \alpha((\beta\gamma)\alpha)$ ("Quasi-gruppe Q^{**}"), where $\alpha, \beta, \gamma \in Q$ [638].

The identities mentioned above are now called left and middle Moufang identities, respectively. G. Bol and D.C. Murdoch proved that in an IP-loop these identities are equivalent ([167, p. 293], [173, p. 115].

We can say that a mathematical theory has been born, if this theory has important definitions and a pair of deep theorems. Ruth Moufang not only gives good definitions (this work has been also done by W. Dörnte), but she also proved probably till now the deepest theorem of quasigroup theory, namely, she proved that any Moufang loop is diassociative. Of course, this theorem was proved with great parallelism with Artin's theorem on the properties of alternative algebras.

Slightly later (1937) A.K. Suschkewitsch defined medial (abelian) quasigroups [864, p. 157]. See also pioneer works of A.A. Albert [10, 11], D.C. Murdoch [645, 646], K. Toyoda [890], R.H. Bruck [166], R. Baer [24, 25], and E.L. Post [701].

We recall that in the 1930s the concept of net (of web) was introduced. The concept of quasigroup in terms of net theory has natural and clear geometrical interpretation [72, 76, 79].

Latin squares are a combinatorial analog of finite binary quasigroups. We notice that the study of Latin squares was started in the frameworks of combinatorics and popular (entertaining) mathematics much earlier than the study of quasigroups in the frameworks of abstract algebra [240, 559, 86]. We only remind readers that Latin squares are used for construction of various magic squares. Orthogonality is one of the most popular among researchers properties of Latin squares, because, probably, this property is one of the most "applicable" properties of Latin squares. The pioneer in the development of Experimental Design was the English statistician R.A. Fisher who worked in the 1930s [324].

Quasigroups as solutions of some functional equations of mathematical logic implicitly (without their name) had appeared in works of German logician Ernst Schröder [437, 552].

The first instance, where a non-associative loop occurs, is probably the multiplicative loop of "Cayley numbers" or "octonions" [548, 80, 512].

At present Quasigroup Theory, Theory of Latin Squares and Net (Web) Theory are developing quite intensively. Also, although each of these theories has its own successes, tasks and problems, they are developing with a great deal of cross-fertilization and are enriching each other.

Quasigroups have various applications in other branches of mathematics, in other sciences and they even have direct applications in practice [559]. For instance, quasigroups have been used in statistics for a long time, for example they (as Latin squares) are used in the theory of design of experiments [324, 597, 527, 304, 26], in the theory of differential equations [201], in differential geometry [727, 201], hyperbolic geometry [893], in physics (for instance, in relativistic mechanics [376, 657, 575]), automata theory [401, 402], coding theory [817, 209, 641, 642], and cryptology [240, 241, 243, 533, 537, 606].

Taking into consideration information accessible to the authors, since the works of Schauffler [748] it is clear that quasigroups have direct applications in cryptography. Latin squares have been used in cryptology, probably from the moment of origin of this science.

V.D. Belousov (1925–1988) obtained fundamental results in the theory of binary and n-ary quasigroups, in the theory on nets and functional equations [72, 80, 77, 76, 78, 85, 121, 293].

A.3 On 20 Belousov problems

Before the conference LOOPS'99, Prof. Hala Orlik Pflugfelder posed a question on the situation with 20 of Belousov's problems. These problems are at the end of V.D. Belousov's book *Foundations of the theory of quasigroups and loops*, Nauka, Moscow, 1967, (in Russian). We give information about these problems in [775]. Here we present update of this survey.

Problem 1a). Find necessary and sufficient conditions that a special loop is isotopic to a left F-quasigroup.

This problem is solved partially in I.A. Florea, M.I. Ursul: *F-quasigroups with an inverse property*. Voprosy teorii kvasigrupp i lup. Kishinev, Shtiintsa, 1970, 145–156 (in Russian).

It is proved that a left F-quasigroup with IP-property is isotopic to an A-loop.

In finite case this problem is also solved. See Chapter 6 of this book.

Problem 1b). Is some identity fulfilled in a special loop?

Yes. L.R. Soikis: *On special loops*. Voprosy teorii kvasigrupp i lup. Kishinev, Shtiintsa, 1970, 122–131 (in Russian).

Problem 1c). To what loops are two-sided F-quasigroups isotopic?

A left (right) F-quasigroup is isotopic to a left (right) M–loop.

V.D. Belousov: *Elements of quasigroup theory. A special course.* Kishinev, 1981, 115 pp., (in Russian).

Kepka posed the following problem. All known examples of two-sided F-quasigroups are isotopic to a Moufang loop. Is it true that every two-sided F-quasigroup is isotopic to a Moufang loop?

T. Kepka, M. Kinyon and J.D. Philips proved that any F-quasigroup is an isotope of a Moufang loop.

T. Kepka, M. Kinyon and J.D. Philips: *The structure of F-quasigroups*, arxiv.org/, abs/math/0510298, 2005; T. Kepka and M.K. Kinyon and J.D. Phillips, *The structure of F-quasigroups*, Journal of Algebra, **317**, 2007, 435–461.

Problem 2. Let $Q(\cdot)$ be a group and let $x \circ y = z_1^{\varepsilon_1} z_2^{\varepsilon_2} \dots z_n^{\varepsilon_n}$, where $z_i = x$ or $z_i = y$, $\varepsilon = \pm 1$ $(i = 1, 2, \dots, n)$. For what sequence of values of ε_i groupoid $Q(\circ)$ is a quasigroup?

Partial results are given in the article

S.V. Larin: *About a quasigroup operation on a group.* Matem. zapiski Krasnoyarskogo gos. ped. instituta, 1970, vyp. 3, 20–26 (in Russian).

Problem 3. A quasigroup $Q(\cdot)$ is called a Stein quasigroup if the identity $x \cdot xy = yx$ holds in the quasigroup $Q(\cdot)$. To what loops are Stein quasigroups isotopic?

A semisymmetric $(x \cdot yx = y)$ Stein quasigroup is isotopic to a loop of exponent 2; see:

G.B. Belyavskaya, A.M. Cheban: *On semisymmetric Stein quasigroups.* Matem. issledov. VII:3(25), 1972, 231–237 (in Russian).

General case: Nothing is known to us.

Problem 4. A quasigroup $Q(A)$ is called *σ–quasigroup* if ${}^{\sigma}A = A$, where ${}^{\sigma}A$ is a parastroph of a quasigroup $Q(A)$. Is a quasigroup A an isotope of some *σ–quasigroup*, if ${}^{\sigma}A = A^T$? (A^T is an isotope of a quasigroup A.)

There are some results in the article A. Sade, Criteres d'isotopie d'un quasigroupe avec un quasigroupe demi-symetrique. Univ. Lisboa Revista Fac. Ci. A (2) 11 1964/1965, 121–136.

Problem 5. Construct and develop a theory of homotopies for quasigroups.

Many authors wrote on this subject. Main: Ja.Ja. Wedel', published in the 1970s and 1980s, A.A. Gvaramija in the 1980s, J.D.H. Smith in the most recent period.

Ja.Ja. Wedel': *On homotopy of quasigroups.* Voprosy teorii kvasigrupp i lup. Kishinev, Shtiintsa, 1970, 30–37 (in Russian).

Ja.Ja. Wedel': *Homotopy of quasigroups.* Sibirsk. Mat. Zh., 1970, II, No. 6, 1236–1246 (in Russian).

Ja.Ja. Wedel': *Homotopy of quasigroups.* Mat. Vesn., 1970, 7, No. 4, 493–506 (in Russian).

I.A. Golovko: *Some remarks on endotopies in quasigroups* Bul. Akad. Ştiinţe RSS Moldoven 1970, no. 1, 12–17 (in Russian).

L.A. Sokolova: *Some properties of homotopical mappings of quasigroups.* Issledov. po obshchei algebre. Kishinev, 1968, vyp. 1, 71–73 (in Russian).

P. Das: *Kernel of homotopy.* Studia Sci. Math. Hungar. 12 (1977), no. 1–2, 89–93.

A.A. Gvaramija, B.I. Plotkin: *The homotopies of quasigroups and universal algebras*, in Universal Algebra and Quasigroup Theory, A. Romanovska, J.D.H. Smith (eds.), Heldermann, Berlin, 1992, pp. 89–99.

J.D.H. Smith: *Homotopy and semisymmetry of quasigroups.* Algebra Universalis 38 (1997), 175–184.

Problem 6. Construct a theory of extensions of quasigroups. Quasigroup extensions were studied by Fenyves:

F. Fenyves: *The Schreier extension of quasigroups.* Publ. Math. Debrecen, v.14, 1967, 35–44 (in Russian).

Extensions of Moufang and Bol loops were studied by:

N. Nishigori: *On skew product of group and loop.* Bull. Sch. Educ. Hiroshima Univ.

Math. and Math. Educ., 1985, 8, No. 2, 113–119, and in his earlier articles. See also the book P.T. Nagy and K. Strambach [654].

G.B. Belyavskaya and A.D. Lumpov defined and researched the so-called crossed product of two systems of quasigroups:

G.B. Belyavskaya, A.D. Lumpov: *A crossed product of two systems of quasigroups and its using for construction partially orthogonal quasigroups.* Issledov. po teorii binarnyh i n–arnyh kvasigrupp, vyp. 83, Kishinev, Shtiintsa, 1985, 26–38 (in Russian).

The singular product of quasigroups was studied in

C. Lindner: *Construction of quasigroups using the singular direct product.* Proc. Amer. Math. Soc., 1972, 29, No. 2, 263–266.

The last two constructions are given in Section 1.6.

Problem 7. Spread (diffuse) theorems on prolongation of groups on loops (in particular, on Moufang loops and other classes of loops).

Prolongations of quasigroups were researched by many authors, for example, from Belousov's school by G.B. Belyavskaya, V.I. Izbash, and I.I. Derienko with W.A. Dudek.

Information about quasigroup prolongations is given in Sections 1.7.5 and 10.2.5.

Problem 8. Does a left distributive quasigroup exist isotopic to a loop and different from a Bol loop?

The positive answer was given by Vasilii Ivanovich Onoi.

V.I. Onoi: *Left distributive quasigroups that are left homogeneous over a quasigroup.* Bul. Akad. Štiince RSS Moldoven. 1970 no. 2, 24–31 (Russian).

Problem 9. If question 8 has the answer "yes," then the problem of studying such loops arises; in particular, it is required to find what identities are true in these loops.

Partial answer: V.D. Belousov and V.I. Onoi defined S-loops, but for S-loops there is no known definition in the language of identities.

V.D. Belousov and V.I. Onoi: *Loops that are isotopic to left-distributive quasigroups.* Matem. issledov., Kishinev, Shtiintsa, 1972, 7, vyp. 3, 135–152.

Problem 10. Find necessary and sufficient conditions for a loop in order that this loop core is a left distributive groupoid (quasigroup).

There exist partial results on this subject: for example, A.A. Gvaramija and P.N. Syrbu wrote on this problem for some classes of loops. But the general case is not solved.

A.A. Gvaramija: *A certain class of loops.* Algebra and number theory. Moskov. Gos. Ped. Inst. Učen. Zap. No. 375 (1971), 25–34 (in Russian).

Problem 11. Find a loop for which a special property is universal.

This problem is solved by:

L. R. Soĭkis: *The special loops.* Questions of the Theory of Quasigroups and Loops, 1970, pp. 122–131, Redakc.-Izdat. Otdel Akad. Nauk Moldav. SSR, Kishinev.

Problem 12. Find classes of G-loops different from classes described in Chapter X, Section 1° of the book by V.D. Belousov: *Foundations of the theory of quasigroups and loops* (in Russian), Nauka, Moscow, 1967. Describe class of all G-loops (for example, using identities).

Many authors wrote on this theme, mainly Alexandr Savel'evich Basarab, D.A. Robinson, Edgar Goodaire, K. Kunen. For example:

A.S. Basarab: *Osborn's G–loops.* Quasigroups Related Systems, v. 3, N. 1, 1996, 1–7.

A.S. Basarab. *Generalized Moufang G–loops.* Quasigroups Related Systems, v. 1, N. 1, 1994, 51–57.

R.L. Wilson, *Isotopy–isomorphy loops of prime order*, J. Algebra, 31, (1974), 117–119.

E.G. Goodaire, D.A. Robinson: *A class of loops which are isomorphic to all loop isotopes.* Canadian J. Math. 34 (1982) 662–672.

K. Kunen: *G-loops and permutation groups.* J. Algebra, 220, (1999), no. 2, 694–708.

There are some other articles about G–loops on K. Kunen's site. He proved that the class of all G-loops does not form a variety and hence it is impossible to characterize the class of G-loops using identities.

Problem 13. Find other closure conditions in nets and construct a general theory of closure conditions.

A very general method of receiving closure conditions for nets was obtained in the article:

V.D. Belousov, V.V. Ryzhkov: *On a method of obtaining closure figures.* Matem. issledov. 1:2, 1966, Kishinev, Shtiintsa, 140–150 (in Russian).

Problem 14. Research loops with the identity $z[(xy)\backslash z] = (z/y)(x\backslash z)$.

A class of middle Bol loops is researched actively. We point out the most important articles.

A.A. Gvaramija: *A certain class of loops.* Algebra and number theory. Moscow. Gos. Ped. Inst. Učen. Zap. No. 375 (1971), 25–34 (in Russian).

I. Grecu, P. Syrbu: *Commutants of middle Bol loops.* Quasigroups Relat. Syst. 22, No. 1, (2014), 81–88.

I. Grecu, P. Syrbu: *On some isostrophy invariants of Bol loops.* Bull. Transilv. Univ. Braşov, Ser. III, Math. Inform. Phys. 5(54), Spec. Iss., (2012), 145–153.

Problem 15. Is infinite anticommutative distributive quasigroup transitive?

Yes. N.I. Sandu: *Medially nilpotent distributive quasigroups and CH–quasigroups.* Sibirsk. Mat. Zh. 28 (1987), no. 2, 159–170 (in Russian).

Problem 16. Let $S(\cdot)$ be a semigroup with cancellation (i.e., from $xy = xz$ or $yx = zx$ follows $y = z$). Is it possible to include this semigroup in a quasigroup (in a loop) with nontrivial identity? (An identity is called nontrivial if there exists a quasigroup of order $q > 1$ in which this identity is not true.)

We do not know articles on this problem.

Problem 17. Find necessary and sufficient conditions in order to have a groupoid $Q(\cdot)$ isomorphically imbedded in a Moufang loop, in a IP–loop, in a distributive quasigroup?

We do not know articles on this problem.

Problem 18. From what identities, that are true in a quasigroup $Q(\cdot)$, does it follow that the quasigroup $Q(\cdot)$ is a loop? (An example of such identity is the identity of associativity).

There are partial answers. If in a quasigroup one of middle Moufang identities holds, then this quasigroup is a loop.

V.M. Galkin: *Quasigroups.* Itogi nauki i tehniki. Ser. Algebra. Topologia. Geometria. Tom 26. Moskva, VINITI, 1988, 3–44 (in Russian).

If in a quasigroup any of the Moufang identities holds, then this quasigroup is a Moufang loop.

K. Kunen: *Moufang quasigroups.* J. Algebra, 183, 1996, N.1, 231–234.

V.A. Shcherbacov, V.I. Izbash: *On quasigroups with Moufang identity.* Bul. Acad. Ştiinţe Repub. Mold. Mat. 1998, no. 2, 109–116.

Quasigroups with identities from the Fenyves list [319] are researched in the article:

K. Kunen. *Quasigroups, loops and associative laws.* J. Algebra, 185, N.1, 1996, 194–204.

In the paper [320] N. C. Fiala has investigated (with computer aid) which quasigroup identities involving at most six variable occurrences imply that the quasigroup is necessarily a non-trivial loop or group. He obtained the following result: There are exactly 35 quasigroup identities connected by the product operation only and with at most six variable occurrences which imply that the quasigroup is necessarily a non-trivial loop or group. Additional information in this direction is given in [483, 694].

Problem 19. Are there loops with type (L, F, R) and with assignment intersections $(L \cap F)$, $(F \cap R)$, $(L \cap R)$, $(L \cap F \cap R)$?

There are some results in Belousov's articles and in articles by E.M. Sadovnikov. General case is not solved.

E.M. Sadovnikov: *The amalgamation of the nuclei of a loop.* Nets and quasigroups. Mat. Issled. Vyp. 39 (1976), 152–171 (in Russian).

Problem 20. What are quasigroups or loops in which all congruences are normal? (In IP–quasigroups all congruences are normal.)

There are some results, for example:

V.A. Shcherbacov: *On congruences of quasigroups.* Kishinev, IM s VTS AN SSRM, 1990, 16 p., Dep. v VINITI 01.08.1990, No. 4413–B90.

V.A. Shcherbacov: *On Bruck-Belousov problem.* Bul. Acad. Stiinte Repub. Mold., Mat., no. 3, 2005, 123–140.

References

[1] A.S. Abashin. Linear recursive MDS-codes of dimensions 2 and 3. *Diskret. Mat.*, 12(2):140–153, 2000 (in Russian).

[2] J. Aczel. On mean values. *Bul. Amer. Math. Soc.*, 54:392–400, 1948.

[3] J. Aczel. Quasigroups-nets-nomograms. *Mat. Lapok*, 15:114–162, 1964.

[4] J. Aczel, V.D. Belousov, and M. Hosszu. Generalized associativity and bisymmetry on quasigroups. *Acta Math. Acad. Sci. Hungar.*, 11:127–136, 1960.

[5] J. Aczel and J. Dhombres. Functional equations in several variables, *Encyclopedia of Mathematics and Its Applications, 31.* Cambridge University Press, Cambridge, 1989.

[6] J.O. Adeniran and Y.T. Oyebo. On the crypto-automorphism of the Buchsteiner loops. *International J.Math. Combin.*, 2:1–6, 2010.

[7] Reza Akhtar, Ashley Arp, Michael Kaminski, Jasmine Van Exel, Davian Vernon, and Cory Washington. The varieties of Bol-Moufang quasigroups defined by a single operation. *Quasigroups Related Systems*, 20(1):1–10, 2012.

[8] M.A. Akivis and V.V. Goldberg. Solution of Belousov's problem, 2000. Arxiv:math.GR/0010175.

[9] M.A. Akivis and V.V. Goldberg. Solution of Belousov's problem. *Discuss. Math. Gen. Algebra Appl.*, 21(1):93–103, 2001.

[10] A.A. Albert. Quasigroups. I. *Trans. Amer. Math. Soc.*, 54:507–519, 1943.

[11] A.A. Albert. Quasigroups. II. *Trans. Amer. Math. Soc.*, 55:401–419, 1944.

[12] A.P. Alferov, A.Yu. Zubov, A.S. Kuz'min, and F.V. Cheremushkin. *Foundations of Cryptography.* Gelios ARV, Moscow, 2005 (in Russian).

[13] J. Arkin. A solution to the classical problem of finding systems of three mutually orthogonal numbers in a cube formed by three superimposed $10 \times 10 \times 10$ cubes. *Fibonacci Quart.*, 11:485–489, 1973.

[14] J. Arkin and E. G. Straus. Latin k-cubes. *Fibonacci Quart.*, 12:288–292, 1974.

[15] V. A. Artamonov, S. Chakrabarti, S. Gangopadhyay, and S. K. Pal. On Latin squares of polynomially complete quasigroups and quasigroups generated by shifts. *Quasigroups and Related Systems*, 21(2):117–130, 2013.

[16] R. Artzy. On loops with a special property. *Proc. Amer. Math. Soc.*, 6:448–453, 1955.

[17] R. Artzy. Crossed-inverse and related loops. *Trans. Amer. Math. Soc.*, 91:480–492, 1959.

[18] R. Artzy. Relation between loop identities. *Proc. Amer. Math. Soc.*, 11:847–851, 1960.

[19] R. Artzy. Cayley diagrams of binary systems. *Duke Math. J.*, 28:491–495, 1961.

[20] R. Artzy. Isotopy and parastrophy of quasigroups. *Proc. Amer. Math. Soc.*, 14:429–431, 1963.

[21] R. Artzy. Net motions and loops. *Arch. Math. (Basel)*, 14:95–101, 1963.

[22] R. Artzy. *Linear geometry.* Addison-Wesley, 1965.

[23] R.E. Atani, Sh.E. Atani, and S. Mirzakuchaki. Public key cryptography based on semimodules over quotient semirings. *International Mathematical Forum*, 2(52):2561–2570, 2007.

[24] R. Baer. Nets and groups, I. *Trans. Amer. Math. Soc.*, 46:110–141, 1939.

[25] R. Baer. Nets and groups, II. *Trans. Amer. Math. Soc.*, 47:435–439, 1940.

[26] R.A. Bailey. *Association Schemes: Designed Experiments, Algebra and Combinatorics.* Cambridge University Press, Cambridge, 2004.

[27] R.A. Bailey and P. Cameron. Encyclopaedia of design theory, 2006. http://designtheory.org/library/encyc/glossary/#top.

[28] R.A. Bailey. Efficient semi-Latin squares. *Statistica Sinica*, 2:413–437, 1992.

[29] H.J. Baker and F. Piper. *Cipher Systems: The Protection of Communications.* Northwood, London, 1982.

[30] V. Bakeva and V. Dimitrova. Some probabilistic properties of quasigroup processed strings useful in cryptanalysis. In *ICT-Innovations 2010*, pages 61–70. Springer, 2010.

[31] V. Bakeva, V. Dimitrova, and A. Popovska-Mitrovikj. Parastrophic quasigroup string processing. In *Proceedings of the 8th Conference on Informatics and Information Technologies with International Participation*, pages 19–21, 2011. http://www.ii.edu.mk/ciit/data/CIITZbornik2011.pdf.

[32] V. Bakeva and N. Ilievska. A probabilistic model of error-detecting codes based on quasigroups. *Quasigroups Related Systems*, 17(2):135–148, 2009.

[33] E. Bannai and T. Ito. *Algebraic Combinatorics. Association Schemes.* Mir, Moscow, 1987 (in Russian).

[34] A. Barlotti and K. Strambach. The geometry of binary systems. *Adv. in Math.*, 49:1–105, 1983.

[35] A.S. Basarab. On a class of WIP-loops. *Mat. Issled.*, 2(2):2–24, 1967.

[36] A.S. Basarab. *Loops with Weak Inverse Property.* PhD thesis, IM AN MSSR, 1968 (in Russian).

[37] A.S. Basarab. A class of LK-loops. *Mat. Issled.*, 120:3–7, 1991 (in Russian).

[38] A.S. Basarab. Osborn's G-loops. *Quasigroups Related Systems*, 1(1):51–56, 1994.

[39] A.S. Basarab. Generalized Moufang G-loops. *Quasigroups Related Systems*, 3(1):1–5, 1996.

[40] A.S. Basarab and A.I. Belioglo. Osborn UAI-loops. *Mat. Issled.*, 51:8–13, 1979.

[41] A.S. Basarab and A.I. Belioglo. Universal-automorphic-inverse G-loops. *Mat. Issled.*, 51:3–7, 1979.

[42] M. Basu, D. K. Ghosh, and S. Bagchi. Pairwise balanced design of order $6n+4$ and 2-fold system of order $3n+2$. *ISRN Discrete Mathematics*, 2012(ID 251457):7 pages, 2012. doi:10.5402/2012/251457.

[43] G.E. Bates and F. Kiokemeister. A note on homomorphic mappings of quasigroups into multiplicative systems. *Bull. Amer. Math. Soc.*, 54:1180–1185, 1948.

[44] R. Bauer. *The Algebra and Geometry of Polyadic Quasigroups and Loops.* PhD thesis, Rutgers State Univ., 1968.

[45] B. Baumeister, G. Stroth, and A. Stein. On Bruck loops of 2-power exponent. *J. Algebra*, 327:316–336, 2011.

[46] B. Baumeister, G. Stroth, and A. Stein. The finite Bruck loops. *J. Algebra*, 330:206–220, 2011.

[47] R. Bean. Critical sets in the elementary abelian 2- and 3-groups. *Util. Math.*, 68:53–61, 2005.

[48] R. Bean, D. Donovan, A. Khodkar, and A.P. Street. Steiner trades that give rise to completely decomposable Latin interchanges. *Int. J. Comput. Math.*, 79(12):1273–1284, 2002.

[49] D.F. Beckley. An optimum systems with modulo 11. *The Computer Bulletin*, 11:213–215, 1967.

[50] D. Bedford. Orthomorphisms and near orthomorphisms of groups and orthogonal latin squares: A survey. *Bull. Inst. Combin. Appl.*, 13:13–33, 1995.

[51] V. A. Beglaryan and K. K. Shchukin. The structure of tri-abelian totally symmetric quasigroups. *Mathematics, Erevan. Univ.*, (3):82–88, 1985 (in Russian).

[52] V.A. Beglaryan. *On the Theory of Homomorphisms in Quasigroups.* PhD thesis, IM AN MSSR, 1982 (in Russian).

[53] A.S. Bektenov. Algebraic (k,n)-nets and orthogonal systems of n-ary quasigroups. *Bul. Akad. Stiince RSS Moldoven*, (1):3–11, 1974 (in Russian).

[54] A.S. Bektenov. Certain properties of a spatial 3-dimensional net and its coordinate ternary quasigroup. *Mat. Issled.*, 39:10–20, 1976 (in Russian).

[55] A.S. Bektenov. Configurations in spatial nets. *Mat. Issled.*, 71:7–21, 1983 (in Russian).

[56] A.S. Bektenov. On a configuration in algebraic spatial nets. *Mat. Issled.*, 83:3–10, 1985 (in Russian).

[57] A.S. Bektenov and T. Yakubov. Systems of orthogonal n-ary operations. *Izv. AN MSSR, Ser. fiz.-teh. i mat. nauk*, (3):7–14, 1974 (in Russian).

[58] V.D. Belousov. On the structure of distributive quasigroups. *Uspekhi Mat. Nauk*, 13(3):235–236, 1958 (in Russian).

[59] V.D. Belousov. Regular permutations in quasigroups. *Uchenye zapiski Bel'tskogo pedinstituta*, 1:39–49, 1958 (in Russian).

[60] V.D. Belousov. Derivative operations and associators in loops. *Matem. sb.*, 45(87):51–70, 1958 (in Russian).

[61] V.D. Belousov. On one class of quasigroups. *Učen. Zap. Bel'ck. Gos. Ped. Inst.*, 5:29–46, 1960 (in Russian).

[62] V.D. Belousov. On properties of binary operations. *Uchenye zapiski Bel'tskogo pedinstituta*, 5:9–28, 1960 (in Russian).

[63] V.D. Belousov. Globally associative systems of quasigroups. *Mat. Sb.*, 55:221–236, 1961 (in Russian).

[64] V.D. Belousov. On a class of left-distributive quasi-groups. *Izv. Vyss. Ucebn. Zaved. Matematika*, (1):16–20, 1963 (in Russian).

[65] V.D. Belousov. Reciprocally-inverse quasi-groups and loops. *Bul. Akad. Štiince RSS Moldoven*, (11):3–10, 1963 (in Russian).

[66] V.D. Belousov. Conjugate operations. *Studies in General Algebra (Sem.)*, 1:37–52, 1965 (in Russian).

[67] V.D. Belousov. Loops with a nucleus of index two. *Studies in Algebra and Math. Anal.*, (1):11–21, 1965 (in Russian).

[68] V.D. Belousov. Two problems of the theory of distributive quasigroups. *Studies in Algebra and Math. Anal.*, (1):109–112, 1965 (in Russian).

[69] V.D. Belousov. Balanced identities on quasigroups. *Mat. Sb.*, 70:55–97, 1966 (in Russian).

[70] V.D. Belousov. Grätzer algebra and the S-systems of quasigroups. *Mat. Issled.*, 1:55–81, 1966 (in Russian).

[71] V.D. Belousov. Extensions of quasigroups. *Bul. Akad. Stiince RSS Moldoven*, (8):3–24, 1967 (in Russian).

[72] V.D. Belousov. *Foundations of the Theory of Quasigroups and Loops.* Nauka, Moscow, 1967 (in Russian).

[73] V.D. Belousov. Some remarks on the functional equation of generalized distributivity. *Aequationes Math.*, 1(1-2):54–65, 1968.

[74] V.D. Belousov. Systems of orthogonal operations. *Mat. Sbornik*, 77 (119)(1):38–58, 1968 (in Russian).

[75] V.D. Belousov. The group associated with a quasigroup. *Mat. Issled.*, 4(3):21–39, 1969 (in Russian).

[76] V.D. Belousov. *Algebraic Nets and Quasigroups.* Stiintsa, Kishinev, 1971 (in Russian).

[77] V.D. Belousov. *n-Ary Quasigroups.* Stiintsa, Kishinev, 1971 (in Russian).

[78] V.D. Belousov. The equation of generalized mediality. *Mat. Issled.*, 39:21–31, 1976 (in Russian).

[79] V.D. Belousov. *Configurations in Algebraic Nets.* Shtiinta, Kishinev, 1979 (in Russian).

[80] V.D. Belousov. *Elements of Quasigroup Theory: a Special Course.* Kishinev State University Printing House, Kishinev, 1981 (in Russian).

[81] V.D. Belousov. *Parastrophic-Orthogonal Quasigroups, Preprint.* Shtiinta, Kishinev, 1983 (in Russian).

[82] V.D. Belousov. Inverse loops. *Mat. Issled.*, 95:3–22, 1987 (in Russian).

[83] V.D. Belousov. Autotopies and anti-autotopies in quasigroups. *Mat. Issled.*, 102:3–25, 1988 (in Russian).

[84] V.D. Belousov. Crossed isotopies of quasigroups. *Mat. Issled.*, 113:14–20, 1990 (in Russian).

[85] V.D. Belousov. Parastrophic-orthogonal quasigroups. Translated from the 1983 Russian original. *Quasigroups Relat. Syst.*, 13(1):25–72, 2005.

[86] V.D. Belousov and G.B. Belyavskaya. *Latin Squares, Quasigroups and Their Applications.* Shtiinta, Kishinev, 1989 (in Russian).

[87] V.D. Belousov and I. A. Florja. On left-distributive quasigroups. *Bul. Akad. Štiince RSS Moldoven*, (7):3–13, 1965 (in Russian).

[88] V.D. Belousov and I. A. Florja. Quasigroups with the inverse property. *Bul. Akad. Ştiine RSS Moldoven.*, (4):3–17, 1966 (in Russian).

[89] V.D. Belousov and A.A. Gvaramiya. Partial identities and nuclei of quasigroups. *Bulletin of the Academy of Sciences of the Georgian SSR*, 65(2):277–279, 1972 (in Russian).

[90] V.D. Belousov and V.I. Onoi. On loops that are isotopic to left distributive quasigroups. *Mat. Issled.*, 3(25):135–152, 1972 (in Russian).

[91] V.D. Belousov and I.B. Raskina. Some remarks on totally symmetric quasigroups and loops that are connected with them. *Izv. Vyssh. Uchebn. Zaved. Mat.*, (2):3–10, 1982 (in Russian).

[92] V.D. Belousov and V.V. Ryzhkov. On a method to obtain closure figures. *Mat. Issled.*, 1(2):140–150, 1966 (in Russian).

[93] V.D. Belousov and M.D. Sandik. n-Ary quasigroups and loops. *Siberian Math. J.*, VII:31–54, 1966 (in Russian).

[94] V.D. Belousov and Z. Stojakovic. Generalized entropy on infinitary quasigroups. *Zbornik rad. Prir.-mat. fak. Univ. u Novom Sadu*, 5:35–42, 1975.

[95] V.D. Belousov and B.V. Tsurkan. Crossed-inverse quasigroups (CI-quasigroups). *Izv. Vyssh. Uchebn. Zaved. Mat.*, 82(3):21–27, 1969 (in Russian).

[96] V.D. Belousov and T. Yakubov. On orthogonal n-ary operations. *Voprosy Kibernetiki*, 16:3–17, 1975 (in Russian).

[97] G.B. Belyavskaya. Contraction of quasigroups. I. *Bul. Akad. Stiince RSS Moldoven*, (1):6–12, 1970 (in Russian).

[98] G.B. Belyavskaya. Contraction of quasigroups. II. *Bul. Akad. Stiince RSS Moldoven*, (3):3–17, 1970 (in Russian).

[99] G.B. Belyavskaya. Generalized extension of quasigroups. *Mat. Issled.*, 5(2):28–48, 1970 (in Russian).

[100] G.B. Belyavskaya. r-orthogonal quasigroups. I. *Mat. Issled.*, 39(1):32–39, 1976 (in Russian).

[101] G.B. Belyavskaya. r-orthogonal quasigroups. II. *Mat. Issled.*, 43(1):39–49, 1976 (in Russian).

[102] G.B. Belyavskaya. On spectrum of partial admissibility of finite quasigroups (Latin squares). *Mat. Zametki*, 32(6):777–788, 1982 (in Russian).

[103] G.B. Belyavskaya. Direct decompositions of quasigroups. *Mat. Issled.*, 95:23–38, 1987 (in Russian).

[104] G.B. Belyavskaya. The nuclei and center of a quasigroup. *Mat. Issled.*, 102:37–52, 1988 (in Russian).

[105] G.B. Belyavskaya. Left, right, middle nuclei and center of a quasigroup. Technical report, Institute of Mathematics of the Academy of Sciences of Republic Moldova, Chisinau, 1988 (in Russian).

[106] G.B. Belyavskaya. T-quasigroups and the center of a quasigroup. *Mat. Issled.*, 111:24–43, 1989 (in Russian).

[107] G.B. Belyavskaya. Full direct decompositions of quasigroups with an idempotent element. *Mat. Issled.*, 113:21–36, 1990 (in Russian).

[108] G.B. Belyavskaya. Abelian quasigroups are T-quasigroups. *Quasigroups Relat. Syst.*, 1:1–7, 1994

[109] G.B. Belyavskaya. Associators, commutators and the linearity of a quasigroup. *Diskret. Mat.*, 7(4):116–125, 1995.

[110] G.B. Belyavskaya. Quasigroup theory: Nuclei, centre, commutants. *Bul. Acad. Stiinte Repub. Mold., Mat.*, (2(21)):47–71, 1996 (in Russian).

[111] G.B. Belyavskaya. Associants and the commutator of quasigroups. *Fundam. Prikl. Mat.*, 3(3):715–737, 1997 (in Russian).

[112] G.B. Belyavskaya. On r-differentiable quasigroups. In *Abstracts of the Int. Conf. on Pure and Applied Math. Dedicated to D.A. Grave*, pages 11–12. Kiev, 2002.

[113] G.B. Belyavskaya. Quasigroup power sets and cyclic S-systems. *Quasigroups Relat. Syst.*, 9:1–17, 2002.

[114] G.B. Belyavskaya. Pairwise orthogonality of n-ary operations. *Bul. Acad. Stiinte Repub. Mold. Mat.*, (3):5–18, 2005.

[115] G.B. Belyavskaya. Identities with permutations associated with quasigroups isotopic to groups. *Bul. Acad. Stiinte Repub. Mold., Mat.*, (2):19–24, 2007.

[116] G.B. Belyavskaya. Power sets of n-ary quasigroups. *Bul. Acad. de Stiinte a Republicii Moldova. Matematica*, (1):37–45, 2007.

[117] G.B. Belyavskaya. Secret-sharing schemes and orthogonal systems of k-ary operations. *Quasigroups and Related Systems*, 17(2):111–130, 2009.

[118] G.B. Belyavskaya. Transformation of orthogonal systems of polynomial n-ary operations. In *VII-th Theoretical and Practical Seminar Combinatorial Configurations and Their Applications, Kirovograd, April 17-18, 2009*, pages 14–17. 2009.

[119] G.B. Belyavskaya. Identities with permutations and quasigroups isotopic to groups and abelian groups. *Discret. Mat.*, 25(2):68–81, 2013 (in Russian).

[120] G.B. Belyavskaya and A. Diordiev. Conjugate-orthogonality and the complete multiplication group of a quasigroup. *Bul. Acad. Stiinte Repub. Mold., Mat.*, (1):22–30, 2009.

[121] G.B. Belyavskaya, W.A. Dudek, and V.A. Shcherbacov. Valentin Danilovich Belousov: His life and work. *Quasigroups Relat. Syst.*, 13:1–7, 2005.

[122] G.B. Belyavskaya, V.I. Izbash, and G.L. Mullen. Check character systems using quasigroups, I. *Des. Codes Cryptogr.*, 37:215–227, 2005.

[123] G.B. Belyavskaya, V.I. Izbash, and V.A. Shcherbacov. Check character systems over quasigroups and loops. *Quasigroups Relat. Syst.*, 10:1–28, 2003.

[124] G.B. Belyavskaya and A.D. Lumpov. Crossed product of two systems of quasigroups and its use for the construction of partially orthogonal quasigroups. *Mat. Issled.*, 83:26–38, 1985.

[125] G.B. Belyavskaya and G. Mullen. Orthogonal hyper-cubes and n-ary operations. *Quasigroups and Related Systems*, 13(1):73–86, 2005.

[126] G.B. Belyavskaya and G.L. Mullen. Strongly orthogonal and uniformly orthogonal many-placed operations. *Algebra and Discrete Mathematics*, (1):1–17, 2006.

[127] G.B. Belyavskaya and T.V. Popovich. Totally conjugate orthogonal quasigroups and complete graphs. *Fundament. i Prikl. Matem.*, 16(8):17–26, 2010 (in Russian).

[128] G.B. Belyavskaya and A.F. Russu. The admissibility of quasigroups. *Mat. Issled.*, 10(1):45–57, 1975. (in Russian).

[129] G.B. Belyavskaya and A.F. Russu. Partial admissibility of quasigroups. *Mat. Issled.*, 43(1):50–58, 1976 (in Russian).

[130] G.B. Belyavskaya and P.N. Syrbu. A class of self-orthogonal n-groupoids. *Izv. Akad. Nauk Moldav. SSR Ser. Fiz.-Tekhn. Mat. Nauk.* (2):25–30, 1989 (in Russian).

[131] G.B. Belyavskaya and A. Kh. Tabarov. Characterization of linear and alinear quasigroups. *Diskret. Mat.*, 4(2):142–147, 1992.

[132] G.B. Belyavskaya and A. Kh. Tabarov. Identities with permutations leading to linearity of quasigroups. *Discrete Mathematics and Applications*, 19(2):173–190, 2009.

[133] G.B. Belyavskaya and A.Kh. Tabarov. The nuclei and center of linear quasigroups. *Bul. Acad. Stiinte Repub. Mold., Mat.*, (3(6)):37–42, 1991 (in Russian).

[134] G.B. Belyavskaya and A.Kh. Tabarov. One-sided T-quasigroups and irreducible balanced identities. *Quasigroups Relat. Syst.*, 1:8–21, 1994.

[135] Josh Benaloh. General linear secret sharing (extended abstract), 1996. https://www.cs.umd.edu/ gasarch/TOPICS/secretsharing/benalohSS.ps.

[136] L. Beneteau. 3-abelian groups and commutative Moufang loops. *European J. Combin.*, 5(3):193–196, 1984.

[137] L. Beneteau and T. Kepka. Theoremes de structure dans certains groupoides localement nilpotents. *C. R. Acad. Sci. Paris Ser. I Math.*, 300(11):327–330, 1985 (in French).

[138] C. H. Bennet and G. Brassard. Quantum cryptography: Public key distribution and coin tossing. In *Proceedings of the IEEE International Conference on Computers, Systems, and Signal Processing*, page 175, Bangalore, 1984.

[139] F.E. Bennett. On a class of $n^2 \times 4$ orthogonal arrays and associated quasigroups. Combinatorics, graph theory and computing, Proc. 14th Southeast. Conf., Boca Raton/Fl, 1983, *Congr. Numerantium* 39:117–122, 1983.

[140] F.E. Bennett. Quasigroup identities and Mendelsohn designs. *Canad. J. Math.*, 41(2):341–368, 1989.

[141] F.E. Bennett. The spectra of a variety of quasigroups and related combinatorial designs. *Discrete Math.*, 77:29–50, 1989.

[142] F.E. Bennett and N.S. Mendelsohn. On the spectrum of Stein quasigroups. *Bull. Aust. Math. Soc.*, 21:47–63, 1980.

[143] F.E. Bennett, Li Sheng Wu, and L. Zhu. Some new conjugate orthogonal Latin squares. *J. Combin. Theory Ser. A*, 46:314–218, 1987.

[144] F.E. Bennett and H. Zhang. Latin squares with self-orthogonal conjugates. *Discrete Math.*, 284:45–55, 2004.

[145] F.E. Bennett and H. Zhang. Schröder quasigroups with a specified number of idempotents. *Discrete Math.*, 312(5):868–882, 2012.

[146] F.E. Bennett and L. Zhu. Conjugate-orthogonal Latin squares and related structures. In J. H. Dinitz and D.R. Stinson, editors, *Contemporary Design Theory, Collect. Surv.*, pages 41–96. Wiley-IEEE, 1992.

[147] Thomas Beth. Eine Bemerkung zur Abschtzung der Anzahl orthogonaler lateinischer Quadrate mittels Siebverfahren. *Abh. Math. Sem. Univ. Hamburg*, 53:283–289, 1983 (in German).

[148] A. Beutelspacher. *Cryptology: An Introduction to the Science of Encoding, Concealing and Hiding.* Vieweg, Wiesbaden, 2002 (in German).

[149] G. Birkhoff. *Lattice Theory*, volume 25. American Mathematical Society Colloquium Publications. Revised edition, New York, 1948.

[150] G. Birkhoff. *Lattice Theory, Third Edition*, volume XXV American Mathematical Society. American Mathematical Society Colloquium Publications, Providence, R.I., 1967.

[151] G. Birkhoff. *Lattice Theory.* Nauka, Moscow, 1984 (in Russian).

[152] V.I. Bityutskov and V.F. Koltchin (Eds.). *Discrete Mathematics. Encyclopedia.* Great Russian Encyclopedia, Moscow, 2004 (in Russian).

[153] Richard E. Blahut. *Theory and Practice of Error Control Codes.* Addison-Wesley Publishing Company, Advanced Book Program, Reading, 1983.

[154] G.R. Blakley. *Proceedings of the National Computer Conference.* 48:313–317, 1979.

[155] M. Blum and S. Goldwasser. An efficient probabilistic public key encryption scheme which hides all partial information. In *Proceedings of Advances in Cryptology CRYPTO '84*, pages 289–299, Springer Verlag, 1985.

[156] N.S. Bolshakova. About one more application of Latin squares. *Vestnik MGTU*, 8:170–173, 2005 (in Russian).

[157] V.V. Borisenko. Irreducible n-quasigroups on finite sets of composite order. *Matem. issledov., Quasigroups and Loops*, 51:38–42, 1979 (in Russian).

[158] R.C. Bose. On the construction of balanced incomplete block designs. *Proc. Nat. Acad. Sci. U.S.A.*, 9:353–399, 1939.

[159] R.C. Bose. Strongly regular graphs, partial geometries and partially balanced designs. *Pac. J. Math.*, 13:389–419, 1963.

[160] R.C. Bose and S.S. Shrikhande. On the falsity of Euler's conjecture about the non-existence of two orthogonal Latin squares of order $4t+2$. *Proc. Nat. Acad. Sci. U.S.A.*, 49:734–737, 1959.

[161] R.C. Bose, S.S. Shrikhande, and E.T. Parker. Further results on the construction of mutually orthogonal Latin squares and the falsity of Euler's conjecture. *Canad. Jour. Math.*, 12:189–203, 1960.

[162] Cyril Branciard, Nicolas Gisin, Barbara Kraus, and Valerio Scarani. Security of two quantum cryptography protocols using the same four qubit states. *Phys. Rev. A*, 72:032301, 2005. arXiv:quant-ph/0505035v2.

[163] R.K. Brayton, Donald Coppersmith, and A.J. Hoffman. Self-orthogonal Latin squares of all orders $n \neq 2, 3, 6$. *Bull. Amer. Math. Soc.*, 80:116–118, 1974.

[164] A.E. Brouwer, A.M. Cohen, and A.Neumaier. *Distance-Regular Graphs.* Springer Verlag, Berlin, 1989.

[165] R.H. Bruck. Simple quasigroups. *Bull. Amer. Math. Soc.*, 50:769–781, 1944.

[166] R.H. Bruck. Some results in the theory of quasigroups. *Trans. Amer. Math. Soc.*, 55:19–52, 1944.

[167] R.H. Bruck. Contribution to the theory of loops. *Trans. Amer. Math. Soc.*, 60:245–354, 1946.

[168] R.H. Bruck. Finite nets I. Numerical invariants. *Canadian J. Math.*, 3:94–107, 1951.

[169] R.H. Bruck. Pseudo-automorphisms and Moufang loops. *Proc. Amer. Math. Soc.*, 3:66–72, 1952.

[170] R.H. Bruck. Normal endomorphisms. *Illinois J. Math.*, 4:38–87, 1960.

[171] R.H. Bruck. Some theorems on Moufang loops. *Math. Z.*, 73:59–78, 1960.

[172] R.H. Bruck. Finite nets II. Uniqueness and embedding. *Pacific J. Math.*, 13:421–457, 1963.

[173] R.H. Bruck. *A Survey of Binary Systems.* Springer Verlag, New York, third printing, corrected edition, 1971.

[174] R.H. Bruck and L.J. Paige. Loops whose inner mappings are automorphisms. *Ann. of Math.*, 63(2):308–332, 1956.

[175] R.H. Bruck and H.J. Ryser. The nonexistence of certain finite projective planes. *Canadian J. Math.*, 1:88–93, 1949.

[176] B.F. Bryant and Hans Schneider. Principal loop-isotopes of quasigroups. *Canad. J. Math.*, 18:120–125, 1966.

[177] D. Bryant, M. Buchanan, and Ian M. Wanless. The spectrum for quasigroups with cyclic automorphisms and additional symmetries. *Discrete Math.*, 309(4):821–833, 2009.

[178] A.A. Buchstab. *Number Theory.* Prosveshchenie, 1966 (in Russian).

[179] H.H. Buchsteiner. O nekotorom klasse binarnych lup. *Mat. Issled.*, 39:54–66, 1976.

[180] I.I. Burdujan. Certain remarks on the geometry of quasigroups. *Matem. issled., Nets and Quasigroups*, 39:40–53, 1976 (in Russian).

[181] V.P. Burichenko. On a special loop, Dixon form and lattice connected with $O_7(3)$. *Mat. Sb.*, 182(10):1408–1429, 1991 (in Russian).

[182] R.P. Burn. Finite Bol loops. *Math. Proc. Camb. Phil. Soc.*, 84(3):377–385, 1978.

[183] R.P. Burn. Finite Bol loops. II. *Math. Proc. Camb. Phil. Soc.*, 89(3):445–455, 1981.

[184] S. Burris and H.P. Sankappanavar. *A Course in Universal Algebra.* Springer-Verlag, 1981.

[185] C. Burstin and W. Mayer. Distributive Gruppen von endliher Ordnung. *J. Reine und Angew. Math.*, 160:111–130, 1929.

[186] R. Capodaglio. On the isotopisms and the pseudo-automorphisms of the loops. *Boll. Un. Mat. Ital. A*, 7(2):199–205, 1993.

[187] E. Cartan. *Lecons sur la Theorie des Spineurs.* Hermann et Cie., Paris, 1938.

[188] Nicholas J. Cavenagh, Ales Drapal, and Carlo Hamalainen. Latin bitrades derived from groups, 2007. arXiv:0704.1730.

[189] N.J. Cavenagh. A superlinear lower bound for the size of a critical set in a Latin square. *J. Combin. Des.*, 15(4):369–282, 2007.

[190] N.J. Cavenagh. The theory and application of Latin bitrades: a survey. *Math. Slovaca*, 58(6):691–718, 2008.

[191] N.J. Cavenagh, D. Donovan, and A. Drapal. Constructing and deconstructing latin trades. *Discrete Math.*, 284:97–105, 2004.

[192] N.J. Cavenagh, D. Donovan, and A. Drapal. 3-homogeneous latin trades. *Discrete Math.*, 300(1-3):57–70, 2005.

[193] D. Ceban and P. Syrbu. On quasigroups with some minimal identities. *Studia Universitatis Moldaviae. Stiinte Exacte si Economice*, 82(2):47–52, 2015.

[194] Sucheta Chakrabarti, Saibal K. Pal, and Sugata Gangopadhyay. An improved 3-quasigroup based encryption scheme. In *ICT Innovations 2012, Secure and Intelligent Systems, 12-15 September, Ohrid, Macedonia, Web Proceedings, S. Markovski, M. Gusev (Editors)*, pages 173–184, 2012.

[195] A.M. Cheban. Loops with identities of length four and of rank three. II. *General Algebra and Discrete Geometry*, pages 117–120, 1980 (in Russian).

[196] A.M. Cheban. The Sushkevich postulates for distributive identities. *Algebraic Structures and Geometry*, 162:152–155, 1991 (in Russian).

[197] A.M. Cheban. WA-quasigroups. *Bulletin of the Transdniestrian University*, 7:31–34, 1997 (in Russian).

[198] O. Chein and E.G. Goodaire. Loops whose loop rings are alternative. *Comm. Algebra*, 14(2):293–310, 1986.

[199] O. Chein and E.G. Goodaire. Subloops of indecomposable RA loops. *Acta Math. Hungar.*, 109(1-2):15–31, 2005.

[200] O. Chein and Edgar Goodaire. A new construction of Bol loops of order 8k. *J. Algebra*, 287(1):103–122, 2005.

[201] O. Chein, H.O. Pflugfelder, and J.D.H. Smith. *Quasigroups and Loops: Theory and Applications*. Heldermann Verlag, 1990.

[202] Orin Chein and Hala Orlik-Pflugfelder. The smallest Moufang loop. *Arch. Math. (Basel)*, 22:573–576, 1971.

[203] A.V. Cheremushkin. Almost all Latin squares have trivial autostrophy group. *Prikladnaya i Discretnaya Matematica*, 1(3):29–32, 2009 (in Russian).

[204] Bill Cherowitzo. Steiner triple systems, 2011. http://math.ucdenver.edu/ wcherowi/courses/m6406/sts.pdf.

[205] L. Chiriac. *Topological Algebraic Systems*. Ştiinţa, Chisinau, 2009.

[206] J.R. Cho. Idempotent medial n-groupoids defined on fields. *Algebra Universalis*, 25:235–246, 1988.

[207] M.M. Choban and L.L. Kiriyak. The medial topological quasigroups with multiple identities. In *Applied and Industrial Mathematics. Oradea, Romania and Chishinau, Moldova, Abstracts*, page 11, Kishinev, Moldova, August 1995.

[208] M.M. Choban and L.L. Kiriyak. The topological quasigroups with multiple identities. *Quasigroups Relat. Syst.*, 9:19–32, 2002.

[209] Dug-Hwan Choi and Jonathan D.H. Smith. Greedy loop transversal codes for correcting error bursts. *Discrete Math.*, 264:37–43, 2003.

[210] S. Chowla and H. J. Ryser. Combinatorial problems. *Canadian J. Math.*, 2:93–99, 1950.

[211] A.H. Clifford and G.B. Preston. *The Algebric Theory of Semigroups*, volume I. American Mathematical Society, Rhode Island, 1961.

[212] P.M. Cohn. *Universal Algebra.* Harper & Row, New York, 1965.

[213] Charles J. Colbourn and Jeffrey H. Dinitz. *Handbook of Combinatorial Designs (*2nd Edition). Chapman & Hall, Boca Raton, 2007.

[214] Charles J. Colbourn and Alexander Rosa. *Triple Systems.* The Clarendon Press, New York, 1999.

[215] J.H. Conway. A simple construction for the Fischer-Griess monster group. *Invent. Math.*, 79(3):513–540, 1985.

[216] J. Cooper, D. Donovan, and J. Seberry. Secret sharing schemes arising from Latin squares. *Bull. Inst. Combin. Appl.*, 12:33–43, 1994.

[217] D. Coppersmith. Weakness in quaternion signatures. *J. Cryptology*, 14:77–85, 2001.

[218] Mariana Cornelissen and Cesar Polcino Milies. Classifying finitely generated indecomposable RA Loops, 2012. arXiv:1204.4277v1.

[219] B. Cote, B. Harvill, M. Huhn, and A. Kirchman. Classification of loops of generalized Bol-Moufang type. *Quasigroups Related Systems*, 19(2):193–206, 2011.

[220] A.V. Covalschi and N.I. Sandu. On the generalized nilpotent and generalized solvable loops. *ROMAI J.*, 7(1):39–62, 2011.

[221] Piroska Csorgo. Every Moufang loop of odd order with nontrivial commutant has nontrivial center. *Archiv der Mathematik*, 100(6):507–519, 2013.

[222] Piroska Csorgo and Ales Drapal. Left conjugacy closed loops of nilpotency class two. *Results Math.*, 47(3-4):242–265, 2005.

[223] Piroska Csorgo, Ales Drapal, and Michael K. Kinyon. Buchsteiner loops. *Internat. J. Algebra Comput.*, 19:1049–1088, 2009.

[224] Piroska Csorgo and Victor Shcherbacov. On some quasigroup cryptographical primitives, 2011. arXiv:1110.6591.

[225] S.B. Damelin, G. Michalski, G.L. Mullen, and D. Stone. The number of linearly independent binary vectors with applications to the construction of hypercubes and orthogonal arrays, pseudo (t, m, s)-nets and linear codes. *Monatshefte für Mathematik*, 141(4):277–288, 2004. DOI: 10.1007/s00605-003-0044-3.

[226] H. Michael Damm. Totally anti-symmetric quasigroups for all orders $n \neq 2, 6$. *Discrete Mathematics*, 307:715–729, 2007.

[227] M. Damm. *Prüfziffersysteme über Quasigruppen.* Master's thesis, Philipps-Universität Marburg, 1998 (in German).

[228] M. Damm. *Total Anti-Symmetrische Quasigruppen.* PhD thesis, Philipps-Universität Marburg, 2004 (in German).

[229] E. Dawson, D. Donowan, and A. Offer. Ouasigroups, isotopisms and authentification schemes. *Australas. J. Combin.*, 13:75–88, 1996.

[230] P. Dehornoy. *Braids and Self-Distributivity*, volume 192 of *Progress in Math.* Birkhäuser, Basel, 2000.

[231] P. Dehornoy. Braid-based cryptography. *Contemp. Math., Group Theory, Statistics, and Cryptography*, 360:5–33, 2004.

[232] P. Delsarte. An algebraic approach to the association schemes of coding theory. *Philips Res. Rep. Suppl.*, No. 10, 1973.

[233] V.I. Dement'eva. Characteristic collineations in k-nets. *Mat. Issled. Vyp.*, 4:161–167, 1969.

[234] V.I. Dement'eva. Paratopies of systems of orthogonal quasigroups. *Questions of the Theory of Quasigroups and Loops, Chisinau, Ştiinţa*, pages 48–62, 1970 (in Russian).

[235] V.I. Dement'eva. Automorphisms of k-nets. *Mat. Issled. Vyp.*, 39:73–81, 1976.

[236] V.I. Dement'eva. Parallel translations in k-nets. *Mat. Issled. Vyp.*, 43:98–107, 1976.

[237] J. Dénes. Latin squares and non-binary encoding. In *Proc. Conf. Information Theory, CNRS*, pages 215–221, Paris, 1979.

[238] J. Dénes. On Latin squares and a digital encrypting communication system. *P.U.M.A., Pure Math. Appl.*, 11(4):559–563, 2000.

[239] J. Dénes and T. Dénes. Non-associative algebraic system in cryptology. Protection against "meet in the middle" attack. *Quasigroups Relat. Syst.*, 8:7–14, 2001.

[240] J. Dénes and A. D. Keedwell. *Latin Squares and Their Applications.* Académiai Kiadó, Budapest, 1974.

[241] J. Dénes and A. D. Keedwell. *Latin Squares. New Development in the Theory and Applications*, volume 46 of *Annals of Discrete Mathematics.* North-Holland, 1991.

[242] J. Dénes and A. D. Keedwell. A new authentication scheme based on Latin squares. *Discrete Math.*, 106/107:157–165, 1992.

[243] J. Dénes and A. D. Keedwell. Some applications of non-associative algebraic systems in cryptology. *P.U.M.A.*, 12(2):147–195, 2002.

[244] J. Dénes and P. Petroczki. A digital encrypting communication systems. *Hungarian Patent*, no. 201437A, 1990.

[245] A.I. Deriyenko, I.I. Deriyenko, and W.A. Dudek. Rigid and super rigid quasigroups. *Quasigroups Relat. Systems*, 17(1):17–28, 2009.

[246] I.I. Deriyenko and W.A. Dudek. Contractions of quasigroups and Latin squares. *Quasigroups Relat. Systems*, 21(2):165–174, 2013.

[247] Ivan I. Deriyenko and Wieslaw A. Dudek. On prolongations of quasigroups. *Quasigroups Relat. Systems*, 16:187–198, 2008.

[248] Chris G. Devillier. Bol loops with non-normal nuclei. *Arch. Math. (Basel)*, 81(4):383–384, 2003.

[249] L.E. Dickson. On semigroups and the general isomorphism between infinite groups. *Trans. Amer. Math. Soc.*, 6:205–208, 1905.

[250] W. Diffie and M.F. Hellman. New directions in cryptography. *IEEE, Transactions of Information Theory*, IT-22:644–654, 1976.

[251] V. Dimitrova, V. Bakeva, A. Popovska-Mitrovikj, and A. Krapez. Classifications of quasigroups of order 4 by parastrophic quasigroups tranformation. In *The International Mathematical Conference on Quasigroups and Loops, LOOPS'11, Booklet of Abstracts*, page 6, Třešt', Czech Republic, July 2011, http://www.karlin.mff.cuni.cz/ loops11/.

[252] V. Dimitrova, S. Markovski, and D. Gligoroski. Classification of quasigroups as Boolean functions, their algebraic complexity and application of Gröbner bases in solving systems of quasigroup equations. *Gröbner, Coding and Cryptography*, pages 415–420, 2009.

[253] Jeffrey H. Dinitz and Douglas R. Stinson. A singular direct product for bicolorable Steiner triple systems. *Codes and designs, Ohio State Univ. Math. Res. Inst. Publ.*, 10:87–97, 2002.

[254] A. Diordiev and V. Shcherbacov. On the existence of rst-inverse loops of small order. In *International conference Loops'03, Praha, August 10-August 17, 2003, Submitted Abstracts*, page 10, Praha, Czech Republic, 2003.

[255] V. Domashev, V. Popov, D. Pravikov, I. Prokof'ev, and A. Shcherbakov. *Programming of Algorithms of Defense of Information.* Nolidge, Moscow, 2000. (in Russian).

[256] D. Donowan. Critical sets for families of Latin squares. *Util. Math.*, 53:3–16, 1998.

[257] D. Donowan. Critical sets in Latin squares of order less than 11. *J. Comb. Math. Comb. Comput.*, 29:223–240, 1999.

[258] D. Donowan. The completion of partial Latin squares. *Australas. J. Combin.*, 22:247–264, 2000.

[259] D. Donowan, J. Fevre, and G.H. John van Rees. On the spectrum of critical sets in Latin squares of order 2^n. *J. Combin. Des.*, 16(1):25–43, 2008.

[260] D. Donowan and A. Howse. Correction to a paper on critical sets. *Australas. J. Combin.*, 21:107–130, 2000.

[261] D. Donowan, J. Lefevre, T. McCourt, N. Cavenagh, and A. Khodkar. Identifying flaws in the security of critical sets in Latin squares via triangulations. *Australas. J. Combin.*, 52:243–268, 2012.

[262] D. Donowan and E.S. Mahmoodian. Correction to a paper on critical sets. *Bull. Inst. Comb. Appl.*, 37:44, 2003.

[263] S.A. Dorichenko and V.V. Yashchenko. *25 Sketches on Ciphers.* Teis, Moscow, 1994 (in Russian).

[264] W. Dornte. Untersuchungen über einen veralgemeinerten Gruppenbegriff. *Math. Z.*, 29:1–19, 1928.

[265] Stephen Doro. Simple Moufang loops. *Math. Proc. Cambridge Philos. Soc.*, 83(3):377–392, 1978.

[266] J. Doyen and R. M. Wilson. Embeddings of Steiner triple systems. *Discrete Math.*, 5:229–239, 1973.

[267] A. Drapal. How far apart can the group multiplication tables be? *Eur. J. Comb.*, 13(5):335–343, 1992.

[268] A. Drapal. Multiplication groups of free loops. I. *Czechoslovak Mathematical Journal*, 46(1):121–131, 1996.

[269] A. Drapal. Multiplication groups of free loops. II. *Czechoslovak Mathematical Journal*, 46(2):201–220, 1996.

[270] A. Drapal. On distances of multiplication tables of groups. *Lond. Math. Soc. Lect. Note Ser.*, 260:248–252, 1999.

[271] A. Drapal. Non-isomorphic 2-groups coincide at most in three quarters of their multiplication table. *Eur. J. Comb.*, 21:301–321, 2000.

[272] A. Drapal. Hamming distances of groups and quasi-groups. *Discrete Math.*, 235(1-3):189–197, 2001.

[273] A. Drapal. On groups that differ in one of four squares. *Eur. J. Comb.*, 23(8):899–918, 2002.

[274] A. Drapal. On multiplication groups of left conjugacy closed loops. *Comment. Math. Univ. Carolin.*, 45(2):223–236, 2004.

[275] A. Drapal and P. Jedlička. On loop identities that can be obtained by a nuclear identification, 2007, http://tf.czu.cz/ jedlickap/a/2007nucid.pdf.

[276] A. Drapal and P. Jedlicka. On loop identities that can be obtained by a nuclear identification. *European J. Combin.*, 31(7):1907–1923, 2010.

[277] A. Drapal and T. Kepka. Exchangeable partial groupoids I. *Acta Univ. Carolin.*, 24:57–72, 1983.

[278] A. Drapal and T. Kepka. Exchangeable partial groupoids II. *Acta Univ. Carolin.*, 26:3–9, 1985.

[279] A. Drapal and T. Kepka. Alternating groups and Latin squares. *European J. Combin.*, 10(2):175–180, 1989.

[280] A. Drapal and T. Kepka. Multiplication groups of quasigroups and loops I. *Acta Univ. Carol.*, 34:85–99, 1993.

[281] A. Drapal and N. Zhukavets. On multiplication tables of groups that agree on half of the columns and half of the rows. *Glasgow Math. J.*, 45:293–308, 2003.

[282] Ales Drapal. Group isotopes and a holomorphic action. *Results Math.*, 54(3-4):253–272, 2009.

[283] Ales Drapal. A simplified proof of Moufang's theorem. *Proc. Amer. Math. Soc.*, 139:93–98, 2011.

[284] Ales Drapal, Terry S. Griggs, and Andrew R. Kozlik. Pure Latin directed triple systems. *Australas. J. Combin.*, 62:59–75, 2015.

[285] Ales Drapal and Victor Shcherbacov. Identities and the group of isostrophisms. *Comment. Math. Univ. Carolin.*, 53:347–374, 2012.

[286] Ales Drapal and Petr Vojtechovsky. Code loops in both parities. *J. Algebraic Combin.*, 31(4):585–611, 2010.

[287] A. Drisko. Loops with transitive automorphisms. *J. Algebra*, 184(1):213–229, 1996.

[288] A. Drisko. On the number of even and odd Latin squares of order $p+1$. *Advances in Mathematics*, 128:20–37, 1997.

[289] W.A. Dudek. Medial n-groups and skew elements. In *Proceedings of the 5-th Symp. Universal and Applied Algebra*, pages 55–80, Turava, Poland, 1988.

[290] W.A. Dudek. On number of transitive distributive quasigroups. *Mat. Issled.*, 120:64–76, 1991 (in Russian).

[291] W.A. Dudek. On some old and new problems in n-ary groups. *Quasigroups Relat. Syst.*, 8:15–36, 2001.

[292] W.A. Dudek and K. Glazek. Around the Hosszu-Gluskin theorem for n-ary groups. *Discrete Mathematics*, 308(21):48614876, 2008.

[293] W.A. Dudek and V.A. Shcherbacov. Remarks to the first publications of V.D. Belousov. *Quasigroups Relat. Syst.*, 13:13–24, 2005.

[294] W.A. Dudek and P.N. Syrbu. About self-orthogonal n-groups. *Bul. Acad. Stiinte Repub. Mold., Mat.*, (3):37–42, 1992. (in Russian).

[295] David S. Dummit. CM elliptic curves and bicolorings of Steiner triple systems. *Bull. London Math. Soc.*, 34(4):403–410, 2002.

[296] David S. Dummit and Richard M. Foote. *Abstract Algebra, Third Edition.* John Wiley & Sons, Inc., Hoboken, NJ, 2004.

[297] J. Dunbar and R. Laskar. Finite nets of dimension d. *Discrete Mathematics*, 22(1):1–24, 1978.

[298] J. Duplak. A parastrophic equivalence in quasigroups. *Quasigroups Relat. Syst.*, 7:7–14, 2000.

[299] J. Dvorsky, E. Ochodkova, and V. Snasel. Hashovaci funkce zalozena na kvazigrupach. In *Workshop Milkulasska kryptobesidka*, Praha, 2000 (in Czech).

[300] J. Dvorsky, E. Ochodkova, and V. Snasel. Hash functions based on large quasigroups. *Velokonocni kryptologie*, pages 1–8, 2002.

[301] A. Ecker and G. Poch. Check character systems. *Computing*, 37/4:277–301, 1986.

[302] A. Ekert. From quantum, code-making to quantum code-breaking. In *Proceedings of the Symposium on Geometric Issues in the Foundations of Science,* Oxford, UK, *June 1996 in Honour of Roger Penrose in His 65th Year*, pages 195–214. Oxford University Press, 1998.

[303] T. ElGamal. A public key cryptosystem and a signature scheme based on discrete logarithms. *IEEE Transactions on Information Theory*, 31(4):469–472, 1985.

[304] S.M. Ermakov, editor. *Mathematical Theory of Design of Experiment.* Nauka, Moscow, 1983. (in Russian).

[305] F. Eugeni and A. Maturo. A new authentication system based on the generalized affine planes. *J. Inf. Optimization Sci.*, 13(2):183–193, 1992.

[306] L. Euler. Recherches sur une nouvelle espece de quarres magiques. *Leonardi Euleri Opera Omnia*, Serie 1(7):291–293, 1923.

[307] T. Evans. Homomorphisms of non-associative systems. *J. London Math. Soc.*, 24:254–260, 1949.

[308] T. Evans. Abstract mean values. *Duke Math. J.*, 30:331–347, 1963.

[309] T. Evans. Latin cubes orthogonal to their transposes: A ternary analogue of Stein quasigroups. *Aequationes Math.*, 9(2/3):296–297, 1973.

[310] T. Evans. Algebraic structures associated with Latin squares and orthogonal arrays. *Congr. Numer.*, 13:31–52, 1975.

[311] T. Evans. The construction of orthogonal k-skeins and Latin k-cubes. *Aequationes Math.*, 14:485–491, 1976.

[312] C. Faith. *Algebra: Rings, Modules and Categories*, volume 1. Springer-Verlag, Berlin, 1973.

[313] R.M. Falcon. Latin squares associated to principal autotopisms of long cycles. Application in Cryptography. In *Proc. Transgressive Computing 2006: A Conference in Honor of Jean Della Dora*, pages 213–230, 2006.

[314] R.M. Falcon. Cycle structures of autotopisms of the Latin squares of order up to 11. http://arxiv.org/, 0709.2973:18 pages, 2007.

[315] R. M. Falcon and J. Martin-Morales. Gröbner bases and the number of Latin squares related to autotopisms of order $\leqslant 7$. *Journal of Symbolic Computation*, 42:1142–1154, 2007.

[316] Horst Feistel. Cryptography and computer privacy. *Scientific American*, 228(5):15–23, 1973.

[317] Walter Feit and John G. Thompson. Solvability of groups of odd order. *Pacific J. Math.*, 13:775–1029, 1963.

[318] F. Fenyves. The Schreier extension of quasigroups. *Publ. Math. Debrecen*, 14:35–44, 1967.

[319] F. Fenyves. Extra loops. II. On loops with identities of Bol-Moufang type. *Publ. Math. Debrecen*, 16:187–192, 1969.

[320] Nick C. Fiala. Short identities implying a quasigroup is a loop or group. *Quasigroups Related Systems*, 15(2):253–271, 2007.

[321] Nick C. Fiala. A shortest single axiom with neutral element for commutative Moufang loops of exponent 3. *Australas. J. Combin.*, 40:167–171, 2008.

[322] Nick C. Fiala. Shortest single axioms for commutative Moufang loops of exponent 3. *Discrete Math.*, 308(15):3381–3385, 2008.

[323] Bernd Fischer. Distributive Quasigruppen endlicher Ordnung. *Math. Z.*, 83:267–303, 1964 (in German).

[324] R.A. Fisher. *Design of Experiments.* Oliver and Boyd, Edinbugh and London, 1935.

[325] L. Fitina, K.G. Russell, and J. Seberry. The power and influence in some Youden squares and secret sharing. *Util. Math.*, 73:143–157, 2007.

[326] I.A. Florja. Bol quasigroups. *Studies in General Algebra*, Ştiinţa, Chişinău, pages 136–154, 1965 (in Russian).

[327] I.A. Florja. Loops with one-sided invertibility. *Bul. Akad. Stiince RSS Moldoven.*, (7):68–79, 1965 (in Russian).

[328] I.A. Florja. *Quasigroups with Inverse Property.* PhD thesis, IM AN MSSR, 1965 (in Russian).

[329] I.A. Florja. F-quasigroups with invertibility property. *Questions of the Theory of Quasigroups and Loops*, Ştiinţa, Chişinău, pages 156–165, 1971 (in Russian).

[330] I.A. Florja. A certain class of F-quasigroups with an invertibility property. *Studies in the theory of quasigroups and loops*, pages 174–186, 1973. (in Russian).

[331] I.A. Florja and N.N. Didurik. About one class of CI-quasigroups. *Bulletin of the Transnistrian University*, (3):20–23, 2012 (in Russian).

[332] I.A. Florja and M.I. Ursul. F-quasigroups with invertibility property (IPF- quasigroups). *Questions of the Theory of Quasigroups and Loops*, Ştiinţa, Chişinău, pages 145–156, 1971 (in Russian).

[333] Tuval Foguel, Michael K. Kinyon, and J.D. Phillips. On twisted subgroups and Bol loops of odd order. *Rocky Mountain J. Math.*, 36(1):183–212, 2006.

[334] John B. Fraleigh. *A First Course in Abstract Algebra.* Addison-Wesley Publishing Company, London, third edition, 1982.

[335] Andrea Frisová. *Quasigroup Based Cryptography.* PhD thesis, Charles University, 2009.

[336] Tsuyoshi Fujiwara. Note on permutability of congruences on algebraic systems. *Proc. Japan Acad.*, 41(9):822–827, 1965.

[337] Stephen M. Gagola. Hall's theorem for Moufang loops. *J. Algebra*, 323(12):3252–3262, 2010.

[338] Stephen M. Gagola. How and why Moufang loops behave like groups. *Quasigroups Related Systems*, 19(1):1–22, 2011.

[339] V.M. Galkin. Finite distributive quasigroups. *Mat. Zametki*, 24(1):39–41, 1978 (in Russian).

[340] V.M. Galkin. Left distributive finite order quasigroups. *Mat. Issled.*, 51:43–54, 1979 (in Russian).

[341] V.M. Galkin. On symmetric quasigroups. *Russ. Math. Surv.*, 39(6):211–212, 1984.

[342] V.M. Galkin. *Quasigroups*, volume 26 of *Algebra, Topology, Geometry*, pages 3–44. VINITI, Moscow, 1988 (in Russian).

[343] V.M. Galkin. *Left Distributive Quasigroups,* dissertation of Doctor of Sciences. Steklov Mathematical Institute, Moscow, 1991 (in Russian).

[344] A.M. Gal'mak. *n-Ary Groups, Part 2.* Publishing Centre of the Belarusian State University, Minsk, 2007 (in Russian).

[345] A.M. Gal'mak and G.N. Vorob'ev. *Ternary Reflection Groups.* Belaruskaya Navuka, Minsk, 1998 (in Russian).

[346] G.N. Garrison. Quasi-groups. *Ann. of Math.*, 41(2):474–487, 1940. (MR0002150 (2,7b)).

[347] Craig Gentry. *A Fully Homomorphic Encryption Scheme.* PhD thesis, The Department of Computer Science of Stanford University, 2009.

[348] Craig Gentry. Computing arbitrary functions of encrypted data. *Communications of the ACM*, 53(3):97–105, 2010.

[349] I. Gessel. Counting Latin rectangles. *Bull. Amer. Math. Soc.*, 16:79–83, 1987.

[350] George Glauberman. On loops of odd order. *J. Algebra*, 1:374–396, 1964.

[351] George Glauberman. On loops of odd order. II. *J. Algebra*, 8:393–414, 1968.

[352] K. Glazek and B. Gleichgewicht. Abelian n-groups. In *Colloq. Math. Soc. J. Bolyai 29, Universal Algebra*, pages 321–329, Esztergom, Hungary, 1977.

[353] D. Gligoroski, S. Markovski, and S. J. Knapskog. A public key block cipher based on multivariate quadratic quasigroups. http://arxiv.org/, 0808.0247:22 pages, 2008.

[354] D. Gligoroski, S. Markovski, and L. Kocarev. Edon-R, An infnite family of cryptographic hash functions, 2006, http://csrc.nist.gov/pki/HashWorkshop/2006/Papers.

[355] D. Gligoroski, R.S. Odegard, M. Mihova, S.J. Knapskog, A. Drapal, V. Klima, J. Amundse, and M. El-Hadedy. Cryptographic Hash Function Edon-R, 2008. SHA-3 Algorithm Submission.

[356] D. Gligoroski, R.S. Odegard, M. Mihova, S.J. Knapskog, L. Kocarev, A. Drapal, and V. Klima. Cryptographic Hash Function EDON-R, 2008. Submission to NIST.

[357] M.M. Glukhov. α-closed classes and α-complete systems of functions of k-valued logic. *Diskretn. Mat.*, 1(1):16–21, 1989 (in Russian).

[358] M.M. Glukhov. T-partitions of quasigroups and groups. *Diskretn. Mat.*, 4(3):47–56, 1992 (in Russian).

[359] M.M. Glukhov. On application of quasigroups in cryptology. *Applied Discrete Mathematics*, 2:28–32, 2008 (in Russian).

[360] M.M. Glukhov. On multiplication groups of free and free commutative quasigroups. *Chebyshevski Sb.*, 11(3):63–77, 2008 (in Russian).

[361] M.M. Glukhov. On multiplication groups of free and free commutative quasigroups. *Fundamental and Applied Mathematics*, 20(1):23–37, 2015 (in Russian).

[362] M.M. Glukhov. On problem of equality of words in free quasigroups of varieties of quasigroups which are isotopic to groups. *Fundamental and Applied Mathematics*, 20(1):39–55, 2015 (in Russian).

[363] M.M. Glukhov and Kh.S. Rasulov. Special C-equivalence of finite one-side invertible quasigroups. *Diskretn. Mat.*, 6(3):3–16, 1994 (in Russian).

[364] L.M. Gluskin. Positional operatives. *Mat. Sb.*, 110:444–472, 1965 (in Russian).

[365] S. Goldwasser and S. Micali. Probabilistic encryption and how to play mental poker keeping secret all partial information. In *Proc. 14th Symposium on Theory of Computing*, pages 365–377, 1982.

[366] S. Goldwasser and S. Micali. Probabilistic encryption. *Journal of Computer and System Sciences*, 28:270–299, 1984.

[367] S. Golomb, L. Welch, and J. Denes. Encryption system based on crossed inverse quasigroups, 2001. US patent, WO0191368.

[368] S.W. Golomb. *Shift Register Sequences.* Holden Day, San Francisco, 1967.

[369] I.A. Golovko. F-quasigroups with idempotent elements. *Mat. Issled.*, 4(2):137–143, 1969 (in Russian).

[370] I.A. Golovko. Loops that are isotopic to F-quasigroups. *Bul. Akad. Štiince RSS Moldoven.*, (2):3–13, 1970 (in Russian).

[371] S. Gonsales, E. Couselo, V.T. Markov, and A.A. Nechaev. Recursive MDS-codes and recursively differentiable quasigroups. *Diskret. Mat.*, 10(2):3–29, 1998, (in Russian).

[372] S. Gonsales, E. Couselo, V.T. Markov, and A.A. Nechaev. Loop codes. *Discrete Mathematics and Applications*, 14(2):163–172, 2004.

[373] E.G. Goodaire, G. Leal, and C.P. Milies. Finite generation of units in alternative loop rings. *Manuscripta Mathematica*, 120(2):233–239, 2006.

[374] E.G. Goodaire and D.A. Robinson. A class of loops which are isomorphic to all loop isotopes. *Canadian J. Math.*, 34:662–672, 1982.

[375] E.G. Goodaire and D.A. Robinson. Semi-direct products and Bol loop. *Demonstratio Math.*, 27(3-4):573–588, 1994.

[376] G.E. Gorelik. *The Dimension of Space.* Moscow State University, Moscow, 1983 (in Russian).

[377] P.V. Gorinchoj and Yu.M. Ryabukhin. About certain algebraic variety of TS-loops. *Mat. Issled.*, 9(4):42–57, 1974 (in Russian).

[378] P.V. Gorincoj. Varieties of nilpotent TS-loops. *Mat. Zametki*, (3):321–334, 1981 (in Russian).

[379] P.V. Gorincoj. *On Varieties of Nilpotent Totally-Symmetric Loops.* PhD thesis, IM AN MSSR, 1982 (in Russian).

[380] P.I. Gramma. On the concept of center in quasigroups. *Studies in General Algebra (Sem.)*, pages 81–88, 1965 (in Russian), MR0206137 (34 5962).

[381] M.J. Grannell, T.S. Griggs, and R.A. Mathon. Steiner systems s(5; 6; v) with v = 72 and 84. *Mathematics of Computation*, 67(221):357–359, 1998.

[382] M.J. Grannell, T.S. Griggs, and K.A.S. Quinn. Mendelsohn directed triple systems. *Discrete Mathematics*, 205(1–3):85–96, 1999.

[383] G. Grätzer. *Universal Algebra.* Springer Verlag, New-York, 1979.

[384] Ion Grecu and Parascovia Syrbu. On some isotrophy invariants of Bol loops. *Bull. Transilv. Univ. Braşov Ser. III*, 54(5):145–153, 2012.

[385] Ion Grecu and Parascovia Syrbu. Commutants of middle Bol loops. *Quasigroups Related Systems*, 22(1):81–88, 2014.

[386] Mark Greer and Michael Kinyon. Pseudoautomorphisms of Bruck loops and their generalizations. *Comment. Math. Univ. Carolin.*, 53(3):383–389, 2012.

[387] A.V. Gribov. *Algebraic Non-Associative Structures and Its Applications in Cryptology.* PhD thesis, Moscov State University, 2015 (in Russian).

[388] A.V. Gribov, P.A. Zolotykh, and A.V. Mikhalev. Construction of algebraic cryptosystem over a quasigroup ring. *Matem. Vopr. Criptogr.*, 1(4):23–32, 2010 (in Russian).

[389] Robert L. Griess. Code loops. *J. Algebra*, 100(1):224–234, 1986.

[390] Robert L. Griess. A Moufang loop, the exceptional Jordan algebra, and a cubic form in 27 variables. *J. Algebra*, 131(1):281–293, 1990.

[391] Robert L. Griess. Codes, loops and p-locals. Groups, difference sets, and the Monster. *Ohio State Univ. Math. Res. Inst. Publ.*, 4:369–375, 1996.

[392] D. Grigoriev and V. Shpilrain. Authentication from matrix conjugation. *Groups-Complexity-Cryptology*, 1(2):199–205, 2009.

[393] A.N. Grishkov and A. V. Zavarnitsine. Lagrange's theorem for Moufang loops. *Math. Proc. Cambridge Philos. Soc.*, 139(1):41–57, 2005.

[394] A.N. Grishkov and A.V. Zavarnitsine. Sylow's theorem for Moufang loops. *J. Algebra*, 321(7):1813–1825, 2009.

[395] O. Grosek and M. Sýs. Isotopy of Latin squares in cryptography. *Tatra Mt. Math. Publ.*, 45:27–36, 2010. DOI: 10.2478/v10127-010-0003-z.

[396] Tony Grubman and Ian M. Wanless. Growth rate of canonical and minimal group embeddings of spherical latin trades. *Journal of Combinatorial Theory, Series A*, 123(1):57–72, 2014.

[397] H. P. Gumm. A new class of check-digit methods for arbitrary number systems. *IEEE Trans. Inf. Th. IT*, 31:102–105, 1985.

[398] A.A. Gvaramiya. A certain class of loops. *Algebra and Number Theory.* Moskov. Gos. Ped. Inst. Ucen. Zap., 375:25–34, 1971 (in Russian).

[399] A.A. Gvaramiya. Nuclei of ternary quasigroups. *Sakharth. SSR Mecn. Akad. Moambe*, 67:537–539, 1972 (in Russian).

[400] A.A. Gvaramiya. *Axiomatized Quasigroup Classes and Multitype Universal Algebra.* Thesis of Doctor of Physical and Mathematical Sciences. Abhaziya State University, Suhumi, 1985 (in Russian).

[401] A.A. Gvaramiya. Quasivarieties of automata. relations with quasigroups. *Sibirsk. Mat. Zh.*, 26(3):11–30, 1985 (in Russian).

[402] A.A. Gvaramiya. Representations of quasigroups, and quasigroup automata. *Fundam. Prikl. Mat.*, 3(3):775–800, 1997 (in Russian).

[403] A.A. Gvaramiya and G.A. Karasev. Diassociative loops with $x \to x^n$ an endomorphism. *Sakharth. SSR Mecn. Akad. Moambe*, 61:541–543, 1971 (in Russian).

[404] A.A. Gvaramiya and M.M. Glukhov. On groups of automorphisms of finitely defined quasigroups from R-varieties. *Topologicheskie Prostranstva Otobrazheniya*, 1985:170–178, 1985.

[405] Jonathan I. Hall. On Mikheev's construction of enveloping groups. *Comment. Math. Univ. Carolin.*, 51(2):245–252, 2010.

[406] Jonathan I. Hall and Gabor P. Nagy. On Moufang 3-nets and groups with triality. *Acta Sci. Math. (Szeged)*, 67(3-4):675–685, 2001.

[407] Marshall Hall. *The Theory of Groups.* The Macmillan Company, New York, 1959.

[408] Marshall Hall. Automorphisms of Steiner triple systems. In *Proc. Sympos. Pure Math.*, volume VI, pages 47–66. American Mathematical Society, Providence, R.I. 50.60, 1962.

[409] Marshall Hall. *Combinatorial Theory.* Blaisdell Publishing Company, Massachusetts, 1967.

[410] Marshall Hall and L. J. Paige. Complete mappings of finite groups. *Pacific Journal of Mathematics*, 5(4):541–549, 1955.

[411] C. Hamalainen. New 2-critical sets in the abelian 2-group. *J. Combin. Math. Combin. Comput.*, 61:193–219, 2007.

[412] Haim Hanani. On quadruple systems. *Canad. J. Math.*, 12:145–157, 1960.

[413] Haim Hanani. The existence and construction of balanced incomplete block designs. *Ann. Math. Statist.*, 32:361–386, 1961.

[414] Haim Hanani. Balanced incomplete block designs and related designs. *Discrete Math.*, 11:255–369, 1975.

[415] Alan Hartman. The fundamental construction for 3-designs. *Discrete Math.*, 124(1-3):107–132, 1994.

[416] M. Hassinen and S. Markovski. Secure SMS messaging using Quasigroup encryption and Java SMS API. In *SPLST'03*, Kuopio, Finland, June 2003.

[417] M. Hassinen and S. Markovski. Differential cryptanalysis of the quasigroup cipher. Definition of the encryption method. In *Differential Cryptanalysis*, Petrozavodsk, June 2004.

[418] B.A. Hausmann and Oystein Ore. Theory of quasi-groups. *Am. J. Math.*, 59:983–1004, 1937.

[419] Aiso Heinze and Mikhail Klin. *Loops, Latin squares and strongly regular graphs: an algorithmic approach via algebraic combinatorics.* In Algorithmic Algebraic Combinatorics and Gröbner Bases, pages 1–65. Springer, 2009.

[420] W. Heise and P. Quattrocci. *Informations- und Codierungstheorie.* Springer, Berlin, 1995.

[421] D. Herbera, T. Kepka, and P. Němec. Hamiltonian self-distributive quasigroups. *J. Algebra*, 289(1):70–104, 2005.

[422] I.N. Herstein. *Abstract Algebra.* Macmillan Publishing Company, New York, second edition, 1990.

[423] Anthony J.W. Hilton and Jerzy Wojciechowski. Amalgamating infinite Latin squares. *Discrete Mathematics*, 292:67–81, 2005.

[424] P. Horak, R. E. L. Aldred, and H. J. Fleischner. Completing Latin squares: Critical sets. I. *J. Combin. Des.*, 10(6):419–432, 2002.

[425] P. Horak and I. J. Dejter. Completing Latin squares: Critical sets. II. *J. Combin. Des.*, 15(1):77–83, 2007.

[426] M. Hosszu. On the explicit form of n-group operations. *Publ. Math. Debrecen*, 10:88–92, 1963.

[427] Miklós Hosszú. Homogeneous groupoids. *Ann. Univ. Sci. Budap. Rolando Etvs, Sect. Math.*, 3–4:95–99, 1961.

[428] D.F. Hsu. *Cyclic Neofields and Combinatorial Designs*, volume 824 of *Lectures Notes in Mathematics*. Springer, Berlin, 1980.

[429] F.L. Hsu, F.A. Hummer, and J.D.H. Smith. Logarithms, syndrome functions, and the information rates of greedy loop transversal codes. *J. Combin. Math. Combin. Comput.*, 22:33–49, 1996.

[430] Tim Hsu. Explicit constructions of code loops as centrally twisted products. *Math. Proc. Cambridge Philos. Soc.*, 128(2):223–232, 2000.

[431] R. Huang and G.C. Rota. On the relations of various conjectures on Latin squares and straitening coefficients. *Discr. Math.*, 128:225–236, 1994.

[432] Lionel Humblot. Sur une extension de la notion de carres latins. *C. R. Acad. Sci. Paris Ser. A-B*, 273:A795–A798, 1971 (in French).

[433] F.A. Hummer and J.D.H. Smith. Greedy loop transversal codes, matrices, and lexicodes. *J. Combin. Math. Combin. Comput.*, 22:143–155, 1996.

[434] Stephen H.Y. Hung and N.S. Mendelsohn. Directed triple systems. *Journal of Combinatorial Theory, Series A*, 14:310–318, 1973.

[435] B. Huppert. *Endliche Gruppen I.* Springer, 1967.

[436] S.G. Ibragimov. On forgotten works of Ernst Schröder lying between algebra and logic. *Istor.-Mat. Issled.*, 17:247–258, 1966.

[437] S.G. Ibragimov. About the logic-algebraic works of Ernest Schröder which have anticipated the theory of quasigroups. In *Kibernetika i logika*, pages 253–313. Nauka, Moscow, 1978 (in Russian).

[438] T. Ihringer. On multiplication groups of quasigroups. *Europ. J. Comb.*, 5(2):137–141, 1984.

[439] A.P. Il'inykh. Classification of finite groupoids with a 2-transitive group of automorphisms. *Mat. Sb.*, 185(6):51–78, 1994.

[440] M. Istinger and H.K. Kaiser. A characterization of polynomially complete algebras. *Journal of Algebra*, 56(6):103–110, 1979.

[441] V.I. Izbash. On quasigroups isotopic to groups, 1989. Reg. in VINITI 29.06.89, No 4228-B89, Moscow (in Russian).

[442] V.I. Izbash. Quasigroups with distributive lattice of subquasigroups. *Mat. Issled.*, 113:42–51, 1990.

[443] V.I. Izbash. Monoquasigroups without congruences and automorphisms. *Bul. Acad. Stiinte Repub. Mold., Mat.*, (4):66–76, 1992 (in Russian).

[444] V.I. Izbash. Isomorphisms of quasigroups isotopic to groups. *Quasigroups Relat. Syst.*, 2:34–50, 1995.

[445] V.I. Izbash and V.A. Shcherbacov. On quasigroups with Moufang identity. In *Abstracts of the Third International Conference in Memory of M.I. Kargapolov (1928–1976)*, pages 134–135, Krasnoyarsk, Russian Federation, August 1993 (in Russian).

[446] V.I. Izbash and P.N. Syrbu. Recursively differentiable quasigroups and complete recursive codes. *Comment. Math. Univ. Carolin.*, 45(2):257–264, 2004.

[447] Vivek Kumar Jain. On the isomorphism classes of transversals III, 2012. arXiv:1112.5530.

[448] Vivek Kumar Jain. A note on transversals, 2013. arXiv:1307.5392.

[449] V.K. Jain and R.P. Shukla. On the isomorphism classes of transversals. *Comm. Alg.*, 36(5):1717–1725, 2008.

[450] V.K. Jain and R.P. Shukla. On the isomorphism classes of transversals II. *Comm. Alg.*, 39(6):2024–2036, 2011.

[451] T.G. Jaiyeola and E. Ilojide. On a group of linear-bivariate polynomials that generate quasigroups over the ring $\mathbb{Z}_n$. *An. Univ. Vest Timis. Ser. Mat.-Inform.*, 50(2):45–53, 2012.

[452] Temitope Gbolahan Jaiyeola, Sunday Peter David, and Yakubu Tunde Oyebo. New algebraic properties of middle Bol loops, 2016. arXiv:1606.09169.

[453] T.G. Jaiyeola and J.O. Adeniran. Not every Osborn loop is universal. *Acta Mathematica Academiae Paedagogiace Nyíregyháziensis*, 25(2):189–190, 2009.

[454] J.C.M. Janssen. On even and odd latin squares. *J.Comb. Th. Ser. A*, 69:173–181, 1995.

[455] Premysl Jedlicka, Michael Kinyon, and Petr Vojtechovsky. The structure of commutative automorphic loops. *Trans. Amer. Math. Soc.*, 363:365–384, 2011.

[456] Premysl Jedlika, David Stanovský, and Petr Vojtchovský. Distributive and trimedial quasigroups of order 243, 2016. arXiv:1603.00608.

[457] E. Jespers, G. Leal, and C. Polcino Milies. Classifying indecomposable RA loops. *J. Algebra*, 176:569–584, 1995.

[458] K.A. Jevlakov, A.M. Slin'ko, I.P. Shestakov, and A.I. Shirshov. *Rings Close to Associative.* Nauka, Moscow, 1978 (in Russian).

[459] J. Ježek and T. Kepka. A note on medial division groupoids. *Proc. Am. Math. Soc.*, 119(2):423–426, 1993.

[460] J. Ježek. Normal subsets of quasigroups. *Comment. Math. Univ. Carolin.*, 16(1):77–85, 1975.

[461] J. Ježek and T. Kepka. Varieties of abelian quasigroups. *Czech. Math. J.*, 27:473–503, 1977.

[462] J. Ježek and T. Kepka. *Medial Groupoids*, volume 93, sešit 2 of *Rozpravy Československe Academie VĚD*. Academia, Praha, 1983.

[463] J. Ježek, T. Kepka, and P. Nemec. *Distributive Groupoids*, volume 91, sešit 3 of *Rozpravy Československe Academie VĚD*. Academia, Praha, 1981.

[464] Jaroslav Jezek and Tomas Kepka. Self-distributive groupoids of small orders. *Czechoslovak Math. J.*, 47(122)(3):463–468, 1997.

[465] L. Ji and R. Wei. The spectrum of 2-idempotent 3-quasigroups with conjugate invariant subgroup. *J. Combin. Des.*, 18(4):292–304, 2010.

[466] D.M. Johnson, A.L. Dulmage, and N.S. Mendelsohn. Orthomorphisms of groups and orthogonal Latin squares, I. *Canad. J. Math.*, 13:356–372, 1961.

[467] K.W. Johnson and B.L. Sharma. A variety of loops. *Ann. Soc. Sci. Bruxelles Ser. I*, 92(1-2):25–41, 1978.

[468] K.W. Johnson and B.L. Sharma. Constructions of weak inverse property loops. *Rocky Mountain J. Math.*, 11(1):1–8, 1981.

[469] D. Kahn. *The Codebreakers: The Story of Secret Writing.* Wiedenfield and Nicolson, London, 1967.

[470] Arkadius Kalka. Non-associative public-key cryptography, 2012. arXiv:1210.8270v1.

[471] M.I. Kargapolov and M. Yu. Merzlyakov. *Foundations of Group Theory.* Nauka, Moscow, 1977 (in Russian).

[472] B.B. Karklin´š and V.B. Karklin´. Inverse loops. *Mat. Issled.*, 39:87–101, 1976 (in Russian).

[473] B.B. Karklin´š and V.B. Karklin´. Universal-automorph-inverse loops. *Mat. Issled.*, 39:82–86, 1976 (in Russian).

[474] S. Yu. Katyshev, V.T. Markov, and A.A. Nechaev. Utilization of nonassociative groupoids for the realization of an open key-distribution procedure. *Diskret. Mat.*, 26:45–64, 2014 (in Russian).

[475] A.D. Keedwell. Critical sets and critical partial Latin squares. In *Graph Theory, Combinatorics, Algorithms and Applications, Proceedings Third China-USA International Conference*, pages 111–124, Singapore, 1994. World Scienti4c Publ. Co.

[476] A.D. Keedwell. Critical sets for Latin squares, graphs and block designs: A survey. *Congressus Numeratium*, 113:231–245, 1996.

[477] A.D. Keedwell. Crossed-inverse quasigroups with long inverse cycles and applications to cryptography. *Australas. J. Combin.*, 20:241–250, 1999.

[478] A.D. Keedwell. Construction, properties and applications of finite neofields. *Comment. Math. Univ. Carolin.*, 41(2):283–297, 2000.

[479] A.D. Keedwell. Critical sets in Latin squares: An intriguing problem. *The Mathematical Gazette*, 85(503):239–244, 2001.

[480] A.D. Keedwell. Critical sets in Latin squares and related matters: An update. *Util. Math.*, 65:97–131, 2004.

[481] A.D. Keedwell. On Sudoku squares. *Bull. Inst. Combin. Appl.*, 50:52–60, 2007.

[482] A.D. Keedwell. The existence of Buchsteiner and conjugacy-closed quasigroups. *European J. Combin.*, 30(5):1382–1385, 2009.

[483] A.D. Keedwell. Realizations of loops and groups defined by short identities. *Comment. Math. Univ. Carolin.*, 50(3):373–383, 2009.

[484] A.D. Keedwell. Constructions of complete sets of orthogonal diagonal Sudoku squares. *Australas. J. Combin.*, 47:227–238, 2010.

[485] A.D. Keedwell. A short note regarding existence of complete sets of orthogonal diagonal Sudoku squares. *Australas. J. Combin.*, 51:271–273, 2011.

[486] A.D. Keedwell and V.A. Shcherbacov. On m-inverse loops and quasigroups with a long inverse cycle. *Australas. J. Combin.*, 26:99–119, 2002.

[487] A.D. Keedwell and V.A. Shcherbacov. Construction and properties of (r,s,t)-inverse quasigroups, I. *Discrete Math.*, 266(1-3):275–291, 2003.

[488] A.D. Keedwell and V.A. Shcherbacov. Construction and properties of (r,s,t)-inverse quasigroups, II. *Discrete Math.*, 288:61–71, 2004.

[489] A.D. Keedwell and V.A. Shcherbacov. Quasigroups with an inverse property and generalized parastrophic identities. *Quasigroups Relat. Syst.*, 13:109–124, 2005.

[490] Donald Keedwell and Jozef Dénes. *Latin Squares and Their Applications, 2nd Edition.* North Holland, Amsterdam, 2015.

[491] T. Kepka. Regular mappings of groupoids. *Acta Univ. Carolin. Math. Phys.*, 12:25–37, 1971.

[492] T. Kepka. Quasigroups which satisfy certain generalized forms of the Abelian identity. *Časopis Pěst. Mat.*, 100:46–60, 1975.

[493] T. Kepka. R-Transitive Groupoids. *Acta Univ. Carolin. Math. Phys.*, 16:71–77, 1975.

[494] T. Kepka. Structure of triabelian quasigroups. *Comment. Math. Univ. Carolin.*, 17:229–240, 1976.

[495] T. Kepka. A note on simple quasigroups. *Acta Univ. Carolin. Math. Phys.*, 19(2):59–60, 1978.

[496] T. Kepka. A note on WA-quasigroups. *Acta Univ. Carolin. Math. Phys.*, 19(2):61–62, 1978.

[497] T. Kepka. Structure of weakly abelian quasigroups. *Czech. Math. J.*, 28:181–188, 1978.

[498] T. Kepka. F-quasigroups isotopic to Moufang loops. *Czech. Math. J.*, 29:62–83, 1979.

[499] T. Kepka, L. Beneteau, and J. Lacaze. Small finite trimedial quasigroups. *Comm. in Algebra*, 14(6):1067–1090, 1986.

[500] T. Kepka, M.K. Kinyon, and J.D. Phillips. The structure of F-quasigroups. http://arxiv.org/, abs/math/0510298:24 pages, 2005.

[501] T. Kepka, M.K. Kinyon, and J.D. Phillips. F-quasigroups isotopic to groups. http://arxiv.org/, math.GR/0601077:11 pages, 2006.

[502] T. Kepka, M.K. Kinyon, and J.D. Phillips. The structure of F-quasigroups. *Journal of Algebra*, 317:435–461, 2007.

[503] T. Kepka, M.K. Kinyon, and J.D. Phillips. F-quasigroups and generalized modules. *Comment. Math. Univ. Carolin.*, 49(2):249–258, 2008.

[504] T. Kepka and P. Nemec. Unipotent quasigroups. *Acta Universitatis Carolinae. Mathematica et Physica*, 18:13–22, 1977.

[505] T. Kepka and P. Nemec. Trimedial quasigroups and generalized modules. *Acta Univ. Carolin. Math. Phys.*, 31(1):3–14, 1990.

[506] T. Kepka and P. Němec. T-quasigroups, II. *Acta Univ. Carolin. Math. Phys.*, 12(2):31–49, 1971.

[507] T. Kepka and P. Němec. Commutative Moufang loops and distributive groupoids of small order. *Czech. Math. J.*, 31:633–670, 1981.

[508] Tomas Kepka and Petr Nemec. Selfdistributive groupoids. Part A1. Non-indempotent left distributive groupoids. *Acta Univ. Carolin. Math. Phys.*, 44(1):3–94, 2003.

[509] Brent L. Kerby and Jonathan D.H. Smith. Quasigroup automorphisms and the Norton-Stein complex. *Proc. Amer. Math. Soc.*, 138(9):3079–3088, 2010.

[510] Brent L. Kerby and Jonathan D.H. Smith. A graph-theoretic approach to quasigroup cycle numbers. *J. Combin. Theory Ser. A*, 118(8):2232–2245, 2011.

[511] A. Kertesz and A. Sade. On nuclei of groupoids. *Publ. Math. Debrecen*, 6:214–233, 1959.

[512] H. Kiechle. *Theory of K-Loops, Habilitationsschrift.* Fachbereich Mathematik der Universität Hamburg, Hamburg, 1998. (Habilitation Dissertation).

[513] Hubert Kiechle and Michael K. Kinyon. Infinite simple Bol loops. *Comment. Math. Univ. Carolin.*, 45(2):275–278, 2004.

[514] M.K. Kinyon and K. Kunen. Power-associative, conjugacy closed loops. http://arxiv.org/, math/0507278v3:30 pages, 2005.

[515] M.K. Kinyon and K. Kunen. Power-associative, conjugacy closed loops. *J. Algebra*, 304(2):679–711, 2006.

[516] M.K. Kinyon, K. Kunen, and J.D. Phillips. Diassociativity in conjugacy closed loops. *Comm. Algebra*, 32(2):767–786, 2004.

[517] M.K. Kinyon, Kenneth Kunen, and J.D. Phillips. A generalization of Moufang and Steiner loops. *Algebra Univers.*, 48:81–101, 2002.

[518] M.K. Kinyon, J.D. Phillips, and P. Vojtechovsky. When is the commutant of a Bol loop a subloop? *Trans. Amer. Math. Soc.*, 360(5):2393–2408, 2008.

[519] M.K. Kinyon and J.D. Phillips. A note on trimedial quasigroups. *Quasigroups Relat. Syst.*, 9:65–66, 2002.

[520] M.K. Kinyon and J.D. Phillips. Axioms for trimedial quasigroups. *Comment. Math. Univ. Carolin.*, 45(2):287–294, 2004.

[521] M.K. Kinyon and P. Vojtechovsky. Primary decompositions in varieties of commutative diassociative loops. *http://arxiv.org/*, math.math.GR/0702874:13 pages, 2007.

[522] Fred Kiokemeister. A theory of normality for quasi-groups. *Am. J. Math.*, 70:99–106, 1948.

[523] O.U. Kirnasovsky. Linear isotopes of small order. *Quasigroups Relat. Syst.*, 2:51–82, 1995.

[524] O.U. Kirnasovsky. The transitive and multitransitive automorphism group of the multiplace quasigroups. *Quasigroups Relat. Syst.*, 4:23–38, 1997.

[525] O.U. Kirnasovsky. *Binary and n-Ary Isotopes of Groups, Main Algebraic Notations and Quantitative Characteristics.* PhD thesis, Taras Shevchenko Kiev State University, Kiev, 2000 (in Ukrainian).

[526] Jenya Kirshtein. Automorphism groups of real Cayley-Dickson loops. http://arxiv.org/, arXiv:1102.5151:15 pages, 2011.

[527] K. Kishen. On the construction of Latin and Hyper-Graceo-Latin cubes and hypercubes. *J. Ind. Soc. Agric. Statist.*, 2:20–48, 1950.

[528] M.D. Kitoroagè. Quasigroups whose inner permutations are automorphisms. *Questions of the Theory of Quasigroups and Loops*, 7:63–76, 1970 (in Russian).

[529] M.D. Kitoroagè. Nuclei in quasigroups. *Mat. Issled.*, 7:60–71, 1972 (in Russian).

[530] A. Klapper. On the existence of secure keystream generators. *J. Cryptology*, 14:1–15, 2001.

[531] Mikhail Klin, Nimrod Kriger, and Andrew Woldar. Classification of highly symmetrical translation loops of order 2 p, p prime. *Beitr Algebra Geom*, 55:253–276, 2014.

[532] N. Koblitz. The uneasy relationship between mathematics and cryptography. *Notices of the AMS*, 54(8):972–979, 2007.

[533] C. Koscielny. A method of constructing quasigroup-based stream ciphers. *Appl. Math. and Comp. Sci.*, 6:109–121, 1996.

[534] C. Koscielny. NLPN Sequences over GF(q). *Quasigroups Relat. Syst.*, 4:89–102, 1997.

[535] C. Koscielny. Generating quasigroups for cryptographic applications. *Int. J. Appl. Math. Comput. Sci.*, 12(4):559–569, 2002.

[536] C. Koscielny. Stegano cryptography with Maple 8. Technical report, Institute of Control and Computation Engineering, University of Zielona Gora, http://www.mapleapps.com/categories/mathematics/Cryptography /html/stegcryp.html, 2003.

[537] C. Koscielny and G.L. Mullen. A quasigroup-based public-key cryptosystem. *Int. J. Appl. Math. Comput. Sci.*, 9(4):955–963, 1999.

[538] A.I. Kostrikin. *Introduction in Algebra.* Nauka, Moscow, 1977. (in Russian).

[539] R. Koval'. On a functional equation with a group isotopy property. *Bul. Acad. Stiinte Repub. Mold. Mat.*, (2):65–71, 2005.

[540] A. Krapez. Generalized associativity on groupoids. *Publ. Inst. Math. (Beograd) (N.S.)*, 28(42):105–112, 1980.

[541] A. Krapez. Cryptographically suitable quasigroups via functional equations. In S. Markovski and M. Gusev (Eds.): *ICT Innovations 2012, AISC 207*, number DOI: 10.1007/978-3-642-37169-1_26, pages 265–274, Berlin, 2012. Springer-Verlag.

[542] A. Krapez and D. Zivkovic. Parastrophically equivalent quasigroup equations. *Publ. Inst. Math. (Beograd) (N.S.)*, 87(101):39–58, 2010.

[543] Denis S. Krotov, Valdimir N. Potapov, and Polina V. Sokolova. On reconstructing reducible n-ary quasigroups and switching subquasigroups. *Quasigroups Related Systems*, 16(1):55–67, 2008.

[544] Jaromy Kuhl and Michael W. Schroeder. Completing partial Latin squares with blocks of non-empty cells. *Graphs and Combinatorics*, 32:241–256, 2016.

[545] K. Kunen. Moufang quasigroups. *J. Algebra*, 183:231–234, 1996.

[546] K. Kunen. G-loops and permutation groups. *J. Algebra*, 220(2):694–708, 1999.

[547] K. Kunen. The structure of conjugacy closed loops. *Transactions of the American Mathematical Society*, 352(6):2889–2911, 2000.

[548] A.G. Kurosh. *Lectures on General Algebra.* Gos. izdatel'stvo fiz.-mat. literatury, Moscow, 1962 (in Russian).

[549] A.G. Kurosh. *Group Theory.* Nauka, Moscow, 1967 (in Russian).

[550] A.G. Kurosh. *Course of Higher Algebra.* Nauka, Moscow, 1968 (in Russian).

[551] A.V. Kuznetsov and A.F. Danilchenko. Functionally complete quasigroups. In *First All-Union Symposium on Quasigroup Theory and Its Applications. Abstracts of Reports and Talks*, pages 17–19, Tbilisi, 1968.

[552] A.V. Kuznetsov and E.A. Kuznetsov. On two generated two homogeneous quasigroups. *Mat. Issled.*, 71:34–53, 1983 (in Russian).

[553] E. Kuznetsov. Transversals in loops. 1. Elementary properties. *Quasigroups Related Systems*, 18(2):43–58, 2010.

[554] E. Kuznetsov. Transversals in loops. 2. Structural theorems. *Quasigroups Related Systems*, 19(2):279–286, 2011.

[555] Xuejia Lai and James L. Massey. A proposal for a new block encryption standard. *Advances in Cryptology - EUROCRYPT 90, LNCS*, 473:389–404, 1991.

[556] Ramji Lal. Transversals in groups. *J. Algebra*, 181:70–81, 1996.

[557] T.Y. Lam. On subgroups of prime index. *The American Mathematical Monthly*, 111(3):256–258, 2004.

[558] R. Laskar. Finite nets of dimension three. I. *J. Algebra*, 32:8–25, 1974.

[559] Charles F. Laywine and Gary L. Mullen. *Discrete Mathematics Using Latin Squares.* John Wiley & Sons, Inc., New York, 1998.

[560] In-Sok Lee, Woo-Hwan Kim, Daesung Kwon, Sangil Nahm, Nam-Seok Kwak, and Yoo-Jin Baek. On the Security of MOR Public Key Cryptosystem, 2004, http://www.iacr.org/cryptodb/archive/2004/ASIACRYPT/326/326.pdf.

[561] Sin-Min Lee. A construction of simple one-element extension of n-groupoids. *Algebra Univers.*, 16:245–249, 1983.

[562] Sin-Min Lee. On finite-element simple extensions of a countable collection of countable groupoids. *Publications de l'Intstitut Mathematique*, 38:65–68, 1985.

[563] Wolfgang Lempken, Spyros S. Magliveras, Tran van Trung, and Wandi Wei. A public key cryptosystem based on non-abelian finite groups. *Journal of Cryptology*, 22:62–74, 2009.

[564] Gaëtan Leurent. Practical key recovery attack against Secret-IV Edon-R. http://www.di.ens.fr/ leurent/files/Edon-RSA10.pdf.

[565] F. Levi and B.L. van der Waerden. Uber eine besondere Klasse von Gruppen. *Abhandlungen Hamburg*, 9:154–158, 1932.

[566] Martin W. Liebeck. The classification of finite simple Moufang loops. *Math. Proc. Cambridge Philos. Soc.*, 102(1):33–47, 1987.

[567] C.C. Lindner and C.A. Rodger. *Design Theory.* CRC Press, New York, 2009.

[568] C.C. Lindner. Construction of quasigroups satisfying the identity x (xy)= yx. *Canad. Math. Bull.*, 14:57–59, 1971.

[569] C.C. Lindner. The generalized singular direct product for quasigroups. *Canad. Math. Bull.*, 14:61–63, 1971.

[570] C.C. Lindner. Identities preserved by the singular direct product. *Algebra Universalis*, 1(1):86–89, 1971.

[571] C.C. Lindner. Quasigroup identities and orthogonal arrays. *London Math. Soc., Lect. Note Ser.*, 82:77–105, 1983.

[572] C.C. Lindner, N.S. Mendelsohn, and S.R. Sun. On the construction of Schroeder quasigroups. *Discrete Math.*, 32(3):271–280, 1980.

[573] Nathan Linial and Zur Luria. An upper bound on the number of high-dimensional permutations, 2011. arXiv:1106.0649v2.

[574] V. Lipski. *Combinatorics for Programmers.* Mir, Moscow, 1988 (in Russian).

[575] Jaak Lohmus, Eugen Paal, and Leo Sorgsepp. About nonassociativity in mathematics and physics. *Acta Appl. Math.*, 50:3–31, 1998.

[576] O. Loos. *Symmetric Spaces. I: General Theory.* W. A. Benjamin, Inc., New York-Amsterdam, 1969.

[577] G.J. Lovegrove. The automorphism groups of Steiner triple systems obtained by the Bose construction. *Journal of Algebraic Combinatorics*, 18:159–170, 2003.

[578] A.D. Lumpov. *C-Equivalence in Quasigroups.* PhD thesis, Institute of Mathematics Academy of Sciences of Republic Moldova, 1990 (in Russian).

[579] I.V. Lyakh. Autotopisms of orthogonal systems of quasigroups. *Izv. Akad. Nauk Moldav. SSR Ser. Fiz.-Tekhn. Mat. Nauk*, (2):6–12, 1985.

[580] I.V. Lyakh. *On Transformations of Orthogonal Systems of Operations and Algebraic Nets.* PhD thesis, Institute of Mathematics Academy of Sciences Republic Moldova, Kishinev, 1986 (in Russian).

[581] I.V. Lyakh. On pseudo-automorphisms of orthogonal systems of loops. *Mat. Issled.*, pages 86–100, 1987.

[582] I.V. Lyakh. On the definition of the concept of the homotopy of orthogonal systems of quasigroups. *Webs and Quasigroups*, pages 77–84, 1988.

[583] E.S. Lyapin. *Semigroups.* Gosudarstv. Izdat. Fiz.-Mat. Lit., Moscow, 1960 (in Russian).

[584] S.S. Magliveras, D.R. Stinson, and Tran van Trung. New approaches to designing public key cryptosystems using one-way functions and trapdoors in finite groups. *J. Cryptology*, 15:285–297, 2002.

[585] Ayan Mahalanobis. A note on using finite non-abelian p-groups in the MOR cryptosystem, 2007. arXiv:cs/0702095v1.

[586] Ayan Mahalanobis. A simple generalization of the El-Gamal cryptosystem to non-abelian groups, 2008. arXiv:cs/0607011v5.

[587] Ayan Mahalanobis. A simple generalization of the Elgamal cryptosystem to non-abelian groups. *Communications in Algebra*, 36:3878–3889, 2008.

[588] Ayan Mahalanobis. A simple generalization of the Elgamal cryptosystem to non-abelian groups II, 2009. arXiv:0706.3305v4.

[589] A.I. Mal'tsev. On the general theory of algebraic systems. *Mat. Sb.*, 35 (77)(1):3–20, 1954 (in Russian).

[590] A.I. Mal'tsev. Identical relations on varieties of quasigroups. *Mat. Sb.*, 69(1):3–12, 1966 (in Russian).

[591] A.I. Mal'tsev. *Algebraic Systems.* Nauka, Moscow, 1976 (in Russian).

[592] Yu. I. Manin. Cubic hypersurfaces. I. Quasigroups of classes of points. *Izv. Akad. Nauk SSSR Ser. Mat.*, 32(6):1223–1244, 1968.

[593] Yu. I. Manin. *Cubic Forms.* Nauka, Moscow, 1972 (in Russian).

[594] H.B. Mann. The construction of orthogonal Latin squares. *Ann. Math. Statist.*, 13:418–423, 1942.

[595] H.B. Mann. On the construction of sets of orthogonal Latin squares. *Ann. Math. Statist.*, 14:401–414, 1943.

[596] H.B. Mann. On orthogonal Latin squares. *Bull. Amer. Math. Soc.*, 50:249–257, 1944.

[597] H.B. Mann. *Analizes and design of experiments.* Dover, New York, 1949.

[598] A. Marini and G. Pirillo. Signs of group Latin squares. *Adv. in Appl. Math.*, 17:117–121, 1995.

[599] A. Marini and V.A. Shcherbacov. About signs of Bol loop translations. *Bul. Acad. Stiinte Repub. Mold., Mat.*, (3):87–92, 1998.

[600] A. Marini and V.A. Shcherbacov. On autotopies and automorphisms of n-ary linear quasigroups. *Algebra and Discrete Math.*, (2):51–75, 2004.

[601] S. Markovski, D. Gligoroski, and S. Andova. Using quasigroups for one-one secure encoding. In *Proc. VIII Conf. Logic and Computer Science "LIRA'97," Novi Sad*, pages 157–167, 1997.

[602] S. Markovski, D. Gligoroski, and V. Bakeva. Quasigroup string processing: Part 1. *Contributions, Sec. Math. Tech. Sci., MANU*, XX(1-2):13–28, 1999.

[603] S. Markovski, D. Gligoroski, and V. Bakeva. Quasigroups and hash functions. In *Res. Math. Comput. Sci.*, volume 6, pages 43–50, South-West Univ., Blagoevgrad, 2002.

[604] S. Markovski, D. Gligoroski, and Lj. Kocarev. Totally Asynchronous Stream Ciphers + Redundancy = Cryptcoding. In *Proceedings of the 2007 International Conference on Security and Management, SAM 2007, June 25-28*, pages 446–451, Las Vegas, 2007. http://www.informatik.uni-trier.de/ ley/db/conf/csreaSAM/csreaSAM2007.html/.../GligoroskiMK07.

[605] S. Markovski, D. Gligoroski, and Lj. Kocarev. Error correcting cryptcodes based on quasigroups. NATO ARW, 6-9 October, 2008, Veliko Tarnovo, Bulgaria, 2008. https://www.cosic.esat.kuleuven.be/.../Markovski_slides_nato08.ppt.

[606] S. Markovski, D. Gligoroski, and B. Stojcevska. Secure two-way on-line communication by using quasigroup enciphering with almost public key. *Novi Sad J. Math.*, 30(2):43–49, 2000.

[607] S. Markovski and A. Mileva. Generating huge quasigroups from small non-linear bijections via extended Feistel function. *Quasigroups Relat. Syst.*, 17(1):91–106, 2009.

[608] Smile Markovski and Aleksandra Mileva. Generating huge quasigroups from small non-linear bijections via extended Feistel function. *Quasigroups Related Systems*, 17(1):91–106, 2009.

[609] S.I. Marnas, L. Angelis, and G.L. Bleris. All-or-nothing transforms using quasigroups. In *Proceedings of 1st Balkan Conference in Informatics*, pages 183–191, Thessaloniki, November 2003.

[610] A. Maturo and M. Zannetti. Redei blocking sets with two Redei lines and quasigroups. *J. Discrete Math. Sci. Cryptography*, 5(1):51–62, 2002.

[611] G. Maze. *Algebraic Methods For Constructing One-Way Trapdoor Functions.* PhD thesis, University of Notre Dame, 2003.

[612] Gerard Maze, Chris Monico, and Joachim Rosenthal. Public key cryptography based on semigroup actions. *Advances in Mathematics of Communications*, 1(4):489–507, 2007.

[613] W. McCune. *OTTER 3.3.* Argonne National Laboratory, www.mcs.anl.gov/AR/otter/, 2004.

[614] W. McCune. *Mace 4.* University of New Mexico, www.cs.unm.edu/mccune/prover9/, 2007.

[615] W. McCune. *Prover 9.* University of New Mexico, www.cs.unm.edu/mccune/prover9/, 2007.

[616] R. McEliece and D. V. Sarwarte. On sharing secrets and reed solomon codes. *Comm. ACM*, 24:583–584, 1981.

[617] Gary McGuire, Bastian Tugemann, and Gilles Civario. There is no 16-clue Sudoku: Solving the Sudoku Minimum Number of Clues Problem, 2013. arxiv: 1201.0749v2.

[618] Brendan D. McKay and Ian M. Wanless. On the number of Latin squares. *Ann. Comb.*, 9(3):335–344, 2005.

[619] B.D. McKay and A. Meynet and W. Myrvold. Small Latin squares, quasigroups and loops. *J. Combin.*, 15(2):98–119, 2007.

[620] R. McKenzie, G. McNulty, and W. Taylor. *Algebras, Lattices, Varieties, volume I.* Wadsworth & Brooks/Cole, Belmont, 1987.

[621] A.J. Menezes, P.C. Van Oorschot, and S.A. Vanstone. *Handbook of Applied Cryptography.* CRC Press, Boca Raton, FL, 1997.

[622] K. A. Meyer. *A New Message Authentication Code Based on the Non-Associativity of Quasigroups.* PhD thesis, Iowa State University, 2006.

[623] P. O. Mikheev. Groups that envelop Moufang loops. *Uspekhi Mat. Nauk*, 48(2):191–192, 1993 (in Russian).

[624] A. Mileva. New developments in quasigroup-based cryptography. In *Multidisciplinary Perspectives in Cryptology and Information Security.* Sattar B. Sadkhan Al Maliky and Nidaa A. Abbas, eds., pages 286–316, 2014.

[625] Aleksandra Mileva and Smile Markovski. Shapeless quasigroups derived by Feistel orthomorphisms. *Glas. Mat. Ser. III*, 47(2):333–349, 2012.

[626] H. Mink. *Permanents.* Mir, Moscow, 1982 (in Russian).

[627] L. Mittenhal. A source of cryptographically strong permutations for use in block ciphers. In *Proc. IEEE, International Sympos. on Information Theory, 1993, IEEE*, pages 17–22, New York, 1993.

[628] L. Mittenhal. Block substitutions using orthomorphic mappings. *Advances in Applied Mathematics*, 16:59–71, 1995.

[629] Mohamed Saied Emam Mohamed, Jintai Ding, and Johannes Buchmann. Algebraic Cryptanalysis of MQQ Public Key Cryptosystem by MutantXL, 2008. eprint.iacr.org/2008/451.pdf.

[630] D.A. Mojdeh and N.J. Rad. Critical sets in latin squares given that they are symmetric. *Univ. Beograd. Publ. Elektrotehn. Fak. Ser. Mat.*, 18:38–45, 2007.

[631] A.A. Moldovyan, N.A. Moldovyan, N.D. Goots, and B.V. Izotov. *Cryptology. Fast codes.* BKhV-Peterburg, St.-Petersburg, 2002 (in Russian).

[632] N.A. Moldovyan. *Problems and Methods of Cryptology.* St.-Petersburg University Press, St.-Petersburg, 1998 (in Russian).

[633] N.A. Moldovyan, A.A. Moldovyan, and M.E. Eremeev. *Cryptology. From Primitives to the Synthesis of Algorithms.* BKhV-Peterburg, St.-Petersburg, 2004 (in Russian).

[634] N.A. Moldovyan and P.A. Moldovyanu. New primitives for digital signature algorithms. *Quasigroups Related Systems*, 17:271–282, 2009.

[635] N.A. Moldovyan, A.V. Shcherbacov, and V.A. Shcherbacov. On some applications of quasigroups in cryptology. In *Workshop on Foundations of Informatics, August 24–29, 2015, Chisinau, Proceedings*, pages 331–341, Chisinau, 2015.

[636] N. Mosina. *Probability on Graphs and Groups: Theory and Applications.* PhD thesis, Columbia University, 2009. Available at http://www.math.columbia.edu/ thaddeus/ theses/2009/mosina.pdf.

[637] N. Mosina and A. Ushakov. Mean-set attack: Cryptanalysis of Sibert and al. authentication protocol. *J. Math. Cryp.*, 4:149–174, 2010.

[638] R. Moufang. Zur Structur von Alternativ Körpern. *Math. Ann.*, 110:416–430, 1935.

[639] Yu. Movsisyan. Hyperidentities in algebras and varieties. *Russ. Math. Surv.*, 53(1):57–108, 1998.

[640] Yu. M. Movsisyan. *Hyperidentities and Hypervarieties in Algebras.* Yerevan State University Press, Yerevan, 1990.

[641] G.L. Mullen and V.A. Shcherbacov. Properties of codes with one check symbol from a quasigroup point of view. *Bul. Acad. Stiinte Repub. Mold., Mat.*, (3):71–86, 2002.

[642] G.L. Mullen and V.A. Shcherbacov. n-T-quasigroup codes with one check symbol and their error detection capabilities. *Comment. Math. Univ. Carolin.*, 45(2):321–340, 2004.

[643] G.L. Mullen and V.A. Shcherbacov. On orthogonality of binary operations and squares. *Bul. Acad. Stiinte Repub. Mold., Mat.*, (2):3–42, 2005.

[644] R.C. Mullin. A generalization of the singular direct product with applications to skew room squares. *J. Combin. Theory Ser. A*, 29(3):306–318, 1980.

[645] D.C. Murdoch. Quasigroups which satisfy certain generalized associative laws. *Amer. J. Math.*, 61:509–522, 1939.

[646] D.C. Murdoch. Structure of abelian quasigroups. *Trans. Amer. Math. Soc.*, 49:392–409, 1941.

[647] William B. Muse and Kevin T. Phelps. Conjugate orthogonal 3-quasigroups. *Proceedings of the Forty-Third Southeastern International Conference on Combinatorics, Graph Theory and Computing. Congr. Numer.*, 213:99–106, 2012.

[648] Gabor P. Nagy. Direct construction of code loops. *Discrete Math.*, 308(23):5349–5357, 2008.

[649] Gabor P. Nagy. A class of simple proper Bol loops. *Manuscripta Math.*, 127(1):81–88, 2008.

[650] Gabor P. Nagy. Some remarks on simple Bol loops. *Comment. Math. Univ. Carolin.*, 49(2):259–270, 2008.

[651] Gabor P. Nagy. A class of finite simple Bol loops of exponent 2. *Transactions of the American Mathematical Society*, 361(10):5331–5343, 2009.

[652] Gabor P. Nagy and Petr Vojtechovsky. Automorphism groups of simple Moufang loops over perfect fields. *Math. Proc. Cambridge Philos. Soc.*, 135(2):193–197, 2003.

[653] Gabor P. Nagy and Petr Vojtechovsky. Octonions, simple Moufang loops and triality. *Quasigroups Related Systems*, 10:65–94, 2003.

[654] Peter T. Nagy and Karl Strambach. *Loops in Group Theory and Lie Theory, de Gruyter Expositions in Mathematics, 35.* Walter de Gruyter & Co., Berlin, 2002.

[655] E. Natale. n-quasigruppi mediali idempotenti commutativi. *Rendiconto Acad. Sci. Fis. e Mat. Napoli*, 46:221–229, 1979.

[656] J.R. Nechvatal. Asymptotic enumeration of generalised Latin rectangles. *Util. Math.*, 20:273–292, 1981.

[657] A.I. Nesterov and L.V. Sabinin. Non-associative geometry and discrete structure of spacetime. *Comment. Math. Univ. Carolin.*, 41(2):347–357, 2000.

[658] Markku Niemenmaa and Tomas Kepka. On multiplication groups of loops. *J. Algebra*, 135(1):112–122, 1990.

[659] Markku Niemenmaa and Miikka Rytty. Centrally nilpotent finite loops. *Quasigroups and Related Systems*, 19:123–132, 2011.

[660] I. Niven and H.S. Zuckerman. *An Introduction to the Theory of Numbers.* Wiley, New York, 1972.

[661] D.A. Norton. Group of orthogonal row-Latin squares. *Pacific J. Math.*, 2:335–341, 1952.

[662] D.A. Norton. Hamiltonian loops. *Proc. Amer. Math. Soc.*, 3:56–65, 1952.

[663] D.A. Norton and Sherman K. Stein. An integer associated with Latin squares. *Proc. Amer. Math. Soc.*, 7:331–334, 1956.

[664] V.A. Nosov and A.E. Pankratiev. Latin squares over abelian groups. *Fundamentalnaya i Prikladnaya Matematika*, 12:65–71, 2006.

[665] P. Němec. Arithmetical forms of quasigroups. *Comment. Math. Univ. Carolin.*, 29(2):295–302, 1988.

[666] P. Němec. Commutative Moufang loops corresponding to linear quasigroups. *Comment. Math. Univ. Carolin.*, 29(2):303–308, 1988.

[667] P. Němec and T. Kepka. T-quasigroups, I. *Acta Univ. Carolin. Math. Phys.*, 12(1):39–49, 1971.

[668] E. Ochadkova and V. Snasel. Using quasigroups for secure encoding of file system. In *Conference Security and Protection of Information, Abstract of Talks*, pages 175–181, Brno, May 2001.

[669] V.I. Onoi. Left distributive quasigroups that are left homogeneous over a quasigroup. *Bul. Akad. Stiince RSS Moldoven*, 2:24–31, 1970 (in Russian).

[670] V.I. Onoi. *Left Distributive Quasigroups and Loops, Which Are Isotopic to These Quasigroups.* PhD thesis, IM AN MSSR, Kishinev, 1972 (in Russian).

[671] V.I. Onoi. Solution of one problem on inverse loops. *Mat. Issled.*, 71:53–58, 1980 (in Russian).

[672] M. Osborn. Loops with the weak inverse property. *Pacific J. Math.*, 10:295–304, 1960.

[673] Seong-Hun Paeng. On the security of cryptosystem using automorphism groups. *Information Processing Letters*, 88:293–298, 2003.

[674] Seong-Hun Paeng, Kil-Chan Ha, Jae Heon Kim, Seongtaek Chee, and Choonsik Park. New public key cryptosystem using finite non-abelian groups. In J. Kilian, editor, *Crypto 2001*, LNCS, vol. 2139, pages 470–485. Springer-Verlag, 2001.

[675] Lowell J. Paige. Neofields. *Duke Math. J.*, 16:39–60, 1949.

[676] Lowell J. Paige. A class of simple Moufang loops. *Proc. Amer. Math. Soc.*, 7:471–482, 1956.

[677] E.T. Parker. Orthogonal Latin squares. *Proc. Nat. Acad. Sci. U.S.A.*, 45:859–862, 1959.

[678] I.I. Parovichenko. *The Theory of Operations over Sets.* Shtiintsa, Kishinev, 1981.

[679] A. Pavlu and A. Vanzurova. On identities of Bol-Moufang type. *Bul. Acad. Stiinte Repub. Mold. Mat.*, (3):68–100, 2005.

[680] M.J. Pelling and D.G. Rogers. Stein quasigroups. I: Combinatorial aspects. *Bull. Aust. Math. Soc.*, 18:221–236, 1978.

[681] M.J. Pelling and D.G. Rogers. Stein quasigroups. II: Algebraic aspects. *Bull. Aust. Math. Soc.*, 20:321–344, 1979.

[682] Ray A. Perlner and David A. Cooper. Quantum resistant public key cryptography: a survey. In *Proceedings of the 8th Symposium on Identity and Trust on the Internet, New York, NY: ACM*, pages 85–93, 2009.

[683] A. Petrescu. n-Quasigroup cryptographic primitives: Stream ciphers. *Studia Univ. Babes-Bolyai, Informatica*, LV(2):27–34, 2010.

[684] Adrian Petrescu. Applications of quasigroups in cryptography. In *Interdisciplinarity in Engineering Scientific International Conference Tg.Mures-Romania, 15–16 November 2007*, 2007. www.upm.ro/InterIng2007/Papers/Section6/16-Petrescu-Quasigroups-pVI- 16-1-5.pdf.

[685] H.O. Pflugfelder. *Quasigroups and Loops: Introduction.* Heldermann Verlag, Berlin, 1990.

[686] H.O. Pflugfelder. Historical notes on loop theory. *Comment. Math. Univ. Carolin.*, 41(2):359–370, 2000.

[687] K.T. Phelps. Conjugate orthogonal quasigroups. *J. Comb. Theory, A*, 25(2):117–127, 1978.

[688] J.D. Phillips. Moufang loop multiplication groups with triality. *Rocky Mountain J. of Math.*, 29(4):1483–1490, 1999.

[689] J.D. Phillips. On Moufang A-loops. *Comment. Math. Univ. Carolin.*, 41(2):371–375, 2000.

[690] J.D. Phillips. See Otter digging for algebraic pearls. *Quasigroups Relat. Syst.*, 10:95–114, 2003.

[691] J.D. Phillips. The Moufang laws, global and local. *J. Algebra Appl.*, 8:477–492, 2009.

[692] J.D. Phillips, D.I. Pushkashu, A.V. Shcherbacov, and V.A. Shcherbacov. On Birkhoff's quasigroup axioms. *J. of Algebra*, 457:7–17, 2016.

[693] J.D. Phillips and V.A. Shcherbacov. Cheban loops. *Journal of Generalized Lie Theory and Applications*, 4:5 pages, 2010. Article ID G100501, doi:10.4303/jglta/G100501.

[694] J.D. Phillips and Petr Vojtechovsky. The varieties of loops of Bol-Moufang type. *Algebra Universalis*, 54(3):259–271, 2005.

[695] J.D. Phillips and V.A. Shcherbacov. Cheban loops, 2010. arxiv.org. 1005.2750v1.

[696] N.C.K. Phillips and W.D. Wallis. All solutions to a tournament problem. *Congressus Numerantium*, 114:193–196, 1996.

[697] David Pointcheval and Jacques Stern. Security arguments for digital signatures and blind signatures. *Journal Cryptology*, 13(3):361–396, 2000.

[698] M. Polonijo. Abelian totally symmetric n-quasigroups. In *Proceedings of the Symposium n-Ary Structures*, pages 185–193, Skopje, 1982.

[699] M. Polonijo. Medial multiquasigroups. *Prilozi, Makedonska Akad. Nauk Umet. Odd. Mat-Tekh. Nauki*, 3:31–36, 1982.

[700] Tatiana Popovich. On conjugate sets of quasigroups. *Bul. Acad. Ştiinţe Repub. Mold. Mat.*, (3):69–76, 2011.

[701] E.L. Post. Polyadic groups. *Trans. Amer. Math. Soc.*, 48:208–350, 1940.

[702] V.N. Potapov and D.S. Krotov. On the number of n-ary quasigroups of finite order. *Diskret. Mat.*, 24(1):60–69, 2012 (in Russian).

[703] V.N. Potapov. *Discrete Functions and Structures in q-Ary Hypercubes*. Thesis of Doctor of Physical and Mathematical Sciences. Sobolev Institute of Mathematics, Novosibirsk, 2013 (in Russian).

[704] G. Grätzer. A theorem on doubly transitive permutation groups with application to universal algebras. *Fund. Math.*, 53(1):25–41, 1963.

[705] R. Pöschel and L. A. Kalužnin. *Funktionen- und Relationenalgebren.* Birkhäuser, Basel, 1979.

[706] D.I. Pushkashu. Para-associative groupoids. *Quasigroups Relat. Syst.*, 18:211–230, 2010.

[707] D.I. Pushkashu. Groupoids with Schröder identity of generalized associativity. In *Proceedings of the Third Conference of Mathematical Society of the Republic of Moldova Dedicated to the 50th Anniversary of the Foundation of the Institute of Mathematics and Computer Science, August 19–23, 2014, Chisinau*, pages 154–157, Chisinau, 2014. Institute of Mathematics and Computer Science.

[708] F. Rado. Generalizarea tesuturilor spatiale pentru structuri algebrice. *Studia Univ. Babes-Bolyai*, 1:41–53, 1960.

[709] F. Rado. On semi-symmetric quasigroups. *Aequationes Math.*, 11:250–255, 1974.

[710] Thomas M. Richardson. Local subgroups of the Monster and odd code loops. *Trans. Amer. Math. Soc.*, 347(5):1453–1531, 1995.

[711] J. Riguet. Relations binares, fermetures, correspondances de Galois. *Bull. Soc. Math. France*, 76:114–155, 1948.

[712] R. Rivest, A. Shamir, and L. Adleman. A method for obtaining digital signatures and public key cryptosystems. *Communications of the ACM*, 21(2):120–126, 1978.

[713] R.L. Rivest. All-or-nothing encryption and the package transform. In *Fast Software Encryption '97*, volume 1267 of *LNCS*. Springer, 1997.

[714] D.A. Robinson. MR1402578. *Mathematical Review*, 97.

[715] D.A. Robinson. *Bol Loops*. PhD thesis, University of Wisconsin, Madison, Wis., 1964.

[716] D.A. Robinson. Bol loops. *Trans. Amer. Math. Soc.*, 123:341–354, 1966.

[717] D.A. Robinson. A special embedding of Bol loops in groups. *Acta Math. Acad. Sci. Hungar.*, 30(1–2):95–103, 1977.

[718] D.A. Robinson and Karl H. Robinson. A class of Bol loops whose nuclei are not normal. *Arch. Math. (Basel)*, 61(6):596–600, 1993.

[719] P. Rowley. Finite groups admitting a fixed-point-free automorphism group. *J. Algebra*, 174:724–727, 1995.

[720] S.A. Rusakov. *Algebraic n-Ary Systems: Sylow Theory of n-Ary Groups*. Navuka i tehnika, Minsk, 1992. (in Russian).

[721] K.A. Rybnikov. *Introduction in Combinatorial Analysis*. Publishing House of Moscow State University, Moscow, 1985 (in Russian).

[722] H.J. Ryser. *Combinatorial Mathematics, The Carus Mathematical Monographs, No. 14*. The Mathematical Association of America, New York, 1963.

[723] L.V. Sabinin. The geometry of loops. *Mat. Zametki*, 12:605–616, 1972 (in Russian).

[724] L.V. Sabinin. Homogeneous spaces and quasigroups. *Izv. Vyssh. Uchebn. Zaved. Mat.*, (7):77–84, 1996 (in Russian).

[725] L.V. Sabinin and Ludmila Sabinina. On the theory of left F-quasigroups. *Algebras Groups Geom.*, Hadronic Press, 12(2):127–137, 1995.

[726] L.V. Sabinin and L.V. Sbitneva. Reductive spaces and left F-quasigroups. *Webs and Quasigroups*,Tver State Univ. Russia. Publication, Tver, pages 123–128, 1991.

[727] Lev V. Sabinin. *Smooth Quasigroups and Loops*, volume 492 of *Mathematics and its Applications*. Kluwer Academic Publishers, Dordrecht, 1999.

[728] L.V. Sabinin. *Analytic Quasigroups and Geometry*. Univ. Druzhby Narodov, Moscow, 1991. (in Russian).

[729] A. Sade. Contribution a la theorie des quasi-groupes: diviseurs singuliers. *C. R. Acad. Sci.*, 237:372–374, 1953.

[730] A. Sade. Quasigroupes obéissant á certaines lois. *Rev. Fac. Sci. Univ. Istambul*, 22:151–184, 1957.

[731] A. Sade. Theorie des nuclei de groupoides et de quasigroupes $\mathfrak{N}\mathfrak{H}, \mathfrak{N}\mathfrak{Q}, \mathfrak{N}\mathfrak{C}$. *Istanbul Univ. Fen Fak. Mec., Ser. A*, 23:73–103, 1958 (in French).

[732] A. Sade. Produit direct-singulier de quasigroupes orthogonaux et anti-abeliens. *Ann. Soc. Sci. Bruxelles*, 74:91–99, 1960.

[733] A. Sade. Paratopie et autoparatopie des quasigroupes. *Ann. Soc. Sci. Bruxelles*, 76(1):88–96, 1962.

[734] A. Sade. Criteres d'isotopie d'un quasigroupe avec un quasigroupe demi-symetrique. *Univ. Lisboa Revista Fac. Ci.*, 11:121–136, 1964/1965. (French).

[735] A. Sade. Quasigroupes demi-symetriques. III. Constructions lineaires, A- maps. *Ann. Soc. Sci. Bruxelles Ser. I*, 81:5–17, 1967.

[736] L.V. Safonova and K.K. Shchukin. Computation of the automorphisms and anti-automorphisms of quasigroups. *Bul. Acad. Stiinte Repub. Mold., Mat.*, (3):49–55, 1990 (in Russian).

[737] Simona Samardjiska. On some cryptographic properties of the polynomial quasigroups. In *ICT Innovations 2010, CCIS 83*, pages 51–60. Springer-Verlag, 20011.

[738] S. Samardziska and S.Markovski. Polynomial n-ary quasigroups. *Mathematica Macedonica*, (5):77–81, 2007.

[739] M.D. Sandik. On identity elements in n-loops. In *Researches in Algebra and Calculus*, pages 140–146. Kartea Moldoveneaske, 1965 (in Russian).

[740] N.I. Sandu. Medial nilpotent distributive quasigroups and CH-quasigroups. *Sib. Math. J.*, 28:307–316, 1987. (in Russian).

[741] D.G. Sarvate and J. Seberry. Encryption methods based on combinatorial designs. *Ars Combinatoria*, 21A:237–245, 1986.

[742] Dinesh G. Sarvate and Alexander L. Strehl. Linear codes through Latin squares modulo n. *Bull. Inst. Combin. Appl.*, 37:73–81, 2003.

[743] M. Satti. A quasigroup based cryptographic system. Technical Report CR/0610017, arxiv.org, 2006.

[744] Larissa V. Sbitneva. Bol loop actions. *Commentationes Mathematicae Universitatis Carolinae*, 41(2):405–408, 2000.

[745] A.V. Scerbacova and V.A. Shcherbacov. About spectrum of T_2-quasigroups. Technical report, arXiv:1509.00796, 2015.

[746] R. D. Schafer. *An Introduction to Nonassociative Algebras.* Dover, New York, 1995.

[747] R. Schauffler. *Eine Anwendung zyklischer Permutationen und ihre Theorie.* PhD thesis, Philipps-Universität Marburg, 1948. (in German).

[748] R. Schauffler. Uber die Bildung von Codewörter. *Arch. Elektr. Übertragung*, 10:303–314, 1956.

[749] E. Schönhardt. Uber Lateinische Quadrate und Unionen. *J. Reine Angew. Math.*, 163:183–229, 1930.

[750] E. Schröder. Über eine eigenthümliche Bestimmung einer Function durch formale Anforderungen. *J. Reine Angew. Math.*, 90:189–220, 1881.

[751] R.H. Schulz. Equivalence of check digit systems over the dicyclic groups of order 8 and 12. In J. Blankenagel and W. Spiegel, editors, *Mathematikdidaktik aus Begeisterung für die Mathematik*, pages 227–237. Klett Verlag, Stuttgart, 2000.

[752] R.H. Schulz. Check character systems and anti-symmetric mappings. In *Computational Discrete Mathematics, LNCS*, volume 2122, pages 136–147. Springer Verlag, 2001.

[753] J.R. Seberry and A.P. Street. Strongbox secured secret sharing schemes. *Util. Math.*, 57:147–163, 2000.

[754] A. Shamir. How to share a secret. *Comm. ACM*, 22:612–613, 1979.

[755] J.-Y. Shao and W.-D. Wei. A formula for the number of Latin squares. *Disc. Math.*, 110:293–296, 1992.

[756] I. G. Shaposhnikov. Congruences of finite multibase universal algebras. *Diskret. Mat.*, 11(3):48–62, 1999.

[757] V.A. Shcherbacov. On generalisation of Markovski cryptoalgorithm. In *Workshop on General Algebra, February 26–March 1, 2015, Technische Universität Dresden, Technical Report,* Technische Universität Dresden, Dresden, 36–37, 2015.

[758] V.A. Shcherbacov. Some properties of full associated group of IP-loop. *Izvestia AN MSSR. Ser. fiz.-techn. i mat. nauk*, (2):51–52, 1984 (in Russian).

[759] V.A. Shcherbacov. About one class of medial quasigroups. *Mat. Issled.*, 102:111–116, 1988 (in Russian).

[760] V.A. Shcherbacov. On left distributive quasigroups isotopic to groups. In *Proceedings of the XI Conference of Young Scientists of Friendship of Nations University*, number 5305-B88, pages 148–149, Moscow, 1988. VINITI.

[761] V.A. Shcherbacov. About automorphism groups of quasigroups isotopic to groups. In *International Conference Universal Algebra, Quasigroups and Related Systems, Abstract of Talks, May 23–28, 1989*, page 31, Jadvisin, Poland, 1989.

[762] V.A. Shcherbacov. On automorphism groups of group isotopes with an idempotent, 1989. Reg. in VINITI 19.04.89, No 3530-B89, Moscow, 14 pages (in Russian).

[763] V.A. Shcherbacov. On automorphism groups of group isotopes with an idempotent. In *International Algebraic Malcev Conference, Novosibirsk, August 21–26, 1989, Abstr. of Commun. on Model Theory and Algebraic Systems*, page 159, Novosibirsk, 1989 (in Russian).

[764] V.A. Shcherbacov. On automorphism groups of quasigroups and linear quasigroups, 1989. Reg. in VINITI 04.11.89, No 6710-B89, Moscow, 32 pages (in Russian).

[765] V.A. Shcherbacov. On automorphism groups of linear quasigroups, 1990. Reg. in VINITI 28.09.90, No 5185-B90, Moscow, 12 pages (in Russian).

[766] V.A. Shcherbacov. On congruences of quasigroups, 1990. Reg. in VINITI 01.08.90, No 4413-B90, Moscow, 16 pages (in Russian).

[767] V.A. Shcherbacov. *On Automorphism Groups and Congruences of Quasigroups.* PhD thesis, Institute of Mathematics Academy of Sciences of Republic Moldova, 1991 (in Russian).

[768] V.A. Shcherbacov. *On Automorphism Groups and Congruences of Quasigroups,* Abstract of Ph.D. thesis. Kishinev, 1991 (in Russian).

[769] V.A. Shcherbacov. On linear quasigroups and their automorphism groups. *Mat. Issled.*, 120:104–113, 1991 (in Russian).

[770] V.A. Shcherbacov. Lattice isomorphic medial distributive quasigroups. In *Algebraic Methods in Geometry. Collection of Scientific Works,* Ryzhkov V. V. (ed.) et al., pages 68–71. Izdatel'stvo Rossijskogo Universiteta Druzhby Narodov, 1992 (in Russian).

[771] V.A. Shcherbacov. On automorphism groups of left distributive quasigroups. In *Collection of Theses of the III International Conference on Algebra, Krasnojarsk, August 23–28, 1993*, pages 371–372, 1993 (in Russian).

[772] V.A. Shcherbacov. On automorphism groups of left distributive quasigroups. *Bul. Acad. Stiinte Repub. Mold., Mat.*, (2):79–86, 1994 (in Russian).

[773] V.A. Shcherbacov. About signs of Moufang loop translations. *Bul. Acad. Stiinte Repub. Mold., Mat.*, (1):17–19, 1996.

[774] V.A. Shcherbacov. On normality of congruences of loops. In *International Conference Loops'99, Submitted Abstracts*, pages 34–35, Prague, August 1999.

[775] V.A. Shcherbacov. On 20 Belousov's problem. Technical Report 99.8, IAMI, Milan, Italy, 1999. http://www.mi.imati.cnr.it/iami/papers/99-08.ps.

[776] V.A. Shcherbacov. On Bruck-Toyoda-Murdoch theorem and isomorphisms of some quasigroups. In *First Conference of the Mathematical Society of the Republic Moldova, Abstracts*, pages 138–139, Chisinau, August 2001.

[777] V.A. Shcherbacov. On loops with the property $(x^{-1})^{-1} = x$. In *Third International Algebraic Conference in Ukraine, Submitted Abstracts*, page 104, Sumy, July 2001.

[778] V.A. Shcherbacov. On n-ary quasigroups and their possible applications in coding theory and cryptology. In *International Congress of Mathematicians. Abstracts of Short Communications and Poster Session*, page 30, Beijing, August 2002.

[779] V.A. Shcherbacov. Elements of quasigroup theory and some its applications in code theory, 2003. www.karlin.mff.cuni.cz/drapal/speccurs.pdf; http://de.wikipedia.org/wiki/Quasigruppe.

[780] V.A. Shcherbacov. On autotopies and automorphisms of n-ary medial quasigroups. In *4th International Algebraic Conference in Ukraine, Abstracts*, pages 205–206, Lviv, August 2003.

[781] V.A. Shcherbacov. On some known possible applications of quasigroups in cryptology, 2003. www.karlin.mff.cuni.cz/ drapal/krypto.pdf.

[782] V.A. Shcherbacov. On Bruck-Belousov problem. *Bul. Acad. Stiinte Repub. Mold., Mat.*, (3):123–140, 2005.

[783] V.A. Shcherbacov. On simple n-ary medial quasigroups. In *Proceedings of Conference Computational Commutative and Non-Commutative Algebraic Geometry*, volume 196 of *NATO Sci. Ser. F Comput. Syst. Sci.*, pages 305–324. IOS Press, 2005.

[784] V.A. Shcherbacov. On structure of finite n-ary medial quasigroups and automorphism groups of these quasigroups. *Quasigroups Relat. Syst.*, 13(1):125–156, 2005.

[785] V.A. Shcherbacov. On the structure of finite medial quasigroups. *Bul. Acad. Stiinte Repub. Mold., Mat.*, (1):11–18, 2005.

[786] V.A. Shcherbacov. On definitions of groupoids closely connected with quasigroups. *Bul. Acad. Stiinte Repub. Mold., Mat.*, (2):43–54, 2007.

[787] V.A. Shcherbacov. *On Linear and Inverse Quasigroups and Their Applications in Code Theory.* Thesis of Doctor of Physical and Mathematical Sciences. Institute of Mathematics and Computer Science of the Academy of Sciences of Moldova, Chişinău, 2008.

[788] V.A. Shcherbacov. On the structure of left and right F-, SM- and E-quasigroups. http://arxiv.org/, arXiv:0811.1725:67, 2008.

[789] V.A. Shcherbacov. On the structure of left and right F-, SM- and E-quasigroups. *J. Gen. Lie Theory Appl.*, 3(3):197–259, 2009.

[790] V.A. Shcherbacov. Quasigroups in cryptology. *Comput. Sci. J. Moldova*, 17(2):193–228, 2009.

[791] V.A. Shcherbacov. A-nuclei and A-centers of a quasigroup, 2011. arXiv:1102.3525.

[792] V.A. Shcherbacov. A-nuclei and A-centers of a quasigroup. Technical report, Central European University, Department of Mathematics and its Applications, February 2011.

[793] V.A. Shcherbacov. Orthogonality of linear (alinear) quasigroups and their parastrophes, 2012. arXiv:1212.1804.

[794] V.A. Shcherbacov and V.I. Izbash. On quasigroups with Moufang identity. *Bul. Acad. Stiinte Repub. Mold., Mat.*, (2):109–116, 1998.

[795] V.A. Shcherbacov and D.I. Pushkashu. On (r, s, t)-inverse loops and totally-anticommutative quasigroups. In *International Algebraic Conference Dedicated to the 100 Anniversary of A.G. Kurosh, Abstracts of Talks, May 27–June 3, Moscow*, pages 265–266, 2008.

[796] V.A. Shcherbacov and D.I. Pushkashu. On the structure of finite paramedial quasigroups. *Comment. Math. Univ. Carolin.*, 51(2):357–370, 2010.

[797] V.A. Shcherbacov, D.I. Pushkashu, and A.V. Shcherbacov. Equational quasigroup definitions, 2010. arXiv:1003.3175.

[798] V.A. Shcherbacov, A.Kh. Tabarov, and D.I. Pushkashu. On congruences of groupoids closely connected with quasigroups. *Fundam. Prikl. Mat.*, 14(1):237–251, 2008.

[799] K.K. Shchukin. *Action of a Group on a Quasigroup.* Kishinev State University, Kishinev, 1985 (in Russian).

[800] K.K. Shchukin. On the structure of trimedial quasigroups. *Investigations in General Algebra, Geometry, and Their Applications*, 162:148–154, 1986.

[801] K.K. Shchukin. On simple medial quasigroups. *Mat. Issled.*, 120:114–117, 1991.

[802] K.K. Shchukin. *Latin $n \times n$-Squares. Information ... and Computers.* Kishinev State University, Kishinev, 2008 (in Russian).

[803] Marlow Sholander. On the existence of the inverse operation in alternation groupoids. *Bull. Am. Math. Soc.*, 55:746–757, 1949.

[804] P.W. Shor. Polynomial-time algorithm for prime factorization and discrete logarithms on a quantum computer. *SIAM J. Comp.*, 26(5):1484–1509, 1997.

[805] P.W. Shor. Quantum computing. In *Proc. Intern. Congress of Mathematicians*, pages 467–486, Berlin, 1998.

[806] V. Shpilrain and A. Ushakov. Thompsons group and public key cryptography. *Lecture Notes in Comput. Sci.*, 3531:151–163, 2005.

[807] G.J. Simmons. *Contemporary Cryptology: The Science of Information Integrity.* IEEE Press, New York, 1992.

[808] Th. Skolem. Some remarks on the triple systems of Steiner. *Mathematica Scandinavica*, 6:273–280, 1958.

[809] L.A. Skornyakov, editor. *General Algebra.* Nauka, Moscow, 1991 (in Russian).

[810] Ivana Slaminkova and Milan Vojvoda. Cryptanalysis of a hash function based on isotopy of quasigroups. *Tatra Mt. Math. Publ.*, 45:137–149, 2010.

[811] Bogdan Smetaniuk. A new construction for Latin squares I. Proof of the Evans conjecture. *Ars Comb.*, 11:155–172, 1981.

[812] O. Yu. Smidt. *Selected works. Mathematics.* Izdat. Akad. Nauk SSSR, Moscow, 1959. (Russian).

[813] J.D.H. Smith. *Mal'cev Varieties*, volume 554 of *Lecture Notes in Math.* Springer Verlag, New York, 1976.

[814] J.D.H. Smith. Finite distributive quasigroups. *Math. Proc. Cambridge Philos. Soc.*, 80(1):37–41, 1976.

[815] J.D.H. Smith. Commutative Moufang loops: The first 50 years. Proceedings of the conference on groups and geometry, Part A (Madison, Wis., 1985). *Algebras Groups Geom.*, 2(3):209–234, 1985.

[816] J.D.H. Smith. *Representation Theory of Infinite Groups and Finite Quasigroups.* Lecture Notes in Mathematics. Universite de Montreal, Montreal, 1986.

[817] J.D.H. Smith. Loop transversals to linear codes. *J. Combin. Inform. System Sci.*, 17:1–8, 1992.

[818] J.D.H. Smith. A class of quasigroups solving a problem of ergodic theory. *Comment. Math. Univ. Carolin.*, 41:409–414, 2000.

[819] J.D.H. Smith. *An Introduction to Quasigroups and Their Representation.* Studies in Advanced Mathematics. Chapman & Hall/CRC, London, 2007.

[820] J.D.H. Smith. Palindromic and sudoku quasigroups. *J. Comb. Math. Comb. Comp.*, 88:85–94, 2014.

[821] V. Snasel, A. Abraham, J. Dvorsky, P. Kromer, and J. Platos. Hash functions based on large quasigroups. In *ICCS 2009, Part I, LNCS 5544*, pages 521–529, Springer-Verlag, Berlin, 2009.

[822] Leonard H. Soicher. On the structure and classification of SOMAs: Generalizations of mutually orthogonal Latin squares. *The Electronic Journal of Combinatorics*, 6(#R32):15 pages, 1999.

[823] L.R. Soikis. The special loops. *Questions of the Theory of Quasigroups and Loops*, pages 122–131, 1970 (in Russian).

[824] F.N. Sokhatskii. About isomorphism of linear quasigroups. In *Abstracts of the International Algebraic Conference*, page 138, Barnaul, Russian Federation, August 1991.

[825] F.N. Sokhatskii. On isotopes of groups, I. *Ukraïn. Mat. Zh.*, 47(10):1387–1398, 1995 (in Ukrainian).

[826] F.N. Sokhatskii. On isotopes of groups, II. *Ukraïn. Mat. Zh.*, 47(12):1692–1703, 1995 (in Ukrainian).

[827] F.N. Sokhatskii. On isotopes of groups, III. *Ukraïn. Mat. Zh.*, 48(2):256–260, 1996 (in Ukrainian).

[828] F.N. Sokhatskii. Some linear conditions and their application to describing group isotopes. *Quasigroups Relat. Syst.*, 6:43–59, 1999.

[829] F.N. Sokhatskii. On the classification of functional equations on quasigroups. *Ukrain. Mat. Zh.*, 56(9):1259–1266, 2004 (in Ukrainian).

[830] F.N. Sokhatskii. *Associates and Decompositions of Multy-Place Operations.* Thesis of Doctor of Physical and Mathematical Sciences. Institute of Mathematics, National Academy of Sciences of Ukraine, 2007 (in Ukrainian).

[831] F.N. Sokhatskii and O. Kirnasovsky. Subquasigroups and normal congruences for multiplace group isotopes. In *Intern. Algebraic Conference Dedicated to the Memory of Prof. L.M.Gluskin (1922–1985), 25–29 August, 1997. Abstracts of Talks*, pages 40–41, Slovyans'k, Ukraine, August 1997.

[832] F.N. Sokhatskii and P. Syvakivskyi. On linear isotopes of cyclic groups. *Quasigroups Relat. Syst.*, 1:66–76, 1994.

[833] Fedir M. Sokhatsky and Iryna V. Fryz. Invertibility criterion of composition of two multiary quasigroups. *Comment. Math. Univ. Carolin.*, 53:429–445, 2012.

[834] F.M. Sokhatsky. On isotopy-isomorphy problem. In *International Conference MITRE-2013, Abstracts*, pages 86–87, Chisinau, Moldova, August 2013.

[835] F.M. Sokhatsky and I.V. Fryz. Invertibility criterion for composition of two quasigroup operations. In *Book of Abstracts of the 8th International Algebraic Conference in Ukraine, July 5–12 (2011), Lugansk, Ukraine*, page 80, 2011.

[836] F.M. Sokhatsky and I.V. Fryz. Invertibility of repetition compositions and its connection with orthogonality. In *The International Mathematical Conference on Quasigroups and Loops, LOOPS'11, Booklet of Abstracts*, page 16, Třešt', Czech Republic, July 2011. http://www.karlin.mff.cuni.cz/ loops11/.

[837] E. I. Sokolov. On the Gluskin-Hosszu theorem for Dörnte n-groups. *Mat. Issled.*, 39:187–189, 1976 (in Russian).

[838] A.R.T. Solarin and L. Sharma. On the construction of Bol loops. *An. Stiint. Univ. "Al. I. Cuza" Iasi Sect. I a Mat. (N.S.)*, 27(1):13–17, 1981.

[839] D. Stanovsky. *Left Distributive Left Quasigroups.* PhD thesis, Charles University in Prague, 2004.

[840] D. Stanovsky. Distributive groupoids are symmetric-by-medial: an elementary proof. *Comment. Math. Univ. Carolin.*, 49(4):541–546, 2008.

[841] D. Stanovsky. Medial quasigroups of prime square order, 2016. arXiv:1604.03347.

[842] D. Stanovsky and P. Vojtechovsky. Commutator theory for loops. *J. Algebra*, 399C:290–322, 2014.

[843] D. Stanovsky and P. Vojtechovsky. Central and medial quasigroups of small order, 2015. arXiv:1511.03534.

[844] Alexander Stein. A conjugacy class as a transversal in a finite group. *J. Algebra*, 239(1):365–390, 2001.

[845] Sh. K. Stein. On the foundations of quasigroups. *Trans. Amer. Math. Soc.*, 85(1):228–256, 1957.

[846] Sh. K. Stein. Homogeneous quasigroups. *Pacific J. Math.*, 14:1091–1102, 1964.

[847] M. Steinberger. On loops with a general weak inverse property. *Mitt. Math. Ges. Hamburg*, 10:573–586, 1979.

[848] O.S. Stepchuk. Non-abelian groups with regular automorphism. In *The XVIth All-Union Algebraic Conference, September, 22–25, 1981, Abstracts, Part II, Leningrad*, page 126, 1981 (in Russian).

[849] E. Stickel. A new method for exchanging secret keys. In *Proceedings of the Third International Conference on Information Technology and Applications*, volume 2, pages 426–430, 2005.

[850] Douglas R. Stinson. *Combinatorial Designs: Constructions and Analysis.* Springer, New York, 2003.

[851] Z. Stojakovic and M. Stojakovic. On sets of orthogonal d-cubes. *Ars Combin.*, 89:21–30, 2008.

[852] Z. Stojakovic and B. Tasic. A generalization of Schröder quasigroups. *Util. Math.*, 58:225–235, 2000.

[853] Zoran Stojakovic. Cyclic n-quasigroups. *Univ. u Novom Sadu Zb. Rad. Prirod.-Mat. Fak. Ser. Mat.*, 12:407415, 1982.

[854] Zoran Stojakovic and Djura Paunic. Self-orthogonal cyclic n-quasigroups. *Aequationes Math.*, 30(2–3):252–257, 1986.

[855] Zoran Stojakovic and Janez Ušan. Orthogonal systems of partial operations. *Aequationes Math.*, 30(2–3):252–257, 1986.

[856] Douglas S. Stones. The many formulae for the number of Latin rectangles. *Electron. J. Combin.*, 17(1):1–46, 2010.

[857] Douglas S. Stones, Petr Vojtechovsky, and Ian M. Wanless. Cycle structure of autotopisms of quasigroups and Latin squares. *J. Combin. Des.*, 20(5):227–263, 2012.

[858] Karl Strambach. Kommutative distributive quasigruppen. *Math.-Phys. Semesterber.*, 24(1):71–83, 1977.

[859] Karl Strambach and Izabella Stuhl. Translation groups of Steiner loops. *Discrete Math.*, 309(13):4225–4227, 2009.

[860] A.P. Street and D.J. Street. *Combinatorics of Experimental Design.* Oxford U. P., Clarendon, 1987.

[861] D.A. Suprunenko. *Groups of Matrixes.* Nauka, Moscow, 1972. (in Russian).

[862] A.K. Suschkewitsch. Über die endlihen Gruppen ohne das Gesetz der eindeutigen Umkehrbarkeit. *Math. Ann.*, 99:30–50, 1928.

[863] A.K. Suschkewitsch. On a generalization of the associative law. *Trans. Amer. Math. Soc.*, 31:204–214, 1929.

[864] A.K. Suschkewitsch. *The Theory of Generalized Groups.* DNTVU, Kiev, 1937 (in Russian).

[865] P.N. Syrbu. *Teoria quasigrupurilor. Introducere.* Universitatea de Stat din Moldova, Chisinau, 2014 (in Romanian).

[866] Parascovia Syrbu and Dina Ceban. On π-quasigroups of type T_1. *Bul. Acad. Ştiinte Repub. Mold. Mat.*, (2):36–43, 2014.

[867] P.N. Syrbu. Self-orthogonal n-ary groups. *Matem. issled.*, 113:99–106, 1990 (in Russian).

[868] P.N. Syrbu. About separable self-orthogonal n-operations. *Bul. Acad. Ştiinţe Repub. Mold., Mat.*, 1994(3(16)):37–42, 1994.

[869] P.N. Syrbu. Loops with universal elasticity. *Quasigroups Relat. Syst.*, 1(1):57–65, 1994.

[870] P.N. Syrbu. On loops with universal elasticity. *Quasigroups Relat. Syst.*, 3(1):41–54, 1996.

[871] P.N. Syrbu. On congruences on n-ary T-quasigroups. *Quasigroups Relat. Syst.*, 6:71–80, 1999.

[872] P.N. Syrbu. On π-quasigroups isotopic to abelian groups. *Bul. Acad. Ştiinţe Repub. Mold. Mat.*, (3):109–117, 2009.

[873] P.N. Syrbu. On middle Bol loops. *ROMAI J.*, 6(2):229–236, 2010.

[874] A.Kh. Tabarov. Groups of regular permutations and nuclei of linear and close to linear quasigroups. *Bul. Acad. Stiinte Repub. Mold., Mat.*, (3):30–36, 1992 (in Russian).

[875] A.Kh. Tabarov. *Nuclei, Linearity and Balanced Identities in Quasigroups.* PhD thesis, IM AN MSSR, 1992 (in Russian).

[876] A.Kh. Tabarov. On the variety of abelian quasigroups. *Diskretn. Mat.*, 10:529–534, 2000 (in Russian).

[877] A.Kh. Tabarov. On endotopisms of linear and alinear quasigroups. In *The XIVth Conference on Applied and Industrial Mathematics, Communications, August 17–19, Chisinau*, pages 319–320, 2006.

[878] A.Kh. Tabarov. Identities and Linearity in Quasigroups, Abstract of D.Sc. thesis. Moscow, 2009 (in Russian), http://vak.ed.gov.ru/common/img/uploaded/files/vak/announcements/-fiz_mat/2009/06-07/TabarovAKH.pdf).

[879] A.Kh. Tabarov. On avtotopies and antiavtotopies quasigroups. In *International Conference on Algebras and Lattices "Jardafest," June 21–25, Prague*, 2010. http://www.karlin.mff.cuni.cz/ ical/print_abstract.php?id=44.

[880] A.Kh. Tabarov. Linear quasigroups. I, 2011. arXiv:1102.5515.

[881] A.Kh. Tabarov. Linear quasigroups. II, 2011. arXiv:1102.5517.

[882] G. Tarry. Le probleme de 36 officiers. *Compte Rendu de l'Association Francaise pour l'Advancement de Science Naturel*, 1:122–123, 1900.

[883] G. Tarry. Le probleme de 36 officiers. *Compte Rendu de l'Association Francaise pour l'Advancement de Science Naturel*, 2:170–203, 1901.

[884] H.A. Thurston. Certain congruences on quasigroups. *Proc. Amer. Math. Soc.*, 3:10–12, 1952.

[885] H.A. Thurston. Equivalences and mappings. *Proc. London Math. Soc.*, 3(2):175–182, 1952.

[886] H.A. Thurston. Noncommuting quasigroup congruences. *Proc. Amer. Math. Soc.*, 3:363–366, 1952.

[887] J. Tits. Sur la trialite et les algebres d'octaves. *Acad. Roy. Belg. Bull. Cl. Sci.*, 5(44):332–350, 1958.

[888] Christian Tobias. Security Analysis of the MOR Cryptosystem. *LNCS*, 2567:175–185, 2003.

[889] D.T. Todorov. Three mutually orthogonal Latin squares of order 14. *Ars Combinatoria*, 20:45–48, 1985.

[890] K. Toyoda. On axioms of linear functions. *Proc. Imp. Acad. Tokyo*, 17:221–227, 1941.

[891] B.V. Tsurkan. Identities with parameters. *General Algebra and Discrete Geometry*, 165:154–157, 1980 (in Russian).

[892] W.T. Tutte. *Graph theory.* Mir, Moscow, 1988 (in Russian).

[893] A. Ungar. The hyperbolic triangle centroid. *Comment. Math. Univ. Carolin.*, 45(2):355–370, 2004.

[894] V.I. Ursu. Quasi-identities of finitely generated commutative Moufang loops. *Algebra i Logika*, 30(6):726–734, 1991 (in Russian).

[895] I.I. Valutse and N.I. Prodan. Structures of congruences on a groupoid with division and on its semigroup of elementary translations. *General Algebra and Discrete Geometry*, 159:18–21, 1980 (in Russian).

[896] J. Verhoeff. *Error Detecting Decimal Codes*, volume 29. Math. Centrum Amsterdam, 1969.

[897] L.R. Vermani. *Elements of Algebraic Coding Theory.* Chapman & Hall, 1996.

[898] P. Vojtechovsky. Distances of groups of prime order. *Contrib. Gen. Algebra*, 11:225–231, 1999.

[899] M. Vojvoda. A survey of security mechanisms in mobile communication systems. *Tatra Mt. Math. Publ.*, 25:109–125, 2002.

[900] M. Vojvoda. Cryptanalysis of one hash function based on quasigroup. *Tatra Mt. Math. Publ.*, 29:173–181, 2004. MR2201663 (2006k:94117).

[901] M. Vojvoda. *Stream Ciphers and Hash Functions: Analysis of Some New Design Approaches.* PhD thesis, Slovak University of Technology, July, 2004.

[902] V.V. Wagner. Generalized groups. *Doklady Akad. Nauk SSSR (N.S.)*, 84:1119–1122, 1952.

[903] Ian M. Wanless. Transversals in Latin squares. *Quasigroups Related Systems*, 15(1):169–190, 2007.

[904] Ian M. Wanless and Bridget S. Webb. The existence of latin squares without orthogonal mates. *Des Codes Crypt.*, 40:131–135, 2006. DOI 10.1007/s10623-006-8168-9.

[905] Ja. Ja. Wedel. Autotopies of loops. *Some Problems of Algebraic Number Theory and Constructive Models*, Alma-Ata, pages 13–21, 1985 (in Russian).

[906] Wikipedia. Heap. http://en.wikipedia.org/wiki/Heap(mathematics).

[907] Wikipedia. Problems in loop theory and quasigroup theory, 2004. http://en.wikipedia.org/wiki.

[908] Wikipedia. Homomorphism, 2007. http://en.wikipedia.org/wiki/Homomorphism.

[909] Wikipedia. Semidirect product, 2007. http://en.wikipedia.org/wiki/Semidirect product.

[910] Wikipedia. Centralizer and normalizer, 2009. http://en.wikipedia.org/wiki.

[911] Wikipedia. Group action, 2009. http://en.wikipedia.org/wiki/Group action.

[912] Wikipedia. Clone, 2011. http://en.wikipedia.org/wiki/Clone.

[913] Wikipedia. Hamming distance, 2011. http://en.wikipedia.org/wiki/Hamming_distance.

[914] Wikipedia. Linear code, 2011. http://en.wikipedia.org/wiki/Linear_code.

[915] Wikipedia. Singleton bound, 2011. http://en.wikipedia.org/wiki/Singleton_bound.

[916] Wikipedia. Systematic code, 2011. http://en.wikipedia.org/wiki/Systematic_code.

[917] Wikipedia. Term, 2011. http://ru.wikipedia.org/wiki/Term.

[918] Wikipedia. Block design, 2013. http://en.wikipedia.org/wiki/Block_design.

[919] Wikipedia. Diophantine equations, 2013. http://en.wikipedia.org/wiki/Diophantine_equation.

[920] Wikipedia. Elementary abelian group, 2013. http://en.wikipedia.org/wiki/Elementary_abelian_group.

[921] Wikipedia. General affine group: Ga(1,5), 2013. http://groupprops.subwiki.org/wiki/General_affine_group:GA(1,5).

[922] Wikipedia. Holomorph, 2013. http://en.wikipedia.org/wiki/Holomorph.

[923] Wikipedia. Monster group, 2013. http://en.wikipedia.org/wiki/Monster_group.

[924] Wikipedia. Dihedral group of order 6, 2014. http://en.wikipedia.org/wiki/Dihedral_group_of_order_6.

[925] Wikipedia. Elgamal encryption, 2014. http://en.wikipedia.org/wiki/ElGamal_encryption.

[926] Wikipedia. Elgamal signature scheme, 2014. http://en.wikipedia.org/wiki/ElGamal_signature_scheme.

[927] Wikipedia. Euler's totient function, 2014. http://en.wikipedia.org/wiki/Euler's_totient_function.

[928] Wikipedia. Finite projective planes, 2014. http://en.wikipedia.org/wiki/Projective_plane #Finite_projective_planes.

[929] Wikipedia. Fisher's inequality, 2014. http://en.wikipedia.org/wiki/Fisher's_inequality.

[930] Wikipedia. Index of a subgroup, 2014. http://en.wikipedia.org/wiki/Index_of_a_subgroup.

[931] Wikipedia. Neighbourhood, 2014. https://en.wikipedia.org/wiki/Neighbourhood_(graph_theory).

[932] Wikipedia. Orthogonal array, 2014. http://en.wikipedia.org/wiki/Orthogonal_array.

[933] Wikipedia. Von Mangoldt function, 2014. http://en.wikipedia.org/wiki/Von_Mangoldt_function.

[934] Wikipedia. Secret sharing, 2014. http://en.wikipedia.org/wiki/Secret_sharing.

[935] Wikipedia. Semantic security, 2014. http://en.wikipedia.org/wiki/Semantic_security.

[936] Wikipedia. Shamir's secret sharing, 2014. http://en.wikipedia.org/wiki/Shamir's_Secret_Sharing.

[937] Wikipedia. Steiner system, 2014. http://en.wikipedia.org/wiki/Steiner_system.

[938] Wikipedia. Strongly regular graph, 2014. http://en.wikipedia.org/wiki/Strongly_regular_graph.

[939] Wikipedia. Variety (universal algebra), 2014. http://en.wikipedia.org/wiki/Variety_(universal_algebra).

[940] Wikipedia. Combinatorial design, 2015. http://en.wikipedia.org/wiki/Combinatorial_design.

[941] Wikipedia. Goldwasser-Micali cryptosystem, 2015. https://en.wikipedia.org/wiki/GoldwasserMicali_cryptosystem.

[942] Eric L. Wilson. A class of loops with the isotopy-isomorphy property. *Canad. J. Math.*, 18:589–592, 1966.

[943] R.L. Wilson. Quasidirect products of quasigroups. *Comm. Algebra*, 3:835–850, 1975.

[944] Richard M. Wilson. Concerning the number of mutually orthogonal Latin squares. *Discrete Math.*, 9:181–198, 1974.

[945] Richard M. Wilson. An existence theory for pairwise balanced designs, III: Proof of the existence conjectures. *Journal of Combinatorial Theory (A)*, 18:71–79, 1975.

[946] Yanping Xu and Yunqing Xu. Self-orthogonal n-cyclic quasigroups and applications in hash functions. *Journal of Information & Computational Science*, 9(15):4525–4530, 2012. http://www.joics.com.

[947] Koichi Yamamoto. Generation principles of Latin squares. *Bull. Inst. Internat. Statist.*, 38:73–76, 1961.

[948] J.W. Young. *Projective Geometry.* Gosizdatel'svo Inostrannaya Literatura, Moscow, 1949 (in Russian).

[949] Fajar Yuliawan. Studi mengenai aplikasi teori quasigroup dalam kriptografi, 2006. Program Studi Teknik Informatika, Institut Teknologi Bandung, www.informatika.org/rinaldi/Kriptografi/2006-2007/Makalah1/Makalah1- 037.pdf.

[950] P. Zappa. Triplets of Latin squares. *Boll. Un. Mat. Ital. A (7)*, 10:63–69, 1996. MR1386246 (97b:05038).

[951] H. Zbingen, N. Gisin, B. Huttner, A. Muller, and W. Tittel. Practical aspects of quantum cryptographical key distributions. *J. Cryptology*, 13:207–220, 2000.

[952] K.A. Zhevlakov, A.M. Slinko, I.P. Shestakov, and A.I. Shirshov. *Rings That Are Nearly Associative.* Nauka, Moscow, 1978.

[953] Lie Zhu. A short disproof of Euler's conjecture concerning orthogonal Latin squares. With editorial comment by A.D. Keedwell. *Ars Combin.*, 14:47–55, 1982.

[954] A.K. Zotov and V.V. Ryzhkov. About the concept of isotopy of n-ry relations and algebraic systems. *Mat. Issled.*, 43:120–128, 1976 (in Russian).

Index

$STS(n)$, 158
m-tuple of maps, 389
    kind of, 390
    product of, 389
n-groupoids
    self-orthogonal cyclic, 429
t-(v, k, λ)-design, 155
t-design, 155
3-CML, 142, 382
3-net, 156
3-quasigroup
    2-idempotent, 163

A-nucleus, 59
    left, 59
    middle, 59
    right, 59
A-pseudo-automorphism
    left, 39
    middle, 39
    right, 39
A. Stein theorem, 149
AIP-loop, 176
algebra
    polynomially complete, 503
    surjective, 15
algebras
    term-equivalent, 260
algorithm
    code-crypt, 491
    Markovski, 20
anti-autotopy, 48
antiautomorphism, 68
antiendomorphism, 68
antihomomorphism, 68
array
    orthogonal, 105
associator, 62, 113, 152
attack
    chosen ciphertext, 482
    chosen plaintext, 482
automorphism, 30
    complete, 148, 290
    fixed point free, 146
    regular, 146
autostrophism, 45
autostrophy, 45
autotopism, 30, 34
autotopy, 30, 34

balanced incomplete block design, 155
Basarab theorem, 113
basic square, 80
Belousov theorem
    on generators, 27
Belousov-Onoi theorem, 148
Belyavskaya theorem, 58
BIBD, 155
binary relation, 274
    admissible, 64
    semi-admissible, 64
    stable, 275
binary relations
    product of, 6, 274
block design, 155
    balanced, 155
    incomplete, 155
    symmetric, 156
    uniform, 154
Bol loop, 24
bound
    Joshi, 459
    Singleton, 459
Bruck theorem, 27, 143
Bruck-Slaby theorem, 144
Buchsteiner loop, 41

Cayley table, 9
CC-loop, 41
centralizer, 216
CH-quasigroup, 114, 382
Cheban loop, 41
CI-quasigroup, 115
cipher, 474
    Caesar's, 473
    quasigroup-based, 474

Scital, 473
stream, 474
clue, 86
code, 440, 459
k-recursive, 460
n-T-quasigroup, 449
n-ary quasigroup, 441
5-n-quasigroup, 450
block-, 459
complete k-recursive, 460
EAN, 373, 453
generalized linear, 459, 465
ISBN, 452
ISSN, 372
linear, 459, 465
MDS-, 459
of German banknotes, 509
quasigroup, 441
Reed-Solomon, 496
systematic, 460
UPC, 372, 453
code theory, 472
commutant, 260
commutator, 62
congruence, 63, 274
left, 274
left normal, 274
normal, 64, 275
regular, 68
right, 274
right normal, 275
core, 147
criteria
Belousov, 166, 233
CGMN, 460
Sokhatskii, 166
critical set, 85
minimal, 85
crossed product, 76
cryptanalysis, 472
cryptcode, 488
crypto-automorphism, 36
cryptography, 472
cryptology, 472
cryptosystem
GM, 500
Goldwasser-Micaly, 500
Rivest-Shamir-Adleman, 500
RSA, 500
semantically secure, 501
derivative
k-recursive, 461
left, 212
left-right, 215
middle, 212
mixed, 215
right, 212
derivative group
n-ary, 329
design
t-, 155
t-$(v, k, 1)$-, 157
t-(v, k, λ)-, 155
balanced, 155
incomplete, 155
Mendelsohn, 161
pairwise balanced, 161
simple, 155
symmetric, 156
transversal, 156
uniform, 154
direct product, 73
external, 73
internal, 73

e-quasigroup, 16
El Gamal
signature scheme, 502
element
(m, n)-, 173
CI-, 260
idempotent, 22, 328
leader, 20
left
Bol, 53
Moufang, 53
middle
Bol, 53
Moufang, 53
right
Bol, 54
Moufang, 53
encryption
homomorphic, 501
encryption system
El Gamal, 499
MOR, 501
RSA, 500
endomorphism, 66
equation
generalized associativity, 166

error, 440
  jump transposition, 440
  jump twin, 440
  phonetic, 440, 451
  single, 440
  transposition, 440
  twin, 440
extension
  Schreier, 74

Fisher theorem, 145
Florja theorem, 124
formula
  Shao-Wei, 101
function
  hash, 493
  one-way, 492

G-loop, 41, 213
G-quasigroup
  left, 213
  right, 213
GA-quasigroup
  left, 42
  middle, 42
  right, 42
Galkin theorem, 148
gisotopy, 431
graph, 101
  degree, 102
  left translation, 103
  middle translation, 103
  of groupoid, 101
  of Latin square, 101
  regular, 103
  right translation, 103
  strongly regular, 103
  valency, 102
grid, 85
group, 24
  of quasiautomorphisms, 35
  AC-, 35
  action, 32
  primitive, 33
  with triality, 177
group isotope
  n-ary, 329
groupoid, 8
  n-ary, 8
  binary, 8
  cancellation, 14
  center, 52
  division, 14
  homogenous
    from the left, 146
    from the right, 146
  left cancellation, 13
  left division, 14
  right cancellation, 13
  right division, 14
  translation, 10
groupoids
  orthogonal, 396

h-nucleus
  left, 57
  middle, 57, 58
  right, 57
Hamiltonian
  group, 29
Hamming distance, 459
  between operations, 507
  of code, 459
hash function
  collision free, 493
heap, 48
holomorph, 375
homomorphism, 66
homotopism, 70

idempotent, 328
identity
  C_3, 405
  T_2, 405
  associative, 23
  balanced, 169
  Belousov I, 169
  Belousov II, 169
  Bol-Moufang type, 108
  Bruck, 113
  CLM, 272
  commutativity, 45
  elasticity, 180
  flexibility, 111
  generalized, 206
    associativity, 152
    parastrophic, 206
  left
    alternative, 111
    Bol, 24, 268
    CLM, 273
    Moufang, 53, 110, 266

semi-medial, 272
semi-symmetric, 149
left semi-symmetric, 46
LWA-, 142
Manin, 152
middle
Moufang, 266
middle Bol, 259
Moufang, 24
middle, 110
Moufang with parameter, 111
parastrophic, 206
retract, 56
right
alternative, 111
CLM, 273
Moufang, 110, 266
semi-medial, 272
right semi-symmetric, 46
Schröder, 108
of generalized associativity, 153
Schröder's 1st, 405
Schröder's 2nd, 405
semi-symmetric, 46
slightly associative, 23
Stein's 1st, 405
Stein's 2nd, 405
Stein's 3rd, 405
triality, 177
Wilson, 205
with permutations, 232
identity element, 21
left, 21
left local, 21
middle, 21
middle local, 21
right, 21
incidence system, 154
index of nucleus, 229
inequality
Fisher's, 155
intercalate, 84
IP-quasigroup, 115
isostrophism, 43
isostrophy, 43
isotopism, 29
nuclear, 212
isotopism (isotopy), 33
isotopy, 29
crossed
left, 435
right, 435
generalized, 431
LP-, 31
nuclear, 212
orthogonal, 434
principal, 31

Ježek-Kepka theorem, 133

k-net, 426
k-transversal, 87
Kepka-Nemec theorem, 135
Kerkhoff's rule, 473
key, 473
non-public, 473
non-symmetric, 473
public, 473
symmetric, 473

Latin bitrade, 84
Latin square, 11, 79
r-semi partial, 84
critical set of, 85
partial, 83
Sudoku, 85
Latin squares
orthogonal, 79
pseudo-orthogonal, 81
Latin trade, 84
lattice, 275
Leakh theorem, 35
left
Bol element, 53
loop, 22
left E-quasigroup, 109, 288
left F-quasigroup, 109, 288
left identity operation, 397
left loop, 21
left neo-field, 98
left quasigroup, 9
left SM-quasigroup, 109, 288
lemma
Izbash, 267
LIP-quasigroup, 115
loop, 21, 23
A-, 109
AAIP-, 47
AI-, 109
ARIF-, 109
automorphic, 109
automorphic inverse, 109

Bol, 24
simple, 113
Bruck, 110
Buchsteiner, 41, 110
Cayley-Dickson, 111
CC-, 41, 110, 182
Cheban, 41, 110
CI-, 189
commutant, 260
diassociative, 112
G-, 41, 213
generalized
Moufang, 110
LCC-, 182
left, 22
Bol, 109
Cheban, 110
M-, 110
left special, 288
left Bol, 99
left M-, 288
M-, 110, 148
middle
Bol, 110
middle Bol, 259
MIP-, 258
Moufang, 24, 99, 110
commutative, 110
Moufang commutative, 142
Osborn, 110
PCC-, 182
RA, 110
RCC-, 182
RIF-, 109
right, 23
Bol, 109
M-, 110
right M-, 288
right special, 288
S-, 148
special, 379
Steiner, 149
tag-, 383
TS-, 149
unipotent, 22
W-, 150
WCIP, 186
WIP, 202
LP-isotopy, 31

M-loop, 148
Manin quasigroup, 114
Mann multiplication, 434
Mann operation
left, 434
right, 434
map, 6
bijective, 7
injective, 6
surjective, 6
mapping
complete, 92
conjugated, 92
quasicomplete, 92
message
authentication, 505
identification, 505
non-repudiation, 505
middle
Moufang
element, 53
middle-left element, 53
MOLS, 80
Moufang identities, 24
Moufang theorem, 112
multiplication group, 25
full, 25
left, 24
middle, 24
right, 24
multiset, 155
Murdoch theorem, 131

n-ary crossed product, 76
n-ary group, 329
n-ary quasigroup, 328
σ-parastrophe, 441
n-operations
orthogonal, 427
n-quasigroup
cyclic, 428
self-orthogonal cyclic, 429
n-T-quasigroup, 114
neo-field, 98
net
(k,n)-, 427
m-dimensional, 427
NLPN sequence, 505
normal
subquasigroup, 67
normal congruence, 64
normalizer, 216

nucleus, 52
σ-A-
left, 48
middle, 48
right, 48
Bol, 54
left, 52
Bol, 54
Moufang, 53
middle, 52
Bol, 54
Moufang, 53
Moufang, 53
right, 52
Bol, 54
Moufang, 53

operation, 8
n-ary, 8, 327
binary, 8
derivative, 212
left derivative, 212
nullary, 8
right derivative, 212
surjective, 15
orbit of element, 32
order of element
left, 173, 174
middle, 173
right, 173, 174
orthogonal array, 105
linear, 106
simple, 106
orthogonality, 395
Belousov criteria, 403
of n-ary operations, 81
of binary groupoids, 80
of Latin squares, 79
strong, 460
orthomorphism, 97

parastrophy, 11
partitions, 5, 6
PBD, 161
permanent, 100
permutation
admissible, 64
semi-admissible, 64
permutation identity, 232
permutation matrix, 100
PI-quasigroup, 185
plane
affine, 159
Fano, 157
projective finite, 157
postulate, 168
Sushkevich A, 168
Sushkevich A*, 168
Sushkevich B, 168
Sushkevich B*, 168
power set, 504
principal isotopism, 31
problem
1a Belousov, 320
Brualdi, 96
Bruck-Belousov, 280
Burmistrovich-Belousov, 265
Jezek-Kepka, 141
Kinyon-Phillips, (1), 326
Kinyon-Phillips, (2), 326
product, 5
Cartesian, 5
crossed, 76
direct, 73
generalized crossed, 76
generalized singular direct, 77
n-ary crossed, 76
quasi-direct, 76
Sabinin, 78
Sade, 76
semidirect, 74
singular direct, 76, 77
prolongation
Belousov, 88
Belyavskaya, 90
Bruck, 88
Derienko-Dudek, 93
generalized Belyavskaya, 90
generalized Derienko-Dudek, 94
Gonsales-Couselo-Markov-Nechaev, 462
Yamamoto, 88
property
AAIP-, 47
WCIP-, 186
pseudo-automorphism
σ-A-
left, 48
middle, 48
right, 48
companion of, 39
left, 39
middle, 39

right, 39
pseudo-Latin square, 396

quasi-identity, 402
quasiautomorphism, 30
quasigroup, 11, 16
(α, β, γ)-inverse, 184
(m, n)-linear, 175
(r, s, t)-inverse, 115
C_3, 109
I-, 185
T_2-, 405
λ-inverse-property, 184
μ-inverse-property, 184
ρ-inverse-property, 184
n-ary, 10, 328
automorphism, 328, 358
autotopism, 358
autotopy, 328, 358
congruence, 331
direct product, 334
homomorphism, 331
isotopy, 328
linear, 114, 329
medial, 330
retract, 345, 441
simple, 344
solvable, 355
translation, 328
unipotently-solvable, 343
n-ary T-, 330
A-, 108
A-center of, 241
A-central, 243
alinear
over a group, 165
alternative, 108
anti-abelian, 422
anti-commutative, 422, 445
anti-symmetric, 422, 445
binary, 10
bundle, 227
center, 58
CH-, 114, 382
CI-, 115
contraction, 97
definition
equational, 16
existential, 10, 19
derivative
left, 212
middle, 212
middle inverse, 212
right, 212
differentiable, 461
distributive, 108, 144
E-, 109, 288
e-, 16
ETS-, 307
F-, 109, 213, 288
flexible, 108
form, 329
functionally complete, 476
G-
left, 213
right, 213
GA-, 42
generalized linear, 164
generated
by left shift, 504
by right shift, 504
half-idempotent, 159
idempotent, 22, 108
inner mapping group of, 25
IP-, 115
LCC-, 182
left
alternative, 108
Bol, 109
CI-, 109
distributive, 108
E-, 109, 288
F-, 109, 288
semi-symmetric, 109
SM-, 109, 288
left alinear
over a group, 165
left derivative, 212
left distributive, 144, 379
minimal, 148
left GA-, 42
left inverse-property, 115
left linear, 190
over a group, 165
over a loop, 190
left-right derivative, 215
linear, 164
over a group, 165
over a quasigroup, 205
over loop, 114
LIP-, 115
LWA-, 288

m-inverse-property, 115
medial, 24, 58
middle GA-, 42
mixed derivative, 215
Moufang, 110
multiplication group of, 25
paramedial, 108
parastrophe, 11
PCC-, 182
PI-, 185
polynomial, 503
polynomially complete, 503
prolongation, 88, 193
pseudo-automorphism
right, 39
RCC-, 182
right
alternative, 108
Bol, 109
distributive, 108
E-, 109, 288
F-, 109, 288
SM-, 109, 288
right CI-, 109
right derivative, 212
right F-, 24
right GA- , 42
RIP-, 115
RWA-, 288
Schröder, 108
self-orthogonal cyclic, 429
semi-symmetric, 108, 428
semicentral, 165
shapeless, 476
simple, 66
SM-, 109, 288
SOC-, 424
solvable, 134
spectrum, 162
Stein, 109
Steiner, 149
Steiner distributive, 383
strongly simple, 66
symmetric, 148
T-, 58
TA-, 93
TAC-, 422
ternary Evans, 428
totally anti-commutative, 422
totally antisymmetric, 93
totCO-, 425
translation, 10
trimedial, 289
TS-, 66, 109, 149
unipotent, 21, 109
unipotently-solvable, 343
WA-, 288
WCIP-, 186
weak totally antisymmetric, 93
weak-inverse-property, 115
WIP-, 202
with elasticity, 108
WTA-, 93

regular permutation, 58
relation, 5, 6
binary, 6
equivalence, 6
n-ary, 6
reflexive, 6
symmetric, 6
transitive, 6
representation, 349
irreducible, 349
retract
of identity, 56
right
Bol element, 54
loop, 23
right E-quasigroup, 109, 288
right identity operation, 397
right loop, 21
right Moufang element, 53
right quasigroup, 9
RIP-quasigroup, 115

S-loop, 148
S-system, 434
S. Stein theorem, 147
Sabinin product, 78
scheme
secret-sharing, 494
secret sharing scheme, 494
selector, 460
binary, 397
k-, 460
semidirect product, 74
semilattice, 275
set, 5
critical, 85
sharply transitive, 25
sets

disjoint, 6
sign function, 99, 466
signature algorithm
Pointcheval-Stern, 503
signature scheme
El Gamal, 502
singular direct product, 77
SOC-quasigroup, 424
SOCQ, 429
SOMA, 81
spectrum, 161
of quasigroup, 162
square, 388
column-Latin, 388
Latin, 388
permutation, 388, 390
row-Latin, 388
Sudoku, 85
Youden, 495
squares
orthogonal, 395
product, 398
stabilizer subgroup, 32
steganography, 472
Steiner
system, 157
triple system, 158
Steiner loop, 149
Steiner quasigroup, 149
Steiner system
(5, 6, 12), 160
generalized, 506
sub-object, 51
subloop
associator, 260
subquasigroup
normal, 67
Sudoku, 85
puzzle, 85
symmetric group S_n, 8
system
incidence, 154
strong orthogonal, 460

T-quasigroup, 58, 190
tactical schema, 155
theorem
n-ary analog of Murdoch, 340
Albert generalized, 32
Basarab, 113
Belousov
on distributive quasigroups, 144
on generators, 27
regular, 230
Belousov-Onoi, 148
Belyavskaya, 58
Belyavskaya-Russu, 87
Belyavskaya-Tabarov, 170
Bruck, 27, 143
Bruck-Ryser-Chowla, 157
Bruck-Slaby, 144
Doyen-Wilson, 159
Fisher B., 145
Fisher R., 155
Florja, 124
Galkin, 148
Gluskin-Hosszu, 329
Grishkov-Zavarnitsine, 112
Izbash-Syrbu, 462
Janssen-Zappa, 100
Ježek-Kepka, 133
Kepka-Kinyon-Phillips, 321
Kepka-Nemec, 135
Leakh, 35
Mann, 399
Mikheev, 178
Moufang, 112
Murdoch, 131
NC-, 216
on four quasigroups, 166
Schreier O., 28
Shchukin, 311
Sokhatsky, 319
Stein A., 149
Stein S., 147
Tabarov, 169
Toyoda, 126
Wilson R.M., 162
Toyoda theorem, 126
translation, 10
left, 10
middle, 12
right, 10
transversal, 26, 86
left, 26
triple system
directed, 161
Mendelsohn, 161
Trojan square, 81
TS-loop, 149
TS-quasigroup, 66, 109

unipotent quasigroup, 21

W-loop, 150
WIP-quasigroup, 115

zero knowledge protocol, 506

Printed in the United States
By Bookmasters